Introductory Algebra

Third Edition

Elayn Martin-Gay

University of New Orleans

PEARSON

Prentice Hall

Upper Saddle River, New Jersey 07458

Library of Congress Cataloging-in-Publication Data

Martin-Gay, K. Elayn
  Introductory algebra/K. Elayn Martin-Gay.—3rd ed.
    p. cm.
  Includes index.
  ISBN 0-13-186843-8 (pbk.: alk. paper) — ISBN 0-13-186839-X (alk. paper)
  1. Algebra.   I. Title.

QA152.3.M37 2007
512.9—dc21

Executive Editor: *Paul Murphy*
Editor in Chief: *Christine Hoag*
Project Manager: *Mary Beckwith*
Production Management: *Elm Street Publishing Services, Inc.*
Senior Managing Editor: *Linda Mihatov Behrens*
Executive Managing Editor: *Kathleen Schiaparelli*
Media Project Manager, Developmental Math: *Audra J. Walsh*
Media Production Editor: *John Cassar*
Managing Editor, Digital Supplements: *Nicole M. Jackson*
Manufacturing Buyer: *Alan Fischer*
Manufacturing Manager: *Alexis Heydt-Long*
Director of Marketing: *Patrice Jones*
Senior Marketing Manager: *Kate Valentine*
Marketing Assistant: *Jennifer de Leeuwerk*
Development Editor: *Laura Wheel*
Editor in Chief, Development: *Carol Trueheart*
Editorial Assistant: *Abigail Rethore*
Art Director: *Maureen Eide*
Interior/Cover Designer: *Suzanne Behnke*
Art Editor: *Thomas Benfatti*
Creative Director: *Juan R. López*
Director of Creative Services: *Paul Belfanti*
Cover Photo: *Digital Vision/Getty Images, Inc.*
Manager, Cover Visual Research & Permissions: *Karen Sanatar*
Director, Image Resource Center: *Melinda Reo*
Manager, Rights and Permissions: *Zina Arabia*
Manager, Visual Research: *Beth Brenzel*
Image Permission Coordinator: *Craig Jones*
Photo Researcher: *Teri Stratford*
Composition: *Interactive Composition Corporation*
Art Studios: *Scientific Illustrators and Laserwords*

© 2007, 2003, 1999 by Prentice-Hall, Inc.
Pearson Prentice Hall
Pearson Education, Inc.
Upper Saddle River, New Jersey 07458

Printed in the United States of America
10 9 8 7 6 5 4 3

ISBN: (paperback) 0-13-186843-8; (case bound) 0-13-186839-X

Pearson Education LTD., *London*
Pearson Education Australia PTY., Limited, *Sydney*
Pearson Education Singapore, Pte. Ltd.
Pearson Education North Asia Ltd., *Hong Kong*
Pearson Education Canada, Ltd., *Canada*
Pearson Educacíon de Mexico, S.A. de C.V.
Pearson Education—Japan, *Tokyo*
Pearson Education Malaysia, Pte. Ltd.

*To my mother, Barbara M. Miller,
and her husband, Leo Miller, and to the memory of
my father, Robert J. Martin*

Contents

1 Real Numbers and Introduction to Algebra   1

2 Equations, Inequalities, and Problem Solving   89

Contents

Tools to Help Students Succeed

Your textbook includes a number of features designed to help you succeed in this math course—as well as the next math course you take. These features include:

Feature	Benefit	Page
Well-crafted Exercise Sets: We learn math by doing math	The exercise sets in your text offer an ample number of exercises carefully ordered so you can master basic mathematical skills and concepts while developing all-important problem solving skills. Exercise sets include Mixed Practice exercises to help you master multiple key concepts, as well as Mental Math, Writing, Applications, Concept Check, Concept Extension, and Review exercises.	298–301
Solutions-to-Selected Exercises: Built-in solutions at the back of the text	If you need to review problems you find difficult, this built-in solutions manual at the back of the text provides the step-by-step solutions to every other odd-numbered exercise in the exercise sets.	A27
Study Skills Builders: Maximize your chances for success	Study Skills Builders reinforce the material in *Section 1.1—Tips for Success in Mathematics*. Study Skills Builders are a great resource for study ideas and self-assessment to maximize your opportunity for success in this course. Take your new study skills with you to help you succeed in your next math course.	278
The Bigger Picture: Succeed in this math course and the next one you take	The Bigger Picture focuses on the key concepts of this course—simplifying expressions and solving equations—and asks you to keep an ongoing outline so you can simplify expressions, solve equations, and recognize the difference between the two. A strong foundation in simplifying expressions and solving equations will help you succeed in this algebra course, as well as the next math course you take.	301
Examples: Step-by-step instruction for you	Examples in the text provide you with clear, concise step-by-step instructions to help you learn. Annotations in the examples provide additional instruction.	258
Helpful Hints: Help where you'll need it most	Helpful Hints provide tips and advice at exact locations where students need it most. Strategically placed where you might have the most difficulty, Helpful Hints will help you work through common trouble spots.	268
Practice Problems: Immediate reinforcement	Practice Problems offer immediate reinforcement after every example. Try each Practice Problem after studying the corresponding example to make sure you have a good working knowledge of the concept.	258
Integrated Review: Mid-chapter progress check	To ensure that you understand the key concepts covered in the first sections of the chapter, work the exercises in the Integrated Review before you continue with the rest of the chapter.	291
Vocabulary Check: Key terms and vocabulary	Make sure you understand key terms and vocabulary in each chapter with the Vocabulary Check.	312
Chapter Highlights: Study smart	Chapter Highlights outline the key concepts of the chapter along with examples to help you focus your studying efforts as you prepare for your test.	312–315
Chapter Test: Take a practice test	In preparation for your classroom test, take this practice test to make sure you understand the key topics in the chapter. Be sure to use the **Chapter Test Prep Video CD** included with this text to see the author present a fully worked-out solution to each exercise in the Chapter Test.	319

Martin-Gay's CD VIDEO RESOURCES Help Students Succeed

Martin-Gay's Chapter Test Prep Video CD (available with this text)

- Provides students with help during their most "teachable moment"—while they are studying for a test.

- Text author Elayn Martin-Gay presents step-by-step solutions to the exact exercises found in each Chapter Test in the book.

- Easy video navigation allows students to instantly access the worked-out solutions to the exercises they want to review.

- A close-captioned option for the hearing impaired is provided.

Martin-Gay's CD Lecture Series (with Tips for Success in Mathematics)

- Text author Elayn Martin-Gay presents the key concepts from every section of the text in 10–15 minute mini-lectures.

- Students can easily review a section or a specific topic before a homework assignment, quiz, or test.

- Includes fully worked-out solutions to exercises marked with a CD Video icon ( ) in each section.

- Includes *Section 1.1, Tips for Success in Mathematics.*

- A close-captioned option for the hearing impaired is provided.

- Ask your bookstore for information about Martin-Gay's, *Introductory Algebra,* Third Edition, CD Lecture Series, or visit www.prenhall.com.

Additional Resources to Help You Succeed

Student Study Pack

A single, easy-to-use package—available bundled with your textbook or by itself—for purchase through your bookstore. This package contains the following resources to help you succeed:

Student Solutions Manual
- Contains worked-out solutions to odd-numbered exercises from each section exercise set, Practice Problems, Mental Math exercises, and all exercises found in the Chapter Review and Chapter Tests.

Prentice Hall Math Tutor Center
- Staffed by qualified math instructors who provide students with tutoring on examples and odd-numbered exercises from the textbook. Tutoring is available via toll-free telephone, toll-free fax, email, or the Internet.

Martin-Gay's CD Lecture Series
- Text author Elayn Martin-Gay presents the key concepts from every section of the text with 10–15 minute mini-lectures. Students can easily review a section or a specific topic before a homework assignment, quiz, or test.
- Includes fully worked-out solutions to exercises marked with a CD Video icon (⊙) in each section. Also includes *Section 1.1, Tips for Success in Mathematics*.
- A close-captioned option for the hearing impaired is provided.

Online Homework and Tutorial Resources

MyMathLab *MyMathLab*

MyMathLab is a series of text specific, easily customizable, online courses for Prentice Hall textbooks in mathematics and statistics. MyMathLab is powered by Course Compass™—Pearson Education's online teaching and learning environment—and by MathXL®—our online homework, tutorial, and assessment system. MyMathLab gives instructors the tools they need to deliver all or a portion of their course online, whether students are in a lab setting or working from home. MyMathLab provides a rich and flexible set of course materials, featuring free-response exercises that are algorithmically generated for unlimited practice and mastery. Students can also use online tools, such as video lectures, animations, and a multimedia textbook, to independently improve their understanding and performance. MyMathLab is available to qualified adopters. For more information, visit our Web site at www.mymathlab.com or contact your Prentice Hall sales representative. (MyMathLab must be set up and assigned by your instructor.)

MathXL® **www.mathxl.com** *MathXL*

MathXL is a powerful online homework, tutorial, and assessment system that accompanies the text. With MathXL, instructors can create, edit, and assign online homework and tests using algorithmically generated exercises correlated to your textbook. All student work is tracked in MathXL's online gradebook. Students can take chapter tests in MathXL and receive personalized study plans based on their test results. The study plan diagnoses weaknesses and links students directly to tutorial exercises for the objectives they need to study and retest. Students can also access supplemental animations and video clips directly from selected exercises. MathXL is available to qualified adopters. For more information, visit our Web site at www.mathxl.com, or contact your Prentice Hall sales representative for a product demonstration. (MathXL must be set up and assigned by your instructor.)

Preface

Introductory Algebra, **Third Edition** was written to provide a solid foundation in algebra for students who might not have previous experience in algebra. Specific care was taken to make sure students have the most up-to-date relevant text preparation for their next mathematics course or for nonmathematical courses that require an understanding of algebraic fundamentals. I have tried to achieve this by writing a user-friendly text that is keyed to objectives and contains many worked-out examples. As suggested by AMATYC and the NCTM Standards (plus Addenda), real-life and real-data applications, data interpretation, conceptual understanding, problem solving, writing, cooperative learning, appropriate use of technology, mental mathematics, number sense, estimation, critical thinking, and geometric concepts are emphasized and integrated throughout the book.

The many factors that contributed to the success of the previous editions have been retained. In preparing the Third Edition, I considered comments and suggestions of colleagues, students, and many users of the prior edition throughout the country.

What's New in the Third Edition?

Enhanced Exercise Sets

- **NEW!** Three forms of mixed sections of exercises have been added to the Third Edition.
    - **Mixed Practice** exercises combining objectives within a section
    - **Mixed Practice** exercises combining previous sections
    - **Mixed Review** exercises included at the end of the Chapter Review

  These exercises require students to determine the problem type and strategy needed in order to solve it. In doing so, students need to think about key concepts to proceed with a correct method of solving—just as they would need to do on a test.

- **NEW! Concept Check exercises** have been added to the section exercise sets. These exercises are related to the Concept Check(s) found within the section. They help students measure their understanding of key concepts by focusing on common trouble areas. These exercises may ask students to identify a common error, and/or provide an explanation.

- **NEW! Concept Extensions** (formerly Combining Concepts) have been revised. These exercises extend the concepts and require students to combine several skills or concepts to solve the exercises in this section.

Increased Emphasis on Study Skills and Student Success

- **NEW! Study Skills Builders** (formerly Study Skill Reminders) Found at the end of many exercise sets, Study Skills Builders allow instructors to assign exercises that will help students improve their study skills and take responsibility for their part of the learning process. Study Skills Builders reinforce the material found in Section 1.1, "Tips for Success in Mathematics" and serve as an excellent tool for self-assessment.

- **NEW! The Bigger Picture** is a recurring feature, starting in Section 1.6, that focuses on the key concepts of the course—simplifying expressions and solving equations. Students develop an outline to help them be able to simplify expressions, solve equations, and to know the difference between the two. By working the exercises and developing this outline throughout the text, students can begin to transition from thinking "section by section" to thinking about how the mathematics in this course is part of the "bigger picture" of mathematics in general. A completed outline is provided in Appendix A so students have a model for their work.

- **NEW! Chapter Test Prep Video CD** provides students with help during their most "teachable moment"—while they are studying for a test. Included with every copy of the student edition of the text, this video CD provides fully worked-out solutions by the author to every exercise from each Chapter Test in the text. The easy video navigation allows students to instantly access the solutions to the exercises they want to review. The problems are solved by the author in the same manner as in the text.
- **NEW! Chapter Test files in TestGen** provide algorithms specific to each exercise from each Chapter Test in the text. Allows for easy replication of Chapter Tests with consistent, algorithmically generated problem types for additional assignments or assessment purposes.

Content Changes in the Third Edition

- Equivalent fractions and rational expressions are presented with more emphasis on multiplying by forms of 1. For example, see fractions on pages R-10, R-11, and rational expressions on pages 326, 327.
- There are additional exercises throughout the text with fractional and decimal coefficients. These help insure a student's full understanding of concepts. For example, Section 3.4, Exercises 9 through 12 and 27 through 30.
- All exercises have been reviewed and updated to ensure that even- and odd-numbered exercises are paired.
- Multiplying and dividing real numbers are now covered in a single Section 1.6. Simplifying expressions is now covered in Section 1.8.
- Material in sections on solving equations (2.1–2.3) has been reorganized for smoother progression and pacing.
- Percent and mixture applications are now covered in Section 2.6.
- Section 5.6 has been revised and now covers problem solving with proportions, problems about numbers, and work and distance problems.
- New Section 6.8, Direct and Inverse Variation.
- New Appendix B, Factoring Sums and Differences of Cubes.
- New Appendix D, Sets.

Key Pedagogical Features

The following key features have been retained and/or updated for the Third Edition of the text:

Problem Solving Process　　This is formally introduced in Chapter 2 with a four-step process that is integrated throughout the text. The four steps are **Understand, Translate, Solve,** and **Interpret.** The repeated use of these steps in a variety of examples shows their wide applicability. Reinforcing the steps can increase students' comfort level and confidence in tackling problems.

Exercise Sets Revised and Updated　　The exercise sets have been carefully examined and extensively revised. Special focus was placed on making sure that even- and odd-numbered exercises are paired.

Examples　　Detailed step-by-step examples were added, deleted, replaced, or updated as needed. Many of these reflect real life. Additional instructional support is provided in the annotated examples.

Practice Problems　　Throughout the text, each worked-out example has a parallel Practice Problem. These invite students to be actively involved in the learning process. Students should try each Practice Problem after finishing the corresponding example. Learning by doing will help students grasp ideas before moving on to other concepts. Answers to the Practice Problems are provided at the bottom of each page.

Helpful Hints　　Helpful Hints contain practical advice on applying mathematical concepts. Strategically placed where students are most likely to need immediate reinforcement, Helpful Hints help students avoid common trouble areas and mistakes.

Concept Checks   This feature allows students to gauge their grasp of an idea as it is being presented in the text. Concept Checks stress conceptual understanding at the point-of-use and help suppress misconceived notions before they start. Answers appear at the bottom of the page. Exercises related to Concept Checks are now included in the exercise sets.

Selected Solutions   Solutions to every-other odd exercise are included in the back of the text. This built-in solutions manual allows students to check their work.

Integrated Reviews   A unique, mid-chapter exercise set that helps students assimilate new skills and concepts that they have learned separately over several sections. These reviews provide yet another opportunity for students to work with "mixed" exercises as they master the topics.

Vocabulary Check   Provides an opportunity for students to become more familiar with the use of mathematical terms as they strengthen their verbal skills. These appear at the end of each chapter before the Chapter Highlights.

Chapter Highlights   Found at the end of every chapter, these contain key definitions and concepts with examples to help students understand and retain what they have learned and help them organize their notes and study for tests.

Chapter Review   The end of every chapter contains a comprehensive review of topics introduced in the chapter. The Chapter Review offers exercises keyed to every section in the chapter, as well as Mixed Review **(NEW!)** exercises that are not keyed to sections.

Chapter Test and Chapter Test Prep Video CD   The Chapter Test is structured to include those problems that involve common student errors. The **Chapter Test Prep Video CD** gives students instant author access to a step-by-step video solution of each exercise in the Chapter Test.

Cumulative Review   Follows every chapter in the text (except Chapters R and 1). Each odd-numbered exercise contained in the Cumulative Review is an earlier worked example in the text that is referenced in the back of the book along with the answer.

Mental Math   Found at the beginning of an exercise set, these mental warm-ups reinforce concepts found in the accompanying section and increase student's confidence before they tackle an exercise set.

Writing Exercises   These exercises occur in almost every exercise set and require students to provide a written response to explain concepts or justify their thinking.

Applications   Real-world and real-data applications have been thoroughly updated and many new applications are included. These exercises occur in almost every exercise set and show the relevance of mathematics and help students gradually, and continuously develop their problem solving skills.

Review Exercises (formerly Review and Preview exercises)   These exercises occur in each exercise set (except in Chapters R and 1) and are keyed to earlier sections. They review concepts learned earlier in the text that will be needed in the next section or chapter.

Exercise Set Resource Icons at the opening of each exercise set remind students of the resources available for extra practice and support:

See Student Resource descriptions pages xviii–xix for details on the individual resources available.

Exercise Icons   These icons facilitate the assignment of specialized exercises and let students know what resources can support them.

 CD Video icon: exercise worked on Martin-Gay's CD Lecture Series.
△ Triangle icon: identifies exercises involving geometric concepts.
✎ Pencil icon: indicates a written response is needed.
🖩 🖩 Calculator icons: optional exercises intended to be solved using a scientific or graphing calculator.

Group Activities   Found at the end of each chapter, these activities are for individual or group completion, and are usually hands-on or data-based activities that extend the concepts found in the chapter allowing students to make decisions and interpretations and to think and write about algebra.

Optional: Calculator Exploration Boxes and Calculator Exercises   The optional Calculator Explorations provide key strokes and exercises at appropriate points to provide an opportunity for students to become familiar with these tools. Section exercises that are best completed by using a calculator are identified by 🖩 or 🖩 for ease of assignment.

A Word about Textbook Design and Student Success

The design of developmental mathematics textbooks has become increasingly important. As students and instructors have told Prentice Hall in focus groups and market research surveys, these textbooks cannot look "cluttered" or "busy." A "busy" design can distract a student from what is most important in the text. It can also heighten math anxiety.

As a result of the conversations and meetings we have had with students and instructors, we concluded the design of this text should be understated and focused on the most important pedagogical elements. Students and instructors helped us to identify the primary elements that are central to student success. These primary elements include:

- Exercise Sets

- Examples and Practice Problems

- Helpful Hints

- Rules, Property, and Definition boxes

As you will notice in this text, these primary features are the most prominent elements in the design. We have made every attempt to make sure these elements are the features the eye is drawn to. The remaining features, the secondary elements in the design, blend into the "fabric" or "grain" of the overall design. These secondary elements complement the primary elements without becoming distractions.

Prentice Hall's thanks goes to all of the students and instructors (as noted by the author in Acknowledgments) who helped us develop the design of this text. At every step in the design process, their feedback proved valuable in helping us to make the right decisions. Thanks to your input, we're confident the design of this text will be both practical and engaging as it serves its educational and learning purposes.

Sincerely,

Paul Murphy

Executive Editor
Developmental Mathematics
Prentice Hall

Instructor and Student Resources

The following resources are available to help instructors and students use this text more effectively.

Instructor Resources

Annotated Instructor's Edition (0-13-186841-1)

- Answers to all exercises printed on the same text page
- Teaching Tips throughout the text placed at key points
- Includes Vocabulary Check at the beginning of relevant sections
- General tips and suggestions for classroom or group activities

Instructor Solutions Manual (0-13-227605-4)

- Solutions to the even- and odd-numbered exercises
- Solutions to every Mental Math exercise
- Solutions to every Practice Problem
- Solutions to every exercise in the Integrated Reviews, Chapter Reviews, Chapter Tests, and Cumulative Reviews

Instructor's Resource Manual with Tests (0-13-227603-8)

- **NEW!** Includes Mini-Lectures for every section from the text
- Group Activities
- Free Response Test Forms, Multiple Choice Test Forms, Cumulative Tests, and Additional Exercises
- Answers to all items

TestGen (0-13-173217-X)

- Enables instructors to build, edit, print, and administer tests
- Features a computerized bank of questions developed to cover all text objectives
- Available on dual-platform Windows/Macintosh CD-Rom

Instructor Adjunct Resource Kit (0-13-188753-X)

The Martin-Gay Instructor/Adjunct Resource Kit (IARK) contains tools and resources to help adjuncts and instructors succeed in the classroom. The IARK includes:

- Instructor-to-Instructor CD Videos that offer tips, suggestions, and strategies for engaging students and presenting key topics
- PDF files of the Instructor Solutions Manual and the Instructor's Resource Manual
- TestGen

MyMathLab Instructor Version (0-13-147898-2)
MyMathLab www.mymathlab.com

MyMathLab is a series of text specific, easily customizable, online courses for Prentice Hall textbooks in mathematics and statistics. MyMathLab is powered by Course Compass™—Pearson Education's online teaching and learning environment—and by MathXL®—our online homework, tutorial, and assessment system. MyMathLab gives instructors the tools they need to deliver all or a portion of their course online, whether students are in a lab setting or working from home. MyMathLab provides a rich and flexible set of course materials, featuring free-response exercises that are algorithmically generated for unlimited practice and mastery. Students can also use online tools, such as video lectures, animations, and a multimedia textbook, to independently improve their understanding and performance. Instructors can use

MyMathLab's homework and test managers to select and assign online exercises correlated directly to the text, and they can import TestGen tests into MyMathLab for added flexibility. MyMathLab's online gradebook—designed specifically for mathematics and statistics—automatically tracks students' homework and test results and gives the instructor control over how to calculate final grades. Instructors can also add offline (paper-and-pencil) grades to the gradebook. MyMathLab is available to qualified adopters. For more information, visit our website at www.mymathlab.com or contact your Prentice Hall sales representative.

MathXL Instructor Version (0-13-147895-8)
MathXL® www.mathxl.com

MathXL is a powerful online homework, tutorial, and assessment system that accompanies the text. With MathXL, instructors can create, edit, and assign online homework and tests using algorithmically generated exercises correlated to your textbook. All student work is tracked in MathXL's online gradebook. Students can take chapter tests in MathXL and receive personalized study plans based on their test results. The study plan diagnoses weaknesses and links students directly to tutorial exercises for the objectives they need to study and retest. Students can also access supplemental animations and video clips directly from selected exercises. MathXL is available to qualified adopters. For more information, visit our Web site at www.mathxl.com, or contact your Prentice Hall sales representative for a product demonstration.

Interact Math® Tutorial Web site www.interactmath.com

Get practice and tutorial help online! This interactive tutorial Web site provides algorithmically generated practice exercises that correlate directly to the exercises in your textbook. You can retry an exercise as many times as you like with new values each time for unlimited practice and mastery. Every exercise is accompanied by an interactive guided solution that gives you helpful feedback if you enter an incorrect answer, and you can also view a worked-out sample problem that steps you through an exercise similar to the one you're working on.

Student Resources

Student Solutions Manual (0-13-227606-2)

- Solutions to the odd-numbered section exercises
- Solutions to the Practice Problems
- Solutions to every Mental Math exercise
- Solutions to every exercise found in the Chapter Reviews and Chapter Tests

Martin-Gay's CD Lecture Series (0-13-173283-8)

- Perfect for review of a section or a specific topic, these mini-lectures by Elayn Martin-Gay cover the key concepts from each section of the text in approximately 10–15 minutes
- Includes fully worked-out solutions to exercises in each section marked with a ⊙
- Includes coverage of Section 1.1, "Tips for Success Mathematics"
- Closed-captioned for the hearing impaired

Prentice Hall Math Tutor Center (0-13-064604-0)

- Staffed by qualified math instructors who provide students with tutoring on examples and odd-numbered exercises from the textbook
- Tutoring is available via toll-free telephone, toll-free fax, e-mail, or the Internet
- Whiteboard technology allows tutors and students to see problems worked while they "talk" in real time over the Internet during tutoring sessions

Introductory Algebra, Third Edition *Student Study Pack (0-13-228613-9)*

The Student Study Pack includes:

- Martin-Gay's CD Lecture Series
- Student Solutions Manual
- Prentice Hall Math Tutor Center access code

Chapter Test Prep Video CD—Standalone (0-13-173277-3)

- Includes fully worked-out solutions to every problem from each Chapter Test in the text.

MathXL Tutorials on CD—Standalone (0-13-186840-3)

- Provides algorithmically generated practice exercises that correlate to exercises at the end of sections.
- Every exercise is accompanied by an example and a guided solution, selected exercises include a video clip.
- The software recognizes student errors and provides feedback. It can also generate printed summaries of students progress.

Interact Math® Tutorial Web Site www.interactmath.com

Get practice and tutorial help online! This interactive tutorial Web site provides algorithmically generated practice exercises that correlate directly to the exercises in your textbook. You can retry an exercise as many times as you like with new values each time for unlimited practice and mastery. Every exercise is accompanied by an interactive guided solution that gives you helpful feedback if you enter an incorrect answer, and you can also view a worked-out sample problem that steps you through an exercise similar to the one you're working on.

Acknowledgments

There are many people who helped me develop this text, and I will attempt to thank some of them here. Cindy Trimble was *invaluable* for contributing to the overall accuracy of the text. Suellen Robinson, Gail Burket, Laura Wheel, and Lori Mancuso were *invaluable* for their many suggestions and contributions during the development and writing of this Third Edition. Ingrid Benson provided guidance throughout the production process. Cheryl Cantwell's contributions throughout production were invaluable.

A special thanks to my editor, Paul Murphy, for all of his assistance, support, and contributions to this project. A very special thank you goes to my project manager, Mary Beckwith, for being there 24/7/365, as my students say. Last, my thanks to the staff at Prentice Hall for all their support: Linda Behrens, Alan Fischer, Patty Burns, Tom Benfatti, Paul Belfanti, Maureen Eide, Suzanne Behnke, Kate Valentine, Patrice Jones, Chris Hoag, Paul Corey, and Tim Bozik.

I would like to thank the following reviewers for their input and suggestions:

Sheila Anderson, *Housatonic Community College*
Tom Blackburn, *Northeastern Illinois University*
Gail Burket, *Palm Beach Community College*
James Butterbach, *Joliet Junior College*
Laura Dyer, *South West Illinois College*
Sharon Edgemon, *Bakersfield College*
Hope Essien, *Olive-Harvey College*
Randa Kress, *Idaho State University*
Ted Lai, *Hudson Community College*
Nicole Lang, *North Hennepin Community College*
Lee LaRue, *Paris Junior College*

Jeri Lee, *Des Moines Area Community College*

Jean McArthur, *Joliet Junior College*

Michael Montano, *Riverside Community College*

Lisa J. Music, *Big Sandy Community and Technical College*

Linda Padilla, *Joliet Junior College*

Scott Perkins, *Lake Sumter Community College*

Marilyn Platt, *Gaston College*

Sandy Spears, *Jefferson Community College*

Ping Charlene Tintera, *Texas A & M University*

Jane Wampler, *Housatonic Community College*

Carol Williams, *Des Moines Area Community College*

Peter Zimmer, *West Cheseter University*

I would also like to thank the following dedicated group of instructors who participated in our focus groups, Martin-Gay Summits, and our design review for this edition of the text. Their feedback and insights have helped to strengthen this edition of the text. These instructors include:

Cedric Atkins, *Mott Community College*

Laurel Berry, *Bryant & Stratton*

Bob Brown, *Community College of Baltimore County–Essex*

Lisa Brown, *Community College of Baltimore County–Essex*

Gail Burkett, *Palm Beach Community College*

Cheryl Cantwell, *Seminole Community College*

Jackie Cohen, *Augusta State College*

Janice Ervin, *Central Piedmont Community College*

Pauline Hall, *Iowa State College*

Sonya Johnson, *Central Piedmont Community College*

Irene Jones, *Fullerton College*

Nancy Lange, *Inver Hills Community College*

Jean McArthur, *Joliet Junior College*

Marica Molle, *Metropolitan Community College*

Linda Padilla, *Joliet Junior College*

Carole Shapero, *Oakton Community College*

Jennifer Strehler, *Oakton Community College*

Tanomo Taguchi, *Fullerton College*

Leigh Ann Wheeler, *Greenville Technical Community College*

Valerie Wright, *Central Piedmont Community College*

A special thank you to those students who participated in our design review: Katherine Browne, Mike Bulfin, Nancy Canipe, Ashley Carpenter, Jeff Chojnachi, Roxanne Davis, Mike Dieter, Amy Dombrowski, Kay Herring, Todd Jaycox, Kaleena Levan, Matt Montgomery, Tony Plese, Abigail Polkinghorn, Harley Price, Eli Robinson, Avery Rosen, Robyn Schott, Cynthia Thomas, and Sherry Ward.

Additional Acknowledgments

As usual, I would like to thank my husband, Clayton, for his constant encouragement. I would also like to thank my children, Eric and Bryan, for providing most of the cooking and humor in our household. I would also like to thank my extended family for their help and wonderful sense of humor. Their contributions are too numerous to list. They are Rod and Karen Pasch; Peter, Michael, Christopher, Matthew, and Jessica Callac; Stuart and Earline Martin; Josh, Mandy, Bailey, Ethan, and Avery Barnes; Mark, Sabrina, and Madison Martin; Leo and Barbara Miller; and Jewett Gay.

Elayn Martin-Gay

About the Author

Elayn Martin-Gay has taught mathematics at the University of New Orleans for more than 25 years. Her numerous teaching awards include the local University Alumni Association's Award for Excellence in Teaching, and Outstanding Developmental Educator at University of New Orleans, presented by the Louisiana Association of Developmental Educators.

Prior to writing textbooks, Elayn Martin-Gay developed an acclaimed series of lecture videos to support developmental mathematics students in their quest for success. These highly successful videos originally served as the foundation material for her texts. Today, the videos are specific to each book in the Martin-Gay series. The author has also created Chapter Test Prep Videos to help students during their most "teachable moment"—as they prepare for a test, along with Instructor-to-Instructor videos that provide teaching tips, hints, and suggestions for each developmental mathematics course, including basic mathematics, prealgebra, beginning algebra, and intermediate algebra.

Elayn is the author of 10 published textbooks as well as multimedia interactive mathematics, all specializing in developmental mathematics courses. She has participated as an author across the broadest range of educational materials: textbooks, videos, tutorial software, and Interactive Math courseware. All of these components are designed to work together. This offers an opportunity of various combinations for an integrated teaching and learning package offering great consistency for the student.

Applications Index

R

Prealgebra Review

This optional review chapter covers basic topics and skills from prealgebra, such as fractions, decimals, and percents. Knowledge of these topics is needed for success in algebra.

Federated is one of the largest department store retailers that you have probably never heard of. The year 2004 marked the seventy-fifth anniversary of this conglomerate, which began with the uniting of the retailing giants who agreed to link their financial interests while maintaining their separate identities. Notice on the bar graph how many of the best-known department stores in this country belong to Federated. In Exercise 89, Exercise Set R.2, you will explore the breakdown of Federated by its franchise companies.

Federated Department Stores

R.1 FACTORS AND THE LEAST COMMON MULTIPLE

Objective **A** Factoring Numbers

In arithmetic we factor numbers, and in algebra we factor expressions containing variables.

To **factor** means to write as a product.

Throughout this text, you will encounter the word *factor* often. Always remember that factoring means writing as a product.

Since $2 \cdot 3 = 6$, we say that 2 and 3 are **factors** of 6. Also, $2 \cdot 3$ is a **factorization** of 6.

PRACTICE PROBLEM 1

List the factors of 10.

EXAMPLE 1 List the factors of 6.

Solution: First we write the different factorizations of 6.

$$6 = 1 \cdot 6, \quad 6 = 2 \cdot 3$$

The factors of 6 are 1, 2, 3, and 6.

▪ **Work Practice Problem 1**

PRACTICE PROBLEM 2

List the factors of 18.

EXAMPLE 2 List the factors of 20.

Solution: $20 = 1 \cdot 20, \quad 20 = 2 \cdot 10, \quad 20 = 4 \cdot 5$

The factors of 20 are 1, 2, 4, 5, 10, and 20.

▪ **Work Practice Problem 2**

In this section, we will concentrate on **natural numbers** only. The natural numbers (also called counting numbers) are

Natural Numbers: 1, 2, 3, 4, 5, 6, 7, and so on

Every natural number except 1 is either a prime number or a composite number.

Prime and Composite Numbers

A **prime number** is a natural number greater than 1 whose only factors are 1 and itself. The first few prime numbers are 2, 3, 5, 7, 11, 13, 17, ... A **composite number** is a natural number greater than 1 that is not prime.

PRACTICE PROBLEM 3

Identify each number as prime or composite: 5, 16, 23, 42.

EXAMPLE 3 Identify each number as prime or composite: 3, 24, 19, 35

Solution:

3 is a prime number. Its factors are 1 and 3 only.
24 is a composite number. Its factors are 1, 2, 3, 4, 6, 8, 12, and 24.
19 is a prime number. Its factors are 1 and 19 only.
35 is a composite number. Its factors are 1, 5, 7, and 35.

▪ **Work Practice Problem 3**

Answers

1. 1, 2, 5, 10, **2.** 1, 2, 3, 6, 9, 18,
3. 5, 23 prime; 16, 42 composite

Objective B Writing Prime Factorizations

When a number is written as a product of primes, this product is called the **prime factorization** of the number. For example, the prime factorization of 12 is $2 \cdot 2 \cdot 3$ since

$$12 = 2 \cdot 2 \cdot 3$$

and all the factors are prime.

EXAMPLE 4 Write the prime factorization of 45.

Solution: We can begin by writing 45 as the product of two numbers, say 9 and 5.

$$45 = 9 \cdot 5$$

The number 5 is prime, but 9 is not. So we write 9 as $3 \cdot 3$.

$$45 = 9 \cdot 5$$
$$= 3 \cdot 3 \cdot 5$$

Each factor is now a prime number, so the prime factorization of 45 is $3 \cdot 3 \cdot 5$.

Work Practice Problem 4

PRACTICE PROBLEM 4

Write the prime factorization of 44.

Helpful Hint

Recall that order is not important when multiplying numbers. For example,

$$3 \cdot 3 \cdot 5 = 3 \cdot 5 \cdot 3 = 5 \cdot 3 \cdot 3 = 45$$

For this reason, any of the products shown can be called *the* prime factorization of 45, and we say that the prime factorization of a number is unique.

EXAMPLE 5 Write the prime factorization of 80.

Solution: We first write 80 as a product of two numbers. We continue this process until all factors are prime.

$$80 = 8 \cdot 10$$
$$4 \cdot 2 \cdot 2 \cdot 5$$
$$= 2 \cdot 2 \cdot 2 \cdot 2 \cdot 5$$

All factors are now prime, so the prime factorization of 80 is

$$2 \cdot 2 \cdot 2 \cdot 2 \cdot 5.$$

Work Practice Problem 5

PRACTICE PROBLEM 5

Write the prime factorization of 60.

✔ **Concept Check** Suppose that you choose $80 = 4 \cdot 20$ as your first step in Example 5 and another student chooses $80 = 5 \cdot 16$. Will you both end up with the same prime factorization as in Example 5? Explain.

Answers
4. $44 = 2 \cdot 2 \cdot 11$, 5. $60 = 2 \cdot 2 \cdot 3 \cdot 5$

✔ **Concept Check Answer**
yes; answers may vary

Helpful Hint

There are a few quick **divisibility tests** to determine if a number is divisible by the primes 2, 3, or 5.

A whole number is divisible by

- **2** if the ones digit is 0, 2, 4, 6, or 8.

132 is divisible by 2
- **3** if the sum of the digits is divisible by 3.

144 is divisible by 3 since $1 + 4 + 4 = 9$ is divisible by 3
- **5** if the ones digit is 0 or 5.

1115 is divisible by 5

When finding the prime factorization of larger numbers, you may want to use the procedure shown in Example 6.

PRACTICE PROBLEM 6

Write the prime factorization of 297.

EXAMPLE 6 Write the prime factorization of 252.

Solution: Since the ones digit of 252 is 2, we know that 252 is divisible by 2.

$$2\overline{)252}$$
$$126$$

126 is divisible by 2 also.

$$2\overline{)252}$$
$$2\overline{)126}$$
$$63$$

63 is not divisible by 2 but is divisible by 3. We divide 63 by 3 and continue in this same manner until the quotient is a prime number.

$$2\overline{)252}$$
$$2\overline{)126}$$
$$3\overline{)63}$$
$$3\overline{)21}$$
$$7$$

The prime factorization of 252 is $2 \cdot 2 \cdot 3 \cdot 3 \cdot 7$.

📙 **Work Practice Problem 6**

Objective C Finding the Least Common Multiple

A **multiple** of a number is the product of that number and any natural number. For example, the multiples of 3 are

$$\underset{3,}{3 \cdot 1} \quad \underset{6,}{3 \cdot 2} \quad \underset{9,}{3 \cdot 3} \quad \underset{12,}{3 \cdot 4} \quad \underset{15,}{3 \cdot 5} \quad \underset{18,}{3 \cdot 6} \quad \underset{21,}{3 \cdot 7} \quad \text{and so on.}$$

The multiples of 2 are

$$\underset{2,}{2 \cdot 1} \quad \underset{4,}{2 \cdot 2} \quad \underset{6,}{2 \cdot 3} \quad \underset{8,}{2 \cdot 4} \quad \underset{10,}{2 \cdot 5} \quad \underset{12,}{2 \cdot 6} \quad \underset{14,}{2 \cdot 7} \quad \text{and so on.}$$

Answer

6. $3 \cdot 3 \cdot 3 \cdot 11$

Notice that 2 and 3 have multiples that are common to both.

Multiples of 2:   2, 4, 6, 8, 10, 12, 14, 16, 18, and so on

Multiples of 3:   3, 6, 9, 12, 15, 18, 21, and so on

Common multiples of 2 and 3: 6, 12, 18, . . .

The least or smallest common multiple of 2 and 3 is 6. The number 6 is called the **least common multiple** or **LCM** of 2 and 3. It is the smallest number that is a multiple of both 2 and 3.

The **least common multiple (LCM)** of a list of numbers is the smallest number that is a multiple of all the numbers in the list.

Finding the LCM by the method above can sometimes be time-consuming. Let's look at another method that uses prime factorization.

To find the LCM of 4 and 10, for example, we write the prime factorization of each.

$$4 = 2 \cdot 2$$
$$10 = 2 \cdot 5$$

If the LCM is to be a multiple of 4, it must contain the factors $2 \cdot 2$. If the LCM is to be a multiple of 10, it must contain the factors $2 \cdot 5$. Since we decide whether the LCM is a multiple of 4 and 10 separately, the LCM does not need to contain three factors of 2. The LCM only needs to contain a factor the greatest number of times that the factor appears in any **one** prime factorization.

The LCM is a
multiple of 4.
The number 2 is a factor twice since that is the
$$LCM = \overbrace{2 \cdot 2} \cdot 5 = 20 \quad \text{greatest number of times that 2 is a factor in either of the prime factorizations.}$$
The LCM is a
multiple of 10.

To Find the LCM of a List of Numbers

Step 1: Write the prime factorization of each number.

Step 2: Write the product containing each different prime factor (from Step 1) the greatest number of times that it appears in any one factorization. This product is the LCM.

EXAMPLE 7 Find the LCM of 18 and 24.

Solution: First we write the prime factorization of each number.

$$18 = 2 \cdot 3 \cdot 3$$
$$24 = 2 \cdot 2 \cdot 2 \cdot 3$$

Now we write each factor the greatest number of times that it appears in any **one** prime factorization.

The greatest number of times that 2 appears is **3** times.
The greatest number of times that 3 appears is **2** times.

$$LCM = \underbrace{2 \cdot 2 \cdot 2}_{\substack{2 \text{ is a factor} \\ 3 \text{ times.}}} \cdot \underbrace{3 \cdot 3}_{\substack{3 \text{ is a factor} \\ 2 \text{ times.}}} = 72$$

□ **Work Practice Problem 7**

PRACTICE PROBLEM 7

Find the LCM of 14 and 35.

Answer
7. 70

PRACTICE PROBLEM 8

Find the LCM of 5 and 9.

EXAMPLE 8 Find the LCM of 11 and 10.

Solution: 11 is a prime number, so we simply rewrite it. Then we write the prime factorization of 10.

$$11 = 11$$
$$10 = 2 \cdot 5$$
$$\text{LCM} = 2 \cdot 5 \cdot 11 = 110$$

☐ **Work Practice Problem 8**

PRACTICE PROBLEM 9

Find the LCM of 4, 15, and 10.

EXAMPLE 9 Find the LCM of 5, 6, and 12.

Solution:

$$5 = 5$$
$$6 = 2 \cdot 3$$
$$12 = 2 \cdot 2 \cdot 3$$
$$\text{LCM} = 2 \cdot 2 \cdot 3 \cdot 5 = 60.$$

☐ **Work Practice Problem 9**

Answers

8. 45, **9.** 60

Objective **A** *List the factors of each number. See Examples 1 and 2.*

1. 9   **2.** 8   **3.** 24   **4.** 36   **5.** 42

6. 63   **7.** 80   **8.** 50   **9.** 19   **10.** 31

Identify each number as prime or composite. See Example 3.

11. 13   **12.** 21   **13.** 39   **14.** 53   **15.** 41

16. 51   **17.** 201   **18.** 307   **19.** 2065   **20.** 1798

Objective **B** *Write each prime factorization. See Examples 4 through 6.*

21. 18   **22.** 28   **23.** 20   **24.** 30

25. 56   **26.** 48   **27.** 81   **28.** 64

29. 300   **30.** 500   **31.** 588   **32.** 315

Multiple choice. Select the best choice to complete each statement.

33. The factors of 48 are
  a. $2 \cdot 2 \cdot 2 \cdot 6$
  b. $2 \cdot 2 \cdot 2 \cdot 3$
  c. $2 \cdot 2 \cdot 2 \cdot 2 \cdot 3$
  d. 1, 2, 3, 4, 6, 8, 12, 16, 24, 48

34. The prime factorization of 63 is
  a. 1, 3, 7, 9, 63
  b. 1, 3, 7, 9, 21, 63
  c. $3 \cdot 3 \cdot 7$
  d. 1, 3, 21, 63

Objective **C** *Find the LCM of each list of numbers. See Examples 7 through 9.*

35. 3, 4   **36.** 4, 5   **37.** 6, 14   **38.** 9, 15

39. 20, 30   **40.** 30, 40   **41.** 5, 7   **42.** 2, 11

43. 9, 12   **44.** 4, 18   **45.** 16, 20   **46.** 18, 30

47. 40, 90   **48.** 50, 70   **49.** 24, 36   **50.** 21, 28

51. 2, 8, 15   **52.** 3, 9, 20   **53.** 2, 3, 7   **54.** 3, 5, 7

55. 8, 24, 48   **56.** 9, 36, 72   **57.** 8, 18, 30   **58.** 4, 14, 35

Concept Extensions

59. Solve. See the concept check in the section.
   a. Write the prime factorization of 40 using 2 and 20 as the first pair of factors.
   b. Write the prime factorization of 40 using 4 and 10 as the first pair of factors.
   c. Explain any similarities or differences found in parts a and b.

60. The LCM of 6 and 7 is 42. In general, describe when the LCM of two numbers is equal to their product.

61. Is the following statement true or false? The number 311 is a prime number.

62. Craig Campanella and Edie Hall both have night jobs. Craig has every fifth night off and Edie has every seventh night off. How often will they have the same night off?

63. Elizabeth Kaster and Lori Sypher are both publishing company representatives in Louisiana. Elizabeth spends a day in New Orleans every 35 days, and Lori spends a day in New Orleans every 20 days. How often are they in New Orleans on the same day?

Find the LCM of each pair of numbers.

64. 315, 504

65. 1000, 1125

R.2 FRACTIONS

Objectives

A Discover Fraction Properties Having to do with 0 and 1.

B Write Equivalent Fractions.

C Write Fractions in Simplest Form.

D Multiply and Divide Fractions.

E Add and Subtract Fractions.

F Perform Operations on Mixed Numbers.

A quotient of two numbers such as $\frac{2}{9}$ is called a **fraction.** The parts of a fraction are:

Fraction bar → $\frac{2 \; ←\text{Numerator}}{9 \; ←\text{Denominator}}$

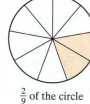

$\frac{2}{9}$ of the circle is shaded.

A fraction may be used to refer to part of a whole. For example, $\frac{2}{9}$ of the circle is shaded. The denominator 9 tells us how many equal parts the whole circle is divided into, and the numerator 2 tells us how many equal parts are shaded.

In this section, we will use numerators that are **whole numbers** and denominators that are nonzero whole numbers. The whole numbers consist of 0 and the natural numbers.

Whole Numbers: 0, 1, 2, 3, 4, 5, and so on

Objective **A** Discovering Fraction Properties with 0 and 1

Before we continue further, don't forget that the fraction bar indicates division. For example,

$$\frac{8}{4} = 8 \div 4 = 2 \quad \text{since} \quad 2 \cdot 4 = 8$$

Thus, we may simplify some fractions by recalling that the fraction bar means division.

$$\frac{6}{6} = 6 \div 6 = 1 \quad \text{and} \quad \frac{3}{1} = 3 \div 1 = 3$$

EXAMPLES Simplify by dividing the numerator by the denominator.

1. $\frac{3}{3} = 1$ Since $3 \div 3 = 1$.

2. $\frac{4}{2} = 2$ Since $4 \div 2 = 2$.

3. $\frac{7}{7} = 1$ Since $7 \div 7 = 1$.

4. $\frac{8}{1} = 8$ Since $8 \div 1 = 8$.

5. $\frac{0}{6} = 0$ Since $0 \cdot 6 = 0$.

6. $\frac{6}{0}$ is undefined because there is no number that when multiplied by 0 gives 6.

🔲 **Work Practice Problems 1–6**

PRACTICE PROBLEMS 1–6

Simplify by dividing the numerator by the denominator.

1. $\frac{4}{4}$ **2.** $\frac{9}{3}$ **3.** $\frac{10}{10}$

4. $\frac{5}{1}$ **5.** $\frac{0}{11}$ **6.** $\frac{11}{0}$

Answers
1. 1, **2.** 3, **3.** 1, **4.** 5, **5.** 0,
6. undefined

From Examples 1 through 6, we can say the following:

Let a be any number other than 0.

$$\frac{a}{a} = 1, \qquad \frac{0}{a} = 0,$$

$$\frac{a}{1} = a, \qquad \frac{a}{0} \text{ is undefined}$$

Objective B Writing Equivalent Fractions

More than one fraction can be used to name the same part of a whole. Such fractions are called **equivalent fractions.**

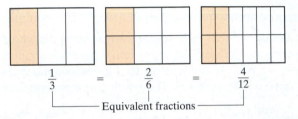

$$\frac{1}{3} = \frac{2}{6} = \frac{4}{12}$$

Equivalent fractions

Equivalent Fractions

Fractions that represent the same portion of a whole are called **equivalent fractions.**

For example, let's write $\frac{1}{3}$ as an equivalent fraction with a denominator of 12. To do so, notice the denominator of 3 multiplied by 4, gives a denominator of 12. Thus let's multiply by 1 in the form of $\frac{4}{4}$.

$$\frac{1}{3} = \frac{1}{3} \cdot 1 = \frac{1}{3} \cdot \frac{4}{4} = \frac{1 \cdot 4}{3 \cdot 4} = \frac{4}{12}$$

$$\frac{4}{4} = 1$$

So $\frac{1}{3} = \frac{4}{12}$.

To Write an Equivalent Fraction

$$\frac{a}{b} = \frac{a}{b} \cdot \frac{c}{c} = \frac{a \cdot c}{b \cdot c}$$

Since $\frac{a}{b} = \frac{a}{b} \cdot 1$

where a, b, and c are nonzero numbers.

EXAMPLE 7 Write $\dfrac{2}{5}$ as an equivalent fraction with a denominator of 15.

Solution: In the denominator, since $5 \cdot 3 = 15$, we multiply the fraction $\dfrac{2}{5}$ by 1 in the form of $\dfrac{3}{3}$.

$$\frac{2}{5} = \frac{2}{5} \cdot \frac{3}{3} = \frac{2 \cdot 3}{5 \cdot 3} = \frac{6}{15}$$

Then $\dfrac{2}{5}$ is equivalent to $\dfrac{6}{15}$. They both represent the same part of a whole.

📖 **Work Practice Problem 7**

Objective **C** Simplifying Fractions

A special equivalent fraction is one that is simplified or in lowest terms. A fraction is said to be **simplified** or in **lowest terms** when the numerator and the denominator have no factors in common other than 1. For example, the fraction $\dfrac{5}{11}$ is in lowest terms since 5 and 11 have no common factors other than 1.

To simplify a fraction, we write an equivalent fraction, but one with no common factors in the numerator and denominator. Since we are writing an equivalent fraction, we use the same method as before, except we are "removing" factors of 1 instead of "inserting" factors of 1.

> ### To Write a Simplified, Equivalent Fraction
>
> $$\frac{a \cdot c}{b \cdot c} = \frac{a}{b} \cdot \frac{c}{c} = \frac{a}{b}$$
>
> Since $\dfrac{a}{b} \cdot 1 = \dfrac{a}{b}$

EXAMPLE 8 Simplify: $\dfrac{42}{49}$

Solution: To help us see common factors in the numerator and denominator, or factors of 1, we write the numerator and the denominator as products of primes.

$$\frac{42}{49} = \frac{2 \cdot 3 \cdot 7}{7 \cdot 7} = \frac{2 \cdot 3}{7} \cdot \frac{7}{7} = \frac{2 \cdot 3}{7} = \frac{6}{7}$$

📖 **Work Practice Problem 8**

✔ **Concept Check** Explain the error in the following steps.

a. $\dfrac{15}{55} = \dfrac{1\cancel{5}}{5\cancel{5}} = \dfrac{1}{5}$ **b.** $\dfrac{6}{7} = \dfrac{5 + 1}{5 + 2} = \dfrac{1}{2}$

EXAMPLES Simplify each fraction.

9. $\dfrac{11}{27} = \dfrac{11}{3 \cdot 3 \cdot 3}$ There are no common factors in the numerator and denominator other than 1, so $\dfrac{11}{27}$ is already simplified.

10. $\dfrac{88}{20} = \dfrac{2 \cdot 2 \cdot 2 \cdot 11}{2 \cdot 2 \cdot 5} = \dfrac{2}{2} \cdot \dfrac{2}{2} \cdot \dfrac{2 \cdot 11}{5} = \dfrac{22}{5}$

📖 **Work Practice Problems 9–10**

PRACTICE PROBLEM 7

Write $\dfrac{1}{4}$ as an equivalent fraction with a denominator of 20.

PRACTICE PROBLEM 8

Simplify: $\dfrac{20}{35}$

PRACTICE PROBLEMS 9–10

Simplify each fraction.

9. $\dfrac{7}{20}$ **10.** $\dfrac{12}{40}$

Answers

7. $\dfrac{5}{20}$, **8.** $\dfrac{4}{7}$, **9.** $\dfrac{7}{20}$, **10.** $\dfrac{3}{10}$

✔ **Concept Check Answer**
answers may vary

Below are two important notes about simplifying fractions.

Note 1: When simplifying, we can use a shortcut notation if desired. From Example 8,

$$\frac{42}{49} = \frac{2\cdot3\cdot\overset{1}{\cancel{7}}}{7\cdot\underset{1}{\cancel{7}}} = \frac{2\cdot3}{7} = \frac{6}{7}$$

Note 2: Also, feel free to save time if you immediately notice common factors. In Example 10, notice that the numerator and denominator of $\frac{88}{20}$ have a common factor of 4.

$$\frac{88}{20} = \frac{\overset{1}{\cancel{4}}\cdot22}{\underset{1}{\cancel{4}}\cdot5} = \frac{22}{5}$$

A **proper fraction** is a fraction whose numerator is less than its denominator. The fraction $\frac{22}{5}$ from Example 10 is called an improper fraction. An **improper fraction** is a fraction whose numerator is greater than or equal to its denominator.

The improper fraction $\frac{22}{5}$ may be written as the mixed number $4\frac{2}{5}$. Notice that a **mixed number** has a whole number part and a fraction part. We review operations on mixed numbers in objective F in this section. First, let's review operations on fractions.

Objective D Multiplying and Dividing Fractions

To multiply two fractions, we multiply numerator times numerator to obtain the numerator of the product. Then we multiply denominator times denominator to obtain the denominator of the product.

Multiplying Fractions

$$\frac{a}{b}\cdot\frac{c}{d} = \frac{a\cdot c}{b\cdot d}\quad\text{if } b\neq0 \text{ and } d\neq0$$

EXAMPLE 11 Multiply: $\frac{2}{15}\cdot\frac{5}{13}$. Simplify the product if possible.

Solution: $\frac{2}{15}\cdot\frac{5}{13} = \frac{2\cdot5}{15\cdot13}$ Multiply numerators. Multiply denominators.

To simplify the product, we divide the numerator and the denominator by any common factors.

$$\frac{2}{15}\cdot\frac{5}{13} = \frac{2\cdot\overset{1}{\cancel{5}}}{3\cdot\underset{1}{\cancel{5}}\cdot13} = \frac{2}{39}$$

Work Practice Problem 11

Before we divide fractions, we first define **reciprocals.** Two numbers are reciprocals of each other if their product is 1.

The reciprocal of $\frac{2}{3}$ is $\frac{3}{2}$ because $\frac{2}{3}\cdot\frac{3}{2} = \frac{6}{6} = 1$.

The reciprocal of 5 is $\frac{1}{5}$ because $5\cdot\frac{1}{5} = \frac{5}{1}\cdot\frac{1}{5} = \frac{5}{5} = 1$.

Helpful Hint
The symbol " $\neq$ " to the right means "is not equal to."

PRACTICE PROBLEM 11

Multiply: $\frac{3}{4}\cdot\frac{8}{9}$. Simplify the product if possible.

Answer
11. $\frac{2}{3}$

To divide fractions, we multiply the first fraction by the reciprocal of the second fraction. For example,

$$\frac{1}{2} \div \underbrace{\frac{5}{7}} = \frac{1}{2} \cdot \overset{\uparrow}{\frac{7}{5}} = \frac{1 \cdot 7}{2 \cdot 5} = \frac{7}{10}$$

To divide, multiply by the reciprocal.

Dividing Fractions

$$\frac{a}{b} \div \frac{c}{d} = \frac{a}{b} \cdot \frac{d}{c}, \qquad \text{if } b \neq 0, d \neq 0, \text{ and } c \neq 0$$

EXAMPLES Divide and simplify.

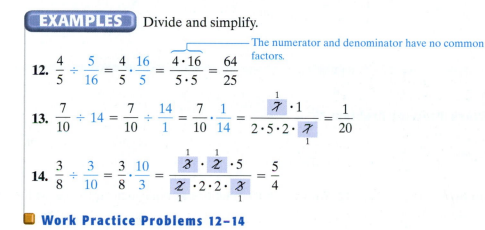

The numerator and denominator have no common factors.

12. $\dfrac{4}{5} \div \dfrac{5}{16} = \dfrac{4}{5} \cdot \dfrac{16}{5} = \dfrac{4 \cdot 16}{5 \cdot 5} = \dfrac{64}{25}$

13. $\dfrac{7}{10} \div 14 = \dfrac{7}{10} \div \dfrac{14}{1} = \dfrac{7}{10} \cdot \dfrac{1}{14} = \dfrac{\overset{1}{\cancel{7}} \cdot 1}{2 \cdot 5 \cdot 2 \cdot \underset{1}{\cancel{7}}} = \dfrac{1}{20}$

14. $\dfrac{3}{8} \div \dfrac{3}{10} = \dfrac{3}{8} \cdot \dfrac{10}{3} = \dfrac{\overset{1}{\cancel{3}} \cdot \overset{1}{\cancel{2}} \cdot 5}{\underset{1}{\cancel{2}} \cdot 2 \cdot 2 \cdot \underset{1}{\cancel{3}}} = \dfrac{5}{4}$

🔲 **Work Practice Problems 12–14**

Objective E Adding and Subtracting Fractions

To add or subtract fractions with the same denominator, we combine numerators and place the sum or difference over the common denominator.

Adding and Subtracting Fractions with the Same Denominator

$$\frac{a}{b} + \frac{c}{b} = \frac{a + c}{b}, \qquad \text{if } b \neq 0$$

$$\frac{a}{b} - \frac{c}{b} = \frac{a - c}{b}, \qquad \text{if } b \neq 0$$

EXAMPLES Add or subtract as indicated. Then simplify if possible.

15. $\dfrac{2}{7} + \dfrac{4}{7} = \dfrac{2 + 4}{7} = \dfrac{6}{7}$ ← Add numerators.
← Keep the common denominator.

16. $\dfrac{3}{10} + \dfrac{2}{10} = \dfrac{3 + 2}{10} = \dfrac{5}{10} = \dfrac{\overset{1}{\cancel{5}}}{2 \cdot \underset{1}{\cancel{5}}} = \dfrac{1}{2}$

17. $\dfrac{5}{3} - \dfrac{1}{3} = \dfrac{5 - 1}{3} = \dfrac{4}{3}$ ← Subtract numerators.
← Keep the common denominator.

18. $\dfrac{9}{7} - \dfrac{2}{7} = \dfrac{9 - 2}{7} = \dfrac{7}{7} = 1$ Subtract numerators.

🔲 **Work Practice Problems 15–18**

PRACTICE PROBLEMS 12–14

Divide and simplify.

12. $\dfrac{2}{9} \div \dfrac{3}{4}$

13. $\dfrac{8}{11} \div 24$

14. $\dfrac{5}{4} \div \dfrac{15}{8}$

PRACTICE PROBLEMS 15–18

Add or subtract as indicated. Then simplify if possible.

15. $\dfrac{2}{11} + \dfrac{5}{11}$ **16.** $\dfrac{1}{8} + \dfrac{3}{8}$

17. $\dfrac{7}{6} - \dfrac{2}{6}$ **18.** $\dfrac{13}{10} - \dfrac{3}{10}$

Answers

12. $\dfrac{8}{27}$, **13.** $\dfrac{1}{33}$, **14.** $\dfrac{2}{3}$, **15.** $\dfrac{7}{11}$,

16. $\dfrac{1}{2}$, **17.** $\dfrac{5}{6}$, **18.** 1

To add or subtract with different denominators, we first write the fractions as **equivalent fractions** with the same denominator. We use the smallest or **least common denominator,** or **LCD.** The LCD is the same as the least common multiple of the denominators (see Section R.1).

PRACTICE PROBLEM 19

Add: $\dfrac{3}{8} + \dfrac{1}{20}$

EXAMPLE 19 Add: $\dfrac{2}{5} + \dfrac{1}{4}$

Solution: We first must find the least common denominator before the fractions can be added. The least common multiple of the denominators 5 and 4 is 20. This is the LCD we will use.

We write both fractions as equivalent fractions with denominators of 20. Since

$$\frac{2}{5} = \frac{2}{5} \cdot 1 = \frac{2}{5} \cdot \frac{4}{4} = \frac{2 \cdot 4}{5 \cdot 4} = \frac{8}{20} \quad \text{and} \quad \frac{1}{4} = \frac{1}{4} \cdot 1 = \frac{1}{4} \cdot \frac{5}{5} = \frac{1 \cdot 5}{4 \cdot 5} = \frac{5}{20}$$

then

$$\frac{2}{5} + \frac{1}{4} = \frac{8}{20} + \frac{5}{20} = \frac{13}{20}$$

🔲 **Work Practice Problem 19**

PRACTICE PROBLEM 20

Subtract and simplify: $\dfrac{8}{15} - \dfrac{1}{3}$

EXAMPLE 20 Subtract and simplify: $\dfrac{19}{6} - \dfrac{23}{12}$

Solution: The LCD is 12. We write both fractions as equivalent fractions with denominators of 12.

$$\frac{19}{6} - \frac{23}{12} = \frac{19}{6} \cdot \frac{2}{2} - \frac{23}{12}$$

$$= \frac{19 \cdot 2}{6 \cdot 2} - \frac{23}{12}$$

$$= \frac{38}{12} - \frac{23}{12}$$

$$= \frac{15}{12} = \frac{\overset{1}{\cancel{3}} \cdot 5}{2 \cdot 2 \cdot \underset{1}{\cancel{3}}} = \frac{5}{4}$$

🔲 **Work Practice Problem 20**

Objective 🇫 Performing Operations on Mixed Numbers

To perform operations on mixed numbers, first write each mixed number as an improper fraction. To recall how this is done, let's write $3\dfrac{1}{5}$ as an improper fraction.

$$3\frac{1}{5} = 3 + \frac{1}{5} = \frac{15}{5} + \frac{1}{5} = \frac{16}{5}$$

Because of the steps above, notice we can use a shortcut process for writing a mixed number as an improper fraction.

$$3\frac{1}{5} = \frac{5 \cdot 3 + 1}{5} = \frac{16}{5}$$

Answers

19. $\dfrac{17}{40}$; **20.** $\dfrac{1}{5}$

EXAMPLE 21 Divide: $2\frac{1}{8} \div 1\frac{2}{3}$

Solution: First write each mixed number as an improper fraction.

$$2\frac{1}{8} = \frac{8 \cdot 2 + 1}{8} = \frac{17}{8}; \qquad 1\frac{2}{3} = \frac{3 \cdot 1 + 2}{3} = \frac{5}{3}$$

Now divide as usual.

$$2\frac{1}{8} \div 1\frac{2}{3} = \frac{17}{8} \div \frac{5}{3} = \frac{17}{8} \cdot \frac{3}{5} = \frac{51}{40}$$

The fraction $\frac{51}{40}$ is improper. To write it as an equivalent mixed number, remember that the fraction bar means division, and divide.

$$\begin{array}{r} 1\frac{11}{40} \\ 40\overline{)51} \\ \underline{-40} \\ 11 \end{array}$$

Thus, the quotient is $\frac{51}{40}$ or $1\frac{11}{40}$.

🔲 **Work Practice Problem 21**

As a general rule, if the original exercise contains mixed numbers, write the result as a mixed number, if possible.

EXAMPLE 22 Add: $2\frac{1}{8} + 1\frac{2}{3}$.

Solution: $2\frac{1}{8} + 1\frac{2}{3} = \frac{17}{8} + \frac{5}{3} = \frac{17 \cdot 3}{8 \cdot 3} + \frac{5 \cdot 8}{3 \cdot 8} = \frac{51}{24} + \frac{40}{24} = \frac{91}{24}$ or $3\frac{19}{24}$

🔲 **Work Practice Problem 22**

When adding or subtracting larger mixed numbers, you might want to use the following method.

EXAMPLE 23 Subtract: $50\frac{1}{6} - 38\frac{1}{3}$

Solution:
$$\begin{array}{r} 50\frac{1}{6} = 50\frac{1}{6} = 49\frac{7}{6} \\ -38\frac{1}{3} = -38\frac{2}{6} = -38\frac{2}{6} \\ \hline 11\frac{5}{6} \end{array}$$

$50\frac{1}{6} = 49 + 1 + \frac{1}{6} = 49\frac{7}{6}$

🔲 **Work Practice Problem 23**

PRACTICE PROBLEM 21

Multiply: $5\frac{1}{6} \cdot 4\frac{2}{5}$

PRACTICE PROBLEM 22

Add: $7\frac{3}{8} + 6\frac{3}{4}$

PRACTICE PROBLEM 23

Subtract: $76\frac{1}{12} - 35\frac{1}{4}$

Answers

21. $22\frac{11}{15}$, **22.** $14\frac{1}{8}$, **23.** $40\frac{5}{6}$

Objective A *Simplify by dividing the numerator by the denominator. See Examples 1 through 6.*

1. $\dfrac{14}{14}$ **2.** $\dfrac{19}{19}$ **3.** $\dfrac{20}{2}$ **4.** $\dfrac{30}{5}$ **5.** $\dfrac{13}{1}$

6. $\dfrac{21}{1}$ **7.** $\dfrac{0}{9}$ **8.** $\dfrac{0}{15}$ **9.** $\dfrac{9}{0}$ **10.** $\dfrac{15}{0}$

Objective B *Write each fraction as an equivalent fraction with the given denominator. See Example 7.*

11. $\dfrac{7}{10}$ with a denominator of 30 **12.** $\dfrac{2}{3}$ with a denominator of 9

13. $\dfrac{2}{9}$ with a denominator of 18 **14.** $\dfrac{8}{7}$ with a denominator of 56

15. $\dfrac{4}{5}$ with a denominator of 20 **16.** $\dfrac{4}{5}$ with a denominator of 25

Objective C *Simplify each fraction. See Examples 8 through 10.*

17. $\dfrac{2}{4}$ **18.** $\dfrac{3}{6}$ **19.** $\dfrac{10}{15}$ **20.** $\dfrac{15}{20}$

21. $\dfrac{3}{7}$ **22.** $\dfrac{5}{9}$ **23.** $\dfrac{18}{30}$ **24.** $\dfrac{42}{45}$

25. $\dfrac{16}{20}$ **26.** $\dfrac{8}{40}$ **27.** $\dfrac{66}{48}$ **28.** $\dfrac{64}{24}$

29. $\dfrac{120}{244}$ **30.** $\dfrac{360}{700}$ **31.** $\dfrac{192}{264}$ **32.** $\dfrac{455}{525}$

Objectives D F Mixed Practice *Multiply or divide as indicated. See Examples 11 through 14 and 21.*

33. $\dfrac{1}{2} \cdot \dfrac{3}{4}$ **34.** $\dfrac{7}{11} \cdot \dfrac{3}{5}$ **35.** $\dfrac{2}{3} \cdot \dfrac{3}{4}$ **36.** $\dfrac{7}{8} \cdot \dfrac{3}{21}$

37. $\dfrac{1}{2} \div \dfrac{7}{12}$ **38.** $\dfrac{7}{12} \div \dfrac{1}{2}$ **39.** $\dfrac{3}{4} \div \dfrac{1}{20}$ **40.** $\dfrac{3}{5} \div \dfrac{9}{10}$

41. $5\dfrac{1}{9} \cdot 3\dfrac{2}{3}$ **42.** $2\dfrac{3}{4} \cdot 1\dfrac{7}{8}$ **43.** $8\dfrac{3}{5} \div 2\dfrac{9}{10}$ **44.** $1\dfrac{7}{8} \div 3\dfrac{8}{9}$

Objectives E F Mixed Practice *Add or subtract as indicated. See Examples 15 through 20, 22, and 23.*

45. $\dfrac{4}{5} + \dfrac{1}{5}$ **46.** $\dfrac{6}{7} + \dfrac{1}{7}$ **47.** $\dfrac{4}{15} - \dfrac{1}{12}$ **48.** $\dfrac{11}{12} - \dfrac{1}{16}$

R-16

49. $\dfrac{2}{3} + \dfrac{3}{7}$

50. $\dfrac{3}{4} + \dfrac{1}{6}$

51. $\dfrac{10}{3} - \dfrac{5}{21}$

52. $\dfrac{11}{7} - \dfrac{3}{35}$

53. $8\dfrac{1}{8} - 6\dfrac{3}{8}$

54. $5\dfrac{2}{5} - 3\dfrac{4}{5}$

55. $9\dfrac{7}{8} + 2\dfrac{3}{10}$

56. $7\dfrac{3}{20} + 2\dfrac{13}{15}$

Objectives **D** **E** **F** **Mixed Practice** *Perform the indicated operations. See Examples 11 through 23.*

57. $\dfrac{23}{105} + \dfrac{4}{105}$

58. $\dfrac{13}{132} + \dfrac{35}{132}$

59. $\dfrac{17}{21} - \dfrac{10}{21}$

60. $\dfrac{18}{35} - \dfrac{11}{35}$

61. $\dfrac{7}{10} \cdot \dfrac{5}{21}$

62. $\dfrac{3}{35} \cdot \dfrac{10}{63}$

63. $\dfrac{9}{20} \div 12$

64. $\dfrac{25}{36} \div 10$

65. $\dfrac{5}{22} - \dfrac{5}{33}$

66. $\dfrac{7}{15} - \dfrac{7}{25}$

67. $17\dfrac{2}{5} + 30\dfrac{2}{3}$

68. $26\dfrac{11}{20} + 40\dfrac{7}{10}$

69. $7\dfrac{2}{5} \div \dfrac{1}{5}$

70. $9\dfrac{5}{6} \div \dfrac{1}{6}$

71. $4\dfrac{2}{11} \cdot 2\dfrac{1}{2}$

72. $6\dfrac{6}{7} \cdot 3\dfrac{1}{2}$

73. $\dfrac{12}{5} - 1$

74. $2 - \dfrac{3}{8}$

75. $8\dfrac{11}{12} - 1\dfrac{5}{6}$

76. $4\dfrac{7}{8} - 2\dfrac{3}{16}$

Concept Extensions

Perform indicated operations.

77. $\dfrac{2}{3} - \dfrac{5}{9} + \dfrac{5}{6}$

78. $\dfrac{8}{11} - \dfrac{1}{4} + \dfrac{1}{2}$

79. Which of the following are correct? See the Concept Check in this section.

a. $\dfrac{12}{24} = \dfrac{2 + 4 + 6}{2 + 4 + 6 + 12} = \dfrac{1}{12}$

b. $\dfrac{12}{24} = \dfrac{2 \cdot 2 \cdot 3}{2 \cdot 2 \cdot 2 \cdot 3} = \dfrac{1}{2}$

c. $\dfrac{12}{24} = \dfrac{2 \cdot 3 + 6 \cdot 1}{2 \cdot 3 + 6 \cdot 3} = \dfrac{1}{3}$

80. In your own words, describe how to add or subtract fractions.

81. In your own words, describe how to divide fractions.

Each circle below represents a whole, or 1. Determine the unknown part of the circle.

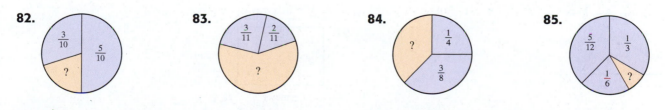

82. $\dfrac{3}{10}$ $\dfrac{5}{10}$?

83. $\dfrac{3}{11}$ $\dfrac{2}{11}$?

84. ? $\dfrac{1}{4}$ $\dfrac{3}{8}$

85. $\dfrac{5}{12}$ $\dfrac{1}{3}$ $\dfrac{1}{6}$?

86. During the 2004 Summer Olympic Games, Natalya Sadova of Russia took the gold medal in the women's discus throw with a distance of $219\frac{11}{12}$ feet. However, the Olympic record for the women's discus throw was set in 1988 by Martina Hellman of East Germany with a distance of $237\frac{1}{6}$ feet. How much longer is the Olympic record discus throw than the gold medal throw in 2004? (*Source:* HickokSports.com)

87. Approximately $\frac{41}{50}$ of all American adults agree that the U.S. federal government should support basic scientific research. What fraction of American adults do *not* agree that the U.S. federal government should support such research? (*Source:* National Science Foundation)

88. The breakdown of science and engineering doctorate degrees awarded in the United States is summarized in the graph shown, called a circle graph or a pie chart. Use the graph to answer the questions. (*Source:* National Science Foundation)

Science and Engineering Doctorates Awarded, by Field of Study

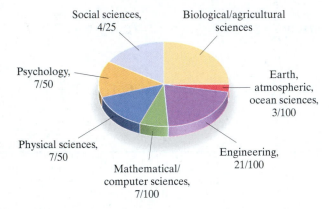

a. What fraction of science and engineering doctorates are awarded in the physical sciences?

b. Engineering doctorates make up what fraction of all science and engineering doctorates awarded in the United States?

c. Social sciences and psychology doctorates together make up what fraction of all science and engineering doctorates awarded in the United States?

d. What fraction of all science and engineering doctorates are awarded in the biological and agricultural sciences?

89. In 2004, Federated Department Stores operated a total of 459 stores throughout the United States, Guam and Puerto Rico through the identity of 8 locally known brand names. The following chart shows the store break down by brand. (*Source:* Federated Department Stores)

Federated Department Stores

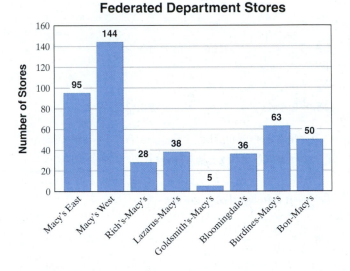

a. What fraction of Federated department stores are Bloomingdale's? Simplify the fraction, if possible.

b. What fraction of the Federated department stores were either Macy's East or Macy's West? Simplify the fraction, if possible.

The area of a plane figure is a measure of the amount of surface of the figure. Find the area of each figure. (The area of a rectangle is the product of its length and width. The area of a triangle is $\frac{1}{2}$ the product of its base and height. Recall that area is measured in square units.)

△ **90.**

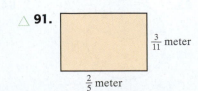

△ **91.**

R.3 DECIMALS AND PERCENTS

Objectives

A Write Decimals as Fractions.

B Add, Subtract, Multiply, and Divide Decimals.

C Round Decimals to a Given Decimal Place.

D Write Fractions as Decimals.

E Write Percents as Decimals and Decimals as Percents.

Objective **A** Writing Decimals as Fractions

Like fractional notation, **decimal notation** is used to denote a part of a whole. Below is a **place value chart** that shows the value of each place.

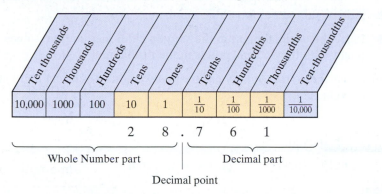

✔ **Concept Check** Fill in the blank: In the number 52.634, the 3 is in the _____ place.

a. Tens **b.** Ones **c.** Tenths

d. Hundredths **e.** Thousandths

The next chart shows decimals written as fractions.

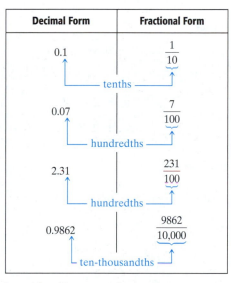

To write a decimal as a fraction, use place values.

EXAMPLES Write each decimal as a fraction. Do not simplify.

1. $0.37 = \dfrac{37}{100}$
2 decimal places 2 zeros

2. $1.3 = \dfrac{13}{10}$
1 decimal place 1 zero

3. $2.649 = \dfrac{2649}{1000}$
3 decimal places 3 zeros

🟧 **Work Practice Problems 1–3**

PRACTICE PROBLEMS 1–3

Write each decimal as a fraction. Do not simplify.

1. 0.27 **2.** 5.1 **3.** 7.685

Answers

1. $\dfrac{27}{100}$, **2.** $\dfrac{51}{10}$, **3.** $\dfrac{7685}{1000}$

✔ **Concept Check Answer**

d

Objective B Adding, Subtracting, Multiplying, and Dividing Decimals

To **add** or **subtract** decimals, follow the steps below.

To Add or Subtract Decimals

Step 1: Write the decimals so that the decimal points line up vertically.

Step 2: Add or subtract as for whole numbers.

Step 3: Place the decimal point in the sum or difference so that it lines up vertically with the decimal points in the problem.

Notice that these steps simply ensure that we add or subtract digits with the same place value.

PRACTICE PROBLEM 4

Add.
a. $7.19 + 19.782 + 1.006$
b. $12 + 0.79 + 0.03$

EXAMPLE 4 Add.

a. $5.87 + 23.279 + 0.003$ **b.** $7 + 0.23 + 0.6$

Solution:

a.
```
    5.87
   23.279
 +  0.003
   29.152
```

b.
```
    7.
    0.23
  +0.6
    7.83
```

☐ **Work Practice Problem 4**

PRACTICE PROBLEM 5

Subtract.
a. $84.23 - 26.982$
b. $90 - 0.19$

EXAMPLE 5 Subtract.

a. $32.15 - 11.237$ **b.** $70 - 0.48$

Solution:

a.
```
      1  11  4  10
    3 2. 1  5  0
  - 1 1. 2  3  7
    2 0. 9  1  3
```

b.
```
        6  9  9  10
      7 0. 0  0
    -   0. 4  8
      6 9. 5  2
```

☐ **Work Practice Problem 5**

Now let's study the following product of decimals. Notice the pattern in the decimal points.

$$0.03 \times 0.6 = \frac{3}{100} \times \frac{6}{10} = \frac{18}{1000} \quad \text{or} \quad 0.018$$

2 decimal places   1 decimal place   3 decimal places

In general, to **multiply** decimals, follow the steps below.

To Multiply Decimals

Step 1: Multiply the decimals as though they are whole numbers.

Step 2: The decimal point in the product is placed so that the number of decimal places in the product is equal to the **sum** of the number of decimal places in the factors.

Answers
4. **a.** 27.978, **b.** 12.82,
5. **a.** 57.248, **b.** 89.81

EXAMPLE 6 Multiply.

a. 0.072×3.5

b. 0.17×0.02

Solution:

a.
```
    0.072    3 decimal places
  ×  3.5     1 decimal place
  ─────
    360
    216
  ─────
  0.2520     4 decimal places
```

b.
```
    0.17     2 decimal places
  × 0.02     2 decimal places
  ──────
  0.0034     4 decimal places
```

Work Practice Problem 6

PRACTICE PROBLEM 6

Multiply.
a. 0.31×4.6
b. 1.26×0.03

To divide a decimal by a whole number using long division, we place the decimal point in the quotient directly above the decimal point in the dividend. For example,

```
      2.47
  3)7.41
   −6
   ──
    1 4
   −1 2
   ────
      21
     −21
     ───
       0
```

To check, see that $2.47 \times 3 = 7.41$

Helpful Hint

Don't forget the names of the numbers in a division problem.

$$\text{divisor} \overline{)\text{dividend}}^{\text{quotient}}$$

In general, to **divide** decimals, use the steps below.

To Divide Decimals

Step 1: Move the decimal point in the divisor to the right until the divisor is a whole number.

Step 2: Move the decimal point in the dividend to the right the **same number of places** as the decimal point was moved in Step 1.

Step 3: Divide. The decimal point in the quotient is directly over the moved decimal point in the dividend.

EXAMPLE 7 Divide.

a. $9.46 \div 0.04$

b. $31.5 \div 0.007$

Solution:

a.
```
        236.5
  004.)946.0      A zero is inserted
   −8             to continue dividing.
   ──
    14
   −12
   ───
     26
    −24
    ───
     20
    −20
    ───
      0
```

b.
```
         4500.
  0007.)31500.     Zeros are inserted
   −28             in order to move the
   ───             decimal point three places
    35             to the right.
   −35
   ───
     0
```

Work Practice Problem 7

PRACTICE PROBLEM 7

Divide.
a. $21.75 \div 0.5$
b. $15.6 \div 0.006$

Answers
6. a. 1.426, **b.** 0.0378, **7. a.** 43.5,
b. 2600

Objective C Rounding Decimals

We **round** the decimal part of a decimal number in nearly the same way as we round the whole numbers. The only difference is that we drop digits to the right of the rounding place, instead of replacing these digits by 0s. For example,

24.954   rounded to the nearest hundredth is   24.95

↑
hundredths place

To Round Decimals to a Place Value to the Right of the Decimal Point

Step 1: Locate the digit to the right of the given place value.

Step 2: • If this digit is 5 or greater, add 1 to the digit in the given place value and drop all digits to its right.

• If this digit is less than 5, drop all digits to the right of the given place.

PRACTICE PROBLEM 8

Round 12.9187 to the nearest hundredth.

EXAMPLE 8 Round 7.8265 to the nearest hundredth.

hundredths place

Solution: 7.8265

Step 1. Locate the digit to the right of the hundredths place.
Step 2. This digit is 5 or greater, so we add 1 to the hundredths place digit and drop all digits to its right.

Thus, 7.8265 rounded to the nearest hundredth is 7.83.

☐ **Work Practice Problem 8**

PRACTICE PROBLEM 9

Round 245.348 to the nearest tenth.

EXAMPLE 9 Round 19.329 to the nearest tenth.

tenths place

Solution: 19.329

Step 1. Locate the digit to the right of the tenths place.
Step 2. This digit is less than 5, so we drop this digit and all digits to its right.

Thus, 19.329 rounded to the nearest tenth is 19.3.

☐ **Work Practice Problem 9**

Objective D Writing Fractions as Decimals

To write fractions as decimals, interpret the fraction bar as division and find the quotient.

To Write a Fraction as a Decimal

Divide the numerator by the denominator.

PRACTICE PROBLEM 10

Write $\frac{2}{5}$ as a decimal.

EXAMPLE 10 Write $\frac{1}{4}$ as a decimal.

Solution:
$$
\begin{array}{r}
0.25 \\
4\overline{)1.00} \\
-8 \\
\hline
20 \\
-20 \\
\hline
0
\end{array}
$$

$\frac{1}{4} = 0.25$

☐ **Work Practice Problem 10**

Answers
8. 12.92, **9.** 245.3, **10.** 0.4

EXAMPLE 11 Write $\frac{2}{3}$ as a decimal.

Solution:

$$
\begin{array}{r}
0.666 \\
3\overline{)2.000} \\
-1\,8 \\
\hline
20 \\
-18 \\
\hline
20 \\
-18 \\
\hline
2
\end{array}
$$

This division pattern will continue so that $\frac{2}{3} = 0.6666\ldots$

A bar can be placed over the digit 6 to indicate that it repeats. We call this a **repeating decimal.**

$$
\frac{2}{3} = 0.666\ldots = 0.\overline{6}
$$

🔲 **Work Practice Problem 11**

We can also write a decimal approximation for $\frac{2}{3}$. For example, $\frac{2}{3}$ rounded to the nearest hundredth is 0.67. This can be written as $\frac{2}{3} \approx 0.67$. The $\approx$ sign means "is approximately equal to."

✔**Concept Check** The notation $0.5\overline{2}$ is the same as

a. $\frac{52}{100}$ **b.** $\frac{52\ldots}{100}$ **c.** $0.52222222\ldots$

EXAMPLE 12 Write $\frac{22}{7}$ as a decimal. Round to the nearest hundredth.

Solution:

$$
\begin{array}{r}
3.142 \approx 3.14 \\
7\overline{)22.000} \\
-21 \\
\hline
1\,0 \\
-7 \\
\hline
30 \\
-28 \\
\hline
20 \\
-14 \\
\hline
6
\end{array}
$$

If rounding to the nearest hundredth, carry the division process out to one more decimal place, the thousandths place.

The fraction $\frac{22}{7}$ in decimal form is approximately 3.14. (The fraction $\frac{22}{7}$ is an approximation for π.)

🔲 **Work Practice Problem 12**

Objective 🅔 Writing Percents as Decimals and Decimals as Percents

The word **percent** comes from the Latin phrase *per centum,* which means **"per 100."** The % symbol is used to denote percent. Thus, 53% means 53 per 100, or

$$
53\% = \frac{53}{100}
$$

PRACTICE PROBLEM 11

Write $\frac{5}{6}$ as a decimal.

PRACTICE PROBLEM 12

Write $\frac{1}{9}$ as a decimal. Round to the nearest thousandth.

Answers

11. $0.8\overline{3}$,   **12.** 0.111

✔ **Concept Check Answer**

c

When solving problems containing percents, it is often necessary to write a percent as a decimal. To see how this is done, study the chart below.

Percent	Fraction	Decimal
7%	$\dfrac{7}{100}$	0.07
63%	$\dfrac{63}{100}$	0.63
109%	$\dfrac{109}{100}$	1.09

To convert directly from a percent to a decimal, notice that

$$7\% = 0.07$$

To Write a Percent as a Decimal

Drop the percent symbol, %, and move the decimal point two places to the left.

PRACTICE PROBLEM 13

Write each percent as a decimal.

a. 20%

b. 1.2%

c. 465%

EXAMPLE 13 Write each percent as a decimal.

a. 25%   **b.** 2.6%   **c.** 195%

Solution: We drop the % and move the decimal point two places to the left. Recall that the decimal point of a whole number is to the right of the ones place digit.

a. $25\% = 25.\% = 0.25$

b. $2.6\% = 02.6\% = 0.026$

c. $195\% = 195.\% = 1.95$

▨ **Work Practice Problem 13**

To write a decimal as a percent, we simply reverse the preceding steps. That is, we move the decimal point two places to the right and attach the percent symbol, %.

To Write a Decimal as a Percent

Move the decimal point two places to the right and attach the percent symbol, %.

PRACTICE PROBLEM 14

Write each decimal as a percent.

a. 0.42

b. 0.003

c. 2.36

d. 0.7

EXAMPLE 14 Write each decimal as a percent.

a. 0.85   **b.** 1.25   **c.** 0.012   **d.** 0.6

Solution: We move the decimal point two places to the right and attach the percent symbol, %.

a. $0.85 = 0.85 = 85\%$

b. $1.25 = 1.25 = 125\%$

c. $0.012 = 0.012 = 1.2\%$

d. $0.6 = 0.60 = 60\%$

▨ **Work Practice Problem 14**

Answers

13. a. 0.20, **b.** 0.012, **c.** 4.65,

14. a. 42%, **b.** 0.3%, **c.** 236%,

d. 70%

R.3 EXERCISE SET

Objective **A** *Write each decimal as a fraction. Do not simplify. See Examples 1 through 3.*

1. 0.6   **2.** 0.9   **3.** 1.86   **4.** 7.23

5. 0.114   **6.** 0.239   **7.** 123.1   **8.** 892.7

Objective **B** *Add or subtract as indicated. See Examples 4 and 5.*

9. 5.7 + 1.13   **10.** 2.31 + 6.4   **11.** 24.6 + 2.39 + 0.0678   **12.** 32.4 + 1.58 + 0.0934

13. 8.8 − 2.3   **14.** 7.6 − 2.1   **15.** 18 − 2.78   **16.** 28 − 3.31

Multiply or divide as indicated. See Examples 6 and 7.

17. $\begin{array}{r} 0.2 \\ \times\ 0.6 \\ \hline \end{array}$ **18.** $\begin{array}{r} 0.7 \\ \times\ 0.9 \\ \hline \end{array}$ **19.** $\begin{array}{r} 0.063 \\ \times\ \ \ 4.2 \\ \hline \end{array}$ **20.** $\begin{array}{r} 0.079 \\ \times\ \ \ 3.6 \\ \hline \end{array}$

21. $5\overline{)8.4}$ **22.** $2\overline{)11.7}$ **23.** $0.82\overline{)4.756}$ **24.** $0.92\overline{)3.312}$

Mixed Practice *Perform the indicated operation. See Examples 4 through 7.*

25. $\begin{array}{r} 45.02 \\ 3.006 \\ +\ 8.405 \\ \hline \end{array}$ **26.** $\begin{array}{r} 65.0028 \\ 5.0903 \\ +\ 6.9 \\ \hline \end{array}$ **27.** $\begin{array}{r} 6.75 \\ \times\ \ \ 10 \\ \hline \end{array}$ **28.** $\begin{array}{r} 8.91 \\ \times\ \ 100 \\ \hline \end{array}$ **29.** $0.6\overline{)42}$

30. $0.9\overline{)36}$ **31.** $\begin{array}{r} 654.9 \\ -\ 56.67 \\ \hline \end{array}$ **32.** $\begin{array}{r} 863.2 \\ -\ 39.45 \\ \hline \end{array}$ **33.** $\begin{array}{r} 5.62 \\ \times\ 7.7 \\ \hline \end{array}$ **34.** $\begin{array}{r} 8.03 \\ \times\ 5.5 \\ \hline \end{array}$

35. $0.063\overline{)52.92}$ **36.** $0.054\overline{)51.84}$ **37.** $\begin{array}{r} 16.003 \\ \times\ 5.31 \\ \hline \end{array}$ **38.** $\begin{array}{r} 31.006 \\ \times\ \ 3.71 \\ \hline \end{array}$

Objective **C** *Round each decimal to the given place value. See Examples 8 and 9.*

39. 0.57, nearest tenth   **40.** 0.75, nearest tenth

41. 0.234, nearest hundredth   **42.** 0.452, nearest hundredth

43. 0.5945, nearest thousandth

44. 63.4529, nearest thousandth

45. 98,207.23, nearest tenth

46. 68,936.543, nearest tenth

47. 12.347, nearest hundredth

48. 42.9878, nearest thousandth

Objective **D** *Write each fraction as a decimal. If the decimal is a repeating decimal, write using the bar notation and then round to the nearest hundredth. See Examples 10 through 12.*

49. $\dfrac{3}{4}$ **50.** $\dfrac{9}{25}$ **51.** $\dfrac{1}{3}$ **52.** $\dfrac{7}{9}$ **53.** $\dfrac{7}{16}$

54. $\dfrac{5}{8}$ **55.** $\dfrac{6}{11}$ **56.** $\dfrac{1}{6}$ **57.** $\dfrac{29}{6}$ **58.** $\dfrac{34}{9}$

Objective **E** *Write each percent as a decimal. See Example 13.*

59. 28%   **60.** 36%   **61.** 3.1%   **62.** 2.2%

63. 135%   **64.** 417%   **65.** 200%   **66.** 700%

67. 96.55%   **68.** 81.49%   **69.** 0.1%   **70.** 0.6%

71. During the 2003–2004 season, grapefruit accounted for 51% of all Florida fresh citrus shipments. Write this percent as a decimal. (*Source:* Citrus Administrative Committee)

72. The average one-year survival rate for a heart transplant recipient is 82.3%. The average one-year survival rate for a liver transplant patient is 81.6%. Write each percent as a decimal. (*Source:* Bureau of Health Resources Development)

Write each decimal as a percent. See Example 14.

73. 0.68   **74.** 0.32   **75.** 0.876   **76.** 0.521

77. 1   **78.** 3   **79.** 0.5   **80.** 0.1

81. 1.92   **82.** 2.15   **83.** 0.004   **84.** 0.005

Concept Extensions

Solve. See the Concept Checks in this section.

85. In the number 3.659, identify the place value of the

   a. 6

   b. 9

   c. 3

86. The notation $0.\overline{67}$ is the same as

   a. 0.6777...

   b. 0.67666...

   c. 0.6767...

87. Write $78.\overline{78}$ as a decimal rounded to the nearest ten-thousandth.

88. In your own words, describe how to add or subtract decimal numbers.

89. In your own words, describe how to multiply decimal numbers.

90. The passenger volume in a 2004 Toyota Prius gas/electric hybrid car is 96.2 cu ft. The passenger volume in a 2004 Honda Civic gas/electric hybrid car is 91.4 cu ft. How much more passenger space is there in the Toyota Prius than the Honda Civic Hybrid? (*Source:* Toyota and Honda)

91. The chart shows the average number of pounds of various dairy products consumed by each U.S. citizen. (*Source:* Dairy Information Center)

Dairy Product	Pounds
Fluid Milk	213.4
Cheese	30.8
Butter	4.4

   a. How much more fluid milk products than cheese products does the average U.S. citizen consume?

   b. What is the total amount of these milk products consumed by the average U.S. citizen annually?

92. An estimated $\dfrac{7}{20}$ of all candy sold in the United States each year is used to give, share, or enjoy during major holidays. What percent of candy is purchased in conjunction with holidays? (*Source:* National Confectioners Association)

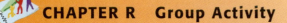

CHAPTER R Group Activity

Interpreting Survey Results

This activity may be completed by working in groups or individually.

Conduct the following survey with 12 students in one of your classes and record the results.

a. What is your age?

Under 20   20s   30s   40s   50s   60 and older

b. What is your gender?

Female   Male

c. How did you arrive on campus today?

Walked   Drove   Bicycled

Took public transportation   Other

1. For each survey question, tally the results for each category.

Age	
Category	**Tally**
Under 20	
20s	
30s	
40s	
50s	
60+	
Total	

Gender	
Category	**Tally**
Female	
Male	
Total	

Mode of Transportation	
Category	**Tally**
Walk	
Drive	
Bicycle	
Public Transit	
Other	
Total	

2. For each survey question, find the fraction of the total number of responses that fall in each answer category. Use the tallies from Question 1 to complete the Fraction columns of the tables at the right.

3. For each survey question, convert the fraction of the total number of responses that fall in each answer category to a decimal number. Use the fractions from Question 2 to complete the Decimal columns of the tables below.

4. For each survey question, find the percent of the total number of responses that fall in each answer category. Complete the Percent columns of the tables below.

5. Study the tables. What may you conclude from them? What do they tell you about your survey respondents? Write a paragraph summarizing your findings.

Age			
Category	**Fraction**	**Decimal**	**Percent**
Under 20			
20s			
30s			
40s			
50s			
60+			

Gender			
Category	**Fraction**	**Decimal**	**Percent**
Female			
Male			

Mode of Transportation			
Category	**Fraction**	**Decimal**	**Percent**
Walk			
Drive			
Bicycle			
Public Transit			
Other			

Chapter R Vocabulary Check

Fill in each blank with one of the words or phrases listed below.

mixed number   factor   improper fraction   percent

multiple   composite number   proper fraction   simplified

prime number   equivalent

1. To _____ means to write as a product.
2. A _____ of a number is the product of that number and any natural number.
3. A _____ is a natural number greater than 1 that is not prime.
4. The word _____ means per 100.
5. Fractions that represent the same portion of a whole are called _____ fractions.
6. An _____ is a fraction whose numerator is greater than or equal to its denominator.
7. A _____ is a natural number greater than 1 whose only factors are 1 and itself.
8. A fraction is _____ when the numerator and the denominator have no factors in common other than 1.
9. A _____ is one whose numerator is less than its denominator.
10. A _____ contains a whole number part and a fraction part.

Helpful Hint

Are you preparing for your test? Don't forget to take the Chapter R Test on page R-34. Then check your answers at the back of the text and use the Chapter Test Prep Video CD to see the fully worked-out solutions to any of the exercises you want to review.

R Chapter Highlights

DEFINITIONS AND CONCEPTS	EXAMPLES
Section R.1 Factors and the Least Common Multiple	

To **factor** means to write as a product.

The factors of 12 are

$$1, 2, 3, 4, 6, 12$$

When a number is written as a product of primes, this product is called the **prime factorization** of a number.

Write the prime factorization of 60.

$$60 = 6 \cdot 10$$
$$= 2 \cdot 3 \cdot 2 \cdot 5$$

The prime factorization of 60 is $2 \cdot 2 \cdot 3 \cdot 5$.

The **least common multiple (LCM)** of a list of numbers is the smallest number that is a multiple of all the numbers in the list.

TO FIND THE LCM OF A LIST OF NUMBERS

Find the LCM of 12 and 40.

Step 1. Write the prime factorization of each number.

Step 2. Write the product containing each different prime factor (from Step 1) the greatest number of times that it appears in any one factorization. This product is the LCM.

$$12 = 2 \cdot 2 \cdot 3$$
$$40 = 2 \cdot 2 \cdot 2 \cdot 5$$
$$\text{LCM} = 2 \cdot 2 \cdot 2 \cdot 3 \cdot 5 = 120$$

DEFINITIONS AND CONCEPTS	EXAMPLES

Section R.2   Fractions

Fractions that represent the same portion of a whole are called **equivalent fractions.**

$$\frac{1}{5} = \frac{1 \cdot 4}{5 \cdot 4} = \frac{4}{20}$$

$\frac{1}{5}$ and $\frac{4}{20}$ are equivalent fractions.

To write an equivalent fraction,

$$\frac{a}{b} = \frac{a}{b} \cdot \frac{c}{c} = \frac{a \cdot c}{b \cdot c}$$

Write $\frac{8}{21}$ as an equivalent fraction with a denominator of 63.

$$\frac{8}{21} = \frac{8}{21} \cdot \frac{3}{3} = \frac{8 \cdot 3}{21 \cdot 3} = \frac{24}{63}$$

A fraction is **simplified** when the numerator and the denominator have no factors in common other than 1.

$\frac{13}{17}$ is simplified.

To simplify a fraction,

$$\frac{a \cdot c}{b \cdot c} = \frac{a}{b} \cdot \frac{c}{c} = \frac{a}{b}$$

Simplify.

$$\frac{6}{14} = \frac{2 \cdot 3}{2 \cdot 7} = \frac{2}{2} \cdot \frac{3}{7} = \frac{3}{7}$$

Two fractions are **reciprocals** if their product is 1. The reciprocal of $\frac{a}{b}$ is $\frac{b}{a}$, as long as a and b are not 0.

The reciprocal of $\frac{6}{25}$ is $\frac{25}{6}$.

To multiply fractions, multiply numerator times numerator to find the numerator of the product and denominator times denominator to find the denominator of the product.

$$\frac{2}{5} \cdot \frac{3}{7} = \frac{6}{35}$$

To divide fractions, multiply the first fraction by the reciprocal of the second fraction.

$$\frac{5}{9} \div \frac{2}{7} = \frac{5}{9} \cdot \frac{7}{2} = \frac{35}{18}$$

To add fractions with the same denominator, add the numerators and place the sum over the common denominator.

$$\frac{5}{11} + \frac{3}{11} = \frac{8}{11}$$

To subtract fractions with the same denominator, subtract the numerators and place the difference over the common denominator.

$$\frac{13}{15} - \frac{3}{15} = \frac{10}{15} = \frac{2}{3}$$

To add or subtract fractions with different denominators, first write each fraction as an equivalent fraction with the LCD as denominator.

$$\frac{2}{9} + \frac{3}{6} = \frac{2 \cdot 2}{9 \cdot 2} + \frac{3 \cdot 3}{6 \cdot 3} = \frac{4 + 9}{18} = \frac{13}{18}$$

Section R.3   Decimals and Percents

To write decimals as fractions, use place values.

$$0.11 = \frac{11}{100}$$

TO ADD OR SUBTRACT DECIMALS

Step 1. Write the decimals so that the decimal points line up vertically.

Step 2. Add or subtract as for whole numbers.

Step 3. Place the decimal point in the sum or difference so that it lines up vertically with the decimal points in the problem.

Subtract: $2.8 - 1.04$ Add: $25 + 0.02$

$$\begin{array}{r} \overset{7\ 10}{2.8\cancel{0}} \\ -1.04 \\ \hline 1.76 \end{array}$$

$$\begin{array}{r} 25. \\ +\ 0.02 \\ \hline 25.02 \end{array}$$

DEFINITIONS AND CONCEPTS	**EXAMPLES**
Section R.3 Decimals and Percents (*continued*)	

TO MULTIPLY DECIMALS

Step 1. Multiply the decimals as though they are whole numbers.

Step 2. The decimal point in the product is placed so that the number of decimal places in the product is equal to the **sum** of the number of decimal places in the factors.

Multiply: 1.48×5.9

$$
\begin{array}{r}
1.4\,8 \quad \leftarrow 2 \text{ decimal places} \\
\times \quad 5.9 \quad \leftarrow 1 \text{ decimal place} \\
\hline
1\,3\,3\,2 \\
7\,4\,0 \\
\hline
8.7\,3\,2 \quad \leftarrow 3 \text{ decimal places}
\end{array}
$$

TO DIVIDE DECIMALS

Step 1. Move the decimal point in the divisor to the right until the divisor is a whole number.

Step 2. Move the decimal point in the dividend to the right the **same number of places** as the decimal point was moved in Step 1.

Step 3. Divide. The decimal point in the quotient is directly over the moved decimal point in the dividend.

Divide: $1.118 \div 2.6$

$$
\begin{array}{r}
0.43 \\
2.6\overline{)1.118} \\
-1\,04 \\
\hline
78 \\
-78 \\
\hline
0
\end{array}
$$

To write fractions as decimals, divide the numerator by the denominator.

Write $\dfrac{3}{8}$ as a decimal.

$$
\begin{array}{r}
0.375 \\
8\overline{)3.000} \\
-2\,4 \\
\hline
60 \\
-56 \\
\hline
40 \\
-40 \\
\hline
0
\end{array}
$$

To write a percent as a decimal, drop the percent symbol, %, and move the decimal point two places to the left.

$25\% = 25.\% = 0.25$

To write a decimal as a percent, move the decimal point two places to the right and attach the percent symbol, %.

$0.7 = 0.70 = 70\%$

CHAPTER REVIEW

(R.1) *Write the prime factorization of each number.*

1. 42

2. 800

Find the least common multiple (LCM) of each list of numbers.

3. 12, 30

4. 7, 42

5. 4, 6, 10

6. 2, 5, 7

(R.2) *Write each fraction as an equivalent fraction with the given denominator.*

7. $\dfrac{5}{8}$ with a denominator of 24

8. $\dfrac{2}{3}$ with a denominator of 60

Simplify each fraction.

9. $\dfrac{8}{20}$

10. $\dfrac{15}{100}$

11. $\dfrac{12}{6}$

12. $\dfrac{8}{8}$

Perform each indicated operation and simplify.

13. $\dfrac{1}{7} \cdot \dfrac{8}{11}$

14. $\dfrac{5}{12} + \dfrac{2}{15}$

15. $\dfrac{3}{10} \div 6$

16. $\dfrac{7}{9} - \dfrac{1}{6}$

17. $3\dfrac{3}{8} \cdot 4\dfrac{1}{4}$

18. $2\dfrac{1}{3} - 1\dfrac{5}{6}$

19. $16\dfrac{9}{10} + 3\dfrac{2}{3}$

20. $6\dfrac{2}{7} \div 2\dfrac{1}{5}$

The area of a plane figure is a measure of the amount of surface of the figure. Find the area of each figure below. (The area of a rectangle is the product of its length and width. The area of a triangle is $\dfrac{1}{2}$ the product of its base and height.)

△ **21.**

△ **22.**

$\dfrac{3}{5}$ mile

$\dfrac{11}{12}$ mile

$\dfrac{1}{2}$ meter

$\dfrac{5}{4}$ meters

(R.3) *Write each decimal as a fraction. Do not simplify.*

23. 1.81

24. 0.035

R-32

Perform each indicated operation.

25. 76.358
      $\underline{+18.76}$

26. $35 + 0.02 + 1.765$

27. $18 - 4.62$

28. 804.062
      $\underline{-112.489}$

29.   7.6
      $\underline{\times\ 12}$

30.   14.63
      $\underline{\times\ \ 3.2}$

31. $27\overline{)772.2}$

32. $0.06\overline{)13.8}$

Round each decimal to the given place value.

33. 0.7652, nearest hundredth

34. 25.6293, nearest tenth

Write each fraction as a decimal. If the decimal is a repeating decimal, write it using the bar notation and then round to the nearest thousandth.

35. $\dfrac{1}{2}$

36. $\dfrac{3}{8}$

37. $\dfrac{4}{11}$

38. $\dfrac{5}{6}$

Write each percent as a decimal.

39. 29%

40. 1.4%

Write each decimal as a percent.

41. 0.39

42. 1.2

43. In 2003, the home ownership rate in the United States was 68.3%. Write this percent as a decimal.

44. Choose the true statement.
   a. $2.3\% = 0.23$
   b. $5 = 500\%$
   c. $40\% = 4$

R CHAPTER TEST

Remember to use the Chapter Test Prep Video CD to see the fully worked-out solutions to any of the exercises you want to review.

1. Write the prime factorization of 72.

2. Find the LCM of 5, 18, 20.

3. Write $\dfrac{5}{12}$ as an equivalent fraction with a denominator of 60.

Simplify each fraction.

4. $\dfrac{15}{20}$

5. $\dfrac{48}{100}$

6. Write 1.3 as a fraction.

Perform each indicated operation and simplify.

7. $\dfrac{5}{8} + \dfrac{7}{10}$

8. $\dfrac{2}{3} \cdot \dfrac{27}{49}$

9. $\dfrac{9}{10} \div 18$

10. $\dfrac{8}{9} - \dfrac{1}{12}$

11. $1\dfrac{2}{9} + 3\dfrac{2}{3}$

12. $5\dfrac{6}{11} - 3\dfrac{7}{22}$

13. $6\dfrac{7}{8} \div \dfrac{1}{8}$

14. $2\dfrac{1}{10} \cdot 6\dfrac{1}{2}$

Perform each indicated operation.

15. $43 + 0.21 + 1.9$

16. $123.6 - 57.72$

17. $\begin{array}{r} 7.93 \\ \times\ 1.6 \\ \hline \end{array}$

18. $0.25\overline{)80}$

19. Round 23.7272 to the nearest hundredth.

20. Write $\dfrac{7}{8}$ as a decimal.

1. _____

2. _____

3. _____

4. _____

5. _____

6. _____

7. _____

8. _____

9. _____

10. _____

11. _____

12. _____

13. _____

14. _____

15. _____

16. _____

17. _____

18. _____

19. _____

20. _____

21. Write $\frac{1}{6}$ as a repeating decimal. Then approximate the result to the nearest thousandth.

22. Write 63.2% as a decimal.

23. Write 0.09 as a percent.

24. Write $\frac{3}{4}$ as a percent. (*Hint:* Write $\frac{3}{4}$ as a decimal, and then write the decimal as a percent.)

Most of the water on Earth is in the form of oceans. Only a small part is fresh water. The graph below is called a circle graph or pie chart. This particular circle graph shows the distribution of fresh water on Earth. Use this graph to answer Exercises 25 through 28. (Source: Philip's World Atlas)

Fresh Water Distribution

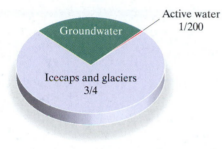

Groundwater

Active water
1/200

Icecaps and glaciers
3/4

25. What fractional part of fresh water is icecaps and glaciers?

26. What fractional part of fresh water is active water?

27. What fractional part of fresh water is groundwater?

28. What fractional part of fresh water is groundwater or icecaps and glaciers?

Find the area of each figure.

△ **29.**

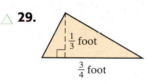

$\frac{1}{3}$ foot

$\frac{3}{4}$ foot

△ **30.**

Rectangle $\frac{7}{8}$ centimeters

$\frac{9}{8}$ centimeters

21. _____

22. _____

23. _____

24. _____

25. _____

26. _____

27. _____

28. _____

29. _____

30. _____

1

Real Numbers and Introduction to Algebra

In this chapter, we begin with a review of the basic symbols—the language—of mathematics. We then introduce algebra by using a variable in place of a number. From there, we translate phrases to algebraic expressions and sentences to equations. This is the beginning of problem solving, which we formally study in Chapter 2.

The apparent magnitude of a star is the measure of its brightness as seen by someone on Earth. The smaller the apparent magnitude, the brighter the star. Below, the apparent magnitudes of some stars are listed.

Star	Apparent Magnitude	Star	Apparent Magnitude
Arcturus	−0.04	Spica	0.98
Sirius	−1.46	Rigel	0.12
Vega	0.03	Regulus	1.35
Antares	0.96	Canopus	−0.72
Sun	−26.7	Hadar	0.61

(*Source: Norton's 2000.0: Star Atlas and Reference Handbook*, 18th ed., Longman Group, UK, 1989)

The stars have been a source of interest to different cultures for centuries. Polaris, the North Star, guided ancient sailors. The Egyptians honored Sirius, the brightest star in the sky, in temples. Around 150 B.C., a Greek astronomer, Hipparchus, devised a system of classifying the brightness of stars. Hipparchus's system is the basis of the apparent magnitude scale used by modern astronomers. In Exercises 81 through 86, Section 1.2, we shall see how this scale is used to describe the brightness of objects such as the sun, the moon, and some planets.

1.1 TIPS FOR SUCCESS IN MATHEMATICS

Before reading this section, remember that your instructor is your best source for information. Please see your instructor for any additional help or information.

Objective A Getting Ready for This Course

Now that you have decided to take this course, remember that a *positive attitude* will make all the difference in the world. Your belief that you can succeed is just as important as your commitment to this course. Make sure that you are ready for this course by having the time and positive attitude that it takes to succeed.

Next, make sure that you have scheduled your math course at a time that will give you the best chance for success. For example, if you are also working, you may want to check with your employer to make sure that your work hours will not conflict with your course schedule.

On the day of your first class period, double-check your schedule and allow yourself extra time to arrive on time in case of traffic problems or difficulty locating your classroom. Make sure that you bring at least your textbook, paper, and a writing instrument. Are you required to have a lab manual, graph paper, calculator, or some other supply besides this text? If so, also bring this material with you.

Objective B General Tips for Success

Below are some general tips that will increase your chance for success in a mathematics class. Many of these tips will also help you in other courses you may be taking.

Exchange names and phone numbers or e-mail addresses with at least one other person in class. This contact person can be a great help if you miss an assignment or want to discuss math concepts or exercises that you find difficult.

Choose to attend all class periods. If possible, sit near the front of the classroom. This way, you will see and hear the presentation better. It may also be easier for you to participate in classroom activities.

Do your homework. You've probably heard the phrase "practice makes perfect" in regard to music and sports. It also applies to mathematics. You will find that the more time you spend solving mathematics exercises, the easier the process becomes. Be sure to schedule enough time to complete your assignments before the next class period.

Check your work. Review the steps you made while working a problem. Learn to check your answers in the original problems. You may also compare your answers with the answers to selected exercises section in the back of the book. If you have made a mistake, try to figure out what went wrong. Then correct your mistake. If you can't find what went wrong, don't erase your work or throw it away. Bring your work to your instructor, a tutor in a math lab, or a classmate. It is easier for someone to find where you had trouble if they look at your original work.

Learn from your mistakes. Everyone, even your instructor, makes mistakes. Use your errors to learn and to become a better math student. The key is finding and understanding your errors. Was your mistake a careless one, or did you make it because you can't read your own math writing? If so, try to work more slowly or write more neatly and make a conscious effort to carefully check your work. Did you make a mistake because you don't understand a concept? If so, take the time to review the concept or ask questions to better understand it.

Know how to get help if you need it. It's all right to ask for help. In fact, it's a good idea to ask for help whenever there is something that you don't understand. Make sure you know when your instructor has office hours and how to find his or her office. Find out whether math tutoring services are available on your campus. Check

out the hours, location, and requirements of the tutoring service. Know whether software is available and how to access this resource.

Organize your class materials, including homework assignments, graded quizzes and tests, and notes from your class or lab. All of these items will make valuable references throughout your course and when studying for upcoming tests and the final exam. Make sure that you can locate these materials when you need them.

Read your textbook before class. Reading a mathematics textbook is unlike reading for fun, such as reading a newspaper. Your pace will be much slower. It is helpful to have paper and a pencil with you when you read. Try to work out examples on your own as you encounter them in your text. You should also write down any questions that you want to ask in class. When you read a mathematics textbook, sometimes some of the information in a section will be unclear. But after you hear a lecture or watch a videotape on that section, you will understand it much more easily than if you had not read your text beforehand.

Don't be afraid to ask questions. You are not the only person in class with questions. Other students are normally grateful that someone has spoken up.

Hand in assignments on time. This way you can be sure that you will not lose points for being late. Show every step of a problem and be neat and organized. Also be sure that you understand which problems are assigned for homework. You can always double-check the assignment with another student in your class.

Objective C Using This Text

There are many helpful resources that are available to you in this text. It is important that you become familiar with and use these resources. They should increase your chances for success in this course.

- *Practice Problems.* Each example in every section has a parallel Practice Problem. As you read a section, try each Practice Problem after you've finished the corresponding example. This "learn-by-doing" approach will help you grasp ideas before you move on to other concepts.

- *Lecture Video CDs.* Exercises marked with a CD are fully worked out on these video CDs by the author.

- *Chapter Test Prep Video CD.* This book contains a video CD. This CD contains all of the Chapter Test exercises worked out by the author.

- *Symbols at the Beginning of an Exercise Set.* If you need help in a particular section, the symbols listed at the beginning of each exercise set will remind you of the numerous supplements available.

- *Objectives.* The main section of exercises in each exercise set is referenced by an objective, such as A or B, and also an example(s). There is also often a section of exercises entitled "Mixed Practice," which is referenced by two or more objectives or sections. These are mixed exercises written to prepare you for your next exam. Use all of this referencing if you have trouble completing an assignment from the exercise set.

- *Icons (Symbols.)* Make sure that you understand the meaning of the icons that are beside many exercises.

    tells you that the corresponding exercise may be viewed on the video segment that corresponds to that section.

    tells you that this exercise is a writing exercise in which you should answer in complete sentences.

    △ tells you that the exercise involves geometry.

- *Integrated Reviews.* Found in the middle of each chapter, these reviews offer you a chance to practice—in one place—the many concepts that you have learned separately over several sections.

- *End-of-Chapter Opportunities.* There are many opportunities at the end of each chapter to help you understand the concepts of the chapter.
  Chapter Highlights contain chapter summaries and examples.
  Chapter Reviews contain review problems. The first part is organized section by section, and the second part contains a set of mixed exercises.
  Chapter Tests are sample tests to help you prepare for an exam. The Chapter Test Prep Video CD, found in this text, contains all the Chapter Test exercises worked by the author.
  Cumulative Reviews are reviews consisting of material from the beginning of the book through the end of that particular chapter.
- *Study Skill Builders.* This feature is found at the end of many exercise sets. In order to increase your chance of success in this course, please read and answer the questions in these Study Skill Builders.
- *The Bigger Picture.* This feature contains the directions for building an outline to be used throughout the course. The purpose of this outline is to help you decide how to solve a problem by making the transition from "what section does this problem come from" to "let me classify this problem based on its appearance."

See the Preface at the beginning of this text for a more thorough explanation of the features of this text.

Objective D Getting Help

If you have trouble completing assignments or understanding the mathematics, get help as soon as you need it! This tip is presented as an objective on its own because it is so important. In mathematics, usually the material presented in one section builds on your understanding of the previous section. This means that if you don't understand the concepts covered during a class period, there is a good chance that you will not understand the concepts covered during the next class period. If this happens to you, get help as soon as you can.

Where can you get help? Many suggestions have been made in this section on where to get help, and now it is up to you to do it. Try your instructor, a tutoring center, or a math lab, or you may want to form a study group with fellow classmates. If you do decide to see your instructor or go to a tutoring center, make sure that you have a neat notebook and are ready with your questions.

Objective E Preparing for and Taking an Exam

Make sure that you allow yourself plenty of time to prepare for a test. If you think that you are a little "math anxious," it may be that you are not preparing for tests in a way that will ensure success. The way that you prepare for a test in mathematics is important. To prepare for a test:

1. Review your previous homework assignments.
2. Review any notes from class and section-level quizzes you have taken. (If this is a final exam, also review chapter tests you have taken.)
3. Review concepts and definitions by reading the Highlights at the end of each chapter.
4. Practice working out exercises by completing the Chapter Review found at the end of each chapter. (If this is a final exam, go through a Cumulative Review. There is one found at the end of each chapter except Chapter 1. Choose the review found at the end of the latest chapter that you have covered in your course.) *Don't stop here!*
5. It is important that you place yourself in conditions similar to test conditions to find out how you will perform. In other words, as soon as you feel that you know the material, get a few blank sheets of paper and take a sample test. There is a Chapter Test available at the end of each chapter, or you can work selected

problems from the Chapter Review. Your instructor may also provide you with a review sheet. During this sample test, do not use your notes or your textbook. Then check your sample test. If you are not satisfied with the results, study the areas that you are weak in and try again.

6. On the day of the test, allow yourself plenty of time to arrive at where you will be taking your exam.

When taking your test:

1. Read the directions on the test carefully.
2. Read each problem carefully as you take the test. Make sure that you answer the question asked.
3. Watch your time and pace yourself so that you can attempt each problem on your test.
4. If you have time, check your work and answers.
5. Do not turn your test in early. If you have extra time, spend it double-checking your work.

Objective F Managing Your Time

As a college student, you know the demands that classes, homework, work, and family place on your time. Some days you probably wonder how you'll ever get everything done. One key to managing your time is developing a schedule. Here are some hints for making a schedule:

1. Make a list of all of your weekly commitments for the term. Include classes, work, regular meetings, extracurricular activities, etc. You may also find it helpful to list such things as laundry, regular workouts, grocery shopping, etc.
2. Next, estimate the time needed for each item on the list. Also make a note of how often you will need to do each item. Don't forget to include time estimates for the reading, studying, and homework you do outside of your classes. You may want to ask your instructor for help estimating the time needed.
3. In the exercise set that follows, you are asked to block out a typical week on the schedule grid given. Start with items with fixed time slots like classes and work.
4. Next, include the items on your list with flexible time slots. Think carefully about how best to schedule some items such as study time.
5. Don't fill up every time slot on the schedule. Remember that you need to allow time for eating, sleeping, and relaxing! You should also allow a little extra time in case some items take longer than planned.
6. If you find that your weekly schedule is too full for you to handle, you may need to make some changes in your workload, classload, or in other areas of your life. You may want to talk to your advisor, manager or supervisor at work, or someone in your college's academic counseling center for help with such decisions.

Note: In this chapter, we begin a feature called Study Skills Builder. The purpose of this feature is to remind you of some of the information given in this section and to further expand on some topics in this section.

Objectives A B C D Mixed Practice

1. What is your instructor's name?

2. What are your instructor's office location and office hours?

3. What is the best way to contact your instructor?

4. Do you have the name and contact information of at least one other student in class? If not, make a point of doing this during your next class gathering.

5. Will your instructor allow you to use a calculator in this class?

6. Is tutorial software available to you? If so, what type and where?

7. Is there a tutoring service available on campus? If so, what are its hours? What services are available?

8. Have you attempted this course before? If so, write down ways that you might improve your chances of success during this second attempt.

9. List some steps that you can take if you begin having trouble understanding the material or completing an assignment.

10. How many hours of studying does your instructor advise for each hour of instruction?

11. What does the ✏ icon in this text mean?

12. What does the 💿 icon in this text mean?

13. What does the △ icon in this text mean?

14. Search the minor columns in your text. What are Practice Problems?

15. When might be the best time to work Practice Problems?

16. Where are the answers to Practice Problems?

17. What answers are contained in this text and where are they?

18. What solutions are contained in this text and where are they?

19. What and where are Integrated Reviews?

20. What video CDs are contained in this book and where are they?

21. Chapter Highlights are found at the end of each chapter. Find the Chapter 1 Highlights and explain how you might use them and how they might be helpful.

22. Chapter Reviews are found at the end of each chapter. Find the Chapter 1 Review and explain how you might use it and how it might be useful.

23. Chapter Tests are found at the end of each chapter. Find the Chapter 1 Test and explain how you might use it and how it might be helpful when preparing for an exam on Chapter 1. Include how the Chapter Test Prep Video may help.

24. Read or reread objective **F** and fill out the schedule grid below.

	Monday	Tuesday	Wednesday	Thursday	Friday	Saturday	Sunday
7:00 a.m.							
8:00 a.m.							
9:00 a.m.							
10:00 a.m.							
11:00 a.m.							
12:00 p.m.							
1:00 p.m.							
2:00 p.m.							
3:00 p.m.							
4:00 p.m.							
5:00 p.m.							
6:00 p.m.							
7:00 p.m.							
8:00 p.m.							
9:00 p.m.							

SYMBOLS AND SETS OF NUMBERS

A Define the Meaning of the Symbols =, ≠, <, >, ≤, and ≥.

B Translate Sentences into Mathematical Statements.

C Identify Integers, Rational Numbers, Irrational Numbers, and Real Numbers.

D Find the Absolute Value of a Real Number.

Helpful Hint

The three dots (an ellipsis) at the end of the list of elements of a set means that the list continues in the same manner indefinitely.

We begin with a review of the set of natural numbers and the set of whole numbers and how we use symbols to compare these numbers. A **set** is a collection of objects, each of which is called a **member** or **element** of the set. A pair of brace symbols { } encloses the list of elements and is translated as "the set of" or "the set containing."

Natural Numbers

$$\{1, 2, 3, 4, 5, 6, \dots\}$$

Whole Numbers

$$\{0, 1, 2, 3, 4, 5, 6, \dots\}$$

These numbers can be pictured on a **number line.** To draw a number line, first draw a line. Choose a point on the line and label it 0. To the right of 0, label any other point 1. Being careful to use the same distance as from 0 to 1, mark off equally spaced distances to the right of 1. Label these points 2, 3, 4, 5, and so on. Since the whole numbers continue indefinitely, it is not possible to show every whole number on the number line. The arrow at the right end of the line indicates that the pattern continues indefinitely.

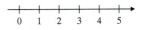

Objective **A** Equality and Inequality Symbols

Picturing natural numbers and whole numbers on a number line helps us to see the order of the numbers. Symbols can be used to describe in writing the order of two quantities. We will use equality symbols and inequality symbols to compare quantities.

Below is a review of these symbols. The letters a and b are used to represent quantities. Letters such as a and b that are used to represent numbers or quantities are called **variables.**

Equality and Inequality Symbols

		Meaning
Equality symbol:	$a = b$	a is equal to b.
Inequality symbols:	$a \neq b$	a is not equal to b.
	$a < b$	a is less than b.
	$a > b$	a is greater than b.
	$a \leq b$	a is less than or equal to b.
	$a \geq b$	a is greater than or equal to b.

These symbols may be used to form **mathematical statements** such as

$$2 = 2 \quad \text{and} \quad 2 \neq 6$$

On the number line, we see that a number **to the right of** another number is **larger.** Similarly, a number **to the left of** another number is **smaller.** For example, 3 is to the left of 5 on the number line, which means that 3 is less than 5, or $3 < 5$. Similarly, 2 is to the right of 0 on the number line, which means 2 is greater than 0, or $2 > 0$. Since 0 is to the left of 2, we can also say that 0 is less than 2, or $0 < 2$.

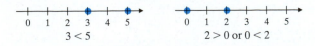

8

Copyright © 2007 Pearson Education, Inc.

Helpful Hint

Notice that $2 > 0$ has exactly the same meaning as $0 < 2$. Switching the order of the numbers and reversing the "direction of the inequality symbol" does not change the meaning of the statement.

$5 > 3$ has the same meaning as $3 < 5$.

Also notice that when the statement is true, the inequality arrow points to the smaller number.

Our discussion above can be generalized in the order property below.

Order Property for Real Numbers

For any two real numbers a and b, a is less than b if a is to the left of b on the number line.

$$a < b \text{ or also } b > a$$

EXAMPLES Determine whether each statement is true or false.

1. $2 < 3$ True.   Since 2 is to the left of 3 on the number line
2. $72 < 27$ False.   72 is to the right of 27 on the number line, so $72 > 27$.
3. $8 \geq 8$ True.   Since $8 = 8$ is true
4. $8 \leq 8$ True.   Since $8 = 8$ is true
5. $23 \leq 0$ False.   Since neither $23 < 0$ nor $23 = 0$ is true
6. $0 \leq 23$ True.   Since $0 < 23$ is true

◻ **Work Practice Problems 1–6**

Objective B Translating Sentences into Mathematical Statements

Now, let's use the symbols discussed above to translate sentences into mathematical statements.

EXAMPLE 7 Translate each sentence into a mathematical statement.

a. Nine is less than or equal to eleven.   b. Eight is greater than one.
c. Three is not equal to four.

Solution:

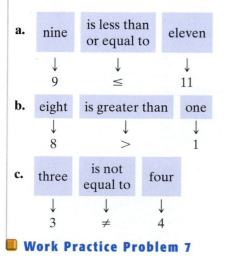

a.

nine	is less than or equal to	eleven
↓	↓	↓
9	$\leq$	11

b.

eight	is greater than	one
↓	↓	↓
8	$>$	1

c.

three	is not equal to	four
↓	↓	↓
3	$\neq$	4

◻ **Work Practice Problem 7**

Objective C Identifying Common Sets of Numbers

Whole numbers are not sufficient to describe many situations in the real world. For example, quantities smaller than zero must sometimes be represented, such as temperatures less than 0 degrees.

We can place numbers less than zero on the number line as follows: Numbers less than 0 are to the left of 0 and are labeled $-1, -2, -3$, and so on. The numbers we have labeled on the number line below are called the set of **integers.**

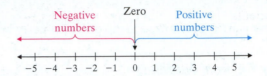

Integers to the left of 0 are called **negative integers;** integers to the right of 0 are called **positive integers.** The integer 0 is neither positive nor negative.

Integers

$$\{\ldots, -3, -2, -1, 0, 1, 2, 3, \ldots\}$$

Helpful Hint

A $-$ sign, such as the one in -2, tells us that the number is to the left of 0 on the number line.

-2 is read "negative two."

A $+$ sign or no sign tells us that a number lies to the right of 0 on the number line. For example, 3 and $+3$ both mean positive three.

PRACTICE PROBLEM 8

Use an integer to express the number in the following. Recently, due to hurricane Katrina the city of New Orleans suffered massive flooding. Under normal conditions, its elevation is 8 feet below sea level. (*Source: The World Almanac,* 2005)

Answer

8. -8

EXAMPLE 8

Use an integer to express the number in the following. "The lowest temperature ever recorded at South Pole Station, Antarctica, occurred during the month of June. The record-low temperature was 117 degrees below zero." (*Source:* The National Oceanic and Atmospheric Administration)

Solution: The integer -117 represents 117 degrees below zero.

Work Practice Problem 8

A problem with integers in real-life settings arises when quantities are smaller than some integer but greater than the next smallest integer. On the number line, these quantities may be visualized by points between integers. Some of these quantities between integers can be represented as a quotient of integers. For example,

The point on the number line halfway between 0 and 1 can be represented by $\frac{1}{2}$, a quotient of integers.

The point on the number line halfway between 0 and −1 can be represented by $-\frac{1}{2}$. Other quotients of integers and their graphs are shown below.

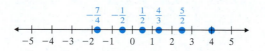

These numbers, each of which can be represented as a quotient of integers, are examples of **rational numbers.** It's not possible to list the set of rational numbers using the notation that we have been using. For this reason, we will use a different notation.

Rational Numbers

$$\left\{ \frac{a}{b} \middle| a \text{ and } b \text{ are integers and } b \neq 0 \right\}$$

We read this set as "the set of numbers $\frac{a}{b}$ such that a and b are integers and **b is not equal to 0.**"

Helpful Hint

We commonly refer to rational numbers as fractions.

Notice that every integer is also a rational number since each integer can be written as a quotient of integers. For example, the integer 5 is also a rational number since $5 = \frac{5}{1}$. For the rational number $\frac{5}{1}$, recall that the top number, 5, is called the numerator and the bottom number, 1, is called the denominator.

Let's practice **graphing** numbers on a number line.

EXAMPLE 9 Graph the numbers on a number line.

$$-\frac{4}{3}, \quad \frac{1}{4}, \quad \frac{3}{2}, \quad -2\frac{1}{8}, \quad 3.5$$

Solution: To help graph the improper fractions in the list, we first write them as mixed numbers.

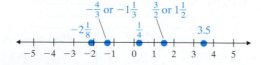

🟨 **Work Practice Problem 9**

Every rational number has a point on the number line that corresponds to it. But not every point on the number line corresponds to a rational number. Those points that do not correspond to rational numbers correspond instead to **irrational numbers.**

PRACTICE PROBLEM 9

Graph the numbers on the number line.

$$-2\frac{1}{2}, \quad -\frac{2}{3}, \quad \frac{1}{5}, \quad \frac{5}{4}, \quad 2.25$$

![number line from -5 to 5]

Answer

9.
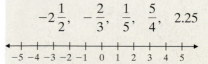

Irrational Numbers

{Nonrational numbers that correspond to points on the number line}

An irrational number that you have probably seen is π. Also, $\sqrt{2}$, the length of the diagonal of the square shown below, is an irrational number.

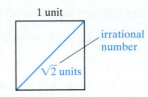

Both rational and irrational numbers can be written as decimal numbers. The decimal equivalent of a rational number will either terminate or repeat in a pattern. For example, upon dividing we find that

$$\frac{3}{4} = 0.75 \qquad \text{(Decimal number terminates or ends.)}$$

$$\frac{2}{3} = 0.66666\ldots \qquad \text{(Decimal number repeats in a pattern.)}$$

The decimal representation of an irrational number will neither terminate nor repeat. (For further review of decimals, see Section R.3.)

The set of numbers, each of which corresponds to a point on the number line, is called the set of **real numbers.** One and only one point on the number line corresponds to each real number.

Real Numbers

{All numbers that correspond to points on the number line}

Several different sets of numbers have been discussed in this section. The following diagram shows the relationships among these sets of real numbers. Notice that, together, the rational numbers and the irrational numbers make up the real numbers.

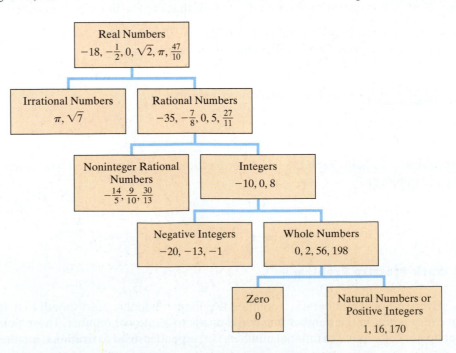

Now that other sets of numbers have been reviewed, let's continue our practice of comparing numbers.

EXAMPLE 10 Insert $<$, $>$, or $=$ between the pairs of numbers to form true statements.

a. -5 -6 b. 3.195 3.2 c. $\dfrac{1}{4}$ $\dfrac{1}{3}$

Solution:

a. $-5 > -6$ since -5 lies to the right of -6 on the number line.

b. By comparing digits in the same place values, we find that $3.195 < 3.2$. Since $0.1 < 0.2$.

c. By dividing, we find that $\dfrac{1}{4} = 0.25$ and $\dfrac{1}{3} = 0.33\ldots$. Since $0.25 < 0.33\ldots$, then $\dfrac{1}{4} < \dfrac{1}{3}$.

■ **Work Practice Problem 10**

PRACTICE PROBLEM 10

Insert $<$, $>$, or $=$ between pairs of numbers to form true statements.

a. -11 -9

b. 4.511 4.151

c. $\dfrac{7}{8}$ $\dfrac{2}{3}$

EXAMPLE 11 Given the set $\left\{-2, 0, \dfrac{1}{4}, 112, -3, 11, \sqrt{2}\right\}$, list the numbers in this set that belong to the set of:

a. Natural numbers   b. Whole numbers   c. Integers

d. Rational numbers   e. Irrational numbers   f. Real numbers

Solution:

a. The natural numbers are 11 and 112.

b. The whole numbers are $0, 11$, and 112.

c. The integers are $-3, -2, 0, 11$, and 112.

d. Recall that integers are rational numbers also. The rational numbers are $-3, -2, 0, \dfrac{1}{4}, 11$, and 112.

e. The irrational number is $\sqrt{2}$.

f. All numbers in the given set are real numbers.

■ **Work Practice Problem 11**

PRACTICE PROBLEM 11

Given the set $\left\{-100, -\dfrac{2}{5}, 0, \pi, 6, 913\right\}$, list the numbers in this set that belong to the set of:

a. Natural numbers

b. Whole numbers

c. Integers

d. Rational numbers

e. Irrational numbers

f. Real numbers

Objective D Finding the Absolute Value of a Number

The number line not only gives us a picture of the real numbers, it also helps us visualize the distance between numbers. The distance between a real number a and 0 is given a special name called the **absolute value** of a. "The absolute value of a" is written in symbols as $|a|$.

Helpful Hint
Since $|a|$ is a distance, $|a|$ is always either positive or 0. It is never negative. That is, **for any real number** a, $|a| \geq 0$.

Absolute Value

The **absolute value** of a real number a, denoted by $|a|$, is the distance between a and 0 on the number line.

For example, $|3| = 3$ and $|-3| = 3$ since both 3 and -3 are a distance of 3 units from 0 on the number line.

Answers

10. a. $<$, b. $>$, c. $>$,
11. a. $6, 913$, b. $0, 6, 913$,
c. $-100, 0, 6, 913$,
d. $-100, -\dfrac{2}{5}, 0, 6, 913$, e. π,
f. all numbers in the given set

PRACTICE PROBLEM 12

Find the absolute value of each number.

a. $|7|$, **b.** $|-8|$, **c.** $\left|\dfrac{2}{3}\right|$, **d.** $|0|$, **e.** $|-3.06|$

EXAMPLE 12 Find the absolute value of each number.

a. $|4|$ **b.** $|-5|$ **c.** $|0|$

d. $\left|-\dfrac{2}{9}\right|$ **e.** $|4.93|$

Solution:

a. $|4| = 4$ since 4 is 4 units from 0 on the number line.

b. $|-5| = 5$ since -5 is 5 units from 0 on the number line.

c. $|0| = 0$ since 0 is 0 units from 0 on the number line.

d. $\left|-\dfrac{2}{9}\right| = \dfrac{2}{9}$

e. $|4.93| = 4.93$

■ **Work Practice Problem 12**

PRACTICE PROBLEM 13

Insert $<$, $>$, or $=$ in the appropriate space to make each statement true.

a. $|-4|$ ___ 4,

b. -3 ___ $|0|$,

c. $|-2.7|$ ___ $|-2|$,

d. $|-6|$ ___ $|-16|$

e. $|10|$ ___ $\left|-10\dfrac{1}{3}\right|$

EXAMPLE 13 Insert $<$, $>$, or $=$ in the appropriate space to make each statement true.

a. $|0|$ ___ 2 **b.** $|-5|$ ___ 5 **c.** $|-3|$ ___ $|-2|$

d. $|-9|$ ___ $|-9.7|$ **e.** $\left|-7\dfrac{1}{6}\right|$ ___ $|7|$

Solution:

a. $|0| < 2$ since $|0| = 0$ and $0 < 2$.

b. $|-5| = 5$.

c. $|-3| > |-2|$ since $3 > 2$.

d. $|-9| < |-9.7|$ since $9 < 9.7$.

e. $\left|-7\dfrac{1}{6}\right| > |7|$ since $7\dfrac{1}{6} > 7$.

■ **Work Practice Problem 13**

Answers

12. a. 7, **b.** 8, **c.** $\dfrac{2}{3}$, **d.** 0, **e.** 3.06,

13. a. $=$, **b.** $<$, **c.** $>$, **d.** $<$,

e. $<$

Objectives Ⓐ Ⓒ **Mixed Practice** *Insert <, >, or = in the space between the paired numbers to make each statement true. See Examples 1 through 6, and 10.*

1. 4   10

2. 8   5

3. 7   3

4. 9   15

5. 6.26   6.26

6. 1.13   1.13

7. 0   7

8. 20   0

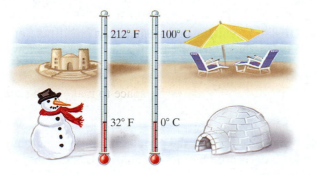

9. The freezing point of water is 32° Fahrenheit. The boiling point of water is 212° Fahrenheit. Write an inequality statement using < or > comparing the numbers 32 and 212.

10. The freezing point of water is 0° Celsius. The boiling point of water is 100° Celsius. Write an inequality statement using < or > comparing the numbers 0 and 100.

△ **11.** An angle measuring 30° and an angle measuring 45° are shown. Use the inequality symbols ≤ or ≥ to write a statement comparing the numbers 30 and 45.

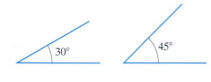

△ **12.** The sum of the measures of the angles of a triangle is 180°. The sum of the measures of the angles of a parallelogram is 360°. Use the inequality symbols ≤ or ≥ to write a statement comparing the numbers 360 and 180.

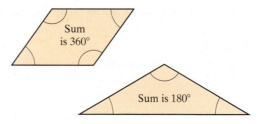

Determine whether each statement is true or false. See Examples 1 through 6 and 10.

13. $11 \leq 11$

14. $8 \geq 9$

15. $-11 > -10$

16. $-16 > -17$

17. $5.092 < 5.902$

18. $1.02 > 1.021$

19. $\dfrac{9}{10} \leq \dfrac{8}{9}$

20. $\dfrac{4}{5} \leq \dfrac{9}{11}$

Rewrite each inequality so that the inequality symbol points in the opposite direction and the resulting statement has the same meaning as the given one.

21. $25 \geq 20$

22. $-13 \leq 13$

23. $0 < 6$

24. $5 > 3$

25. $-10 > -12$

26. $-4 < -2$

Objectives **B** **C** **Mixed Practice** *Write each sentence as a mathematical statement. See Example 7, 10, and 11.*

27. Seven is less than eleven.

28. Twenty is greater than two.

29. Five is greater than or equal to four.

30. Negative ten is less than or equal to thirty-seven.

31. Fifteen is not equal to negative two.

32. Negative seven is not equal to seven.

Use integers to represent the values in each statement. See Example 8.

33. The highest elevation in California is Mt. Whitney with an altitude of 14,494 feet. The lowest elevation in California is Death Valley with an altitude of 282 feet below sea level. (*Source:* U.S. Geological Survey)

34. Driskill Mountain, in Louisiana, has an altitude of 535 feet. New Orleans, Louisiana, lies 8 feet below sea level. (*Source:* U.S. Geological Survey)

35. The number of students admitted to the Class of 2008 at UCLA was 43,413 fewer students than the number that applied. (*Source:* UCLA)

36. From 1990 to 2000, the population of Washington, D.C., decreased by 34,841. (*Source:* U.S. Census Bureau)

37. Gretchen Bertani deposited $475 in her savings account. She later withdrew $195.

38. David Lopez was deep-sea diving. During his dive, he ascended 17 feet and later descended 15 feet.

Graph each set of numbers on the number line. See Example 9.

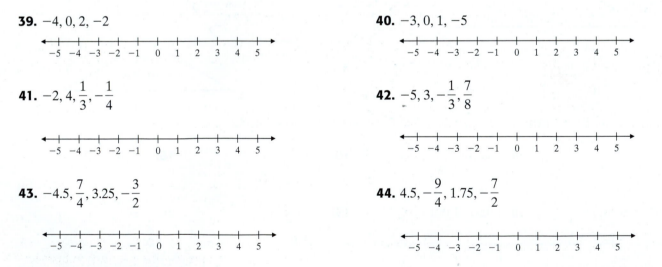

39. $-4, 0, 2, -2$

40. $-3, 0, 1, -5$

41. $-2, 4, \dfrac{1}{3}, -\dfrac{1}{4}$

42. $-5, 3, -\dfrac{1}{3}, \dfrac{7}{8}$

43. $-4.5, \dfrac{7}{4}, 3.25, -\dfrac{3}{2}$

44. $4.5, -\dfrac{9}{4}, 1.75, -\dfrac{7}{2}$

Tell which set or sets each number belongs to: natural numbers, whole numbers, integers, rational numbers, irrational numbers, and real numbers. See Example 11.

45. 0

46. $\dfrac{1}{4}$

47. -7

48. $-\dfrac{1}{7}$

49. 265

50. 7941

51. $\dfrac{2}{3}$

52. $\sqrt{3}$

Determine whether each statement is true or false.

53. Every rational number is also an integer.

54. Every natural number is positive.

55. 0 is a real number.

56. $\frac{1}{2}$ is an integer.

57. Every negative number is also a rational number.

58. Every rational number is also a real number.

59. Every real number is also a rational number.

60. Every whole number is an integer.

Objective **D** *Find each absolute value. See Example 12.*

61. $|8.9|$

62. $|11.2|$

63. $|-20|$

64. $|-17|$

65. $\left|\frac{9}{2}\right|$

66. $\left|\frac{10}{7}\right|$

67. $\left|-\frac{12}{13}\right|$

68. $\left|-\frac{1}{15}\right|$

Insert $<$, $>$, or $=$ in the appropriate space to make each statement true. See Examples 12 and 13.

69. $|-5|$ -4

70. $|-12|$ $|0|$

71. $\left|-\frac{5}{8}\right|$ $\left|\frac{5}{8}\right|$

72. $\left|\frac{2}{5}\right|$ $\left|-\frac{2}{5}\right|$

73. $|-2|$ $|-2.7|$

74. $|-5.01|$ $|-5|$

75. $|0|$ $|-8|$

76. $|-12|$ $\frac{-24}{2}$

Concept Extensions

The graph below is called a bar graph. This graph shows apple production in Massachusetts from 1994 through 2003. Each bar represents a different year, and the height of each bar represents the apple production for that year in thousands of bushels. (A bushel is 42 lb.)

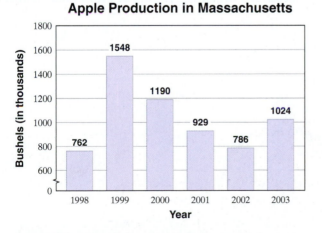

Apple Production in Massachusetts

(Note: The ⌇ symbol means that some numbers are missing. Along the vertical data line, notice the numbers between 0 and 600 are missing or not shown.) (Source: New England Agriculture Statistical Service.)

77. Write an inequality comparing the apple production in 1998 with the apple production in 1999.

78. Write an inequality comparing the apple production in 2003 with the apple production in 2002.

79. Determine the change in apple production between 2000 and 2001.

80. According to the bar graph, which year produced the largest crops?

The apparent magnitude of a star is the measure of its brightness as seen by someone on Earth. The smaller the apparent magnitude, the brighter the star. Below, the apparent magnitudes of some stars are listed. Use this table to answer Exercises 81 through 86.

Star	Apparent Magnitude	Star	Apparent Magnitude
Arcturus	−0.04	Spica	0.98
Sirius	−1.46	Rigel	0.12
Vega	0.03	Regulus	1.35
Antares	0.96	Canopus	−0.72
Sun	−26.7	Hadar	0.61

(*Source: Norton's 2000.0: Star Atlas and Reference Handbook,* 18th ed., Longman Group, UK, 1989)

81. The apparent magnitude of the sun is −26.7. The apparent magnitude of the star Arcturus is −0.04. Write an inequality statement comparing the numbers −0.04 and −26.7.

82. The apparent magnitude of Antares is 0.96. The apparent magnitude of Spica is 0.98. Write an inequality statement comparing the numbers 0.96 and 0.98.

83. Which is brighter, the sun or Arcturus?

84. Which is dimmer, Antares or Spica?

85. Which star listed is the brightest?

86. Which star listed is the dimmest?

87. In your own words, explain how to find the absolute value of a number.

88. Give an example of a real-life situation that can be described with integers but not with whole numbers.

1.3 EXPONENTS, ORDER OF OPERATIONS, AND VARIABLE EXPRESSIONS

Objectives

A Define and Use Exponents and the Order of Operations.

B Evaluate Algebraic Expressions, Given Replacement Values for Variables.

C Determine Whether a Number Is a Solution of a Given Equation.

D Translate Phrases into Expressions and Sentences into Equations.

Objective **A** Exponents and the Order of Operations

Frequently in algebra, products occur that contain repeated multiplication of the same factor. For example, the volume of a cube whose sides each measure 2 centimeters is $(2 \cdot 2 \cdot 2)$ cubic centimeters. We may use **exponential notation** to write such products in a more compact form. For example,

$2 \cdot 2 \cdot 2$ may be written as 2^3.

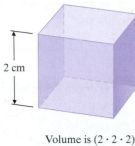

2 cm

Volume is $(2 \cdot 2 \cdot 2)$ cubic centimeters.

The 2 in 2^3 is called the **base**; it is the repeated factor. The 3 in 2^3 is called the **exponent** and is the number of times the base is used as a factor. The expression 2^3 is called an **exponential expression.**

$$\overset{\text{exponent}}{2^3} = 2 \cdot 2 \cdot 2 = 8$$
base $\qquad$ 2 is a factor 3 times.

EXAMPLE 1 Evaluate (find the value of) each expression.

a. 3^2 [read as "3 squared" or as "3 to the second power"]
b. 5^3 [read as "5 cubed" or as "5 to the third power"]
c. 2^4 [read as "2 to the fourth power"]
d. 7^1
e. $\left(\dfrac{3}{7}\right)^2$

Solution:

a. $3^2 = 3 \cdot 3 = 9$
b. $5^3 = 5 \cdot 5 \cdot 5 = 125$
c. $2^4 = 2 \cdot 2 \cdot 2 \cdot 2 = 16$
d. $7^1 = 7$
e. $\left(\dfrac{3}{7}\right)^2 = \left(\dfrac{3}{7}\right)\left(\dfrac{3}{7}\right) = \dfrac{3 \cdot 3}{7 \cdot 7} = \dfrac{9}{49}$

🟨 **Work Practice Problem 1**

PRACTICE PROBLEM 1

Evaluate each expression.
a. 4^2
b. 2^2
c. 3^4
d. 9^1
e. $\left(\dfrac{2}{5}\right)^3$

Helpful Hint

$2^3 \neq 2 \cdot 3$ since 2^3 indicates **repeated multiplication of the same factor.**

$2^3 = 2 \cdot 2 \cdot 2 = 8$, whereas $2 \cdot 3 = 6$

Answers
1. a. 16, **b.** 4, **c.** 81, **d.** 9, **e.** $\dfrac{8}{125}$

Using symbols for mathematical operations is a great convenience. The more operation symbols presented in an expression, the more careful we must be when performing the indicated operation. For example, in the expression $2 + 3 \cdot 7$, do we add first or multiply first? To eliminate confusion, **grouping symbols** are used. Examples of grouping symbols are parentheses ( ), brackets [ ], braces { }, absolute value bars | |, and the fraction bar. If we wish $2 + 3 \cdot 7$ to be simplified by adding first, we enclose $2 + 3$ in parentheses.

$$(2 + 3) \cdot 7 = 5 \cdot 7 = 35$$

If we wish to multiply first, $3 \cdot 7$ may be enclosed in parentheses.

$$2 + (3 \cdot 7) = 2 + 21 = 23$$

To eliminate confusion when no grouping symbols are present, we use the following agreed-upon order of operations.

Order of Operations

1. Perform all operations within grouping symbols first, starting with the innermost set.
2. Evaluate exponential expressions.
3. Multiply or divide in order from left to right.
4. Add or subtract in order from left to right.

Using this order of operations, we now simplify $2 + 3 \cdot 7$. There are no grouping symbols and no exponents, so we multiply and then add.

$$2 + 3 \cdot 7 = 2 + 21 \qquad \text{Multiply.}$$
$$= 23 \qquad \text{Add.}$$

PRACTICE PROBLEMS 2–5

Simplify each expression.

2. $3 \cdot 2 + 4^2$

3. $28 \div 7 \cdot 2$

4. $\dfrac{9}{5} \cdot \dfrac{1}{3} - \dfrac{1}{3}$

5. $5 + 3[2(3 \cdot 4 + 1) - 20]$

EXAMPLES Simplify each expression.

2. $6 \div 3 + 5^2 = 6 \div 3 + 25$ Evaluate 5^2
$$= 2 + 25 \qquad \text{Divide.}$$
$$= 27 \qquad \text{Add.}$$

3. $20 \div 5 \cdot 4 = 4 \cdot 4$
$$= 16$$

> **Helpful Hint**
> Remember to multiply or divide in order from left to right.

4. $\dfrac{3}{2} \cdot \dfrac{1}{2} - \dfrac{1}{2} = \dfrac{3}{4} - \dfrac{1}{2}$ Multiply.
$$= \dfrac{3}{4} - \dfrac{2}{4} \qquad \text{The least common denominator is 4.}$$
$$= \dfrac{1}{4} \qquad \text{Subtract.}$$

5. $1 + 2[5(2 \cdot 3 + 1) - 10] = 1 + 2[5(7) - 10]$ Simplify the expression in the innermost set of parentheses. $2 \cdot 3 + 1 = 6 + 1 = 7$.
$$= 1 + 2[35 - 10] \qquad \text{Multiply 5 and 7.}$$
$$= 1 + 2[25] \qquad \text{Subtract inside the brackets.}$$
$$= 1 + 50 \qquad \text{Multiply 2 and 25.}$$
$$= 51 \qquad \text{Add.}$$

Work Practice Problems 2–5

In the next example, the fraction bar serves as a grouping symbol and separates the numerator and denominator. Simplify each separately.

Answers

2. 22,   **3.** 8,   **4.** $\dfrac{4}{15}$,   **5.** 23

EXAMPLE 6 Simplify: $\dfrac{3 + |4 - 3| + 2^2}{6 - 3}$

Solution:

$$\dfrac{3 + |4 - 3| + 2^2}{6 - 3} = \dfrac{3 + |1| + 2^2}{6 - 3}$$ Simplify the expression inside the absolute value bars.

$$= \dfrac{3 + 1 + 2^2}{3}$$ Find the absolute value and simplify the denominator.

$$= \dfrac{3 + 1 + 4}{3}$$ Evaluate the exponential expression.

$$= \dfrac{8}{3}$$ Simplify the numerator.

🔲 **Work Practice Problem 6**

PRACTICE PROBLEM 6

Simplify: $\dfrac{1 + |7 - 4| + 3^2}{8 - 5}$

Helpful Hint

Be careful when evaluating an exponential expression.

$3 \cdot 4^2 = 3 \cdot 16 = 48$ $(3 \cdot 4)^2 = (12)^2 = 144$

    ↑   ↑

Base is 4.   Base is $3 \cdot 4$.

Objective B Evaluating Algebraic Expressions

Recall that letters used to represent quantities are called **variables.** An **algebraic expression** is a collection of numbers, variables, operation symbols, and grouping symbols. For example,

$$2x, \quad -3, \quad 2x - 10, \quad 5(p^2 + 1), \quad xy, \quad \text{and} \quad \dfrac{3y^2 - 6y + 1}{5}$$

are algebraic expressions.

Expressions	Meaning
$2x$	$2 \cdot x$
$5(p^2 + 1)$	$5 \cdot (p^2 + 1)$
$3y^2$	$3 \cdot y^2$
xy	$x \cdot y$

If we give a specific value to a variable, we can **evaluate an algebraic expression.** To evaluate an algebraic expression means to find its numerical value once we know the values of the variables.

Algebraic expressions are often used in problem solving. For example, the expression

$$16t^2$$

gives the distance in feet (neglecting air resistance) that an object will fall in t seconds.

Answer

6. $\dfrac{13}{3}$

PRACTICE PROBLEM 7

Evaluate each expression when $x = 1$ and $y = 4$.

a. $3y^2$

b. $2y - x$

c. $\dfrac{11x}{3y}$

d. $\dfrac{x}{y} + \dfrac{6}{y}$

e. $y^2 - x^2$

EXAMPLE 7 Evaluate each expression when $x = 3$ and $y = 2$.

a. $5x^2$ **b.** $2x - y$ **c.** $\dfrac{3x}{2y}$ **d.** $\dfrac{x}{y} + \dfrac{y}{2}$ **e.** $x^2 - y^2$

Solution:

a. Replace x with 3. Then simplify.

$$5x^2 = 5 \cdot (3)^2 = 5 \cdot 9 = 45$$

b. Replace x with 3 and y with 2.

$$2x - y = 2(3) - 2 \quad \text{Let } x = 3 \text{ and } y = 2.$$
$$= 6 - 2 \quad \text{Multiply.}$$
$$= 4 \quad \text{Subtract.}$$

c. $\dfrac{3x}{2y} = \dfrac{3 \cdot 3}{2 \cdot 2} = \dfrac{9}{4} \quad \text{Let } x = 3 \text{ and } y = 2.$

d. Replace x with 3 and y with 2. Then simplify.

$$\dfrac{x}{y} + \dfrac{y}{2} = \dfrac{3}{2} + \dfrac{2}{2} = \dfrac{5}{2}$$

e. Replace x with 3 and y with 2.

$$x^2 - y^2 = 3^2 - 2^2 = 9 - 4 = 5$$

Work Practice Problem 7

Objective C Solutions of Equations

Many times a problem-solving situation is modeled by an equation. An **equation** is a mathematical statement that two expressions have equal value. The equal symbol "=" is used to equate the two expressions. For example, $3 + 2 = 5$, $7x = 35$, $\dfrac{2(x - 1)}{3} = 0$, and $I = PRT$ are all equations.

> **Helpful Hint**
>
> An equation contains the equal symbol "=". An algebraic expression does not.

✔**Concept Check** Which of the following are equations? Which are expressions?

a. $5x = 8$ **b.** $5x - 8$ **c.** $12y + 3x$ **d.** $12y = 3x$

When an equation contains a variable, deciding which value(s) of the variable make the equation a true statement is called **solving** the equation for the variable. A **solution** of an equation is a value for the variable that makes the equation a true statement. For example, 3 is a solution of the equation $x + 4 = 7$, because if x is replaced with 3 the statement is true.

$$x + 4 = 7$$
$$\downarrow$$
$$3 + 4 \overset{?}{=} 7 \quad \text{Replace } x \text{ with 3.}$$
$$7 = 7 \quad \text{True}$$

Similarly, 1 is not a solution of the equation $x + 4 = 7$, because $1 + 4 = 7$ is **not** a true statement.

Answers

7. **a.** 48, **b.** 7, **c.** $\dfrac{11}{12}$, **d.** $\dfrac{7}{4}$, **e.** 15

✔ **Concept Check Answer**

equations: **a, d**; expressions: **b, c**

EXAMPLE 8 Decide whether 2 is a solution of $3x + 10 = 8x$.

Solution: Replace x with 2 and see if a true statement results.

$$3x + 10 = 8x \qquad \text{Original equation}$$
$$3(2) + 10 \stackrel{?}{=} 8(2) \qquad \text{Replace } x \text{ with 2.}$$
$$6 + 10 \stackrel{?}{=} 16 \qquad \text{Simplify each side.}$$
$$16 = 16 \qquad \text{True}$$

Since we arrived at a true statement after replacing x with 2 and simplifying both sides of the equation, 2 is a solution of the equation.

🔲 **Work Practice Problem 8**

Objective Ⓓ Translating Words to Symbols

Now that we know how to represent an unknown number by a variable, let's practice translating phrases into algebraic expressions and sentences into equations. Oftentimes solving problems involves the ability to translate word phrases and sentences into symbols. Below is a list of key words and phrases to help us translate.

> **Helpful Hint**
>
> Order matters when subtracting and also dividing, so be especially careful with these translations.

Addition (+)	Subtraction (−)	Multiplication (·)	Division (÷)	Equality (=)
Sum	Difference of	Product	Quotient	Equals
Plus	Minus	Times	Divide	Gives
Added to	Subtracted from	Multiply	Into	Is/was/ should be
More than	Less than	Twice	Ratio	Yields
Increased by	Decreased by	Of	Divided by	Amounts to
Total	Less			Represents Is the same as

EXAMPLE 9 Write an algebraic expression that represents each phrase. Let the variable x represent the unknown number.

a. The sum of a number and 3
b. The product of 3 and a number
c. The quotient of 7.3 and a number
d. 10 decreased by a number
e. 5 times a number, increased by 7

Solution:

a. $x + 3$ since "sum" means to add
b. $3 \cdot x$ and $3x$ are both ways to denote the product of 3 and x
c. $7.3 \div x$ or $\dfrac{7.3}{x}$
d. $10 - x$ because "decreased by" means to subtract
e. $\underbrace{5x}_{\substack{5 \text{ times} \\ \text{a number}}} + 7$

🔲 **Work Practice Problem 9**

PRACTICE PROBLEM 8

Decide whether 3 is a solution of $5x - 10 = x + 2$.

PRACTICE PROBLEM 9

Write an algebraic expression that represents each phrase. Let the variable x represent the unknown number.

a. The product of 5 and a number
b. A number added to 7
c. A number divided by 11.2
d. A number subtracted from 8
e. Twice a number, plus 1

Answers

8. It is a solution.,
9. a. $5 \cdot x$ or $5x$, b. $7 + x$,
c. $x \div 11.2$ or $\dfrac{x}{11.2}$, d. $8 - x$,
e. $2x + 1$

Helpful
Hint

Make sure you understand the difference when translating phrases containing "decreased by," "subtracted from," and "less than."

Phrase	Translation
A number decreased by 10	$x - 10$
A number subtracted from 10	$10 - x$
10 less than a number	$x - 10$
A number less 10	$x - 10$

Notice the order.

Now let's practice translating sentences into equations.

PRACTICE PROBLEM 10

Write each sentence as an equation. Let x represent the unknown number.

a. The ratio of a number and 6 is 24.

b. The difference of 10 and a number is 18.

c. One less than twice a number is 99.

EXAMPLE 10 Write each sentence as an equation. Let x represent the unknown number.

a. The quotient of 15 and a number is 4.

b. Three subtracted from 12 is a number.

c. 17 added to four times a number is 21.

Solution:

a. In words: the quotient of 15 and a number   is   4

Translate: $\dfrac{15}{x}$ = 4

b. In words: three subtracted **from** 12   is   a number

Translate: $12 - 3$ = x

Care must be taken when the operation is subtraction. The expression $3 - 12$ would be incorrect. Notice that $3 - 12 \neq 12 - 3$.

c. In words: 17   added to   four times a number   is   21

Translate: 17 + $4x$ = 21

🔲 **Work Practice Problem 10**

Answers

10. a. $\dfrac{x}{6} = 24$, **b.** $10 - x = 18$,

c. $2x - 1 = 99$

CALCULATOR EXPLORATIONS

Exponents

To evaluate exponential expressions on a calculator, find the key marked $\boxed{y^x}$ or $\boxed{\wedge}$. To evaluate, for example, 6^5, press the following keys: $\boxed{6}$ $\boxed{y^x}$ $\boxed{5}$ $\boxed{=}$ or $\boxed{6}$ $\boxed{\wedge}$ $\boxed{5}$ $\boxed{=}$.

↕ or
$\boxed{\text{ENTER}}$

The display should read $\boxed{7776}$.

Order of Operations

Some calculators follow the order of operations, and others do not. To see whether or not your calculator has the order of operations built in, use your calculator to find $2 + 3 \cdot 4$. To do this, press the following sequence of keys:

$\boxed{2}$ $\boxed{+}$ $\boxed{3}$ $\boxed{\times}$ $\boxed{4}$ $\boxed{=}$.

↕ or
$\boxed{\text{ENTER}}$

The correct answer is 14 because the order of operations is to multiply before we add. If the calculator displays $\boxed{14}$, then it has the order of operations built in.

Even if the order of operations is built in, parentheses must sometimes be inserted. For example, to simplify $\dfrac{5}{12 - 7}$, press the keys

$\boxed{5}$ $\boxed{\div}$ $\boxed{(}$ $\boxed{1}$ $\boxed{2}$ $\boxed{-}$ $\boxed{7}$ $\boxed{)}$ $\boxed{=}$.

↕ or
$\boxed{\text{ENTER}}$

The display should read $\boxed{1}$.

Use a calculator to evaluate each expression.

1. 5^3 　　　　　　　　 **2.** 7^4

3. 9^5 　　　　　　　　 **4.** 8^6

5. $2(20 - 5)$ 　　　　　 **6.** $3(14 - 7) + 21$

7. $24(862 - 455) + 89$

8. $99 + (401 + 962)$

9. $\dfrac{4623 + 129}{36 - 34}$

10. $\dfrac{956 - 452}{89 - 86}$

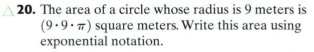

1.3 EXERCISE SET

Objective A *Evaluate. See Example 1.*

1. 3^5 　　　 **2.** 5^4 　　　 ⊙ **3.** 3^3 　　　 **4.** 4^4 　　　 **5.** 1^5 　　　 **6.** 1^8

7. 5^1 　　　 **8.** 8^1 　　　 **9.** 7^2 　　　 **10.** 9^2 　　　 ⊙ **11.** $\left(\dfrac{2}{3}\right)^4$ 　　　 **12.** $\left(\dfrac{6}{11}\right)^2$

13. $\left(\dfrac{1}{5}\right)^3$ 　　　 **14.** $\left(\dfrac{1}{2}\right)^5$ 　　　 **15.** $(1.2)^2$ 　　　 **16.** $(1.5)^2$ 　　　 **17.** $(0.7)^3$ 　　　 **18.** $(0.4)^3$

△ **19.** The area of a square whose sides each measure 5 meters is $(5 \cdot 5)$ square meters. Write this area using exponential notation.

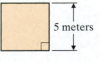

5 meters

△ **20.** The area of a circle whose radius is 9 meters is $(9 \cdot 9 \cdot \pi)$ square meters. Write this area using exponential notation.

9 m

Simplify each expression. See Examples 2 through 6.

21. $5 + 6 \cdot 2$

22. $8 + 5 \cdot 3$

23. $4 \cdot 8 - 6 \cdot 2$

24. $12 \cdot 5 - 3 \cdot 6$

25. $18 \div 3 \cdot 2$

26. $48 \div 6 \cdot 2$

27. $2 + (5 - 2) + 4^2$

28. $6 - 2 \cdot 2 + 2^5$

29. $5 \cdot 3^2$

30. $2 \cdot 5^2$

31. $\dfrac{1}{4} \cdot \dfrac{2}{3} - \dfrac{1}{6}$

32. $\dfrac{3}{4} \cdot \dfrac{1}{2} + \dfrac{2}{3}$

33. $\dfrac{6 - 4}{9 - 2}$

34. $\dfrac{8 - 5}{24 - 20}$

35. $2[5 + 2(8 - 3)]$

36. $3[4 + 3(6 - 4)]$

37. $\dfrac{19 - 3 \cdot 5}{6 - 4}$

38. $\dfrac{14 - 2 \cdot 3}{12 - 8}$

39. $\dfrac{|6 - 2| + 3}{8 + 2 \cdot 5}$

40. $\dfrac{15 - |3 - 1|}{12 - 3 \cdot 2}$

41. $\dfrac{3 + 3(5 + 3)}{3^2 + 1}$

42. $\dfrac{3 + 6(8 - 5)}{4^2 + 2}$

43. $\dfrac{6 + |8 - 2| + 3^2}{18 - 3}$

44. $\dfrac{16 + |13 - 5| + 4^2}{17 - 5}$

45. $2 + 3[10(4 \cdot 5 - 16) - 30]$

46. $3 + 4[8(5 \cdot 5 - 20) - 41]$

47. $\left(\dfrac{2}{3}\right)^3 + \dfrac{1}{9} + \dfrac{1}{3} \cdot \dfrac{4}{3}$

48. $\left(\dfrac{3}{8}\right)^2 + \dfrac{1}{4} + \dfrac{1}{8} \cdot \dfrac{3}{2}$

Objective **B** *Evaluate each expression when $x = 1$, $y = 3$, and $z = 5$. See Example 7.*

49. $3y$

50. $4x$

51. $\dfrac{z}{5x}$

52. $\dfrac{y}{2z}$

53. $3x - 2$

54. $6y - 8$

55. $|2x + 3y|$

56. $|5z - 2y|$

57. $xy + z$

58. $yz - x$

59. $5y^2$

60. $2z^2$

Evaluate each expression when $x = 2$, $y = 6$, and $z = 3$. See Example 7.

61. $5z$

62. $7x$

63. $\dfrac{z}{xy}$

64. $\dfrac{x}{yz}$

65. $\dfrac{y}{x} + \dfrac{y}{x}$

66. $\dfrac{9}{z} + \dfrac{4z}{y}$

Objective C *Decide whether the given number is a solution of the given equation. See Example 8.*

67. $3x - 6 = 9$; 5

68. $2x + 7 = 3x$; 6

69. $2x + 6 = 5x - 1$; 0

70. $4x + 2 = x + 8$; 2

71. $2x - 5 = 5$; 8

72. $3x - 10 = 8$; 6

73. $x + 6 = x + 6$; 2

74. $x + 6 = x + 6$; 10

75. $x = 5x + 15$; 0

76. $4 = 1 - x$; 1

77. $\frac{1}{3}x = 9$; 27

78. $\frac{2}{7}x = \frac{3}{14}$; 6

Objective D *Write each phrase as an algebraic expression. Let x represent the unknown number. See Example 9.*

79. Fifteen more than a number

80. A number increased by 9

81. Five subtracted from a number

82. Five decreased by a number

83. The ratio of a number and 4

84. The quotient of a number and 9

85. Three times a number, increased by 22

86. Twice a number, decreased by 72

Write each sentence as an equation. Use x to represent any unknown number. See Example 10.

87. One increased by two equals the quotient of nine and three.

88. Four subtracted from eight is equal to two squared.

89. Three is not equal to four divided by two.

90. The difference of sixteen and four is greater than ten.

91. The sum of 5 and a number is 20.

92. Seven subtracted from a number is 0.

93. The product of 7.6 and a number is 17.

94. 9.1 times a number equals 4

95. Thirteen minus three times a number is 13.

96. Eight added to twice a number is 42.

Concept Extensions

97. Are parentheses necessary in the expression $2 + (3 \cdot 5)$? Explain your answer.

98. Are parentheses necessary in the expression $(2 + 3) \cdot 5$? Explain your answer.

For Exercises 99 and 100, match each expression in the first column with its value in the second column.

99.
 a. $(6 + 2) \cdot (5 + 3)$ 19
 b. $(6 + 2) \cdot 5 + 3$ 22
 c. $6 + 2 \cdot 5 + 3$ 64
 d. $6 + 2 \cdot (5 + 3)$ 43

100.
 a. $(1 + 4) \cdot 6 - 3$ 15
 b. $1 + 4 \cdot (6 - 3)$ 13
 c. $1 + 4 \cdot 6 - 3$ 27
 d. $(1 + 4) \cdot (6 - 3)$ 22

Recall that perimeter measures the distance around a plane figure and area measures the amount of surface of a plane figure. The expression $2l + 2w$ gives the perimeter of the rectangle below, and the expression lw gives its area (measured in square units).

101. Complete the chart below for the given lengths and widths. Be sure to include units.

Length: l	Width: w	Perimeter of Rectangle: 2l + 2w	Area of Rectangle: lw
4 in.	3 in.		
6 in.	1 in.		
5 in.	2 in.		

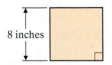

102. Study the perimeters and areas found in the chart to the left. Do you notice any trends?

103. Insert one set of parentheses so that the following expression simplifies to 32.

$$20 - 4 \cdot 4 \div 2$$

104. Insert parentheses so that the following expression simplifies to 28.

$$2 \cdot 5 + 3^2$$

105. In your own words, explain the difference between an expression and an equation.

Determine whether each is an expression or an equation. See the Concept Check in this section.

106.
 a. $3x^2 - 26$
 b. $3x^2 - 26 = 1$
 c. $2x - 5 = 7x - 5$
 d. $9y + x - 8$
 e. $3^2 - 4(5 - 3)$

107.
 a. $5x + 6$
 b. $2a = 7$
 c. $3a + 2 = 9$
 d. $4x + 3y - 8z$
 e. $5^2 - 2(6 - 2)$

108. Why is 8^2 usually read as "eight squared"? (*Hint:* What is the area of the **square** below?)

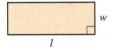

8 inches

109. Why is 4^3 usually read as "four cubed"? (*Hint:* What is the volume of the **cube** below?)

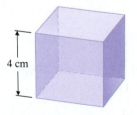

4 cm

110. Write any expression, using 4 or more numbers, that simplifies to 7.

1.4 ADDING REAL NUMBERS

Objectives

A Add Real Numbers.

B Find the Opposite of a Number.

C Solve Problems That Involve Addition of Real Numbers.

Real numbers can be added, subtracted, multiplied, divided, and raised to powers, just as whole numbers can.

Objective A Adding Real Numbers

Adding real numbers can be visualized by using a number line. A positive number can be represented on the number line by an arrow of appropriate length pointing to the right, and a negative number by an arrow of appropriate length pointing to the left.

Both arrows represent 2 or +2.

They both point to the right, and they are both 2 units long.

Both arrows represent −3.

They both point to the left, and they are both 3 units long.

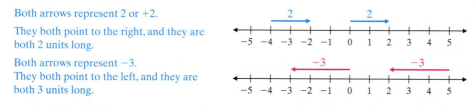

To add signed numbers such as $5 + (-2)$ on a number line, we start at 0 on the number line and draw an arrow representing 5. From the tip of this arrow, we draw another arrow representing −2. The tip of the second arrow ends at their sum, 3.

$$5 + (-2) = 3$$

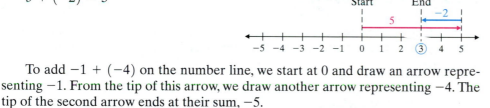

To add $-1 + (-4)$ on the number line, we start at 0 and draw an arrow representing −1. From the tip of this arrow, we draw another arrow representing −4. The tip of the second arrow ends at their sum, −5.

$$-1 + (-4) = -5$$

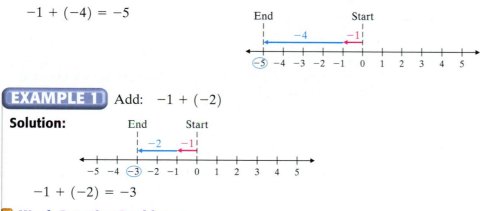

EXAMPLE 1 Add: $-1 + (-2)$

Solution:

$$-1 + (-2) = -3$$

▢ **Work Practice Problem 1**

PRACTICE PROBLEM 1

Add using a number line:
$-2 + (-4)$

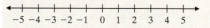

Thinking of integers as money earned or lost might help make addition more meaningful. Earnings can be thought of as positive numbers. If $1 is earned and later another $3 is earned, the total amount earned is $4. In other words, $1 + 3 = 4$.

On the other hand, losses can be thought of as negative numbers. If $1 is lost and later another $3 is lost, a total of $4 is lost. In other words, $(-1) + (-3) = -4$.

In Example 1, we added numbers with the same sign. Adding numbers whose signs are not the same can be pictured on a number line also.

Answer

1. −6

PRACTICE PROBLEM 2

Add using a number line:
$-5 + 8$

EXAMPLE 2 Add: $-4 + 6$

Solution:

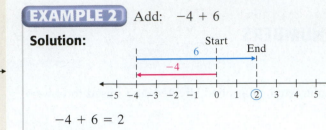

$-4 + 6 = 2$

■ **Work Practice Problem 2**

Let's use temperature as an example. If the thermometer registers 4 degrees below 0 degrees and then rises 6 degrees, the new temperature is 2 degrees above 0 degrees. Thus, it is reasonable that $-4 + 6 = 2$. (See the diagram in the margin.)

EXAMPLE 3 Add: $4 + (-6)$

Solution:

$4 + (-6) = -2$

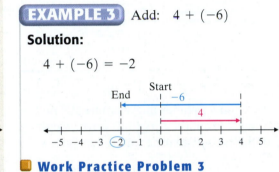

■ **Work Practice Problem 3**

PRACTICE PROBLEM 3

Add using a number line:
$5 + (-4)$

Using a number line each time we add two numbers can be time consuming. Instead, we can notice patterns in the previous examples and write rules for adding real numbers.

Adding Real Numbers

To add two real numbers

1. with the *same sign,* add their absolute values. Use their common sign as the sign of the answer.
2. with *different signs,* subtract their absolute values. Give the answer the same sign as the number with the larger absolute value.

PRACTICE PROBLEM 4

Add without using a number line: $(-8) + (-5)$

EXAMPLE 4 Add without using a number line: $(-7) + (-6)$

Solution: Here, we are adding two numbers with the same sign.

$(-7) + (-6) = -13$

↑ ↖ sum of absolute values ($|-7| = 7, |-6| = 6, 7 + 6 = 13$)
same sign

■ **Work Practice Problem 4**

PRACTICE PROBLEM 5

Add without using a number line: $(-14) + 6$

EXAMPLE 5 Add without using a number line: $(-10) + 4$

Solution: Here, we are adding two numbers with different signs.

$(-10) + 4 = -6$

↑ ↖ difference of absolute values ($|-10| = 10, |4| = 4, 10 - 4 = 6$)
sign of number with larger absolute value, -10

■ **Work Practice Problem 5**

Answers

2. 3, **3.** 1, **4.** -13, **5.** -8

EXAMPLES Add without using a number line.

6. $(-8) + (-11) = -19$
7. $(-2) + 10 = 8$
8. $0.2 + (-0.5) = -0.3$
9. $-\dfrac{7}{10} + \left(-\dfrac{1}{10}\right) = -\dfrac{8}{10} = -\dfrac{\cancel{2} \cdot 4}{\cancel{2} \cdot 5} = -\dfrac{4}{5}$
10. $11.4 + (-4.7) = 6.7$
11. $-\dfrac{3}{8} + \dfrac{2}{5} = -\dfrac{15}{40} + \dfrac{16}{40} = \dfrac{1}{40}$

▪ **Work Practice Problems 6–11**

EXAMPLE 12 Find each sum.

a. $3 + (-7) + (-8)$
b. $[7 + (-10)] + [-2 + (-4)]$

Solution:

a. Perform the additions from left to right.

$3 + (-7) + (-8) = -4 + (-8)$ Adding numbers with different signs
$\qquad\qquad\qquad\quad = -12$ Adding numbers with like signs

b. Simplify inside the brackets first.

$[7 + (-10)] + [-2 + (-4)] = [-3] + [-6]$
$\qquad\qquad\qquad\qquad\qquad = -9$ Add.

▪ **Work Practice Problem 12**

Objective B Finding Opposites

To help us subtract real numbers in the next section, we first review what we mean by opposites. The graphs of 4 and −4 are shown on the number line below.

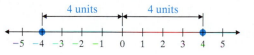

Notice that the graph of 4 and −4 lie on opposite sides of 0, and each is 4 units away from 0. Such numbers are known as **opposites** or **additive inverses** of each other.

Opposite or Additive Inverse

Two numbers that are the same distance from 0 but lie on opposite sides of 0 are called **opposites** or **additive inverses** of each other.

EXAMPLES Find the opposite of each number.

13. 10   The opposite of 10 is −10.
14. −3   The opposite of −3 is 3.
15. $\dfrac{1}{2}$ The opposite of $\dfrac{1}{2}$ is $-\dfrac{1}{2}$.
16. −4.5   The opposite of −4.5 is 4.5.

▪ **Work Practice Problems 13–16**

PRACTICE PROBLEMS 6–11

Add without using a number line.

6. $(-17) + (-10)$
7. $(-4) + 12$
8. $1.5 + (-3.2)$
9. $-\dfrac{6}{12} + \left(-\dfrac{3}{12}\right)$
10. $12.1 + (-3.6)$
11. $-\dfrac{4}{5} + \dfrac{2}{3}$

PRACTICE PROBLEM 12

Find each sum.

a. $16 + (-9) + (-9)$
b. $[3 + (-13)] + [-4 + (-7)]$

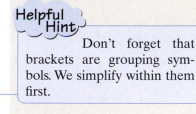

Helpful Hint

Don't forget that brackets are grouping symbols. We simplify within them first.

PRACTICE PROBLEMS 13–16

Find the opposite of each number.

13. −35   14. 12
15. $-\dfrac{3}{11}$ 16. 1.9

Answers

6. −27,   7. 8,   8. −1.7,   9. $-\dfrac{3}{4}$,
10. 8.5,   11. $-\dfrac{2}{15}$,   12. a. −2,
b. −21,   13. 35,   14. −12,   15. $\dfrac{3}{11}$,
16. −1.9

We use the symbol "−" to represent the phrase "the opposite of" or "the additive inverse of." In general, if a is a number, we write the opposite or additive inverse of a as $-a$. We know that the opposite of -3 is 3. Notice that this translates as

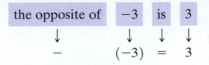

	the opposite of	−3	is	3
	↓	↓	↓	↓
	−	(−3)	=	3

This is true in general.

> If a is a number, then $-(-a) = a$.

PRACTICE PROBLEM 17

Simplify each expression.

a. $-(-22)$

b. $-\left(-\dfrac{2}{7}\right)$

c. $-(-x)$

d. $-|-14|$

e. $-|2.3|$

EXAMPLE 17 Simplify each expression.

a. $-(-10)$ **b.** $-\left(-\dfrac{1}{2}\right)$ **c.** $-(-2x)$

d. $-|-6|$ **e.** $-|4.1|$

Solution:

a. $-(-10) = 10$ **b.** $-\left(-\dfrac{1}{2}\right) = \dfrac{1}{2}$

c. $-(-2x) = 2x$

d. $-|-6| = -6$ Since $|-6| = 6$.

e. $-|4.1| = -4.1$ Since $|4.1| = 4.1$.

■ **Work Practice Problem 17**

Let's discover another characteristic about opposites. Notice that the sum of a number and its opposite is 0.

$$10 + (-10) = 0$$
$$-3 + 3 = 0$$
$$\frac{1}{2} + \left(-\frac{1}{2}\right) = 0$$

In general, we can write the following:

> The sum of a number a and its opposite $-a$ is 0.
>
> $$a + (-a) = 0 \qquad \text{Also,} \qquad -a + a = 0.$$

Notice that this means that the opposite of 0 is then 0 since $0 + 0 = 0$.

Answers

17. a. 22, **b.** $\dfrac{2}{7}$, **c.** x, **d.** -14, **e.** -2.3

Objective C Solving Problems That Involve Addition

Positive and negative numbers are used in everyday life. Stock market returns show gains and losses as positive and negative numbers. Temperatures in cold climates often dip into the negative range, commonly referred to as "below zero" temperatures. Bank statements report deposits and withdrawals as positive and negative numbers.

EXAMPLE 18 Calculating Gain or Loss

During a three-day period, a share of Lamplighter's International stock recorded the following gains and losses:

Monday	Tuesday	Wednesday
a gain of $2	a loss of $1	a loss of $3

Find the overall gain or loss for the stock for the three days.

Solution: Gains can be represented by positive numbers. Losses can be represented by negative numbers. The overall gain or loss is the sum of the gains and losses.

In words: gain plus loss plus loss

Translate: $2 + (-1) + (-3) = -2$

The overall loss is $2.

🟧 **Work Practice Problem 18**

PRACTICE PROBLEM 18

During a four-day period, a share of Walco stock recorded the following gains and losses:

Tuesday	Wednesday
a loss of $2	a loss of $1
Thursday	**Friday**
a gain of $3	a gain of $3

Find the overall gain or loss for the stock for the four days.

Answer

18. a gain of $3

Objective A *Add. See Examples 1 through 12.*

1. $6 + (-3)$

2. $9 + (-12)$

3. $-6 + (-8)$

4. $-6 + (-14)$

5. $8 + (-7)$

6. $16 + (-4)$

7. $-14 + 2$

8. $-10 + 5$

9. $-2 + (-3)$

10. $-7 + (-4)$

11. $-9 + (-3)$

12. $-11 + (-5)$

13. $-7 + 3$

14. $-5 + 9$

15. $10 + (-3)$

16. $8 + (-6)$

17. $5 + (-7)$

18. $3 + (-6)$

19. $-16 + 16$

20. $23 + (-23)$

21. $27 + (-46)$

22. $53 + (-37)$

23. $-18 + 49$

24. $-26 + 14$

25. $-33 + (-14)$

26. $-18 + (-26)$

27. $6.3 + (-8.4)$

28. $9.2 + (-11.4)$

29. $117 + (-79)$

30. $144 + (-88)$

31. $-9.6 + (-3.5)$

32. $-6.7 + (-7.6)$

33. $-\dfrac{3}{8} + \dfrac{5}{8}$

34. $-\dfrac{5}{12} + \dfrac{7}{12}$

35. $-\dfrac{7}{16} + \dfrac{1}{4}$

36. $-\dfrac{5}{9} + \dfrac{1}{3}$

37. $-\dfrac{7}{10} + \left(-\dfrac{3}{5}\right)$

38. $-\dfrac{5}{6} + \left(-\dfrac{2}{3}\right)$

39. $|-8| + (-16)$

40. $|-6| + (-61)$

41. $-15 + 9 + (-2)$

42. $-9 + 15 + (-5)$

43. $-21 + (-16) + (-22)$

44. $-18 + (-6) + (-40)$

45. $-23 + 16 + (-2)$

46. $-14 + (-3) + 11$

47. $|5 + (-10)|$

48. $|7 + (-17)|$

49. $6 + (-4) + 9$

50. $8 + (-2) + 7$

51. $[-17 + (-4)] + [-12 + 15]$

52. $[-2 + (-7)] + [-11 + 22]$

53. $|9 + (-12)| + |-16|$

54. $|43 + (-73)| + |-20|$

55. $-13 + [5 + (-3) + 4]$

56. $-30 + [1 + (-6) + 8]$

57. Find the sum of -38 and 12.

58. Find the sum of -44 and 16.

Objective **B** *Find each additive inverse or opposite. See Examples 13 through 17.*

59. 6   **60.** 4   **61.** -2 **62.** -8

63. 0   **64.** $-\dfrac{1}{4}$ **65.** $|-6|$ **66.** $|-11|$

Simplify each of the following. See Example 17.

67. $-|-2|$ **68.** $-|-5|$ **69.** $-(-7)$ **70.** $-(-14)$ **71.** $-(-7.9)$

72. $-(-8.4)$ **73.** $-(-5z)$ **74.** $-(-7m)$ **75.** $\left|-\dfrac{2}{3}\right|$ **76.** $-\left|-\dfrac{2}{3}\right|$

Objective **C** *Solve each of the following. See Example 18.*

77. The lowest temperature ever recorded in Massachusetts was $-35°$F. The highest recorded temperature in Massachusetts was $142°$ higher than the record low temperature. Find Massachusetts' highest recorded temperature. (*Source:* National Climatic Data Center)

78. On January 2, 1943, the temperature was $-4°$ at 7:30 a.m. in Spearfish, South Dakota. Incredibly, it got $49°$ warmer in the next 2 minutes. To what temperature did it rise by 7:32?

79. The lowest elevation on Earth is -411 meters (that is, 411 meters below sea level) at the Dead Sea. If you are standing 316 meters above the Dead Sea, what is your elevation? (*Source:* National Geographic Society)

80. The lowest elevation in Australia is -52 feet at Lake Eyre. If you are standing at a point 439 feet above Lake Eyre, what is your elevation? (*Source:* National Geographic Society)

81. When checking the stock listing in the newspaper, LaTonda finds that one of her stocks posted net changes of -2.50 points and -0.86 point over the last two days. What is the combined change?

82. Yesterday your stock posted a net change of $+0.93$ point, but today it showed a loss of -1.25 points. Find the overall change for two days.

83. In golf, scores that are under par for the entire round are shown as negative scores; scores that are over par are positive, and par is 0. Vijay Singh won the 2004 PGA championship with round scores of -5, -4, -3, $+4$. What was his total overall score? (*Source:* PGA of America)

84. Annika Sorenstam won the 2004 LPGA Samsung World Championship with the following round scores: -6, -4, -3, and -5. What was her total overall score? (*Source:* LPGA of America)

85. A negative net income results when a company spends more money than it brings in. KMart had the following quarterly net incomes during its 2004 fiscal year. (*Source:* Yahoo finance)

Quarter of Fiscal 2004	Net Income (in millions)
First	-23
Second	-581
Third	93
Fourth	155

What was the total net income for fiscal year 2004?

86. Northwest Airlines had the following quarterly net incomes during its 2004 fiscal year. (*Source:* Yahoo finance)

Quarter of Fiscal 2004	Net Income (in millions)
First	13
Second	-1724
Third	-177
Fourth	34

What was the total net income for fiscal year 2004?

Concept Extensions

The following bar graph shows each month's average daily low temperature in degrees Fahrenheit for Barrow, Alaska. Use this graph to answer Exercises 87 through 90.

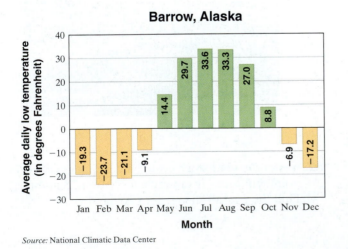

Source: National Climatic Data Center

87. Give any two numbers whose sum is -5.

88. Give two numbers, one positive and one negative whose sum is -10.

89. For what month is the graphed temperature the highest?

90. For what month is the graphed temperature the lowest?

91. For what month is the graphed temperature positive *and* closest to 0°?

92. For what month is the graphed temperature negative *and* closest to 0°?

93. Find the average of the temperatures shown for the months of April, May, and October. (To find the average of three temperatures, find their sum and divide by 3.)

94. Find the average of the temperatures shown for the months of January, September, and October.

If p is a positive number and n is a negative number, fill in the blanks with the words positive or negative.

95. $-p$ is a _____ number.

96. $-n$ is a _____ number.

97. $p + p$ is a _____ number.

98. $n + n$ is a _____ number.

99. Explain why adding a negative number to another negative number always gives a negative sum.

100. When a positive and a negative number are added, sometimes the sum is positive, sometimes it is zero, and sometimes it is negative. Explain why this happens.

101. In your own words, explain how to find the opposite of a number.

102. In your own words, explain why 0 is the only number that is its own opposite.

A Subtract Real Numbers.

B Evaluate Algebraic Expressions Using Real Numbers.

C Determine Whether a Number Is a Solution of a Given Equation.

D Solve Problems That Involve Subtraction of Real Numbers.

E Find Complementary and Supplementary Angles.

1.5 SUBTRACTING REAL NUMBERS

Objective **A** Subtracting Real Numbers

Now that addition of real numbers has been discussed, we can explore subtraction. We know that $9 - 7 = 2$. Notice that $9 + (-7) = 2$, also. This means that

$$9 - 7 = 9 + (-7)$$

Notice that the *difference* of 9 and 7 is the same as the *sum* of 9 and the opposite of 7. This is how we can subtract real numbers.

Subtracting Real Numbers

If a and b are real numbers, then $a - b = a + (-b)$.

In other words, to find the difference of two numbers, we add the opposite of the number being subtracted.

PRACTICE PROBLEM 1

Subtract.
a. $-20 - 6$
b. $3 - (-5)$
c. $7 - 17$
d. $-4 - (-9)$

EXAMPLE 1 Subtract.

a. $-13 - 4$ **b.** $5 - (-6)$ **c.** $3 - 6$ **d.** $-1 - (-7)$

Solution:

a. $-13 - 4 = -13 + (-4)$ Add -13 to the opposite of 4, which is -4.

     add $\longrightarrow$
     $\longleftarrow$ opposite $\longrightarrow$

       $= -17$

b. $5 - (-6) = 5 + (6)$ Add 5 to the opposite of -6, which is 6.

     add $\longrightarrow$
     $\longleftarrow$ opposite $\longrightarrow$

       $= 11$

c. $3 - 6 = 3 + (-6)$ Add 3 to the opposite of 6, which is -6.

       $= -3$

d. $-1 - (-7) = -1 + (7) = 6$

⬛ **Work Practice Problem 1**

> **Helpful Hint**
>
> Study the patterns indicated.
>
> No change $\longrightarrow$ Change to addition.
>   Change to opposite.
>
> $5 - 11 = 5 + (-11) = -6$
>
> $-3 - 4 = -3 + (-4) = -7$
>
> $7 - (-1) = 7 + (1) = 8$

PRACTICE PROBLEMS 2–4

Subtract.
2. $9.6 - (-5.7)$

3. $-\dfrac{4}{9} - \dfrac{2}{9}$

4. $-\dfrac{1}{4} - \left(-\dfrac{2}{5}\right)$

EXAMPLES Subtract.

2. $5.3 - (-4.6) = 5.3 + (4.6) = 9.9$

3. $-\dfrac{3}{10} - \dfrac{5}{10} = -\dfrac{3}{10} + \left(-\dfrac{5}{10}\right) = -\dfrac{8}{10} = -\dfrac{4}{5}$

4. $-\dfrac{2}{3} - \left(-\dfrac{4}{5}\right) = -\dfrac{2}{3} + \left(\dfrac{4}{5}\right) = -\dfrac{10}{15} + \dfrac{12}{15} = \dfrac{2}{15}$

⬛ **Work Practice Problems 2–4**

Answers

1. a. -26, **b.** 8, **c.** -10, **d.** 5,

2. 15.3, **3.** $-\dfrac{2}{3}$, **4.** $\dfrac{3}{20}$

EXAMPLE 5 Write each phrase as an expression and simplify.

a. Subtract 8 from −4.   **b.** Decrease 10 by −20.

Solution: Be careful when interpreting these. The order of numbers in subtraction is important.

a. 8 is to be subtracted **from** −4.

$$-4 - 8 = -4 + (-8) = -12$$

b. To decrease 10 by −20, we find 10 **minus** −20.

$$10 - (-20) = 10 + 20 = 30$$

□ **Work Practice Problem 5**

If an expression contains additions and subtractions, just write the subtractions as equivalent additions. Then simplify from left to right.

EXAMPLE 6 Simplify each expression.

a. $-14 - 8 + 10 - (-6)$ **b.** $1.6 - (-10.3) + (-5.6)$

Solution:

a. $-14 - 8 + 10 - (-6) = -14 + (-8) + 10 + 6 = -6$

b. $1.6 - (-10.3) + (-5.6) = 1.6 + 10.3 + (-5.6) = 6.3$

□ **Work Practice Problem 6**

When an expression contains parentheses and brackets, remember the order of operations. Start with the innermost set of parentheses or brackets and work your way outward.

EXAMPLE 7 Simplify each expression.

a. $-3 + [(-2 - 5) - 2]$ **b.** $2^3 - 10 + [-6 - (-5)]$

Solution:

a. Start with the innermost set of parentheses. Rewrite −2 − 5 as an addition.

$$
\begin{aligned}
-3 + [(-2 - 5) - 2] &= -3 + [(-2 + (-5)) - 2] \\
&= -3 + [(-7) - 2] &&\text{Add: } -2 + (-5). \\
&= -3 + [-7 + (-2)] &&\text{Write } -7 - 2 \text{ as an addition.} \\
&= -3 + [-9] &&\text{Add.} \\
&= -12 &&\text{Add.}
\end{aligned}
$$

b. Start simplifying the expression inside the brackets by writing −6 − (−5) as an addition.

$$
\begin{aligned}
2^3 - 10 + [-6 - (-5)] &= 2^3 - 10 + [-6 + 5] \\
&= 2^3 - 10 + [-1] &&\text{Add.} \\
&= 8 - 10 + (-1) &&\text{Evaluate } 2^3. \\
&= 8 + (-10) + (-1) &&\text{Write } 8 - 10 \text{ as an addition.} \\
&= -2 + (-1) &&\text{Add.} \\
&= -3 &&\text{Add.}
\end{aligned}
$$

□ **Work Practice Problem 7**

Objective B Evaluating Algebraic Expressions

It is important to be able to evaluate expressions for given replacement values. This helps, for example, when checking solutions of equations.

PRACTICE PROBLEM 5
Write each phrase as an expression and simplify.
a. Subtract 7 from −11.
b. Decrease 35 by −25.

PRACTICE PROBLEM 6
Simplify each expression.
a. $-20 - 5 + 12 - (-3)$
b. $5.2 - (-4.4) + (-8.8)$

PRACTICE PROBLEM 7
Simplify each expression.
a. $-9 + [(-4 - 1) - 10]$
b. $5^2 - 20 + [-11 - (-3)]$

Answers
5. a. −18, b. 60, 6. a. −10,
b. 0.8, 7. a. −24, b. −3

PRACTICE PROBLEM 8

Find the value of each expression when $x = 1$ and $y = -4$.

a. $\dfrac{x - y}{14 + x}$

b. $x^2 - y$

EXAMPLE 8 Find the value of each expression when $x = 2$ and $y = -5$.

a. $\dfrac{x - y}{12 + x}$ b. $x^2 - y$

Solution:

a. Replace x with 2 and y with -5. Be sure to put parentheses around -5 to separate signs. Then simplify the resulting expression.

$$\frac{x - y}{12 + x} = \frac{2 - (-5)}{12 + 2} = \frac{2 + 5}{14} = \frac{7}{14} = \frac{1}{2}$$

b. Replace the x with 2 and y with -5 and simplify.

$$x^2 - y = 2^2 - (-5) = 4 - (-5) = 4 + 5 = 9$$

Work Practice Problem 8

Helpful Hint

For additional help when replacing variables with replacement values, first place parentheses about any variables.

For Example 8b above, we have

$$x^2 - y = \underbrace{(x)^2 - (y)}_{\substack{\text{Place parentheses}\\ \text{about variables}}} = \underbrace{(2)^2 - (-5)}_{\substack{\text{Replace variables}\\ \text{with values}}} = 4 - (-5) = 4 + 5 = 9$$

Objective C Solutions of Equations

Recall from Section 1.3 that a solution of an equation is a value for the variable that makes the equation true.

PRACTICE PROBLEM 9

Determine whether -2 is a solution of $-1 + x = 1$.

EXAMPLE 9 Determine whether -4 is a solution of $x - 5 = -9$.

Solution: Replace x with -4 and see if a true statement results.

$$x - 5 = -9 \quad \text{Original equation}$$
$$-4 - 5 \stackrel{?}{=} -9 \quad \text{Replace } x \text{ with } -4.$$
$$-4 + (-5) \stackrel{?}{=} -9$$
$$-9 = -9 \quad \text{True}$$

Thus -4 is a solution of $x - 5 = -9$.

Work Practice Problem 9

Objective D Solving Problems That Involve Subtraction

Another use of real numbers is in recording altitudes above and below sea level, as shown in the next example.

PRACTICE PROBLEM 10

At 6:00 p.m., the temperature at the Winter Olympics was 14°; by morning the temperature dropped to $-23°$. Find the overall change in temperature.

EXAMPLE 10 Finding the Difference in Elevations

The lowest point on the surface of the Earth is the Dead Sea, at an elevation of 1349 feet below sea level. The highest point is Mt. Everest, at an elevation of 29,035 feet. What is the difference in elevation between these two world extremes? (*Source:* National Geographic Society)

Solution: To find the difference in elevation between the two heights, find the difference of the high point and the low point.

Answers

8. a. $\dfrac{1}{3}$, b. 5, 9. -2 is not a solution., 10. $-37°$

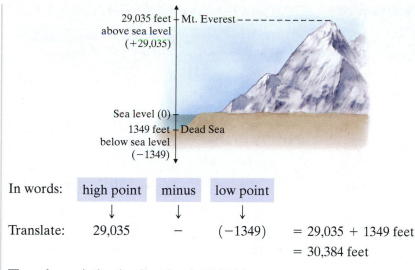

29,035 feet — Mt. Everest
above sea level
(+29,035)

Sea level (0)
1349 feet — Dead Sea
below sea level
(−1349)

In words: | high point | minus | low point |

Translate: 29,035 − (−1349) = 29,035 + 1349 feet
= 30,384 feet

Thus, the variation in elevation is 30,384 feet.

📙 **Work Practice Problem 10**

Objective E Finding Complementary and Supplementary Angles

A knowledge of geometric concepts is needed by many professionals, such as doctors, carpenters, electronic technicians, gardeners, machinists, and pilots, just to name a few. With this in mind, we review the geometric concepts of **complementary** and **supplementary angles.**

Complementary and Supplementary Angles

Two angles are **complementary** if the sum of their measures is 90°.

Two angles are **supplementary** if the sum of their measures is 180°.

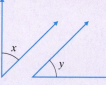

$m\angle x + m\angle y = 90°$

$m\angle x + m\angle y = 180°$

△ **EXAMPLE 11** Find the measure of each unknown complementary or supplementary angle.

a.

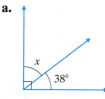

b.

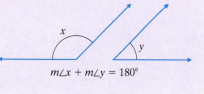

Solution:

a. These angles are complementary, so their sum is 90°. This means that the measure of angle x, $m\angle x$, is $90° − 38°$.

$m\angle x = 90° − 38° = 52°$

b. These angles are supplementary, so their sum is 180°. This means that $m\angle y$ is $180° − 62°$.

$m\angle y = 180° − 62° = 118°$

📙 **Work Practice Problem 11**

PRACTICE PROBLEM 11

Find the measure of each unknown complementary or supplementary angle.

a.

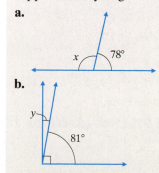

b.

Answers
11. a. 102°, **b.** 9°

Objective **A** *Subtract. See Examples 1 through 4.*

1. $-6 - 4$ **2.** $-12 - 8$ **3.** $4 - 9$ **4.** $8 - 11$ **5.** $16 - (-3)$

6. $12 - (-5)$ **7.** $7 - (-4)$ **8.** $3 - (-6)$ **9.** $-26 - (-18)$ **10.** $-60 - (-48)$

11. $-6 - 5$ **12.** $-8 - 4$ **13.** $16 - (-21)$ **14.** $15 - (-33)$ **15.** $-6 - (-11)$

16. $-4 - (-16)$ **17.** $-44 - 27$ **18.** $-36 - 51$ **19.** $-21 - (-21)$ **20.** $-17 - (-17)$

21. $-\dfrac{3}{11} - \left(-\dfrac{5}{11}\right)$ **22.** $-\dfrac{4}{7} - \left(-\dfrac{1}{7}\right)$ **23.** $9.7 - 16.1$ **24.** $8.3 - 11.2$ **25.** $-2.6 - (-6.7)$

26. $-6.1 - (-5.3)$ **27.** $\dfrac{1}{2} - \dfrac{2}{3}$ **28.** $\dfrac{3}{4} - \dfrac{7}{8}$ **29.** $-\dfrac{1}{6} - \dfrac{3}{4}$ **30.** $-\dfrac{1}{10} - \dfrac{7}{8}$

31. $8.3 - (-0.62)$ **32.** $4.3 - (-0.87)$ **33.** $0 - 8.92$ **34.** $0 - (-4.21)$

Write each phrase as an expression and simplify. See Example 5.

35. Subtract -5 from 8.

36. Subtract -2 from 3.

37. Find the difference between -6 and -1.

38. Find the difference between -17 and -1.

39. Subtract 8 from 7.

40. Subtract 9 from -4.

41. Decrease -8 by 15.

42. Decrease 11 by -14.

Mixed Practice (*Sections 1.3, 1.4, 1.5*) *Simplify each expression. (Remember the order of operations.) See Examples 6 and 7.*

43. $-10 - (-8) + (-4) - 20$

44. $-16 - (-3) + (-11) - 14$

45. $5 - 9 + (-4) - 8 - 8$

46. $7 - 12 + (-5) - 2 + (-2)$

47. $-6 - (2 - 11)$

48. $-9 - (3 - 8)$

49. $3^3 - 8 \cdot 9$

50. $2^3 - 6 \cdot 3$

51. $2 - 3(8 - 6)$

52. $4 - 6(7 - 3)$

53. $(3 - 6) + 4^2$

54. $(2 - 3) + 5^2$

55. $-2 + [(8 - 11) - (-2 - 9)]$

56. $-5 + [(4 - 15) - (-6) - 8]$

57. $|-3| + 2^2 + [-4 - (-6)]$

58. $|-2| + 6^2 + (-3 - 8)$

Objective B *Evaluate each expression when $x = -5$, $y = 4$, and $t = 10$. See Example 8.*

59. $x - y$

60. $y - x$

61. $\dfrac{9 - x}{y + 6}$

62. $\dfrac{15 - x}{y + 2}$

63. $|x| + 2t - 8y$

64. $|y| + 3x - 2t$

65. $y^2 - x$

66. $t^2 - x$

67. $\dfrac{|x - (-10)|}{2t}$

68. $\dfrac{|5y - x|}{6t}$

Objective C *Decide whether the given number is a solution of the given equation. See Example 9.*

69. $x - 9 = 5$; -4

70. $x - 10 = -7$; 3

71. $-x + 6 = -x - 1$; -2

72. $-x - 6 = -x - 1$; -10

73. $-x - 13 = -15$; 2

74. $4 = 1 - x$; 5

Objectives D E **Mixed Practice** *Solve. See Examples 10 and 11.*

75. Within 24 hours in 1916, the temperature in Browning, Montana, fell from 44° to −56°. How large a drop in temperature was this?

76. The coldest temperature ever recorded in Louisiana was −16°F. The hottest temperature ever recorded in Louisiana was 114°F. How much of a difference in temperature is there between these two extremes? (*Source:* National Climatic Data Center)

77. In a series of plays, the San Francisco 49ers gain 2 yards, lose 5 yards, and then lose another 20 yards. What is their total gain or loss of yardage?

78. In some card games, it is possible to have a negative score. Lavonne Schultz currently has a score of 15 points. She then loses 24 points. What is her new score?

79. Pythagoras died in the year −475 (or 475 B.C.). When was he born, if he was 94 years old when he died?

80. The Greek astronomer and mathematician Geminus died in 60 A.D. at the age of 70. When was he born?

81. Find *x* if the angles below are complementary angles.

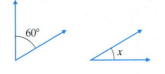

82. Find *x* if the angles below are supplementary angles.

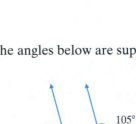

83. A commercial jet liner hits an air pocket and drops 250 feet. After climbing 120 feet, it drops another 178 feet. What is its overall vertical change?

84. Tyson Industries stock posted a loss of 1.625 points yesterday. If it drops another 0.75 point today, find its overall change for the two days.

85. The highest point in Africa is Mt. Kilimanjaro, Tanzania, at an elevation of 19,340 feet. The lowest point is Lake Assal, Djibouti, at 512 feet below sea level. How much higher is Mt. Kilimanjaro than Lake Assal? (*Source:* National Geographic Society)

86. The airport in Bishop, California, is at an elevation of 4101 feet above sea level. The nearby Furnace Creek Airport in Death Valley, California, is at an elevation of 226 feet below sea level. How much higher in elevation is the Bishop Airport than the Furnace Creek Airport? (*Source:* National Climatic Data Center)

Find each unknown complementary or supplementary angle.

87.

88.

Concept Extensions

The following bar graph shows each month's average daily low temperature in degrees Fahrenheit for Barrow, Alaska. Use this graph to answer Exercises 89 through 91.

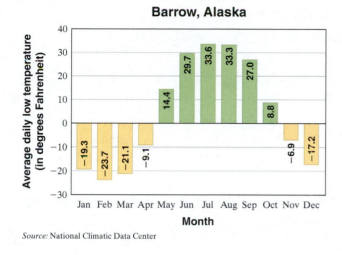

Barrow, Alaska

Source: National Climatic Data Center

89. Record the monthly increases and decreases in the low temperature from the previous month.

Month	Monthly Increase or Decrease
February	
March	
April	
May	
June	
July	
August	
September	
October	
November	
December	

90. Which month had the greatest increase in temperature?

91. Which month had the greatest decrease in temperature?

92. Find two numbers whose difference is -5.

93. Find two numbers whose difference is -9.

If p is a positive number and n is a negative number, determine whether each statement is true or false. Explain your answer.

94. $p - n$ is always a positive number.

95. $n - p$ is always a negative number.

96. $|n| - |p|$ is always a positive number.

97. $|n - p|$ is always a positive number.

Without calculating, determine whether each answer is positive or negative. Then use a calculator to find the exact difference.

98. $56{,}875 - 87{,}262$

99. $4.362 - 7.0086$

100. If a and b are positive numbers, then is $a - b$ always positive, always negative, or sometimes positive and sometimes negative?

101. If a and b are negative numbers, then is $a - b$ always positive, always negative, or sometimes positive and sometimes negative?

Operations on Real Numbers

Answer the following with positive, negative, or 0.

1. The opposite of a positive number is a _____ number.

2. The sum of two negative numbers is a _____ number.

3. The absolute value of a negative number is a _____ number.

4. The absolute value of zero is _____.

5. The reciprocal of a positive number is a _____ number.

6. The sum of a number and its opposite is _____.

7. The absolute value of a positive number is a _____ number.

8. The opposite of a negative number is a _____ number.

Fill in the chart:

	Number	Opposite	Absolute Value
9.	$\frac{1}{7}$		
10.	$-\frac{12}{5}$		
11.		-3	
12.		$\frac{9}{11}$	

Perform each indicated operation and simplify.

13. $-19 + (-23)$ **14.** $7 - (-3)$ **15.** $-15 + 17$ **16.** $-8 - 10$

17. $18 + (-25)$ **18.** $-2 + (-37)$ **19.** $-14 - (-12)$ **20.** $5 - 14$

21. $4.5 - 7.9$ **22.** $-8.6 - 1.2$ **23.** $-\dfrac{3}{4} - \dfrac{1}{7}$ **24.** $\dfrac{2}{3} - \dfrac{7}{8}$

25. $-9 - (-7) + 4 - 6$ **26.** $11 - 20 + (-3) - 12$ **27.** $24 - 6(14 - 11)$

28. $30 - 5(10 - 8)$ **29.** $(7 - 17) + 4^2$ **30.** $9^2 + (10 - 30)$

31. $|-9| + 3^2 + (-4 - 20)$ **32.** $|-4 - 5| + 5^2 + (-50)$

33. $-7 + [(1 - 2) + (-2 - 9)]$ **34.** $-6 + [(-3 + 7) + (4 - 15)]$

35. Subtract 5 from 1.   **36.** Subtract -2 from -3.

37. Subtract $-\dfrac{2}{5}$ from $\dfrac{1}{4}$.   **38.** Subtract $\dfrac{1}{10}$ from $-\dfrac{5}{8}$.

39. $2(19 - 17)^3 - 3(-7 + 9)^2$ **40.** $3(10 - 9)^2 + 6(20 - 19)^3$

Evaluate each expression when $x = -2$, $y = -1$, and $z = 9$.

41. $x - y$ **42.** $x + y$

43. $y + z$ **44.** $z - y$

45. $\dfrac{|5z - x|}{y - x}$ **46.** $\dfrac{|-x - y + z|}{2z}$

25. _____

26. _____

27. _____

28. _____

29. _____

30. _____

31. _____

32. _____

33. _____

34. _____

35. _____

36. _____

37. _____

38. _____

39. _____

40. _____

41. _____

42. _____

43. _____

44. _____

45. _____

46. _____

A Multiply Real Numbers.

B Find the Reciprocal of a Real Number.

C Divide Real Numbers.

D Evaluate Expressions Using Real Numbers.

E Determine Whether a Number is a Solution of a Given Equation.

1.6 MULTIPLYING AND DIVIDING REAL NUMBERS

Objective A Multiplying Real Numbers

Multiplication of real numbers is similar to multiplication of whole numbers. We just need to determine when the answer is positive, when it is negative, and when it is zero. To discover sign patterns for multiplication, recall that multiplication is repeated addition. For example, 3(2) means that 2 is added to itself three times, or

$$3(2) = 2 + 2 + 2 = 6$$

Also,

$$3(-2) = (-2) + (-2) + (-2) = -6$$

Since $3(-2) = -6$, this suggests that the product of a positive number and a negative number is a negative number.

What about the product of two negative numbers? To find out, consider the following pattern.

Factor decreases by 1 each time.

$$-3 \cdot 2 = -6$$
$$-3 \cdot 1 = -3 \quad \text{Product increases by 3 each time.}$$
$$-3 \cdot 0 = 0$$
$$-3 \cdot -1 = 3$$
$$-3 \cdot -2 = 6$$

This suggests that the product of two negative numbers is a positive number. Our results are given below.

Multiplying Real Numbers

1. The product of two numbers with the *same* sign is a positive number.
2. The product of two numbers with *different* signs is a negative number.

EXAMPLES Multiply.

1. $-7(6) = -42$ Different signs, so the product is negative.
2. $2(-10) = -20$
3. $-2(-14) = 28$ Same sign, so the product is positive.
4. $-\dfrac{2}{3} \cdot \dfrac{4}{7} = -\dfrac{2 \cdot 4}{3 \cdot 7} = -\dfrac{8}{21}$
5. $5(-1.7) = -8.5$
6. $-18(-3) = 54$

Work Practice Problems 1–6

We already know that the product of 0 and any whole number is 0. This is true of all real numbers.

Products Involving Zero

If b is a real number, then $b \cdot 0 = 0$. Also $0 \cdot b = 0$.

EXAMPLE 7 Multiply.

a. $7(0)(-6)$ **b.** $(-2)(-3)(-4)$ **c.** $(-1)(5)(-9)$

Solution:

a. By the order of operations, we multiply from left to right. Notice that because one of the factors is 0, the product is 0.

$$7(0)(-6) = 0(-6) = 0$$

b. Multiply two factors at a time, from left to right.

$$(-2)(-3)(-4) = (6)(-4) \quad \text{Multiply } (-2)(-3).$$
$$= -24$$

c. Multiply from left to right.

$$(-1)(5)(-9) = (-5)(-9) \quad \text{Multiply } (-1)(5).$$
$$= 45$$

🔲 **Work Practice Problem 7**

Helpful Hint

You may have noticed from the example that if we multiply:

- an *even* number of negative numbers, the product is *positive*.
- an *odd* number of negative numbers, the product is *negative*.

Now that we know how to multiply positive and negative numbers, let's see how we find the values of $(-5)^2$ and -5^2, for example. Although these two expressions look similar, the difference between the two is the parentheses. In $(-5)^2$, the parentheses tell us that the base, or repeated factor, is -5. In -5^2, only 5 is the base. Thus,

$$(-5)^2 = (-5)(-5) = 25 \quad \text{The base is } -5.$$
$$-5^2 = -(5 \cdot 5) = -25 \quad \text{The base is } 5.$$

EXAMPLE 8 Evaluate.

a. $(-2)^3$ **b.** -2^3 **c.** $(-3)^2$ **d.** -3^2 **e.** $\left(-\dfrac{2}{3}\right)^2$

Solution:

a. $(-2)^3 = (-2)(-2)(-2) = -8$ The base is -2.
b. $-2^3 = -(2 \cdot 2 \cdot 2) = -8$ The base is 2.
c. $(-3)^2 = (-3)(-3) = 9$ The base is -3.
d. $-3^2 = -(3 \cdot 3) = -9$ The base is 3.
e. $\left(-\dfrac{2}{3}\right)^2 = \left(-\dfrac{2}{3}\right)\left(-\dfrac{2}{3}\right) = \dfrac{4}{9}$ The base is $-\dfrac{2}{3}$.

🔲 **Work Practice Problem 8**

Helpful Hint

Be careful when identifying the base of an exponential expression.

$$(-3)^2 \qquad\qquad\qquad -3^2$$
Base is -3 Base is 3
$$(-3)^2 = (-3)(-3) = 9 \qquad -3^2 = -(3 \cdot 3) = -9$$

PRACTICE PROBLEM 7

Multiply.
a. $5(0)(-3)$
b. $(-1)(-6)(-7)$
c. $(-2)(4)(-8)$

PRACTICE PROBLEM 8

Evaluate.
a. $(-2)^4$ **b.** -2^4
c. $(-1)^5$ **d.** -1^5
e. $\left(-\dfrac{7}{9}\right)^2$

Answers
7. a. 0, **b.** -42, **c.** 64, **8. a.** 16
b. -16, **c.** -1, **d.** -1 **e.** $\dfrac{49}{81}$

Objective B Finding Reciprocals

Addition and subtraction are related. Every difference of two numbers $a - b$ can be written as the sum $a + (-b)$. Multiplication and division are related also. For example, the quotient $6 \div 3$ can be written as the product $6 \cdot \dfrac{1}{3}$. Recall that the pair of numbers 3 and $\dfrac{1}{3}$ has a special relationship. Their product is 1 and they are called **reciprocals** or **multiplicative inverses** of each other.

> ### Reciprocal or Multiplicative Inverse
>
> Two numbers whose product is 1 are called **reciprocals** or **multiplicative inverses** of each other.

PRACTICE PROBLEM 9

Find the reciprocal of each number.

a. 13   **b.** $\dfrac{7}{15}$

c. -5 **d.** $-\dfrac{8}{11}$

e. 7.9

EXAMPLE 9 Find the reciprocal of each number.

a. 22   Reciprocal is $\dfrac{1}{22}$ since $22 \cdot \dfrac{1}{22} = 1$.

b. $\dfrac{3}{16}$ Reciprocal is $\dfrac{16}{3}$ since $\dfrac{3}{16} \cdot \dfrac{16}{3} = 1$.

c. -10 Reciprocal is $-\dfrac{1}{10}$ since $-10 \cdot -\dfrac{1}{10} = 1$.

d. $-\dfrac{9}{13}$ Reciprocal is $-\dfrac{13}{9}$ since $-\dfrac{9}{13} \cdot -\dfrac{13}{9} = 1$.

e. 1.7   Reciprocal is $\dfrac{1}{1.7}$ since $1.7 \cdot \dfrac{1}{1.7} = 1$.

🔶 **Work Practice Problem 9**

☁ **Helpful Hint**

The fraction $\dfrac{1}{1.7}$ is not simplified since the denominator is a decimal number. For the purpose of finding a reciprocal, we will leave the fraction as is.

Does the number 0 have a reciprocal? If it does, it is a number n such that $0 \cdot n = 1$. Notice that this can never be true since $0 \cdot n = 0$. This means that 0 has no reciprocal.

> ### Quotients Involving Zero
>
> The number 0 does not have a reciprocal.

Objective C Dividing Real Numbers

We may now write a quotient as an equivalent product.

> ### Quotient of Two Real Numbers
>
> If a and b are real numbers and b is not 0, then
>
> $$a \div b = \frac{a}{b} = a \cdot \frac{1}{b}$$

Answers

9. a. $\dfrac{1}{13}$, **b.** $\dfrac{15}{7}$, **c.** $-\dfrac{1}{5}$,

d. $-\dfrac{11}{8}$, **e.** $\dfrac{1}{7.9}$

In other words, the quotient of two real numbers is the product of the first number and the multiplicative inverse or reciprocal of the second number.

EXAMPLE 10 Use the definition of the quotient of two numbers to find each quotient.

a. $-18 \div 3$ **b.** $\dfrac{-14}{-2}$ **c.** $\dfrac{20}{-4}$

Solution:

a. $-18 \div 3 = -18 \cdot \dfrac{1}{3} = -6$

b. $\dfrac{-14}{-2} = -14 \cdot -\dfrac{1}{2} = 7$

c. $\dfrac{20}{-4} = 20 \cdot -\dfrac{1}{4} = -5$

🔲 **Work Practice Problem 10**

Since the quotient $a \div b$ can be written as the product $a \cdot \dfrac{1}{b}$, it follows that sign patterns for dividing two real numbers are the same as sign patterns for multiplying two real numbers.

Dividing Real Numbers

1. The quotient of two numbers with the same sign is a positive number.
2. The quotient of two numbers with different signs is a negative number.

EXAMPLE 11 Divide.

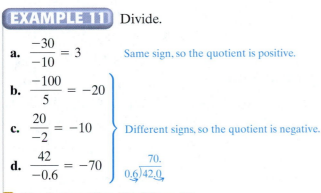

a. $\dfrac{-30}{-10} = 3$ Same sign, so the quotient is positive.

b. $\dfrac{-100}{5} = -20$

c. $\dfrac{20}{-2} = -10$ Different signs, so the quotient is negative.

d. $\dfrac{42}{-0.6} = -70$ $0.6\overline{)42.0}\,\,^{70.}$

🔲 **Work Practice Problem 11**

In the examples above, we divided mentally or by long division. When we divide by a fraction, it is usually easier to multiply by its reciprocal.

EXAMPLES Divide.

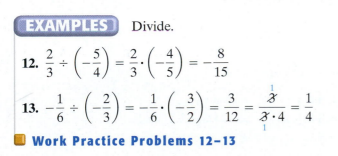

12. $\dfrac{2}{3} \div \left(-\dfrac{5}{4}\right) = \dfrac{2}{3} \cdot \left(-\dfrac{4}{5}\right) = -\dfrac{8}{15}$

13. $-\dfrac{1}{6} \div \left(-\dfrac{2}{3}\right) = -\dfrac{1}{6} \cdot \left(-\dfrac{3}{2}\right) = \dfrac{3}{12} = \dfrac{\cancel{3}^{1}}{\cancel{3} \cdot 4} = \dfrac{1}{4}$

🔲 **Work Practice Problems 12–13**

PRACTICE PROBLEM 10

Use the definition of the quotient of two numbers to find each quotient.

a. $-12 \div 4$ **b.** $\dfrac{-20}{-10}$

c. $\dfrac{36}{-4}$

PRACTICE PROBLEM 11

Divide.

a. $\dfrac{-25}{5}$ **b.** $\dfrac{-48}{-6}$

c. $\dfrac{50}{-2}$ **d.** $\dfrac{-72}{0.2}$

PRACTICE PROBLEMS 12–13

Divide.

12 $-\dfrac{5}{9} \div \dfrac{2}{3}$ **13** $-\dfrac{2}{7} \div \left(-\dfrac{1}{5}\right)$

Answers
10. a. -3, **b.** 2, **c.** -9,
11. a. -5, **b.** 8, **c.** -25, **d.** -360,
12. $-\dfrac{5}{6}$, **13.** $\dfrac{10}{7}$

Our definition of the quotient of two real numbers does not allow for division by 0 because 0 does not have a reciprocal. How then do we interpret $\frac{3}{0}$? We say that an expression such as this one is **undefined.** Can we divide 0 by a number other than 0? Yes; for example,

$$\frac{0}{3} = 0 \cdot \frac{1}{3} = 0$$

Division Involving Zero

If a is a nonzero number, then $\frac{0}{a} = 0$ and $\frac{a}{0}$ is undefined.

PRACTICE PROBLEM 14

Divide if possible.

a. $\dfrac{-7}{0}$ **b.** $\dfrac{0}{-2}$

EXAMPLE 14   Divide, if possible.

a. $\dfrac{1}{0}$ is undefined.   **b.** $\dfrac{0}{-3} = 0$

■ **Work Practice Problem 14**

Notice that $\dfrac{12}{-2} = -6, -\dfrac{12}{2} = -6,$ and $\dfrac{-12}{2} = -6.$ This means that

$$\frac{12}{-2} = -\frac{12}{2} = \frac{-12}{2}$$

In other words, a single negative sign in a fraction can be written in the denominator, in the numerator, or in front of the fraction without changing the value of the fraction.

If a and b are real numbers, and $b \neq 0,$ then $\dfrac{a}{-b} = \dfrac{-a}{b} = -\dfrac{a}{b}.$

Objective D Evaluating Expressions

Examples combining basic arithmetic operations along with the principles of the order of operations help us to review these concepts of multiplying and dividing real numbers.

PRACTICE PROBLEM 15

Use order of operations to evaluate each expression.

a. $\dfrac{0(-5)}{3}$

b. $-3(-9) - 4(-4)$

c. $(-3)^2 + 2[(5 - 15) - |-4 - 1|]$

d. $\dfrac{-7(-4) + 2}{-10 - (-5)}$

e. $\dfrac{5(-2)^3 + 52}{-4 + 1}$

EXAMPLE 15   Use order of operations to evaluate each expression.

a. $\dfrac{0(-8)}{2}$

b. $-4(-11) - 5(-2)$

c. $(-2)^2 + 3[(-3 - 2) - |4 - 6|]$

d. $\dfrac{(-12)(-3) + 4}{-7 - (-2)}$

e. $\dfrac{2(-3)^2 - 20}{|-5| + 4}$

Solution:

a. $\dfrac{0(-8)}{2} = \dfrac{0}{2} = 0$

b. $(-4)(-11) - 5(-2) = 44 - (-10)$ Find the products.

$\qquad\qquad\qquad\qquad\quad = 44 + 10$ Add 44 to the opposite of -10.

$\qquad\qquad\qquad\qquad\quad = 54$ Add.

Answers

14. a. undefined, **b.** 0, **15. a.** 0,
b. 43, **c.** −21, **d.** −6, **e.** −4

c. $(-2)^2 + 3[(-3 - 2) - |4 - 6|] = (-2)^2 + 3[(-5) - |-2|]$ Simplify within innermost sets of grouping symbols.

$$= (-2)^2 + 3[-5 - 2] \quad \text{Write } |-2| \text{ as 2.}$$
$$= (-2)^2 + 3(-7) \quad \text{Combine.}$$
$$= 4 + (-21) \quad \text{Evaluate } (-2)^2 \text{ and then multiply } 3(-7).$$
$$= -17 \quad \text{Add.}$$

For parts d and e, first simplify the numerator and denominator separately; then divide.

d. $\dfrac{(-12)(-3) + 4}{-7 - (-2)} = \dfrac{36 + 4}{-7 + 2}$

$$= \dfrac{40}{-5}$$
$$= -8 \quad \text{Divide.}$$

e. $\dfrac{2(-3)^2 - 20}{|-5| + 4} = \dfrac{2 \cdot 9 - 20}{5 + 4} = \dfrac{18 - 20}{9} = \dfrac{-2}{9} = -\dfrac{2}{9}$

🔲 **Work Practice Problem 15**

Using what we have learned about multiplying and dividing real numbers, we continue to practice evaluating algebraic expressions.

EXAMPLE 16 Evaluate each expression when $x = -2$ and $y = -4$.

a. $\dfrac{3x}{2y}$ **b.** $x^3 - y^2$ **c.** $\dfrac{x - y}{-x}$

Solution: Replace x with -2 and y with -4 and simplify.

a. $\dfrac{3x}{2y} = \dfrac{3(-2)}{2(-4)} = \dfrac{-6}{-8} = \dfrac{6}{8} = \dfrac{\overset{1}{\cancel{2}} \cdot 3}{\underset{1}{\cancel{2}} \cdot 4} = \dfrac{3}{4}$

b. Replace x with -2 and y with -4.

$$x^3 - y^2 = (-2)^3 - (-4)^2 \quad \text{Substitute the given values for the variables.}$$
$$= -8 - (16) \quad \text{Evaluate } (-2)^3 \text{ and } (-4)^2.$$
$$= -8 + (-16) \quad \text{Write as a sum.}$$
$$= -24 \quad \text{Add.}$$

c. $\dfrac{x - y}{-x} = \dfrac{-2 - (-4)}{-(-2)} = \dfrac{-2 + 4}{2} = \dfrac{2}{2} = 1$

🔲 **Work Practice Problem 16**

Helpful Hint

Remember: For additional help when replacing variables with replacement values, first place parentheses about any variables.
Evaluate $3x - y^2$ when $x = 5$ and $y = -4$.

$$3x - y^2 = 3(x) - (y)^2 \quad \text{Place parentheses about variables only.}$$
$$= 3(5) - (-4)^2 \quad \text{Replace variables with values.}$$
$$= 15 - 16 \quad \text{Simplify.}$$
$$= -1$$

PRACTICE PROBLEM 16

Evaluate each expression when $x = -1$ and $y = -5$.

a. $\dfrac{3y}{45x}$

b. $x^2 - y^3$

c. $\dfrac{x + y}{3x}$

Answers

16. **a.** $\dfrac{1}{3}$, **b.** 126, **c.** 2

Objective E Solutions of Equations

We use our skills in multiplying and dividing real numbers to check possible solutions of an equation.

PRACTICE PROBLEM 17

Determine whether -8 is a solution of $\dfrac{x}{4} - 3 = x + 3$.

EXAMPLE 17 Determine whether -10 is a solution of $\dfrac{-20}{x} + 15 = 2x$.

Solution:

$$\dfrac{-20}{x} + 15 = 2x \qquad \text{Original equation}$$

$$\dfrac{-20}{-10} + 15 \overset{?}{=} 2(-10) \qquad \text{Replace } x \text{ with } -10.$$

$$2 + 15 \overset{?}{=} -20 \qquad \text{Divide and multiply.}$$

$$17 = -20 \qquad \text{False}$$

Since we have a false statement, -10 is *not* a solution of the equation.

☐ **Work Practice Problem 17**

Answer

17. -8 is a solution.

![calculator icon] **CALCULATOR EXPLORATIONS**

Entering Negative Numbers on a Scientific Calculator

To enter a negative number on a scientific calculator, find a key marked $\boxed{+/-}$. (On some calculators, this key is marked $\boxed{\text{CHS}}$ for "change sign.") To enter -8, for example, press the keys $\boxed{8}\ \boxed{+/-}$. The display will read $\boxed{-8}$.

Entering Negative Numbers on a Graphing Calculator

To enter a negative number on a graphing calculator, find a key marked $\boxed{(-)}$. Do not confuse this key with the key $\boxed{-}$, which is used for subtraction. To enter -8, for example, press the keys $\boxed{(-)}\ \boxed{8}$. The display will read $\boxed{-8}$.

Operations with Real Numbers

To evaluate $-2(7 - 9) - 20$ on a calculator, press the keys

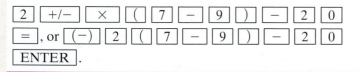

$\boxed{=}$, or $\boxed{(-)}\ \boxed{2}\ \boxed{(}\ \boxed{7}\ \boxed{-}\ \boxed{9}\ \boxed{)}\ \boxed{-}\ \boxed{2}\ \boxed{0}$

$\boxed{\text{ENTER}}$.

The display will read $\boxed{-16}$ or $\boxed{\begin{array}{r} -2(7-9)-20 \\ -16 \end{array}}$

Use a calculator to simplify each expression.

1. $-38(26 - 27)$

2. $-59(-8) + 1726$

3. $134 + 25(68 - 91)$

4. $45(32) - 8(218)$

5. $\dfrac{-50(294)}{175 - 205}$

6. $\dfrac{-444 - 444.8}{-181 - (-181)}$

7. $9^5 - 4550$

8. $5^8 - 6259$

9. $(-125)^2$ (Be careful.)

10. -125^2 (Be careful.)

Mental Math

Answer the following with positive or negative.

1. The product of two negative numbers is a _____ number.

2. The quotient of two negative numbers is a _____ number.

3. The quotient of a positive number and a negative number is a _____ number.

4. The product of a positive number and a negative number is a _____ number.

5. The reciprocal of a positive number is a _____ number.

6. The opposite of a positive number is a _____ number.

FOR EXTRA HELP

1.6 EXERCISE SET

Student Solutions Manual   PH Math/Tutor Center   CD/Video for Review   Math XL MathXL®   MyMathLab MyMathLab

Objective A *Multiply. See Examples 1 through 6.*

1. $-6(4)$

2. $-8(5)$

3. $2(-1)$

4. $7(-4)$

5. $-5(-10)$

6. $-6(-11)$

7. $-3 \cdot 15$

8. $-2 \cdot 37$

9. $-\dfrac{1}{2}\left(-\dfrac{3}{5}\right)$

10. $-\dfrac{1}{8}\left(-\dfrac{1}{3}\right)$

11. $5(-1.4)$

12. $6(-2.5)$

Evaluate. See Examples 7 and 8.

13. $(-1)(-3)(-5)$

14. $(-2)(-3)(-4)$

15. $(2)(-1)(-3)(0)$

16. $(3)(-5)(-2)(0)$

17. $(-4)^2$

18. $(-3)^3$

19. -4^2

20. -6^2

21. $\left(-\dfrac{3}{4}\right)^2$

22. $\left(-\dfrac{2}{7}\right)^2$

23. -0.7^2

24. -0.8^2

Objective B *Find each reciprocal. See Example 9.*

25. $\dfrac{2}{3}$

26. $\dfrac{1}{7}$

27. -14

28. -8

29. $-\dfrac{3}{11}$

30. $-\dfrac{6}{13}$

31. 0.2

32. 1.5

Objective **C** *Divide. See Examples 10 through 14.*

33. $\dfrac{18}{-2}$

34. $\dfrac{36}{-9}$

35. $-48 \div 12$

36. $-60 \div 5$

37. $\dfrac{0}{-4}$

38. $\dfrac{0}{-9}$

39. $\dfrac{5}{0}$

40. $\dfrac{8}{0}$

41. $\dfrac{6}{7} \div \left(-\dfrac{1}{3}\right)$

42. $\dfrac{4}{5} \div \left(-\dfrac{1}{2}\right)$

43. $-3.2 \div -0.02$

44. $-4.9 \div -0.07$

Objectives **A** **C** **Mixed Practice** *Perform the indicated operation. See Examples 1–14.*

45. $(-8)(-8)$

46. $(-7)(-7)$

47. $\dfrac{2}{3}\left(-\dfrac{4}{9}\right)$

48. $\dfrac{2}{7}\left(-\dfrac{2}{11}\right)$

49. $\dfrac{-12}{-4}$

50. $\dfrac{-45}{-9}$

51. $\dfrac{30}{-2}$

52. $\dfrac{14}{-2}$

53. $(-5)^3$

54. $(-2)^5$

55. $(-0.2)^3$

56. $(-0.3)^3$

57. $-\dfrac{3}{4}\left(-\dfrac{8}{9}\right)$

58. $-\dfrac{5}{6}\left(-\dfrac{3}{10}\right)$

59. $-\dfrac{5}{9} \div \left(-\dfrac{3}{4}\right)$

60. $-\dfrac{1}{10} \div \left(-\dfrac{8}{11}\right)$

61. $-2.1(-0.4)$

62. $-1.3(-0.6)$

63. $\dfrac{-48}{1.2}$

64. $\dfrac{-86}{2.5}$

65. $(-3)^4$

66. -3^4

67. -1^7

68. $(-1)^7$

69. Multiply -11 by 11.

70. Multiply -12 by 12.

71. Find the quotient of $-\dfrac{4}{9}$ and $\dfrac{4}{9}$.

72. Find the quotient of $-\dfrac{5}{12}$ and $\dfrac{5}{12}$.

Mixed Practice (*Sections 1.4, 1.5, 1.6*) *Perform the indicated operation.*

73. $-9 - 10$

74. $-8 - 11$

75. $-9(-10)$

76. $-8(-11)$

77. $7(-12)$

78. $6(-15)$

79. $7 + (-12)$

80. $6 + (-15)$

Objective D *Evaluate each expression. See Example 15.*

81. $\dfrac{-9(-3)}{-6}$

82. $\dfrac{-6(-3)}{-4}$

83. $-3(2-8)$

84. $-4(3-9)$

85. $-7(-2)-3(-1)$

86. $-8(-3)-4(-1)$

87. $2^2-3[(2-8)-(-6-8)]$

88. $3^2-2[(3-5)-(2-9)]$

89. $\dfrac{-6^2+4}{-2}$

90. $\dfrac{3^2+4}{5}$

91. $\dfrac{-3-5^2}{2(-7)}$

92. $\dfrac{-2-4^2}{3(-6)}$

93. $\dfrac{22+(3)(-2)^2}{-5-2}$

94. $\dfrac{-20+(-4)^2(3)}{1-5}$

95. $\dfrac{(-4)^2-16}{4-12}$

96. $\dfrac{(-2)^2-4}{4-9}$

97. $\dfrac{6-2(-3)}{4-3(-2)}$

98. $\dfrac{8-3(-2)}{2-5(-4)}$

99. $\dfrac{|5-9|+|10-15|}{|2(-3)|}$

100. $\dfrac{|-3+6|+|-2+7|}{|-2\cdot 2|}$

101. $\dfrac{-7(-1)+(-3)4}{(-2)(5)+(-6)(-8)}$

102. $\dfrac{8(-7)+(-2)(-6)}{(-9)(3)+(-10)(-11)}$

Evaluate each expression when $x = -5$ and $y = -3$. See Example 16.

103. $\dfrac{2x-5}{y-2}$

104. $\dfrac{2y-12}{x-4}$

105. $\dfrac{6-y}{x-4}$

106. $\dfrac{10-y}{x-8}$

107. $\dfrac{4-2x}{y+3}$

108. $\dfrac{2y+3}{-5-x}$

109. $\dfrac{x^2+y}{3y}$

110. $\dfrac{y^2-x}{2x}$

Objective E *Decide whether the given number is a solution of the given equation. See Example 17.*

111. $-3x-5=-20;\ \ 5$

112. $17-4x=x+27;\ \ -2$

113. $\dfrac{x}{5}+2=-1;\ \ 15$

114. $\dfrac{x}{6}-3=5;\ \ 48$

115. $\dfrac{x-3}{7}=-2;\ \ -11$

116. $\dfrac{x+4}{5}=-6;\ \ -30$

Concept Extensions

State whether each statement is true or false.

117. The product of three negative integers is negative.

118. The product of three positive integers is positive.

119. The product of four negative integers is negative.

120. The product of four positive integers is positive.

Study the bar graph below showing the average surface temperatures of planets. Use Exercises 121 and 122 to complete the planet temperatures on the graph.

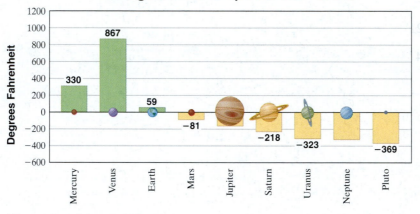

Average Surface Temperature of Planets*

*(For some planets, the temperature given is the temperature where the atmosphere pressure equals 1 Earth atmosphere; Source: *The World Almanac*, 2005)

121. The surface temperature of Jupiter is twice the temperature of Mars. Find this temperature.

122. The surface temperature of Neptune is equal to the temperature of Mercury divided by −1. Find this temperature.

123. Explain why the product of an even number of negative numbers is a positive number.

124. If a and b are any real numbers, is the statement $a \cdot b = b \cdot a$ always true? Why or why not?

125. Find any real numbers that are their own reciprocal.

126. Explain why 0 has no reciprocal.

Write each as an algebraic expression. Then simplify the expression.

127. 7 subtracted from the quotient of 0 and 5

128. Twice the sum of −3 and −4

129. −1 added to the product of −8 and −5

130. The difference of −9 and the product of −4 and −6

This is a special feature that we introduce in this section. Among other concepts introduced later in this text, it is very important for you to be able to simplify expressions and solve equations and to know the difference between the two. To help with this, we begin an outline below and expand this outline throughout the text. Although suggestions are given, this outline should be in your own words. Once you complete the new portion of your outline, try the exercises below. Remember: Study your outline often as you proceed through this text.

I. Simplifying Expressions

   A. Real Numbers

      1. Add:

               Adding like signs.

        $-1.7 + (-0.21) = -1.91$ Add absolute values.

               Attach the common sign.

        $-7 + 3 = -4$ Adding unlike signs.

               Subtract absolute values. Attach the sign of the number with the larger absolute value.

      2. Subtract: Add the first number to the opposite of the second number.

$$\frac{1}{7} - \frac{1}{3} = \frac{3}{21} + \left(-\frac{7}{21}\right) = -\frac{4}{21}$$

      3. Multiply or Divide: Multiply or divide as usual. If the signs of the two numbers are the same, the answer is positive. If the signs of the two numbers are different, the answer is negative.

$$-\frac{3}{8} \cdot \frac{7}{11} = -\frac{21}{88}, \qquad -42 \div (-10) = 4.2$$

Perform the indicated operations.

1. $-0.2(25)$

2. $86 - 100$

3. $-\frac{1}{7} + \left(-\frac{3}{5}\right)$

4. $\dfrac{-40}{-5}$

5. $(-7)^2$

6. -7^2

7. $\dfrac{|-42|}{-|-2|}$

8. $\dfrac{8.6}{0}$

9. $\dfrac{0}{8.6}$

10. $-25 - (-13)$

11. $-8.3 - 8.3$

12. $-\frac{8}{9}\left(-\frac{3}{16}\right)$

13. $2 + 3(8 - 11)^3$

14. $-2\frac{1}{2} \div \left(-3\frac{1}{4}\right)$

15. $20 \div 2 \cdot 5$

16. $-2[(1 - 5) - (7 - 17)]$

1.7 PROPERTIES OF REAL NUMBERS

Objective A Using the Commutative and Associative Properties

In this section we give names to properties of real numbers with which we are already familiar. Throughout this section, the variables a, b, and c represent real numbers.

Recall that order does not matter when adding numbers. For example, we know that $7 + 5$ is the same as $5 + 7$. This property is given a special name—the **commutative property of addition.** We also know that order does not matter when multiplying numbers. For example, we know that $-5(6) = 6(-5)$. This property means that multiplication is commutative also and is called the **commutative property of multiplication.**

Commutative Properties

Addition:	$a + b = b + a$
Multiplication:	$a \cdot b = b \cdot a$

These properties state that the *order* in which any two real numbers are added or multiplied does not change their sum or product. For example, if we let $a = 3$ and $b = 5$, then the commutative properties guarantee that

$$3 + 5 = 5 + 3 \quad \text{and} \quad 3 \cdot 5 = 5 \cdot 3$$

Helpful Hint

Is subtraction also commutative? Try an example. Is $3 - 2 = 2 - 3$? **No!** The left side of this statement equals 1; the right side equals -1. There is no commutative property of subtraction. Similarly, there is no commutative property of division. For example, $10 \div 2$ does not equal $2 \div 10$.

PRACTICE PROBLEM 1

Use a commutative property to complete each statement.

a. $7 \cdot y = $ _____

b. $4 + x = $ _____

EXAMPLE 1 Use a commutative property to complete each statement.

a. $x + 5 = $ _____   **b.** $3 \cdot x = $ _____

Solution:

a. $x + 5 = 5 + x$ By the commutative property of addition

b. $3 \cdot x = x \cdot 3$ By the commutative property of multiplication

■ **Work Practice Problem 1**

✔ **Concept Check** Which of the following pairs of actions are commutative?

a. "raking the leaves" and "bagging the leaves"

b. "putting on your left glove" and "putting on your right glove"

c. "putting on your coat" and "putting on your shirt"

d. "reading a novel" and "reading a newspaper"

Answers

1. **a.** $y \cdot 7$, **b.** $x + 4$

✔ **Concept Check Answer**

b, d

Let's now discuss grouping numbers. When we add three numbers, the way in which they are grouped or associated does not change their sum. For example, we know that $2 + (3 + 4) = 2 + 7 = 9$. This result is the same if we group the numbers differently. In other words, $(2 + 3) + 4 = 5 + 4 = 9$, also. Thus, $2 + (3 + 4) = (2 + 3) + 4$. This property is called the **associative property of addition**.

In the same way, changing the grouping of numbers when multiplying does not change their product. For example, $2 \cdot (3 \cdot 4) = (2 \cdot 3) \cdot 4$ (check it). This is the **associative property of multiplication.**

Associative Properties

Addition:	$(a + b) + c = a + (b + c)$
Multiplication:	$(a \cdot b) \cdot c = a \cdot (b \cdot c)$

These properties state that the way in which three numbers are *grouped* does not change their sum or their product.

EXAMPLE 2 Use an associative property to complete each statement.

a. $5 + (4 + 6) =$ _____

b. $(-1 \cdot 2) \cdot 5 =$ _____

c. $(m + n) + 9 =$ _____

d. $(xy) \cdot 12 =$ _____

Solution:

a. $5 + (4 + 6) = (5 + 4) + 6$ By the associative property of addition

b. $(-1 \cdot 2) \cdot 5 = -1 \cdot (2 \cdot 5)$ By the associative property of multiplication

c. $(m + n) + 9 = m + (n + 9)$ By the associative property of addition

d. $(xy) \cdot 12 = x \cdot (y \cdot 12)$ Recall that xy means $x \cdot y$.

☐ **Work Practice Problem 2**

Helpful Hint

Remember the difference between the commutative properties and the associative properties. The commutative properties have to do with the *order* of numbers and the associative properties have to do with the *grouping* of numbers.

EXAMPLES

Determine whether each statement is true by an associative property or a commutative property.

3. $(7 + 10) + 4 = (10 + 7) + 4$ Since the order of two numbers was changed and their grouping was not, this is true by the commutative property of addition.

4. $2 \cdot (3 \cdot 1) = (2 \cdot 3) \cdot 1$ Since the grouping of the numbers was changed and their order was not, this is true by the associative property of multiplication.

☐ **Work Practice Problems 3–4**

Let's now illustrate how these properties can help us simplify expressions.

PRACTICE PROBLEM 2

Use an associative property to complete each statement.

a. $5 \cdot (-3 \cdot 6) =$ _____

b. $(-2 + 7) + 3 =$ _____

c. $(q + r) + 17 =$ _____

d. $(ab) \cdot 21 =$ _____

PRACTICE PROBLEMS 3–4

Determine whether each statement is true by an associative property or a commutative property.

3. $5 \cdot (4 \cdot 7) = 5 \cdot (7 \cdot 4)$

4. $-2 + (4 + 9)$
$= (-2 + 4) + 9$

Answers

2. a. $(5 \cdot -3) \cdot 6$,   **b.** $-2 + (7 + 3)$,
c. $q + (r + 17)$,   **d.** $a \cdot (b \cdot 21)$,
3. commutative,   **4.** associative

Simplify each expression.

5. $(-3 + x) + 17$

6. $4(5x)$

EXAMPLES Simplify each expression.

5. $10 + (x + 12) = 10 + (12 + x)$ By the commutative property of addition

$$= (10 + 12) + x$$ By the associative property of addition

$$= 22 + x$$ Add.

6. $-3(7x) = (-3 \cdot 7)x$ By the associative property of multiplication

$$= -21x$$ Multiply.

◻ **Work Practice Problems 5–6**

Objective B Using the Distributive Property

The **distributive property of multiplication over addition** is used repeatedly throughout algebra. It is useful because it allows us to write a product as a sum or a sum as a product.

We know that $7(2 + 4) = 7(6) = 42$. Compare that with

$$7(2) + 7(4) = 14 + 28 = 42$$

Since both original expressions equal 42, they must equal each other, or

$$7(2 + 4) = 7(2) + 7(4)$$

This is an example of the distributive property. The product on the left side of the equal sign is equal to the sum on the right side. We can think of the 7 as being distributed to each number inside the parentheses.

Distributive Property of Multiplication Over Addition

$$a(b + c) = ab + ac$$

Since multiplication is commutative, this property can also be written as

$$(b + c)a = ba + ca$$

The distributive property can also be extended to more than two numbers inside the parentheses. For example,

$$3(x + y + z) = 3(x) + 3(y) + 3(z)$$
$$= 3x + 3y + 3z$$

Since we define subtraction in terms of addition, the distributive property is also true for subtraction. For example,

$$2(x - y) = 2(x) - 2(y)$$
$$= 2x - 2y$$

Use the distributive property to write each expression without parentheses. Then simplify the result.

7. $5(x + y)$

8. $-3(2 + 7x)$

9. $4(x + 6y - 2z)$

10. $-1(3 - a)$

11. $-(8 + a - b)$

12. $\dfrac{1}{2}(2x + 4) + 9$

EXAMPLES Use the distributive property to write each expression without parentheses. Then simplify the result.

7. $2(x + y) = 2(x) + 2(y)$
$$= 2x + 2y$$

8. $-5(-3 + 2z) = -5(-3) + (-5)(2z)$
$$= 15 - 10z$$

9. $5(x + 3y - z) = 5(x) + 5(3y) - 5(z)$
$$= 5x + 15y - 5z$$

Answers

5. $14 + x$,   **6.** $20x$,   **7.** $5x + 5y$,
8. $-6 - 21x$,   **9.** $4x + 24y - 8z$,
10. $-3 + a$,   **11.** $-8 - a + b$,
12. $x + 11$

10. $-1(2 - y) = (-1)(2) - (-1)(y)$
$$= -2 + y$$

11. $-(3 + x - w) = -1(3 + x - w)$
$$= (-1)(3) + (-1)(x) - (-1)(w)$$
$$= -3 - x + w$$

> **Helpful Hint**
> Notice in Example 11 that $-(3 + x - w)$ is first rewritten as $-1(3 + x - w)$.

12. $\dfrac{1}{2}(6x + 14) + 10 = \dfrac{1}{2}(6x) + \dfrac{1}{2}(14) + 10$ Apply the distributive property.
$$= 3x + 7 + 10 \qquad\qquad \text{Multiply.}$$
$$= 3x + 17 \qquad\qquad\quad \text{Add.}$$

🔲 **Work Practice Problems 7–12**

The distributive property can also be used to write a sum as a product.

EXAMPLES Use the distributive property to write each sum as a product.

13. $8 \cdot 2 + 8 \cdot x = 8(2 + x)$

14. $7s + 7t = 7(s + t)$

🔲 **Work Practice Problems 13–14**

Objective C Using the Identity and Inverse Properties

Next, we look at the **identity properties.**

The number 0 is called the identity for addition because when 0 is added to any real number, the result is the same real number. In other words, the *identity* of the real number is not changed.

The number 1 is called the identity for multiplication because when a real number is multiplied by 1, the result is the same real number. In other words, the *identity* of the real number is not changed.

Identities for Addition and Multiplication

0 is the identity element for addition.

$$a + 0 = a \quad \text{and} \quad 0 + a = a$$

1 is the identity element for multiplication.

$$a \cdot 1 = a \quad \text{and} \quad 1 \cdot a = a$$

Notice that 0 is the *only* number that can be added to any real number with the result that the sum is the same real number. Also, 1 is the *only* number that can be multiplied by any real number with the result that the product is the same real number.

Additive inverses or **opposites** were introduced in Section 1.4. Two numbers are called additive inverses or opposites if their sum is 0. The additive inverse or opposite of 6 is -6 because $6 + (-6) = 0$. The additive inverse or opposite of -5 is 5 because $-5 + 5 = 0$.

Reciprocals or **multiplicative inverses** were introduced in Section R.2. Two nonzero numbers are called reciprocals or multiplicative inverses if their product is 1. The reciprocal or multiplicative inverse of $\dfrac{2}{3}$ is $\dfrac{3}{2}$ because $\dfrac{2}{3} \cdot \dfrac{3}{2} = 1$. Likewise, the reciprocal of -5 is $-\dfrac{1}{5}$ because $-5\left(-\dfrac{1}{5}\right) = 1$.

PRACTICE PROBLEMS 13–14

Use the distributive property to write each sum as a product.

13. $9 \cdot 3 + 9 \cdot y$

14. $4x + 4y$

Answers

13. $9(3 + y)$,   **14.** $4(x + y)$

Additive or Multiplicative Inverses

The numbers a and $-a$ are additive inverses or opposites of each other because their sum is 0; that is,

$$a + (-a) = 0$$

The numbers b and $\dfrac{1}{b}$ (for $b \neq 0$) are reciprocals or multiplicative inverses of each other because their product is 1; that is,

$$b \cdot \dfrac{1}{b} = 1$$

✔ **Concept Check** Which of the following is the

a. opposite of $-\dfrac{3}{10}$, and which is the

b. reciprocal of $-\dfrac{3}{10}$?

$$1, -\dfrac{10}{3}, \dfrac{3}{10}, 0, \dfrac{10}{3}, -\dfrac{3}{10}$$

PRACTICE PROBLEMS 15–21

Name the property illustrated by each true statement.

15. $7(a + b) = 7 \cdot a + 7 \cdot b$

16. $12 + y = y + 12$

17. $-4 \cdot (6 \cdot x) = (-4 \cdot 6) \cdot x$

18. $6 + (z + 2) = 6 + (2 + z)$

19. $3\left(\dfrac{1}{3}\right) = 1$

20. $(x + 0) + 23 = x + 23$

21. $(7 \cdot y) \cdot 10 = y \cdot (7 \cdot 10)$

EXAMPLES Name the property illustrated by each true statement.

15. $3(x + y) = 3 \cdot x + 3 \cdot y$ Distributive property

16. $(x + 7) + 9 = x + (7 + 9)$ Associative property of addition (grouping changed)

17. $(b + 0) + 3 = b + 3$ Identity element for addition

18. $2 \cdot (z \cdot 5) = 2 \cdot (5 \cdot z)$ Commutative property of multiplication (order changed)

19. $-2 \cdot \left(-\dfrac{1}{2}\right) = 1$ Multiplicative inverse property

20. $-2 + 2 = 0$ Additive inverse property

21. $-6 \cdot (y \cdot 2) = (-6 \cdot 2) \cdot y$ Commutative and associative properties of multiplication (order and grouping changed)

■ **Work Practice Problems 15–21**

Answers

15. distributive property,
16. commutative property of addition,
17. associative property of multiplication, **18.** commutative property of addition,
19. multiplicative inverse property,
20. identity element for addition,
21. commutative and associative properties of multiplication

✔ **Concept Check Answers**

a. $\dfrac{3}{10}$, **b.** $-\dfrac{10}{3}$

Objective A *Use a commutative property to complete each statement. See Examples 1 and 3.*

1. $x + 16 =$ _____

2. $8 + y =$ _____

3. $-4 \cdot y =$ _____

4. $-2 \cdot x =$ _____

5. $xy =$ _____

6. $ab =$ _____

7. $2x + 13 =$ _____

8. $19 + 3y =$ _____

Use an associative property to complete each statement. See Examples 2 and 4.

9. $(xy) \cdot z =$ _____

10. $3 \cdot (x \cdot y) =$ _____

11. $2 + (a + b) =$ _____

12. $(y + 4) + z =$ _____

13. $4 \cdot (ab) =$ _____

14. $(-3y) \cdot z =$ _____

15. $(a + b) + c =$ _____

16. $6 + (r + s) =$ _____

Use the commutative and associative properties to simplify each expression. See Examples 5 and 6.

17. $8 + (9 + b)$

18. $(r + 3) + 11$

19. $4(6y)$

20. $2(42x)$

21. $\frac{1}{5}(5y)$

22. $\frac{1}{8}(8z)$

23. $(13 + a) + 13$

24. $7 + (x + 4)$

25. $-9(8x)$

26. $-3(12y)$

27. $\frac{3}{4}\left(\frac{4}{3}s\right)$

28. $\frac{2}{7}\left(\frac{7}{2}r\right)$

29. $-\frac{1}{2}(5x)$

30. $-\frac{1}{3}(7x)$

Objective B *Use the distributive property to write each expression without parentheses. Then simplify the result, if possible. See Examples 7 through 12.*

31. $4(x + y)$

32. $7(a + b)$

33. $9(x - 6)$

34. $11(y - 4)$

35. $2(3x + 5)$

36. $5(7 + 8y)$

37. $7(4x - 3)$

38. $3(8x - 1)$

39. $3(6 + x)$

40. $2(x + 5)$

41. $-2(y - z)$

42. $-3(z - y)$

43. $-\frac{1}{3}(3y + 5)$

44. $-\frac{1}{2}(2r + 11)$

45. $5(x + 4m + 2)$

46. $8(3y + z - 6)$

47. $-4(1 - 2m + n) + 4$

48. $-4(4 + 2p + 5) + 16$

49. $-(5x + 2)$

50. $-(9r + 5)$

51. $-(r - 3 - 7p) + 3$

52. $-(q - 2 + 6r) + 2$

53. $\frac{1}{2}(6x + 7) + \frac{1}{2}$

54. $\frac{1}{4}(4x - 2) - \frac{7}{2}$

55. $-\dfrac{1}{3}(3x - 9y)$

56. $-\dfrac{1}{5}(10a - 25b)$

57. $3(2r + 5) - 7$

58. $10(4s + 6) - 40$

59. $-9(4x + 8) + 2$

60. $-11(5x + 3) + 10$

61. $-0.4(4x + 5) - 0.5$

62. $-0.6(2x + 1) - 0.1$

Use the distributive property to write each sum as a product. See Examples 13 and 14.

63. $4 \cdot 1 + 4 \cdot y$

64. $14 \cdot z + 14 \cdot 5$

65. $11x + 11y$

66. $9a + 9b$

67. $(-1) \cdot 5 + (-1) \cdot x$

68. $(-3)a + (-3)y$

69. $30a + 30b$

70. $25x + 25y$

Objectives **A** **C** *Name the properties illustrated by each true statement. See Examples 15 through 21.*

71. $3 \cdot 5 = 5 \cdot 3$

72. $4(3 + 8) = 4 \cdot 3 + 4 \cdot 8$

73. $2 + (x + 5) = (2 + x) + 5$

74. $9 \cdot (x \cdot 7) = (9 \cdot x) \cdot 7$

75. $(x + 9) + 3 = (9 + x) + 3$

76. $1 \cdot 9 = 9$

77. $(4 \cdot y) \cdot 9 = 4 \cdot (y \cdot 9)$

78. $-4 \cdot (8 \cdot 3) = (8 \cdot 3) \cdot (-4)$

79. $0 + 6 = 6$

80. $(a + 9) + 6 = a + (9 + 6)$

81. $-4(y + 7) = -4 \cdot y + (-4) \cdot 7$

82. $(11 + r) + 8 = (r + 11) + 8$

83. $6 \cdot \dfrac{1}{6} = 1$

84. $r + 0 = r$

85. $-6 \cdot 1 = -6$

86. $-\dfrac{3}{4}\left(-\dfrac{4}{3}\right) = 1$

Concept Extensions

Fill in the table with the opposite (additive inverse), the reciprocal (multiplicative inverse), or the expression. Assume that the value of each expression is not 0.

	87.	88.	89.	90.	91.	92.
Expression	8	$-\dfrac{2}{3}$	x	$4y$		
Opposite						$7x$
Reciprocal					$\dfrac{1}{2x}$	

Decide whether each statement is true or false. See the Concept Check in this section.

93. The opposite of $-\dfrac{a}{2}$ is $-\dfrac{2}{a}$.

94. The reciprocal of $-\dfrac{a}{2}$ is $\dfrac{a}{2}$.

Determine which pairs of actions are commutative. See the Concept Check in this section.

95. "taking a test" and "studying for the test"

96. "putting on your shoes" and "putting on your socks"

97. "putting on your left shoe" and "putting on your right shoe"

98. "reading the sports section" and "reading the comics section"

99. "mowing the lawn" and "trimming the hedges"

100. "baking a cake" and "eating the cake"

101. "feeding the dog" and "feeding the cat"

102. "dialing a number" and "turning on the cell phone"

Name the property illustrated by each step.

103. **a.** $\triangle + (\square + \bigcirc) = (\square + \bigcirc) + \triangle$

    b. $= (\bigcirc + \square) + \triangle$

    c. $= \bigcirc + (\square + \triangle)$

104. **a.** $(x + y) + z = x + (y + z)$

    b. $= (y + z) + x$

    c. $= (z + y) + x$

105. Explain why 0 is called the identity element for addition.

106. Explain why 1 is called the identity element for multiplication.

 107. Write an example that shows that division is not commutative.

108. Write an example that shows that subtraction is not commutative.

STUDY SKILLS BUILDER

Are You Familiar with Your Textbook Supplements?

There are many student supplements available for additional study. Below, I have listed some of these. See the preface of this text or your instructor for further information.

Chapter Test Prep Video CD. This material is found in your textbook and is fully explained there. The CD contains video clips of solutions to the Chapter Test exercises in this text and is excellent help when studying for chapter tests.

Lecture Video CDs. These video segments are keyed to each section of the text. The material is presented by me, Elayn Martin-Gay, and I have placed a video icon by the exercises in the text that I have worked on the video.

The Student Solutions Manual. This contains worked out solutions to odd-numbered exercises as well as every exercise in the Integrated Reviews, Chapter Reviews, Chapter Tests, and Cumulative Reviews.

Prentice Hall Tutor Center. Mathematics questions may be phoned, faxed, or emailed to this center.

MyMathLab, MathXL, and Interact Math. These are computer and Internet tutorials. This supplement may already be available to you somewhere on campus, for example at your local learning resource lab. Take a moment and find the name and location of any such lab on campus.

As usual, your instructor is your best source of information.

Let's see how you are doing with textbook supplements:

1. Name one way the Chapter Test Prep Video can help you prepare for a chapter test.

2. List any textbook supplements that you have found useful.

3. Have you located and visited a learning resource lab located on your campus?

4. List the textbook supplements that are currently housed in your campus' learning resource lab.

1.8 SIMPLIFYING EXPRESSIONS

As we explore in this section, we will see that an expression such as $3x + 2x$ is not written as simply as possible. This is because—even without replacing x by a value—we can perform the indicated addition.

Objective **A** Identifying Terms, Like Terms, and Unlike Terms

Before we practice simplifying expressions, we must learn some new language. A **term** is a number or the product of a number and variables raised to powers.

Terms

$$-y, \quad 2x^3, \quad -5, \quad 3xz^2, \quad \frac{2}{y}, \quad 0.8z$$

The **numerical coefficient** of a term is the numerical factor. The numerical coefficient of $3x$ is 3. Recall that $3x$ means $3 \cdot x$.

Term	Numerical Coefficient	
$3x$	3	
$\dfrac{y^3}{5}$	$\dfrac{1}{5}$	since $\dfrac{y^3}{5}$ means $\dfrac{1}{5} \cdot y^3$
$-0.7ab^3c^5$	-0.7	
z	1	
$-y$	-1	
-5	-5	

Helpful Hint

The term z means $1z$ and thus has a numerical coefficient of 1.
The term $-y$ means $-1y$ and thus has a numerical coefficient of -1.

Identify the numerical coefficient of each term.

a. $-4x$ **b.** $15y^3$ **c.** x

d. $-y$ **e.** $\dfrac{z}{4}$

EXAMPLE 1 Identify the numerical coefficient of each term.

a. $-3y$ **b.** $22z^4$ **c.** y **d.** $-x$ **e.** $\dfrac{x}{7}$

Solution:

a. The numerical coefficient of $-3y$ is -3.

b. The numerical coefficient of $22z^4$ is 22.

c. The numerical coefficient of y is 1, since y is $1y$.

d. The numerical coefficient of $-x$ is -1, since $-x$ is $-1x$.

e. The numerical coefficient of $\dfrac{x}{7}$ is $\dfrac{1}{7}$, since $\dfrac{x}{7}$ is $\dfrac{1}{7} \cdot x$.

Answers

1. **a.** -4, **b.** 15, **c.** 1,
d. -1, **e.** $\dfrac{1}{4}$

■ **Work Practice Problem 1**

Terms with the same variables raised to exactly the same powers are called **like terms**. Terms that aren't like terms are called **unlike terms**.

Like Terms	Unlike Terms	Reason Why
$3x, 2x$	$5x, 5x^2$	Why? Same variable x, but different powers of x and x^2
$-6x^2y, 2x^2y, 4x^2y$	$7y, 3z, 8x^2$	Why? Different variables
$2ab^2c^3, ac^3b^2$	$6abc^3, 6ab^2$	Why? Different variables and different powers

Helpful Hint

In like terms, each variable and its exponent must match exactly, but these factors don't need to be in the same order.

$2x^2y$ and $3yx^2$ are like terms.

EXAMPLE 2 Determine whether the terms are like or unlike.

a. $2x, 3x^2$ **b.** $4x^2y, x^2y, -2x^2y$ **c.** $-2yz, -3zy$
d. $-x^4, x^4$ **e.** $-8a^5, 8a^5$

Solution:

a. Unlike terms, since the exponents on x are not the same.
b. Like terms, since each variable and its exponent match.
c. Like terms, since $zy = yz$ by the commutative property.
d. Like terms. The variable and its exponent match.
e. Like terms. The variable and its exponent match.

📙 **Work Practice Problem 2**

PRACTICE PROBLEM 2

Determine whether the terms are like or unlike.

a. $7x^2, -6x^3$
b. $3x^2y^2, -x^2y^2, 4x^2y^2$
c. $-5ab, 3ba$
d. $2x^3, 4y^3$
e. $-7m^4, 7m^4$

Objective B Combining Like Terms

An algebraic expression containing the sum or difference of like terms can be simplified by applying the distributive property. For example, by the distributive property, we rewrite the sum of the like terms $6x + 2x$ as

$$6x + 2x = (6 + 2)x = 8x$$

Also,

$$-y^2 + 5y^2 = (-1 + 5)y^2 = 4y^2$$

Simplifying the sum or difference of like terms is called **combining like terms**.

EXAMPLE 3 Simplify each expression by combining like terms.

a. $7x - 3x$ **b.** $10y^2 + y^2$
c. $8x^2 + 2x - 3x$ **d.** $9n^2 - 5n^2 + n^2$

Solution:

a. $7x - 3x = (7 - 3)x = 4x$
b. $10y^2 + y^2 = (10 + 1)y^2 = 11y^2$
c. $8x^2 + 2x - 3x = 8x^2 + (2 - 3)x = 8x^2 - 1x$ or $8x^2 - x$
d. $9n^2 - 5n^2 + n^2 = (9 - 5 + 1)n^2 = 5n^2$

📙 **Work Practice Problem 3**

PRACTICE PROBLEM 3

Simplify each expression by combining like terms.

a. $9y - 4y$
b. $11x^2 + x^2$
c. $5y - 3x + 4x$
d. $14m^2 - m^2 + 3m^2$

Answers

2. **a.** unlike, **b.** like, **c.** like,
d. unlike, **e.** like, 3. **a.** $5y$, **b.** $12x^2$,
c. $5y + x$, **d.** $16m^2$

The preceding examples suggest the following.

Combining Like Terms

To **combine like terms,** combine the numerical coefficients and multiply the result by the common variable factors.

EXAMPLES Simplify each expression by combining like terms.

4. $2x + 3x + 5 + 2 = (2 + 3)x + (5 + 2)$
$$= 5x + 7$$

5. $-5a - 3 + a + 2 = -5a + 1a + (-3 + 2)$
$$= (-5 + 1)a + (-3 + 2)$$
$$= -4a - 1$$

6. $4y - 3y^2$ These two terms cannot be combined because they are unlike terms.

7. $2.3x + 5x - 6 = (2.3 + 5)x - 6$
$$= 7.3x - 6$$

■ **Work Practice Problems 4–7**

Objective C Simplifying Expressions Containing Parentheses

In simplifying expressions we make frequent use of the distributive property to remove parentheses.

It may be helpful to study the examples below.

$$+(3a + 2) = +1(3a + 2) = +1(3a) + (+1)(2) = 3a + 2$$
$$\text{means}$$

$$-(3a + 2) = -1(3a + 2) = -1(3a) + (-1)(2) = -3a - 2$$
$$\text{means}$$

EXAMPLES Find each product by using the distributive property to remove parentheses.

8. $5(3x + 2) = 5(3x) + 5(2)$ Apply the distributive property.
$$= 15x + 10$$ Multiply.

9. $-2(y + 0.3z - 1) = -2(y) + (-2)(0.3z)$ Apply the distributive property.
$$- (-2)(1)$$
$$= -2y - 0.6z + 2$$ Multiply.

10. $-(9x + y - 2z + 6) = -1(9x + y - 2z + 6)$ Distribute -1 over each term.
$$= -1(9x) + (-1)(y) - (-1)(2z) + (-1)(6)$$
$$= -9x - y + 2z - 6$$

■ **Work Practice Problems 8–10**

Helpful Hint

If a "−" sign precedes parentheses, the sign of each term inside the parentheses is changed when the distributive property is applied to remove the parentheses.

Examples:

$$-(2x + 1) = -2x - 1$$
$$-(x - 2y) = -x + 2y$$
$$-(-5x + y - z) = 5x - y + z$$
$$-(-3x - 4y - 1) = 3x + 4y + 1$$

When simplifying an expression containing parentheses, we often use the distributive property first to remove parentheses and then again to combine any like terms.

EXAMPLES Simplify each expression.

11. $3(2x - 5) + 1 = 6x - 15 + 1$ Apply the distributive property.
        $= 6x - 14$ Combine like terms.

12. $8 - (7x + 2) + 3x = 8 - 7x - 2 + 3x$ Apply the distributive property.
          $= -7x + 3x + 8 - 2$
          $= -4x + 6$ Combine like terms.

13. $-2(4x + 7) - (3x - 1) = -8x - 14 - 3x + 1$ Apply the distributive property.
          $= -11x - 13$ Combine like terms.

14. $9 + 3(4x - 10) = 9 + 12x - 30$ Apply the distributive property.
          $= -21 + 12x$ Combine like terms.
          or $12x - 21$

▢ **Work Practice Problems 11–14**

PRACTICE PROBLEMS 11–14

Simplify each expression.
11. $4(4x - 6) + 20$
12. $5 - (3x + 9) + 6x$
13. $-3(7x + 1) - (4x - 2)$
14. $8 + 11(2y - 9)$

Helpful Hint Don't forget to use the distributive property and multiply before adding or subtracting like terms.

EXAMPLE 15 Subtract $4x - 2$ from $2x - 3$.

Solution: We first note that "subtract $4x - 2$ **from** $2x - 3$" translates to $(2x - 3) - (4x - 2)$. Notice that parentheses were placed around each given expression. This is to ensure that the entire expression after the subtraction sign is subtracted. Next, we simplify the algebraic expression.

$$(2x - 3) - (4x - 2) = 2x - 3 - 4x + 2$$ Apply the distributive property.
           $= -2x - 1$ Combine like terms.

▢ **Work Practice Problem 15**

PRACTICE PROBLEM 15

Subtract $9x - 10$ from $4x - 3$.

Objective D Writing Algebraic Expressions

To prepare for problem solving, we next practice writing word phrases as algebraic expressions.

Answers
11. $16x - 4$,   **12.** $3x - 4$,
13. $-25x - 1$,   **14.** $-91 + 22y$,
15. $-5x + 7$

PRACTICE PROBLEMS 16–19

Write each phrase as an algebraic expression and simplify if possible. Let x represent the unknown number.

16. Three times a number, subtracted from 10

17. The sum of a number and 2, divided by 5

18. Three times a number, added to the sum of a number and 6

19. Seven times the difference of a number and 4.

 Write each phrase as an algebraic expression and simplify if possible. Let x represent the unknown number.

16. Twice a number, plus 6

$$2x \qquad + \quad 6$$

This expression cannot be simplified.

17. The difference of a number and 4, divided by 7

$$(x - 4) \qquad\qquad \div \quad 7 \quad \text{or} \quad \frac{x - 4}{7}$$

This expression cannot be simplified.

18. Five plus the sum of a number and 1

$$5 \quad + \qquad\qquad (x + 1)$$

We can simplify this expression.

$$5 + (x + 1) = 5 + x + 1$$
$$= 6 + x$$

19. Four times the sum of a number and 3

$$4 \quad \cdot \qquad\qquad (x + 3)$$

Use the distributive property to simplify the expression.

$$4 \cdot (x + 3) = 4(x + 3)$$
$$= 4 \cdot x + 4 \cdot 3$$
$$= 4x + 12$$

▢ **Work Practice Problems 16–19**

Answers

16. $10 - 3x$,   **17.** $(x + 2) \div 5$ or $\dfrac{x + 2}{5}$,

18. $4x + 6$,   **19.** $7x - 28$

Mental Math

Objective A *Identify the numerical coefficient of each term. See Example 1.*

1. $-7y$

2. $3x$

3. x

4. $-y$

5. $17x^2y$

6. $1.2xyz$

Indicate whether the terms in each list are like or unlike. See Example 2.

7. $5y, -y$

8. $-2x^2y, 6xy$

9. $2z, 3z^2$

10. $ab^2, -7ab^2$

11. $8wz, \dfrac{1}{7}zw$

12. $7.4p^3q^2, 6.2p^3q^2r$

1.8 EXERCISE SET

FOR EXTRA HELP

Student Solutions Manual   PH Math/Tutor Center   CD/Video for Review   Math XL   MathXL®   MyMathLab   MyMathLab

Objective B *Simplify each expression by combining any like terms. See Examples 3 through 7.*

1. $7y + 8y$

2. $3x + 2x$

3. $8w - w + 6w$

4. $c - 7c + 2c$

5. $3b - 5 - 10b - 4$

6. $6g + 5 - 3g - 7$

7. $m - 4m + 2m - 6$

8. $a + 3a - 2 - 7a$

9. $5g - 3 - 5 - 5g$

10. $8p + 4 - 8p - 15$

11. $6.2x - 4 + x - 1.2$

12. $7.9y - 0.7 - y + 0.2$

13. $2k - k - 6$

14. $7c - 8 - c$

15. $-9x + 4x + 18 - 10x$

16. $5y - 14 + 7y - 20y$

17. $6x - 5x + x - 3 + 2x$

18. $8h + 13h - 6 + 7h - h$

19. $7x^2 + 8x^2 - 10x^2$

20. $8x^3 + x^3 - 11x^3$

21. $3.4m - 4 - 3.4m - 7$

22. $2.8w - 0.9 - 0.5 - 2.8w$

23. $6x + 0.5 - 4.3x - 0.4x + 3$

24. $0.4y - 6.7 + y - 0.3 - 2.6y$

Objective C *Simplify each expression. Use the distributive property to remove any parentheses. See Examples 8 through 10.*

25. $5(y + 4)$

26. $7(r + 3)$

27. $-2(x + 2)$

28. $-4(y + 6)$

29. $-5(2x - 3y + 6)$

30. $-2(4x - 3z - 1)$

31. $-(3x - 2y + 1)$

32. $-(y + 5z - 7)$

Objectives **B** **C** **Mixed Practice** *Remove parentheses and simplify each expression. See Examples 8 through 14.*

33. $7(d - 3) + 10$

34. $9(z + 7) - 15$

35. $-4(3y - 4) + 12y$

36. $-3(2x + 5) - 6x$

37. $3(2x - 5) - 5(x - 4)$

38. $2(6x - 1) - (x - 7)$

39. $-2(3x - 4) + 7x - 6$

40. $8y - 2 - 3(y + 4)$

41. $5k - (3k - 10)$

42. $-11c - (4 - 2c)$

43. Subtract $6x - 1$ from $3x + 4$

44. Subtract $4 + 3y$ from $8 - 5y$

45. $5(x + 2) - (3x - 4)$

46. $4(2x - 3) - (x + 1)$

47. $\frac{1}{3}(7y - 1) + \frac{1}{6}(4y + 7)$

48. $\frac{1}{5}(9y + 2) + \frac{1}{10}(2y - 1)$

49. $2 + 4(6x - 6)$

50. $8 + 4(3x - 4)$

51. $0.5(m + 2) + 0.4m$

52. $0.2(k + 8) - 0.1k$

53. $10 - 3(2x + 3y)$

54. $14 - 11(5m + 3n)$

55. $6(3x - 6) - 2(x + 1) - 17x$

56. $7(2x + 5) - 4(x + 2) - 20x$

57. $\frac{1}{2}(12x - 4) - (x + 5)$

58. $\frac{1}{3}(9x - 6) - (x - 2)$

Perform each indicated operation. Don't forget to simplify if possible. See Example 15.

59. Add $6x + 7$ to $4x - 10$.

60. Add $3y - 5$ to $y + 16$.

61. Subtract $7x + 1$ from $3x - 8$.

62. Subtract $4x - 7$ from $12 + x$.

63. Subtract $5m - 6$ from $m - 9$.

64. Subtract $m - 3$ from $2m - 6$.

Objective **D** *Write each phrase as an algebraic expression and simplify if possible. Let x represent the unknown number. See Examples 16 through 19.*

65. Twice a number, decreased by four

66. The difference of a number and two, divided by five

67. Three-fourths of a number, increased by twelve

68. Eight more than triple a number

69. The sum of 5 times a number and -2, added to 7 times the number

70. The sum of 3 times a number and 10, **subtracted from** 9 times the number

71. Eight times the sum of a number and six

72. Six times the difference of a number and five

73. Double a number minus the sum of the number and ten

74. Half a number minus the product of the number and eight

Review

Evaluate each expression for the given values. See Section 1.5.

75. If $x = -1$ and $y = 3$, find $y - x^2$

76. If $g = 0$ and $h = -4$, find $gh - h^2$

77. If $a = 2$ and $b = -5$, find $a - b^2$

78. If $x = -3$, find $x^3 - x^2 + 4$

79. If $y = -5$ and $z = 0$, find $yz - y^2$

80. If $x = -2$, find $x^3 - x^2 - x$

Concept Extensions

Given the following information, determine whether each scale is balanced or not.

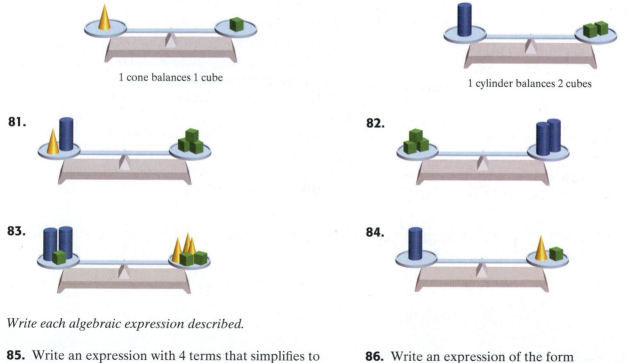

1 cone balances 1 cube

1 cylinder balances 2 cubes

81.

82.

83.

84.

Write each algebraic expression described.

85. Write an expression with 4 terms that simplifies to $3x - 4$.

86. Write an expression of the form

____ (___ + ___) whose product is $6x + 24$.

△ **87.** Recall that the perimeter of a figure is the total distance around the figure. Given the following rectangle, express the perimeter as an algebraic expression containing the variable *x*.

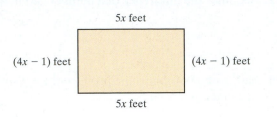

5x feet

(4x − 1) feet   (4x − 1) feet

5x feet

△ **88.** Given the following triangle, express its perimeter as an algebraic expression containing the variable *x*.

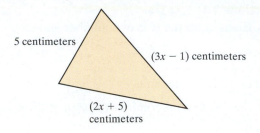

5 centimeters

(3x − 1) centimeters

(2x + 5) centimeters

△ **89.** To convert from feet to inches, we multiply by 12. For example, the number of inches in 2 feet is 12 · 2 inches. If one board has a length of (*x* + 2) *feet* and a second board has a length of (3*x* − 1) *inches,* express their total length in inches as an algebraic expression.

90. The value of 7 nickels is 5 · 7 cents. Likewise, the value of *x* nickels is 5*x* cents. If the money box in a drink machine contains *x nickels,* 3*x dimes,* and (30*x* − 1) *quarters,* express their total value in cents as an algebraic expression.

91. In your own words, explain how to combine like terms.

92. Do like terms contain the same numerical coefficients? Explain your answer.

STUDY SKILLS BUILDER

What to Do the Day of an Exam

Your first exam may be soon. On the day of an exam, don't forget to try the following:

- Allow yourself plenty of time to arrive.
- Read the directions on the test carefully.
- Read each problem carefully as you take your test. Make sure that you answer the question asked.
- Watch your time and pace yourself so that you may attempt each problem on your test.
- Check your work and answers.
- ***Do not turn your test in early.*** If you have extra time, spend it double-checking your work.

Good luck!

Answer the following questions based on your most recent mathematics exam, whenever that was.

1. How soon before class did you arrive?
2. Did you read the directions on the test carefully?
3. Did you make sure you answered the question asked for each problem on the exam?
4. Were you able to attempt each problem on your exam?
5. If your answer to question 4 is no, list reasons why.
6. Did you have extra time on your exam?
7. If your answer to question 6 is yes, describe how you spent that extra time.

CHAPTER 1   Group Activity

Sections 1.3, 1.4, 1.5

Magic Squares

A magic square is a set of numbers arranged in a square table so that the sum of the numbers in each column, row, and diagonal is the same. For instance, in the magic square below, the sum of each column, row, and diagonal is 15. Notice that no number is used more than once in the magic square.

2	9	4
7	5	3
6	1	8

The properties of magic squares have been known for a very long time and once were thought to be good luck charms. The ancient Egyptians and Greeks understood their patterns. A magic square even made it into a famous work of art. The engraving titled *Melencolia I,* created by German artist Albrecht Dürer in 1514, features the following four-by-four magic square on the building behind the central figure.

16	3	2	13
5	10	11	8
9	6	7	12
4	15	14	1

Group Exercises

1. Verify that what is shown in the Dürer engraving is, in fact, a magic square. What is the common sum of the columns, rows, and diagonals?

2. Negative numbers can also be used in magic squares. Complete the following magic square:

		−2
	−1	
0		−4

3. Use the numbers −12, −9, −6, −3, 0, 3, 6, 9, and 12 to form a magic square.

Chapter 1 Vocabulary Check

Fill in each blank with one of the words or phrases listed below.

inequality symbols   exponent   term   numerical coefficient

grouping symbols   solution   like terms   unlike terms

equation   absolute value   numerator   denominator

opposites   base   reciprocals   variable

1. The symbols $\neq$, $<$, and $>$ are called _____.
2. A mathematical statement that two expressions are equal is called an _____.
3. The _____ of a number is the distance between that number and 0 on the number line.
4. A symbol used to represent a number is called a _____.
5. Two numbers that are the same distance from 0 but lie on opposite sides of 0 are called _____.
6. The number in a fraction above the fraction bar is called the _____.
7. A _____ of an equation is a value for the variable that makes the equation a true statement.
8. Two numbers whose product is 1 are called _____.
9. In 2^3, the 2 is called the _____ and the 3 is called the _____.
10. The _____ of a term is its numerical factor.
11. The number in a fraction below the fraction bar is called the _____.
12. Parentheses and brackets are examples of _____.
13. A _____ is a number or the product of a number and variables raised to powers.
14. Terms with the same variables raised to the same powers are called _____.
15. If terms are not like terms, then they are _____.

> **Helpful Hint**
>
> Are you preparing for your test? Don't forget to take the Chapter 1 Test on page 87. Then check your answers at the back of the text and use the Chapter Test Prep Video CD to see the fully worked-out solutions to any of the exercises you want to review.

1 Chapter Highlights

DEFINITIONS AND CONCEPTS	**EXAMPLES**
Section 1.2 Symbols and Sets of Numbers	

A **set** is a collection of objects, called **elements,** enclosed in braces.	$\{a, c, e\}$
Natural numbers: $\{1, 2, 3, 4, \dots\}$	Given the set $\left\{-3.4, \sqrt{3}, 0, \frac{2}{3}, 5, -4\right\}$ list the numbers that belong to the set of
Whole numbers: $\{0, 1, 2, 3, 4, \dots\}$	Natural numbers: 5
Integers: $\{\dots, -3, -2, -1, 0, 1, 2, 3, \dots\}$	Whole numbers: $0, 5$
Rational numbers: {real numbers that can be expressed as a quotient of integers}	Integers: $-4, 0, 5$
Irrational numbers: {real numbers that cannot be expressed as a quotient of integers}	Rational numbers: $-3.4, 0, \frac{2}{3}, 5, -4$
A line used to picture numbers is called a **number line.**	Irrational numbers: $\sqrt{3}$
Real numbers: {all numbers that correspond to a point on the number line}	Real numbers: $-3.4, \sqrt{3}, 0, \frac{2}{3}, 5, -4$

Number line: $-5 \ -4 \ -3 \ -2 \ -1 \ \ 0 \ \ 1 \ \ 2 \ \ 3 \ \ 4 \ \ 5$

DEFINITIONS AND CONCEPTS	EXAMPLES

Section 1.2   Symbols and Sets of Numbers (*continued*)

The **absolute value** of a real number a denoted by $|a|$ is the distance between a and 0 on the number line.

$$|5| = 5 \quad |0| = 0 \quad |-2| = 2$$

SYMBOLS: = is equal to
$\neq$ is not equal to
$>$ is greater than
$<$ is less than
$\leq$ is less than or equal to
$\geq$ is greater than or equal to

$-7 = -7$
$3 \neq -3$
$4 > 1$
$1 < 4$
$6 \leq 6$
$18 \geq -\dfrac{1}{3}$

ORDER PROPERTY FOR REAL NUMBERS

For any two real numbers a and b, a is less than b if a is to the left of b on the number line.

$$0 > -3$$
$$-3 < 0 \quad 0 < 2.5 \quad 2.5 > 0$$

```
<-+--+--●--+--+--●--+--+--●--+--+->
 -5 -4 -3 -2 -1  0  1  2  3  4  5
```

Section 1.3   Exponents, Order of Operations, and Variable Expressions

The expression a^n is an **exponential expression.** The number a is called the **base;** it is the repeated factor. The number n is called the **exponent;** it is the number of times that the base is a factor.

$$4^3 = 4 \cdot 4 \cdot 4 = 64$$
$$7^2 = 7 \cdot 7 = 49$$

ORDER OF OPERATIONS

1. Perform all operations within grouping symbols first, starting with the innermost set.
2. Evaluate exponential expressions.
3. Multiply or divide in order from left to right.
4. Add or subtract in order from left to right.

$$\frac{8^2 + 5(7 - 3)}{3 \cdot 7} = \frac{8^2 + 5(4)}{21}$$
$$= \frac{64 + 5(4)}{21}$$
$$= \frac{64 + 20}{21}$$
$$= \frac{84}{21}$$
$$= 4$$

A symbol used to represent a number is called a **variable.**

Examples of variables are

$$q, x, z$$

An **algebraic expression** is a collection of numbers, variables, operation symbols, and grouping symbols.

Examples of algebraic expressions are

$$5x, \quad 2(y - 6), \quad \frac{q^2 - 3q + 1}{6}$$

To **evaluate an algebraic expression** containing a variable, substitute a given number for the variable and simplify.

Evaluate $x^2 - y^2$ when $x = 5$ and $y = 3$.

$$x^2 - y^2 = (5)^2 - 3^2$$
$$= 25 - 9$$
$$= 16$$

A mathematical statement that two expressions are equal is called an **equation.**

Equations:
$$3x - 9 = 20$$
$$A = \pi r^2$$

continued

DEFINITIONS AND CONCEPTS	**EXAMPLES**

Section 1.3   Exponents, Order of Operations, and Variable Expressions (*continued*)

A **solution** of an equation is a value for the variable that makes the equation a true statement.	Determine whether 4 is a solution of $5x + 7 = 27$. $$5x + 7 = 27$$ $$5(4) + 7 \overset{?}{=} 27$$ $$20 + 7 \overset{?}{=} 27$$ $$27 = 27 \text{ True}$$ 4 is a solution.

Section 1.4   Adding Real Numbers

TO ADD TWO NUMBERS WITH THE SAME SIGN **1.** Add their absolute values. **2.** Use their common sign as the sign of the sum.	Add. $$10 + 7 = 17$$ $$-3 + (-8) = -11$$
TO ADD TWO NUMBERS WITH DIFFERENT SIGNS **1.** Subtract their absolute values. **2.** Use the sign of the number whose absolute value is larger as the sign of the sum.	$$-25 + 5 = -20$$ $$14 + (-9) = 5$$
Two numbers that are the same distance from 0 but lie on opposite sides of 0 are called **opposites** or **additive inverses.** The opposite of a number a is denoted by $-a$.	The opposite of -7 is 7. The opposite of 123 is -123.

Section 1.5   Subtracting Real Numbers

To subtract two numbers a and b, add the first number a to the opposite of the second number, b. $$a - b = a + (-b)$$	Subtract. $$3 - (-44) = 3 + 44 = 47$$ $$-5 - 22 = -5 + (-22) = -27$$ $$-30 - (-30) = -30 + 30 = 0$$

Section 1.6   Multiplying and Dividing Real Numbers

MULTIPLYING REAL NUMBERS The product of two numbers with the same sign is a positive number. The product of two numbers with different signs is a negative number.	Multiply. $$7 \cdot 8 = 56 \qquad -7 \cdot (-8) = 56$$ $$-2 \cdot 4 = -8 \qquad 2 \cdot (-4) = -8$$
PRODUCTS INVOLVING ZERO The product of 0 and any number is 0. $$b \cdot 0 = 0 \quad \text{and} \quad 0 \cdot b = 0$$	$$-4 \cdot 0 = 0 \qquad 0 \cdot \left(-\frac{3}{4}\right) = 0$$
QUOTIENT OF TWO REAL NUMBERS $$\frac{a}{b} = a \cdot \frac{1}{b}$$	Divide. $$\frac{42}{2} = 42 \cdot \frac{1}{2} = 21$$

DEFINITIONS AND CONCEPTS	**EXAMPLES**

Section 1.6   Multiplying and Dividing Real Numbers (*continued*)

DIVIDING REAL NUMBERS

The quotient of two numbers with the same sign is a positive number. The quotient of two numbers with different signs is a negative number.

$$\frac{90}{10} = 9 \qquad \frac{-90}{-10} = 9$$

$$\frac{42}{-6} = -7 \qquad \frac{-42}{6} = -7$$

QUOTIENTS INVOLVING ZERO

Let a be a nonzero number. $\dfrac{0}{a} = 0$ and $\dfrac{a}{0}$ is undefined.

$$\frac{0}{18} = 0 \qquad \frac{0}{-47} = 0 \qquad \frac{-85}{0} \text{ is undefined.}$$

Section 1.7   Properties of Real Numbers

COMMUTATIVE PROPERTIES

Addition: $a + b = b + a$
Multiplication: $a \cdot b = b \cdot a$

$3 + (-7) = -7 + 3$
$-8 \cdot 5 = 5 \cdot (-8)$

ASSOCIATIVE PROPERTIES

Addition: $(a + b) + c = a + (b + c)$
Multiplication: $(a \cdot b) \cdot c = a \cdot (b \cdot c)$

$(5 + 10) + 20 = 5 + (10 + 20)$
$(-3 \cdot 2) \cdot 11 = -3 \cdot (2 \cdot 11)$

Two numbers whose product is 1 are called **multiplicative inverses** or **reciprocals.** The reciprocal of a nonzero number a is $\dfrac{1}{a}$ because $a \cdot \dfrac{1}{a} = 1$.

The reciprocal of 3 is $\dfrac{1}{3}$.

The reciprocal of $-\dfrac{2}{5}$ is $-\dfrac{5}{2}$.

DISTRIBUTIVE PROPERTY

$$a(b + c) = a \cdot b + a \cdot c$$

$5(6 + 10) = 5 \cdot 6 + 5 \cdot 10$
$-2(3 + x) = -2 \cdot 3 + (-2)(x)$

IDENTITIES

$$a + 0 = a \qquad 0 + a = a$$
$$a \cdot 1 = a \qquad 1 \cdot a = a$$

$5 + 0 = 5 \qquad 0 + (-2) = -2$
$-14 \cdot 1 = -14 \qquad 1 \cdot 27 = 27$

INVERSES

Additive or opposite: $a + (-a) = 0$

Multiplicative or reciprocal: $b \cdot \dfrac{1}{b} = 1, \qquad b \neq 0$

$7 + (-7) = 0$

$3 \cdot \dfrac{1}{3} = 1$

Section 1.8   Simplifying Expressions

The **numerical coefficient** of a **term** is its numerical factor.

TERM	NUMERICAL COEFFICIENT
$-7y$	-7
x	1
$\dfrac{1}{5}a^2b$	$\dfrac{1}{5}$

continued

DEFINITIONS AND CONCEPTS	**EXAMPLES**

Section 1.8   Simplifying Expressions (*continued*)

Terms with the same variables raised to exactly the same powers are **like terms**.	**LIKE TERMS** **UNLIKE TERMS** $12x, -x$ $3y, 3y^2$ $-2xy, 5yx$ $7a^2b, -2ab^2$
To combine like terms, add the numerical coefficients and multiply the result by the common variable factor.	$9y + 3y = 12y$ $-4z^2 + 5z^2 - 6z^2 = -5z^2$
To remove parentheses, apply the distributive property.	$-4(x + 7) + 10(3x - 1)$ $\quad = -4x - 28 + 30x - 10$ $\quad = 26x - 38$

STUDY SKILLS BUILDER

Are You Prepared for a Test on Chapter 1?

Below I have listed some *common trouble areas* for students in Chapter 1. After studying for your test—but before taking your test—read these.

- Do you know the difference between $|-3|$, $-|-3|$, and $-(-3)$?

    $|-3| = 3;\quad -|-3| = -3;\quad$ and $\quad -(-3) = 3$

    (Section 1.2)

- Evaluate $x - y$ if $x = 7$ and $y = -3$.

    $x - y = 7 - (-3) = 10$ (Section 1.5)

- Make sure you are familiar with order of operations. Sometimes the simplest-looking expressions can give you the most trouble.

    $1 + 2(3 + 6) = 1 + 2(9) = 1 + 18 = 19$

    (Section 1.3)

- Do you know the difference between $(-3)^2$ and -3^2?

    $(-3)^2 = 9$ and $-3^2 = -9$ (Section 1.6)

- Do you know that these fractions are equivalent?

    $$-\frac{1}{3} = \frac{-1}{3} = \frac{1}{-3}$$ (Section 1.6)

Remember: This is simply a checklist of selected topics given to check your understanding. For a review of Chapter 1 in the text, see the material at the end of Chapter 1.

1 CHAPTER REVIEW

(1.2) *Insert $<$, $>$, or $=$ in the appropriate space to make each statement true.*

1. 8   10

2. 7   2

3. -4 -5

4. $\dfrac{12}{2}$ -8

5. $|-7|$ $|-8|$

6. $|-9|$ -9

7. $-|-1|$ -1

8. $|-14|$ $-(-14)$

9. 1.2   1.02

10. $-\dfrac{3}{2}$ $-\dfrac{3}{4}$

Translate each statement into symbols.

11. Four is greater than or equal to negative three.

12. Six is not equal to five.

13. 0.03 is less than 0.3.

14. New York City has 155 museums and 400 art galleries. Write an inequality comparing the numbers 155 and 400. (*Source:* Absolute Trivia.com)

Given the sets of numbers below, list the numbers in each set that also belong to the set of:

a. Natural numbers

b. Whole numbers

c. Integers

d. Rational numbers

e. Irrational numbers

f. Real numbers

15. $\left\{-6, 0, 1, 1\dfrac{1}{2}, 3, \pi, 9.62\right\}$

16. $\left\{-3, -1.6, 2, 5, \dfrac{11}{2}, 15.1, \sqrt{5}, 2\pi\right\}$

The following chart shows the gains and losses in dollars of Density Oil and Gas stock for a particular week. Use this chart to answer Exercises 17 and 18.

Day	Gain or Loss (in dollars)
Monday	+1
Tuesday	−2
Wednesday	+5
Thursday	+1
Friday	−4

17. Which day showed the greatest loss?

18. Which day showed the greatest gain?

(1.3) *Choose the correct answer for each statement.*

19. The expression $6 \cdot 3^2 + 2 \cdot 8$ simplifies to

    a. -52 **b.** 440   **c.** 70   **d.** 64

20. The expression $68 - 5 \cdot 2^3$ simplifies to

    a. -232 **b.** 28   **c.** 38   **d.** 504

Simplify each expression.

21. $3(1 + 2 \cdot 5) + 4$

22. $8 + 3(2 \cdot 6 - 1)$

23. $\dfrac{4 + |6 - 2| + 8^2}{4 + 6 \cdot 4}$

24. $5[3(2 + 5) - 5]$

Translate each word statement to symbols.

25. The difference of twenty and twelve is equal to the product of two and four.

26. The quotient of nine and two is greater than negative five.

Evaluate each expression when $x = 6$, $y = 2$, and $z = 8$.

27. $2x + 3y$

28. $x(y + 2z)$

29. $\dfrac{x}{y} + \dfrac{z}{2y}$

30. $x^2 - 3y^2$

△ **31.** The expression $180 - a - b$ represents the measure of the unknown angle of the given triangle. Replace a with 37 and b with 80 to find the measure of the unknown angle.

Decide whether the given number is a solution to the given equation.

32. $7x - 3 = 18$;   3

33. $3x^2 + 4 = x - 1$;   1

(1.4) *Find the additive inverse or opposite of each number.*

34. -9

35. $\dfrac{2}{3}$

36. $|-2|$

37. $-|-7|$

Add.

38. $-15 + 4$

39. $-6 + (-11)$

40. $\dfrac{1}{16} + \left(-\dfrac{1}{4}\right)$

41. $-8 + |-3|$

42. $-4.6 + (-9.3)$

43. $-2.8 + 6.7$

(1.5) *Perform each indicated operation.*

44. $6 - 20$

45. $-3.1 - 8.4$

46. $-6 - (-11)$

47. $4 - 15$

48. $-21 - 16 + 3(8 - 2)$

49. $\dfrac{11 - (-9) + 6(8 - 2)}{2 + 3 \cdot 4}$

Evaluate each expression for $x = 3$, $y = -6$, and $z = -9$. Then choose the correct evaluation.

50. $2x^2 - y + z$

  a. 15 **b.** 3 **c.** 27 **d.** -3

51. $\dfrac{|y - 4x|}{2x}$

  a. 3 **b.** 1 **c.** -1 **d.** -3

52. At the beginning of the week the price of Density Oil and Gas stock from Exercises 17 and 18 is $50 per share. Find the price of a share of stock at the end of the week.

Find each multiplicative inverse or reciprocal.

53. -6

54. $\dfrac{3}{5}$

(1.6) *Simplify each expression.*

55. $6(-8)$

56. $(-2)(-14)$

57. $\dfrac{-18}{-6}$

58. $\dfrac{42}{-3}$

59. $-3(-6)(-2)$

60. $(-4)(-3)(0)(-6)$

61. $\dfrac{4(-3) + (-8)}{2 + (-2)}$

62. $\dfrac{3(-2)^2 - 5}{-14}$

(1.7) *Name the property illustrated in each equation.*

63. $-6 + 5 = 5 + (-6)$

64. $6 \cdot 1 = 6$

65. $3(8 - 5) = 3 \cdot 8 + 3 \cdot (-5)$

66. $4 + (-4) = 0$

67. $2 + (3 + 9) = (2 + 3) + 9$

68. $2 \cdot 8 = 8 \cdot 2$

69. $6(8 + 5) = 6 \cdot 8 + 6 \cdot 5$

70. $(3 \cdot 8) \cdot 4 = 3 \cdot (8 \cdot 4)$

71. $4 \cdot \dfrac{1}{4} = 1$

72. $8 + 0 = 8$

73. $4(8 + 3) = 4(3 + 8)$

74. $5(2 + 1) = 5 \cdot 2 + 5 \cdot 1$

(1.8) *Simplify each expression.*

75. $5x - x + 2x$

76. $0.2z - 4.6z - 7.4z$

77. $\dfrac{1}{2}x + 3 + \dfrac{7}{2}x - 5$

78. $\dfrac{4}{5}y + 1 + \dfrac{6}{5}y + 2$ **79.** $2(n - 4) + n - 10$ **80.** $3(w + 2) - (12 - w)$

81. Subtract $7x - 2$ from $x + 5$.   **82.** Subtract $1.4y - 3$ from $y - 0.7$.

Write each phrase as an algebraic expression. Simplify if possible.

83. Three times a number decreased by 7

84. Twice the sum of a number and 2.8, added to 3 times the number

Mixed Review

Insert $<$, $>$, or $=$ in the space between each pair of numbers.

85. $-|-11|$ $|11.4|$

86. $-1\dfrac{1}{2}$ $-2\dfrac{1}{2}$

Perform the indicated operations.

87. $-7.2 + (-8.1)$ **88.** $14 - 20$ **89.** $4(-20)$

90. $\dfrac{-20}{4}$ **91.** $-\dfrac{4}{5}\left(\dfrac{5}{16}\right)$ **92.** $-0.5(-0.3)$

93. $8 \div 2 \cdot 4$ **94.** $(-2)^4$ **95.** $\dfrac{-3 - 2(-9)}{-15 - 3(-4)}$

96. $5 + 2[(7 - 5)^2 + (1 - 3)]$ **97.** $-\dfrac{5}{8} \div \dfrac{3}{4}$ **98.** $\dfrac{-15 + (-4)^2 + |-9|}{10 - 2 \cdot 5}$

Remove parentheses and simplify each expression.

99. $7(3x - 3) - 5(x + 4)$ **100.** $8 + 2(9x - 10)$

1 CHAPTER TEST

 Remember to use the Chapter Test Prep Video CD to see the fully worked-out solutions to any of the exercises you want to review.

Translate each statement into symbols.

1. The absolute value of negative seven is greater than five.

2. The sum of nine and five is greater than or equal to four.

Simplify each expression.

3. $-13 + 8$

4. $-13 - (-2)$

5. $6 \cdot 3 - 8 \cdot 4$

6. $13(-3)$

7. $(-6)(-2)$

8. $\dfrac{|-16|}{-8}$

9. $\dfrac{-8}{0}$

10. $\dfrac{|-6| + 2}{5 - 6}$

11. $\dfrac{1}{2} - \dfrac{5}{6}$

12. $-1\dfrac{1}{8} + 5\dfrac{3}{4}$

13. $-\dfrac{3}{5} + \dfrac{15}{8}$

14. $3(-4)^2 - 80$

15. $6[5 + 2(3 - 8) - 3]$

16. $\dfrac{-12 + 3 \cdot 8}{4}$

17. $\dfrac{(-2)(0)(-3)}{-6}$

Insert $<$, $>$, or $=$ in the appropriate space to make each statement true.

18. $-3 \quad -7$

19. $4 \quad -8$

20. $|-3| \quad 2$

21. $|-2| \quad -1 - (-3)$

1. _____

2. _____

3. _____

4. _____

5. _____

6. _____

7. _____

8. _____

9. _____

10. _____

11. _____

12. _____

13. _____

14. _____

15. _____

16. _____

17. _____

18. _____

19. _____

20. _____

21. _____

22. a. _____

b. _____

c. _____

d. _____

e. _____

f. _____

23. _____

24. _____

25. _____

26. _____

27. _____

28. _____

29. _____

30. _____

31. _____

32. _____

33. _____

34. _____

35. _____

36. _____

37. _____

38. _____

39. _____

40. _____

22. Given $\left\{-5, -1, \frac{1}{4}, 0, 1, 7, 11.6, \sqrt{7}, 3\pi\right\}$, list the numbers in this set that also belong to the set of:

  a. Natural numbers   **b.** Whole numbers
  c. Integers   **d.** Rational numbers
  e. Irrational numbers   **f.** Real numbers

Evaluate each expression when $x = 6$, $y = -2$, and $z = -3$.

23. $x^2 + y^2$ **24.** $x + yz$ **25.** $2 + 3x - y$ **26.** $\dfrac{y + z - 1}{x}$

Identify the property illustrated by each expression.

27. $8 + (9 + 3) = (8 + 9) + 3$ **28.** $6 \cdot 8 = 8 \cdot 6$

29. $-6(2 + 4) = -6 \cdot 2 + (-6) \cdot 4$ **30.** $\dfrac{1}{6}(6) = 1$

31. Find the opposite of -9.   **32.** Find the reciprocal of $-\dfrac{1}{3}$.

The New Orleans Saints were 22 yards from the goal when the series of gains and losses shown in the chart occurred. Use this chart to answer Exercises 33 and 34.

	Gains and Losses (in yards)
First down	5
Second down	−10
Third down	−2
Fourth down	29

33. During which down did the greatest loss of yardage occur?

34. Was a touchdown scored?

35. The temperature at the Winter Olympics was a frigid 14° below zero in the morning, but by noon it had risen 31°. What was the temperature at noon?

36. Jean Avarez decided to sell 280 shares of stock, which decreased in value by $1.50 per share yesterday. How much money did she lose?

Simplify each expression.

37. $2y - 6 - y - 4$ **38.** $2.7x + 6.1 + 3.2x - 4.9$

39. $4(x - 2) - 3(2x - 6)$ **40.** $-5(y + 1) + 2(3 - 5y)$

2

Equations, Inequalities, and Problem Solving

In this chapter, we solve equations and inequalities. Once we know how to solve equations and inequalities, we may solve word problems. Of course, problem solving is an integral topic in algebra and its discussion is continued throughout this text.

Since 1948, when NASCAR began, the cars have been transformed from the original "stock" cars, or road models, into the technologically advanced racing machines on the tracks today. In fact, auto manufacturers are creating more advanced street vehicles that can also be used for racing. NASCAR is an increasingly popular sport, with the audience growing daily. In Exercise 35 on page 127 (Section 2.4), you will find the number of points accumulated by the top two finishers for a recent Winston Cup.

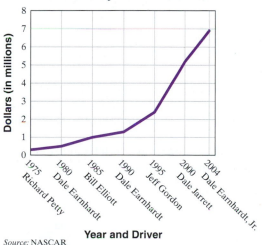

Most Money Won Each Year

Source: NASCAR

A Use the Addition Property of Equality to Solve Linear Equations.

B Simplify an Equation and Then Use the Addition Property of Equality.

C Write Word Phrases as Algebraic Expressions.

2.1 THE ADDITION PROPERTY OF EQUALITY

Objective **A** Using the Addition Property

Recall from Section 1.3 that an equation is a statement in which two expressions have the same value. Also, a value of the variable that makes an equation a true statement is called a solution or root of the equation. The process of finding the solution of an equation is called **solving** the equation for the variable. In this section, we concentrate on solving *linear equations* in one variable.

> ### Linear Equation in One Variable
>
> A **linear equation in one variable** can be written in the form
>
> $$Ax + B = C$$
>
> where A, B, and C are real numbers and $A \neq 0$.

Evaluating a linear equation for a given value of the variable, as we did in Section 1.3, can tell us whether that value is a solution. But we can't rely on evaluating an equation as our method of solving it—with what value would we start?

Instead, to solve a linear equation in x, we write a series of simpler equations, all *equivalent* to the original equation, so that the final equation has the form

$$x = \text{number} \qquad \text{or} \qquad \text{number} = x$$

Equivalent equations are equations that have the same solution. This means that the "number" above is the solution to the original equation.

The first property of equality that helps us write simpler equivalent equations is the **addition property of equality.**

> ### Addition Property of Equality
>
> If a, b, and c are real numbers, then
>
> $$a = b \qquad \text{and} \qquad a + c = b + c$$
>
> are equivalent equations.

This property guarantees that adding the same number to both sides of an equation does not change the solution of the equation. Since subtraction is defined in terms of addition, we may also **subtract the same number from both sides** without changing the solution.

A good way to picture a true equation is as a balanced scale. Since it is balanced, each side of the scale weighs the same amount.

$x - 2 \qquad\qquad 5$

90

If the same weight is added to or subtracted from each side, the scale remains balanced.

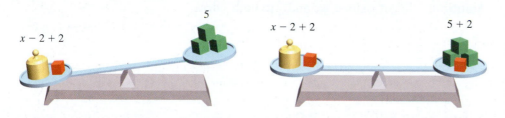

We use the addition property of equality to write equivalent equations until the variable is alone (by itself on one side of the equation) and the equation looks like "$x = $ number" or "number $= x$."

✔ **Concept Check** Use the addition property to fill in the blanks so that the middle equation simplifies to the last equation.

$$x - 5 = 3$$
$$x - 5 + \underline{} = 3 + \underline{}$$
$$x = 8$$

EXAMPLE 1 Solve $x - 7 = 10$ for x.

Solution: To solve for x, we first get x alone on one side of the equation. To do this, we add 7 to both sides of the equation.

$$x - 7 = 10$$
$$x - 7 + 7 = 10 + 7 \qquad \text{Add 7 to both sides.}$$
$$x = 17 \qquad \text{Simplify.}$$

The solution of the equation $x = 17$ is obviously 17.
Since we are writing equivalent equations, the solution of the equation $x - 7 = 10$ is also 17.

Check: To check, replace x with 17 in the original equation.

$$x - 7 = 10 \qquad \text{Original equation.}$$
$$17 - 7 \stackrel{?}{=} 10 \qquad \text{Replace } x \text{ with 17.}$$
$$10 = 10 \qquad \text{True}$$

Since the statement is true, 17 is the solution.

▢ **Work Practice Problem 1**

EXAMPLE 2 Solve: $y + 0.6 = -1.0$

Solution: To solve for y (get y alone on one side of the equation), we subtract 0.6 from both sides of the equation.

$$y + 0.6 = -1.0$$
$$y + 0.6 - 0.6 = -1.0 - 0.6 \qquad \text{Subtract 0.6 from both sides.}$$
$$y = -1.6 \qquad \text{Combine like terms.}$$

Check: $y + 0.6 = -1.0 \qquad \text{Original equation.}$
$$-1.6 + 0.6 \stackrel{?}{=} -1.0 \qquad \text{Replace } y \text{ with } -1.6.$$
$$-1.0 = -1.0 \qquad \text{True}$$

The solution is -1.6.

▢ **Work Practice Problem 2**

PRACTICE PROBLEM 1

Solve $x - 5 = 8$ for x.

PRACTICE PROBLEM 2

Solve: $y + 1.7 = 0.3$

Answers
1. $x = 13$, **2.** $y = -1.4$

✔ **Concept Check Answer**
5

PRACTICE PROBLEM 3

Solve: $\dfrac{7}{8} = y - \dfrac{1}{3}$

EXAMPLE 3 Solve: $\dfrac{1}{2} = x - \dfrac{3}{4}$

Solution: To get x alone, we add $\dfrac{3}{4}$ to both sides.

$$\dfrac{1}{2} = x - \dfrac{3}{4}$$

$$\dfrac{1}{2} + \dfrac{3}{4} = x - \dfrac{3}{4} + \dfrac{3}{4} \qquad \text{Add } \dfrac{3}{4} \text{ to both sides.}$$

$$\dfrac{1}{2}\cdot\dfrac{2}{2} + \dfrac{3}{4} = x \qquad \text{The LCD is 4.}$$

$$\dfrac{2}{4} + \dfrac{3}{4} = x \qquad \text{Add the fractions.}$$

$$\dfrac{5}{4} = x$$

Check: $\dfrac{1}{2} = x - \dfrac{3}{4} \qquad$ Original equation.

$$\dfrac{1}{2} \overset{?}{=} \dfrac{5}{4} - \dfrac{3}{4} \qquad \text{Replace } x \text{ with } \dfrac{5}{4}.$$

$$\dfrac{1}{2} \overset{?}{=} \dfrac{2}{4} \qquad \text{Subtract.}$$

$$\dfrac{1}{2} = \dfrac{1}{2} \qquad \text{True}$$

The solution is $\dfrac{5}{4}$.

🔲 **Work Practice Problem 3**

Helpful Hint

We may solve an equation so that the variable is alone on *either* side of the equation. For example, $\dfrac{5}{4} = x$ is equivalent to $x = \dfrac{5}{4}$.

PRACTICE PROBLEM 4

Solve: $3x + 10 = 4x$

Helpful Hint

For Example 4, why not subtract $6t$ from both sides? The addition property allows us to do this, and we would have $-t - 5 = 0$. We are just no closer to our goal of having variable terms on one side of the equation and numbers on the other.

EXAMPLE 4 Solve: $5t - 5 = 6t$

Solution: To solve for t, we first want all terms containing t on one side of the equation and numbers on the other side. Notice that if we subtract $5t$ from both sides of the equation, then variable terms will be on one side of the equation and the number -5 will be alone on the other side.

$$5t - 5 = 6t$$

$$5t - 5 - 5t = 6t - 5t \qquad \text{Subtract } 5t \text{ from both sides.}$$

$$-5 = t \qquad \text{Combine like terms.}$$

Check: $5t - 5 = 6t \qquad$ Original equation.

$$5(-5) - 5 \overset{?}{=} 6(-5) \qquad \text{Replace } t \text{ with } -5.$$

$$-25 - 5 \overset{?}{=} -30$$

$$-30 = -30 \qquad \text{True}$$

The solution is -5.

🔲 **Work Practice Problem 4**

Answers

3. $y = \dfrac{29}{24}$, **4.** $x = 10$

Objective B Simplifying Equations

Many times, it is best to simplify one or both sides of an equation before applying the addition property of equality.

EXAMPLE 5 Solve: $2x + 3x - 5 + 7 = 10x + 3 - 6x - 4$

Solution: First we simplify both sides of the equation.

$$2x + 3x - 5 + 7 = 10x + 3 - 6x - 4$$
$$5x + 2 = 4x - 1 \qquad \text{Combine like terms on each side of the equation.}$$

Next, we want all terms with a variable on one side of the equation and all numbers on the other side.

$$5x + 2 - 4x = 4x - 1 - 4x \qquad \text{Subtract } 4x \text{ from both sides.}$$
$$x + 2 = -1 \qquad \text{Combine like terms.}$$
$$x + 2 - 2 = -1 - 2 \qquad \text{Subtract 2 from both sides to get } x \text{ alone.}$$
$$x = -3 \qquad \text{Combine like terms.}$$

Check:
$$2x + 3x - 5 + 7 = 10x + 3 - 6x - 4 \qquad \text{Original equation.}$$
$$2(-3) + 3(-3) - 5 + 7 \stackrel{?}{=} 10(-3) + 3 - 6(-3) - 4 \qquad \text{Replace } x \text{ with } -3.$$
$$-6 - 9 - 5 + 7 \stackrel{?}{=} -30 + 3 + 18 - 4 \qquad \text{Multiply.}$$
$$-13 = -13 \qquad \text{True}$$

The solution is -3.

■ **Work Practice Problem 5**

If an equation contains parentheses, we use the distributive property to remove them, as before. Then we combine any like terms.

EXAMPLE 6 Solve: $6(2a - 1) - (11a + 6) = 7$

Solution: $6(2a - 1) - 1(11a + 6) = 7$

$$6(2a) + 6(-1) - 1(11a) - 1(6) = 7 \qquad \text{Apply the distributive property.}$$
$$12a - 6 - 11a - 6 = 7 \qquad \text{Multiply.}$$
$$a - 12 = 7 \qquad \text{Combine like terms.}$$
$$a - 12 + 12 = 7 + 12 \qquad \text{Add 12 to both sides.}$$
$$a = 19 \qquad \text{Simplify.}$$

Check: Check by replacing a with 19 in the original equation.

■ **Work Practice Problem 6**

EXAMPLE 7 Solve: $3 - x = 7$

Solution: First we subtract 3 from both sides.

$$3 - x = 7$$
$$3 - x - 3 = 7 - 3 \qquad \text{Subtract 3 from both sides.}$$
$$-x = 4 \qquad \text{Simplify.}$$

We have not yet solved for x since x is not alone. However, this equation does say that the opposite of x is 4. If the opposite of x is 4, then x is the opposite of 4, or $x = -4$.

If $\quad -x = 4$,

then $\quad x = -4$.

Continued on next page

PRACTICE PROBLEM 5

Solve:
$$10w + 3 - 4w + 4 = -2w + 3 + 7w$$

PRACTICE PROBLEM 6

Solve:
$$3(2w - 5) - (5w + 1) = -3$$

PRACTICE PROBLEM 7

Solve: $12 - y = 9$

Answers

5. $w = -4$, **6.** $w = 13$, **7.** $y = 3$

Check: $3 - x = 7$ Original equation.

$3 - (-4) \overset{?}{=} 7$ Replace x with -4.

$3 + 4 \overset{?}{=} 7$ Add.

$7 = 7$ True

The solution is -4.

☐ **Work Practice Problem 7**

Objective C Writing Algebraic Expressions

In this section, we continue to practice writing algebraic expressions.

PRACTICE PROBLEM 8

a. The sum of two numbers is 11. If one number is 4, find the other number.

b. The sum of two numbers is 11. If one number is x, write an expression representing the other number.

c. The sum of two numbers is 56. If one number is a, write an expression representing the other number.

EXAMPLE 8

a. The sum of two numbers is 8. If one number is 3, find the other number.

b. The sum of two numbers is 8. If one number is x, write an expression representing the other number.

Solution:

a. If the sum of two numbers is 8 and one number is 3, we find the other number by subtracting 3 from 8. The other number is $8 - 3$, or 5.

b. If the sum of two numbers is 8 and one number is x, we find the other number by subtracting x from 8. The other number is represented by $8 - x$.

☐ **Work Practice Problem 8**

PRACTICE PROBLEM 9

In a recent House of Representatives race in California, Lucille Roybal-Allard received 49,489 more votes than Wayne Miller. If Wayne received n votes, how many did Lucille receive? (*Source:* Voter News Service)

EXAMPLE 9 The Verrazano-Narrows Bridge in New York City is the longest suspension bridge in North America. The Golden Gate Bridge in San Francisco is 60 feet shorter than the Verrazano-Narrows Bridge. If the length of the Verrazano-Narrows Bridge is m feet, express the length of the Golden Gate Bridge as an algebraic expression in m. (*Source:* Survey of State Highway Engineers)

Solution: Since the Golden Gate is 60 feet shorter than the Verrazano-Narrows Bridge, we have that its length is

In words:	Length of Verrazano-Narrows Bridge	minus	60
Translate:	m	$-$	60

The Golden Gate Bridge is $(m - 60)$ feet long.

☐ **Work Practice Problem 9**

Answers

8. a. $11 - 4$ or 7,   **b.** $11 - x$,
c. $56 - a$,   **9.** $(n + 49{,}489)$ votes

Mental Math

Solve each equation mentally. See Examples 1 and 2.

1. $x + 4 = 6$

2. $x + 7 = 17$

3. $n + 18 = 30$

4. $z + 22 = 40$

5. $b - 11 = 6$

6. $d - 16 = 5$

2.1 EXERCISE SET

FOR EXTRA HELP

Student Solutions Manual   PH Math/Tutor Center   CD/Video for Review   Math XL
MathXL®   MyMathLab
MyMathLab

Objective A *Solve each equation. Check each solution. See Examples 1 through 4.*

1. $x + 7 = 10$

2. $x + 14 = 25$

3. $x - 2 = -4$

4. $y - 9 = 1$

5. $-11 = 3 + x$

6. $-8 = 8 + z$

7. $r - 8.6 = -8.1$

8. $t - 9.2 = -6.8$

9. $x - \dfrac{2}{5} = -\dfrac{3}{20}$

10. $y - \dfrac{4}{7} = -\dfrac{3}{14}$

11. $\dfrac{1}{3} + f = \dfrac{3}{4}$

12. $c + \dfrac{1}{6} = \dfrac{3}{8}$

Objective B *Solve each equation. Don't forget to first simplify each side of the equation, if possible. Check each solution. See Examples 5 through 7.*

13. $7x + 2x = 8x - 3$

14. $3n + 2n = 7 + 4n$

15. $\dfrac{5}{6}x + \dfrac{1}{6}x = -9$

16. $\dfrac{13}{11}y - \dfrac{2}{11}y = -3$

17. $2y + 10 = 5y - 4y$

18. $4x - 4 = 10x - 7x$

19. $-5(n - 2) = 8 - 4n$

20. $-4(z - 3) = 2 - 3z$

21. $\dfrac{3}{7}x + 2 = -\dfrac{4}{7}x - 5$

22. $\dfrac{1}{5}x - 1 = -\dfrac{4}{5}x - 13$

23. $5x - 6 = 6x - 5$

24. $2x + 7 = x - 10$

25. $8y + 2 - 6y = 3 + y - 10$

26. $4p - 11 - p = 2 + 2p - 20$

27. $-3(x - 4) = -4x$

28. $-2(x - 1) = -3x$

29. $\dfrac{3}{8}x - \dfrac{1}{6} = -\dfrac{5}{8}x - \dfrac{2}{3}$

30. $\dfrac{2}{5}x - \dfrac{1}{12} = -\dfrac{3}{5}x - \dfrac{3}{4}$

31. $2(x - 4) = x + 3$

32. $3(y + 7) = 2y - 5$

33. $3(n - 5) - (6 - 2n) = 4n$

34. $5(3 + z) - (8z + 9) = -4z$

35. $-2(x + 6) + 3(2x - 5) = 3(x - 4) + 10$

36. $-5(x + 1) + 4(2x - 3) = 2(x + 2) - 8$

Objectives **A** **B** **Mixed Practice** *Solve. See Examples 1 through 7.*

37. $13x - 3 = 14x$

38. $18x - 9 = 19x$

39. $5b - 0.7 = 6b$

40. $9x + 5.5 = 10x$

41. $3x - 6 = 2x + 5$

42. $7y + 2 = 6y + 2$

43. $13x - 9 + 2x - 5 = 12x - 1 + 2x$

44. $15x + 20 - 10x - 9 = 25x + 8 - 21x - 7$

45. $7(6 + w) = 6(2 + w)$

46. $6(5 + c) = 5(c - 4)$

47. $n + 4 = 3.6$

48. $m + 2 = 7.1$

49. $10 - (2x - 4) = 7 - 3x$

50. $15 - (6 - 7k) = 2 + 6k$

51. $\dfrac{1}{3} = x + \dfrac{2}{3}$

52. $\dfrac{1}{11} = y + \dfrac{10}{11}$

53. $-6.5 - 4x - 1.6 - 3x = -6x + 9.8$

54. $-1.4 - 7x - 3.6 - 2x = -8x + 4.4$

Objective **C** *Write each algebraic expression described. See Examples 8 and 9.*

55. A 10-foot board is cut into two pieces. If one piece is x feet long, express the other length in terms of x.

56. A 5-foot piece of string is cut into two pieces. If one piece is x feet long, express the other length in terms of x.

57. Recall that two angles are *supplementary* if their sum is 180°. If one angle measures $x°$, express the measure of its supplement in terms of x.

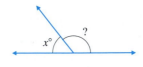

58. Recall that two angles are *complementary* if their sum is 90°. If one angle measures $x°$, express the measure of its complement in terms of x.

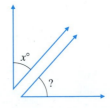

59. In 2004, the number of graduate students at the University of Texas at Austin was approximately 28,000 fewer than the number of undergraduate students. If the number of undergraduate students was n, how many graduate students attend UT Austin? (*Source:* www.utexas.edu)

60. The longest interstate highway in the U.S. is I-90, which connects Seattle, Washington, and Boston, Massachusetts. The second longest interstate highway, I-80 (connecting San Francisco, California, and Teaneck, New Jersey), is 178.5 miles shorter than I-90. If the length of I-80 is m miles, express the length of I-90 as an algebraic expression in m. (*Source:* U.S. Department of Transportation—Federal Highway Administration)

61. The area of the Sahara Desert in Africa is 7 times the area of the Gobi Desert in Asia. If the area of the Gobi Desert is x square miles, express the area of the Sahara Desert as an algebraic expression in x.

62. The largest meteorite in the world is the Hoba West located in Namibia. Its weight is 3 times the weight of the Armanty meteorite located in Outer Mongolia. If the weight of the Armanty meteorite is y kilograms, express the weight of the Hoba West meteorite as an algebraic expression in y.

Review

Find each multiplicative inverse or reciprocal. See Section 1.7.

63. $\dfrac{5}{8}$ **64.** $\dfrac{7}{6}$ **65.** 2   **66.** 5   **67.** $-\dfrac{1}{9}$ **68.** $-\dfrac{3}{5}$

Perform each indicated operation and simplify. See Sections 1.5 and 1.6.

69. $\dfrac{3x}{3}$ **70.** $\dfrac{-2y}{-2}$ **71.** $-5\left(-\dfrac{1}{5}y\right)$ **72.** $7\left(\dfrac{1}{7}r\right)$ **73.** $\dfrac{3}{5}\left(\dfrac{5}{3}x\right)$ **74.** $\dfrac{9}{2}\left(\dfrac{2}{9}x\right)$

Concept Extensions

75. Write two terms whose sum is $-3x$.

76. Write four terms whose sum is $2y - 6$.

Use the addition property to fill in the blank so that the middle equation simplifies to the last equation. See the Concept Check in this section.

77. $x - 4 = -9$
$x - 4 + (\ \ ) = -9 + (\ \ )$
$x = -5$

78. $a + 9 = 15$
$a + 9 + (\ \ ) = 15 + (\ \ )$
$a = 6$

Fill in the blanks with numbers of your choice so that each equation has the given solution. Note: Each blank may be replaced with a different number.

79. ____ $+ x =$ ____ ; Solution: -3

80. $x -$ ____ $=$ ____ ; Solution: -10

Solve.

△ **81.** The sum of the angles of a triangle is 180°. If one angle of a triangle measures $x°$ and a second angle measures $(2x + 7)°$, express the measure of the third angle in terms of x. Simplify the expression.

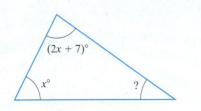

△ **82.** A quadrilateral is a four-sided figure (like the one shown in the figure) whose angle sum is 360°. If one angle measures $x°$, a second angle measures $3x°$, and a third angle measures $5x°$, express the measure of the fourth angle in terms of x. Simplify the expression.

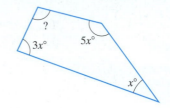

✏ **83.** In your own words, explain what is meant by the solution of an equation.

✏ **84.** In your own words, explain how to check a solution of an equation.

Use a calculator to determine the solution of each equation.

🖩 **85.** $36.766 + x = -108.712$

🖩 **86.** $-85.325 = x - 97.985$

STUDY SKILLS BUILDER

Have You Decided to Complete This Course Successfully?

Ask yourself if one of your current goals is to complete this course successfully.

If it is not a goal of yours, ask yourself why? One common reason is fear of failure. Amazingly enough, fear of failure alone can be strong enough to keep many of us from doing our best in any endeavor.

Another common reason is that you simply haven't taken the time to make successfully completing this course one of your goals. How do you do this? Start by writing this goal in your mathematics notebook. Then list steps you will take to ensure success. A great first step is to read or reread Section 1.1 and make a commitment to try the suggestions in that section.

Good luck, and don't forget that a positive attitude will make a big difference.

Let's see how you are doing.

1. Have you decided to make "successfully completing this course" a goal of yours? If no, please list reasons why this has not happened. Study your list and talk to your instructor about this.

2. If your answer to question 1 is yes, take a moment and list in your notebook further specific goals that will help you achieve this major goal of successfully completing this course. (For example, "My goal this semester is not to miss any of my mathematics classes.")

3. Rate your commitment to this course with a number between 1 and 5. Use the diagram below to help.

High Commitment		Average Commitment		Not committed at all
5	4	3	2	1

4. If you have rated your personal commitment level (from the exercise above) as a 1, 2, or 3, list the reasons why this is so. Then determine whether it is possible to increase your commitment level to a 4 or 5.

2.2 THE MULTIPLICATION PROPERTY OF EQUALITY

Objectives

A Use the Multiplication Property of Equality to Solve Linear Equations.

B Use Both the Addition and Multiplication Properties of Equality to Solve Linear Equations.

C Write Word Phrases as Algebraic Expressions.

Objective A Using the Multiplication Property

As useful as the addition property of equality is, it cannot help us solve every type of linear equation in one variable. For example, adding or subtracting a value on both sides of the equation does not help solve

$$\frac{5}{2}x = 15$$

because the variable x is being multiplied by a number (other than 1). Instead, we apply another important property of equality, the **multiplication property of equality.**

> **Multiplication Property of Equality**
>
> If $a, b,$ and c are real numbers and $c \neq 0$, then
>
> $$a = b \quad \text{and} \quad ac = bc$$
>
> are equivalent equations.

This property guarantees that multiplying both sides of an equation by the same nonzero number does not change the solution of the equation. Since division is defined in terms of multiplication, we may also **divide both sides of the equation by the same nonzero number** without changing the solution.

EXAMPLE 1 Solve: $\dfrac{5}{2}x = 15$

Solution: To get x alone, we multiply both sides of the equation by the reciprocal (or multiplicative inverse) of $\dfrac{5}{2}$, which is $\dfrac{2}{5}$.

$$\frac{5}{2}x = 15$$

$$\frac{2}{5} \cdot \left(\frac{5}{2}x\right) = \frac{2}{5} \cdot 15 \quad \text{Multiply both sides by } \frac{2}{5}.$$

$$\left(\frac{2}{5} \cdot \frac{5}{2}\right)x = \frac{2}{5} \cdot 15 \quad \text{Apply the associative property.}$$

$$1x = 6 \quad \text{Simplify.}$$

or

$$x = 6$$

Check: Replace x with 6 in the original equation.

$$\frac{5}{2}x = 15 \quad \text{Original equation.}$$

$$\frac{5}{2}(6) \stackrel{?}{=} 15 \quad \text{Replace } x \text{ with 6.}$$

$$15 = 15 \quad \text{True}$$

The solution is 6.

Work Practice Problem 1

PRACTICE PROBLEM 1

Solve: $\dfrac{3}{7}x = 9$

Answer

1. $x = 21$

99

In the equation $\frac{5}{2}x = 15$, $\frac{5}{2}$ is the coefficient of x. When the coefficient of x is a *fraction*, we will get x alone by multiplying by the reciprocal. When the coefficient of x is an integer or a decimal, it is usually more convenient to divide both sides by the coefficient. (Dividing by a number is, of course, the same as multiplying by the reciprocal of the number.)

PRACTICE PROBLEM 2

Solve: $7x = 42$

EXAMPLE 2 Solve: $5x = 30$

Solution: To get x alone, we divide both sides of the equation by 5, the coefficient of x.

$$5x = 30$$

$$\frac{5x}{5} = \frac{30}{5} \quad \text{Divide both sides by 5.}$$

$$1 \cdot x = 6 \quad \text{Simplify.}$$

$$x = 6$$

Check: $5x = 30$ Original equation.

$5 \cdot 6 \stackrel{?}{=} 30$ Replace x with 6.

$30 = 30$ True

The solution is 6.

■ **Work Practice Problem 2**

PRACTICE PROBLEM 3

Solve: $-4x = 52$

EXAMPLE 3 Solve: $-3x = 33$

Solution: Recall that $-3x$ means $-3 \cdot x$. To get x alone, we divide both sides by the coefficient of x, that is, -3.

$$-3x = 33$$

$$\frac{-3x}{-3} = \frac{33}{-3} \quad \text{Divide both sides by } -3.$$

$$1x = -11 \quad \text{Simplify.}$$

$$x = -11$$

Check: $-3x = 33$ Original equation.

$-3(-11) \stackrel{?}{=} 33$ Replace x with -11.

$33 = 33$ True

The solution is -11.

■ **Work Practice Problem 3**

PRACTICE PROBLEM 4

Solve: $\frac{y}{5} = 13$

EXAMPLE 4 Solve: $\frac{y}{7} = 20$

Solution: Recall that $\frac{y}{7} = \frac{1}{7}y$. To get y alone, we multiply both sides of the equation by 7, the reciprocal of $\frac{1}{7}$.

$$\frac{y}{7} = 20$$

$$\frac{1}{7}y = 20$$

$$7 \cdot \frac{1}{7}y = 7 \cdot 20 \quad \text{Multiply both sides by 7.}$$

$$1y = 140 \quad \text{Simplify.}$$

$$y = 140$$

Answers

2. $x = 6$, **3.** $x = -13$, **4.** $y = 65$

Check: $\dfrac{y}{7} = 20$ Original equation.

$\dfrac{140}{7} \overset{?}{=} 20$ Replace y with 140.

$20 = 20$ True

The solution is 140.

🔲 **Work Practice Problem 4**

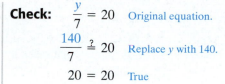 **EXAMPLE 5** Solve: $3.1x = 4.96$

Solution: $3.1x = 4.96$

$\dfrac{3.1x}{3.1} = \dfrac{4.96}{3.1}$ Divide both sides by 3.1.

$1x = 1.6$ Simplify.

$x = 1.6$

Check: Check by replacing x with 1.6 in the original equation. The solution is 1.6.

🔲 **Work Practice Problem 5**

EXAMPLE 6 Solve: $-\dfrac{2}{3}x = -\dfrac{5}{2}$

Solution: To get x alone, we multiply both sides of the equation by $-\dfrac{3}{2}$, the reciprocal of the coefficient of x.

$-\dfrac{2}{3}x = -\dfrac{5}{2}$

$-\dfrac{3}{2} \cdot -\dfrac{2}{3}x = -\dfrac{3}{2} \cdot -\dfrac{5}{2}$ Multiply both sides by $-\dfrac{3}{2}$, the reciprocal of $-\dfrac{2}{3}$.

$x = \dfrac{15}{4}$ Simplify.

Check: Check by replacing x with $\dfrac{15}{4}$ in the original equation. The solution is $\dfrac{15}{4}$.

🔲 **Work Practice Problem 6**

Objective B Using Both the Addition and Multiplication Properties

We are now ready to combine the skills learned in the last section with the skills learned from this section to solve equations by applying more than one property.

EXAMPLE 7 Solve: $-z - 4 = 6$

Solution: First, let's get $-z$, the term containing the variable, alone. To do so, we add 4 to both sides of the equation.

$-z - 4 + 4 = 6 + 4$ Add 4 to both sides.

$-z = 10$ Simplify.

Next, recall that $-z$ means $-1 \cdot z$. Thus to get z alone, we either multiply or divide both sides of the equation by -1. In this example, we divide.

$-z = 10$

$\dfrac{-z}{-1} = \dfrac{10}{-1}$ Divide both sides by the coefficient -1.

$1z = -10$ Simplify.

$z = -10$

Continued on next page

PRACTICE PROBLEM 5
Solve: $2.6x = 13.52$

PRACTICE PROBLEM 6
Solve: $-\dfrac{5}{6}y = -\dfrac{3}{5}$

PRACTICE PROBLEM 7
Solve: $-x + 7 = -12$

Answers
5. $x = 5.2$,   **6.** $y = \dfrac{18}{25}$,   **7.** $x = 19$

Check:

$$
\begin{aligned}
-z - 4 &= 6 \qquad &\text{Original equation.} \\
-(-10) - 4 &\overset{?}{=} 6 \qquad &\text{Replace } z \text{ with } -10. \\
10 - 4 &\overset{?}{=} 6 \\
6 &= 6 \qquad &\text{True}
\end{aligned}
$$

The solution is -10.

■ **Work Practice Problem 7**

Don't forget to first simplify one or both sides of an equation, if possible.

PRACTICE PROBLEM 8

Solve:
$-7x + 2x + 3 - 20 = -2$

EXAMPLE 8 Solve: $a + a - 10 + 7 = -13$

Solution: First, we simplify both sides of the equation by combining like terms.

$$
\begin{aligned}
a + a - 10 + 7 &= -13 \\
2a - 3 &= -13 \qquad &\text{Combine like terms.} \\
2a - 3 + 3 &= -13 + 3 \qquad &\text{Add 3 to both sides.} \\
2a &= -10 \qquad &\text{Simplify.} \\
\frac{2a}{2} &= \frac{-10}{2} \qquad &\text{Divide both sides by 2.} \\
a &= -5 \qquad &\text{Simplify.}
\end{aligned}
$$

Check: To check, replace a with -5 in the original equation. The solution is -5.

■ **Work Practice Problem 8**

PRACTICE PROBLEM 9

Solve: $10x - 4 = 7x + 14$

EXAMPLE 9 Solve: $7x - 3 = 5x + 9$

Solution: To get x alone, let's first use the addition property to get variable terms on one side of the equation and numbers on the other side. One way to get variable terms on one side is to subtract $5x$ from both sides.

$$
\begin{aligned}
7x - 3 &= 5x + 9 \\
7x - 3 - 5x &= 5x + 9 - 5x \qquad &\text{Subtract } 5x \text{ from both sides.} \\
2x - 3 &= 9 \qquad &\text{Simplify.}
\end{aligned}
$$

Now, to get numbers on the other side, let's add 3 to both sides.

$$
\begin{aligned}
2x - 3 + 3 &= 9 + 3 \qquad &\text{Add 3 to both sides.} \\
2x &= 12 \qquad &\text{Simplify.}
\end{aligned}
$$

Use the multiplication property to get x alone.

$$
\begin{aligned}
\frac{2x}{2} &= \frac{12}{2} \qquad &\text{Divide both sides by 2.} \\
x &= 6 \qquad &\text{Simplify.}
\end{aligned}
$$

Check: To check, replace x with 6 in the original equation to see that a true statement results. The solution is 6.

■ **Work Practice Problem 9**

If an equation has parentheses, don't forget to use the distributive property to remove them. Then combine any like terms.

Answers

8. $x = -3$, **9.** $x = 6$

EXAMPLE 10 Solve: $5(2x + 3) = -1 + 7$

Solution:

$$5(2x + 3) = -1 + 7$$
$$5(2x) + 5(3) = -1 + 7 \quad \text{Apply the distributive property.}$$
$$10x + 15 = 6 \quad \text{Multiply and write } -1 + 7 \text{ as } 6.$$
$$10x + 15 - 15 = 6 - 15 \quad \text{Subtract 15 from both sides.}$$
$$10x = -9 \quad \text{Simplify.}$$
$$\frac{10x}{10} = -\frac{9}{10} \quad \text{Divide both sides by 10.}$$
$$x = -\frac{9}{10} \quad \text{Simplify.}$$

Check: To check, replace x with $-\dfrac{9}{10}$ in the original equation to see that a true statement results. The solution is $-\dfrac{9}{10}$.

🔲 **Work Practice Problem 10**

PRACTICE PROBLEM 10

Solve: $4(3x - 2) = -1 + 4$

Objective C Writing Algebraic Expressions

We continue to sharpen our problem-solving skills by writing algebraic expressions.

EXAMPLE 11 **Writing an Expression for Consecutive Integers**

If x is the first of three consecutive integers, express the sum of the three integers in terms of x. Simplify if possible.

Solution: An example of three consecutive integers is 7, 8, and 9.

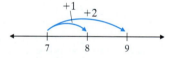

The second consecutive integer is always 1 more than the first, and the third consecutive integer is 2 more than the first. If x is the first of three consecutive integers, the three consecutive integers are $x, x + 1$, and $x + 2$.

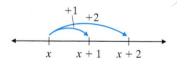

Their sum is shown below.

In words:
| first integer | + | second integer | + | third integer |

Translate: $\quad x \quad + \quad (x + 1) \quad + \quad (x + 2)$

This simplifies to $3x + 3$.

🔲 **Work Practice Problem 11**

PRACTICE PROBLEM 11

a. If x is the first of two consecutive integers, express the sum of the first and the second integer in terms of x. Simplify if possible.

b. If x is the first of two consecutive even integers (see next page), express the sum of the first and second integer in terms of x. Simplify if possible.

Answers

10. $x = \dfrac{11}{12}$, **11. a.** $2x + 1$, **b.** $2x + 2$

Study these examples of consecutive even and consecutive odd integers.

Consecutive even integers:

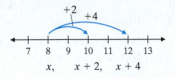

$$x, \quad x+2, \quad x+4$$

Consecutive odd integers:

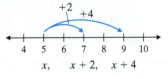

$$x, \quad x+2, \quad x+4$$

If x is an odd integer, then $x + 2$ is the next odd integer. This 2 simply means that odd integers are always 2 units from each other.

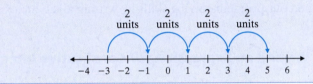

Mental Math

Solve each equation mentally. See Examples 2 and 3.

1. $3a = 27$ **2.** $9c = 54$ **3.** $5b = 10$ **4.** $7t = 14$ **5.** $6x = -30$ **6.** $8r = -64$

2.2 EXERCISE SET

FOR EXTRA HELP

Student Solutions Manual   PH Math/Tutor Center   CD/Video for Review   Math XL   MathXL®   MyMathLab   MyMathLab

Objective A *Solve each equation. Check each solution. See Examples 1 through 6.*

1. $-5x = -20$

2. $-7x = -49$

3. $3x = 0$

4. $2x = 0$

5. $-x = -12$

6. $-y = 8$

7. $\dfrac{2}{3}x = -8$

8. $\dfrac{3}{4}n = -15$

9. $\dfrac{1}{6}d = \dfrac{1}{2}$

10. $\dfrac{1}{8}v = \dfrac{1}{4}$

11. $\dfrac{a}{2} = 1$

12. $\dfrac{d}{15} = 2$

13. $\dfrac{k}{-7} = 0$

14. $\dfrac{f}{-5} = 0$

15. $1.7x = 10.71$

16. $8.5y = 19.55$

Objective B *Solve each equation. Check each solution. See Examples 7 and 8.*

17. $2x - 4 = 16$

18. $3x - 1 = 26$

19. $-x + 2 = 22$

20. $-x + 4 = -24$

21. $6a + 3 = 3$

22. $8t + 5 = 5$

23. $\dfrac{x}{3} - 2 = -5$

24. $\dfrac{b}{4} - 1 = -7$

25. $6z - 8 - z + 3 = 0$

26. $4a + 1 + a - 11 = 0$

27. $1 = 0.4x - 0.6x - 5$

28. $19 = 0.4x - 0.9x - 6$

29. $\dfrac{2}{3}y - 11 = -9$

30. $\dfrac{3}{5}x - 14 = -8$

31. $\dfrac{3}{4}t - \dfrac{1}{2} = \dfrac{1}{3}$

32. $\dfrac{2}{7}z - \dfrac{1}{5} = \dfrac{1}{2}$

Solve each equation. See Examples 9 and 10.

33. $8x + 20 = 6x + 18$

34. $11x + 13 = 9x + 9$

35. $3(2x + 5) = -18 + 9$

36. $2(4x + 1) = -12 + 6$

37. $2x - 5 = 20x + 4$

38. $6x - 4 = -2x - 10$

39. $2 + 14 = -4(3x - 4)$

40. $8 + 4 = -6(5x - 2)$

41. $-6y - 3 = -5y - 7$

42. $-17z - 4 = -16z - 20$

43. $\frac{1}{2}(2x - 1) = -\frac{1}{7} - \frac{3}{7}$

44. $\frac{1}{3}(3x - 1) = -\frac{1}{10} - \frac{2}{10}$

45. $-10z - 0.5 = -20z + 1.6$

46. $-14y - 1.8 = -24y + 3.9$

47. $-4x + 20 = 4x - 20$

48. $-3x + 15 = 3x - 15$

Objectives Ⓐ Ⓑ **Mixed Practice** *See Examples 1 through 10.*

49. $42 = 7x$

50. $81 = 3x$

51. $4.4 = -0.8x$

52. $6.3 = -0.6x$

53. $6x + 10 = -20$

54. $10y + 15 = -5$

55. $5 - 0.3k = 5$

56. $2 - 0.4p = 2$

57. $13x - 5 = 11x - 11$

58. $20x - 20 = 16x - 40$

59. $9(3x + 1) = 4x - 5x$

60. $7(2x + 1) = 18x - 19x$

61. $-\frac{3}{7}p = -2$

62. $-\frac{4}{5}r = -5$

63. $-\frac{4}{3}x = 12$

64. $-\frac{10}{3}x = 30$

65. $-2x - \frac{1}{2} = \frac{7}{2}$

66. $-3n - \frac{1}{3} = \frac{8}{3}$

67. $10 = 2x - 1$

68. $12 = 3j - 4$

69. $10 - 3x - 6 - 9x = 7$

70. $12x + 30 + 8x - 6 = 10$

🔘 **71.** $z - 5z = 7z - 9 - z$

72. $t - 6t = -13 + t - 3t$

73. $-x - \frac{4}{5} = x + \frac{1}{2} + \frac{2}{5}$

74. $x + \frac{3}{7} = -x + \frac{1}{3} + \frac{4}{7}$

75. $-15 + 37 = -2(x + 5)$

76. $-19 + 74 = -5(x + 3)$

Objective Ⓒ *Write each algebraic expression described. Simplify if possible. See Example 11.*

77. If x represents the first of two consecutive odd integers, express the sum of the two integers in terms of x.

78. If x is the first of three consecutive even integers, write their sum as an algebraic expression in x.

79. If x is the first of four consecutive integers, express the sum of the first integer and the third integer as an algebraic expression containing the variable x.

80. If x is the first of two consecutive integers, express the sum of 20 and the second consecutive integer as an algebraic expression containing the variable x.

81. Classrooms on one side of the science building are all numbered with consecutive even integers. If the first room on this side of the building is numbered x, write an expression in x for the sum of five classroom numbers in a row. Then simplify this expression.

82. Two sides of a quadrilateral have the same length, x, while the other two sides have the same length, both being the next consecutive odd integer. Write the sum of these lengths. Then simplify this expression.

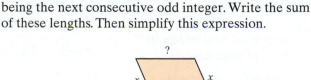

Review

Simplify each expression. See Section 1.8.

83. $5x + 2(x - 6)$

84. $-7y + 2y - 3(y + 1)$

85. $6(2z + 4) + 20$

86. $-(3a - 3) + 2a - 6$

87. $-(x - 1) + x$

88. $8(z - 6) + 7z - 1$

Concept Extensions

Fill in the blank with a number of your choice so that each equation has the given solution.

89. $6x =$ _____; solution: -8

90. _____ $x = 10$; solution: $\dfrac{1}{2}$

91. The equation $3x + 6 = 2x + 10 + x - 4$ is true for all real numbers. Substitute a few real numbers for x to see that this is so and then try solving the equation. Describe what happens.

92. The equation $6x + 2 - 2x = 4x + 1$ has no solution. Try solving this equation for x and describe what happens.

93. From the results of Exercises 91 and 92, when do you think an equation has all real numbers as its solutions?

94. From the results of Exercises 91 and 92, when do you think an equation has no solution?

Solve.

95. $0.07x - 5.06 = -4.92$

96. $0.06y + 2.63 = 2.5562$

STUDY SKILLS BUILDER

How Are Your Homework Assignments Going?

It is very important in mathematics to keep up with homework. Why? Many concepts build on each other. Often your understanding of a day's concepts depends on an understanding of the previous day's material.

Remember that completing your homework assignment involves a lot more than attempting a few of the problems assigned.

To complete a homework assignment, remember these four things:

- Attempt all of it.
- Check it.
- Correct it.
- If needed, ask questions about it.

Take a moment and review your completed homework assignments. Answer the questions below based on this review.

1. Approximate the fraction of your homework you have attempted.

2. Approximate the fraction of your homework you have checked (if possible).

3. If you are able to check your homework, have you corrected it when errors have been found?

4. When working homework, if you do not understand a concept, what do you do?

A Apply the General Strategy for Solving a Linear Equation.

B Solve Equations Containing Fractions or Decimals.

C Recognize Identities and Equations with No Solution.

2.3 FURTHER SOLVING LINEAR EQUATIONS

Objective **A** Solving Linear Equations

We now combine our knowledge from the previous sections into a general strategy for solving linear equations. One new piece in this strategy is a suggestion to "clear an equation of fractions" as a first step. Doing so makes the equation more manageable, since working with integers is more convenient than working with fractions. We will discuss this further in Example 3.

To Solve Linear Equations in One Variable

Step 1: If an equation contains fractions, multiply both sides by the LCD to clear the equation of fractions.

Step 2: Use the distributive property to remove parentheses if they occur.

Step 3: Simplify each side of the equation by combining like terms.

Step 4: Get all variable terms on one side and all numbers on the other side by using the addition property of equality.

Step 5: Get the variable alone by using the multiplication property of equality.

Step 6: Check the solution by substituting it into the original equation.

PRACTICE PROBLEM 1

Solve:

$5(3x - 1) + 2 = 12x + 6$

EXAMPLE 1 Solve: $4(2x - 3) + 7 = 3x + 5$

Solution: There are no fractions, so we begin with Step 2.

$$4(2x - 3) + 7 = 3x + 5$$

Step 2: $8x - 12 + 7 = 3x + 5$ Use the distributive property.

Step 3: $8x - 5 = 3x + 5$ Combine like terms.

Step 4: Get all variable terms on one side of the equation and all numbers on the other side. One way to do this is by subtracting $3x$ from both sides and then adding 5 to both sides.

$$8x - 5 - 3x = 3x + 5 - 3x$$ Subtract $3x$ from both sides.

$$5x - 5 = 5$$ Simplify.

$$5x - 5 + 5 = 5 + 5$$ Add 5 to both sides.

$$5x = 10$$ Simplify.

Step 5: Use the multiplication property of equality to get x alone.

$$\frac{5x}{5} = \frac{10}{5}$$ Divide both sides by 5.

$$x = 2$$ Simplify.

Step 6: Check.

$$4(2x - 3) + 7 = 3x + 5$$ Original equation

$$4[2(2) - 3] + 7 \stackrel{?}{=} 3(2) + 5$$ Replace x with 2.

$$4(4 - 3) + 7 \stackrel{?}{=} 6 + 5$$

$$4(1) + 7 \stackrel{?}{=} 11$$

$$4 + 7 \stackrel{?}{=} 11$$

$$11 = 11$$ True

The solution is 2.

Answer

1. $x = 3$

■ Work Practice Problem 1

EXAMPLE 2 Solve: $8(2 - t) = -5t$

Solution: First, we apply the distributive property.

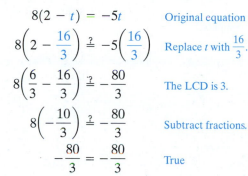

$$8(2 - t) = -5t$$

Step 2:	$16 - 8t = -5t$	Use the distributive property.
Step 4:	$16 - 8t + 8t = -5t + 8t$	Add $8t$ to both sides.
	$16 = 3t$	Combine like terms.
Step 5:	$\dfrac{16}{3} = \dfrac{3t}{3}$	Divide both sides by 3.
	$\dfrac{16}{3} = t$	Simplify.

Step 6: Check.

$8(2 - t) = -5t$	Original equation
$8\left(2 - \dfrac{16}{3}\right) \stackrel{?}{=} -5\left(\dfrac{16}{3}\right)$	Replace t with $\dfrac{16}{3}$.
$8\left(\dfrac{6}{3} - \dfrac{16}{3}\right) \stackrel{?}{=} -\dfrac{80}{3}$	The LCD is 3.
$8\left(-\dfrac{10}{3}\right) \stackrel{?}{=} -\dfrac{80}{3}$	Subtract fractions.
$-\dfrac{80}{3} = -\dfrac{80}{3}$	True

The solution is $\dfrac{16}{3}$.

🔲 **Work Practice Problem 2**

Objective B Solving Equations Containing Fractions or Decimals

If an equation contains fractions, we can clear the equation of fractions by multiplying both sides by the LCD of all denominators. By doing this, we avoid working with time-consuming fractions.

EXAMPLE 3 Solve: $\dfrac{x}{2} - 1 = \dfrac{2}{3}x - 3$

Solution: We begin by clearing fractions. To do this, we multiply both sides of the equation by the LCD of 2 and 3, which is 6.

$$\frac{x}{2} - 1 = \frac{2}{3}x - 3$$

Step 1:	$6\left(\dfrac{x}{2} - 1\right) = 6\left(\dfrac{2}{3}x - 3\right)$	Multiply both sides by the LCD, 6.
Step 2:	$6\left(\dfrac{x}{2}\right) - 6(1) = 6\left(\dfrac{2}{3}x\right) - 6(3)$	Use the distributive property.
	$3x - 6 = 4x - 18$	Simplify.

There are no longer grouping symbols and no like terms on either side of the equation, so we continue with Step 4.

Continued on next page

PRACTICE PROBLEM 2

Solve: $9(5 - x) = -3x$

> **Helpful Hint**
>
> When checking solutions, use the original equation.

PRACTICE PROBLEM 3

Solve: $\dfrac{5}{2}x - 1 = \dfrac{3}{2}x - 4$

> **Helpful Hint**
>
> Don't forget to multiply *each* term by the LCD.

Answers

2. $x = \dfrac{15}{2}$, **3.** $x = -3$

$$3x - 6 = 4x - 18$$

Step 4: $3x - 6 - 3x = 4x - 18 - 3x$ Subtract $3x$ from both sides.

$$-6 = x - 18$$ Simplify.

$$-6 + 18 = x - 18 + 18$$ Add 18 to both sides.

$$12 = x$$ Simplify.

Step 5: The variable is now alone, so there is no need to apply the multiplication property of equality.

Step 6: Check.

$$\frac{x}{2} - 1 = \frac{2}{3}x - 3$$ Original equation

$$\frac{12}{2} - 1 \stackrel{?}{=} \frac{2}{3} \cdot 12 - 3$$ Replace x with 12.

$$6 - 1 \stackrel{?}{=} 8 - 3$$ Simplify.

$$5 = 5$$ True

The solution is 12.

■ **Work Practice Problem 3**

PRACTICE PROBLEM 4

Solve: $\dfrac{3(x - 2)}{5} = 3x + 6$

EXAMPLE 4 Solve: $\dfrac{2(a + 3)}{3} = 6a + 2$

Solution: We clear the equation of fractions first.

$$\frac{2(a + 3)}{3} = 6a + 2$$

Step 1: $3 \cdot \dfrac{2(a + 3)}{3} = 3(6a + 2)$ Clear the fraction by multiplying both sides by the LCD, 3.

$$2(a + 3) = 3(6a + 2)$$ Simplify.

Step 2: Next, we use the distributive property to remove parentheses.

$$2a + 6 = 18a + 6$$ Use the distributive property.

Step 4: $2a + 6 - 18a = 18a + 6 - 18a$ Subtract $18a$ from both sides.

$$-16a + 6 = 6$$ Simplify.

$$-16a + 6 - 6 = 6 - 6$$ Subtract 6 from both sides.

$$-16a = 0$$

Step 5: $\dfrac{-16a}{-16} = \dfrac{0}{-16}$ Divide both sides by -16.

$$a = 0$$ Simplify.

Step 6: To check, replace a with 0 in the original equation. The solution is 0.

■ **Work Practice Problem 4**

Helpful Hint

Remember: When solving an equation, it makes no difference on which side of the equation variable terms lie. Just make sure that constant terms lie on the other side.

When solving a problem about money, you may need to solve an equation containing decimals. If you choose, you may multiply to clear the equation of decimals.

Answer

4. $x = -3$

EXAMPLE 5 Solve: $0.25x + 0.10(x - 3) = 1.1$

Solution: First we clear this equation of decimals by multiplying both sides of the equation by 100. Recall that multiplying a decimal number by 100 has the effect of moving the decimal point 2 places to the right.

$$0.25x + 0.10(x - 3) = 1.1$$

Step 1: $0.25x + 0.10(x - 3) = 1.10$ Multiply both sides by 100

$$25x + 10(x - 3) = 110$$

Step 2: $\qquad\quad 25x + 10x - 30 = 110$ Apply the distributive property.

Step 3: $\qquad\qquad\quad 35x - 30 = 110$ Combine like terms.

Step 4: $\quad 35x - 30 + 30 = 110 + 30$ Add 30 to both sides.

$$35x = 140$$ Combine like terms.

Step 5: $\qquad\qquad\quad \dfrac{35x}{35} = \dfrac{140}{35}$ Divide both sides by 35.

$$x = 4$$

Step 6: To check, replace x with 4 in the original equation. The solution is 4.

■ **Work Practice Problem 5**

Objective **C** Recognizing Identities and Equations with No Solution

So far, each equation that we have solved has had a single solution. However, not every equation in one variable has a single solution. Some equations have no solution, while others have an infinite number of solutions. For example,

$$x + 5 = x + 7$$

has **no solution** since no matter which real number we replace x with, the equation is false.

    real number $+ 5 =$ same real number $+ 7$ FALSE

On the other hand,

$$x + 6 = x + 6$$

has infinitely many solutions since x can be replaced by any real number and the equation is always true.

    real number $+ 6 =$ same real number $+ 6$ TRUE

The equation $x + 6 = x + 6$ is called an **identity.** The next few examples illustrate special equations like these.

EXAMPLE 6 Solve: $-2(x - 5) + 10 = -3(x + 2) + x$

Solution:

$$-2(x - 5) + 10 = -3(x + 2) + x$$

$$-2x + 10 + 10 = -3x - 6 + x$$ Apply the distributive property on both sides.

$$-2x + 20 = -2x - 6$$ Combine like terms.

$$-2x + 20 + 2x = -2x - 6 + 2x$$ Add $2x$ to both sides.

$$20 = -6$$ Combine like terms.

The final equation contains no variable terms, and the result is the false statement $20 = -6$. This means that there is no value for x that makes $20 = -6$ a true equation. Thus, we conclude that there is **no solution** to this equation.

■ **Work Practice Problem 6**

PRACTICE PROBLEM 5

Solve:

$$0.06x - 0.10(x - 2) = -0.16$$

Helpful Hint

If you have trouble with this step, try removing parentheses first.

$$0.25x + 0.10(x - 3) = 1.1$$
$$0.25x + 0.10x - 0.3 = 1.1$$
$$0.25x + 0.10x - 0.30 = 1.10$$
$$25x + 10x - 30 = 110$$

Then continue.

PRACTICE PROBLEM 6

Solve:

$$5(2 - x) + 8x = 3(x - 6)$$

Answers

5. $x = 9$,   **6.** no solution

PRACTICE PROBLEM 7

Solve:

$-6(2x + 1) - 14$
$= -10(x + 2) - 2x$

EXAMPLE 7 Solve: $3(x - 4) = 3x - 12$

Solution: $3(x - 4) = 3x - 12$

$$3x - 12 = 3x - 12 \quad \text{Apply the distributive property.}$$

The left side of the equation is now identical to the right side. Every real number may be substituted for x and a true statement will result. We arrive at the same conclusion if we continue.

$$3x - 12 = 3x - 12$$
$$3x - 12 - 3x = 3x - 12 - 3x \quad \text{Subtract } 3x \text{ from both sides.}$$
$$-12 = -12 \quad \text{Combine like terms.}$$

Again, the final equation contains no variables, but this time the result is the true statement $-12 = -12$. This means that one side of the equation is identical to the other side. Thus, $3(x - 4) = 3x - 12$ is an **identity** and **every real number** is a solution.

Answer

7. Every real number is a solution.

✔ **Concept Check Answer**

a. Every real number is a solution.
b. The solution is 0.
c. There is no solution.

■ **Work Practice Problem 7**

✔**Concept Check** Suppose you have simplified several equations and obtain the following results. What can you conclude about the solutions to the original equation?

a. $7 = 7$ **b.** $x = 0$ **c.** $7 = -4$

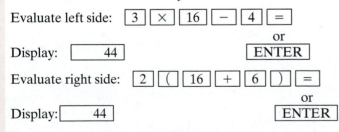 **CALCULATOR EXPLORATIONS**   Checking Equations

We can use a calculator to check possible solutions of equations. To do this, replace the variable by the possible solution and evaluate both sides of the equation separately.

Equation: $3x - 4 = 2(x + 6)$ Solution: $x = 16$

$$3x - 4 = 2(x + 6) \quad \text{Original equation}$$
$$3(16) - 4 \overset{?}{=} 2(16 + 6) \quad \text{Replace } x \text{ with 16.}$$

Now evaluate each side with your calculator.

Evaluate left side: $\boxed{3}\ \boxed{\times}\ \boxed{16}\ \boxed{-}\ \boxed{4}\ \boxed{=}$

                   or

Display: $\boxed{44}$ $\boxed{\text{ENTER}}$

Evaluate right side: $\boxed{2}\ \boxed{(}\ \boxed{16}\ \boxed{+}\ \boxed{6}\ \boxed{)}\ \boxed{=}$

                   or

Display: $\boxed{44}$ $\boxed{\text{ENTER}}$

Since the left side equals the right side, the equation checks.

Use a calculator to check the possible solutions to each equation.

1. $2x = 48 + 6x; \quad x = -12$

2. $-3x - 7 = 3x - 1; \quad x = -1$

3. $5x - 2.6 = 2(x + 0.8); \quad x = 4.4$

4. $-1.6x - 3.9 = -6.9x - 25.6; \quad x = 5$

5. $\dfrac{564x}{4} = 200x - 11(649); \quad x = 121$

6. $20(x - 39) = 5x - 432; \quad x = 23.2$

2.3 EXERCISE SET

Objective A *Solve each equation. See Examples 1 and 2.*

1. $-4y + 10 = -2(3y + 1)$

2. $-3x + 1 = -2(4x + 2)$

3. $15x - 8 = 10 + 9x$

4. $15x - 5 = 7 + 12x$

5. $-2(3x - 4) = 2x$

6. $-(5x - 10) = 5x$

7. $5(2x - 1) - 2(3x) = 1$

8. $3(2 - 5x) + 4(6x) = 12$

9. $-6(x - 3) - 26 = -8$

10. $-4(n - 4) - 23 = -7$

11. $8 - 2(a + 1) = 9 + a$

12. $5 - 6(2 + b) = b - 14$

13. $4x + 3 = -3 + 2x + 14$

14. $6y - 8 = -6 + 3y + 13$

15. $-2y - 10 = 5y + 18$

16. $-7n + 5 = 8n - 10$

Objective B *Solve each equation. See Examples 3 through 5.*

17. $\dfrac{2}{3}x + \dfrac{4}{3} = -\dfrac{2}{3}$

18. $\dfrac{4}{5}x - \dfrac{8}{5} = -\dfrac{16}{5}$

19. $\dfrac{3}{4}x - \dfrac{1}{2} = 1$

20. $\dfrac{2}{9}x - \dfrac{1}{3} = 1$

21. $0.50x + 0.15(70) = 35.5$

22. $0.40x + 0.06(30) = 9.8$

23. $\dfrac{2(x + 1)}{4} = 3x - 2$

24. $\dfrac{3(y + 3)}{5} = 2y + 6$

25. $x + \dfrac{7}{6} = 2x - \dfrac{7}{6}$

26. $\dfrac{5}{2}x - 1 = x + \dfrac{1}{4}$

27. $0.12(y - 6) + 0.06y = 0.08y - 0.7$

28. $0.60(z - 300) + 0.05z = 0.70z - 205$

Objective C *Solve each equation. See Examples 6 and 7.*

29. $4(3x + 2) = 12x + 8$

30. $14x + 7 = 7(2x + 1)$

31. $\dfrac{x}{4} + 1 = \dfrac{x}{4}$

32. $\dfrac{x}{3} - 2 = \dfrac{x}{3}$

33. $3x - 7 = 3(x + 1)$

34. $2(x - 5) = 2x + 10$

35. $-2(6x - 5) + 4 = -12x + 14$

36. $-5(4y - 3) + 2 = -20y + 17$

Objectives Ⓐ Ⓑ Ⓒ **Mixed Practice** *Solve. See Examples 1 through 7.*

37. $\dfrac{6(3 - z)}{5} = -z$

38. $\dfrac{4(5 - w)}{3} = -w$

39. $-3(2t - 5) + 2t = 5t - 4$

40. $-(4a - 7) - 5a = 10 + a$

41. $5y + 2(y - 6) = 4(y + 1) - 2$

42. $9x + 3(x - 4) = 10(x - 5) + 7$

43. $\dfrac{3(x - 5)}{2} = \dfrac{2(x + 5)}{3}$

44. $\dfrac{5(x - 1)}{4} = \dfrac{3(x + 1)}{2}$

45. $0.7x - 2.3 = 0.5$

46. $0.9x - 4.1 = 0.4$

47. $5x - 5 = 2(x + 1) + 3x - 7$

48. $3(2x - 1) + 5 = 6x + 2$

49. $4(2n + 1) = 3(6n + 3) + 1$

50. $4(4y + 2) = 2(1 + 6y) + 8$

51. $x + \dfrac{5}{4} = \dfrac{3}{4}x$

52. $\dfrac{7}{8}x + \dfrac{1}{4} = \dfrac{3}{4}x$

53. $\dfrac{x}{2} - 1 = \dfrac{x}{5} + 2$

54. $\dfrac{x}{5} - 7 = \dfrac{x}{3} - 5$

55. $2(x + 3) - 5 = 5x - 3(1 + x)$

56. $4(2 + x) + 1 = 7x - 3(x - 2)$

57. $0.06 - 0.01(x + 1) = -0.02(2 - x)$

58. $-0.01(5x + 4) = 0.04 - 0.01(x + 4)$

59. $\dfrac{9}{2} + \dfrac{5}{2}y = 2y - 4$

60. $3 - \dfrac{1}{2}x = 5x - 8$

Review

Write each algebraic expression described. See Section 2.1.

△ **61.** A plot of land is in the shape of a triangle. If one side is x meters, a second side is $(2x - 3)$ meters, and a third side is $(3x - 5)$ meters, express the perimeter of the lot as a simplified expression in x.

62. A portion of a board has length x feet. The other part has length $(7x - 9)$ feet. Express the total length of the board as a simplified expression in x.

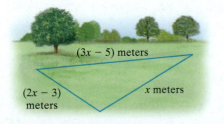

(3x − 5) meters
(2x − 3) meters   x meters

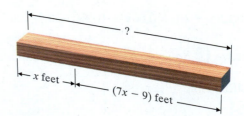

— ? —
x feet   (7x − 9) feet

Write each phrase as an algebraic expression. Use x for the unknown number. See Section 2.1.

63. A number subtracted from -8

64. Three times a number

65. The sum of -3 and twice a number

66. The difference of 8 and twice a number

67. The product of 9 and the sum of a number and 20

68. The quotient of -12 and the difference of a number and 3

Concept Extensions

See the Concept Check in this section.

69. a. Solve: $x + 3 = x + 3$
   b. If you simplify an equation and get $0 = 0$, what can you conclude about the solution(s) of the original equation?
   c. On your own, construct an equation for which every real number is a solution.

70. a. Solve: $x + 3 = x + 5$
   b. If you simplify an equation and get $3 = 5$, what can you conclude about the solution(s) of the original equation?
   c. On your own, construct an equation that has no solution.

Match each equation in the first column with its solution in the second column. Items in the second column may be used more than once.

71. $5x + 1 = 5x + 1$

72. $3x + 1 = 3x + 2$

73. $2x - 6x - 10 = -4x + 3 - 10$

74. $x - 11x - 3 = -10x - 1 - 2$

75. $9x - 20 = 8x - 20$

76. $-x + 15 = x + 15$

a. all real numbers
b. no solution
c. 0

77. Explain the difference between simplifying an expression and solving an equation.

78. On your own, write an expression and then an equation. Label each.

For Exercises 79 and 80, **a.** *Write an equation for perimeter.* **b.** *Solve the equation in part (a).* **c.** *Find the length of each side.*

79. The perimeter of a geometric figure is the sum of the lengths of its sides. If the perimeter of the following pentagon (five-sided figure) is 28 centimeters, find the length of each side.

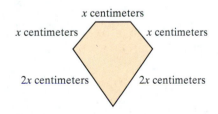

x centimeters
x centimeters   *x* centimeters
2*x* centimeters   2*x* centimeters

80. The perimeter of the following triangle is 35 meters. Find the length of each side.

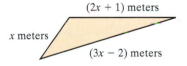

(2*x* + 1) meters
x meters
(3*x* − 2) meters

Fill in the blanks with numbers of your choice so that each equation has the given solution. Note: *Each blank may be replaced by a different number.*

81. $x +$ ____ $= 2x -$ ____ ; solution: 9

82. $-5x -$ ____ $=$ ____ ; solution: 2

Solve.

83. $1000(7x - 10) = 50(412 + 100x)$

84. $1000(x + 40) = 100(16 + 7x)$

85. $0.035x + 5.112 = 0.010x + 5.107$

86. $0.127x - 2.685 = 0.027x - 2.38$

INTEGRATED REVIEW | Sections 2.1–2.3

Solving Linear Equations

Solve. Feel free to use the steps given in Section 2.3.

1. $x - 10 = -4$

2. $y + 14 = -3$

3. $9y = 108$

4. $-3x = 78$

5. $-6x + 7 = 25$

6. $5y - 42 = -47$

7. $\dfrac{2}{3}x = 9$

8. $\dfrac{4}{5}z = 10$

9. $\dfrac{r}{-4} = -2$

10. $\dfrac{y}{-8} = 8$

11. $6 - 2x + 8 = 10$

12. $-5 - 6y + 6 = 19$

13. $2x - 7 = 6x - 27$

14. $3 + 8y = 3y - 2$

15. $9(3x - 1) = -4 + 49$

16. $12(2x + 1) = -6 + 66$

17. $-3a + 6 + 5a = 7a - 8a$

18. $4b - 8 - b = 10b - 3b$

19. $-\dfrac{2}{3}x = \dfrac{5}{9}$

20. $-\dfrac{3}{8}y = -\dfrac{1}{16}$

21. $10 = -6n + 16$

22. $-5 = -2m + 7$

23. $3(5c - 1) - 2 = 13c + 3$

24. $4(3t + 4) - 20 = 3 + 5t$

25. $\dfrac{2(z + 3)}{3} = 5 - z$

26. $\dfrac{3(w + 2)}{4} = 2w + 3$

27. $-2(2x - 5) = -3x + 7 - x + 3$

28. $-4(5x - 2) = -12x + 4 - 8x + 4$

29. $0.02(6t - 3) = 0.04(t - 2) + 0.02$

30. $0.03(m + 7) = 0.02(5 - m) + 0.03$

31. $-3y = \dfrac{4(y - 1)}{5}$

32. $-4x = \dfrac{5(1 - x)}{6}$

33. $\dfrac{5}{3}x - \dfrac{7}{3} = x$

34. $\dfrac{7}{5}n + \dfrac{3}{5} = -n$

35. $\dfrac{1}{10}(3x - 7) = \dfrac{3}{10}x + 5$

36. $\dfrac{1}{7}(2x - 5) = \dfrac{2}{7}x + 1$

37. $5 + 2(3x - 6) = -4(6x - 7)$

38. $3 + 5(2x - 4) = -7(5x + 2)$

23. _____

24. _____

25. _____

26. _____

27. _____

28. _____

29. _____

30. _____

31. _____

32. _____

33. _____

34. _____

35. _____

36. _____

37. _____

38. _____

A Translate a Problem to an
Equation, Then Use the Equation
to Solve the Problem.

2.4 AN INTRODUCTION TO PROBLEM SOLVING

In the preceding sections, we practiced translating phrases into expressions and sentences into equations as well as solving linear equations. We are now ready to put our skills to practical use. To begin, we present a general strategy for problem solving.

General Strategy for Problem Solving

1. **UNDERSTAND** the problem. During this step, become comfortable with the problem. Some ways of doing this are:

   Read and reread the problem.

   Choose a variable to represent the unknown.

   Construct a drawing.

   Propose a solution and check. Pay careful attention to how you check your proposed solution. This will help when writing an equation to model the problem.

2. **TRANSLATE** the problem into an equation.

3. **SOLVE** the equation.

4. **INTERPRET** the results: *Check* the proposed solution in the stated problem and *state* your conclusion.

Objective A Translating and Solving Problems

Much of problem solving involves a direct translation from a sentence to an equation.

EXAMPLE 1 Finding an Unknown Number

Twice the sum of a number and 4 is the same as four times the number, decreased by 12. Find the number.

Solution:

1. **UNDERSTAND.** Read and reread the problem. If we let x = the unknown number, then
   "the sum of a number and 4" translates to "$x + 4$" and
   "four times the number" translates to "$4x$"

2. **TRANSLATE.**

twice	sum of a number and 4	is the same as	four times the number	decreased by	12
↓	↓	↓	↓	↓	↓
2	$(x + 4)$	=	$4x$	−	12

3. **SOLVE**

$$2(x + 4) = 4x - 12$$
$$2x + 8 = 4x - 12 \qquad \text{Apply the distributive property.}$$
$$2x + 8 - 4x = 4x - 12 - 4x \qquad \text{Subtract } 4x \text{ from both sides.}$$
$$-2x + 8 = -12$$
$$-2x + 8 - 8 = -12 - 8 \qquad \text{Subtract 8 from both sides.}$$
$$-2x = -20$$
$$\frac{-2x}{-2} = \frac{-20}{-2} \qquad \text{Divide both sides by } -2.$$
$$x = 10$$

PRACTICE PROBLEM 1

Three times the difference of a number and 5 is the same as twice the number decreased by 3. Find the number.

4. INTERPRET.

Check: Check this solution in the problem as it was originally stated. To do so, replace "number" with 10. Twice the sum of "10" and 4 is 28, which is the same as 4 times "10" decreased by 12.

State: The number is 10.

☐ **Work Practice Problem 1**

The next example has to do with consecutive integers. For a review, see Section 2.2.

EXAMPLE 2

Some states have a single area code for the entire state. Two such states have area codes that are consecutive odd integers. If the sum of these integers is 1208, find the two area codes. (*Source: World Almanac*)

Solution:

1. **UNDERSTAND.** Read and reread the problem. If we let

   x = the first odd integer, then

   $x + 2$ = the next odd integer

2. **TRANSLATE.**

first odd integer	the sum of ↓	next odd integer	is	1208
↓		↓		
x	$+$	$(x + 2)$	$=$	1208

3. **SOLVE.**

$$x + x + 2 = 1208$$
$$2x + 2 = 1208$$
$$2x + 2 - 2 = 1208 - 2$$
$$2x = 1206$$
$$\frac{2x}{2} = \frac{1206}{2}$$
$$x = 603$$

4. **INTERPRET.**

Check: If $x = 603$, then the next odd integer $x + 2 = 603 + 2 = 605$. Notice their sum, $603 + 605 = 1208$, as needed.

State: The area codes are 603 and 605.

Note: New Hampshire's area code is 603 and South Dakota's area code is 605.

☐ **Work Practice Problem 2**

During the next example, we expand our discussion of the UNDERSTAND part of the problem-solving process.

PRACTICE PROBLEM 2

The sum of three consecutive even integers is 144. Find the integers.

> **Helpful Hint**
>
> Remember, the 2 here means that odd integers are 2 units apart, for example, the odd integers 13 and $13 + 2 = 15$.

Answer

2. 46, 48, 50

PRACTICE PROBLEM 3

An 18-foot wire is to be cut so that the longer piece is 5 times longer than the shorter piece. Find the length of each piece.

EXAMPLE 3 **Finding the Length of a Board**

A 10-foot board is to be cut into two pieces so that the longer piece is 4 times the shorter. Find the length of each piece.

Solution:

1. UNDERSTAND the problem. To do so, read and reread the problem. You may also want to propose a solution. For example, if 3 feet represents the length of the shorter piece, then $4(3) = 12$ feet is the length of the longer piece, since it is 4 times the length of the shorter piece. This guess gives a total board length of 3 feet + 12 feet = 15 feet, which is too long. However, the purpose of proposing a solution is not to guess correctly, but to help better understand the problem and how to model it.

   In general, if we let

   x = length of shorter piece, then
   $4x$ = length of longer piece

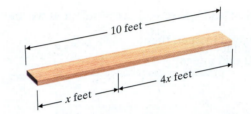

2. TRANSLATE the problem. First, we write the equation in words.

length of shorter piece	added to	length of longer piece	equals	total length of board
↓	↓	↓	↓	↓
x	$+$	$4x$	$=$	10

3. SOLVE.

$$x + 4x = 10$$
$$5x = 10 \quad \text{Combine like terms.}$$
$$\frac{5x}{5} = \frac{10}{5} \quad \text{Divide both sides by 5.}$$
$$x = 2$$

4. INTERPRET.

Check: Check the solution in the stated problem. If the shorter piece of board is 2 feet, the longer piece is $4 \cdot (2 \text{ feet}) = 8$ feet and the sum of the two pieces is 2 feet + 8 feet = 10 feet.

State: The shorter piece of board is 2 feet and the longer piece of board is 8 feet.

■ **Work Practice Problem 3**

Helpful Hint

Make sure that units are included in your answer, if appropriate.

Answer

3. shorter piece: 3 feet; longer piece: 15 feet

EXAMPLE 4 **Finding the Number of Republican and Democratic Senators**

In a recent year, the U.S. House of Representatives had a total of 431 Democrats and Republicans. There were 15 more Republican representatives than Democratic. Find the number of representatives from each party. (*Source:* Office of the Clerk of the U.S. House of Representatives)

Solution:

1. UNDERSTAND the problem. Read and reread the problem. Let's suppose that there are 200 Democratic representatives. Since there are 15 more Republicans than Democrats, there must be $200 + 15 = 215$ Republicans. The total number of Democrats and Republicans is then $200 + 215 = 415$. This is incorrect since the total should be 431, but we now have a better understanding of the problem.
    In general, if we let

    x = number of Democrats, then

    $x + 15$ = number of Republicans

2. TRANSLATE the problem. First, we write the equation in words.

number of Democrats	added to	number of Republicans	equals	431
↓	↓	↓	↓	↓
x	$+$	$(x + 15)$	$=$	431

3. SOLVE.

$$x + (x + 15) = 431$$
$$2x + 15 = 431 \qquad \text{Combine like terms.}$$
$$2x + 15 - 15 = 431 - 15 \qquad \text{Subtract 15 from both sides.}$$
$$2x = 416$$
$$\frac{2x}{2} = \frac{416}{2} \qquad \text{Divide both sides by 2.}$$
$$x = 208$$

4. INTERPRET.

Check: If there were 208 Democratic representatives, then there were $208 + 15 = 223$ Republican representatives. The total number of representatives is then $208 + 223 = 431$. The results check.

State: There were 208 Democratic and 223 Republican representatives in Congress.

🔲 **Work Practice Problem 4**

EXAMPLE 5 **Calculating Hours on Job**

A computer science major at a local university has a part-time job working on computers for his clients. He charges $20 to come to your home or office and then $25 per hour. During one month he visited 10 homes or offices and his total income was $575. How many hours did he spend working on computers?

Solution:

1. UNDERSTAND. Read and reread the problem. Let's propose that the student spent 20 hours working on computers. Pay careful attention as to how his income is calculated. For 20 hours and 10 visits, his income is $20(\$25) + 10(\$20) = \$700$, more than $575. We now have a better understanding of the problem and know that the time working on computers is less than 20 hours.

Continued on next page

PRACTICE PROBLEM 4

Through the year 2010, the state of California will have 21 more electoral votes for president than the state of Texas. If the total electoral votes for these two states is 89, find the number of electoral votes for each state.

PRACTICE PROBLEM 5

A car rental agency charges $28 a day and $0.15 a mile. If you rent a car for a day and your bill (before taxes) is $52, how many miles did you drive?

Answers
4. Texas: 34 electoral votes; California: 55 electoral votes,
5. 160 miles

Let's let

x = hours working on computers. Then

$25x$ = amount of money made while working on computers

2. TRANSLATE.

money made while working on computers	plus	money made for visits	is equal to	575
↓	↓	↓	↓	↓
$25x$	$+$	$10(20)$	$=$	575

3. SOLVE.

$$25x + 200 = 575$$
$$25x + 200 - 200 = 575 - 200 \quad \text{Subtract 200 from both sides.}$$
$$25x = 375 \quad \text{Simplify.}$$
$$\frac{25x}{25} = \frac{375}{25} \quad \text{Divide both sides by 25.}$$
$$x = 15 \quad \text{Simplify.}$$

4. INTERPRET.

Check: If the student works 15 hours and makes 10 visits, his income is $15(\$25) + 10(\$20) = \$575$.

State: The student spent 15 hours working on computers.

🔲 **Work Practice Problem 5**

PRACTICE PROBLEM 6

The measure of the second angle of a triangle is twice the measure of the smallest angle. The measure of the third angle of the triangle is three times the measure of the smallest angle. Find the measures of the angles.

△ **EXAMPLE 6** **Finding Angle Measures**

If the two walls of the Vietnam Veterans Memorial in Washington, D.C., were connected, an isosceles triangle would be formed. The measure of the third angle is 97.5° more than the measure of either of the two equal angles. Find the measure of the third angle. (*Source:* National Park Service)

Solution:

1. UNDERSTAND. Read and reread the problem. We then draw a diagram (recall that an isosceles triangle has two angles with the same measure) and let

x = degree measure of one angle

x = degree measure of the second equal angle

$x + 97.5$ = degree measure of the third angle

Answer

6. smallest: 30°;   second: 60°; third: 90°

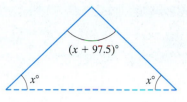

2. **TRANSLATE.** Recall that the sum of the measures of the angles of a triangle equals 180.

measure of first angle	+	measure of second angle	+	measure of third angle	equals	180
↓		↓		↓	↓	↓
x	$+$	x	$+ (x + 97.5)$	$=$		180

3. **SOLVE.**

$$x + x + (x + 97.5) = 180$$

$$3x + 97.5 = 180 \qquad \text{Combine like terms.}$$

$$3x + 97.5 - 97.5 = 180 - 97.5 \qquad \text{Subtract 97.5 from both sides.}$$

$$3x = 82.5$$

$$\frac{3x}{3} = \frac{82.5}{3} \qquad \text{Divide both sides by 3.}$$

$$x = 27.5$$

4. **INTERPRET.**

Check: If $x = 27.5$, then the measure of the third angle is $x + 97.5 = 125$. The sum of the angles is then $27.5 + 27.5 + 125 = 180$, the correct sum.

State: The third angle measures $125°$.*

🔲 **Work Practice Problem 6**

---

*The two walls actually meet at an angle of 125 degrees 12 minutes. The measurement of 97.5° given in the problem is an approximation.

Objective Ⓐ *Solve. See Example 1.*

1. Twice the difference of a number and 8 is equal to three times the sum of the number and 3. Find the number.

2. Five times the sum of a number and −1 is the same as 6 times the number. Find the number.

3. The product of twice a number and three is the same as the difference of five times the number and $\frac{3}{4}$. Find the number.

4. If the difference of a number and four is doubled, the result is $\frac{1}{4}$ less than the number. Find the number.

Solve. See Example 2.

5. The left and right page numbers of an open book are two consecutive integers whose sum is 469. Find these page numbers.

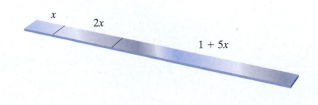

6. The room numbers of two adjacent classrooms are two consecutive even numbers. If their sum is 654, find the classroom numbers.

7. To make an international telephone call, you need the code for the country you are calling. The codes for Belgium, France, and Spain are three consecutive integers whose sum is 99. Find the code for each country. (*Source: The World Almanac and Book of Facts*)

8. The code to unlock a student's combination lock happens to be three consecutive odd integers whose sum is 51. Find the integers.

Solve. See Examples 3 and 4.

9. A 25-inch piece of steel is cut into three pieces so that the second piece is twice as long as the first piece, and the third piece is one inch more than five times the length of the first piece. Find the lengths of the pieces.

10. A 46-foot piece of rope is cut into three pieces so that the second piece is three times as long as the first piece, and the third piece is two feet more than seven times the length of the first piece. Find the lengths of the pieces.

11. A 40-inch board is to be cut into three pieces so that the second piece is twice as long as the first piece and the third piece is 5 times as long as the first piece. If x represents the length of the first piece, find the lengths of all three pieces.

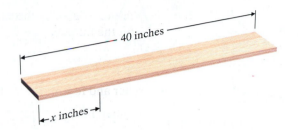

12. A 21-foot beam is to be divided so that the longer piece is 1 foot more than 3 times the shorter piece. If x represents the length of the shorter piece, find the lengths of both pieces.

13. The governor of California earns $50,425 more than the governor of Florida. If the total of their salaries is $299,575, find the salaries of each. (*Source: The World Almanac*, 2005)

14. In the 2004 Summer Olympics, the United States team won 3 more gold medals than the Chinese team. If the total number of gold medals won by both teams was 67, find the number of gold medals won by each team. (*Source:* Wikipedia)

Solve. See Example 5.

15. A car rental agency advertised renting a Buick Century for $24.95 per day and $0.29 per mile. If you rent this car for 2 days, how many whole miles can you drive on a $100 budget?

16. A plumber gave an estimate for the renovation of a kitchen. Her hourly pay is $27 per hour and the plumbing parts will cost $80. If her total estimate is $404, how many hours does she expect this job to take?

17. In one U.S. city, the taxi cost is $3 plus $0.80 per mile. If you are traveling from the airport, there is an additional charge of $4.50 for tolls. How far can you travel from the airport by taxi for $27.50?

18. A professional carpet cleaning service charges $30 plus $25.50 per hour to come to your home. If your total bill from this company is $119.25 before taxes, for how many hours were you charged?

Solve. See Example 6.

△ **19.** The flag of Equatorial Guinea contains an isosceles triangle. (Recall that an isosceles triangle contains two angles with the same measure.) If the measure of the third angle of the triangle is 30° more than twice the measure of either of the other two angles, find the measure of each angle of the triangle. (*Hint:* Recall that the sum of the measures of the angles of a triangle is 180°.)

△ **20.** The flag of Brazil contains a parallelogram. One angle of the parallelogram is 15° less than twice the measure of the angle next to it. Find the measure of each angle of the parallelogram. (*Hint:* Recall that opposite angles of a parallelogram have the same measure and that the sum of the measures of the angles is 360°.)

21. The sum of the measures of the angles of a parallelogram is 360°. In the parallelogram below, angles A and D have the same measure as well as angles C and B. If the measure of angle C is twice the measure of angle A, find the measure of each angle.

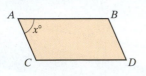

22. Recall that the sum of the measures of the angles of a triangle is 180°. In the triangle below, angle C has the same measure as angle B, and angle A measures 42° less than angle B. Find the measure of each angle.

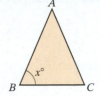

Mixed Practice

23. A 17-foot piece of string is cut into two pieces so that the longer piece is 2 feet longer than twice the shorter piece. Find the lengths of both pieces.

24. A 25-foot wire is to be cut so that the longer piece is one foot longer than 5 times the shorter piece. Find the length of each piece.

25. From 1997 to 2001, the number of prescriptions written for ADHD drugs increased by 5.5 million. If the sum of the number of prescriptions for these two years is 35.7 million, find the number of prescriptions for each year. Check to see that your results agree with the heights of the bars in the graph.

26. The Pentagon Building in Washington, D.C., is the headquarters for the U.S. Department of Defense. The Pentagon is also the world's largest office building in terms of ground space with a floor area of over 6.5 million square feet. This is three times the floor area of the Empire State Building. About how much floor space does the Empire State Building have? Round to the nearest tenth.

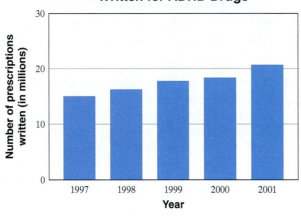

Source: IMS Health

27. Two angles are supplementary if their sum is 180°. One angle measures three times the measure of a smaller angle. If x represents the measure of the smaller angle and these two angles are supplementary, find the measure of each angle.

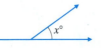

28. Two angles are complementary if their sum is 90°. Given the measures of the complementary angles shown, find the measure of each angle.

29. The measures of the angles of a triangle are 3 consecutive even integers. Find the measure of each angle.

30. A quadrilateral is a polygon with 4 sides. The sum of the measures of the 4 angles in a quadrilateral is 360°. If the measures of the angles of a quadrilateral are consecutive odd integers, find the measures.

31. The sum of $\frac{1}{5}$ and twice a number is equal to $\frac{4}{5}$ subtracted from three times the number. Find the number.

32. The sum of $\frac{2}{3}$ and four times a number is equal to $\frac{5}{6}$ subtracted from five times the number. Find the number.

33. Hertz Car Rental charges a daily rate of $39 plus $0.20 per mile for a certain car. Suppose that you rent that car for a day and your bill (before taxes) is $95. How many miles did you drive?

34. A woman's $15,000 estate is to be divided so that her husband receives twice as much as her son. Find the amount of money that her husband receives and the amount of money that her son receives.

35. The winner of the NASCAR Winston Cup in 2003 was Matt Kenseth. Kenseth earned 90 more points than his closest rival, Jimmie Johnson. Together they earned 9954 points. How many points did each driver accumulate during the 2003 Winston Cup Series?

36. During the 2004 Houston Bowl, University of Colorado beat University of Texas–El Paso by 5 points. If their combined scores totaled 61, find the individual team scores. (*Source:* ESPN)

37. The number of counties in California and the number of counties in Montana are consecutive even integers whose sum is 114. If California has more counties than Montana, how many counties does each state have? (*Source: The World Almanac and Book of Facts,* 2005)

38. After a recent election, there were 2 more Republican governors than Democratic governors in the United States. How many Democrats and how many Republicans held governor's offices after this election? (*Source: The World Almanac and Book of Facts*)

39. Over the past few years the satellite Voyager II has passed by the planets Saturn, Uranus, and Neptune, continually updating information about these planets, including the number of moons for each. Uranus is now believed to have 13 more moons than Neptune. Also, Saturn is now believed to have 2 more than twice the number of moons of Neptune. If the total number of moons for these planets is 47, find the number of moons for each planet. (*Source:* National Space Science Data Center)

40. On April 7, 2001, the Mars Odyssey spacecraft was launched, beginning a multi-year mission to observe and map the planet Mars. Mars Odyssey was launched on Boeing's Delta II 7925 launch vehicle using nine strap-on solid rocket motors. Each solid rocket motor has a height that is 8 meters more than 5 times its diameter. If the sum of the height and the diameter for a single solid rocket motor is 14 meters, find each dimension. (*Source:* NASA)

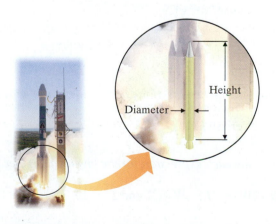

41. If the sum of a number and five is tripled, the result is one less than twice the number. Find the number.

42. Twice the sum of a number and six equals three times the sum of the number and four. Find the number.

43. The area of the Sahara Desert is 7 times the area of the Gobi Desert. If the sum of their areas is 4,000,000 square miles, find the area of each desert.

44. The largest meteorite in the world is the Hoba West located in Namibia. Its weight is 3 times the weight of the Armanty meteorite located in Outer Mongolia. If the sum of their weights is 88 tons, find the weight of each.

45. In the 2004 summer Olympics, France won more gold medals than Italy, who won more gold medals than Korea. If the total number of gold medals won by these three countries is three consecutive integers whose sum is 30, find the number of gold medals won by each. (*Source: The World Almanac, 2005*)

46. To make an international telephone call, you need the code for the country you are calling. The codes for Mali Republic, Côte d'Ivoire, and Niger are three consecutive odd integers whose sum is 675. Find the code for each country.

47. In a recent election in Florida for a seat in the United States House of Representatives, Corrine Brown received 13,288 more votes than Bill Randall. If the total number of votes was 119,436, find the number of votes for each candidate.

48. In a recent election in Texas for a seat in the United States House of Representatives, Max Sandlin received 25,557 more votes than opponent Dennis Boerner. If the total number of votes was 135,821, find the number of votes for each candidate. (*Source: Voter News Service*)

The graph below shows the states with the highest tourism budgets. Use the graph for Exercises 49 through 52.

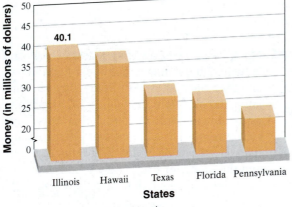

Source: Travel Industry Association of America

49. Which state spends the most money on tourism?

50. Which states spend between $25 and $30 million on tourism?

51. The states of Texas and Florida spend a total of $56.6 million for tourism. The state of Texas spends $2.2 million more than the state of Florida. Find the amount that each state spends on tourism.

52. The states of Hawaii and Pennsylvania spend a total of $60.9 million for tourism. The state of Hawaii spends $8.1 million less than twice the amount of money that the state of Pennsylvania spends. Find the amount that each state spends on tourism.

Compare the heights of the bars in the graph with your results of the exercises below. Are your answers reasonable?

53. Exercise 51.

54. Exercise 52

Evaluate each expression for the given values. See Section 1.3.

55. $2W + 2L$; $W = 7$ and $L = 10$

56. $\frac{1}{2}Bh$; $B = 14$ and $h = 22$

57. πr^2; $r = 15$

58. $r \cdot t$; $r = 15$ and $t = 2$

Concept Extensions

△ **59.** A golden rectangle is a rectangle whose length is approximately 1.6 times its width. The early Greeks thought that a rectangle with these dimensions was the most pleasing to the eye and examples of the golden rectangle are found in many early works of art. For example, the Parthenon in Athens contains many examples of golden rectangles.

Mike Hallahan would like to plant a rectangular garden in the shape of a golden rectangle. If he has 78 feet of fencing available, find the dimensions of the garden.

60. Dr. Dorothy Smith gave the students in her geometry class at the University of New Orleans the following question. Is it possible to construct a triangle such that the second angle of the triangle has a measure that is twice the measure of the first angle and the measure of the third angle is 3 times the measure of the first? If so, find the measure of each angle. (*Hint:* Recall that the sum of the measures of the angles of a triangle is 180°.)

61. The human eye blinks once every 5 seconds on average. How many times does the average eye blink in one hour? In one 16-hour day while awake? In one year?

62. Give an example of how you recently solved a problem using mathematics.

63. In your own words, explain why a solution of a word problem should be checked using the original wording of the problem and not the equation written from the wording.

Recall from Exercise 59 that a golden rectangle is a rectangle whose length is approximately 1.6 times its width.

△ **64.** It is thought that for about 75% of adults, a rectangle in the shape of the golden rectangle is the most pleasing to the eye. Draw three rectangles, one in the shape of the golden rectangle, and poll your class. Do the results agree with the percentage given above?

△ **65.** Examples of golden rectangles can be found today in architecture and manufacturing packaging. Find an example of a golden rectangle in your home. A few suggestions: the front face of a book, the floor of a room, the front of a box of food.

△ **66.** Measure the dimensions of each rectangle and decide which one best approximates the shape of a golden rectangle.

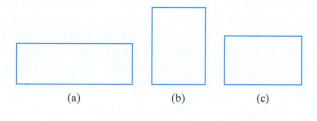

(a)   (b)   (c)

2.5 FORMULAS AND PROBLEM SOLVING

Objective **A** Using Formulas to Solve Problems

A **formula** describes a known relationship among quantities. Many formulas are given as equations. For example, the formula

$$d = r \cdot t$$

stands for the relationship

$$\text{distance} = \text{rate} \cdot \text{time}$$

Let's look at one way that we can use this formula.

If we know we traveled a distance of 100 miles at a rate of 40 miles per hour, we can replace the variables d and r in the formula $d = rt$ and find our travel time, t.

$$d = rt \qquad \text{Formula}$$
$$100 = 40t \qquad \text{Replace } d \text{ with 100 and } r \text{ with 40.}$$

To solve for t, we divide both sides of the equation by 40.

$$\frac{100}{40} = \frac{40t}{40} \qquad \text{Divide both sides by 40.}$$

$$\frac{5}{2} = t \qquad \text{Simplify.}$$

The travel times was $\frac{5}{2}$ hours, or $2\frac{1}{2}$ hours, or 2.5 hours.

In this section, we solve problems that can be modeled by known formulas. We use the same problem-solving strategy that was introduced in the previous section.

PRACTICE PROBLEM 1

A family is planning their vacation to visit relatives. They will drive from Cincinnati, Ohio to Rapid City, South Dakota, a distance of 1180 miles. They plan to average a rate of 50 miles per hour. How much time will they spend driving?

EXAMPLE 1 **Finding Time Given Rate and Distance**

A glacier is a giant mass of rocks and ice that flows downhill like a river. Portage Glacier in Alaska is about 6 miles, or 31,680 *feet,* long and moves 400 *feet* per year. Icebergs are created when the front end of the glacier flows into Portage Lake. How long does it take for ice at the head (beginning) of the glacier to reach the lake?

Solution:

1. UNDERSTAND. Read and reread the problem. The appropriate formula needed to solve this problem is the distance formula, $d = rt$. To become familiar with this formula, let's find the distance that ice traveling at a rate of 400 feet per year travels in 100 years. To do so, we let time t be 100 years and rate r be the given 400 feet per year, and substitute these values into the formula $d = rt$. We then have that distance $d = 400(100) = 40,000$ feet. Since we are interested in finding how long it takes ice to travel 31,680 feet, we now know that it is less than 100 years.

Answer
1. 23.6 hours

Since we are using the formula $d = rt$, we let

t = the time in years for ice to reach the lake

r = rate or speed of ice

d = distance from beginning of glacier to lake

2. TRANSLATE. To translate to an equation, we use the formula $d = rt$ and let distance $d = 31{,}680$ feet and rate $r = 400$ feet per year.

$$d = r \cdot t$$
$$31{,}680 = 400 \cdot t \quad \text{Let } d = 31{,}680 \text{ and } r = 400.$$

3. SOLVE. Solve the equation for t. To solve for t, divide both sides by 400.

$$\frac{31{,}680}{400} = \frac{400 \cdot t}{400} \quad \text{Divide both sides by 400.}$$
$$79.2 = t \quad \text{Simplify.}$$

4. INTERPRET.

Check: To check, substitute 79.2 for t and 400 for r in the distance formula and check to see that the distance is 31,680 feet.

State: It takes 79.2 years for the ice at the head of Portage Glacier to reach the lake.

▢ **Work Practice Problem 1**

Helpful Hint

Don't forget to include units, if appropriate.

△ **EXAMPLE 2 Calculating the Length of a Garden**

Charles Pecot can afford enough fencing to enclose a rectangular garden with a perimeter of 140 feet. If the width of his garden is to be 30 feet, find the length.

$w = 30$ feet
l

△ **PRACTICE PROBLEM 2**

A wood deck is being built behind a house. The width of the deck must be 18 feet because of the shape of the house. If there is 450 square feet of decking material, find the length of the deck.

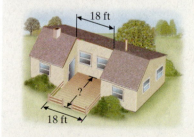

18 ft
?
18 ft

Solution:

1. UNDERSTAND. Read and reread the problem. The formula needed to solve this problem is the formula for the perimeter of a rectangle, $P = 2l + 2w$. Before continuing, let's become familar with this formula.

   $l =$, the length of the rectangular garden

   w = the width of the rectangular garden

   P = perimeter of the garden

2. TRANSLATE. To translate to an equation, we use the formula $P = 2l + 2w$ and let perimeter $P = 140$ feet and width $w = 30$ feet.

   $$P = 2l + 2w \quad \text{Let } P = 140 \text{ and } w = 30.$$
   $$140 = 2l + 2(30)$$

Continued on next page

Answer
2. 25 feet

3. SOLVE.

$$140 = 2l + 2(30)$$
$$140 = 2l + 60 \qquad \text{Multiply } 2(30).$$
$$140 - 60 = 2l + 60 - 60 \qquad \text{Subtract 60 from both sides.}$$
$$80 = 2l \qquad \text{Combine like terms.}$$
$$40 = l \qquad \text{Divide both sides by 2.}$$

4. INTERPRET.

Check: Substitute 40 for l and 30 for w in the perimeter formula and check to see that the perimeter is 140 feet.

State: The length of the rectangular garden is 40 feet.

◻ **Work Practice Problem 2**

PRACTICE PROBLEM 3

Convert the temperature 5°C to Fahrenheit.

EXAMPLE 3 **Finding an Equivalent Temperature**

The average maximum temperature for January in Algerias, Algeria, is 59° Fahrenheit. Find the equivalent temperature in degrees Celsius.

Solution:

1. UNDERSTAND. Read and reread the problem. A formula that can be used to solve this problem is the formula for converting degrees Celsius to degrees Fahrenheit, $F = \dfrac{9}{5}C + 32$. Before continuing, become familiar with this formula. Using this formula, we let

   C = temperature in degrees Celsius, and

   F = temperature in degrees Fahrenheit.

2. TRANSLATE. To translate to an equation, we use the formula $F = \dfrac{9}{5}C + 32$ and let degrees Fahrenheit $F = 59$.

   Formula: $\qquad F = \dfrac{9}{5}C + 32$

   Substitute: $\qquad 59 = \dfrac{9}{5}C + 32 \qquad$ Let $F = 59$.

3. SOLVE.

   $$59 = \frac{9}{5}C + 32$$

   $$59 - 32 = \frac{9}{5}C + 32 - 32 \qquad \text{Subtract 32 from both sides.}$$

   $$27 = \frac{9}{5}C \qquad \text{Combine like terms.}$$

   $$\frac{5}{9} \cdot 27 = \frac{5}{9} \cdot \frac{9}{5}C \qquad \text{Multiply both sides by } \tfrac{5}{9}.$$

   $$15 = C \qquad \text{Simplify.}$$

4. INTERPRET.

Check: To check, replace C with 15 and F with 59 in the formula and see that a true statement results.

State: Thus, 59° Fahrenheit is equivalent to 15° Celsius.

◻ **Work Practice Problem 3**

Answer

3. 41°F

In the next example, we again use the formula for perimeter of a rectangle as in Example 2. In Example 2, we knew the width of the rectangle. In this example, both the length and width are unknown.

EXAMPLE 4 Finding Road Sign Dimensions

The length of a rectangular road sign is 2 feet less than three times its width. Find the dimensions if the perimeter is 28 feet.

PRACTICE PROBLEM 4

The length of a rectangle is one more meter than 4 times its width. Find the dimensions if the perimeter is 52 meters.

Solution:

1. UNDERSTAND. Read and reread the problem. Recall that the formula for the perimeter of a rectangle is $P = 2l + 2w$. Draw a rectangle and guess the solution. If the width of the rectangular sign is 5 feet, its length is 2 feet less than 3 times the width or $3(5 \text{ feet}) - 2 \text{ feet} = 13 \text{ feet}$. The perimeter P of the rectangle is then $2(13 \text{ feet}) + 2(5 \text{ feet}) = 36 \text{ feet}$, too much. We now know that the width is less than 5 feet.

   Proposed rectangle:

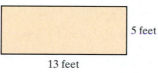

5 feet

13 feet

Let

$\qquad w = $ the width of the rectangular sign; then

$3w - 2 = $ the length of the sign.

w

$3w - 2$

Draw a rectangle and label it with the assigned variables.

2. TRANSLATE.

   Formula: $P = 2l + 2w$

   Substitute: $28 = 2(3w - 2) + 2w.$

3. SOLVE.

$$28 = 2(3w - 2) + 2w$$
$$28 = 6w - 4 + 2w \qquad \text{Apply the distributive property.}$$
$$28 = 8w - 4$$
$$28 + 4 = 8w - 4 + 4 \qquad \text{Add 4 to both sides.}$$
$$32 = 8w$$
$$\frac{32}{8} = \frac{8w}{8} \qquad \text{Divide both sides by 8.}$$
$$4 = w$$

4. INTERPRET.

Check: If the width of the sign is 4 feet, the length of the sign is $3(4 \text{ feet}) - 2 \text{ feet} = 10 \text{ feet}$. This gives a perimeter of $P = 2(4 \text{ feet}) + 2(10 \text{ feet}) = 28 \text{ feet}$, the correct perimeter.

State: The width of the sign is 4 feet and the length of the sign is 10 feet.

■ **Work Practice Problem 4**

Answer

4. length: 21m; width: 5m

Objective B Solving a Formula for a Variable

We say that the formula

$$d = rt$$

is solved for d because d is alone on one side of the equation and the other side contains no d's. Suppose that we have a large number of problems to solve where we are given distance d and rate r and asked to find time t. In this case, it may be easier to first solve the formula $d = rt$ for t. To solve for t, we divide both sides of the equation by r.

$$d = rt$$

$$\frac{d}{r} = \frac{rt}{r} \quad \text{Divide both sides by } r.$$

$$\frac{d}{r} = t \quad \text{Simplify.}$$

To solve a formula or an equation for a specified variable, we use the same steps as for solving a linear equation except that we treat the specified variable as the only variable in the equation. These steps are listed next.

Solving Equations for a Specified Variable

Step 1: Multiply on both sides to clear the equation of fractions if they occur.

Step 2: Use the distributive property to remove parentheses if they occur.

Step 3: Simplify each side of the equation by combining like terms.

Step 4: Get all terms containing the specified variable on one side and all other terms on the other side by using the addition property of equality.

Step 5: Get the specified variable alone by using the multiplication property of equality.

PRACTICE PROBLEM 5

Solve $C = 2\pi r$ for r. (This formula is used to find the circumference C of a circle given its radius r.)

EXAMPLE 5 Solve $V = lwh$ for l.

Solution: This formula is used to find the volume of a box. To solve for l, we divide both sides by wh.

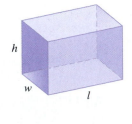

$$V = lwh$$

$$\frac{V}{wh} = \frac{lwh}{wh} \quad \text{Divide both sides by } wh.$$

$$\frac{V}{wh} = l \quad \text{Simplify.}$$

Since we have l alone on one side of the equation, we have solved for l in terms of $V, w,$ and h. Remember that it does not matter on which side of the equation we get the variable alone.

■ **Work Practice Problem 5**

Answer

5. $r = \dfrac{C}{2\pi}$

EXAMPLE 6 Solve $y = mx + b$ for x.

Solution: First we get mx alone by subtracting b from both sides.

$$y = mx + b$$
$$y - b = mx + b - b \quad \text{Subtract } b \text{ from both sides.}$$
$$y - b = mx \quad \text{Combine like terms.}$$

Next we solve for x by dividing both sides by m.

$$\frac{y - b}{m} = \frac{mx}{m}$$
$$\frac{y - b}{m} = x \quad \text{Simplify.}$$

Work Practice Problem 6

✔ Concept Check Solve:

a.

b. ⬤ = 🟥 · 🔺 − 🟩 for 🟥

EXAMPLE 7 Solve $P = 2l + 2w$ for w.

Solution: This formula relates the perimeter of a rectangle to its length and width. Find the term containing the variable w. To get this term, $2w$, alone subtract $2l$ from both sides.

$$P = 2l + 2w$$
$$P - 2l = 2l + 2w - 2l \quad \text{Subtract } 2l \text{ from both sides.}$$
$$P - 2l = 2w \quad \text{Combine like terms.}$$
$$\frac{P - 2l}{2} = \frac{2w}{2} \quad \text{Divide both sides by 2.}$$
$$\frac{P - 2l}{2} = w \quad \text{Simplify.}$$

Work Practice Problem 7

The next example has an equation containing a fraction. We will first clear the equation of fractions and then solve for the specified variable.

EXAMPLE 8 Solve $F = \frac{9}{5}C + 32$ for C.

Solution:
$$F = \frac{9}{5}C + 32$$
$$5(F) = 5\left(\frac{9}{5}C + 32\right) \quad \text{Clear the fraction by multiplying both sides by the LCD.}$$
$$5F = 9C + 160 \quad \text{Distribute the 5.}$$
$$5F - 160 = 9C + 160 - 160 \quad \text{To get the term containing the variable } C \text{ alone, subtract 160 from both sides.}$$
$$5F - 160 = 9C \quad \text{Combine like terms.}$$
$$\frac{5F - 160}{9} = \frac{9C}{9} \quad \text{Divide both sides by 9.}$$
$$\frac{5F - 160}{9} = C \quad \text{Simplify.}$$

Work Practice Problem 8

PRACTICE PROBLEM 6

Solve $P = 2l + 2w$ for l.

PRACTICE PROBLEM 7

Solve: $P = 2a + b - c$ for a.

Helpful Hint
The 2's may *not* be divided out here. Although 2 is a factor of the denominator, 2 is *not* a factor of the numerator since it is not a factor of both terms in the numerator.

PRACTICE PROBLEM 8

Solve $A = \frac{a + b}{2}$ for b.

Answers
6. $l = \frac{P - 2w}{2}$, **7.** $a = \frac{P - b + c}{2}$,
8. $b = 2A - a$

✔ Concept Check Answer

a. ⬤ + 🟩 **b.** $\frac{⬤ + 🟩}{🔺}$

2.5 EXERCISE SET

Objective **A** *Substitute the given values into each given formula and solve for the unknown variable. If necessary, round to one decimal place. See Examples 1 through 4.*

△ **1.** $A = bh$; $A = 45, b = 15$ (Area of a parallelogram)

2. $d = rt$; $d = 195, t = 3$ (Distance formula)

△ **3.** $S = 4lw + 2wh$; $S = 102, l = 7, w = 3$ (Surface area of a special rectangular box)

△ **4.** $V = lwh$; $l = 14, w = 8, h = 3$ (Volume of a rectangular box)

△ **5.** $A = \frac{1}{2}h(B + b)$; $A = 180, B = 11, b = 7$ (Area of a trapezoid)

△ **6.** $A = \frac{1}{2}h(B + b)$; $A = 60, B = 7, b = 3$ (Area of a trapezoid)

△ **7.** $P = a + b + c$; $P = 30, a = 8, b = 10$ (Perimeter of a triangle)

△ **8.** $V = \frac{1}{3}Ah$; $V = 45, h = 5$ (Volume of a pyramid)

△ **9.** $C = 2\pi r$; $C = 15.7$ (Circumference of a circle) (use the approximation 3.14 for π)

△ **10.** $A = \pi r^2$; $r = 4$ (Area of a circle) (use the approximation 3.14 for π)

Objective **B** *Solve each formula for the specified variable. See Examples 5 through 8.*

11. $f = 5gh$ for h

△ **12.** $C = 2\pi r$ for r

13. $V = lwh$ for w

14. $T = mnr$ for n

15. $3x + y = 7$ for y

16. $-x + y = 13$ for y

17. $A = P + PRT$ for R

18. $A = P + PRT$ for T

19. $V = \frac{1}{3}Ah$ for A

20. $D = \frac{1}{4}fk$ for k

21. $P = a + b + c$ for a

22. $PR = x + y + z + w$ for z

23. $S = 2\pi rh + 2\pi r^2$ for h

△ **24.** $S = 4lw + 2wh$ for h

Solve. See Examples 1 through 4.

25. For the purpose of purchasing new baseboard and carpet,
   a. Find the area and perimeter of the room below (neglecting doors).
   b. Identify whether baseboard has to do with area or perimeter and the same with carpet.

11.5 ft   9 ft

26. For the purpose of purchasing lumber for a new fence and seed to plant grass,
   a. Find the area and perimeter of the yard below.
   b. Identify whether a fence has to do with area or perimeter and the same with grass seed.

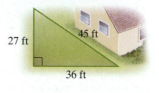

27 ft   45 ft   36 ft

27. A frame shop charges according to both the amount of framing needed to surround the picture and the amount of glass needed to cover the picture.
   a. Find the area and perimeter of the picture below.
   b. Identify whether the frame has to do with perimeter or area and the same with the glass.

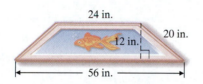

24 in.   12 in.   20 in.   56 in.

28. A decorator is painting and placing a border completely around the parallelogram-shaped wall.
   a. Find the area and perimeter of the wall below.
   b. Identify whether the border has to do with perimeter or area and the same with paint.

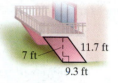

7 ft   11.7 ft   9.3 ft

29. The world's largest pink ribbon, the sign of the fight against breast cancer, was erected out of pink post-it notes on a billboard in New York City in October, 2004. If the area of the rectangular billboard covered by the ribbon is approximately 3990 sq ft, and the width of the ribbon was approximately 57 ft, what was the height of this gigantic symbol?

△ **30.** The world's largest sign for Coca-Cola is located in Arica, Chile. The rectangular sign has a length of 400 feet and has an area of 52,400 square feet. Find the width of the sign. (*Source:* Fabulous Facts about Coca-Cola, Atlanta, GA)

31. Convert Nome, Alaska's 14°F high temperature to Celsius.

32. Convert Paris, France's low temperature of −5°C to Fahrenheit.

33. The X-30 is a "space plane" that skims the edge of space at 4000 miles per hour. Neglecting altitude, if the circumference of the Earth is approximately 25,000 miles, how long will it take for the X-30 to travel around the Earth?

34. In the United States, a notable hang glider flight was a 303-mile, $8\frac{1}{2}$ hour flight from New Mexico to Kansas. What was the average rate during this flight?

△ **35.** An architect designs a rectangular flower garden such that the width is exactly two-thirds of the length. If 260 feet of antique picket fencing are to be used to enclose the garden, find the dimensions of the garden.

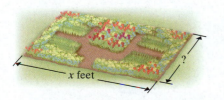

x feet

△ **36.** If the length of a rectangular parking lot is 10 meters less than twice its width, and the perimeter is 400 meters, find the length of the parking lot.

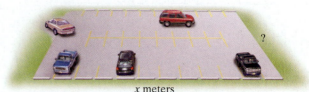

x meters

△ **37.** A flower bed is in the shape of a triangle with one side twice the length of the shortest side, and the third side is 30 feet more than the length of the shortest side. Find the dimensions if the perimeter is 102 feet.

△ **38.** The perimeter of a yield sign in the shape of an isosceles triangle is 22 feet. If the shortest side is 2 feet less than the other two sides, find the length of the shortest side. (*Hint:* An isosceles triangle has two sides the same length.)

39. The Cat is a high-speed catamaran auto ferry that operates between Bar Harbor, Maine, and Yarmouth, Nova Scotia. The Cat can make the trip in about $2\frac{1}{2}$ hours at a speed of 55 mph. About how far apart are Bar Harbor and Yarmouth? (*Source:* Bay Ferries)

40. A family is planning their vacation to Disney World. They will drive from a small town outside New Orleans, Louisiana, to Orlando, Florida, a distance of 700 miles. They plan to average a rate of 55 mph. How long will this trip take?

△ **41.** Piranha fish require 1.5 cubic feet of water per fish to maintain a healthy environment. Find the maximum number of piranhas you could put in a tank measuring 8 feet by 3 feet by 6 feet.

△ **42.** Find the maximum number of goldfish you can put in a cylindrical tank whose diameter is 8 meters and whose height is 3 meters if each goldfish needs 2 cubic meters of water.

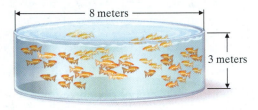

△ **43.** A lawn is in the shape of a trapezoid with a height of 60 feet and bases of 70 feet and 130 feet. How many bags of fertilizer must be purchased to cover the lawn if each bag covers 4000 square feet?

△ **44.** If the area of a right-triangularly shaped sail is 20 square feet and its base is 5 feet, find the height of the sail.

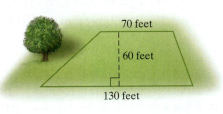

△ **45.** Maria's Pizza sells one 16-inch cheese pizza or two 10-inch cheese pizzas for $9.99. Determine which size gives more pizza.

△ **46.** Find how much rope is needed to wrap around the Earth at the equator, if the radius of the Earth is 4000 miles. (*Hint:* Use 3.14 for π and the formula for circumference.)

47. A Japanese "bullet" train set a new world record for train speed at 552 kilometers per hour during a manned test run on the Yamanashi Maglev Test Line in April 1999. The Yamanashi Maglev Test Line is 42.8 kilometers long. How many *minutes* would a test run on the Yamanashi Line last at this record-setting speed? Round to the nearest hundredth of a minute. (*Source:* Japan Railways Central Co.)

48. In 1983, the Hawaiian volcano Kilauea began erupting in a series of episodes still occurring at the time of this writing. At times, the lava flows advanced at speeds of up to 0.5 kilometer per hour. In 1983 and 1984 lava flows destroyed 16 homes in the Royal Gardens subdivision, about 6 km away from the eruption site. Roughly how long did it take the lava to reach Royal Gardens? (*Source:* U.S. Geological Survey Hawaiian Volcano Observatory)

△ **49.** The perimeter of an equilateral triangle is 7 inches more than the perimeter of a square, and the side of the triangle is 5 inches longer than the side of the square. Find the side of the triangle. (*Hint:* An equilateral triangle has three sides the same length.)

△ **50.** A square animal pen and a pen shaped like an equilateral triangle have equal perimeters. Find the length of the sides of each pen if the sides of the triangular pen are fifteen less than twice a side of the square pen.

51. Find how long it takes Tran Nguyen to drive 135 miles on I-10 if he merges onto I-10 at 10 A.M. and drives nonstop with his cruise control set on 60 mph.

52. Beaumont, Texas, is about 150 miles from Toledo Bend. If Leo Miller leaves Beaumont at 4 A.M. and averages 45 mph, when should he arrive at Toledo Bend?

△ **53.** The longest runway at Los Angeles International Airport has the shape of a rectangle and an area of 1,813,500 square feet. This runway is 150 feet wide. How long is the runway? (*Source:* Los Angeles World Airports)

54. Bolts of lightning can travel at the speed of 270,000 miles per second. How many times can a lightning bolt travel around the world in one second? (See Exercise 46. Round to the nearest tenth.)

55. The highest temperature ever recorded in Europe was 122°F in Seville, Spain, in August of 1881. Convert this record high temperature to Celsius. (*Source:* National Climatic Data Center)

56. The lowest temperature ever recorded in Oceania was −10°C at the Haleakala Summit in Maui, Hawaii, in January 1961. Convert this record low temperature to Fahrenheit. (*Source:* National Climatic Data Center)

△ **57.** The CART FedEx Championship Series is an open-wheeled race car competition based in the United States. A CART car has a maximum length of 199 inches, a maximum width of 78.5 inches, and a maximum height of 33 inches. When the CART series travels to another country for a grand prix, teams must ship their cars. Find the volume of the smallest shipping crate needed to ship a CART car of maximum dimensions. (*Source:* Championship Auto Racing Teams, Inc.)

58. On a road course, a CART car's speed can average up to around 105 mph. Based on this speed, how long would it take a CART driver to travel from Los Angeles to New York City, a distance of about 2810 miles by road, without stopping? Round to the nearest tenth of an hour.

CART Racing Car

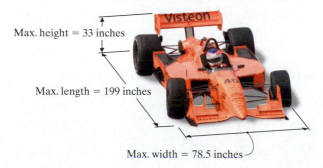

Max. height = 33 inches

Max. length = 199 inches

Max. width = 78.5 inches

△ **59.** The Hoberman Sphere is a toy ball that expands and contracts. When it is completely closed, it has a diameter of 9.5 inches. Find the volume of the Hoberman Sphere when it is completely closed. Use 3.14 for π. Round to the nearest whole cubic inch. (*Hint:* volume of a sphere $= \dfrac{4}{3}\pi r^3$. *Source:* Hoberman Designs, Inc.)

△ **60.** When the Hoberman Sphere (see Exercise 59) is completely expanded, its diameter is 30 inches. Find the volume of the Hoberman Sphere when it is completely expanded. Use 3.14 for π. Round to the nearest whole cubic inch. (*Source:* Hoberman Designs, Inc.)

61. The average temperature on the planet Mercury is 167°C. Convert this temperature to degrees Fahrenheit. Round to the nearest degree. (*Source:* National Space Science Data Center)

62. The average temperature on the planet Jupiter is −227°F. Convert this temperature to degrees Celsius. Round to the nearest degree. (*Source:* National Space Science Data Center)

Review

Write each percent as a decimal. See Section R.3.

63. 32%   **64.** 8%   **65.** 200%   **66.** 0.5%

Write each decimal as a percent. See Section R.3.

67. 0.17   **68.** 0.03   **69.** 7.2   **70.** 5

Concept Extensions

Solve.

71. $N = R + \dfrac{V}{G}$ for V (Urban forestry: tree plantings per year)

72. $B = \dfrac{F}{P - V}$ for V (Business: break-even point)

73. The formula $V = lwh$ is used to find the volume of a box. If the length of a box is doubled, the width is doubled, and the height is doubled, how does this affect the volume? Explain your answer.

74. The formula $A = bh$ is used to find the area of a parallelogram. If the base of a parallelogram is doubled and its height is doubled, how does this affect the area? Explain your answer.

75. Find the temperature at which the Celsius measurement and Fahrenheit measurement are the same number.

Solve. See the Concept Check in this section.

76.  ⬠ · ◼ + △ = ● for ◼

77. ◼ − ● · ▮ = △ for ●

78. A glacier is a giant mass of rocks and ice that flows downhill like a river. Exit Glacier, near Seward, Alaska, moves at a rate of 20 inches a day. Find the distance in feet the glacier moves in a year. (Assume 365 days a year. Round to two decimal places.)

79. Flying fish do not *actually* fly, but glide. They have been known to travel a distance of 1300 feet at a rate of 20 miles per hour. How many seconds did it take to travel this distance? (*Hint:* First convert miles per hour to feet per second. Recall that 1 mile = 5280 feet. Round to the nearest tenth of a second.)

Substitute the given values into each given formula and solve for the unknown variable. If necessary, round to one decimal place.

80. $I = PRT$; $I = 3750, P = 25,000, R = 0.05$ (Simple interest formula)

81. $I = PRT$; $I = 1,056,000, R = 0.055, T = 6$ (Simple interest formula)

82. $V = \frac{1}{3}\pi r^2 h$; $V = 565.2, r = 6$ (use a calculator approximation for π)(Volume of a cone)

83. $V = \frac{4}{3}\pi r^3$; $r = 3$ (use a calculator approximation for π) (Volume of a sphere)

STUDY SKILLS BUILDER

Organizing a Notebook

It's never too late to get organized. If you need ideas about organizing a notebook for your mathematics course, try some of these:

- Use a spiral or ring binder notebook with pockets and use it for mathematics only.
- Start each page by writing the book's section number you are working on at the top.
- When your instructor is lecturing, take notes. *Always* include any examples your instructor works for you.
- Place your worked-out homework exercises in your notebook immediately after the lecture notes from that section. This way, a section's worth of material is together.
- Homework exercises: Attempt all assigned homework. For odd-numbered exercises, you are not through until you check your answers against the back of the book. Correct any exercises with incorrect answers. You may want to place a "?" by any homework exercises or notes that you need to ask questions about. Also, consider placing a "!" by any notes or exercises you feel are important.
- Place graded quizzes in the pockets of your notebook. If you are using a binder, you can place your quizzes in a special section of your binder.

Let's check your notebook organization by answering the following questions.

1. Do you have a spiral or ring binder notebook for your mathematics course only?
2. Have you ever had to flip through several sheets of notes and work in your mathematics notebook to determine what section's work you are in?
3. Are you now writing the textbook's section number at the top of each notebook page?
4. Have you ever lost or had trouble finding a graded quiz or test?
5. Are you now placing all your graded work in a dedicated place in your notebook?
6. Are you attempting all of your homework and placing all of your work in your notebook?
7. Are you checking and correcting your homework in your notebook? If not, why not?
8. Are you writing in your notebook the examples your instructor works for you in class?

2.6 PERCENT AND MIXTURE PROBLEM SOLVING

This section is devoted to solving problems in the categories listed. The same problem-solving steps used in previous sections are also followed in this section. They are listed below for review.

General Strategy for Problem Solving

1. UNDERSTAND the problem. During this step, become comfortable with the problem. Some ways of doing this are as follows:

   Read and reread the problem.

   Choose a variable to represent the unknown.

   Construct a drawing, whenever possible.

   Propose a solution and check. Pay careful attention to how you check your proposed solution. This will help writing an equation to model the problem.

2. TRANSLATE the problem into an equation.

3. SOLVE the equation.

4. INTERPRET the results: *Check* the proposed solution in the stated problem and *state* your conclusion.

Objective **A** Solving Percent Equations

Many of today's statistics are given in terms of percent: a basketball player's free throw percent, current interest rates, stock market trends, and nutrition labeling, just to name a few. In this section, we first explore percent, percent equations, and applications involving percents. See Section R.3 if a further review of percents is needed.

PRACTICE PROBLEM 1

The number 22 is what percent of 40?

EXAMPLE 1 The number 63 is what percent of 72?

Solution:

1. UNDERSTAND. Read and reread the problem. Next, let's suppose that the percent is 80%. To check, we find 80% of 72.

   $$80\% \text{ of } 72 = 0.80(72) = 57.6$$

   This is close, but not 63. At this point, though, we have a better understanding of the problem, we know the correct answer is close to and greater than 80%, and we know how to check our proposed solution later.

   Let x = the unknown percent.

2. TRANSLATE. Recall that "is" means "equals" and "of" signifies multiplying. Let's translate the sentence directly.

the number 63	is	what percent	of	72
↓	↓	↓	↓	↓
63	=	x	·	72

3. SOLVE.

$$63 = 72x$$
$$0.875 = x \quad \text{Divide both sides by 72.}$$
$$87.5\% = x \quad \text{Write as a percent.}$$

Answer

1. 55%

4. INTERPRET.

Check: Verify that 87.5% of 72 is 63.

State: The number 63 is 87.5% of 72.

 **Work Practice Problem 1**

EXAMPLE 2 The number 120 is 15% of what number?

Solution:

1. UNDERSTAND. Read and reread the problem.

   Let x = the unknown number.

2. TRANSLATE.

the number 120	is	15%	of	what number
↓	↓	↓	↓	↓
120	=	15%	·	x

3. SOLVE.

   $120 = 0.15x$ Write 15% as 0.15.

   $800 = x$ Divide both sides by 0.15.

4. INTERPRET.

Check: Check the proposed solution by finding 15% of 800 and verifying that the result is 120.

State: Thus, 120 is 15% of 800.

 **Work Practice Problem 2**

PRACTICE PROBLEM 2

The number 150 is 40% of what number?

EXAMPLE 3 The circle graph below shows the purpose of trips made by American travelers. Use this graph to answer the questions below.

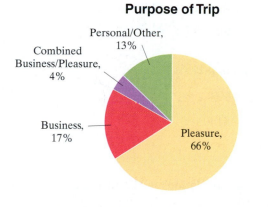

Purpose of Trip

Personal/Other, 13%

Combined Business/Pleasure, 4%

Business, 17%

Pleasure, 66%

Source: Travel Industry Association of America

a. What percent of trips made by American travelers are solely for the purpose of business?

b. What percent of trips made by American travelers are for the purpose of business or combined business/pleasure?

c. On an airplane flight of 253 Americans, how many of these people might we expect to be traveling solely for business?

Solution:

a. From the circle graph, we see that 17% of trips made by American travelers are solely for the purpose of business.

Continued on next page

PRACTICE PROBLEM 3

Use the circle graph to answer each question.

a. What percent of trips made by American travelers are solely for pleasure?

b. What percent of trips made by American travelers are for the purpose of pleasure or combined business/ pleasure?

c. On an airplane flight of 250 Americans, how many of these people might we expect to be traveling solely for pleasure?

Answers

2. 375, **3. a.** 66%, **b.** 70%, **c.** 165 people

b. From the circle graph, we know that 17% of trips are solely for business and 4% of trips are for combined business/pleasure. The sum 17% + 4% or 21% of trips made by American travelers are for the purpose of business or combined business/pleasure.

c. Since 17% of trips made by American travelers are for business, we find 17% of 253. Remember that "of" translates to "multiplication."

17% of 253 = 0.17(253)　Replace "of" with the operation of multiplication.

= 43.01

We might then expect that about 43 American travelers on the flight are traveling solely for business.

■ **Work Practice Problem 3**

Objective B Solving Discount and Mark-Up Problems

The next example has to do with discounting the price of a cell phone.

PRACTICE PROBLEM 4

A surfboard, originally purchased for $400, was sold on eBay at a discount of 40% of the original price. What is the discount and the new price?

EXAMPLE 4 Cell Phones Unlimited recently reduced the price of a $140 phone by 20%. What is the discount and the new price?

Solution:

1. UNDERSTAND. Read and reread the problem. Make sure you understand the meaning of the word "discount." Discount is the amount of money by which an item has been decreased. To find the discount, we simply find 20% of $140. In other words, we have the formulas,

discount = percent · original price　Then

new price = original price − discount

2, 3. TRANSLATE and SOLVE.

discount = percent · original price

= 20% · $140

= 0.20 · $140

= $28

Thus, the discount in price is $28.

new price = original price − discount

= $140 − $28

= $112

4. INTERPRET.

Check: Check your calculations in the formulas, and also see if our results are reasonable. They are.

State: The discount in price is $28 and the new price is $112.

■ **Work Practice Problem 4**

A concept similar to discount is mark-up. What is the difference between the two? A discount is subtracted from the original price while a mark-up is added to the original price. For mark-ups,

mark-up = percent · original price

new price = original price + mark-up

Mark-up exercises can be found in Exercise Set 2.6.

Answer

4. discount: $160,　new price: $240

Objective **C** Solving Percent Increase and Percent Decrease Problems

Percent increase or percent decrease is a common way to describe how some measurement has increased or decreased. For example, crime increased by 8%, teachers received a 5.5% increase in salary, or a company decreased its employees by 10%. The next example is a review of percent increase.

EXAMPLE 5 **Calculating the Percent Increase of Attending College**

The tuition and fees cost of attending a public college rose from $1454 in 1990 to $2928 in 2003. Find the percent increase. (*Source:* National Center for Education Statistics and U.S. Department of Education) *Note:* These costs are an average of two-year and four-year colleges.

PRACTICE PROBLEM 5

If a number increases from 120 to 200, find the percent increase. Round to the nearest tenth of a percent.

Solution:

1. UNDERSTAND. Read and reread the problem. Notice that the new tuition, $2928, is over double the old tuition of $1454. Because of that, we know that the percent increase is greater than 100%. To see this, let's guess that the percent increase is 100%. To check, we find 100% of $1454 to find the *increase* in cost. Then we add this increase to $1454 to find the *new cost*. In other words, 100% ($1454) = 1.00($1454) = $1454, the *increase* in cost. The *new cost* would be old cost + increase = $1454 + $1454 = $2908, very close to the actual new cost of $2928. We now know that the increase is close to, but greater than 100% and we know how to check our proposed solution.

   Let x = the percent increase.

2. TRANSLATE. First, find the **increase,** and then the **percent increase.** The increase in cost is found by:

   In words: increase   = | new cost | − | old cost | or

   Translate: increase   = $2928   − $1454

                            = $1474

   Next, find the percent increase. The percent increase or percent decrease is always a percent of the original number or in this case, the old cost.

   In words: | increase | | is | | what percent increase | | of | | old cost |

   Translate: $1474   = x · $1454

3. SOLVE.

   $$1474 = 1454x$$

   $$1.014 \approx x \qquad \text{Divide both sides by 1454 and round to 3 decimal places.}$$

   $$101.4\% \approx x \qquad \text{Write as a percent.}$$

4. INTERPRET.

Check: Check the proposed solution.

State: The percent increase in cost is approximately 101.4%.

▢ Work Practice Problem 5

   Percent decrease is found using a similar method. First find the decrease, then determine what percent of the original or first amount is that decrease.

   Read the next example carefully. For Example 5, we were asked to find percent increase. In Example 6, we are given the percent increase and asked to find the number before the increase.

PRACTICE PROBLEM 6

Find the original price of a suit if the sale price is $46 after a 20% discount.

EXAMPLE 6 Most of the movie screens in the United States project analog films, but the number of cinemas using digital are increasing. Find the number of digital screens last year if after a 175% increase, the number this year is 124. Round to the nearest whole.

Solution:

1. UNDERSTAND. Read and reread the problem. Let's guess a solution and see how we would check our guess. If the number of digital screens last year was 50, we would see if 50 plus the increase is 124; that is,

$$50 + 175\%(50) = 50 + 1.75(50) = 50 + 87.5 = 137.50$$

Since 137.5 is too large, we know that our guess of 50 is too large. We also have a better understanding of the problem. Let

x = number of digital screens last year

2. TRANSLATE. To translate to an equation, we remember that

In words:	number of digital screen last year	plus	increase	equals	number of digital screens this year
Translate:	x	+	$1.75x$	=	124

3. SOLVE.

$$2.75x = 124$$
$$x = \frac{124}{2.75}$$
$$x \approx 45$$

4. INTERPRET.

Check: Recall that x represents the number of digital screens last year. If this number is approximately 45, let's see if 45 plus the increase is close to 124. (We use the word "close" since 45 is rounded.)

$$45 + 175\%(45) = 45 + 1.75(45) = 45 + 78.75 = 123.75,$$

which is close to 124.

State: There were approximately 45 digital screens last year.

☐ **Work Practice Problem 6**

Objective D Solving Mixture Problems

Mixture problems involve two or more different quantities being combined to form a new mixture. These applications range from Dow Chemical's need to form a chemical mixture of a required strength to Planter's Peanut Company's need to find the correct mixture of peanuts and cashews, given taste and price constraints.

PRACTICE PROBLEM 7

How much 20% dye solution and 50% dye solution should be mixed to obtain 6 liters of a 40% solution?

EXAMPLE 7 **Calculating Percent for a Lab Experiment**

A chemist working on his doctoral degree at Massachusetts Institute of Technology needs 12 liters of a 50% acid solution for a lab experiment. The stockroom has only 40% and 70% solutions. How much of each solution should be mixed together to form 12 liters of a 50% solution?

Solution:

1. UNDERSTAND. First, read and reread the problem a few times. Next, guess a solution. Suppose that we need 7 liters of the 40% solution. Then we need $12 - 7 = 5$ liters of the 70% solution. To see if this is indeed the solution, find

Answers

6. $57.50, 7. 2 liters of the 20% solution; 4 liters of the 50% solution

the amount of pure acid in 7 liters of the 40% solution, in 5 liters of the 70% solution, and in 12 liters of a 50% solution, the required amount and strength.

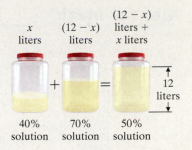

number of liters	×	acid strength	=	amount of pure acid
7 liters	×	40%	=	7(0.40) or 2.8 liters
5 liters	×	70%	=	5(0.70) or 3.5 liters
12 liters	×	50%	=	12(0.50) or 6 liters

Since 2.8 liters + 3.5 liters = 6.3 liters and not 6, our guess is incorrect, but we have gained some valuable insight into how to model and check this problem.
   Let

$$x = \text{number of liters of 40\% solution; then}$$

$12 - x = $ number of liters of 70% solution.

2. **TRANSLATE.** To help us translate to an equation, the following table summarizes the information given. Recall that the amount of acid in each solution is found by multiplying the acid strength of each solution by the number of liters.

	No. of Liters	·	Acid Strength	=	Amount of Acid
40% Solution	x		40%		$0.40x$
70% Solution	$12 - x$		70%		$0.70(12 - x)$
50% Solution Needed	12		50%		$0.50(12)$

The amount of acid in the final solution is the sum of the amounts of acid in the two beginning solutions.

In words: | acid in 40% solution | + | acid in 70% solution | = | acid in 50% mixture |
|---|---|---|---|---|

Translate: $0.40x$ + $0.70(12 - x)$ = $0.50(12)$

3. **SOLVE.**

$$0.40x + 0.70(12 - x) = 0.50(12)$$

$0.4x + 8.4 - 0.7x = 6$	Apply the distributive property.
$-0.3x + 8.4 = 6$	Combine like terms.
$-0.3x = -2.4$	Subtract 8.4 from both sides.
$x = 8$	Divide both sides by -0.3.

4. **INTERPRET.**

Check: To check, recall how we checked our guess.

State: If 8 liters of the 40% solution are mixed with $12 - 8$ or 4 liters of the 70% solution, the result is 12 liters of a 50% solution.

□ **Work Practice Problem 7**

Mental Math

Tell whether the percent labels in the circle graphs are correct.

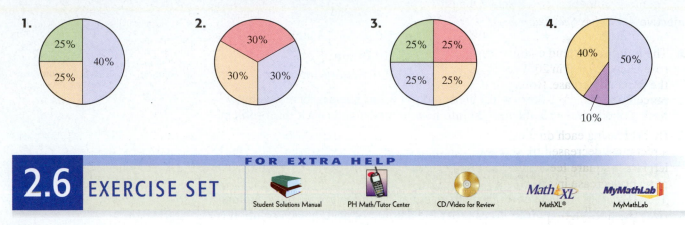

1. 25% 40% 25%

2. 30% 30% 30%

3. 25% 25% 25% 25%

4. 40% 50% 10%

FOR EXTRA HELP

2.6 EXERCISE SET

Student Solutions Manual   PH Math/Tutor Center   CD/Video for Review   MathXL   MyMathLab

Objective Ⓐ *Find each number described. See Examples 1 and 2.*

1. What number is 16% of 70?

2. What number is 88% of 1000?

3. The number 28.6 is what percent of 52?

4. The number 87.2 is what percent of 436?

5. The number 45 is 25% of what number?

6. The number 126 is 35% of what number?

The circle graph below shows the number of minutes that adults spend on their home phone each day. Use this graph for Exercises 7 through 10. See Example 3.

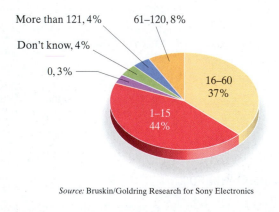

More than 121, 4%   61–120, 8%
Don't know, 4%
0, 3%
16–60 37%
1–15 44%

Source: Bruskin/Goldring Research for Sony Electronics

7. What percent of adults spend more than 121 minutes on the phone each day?

8. What percent of adults spend no time on the phone each day?

9. Florence is a town in Alabama whose adult population is approximately 27,000. How many of these adults might you expect to talk 16–60 minutes on the phone each day?

10. Columbus is a town in Indiana whose adult population is approximately 29,250. How many of these adults might you expect to talk 61–120 minutes on the phone each day?

Objective Ⓑ *Solve. If needed, round answers to the nearest cent. See Example 4.*

11. A used automobile dealership recently reduced the price of a used sports car by 8%. If the price of the car before discount was $18,500, find the discount and the new price.

12. A music store is advertising a 25%-off sale on all new releases. Find the discount and the sale price of a newly released CD that regularly sells for $12.50.

148

13. A birthday celebration meal is $40.50 including tax. Find the total cost if a 15% tip is added to the cost.

14. A retirement dinner for two is $65.40 including tax. Find the total cost if a 20% tip is added to the cost.

Objective **C** *Solve. See Example 5.*

15. The number of fraud complaints (usually ID theft) rose from 220,000 in 2001 to 380,000 in 2002. Find the percent increase. Round to the nearest whole percent.

16. The number of text messages rose from 996 million in June to 1100 million in December. Find the percent increase. Round to the nearest whole percent.

17. By decreasing each dimension by 1 unit, the area of a rectangle decreased from 40 square feet (on the left) to 28 square feet (on the right). Find the percent decrease in area.

18. By decreasing the length of the side by one unit, the area of a square decreased from 100 square meters to 81 square meters. Find the percent decrease in area.

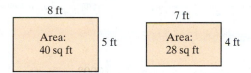

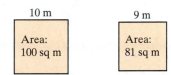

Solve. See Example 6.

19. Find the original price of a pair of shoes if the sale price is $78 after a 25% discount.

20. Find the original price of a popular pair of shoes if the increased price is $80 after a 25% increase.

21. Find last year's salary if after a 4% pay raise, this year's salary is $44,200.

22. Find last year's salary if after a 3% pay raise, this year's salary is $55,620.

Objective **D** *Solve. For each exercise, a table is given for you to complete and use to write an equation that models the situation. See Example 7.*

23. How much pure acid should be mixed with 2 gallons of a 40% acid solution in order to get a 70% acid solution?

	Number of Gallons ·	Acid Strength =	Amount of Acid
Pure Acid		100%	
40% Acid Solution			
70% Acid Solution Needed			

24. How many cubic centimeters (cc) of a 25% antibiotic solution should be added to 10 cubic centimeters of a 60% antibiotic solution in order to get a 30% antibiotic solution?

	Number of Cubic cm ·	Antibiotic Strength =	Amount of Antibiotic
25% Antibiotic Solution			
60% Antibiotic Solution			
30% Antibiotic Solution Needed			

25. Community Coffee Company wants a new flavor of Cajun coffee. How many pounds of coffee worth $7 a pound should be added to 14 pounds of coffee worth $4 a pound to get a mixture worth $5 a pound?

	Number of Pounds ·	Cost per Pound =	Value
$7 per lb Coffee			
$4 per lb Coffee			
$5 per lb Coffee Wanted			

26. Planter's Peanut Company wants to mix 20 pounds of peanuts worth $3 a pound with cashews worth $5 a pound in order to make an experimental mix worth $3.50 a pound. How many pounds of cashews should be added to the peanuts?

	Number of Pounds ·	Cost per Pound =	Value
$3 per lb Peanuts			
$5 per lb Cashews			
$3.50 per lb Mixture Wanted			

Objectives Ⓐ Ⓑ Ⓒ **Mixed Practice** *Solve. If needed, round money amounts to two decimal places and all other amounts to one decimal place. See Examples 1 through 6.*

27. Find 23% of 20.

28. Find 140% of 86.

29. The number 40 is 80% of what number?

30. The number 56.25 is 45% of what number?

31. The number 144 is what percent of 480?

32. The number 42 is what percent of 35?

The graph shows the communities in the United States that have the highest percentages of citizens that shop by catalog. Use the graph to answer Exercises 33 through 36.

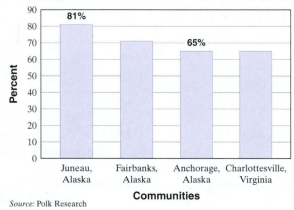

Highest Percent that Shop by Category

Source: Polk Research

33. Estimate the percent of the population in Fairbanks, Alaska, who shops by catalog?

34. Estimate the percent of the population in Charlottesville, Virginia, who shops by catalog?

35. According to the 2005 *World Almanac*, Anchorage has a population of 270,951. How many catalog shoppers might we predict live in Anchorage? Round to the nearest whole number.

36. According to the 2005 *World Almanac*, Juneau has a population of 31,187. How many catalog shoppers might we predict live in Juneau? Round to the nearest whole number.

For Exercises 37 and 38, fill in the percent column in each table. Each table contains a worked-out example.

37.

Ford Motor Company Model Year 2004 Vehicle Sales Worldwide		
	Thousands of Vehicles	**Percent of Total** (Rounded to Nearest Percent)
North America	3277	
Europe	1474	
Asia-Pacific	328	
Rest of the World	383	Example: $\frac{383}{5462} \approx 7\%$
Total	5462	

Source: Ford Motor Company

38.

Kraft Foods North America Year 2003 Volume Food Produced		
Food Group	**Volume** (in pounds)	**Percent** (Round to Nearest Percent)
Cheese, Meals, and Enhancers	6183	
Biscuits, Snacks, and Confectionaries	2083	Example: $\frac{2083}{13,741} \approx 15\%$
Beverages, Desserts, and Cereals	3905	
Oscar Mayer and Pizza	1570	
Total	13,741	

Source: Kraft Foods, North America

39. Iceberg lettuce is grown and shipped to stores for about 40 cents a head, and consumers purchase it for about 70 cents a head. Find the percent increase.

40. The lettuce consumption per capita in 1980 was about 25.6 pounds, and in 2002 the consumption dropped to 22.4 pounds. Find the percent decrease.

41. A student at the University of New Orleans makes money by buying and selling used cars. Charles bought a used car and later sold it for a 20% profit. If he sold it for $4680, how much did Charles pay for the car?

42. Smart Cards (cards with an embedded computer chip) have been growing in popularity in recent years. In 2006, 500 million Smart Cards are expected to be issued. This represents a 117% increase from the number of cards that were issued in 2001. How many Smart Cards were issued in 2001? Round to the nearest million. (*Source:* The Freedonia Group)

43. By doubling each dimension, the area of a parallelogram increased from 36 square centimeters to 144 square centimeters. Find the percent increase in area.

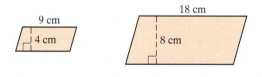

44. By doubling each dimension, the area of a triangle increased from 6 square miles to 24 square miles. Find the percent increase in area.

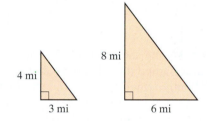

45. A gasoline station recently increased the price of one grade of gasoline by 5%. If this gasoline originally cost $2.20 per gallon, find the mark-up and the new price.

46. The price of a biology book recently increased by 10%. If this book originally cost $89.90, find the mark-up and the new price.

47. How much of an alloy that is 20% copper should be mixed with 200 ounces of an alloy that is 50% copper in order to get an alloy that is 30% copper?

48. How much water should be added to 30 gallons of a solution that is 70% antifreeze in order to get a mixture that is 60% antifreeze?

49. During the 1982–1983 term, the Supreme Court made 151 decisions while during the 2003–2004 term, they only made 73. Find the percent decrease in number of decisions. Round to the nearest tenth of a percent.

50. The number of farms in the United States is decreasing. In 1940, there were approximately 6.3 million farms, while in 2003 there were only 2.1 million. Find the percent decrease in the number of farms. Round to the nearest tenth of a percent.

51. A company recently downsized its number of employees by 35%. If there are still 78 employees, how many employees were there prior to the layoffs?

52. The average number of children born to each U.S. woman has decreased by 44% since 1920. If this average is now 1.9, find the average in 1920. Round to the nearest tenth.

53. Nordstrom advertised a 25%-off sale. If a London Fog coat originally sold for $256, find the decrease in price and the sale price.

54. A gasoline station decreased the price of a $0.95 cola by 15%. Find the decrease in price and the new price.

55. Scoville units are used to measure the hotness of a pepper. Measuring 577 thousand Scoville units, the "Red Savina" habañero pepper was known as the hottest chili pepper. That has recently changed with the discovery of Naga Jolokia pepper from India. It measures 48% hotter than the habañero. Find the measure of the Naga Jolokia pepper. Round to the nearest thousand units.

56. At this writing, the women's world record for throwing a disc (like a heavy Frisbee) was set by Jennifer Griffin of the United States in 2000. Her throw was 138.56 meters. The men's world record was set by Christian Sandstrom of Sweden in 2002. His throw was 80.4% farther than Jennifer's. Find the distance of his throw. Round to the nearest meter. (*Source:* World Flying Disc Federation)

57. A recent survey showed that 42% of recent college graduates named flexible hours as their most desired employment benefit. In a graduating class of 860 college students, how many would you expect to rank flexible hours as their top priority in job benefits? (Round to the nearest whole.) (*Source:* JobTrak.com)

58. A recent survey showed that 64% of U.S. colleges have Internet access in their classrooms. There are approximately 9800 post-secondary institutions in the United States. How many of these would you expect to have Internet access in their classrooms? (*Source:* Market Data Retrieval, National Center for Education Statistics)

59. A new self-tanning lotion for everyday use is to be sold. First, an experimental lotion mixture is made by mixing 800 ounces of everyday moisturizing lotion worth $0.30 an ounce with self-tanning lotion worth $3 per ounce. If the experimental lotion is to cost $1.20 per ounce, how many ounces of the self-tanning lotion should be in the mixture?

60. The owner of a local chocolate shop wants to develop a new trail mix. How many pounds of chocolate-covered peanuts worth $5 a pound should be mixed with 10 pounds of granola bites worth $2 a pound to get a mixture worth $3 per pound?

Review

Place $<$, $>$, or $=$ in the appropriate space to make each a true statement. See Sections 1.2, 1.3, and 1.6.

61. -5 -7

62. $\dfrac{12}{3}$ 2^2

63. $|-5|$ $-(-5)$

64. -3^3 $(-3)^3$

65. $(-3)^2$ -3^2

66. $|-2|$ $-|-2|$

Concept Extensions

67. Is it possible to mix a 10% acid solution and a 40% acid solution to obtain a 60% acid solution? Why or why not?

68. Must the percents in a circle graph have a sum of 100%? Why or why not?

Standardized nutrition labels like the one below have been displayed on food items since 1994. The percent column on the right shows the percent of daily values (based on a 2000-calorie diet) shown at the bottom of the label. For example, a serving of this food contains 4 grams of total fat, where the recommended daily fat based on a 2000-calorie diet is less than 65 grams of fat. This means that $\frac{4}{65}$ or approximately 6% (as shown) of your daily recommended fat is taken in by eating a serving of this food. Use this nutrition label to answer Exercises 69 through 71.

Nutrition Facts

Serving Size   18 Crackers (31g)
Servings Per Container   About 9

Amount Per Serving

Calories 130   Calories from Fat 35

	% Daily Value*
Total Fat 4g	**6%**
Saturated Fat 0.5g	**3%**
Polyunsaturated Fat 0g	
Monounsaturated Fat 1.5g	
Cholesterol 0mg	**0%**
Sodium 230mg	*x*
Total Carbohydrate 23g	*y*
Dietary Fiber 2g	**8%**
Sugars 3g	
Protein 2g	

Vitamin A 0%	•	Vitamin C 0%
Calcium 2%	•	Iron 6%

* Percent Daily Values are based on a 2,000 calorie diet. Your daily values may be higher or lower depending on your calorie needs.

		Calories	2,000	2,500
Total Fat	Less than		65g	80g
Sat. Fat	Less than		20g	25g
Cholesterol	Less than		300mg	300mg
Sodium	Less than		2400mg	2400mg
Total Carbohydrate			300g	375g
Dietary Fiber			25g	30g

69. Based on a 2000-calorie diet, what percent of daily value of sodium is contained in a serving of this food? In other words, find *x* in the label. (Round to the nearest tenth of a percent.)

70. Based on a 2000-calorie diet, what percent of daily value of total carbohydrate is contained in a serving of this food? In other words, find *y* in the label. (Round to the nearest tenth of a percent.)

71. Notice on the nutrition label that one serving of this food contains 130 calories and 35 of these calories are from fat. Find the percent of calories from fat. (Round to the nearest tenth of a percent.) It is recommended that no more than 30% of calorie intake come from fat. Does this food satisfy this recommendation?

Use the nutrition label below to answer Exercises 72 through 74.

NUTRITIONAL INFORMATION PER SERVING

Serving Size: 9.8 oz.   **Servings Per Container: 1**

Calories 280	Polyunsaturated Fat 1g
Protein 12g	Saturated Fat 3g
Carbohydrate 45g	Cholesterol 20mg
Fat 6g	Sodium 520mg
Percent of Calories from Fat....?	Potassium 220mg

72. If fat contains approximately 9 calories per gram, find the percent of calories from fat in one serving of this food. (Round to the nearest tenth of a percent.)

73. If protein contains approximately 4 calories per gram, find the percent of calories from protein from one serving of this food. (Round to the nearest tenth of a percent.)

74. Find a food that contains more than 30% of its calories per serving from fat. Analyze the nutrition label and verify that the percents shown are correct.

A Graph Inequalities on a Number Line.

B Use the Addition Property of Inequality to Solve Inequalities.

C Use the Multiplication Property of Inequality to Solve Inequalities.

D Use Both Properties to Solve Inequalities.

E Solve Problems Modeled by Inequalities.

2.7 SOLVING LINEAR INEQUALITIES

In Chapter 1, we reviewed these inequality symbols and their meanings:

    < means "is less than"   ≤ means "is less than or equal to"
    > means "is greater than"   ≥ means "is greater than or equal to"

An **inequality** is a statement that contains one of the symbols above.

Equations	Inequalities
$x = 3$	$x \leq 3$
$5n - 6 = 14$	$5n - 6 > 14$
$12 = 7 - 3y$	$12 \leq 7 - 3y$
$\dfrac{x}{4} - 6 = 1$	$\dfrac{x}{4} - 6 > 1$

Objective **A** Graphing Inequalities on a Number Line

Recall that the single solution to the equation $x = 3$ is 3. The solutions of the inequality $x \leq 3$ include 3 and *all real numbers less than 3* (for example, $-10, \frac{1}{2}, 2,$ and 2.9). Because we can't list all numbers less than 3, we show instead a picture of the solutions by graphing them on a number line.

To graph the solutions of $x \leq 3$, we shade the numbers to the left of 3 since they are less than 3. Then we place a closed circle on the point representing 3. The closed circle indicates that 3 *is* a solution: 3 *is* less than or equal to 3.

To graph the solutions of $x < 3$, we shade the numbers to the left of 3. Then we place an open circle on the point representing 3. The open circle indicates that 3 *is not* a solution: 3 *is not* less than 3.

PRACTICE PROBLEM 1

Graph: $x \geq -2$

PRACTICE PROBLEM 2

Graph: $5 > x$

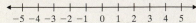

Answers

1.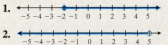

2.

154

EXAMPLE 1 Graph: $x \geq -1$

Solution: To graph the solutions of $x \geq -1$, we place a closed circle at -1 since the inequality symbol is $\geq$ and -1 is greater than or equal to -1. Then we shade to the right of -1.

◼ **Work Practice Problem 1**

EXAMPLE 2 Graph: $-1 > x$

Solution: Recall from Chapter 1 that $-1 > x$ means the same as $x < -1$. The graph of the solutions of $x < -1$ is shown below.

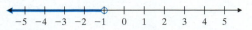

◼ **Work Practice Problem 2**

EXAMPLE 3 Graph: $-4 < x \le 2$

Solution: We read $-4 < x \le 2$ as "-4 is less than x and x is less than or equal to 2," or as "x is greater than -4 and x is less than or equal to 2." To graph the solutions of this inequality, we place an open circle at -4 (-4 is not part of the graph), a closed circle at 2 (2 is part of the graph), and we shade all numbers between -4 and 2. Why? All numbers between -4 and 2 are greater than -4 *and* also less than 2.

□ **Work Practice Problem 3**

Objective B Using the Addition Property

When solutions of a linear inequality are not immediately obvious, they are found through a process similar to the one used to solve a linear equation. Our goal is to get the variable alone on one side of the inequality. We use properties of inequality similar to properties of equality.

Addition Property of Inequality

If a, b, and c are real numbers, then

$$a < b \quad \text{and} \quad a + c < b + c$$

are equivalent inequalities.

This property also holds true for subtracting values, since subtraction is defined in terms of addition. In other words, adding or subtracting the same quantity from both sides of an inequality does not change the solutions of the inequality.

EXAMPLE 4 Solve $x + 4 \le -6$. Graph the solutions.

Solution: To solve for x, subtract 4 from both sides of the inequality.

$$x + 4 \le -6 \qquad \text{Original inequality}$$
$$x + 4 - 4 \le -6 - 4 \qquad \text{Subtract 4 from both sides.}$$
$$x \le -10 \qquad \text{Simplify.}$$

The graph of the solutions is shown below.

□ **Work Practice Problem 4**

Helpful Hint

Notice that any number less than or equal to -10 is a solution to $x \le -10$. For example, solutions include

$$-10, \quad -200, \quad -11\frac{1}{2}, \quad -\sqrt{130}, \quad \text{and} \quad -50.3$$

Objective C Using the Multiplication Property

An important difference between solving linear equations and solving linear inequalities is shown when we multiply or divide both sides of an inequality by a nonzero real number. For example, start with the true statement $6 < 8$ and multiply both sides by 2. As we see below, the resulting inequality is also true.

$$6 < 8 \qquad \text{True}$$
$$2(6) < 2(8) \qquad \text{Multiply both sides by 2.}$$
$$12 < 16 \qquad \text{True}$$

PRACTICE PROBLEM 3

Graph: $-3 \le x < 1$

PRACTICE PROBLEM 4

Solve $x - 6 \ge -11$. Graph the solutions.

Answers

3.

4. $x \ge -5$,

But if we start with the same true statement $6 < 8$ and multiply both sides by -2, the resulting inequality is not a true statement.

$$6 < 8 \qquad \text{True}$$
$$-2(6) < -2(8) \qquad \text{Multiply both sides by } -2.$$
$$-12 < -16 \qquad \text{False}$$

Notice, however, that if we reverse the direction of the inequality symbol, the resulting inequality is true.

$$-12 < -16 \qquad \text{False}$$
$$-12 > -16 \qquad \text{True}$$

This demonstrates the multiplication property of inequality.

Multiplication Property of Inequality

1. If a, b, and c are real numbers, and c is **positive,** then

$$a < b \qquad \text{and} \qquad ac < bc$$

are equivalent inequalities.

2. If a, b, and c are real numbers, and c is **negative,** then

$$a < b \quad \text{and} \quad ac > bc$$

are equivalent inequalities.

Because division is defined in terms of multiplication, this property also holds true when dividing both sides of an inequality by a nonzero number: If we multiply or divide both sides of an inequality by a negative number, **the direction of the inequality sign must be reversed for the inequalities to remain equivalent.**

✔ **Concept Check** Fill in the box with $<$, $>$, $\leq$, or $\geq$.

a. Since $-8 < -4$, then $3(-8) \; \square \; 3(-4)$.

b. Since $5 \geq -2$, then $\dfrac{5}{-7} \; \square \; \dfrac{-2}{-7}$.

c. If $a < b$, then $2a \; \square \; 2b$.

d. If $a \geq b$, then $\dfrac{a}{-3} \; \square \; \dfrac{b}{-3}$.

PRACTICE PROBLEM 5

Solve $-3x \leq 12$. Graph the solutions.

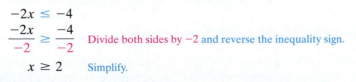

EXAMPLE 5 Solve $-2x \leq -4$. Graph the solutions.

Solution: Remember to reverse the direction of the inequality symbol when dividing by a negative number.

$$-2x \leq -4$$
$$\frac{-2x}{-2} \geq \frac{-4}{-2} \qquad \text{\color{red}Divide both sides by } -2 \text{ and reverse the inequality sign.}$$
$$x \geq 2 \qquad \text{\color{red}Simplify.}$$

The graph of the solutions is shown.

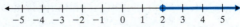

🟧 **Work Practice Problem 5**

Answer

5. $x \geq -4$

✔ **Concept Check Answer**

a. $<$, **b.** $\leq$, **c.** $<$, **d.** $\leq$

EXAMPLE 6 Solve $2x < -4$. Graph the solutions.

Solution: $2x < -4$

$$\frac{2x}{2} < \frac{-4}{2}$$ Divide both sides by 2. Do not reverse the inequality sign.

$$x < -2$$ Simplify.

The graph of the solutions is shown.

▢ Work Practice Problem 6

Since we cannot list all solutions to an inequality such as $x < -2$, we will use the set notation $\{x \mid x < -2\}$. Recall from Section 1.2 that this is read "the set of all x such that x is less than -2." We will use this notation when solving inequalities.

Objective D Using Both Properties of Inequality

The following steps may be helpful when solving inequalities in one variable. Notice that these steps are similar to the ones given in Section 2.3 for solving equations.

To Solve Linear Inequalities in One Variable

Step 1: If an inequality contains fractions, multiply both sides by the LCD to clear the inequality of fractions.

Step 2: Use the distributive property to remove parentheses if they occur.

Step 3: Simplify each side of the inequality by combining like terms.

Step 4: Get all variable terms on one side and all numbers on the other side by using the addition property of inequality.

Step 5: Get the variable alone by using the multiplication property of inequality.

Helpful Hint

Don't forget that if both sides of an inequality are multiplied or divided by a negative number, the direction of the inequality sign must be reversed.

EXAMPLE 7 Solve $-4x + 7 \geq -9$. Graph the solution set.

Solution: $-4x + 7 \geq -9$

$$-4x + 7 - 7 \geq -9 - 7$$ Subtract 7 from both sides.

$$-4x \geq -16$$ Simplify.

$$\frac{-4x}{-4} \leq \frac{-16}{-4}$$ Divide both sides by -4 and reverse the direction of the inequality sign.

$$x \leq 4$$ Simplify.

The graph of the solution set $\{x \mid x \leq 4\}$ is shown.

▢ Work Practice Problem 7

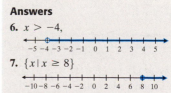

PRACTICE PROBLEM 8

Solve $2x - 3 > 4(x - 1)$.
Graph the solution set.

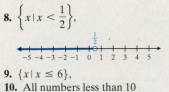

EXAMPLE 8 Solve $-5x + 7 < 2(x - 3)$. Graph the solution set.

Solution: $-5x + 7 < 2(x - 3)$

$-5x + 7 < 2x - 6$	Apply the distributive property.
$-5x + 7 - 2x < 2x - 6 - 2x$	Subtract $2x$ from both sides.
$-7x + 7 < -6$	Combine like terms.
$-7x + 7 - 7 < -6 - 7$	Subtract 7 from both sides.
$-7x < -13$	Combine like terms.
$\dfrac{-7x}{-7} > \dfrac{-13}{-7}$	Divide both sides by -7 and reverse the direction of the inequality sign.
$x > \dfrac{13}{7}$	Simplify.

The graph of the solution set $\left\{ x \,\middle|\, x > \dfrac{13}{7} \right\}$ is shown.

![number line graph with open circle at 13/7, shaded to the right, labeled -5 -4 -3 -2 -1 0 1 2 3 4 5]

🔲 **Work Practice Problem 8**

PRACTICE PROBLEM 9

Solve:
$3(x + 5) - 1 \geq 5(x - 1) + 7$

EXAMPLE 9 Solve: $2(x - 3) - 5 \leq 3(x + 2) - 18$

Solution: $2(x - 3) - 5 \leq 3(x + 2) - 18$

$2x - 6 - 5 \leq 3x + 6 - 18$	Apply the distributive property.
$2x - 11 \leq 3x - 12$	Combine like terms.
$-x - 11 \leq -12$	Subtract $3x$ from both sides.
$-x \leq -1$	Add 11 to both sides.
$\dfrac{-x}{-1} \geq \dfrac{-1}{-1}$	Divide both sides by -1 and reverse the direction of the inequality sign.
$x \geq 1$	Simplify.

The solution set is $\{ x \,|\, x \geq 1 \}$.

🔲 **Work Practice Problem 9**

Objective E Solving Problems Modeled by Inequalities

Problems containing words such as "at least," "at most," "between," "no more than," and "no less than" usually indicate that an inequality should be solved instead of an equation. In solving applications involving linear inequalities, we use the same procedure we used to solve applications involving linear equations.

Some Inequality Translations			
≥	≤	<	>
at least	at most	is less than	is greater than
no less than	no more than		

PRACTICE PROBLEM 10

Twice a number, subtracted from 35 is greater than 15. Find all numbers that make this true.

EXAMPLE 10 12 subtracted from 3 times a number is less than 21. Find all numbers that make this true.

Solution:

1. UNDERSTAND. Read and reread the problem. This is a direct translation problem, and let's let

    x = the unknown number

Answers

8. $\left\{ x \,\middle|\, x < \dfrac{1}{2} \right\}$,

![number line graph with open circle at 1/2, shaded to the left, labeled -5 -4 -3 -2 -1 0 1 2 3 4 5]

9. $\{ x \,|\, x \leq 6 \}$,

10. All numbers less than 10

2. TRANSLATE.

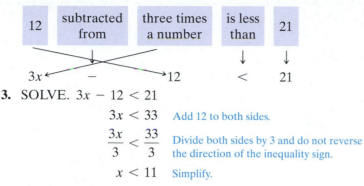

12	subtracted from	three times a number	is less than	21

$$3x \qquad - \qquad 12 \qquad < \qquad 21$$

3. SOLVE. $3x - 12 < 21$

$$3x < 33 \qquad \text{Add 12 to both sides.}$$

$$\frac{3x}{3} < \frac{33}{3} \qquad \text{Divide both sides by 3 and do not reverse the direction of the inequality sign.}$$

$$x < 11 \qquad \text{Simplify.}$$

4. INTERPRET.

Check: Check the translation; then let's choose a number less than 11 to see if it checks. For example, let's check 10. 12 subtracted from 3 times 10 is 12 subtracted from 30, or 18. Since 18 is less than 21, the number 10 checks.

State: All numbers less than 11 make the original statement true.

▢ **Work Practice Problem 10**

EXAMPLE 11 **Budgeting for a Wedding**

Marie Chase and Jonathan Edwards are having their wedding reception at the Gallery reception hall. They may spend at most $1000 for the reception. If the reception hall charges a $100 cleanup fee plus $14 per person, find the greatest number of people that they can invite and still stay within their budget.

Solution:

1. UNDERSTAND. Read and reread the problem. Suppose that 50 people attend the reception. The cost is then $100 + \$14(50) = \$100 + \$700 = \$800$.

Let x = the number of people who attend the reception.

2. TRANSLATE.

cleanup fee	+	cost per person	times	number of people	must be less than or equal to	$1000
100	+	14	·	x	≤	1000

3. SOLVE.

$$100 + 14x \leq 1000$$

$$14x \leq 900 \qquad \text{Subtract 100 from both sides.}$$

$$x \leq 64\frac{2}{7} \qquad \text{Divide both sides by 14.}$$

4. INTERPRET.

Check: Since x represents the number of people, we round down to the nearest whole, or 64. Notice that if 64 people attend, the cost is $100 + \$14(64) = \$996$. If 65 people attend, the cost is $100 + \$14(65) = \$1010$, which is more than the given $1000.

State: Marie Chase and Jonathan Edwards can invite at most 64 people to the reception.

▢ **Work Practice Problem 11**

Mental Math

Solve each inequality.

1. $5x > 10$ **2.** $4x < 20$ **3.** $2x \geq 16$ **4.** $9x \leq 63$

Decide which number listed is not a solution to each given inequality.

5. $x \geq -3; \quad -3, 0, -5, \pi$

6. $x < 6; \quad -6, |-6|, 0, -3.2$

7. $x < 4.01; \quad 4, -4.01, 4.1, -4.1$

8. $x \geq -3; \quad -4, -3, -2, -(-2)$

2.7 EXERCISE SET

Objective A *Graph each inequality on a number line. See Examples 1 and 2.*

1. $x \leq -1$

-5 -4 -3 -2 -1 0 1 2 3 4 5

2. $y < 0$

-5 -4 -3 -2 -1 0 1 2 3 4 5

3. $x > \dfrac{1}{2}$

-5 -4 -3 -2 -1 0 1 2 3 4 5

4. $z \geq -\dfrac{2}{3}$

-5 -4 -3 -2 -1 0 1 2 3 4 5

5. $y < 4$

-5 -4 -3 -2 -1 0 1 2 3 4 5

6. $x > 3$

-5 -4 -3 -2 -1 0 1 2 3 4 5

7. $-2 \leq m$

-5 -4 -3 -2 -1 0 1 2 3 4 5

8. $-5 \geq x$

-5 -4 -3 -2 -1 0 1 2 3 4 5

Graph each inequality on a number line. See Example 3.

9. $-1 < x < 3$

-5 -4 -3 -2 -1 0 1 2 3 4 5

10. $-2 \leq x \leq 3$

-5 -4 -3 -2 -1 0 1 2 3 4 5

11. $0 \leq y < 2$

-5 -4 -3 -2 -1 0 1 2 3 4 5

12. $-4 < x \leq 0$

-5 -4 -3 -2 -1 0 1 2 3 4 5

Objective B *Solve each inequality. Graph the solution set. See Example 4.*

13. $x - 2 \geq -7$

-5 -4 -3 -2 -1 0 1 2 3 4 5

14. $x + 4 \leq 1$

-5 -4 -3 -2 -1 0 1 2 3 4 5

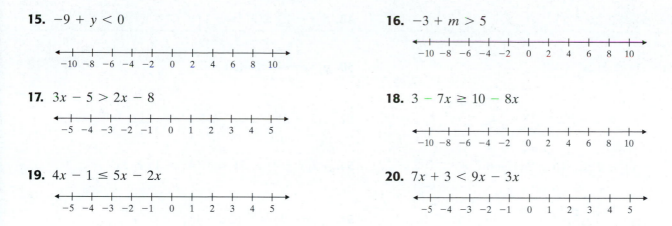

15. $-9 + y < 0$

16. $-3 + m > 5$

17. $3x - 5 > 2x - 8$

18. $3 - 7x \geq 10 - 8x$

19. $4x - 1 \leq 5x - 2x$

20. $7x + 3 < 9x - 3x$

Objective **C** *Solve each inequality. Graph the solution set. See Examples 5 and 6.*

21. $2x < -6$

22. $3x > -9$

23. $-8x \leq 16$

24. $-5x < 20$

25. $-x > 0$

26. $-y \geq 0$

27. $\dfrac{3}{4}y \geq -2$

28. $\dfrac{5}{6}x \leq -8$

29. $-0.6y < -1.8$

30. $-0.3x > -2.4$

Objectives **B** **C** **D** **Mixed Practice** *Solve each inequality. See Examples 4 through 9.*

31. $-8 < x + 7$

32. $-11 > x + 4$

33. $7(x + 1) - 6x \geq -4$

34. $10(x + 2) - 9x \leq -1$

35. $4x > 1$

36. $6x < 5$

37. $-\dfrac{2}{3}y \leq 8$

38. $-\dfrac{3}{4}y \geq 9$

39. $4(2z + 1) < 4$

40. $6(2 - z) \geq 12$

41. $3x - 7 < 6x + 2$

42. $2x - 1 \geq 4x - 5$

43. $5x - 7x \leq x + 2$

44. $4 - x < 8x + 2x$

45. $-6x + 2 \geq 2(5 - x)$

46. $-7x + 4 > 3(4 - x)$

47. $3(x - 5) < 2(2x - 1)$

48. $5(x - 2) \le 3(2x - 1)$

49. $4(3x - 1) \le 5(2x - 4)$

50. $3(5x - 4) \le 4(3x - 2)$

51. $3(x + 2) - 6 > -2(x - 3) + 14$

52. $7(x - 2) + x \le -4(5 - x) - 12$

53. $-5(1 - x) + x \le -(6 - 2x) + 6$

54. $-2(x - 4) - 3x < -(4x + 1) + 2x$

55. $\dfrac{1}{4}(x + 4) < \dfrac{1}{5}(2x + 3)$

56. $\dfrac{1}{2}(x - 5) < \dfrac{1}{3}(2x - 1)$

57. $-5x + 4 \le -4(x - 1)$

58. $-6x + 2 < -3(x + 4)$

Objective **E** *Solve the following. See Examples 10 and 11.*

59. Six more than twice a number is greater than negative fourteen. Find all numbers that make this statement true.

60. One more than five times a number is less than or equal to ten. Find all such numbers.

61. The perimeter of a rectangle is to be no greater than 100 centimeters and the width must be 15 centimeters. Find the maximum length of the rectangle.

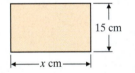

62. One side of a triangle is four times as long as another side, and the third side is 12 inches long. If the perimeter can be no longer than 87 inches, find the maximum lengths of the other two sides.

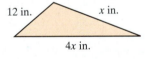

63. Ben Holladay bowled 146 and 201 in his first two games. What must he bowl in his third game to have an average of at least 180? (*Hint:* The average of a list of numbers is their sum divided by the number of numbers in the list.)

64. On an NBA team the two forwards measure 6'8" and 6'6" tall and the two guards measure 6'0" and 5'9" tall. How tall a center should they hire if they wish to have a starting team average height of at least 6'5"?

65. Dennis and Nancy Wood are celebrating their 30th wedding anniversary by having a reception at Tiffany Oaks reception hall. They have budgeted $3000 for their reception. If the reception hall charges a $50.00 cleanup fee plus $34 per person, find the greatest number of people that they may invite and still stay within their budget.

66. A surprise retirement party is being planned for Pratap Puri. A total of $860 has been collected for the event, which is to be held at a local reception hall. This reception hall charges a cleanup fee of $40 and $15 per person for drinks and light snacks. Find the greatest number of people that may be invited and still stay within the $860 budget.

67. A 150-pound person uses 5.8 calories per minute when walking at a speed of 4 mph. How long must a person walk at this speed to use at least 200 calories? (Round up to the nearest minute.) (*Source:* Home & Garden Bulletin No. 72)

68. A 170-pound person uses 5.3 calories per minute when bicycling at a speed of 5.5 mph. How long must a person ride a bike at this speed in order to use at least 200 calories? (Round up to the nearest minute.) (*Source:* Same as Exercise 67.)

Review

Evaluate each expression. See Section 1.3.

69. 3^4 **70.** 4^3 **71.** 1^8 **72.** 0^7 **73.** $\left(\dfrac{7}{8}\right)^2$ **74.** $\left(\dfrac{2}{3}\right)^3$

The graph shows the number of Krispy Kreme Doughnut locations from 1996 to 2004. The height of the graph for each year shown corresponds to the number of Krispy Kreme locations. Use this graph to answer Exercises 75 through 80.

Krispy Kreme Doughnut Locations

75. How many Krispy Kreme locations were there in 1998?

76. How many Krispy Kreme locations were there in 2003?

77. Between which two years did the greatest increase in the number of Krispy Kreme locations occur?

78. In what year were there approximately 150 Krispy Kreme locations?

79. During which year did the number of Krispy Kreme locations rise above 200?

80. During which year did the number of Krispy Kreme locations rise above 300?

Concept Extensions

Fill in the box with $<$, $>$, $\leq$, or $\geq$. See the Concept Check in this Section.

81. Since $3 < 5$, then $3(-4) \ \square \ 5(-4)$.

82. If $m \leq n$, then $2m \ \square \ 2n$.

83. If $m \leq n$, then $-2m \ \square \ -2n$.

84. If $-x < y$, then $x \ \square \ -y$.

85. When solving an inequality, when must you reverse the direction of an inequality symbol?

86. If both sides of the inequality $-3x < -30$ are divided by 3, do you reverse the direction of the inequality symbol? Why or why not?

Solve.

87. Eric Daly has scores of 75, 83, and 85 on his history tests. Use an inequality to find the scores he can make on his final exam to receive a B in the class. The final exam counts as **two** tests, and a B is received if the final course average is greater than or equal to 80.

88. Maria Lipco has scores of 85, 95, and 92 on her algebra tests. Use an inequality to find the scores she can make on her final exam to receive an A in the course. The final exam counts as three tests, and an A is received if the final course average is greater than or equal to 90. Round to one decimal place.

THE BIGGER PICTURE Simplifying Expressions and Solving Equations

Now we continue our outline started in Section 1.6. Although suggestions are given, this outline should be in your own words. Once you complete this new portion, try the exercises below.

I. Simplifying Expressions

  A. Real Numbers

    1. Add (Section 1.4)

    2. Subtract (Section 1.5)

    3. Multiply or Divide (Section 1.6)

II. Solving Equations and Inequalities

  A. Linear Equations: power on variable is 1 and there are no variables in the denominator

$$7(x - 3) = 4x + 6 \quad \text{Linear equation. Simplify both sides, then get variable terms on one side, numbers on the other side.}$$

$$7x - 21 = 4x + 6 \quad \text{Use the distributive property.}$$

$$7x = 4x + 27 \quad \text{Add 21 to both sides.}$$

$$3x = 27 \quad \text{Subtract } 4x \text{ from both sides.}$$

$$x = 9 \quad \text{Divide both sides by 3.}$$

  B. Linear Inequalities: same as linear equation, except there are inequality symbols, $\leq, <, \geq, >$ Remember, if you multiply or divide by a negative number,

then reverse the direction of the inequality symbol.

$$-4x - 11 \leq 1 \quad \text{Linear inequality.}$$

$$-4x \leq 12 \quad \text{Add 11 to both sides.}$$

$$\frac{-4x}{-4} \geq \frac{12}{-4} \quad \text{Divide both sides by } -4 \text{ and reverse the direction of the inequality symbol.}$$

$$x \geq -3 \quad \text{Simplify.}$$

Solve each equation or inequality.

1. $-5x = 15$

2. $-5x > 15$

3. $9y - 14 = -12$

4. $9x - 3 = 5x - 4$

5. $4(x - 2) \leq 5x + 7$

6. $5(4x - 1) = 2(10x - 1)$

7. $-5.4 = 0.6x - 9.6$

8. $\frac{1}{3}(x - 4) < \frac{1}{4}(x + 7)$

9. $3y - 5(y - 4) = -2(y - 10)$

10. $\frac{7(x - 1)}{3} = \frac{2(x + 1)}{5}$

CHAPTER 2   Group Activity

Investigating Averages

Sections 2.1–2.7

Materials:

- small rubber ball or crumpled paper ball
- bucket or waste can

This activity may be completed by working in groups or individually.

1. Try shooting the ball into the bucket or waste can 5 times. Record your results below.

    Shots Made   **Shots Missed**

2. Find your shooting percent for the 5 shots (that is, the percent of the shots you actually made out of the number you tried).

3. Suppose you are going to try an additional 5 shots. How many of the next 5 shots will you have to make to have

a 50% shooting percent for all 10 shots? An 80% shooting percent?

4. Did you solve an equation in Question 3? If so, explain what you did. If not, explain how you could use an equation to find the answers.

5. Now suppose you are going to try an additional 22 shots. How many of the next 22 shots will you have to make to have at least a 50% shooting percent for all 27 shots? At least a 70% shooting percent?

6. Choose one of the sports played at your college that is currently in season. How many regular-season games are scheduled? What is the team's current percent of games won?

7. Suppose the team has a goal of finishing the season with a winning percent better than 110% of their current wins. At least how many of the remaining games must they win to achieve their goal?

Chapter 2 Vocabulary Check

Fill in each blank with one of the words or phrases listed below.

no solution	all real numbers	linear equation in one variable
equivalent equations	formula	reversed
linear inequality in one variable	the same	

1. A _____ can be written in the form $ax + b = c$.
2. Equations that have the same solution are called _____.
3. An equation that describes a known relationship among quantities is called a _____.
4. A _____ can be written in the form $ax + b < c$, (or $>$, $\leq$, $\geq$).
5. The solution(s) to the equation $x + 5 = x + 5$ is/are _____.
6. The solution(s) to the equation $x + 5 = x + 4$ is/are _____.
7. If both sides of an inequality are multiplied or divided by the same positive number, the direction of the inequality symbol is _____.
8. If both sides of an inequality are multiplied by the same negative number, the direction of the inequality symbol is _____.

Helpful Hint

Are you preparing for your test? Don't forget to take the Chapter 2 Test on page 173. Then check your answers at the back of the text and use the Chapter Test Prep Video CD to see the fully worked-out solutions to any of the exercises you want to review.

2 Chapter Highlights

DEFINITIONS AND CONCEPTS	EXAMPLES
Section 2.1 The Addition Property of Equality	
A **linear equation in one variable** can be written in the form $Ax + B = C$ where $A, B,$ and C are real numbers and $A \neq 0$. **Equivalent equations** are equations that have the same solution.	$-3x + 7 = 2$ $3(x - 1) = -8(x + 5) + 4$ $x - 7 = 10$ and $x = 17$ are equivalent equations.
ADDITION PROPERTY OF EQUALITY Adding the same number to or subtracting the same number from both sides of an equation does not change its solution.	$y + 9 = 3$ $y + 9 - 9 = 3 - 9$ $y = -6$
Section 2.2 The Multiplication Property of Equality	
MULTIPLICATION PROPERTY OF EQUALITY Multiplying both sides or dividing both sides of an equation by the same nonzero number does not change its solution.	$\dfrac{2}{3}a = 18$ $\dfrac{3}{2}\left(\dfrac{2}{3}a\right) = \dfrac{3}{2}(18)$ $a = 27$

DEFINITIONS AND CONCEPTS	EXAMPLES

Section 2.3   Further Solving Linear Equations

To Solve Linear Equations

$$Solve: \quad \frac{5(-2x + 9)}{6} + 3 = \frac{1}{2}$$

1. Clear the equation of fractions.

1. $6 \cdot \dfrac{5(-2x + 9)}{6} + 6 \cdot 3 = 6 \cdot \dfrac{1}{2}$

2. Remove any grouping symbols such as parentheses.

2. $5(-2x + 9) + 18 = 3$ Apply the distributive property.
$-10x + 45 + 18 = 3$

3. Simplify each side by combining like terms.

3. $-10x + 63 = 3$ Combine like terms.

4. Get all variable terms on one side and all numbers on the other side by using the addition property of equality.

4. $-10x + 63 - 63 = 3 - 63$ Subtract 63.
$-10x = -60$

5. Get the variable alone by using the multiplication property of equality.

5. $\dfrac{-10x}{-10} = \dfrac{-60}{-10}$ Divide by -10.
$x = 6$

6. Check the solution by substituting it into the original equation.

Section 2.4   An Introduction to Problem Solving

Problem-Solving Steps

The height of the Hudson volcano in Chile is twice the height of the Kiska volcano in the Aleutian Islands. If the sum of their heights is 12,870 feet, find the height of each.

1. UNDERSTAND the problem.

1. Read and reread the problem. Guess a solution and check your guess.
Let x be the height of the Kiska volcano. Then $2x$ is the height of the Hudson volcano.

x $2x$

2. TRANSLATE the problem.

2.

height of Kiska	added to	height of Hudson	is	12,870
↓	↓	↓	↓	↓
x	$+$	$2x$	$=$	12,870

3. SOLVE the equation.

3. $x + 2x = 12{,}870$
$3x = 12{,}870$
$x = 4290$

4. INTERPRET the results.

4. *Check:* If x is 4290, then $2x$ is 2(4290) or 8580. Their sum is $4290 + 8580$ or 12,870, the required amount.

State: The Kiska volcano is 4290 feet high, and the Hudson volcano is 8580 feet high.

DEFINITIONS AND CONCEPTS	**EXAMPLES**

Section 2.5   Formulas and Problem Solving

An equation that describes a known relationship among quantities is called a **formula.**	$A = lw$ (area of a rectangle) $I = PRT$ (simple interest)
To solve a formula for a specified variable, use the same steps as for solving a linear equation. Treat the specified variable as the only variable of the equation.	*Solve:* $P = 2l + 2w$ for l. $P = 2l + 2w$ $P - 2w = 2l + 2w - 2w$ Subtract $2w$. $P - 2w = 2l$ $\dfrac{P - 2w}{2} = \dfrac{2l}{2}$ Divide by 2. $\dfrac{P - 2w}{2} = l$

Section 2.6   Percent and Mixture Problem Solving

Use the same problem-solving steps to solve a problem containing percents.

32% of what number is 36.8?

1.   UNDERSTAND.

1.   Read and reread. Propose a solution and check. Let x = the unknown number.

2.   TRANSLATE.

2.

32%	of	what number	is	36.8
↓	↓	↓	↓	↓
32%	·	x	=	36.8

3.   SOLVE.

3.   *Solve:* $32\% \cdot x = 36.8$

$$0.32x = 36.8$$
$$\frac{0.32x}{0.32} = \frac{36.8}{0.32} \quad \text{Divide by 0.32.}$$
$$x = 115 \quad \text{Simplify.}$$

4.   INTERPRET.

4.   *Check, then state:* 32% of 115 is 36.8.

How many liters of a 20% acid solution must be mixed with a 50% acid solution in order to obtain 12 liters of a 30% solution?

1.   UNDERSTAND.

1.   Read and reread. Guess a solution and check. Let x = number of liters of 20% solution. Then $12 - x$ = number of liters of 50% solution.

2.   TRANSLATE.

2.

	No. of Liters ·	Acid Strength =	Amount of Acid
20% Solution	x	20%	$0.20x$
50% Solution	$12 - x$	50%	$0.50(12 - x)$
30% Solution Needed	12	30%	$0.30(12)$

	acid in 20% solution		acid in 50% solution		acid in 30% solution
In words:	acid in 20% solution	+	acid in 50% solution	=	acid in 30% solution
Translate:	$0.20x$	+	$0.50(12 - x)$	=	$0.30(12)$

continued

DEFINITIONS AND CONCEPTS	**EXAMPLES**
Section 2.6 Percent and Mixture Problem Solving (*continued*)	

3. SOLVE.	**3.** *Solve:* $0.20x + 0.50(12 - x) = 0.30(12)$
	$0.20x + 6 - 0.50x = 3.6$ Apply the distributive property.
	$-0.30x + 6 = 3.6$ Combine like terms.
	$-0.30x = -2.4$ Subtract 6.
	$x = 8$ Divide by -0.30.
4. INTERPRET.	**4.** *Check, then state:* If 8 liters of a 20% acid solution are mixed with $12 - 8$ or 4 liters of a 50% acid solution, the result is 12 liters of a 30% solution.

Section 2.7 Solving Linear Inequalities	

Properties of inequalities are similar to properties of equations. However, if you multiply or divide both sides of an inequality by the same *negative* number, you must reverse the direction of the inequality symbol.	$-2x \le 4$ $\dfrac{-2x}{-2} \ge \dfrac{4}{-2}$ Divide by -2; reverse the inequality symbol. $x \ge -2$ $\begin{array}{c} \leftarrow\!\!+\!\!\!+\!\!\!+\!\!\!+\!\!\!+\!\!\!\bullet\!\!\!+\!\!\!+\!\!\!+\!\!\!+\!\!\!+\!\!\!+\!\!\!\rightarrow \\ {\scriptstyle -5\ -4\ -3\ -2\ -1\ \ 0\ \ 1\ \ 2\ \ 3\ \ 4\ \ 5} \end{array}$
TO SOLVE LINEAR INEQUALITIES	*Solve:* $3(x + 2) \le -2 + 8$
1. Clear the inequality of fractions.	**1.** $3(x + 2) \le -2 + 8$ No fractions to clear.
2. Remove grouping symbols.	**2.** $3x + 6 \le -2 + 8$ Apply the distributive property.
3. Simplify each side by combining like terms.	**3.** $3x + 6 \le 6$ Combine like terms.
4. Write all variable terms on one side and all numbers on the other side using the addition property of inequality.	**4.** $3x + 6 - 6 \le 6 - 6$ Subtract 6. $\qquad\ 3x \le 0$
5. Get the variable alone by using the multiplication property of inequality.	**5.** $\dfrac{3x}{3} \le \dfrac{0}{3}$ Divide by 3. $x \le 0$ The solution set is $\{x \mid x \le 0\}$. $\begin{array}{c} \leftarrow\!\!+\!\!\!+\!\!\!+\!\!\!+\!\!\!+\!\!\!\bullet\!\!\!+\!\!\!+\!\!\!+\!\!\!+\!\!\!+\!\!\!+\!\!\!\rightarrow \\ {\scriptstyle -5\ -4\ -3\ -2\ -1\ \ 0\ \ 1\ \ 2\ \ 3\ \ 4\ \ 5} \end{array}$

2 CHAPTER REVIEW

(2.1) *Solve each equation.*

1. $8x + 4 = 9x$

2. $5y - 3 = 6y$

3. $\dfrac{2}{7}x + \dfrac{5}{7}x = 6$

4. $3x - 5 = 4x + 1$

5. $2x - 6 = x - 6$

6. $4(x + 3) = 3(1 + x)$

7. $6(3 + n) = 5(n - 1)$

8. $5(2 + x) - 3(3x + 2) = -5(x - 6) + 2$

Choose the correct algebraic expression.

9. The sum of two numbers is 10. If one number is x, express the other number in terms of x.

a. $x - 10$

b. $10 - x$

c. $10 + x$

d. $10x$

10. Mandy is 5 inches taller than Melissa. If x inches represents the height of Mandy, express Melissa's height in terms of x.

a. $x - 5$

b. $5 - x$

c. $5 + x$

d. $5x$

△ **11.** If one angle measures $x°$, express the measure of its complement in terms of x.

a. $(180 - x)°$

b. $(90 - x)°$

c. $(x - 180)°$

d. $(x - 90)°$

△ **12.** If one angle measures $(x + 5)°$, express the measure of its supplement in terms of x.

a. $(185 + x)°$

b. $(95 + x)°$

c. $(175 - x)°$

d. $(x - 170)°$

(2.2) *Solve each equation.*

13. $\dfrac{3}{4}x = -9$

14. $\dfrac{x}{6} = \dfrac{2}{3}$

15. $-5x = 0$

16. $-y = 7$

17. $0.2x = 0.15$

18. $\dfrac{-x}{3} = 1$

19. $-3x + 1 = 19$

20. $5x + 25 = 20$

21. $7(x - 1) + 9 = 5x$

22. $7x - 6 = 5x - 3$

23. $-5x + \dfrac{3}{7} = \dfrac{10}{7}$

24. $5x + x = 9 + 4x - 1 + 6$

25. Write the sum of three consecutive integers as an expression in x. Let x be the first integer.

26. Write the sum of the first and fourth of four consecutive even integers. Let x be the first even integer.

(2.3) *Solve each equation.*

27. $\dfrac{5}{3}x + 4 = \dfrac{2}{3}x$

28. $\dfrac{7}{8}x + 1 = \dfrac{5}{8}x$

29. $-(5x + 1) = -7x + 3$

30. $-4(2x + 1) = -5x + 5$

31. $-6(2x - 5) = -3(9 + 4x)$

32. $3(8y - 1) = 6(5 + 4y)$

33. $\dfrac{3(2 - z)}{5} = z$

34. $\dfrac{4(n + 2)}{5} = -n$

35. $0.5(2n - 3) - 0.1 = 0.4(6 + 2n)$

36. $-9 - 5a = 3(6a - 1)$

37. $\dfrac{5(c + 1)}{6} = 2c - 3$

38. $\dfrac{2(8 - a)}{3} = 4 - 4a$

▦ **39.** $200(70x - 3560) = -179(150x - 19,300)$

40. $1.72y - 0.04y = 0.42$

(2.4) *Solve each of the following.*

41. The height of the Washington Monument is 50.5 inches more than 10 times the length of a side of its square base. If the sum of these two dimensions is 7327 inches, find the height of the Washington Monument. (*Source:* National Park Service)

42. A 12-foot board is to be divided into two pieces so that one piece is twice as long as the other. If *x* represents the length of the shorter piece, find the length of each piece.

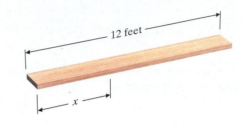

43. In a recent year, Kellogg Company acquired Keebler Foods Company. After the merger, the total number of Kellogg and Keebler manufacturing plants was 53. The number of Kellogg plants was one less than twice the number of Keebler plants. How many of each type of plant were there? (*Source: Kellogg Company 2000 Annual Report*)

44. Find three consecutive integers whose sum is −114.

45. The quotient of a number and 3 is the same as the difference of the number and two. Find the number.

46. Double the sum of a number and 6 is the opposite of the number. Find the number.

(2.5) *Substitute the given values into the given formulas and solve for the unknown variable.*

47. $P = 2l + 2w$; $P = 46, l = 14$

48. $V = lwh$; $V = 192, l = 8, w = 6$

Solve each equation for the indicated variable.

49. $y = mx + b$ for m

50. $r = vst - 5$ for s

51. $2y - 5x = 7$ for x

52. $3x - 6y = -2$ for y

△ **53.** $C = \pi D$ for π

△ **54.** $C = 2\pi r$ for π

△ **55.** A swimming pool holds 900 cubic meters of water. If its length is 20 meters and its height is 3 meters, find its width.

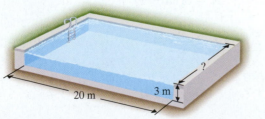

56. The perimeter of a rectangular billboard is 60 feet and has a length 6 feet longer than its width. Find the dimensions of the billboard.

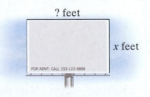

57. A charity 10K race is given annually to benefit a local hospice organization. How long will it take to run/walk a 10K race (10 kilometers or 10,000 meters) if your average pace is 125 **meters** per minute? Give your time in hours and minutes.

58. On April 28, 2001, the highest temperature recorded in the United States was 104°F, which occurred in Death Valley, California. Convert this temperature to degrees Celsius. (*Source:* National Weather Service)

(2.6) *Find each of the following.*

59. The number 9 is what percent of 45?

60. The number 59.5 is what percent of 85?

61. The number 137.5 is 125% of what number?

62. The number 768 is 60% of what number?

63. The price of a small diamond ring was recently increased by 11%. If the ring originally cost $1900, find the mark-up and the new price of the ring.

64. A recent survey found that 66.9% of Americans use the Internet. If a city has a population of 76,000 how many people in that city would you expect to use the Internet? (*Source:* UCLA Center for Communication Policy)

65. Thirty gallons of a 20% acid solution is needed for an experiment. Only 40% and 10% acid solutions are available. How much of each should be mixed to form the needed solution?

66. The ACT Assessment is a college entrance exam taken by about 60% of college-bound students. The national average score was 20.7 in 1993 and rose to 21.0 in 2001. Find the percent increase. (Round to the nearest hundredth of a percent.)

The graph below shows the percent(s) of cell phone users who have engaged in various behaviors while driving and talking on their cell phones. Use this graph to answer Exercises 67 through 70.

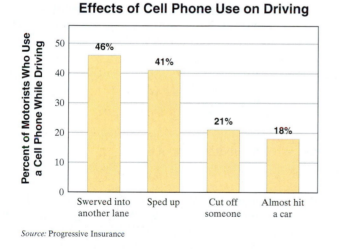

Effects of Cell Phone Use on Driving

Source: Progressive Insurance

67. What percent of motorists who use a cell phone while driving have almost hit another car?

68. What is the most common effect of cell phone use on driving?

69. If a cell-phone service has an estimated 4600 customers who use their cell phones while driving, how many of these customers would you expect to have cut someone off while driving and talking on their cell phones?

70. Do the percents in the graph to the left have a sum of 100%? Why or why not?

(2.7) *Graph on a number line.*

71. $x \leq -2$

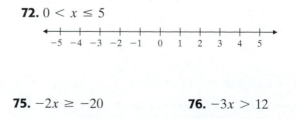

72. $0 < x \leq 5$

Solve each inequality.

73. $x - 5 \leq -4$

74. $x + 7 > 2$

75. $-2x \geq -20$

76. $-3x > 12$

77. $5x - 7 > 8x + 5$ **78.** $x + 4 \geq 6x - 16$ **79.** $\frac{2}{3}y > 6$

80. $-0.5y \leq 7.5$ **81.** $-2(x - 5) > 2(3x - 2)$ **82.** $4(2x - 5) \leq 5x - 1$

83. Carol Abolafia earns $175 per week plus a 5% commission on all her sales. Find the minimum amount of sales she must make to ensure that she earns at least $300 per week.

84. Joseph Barrow shot rounds of 76, 82, and 79 golfing. What must he shoot on his next round so that his average will be below 80?

Mixed Review

Solve each equation.

85. $6x + 2x - 1 = 5x + 11$ **86.** $2(3y - 4) = 6 + 7y$ **87.** $4(3 - a) - (6a + 9) = -12a$

88. $\frac{x}{3} - 2 = 5$ **89.** $2(y + 5) = 2y + 10$ **90.** $7x - 3x + 2 = 2(2x - 1)$

Solve.

91. The sum of six and twice a number is equal to seven less than the number. Find the number.

92. A 23-inch piece of string is to be cut into two pieces so that the length of the longer piece is three more than four times the shorter piece. If x represents the length of the shorter piece, find the lengths of both pieces.

Solve for the specified variable.

93. $V = \frac{1}{3}Ah$ for h

94. What number is 26% of 85?

95. The number 72 is 45% of what number?

96. A company recently increased their number of employees from 235 to 282. Find the percent increase.

Solve each inequality. Graph the solution set.

97. $4x - 7 > 3x + 2$

-10 -8 -6 -4 -2 0 2 4 6 8 10

98. $-5x < 20$

-5 -4 -3 -2 -1 0 1 2 3 4 5

99. $-3(1 + 2x) + x \geq -(3 - x)$

-5 -4 -3 -2 -1 0 1 2 3 4 5

2 CHAPTER TEST

Remember to use the Chapter Test Prep Video CD to see the fully worked-out solutions to any of the exercises you want to review.

Solve each equation.

1. $-\dfrac{4}{5}x = 4$

2. $4(n - 5) = -(4 - 2n)$

3. $5y - 7 + y = -(y + 3y)$

4. $4z + 1 - z = 1 + z$

5. $\dfrac{2(x + 6)}{3} = x - 5$

6. $\dfrac{4(y - 1)}{5} = 2y + 3$

7. $\dfrac{1}{2} - x + \dfrac{3}{2} = x - 4$

8. $\dfrac{1}{3}(y + 3) = 4y$

9. $-0.3(x - 4) + x = 0.5(3 - x)$

10. $-4(a + 1) - 3a = -7(2a - 3)$

11. $-2(x - 3) = x + 5 - 3x$

Solve each application.

12. A number increased by two-thirds of the number is 35. Find the number.

△ **13.** A gallon of water seal covers 200 square feet. How many gallons are needed to paint two coats of water seal on a deck that measures 20 feet by 35 feet?

20 feet   35 feet

14. Find the value of x if $y = -14$, $m = -2$, and $b = -2$ in the formula $y = mx + b$.

Solve each equation for the indicated variable.

15. $V = \pi r^2 h$ for h

16. $3x - 4y = 10$ for y

Answers

1. _____

2. _____

3. _____

4. _____

5. _____

6. _____

7. _____

8. _____

9. _____

10. _____

11. _____

12. _____

13. _____

14. _____

15. _____

16. _____

17. _____

18. _____

19. _____

20. _____

21. _____

22. _____

23. _____

24. _____

25. _____

Solve each inequality. Graph the solution set.

17. $3x - 5 > 7x + 3$

18. $x + 6 > 4x - 6$

Solve each inequality.

19. $-0.3x \geq 2.4$

20. $-5(x - 1) + 6 \leq -3(x + 4) + 1$

21. $\dfrac{2(5x + 1)}{3} > 2$

The following graph shows the breakdown of tornadoes occurring in the United States by strength. The corresponding Fujita Tornado Scale categories are shown in parentheses. Use this graph to answer Exercises 22 and 23.

Violent tornadoes (F4–F5) 2%

Strong tornadoes (F2–F3) 29%

Weak tornadoes (F0–F1) 69%

Source: National Climatic Data Center

22. What percent of tornadoes occurring in the United States are classified as "strong," that is, F2 or F3 on the Fujita Scale?

23. According to the National Climatic Data Center, in an average year, about 800 tornadoes are reported in the United States. How many of these would you expect to be classified as "weak" tornadoes?

24. The number 72 is what percent of 180?

25. New York State has more public libraries than any other state. It has 650 more public libraries than Indiana does. If the total number of public libraries for these states is 1504, find the number of public libraries in New York and the number in Indiana. (*Source: The World Almanac and Book of Facts,* 2001)

Chapters 1–2

Determine whether each statement is true or false.

1. $8 \geq 8$

2. $-4 < -6$

3. $8 \leq 8$

4. $3 > -3$

5. $23 \leq 0$

6. $-8 \geq -8$

7. $0 \leq 23$

8. $-8 \leq -8$

9. Insert $<$, $>$, or $=$ in the appropriate space to make each statement true.

 a. $|0|$ 2
 b. $|-5|$ 5
 c. $|-3|$ $|-2|$
 d. $|-9|$ $|-9.7|$
 e. $\left|-7\frac{1}{6}\right|$ $|7|$

10. Find the absolute value of each number.

 a. $|5|$
 b. $|-8|$
 c. $\left|-\frac{2}{3}\right|$

11. Simplify the expression

$$\frac{3 + |4 - 3| + 2^2}{6 - 3}.$$

12. $1 + 2(9 - 7)^3 + 4^2$

Add without using number lines.

13. $(-8) + (-11)$

14. $-2 + (-8)$

15. $(-2) + 10$

16. $-10 + 20$

17. $0.2 + (-0.5)$

18. $1.2 + (-1.2)$

Answers

1. _____

2. _____

3. _____

4. _____

5. _____

6. _____

7. _____

8. _____

9. a. _____

 b. _____

 c. _____

 d. _____

 e. _____

10. a. _____

 b. _____

 c. _____

11. _____

12. _____

13. _____

14. _____

15. _____

16. _____

17. _____

18. _____

175

19. a. _____

b. _____

20. a. _____

b. _____

c. _____

d. _____

21. a. _____

b. _____

c. _____

22. a. _____

b. _____

c. _____

23. a. _____

b. _____

c. _____

24. a. _____

b. _____

25. _____

26. _____

27. _____

28. _____

29. a. _____

b. _____

c. _____

d. _____

e. _____

19. Simplify each expression.

   a. $-3 + [(-2 - 5) - 2]$

   b. $2^3 - 10 + [-6 - (-5)]$

20. Simplify each expression.

   a. $-(-5)$ **c.** $-(-a)$

   b. $-\left(-\dfrac{2}{3}\right)$ **d.** $-|-3|$

21. Use order of operations and simplify each expression.

   a. $7(0)(-6)$

   b. $(-2)(-3)(-4)$

   c. $(-1)(5)(-9)$

22. Subtract

   a. $-2.7 - 8.4$

   b. $-\dfrac{4}{5} - \left(-\dfrac{3}{5}\right)$

   c. $\dfrac{1}{4} - \left(-\dfrac{1}{2}\right)$

23. Use the definition of the quotient of two numbers to find each quotient.

   a. $-18 \div 3$

   b. $\dfrac{-14}{-2}$

   c. $\dfrac{20}{-4}$

24. Find each product.

   a. $(4.5)(-0.08)$

   b. $-\dfrac{3}{4} \cdot -\dfrac{8}{17}$

Use the distributive property to write each expression without parentheses. Then simplify the result.

25. $-5(-3 + 2z)$

26. $2x(x^2 - 3x + 4)$

27. $\dfrac{1}{2}(6x + 14) + 10$

28. $-(x + 4) + 3(x + 4)$

29. Tell whether the terms are like or unlike.

   a. $2x, 3x^2$

   b. $4x^2y, x^2y, -2x^2y$

   c. $-2yz, -3zy$

   d. $-x^4, x^4$

   e. $-8a^5, 8a^5$

30. Find each quotient.

a. $\dfrac{-32}{8}$ **b.** $\dfrac{-108}{-12}$

c. $-\dfrac{5}{7} \div \left(-\dfrac{9}{2}\right)$

31. Subtract $4x - 2$ from $2x - 3$.

32. Subtract $10x + 3$ from $-5x + 1$.

33. Solve: $x - 7 = 10$

Solve.

34. $\dfrac{5}{6} + x = \dfrac{2}{3}$

35. Solve: $-z - 4 = 6$

36. $-3x + 1 - (-4x - 6) = 10$

37. Solve: $\dfrac{2(a + 3)}{3} = 6a + 2$

38. $\dfrac{x}{4} = 18$

39. In a recent year, the U.S. House of Representatives had a total of 431 Democrats and Republicans. There were 15 more Republican representatives than Democratic. Find the number of representatives from each party. (*Source:* Office of the Clerk of the U.S. House of Representatives)

40. $6x + 5 = 4(x + 4) - 1$

41. A glacier is a giant mass of rocks and ice that flows downhill like a river. Portage Glacier in Alaska is about 6 miles, or 31,680 feet, long and moves 400 feet per year. Icebergs are created when the front end of the glacier flows into Portage Lake. How long does it take for ice at the head (beginning) of the glacier to reach the lake?

42. A number increased by 4 is the same as 3 times the number decreased by 8. Find the number.

43. The number 63 is what percent of 72?

44. Solve $C = 2\pi r$ for r.

45. Solve: $5(2x + 3) = -1 + 7$

46. Solve: $x - 3 > 2$

47. Graph $-1 > x$.

48. Solve: $3x - 4 \le 2x - 14$

49. Solve: $2(x - 3) - 5 \le 3(x + 2) - 18$

50. Solve: $-3x \ge 9$

30. a. _____ b. _____ c. _____
31. _____
32. _____
33. _____
34. _____
35. _____
36. _____
37. _____
38. _____
39. _____
40. _____
41. _____
42. _____
43. _____
44. _____
45. _____
46. _____
47. see graph
48. _____
49. _____
50. _____

3

Exponents and Polynomials

Recall from Chapter 1 that an exponent is a shorthand notation for repeated factors. This chapter explores additional concepts about exponents and exponential expressions. An especially useful type of exponential expression is a polynomial. Polynomials model many real-world phenomena. Our goal in this chapter is to become proficient with operations on polynomials.

A popular use of the Internet is the World Wide Web. The World Wide Web was invented in 1989–1990 as an environment originally by which scientists could share information. It has grown into a medium containing text, graphics, audio, animation, and video. Each of the locations, or Web sites below, has an address and can be used to locate other Web sites. In Section 3.2, Exercise 95, you will have the opportunity to estimate the number of visitors of the most popular Web sites.

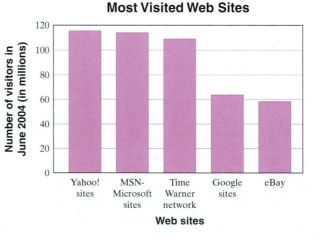

Most Visited Web Sites

3.1 EXPONENTS

Objective A Evaluating Exponential Expressions

In this section, we continue our work with integer exponents. Recall from Section 1.3 that repeated multiplication of the same factor can be written using exponents. For example,

$$2 \cdot 2 \cdot 2 \cdot 2 \cdot 2 = 2^5$$

The exponent 5 tells us how many times that 2 is a factor. The expression 2^5 is called an **exponential expression.** It is also called the fifth **power** of 2, or we can say that 2 is **raised** to the fifth power.

$$5^6 = \underbrace{5 \cdot 5 \cdot 5 \cdot 5 \cdot 5 \cdot 5}_{\text{6 factors; each factor is 5}} \quad \text{and} \quad (-3)^4 = \underbrace{(-3) \cdot (-3) \cdot (-3) \cdot (-3)}_{\text{4 factors; each factor is } -3}$$

The base of an exponential expression is the repeated factor. The exponent is the number of times that the base is used as a factor.

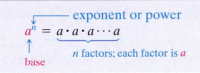

$$a^n = \underbrace{a \cdot a \cdot a \cdots a}_{n \text{ factors; each factor is } a}$$

exponent or power
base

EXAMPLES Evaluate each expression.

1. $2^3 = 2 \cdot 2 \cdot 2 = 8$
2. $3^1 = 3$. To raise 3 to the first power means to use 3 as a factor only once. When no exponent is shown, the exponent is assumed to be 1.
3. $(-4)^2 = (-4)(-4) = 16$
4. $-4^2 = -(4 \cdot 4) = -16$
5. $\left(\dfrac{1}{2}\right)^4 = \dfrac{1}{2} \cdot \dfrac{1}{2} \cdot \dfrac{1}{2} \cdot \dfrac{1}{2} = \dfrac{1}{16}$
6. $4 \cdot 3^2 = 4 \cdot 9 = 36$

▢ Work Practice Problems 1–6

Notice how similar -4^2 is to $(-4)^2$ in the examples above. The difference between the two is the parentheses. In $(-4)^2$, the parentheses tell us that the base, or the repeated factor, is -4. In -4^2, only 4 is the base.

Helpful Hint

Be careful when identifying the base of an exponential expression. Pay close attention to the use of parentheses.

$(-3)^2$	-3^2	$2 \cdot 3^2$
The base is -3.	The base is 3.	The base is 3.
$(-3)^2 = (-3)(-3) = 9$	$-3^2 = -(3 \cdot 3) = -9$	$2 \cdot 3^2 = 2 \cdot 3 \cdot 3 = 18$

An exponent has the same meaning whether the base is a number or a variable. If x is a real number and n is a positive integer, then x^n is the product of n factors, each of which is x.

$$x^n = \underbrace{x \cdot x \cdot x \cdot x \cdot x \cdots x}_{n \text{ factors; each factor is } x}$$

Objectives

A Evaluate Exponential Expressions.

B Use the Product Rule for Exponents.

C Use the Power Rule for Exponents.

D Use the Power Rules for Products and Quotients.

E Use the Quotient Rule for Exponents, and Define a Number Raised to the 0 Power.

F Decide Which Rule(s) to Use to Simplify an Expression.

PRACTICE PROBLEMS 1–6

Evaluate each expression.

1. 3^4
2. 7^1
3. $(-2)^3$
4. -2^3
5. $\left(\dfrac{2}{3}\right)^2$
6. $5 \cdot 6^2$

Answers

1. 81, **2.** 7, **3.** -8, **4.** -8, **5.** $\dfrac{4}{9}$, **6.** 180

PRACTICE PROBLEM 7

Evaluate each expression for the given value of x.

a. $3x^2$ when x is 4

b. $\dfrac{x^4}{-8}$ when x is -2

EXAMPLE 7 Evaluate each expression for the given value of x.

a. $2x^3$ when x is 5   **b.** $\dfrac{9}{x^2}$ when x is -3

Solution:

a. When x is 5,

$$2x^3 = 2 \cdot 5^3$$
$$= 2 \cdot (5 \cdot 5 \cdot 5)$$
$$= 2 \cdot 125$$
$$= 250$$

b. When x is -3,

$$\frac{9}{x^2} = \frac{9}{(-3)^2}$$
$$= \frac{9}{(-3)(-3)}$$
$$= \frac{9}{9} = 1$$

🟧 **Work Practice Problem 7**

Objective B Using the Product Rule

Exponential expressions can be multiplied, divided, added, subtracted, and themselves raised to powers. Let's see if we can discover a shortcut method for multiplying exponential expressions with the same base. By our definition of an exponent,

$$5^4 \cdot 5^3 = \underbrace{(5 \cdot 5 \cdot 5 \cdot 5)}_{4 \text{ factors of } 5} \cdot \underbrace{(5 \cdot 5 \cdot 5)}_{3 \text{ factors of } 5}$$

$$= \underbrace{5 \cdot 5 \cdot 5 \cdot 5 \cdot 5 \cdot 5 \cdot 5}_{7 \text{ factors of } 5}$$

$$= 5^7$$

Also,

$$x^2 \cdot x^3 = (x \cdot x) \cdot (x \cdot x \cdot x)$$
$$= x \cdot x \cdot x \cdot x \cdot x$$
$$= x^5$$

In both cases, notice that the result is exactly the same if the exponents are added.

$$5^4 \cdot 5^3 = 5^{4+3} = 5^7 \quad \text{and} \quad x^2 \cdot x^3 = x^{2+3} = x^5$$

This suggests the following rule.

Product Rule for Exponents

If m and n are positive integers and a is a real number, then

$$a^m \cdot a^n = a^{m+n} \;\leftarrow \text{Add exponents.}$$
$$\uparrow \text{————— Keep common base.}$$

For example,

$$3^5 \cdot 3^7 = 3^{5+7} = 3^{12} \;\leftarrow \text{Add exponents.}$$
$$\uparrow \text{————— Keep common base.}$$

Answers

7. a. 48,   **b.** -2

Helpful
Hint

> **Helpful Hint**
>
> Don't forget that
>
> $3^5 \cdot 3^7 \neq 9^{12}$ ← Add exponents.
>   └── Common base *not* kept.
>
> $3^5 \cdot 3^7 = \underbrace{3 \cdot 3 \cdot 3 \cdot 3 \cdot 3}_{\text{5 factors of 3}} \cdot \underbrace{3 \cdot 3 \cdot 3 \cdot 3 \cdot 3 \cdot 3 \cdot 3}_{\text{7 factors of 3}}$
>
> $= 3^{12}$ 12 factors of 3, *not* 9.

In other words, to multiply two exponential expressions with the **same base,** we keep the base and add the exponents. We call this **simplifying** the exponential expression.

EXAMPLES Use the product rule to simplify each expression.

8. $4^2 \cdot 4^5 = 4^{2+5} = 4^7$ ← Add exponents.
                              └── Keep common base.

9. $x^2 \cdot x^5 = x^{2+5} = x^7$

10. $y^3 \cdot y = y^3 \cdot y^1$
$= y^{3+1}$
$= y^4$

> **Helpful Hint**
>
> Don't forget that if no exponent is written, it is assumed to be 1.

11. $y^3 \cdot y^2 \cdot y^7 = y^{3+2+7} = y^{12}$

12. $(-5)^7 \cdot (-5)^8 = (-5)^{7+8} = (-5)^{15}$

🔲 **Work Practice Problems 8–12**

✔ **Concept Check** Where possible, use the product rule to simplify the expression.

a. $z^2 \cdot z^{14}$ **b.** $x^2 \cdot z^{14}$ **c.** $9^8 \cdot 9^3$ **d.** $9^8 \cdot 2^7$

EXAMPLE 13 Use the product rule to simplify $(2x^2)(-3x^5)$.

Solution: Recall that $2x^2$ means $2 \cdot x^2$ and $-3x^5$ means $-3 \cdot x^5$.

$(2x^2)(-3x^5) = (2 \cdot x^2) \cdot (-3 \cdot x^5)$

$= (2 \cdot -3) \cdot (x^2 \cdot x^5)$ Group factors with common bases (using commutative and associative properties.)

$= -6x^7$ Simplify.

🔲 **Work Practice Problem 13**

EXAMPLES Simplify.

14. $(x^2y)(x^3y^2) = (x^2 \cdot x^3) \cdot (y^1 \cdot y^2)$ Group like bases and write y as y^1.
$= x^5 \cdot y^3$ or x^5y^3 Multiply.

15. $(-a^7b^4)(3ab^9) = (-1 \cdot 3) \cdot (a^7 \cdot a^1) \cdot (b^4 \cdot b^9)$
$= -3a^8b^{13}$

🔲 **Work Practice Problems 14–15**

PRACTICE PROBLEMS 8–12

Use the product rule to simplify each expression.

8. $7^3 \cdot 7^2$ **9.** $x^4 \cdot x^9$
10. $r^5 \cdot r$ **11.** $s^6 \cdot s^2 \cdot s^3$
12. $(-3)^9 \cdot (-3)$

PRACTICE PROBLEM 13

Use the product rule to simplify $(6x^3)(-2x^9)$.

PRACTICE PROBLEMS 14–15

Simplify.
14. $(m^5n^{10})(mn^8)$
15. $(-x^9y)(4x^2y^{11})$

Answers
8. 7^5, **9.** x^{13}, **10.** r^6, **11.** s^{11},
12. $(-3)^{10}$, **13.** $-12x^{12}$,
14. m^6n^{18}, **15.** $-4x^{11}y^{12}$

✔ **Concept Check Answers**

a. z^{16}, **b.** cannot be simplified,
c. 9^{11}, **d.** cannot be simplified

These examples will remind you of the difference between adding and multiplying terms.

Addition

$$5x^3 + 3x^3 = (5 + 3)x^3 = 8x^3 \qquad \text{By the distributive property.}$$
$$7x + 4x^2 = 7x + 4x^2 \qquad \text{Cannot be combined.}$$

Multiplication

$$(5x^3)(3x^3) = 5 \cdot 3 \cdot x^3 \cdot x^3 = 15x^{3+3} = 15x^6 \qquad \text{By the product rule.}$$
$$(7x)(4x^2) = 7 \cdot 4 \cdot x \cdot x^2 = 28x^{1+2} = 28x^3 \qquad \text{By the product rule.}$$

Objective C Using the Power Rule

Exponential expressions can themselves be raised to powers. Let's try to discover a rule that simplifies an expression like $(x^2)^3$. By the definition of a^n,

$$(x^2)^3 = (x^2)(x^2)(x^2) \qquad (x^2)^3 \text{ means 3 factors of } (x^2).$$

which can be simplified by the product rule for exponents.

$$(x^2)^3 = (x^2)(x^2)(x^2) = x^{2+2+2} = x^6$$

Notice that the result is exactly the same if we multiply the exponents.

$$(x^2)^3 = x^{2 \cdot 3} = x^6$$

The following rule states this result.

Power Rule for Exponents

If m and n are positive integers and a is a real number, then

$$(a^m)^n = a^{mn} \quad \leftarrow \text{Multiply exponents.}$$
$$\qquad\qquad\qquad \leftarrow \text{Keep common base.}$$

For example,

$$(7^2)^5 = 7^{2 \cdot 5} = 7^{10} \quad \leftarrow \text{Multiply exponents.}$$
$$\qquad\qquad\qquad\qquad \leftarrow \text{Keep common base.}$$

In other words, to raise an exponential expression to a power, we keep the base and multiply the exponents.

PRACTICE PROBLEMS 16–17

Use the power rule to simplify each expression.

16. $(9^4)^{10}$ **17.** $(z^6)^3$

EXAMPLES Use the power rule to simplify each expression.

16. $(5^3)^6 = 5^{3 \cdot 6} = 5^{18}$
17. $(y^8)^2 = y^{8 \cdot 2} = y^{16}$

Work Practice Problems 16–17

Take a moment to make sure that you understand when to apply the product rule and when to apply the power rule.

Product Rule → Add Exponents	Power Rule → Multiply Exponents
$x^5 \cdot x^7 = x^{5+7} = x^{12}$	$(x^5)^7 = x^{5 \cdot 7} = x^{35}$
$y^6 \cdot y^2 = y^{6+2} = y^8$	$(y^6)^2 = y^{6 \cdot 2} = y^{12}$

Answers

16. 9^{40},   **17.** z^{18}

Objective D Using the Power Rules for Products and Quotients

When the base of an exponential expression is a product, the definition of a^n still applies. For example, simplify $(xy)^3$ as follows.

$$(xy)^3 = (xy)(xy)(xy) \qquad \text{\color{blue}$(xy)^3$ means 3 factors of (xy).}$$
$$= x \cdot x \cdot x \cdot y \cdot y \cdot y \qquad \text{\color{blue}Group factors with common bases.}$$
$$= x^3 y^3 \qquad \text{\color{blue}Simplify.}$$

Notice that to simplify the expression $(xy)^3$, we raise each factor within the parentheses to a power of 3.

$$(xy)^3 = x^3 y^3$$

In general, we have the following rule.

Power of a Product Rule

If n is a positive integer and a and b are real numbers, then

$$(ab)^n = a^n b^n$$

For example,

$$(3x)^5 = 3^5 x^5$$

In other words, to raise a product to a power, we raise each factor to the power.

EXAMPLES Simplify each expression.

18. $(st)^4 = s^4 \cdot t^4 = s^4 t^4$ {\color{blue}Use the power of a product rule.}

19. $(2a)^3 = 2^3 \cdot a^3 = 8a^3$ {\color{blue}Use the power of a product rule.}

20. $(-5x^2 y^3 z)^2 = (-5)^2 \cdot (x^2)^2 \cdot (y^3)^2 \cdot (z^1)^2$ {\color{blue}Use the power of a product rule.}
$$= 25x^4 y^6 z^2 \qquad \text{\color{blue}Use the power rule for exponents.}$$

21. $(-xy^3)^5 = (-1xy^3)^5 = (-1)^5 \cdot x^5 \cdot (y^3)^5$
$$= -1x^5 y^{15} \quad \text{or} \quad -x^5 y^{15}$$

🔲 **Work Practice Problems 18–21**

PRACTICE PROBLEMS 18–21

Simplify each expression.
18. $(xy)^7$ **19.** $(3y)^4$
20. $(-2p^4 q^2 r)^3$ **21.** $(-a^4 b)^7$

Let's see what happens when we raise a quotient to a power. For example, we simplify $\left(\dfrac{x}{y}\right)^3$ as follows.

$$\left(\frac{x}{y}\right)^3 = \left(\frac{x}{y}\right)\left(\frac{x}{y}\right)\left(\frac{x}{y}\right) \qquad \text{\color{blue}$\left(\frac{x}{y}\right)^3$ means 3 factors of $\left(\frac{x}{y}\right)$.}$$

$$= \frac{x \cdot x \cdot x}{y \cdot y \cdot y} \qquad \text{\color{blue}Multiply fractions.}$$

$$= \frac{x^3}{y^3} \qquad \text{\color{blue}Simplify.}$$

Notice that to simplify the expression, $\left(\dfrac{x}{y}\right)^3$, we raise both the numerator and the denominator to a power of 3.

$$\left(\frac{x}{y}\right)^3 = \frac{x^3}{y^3}$$

In general, we have the following rule.

Answers
18. $x^7 y^7$,   **19.** $81y^4$,   **20.** $-8p^{12} q^6 r^3$,
21. $-a^{28} b^7$

Power of a Quotient Rule

If n is a positive integer and a and c are real numbers, then

$$\left(\frac{a}{c}\right)^n = \frac{a^n}{c^n}, \quad c \neq 0$$

For example,

$$\left(\frac{y}{7}\right)^3 = \frac{y^3}{7^3}$$

In other words, to raise a quotient to a power, we raise both the numerator and the denominator to the power.

PRACTICE PROBLEMS 22–23

Simplify each expression.

22. $\left(\dfrac{r}{s}\right)^6$ **23.** $\left(\dfrac{5x^6}{9y^3}\right)^2$

EXAMPLES Simplify each expression.

22. $\left(\dfrac{m}{n}\right)^7 = \dfrac{m^7}{n^7}, \quad n \neq 0$ Use the power of a quotient rule.

23. $\left(\dfrac{2x^4}{3y^5}\right)^4 = \dfrac{2^4 \cdot (x^4)^4}{3^4 \cdot (y^5)^4}$ Use the power of a quotient rule.

$\qquad\qquad = \dfrac{16x^{16}}{81y^{20}}, \quad y \neq 0$ Use the power rule for exponents.

🔶 **Work Practice Problems 22–23**

Objective 🄴 Using the Quotient Rule and Defining the Zero Exponent

Another pattern for simplifying exponential expressions involves quotients.

$$\frac{x^5}{x^3} = \frac{x \cdot x \cdot x \cdot x \cdot x}{x \cdot x \cdot x}$$

$$= \frac{x \cdot x \cdot x \cdot x \cdot x}{x \cdot x \cdot x}$$

$$= 1 \cdot 1 \cdot 1 \cdot x \cdot x$$

$$= x \cdot x$$

$$= x^2$$

Notice that the result is exactly the same if we subtract exponents of the common bases.

$$\frac{x^5}{x^3} = x^{5-3} = x^2$$

The following rule states this result in a general way.

Quotient Rule for Exponents

If m and n are positive integers and a is a real number, then

$$\frac{a^m}{a^n} = a^{m-n}, \quad a \neq 0$$

For example,

$$\frac{x^6}{x^2} = x^{6-2} = x^4, \quad x \neq 0$$

Answers

22. $\dfrac{r^6}{s^6}, \quad s \neq 0$, **23.** $\dfrac{25x^{12}}{81y^6}, \quad y \neq 0$

In other words, to divide one exponential expression by another with a common base, we keep the base and subtract the exponents.

EXAMPLES Simplify each quotient.

24. $\dfrac{x^5}{x^2} = x^{5-2} = x^3$ Use the quotient rule.

25. $\dfrac{4^7}{4^3} = 4^{7-3} = 4^4 = 256$ Use the quotient rule.

26. $\dfrac{(-3)^5}{(-3)^2} = (-3)^3 = -27$ Use the quotient rule.

27. $\dfrac{2x^5y^2}{xy} = 2 \cdot \dfrac{x^5}{x^1} \cdot \dfrac{y^2}{y^1}$

$\qquad = 2 \cdot (x^{5-1}) \cdot (y^{2-1})$ Use the quotient rule.

$\qquad = 2x^4y^1 \quad \text{or} \quad 2x^4y$

⬛ **Work Practice Problems 24–27**

PRACTICE PROBLEMS 24-27

Simplify each quotient.

24. $\dfrac{y^7}{y^3}$ **25.** $\dfrac{5^9}{5^6}$

26. $\dfrac{(-2)^{14}}{(-2)^{10}}$ **27.** $\dfrac{7a^4b^{11}}{ab}$

Let's now give meaning to an expression such as x^0. To do so, we will simplify $\dfrac{x^3}{x^3}$ in two ways and compare the results.

$\dfrac{x^3}{x^3} = x^{3-3} = x^0$ Apply the quotient rule.

$\dfrac{x^3}{x^3} = \dfrac{x \cdot x \cdot x}{x \cdot x \cdot x} = 1$ Apply the fundamental principle for fractions.

Since $\dfrac{x^3}{x^3} = x^0$ and $\dfrac{x^3}{x^3} = 1$, we define that $x^0 = 1$ as long as x is not 0.

Zero Exponent

$a^0 = 1$, as long as a is not 0.

For example, $5^0 = 1$.

In other words, a base raised to the 0 power is 1, as long as the base is not 0.

EXAMPLES Simplify each expression.

28. $3^0 = 1$

29. $(5x^3y^2)^0 = 1$

30. $(-4)^0 = 1$

31. $-4^0 = -1 \cdot 4^0 = -1 \cdot 1 = -1$

32. $5x^0 = 5 \cdot x^0 = 5 \cdot 1 = 5$

⬛ **Work Practice Problems 28–32**

PRACTICE PROBLEMS 28-32

Simplify each expression.

28. 8^0 **29.** $(2r^2s)^0$

30. $(-7)^0$ **31.** -7^0

32. $7y^0$

Answers

24. y^4, **25.** 125, **26.** 16, **27.** $7a^3b^{10}$,
28. 1, **29.** 1, **30.** 1, **31.** −1,
32. 7

✔**Concept Check** Suppose you are simplifying each expression. Tell whether you would *add* the exponents, *subtract* the exponents, *multiply* the exponents, *divide* the exponents, or *none of these*.

a. $(x^{63})^{21}$ **b.** $\dfrac{y^{15}}{y^3}$ **c.** $z^{16} + z^8$ **d.** $w^{45} \cdot w^9$

Objective **F** Deciding Which Rule to Use

Let's practice deciding which rule to use to simplify. We will continue this discussion with more examples in the next section.

PRACTICE PROBLEM 33

Simplify each expression.

a. $\dfrac{x^7}{x^4}$ **b.** $(3y^4)^4$ **c.** $\left(\dfrac{x}{4}\right)^3$

EXAMPLE 33 Simplify each expression.

a. $x^7 \cdot x^4$ **b.** $\left(\dfrac{t}{2}\right)^4$ **c.** $(9y^5)^2$

Solution:

a. Here, we have a product, so we use the product rule to simplify.

$$x^7 \cdot x^4 = x^{7+4} = x^{11}$$

b. This is a quotient raised to a power, so we use the power of a quotient rule.

$$\left(\frac{t}{2}\right)^4 = \frac{t^4}{2^4} = \frac{t^4}{16}$$

c. This is a product raised to a power, so we use the power of a product rule.

$$(9y^5)^2 = 9^2(y^5)^2 = 81y^{10}$$

🔲 **Work Practice Problem 33**

Answers

33. a. x^3,   **b.** $81y^{16}$,   **c.** $\dfrac{x^3}{64}$

✔ **Concept Check Answers**

a. multiply,   **b.** subtract,
c. none of these,   **d.** add

Mental Math

State the bases and the exponents for each expression.

1. 3^2 **2.** 5^4 **3.** $(-3)^6$ **4.** -3^7 **5.** -4^2

6. $(-4)^3$ **7.** $5 \cdot 3^4$ **8.** $9 \cdot 7^6$ **9.** $5x^2$ **10.** $(5x)^2$

3.1 EXERCISE SET

FOR EXTRA HELP

Student Solutions Manual   PH Math/Tutor Center   CD/Video for Review   Math XL MathXL®   MyMathLab MyMathLab

Objective A *Evaluate each expression. See Examples 1 through 6.*

1. 7^2 **2.** -3^2 **3.** $(-5)^1$ **4.** $(-3)^2$ **5.** -2^4 **6.** -4^3

7. $(-2)^4$ **8.** $(-4)^3$ **9.** $\left(\dfrac{1}{3}\right)^3$ **10.** $\left(-\dfrac{1}{9}\right)^2$ **11.** $7 \cdot 2^4$ **12.** $9 \cdot 2^2$

Evaluate each expression with the given replacement values. See Example 7.

13. x^2 when $x = -2$ **14.** x^3 when $x = -2$ **15.** $5x^3$ when $x = 3$

16. $4x^2$ when $x = 5$ **17.** $2xy^2$ when $x = 3$ and $y = -5$ **18.** $-4x^2y^3$ when $x = 2$ and $y = -1$

19. $\dfrac{5z^4}{7}$ when $z = -2$ **20.** $\dfrac{10}{3y^3}$ when $y = -3$

Objective B *Use the product rule to simplify each expression. Write the results using exponents. See Examples 8 through 13.*

21. $x^2 \cdot x^5$ **22.** $y^2 \cdot y$ **23.** $(-3)^3 \cdot (-3)^9$ **24.** $(-5)^7 \cdot (-5)^6$

25. $(5y^4)(3y)$ **26.** $(-2z^3)(-2z^2)$ **27.** $(x^9y)(x^{10}y^5)$ **28.** $(a^2b)(a^{13}b^{17})$

29. $(-8mn^6)(9m^2n^2)$ **30.** $(-7a^3b^3)(7a^{19}b)$ **31.** $(4z^{10})(-6z^7)(z^3)$ **32.** $(12x^5)(-x^6)(x^4)$

△ **33.** The rectangle below has width $4x^2$ feet and length $5x^3$ feet. Find its area as an expression in x.

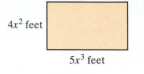

$4x^2$ feet

$5x^3$ feet

△ **34.** The parallelogram below has base length $9y^7$ meters and height $2y^{10}$ meters. Find its area as an expression in y.

$2y^{10}$ meters

$9y^7$ meters

Objectives C D **Mixed Practice** *Use the power rule and the power of a product or quotient rule to simplify each expression. See Examples 16 through 23.*

35. $(x^9)^4$

36. $(y^7)^5$

37. $(pq)^8$

38. $(ab)^6$

39. $(2a^5)^3$

40. $(4x^6)^2$

41. $(x^2y^3)^5$

42. $(a^4b)^7$

43. $(-7a^2b^5c)^2$

44. $(-3x^7yz^2)^3$

45. $\left(\dfrac{r}{s}\right)^9$

46. $\left(\dfrac{q}{t}\right)^{11}$

47. $\left(\dfrac{mp}{n}\right)^9$

48. $\left(\dfrac{xy}{7}\right)^2$

49. $\left(\dfrac{-2xz}{y^5}\right)^2$

50. $\left(\dfrac{xy^4}{-3z^3}\right)^3$

△ **51.** The square shown has sides of length $8z^5$ decimeters. Find its area.

$8z^5$
decimeters

△ **52.** Given the circle below with radius $5y$ centimeters, find its area. Do not approximate π.

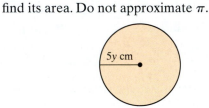

$5y$ cm

△ **53.** The vault below is in the shape of a cube. If each side is $3y^4$ feet, find its volume.

$3y^4$ feet $3y^4$ feet

$3y^4$ feet

△ **54.** The silo shown is in the shape of a cylinder. If its radius is $4x$ meters and its height is $5x^3$ meters, find its volume. Do not approximate π.

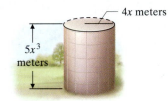

$4x$ meters

$5x^3$
meters

Objective E *Use the quotient rule and simplify each expression. See Examples 24 through 28.*

55. $\dfrac{x^3}{x}$

56. $\dfrac{y^{10}}{y^9}$

57. $\dfrac{(-4)^6}{(-4)^3}$

58. $\dfrac{(-6)^{13}}{(-6)^{11}}$

59. $\dfrac{p^7q^{20}}{pq^{15}}$

60. $\dfrac{x^8y^6}{xy^5}$

61. $\dfrac{7x^2y^6}{14x^2y^3}$

62. $\dfrac{9a^4b^7}{27ab^2}$

Simplify each expression. See Examples 28 through 32.

63. 7^0

64. 23^0

65. $(2x)^0$

66. $(4y)^0$

67. $-7x^0$

68. $-2x^0$

69. $5^0 + y^0$

70. $-3^0 + 4^0$

Objectives Ⓐ Ⓑ Ⓒ Ⓓ Ⓔ Ⓕ **Mixed Practice** *Simplify each expression. See Examples 1 through 6, and 8 through 33.*

71. -9^2

72. $(-9)^2$

73. $\left(\frac{1}{4}\right)^3$

74. $\left(\frac{2}{3}\right)^3$

75. $b^4 b^2$

76. $y^4 y$

77. $a^2 a^3 a^4$

78. $x^2 x^{15} x^9$

79. $(2x^3)(-8x^4)$

80. $(3y^4)(-5y)$

81. $(a^7 b^{12})(a^4 b^8)$

82. $(y^2 z^2)(y^{15} z^{13})$

83. $(-2mn^6)(-13m^8 n)$

84. $(-3s^5 t)(-7st^{10})$

85. $(z^4)^{10}$

86. $(t^5)^{11}$

87. $(4ab)^3$

88. $(2ab)^4$

89. $(-6xyz^3)^2$

90. $(-3xy^2 a^3)^3$

91. $\dfrac{z^{12}}{z^4}$

92. $\dfrac{b^4}{b}$

93. $\dfrac{3x^5}{x^4}$

94. $\dfrac{5x^9}{x^3}$

95. $(6b)^0$

96. $(5ab)^0$

97. $(9xy)^2$

98. $(2ab)^5$

99. $2^3 + 2^5$

100. $7^2 - 7^0$

101. $\left(\dfrac{3y^5}{6x^4}\right)^3$

102. $\left(\dfrac{2ab}{6yz}\right)^4$

103. $\dfrac{2x^3 y^2 z}{xyz}$

104. $\dfrac{x^{12} y^{13}}{x^5 y^7}$

Review

Subtract.

105. $5 - 7$

106. $9 - 12$

107. $3 - (-2)$

108. $5 - (-10)$

109. $-11 - (-4)$

110. $-15 - (-21)$

Solve. See the Concept Checks in this section. For Exercises 111 through 114, match the expression with the operation needed to simplify each. A letter may be used more than once and a letter may not be used at all.

111. $(x^{14})^{23}$

112. $x^{14} \cdot x^{23}$

113. $x^{14} + x^{23}$

114. $\dfrac{x^{35}}{x^{17}}$

a. Add the exponents
b. Subtract the exponents
c. Multiply the exponents
d. Divide the exponents
e. None of these

Fill in the boxes so that each statement is true. (More than one answer is possible for each exercise.)

115. $x^{\square} \cdot x^{\square} = x^{12}$

116. $(x^{\square})^{\square} = x^{20}$

117. $\dfrac{y^{\square}}{y^{\square}} = y^7$

118. $(y^{\square})^{\square} \cdot (y^{\square})^{\square} = y^{30}$

Concept Extensions

△ **119.** The formula $V = x^3$ can be used to find the volume V of a cube with side length x. Find the volume of a cube with side length 7 meters. (Volume is measured in cubic units.)

x

△ **120.** The formula $S = 6x^2$ can be used to find the surface area S of a cube with side length x. Find the surface area of a cube with side length 5 meters. (Surface area is measured in square units.)

△ **121.** To find the amount of water that a swimming pool in the shape of a cube can hold, do we use the formula for volume of the cube or surface area of the cube? (See Exercises 119 and 120.)

△ **122.** To find the amount of material needed to cover an ottoman in the shape of a cube, do we use the formula for volume of the cube or surface area of the cube? (See Exercises 119 and 120.)

123. Explain why $(-5)^4 = 625$, while $-5^4 = -625$.

124. Explain why $5 \cdot 4^2 = 80$, while $(5 \cdot 4)^2 = 400$.

125. In your own words, explain why $5^0 = 1$.

126. In your own words, explain when $(-3)^n$ is positive and when it is negative.

Simplify each expression. Assume that variables represent positive integers.

127. $x^{5a}x^{4a}$

128. $b^{9a}b^{4a}$

129. $(a^b)^5$

130. $(2a^{4b})^4$

131. $\dfrac{x^{9a}}{x^{4a}}$

132. $\dfrac{y^{15b}}{y^{6b}}$

 STUDY SKILLS BUILDER

How Well Do You Know Your Textbook?

The questions below will determine whether you are familiar with your textbook. For help, see Section 1.1 in this text.

1. What does the 💿 icon mean?

2. What does the ✏ icon mean?

3. What does the △ icon mean?

4. Where can you find a review for each chapter? What answers to this review can be found in the back of your text?

5. Each chapter contains an overview of the chapter along with examples. What is this feature called?

6. Each chapter contains a review of vocabulary. What is this feature called?

7. There is a CD in your text. What content is contained on this CD?

8. What is the location of the section that is entirely devoted to study skills?

9. There are Practice Problems that are contained in the margin of the text. What are they and how can they be used?

3.2 NEGATIVE EXPONENTS AND SCIENTIFIC NOTATION

Objectives

A Simplify Expressions Containing Negative Exponents.

B Use the Rules and Definitions for Exponents to Simplify Exponential Expressions.

C Write Numbers in Scientific Notation.

D Convert Numbers in Scientific Notation to Standard Form.

Objective **A** Simplifying Expressions Containing Negative Exponents

Our work with exponential expressions so far has been limited to exponents that are positive integers or 0. Here we will also give meaning to an expression like x^{-3}.

Suppose that we wish to simplify the expression $\dfrac{x^2}{x^5}$. If we use the quotient rule for exponents, we subtract exponents:

$$\frac{x^2}{x^5} = x^{2-5} = x^{-3}, \quad x \neq 0$$

But what does x^{-3} mean? Let's simplify $\dfrac{x^2}{x^5}$ using the definition of a^n.

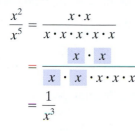

$$\frac{x^2}{x^5} = \frac{x \cdot x}{x \cdot x \cdot x \cdot x \cdot x}$$

$$= \frac{x \cdot x}{x \cdot x \cdot x \cdot x \cdot x}$$ Divide numerator and denominator by common factors by applying the fundamental principle for fractions.

$$= \frac{1}{x^3}$$

If the quotient rule is to hold true for negative exponents, then x^{-3} must equal $\dfrac{1}{x^3}$. From this example, we state the definition for negative exponents.

Negative Exponents

If a is a real number other than 0 and n is an integer, then

$$a^{-n} = \frac{1}{a^n}$$

For example,

$$x^{-3} = \frac{1}{x^3}$$

In other words, another way to write a^{-n} is to take its reciprocal and change the sign of its exponent.

EXAMPLES Simplify by writing each expression with positive exponents only.

1. $3^{-2} = \dfrac{1}{3^2} = \dfrac{1}{9}$ Use the definition of negative exponents.

2. $2x^{-3} = 2^1 \cdot \dfrac{1}{x^3} = \dfrac{2^1}{x^3}$ or $\dfrac{2}{x^3}$ Use the definition of negative exponents.

3. $2^{-1} + 4^{-1} = \dfrac{1}{2} + \dfrac{1}{4} = \dfrac{2}{4} + \dfrac{1}{4} = \dfrac{3}{4}$

4. $(-2)^{-4} = \dfrac{1}{(-2)^4} = \dfrac{1}{(-2)(-2)(-2)(-2)} = \dfrac{1}{16}$

> **Helpful Hint**
> Don't forget that since there are no parentheses, only x is the base for the exponent -3.

Work Practice Problems 1–4

PRACTICE PROBLEMS 1–4

Simplify by writing each expression with positive exponents only.

1. 5^{-3} 2. $7x^{-4}$

3. $5^{-1} + 3^{-1}$ 4. $(-3)^{-4}$

Answers

1. $\dfrac{1}{125}$, 2. $\dfrac{7}{x^4}$, 3. $\dfrac{8}{15}$, 4. $\dfrac{1}{81}$

191

> **Helpful Hint**
>
> A negative exponent *does not affect* the sign of its base.
> Remember: Another way to write a^{-n} is to take its reciprocal and change the sign of its exponent: $a^{-n} = \dfrac{1}{a^n}$. For example,
>
> $$x^{-2} = \frac{1}{x^2}, \qquad 2^{-3} = \frac{1}{2^3} \ \text{ or } \ \frac{1}{8}$$
>
> $$\frac{1}{y^{-4}} = \frac{1}{\frac{1}{y^4}} = y^4, \qquad \frac{1}{5^{-2}} = 5^2 \ \text{ or } \ 25$$

From the preceding Helpful Hint, we know that $x^{-2} = \dfrac{1}{x^2}$ and $\dfrac{1}{y^{-4}} = y^4$. We can use this to include another statement in our definition of negative exponents.

Negative Exponents

If a is a real number other than 0 and n is an integer, then

$$a^{-n} = \frac{1}{a^n} \quad \text{and} \quad \frac{1}{a^{-n}} = a^n$$

PRACTICE PROBLEMS 5–8

Simplify each expression. Write each result using positive exponents only.

5. $\left(\dfrac{6}{7}\right)^{-2}$ **6.** $\dfrac{x}{x^{-4}}$

7. $\dfrac{y^{-9}}{z^{-5}}$ **8.** $\dfrac{y^{-4}}{y^6}$

EXAMPLES Simplify each expression. Write each result using positive exponents only.

5. $\left(\dfrac{2}{x}\right)^{-3} = \dfrac{2^{-3}}{x^{-3}} = \dfrac{2^{-3}}{1} \cdot \dfrac{1}{x^{-3}} = \dfrac{1}{2^3} \cdot \dfrac{x^3}{1} = \dfrac{x^3}{2^3} = \dfrac{x^3}{8}$ Use the negative exponents rule.

6. $\dfrac{y}{y^{-2}} = \dfrac{y^1}{y^{-2}} = y^{1-(-2)} = y^3$ Use the quotient rule.

7. $\dfrac{p^{-4}}{q^{-9}} = p^{-4} \cdot \dfrac{1}{q^{-9}} = \dfrac{1}{p^4} \cdot q^9 = \dfrac{q^9}{p^4}$ Use the negative exponents rule.

8. $\dfrac{x^{-5}}{x^7} = x^{-5-7} = x^{-12} = \dfrac{1}{x^{12}}$

🔲 **Work Practice Problems 5–8**

Objective B Simplifying Exponential Expressions

All the previously stated rules for exponents apply for negative exponents also. Here is a summary of the rules and definitions for exponents.

Summary of Exponent Rules

If m and n are integers and a, b, and c are real numbers, then

Product rule for exponents:	$a^m \cdot a^n = a^{m+n}$
Power rule for exponents:	$(a^m)^n = a^{m \cdot n}$
Power of a product:	$(ab)^n = a^n b^n$
Power of a quotient:	$\left(\dfrac{a}{c}\right)^n = \dfrac{a^n}{c^n}, \quad c \neq 0$
Quotient rule for exponents:	$\dfrac{a^m}{a^n} = a^{m-n}, \quad a \neq 0$
Zero exponent:	$a^0 = 1, \quad a \neq 0$
Negative exponent:	$a^{-n} = \dfrac{1}{a^n}, \quad a \neq 0$

Answers

5. $\dfrac{49}{36}$, **6.** x^5, **7.** $\dfrac{z^5}{y^9}$, **8.** $\dfrac{1}{y^{10}}$

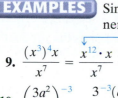 **EXAMPLES** Simplify each expression. Write each result using positive exponents only.

9. $\dfrac{(x^3)^4 x}{x^7} = \dfrac{x^{12} \cdot x}{x^7} = \dfrac{x^{12+1}}{x^7} = \dfrac{x^{13}}{x^7} = x^{13-7} = x^6$ Use the power rule.

10. $\left(\dfrac{3a^2}{b}\right)^{-3} = \dfrac{3^{-3}(a^2)^{-3}}{b^{-3}}$ Raise each factor in the numerator and the denominator to the -3 power.

$\quad = \dfrac{3^{-3}a^{-6}}{b^{-3}}$ Use the power rule.

$\quad = \dfrac{b^3}{3^3 a^6}$ Use the negative exponent rule.

$\quad = \dfrac{b^3}{27a^6}$ Write 3^3 as 27.

11. $(y^{-3}z^6)^{-6} = (y^{-3})^{-6}(z^6)^{-6}$ Raise each factor to the -6 power.

$\quad = y^{18}z^{-36} = \dfrac{y^{18}}{z^{36}}$

12. $\dfrac{(2x)^5}{x^3} = \dfrac{2^5 \cdot x^5}{x^3} = 2^5 \cdot x^{5-3} = 32x^2$ Raise each factor in the numerator to the fifth power.

13. $\dfrac{x^{-7}}{(x^4)^3} = \dfrac{x^{-7}}{x^{12}} = x^{-7-12} = x^{-19} = \dfrac{1}{x^{19}}$

14. $(5y^3)^{-2} = 5^{-2}(y^3)^{-2} = 5^{-2}y^{-6} = \dfrac{1}{5^2 y^6} = \dfrac{1}{25y^6}$

15. $-\dfrac{22a^7 b^{-5}}{11a^{-2}b^3} = -\dfrac{22}{11} \cdot a^{7-(-2)}b^{-5-3} = -2a^9 b^{-8} = -\dfrac{2a^9}{b^8}$

16. $\dfrac{(2xy)^{-3}}{(x^2y^3)^2} = \dfrac{2^{-3}x^{-3}y^{-3}}{(x^2)^2(y^3)^2} = \dfrac{2^{-3}x^{-3}y^{-3}}{x^4 y^6} = 2^{-3}x^{-3-4}y^{-3-6}$

$\quad = 2^{-3}x^{-7}y^{-9} = \dfrac{1}{2^3 x^7 y^9}$ or $\dfrac{1}{8x^7 y^9}$

🔲 **Work Practice Problems 9–16**

Objective **C** Writing Numbers in Scientific Notation

Both very large and very small numbers frequently occur in many fields of science. For example, the distance between the sun and the planet Pluto is approximately 5,906,000,000 kilometers, and the mass of a proton is approximately 0.000000000000000000000000165 gram. It can be tedious to write these numbers in this standard decimal notation, so **scientific notation** is used as a convenient shorthand for expressing very large and very small numbers.

5,906,000,000 kilometers

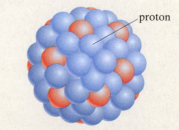
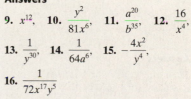

Scientific Notation

A positive number is written in scientific notation if it is written as the product of a number a, where $1 \leq a < 10$, and an integer power r of 10: $a \times 10^r$

The following numbers are written in scientific notation. The $\times$ sign for multiplication is used as part of the notation.

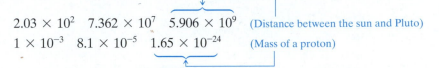

$2.03 \times 10^2 \quad 7.362 \times 10^7 \quad 5.906 \times 10^9$ (Distance between the sun and Pluto)

$1 \times 10^{-3} \quad 8.1 \times 10^{-5} \quad 1.65 \times 10^{-24}$ (Mass of a proton)

The following steps are useful when writing numbers in scientific notation.

To Write a Number in Scientific Notation

Step 1: Move the decimal point in the original number so that the new number has a value between 1 and 10.

Step 2: Count the number of decimal places the decimal point is moved in Step 1. If the original number is 10 or greater, the count is positive. If the original number is less than 1, the count is negative.

Step 3: Multiply the new number in Step 1 by 10 raised to an exponent equal to the count found in Step 2.

PRACTICE PROBLEM 17

Write each number in scientific notation.

a. 420,000   **b.** 0.00017

c. 9,060,000,000   **d.** 0.000007

EXAMPLE 17   Write each number in scientific notation.

a. 367,000,000   **c.** 20,520,000,000

b. 0.000003   **d.** 0.00085

Solution:

a. Step 1: Move the decimal point until the number is between 1 and 10.

367,000,000.

8 places

Step 2: The decimal point is moved 8 places and the original number is 10 or greater, so the count is positive 8.

Step 3: $367,000,000 = 3.67 \times 10^8$.

b. Step 1: Move the decimal point until the number is between 1 and 10.

0.000003

6 places

Step 2: The decimal point is moved 6 places and the original number is less than 1, so the count is −6.

Step 3: $0.000003 = 3.0 \times 10^{-6}$

c. $20,520,000,000 = 2.052 \times 10^{10}$

d. $0.00085 = 8.5 \times 10^{-4}$

☐ **Work Practice Problem 17**

Objective D Converting Numbers to Standard Form

A number written in scientific notation can be rewritten in standard form. For example, to write 8.63×10^3 in standard form, recall that $10^3 = 1000$.

$$8.63 \times 10^3 = 8.63(1000) = 8630$$

Notice that the exponent on the 10 is positive 3, and we moved the decimal point 3 places to the right.

To write 7.29×10^{-3} in standard form, recall that $10^{-3} = \dfrac{1}{10^3} = \dfrac{1}{1000}$.

$$7.29 \times 10^{-3} = 7.29\left(\dfrac{1}{1000}\right) = \dfrac{7.29}{1000} = 0.00729$$

The exponent on the 10 is negative 3, and we moved the decimal to the left 3 places.

In general, **to write a scientific notation number in standard form,** move the decimal point the same number of places as the exponent on 10. If the exponent is positive, move the decimal point to the right; if the exponent is negative, move the decimal point to the left.

✔ **Concept Check** Which number in each pair is larger?

a. 7.8×10^3 or 2.1×10^5
b. 9.2×10^{-2} or 2.7×10^4
c. 5.6×10^{-4} or 6.3×10^{-5}

EXAMPLE 18 Write each number in standard notation, without exponents.

a. 1.02×10^5 **c.** 8.4×10^7
b. 7.358×10^{-3} **d.** 3.007×10^{-5}

Solution:

a. Move the decimal point 5 places to the right.

$$1.02 \times 10^5 = 102{,}000.$$

b. Move the decimal point 3 places to the left.

$$7.358 \times 10^{-3} = 0.007358$$

c. $8.4 \times 10^7 = 84{,}000{,}000.$ 7 places to the right

d. $3.007 \times 10^{-5} = 0.00003007$ 5 places to the left

▢ **Work Practice Problem 18**

Performing operations on numbers written in scientific notation makes use of the rules and definitions for exponents.

EXAMPLE 19 Perform each indicated operation. Write each result in standard decimal notation.

a. $(8 \times 10^{-6})(7 \times 10^3)$
b. $\dfrac{12 \times 10^2}{6 \times 10^{-3}}$

Solution:

a. $(8 \times 10^{-6})(7 \times 10^3) = 8 \cdot 7 \cdot 10^{-6} \cdot 10^3$
$= 56 \times 10^{-3}$
$= 0.056$

b. $\dfrac{12 \times 10^2}{6 \times 10^{-3}} = \dfrac{12}{6} \times 10^{2-(-3)} = 2 \times 10^5 = 200{,}000$

▢ **Work Practice Problem 19**

PRACTICE PROBLEM 18

Write the numbers in standard notation, without exponents.

a. 3.062×10^{-4}
b. 5.21×10^4
c. 9.6×10^{-5}
d. 6.002×10^6

PRACTICE PROBLEM 19

Perform each indicated operation. Write each result in standard decimal notation.

a. $(9 \times 10^7)(4 \times 10^{-9})$
b. $\dfrac{8 \times 10^4}{2 \times 10^{-3}}$

Answers
18. a. 0.0003062, **b.** 52,100,
c. 0.000096, **d.** 6,002,000,
19. a. 0.36, **b.** 40,000,000

✔ **Concept Check Answer**
a. 2.1×10^5, **b.** 2.7×10^4,
c. 5.6×10^{-4}

To enter a number written in scientific notation on a scientific calculator, locate the scientific notation key, which may be marked $\boxed{\text{EE}}$ or $\boxed{\text{EXP}}$. To enter 3.1×10^7, press $\boxed{3.1}$ $\boxed{\text{EE}}$ $\boxed{7}$. The display should read $\boxed{3.1 \quad 07}$.

Enter each number written in scientific notation on your calculator.

1. 5.31×10^3

2. -4.8×10^{14}

3. 6.6×10^{-9}

4. -9.9811×10^{-2}

Multiply each of the following on your calculator. Notice the form of the result.

5. $3,000,000 \times 5,000,000$

6. $230,000 \times 1,000$

Multiply each of the following on your calculator. Write the product in scientific notation.

7. $(3.26 \times 10^6)(2.5 \times 10^{13})$

8. $(8.76 \times 10^{-4})(1.237 \times 10^9)$

Mental Math

Write each expression using positive exponents only.

1. $5x^{-2}$ **2.** $3x^{-3}$ **3.** $\dfrac{1}{y^{-6}}$ **4.** $\dfrac{1}{x^{-3}}$ **5.** $\dfrac{4}{y^{-3}}$ **6.** $\dfrac{16}{y^{-7}}$

3.2 EXERCISE SET

Objective **A** *Simplify each expression. Write each result using positive exponents only. See Examples 1 through 8.*

1. 4^{-3} **2.** 6^{-2} **3.** $7x^{-3}$ **4.** $(7x)^{-3}$ **5.** $\left(-\dfrac{1}{4}\right)^{-3}$ **6.** $\left(-\dfrac{1}{8}\right)^{-2}$

7. $3^{-1} + 2^{-1}$ **8.** $4^{-1} + 4^{-2}$ **9.** $\dfrac{1}{p^{-3}}$ **10.** $\dfrac{1}{q^{-5}}$ **11.** $\dfrac{p^{-5}}{q^{-4}}$ **12.** $\dfrac{r^{-5}}{s^{-2}}$

13. $\dfrac{x^{-2}}{x}$ **14.** $\dfrac{y}{y^{-3}}$ **15.** $\dfrac{z^{-4}}{z^{-7}}$ **16.** $\dfrac{x^{-4}}{x^{-1}}$ **17.** $3^{-2} + 3^{-1}$ **18.** $4^{-2} - 4^{-3}$

19. $(-3)^{-2}$ **20.** $(-2)^{-6}$ **21.** $\dfrac{-1}{p^{-4}}$ **22.** $\dfrac{-1}{y^{-6}}$ **23.** $-2^0 - 3^0$ **24.** $5^0 + (-5)^0$

Objective **B** *Simplify each expression. Write each result using positive exponents only. See Examples 9 through 16.*

25. $\dfrac{x^2 x^5}{x^3}$ **26.** $\dfrac{y^4 y^5}{y^6}$ **27.** $\dfrac{p^2 p}{p^{-1}}$ **28.** $\dfrac{y^3 y}{y^{-2}}$ **29.** $\dfrac{(m^5)^4 m}{m^{10}}$ **30.** $\dfrac{(x^2)^8 x}{x^9}$

31. $\dfrac{r}{r^{-3}r^{-2}}$

32. $\dfrac{p}{p^{-3}q^{-5}}$

33. $(x^5y^3)^{-3}$

34. $(z^5x^5)^{-3}$

35. $\dfrac{(x^2)^3}{x^{10}}$

36. $\dfrac{(y^4)^2}{y^{12}}$

37. $\dfrac{(a^5)^2}{(a^3)^4}$

38. $\dfrac{(x^2)^5}{(x^4)^3}$

39. $\dfrac{8k^4}{2k}$

40. $\dfrac{27r^6}{3r^4}$

41. $\dfrac{-6m^4}{-2m^3}$

42. $\dfrac{15a^4}{-15a^5}$

43. $\dfrac{-24a^6b}{6ab^2}$

44. $\dfrac{-5x^4y^5}{15x^4y^2}$

45. $\dfrac{6x^2y^3}{-7x^2y^5}$

46. $\dfrac{-8xa^2b}{-5xa^5b}$

47. $(3a^2b^{-4})^3$

48. $(5x^3y^{-2})^2$

49. $(a^{-5}b^2)^{-6}$

50. $(4^{-1}x^5)^{-2}$

51. $\left(\dfrac{x^{-2}y^4}{x^3y^7}\right)^2$

52. $\left(\dfrac{a^5b}{a^7b^{-2}}\right)^{-3}$

53. $\dfrac{4^2z^{-3}}{4^3z^{-5}}$

54. $\dfrac{5^{-1}z^7}{5^{-2}z^9}$

55. $\dfrac{3^{-1}x^4}{3^3x^{-7}}$

56. $\dfrac{2^{-3}x^{-4}}{2^2x}$

57. $\dfrac{7ab^{-4}}{7^{-1}a^{-3}b^2}$

58. $\dfrac{6^{-5}x^{-1}y^2}{6^{-2}x^{-4}y^4}$

59. $\dfrac{-12m^5n^{-7}}{4m^{-2}n^{-3}}$

60. $\dfrac{-15r^{-6}s}{5r^{-4}s^{-3}}$

61. $\left(\dfrac{a^{-5}b}{ab^3}\right)^{-4}$

62. $\left(\dfrac{r^{-2}s^{-3}}{r^{-4}s^{-3}}\right)^{-3}$

63. $(5^2)(8)(2^0)$

64. $(3^4)(7^0)(2)$

65. $\dfrac{(xy^3)^5}{(xy)^{-4}}$

66. $\dfrac{(rs)^{-3}}{(r^2s^3)^2}$

67. $\dfrac{(-2xy^{-3})^{-3}}{(xy^{-1})^{-1}}$

68. $\dfrac{(-3x^2y^2)^{-2}}{(xyz)^{-2}}$

69. $\dfrac{(a^4b^{-7})^{-5}}{(5a^2b^{-1})^{-2}}$

70. $\dfrac{(a^6b^{-2})^4}{(4a^{-3}b^{-3})^3}$

△ **71.** Find the volume of the cube.

$\dfrac{3x^{-2}}{z}$ inches

△ **72.** Find the area of the triangle.

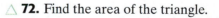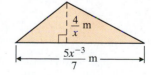

$\dfrac{4}{x}$ m

$\dfrac{5x^{-3}}{7}$ m

Objective C *Write each number in scientific notation. See Example 17.*

73. 78,000

74. 9,300,000,000

75. 0.00000167

76. 0.00000017

77. 0.00635

78. 0.00194

79. 1,160,000

80. 700,000

81. At this writing, the world's largest optical telescopes are the twin Keck Telescopes located near the summit of Mauna Kea in Hawaii. The elevation of the Keck Telescopes is about 13,600 feet above sea level. Write 13,600 in scientific notation. (*Source:* W.M. Keck Observatory)

82. After more than 30 years, the *Pioneer 10* spacecraft sent its last signal to Earth. Launched on March 2, 1972, it became the first spacecraft to leave our solar system. When it transmitted its last signal, in January 2003, it was approximately 8,000,000,000 miles from Earth. Write 8,000,000,000 in scientific notation. (*Source:* NASA Ames Research Center)

Objective **D** *Write each number in standard notation. See Example 18.*

83. 8.673×10^{-10}

84. 9.056×10^{-4}

85. 3.3×10^{-2}

86. 4.8×10^{-6}

87. 2.032×10^{4}

88. 9.07×10^{10}

89. Each second, the Sun converts 7.0×10^{8} tons of hydrogen into helium and energy in the form of gamma rays. Write this number in standard notation. (*Source:* Students for the Exploration and Development of Space)

90. In chemistry, Avogadro's number is the number of atoms in one mole of an element. Avogadro's number is $6.02214199 \times 10^{23}$. Write this number in standard notation. (*Source:* National Institute of Standards and Technology)

Objectives **C** **D** **Mixed Practice** *See Examples 17 and 18. Below are some interesting facts about the Internet. If a number is written in standard form, write it in scientific notation. If a number is written in scientific notation, write it in standard form.*

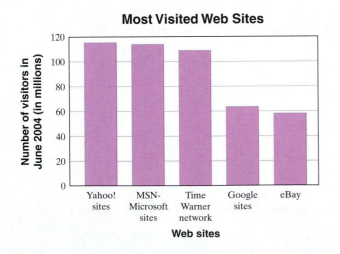

Most Visited Web Sites

91. The total number of Internet users is 940,000,000.

92. The total number of Internet hosts (sites) is 233,000,000.

93. In a recent year, the revenue generated by the Internet was 1.23×10^{12} dollars.

94. The estimated number of e-mail boxes is 1.2×10^{9}.

95. The bar graph above shows the most visited Web sites on the computer. Estimate the height of the tallest bar and the shortest bar. Then write each number in scientific notation.

96. Junk e-mail (SPAM) costs consumers and businesses an estimated $23,000,000,000.

Objective D *Evaluate each expression using exponential rules. Write each result in standard notation. See Example 19.*

97. $(1.2 \times 10^{-3})(3 \times 10^{-2})$

98. $(2.5 \times 10^{6})(2 \times 10^{-6})$

99. $(4 \times 10^{-10})(7 \times 10^{-9})$

100. $(5 \times 10^{6})(4 \times 10^{-8})$

101. $\dfrac{8 \times 10^{-1}}{16 \times 10^{5}}$

102. $\dfrac{25 \times 10^{-4}}{5 \times 10^{-9}}$

103. $\dfrac{1.4 \times 10^{-2}}{7 \times 10^{-8}}$

104. $\dfrac{0.4 \times 10^{5}}{0.2 \times 10^{11}}$

105. Although the actual amount varies by season and time of day, the average volume of water that flows over Niagara Falls (the American and Canadian falls combined) each second is 7.5×10^{5} gallons. How much water flows over Niagara Falls in an hour? Write the result in scientific notation. (*Hint:* 1 hour equals 3600 seconds) (*Source:* niagarafallslive.com)

106. A beam of light travels 9.460×10^{12} kilometers per year. How far does light travel in 10,000 years? Write the result in scientific notation.

Review

Simplify each expression by combining any like terms. See Section 2.1.

107. $3x - 5x + 7$

108. $7w + w - 2w$

109. $y - 10 + y$

110. $-6z + 20 - 3z$

111. $7x + 2 - 8x - 6$

112. $10y - 14 - y - 14$

Concept Extensions

Simplify.

113. $(2a^{3})^{3}a^{4} + a^{5}a^{8}$

114. $(2a^{3})^{3}a^{-3} + a^{11}a^{-5}$

Fill in the boxes so that each statement is true. (More than one answer is possible for these exercises.)

115. $x^{\square} = \dfrac{1}{x^{5}}$

116. $7^{\square} = \dfrac{1}{49}$

117. $z^{\square} \cdot z^{\square} = z^{-10}$

118. $(x^{\square})^{\square} = x^{-15}$

119. Which is larger? See the Concept Check in this section.
   a. 9.7×10^{-2} or 1.3×10^{1}
   b. 8.6×10^{5} or 4.4×10^{7}
   c. 6.1×10^{-2} or 5.6×10^{-4}

120. It was stated earlier that for an integer n,
$$x^{-n} = \dfrac{1}{x^{n}}, \quad x \neq 0$$
Explain why x may not equal 0.

121. Determine whether each statement is true or false.
   a. $5^{-1} < 5^{-2}$
   b. $\left(\dfrac{1}{5}\right)^{-1} < \left(\dfrac{1}{5}\right)^{-2}$
   c. $a^{-1} < a^{-2}$ for all nonzero numbers.

Simplify each expression. Assume that variables represent positive integers.

122. $a^{-4m} \cdot a^{5m}$

123. $(x^{-3s})^{3}$

124. $(3y^{2z})^{3}$

125. $a^{4m+1} \cdot a^{4}$

A Define Term and Coefficient of a Term.

B Define Polynomial, Monomial, Binomial, Trinomial, and Degree.

C Evaluate Polynomials for Given Replacement Values.

D Simplify a Polynomial by Combining Like Terms.

E Simplify a Polynomial in Several Variables.

F Write a Polynomial in Descending Powers of the Variable and with No Missing Powers of the Variable.

3.3 INTRODUCTION TO POLYNOMIALS

Objective **A** Defining Term and Coefficient

In this section, we introduce a special algebraic expression called a polynomial. Let's first review some definitions presented in Section 1.8.

Recall that a term is a number or the product of a number and variables raised to powers. The terms of an expression are separated by plus signs. The terms of the expression $4x^2 + 3x$ are $4x^2$ and $3x$. The terms of the expression $9x^4 - 7x - 1$, or $9x^4 + (-7x) + (-1)$, are $9x^4$, $-7x$, and -1.

Expression	Terms
$4x^2 + 3x$	$4x^2, 3x$
$9x^4 - 7x - 1$	$9x^4, -7x, -1$
$7y^3$	$7y^3$

The **numerical coefficient** of a term, or simply the **coefficient**, is the numerical factor of each term. If no numerical factor appears in the term, then the coefficient is understood to be 1. If the term is a number only, it is called a **constant term** or simply a **constant.**

Term	Coefficient
x^5	1
$3x^2$	3
$-4x$	-4
$-x^2y$	-1
3 (constant)	3

PRACTICE PROBLEM 1

Complete the table for the expression
$-6x^6 + 4x^5 + 7x^3 - 9x^2 - 1$.

Term	Coefficient
$7x^3$	
	-9
$-6x^6$	
	4
-1	

EXAMPLE 1

Complete the table for the expression $7x^5 - 8x^4 + x^2 - 3x + 5$.

Term	Coefficient
x^2	
	-8
$-3x$	
	7
5	

Solution: The completed table is shown below.

Term	Coefficient
x^2	1
$-8x^4$	-8
$-3x$	-3
$7x^5$	7
5	5

□ **Work Practice Problem 1**

Objective **B** **Defining Polynomial, Monomial, Binomial, Trinomial, and Degree**

Now we are ready to define what we mean by a polynomial.

Polynomial

A **polynomial in x** is a finite sum of terms of the form ax^n, where a is a real number and n is a whole number.

For example,

$$x^5 - 3x^3 + 2x^2 - 5x + 1$$

is a polynomial in x. Notice that this polynomial is written in **descending powers** of x because the powers of x decrease from left to right. (Recall that the term 1 can be thought of as $1x^0$.)

On the other hand,

$$x^{-5} + 2x - 3$$

is **not** a polynomial because one of its terms contains a variable with an exponent, -5, that is not a whole number.

Types of Polynomials

A **monomial** is a polynomial with exactly one term.

A **binomial** is a polynomial with exactly two terms.

A **trinomial** is a polynomial with exactly three terms.

The following are examples of monomials, binomials, and trinomials. Each of these examples is also a polynomial.

Polynomials			
Monomials	**Binomials**	**Trinomials**	**More Than Three Terms**
ax^2	$x + y$	$x^2 + 4xy + y^2$	$5x^3 - 6x^2 + 3x - 6$
$-3z$	$3p + 2$	$x^5 + 7x^2 - x$	$-y^5 + y^4 - 3y^3 - y^2 + y$
4	$4x^2 - 7$	$-q^4 + q^3 - 2q$	$x^6 + x^4 - x^3 + 1$

Each term of a polynomial has a degree. The **degree of a term in one variable** is the exponent on the variable.

EXAMPLE 2 Identify the degree of each term of the trinomial $12x^4 - 7x + 3$.

Solution: The term $12x^4$ has degree 4.

The term $-7x$ has degree 1 since $-7x$ is $-7x^1$.

The term 3 has degree 0 since 3 is $3x^0$.

■ **Work Practice Problem 2**

Each polynomial also has a degree.

Degree of a Polynomial

The **degree of a polynomial** is the greatest degree of any term of the polynomial.

PRACTICE PROBLEM 2

Identify the degree of each term of the trinomial $-15x^3 + 2x^2 - 5$.

Answer

2. $3; 2; 0$

PRACTICE PROBLEM 3

Find the degree of each polynomial and tell whether the polynomial is a monomial, binomial, trinomial, or none of these.

a. $-6x + 14$

b. $9x - 3x^6 + 5x^4 + 2$

c. $10x^2 - 6x - 6$

EXAMPLE 3 Find the degree of each polynomial and tell whether the polynomial is a monomial, binomial, trinomial, or none of these.

a. $-2t^2 + 3t + 6$ **b.** $15x - 10$ **c.** $7x + 3x^3 + 2x^2 - 1$

Solution:

a. The degree of the trinomial $-2t^2 + 3t + 6$ is 2, the greatest degree of any of its terms.

b. The degree of the binomial $15x - 10$ or $15x^1 - 10$ is 1.

c. The degree of the polynomial $7x + 3x^3 + 2x^2 - 1$ is 3. The polynomial is neither a monomial, binomial, nor trinomial.

◻ **Work Practice Problem 3**

Objective C Evaluating Polynomials

Polynomials have different values depending on the replacement values for the variables. When we find the value of a polynomial for a given replacement value, we are evaluating the polynomial for that value.

PRACTICE PROBLEM 4

Evaluate each polynomial when $x = -1$.

a. $-2x + 10$

b. $6x^2 + 11x - 20$

EXAMPLE 4 Evaluate each polynomial when $x = -2$.

a. $-5x + 6$ **b.** $3x^2 - 2x + 1$

Solution:

a. $\begin{aligned} -5x + 6 &= -5(-2) + 6 \quad \text{Replace } x \text{ with } -2. \\ &= 10 + 6 \\ &= 16 \end{aligned}$

b. $\begin{aligned} 3x^2 - 2x + 1 &= 3(-2)^2 - 2(-2) + 1 \quad \text{Replace } x \text{ with } -2. \\ &= 3(4) + 4 + 1 \\ &= 12 + 4 + 1 \\ &= 17 \end{aligned}$

◻ **Work Practice Problem 4**

Many physical phenomena can be modeled by polynomials.

PRACTICE PROBLEM 5

Find the height of the object in example 5 when $t = 2$ seconds and $t = 4$ seconds.

EXAMPLE 5 **Finding Free-Fall Time**

The Swiss Re Building, completed in London in 2003, is a unique building. Londoners often refer to it as the "pickle building." The building is 592.1 feet tall. An object is dropped from the highest point of this building. Neglecting air resistance, the height in feet of the object above ground at time t seconds is given by the polynomial $-16t^2 + 592.1$. Find the height of the object when $t = 1$ second, and when $t = 6$ seconds.

Solution: To find each height, we evaluate the polynomial when $t = 1$ and when $t = 6$.

$$\begin{aligned} -16t^2 + 592.1 &= -16(1)^2 + 592.1 \quad \text{Replace } t \text{ with } 1. \\ &= -16(1) + 592.1 \\ &= -16 + 592.1 \\ &= 576.1 \end{aligned}$$

The height of the object at 1 second is 576.1 feet.

$$\begin{aligned} -16t^2 + 592.1 &= -16(6)^2 + 592.1 \quad \text{Replace } t \text{ with } 6. \\ &= -16(36) + 592.1 \\ &= -576 + 592.1 = 16.1 \end{aligned}$$

Answers

3. a. binomial, 1, **b.** none of these, 6, **c.** trinomial, 2, **4. a.** 12, **b.** −25, **5.** 528.1 feet, 336.1 feet

The height of the object at 6 seconds is 16.1 feet.

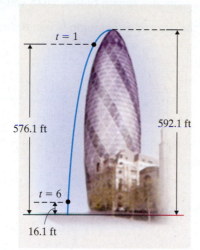

576.1 ft
592.1 ft
16.1 ft
$t = 1$
$t = 6$

🔲 **Work Practice Problem 5**

Objective D Simplifying Polynomials by Combining Like Terms

We can simplify polynomials with like terms by combining the like terms. Recall from Section 1.8 that like terms are terms that contain exactly the same variables raised to exactly the same powers.

Like Terms	Unlike Terms
$5x^2, -7x^2$	$3x, 3y$
$y, 2y$	$-2x^2, -5x$
$\frac{1}{2}a^2b, -a^2b$	$6st^2, 4s^2t$

Only like terms can be combined. We combine like terms by applying the distributive property.

EXAMPLES Simplify each polynomial by combining any like terms.

6. $-3x + 7x = (-3 + 7)x = 4x$

7. $11x^2 + 5 + 2x^2 - 7 = 11x^2 + 2x^2 + 5 - 7$
$$= 13x^2 - 2$$

8. $9x^3 + x^3 = 9x^3 + 1x^3$ Write x^3 as $1x^3$.
$$= 10x^3$$

9. $5x^2 + 6x - 9x - 3 = 5x^2 - 3x - 3$ Combine like terms $6x$ and $-9x$.

10. $\frac{2}{5}x^4 + \frac{2}{3}x^3 - x^2 + \frac{1}{10}x^4 - \frac{1}{6}x^3$

$$= \left(\frac{2}{5} + \frac{1}{10}\right)x^4 + \left(\frac{2}{3} - \frac{1}{6}\right)x^3 - x^2$$

$$= \left(\frac{4}{10} + \frac{1}{10}\right)x^4 + \left(\frac{4}{6} - \frac{1}{6}\right)x^3 - x^2$$

$$= \frac{5}{10}x^4 + \frac{3}{6}x^3 - x^2$$

$$= \frac{1}{2}x^4 + \frac{1}{2}x^3 - x^2$$

🔲 **Work Practice Problems 6–10**

PRACTICE PROBLEMS 6–10

Simplify each polynomial by combining any like terms.

6. $-6y + 8y$

7. $14y^2 + 3 - 10y^2 - 9$

8. $7x^3 + x^3$

9. $23x^2 - 6x - x - 15$

10. $\frac{2}{7}x^3 - \frac{1}{4}x + 2 - \frac{1}{2}x^3 + \frac{3}{8}x$

Answers
6. $2y$, **7.** $4y^2 - 6$, **8.** $8x^3$,
9. $23x^2 - 7x - 15$,
10. $-\frac{3}{14}x^3 + \frac{1}{8}x + 2$

PRACTICE PROBLEM 11 △

Write a polynomial that describes the total area of the squares and rectangles shown below. Then simplify the polynomial.

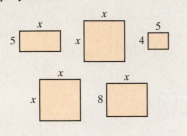

△ **EXAMPLE 11** Write a polynomial that describes the total area of the squares and rectangles shown below. Then simplify the polynomial.

Solution: Recall that the area of a rectangle is length times width.

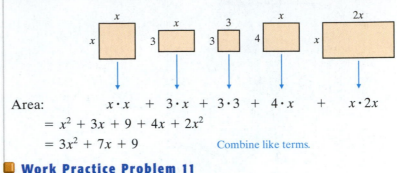

Area: $x \cdot x$ + $3 \cdot x$ + $3 \cdot 3$ + $4 \cdot x$ + $x \cdot 2x$

$= x^2 + 3x + 9 + 4x + 2x^2$

$= 3x^2 + 7x + 9$ Combine like terms.

Work Practice Problem 11

Objective E Simplifying Polynomials Containing Several Variables

A polynomial may contain more than one variable. One example is

$$5x + 3xy^2 - 6x^2y^2 + x^2y - 2y + 1$$

We call this expression a polynomial in several variables.

The **degree of a term** with more than one variable is the sum of the exponents on the variables. The **degree of the polynomial** in several variables is still the greatest degree of the terms of the polynomial.

PRACTICE PROBLEM 12

Identify the degrees of the terms and the degree of the polynomial $-2x^3y^2 + 4 - 8xy + 3x^3y + 5xy^2$.

EXAMPLE 12 Identify the degrees of the terms and the degree of the polynomial $5x + 3xy^2 - 6x^2y^2 + x^2y - 2y + 1$.

Solution: To organize our work, we use a table.

Terms of Polynomial	Degree of Term	Degree of Polynomial
$5x$	1	
$3xy^2$	1 + 2 or 3	
$-6x^2y^2$	2 + 2 or 4	4 (greatest degree)
x^2y	2 + 1 or 3	
$-2y$	1	
1	0	

Work Practice Problem 12

To simplify a polynomial containing several variables, we combine any like terms.

PRACTICE PROBLEMS 13–14

Simplify each polynomial by combining any like terms.

13. $11ab - 6a^2 - ba + 8b^2$
14. $7x^2y^2 + 2y^2 - 4y^2x^2 + x^2 - y^2 + 5x^2$

EXAMPLES Simplify each polynomial by combining any like terms.

13. $3xy - 5y^2 + 7yx - 9x^2 = (3 + 7)xy - 5y^2 - 9x^2$

$= 10xy - 5y^2 - 9x^2$

14. $9a^2b - 6a^2 + 5b^2 + a^2b - 11a^2 + 2b^2$

$= 10a^2b - 17a^2 + 7b^2$

Helpful Hint
This term can be written as $7yx$ or $7xy$.

Work Practice Problems 13–14

Answers

11. $2x^2 + 13x + 20$,
12. $5, 0, 2, 4, 3; 5$,
13. $10ab - 6a^2 + 8b^2$,
14. $3x^2y^2 + y^2 + 6x^2$

Objective F Inserting "Missing" Terms

To prepare for dividing polynomials in Section 3.7, let's practice writing a polynomial in descending powers of the variable and with no "missing" powers.

Recall from Objective B that a polynomial such as

$$x^5 - 3x^3 + 2x^2 - 5x + 1$$

is written in descending powers of x because the powers of x decrease from left to right. Study the decreasing powers of x and notice that there is a "missing" power of x. This missing power is x^4. Writing a polynomial in decreasing powers of the variable helps you immediately determine important features of the polynomial, such as its degree. It is also sometimes helpful to write a polynomial so that there are no "missing" powers of x. For our polynomial above, if we simply insert a term of $0x^4$, which equals 0, we have an equivalent polynomial with no missing powers of x.

$$x^5 - 3x^3 + 2x^2 - 5x + 1 = x^5 + 0x^4 - 3x^3 + 2x^2 - 5x + 1$$

EXAMPLE 15 Write each polynomial in descending powers of the variable with no missing powers.

a. $x^2 - 4$

b. $3m^3 - m + 1$

c. $2x + x^4$

Solution:

a. $x^2 - 4 = x^2 + 0x^1 - 4$ or $x^2 + 0x - 4$ Insert a missing term of $0x^1$ or $0x$.

b. $3m^3 - m + 1 = 3m^3 + 0m^2 - m + 1$ Insert a missing term of $0m^2$.

c. $2x + x^4 = x^4 + 2x$ Write in descending power of variable.

$\qquad = x^4 + 0x^3 + 0x^2 + 2x + 0x^0$ Insert missing terms of $0x^3, 0x^2$, and $0x^0$ (or 0).

◻ **Work Practice Problem 15**

Helpful Hint

Since there is no constant as a last term, we insert a $0x^0$. This $0x^0$ (or 0) is the final power of x in our polynomial.

PRACTICE PROBLEM 15

Write each polynomial in descending powers of the variable with no missing powers.

a. $x^2 + 9$

b. $9m^3 + m^2 - 5$

c. $-3a^3 + a^4$

Objective A *Complete each table for each polynomial. See Example 1.*

1. $x^2 - 3x + 5$

Term	Coefficient
x^2	
	-3
5	

2. $2x^3 - x + 4$

Term	Coefficient
	2
$-x$	
4	

3. $-5x^4 + 3.2x^2 + x - 5$

Term	Coefficient
$-5x^4$	
$3.2x^2$	
x	
-5	

4. $9.7x^7 - 3x^5 + x^3 - \dfrac{1}{4}x^2$

Term	Coefficient
$9.7x^7$	
$-3x^5$	
x^3	
$-\dfrac{1}{4}x^2$	

Objective B *Find the degree of each polynomial and determine whether it is a monomial, binomial, trinomial, or none of these. See Examples 2 and 3.*

5. $x + 2$

6. $-6y + 4$

7. $9m^3 - 5m^2 + 4m - 8$

8. $a + 5a^2 + 3a^3 - 4a^4$

9. $12x^4 - x^6 - 12x^2$

10. $7r^2 + 2r - 3r^5$

11. $3z - 5z^4$

12. $5y^6 + 2$

Objective C *Evaluate each polynomial when* **(a)** $x = 0$ *and* **(b)** $x = -1$. *See Examples 4 and 5.*

13. $5x - 6$

14. $2x - 10$

15. $x^2 - 5x - 2$

16. $x^2 + 3x - 4$

17. $-x^3 + 4x^2 - 15$

18. $-2x^3 + 3x^2 - 6$

A rocket is fired upward from the ground with an initial velocity of 200 feet per second. Neglecting air resistance, the height of the rocket at any time t can be described in feet by the polynomial $-16t^2 + 200t$. *Find the height of the rocket at the time given in Exercises 19 through 22. See Example 5.*

	Time, t (in seconds)	Height $-16t^2 + 200t$
19.	1	
20.	5	
21.	7.6	
22.	10.3	

23. The polynomial $-24x^2 + 336x - 132$ represents the average number of visitors (in thousands) per day to National Park Service areas, where x represents the month of the year. Use this model to predict the average daily attendance at our national parks for the month of July. (*Hint:* July is the seventh month.) (*Source:* Based on data from the National Park Service)

24. The number of wireless telephone subscribers (in millions) x years after 1994 is given by the polynomial $0.56x^2 + 10x + 15.25$ for 1994 through 2004. Use this model to predict the number of wireless telephone subscribers in 2010 ($x = 16$). (*Source:* Based on data from Cellular Telecommunications & Internet Association)

Objective D *Simplify each expression by combining like terms. See Examples 6 through 10.*

25. $9x - 20x$

26. $14y - 30y$

27. $14x^3 + 9x^3$

28. $18x^3 + 4x^3$

29. $7x^2 + 3 + 9x^2 - 10$

30. $8x^2 + 4 + 11x^2 - 20$

31. $15x^2 - 3x^2 - 13$

32. $12k^3 - 9k^3 + 11$

33. $8s - 5s + 4s$

34. $5y + 7y - 6y$

35. $0.1y^2 - 1.2y^2 + 6.7 - 1.9$

36. $7.6y + 3.2y^2 - 8y - 2.5y^2$

37. $\frac{2}{3}x^4 + 12x^3 + \frac{1}{6}x^4 - 19x^3 - 19$

38. $\frac{2}{5}x^4 - 23x^2 + \frac{1}{15}x^4 + 5x^2 - 5$

39. $\frac{3}{20}x^3 + \frac{1}{10} - \frac{3}{10}x - \frac{1}{5} - \frac{7}{20}x + 6x^2$

40. $\frac{5}{16}x^3 - \frac{1}{8} + \frac{3}{8}x + \frac{1}{4} - \frac{9}{16}x - 14x^2$

Objective E *Identify the degrees of the terms and the degree of the polynomial. See Example 12.*

41. $9ab - 6a + 5b - 3$

42. $y^4 - 6y^3x + 2x^2y^2 - 5y^2 + 3$

43. $x^3y - 6 + 2x^2y^2 + 5y^3$

44. $2a^2b + 10a^4b - 9ab + 6$

Simplify each polynomial by combining any like terms. See Examples 13 and 14.

45. $3ab - 4a + 6ab - 7a$

46. $-9xy + 7y - xy - 6y$

47. $4x^2 - 6xy + 3y^2 - xy$

48. $3a^2 - 9ab + 4b^2 - 7ab$

49. $5x^2y + 6xy^2 - 5yx^2 + 4 - 9y^2x$

50. $17a^2b - 16ab^2 + 3a^3 + 4ba^3 - b^2a$

51. $14y^3 - 9 + 3a^2b^2 - 10 - 19b^2a^2$

52. $18x^4 + 2x^3y^3 - 1 - 2y^3x^3 - 17x^4$

Objective ⬛F *Write each polynomial in descending powers of the variable and with no missing powers. See Example 15.*

53. $7x^2 + 3$

54. $5x^2 - 2$

55. $x^3 - 64$

56. $x^3 - 8$

57. $5y^3 + 2y - 10$

58. $6m^3 - 3m + 4$

59. $8y + 2y^4$

60. $11z + 4z^4$

61. $6x^5 + x^3 - 3x + 15$

62. $9y^5 - y^2 + 2y - 11$

Objective ⬛D *Write a polynomial that describes the total area of each set of rectangles and squares shown in Exercises 63 and 64. Then simplify the polynomial. See Example 11.*

△ **63.**

△ **64.**

Recall that the perimeter of a figure such as the ones shown in Exercises 65 and 66 is the sum of the lengths of its sides. Write each perimeter as a polynomial. Then simplify the polynomial.

△ **65.**

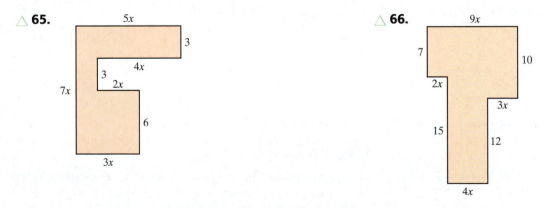

△ **66.**

Review

Simplify each expression. See Section 1.8.

67. $4 + 5(2x + 3)$ **68.** $9 - 6(5x + 1)$ **69.** $2(x - 5) + 3(5 - x)$ **70.** $-3(w + 7) + 5(w + 1)$

Concept Extensions

71. Describe how to find the degree of a term.

72. Describe how to find the degree of a polynomial.

73. Explain why xyz is a monomial while $x + y + z$ is a trinomial.

74. Explain why the degree of the term $5y^3$ is 3 and the degree of the polynomial $2y + y + 2y$ is 1.

Simplify, if possible.

75. $x^4 \cdot x^9$

76. $x^4 + x^9$

77. $a \cdot b^3 \cdot a^2 \cdot b^7$

78. $a + b^3 + a^2 + b^7$

79. $(y^5)^4 + (y^2)^{10}$

80. $x^5 y^2 + y^2 x^5$

Fill in the boxes so that the terms in each expression can be combined. Then simplify. Each exercise has more than one solution.

81. $7x^{\square} + 2x^{\square}$

82. $(3y^2)^{\square} + (4y^3)^{\square}$

83. Explain why the height of the rocket in Exercises 19 through 22 increases and then decreases as time passes.

84. Approximate (to the nearest tenth of a second) how long before the rocket in Exercises 19 through 22 hits the ground.

Simplify each polynomial by combining like terms.

85. $1.85x^2 - 3.76x + 9.25x^2 + 10.76 - 4.21x$

86. $7.75x + 9.16x^2 - 1.27 - 14.58x^2 - 18.34$

STUDY SKILLS BUILDER

Are You Organized?

Have you ever had trouble finding a completed assignment? When it's time to study for a test, are your notes neat and organized? Have you ever had trouble reading your own mathematics handwriting? (Be honest—I have.)

When any of these things happen, it's time to get organized. Here are a few suggestions:

Write your notes and complete your homework assignment in a notebook with pockets (spiral or ring binder.) Take class notes in this notebook, and then follow the notes with your completed homework assignment. When you receive graded papers or handouts, place them in the notebook pocket so that you will not lose them.

Remember to mark (possibly with an exclamation point) any note(s) that seem extra important to you. Also remember to mark (possibly with a question mark) any notes or homework that you are having trouble with. Don't forget to see your instructor or a math tutor to help you with the concepts or exercises that you are having trouble understanding.

Also, if you are having trouble reading your own handwriting, *slow down* and write your mathematics work clearly!

Exercises

1. Have you been completing your assignments on time?

2. Have you been correcting any exercises you may be having difficulty with?

3. If you are having trouble with a mathematical concept or correcting any homework exercises, have you visited your instructor, a tutor, or your campus math lab?

4. Are you taking lecture notes in your mathematics course? (By the way, these notes should include worked-out examples solved by your instructor.)

5. Is your mathematics course material (handouts, graded papers, lecture notes) organized?

6. If your answer to Exercise 5 is no, take a moment and review your course material. List at least two ways that you might better organize it. Then read the Study Skills Builder on organizing a notebook in Chapter 2.

3.4 ADDING AND SUBTRACTING POLYNOMIALS

Objective **A** Adding Polynomials

To add polynomials, we use commutative and associative properties and then combine like terms. To see if you are ready to add polynomials, try the Concept Check.

✔ **Concept Check** When combining like terms in the expression $5x - 8x^2 - 8x$, which of the following is the proper result?

a. $-11x^2$ **b.** $-3x - 8x^2$ **c.** $-11x$ **d.** $-11x^4$

> ### To Add Polynomials
>
> To add polynomials, combine all like terms.

PRACTICE PROBLEMS 1–2

Add.

1. $(3x^5 - 7x^3 + 2x - 1)$
$+ (3x^3 - 2x)$

2. $(5x^2 - 2x + 1)$
$+ (-6x^2 + x - 1)$

EXAMPLES Add.

1. $(4x^3 - 6x^2 + 2x + 7) + (5x^2 - 2x)$

$= 4x^3 - 6x^2 + 2x + 7 + 5x^2 - 2x$ Remove parentheses.

$= 4x^3 + (-6x^2 + 5x^2) + (2x - 2x) + 7$ Combine like terms.

$= 4x^3 - x^2 + 7$ Simplify.

2. $(-2x^2 + 5x - 1) + (-2x^2 + x + 3)$

$= -2x^2 + 5x - 1 - 2x^2 + x + 3$ Remove parentheses.

$= (-2x^2 - 2x^2) + (5x + 1x) + (-1 + 3)$ Combine like terms.

$= -4x^2 + 6x + 2$ Simplify.

▢ **Work Practice Problems 1–2**

Just as we can add numbers vertically, polynomials can be added vertically if we line up like terms underneath one another.

PRACTICE PROBLEM 3

Add $(9y^2 - 6y + 5)$ and $(4y + 3)$ using a vertical format.

EXAMPLE 3 Add $(7y^3 - 2y^2 + 7)$ and $(6y^2 + 1)$ using a vertical format.

Solution: Vertically line up like terms and add.

$$
\begin{array}{r}
7y^3 - 2y^2 + 7 \\
6y^2 + 1 \\
\hline
7y^3 + 4y^2 + 8
\end{array}
$$

▢ **Work Practice Problem 3**

Objective **B** Subtracting Polynomials

To subtract one polynomial from another, recall the definition of subtraction. To subtract a number, we add its opposite: $a - b = a + (-b)$. To subtract a polynomial, we also add its opposite. Just as $-b$ is the opposite of b, $-(x^2 + 5)$ is the opposite of $(x^2 + 5)$.

Answers

1. $3x^5 - 4x^3 - 1$,   **2.** $-x^2 - x$,

3. $9y^2 - 2y + 8$

✔ **Concept Check Answer**

b

> ### To Subtract Polynomials
>
> To subtract two polynomials, change the signs of the terms of the polynomial being subtracted and then add.

EXAMPLE 4 Subtract: $(5x - 3) - (2x - 11)$

Solution: From the definition of subtraction, we have

$$(5x - 3) - (2x - 11) = (5x - 3) + [-(2x - 11)] \quad \text{Add the opposite.}$$
$$= (5x - 3) + (-2x + 11) \quad \text{Apply the distributive property.}$$
$$= 5x - 3 - 2x + 11 \quad \text{Remove parentheses.}$$
$$= 3x + 8 \quad \text{Combine like terms.}$$

🔲 **Work Practice Problem 4**

EXAMPLE 5 Subtract: $(2x^3 + 8x^2 - 6x) - (2x^3 - x^2 + 1)$

Solution: First, we change the sign of each term of the second polynomial; then we add.

$$(2x^3 + 8x^2 - 6x) - (2x^3 - x^2 + 1)$$
$$= (2x^3 + 8x^2 - 6x) + (-2x^3 + x^2 - 1)$$
$$= 2x^3 + 8x^2 - 6x - 2x^3 + x^2 - 1$$
$$= 2x^3 - 2x^3 + 8x^2 + x^2 - 6x - 1$$
$$= 9x^2 - 6x - 1 \quad \text{Combine like terms.}$$

🔲 **Work Practice Problem 5**

Just as polynomials can be added vertically, so can they be subtracted vertically.

EXAMPLE 6 Subtract $(5y^2 + 2y - 6)$ from $(-3y^2 - 2y + 11)$ using a vertical format.

Solution: Arrange the polynomials in a vertical format, lining up like terms.

$$
\begin{array}{ll}
-3y^2 - 2y + 11 & \quad -3y^2 - 2y + 11 \\
\underline{-(5y^2 + 2y - 6)} & \quad \underline{-5y^2 - 2y + 6} \\
 & \quad -8y^2 - 4y + 17
\end{array}
$$

🔲 **Work Practice Problem 6**

Helpful Hint
Don't forget to change the sign of each term in the polynomial being subtracted.

Objective C Adding and Subtracting Polynomials in One Variable

Let's practice adding and subtracting polynomials in one variable.

EXAMPLE 7 Subtract $(5z - 7)$ from the sum of $(8z + 11)$ and $(9z - 2)$.

Solution: Notice that $(5z - 7)$ is to be subtracted **from** a sum. The translation is

$$[(8z + 11) + (9z - 2)] - (5z - 7)$$
$$= 8z + 11 + 9z - 2 - 5z + 7 \quad \text{Remove grouping symbols.}$$
$$= 8z + 9z - 5z + 11 - 2 + 7 \quad \text{Group like terms.}$$
$$= 12z + 16 \quad \text{Combine like terms.}$$

🔲 **Work Practice Problem 7**

PRACTICE PROBLEM 4
Subtract:
$(9x + 5) - (4x - 3)$

PRACTICE PROBLEM 5
Subtract:
$(4x^3 - 10x^2 + 1)$
$\quad - (-4x^3 + x^2 - 11)$

PRACTICE PROBLEM 6
Subtract $(6y^2 - 3y + 2)$ from $(2y^2 - 2y + 7)$ using a vertical format.

PRACTICE PROBLEM 7
Subtract $(3x + 1)$ from the sum of $(4x - 3)$ and $(12x - 5)$.

Answers
4. $5x + 8$, **5.** $8x^3 - 11x^2 + 12$,
6. $-4y^2 + y + 5$, **7.** $13x - 9$

Objective D Adding and Subtracting Polynomials in Several Variables

Now that we know how to add or subtract polynomials in one variable, we can also add and subtract polynomials in several variables.

Add or subtract as indicated.

8. $(2a^2 - ab + 6b^2)$
   $+ (-3a^2 + ab - 7b^2)$

9. $(5x^2y^2 + 3 - 9x^2y + y^2)$
   $- (-x^2y^2 + 7 - 8xy^2 + 2y^2)$

EXAMPLES Add or subtract as indicated.

8. $(3x^2 - 6xy + 5y^2) + (-2x^2 + 8xy - y^2)$

   $= 3x^2 - 6xy + 5y^2 - 2x^2 + 8xy - y^2$

   $= x^2 + 2xy + 4y^2$ Combine like terms.

9. $(9a^2b^2 + 6ab - 3ab^2) - (5b^2a + 2ab - 3 - 9b^2)$

   $= 9a^2b^2 + 6ab - 3ab^2 - 5b^2a - 2ab + 3 + 9b^2$

   $= 9a^2b^2 + 4ab - 8ab^2 + 9b^2 + 3$ Combine like terms.

□ **Work Practice Problems 8–9**

✔ **Concept Check** If possible, simplify each expression by performing the indicated operation.

a. $2y + y$

b. $2y \cdot y$

c. $-2y - y$

d. $(-2y)(-y)$

e. $2x + y$

Answers

8. $-a^2 - b^2$,

9. $6x^2y^2 - 4 - 9x^2y + 8xy^2 - y^2$

✔ Concept Check Answers

a. $3y$, b. $2y^2$, c. $-3y$, d. $2y^2$,

e. cannot be simplified

Objective A *Add. See Examples 1 and 2.*

1. $(3x + 7) + (9x + 5)$

2. $(-y - 2) + (3y + 5)$

3. $(-7x + 5) + (-3x^2 + 7x + 5)$

4. $(3x - 8) + (4x^2 - 3x + 3)$

5. $(-5x^2 + 3) + (2x^2 + 1)$

6. $(3x^2 + 7) + (3x^2 + 9)$

7. $(-3y^2 - 4y) + (2y^2 + y - 1)$

8. $(7x^2 + 2x - 9) + (-3x^2 + 5)$

9. $(1.2x^3 - 3.4x + 7.9) + (6.7x^3 + 4.4x^2 - 10.9)$

10. $(9.6y^3 + 2.7y^2 - 8.6) + (1.1y^3 - 8.8y + 11.6)$

11. $\left(\dfrac{3}{4}m^2 - \dfrac{2}{5}m + \dfrac{1}{8}\right) + \left(-\dfrac{1}{4}m^2 - \dfrac{3}{10}m + \dfrac{11}{16}\right)$

12. $\left(-\dfrac{4}{7}n^2 + \dfrac{5}{6}m - \dfrac{1}{20}\right) + \left(\dfrac{3}{7}n^2 - \dfrac{5}{12}m - \dfrac{3}{10}\right)$

Add using a vertical format. See Example 3.

13. $\begin{aligned}3t^2 + 4\\ \underline{5t^2 - 8}\end{aligned}$

14. $\begin{aligned}7x^3 + 3\\ \underline{2x^3 + 1}\end{aligned}$

15. $\begin{aligned}10a^3 - 8a^2 + 4a + 9\\ \underline{5a^3 + 9a^2 - 7a + 7}\end{aligned}$

16. $\begin{aligned}2x^3 - 3x^2 + \ x - 4\\ \underline{5x^3 + 2x^2 - 3x + 2}\end{aligned}$

Objective B *Subtract. See Examples 4 and 5.*

17. $(2x + 5) - (3x - 9)$

18. $(4 + 5a) - (-a - 5)$

19. $(5x^2 + 4) - (-2y^2 + 4)$

20. $(-7y^2 + 5) - (-8y^2 + 12)$

21. $3x - (5x - 9)$

22. $4 - (-y - 4)$

23. $(2x^2 + 3x - 9) - (-4x + 7)$

24. $(-7x^2 + 4x + 7) - (-8x + 2)$

25. $(5x + 8) - (-2x^2 - 6x + 8)$

26. $(-6y^2 + 3y - 4) - (9y^2 - 3y)$

27. $(0.7x^2 + 0.2x - 0.8) - (0.9x^2 + 1.4)$

28. $(-0.3y^2 + 0.6y - 0.3) - (0.5y^2 + 0.3)$

29. $\left(\dfrac{1}{4}z^2 - \dfrac{1}{5}z\right) - \left(-\dfrac{3}{20}z^2 + \dfrac{1}{10}z - \dfrac{7}{20}\right)$

30. $\left(\dfrac{1}{3}x^2 - \dfrac{2}{7}x\right) - \left(\dfrac{4}{21}x^2 + \dfrac{1}{21}x - \dfrac{2}{3}\right)$

Subtract using a vertical format. See Example 6.

31.
$$\begin{array}{r} 4z^2 - 8z + 3 \\ -(6z^2 + 8z - 3) \\ \hline \end{array}$$

32.
$$\begin{array}{r} 7a^2 - 9a + 6 \\ -(11a^2 - 4a + 2) \\ \hline \end{array}$$

33.
$$\begin{array}{r} 5u^5 - 4u^2 + 3u - 7 \\ -(3u^5 + 6u^2 - 8u + 2) \\ \hline \end{array}$$

34.
$$\begin{array}{r} 5x^3 - 4x^2 + 6x - 2 \\ -(3x^3 - 2x^2 - \ x - 4) \\ \hline \end{array}$$

Objectives **A** **B** **C** **Mixed Practice** *Add or subtract as indicated. See Examples 1 through 7.*

35. $(3x + 5) + (2x - 14)$

36. $(2y + 20) + (5y - 30)$

37. $(9x - 1) - (5x + 2)$

38. $(7y + 7) - (y - 6)$

39. $(14y + 12) + (-3y - 5)$

40. $(26y + 17) + (-20y - 10)$

41. $(x^2 + 2x + 1) - (3x^2 - 6x + 2)$

42. $(5y^2 - 3y - 1) - (2y^2 + y + 1)$

43. $(3x^2 + 5x - 8) + (5x^2 + 9x + 12) - (8x^2 - 14)$

44. $(2x^2 + 7x - 9) + (x^2 - x + 10) - (3x^2 - 30)$

45. $(-a^2 + 1) - (a^2 - 3) + (5a^2 - 6a + 7)$

46. $(-m^2 + 3) - (m^2 - 13) + (6m^2 - m + 1)$

Perform each indicated operation. See Examples 3, 6, and 7.

47. Subtract $4x$ from $7x - 3$.

48. Subtract y from $y^2 - 4y + 1$.

49. Add $(4x^2 - 6x + 1)$ and $(3x^2 + 2x + 1)$.

50. Add $(-3x^2 - 5x + 2)$ and $(x^2 - 6x + 9)$.

51. Subtract $(5x + 7)$ from $(7x^2 + 3x + 9)$.

52. Subtract $(5y^2 + 8y + 2)$ from $(7y^2 + 9y - 8)$.

53. Subtract $(4y^2 - 6y - 3)$ from the sum of $(8y^2 + 7)$ and $(6y + 9)$.

54. Subtract $(4x^2 - 2x + 2)$ from the sum of $(x^2 + 7x + 1)$ and $(7x + 5)$.

55. Subtract $(3x^2 - 4)$ from the sum of $(x^2 - 9x + 2)$ and $(2x^2 - 6x + 1)$.

56. Subtract $(y^2 - 9)$ from the sum of $(3y^2 + y + 4)$ and $(2y^2 - 6y - 10)$.

Objective **D** *Add or subtract as indicated. See Examples 8 and 9.*

57. $(9a + 6b - 5) + (-11a - 7b + 6)$

58. $(3x - 2 + 6y) + (7x - 2 - y)$

59. $(4x^2 + y^2 + 3) - (x^2 + y^2 - 2)$

60. $(7a^2 - 3b^2 + 10) - (-2a^2 + b^2 - 12)$

61. $(x^2 + 2xy - y^2) + (5x^2 - 4xy + 20y^2)$

62. $(a^2 - ab + 4b^2) + (6a^2 + 8ab - b^2)$

63. $(11r^2s + 16rs - 3 - 2r^2s^2) - (3sr^2 + 5 - 9r^2s^2)$

64. $(3x^2y - 6xy + x^2y^2 - 5) - (11x^2y^2 - 1 + 5yx^2)$

For Exercises 65 through 68, find the perimeter of each figure.

65.

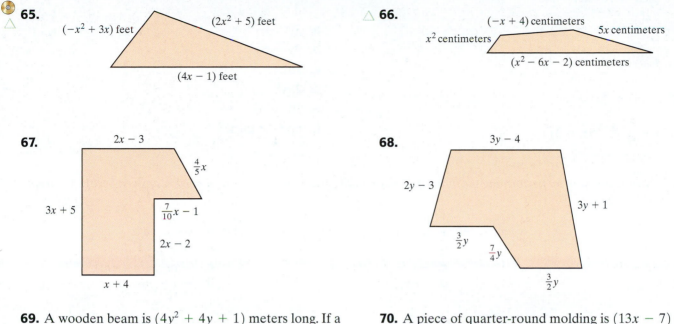

$(-x^2 + 3x)$ feet $(2x^2 + 5)$ feet

$(4x - 1)$ feet

66.

$(-x + 4)$ centimeters $5x$ centimeters

x^2 centimeters

$(x^2 - 6x - 2)$ centimeters

67.

$2x - 3$

$\frac{4}{5}x$

$3x + 5$

$\frac{7}{10}x - 1$

$2x - 2$

$x + 4$

68.

$3y - 4$

$2y - 3$

$3y + 1$

$\frac{3}{2}y$

$\frac{7}{4}y$

$\frac{3}{2}y$

69. A wooden beam is $(4y^2 + 4y + 1)$ meters long. If a piece $(y^2 - 10)$ meters is cut, express the length of the remaining piece of beam as a polynomial in y.

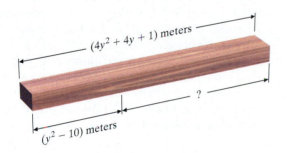

$(4y^2 + 4y + 1)$ meters

?

$(y^2 - 10)$ meters

70. A piece of quarter-round molding is $(13x - 7)$ inches long. If a piece $(2x + 2)$ inches is removed, express the length of the remaining piece of molding as a polynomial in x.

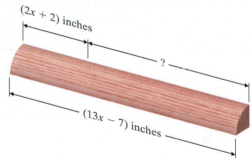

$(2x + 2)$ inches

?

$(13x - 7)$ inches

Perform each indicated operation.

71. $[(1.2x^2 - 3x + 9.1) - (7.8x^2 - 3.1 + 8)] + (1.2x - 6)$

72. $[(7.9y^4 - 6.8y^3 + 3.3y) + (6.1y^3 - 5)] - (4.2y^4 + 1.1y - 1)$

Review

Multiply. See Section 3.1.

73. $3x(2x)$ **74.** $-7x(x)$ **75.** $(12x^3)(-x^5)$ **76.** $6r^3(7r^{10})$ **77.** $10x^2(20xy^2)$ **78.** $-z^2y(11zy)$

Fill in the squares so that each is a true statement.

79. $3x^{\square} + 4x^2 = 7x^{\square}$

80. $9y^7 + 3y^{\square} = 12y^7$

81. $2x^{\square} + 3x^{\square} - 5x^{\square} + 4x^{\square} = 6x^4 - 2x^3$

82. $3y^{\square} + 7y^{\square} - 2y^{\square} - y^{\square} = 10y^5 - 3y^2$

Match each expression on the left with its simplification on the right. Not all letters on the right must be used and a letter may be used more than once.

83. $10y - 6y^2 - y$

84. $5x + 5x$

85. $(5x - 3) + (5x - 3)$

86. $(15x - 3) - (5x - 3)$

a. $3y$

b. $9y - 6y^2$

c. $10x$

d. $25x^2$

e. $10x - 6$

f. none of these

Simplify each expression by performing the indicated operation. Explain how you arrived at each answer. See the Concept Check in this section.

87. a. $z + 3z$
  b. $z \cdot 3z$
  c. $-z - 3z$
  d. $(-z)(-3z)$

88. a. $x + x$
  b. $x \cdot x$
  c. $-x - x$
  d. $(-x)(-x)$

89. a. $m \cdot m \cdot m$
  b. $m + m + m$
  c. $(-m)(-m)(-m)$
  d. $-m - m - m$

90. The polynomial $0.0005x^2 + 0.0303x + 1.156$ represents the sale of electricity (in trillion kilowatt-hours) in the U.S. residential sector during 1999–2003. The polynomial $0.0215x^2 - 0.1073x + 2.31$ represents the sale of electricity (in trillion kilowatt-hours) in all other U.S. sectors during 1999–2003. In both polynomials, x represents the number of years after 1999. Find a polynomial for the total sales of electricity (in trillion kilowatt hours) to all sectors in the United States during this period. (*Source:* Based on data from the Energy Information Administration)

91. The polynomial $-0.35x^2 + 0.49x + 71.75$ represents the percent of Americans under age 65 covered by private health insurance during 1999–2003. The polynomial $0.025x^2 + 9.65x + 11.83$ represents the percent of Americans under age 65 covered by public health programs during 1999–2003. In both polynomials, x represents the number of years since 1999. Find a polynomial for the total percent of Americans under age 65 with some form of health coverage during this period. (*Source:* Based on data from the Public Health Service)

3.5 MULTIPLYING POLYNOMIALS

Objectives

A Multiply Monomials.

B Multiply a Monomial by a Polynomial.

C Multiply Two Polynomials.

D Multiply Polynomials Vertically.

Objective A Multiplying Monomials

Recall from Section 3.1 that to multiply two monomials such as $(-5x^3)$ and $(-2x^4)$, we use the associative and commutative properties and regroup. Remember also that to multiply exponential expressions with a common base, we use the product rule for exponents and add exponents.

$$(-5x^3)(-2x^4) = (-5)(-2)(x^3 \cdot x^4) \quad \text{Use the commutative and associative properties.}$$
$$= 10x^7 \quad \text{Multiply.}$$

EXAMPLES Multiply.

1. $6x \cdot 4x = (6 \cdot 4)(x \cdot x)$ Use the commutative and associative properties.
$$= 24x^2 \quad \text{Multiply.}$$

2. $-7x^2 \cdot 2x^5 = (-7 \cdot 2)(x^2 \cdot x^5)$
$$= -14x^7$$

3. $(-12x^5)(-x) = (-12x^5)(-1x)$
$$= (-12)(-1)(x^5 \cdot x)$$
$$= 12x^6$$

◻ **Work Practice Problems 1–3**

✔ **Concept Check** Simplify.

a. $3x \cdot 2x$ **b.** $3x + 2x$

Objective B Multiplying Monomials by Polynomials

To multiply a monomial such as $7x$ by a trinomial such as $x^3 + 2x + 5$, we use the distributive property.

EXAMPLES Multiply.

4. $7x(x^2 + 2x + 5) = 7x(x^2) + 7x(2x) + 7x(5)$ Apply the distributive property.
$$= 7x^3 + 14x^2 + 35x \quad \text{Multiply.}$$

5. $5x(2x^3 + 6) = 5x(2x^3) + 5x(6)$ Apply the distributive property.
$$= 10x^4 + 30x \quad \text{Multiply.}$$

6. $-3x^2(5x^2 + 6x - 1)$
$$= (-3x^2)(5x^2) + (-3x^2)(6x) + (-3x^2)(-1) \quad \text{Apply the distributive property.}$$
$$= -15x^4 - 18x^3 + 3x^2 \quad \text{Multiply.}$$

◻ **Work Practice Problems 4–6**

PRACTICE PROBLEMS 1–3

Multiply.

1. $10x \cdot 9x$

2. $8x^3(-11x^7)$

3. $(-5x^4)(-x)$

PRACTICE PROBLEMS 4–6

Multiply.

4. $4x(x^2 + 4x + 3)$

5. $8x(7x^4 + 1)$

6. $-2x^3(3x^2 - x + 2)$

Answers

1. $90x^2$, **2.** $-88x^{10}$, **3.** $5x^5$,
4. $4x^3 + 16x^2 + 12x$, **5.** $56x^5 + 8x$,
6. $-6x^5 + 2x^4 - 4x^3$

✔ **Concept Check Answers**

a. $6x^2$, **b.** $5x$

Objective C Multiplying Two Polynomials

We also use the distributive property to multiply two binomials.

EXAMPLE 7 Multiply.

a. $(m + 4)(m + 6)$ **b.** $(3x + 2)(2x - 5)$

Solution:

a. $(m + 4)(m + 6) = m(m + 6) + 4(m + 6)$ Use the distributive property.

$\qquad\qquad\qquad\quad = m \cdot m + m \cdot 6 + 4 \cdot m + 4 \cdot 6$ Use the distributive property.

$\qquad\qquad\qquad\quad = m^2 + 6m + 4m + 24$ Multiply.

$\qquad\qquad\qquad\quad = m^2 + 10m + 24$ Combine like terms.

b. $(3x + 2)(2x - 5) = 3x(2x - 5) + 2(2x - 5)$ Use the distributive property.

$\qquad\qquad\qquad\quad = 3x(2x) + 3x(-5) + 2(2x) + 2(-5)$

$\qquad\qquad\qquad\quad = 6x^2 - 15x + 4x - 10$ Multiply.

$\qquad\qquad\qquad\quad = 6x^2 - 11x - 10$ Combine like terms.

◻ **Work Practice Problem 7**

This idea can be expanded so that we can multiply any two polynomials.

> ### To Multiply Two Polynomials
>
> Multiply each term of the first polynomial by each term of the second polynomial, and then combine like terms.

EXAMPLES Multiply.

8. $(2x - y)^2$

$\quad = (2x - y)(2x - y)$ Using the meaning of an exponent, we have 2 factors of $(2x - y)$.

$\quad = 2x(2x) + 2x(-y) + (-y)(2x) + (-y)(-y)$

$\quad = 4x^2 - 2xy - 2xy + y^2$ Multiply.

$\quad = 4x^2 - 4xy + y^2$ Combine like terms.

9. $(t + 2)(3t^2 - 4t + 2)$

$\quad = t(3t^2) + t(-4t) + t(2) + 2(3t^2) + 2(-4t) + 2(2)$

$\quad = 3t^3 - 4t^2 + 2t + 6t^2 - 8t + 4$

$\quad = 3t^3 + 2t^2 - 6t + 4$ Combine like terms.

◻ **Work Practice Problems 8–9**

✔ **Concept Check**   Square where indicated. Simplify if possible.

a. $(4a)^2 + (3b)^2$ **b.** $(4a + 3b)^2$

Objective D Multiplying Polynomials Vertically

Another convenient method for multiplying polynomials is to multiply vertically, similar to the way we multiply real numbers. This method is shown in the next examples.

EXAMPLE 10 Multiply vertically: $(2y^2 + 5)(y^2 - 3y + 4)$

Solution:
$$
\begin{array}{r}
y^2 - 3y + 4 \\
2y^2 + 5 \\
\hline
5y^2 - 15y + 20 \\
2y^4 - 6y^3 + 8y^2 \\
\hline
2y^4 - 6y^3 + 13y^2 - 15y + 20
\end{array}
$$

Multiply $y^2 - 3y + 4$ by 5

Multiply $y^2 - 3y + 4$ by $2y^2$

Combine like terms.

🔲 **Work Practice Problem 10**

EXAMPLE 11 Find the product of $(2x^2 - 3x + 4)$ and $(x^2 + 5x - 2)$ using a vertical format.

Solution: First, we arrange the polynomials in a vertical format. Then we multiply each term of the second polynomial by each term of the first polynomial.

$$
\begin{array}{r}
2x^2 - 3x + 4 \\
x^2 + 5x - 2 \\
\hline
-4x^2 + 6x - 8 \\
10x^3 - 15x^2 + 20x \\
2x^4 - 3x^3 + 4x^2 \\
\hline
2x^4 + 7x^3 - 15x^2 + 26x - 8
\end{array}
$$

Multiply $2x^2 - 3x + 4$ by -2.

Multiply $2x^2 - 3x + 4$ by $5x$.

Multiply $2x^2 - 3x + 4$ by x^2.

Combine like terms.

🔲 **Work Practice Problem 11**

PRACTICE PROBLEM 10

Multiply vertically:
$(3y^2 + 1)(y^2 - 4y + 5)$

PRACTICE PROBLEM 11

Find the product of
$(4x^2 - x - 1)$ and
$(3x^2 + 6x - 2)$ using a vertical format.

Answers
10. $3y^4 - 12y^3 + 16y^2 - 4y + 5$,
11. $12x^4 + 21x^3 - 17x^2 - 4x + 2$

Mental Math

Perform the indicated operation, if possible.

1. $x^3 \cdot x^5$

2. $x^2 \cdot x^6$

3. $x^3 + x^5$

4. $x^2 + x^6$

5. $y^4 \cdot y$

6. $y^9 \cdot y$

7. $x^7 \cdot x^7$

8. $x^{11} \cdot x^{11}$

9. $x^7 + x^7$

10. $x^{11} + x^{11}$

3.5 EXERCISE SET

Objective A *Multiply. See Examples 1 through 3.*

1. $8x^2 \cdot 3x$

2. $6x \cdot 3x^2$

3. $(-x^3)(-x)$

4. $(-x^6)(-x)$

5. $-4n^3 \cdot 7n^7$

6. $9t^6(-3t^5)$

7. $(-3.1x^3)(4x^9)$

8. $(-5.2x^4)(3x^4)$

9. $\left(-\frac{1}{3}y^2\right)\left(\frac{2}{5}y\right)$

10. $\left(-\frac{3}{4}y^7\right)\left(\frac{1}{7}y^4\right)$

11. $(2x)(-3x^2)(4x^5)$

12. $(x)(5x^4)(-6x^7)$

Objective B *Multiply. See Examples 4 through 6.*

13. $3x(2x + 5)$

14. $2x(6x + 3)$

15. $7x(x^2 + 2x - 1)$

16. $5y(y^2 + y - 10)$

17. $-2a(a + 4)$

18. $-3a(2a + 7)$

19. $3x(2x^2 - 3x + 4)$

20. $4x(5x^2 - 6x - 10)$

21. $3a^2(4a^3 + 15)$

22. $9x^3(5x^2 + 12)$

23. $-2a^2(3a^2 - 2a + 3)$

24. $-4b^2(3b^3 - 12b^2 - 6)$

25. $3x^2y(2x^3 - x^2y^2 + 8y^3)$

26. $4xy^2(7x^3 + 3x^2y^2 - 9y^3)$

27. $-y(4x^3 - 7x^2y + xy^2 + 3y^3)$

28. $-x(6y^3 - 5xy^2 + x^2y - 5x^3)$

29. $\frac{1}{2}x^2(8x^2 - 6x + 1)$

30. $\frac{1}{3}y^2(9y^2 - 6y + 1)$

Objective C *Multiply. See Examples 7 through 9.*

31. $(x + 4)(x + 3)$

32. $(x + 2)(x + 9)$

33. $(a + 7)(a - 2)$

34. $(y - 10)(y + 11)$

35. $\left(x + \dfrac{2}{3}\right)\left(x - \dfrac{1}{3}\right)$ **36.** $\left(x + \dfrac{3}{5}\right)\left(x - \dfrac{2}{5}\right)$ **37.** $(3x^2 + 1)(4x^2 + 7)$ **38.** $(5x^2 + 2)(6x^2 + 2)$

39. $(4x - 3)(3x - 5)$ **40.** $(8x - 3)(2x - 4)$ **41.** $(1 - 3a)(1 - 4a)$ **42.** $(3 - 2a)(2 - a)$

43. $(2y - 4)^2$ **44.** $(6x - 7)^2$ **45.** $(x - 2)(x^2 - 3x + 7)$ **46.** $(x + 3)(x^2 + 5x - 8)$

47. $(x + 5)(x^3 - 3x + 4)$ **48.** $(a + 2)(a^3 - 3a^2 + 7)$ **49.** $(2a - 3)(5a^2 - 6a + 4)$

50. $(3 + b)(2 - 5b - 3b^2)$ **51.** $(7xy - y)^2$ **52.** $(x^2 - 4)^2$

Objective **D** *Multiply vertically. See Examples 10 and 11.*

53. $(2x - 11)(6x + 1)$ **54.** $(4x - 7)(5x + 1)$ **55.** $(x + 3)(2x^2 + 4x - 1)$

56. $(4x - 5)(8x^2 + 2x - 4)$ **57.** $(x^2 + 5x - 7)(2x^2 - 7x - 9)$ **58.** $(3x^2 - x + 2)(x^2 + 2x + 1)$

Objectives **A** **B** **C** **D** **Mixed Practice** *Multiply. See Examples 1 through 11.*

59. $-1.2y(-7y^6)$ **60.** $-4.2x(-2x^5)$ **61.** $-3x(x^2 + 2x - 8)$ **62.** $-5x(x^2 - 3x + 10)$

63. $(x + 19)(2x + 1)$ **64.** $(3y + 4)(y + 11)$ **65.** $\left(x + \dfrac{1}{7}\right)\left(x - \dfrac{3}{7}\right)$ **66.** $\left(m + \dfrac{2}{9}\right)\left(m - \dfrac{1}{9}\right)$

67. $(3y + 5)^2$ **68.** $(7y + 2)^2$ **69.** $(a + 4)(a^2 - 6a + 6)$ **70.** $(t + 3)(t^2 - 5t + 5)$

Express as the product of polynomials. Then multiply.

△ **71.** Find the area of the rectangle.

$(2x + 5)$ yards

$(2x - 5)$ yards

△ **72.** Find the area of the square field.

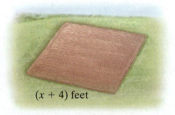

$(x + 4)$ feet

△ **73.** Find the area of the triangle.

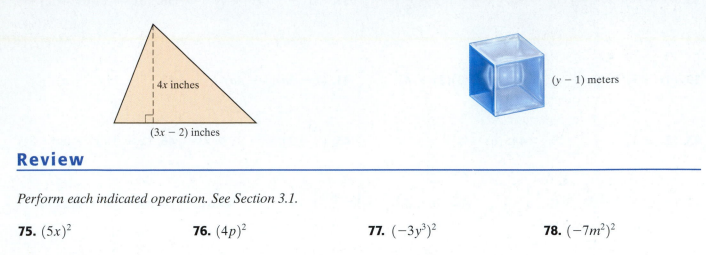

4x inches

(3x − 2) inches

△ **74.** Find the volume of the cube-shaped glass block.

(y − 1) meters

Review

Perform each indicated operation. See Section 3.1.

75. $(5x)^2$

76. $(4p)^2$

77. $(-3y^3)^2$

78. $(-7m^2)^2$

Concept Extensions

79. Perform each indicated operation. Explain the difference between the two expressions.
   a. $(3x + 5) + (3x + 7)$
   b. $(3x + 5)(3x + 7)$

80. Perform each operation. Explain the difference between the two expressions.
   a. $(8x − 3) − (5x − 2)$
   b. $(8x − 3)(5x − 2)$

Mixed Practice *See Sections 3.4, 3.5. Perform the indicated operations.*

81. $(3x − 1) + (10x − 6)$

82. $(2x − 1) + (10x − 7)$

83. $(3x − 1)(10x − 6)$

84. $(2x − 1)(10x − 7)$

85. $(3x − 1) − (10x − 6)$

86. $(2x − 1) − (10x − 7)$

△ **87.** The area of the largest rectangle below is $x(x + 3)$. Find another expression for this area by finding the sum of the areas of the smaller rectangles.

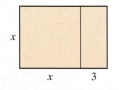

x

x   3

△ **88.** Write an expression for the area of the largest rectangle below in two different ways.

x

1   2x

△ **89.** The area of the figure below is $(x + 2)(x + 3)$. Find another expression for this area by finding the sum of the areas of the smaller rectangles.

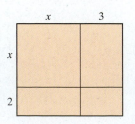

x   3

x

2

△ **90.** Write an expression for the area of the figure below in two different ways

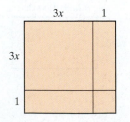

3x   1

3x

1

Simplify. See the Concept Checks in this section.

91. $5a + 6a$

92. $5a \cdot 6a$

Square where indicated. Simplify if possible.

93. $(5x)^2 + (2y)^2$

94. $(5x + 2y)^2$

 95. Multiply each of the following polynomials.
- **a.** $(a + b)(a - b)$
- **b.** $(2x + 3y)(2x - 3y)$
- **c.** $(4x + 7)(4x - 7)$
- **d.** Can you make a general statement about all products of the form $(x + y)(x - y)$?

96. Evaluate each of the following.
- **a.** $(2 + 3)^2$; $2^2 + 3^2$
- **b.** $(8 + 10)^2$; $8^2 + 10^2$
  Does $(a + b)^2 = a^2 + b^2$ no matter what the values of a and b are? Why or why not?

STUDY SKILLS BUILDER

Tips for Studying for an Exam

To prepare for an exam, try the following study techniques:

- Start the study process days before your exam.
- Make sure that you are up-to-date on your assignments.
- If there is a topic that you are unsure of, use one of the many resources that are available to you. For example,

  See your instructor.
  Visit a learning resource center on campus.
  Read the textbook material and examples on the topic.
  View a video on the topic.

- Reread your notes and carefully review the Chapter Highlights at the end of the chapter.
- Work the review exercises at the end of the chapter. Check your answers and correct any mistakes. If you have trouble, use a resource listed above.
- Find a quiet place to take the Chapter Test found at the end of the chapter. Do not use any resources when taking this sample test. This way, you will have a clear indication of how prepared you are for your exam.

- Check your answers and make sure that you correct any missed exercises.
- Get lots of rest the night before the exam. It's hard to show how well you know the material if your brain is foggy from lack of sleep.

Good luck and keep a positive attitude.

Let's see how you did on your last exam.

1. How many days before your last exam did you start studying for that exam?
2. Were you up-to-date on your assignments at that time or did you need to catch up on assignments?
3. List the most helpful text supplement (if you used one).
4. List the most helpful campus supplement (if you used one).
5. List your process for preparing for a mathematics test.
6. Was this process helpful? In other words, were you satisfied with your performance on your exam?
7. If not, what changes can you make in your process that will make it more helpful to you?

3.6 SPECIAL PRODUCTS

Objective A Using the FOIL Method

In this section, we multiply binomials using special products. First, we introduce a special order for multiplying binomials called the FOIL order or method. This order, or pattern, is a result of the distributive property. We demonstrate by multiplying $(3x + 1)$ by $(2x + 5)$.

The FOIL Method

F stands for the
product of the **First** terms. $(3x + 1)(2x + 5)$

$(3x)(2x) = 6x^2$ F

O stands for the
product of the **Outer** terms. $(3x + 1)(2x + 5)$

$(3x)(5) = 15x$ O

I stands for the
product of the **Inner** terms. $(3x + 1)(2x + 5)$

$(1)(2x) = 2x$ I

L stands for the
product of the **Last** terms. $(3x + 1)(2x + 5)$

$(1)(5) = 5$ L

$$\begin{array}{cccc} \text{F} & \text{O} & \text{I} & \text{L} \end{array}$$
$$(3x + 1)(2x + 5) = 6x^2 + 15x + 2x + 5$$
$$= 6x^2 + 17x + 5 \qquad \text{Combine like terms.}$$

Let's practice multiplying binomials using the FOIL method.

PRACTICE PROBLEM 1

Multiply: $(x + 7)(x - 5)$

EXAMPLE 1 Multiply: $(x - 3)(x + 4)$

Solution:

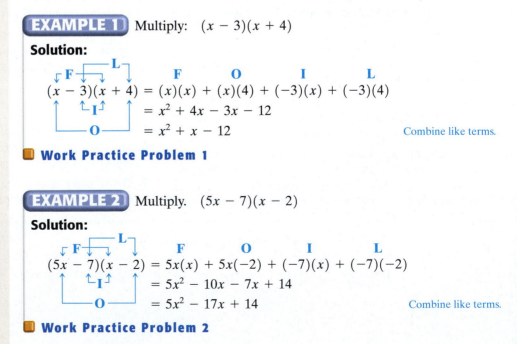

$$(x - 3)(x + 4) = (x)(x) + (x)(4) + (-3)(x) + (-3)(4)$$
$$= x^2 + 4x - 3x - 12$$
$$= x^2 + x - 12 \qquad \text{Combine like terms.}$$

☐ **Work Practice Problem 1**

PRACTICE PROBLEM 2

Multiply: $(6x - 1)(x - 4)$

EXAMPLE 2 Multiply. $(5x - 7)(x - 2)$

Solution:

$$(5x - 7)(x - 2) = 5x(x) + 5x(-2) + (-7)(x) + (-7)(-2)$$
$$= 5x^2 - 10x - 7x + 14$$
$$= 5x^2 - 17x + 14 \qquad \text{Combine like terms.}$$

☐ **Work Practice Problem 2**

Answers

1. $x^2 + 2x - 35$,   **2.** $6x^2 - 25x + 4$

EXAMPLE 3 Multiply: $(y^2 + 6)(2y - 1)$

Solution: F   O   I   L
$(y^2 + 6)(2y - 1) = 2y^3 - 1y^2 + 12y - 6$

Notice in this example that there are no like terms that can be combined, so the product is $2y^3 - y^2 + 12y - 6$.

Work Practice Problem 3

PRACTICE PROBLEM 3
Multiply: $(2y^2 + 3)(y - 4)$

Objective B Squaring Binomials

An expression such as $(3y + 1)^2$ is called the square of a binomial. Since $(3y + 1)^2 = (3y + 1)(3y + 1)$, we can use the FOIL method to find this product.

EXAMPLE 4 Multiply: $(3y + 1)^2$

Solution: $(3y + 1)^2 = (3y + 1)(3y + 1)$

$$\qquad\qquad\quad F \qquad O \qquad I \qquad L$$
$$= (3y)(3y) + (3y)(1) + 1(3y) + 1(1)$$
$$= 9y^2 + 3y + 3y + 1$$
$$= 9y^2 + 6y + 1$$

Work Practice Problem 4

PRACTICE PROBLEM 4
Multiply: $(2x + 9)^2$

Notice the pattern that appears in Example 4.

$(3y + 1)^2 = 9y^2 + 6y + 1$

→ $9y^2$ is the first term of the binomial squared: $(3y)^2 = 9y^2$.

→ $6y$ is 2 times the product of both terms of the binomial: $(2)(3y)(1) = 6y$.

→ 1 is the second term of the binomial squared: $(1)^2 = 1$.

This pattern leads to the formulas below, which can be used when squaring a binomial. We call these **special products.**

Squaring a Binomial

A binomial squared is equal to the square of the first term plus or minus twice the product of both terms plus the square of the second term.

$$(a + b)^2 = a^2 + 2ab + b^2$$
$$(a - b)^2 = a^2 - 2ab + b^2$$

This product can be visualized geometrically.

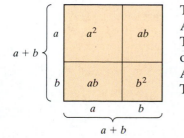

The area of the large square is side · side.
Area $= (a + b)(a + b) = (a + b)^2$
The area of the large square is also the sum of the areas of the smaller rectangles.
Area $= a^2 + ab + ab + b^2 = a^2 + 2ab + b^2$
Thus, $(a + b)^2 = a^2 + 2ab + b^2$.

Answers
3. $2y^3 - 8y^2 + 3y - 12$,
4. $4x^2 + 36x + 81$

PRACTICE PROBLEMS 5–8

Use a special product to square each binomial.

5. $(y + 3)^2$

6. $(r - s)^2$

7. $(6x + 5)^2$

8. $(x^2 - 3y)^2$

EXAMPLES   Use a special product to square each binomial.

first term squared	plus or minus	twice the product of the terms	plus	second term squared

5. $(t + 2)^2 = t^2 + 2(t)(2) + 2^2 = t^2 + 4t + 4$

6. $(p - q)^2 = p^2 - 2(p)(q) + q^2 = p^2 - 2pq + q^2$

7. $(2x + 5)^2 = (2x)^2 + 2(2x)(5) + 5^2 = 4x^2 + 20x + 25$

8. $(x^2 - 7y)^2 = (x^2)^2 - 2(x^2)(7y) + (7y)^2 = x^4 - 14x^2y + 49y^2$

◻ **Work Practice Problems 5–8**

Helpful Hint

Notice that

$(a + b)^2 \neq a^2 + b^2$ The middle term $2ab$ is missing.

$(a + b)^2 = (a + b)(a + b) = a^2 + 2ab + b^2$

Likewise,

$(a - b)^2 \neq a^2 - b^2$

$(a - b)^2 = (a - b)(a - b) = a^2 - 2ab + b^2$

Objective C Multiplying the Sum and Difference of Two Terms

Another special product is the product of the sum and difference of the same two terms, such as $(x + y)(x - y)$. Finding this product by the FOIL method, we see a pattern emerge.

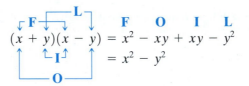

$$(x + y)(x - y) = x^2 - xy + xy - y^2$$
$$= x^2 - y^2$$

Notice that the two middle terms subtract out. This is because the **O**uter product is the opposite of the **I**nner product. Only the **difference of squares** remains.

Multiplying the Sum and Difference of Two Terms

The product of the sum and difference of two terms is the square of the first term minus the square of the second term.

$$(a + b)(a - b) = a^2 - b^2$$

EXAMPLES Use a special product to multiply.

first term squared	minus	second term squared
↓	↓	↓

9. $(x + 4)(x - 4) = x^2 \qquad - \qquad 4^2 = x^2 - 16$

10. $(6t + 7)(6t - 7) = (6t)^2 \qquad - \qquad 7^2 = 36t^2 - 49$

11. $\left(x - \dfrac{1}{4}\right)\left(x + \dfrac{1}{4}\right) = x^2 \qquad - \qquad \left(\dfrac{1}{4}\right)^2 = x^2 - \dfrac{1}{16}$

12. $(2p - q)(2p + q) = (2p)^2 - q^2 = 4p^2 - q^2$

13. $(3x^2 - 5y)(3x^2 + 5y) = (3x^2)^2 - (5y)^2 = 9x^4 - 25y^2$

Work Practice Problems 9–13

✔**Concept Check** Match each expression on the left to the equivalent expression or expressions in the list on the right.

$(a + b)^2$ **a.** $(a + b)(a + b)$

$(a + b)(a - b)$ **b.** $a^2 - b^2$

                     c. $a^2 + b^2$

                     d. $a^2 - 2ab + b^2$

                     e. $a^2 + 2ab + b^2$

Objective D Using Special Products

Let's now practice using our special products on a variety of multiplication problems. This practice will help us recognize when to apply what special product formula.

EXAMPLES Use a special product to multiply, if possible.

14. $(4x - 9)(4x + 9)$ *This is the sum and difference of the same two terms.*

$= (4x)^2 - 9^2 = 16x^2 - 81$

15. $(3y + 2)^2$ *This is a binomial squared.*

$= (3y)^2 + 2(3y)(2) + 2^2$

$= 9y^2 + 12y + 4$

16. $(6a + 1)(a - 7)$ *No special product applies.*

       F   O   I   L   *Use the FOIL method.*

$= 6a \cdot a + 6a(-7) + 1 \cdot a + 1(-7)$

$= 6a^2 - 42a + a - 7$

$= 6a^2 - 41a - 7$

17. $\left(4x - \dfrac{1}{11}\right)^2$ *This is a binomial squared.*

$= (4x)^2 - 2(4x)\left(\dfrac{1}{11}\right) + \left(\dfrac{1}{11}\right)^2$

$= 16x^2 - \dfrac{8}{11}x + \dfrac{1}{121}$

Work Practice Problems 14–17

Helpful Hint

- When multiplying two binomials, you may always use the FOIL order or method.
- When multiplying any two polynomials, you may always use the distributive property to find the product.

Objective A *Multiply using the FOIL method. See Examples 1 through 3.*

1. $(x + 3)(x + 4)$

2. $(x + 5)(x + 1)$

3. $(x - 5)(x + 10)$

4. $(y - 12)(y + 4)$

5. $(5x - 6)(x + 2)$

6. $(3y - 5)(2y - 7)$

7. $(y - 6)(4y - 1)$

8. $(2x - 9)(x - 11)$

9. $(2x + 5)(3x - 1)$

10. $(6x + 2)(x - 2)$

11. $(y^2 + 7)(6y + 4)$

12. $(y^2 + 3)(5y + 6)$

13. $\left(x - \dfrac{1}{3}\right)\left(x + \dfrac{2}{3}\right)$

14. $\left(x - \dfrac{2}{5}\right)\left(x + \dfrac{1}{5}\right)$

15. $(0.4 - 3a)(0.2 - 5a)$

16. $(0.3 - 2a)(0.6 - 5a)$

17. $(x + 5y)(2x - y)$

18. $(x + 4y)(3x - y)$

Objective B *Multiply. See Examples 4 through 8.*

19. $(x + 2)^2$

20. $(x + 7)^2$

21. $(2x - 1)^2$

22. $(7x - 3)^2$

23. $(3a - 5)^2$

24. $(5a + 2)^2$

25. $(x^2 + 0.5)^2$

26. $(x^2 + 0.3)^2$

27. $\left(y - \dfrac{2}{7}\right)^2$

28. $\left(y - \dfrac{3}{4}\right)^2$

29. $(2a - 3)^2$

30. $(5b - 4)^2$

31. $(5x + 9)^2$

32. $(6s + 2)^2$

33. $(3x - 7y)^2$

34. $(4s - 2y)^2$

35. $(4m + 5n)^2$

36. $(3n + 5m)^2$

37. $(5x^4 - 3)^2$

38. $(7x^3 - 6)^2$

Objective C *Multiply. See Examples 9 through 13.*

39. $(a - 7)(a + 7)$

40. $(b + 3)(b - 3)$

41. $(x + 6)(x - 6)$

42. $(x - 8)(x + 8)$

43. $(3x - 1)(3x + 1)$

44. $(4x - 5)(4x + 5)$

45. $(x^2 + 5)(x^2 - 5)$

46. $(a^2 + 6)(a^2 - 6)$

47. $(2y^2 - 1)(2y^2 + 1)$ **48.** $(3x^2 + 1)(3x^2 - 1)$ **49.** $(4 - 7x)(4 + 7x)$ **50.** $(8 - 7x)(8 + 7x)$

51. $\left(3x - \dfrac{1}{2}\right)\left(3x + \dfrac{1}{2}\right)$ **52.** $\left(10x + \dfrac{2}{7}\right)\left(10x - \dfrac{2}{7}\right)$ **53.** $(9x + y)(9x - y)$ **54.** $(2x - y)(2x + y)$

55. $(2m + 5n)(2m - 5n)$ **56.** $(5m + 4n)(5m - 4n)$

Objective Ⓓ **Mixed Practice** *Multiply. See Examples 14 through 17.*

57. $(a + 5)(a + 4)$ **58.** $(a + 5)(a + 7)$ **59.** $(a - 7)^2$ **60.** $(b - 2)^2$

61. $(4a + 1)(3a - 1)$ **62.** $(6a + 7)(6a + 5)$ **63.** $(x + 2)(x - 2)$ **64.** $(x - 10)(x + 10)$

65. $(3a + 1)^2$ **66.** $(4a + 2)^2$ **67.** $(x + y)(4x - y)$ **68.** $(3x + 2)(4x - 2)$

⊙ **69.** $\left(a - \dfrac{1}{2}y\right)\left(a + \dfrac{1}{2}y\right)$ **70.** $\left(\dfrac{a}{2} + 4y\right)\left(\dfrac{a}{2} - 4y\right)$ ⊙ **71.** $(3b + 7)(2b - 5)$ **72.** $(3y - 13)(y - 3)$

73. $(x^2 + 10)(x^2 - 10)$ **74.** $(x^2 + 8)(x^2 - 8)$ ⊙ **75.** $(4x + 5)(4x - 5)$ **76.** $(3x + 5)(3x - 5)$

77. $(5x - 6y)^2$ **78.** $(4x - 9y)^2$ **79.** $(2r - 3s)(2r + 3s)$ **80.** $(6r - 2x)(6r + 2x)$

Express each as a product of polynomials in x. Then multiply and simplify.

△ **81.** Find the area of the square rug if its side is $(2x + 1)$ feet.

$(2x + 1)$ feet

$(2x + 1)$ feet

△ **82.** Find the area of the rectangular canvas if its length is $(3x - 2)$ inches and its width is $(x - 4)$ inches.

$(x - 4)$ inches

$(3x - 2)$ inches

Review

Simplify each expression. See Sections 3.1 and 3.2.

83. $\dfrac{50b^{10}}{70b^5}$ **84.** $\dfrac{60y^6}{80y^2}$ **85.** $\dfrac{8a^{17}b^5}{-4a^7b^{10}}$ **86.** $\dfrac{-6a^8y}{3a^4y}$ **87.** $\dfrac{2x^4y^{12}}{3x^4y^4}$ **88.** $\dfrac{-48ab^6}{32ab^3}$

Concept Extensions

Match each expression on the left to the equivalent expression on the right. See the Concept Check in this section.

89. $(a - b)^2$

90. $(a - b)(a + b)$

91. $(a + b)^2$

92. $(a + b)^2(a - b)^2$

a. $a^2 - b^2$

b. $a^2 + b^2$

c. $a^2 - 2ab + b^2$

d. $a^2 + 2ab + b^2$

e. none of these

Fill in the squares so that a true statement forms.

93. $(x^\square + 7)(x^\square + 3) = x^4 + 10x^2 + 21$

94. $(5x^\square - 2)^2 = 25x^6 - 20x^3 + 4$

△ **95.** Find the area of the shaded region.

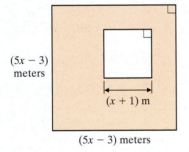

(5x − 3) meters

(5x − 3) meters

(x + 1) m

△ **96.** Find the area of the shaded region.

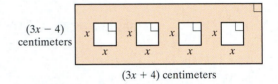

(3x − 4) centimeters

(3x + 4) centimeters

97. In your own words, describe the different methods that can be used to find the product: $(2x - 5)(3x + 1)$.

98. In your own words, describe the different methods that can be used to find the product: $(5x + 1)^2$.

Sections 3.1–3.6

Exponents and Operations on Polynomials

Perform operations and simplify.

1. $(5x^2)(7x^3)$

2. $(4y^2)(-8y^7)$

3. -4^2

4. $(-4)^2$

5. $(x - 5)(2x + 1)$

6. $(3x - 2)(x + 5)$

7. $(x - 5) + (2x + 1)$

8. $(3x - 2) + (x + 5)$

9. $\dfrac{7x^9 y^{12}}{x^3 y^{10}}$

10. $\dfrac{20a^2 b^8}{14a^2 b^2}$

11. $(12m^7 n^6)^2$

12. $(4y^9 z^{10})^3$

13. $(4y - 3)(4y + 3)$

14. $(7x - 1)(7x + 1)$

15. $(x^{-7} y^5)^9$

16. 8^{-2}

17. $(3^{-1} x^9)^3$

18. $\dfrac{(r^7 s^{-5})^6}{(2r^{-4} s^{-4})^4}$

19. $(7x^2 - 2x + 3) - (5x^2 + 9)$

20. $(10x^2 + 7x - 9) - (4x^2 - 6x + 2)$

1. _____
2. _____
3. _____
4. _____
5. _____
6. _____
7. _____
8. _____
9. _____
10. _____
11. _____
12. _____
13. _____
14. _____
15. _____
16. _____
17. _____
18. _____
19. _____
20. _____

21. $0.7y^2 - 1.2 + 1.8y^2 - 6y + 1$

22. $7.8x^2 - 6.8x - 3.3 + 0.6x^2 - 0.9$

23. Subtract $y^2 + 2$ from $3y^2 - 6y + 1$

24. $(z^2 + 5) - (3z^2 - 1) + \left(8z^2 + 2z - \dfrac{1}{2}\right)$

25. $(x + 4)^2$

26. $(y - 9)^2$

27. $(x + 4) + (x + 4)$

28. $(y - 9) + (y - 9)$

29. $7x^2 - 6xy + 4(y^2 - xy)$

30. $5a^2 - 3ab + 6(b^2 - a^2)$

31. $(x - 3)(x^2 + 5x - 1)$

32. $(x + 1)(x^2 - 3x - 2)$

33. $(2x - 7)(3x + 10)$

34. $(5x - 1)(4x + 5)$

35. $(2x - 7)(x^2 - 6x + 1)$

36. $(5x - 1)(x^2 + 2x - 3)$

37. $\left(2x + \dfrac{5}{9}\right)\left(2x - \dfrac{5}{9}\right)$

38. $\left(12y + \dfrac{3}{7}\right)\left(12y - \dfrac{3}{7}\right)$

21. _____

22. _____

23. _____

24. _____

25. _____

26. _____

27. _____

28. _____

29. _____

30. _____

31. _____

32. _____

33. _____

34. _____

35. _____

36. _____

37. _____

38. _____

3.7 DIVIDING POLYNOMIALS

Objectives

A Divide a Polynomial by a Monomial.

B Use Long Division to Divide a Polynomial by a Polynomial Other Than a Monomial.

Objective **A** Dividing by a Monomial

To divide a polynomial by a monomial, recall addition of fractions. Fractions that have a common denominator are added by adding the numerators:

$$\frac{a}{c} + \frac{b}{c} = \frac{a + b}{c}$$

If we read this equation from right to left and let a, b, and c be monomials, $c \neq 0$, we have the following.

To Divide a Polynomial by a Monomial

Divide each term of the polynomial by the monomial.

$$\frac{a + b}{c} = \frac{a}{c} + \frac{b}{c}, \quad c \neq 0$$

Throughout this section, we assume that denominators are not 0.

EXAMPLE 1 Divide: $(6m^2 + 2m) \div 2m$

Solution: We begin by writing the quotient in fraction form. Then we divide each term of the polynomial $6m^2 + 2m$ by the monomial $2m$ and use the quotient rule for exponents to simplify.

$$\frac{6m^2 + 2m}{2m} = \frac{6m^2}{2m} + \frac{2m}{2m}$$
$$= 3m + 1 \qquad \text{Simplify.}$$

Check: To check, we multiply.

$$2m(3m + 1) = 2m(3m) + 2m(1) = 6m^2 + 2m$$

The quotient $3m + 1$ checks.

 Work Practice Problem 1

✔ **Concept Check** In which of the following is $\dfrac{x + 5}{5}$ simplified correctly?

a. $\dfrac{x}{5} + 1$ **b.** x **c.** $x + 1$

EXAMPLE 2 Divide: $\dfrac{9x^5 - 12x^2 + 3x}{3x^2}$

Solution: $\dfrac{9x^5 - 12x^2 + 3x}{3x^2} = \dfrac{9x^5}{3x^2} - \dfrac{12x^2}{3x^2} + \dfrac{3x}{3x^2}$ Divide each term by $3x^2$.

$$= 3x^3 - 4 + \frac{1}{x} \qquad \text{Simplify.}$$

Notice that the quotient is not a polynomial because of the term $\dfrac{1}{x}$. This expression is called a rational expression—we will study rational expressions in Chapter 5. Although the quotient of two polynomials is not always a polynomial, we may still check by multiplying.

PRACTICE PROBLEM 1

Divide: $(25x^3 + 5x^2) \div 5x^2$

PRACTICE PROBLEM 2

Divide: $\dfrac{24x^7 + 12x^2 - 4x}{4x^2}$

Answers

1. $5x + 1$,   **2.** $6x^5 + 3 - \dfrac{1}{x}$

✔ **Concept Check Answer**

a

Continued on next page

Check: $3x^2\left(3x^3 - 4 + \dfrac{1}{x}\right) = 3x^2(3x^3) - 3x^2(4) + 3x^2\left(\dfrac{1}{x}\right)$

$$= 9x^5 - 12x^2 + 3x$$

🟧 **Work Practice Problem 2**

PRACTICE PROBLEM 3

Divide: $\dfrac{12x^3y^3 - 18xy + 6y}{3xy}$

EXAMPLE 3 Divide: $\dfrac{8x^2y^2 - 16xy + 2x}{4xy}$

Solution: $\dfrac{8x^2y^2 - 16xy + 2x}{4xy} = \dfrac{8x^2y^2}{4xy} - \dfrac{16xy}{4xy} + \dfrac{2x}{4xy}$ Divide each term by 4xy.

$$= 2xy - 4 + \dfrac{1}{2y}$$ Simplify.

Check: $4xy\left(2xy - 4 + \dfrac{1}{2y}\right) = 4xy(2xy) - 4xy(4) + 4xy\left(\dfrac{1}{2y}\right)$

$$= 8x^2y^2 - 16xy + 2x$$

🟧 **Work Practice Problem 3**

Objective 🅑 Dividing by a Polynomial Other Than a Monomial

To divide a polynomial by a polynomial other than a monomial, we use a process known as long division. Polynomial long division is similar to number long division, so we review long division by dividing 13 into 3660.

$$
\begin{array}{r}
281 \\
13\overline{)3660} \\
\end{array}
$$

$26\downarrow\downarrow$	$2 \cdot 13 = 26$
$\overline{106}$	Subtract and bring down the next digit in the dividend.
$104\downarrow$	$8 \cdot 13 = 104$
$\overline{20}$	Subtract and bring down the next digit in the dividend.
13	$1 \cdot 13 = 13$
$\overline{7}$	Subtract. There are no more digits to bring down, so the remainder is 7.

> **Helpful Hint**
> Recall that 3660 is called the dividend.

The quotient is 281 R 7, which can be written as $281\dfrac{7 \;\leftarrow \text{remainder}}{13 \;\leftarrow \text{divisor}}$

Recall that division can be checked by multiplication. To check this division problem, we see that

$13 \cdot 281 + 7 = 3660$, the dividend.

Now we demonstrate long division of polynomials.

PRACTICE PROBLEM 4

Divide $x^2 + 12x + 35$ by $x + 5$ using long division.

EXAMPLE 4 Divide $x^2 + 7x + 12$ by $x + 3$ using long division.

Solution:

> To subtract, change the signs of these terms and add.

$$
\begin{array}{r}
x \\
x + 3\overline{)x^2 + 7x + 12} \\
\underline{x^2 + 3x} \downarrow \\
4x + 12
\end{array}
$$

How many times does x divide x^2?
$\dfrac{x^2}{x} = x$.

Multiply: $x(x + 3)$.

Subtract and bring down the next term.

Answers

3. $4x^2y^2 - 6 + \dfrac{2}{x}$,   4. $x + 7$

Now we repeat this process.

$$\begin{array}{r} x + 4 \\ x + 3\overline{)x^2 + 7x + 12} \\ \underline{x^2 + 3x} \\ 4x + 12 \\ \underline{4x + 12} \\ 0 \end{array}$$

How many times does x divide $4x$? $\dfrac{4x}{x} = 4$.

Multiply: $4(x + 3)$.

Subtract. The remainder is 0.

To subtract, change the signs of these terms and add.

The quotient is $x + 4$.

Check: We check by multiplying.

divisor $\cdot$ quotient $+$ remainder $=$ dividend

or

$(x + 3) \cdot (x + 4) + 0 = x^2 + 7x + 12$

The quotient checks.

📘 **Work Practice Problem 4**

EXAMPLE 5 Divide $6x^2 + 10x - 5$ by $3x - 1$ using long division.

Solution:

$$\begin{array}{r} 2x + 4 \\ 3x - 1\overline{)6x^2 + 10x - 5} \\ \underline{6x^2 - 2x} \\ 12x - 5 \\ \underline{12x - 4} \\ -1 \end{array}$$

$\dfrac{6x^2}{3x} = 2x$, so $2x$ is a term of the quotient.

Multiply: $2x(3x - 1)$.

Subtract and bring down the next term.

$\dfrac{12x}{3x} = 4$. Multiply: $4(3x - 1)$.

Subtract. The remainder is -1.

Thus $(6x^2 + 10x - 5)$ divided by $(3x - 1)$ is $(2x + 4)$ with a remainder of -1. This can be written as follows.

$$\dfrac{6x^2 + 10x - 5}{3x - 1} = 2x + 4 + \dfrac{-1}{3x - 1} \quad \leftarrow \text{remainder}$$
$$\phantom{\dfrac{6x^2 + 10x - 5}{3x - 1} = 2x + 4 + \dfrac{-1}{3x - 1}} \leftarrow \text{divisor}$$
$$\text{or } 2x + 4 - \dfrac{1}{3x - 1}$$

Check: To check, we multiply $(3x - 1)(2x + 4)$. Then we add the remainder, -1, to this product.

$$(3x - 1)(2x + 4) + (-1) = (6x^2 + 12x - 2x - 4) - 1$$
$$= 6x^2 + 10x - 5$$

The quotient checks.

📘 **Work Practice Problem 5**

Notice that the division process is continued until the degree of the remainder polynomial is less than the degree of the divisor polynomial.

Recall in Section 3.3 that we practiced writing polynomials in descending order of powers and with no missing terms. For example, $2 - 4x^2$ written in this form is $-4x^2 + 0x + 2$. Writing the dividend and divisor in this form is helpful when dividing polynomials.

PRACTICE PROBLEM 5

Divide: $8x^2 + 2x - 7$ by $2x - 1$

Answer

5. $4x + 3 + \dfrac{-4}{2x - 1}$ or $4x + 3 - \dfrac{4}{2x - 1}$

PRACTICE PROBLEM 6

Divide: $(15 - 2x^2) \div (x - 3)$

EXAMPLE 6 Divide: $(2 - 4x^2) \div (x + 1)$

Solution: We use the rewritten form of $2 - 4x^2$ from the previous page.

$$
\begin{array}{r}
-4x + 4 \\
x + 1\overline{)-4x^2 + 0x + 2} \\
-4x^2 - 4x \\
\hline
4x + 2 \\
4x + 4 \\
\hline
-2
\end{array}
$$

$\dfrac{-4x^2}{x} = -4x$, so $-4x$ is a term of the quotient.

Multiply: $-4x(x + 1)$.

Subtract and bring down the next term.

$\dfrac{4x}{x} = 4$. Multiply: $4(x + 1)$.

Remainder.

Thus, $\dfrac{-4x^2 + 0x + 2}{x + 1}$ or $\dfrac{2 - 4x^2}{x + 1} = -4x + 4 + \dfrac{-2}{x + 1}$ or $-4x + 4 - \dfrac{2}{x + 1}$.

Check: To check, see that $(x + 1)(-4x + 4) + (-2) = 2 - 4x^2$.

🟧 **Work Practice Problem 6**

PRACTICE PROBLEM 7

Divide: $\dfrac{5 - x + 9x^3}{3x + 2}$

EXAMPLE 7 Divide: $\dfrac{4x^2 + 7 + 8x^3}{2x + 3}$

Solution: Before we begin the division process, we rewrite $4x^2 + 7 + 8x^3$ as $8x^3 + 4x^2 + 0x + 7$. Notice that we have written the polynomial in descending order and have represented the missing x term by $0x$.

$$
\begin{array}{r}
4x^2 - 4x + 6 \\
2x + 3\overline{)8x^3 + 4x^2 + 0x + 7} \\
8x^3 + 12x^2 \\
\hline
-8x^2 + 0x \\
-8x^2 - 12x \\
\hline
12x + 7 \\
12x + 18 \\
\hline
-11
\end{array}
$$

Remainder.

Thus, $\dfrac{4x^2 + 7 + 8x^3}{2x + 3} = 4x^2 - 4x + 6 + \dfrac{-11}{2x + 3}$ or $4x^2 - 4x + 6 - \dfrac{11}{2x + 3}$.

🟧 **Work Practice Problem 7**

PRACTICE PROBLEM 8

Divide: $x^3 - 1$ by $x - 1$

EXAMPLE 8 Divide $x^3 - 8$ by $x - 2$.

Solution: Notice that the polynomial $x^3 - 8$ is missing an x^2 term and an x term. We'll represent these terms by inserting $0x^2$ and $0x$.

$$
\begin{array}{r}
x^2 + 2x + 4 \\
x - 2\overline{)x^3 + 0x^2 + 0x - 8} \\
x^3 - 2x^2 \\
\hline
2x^2 + 0x \\
2x^2 - 4x \\
\hline
4x - 8 \\
4x - 8 \\
\hline
0
\end{array}
$$

Thus, $\dfrac{x^3 - 8}{x - 2} = x^2 + 2x + 4$.

Check: To check, see that $(x^2 + 2x + 4)(x - 2) = x^3 - 8$.

🟧 **Work Practice Problem 8**

Answers

6. $-2x - 6 + \dfrac{-3}{x - 3}$

or $-2x - 6 - \dfrac{3}{x - 3}$,

7. $3x^2 - 2x + 1 + \dfrac{3}{3x + 2}$,

8. $x^2 + x + 1$

Mental Math

Simplify each expression.

1. $\dfrac{a^6}{a^4}$

2. $\dfrac{y^2}{y}$

3. $\dfrac{a^3}{a}$

4. $\dfrac{p^8}{p^3}$

5. $\dfrac{k^5}{k^2}$

6. $\dfrac{k^7}{k^5}$

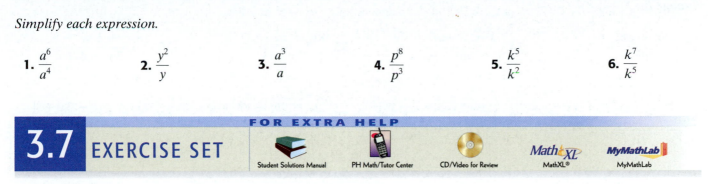

3.7 EXERCISE SET

FOR EXTRA HELP

Student Solutions Manual   PH Math/Tutor Center   CD/Video for Review   MathXL®   MyMathLab

Objective A *Perform each division. See Examples 1 through 3.*

1. $\dfrac{12x^4 + 3x^2}{x}$

2. $\dfrac{15x^2 - 9x^5}{x}$

3. $\dfrac{20x^3 - 30x^2 + 5x + 5}{5}$

4. $\dfrac{8x^3 - 4x^2 + 6x + 2}{2}$

5. $\dfrac{15p^3 + 18p^2}{3p}$

6. $\dfrac{14m^2 - 27m^3}{7m}$

7. $\dfrac{-9x^4 + 18x^5}{6x^5}$

8. $\dfrac{6x^5 + 3x^4}{3x^4}$

9. $\dfrac{-9x^5 + 3x^4 - 12}{3x^3}$

10. $\dfrac{6a^2 - 4a + 12}{-2a^2}$

11. $\dfrac{4x^4 - 6x^3 + 7}{-4x^4}$

12. $\dfrac{-12a^3 + 36a - 15}{3a}$

Objective B *Find each quotient using long division. See Examples 4 and 5.*

13. $\dfrac{x^2 + 4x + 3}{x + 3}$

14. $\dfrac{x^2 + 7x + 10}{x + 5}$

15. $\dfrac{2x^2 + 13x + 15}{x + 5}$

16. $\dfrac{3x^2 + 8x + 4}{x + 2}$

17. $\dfrac{2x^2 - 7x + 3}{x - 4}$

18. $\dfrac{3x^2 - x - 4}{x - 1}$

19. $\dfrac{9a^3 - 3a^2 - 3a + 4}{3a + 2}$

20. $\dfrac{4x^3 + 12x^2 + x - 14}{2x + 3}$

21. $\dfrac{8x^2 + 10x + 1}{2x + 1}$

22. $\dfrac{3x^2 + 17x + 7}{3x + 2}$

23. $\dfrac{2x^3 + 2x^2 - 17x + 8}{x - 2}$

24. $\dfrac{4x^3 + 11x^2 - 8x - 10}{x + 3}$

Find each quotient using long division. Don't forget to write the polynomials in descending order and fill in any missing terms. See Examples 6 through 8.

25. $\dfrac{x^2 - 36}{x - 6}$

26. $\dfrac{a^2 - 49}{a - 7}$

27. $\dfrac{x^3 - 27}{x - 3}$

28. $\dfrac{x^3 + 64}{x + 4}$

29. $\dfrac{1 - 3x^2}{x + 2}$

30. $\dfrac{7 - 5x^2}{x + 3}$

31. $\dfrac{-4b + 4b^2 - 5}{2b - 1}$

32. $\dfrac{-3y + 2y^2 - 15}{2y + 5}$

Objectives Ⓐ Ⓑ **Mixed Practice** *Divide. If the divisor contains 2 or more terms, use long division. See Examples 1 through 8.*

33. $\dfrac{a^2b^2 - ab^3}{ab}$

34. $\dfrac{m^3n^2 - mn^4}{mn}$

35. $\dfrac{8x^2 + 6x - 27}{2x - 3}$

36. $\dfrac{18w^2 + 18w - 8}{3w + 4}$

37. $\dfrac{2x^2y + 8x^2y^2 - xy^2}{2xy}$

38. $\dfrac{11x^3y^3 - 33xy + x^2y^2}{11xy}$

39. $\dfrac{2b^3 + 9b^2 + 6b - 4}{b + 4}$

40. $\dfrac{2x^3 + 3x^2 - 3x + 4}{x + 2}$

41. $\dfrac{y^3 + 3y^2 + 4}{y - 2}$

42. $\dfrac{3x^3 + 11x + 12}{x + 4}$

43. $\dfrac{5 - 6x^2}{x - 2}$

44. $\dfrac{3 - 7x^2}{x - 3}$

Divide.

45. $\dfrac{x^5 + x^2}{x^2 + x}$

46. $\dfrac{x^6 - x^4}{x^3 + 1}$

Review

Fill in each blank. See Sections 3.1 and 3.2.

47. $12 = 4 \cdot$ _____

48. $12 = 2 \cdot$ _____

49. $20 = -5 \cdot$ _____

50. $20 = -4 \cdot$ _____

51. $9x^2 = 3x \cdot$ _____

52. $9x^2 = 9x \cdot$ _____

53. $36x^2 = 4x \cdot$ _____

54. $36x^2 = 2x \cdot$ _____

Concept Extensions

Solve.

△ **55.** The perimeter of a square is $(12x^3 + 4x - 16)$ feet. Find the length of its side.

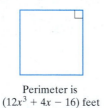

Perimeter is
$(12x^3 + 4x - 16)$ feet

△ **56.** The volume of the swimming pool shown is $(36x^5 - 12x^3 + 6x^2)$ cubic feet. If its height is $2x$ feet and its width is $3x$ feet, find its length.

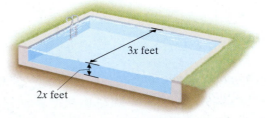

3x feet

2x feet

△ **57.** The area of the parallelogram shown is $(10x^2 + 31x + 15)$ square meters. If its base is $(5x + 3)$ meters, find its height.

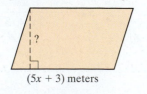

$(5x + 3)$ meters

△ **58.** The area of the top of the Ping-Pong table shown is $(49x^2 + 70x - 200)$ square inches. If its length is $(7x + 20)$ inches, find its width.

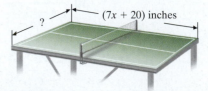

$(7x + 20)$ inches

59. Explain how to check a polynomial long division result when the remainder is 0.

60. Explain how to check a polynomial long division result when the remainder is not 0.

61. In which of the following is $\dfrac{a+7}{7}$ simplified correctly? See the Concept Check in this section.

   a. $a + 1$

   b. a

   c. $\dfrac{a}{7} + 1$

THE BIGGER PICTURE Simplifying Expressions and Solving Equations

Now we continue our outline from Sections 1.6 and 2.7. Although suggestions are given, this outline should be in your own words. Once you complete this new portion, try the exercises below.

I. Simplifying Expressions

  A. Real Numbers

    1. Add (Section 1.4)

    2. Subtract (Section 1.5)

    3. Multiply or Divide (Section 1.6)

  B. Exponents — $x^7 \cdot x^5 = x^{12}$; $(x^7)^5 = x^{35}$; $\dfrac{x^7}{x^5} = x^2$;

    $x^0 = 1$; $8^{-2} = \dfrac{1}{8^2} = \dfrac{1}{64}$

  C. Polynomials

    1. Add: Combine like terms.

      $(3y^2 + 6y + 7) + (9y^2 - 11y - 15)$

      $= 3y^2 + 6y + 7 + 9y^2 - 11y - 15$

      $= 12y^2 - 5y - 8$

    2. Subtract: Change the sign of the terms of the polynomial being subtracted, then add.

      $(3y^2 + 6y + 7) - (9y^2 - 11y - 15)$

      $= 3y^2 + 6y + 7 - 9y^2 + 11y + 15$

      $= -6y^2 + 17y + 22$

    3. Multiply: Multiply each term of one polynomial by each term of the other polynomial.

      $(x + 5)(2x^2 - 3x + 4)$

      $= x(2x^2 - 3x + 4) + 5(2x^2 - 3x + 4)$

      $= 2x^3 - 3x^2 + 4x + 10x^2 - 15x + 20$

      $= 2x^3 + 7x^2 - 11x + 20$

    4. Divide:

      a. To divide by a monomial, divide each term of the polynomial by the monomial.

        $\dfrac{8x^2 + 2x - 6}{2x} = \dfrac{8x^2}{2x} + \dfrac{2x}{2x} - \dfrac{6}{2x}$

        $= 4x + 1 - \dfrac{3}{x}$

      b. To divide by a polynomial other than a monomial, use long division.

$$x - 6 + \frac{40}{2x + 5}$$

$$2x + 5 \overline{\smash{)}\,2x^2 - 7x + 10}$$
$$\underline{2x^2 + 5x}$$
$$-12x + 10$$
$$\underline{-12x - 30}$$
$$40$$

II. Solving Equations and Inequalities

  A. Linear Equations (Section 2.3)

  B. Linear Inequalities (Section 2.7)

Simplify the expressions.

1. $-5.7 + (-0.23)$

2. $\dfrac{1}{2} - \dfrac{9}{10}$

3. $(-5x^2y^3)(-x^7y)$

4. $2^{-3}a^{-7}a^3$

5. $(7y^3 - 6y + 2) - (y^3 + 2y^2 + 2)$

6. Subtract $(y^2 + 7)$ from $(9y^2 - 3y)$

7. Multiply: $(x - 3)(4x^2 - x + 7)$

8. Multiply: $(6m - 5)^2$

9. Divide: $\dfrac{20n^2 - 5n + 10}{5n}$

10. Divide: $\dfrac{6x^2 - 20x + 20}{3x - 1}$

Solve the equations or inequalities.

11. $-6x = 3.6$

12. $-6x < 3.6$

13. $6x + 6 \geq 8x + 2$

14. $7y + 3(y - 1) = 4(y + 1) - 3$

CHAPTER 3 Group Activity

Modeling with Polynomials

Materials

Calculator

This activity may be completed by working in groups or individually.

The polynomial model $-13x^2 + 221x + 8476$ gives the average daily total supply of motor gasoline (in thousand barrels per day) in the United States for the period 2000–2003. The polynomial model $-23x^2 + 192x + 7825$ gives the average daily supply of domestically produced motor gasoline (in thousand barrels per day) in the United States for the same period. In both models, x is the number of years after 2000. The other source of motor gasoline in the United States, contributing to the total supply, is imported motor gasoline. (*Source:* Based on data from the Energy Information Administration)

1. Use the given polynomials to complete the following table showing the average daily supply (both total and domestic) over the period 2000–2003 by evaluating each polynomial at the given values of x. Then subtract each value in the fourth column from the corresponding value in the third column. Record the result in the last column, titled "Difference." What do you think these values represent?

Year	x	Average Daily Total Supply (thousand barrels per day)	Average Daily Domestic Supply (thousand barrels per day)	Difference
2000	0			
2001	1			
2002	2			
2003	3			

2. Use the polynomial models to find a new polynomial model representing the average daily supply of imported motor gasoline. Then evaluate your new polynomial model to complete the accompanying table.

Year	x	Average Daily Imported Supply (thousand barrels per day)
2000	0	
2001	1	
2002	2	
2003	3	

3. Compare the values in the last column of the table in question 1 to the values in the last column of the table in question 2. What do you notice? What can you conclude?

4. Make a bar graph of the data in the table in question 2. Describe what you see.

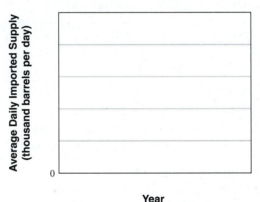

Chapter 3 Vocabulary Check

Fill in each blank with one of the words or phrases listed below.

term	coefficient	monomial	binomial	trinomial

polynomials	degree of a term	degree of a polynomial	FOIL

1. A _____ is a number or the product of a number and variables raised to powers.
2. The _____ method may be used when multiplying two binomials.
3. A polynomial with exactly 3 terms is called a _____.
4. The _____ is the greatest degree of any term of the polynomial.
5. A polynomial with exactly 2 terms is called a _____.
6. The _____ of a term is its numerical factor.
7. The _____ is the sum of the exponents on the variables in the term.
8. A polynomial with exactly 1 term is called a _____.
9. Monomials, binomials, and trinomials are all examples of _____.

Helpful Hint

Are you preparing for your test? Don't forget to take the Chapter 3 Test on page 249. Then check your answers at the back of the text and use the Chapter Test Prep Video CD to see the fully worked-out solutions to any of the exercises you want to review.

3 Chapter Highlights

DEFINITIONS AND CONCEPTS	EXAMPLES

Section 3.1 Exponents

a^n means the product of n factors, each of which is a.

$$3^2 = 3 \cdot 3 = 9$$
$$(-5)^3 = (-5)(-5)(-5) = -125$$
$$\left(\frac{1}{2}\right)^4 = \frac{1}{2} \cdot \frac{1}{2} \cdot \frac{1}{2} \cdot \frac{1}{2} = \frac{1}{16}$$

Let m and n be integers and no denominators be 0.

Product Rule: $a^m \cdot a^n = a^{m+n}$

Power Rule: $(a^m)^n = a^{mn}$

Power of a Product Rule: $(ab)^n = a^n b^n$

Power of a Quotient Rule: $\left(\dfrac{a}{b}\right)^n = \dfrac{a^n}{b^n}$

Quotient Rule: $\dfrac{a^m}{a^n} = a^{m-n}$

Zero Exponent: $a^0 = 1, a \neq 0$

$$x^2 \cdot x^7 = x^{2+7} = x^9$$
$$(5^3)^8 = 5^{3 \cdot 8} = 5^{24}$$
$$(7y)^4 = 7^4 y^4$$
$$\left(\frac{x}{8}\right)^3 = \frac{x^3}{8^3}$$
$$\frac{x^9}{x^4} = x^{9-4} = x^5$$
$$5^0 = 1; \ x^0 = 1, x \neq 0$$

DEFINITIONS AND CONCEPTS	**EXAMPLES**

Section 3.2   Negative Exponents and Scientific Notation

If $a \neq 0$ and n is an integer,

$$a^{-n} = \frac{1}{a^n}$$

$$3^{-2} = \frac{1}{3^2} = \frac{1}{9}; 5x^{-2} = \frac{5}{x^2}$$

Simplify:
$$\left(\frac{x^{-2}y}{x^5}\right)^{-2} = \frac{x^4 y^{-2}}{x^{-10}}$$
$$= x^{4-(-10)}y^{-2}$$
$$= \frac{x^{14}}{y^2}$$

A positive number is written in scientific notation if it is written as the product of a number a, where $1 \leq a < 10$, and an integer power r of 10.

$$a \times 10^r$$

$$1200 = 1.2 \times 10^3$$

$$0.000000568 = 5.68 \times 10^{-7}$$

Section 3.3   Introduction to Polynomials

A **term** is a number or the product of a number and variables raised to powers.

$$-5x, 7a^2b, \frac{1}{4}y^4, 0.2$$

The **numerical coefficient** or **coefficient** of a term is its numerical factor.

Term	Coefficient
$7x^2$	7
y	1
$-a^2b$	-1

A **polynomial** is a finite sum of terms of the form ax^n where a is a real number and n is a whole number.

$5x^3 - 6x^2 + 3x - 6$ (Polynomial)

A **monomial** is a polynomial with exactly 1 term.

$\frac{5}{6}y^3$ (Monomial)

A **binomial** is a polynomial with exactly 2 terms.

$-0.2a^2b - 5b^2$ (Binomial)

A **trinomial** is a polynomial with exactly 3 terms.

$3x^2 - 2x + 1$ (Trinomial)

The **degree of a polynomial** is the greatest degree of any term of the polynomial.

Polynomial	Degree
$5x^2 - 3x + 2$	2
$7y + 8y^2z^3 - 12$	$2 + 3 = 5$

Section 3.4   Adding and Subtracting Polynomials

To add polynomials, combine like terms.

Add.

$$(7x^2 - 3x + 2) + (-5x - 6)$$
$$= 7x^2 - 3x + 2 - 5x - 6$$
$$= 7x^2 - 8x - 4$$

To subtract two polynomials, change the signs of the terms of the second polynomial, and then add.

Subtract.

$$(17y^2 - 2y + 1) - (-3y^3 + 5y - 6)$$
$$= (17y^2 - 2y + 1) + (3y^3 - 5y + 6)$$
$$= 17y^2 - 2y + 1 + 3y^3 - 5y + 6$$
$$= 3y^3 + 17y^2 - 7y + 7$$

DEFINITIONS AND CONCEPTS	**EXAMPLES**

Section 3.5   Multiplying Polynomials

To multiply two polynomials, multiply each term of one polynomial by each term of the other polynomial, and then combine like terms.	Multiply. $$(2x + 1)(5x^2 - 6x + 2)$$ $$= 2x(5x^2 - 6x + 2) + 1(5x^2 - 6x + 2)$$ $$= 10x^3 - 12x^2 + 4x + 5x^2 - 6x + 2$$ $$= 10x^3 - 7x^2 - 2x + 2$$

Section 3.6   Special Products

The **FOIL method** may be used when multiplying two binomials.	Multiply: $(5x - 3)(2x + 3)$ First └─── Last ─┐ $(5x - 3)(2x + 3)$ └── Inner ──┘ └── Outer ──┘ $\qquad$ F $\qquad$ O $\qquad$ I $\qquad$ L $= (5x)(2x) + (5x)(3) + (-3)(2x) + (-3)(3)$ $= 10x^2 + 15x - 6x - 9$ $= 10x^2 + 9x - 9$
Squaring a Binomial $$(a + b)^2 = a^2 + 2ab + b^2$$ $$(a - b)^2 = a^2 - 2ab + b^2$$	Square each binomial. $$(x + 5)^2 = x^2 + 2(x)(5) + 5^2$$ $$= x^2 + 10x + 25$$ $$(3x - 2y)^2 = (3x)^2 - 2(3x)(2y) + (2y)^2$$ $$= 9x^2 - 12xy + 4y^2$$
Multiplying the Sum and Difference of Two Terms $$(a + b)(a - b) = a^2 - b^2$$	Multiply. $$(6y + 5)(6y - 5) = (6y)^2 - 5^2$$ $$= 36y^2 - 25$$

Section 3.7   Dividing Polynomials

To divide a polynomial by a monomial, $$\frac{a + b}{c} = \frac{a}{c} + \frac{b}{c}, c \neq 0$$	Divide. $$\frac{15x^5 - 10x^3 + 5x^2 - 2x}{5x^2}$$ $$= \frac{15x^5}{5x^2} - \frac{10x^3}{5x^2} + \frac{5x^2}{5x^2} - \frac{2x}{5x^2}$$ $$= 3x^3 - 2x + 1 - \frac{2}{5x}$$
To divide a polynomial by a polynomial other than a monomial, use long division.	$$\begin{array}{r} 5x - 1 + \dfrac{-4}{2x + 3} \\ 2x + 3 \overline{)10x^2 + 13x - 7} \\ \underline{10x^2 + 15x} \\ -2x - 7 \\ \underline{-2x - 3} \\ -4 \end{array}$$ or $5x - 1 - \dfrac{4}{2x + 3}$

Are You Prepared for a Test on Chapter 3?

Below is a list of some *common trouble areas* for students in Chapter 3. After studying for your test—but before taking your test—read these.

- Do you know that a negative exponent does not make the base a negative number? For example,

$$3^{-2} = \frac{1}{3^2} = \frac{1}{9}$$

- Make sure you remember that x has an understood coefficient of 1 and an understood exponent of 1. For example,

$$2x + x = 2x + 1x = 3x; \quad x^5 \cdot x = x^5 \cdot x^1 = x^6$$

- Do you know the difference between $5x^2$ and $(5x)^2$?

$$5x^2 \text{ is } 5 \cdot x^2; \quad (5x)^2 = 5^2 \cdot x^2 \text{ or } 25 \cdot x^2$$

- Can you evaluate $x^2 - x$ when $x = -2$?

$$x^2 - x = (-2)^2 - (-2) = 4 - (-2) = 4 + 2 = 6$$

- Can you subtract $5x^2 + 1$ from $3x^2 - 6$?

$$(3x^2 - 6) - (5x^2 + 1) = 3x^2 - 6 - 5x^2 - 1$$
$$= -2x^2 - 7$$

- Make sure you are familiar with squaring a binomial.

$$(3x - 4)^2 = (3x)^2 - 2(3x)(4) + 4^2$$
$$= 9x^2 - 24x + 16$$

or

$$(3x - 4)^2 = (3x - 4)(3x - 4)$$
$$= 9x^2 - 24x + 16$$

Remember: This is simply a checklist of common trouble areas. For a review of Chapter 3, see the Highlights and Chapter Review.

3 CHAPTER REVIEW

(3.1) *State the base and the exponent for each expression.*

1. 3^2

2. $(-5)^4$

3. -5^4

4. x^6

Evaluate each expression.

5. 8^3

6. $(-6)^2$

7. -6^2

8. $-4^3 - 4^0$

9. $(3b)^0$

10. $\dfrac{8b}{8b}$

Simplify each expression.

11. $y^2 \cdot y^7$

12. $x^9 \cdot x^5$

13. $(2x^5)(-3x^6)$

14. $(-5y^3)(4y^4)$

15. $(x^4)^2$

16. $(y^3)^5$

17. $(3y^6)^4$

18. $(2x^3)^3$

19. $\dfrac{x^9}{x^4}$

20. $\dfrac{z^{12}}{z^5}$

21. $\dfrac{a^5b^4}{ab}$

22. $\dfrac{x^4y^6}{xy}$

23. $\dfrac{12xy^6}{3x^4y^{10}}$

24. $\dfrac{2x^7y^8}{8xy^2}$

25. $5a^7(2a^4)^3$

26. $(2x)^2(9x)$

27. $(-5a)^0 + 7^0 + 8^0$

28. $8x^0 + 9^0$

244

Simplify the given expression and choose the correct result.

29. $\left(\dfrac{3x^4}{4y}\right)^3$

 a. $\dfrac{27x^{64}}{64y^3}$ **c.** $\dfrac{9x^{12}}{12y^3}$

 b. $\dfrac{27x^{12}}{64y^3}$ **d.** $\dfrac{3x^{12}}{4y^3}$

30. $\left(\dfrac{5a^6}{b^3}\right)^2$

 a. $\dfrac{10a^{12}}{b^6}$ **c.** $\dfrac{25a^{12}}{b^6}$

 b. $\dfrac{25a^{36}}{b^9}$ **d.** $25a^{12}b^6$

(3.2) *Simplify each expression.*

31. 7^{-2} **32.** -7^{-2} **33.** $2x^{-4}$ **34.** $(2x)^{-4}$

35. $\left(\dfrac{1}{5}\right)^{-3}$ **36.** $\left(\dfrac{-2}{3}\right)^{-2}$ **37.** $2^0 + 2^{-4}$ **38.** $6^{-1} - 7^{-1}$

Simplify each expression. Write each answer using positive exponents only.

39. $\dfrac{x^5}{x^{-3}}$ **40.** $\dfrac{z^4}{z^{-4}}$ **41.** $\dfrac{r^{-3}}{r^{-4}}$ **42.** $\dfrac{y^{-2}}{y^{-5}}$

43. $\left(\dfrac{bc^{-2}}{bc^{-3}}\right)^4$ **44.** $\left(\dfrac{x^{-3}y^{-4}}{x^{-2}y^{-5}}\right)^{-3}$ **45.** $\dfrac{x^{-4}y^{-6}}{x^2y^7}$ **46.** $\dfrac{a^5b^{-5}}{a^{-5}b^5}$

Write each number in scientific notation.

47. 0.00027 **48.** 0.8868 **49.** 80,800,000 **50.** 868,000

51. In August 2004, the United States imported approximately 112,400,000 kilograms of coffee. Write this number in scientific notation. (*Source:* International Coffee Organization)

52. The approximate diameter of the Milky Way galaxy is 150,000 light years. Write this number in scientific notation. (*Source:* NASA IMAGE/POETRY Education and Public Outreach Program)

150,000 light years

Write each number in standard form.

53. 8.67×10^5 **54.** 3.86×10^{-3} **55.** 8.6×10^{-4} **56.** 8.936×10^5

57. The volume of the planet Jupiter is 1.43128×10^{15} cubic kilometers. Write this number in standard notation. (*Source:* National Space Science Data Center)

58. An angstrom is a unit of measure, equal to 1×10^{-10} meter, used for measuring wavelengths or the diameters of atoms. Write this number in standard notation. (*Source:* National Institute of Standards and Technology)

Simplify. Express each result in standard form.

59. $(8 \times 10^4)(2 \times 10^{-7})$

60. $\dfrac{8 \times 10^4}{2 \times 10^{-7}}$

(3.3) *Find the degree of each polynomial.*

61. $y^5 + 7x - 8x^4$

62. $9y^2 + 30y + 25$

63. $-14x^2y - 28x^2y^3 - 42x^2y^2$

64. $6x^2y^2z^2 + 5x^2y^3 - 12xyz$

△ **65.** The surface area of a box with a square base and a height of 5 units is given by the polynomial $2x^2 + 20x$. Fill in the table below by evaluating $2x^2 + 20x$ for the given values of x.

x	1	3	5.1	10
$2x^2 + 20x$				

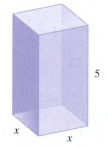

Combine like terms in each expression.

66. $7a^2 - 4a^2 - a^2$

67. $9y + y - 14y$

68. $6a^2 + 4a + 9a^2$

69. $21x^2 + 3x + x^2 + 6$

70. $4a^2b - 3b^2 - 8q^2 - 10a^2b + 7q^2$

71. $2s^{14} + 3s^{13} + 12s^{12} - s^{10}$

(3.4) *Add or subtract as indicated.*

72. $(3x^2 + 2x + 6) + (5x^2 + x)$

73. $(2x^5 + 3x^4 + 4x^3 + 5x^2) + (4x^2 + 7x + 6)$

74. $(-5y^2 + 3) - (2y^2 + 4)$

75. $(2m^7 + 3x^4 + 7m^6) - (8m^7 + 4m^2 + 6x^4)$

76. $(3x^2 - 7xy + 7y^2) - (4x^2 - xy + 9y^2)$

77. Add $(-9x^2 + 6x + 2)$ and $(4x^2 - x - 1)$.

78. Subtract $(4x^2 + 8x - 7)$ from the sum of $(x^2 + 7x + 9)$ and $(x^2 + 4)$.

(3.5) *Multiply each expression.*

79. $6(x + 5)$ **80.** $9(x - 7)$ **81.** $4(2a + 7)$ **82.** $9(6a - 3)$

83. $-7x(x^2 + 5)$ **84.** $-8y(4y^2 - 6)$ **85.** $-2(x^3 - 9x^2 + x)$ **86.** $-3a(a^2b + ab + b^2)$

87. $(3a^3 - 4a + 1)(-2a)$ **88.** $(6b^3 - 4b + 2)(7b)$ **89.** $(2x + 2)(x - 7)$

90. $(2x - 5)(3x + 2)$ **91.** $(4a - 1)(a + 7)$ **92.** $(6a - 1)(7a + 3)$

93. $(x + 7)(x^3 + 4x - 5)$ **94.** $(x + 2)(x^5 + x + 1)$ **95.** $(x^2 + 2x + 4)(x^2 + 2x - 4)$

96. $(x^3 + 4x + 4)(x^3 + 4x - 4)$ **97.** $(x + 7)^3$ **98.** $(2x - 5)^3$

(3.6) *Use special products to multiply each of the following.*

99. $(x + 7)^2$ **100.** $(x - 5)^2$ **101.** $(3x - 7)^2$ **102.** $(4x + 2)^2$

103. $(5x - 9)^2$ **104.** $(5x + 1)(5x - 1)$ **105.** $(7x + 4)(7x - 4)$ **106.** $(a + 2b)(a - 2b)$

107. $(2x - 6)(2x + 6)$ **108.** $(4a^2 - 2b)(4a^2 + 2b)$

Express each as a product of polynomials in x. Then multiply and simplify.

△ **109.** Find the area of the square if its side is $(3x - 1)$ meters.

$(3x - 1)$ meters

△ **110.** Find the area of the rectangle.

$(x - 1)$ miles

$(5x + 2)$ miles

(3.7) *Divide.*

111. $\dfrac{x^2 + 21x + 49}{7x^2}$

112. $\dfrac{5a^3b - 15ab^2 + 20ab}{-5ab}$

113. $(a^2 - a + 4) \div (a - 2)$

114. $(4x^2 + 20x + 7) \div (x + 5)$

115. $\dfrac{a^3 + a^2 + 2a + 6}{a - 2}$

116. $\dfrac{9b^3 - 18b^2 + 8b - 1}{3b - 2}$

117. $\dfrac{4x^4 - 4x^3 + x^2 + 4x - 3}{2x - 1}$

118. $\dfrac{-10x^2 - x^3 - 21x + 18}{x - 6}$

△ **119.** The area of the rectangle below is $(15x^3 - 3x^2 + 60)$ square feet. If its length is $3x^2$ feet, find its width.

Area is $(15x^3 - 3x^2 + 60)$ sq feet

△ **120.** The perimeter of the equilateral triangle below is $(21a^3b^6 + 3a - 3)$ units. Find the length of a side.

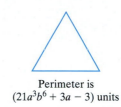

Perimeter is
$(21a^3b^6 + 3a - 3)$ units

Mixed Review

Evaluate.

121. $\left(-\dfrac{1}{2}\right)^3$

Simplify each expression. Write each answer using positive exponents only.

122. $(4xy^2)(x^3y^5)$

123. $\dfrac{18x^9}{27x^3}$

124. $\left(\dfrac{3a^4}{b^2}\right)^3$

125. $(2x^{-4}y^3)^{-4}$

126. $\dfrac{a^{-3}b^6}{9^{-1}a^{-5}b^{-2}}$

Perform the indicated operations and simplify.

127. $(6x + 2) + (5x - 7)$

128. $(-y^2 - 4) + (3y^2 - 6)$

129. $(8y^2 - 3y + 1) - (3y^2 + 2)$

130. $(5x^2 + 2x - 6) - (-x - 4)$

131. $4x(7x^2 + 3)$

132. $(2x + 5)(3x - 2)$

133. $(x - 3)(x^2 + 4x - 6)$

134. $(7x - 2)(4x - 9)$

Use special products to multiply.

135. $(5x + 4)^2$

136. $(6x + 3)(6x - 3)$

Divide.

137. $\dfrac{8a^4 - 2a^3 + 4a - 5}{2a^3}$

138. $\dfrac{x^2 + 2x + 10}{x + 5}$

139. $\dfrac{4x^3 + 8x^2 - 11x + 4}{2x - 3}$

3 CHAPTER TEST

Remember to use the Chapter Test Prep Video CD to see the fully worked-out solutions to any of the exercises you want to review.

Answers

Evaluate each expression.

1. 2^5 **2.** $(-3)^4$ **3.** -3^4 **4.** 4^{-3}

Simplify each exponential expression.

5. $(3x^2)(-5x^9)$ **6.** $\dfrac{y^7}{y^2}$ **7.** $\dfrac{r^{-8}}{r^{-3}}$

Simplify each expression. Write the result using only positive exponents.

8. $\left(\dfrac{x^2 y^3}{x^3 y^{-4}}\right)^2$ **9.** $\dfrac{6^2 x^{-4} y^{-1}}{6^3 x^{-3} y^7}$

Express each number in scientific notation.

10. 563,000   **11.** 0.0000863

Write each number in standard form.

12. 1.5×10^{-3} **13.** 6.23×10^4

14. Simplify. Write the answer in standard form.

$(1.2 \times 10^5)(3 \times 10^{-7})$

15. a. Complete the table for the polynomial $4xy^2 + 7xyz + x^3y - 2$.

Term	Numerical Coefficient	Degree of Term
$4xy^2$		
$7xyz$		
x^3y		
-2		

b. What is the degree of the polynomial?

16. Simplify by combining like terms.

$5x^2 + 4x - 7x^2 + 11 + 8x$

Perform each indicated operation.

17. $(8x^3 + 7x^2 + 4x - 7) + (8x^3 - 7x - 6)$

18. $\begin{array}{r} 5x^3 + x^2 + 5x - 2 \\ -(8x^3 - 4x^2 + x - 7) \\ \hline \end{array}$

19. Subtract $(4x + 2)$ from the sum of $(8x^2 + 7x + 5)$ and $(x^3 - 8)$.

Answers

1. _____

2. _____

3. _____

4. _____

5. _____

6. _____

7. _____

8. _____

9. _____

10. _____

11. _____

12. _____

13. _____

14. _____

15. **a.** see table

b. _____

16. _____

17. _____

18. _____

19. _____

249

20. _____

21. _____

22. _____

23. _____

24. _____

25. _____

26. _____

27. see table _____

28. _____

29. _____

30. _____

31. _____

Multiply. See Exercises 20 through 26.

20. $(3x + 7)(x^2 + 5x + 2)$

21. $3x^2(2x^2 - 3x + 7)$

22. $(x + 7)(3x - 5)$

23. $\left(3x - \dfrac{1}{5}\right)\left(3x + \dfrac{1}{5}\right)$

24. $(4x - 2)^2$

25. $(8x + 3)^2$

26. $(x^2 - 9b)(x^2 + 9b)$

27. The height of the Bank of China in Hong Kong is 1001 feet. Neglecting air resistance, the height of an object dropped from this building at time t seconds is given by the polynomial $-16t^2 + 1001$. Find the height of the object at the given times below.

t	0 seconds	1 second	3 seconds	5 seconds
$-16t^2 + 1001$				

△ **28.** Find the area of the top of the table. Express the area as a product, then multiply and simplify.

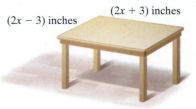

(2x − 3) inches   (2x + 3) inches

Divide.

29. $\dfrac{4x^2 + 2xy - 7x}{8xy}$

30. $(x^2 + 7x + 10) \div (x + 5)$

31. $\dfrac{27x^3 - 8}{3x + 2}$

Chapters 1–3

1. Given the set
$$\left\{-2, 0, \frac{1}{4}, 112, -3, 11, \sqrt{2}\right\}, \text{ list the}$$
numbers in this set that belong to the set of:

  a. Natural numbers
  b. Whole numbers
  c. Integers
  d. Rational numbers
  e. Irrational numbers
  f. Real numbers

2. Find the absolute value of each number.

  a. $|-7.2|$
  b. $|0|$
  c. $\left|-\dfrac{1}{2}\right|$

3. Evaluate (find the value of) the following:

  a. 3^2
  b. 5^3
  c. 2^4
  d. 7^1
  e. $\left(\dfrac{3}{7}\right)^2$

4. Multiply. Write products in lowest terms.

  a. $\dfrac{3}{4} \cdot \dfrac{7}{21}$
  b. $\dfrac{1}{2} \cdot 4\dfrac{5}{6}$

5. Simplify: $\dfrac{3}{2} \cdot \dfrac{1}{2} - \dfrac{1}{2}$

6. Evaluate $\dfrac{2x - 7y}{x^2}$ for $x = 5$ and $y = 1$.

7. Write an algebraic expression that represents each phrase. Let the variable x represent the unknown number.

  a. The sum of a number and 3
  b. The product of 3 and a number
  c. The quotient of 7.3 and a number
  d. 10 decreased by a number
  e. 5 times a number, increased by 7

8. Simplify: $8 + 3(2 \cdot 6 - 1)$

Answers

1. a. _____
  b. _____
  c. _____
  d. _____
  e. _____
  f. _____
2. a. _____
  b. _____
  c. _____
3. a. _____
  b. _____
  c. _____
  d. _____
  e. _____
4. a. _____
  b. _____
5. _____
6. _____
7. a. _____
  b. _____
  c. _____
  d. _____
  e. _____
8. _____

9. Add: $11.4 + (-4.7)$

10. Is $x = 1$ a solution of $5x^2 + 2 = x - 8$.

11. If $x = 2$ and $y = -5$, find the value of each expression.

  a. $\dfrac{x - y}{12 + x}$ **b.** $x^2 - y$

12. Subtract:

  a. $7 - 40$

  b. $-5 - (-10)$

Divide.

13. $\dfrac{-30}{-10}$

14. $\dfrac{-48}{6}$

15. $\dfrac{42}{-0.6}$

16. $\dfrac{-30}{-0.2}$

Find each product by using the distributive property to remove parentheses.

17. $5(3x + 2)$

18. $-3(2x - 3)$

19. $-2(y + 0.3z - 1)$

20. $4x(-x^2 + 6x - 1)$

21. $-(9x + y - 2z + 6)$

22. $-(-4xy + 6y - 2)$

23. Solve: $6(2a - 1) - (11a + 6) = 7$

24. Solve: $2x + \dfrac{1}{8} = x - \dfrac{3}{8}$

25. Solve: $\dfrac{y}{7} = 20$

26. Solve: $10 = 5j - 2$

27. Solve: $0.25x + 0.10(x - 3) = 1.1$

28. Solve: $\dfrac{7x + 5}{3} = x + 3$

29. Twice the sum of a number and 4 is the same as four times the number decreased by 12. Find the number.

30. Write the phrase as an algebraic expression and simplify if possible. Double a number, subtracted from the sum of a number and seven.

△ **31.** Charles Pecot can afford enough fencing to enclose a rectangular garden with a perimeter of 140 feet. If the width of his garden is to be 30 feet, find the length.

32. Simplify: $\dfrac{4(-3) + (-8)}{5 + (-5)}$

9. _____

10. _____

11. a. _____

   b. _____

12. a. _____

   b. _____

13. _____

14. _____

15. _____

16. _____

17. _____

18. _____

19. _____

20. _____

21. _____

22. _____

23. _____

24. _____

25. _____

26. _____

27. _____

28. _____

29. _____

30. _____

31. _____

32. _____

33. The number 120 is 15% of what number?

34. Graph $x < 5$.

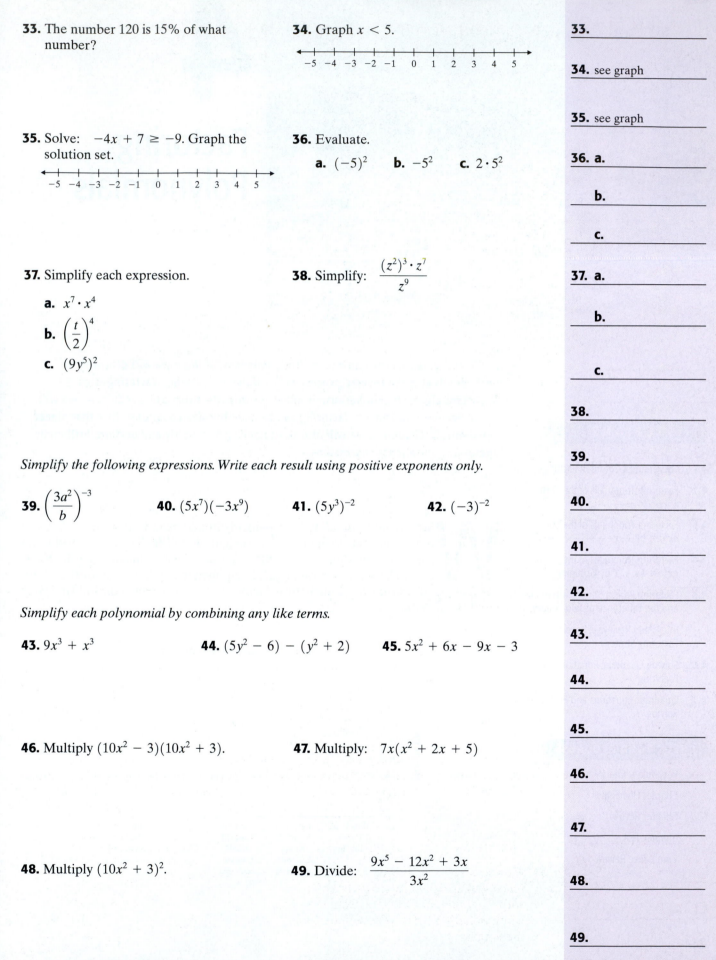

35. Solve: $-4x + 7 \geq -9$. Graph the solution set.

36. Evaluate.

    a. $(-5)^2$ **b.** -5^2 **c.** $2 \cdot 5^2$

37. Simplify each expression.

  a. $x^7 \cdot x^4$

  b. $\left(\dfrac{t}{2}\right)^4$

  c. $(9y^5)^2$

38. Simplify: $\dfrac{(z^2)^3 \cdot z^7}{z^9}$

Simplify the following expressions. Write each result using positive exponents only.

39. $\left(\dfrac{3a^2}{b}\right)^{-3}$ **40.** $(5x^7)(-3x^9)$ **41.** $(5y^3)^{-2}$ **42.** $(-3)^{-2}$

Simplify each polynomial by combining any like terms.

43. $9x^3 + x^3$ **44.** $(5y^2 - 6) - (y^2 + 2)$ **45.** $5x^2 + 6x - 9x - 3$

46. Multiply $(10x^2 - 3)(10x^2 + 3)$.

47. Multiply: $7x(x^2 + 2x + 5)$

48. Multiply $(10x^2 + 3)^2$.

49. Divide: $\dfrac{9x^5 - 12x^2 + 3x}{3x^2}$

33. _____

34. see graph _____

35. see graph _____

36. a. _____

    b. _____

    c. _____

37. a. _____

    b. _____

    c. _____

38. _____

39. _____

40. _____

41. _____

42. _____

43. _____

44. _____

45. _____

46. _____

47. _____

48. _____

49. _____

4

Factoring Polynomials

In Chapter 3, we learned how to multiply polynomials. Now we will deal with an operation that is the reverse process of multiplying—factoring. Factoring is an important algebraic skill because it allows us to write a sum as a product. As we will see in Sections 4.6 and 4.7, factoring can be used to solve equations other than linear equations. In Chapter 5, we will also use factoring to simplify and perform arithmetic operations on rational expressions.

When recently completed, the Taipei 101 building in Taipei, Taiwan became the world's tallest building. At a height of 1671 feet, it is the world's first super tall building to be built in an active earthquake zone. In Exercise 107, Section 4.5, on page 290, a polynomial expression for the height of an object dropped from Taipei 101 is factored. (*Source:* Council on Tall Buildings and Urban Habitats)

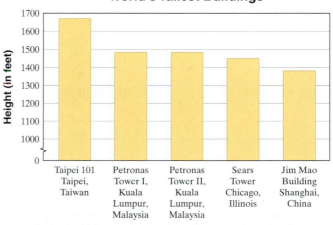

World's Tallest Buildings

4.1 THE GREATEST COMMON FACTOR

Objectives

A Find the Greatest Common Factor of a List of Numbers.

B Find the Greatest Common Factor of a List of Terms.

C Factor Out the Greatest Common Factor from the Terms of a Polynomial.

D Factor by Grouping.

In the product $2 \cdot 3 = 6$, the numbers 2 and 3 are called **factors** of 6 and $2 \cdot 3$ is a **factored form** of 6. This is true of polynomials also. Since $(x + 2)(x + 3) = x^2 + 5x + 6$, then $(x + 2)$ and $(x + 3)$ are factors of $x^2 + 5x + 6$, and $(x + 2)(x + 3)$ is a factored form of the polynomial.

> The process of writing a polynomial as a product is called **factoring** the polynomial.

Study the examples below and look for a pattern.

Multiplying: $5(x^2 + 3) = 5x^2 + 15 \qquad 2x(x - 7) = 2x^2 - 14x$

Factoring: $5x^2 + 15 = 5(x^2 + 3) \qquad 2x^2 - 14x = 2x(x - 7)$

Do you see that factoring is the reverse process of multiplying?

factoring
$x^2 + 5x + 6 = (x + 2)(x + 3)$
multiplying

✔ **Concept Check** Multiply: $2(x - 4)$
What do you think the result of factoring $2x - 8$ would be? Why?

Objective A Finding the Greatest Common Factor of a List of Numbers

The first step in factoring a polynomial is to see whether the terms of the polynomial have a common factor. If there is one, we can write the polynomial as a product by **factoring out** the common factor. We will usually factor out the *greatest* common factor (GCF).

The GCF of a list of integers is the largest integer that is a factor of all the integers in the list. For example, the GCF of 12 and 20 is 4 because 4 is the largest integer that is a factor of both 12 and 20. With large integers, the GCF may not be easily found by inspection. When this happens, we will write each integer as a product of prime numbers. Recall from Section R.1 that a prime number is a whole number other than 1, whose only factors are 1 and itself.

EXAMPLE 1 Find the GCF of each list of numbers.

a. 28 and 40   **b.** 55 and 21   **c.** 15, 18, and 66

Solution:

a. Write each number as a product of primes.

$28 = 2 \cdot 2 \cdot 7 = 2^2 \cdot 7$

$40 = 2 \cdot 2 \cdot 2 \cdot 5 = 2^3 \cdot 5$

There are two common factors, each of which is 2, so the GCF is

$GCF = 2 \cdot 2 = 4$

Continued on next page

PRACTICE PROBLEM 1

Find the GCF of each list of numbers.

a. 45 and 75   **b.** 32 and 33

c. 14, 24, and 60

Answers

1. **a.** 15, **b.** 1, **c.** 2

✔ **Concept Check Answer**

$2x - 8$; The result would be $2(x - 4)$ because factoring is the reverse process of multiplying.

b. $55 = 5 \cdot 11$

$21 = 3 \cdot 7$

There are no common prime factors; thus, the GCF is 1.

c. $15 = 3 \cdot 5$

$18 = 2 \cdot 3 \cdot 3 = 2 \cdot 3^2$

$66 = 2 \cdot 3 \cdot 11$

The only prime factor common to all three numbers is 3, so the GCF is

$GCF = 3$

▣ **Work Practice Problem 1**

Objective **B** Finding the Greatest Common Factor of a List of Terms

The greatest common factor of a list of variables raised to powers is found in a similar way. For example, the GCF of x^2, x^3, and x^5 is x^2 because each term contains a factor of x^2 and no higher power of x is a factor of each term.

$$x^2 = x \cdot x$$
$$x^3 = x \cdot x \cdot x$$
$$x^5 = x \cdot x \cdot x \cdot x \cdot x$$

There are two common factors, each of which is x, so the GCF $= x \cdot x$ or x^2. From this example, we see that **the GCF of a list of common variables raised to powers is the variable raised to the smallest exponent in the list.**

PRACTICE PROBLEM 2

Find the GCF of each list of terms.

a. y^4, y^5, and y^8

b. x and x^{10}

EXAMPLE 2 Find the GCF of each list of terms.

a. x^3, x^7, and x^5

b. y, y^4, and y^7

Solution:

a. The GCF is x^3, since 3 is the smallest exponent to which x is raised.

b. The GCF is y^1 or y, since 1 is the smallest exponent on y.

▣ **Work Practice Problem 2**

The **greatest common factor (GCF) of a list of terms** is the product of the GCF of the numerical coefficients and the GCF of the variable factors.

$$20x^2y^2 = 2 \cdot 2 \cdot 5 \cdot x \cdot x \cdot y \cdot y$$
$$6xy^3 = 2 \cdot 3 \cdot x \cdot y \cdot y \cdot y$$
$$GCF = 2 \cdot x \cdot y \cdot y = 2xy^2$$

Helpful Hint

Remember that the GCF of a list of terms contains the smallest exponent on each common variable.

Smallest exponent on x.

The GCF of x^5y^6, x^2y^7 and x^3y^4 is x^2y^4. ——— Smallest exponent on y.

Answers

2. a. y^4, **b.** x

EXAMPLE 3 Find the greatest common factor of each list of terms.

a. $6x^2$, $10x^3$, and $-8x$
b. $-18y^2$, $-63y^3$, and $27y^4$
c. a^3b^2, a^5b, and a^6b^2

Solution:

a. $6x^2 = 2 \cdot 3 \cdot x^2$
   $10x^3 = 2 \cdot 5 \cdot x^3$ → The GCF of x^2, x^3, and x^1 is x^1 or x.
   $-8x = -1 \cdot 2 \cdot 2 \cdot 2 \cdot x^1$
   GCF $= 2 \cdot x^1$ or $2x$

b. $-18y^2 = -1 \cdot 2 \cdot 3 \cdot 3 \cdot y^2$
   $-63y^3 = -1 \cdot 3 \cdot 3 \cdot 7 \cdot y^3$ → The GCF of y^2, y^3, and y^4 is y^2.
   $27y^4 = 3 \cdot 3 \cdot 3 \cdot y^4$
   GCF $= 3 \cdot 3 \cdot y^2$ or $9y^2$

c. The GCF of a^3, a^5, and a^6 is a^3.
   The GCF of b^2, b, and b^2 is b. Thus,
   the GCF of a^3b^2, a^5b, and a^6b^2 is a^3b.

🔲 **Work Practice Problem 3**

Objective C Factoring Out the Greatest Common Factor

To factor a polynomial such as $8x + 14$, we first see whether the terms have a greatest common factor other than 1. In this case, they do: The GCF of $8x$ and 14 is 2.

We factor out 2 from each term by writing each term as the product of 2 and the term's remaining factors.

$8x + 14 = 2 \cdot 4x + 2 \cdot 7$

Using the distributive property, we can write

$8x + 14 = 2 \cdot 4x + 2 \cdot 7$
$ = 2(4x + 7)$

Thus, a factored form of $8x + 14$ is $2(4x + 7)$. We can check by multiplying:

$2(4x + 7) = 2 \cdot 4x + 2 \cdot 7 = 8x + 14.$

Helpful Hint

A factored form of $8x + 14$ is *not*

$2 \cdot 4x + 2 \cdot 7$

Although the *terms* have been factored (written as products), the *polynomial* $8x + 14$ has not been factored. A factored form of $8x + 14$ is the *product* $2(4x + 7)$.

✔ **Concept Check** Which of the following is/are factored form(s) of $6t + 18$?

a. 6
b. $6 \cdot t + 6 \cdot 3$
c. $6(t + 3)$
d. $3(t + 6)$

PRACTICE PROBLEM 3

Find the greatest common factor of each list of terms.
a. $6x^2$, $9x^4$, and $-12x^5$
b. $-16y$, $-20y^6$, and $40y^4$
c. a^5b^4, ab^3, and a^3b^2

Answers
3. a. $3x^2$, b. $4y$, c. ab^2

✔ **Concept Check Answer**

c

PRACTICE PROBLEM 4

Factor each polynomial by factoring out the greatest common factor (GCF).

a. $10y + 25$

b. $x^4 - x^9$

EXAMPLE 4 Factor each polynomial by factoring out the greatest common factor (GCF).

a. $5ab + 10a$ **b.** $y^5 - y^{12}$

Solution:

a. The GCF of terms $5ab$ and $10a$ is $5a$. Thus,

$$5ab + 10a = 5a \cdot b + 5a \cdot 2$$
$$= 5a(b + 2) \qquad \text{Apply the distributive property.}$$

We can check our work by multiplying $5a$ and $(b + 2)$.
$5a(b + 2) = 5a \cdot b + 5a \cdot 2 = 5ab + 10a$, the original polynomial.

b. The GCF of y^5 and y^{12} is y^5. Thus,

$$y^5 - y^{12} = y^5(1) - y^5(y^7)$$
$$= y^5(1 - y^7)$$

Helpful Hint
Don't forget the 1.

■ **Work Practice Problem 4**

PRACTICE PROBLEM 5

Factor: $-10x^3 + 8x^2 - 2x$

EXAMPLE 5 Factor: $-9a^5 + 18a^2 - 3a$

Solution:

$$-9a^5 + 18a^2 - 3a = 3a(-3a^4) + 3a(6a) + 3a(-1)$$
$$= 3a(-3a^4 + 6a - 1)$$

Helpful Hint
Don't forget the -1.

■ **Work Practice Problem 5**

In Example 5, we could have chosen to factor out $-3a$ instead of $3a$. If we factor out $-3a$, we have

$$-9a^5 + 18a^2 - 3a = (-3a)(3a^4) + (-3a)(-6a) + (-3a)(1)$$
$$= -3a(3a^4 - 6a + 1)$$

Helpful Hint
Notice the changes in signs when factoring out $-3a$.

PRACTICE PROBLEMS 6–8

Factor.

6. $4x^3 + 12x$

7. $\dfrac{2}{5}a^5 - \dfrac{4}{5}a^3 + \dfrac{1}{5}a^2$

8. $6a^3b + 3a^3b^2 + 9a^2b^4$

EXAMPLES Factor.

6. $6a^4 - 12a = 6a(a^3 - 2)$

7. $\dfrac{3}{7}x^4 + \dfrac{1}{7}x^3 - \dfrac{5}{7}x^2 = \dfrac{1}{7}x^2(3x^2 + x - 5)$

8. $15p^2q^4 + 20p^3q^5 + 5p^3q^3 = 5p^2q^3(3q + 4pq^2 + p)$

■ **Work Practice Problems 6–8**

PRACTICE PROBLEM 9

Factor: $7(p + 2) + q(p + 2)$

EXAMPLE 9 Factor: $5(x + 3) + y(x + 3)$

Solution: The binomial $(x + 3)$ is present in both terms and is the greatest common factor. We use the distributive property to factor out $(x + 3)$.

$$5(x + 3) + y(x + 3) = (x + 3)(5 + y)$$

■ **Work Practice Problem 9**

Answers

4. a. $5(2y + 5)$, **b.** $x^4(1 - x^5)$,

5. $2x(-5x^2 + 4x - 1)$,

6. $4x(x^2 + 3)$,

7. $\dfrac{1}{5}a^2(2a^3 - 4a + 1)$,

8. $3a^2b(2a + ab + 3b^3)$,

9. $(p + 2)(7 + q)$

Objective D Factoring by Grouping

Once the GCF is factored out, we can often continue to factor the polynomial, using a variety of techniques. We discuss here a technique called **factoring by grouping.** This technique can be used to factor some polynomials with four terms.

EXAMPLE 10   Factor $xy + 2x + 3y + 6$ by grouping.

Solution: Notice that the first two terms of this polynomial have a common factor of x and the second two terms have a common factor of 3. Because of this, group the first two terms, then the last two terms, and then factor out these common factors.

$$xy + 2x + 3y + 6 = (xy + 2x) + (3y + 6) \quad \text{Group terms.}$$
$$= x(y + 2) + 3(y + 2) \quad \text{Factor out GCF from each grouping.}$$

Next we factor out the common binomial factor, $(y + 2)$.

$$x(y + 2) + 3(y + 2) = (y + 2)(x + 3)$$

Now the result is a factored form because it is a product. We were able to write the polynomial as a product because of the common binomial factor, $(y + 2)$, that appeared. If this does not happen, try rearranging the terms of the original polynomial.

Check: Multiply $(y + 2)$ by $(x + 3)$.

$$(y + 2)(x + 3) = xy + 2x + 3y + 6,$$

the original polynomial.
Thus, the factored form of $xy + 2x + 3y + 6$ is the product $(y + 2)(x + 3)$.

☐ **Work Practice Problem 10**

You may want to try these steps when factoring by grouping.

To Factor by Grouping

Step 1: Group the terms in two groups so that each group has a common factor.

Step 2: Factor out the GCF from each group.

Step 3: If there is a common binomial factor, factor it out.

Step 4: If not, rearrange the terms and try these steps again.

EXAMPLES   Factor by grouping.

11. $15x^3 - 10x^2 + 6x - 4$
$$= (15x^3 - 10x^2) + (6x - 4) \quad \text{Group the terms.}$$
$$= 5x^2(3x - 2) + 2(3x - 2) \quad \text{Factor each group.}$$
$$= (3x - 2)(5x^2 + 2) \quad \text{Factor out the common factor, } (3x - 2).$$

12. $3x^2 + 4xy - 3x - 4y$
$$= (3x^2 + 4xy) + (-3x - 4y)$$
$$= x(3x + 4y) - 1(3x + 4y) \quad \text{Factor each group. A } -1 \text{ is factored from the second pair of terms so that there is a common factor, } (3x + 4y).$$
$$= (3x + 4y)(x - 1) \quad \text{Factor out the common factor, } (3x + 4y).$$

Continued on next page

PRACTICE PROBLEM 10

Factor $ab + 7a + 2b + 14$ by grouping.

Helpful Hint

Notice that this form, $x(y + 2) + 3(y + 2)$, is *not* a factored form of the original polynomial. It is a sum, not a product.

PRACTICE PROBLEMS 11–13

Factor by grouping.
11. $28x^3 - 7x^2 + 12x - 3$
12. $2xy + 5y^2 - 4x - 10y$
13. $3x^2 + 4xy + 3x + 4y$

Answers
10. $(b + 7)(a + 2)$,
11. $(4x - 1)(7x^2 + 3)$,
12. $(2x + 5y)(y - 2)$,
13. $(3x + 4y)(x + 1)$

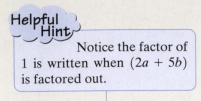

Helpful Hint

Notice the factor of 1 is written when $(2a + 5b)$ is factored out.

PRACTICE PROBLEMS 14–16

Factor by grouping.

14. $4x^3 + x - 20x^2 - 5$

15. $3xy - 4 + x - 12y$

16. $2x - 2 + x^3 - 3x^2$

13. $2a^2 + 5ab + 2a + 5b$

$\quad = (2a^2 + 5ab) + (2a + 5b)$ Factor each group. An understood 1 is written before

$\quad = a(2a + 5b) + 1(2a + 5b)$ $(2a + 5b)$ to help remember that $(2a + 5b)$ is $1(2a + 5b)$.

$\quad = (2a + 5b)(a + 1)$ Factor out the common factor, $(2a + 5b)$.

🟧 **Work Practice Problems 11–13**

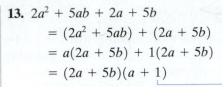 Factor by grouping.

14. $3x^3 - 2x - 9x^2 + 6$

$\quad = x(3x^2 - 2) - 3(3x^2 - 2)$ Factor each group. A -3 is factored from the second

$\quad = (3x^2 - 2)(x - 3)$ pair of terms so that there is a common factor, $(3x^2 - 2)$. Factor out the common factor, $(3x^2 - 2)$.

15. $3xy + 2 - 3x - 2y$

Notice that the first two terms have no common factor other than 1. However, if we rearrange these terms, a grouping emerges that does lead to a common factor.

$3xy + 2 - 3x - 2y$

$\quad = (3xy - 3x) + (-2y + 2)$

$\quad = 3x(y - 1) - 2(y - 1)$ Factor -2 from the second group.

$\quad = (y - 1)(3x - 2)$ Factor out the common factor, $(y - 1)$.

16. $5x - 10 + x^3 - x^2 = 5(x - 2) + x^2(x - 1)$

There is no common binomial factor that can now be factored out. No matter how we rearrange the terms, no grouping will lead to a common factor. Thus, this polynomial is not factorable by grouping.

🟧 **Work Practice Problems 14–16**

Helpful Hint

Throughout this chapter, we will be factoring polynomials. Even when the instructions do not so state, it is always a good idea to check your answers by multiplying.

Mental Math

Find the GCF of each pair of integers.

1. 2, 16
2. 3, 18
3. 6, 7
4. 9, 11
5. 14, 35
6. 33, 55

4.1 EXERCISE SET

Objectives **A** **B** **Mixed Practice** *Find the GCF for each list. See Examples 1 through 3.*

1. 32, 36
2. 36, 90
3. 18, 42, 84
4. 30, 75, 135

5. 24, 14, 21
6. 15, 25, 27
7. y^2, y^4, y^7
8. x^3, x^2, x^5

9. z^7, z^9, z^{11}
10. y^8, y^{10}, y^{12}
11. $x^{10}y^2, xy^2, x^3y^3$
12. p^7q, p^8q^2, p^9q^3

13. $14x, 21$
14. $20y, 15$
15. $12y^4, 20y^3$
16. $32x^5, 18x^2$

17. $-10x^2, 15x^3$
18. $-21x^3, 14x$
19. $12x^3, -6x^4, 3x^5$
20. $15y^2, 5y^7, -20y^3$

21. $-18x^2y, 9x^3y^3, 36x^3y$
22. $7x^3y^3, -21x^2y^2, 14xy^4$
23. $20a^6b^2c^8, 50a^7b$
24. $40x^7y^2z, 64x^9y$

Objective **C** *Factor out the GCF from each polynomial. See Examples 4 through 9.*

25. $3a + 6$
26. $18a + 12$
27. $30x - 15$
28. $42x - 7$
29. $x^3 + 5x^2$

30. $y^5 + 6y^4$
31. $6y^4 + 2y^3$
32. $5x^2 + 10x^6$
33. $32xy - 18x^2$
34. $10xy - 15x^2$

35. $4x - 8y + 4$
36. $7x + 21y - 7$
37. $6x^3 - 9x^2 + 12x$
38. $12x^3 + 16x^2 - 8x$

39. $a^7b^6 - a^3b^2 + a^2b^5 - a^2b^2$
40. $x^9y^6 + x^3y^5 - x^4y^3 + x^3y^3$
41. $5x^3y - 15x^2y + 10xy$

42. $14x^3y + 7x^2y - 7xy$
43. $8x^5 + 16x^4 - 20x^3 + 12$
44. $9y^6 - 27y^4 + 18y^2 + 6$

45. $\frac{1}{3}x^4 + \frac{2}{3}x^3 - \frac{4}{3}x^5 + \frac{1}{3}x$

46. $\frac{2}{5}y^7 - \frac{4}{5}y^5 + \frac{3}{5}y^2 - \frac{2}{5}y$

47. $y(x^2 + 2) + 3(x^2 + 2)$

48. $x(y^2 + 1) - 3(y^2 + 1)$

49. $z(y + 4) + 3(y + 4)$

50. $8(x + 2) - y(x + 2)$

51. $r(z^2 - 6) + (z^2 - 6)$

52. $q(b^3 - 5) + (b^3 - 5)$

Factor a −1 from each polynomial. See Example 5.

53. $-x - 7$

54. $-y - 3$

55. $-2 + z$

56. $-5 + y$

57. $3a - b + 2$

58. $2y - z - 11$

Objective D *Factor each four-term polynomial by grouping. See Examples 10 through 16.*

59. $x^3 + 2x^2 + 5x + 10$

60. $x^3 + 4x^2 + 3x + 12$

61. $5x + 15 + xy + 3y$

62. $xy + y + 2x + 2$

63. $6x^3 - 4x^2 + 15x - 10$

64. $16x^3 - 28x^2 + 12x - 21$

65. $5m^3 + 6mn + 5m^2 + 6n$

66. $8w^2 + 7wv + 8w + 7v$

67. $2y - 8 + xy - 4x$

68. $6x - 42 + xy - 7y$

69. $2x^3 + x^2 + 8x + 4$

70. $2x^3 - x^2 - 10x + 5$

71. $4x^2 - 8xy - 3x + 6y$

72. $5xy - 15x - 6y + 18$

73. $5q^2 - 4pq - 5q + 4p$

74. $6m^2 - 5mn - 6m + 5n$

Factor out the GCF from each polynomial. Then factor by grouping.

75. $12x^2y - 42x^2 - 4y + 14$

76. $90 + 15y^2 - 18x - 3xy^2$

Review

Multiply. See Section 3.5.

77. $(x + 2)(x + 5)$

78. $(y + 3)(y + 6)$

79. $(b + 1)(b - 4)$

80. $(x - 5)(x + 10)$

Fill in the chart by finding two numbers that have the given product and sum. The first column is filled in for you.

		81.	**82.**	**83.**	**84.**	**85.**	**86.**	**87.**	**88.**
Two Numbers	4, 7								
Their Product	28	12	20	8	16	−10	−9	−24	−36
Their Sum	11	8	9	−9	−10	3	0	−5	−5

Concept Extensions

See the Concept Checks in this section.

89. Which of the following is/are factored form(s) of $8a - 24$?

   a. $8 \cdot a - 24$ **b.** $8(a - 3)$ **c.** $4(2a - 12)$ **d.** $8 \cdot a - 2 \cdot 12$

Which of the following expressions are factored?

90. $(a + 6)(a + 2)$

91. $(x + 5)(x + y)$

92. $5(2y + z) - b(2y + z)$

93. $3x(a + 2b) + 2(a + 2b)$

94. The polynomial $-24x^2 + 336x - 132$ represents the average number of visitors (in thousands) per day to National Park Service areas, where x represents the month of the year. (*Source:* Based on data from National Park Service)

   a. Find the average daily number of visitors to National Park Service areas during the month of August. To do so, let $x = 8$ and evaluate $-24x^2 + 336x - 132$.

   b. Find the average daily number of visitors in May.

   c. Factor the polynomial $-24x^2 + 336x - 132$.

95. The average total daily supply of motor gasoline (in thousands of barrels per day) in the United States for the period 2000–2003 can be approximated by the polynomial $-13x^2 + 221x + 8476$, where x is the number of years after 2000. (*Source:* Based on data from Energy Information Administration)

   a. Find the average daily total supply of motor gasoline in 2001. To do so, let $x = 1$ and evaluate $-13x^2 + 221x + 8476$.

   b. Find the average daily total supply of motor gasoline in 2003.

   c. Factor the polynomial $-13x^2 + 221x + 8476$.

Write an expression for the area of each shaded region. Then write the expression as a factored polynomial.

△ **96.**

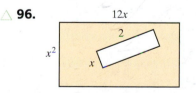

△ **97.**

Write an expression for the length of each rectangle. (Hint: Factor the area binomial and recall that Area = width · length.)

△ **98.**

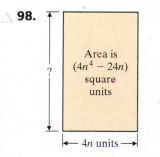

Area is $(4n^4 - 24n)$ square units

$?$

$\leftarrow 4n$ units $\rightarrow$

△ **99.**

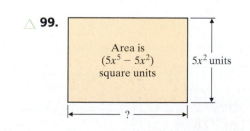

Area is $(5x^5 - 5x^2)$ square units

$5x^2$ units

$?$

100. Construct a binomial whose greatest common factor is $5a^3$. (*Hint:* Multiply $5a^3$ by a binomial whose terms contain no common factor other than 1. $5a^3(\square + \square)$.)

101. Construct a trinomial whose greatest common factor is $2x^2$. See the hint for Exercise 100.

 102. Explain how you can tell whether a polynomial is written in factored form.

103. Construct a four-term polynomial that can be factored by grouping.

STUDY SKILLS BUILDER

Are You Getting All the Mathematics Help That You Need?

Remember that, in addition to your instructor, there are many places to get help with your mathematics course. For example.

- This text has an accompanying video lesson for every section and worked out solutions to every Chapter Test exercise on video.

- The back of the book contains answers to odd-numbered exercises and selected solutions.

- A student *Solutions Manual* is available that contains worked-out solutions to odd-numbered exercises as well as solutions to every exercise in the Integrated Reviews, Chapter Reviews, Chapter Tests, and Cumulative Reviews.

- Don't forget to check with your instructor for other local resources available to you, such as a tutor center.

Exercises

1. List items you find helpful in the text and all student supplements to this text.

2. List all the campus help that is available to you for this course.

3. List any help (besides the textbook) from Exercises 1 and 2 above that you are using.

4. List any help (besides the textbook) that you feel you should try.

5. Write a goal for yourself that includes trying anything you listed in Exercise 4 during the next week.

FACTORING TRINOMIALS OF THE FORM $x^2 + bx + c$

A Factor Trinomials of the Form $x^2 + bx + c$.

B Factor Out the Greatest Common Factor and Then Factor a Trinomial of the Form $x^2 + bx + c$.

Objective **A** Factoring Trinomials of the Form $x^2 + bx + c$

In this section, we factor trinomials of the form $x^2 + bx + c$, such as

$$x^2 + 7x + 12, \quad x^2 - 12x + 35, \quad x^2 + 4x - 12, \quad \text{and} \quad r^2 - r - 42$$

Notice that for these trinomials, the coefficient of the squared variable is 1.

   Recall that factoring means to write as a product and that factoring and multiplying are reverse processes. Using the FOIL method of multiplying binomials, we have the following.

$$
\begin{array}{cccc}
\text{F} & \text{O} & \text{I} & \text{L}
\end{array}
$$
$$(x + 3)(x + 1) = x^2 + 1x + 3x + 3$$
$$= x^2 + 4x + 3$$

Thus, a factored form of $x^2 + 4x + 3$ is $(x + 3)(x + 1)$.

   Notice that the product of the first terms of the binomials is $x \cdot x = x^2$, the first term of the trinomial. Also, the product of the last two terms of the binomials is $3 \cdot 1 = 3$, the third term of the trinomial. The sum of these same terms is $3 + 1 = 4$, the coefficient of the middle, x, term of the trinomial.

The product of these numbers is 3.

$$x^2 + 4x + 3 = (x + 3)(x + 1)$$

The sum of these numbers is 4.

   Many trinomials, such as the one above, factor into two binomials. To factor $x^2 + 7x + 10$, let's assume that it factors into two binomials and begin by writing two pairs of parentheses. The first term of the trinomial is x^2, so we use x and x as the first terms of the binomial factors.

$$x^2 + 7x + 10 = (x + \square)(x + \square)$$

To determine the last term of each binomial factor, we look for two integers whose product is 10 and whose sum is 7. The integers are 2 and 5. Thus,

$$x^2 + 7x + 10 = (x + 2)(x + 5)$$

Check: To see if we have factored correctly, we multiply.

$$(x + 2)(x + 5) = x^2 + 5x + 2x + 10$$
$$= x^2 + 7x + 10 \qquad \text{Combine like terms.}$$

Helpful Hint

   Since multiplication is commutative, the factored form of $x^2 + 7x + 10$ can be written as either $(x + 2)(x + 5)$ or $(x + 5)(x + 2)$.

To Factor a Trinomial of the Form $x^2 + bx + c$

The product of these numbers is c.

$$x^2 + bx + c = (x + \square)(x + \square)$$

The sum of these numbers is b.

PRACTICE PROBLEM 1

Factor: $x^2 + 12x + 20$

EXAMPLE 1 Factor: $x^2 + 7x + 12$

Solution: We begin by writing the first terms of the binomial factors.

$$(x + \square)(x + \square)$$

Next we look for two numbers whose product is 12 and whose sum is 7. Since our numbers must have a positive product and a positive sum, we look at pairs of positive factors of 12 only.

Factors of 12	Sum of Factors
1, 12	13
2, 6	8
3, 4	7

Correct sum, so the numbers are 3 and 4.

Thus, $x^2 + 7x + 12 = (x + 3)(x + 4)$

Check: $(x + 3)(x + 4) = x^2 + 4x + 3x + 12 = x^2 + 7x + 12$.

◻ **Work Practice Problem 1**

PRACTICE PROBLEM 2

Factor each trinomial.
a. $x^2 - 23x + 22$
b. $x^2 - 27x + 50$

EXAMPLE 2 Factor: $x^2 - 12x + 35$

Solution: Again, we begin by writing the first terms of the binomials.

$$(x + \square)(x + \square)$$

Now we look for two numbers whose product is 35 and whose sum is -12. Since our numbers must have a positive product and a negative sum, we look at pairs of negative factors of 35 only.

Factors of 35	Sum of Factors
$-1, -35$	-36
$-5, -7$	-12

Correct sum, so the numbers are -5 and -7.

$$x^2 - 12x + 35 = (x - 5)(x - 7)$$

Check: To check, multiply $(x - 5)(x - 7)$.

◻ **Work Practice Problem 2**

PRACTICE PROBLEM 3

Factor: $x^2 + 5x - 36$

EXAMPLE 3 Factor: $x^2 + 4x - 12$

Solution: $x^2 + 4x - 12 = (x + \square)(x + \square)$

We look for two numbers whose product is -12 and whose sum is 4. Since our numbers must have a negative product, we look at pairs of factors with opposite signs.

Factors of -12	Sum of Factors
$-1, 12$	11
$1, -12$	-11
$-2, 6$	4
$2, -6$	-4
$-3, 4$	1
$3, -4$	-1

Correct sum, so the numbers are -2 and 6.

$$x^2 + 4x - 12 = (x - 2)(x + 6)$$

◻ **Work Practice Problem 3**

Answers
1. $(x + 10)(x + 2)$,
2. a. $(x - 1)(x - 22)$,
b. $(x - 2)(x - 25)$,
3. $(x + 9)(x - 4)$

EXAMPLE 4 Factor: $r^2 - r - 42$

Solution: Because the variable in this trinomial is r, the first term of each binomial factor is r.

$$r^2 - r - 42 = (r + \Box)(r + \Box)$$

Now we look for two numbers whose product is -42 and whose sum is -1, the numerical coefficient of r. The numbers are 6 and -7. Therefore,

$$r^2 - r - 42 = (r + 6)(r - 7)$$

◻ **Work Practice Problem 4**

PRACTICE PROBLEM 4

Factor each trinomial.

a. $q^2 - 3q - 40$

b. $y^2 + 2y - 48$

EXAMPLE 5 Factor: $a^2 + 2a + 10$

Solution: Look for two numbers whose product is 10 and whose sum is 2. Neither 1 and 10 nor 2 and 5 give the required sum, 2. We conclude that $a^2 + 2a + 10$ is not factorable with integers. A polynomial such as $a^2 + 2a + 10$ is called a **prime polynomial.**

◻ **Work Practice Problem 5**

PRACTICE PROBLEM 5

Factor: $x^2 + 6x + 15$

EXAMPLE 6 Factor: $x^2 + 5xy + 6y^2$

Solution: $x^2 + 5xy + 6y^2 = (x + \Box)(x + \Box)$

Recall that the middle term $5xy$ is the same as $5yx$. Thus, we can see that $5y$ is the "coefficient" of x. We then look for two terms whose product is $6y^2$ and whose sum is $5y$. The terms are $2y$ and $3y$ because $2y \cdot 3y = 6y^2$ and $2y + 3y = 5y$. Therefore,

$$x^2 + 5xy + 6y^2 = (x + 2y)(x + 3y)$$

◻ **Work Practice Problem 6**

PRACTICE PROBLEM 6

Factor each trinomial.

a. $x^2 + 9xy + 14y^2$

b. $a^2 - 13ab + 30b^2$

EXAMPLE 7 Factor: $x^4 + 5x^2 + 6$

Solution: As usual, we begin by writing the first terms of the binomials. Since the greatest power of x in this polynomial is x^4, we write

$$(x^2 + \Box)(x^2 + \Box) \quad \text{since } x^2 \cdot x^2 = x^4$$

Now we look for two factors of 6 whose sum is 5. The numbers are 2 and 3. Thus,

$$x^4 + 5x^2 + 6 = (x^2 + 2)(x^2 + 3)$$

◻ **Work Practice Problem 7**

PRACTICE PROBLEM 7

Factor: $x^4 + 8x^2 + 12$

If the terms of a polynomial are not written in descending powers of the variable, you may want to do so before factoring.

EXAMPLE 8 Factor: $40 - 13t + t^2$

Solution: First, we rearrange terms so that the trinomial is written in descending powers of t.

$$40 - 13t + t^2 = t^2 - 13t + 40$$

Next, try to factor.

$$t^2 - 13t + 40 = (t + \Box)(t + \Box)$$

Now we look for two factors of 40 whose sum is -13. The numbers are -8 and -5. Thus,

$$t^2 - 13t + 40 = (t - 8)(t - 5)$$

◻ **Work Practice Problem 8**

PRACTICE PROBLEM 8

Factor: $48 - 14x + x^2$

Answers

4. a. $(q - 8)(q + 5)$,
b. $(y + 8)(y - 6)$,
5. prime polynomial,
6. a. $(x + 2y)(x + 7y)$,
b. $(a - 3b)(a - 10b)$,
7. $(x^2 + 6)(x^2 + 2)$,
8. $(x - 6)(x - 8)$

The following sign patterns may be useful when factoring trinomials.

> **Helpful Hint**
>
> A positive constant in a trinomial tells us to look for two numbers with the same sign. The sign of the coefficient of the middle term tells us whether the signs are both positive or both negative.
>
> both positive   same sign   both negative   same sign
>
> $$x^2 + 10x + 16 = (x + 2)(x + 8) \qquad x^2 - 10x + 16 = (x - 2)(x - 8)$$
>
> A negative constant in a trinomial tells us to look for two numbers with opposite signs.
>
> opposite signs   opposite signs
>
> $$x^2 + 6x - 16 = (x + 8)(x - 2) \qquad x^2 - 6x - 16 = (x - 8)(x + 2)$$

Objective B Factoring Out the Greatest Common Factor

Remember that the first step in factoring any polynomial is to factor out the greatest common factor (if there is one other than 1 or -1).

PRACTICE PROBLEM 9

Factor each trinomial.
a. $4x^2 - 24x + 36$
b. $x^3 + 3x^2 - 4x$

EXAMPLE 9 Factor: $3m^2 - 24m - 60$

Solution: First we factor out the greatest common factor, 3, from each term.

$$3m^2 - 24m - 60 = 3(m^2 - 8m - 20)$$

Now we factor $m^2 - 8m - 20$ by looking for two factors of -20 whose sum is -8. The factors are -10 and 2. Therefore, the complete factored form is

$$3m^2 - 24m - 60 = 3(m + 2)(m - 10)$$

🟧 **Work Practice Problem 9**

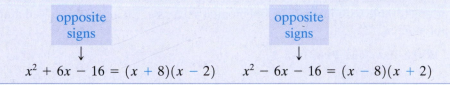

> **Helpful Hint**
>
> Remember to write the common factor 3 as part of the factored form.

PRACTICE PROBLEM 10

Factor: $5x^5 - 25x^4 - 30x^3$

EXAMPLE 10 Factor: $2x^4 - 26x^3 + 84x^2$

Solution:

$$2x^4 - 26x^3 + 84x^2 = 2x^2(x^2 - 13x + 42) \qquad \text{Factor out common factor, } 2x^2.$$
$$= 2x^2(x - 6)(x - 7) \qquad \text{Factor } x^2 - 13x + 42.$$

🟧 **Work Practice Problem 10**

Answers
9. a. $4(x - 3)(x - 3)$,
b. $x(x + 4)(x - 1)$,
10. $5x^3(x + 1)(x - 6)$

Mental Math

Complete each factored form.

1. $x^2 + 9x + 20 = (x + 4)(x \quad)$ **2.** $x^2 + 12x + 35 = (x + 5)(x \quad)$ **3.** $x^2 - 7x + 12 = (x - 4)(x \quad)$

4. $x^2 - 13x + 22 = (x - 2)(x \quad)$ **5.** $x^2 + 4x + 4 = (x + 2)(x \quad)$ **6.** $x^2 + 10x + 24 = (x + 6)(x \quad)$

4.2 EXERCISE SET

Objective **A** *Factor each trinomial completely. If a polynomial can't be factored, write "prime." See Examples 1 through 8.*

1. $x^2 + 7x + 6$ **2.** $x^2 + 6x + 8$ **3.** $y^2 - 10y + 9$ **4.** $y^2 - 12y + 11$

5. $x^2 - 6x + 9$ **6.** $x^2 - 10x + 25$ **7.** $x^2 - 3x - 18$ **8.** $x^2 - x - 30$

9. $x^2 + 3x - 70$ **10.** $x^2 + 4x - 32$ **11.** $x^2 + 5x + 2$ **12.** $x^2 - 7x + 5$

13. $x^2 + 8xy + 15y^2$ **14.** $x^2 + 6xy + 8y^2$ **15.** $a^4 - 2a^2 - 15$ **16.** $y^4 - 3y^2 - 70$

17. $13 + 14m + m^2$ **18.** $17 + 18n + n^2$ **19.** $10t - 24 + t^2$ **20.** $6q - 27 + q^2$

21. $a^2 - 10ab + 16b^2$ **22.** $a^2 - 9ab + 18b^2$

Objectives **A** **B** **Mixed Practice** *Factor each trinomial completely. Some of these trinomials contain a greatest common factor (other than 1). Don't forget to factor out the GCF first. See Examples 1 through 10.*

23. $2z^2 + 20z + 32$ **24.** $3x^2 + 30x + 63$ **25.** $2x^3 - 18x^2 + 40x$ **26.** $3x^3 - 12x^2 - 36x$

27. $x^2 - 3xy - 4y^2$ **28.** $x^2 - 4xy - 77y^2$ **29.** $x^2 + 15x + 36$ **30.** $x^2 + 19x + 60$

31. $x^2 - x - 2$ **32.** $x^2 - 5x - 14$ **33.** $r^2 - 16r + 48$ **34.** $r^2 - 10r + 21$

35. $x^2 + xy - 2y^2$ **36.** $x^2 - xy - 6y^2$ **37.** $3x^2 + 9x - 30$ **38.** $4x^2 - 4x - 48$

269

39. $3x^2 - 60x + 108$ **40.** $2x^2 - 24x + 70$ **41.** $x^2 - 18x - 144$ **42.** $x^2 + x - 42$

43. $r^2 - 3r + 6$ **44.** $x^2 + 4x - 10$ **45.** $x^2 - 8x + 15$ **46.** $x^2 - 9x + 14$

47. $6x^3 + 54x^2 + 120x$ **48.** $3x^3 + 3x^2 - 126x$ **49.** $4x^2y + 4xy - 12y$ **50.** $3x^2y - 9xy + 45y$

51. $x^2 - 4x - 21$ **52.** $x^2 - 4x - 32$ **53.** $x^2 + 7xy + 10y^2$ **54.** $x^2 - 3xy - 4y^2$

55. $64 + 24t + 2t^2$ **56.** $50 + 20t + 2t^2$ **57.** $x^3 - 2x^2 - 24x$ **58.** $x^3 - 3x^2 - 28x$

59. $2t^5 - 14t^4 + 24t^3$ **60.** $3x^6 + 30x^5 + 72x^4$ **61.** $5x^3y - 25x^2y^2 - 120xy^3$ **62.** $7a^3b - 35a^2b^2 + 42ab^3$

63. $162 - 45m + 3m^2$ **64.** $48 - 20n + 2n^2$ **65.** $-x^2 + 12x - 11$
(Factor out -1 first.)   **66.** $-x^2 + 8x - 7$
(Factor out -1 first.)

67. $\dfrac{1}{2}y^2 - \dfrac{9}{2}y - 11$
(Factor out $\dfrac{1}{2}$ first.)   **68.** $\dfrac{1}{3}y^2 - \dfrac{5}{3}y - 8$
(Factor out $\dfrac{1}{3}$ first.)   **69.** $x^3y^2 + x^2y - 20x$ **70.** $a^2b^3 + ab^2 - 30b$

Review

Multiply. See Section 3.5.

71. $(2x + 1)(x + 5)$ **72.** $(3x + 2)(x + 4)$ **73.** $(5y - 4)(3y - 1)$

74. $(4z - 7)(7z - 1)$ **75.** $(a + 3b)(9a - 4b)$ **76.** $(y - 5x)(6y + 5x)$

Concept Extensions

77. Write a polynomial that factors as $(x - 3)(x + 8)$.

78. To factor $x^2 + 13x + 42$, think of two numbers whose _____ is 42 and whose _____ is 13.

Complete each sentence in your own words.

79. If $x^2 + bx + c$ is factorable and c is negative, then the signs of the last-term factors of the binomials are opposite because . . . .

80. If $x^2 + bx + c$ is factorable and c is positive, then the signs of the last-term factors of the binomials are the same because . . . .

Remember that perimeter means distance around. Write the perimeter of each rectangle as a simplified polynomial. Then factor the polynomial.

△ **81.**

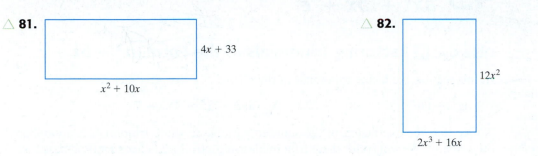

$4x + 33$

$x^2 + 10x$

△ **82.**

$12x^2$

$2x^3 + 16x$

83. An object is thrown upward from the top of an 80-foot building with an initial velocity of 64 feet per second. The height of the object after t seconds is given by $-16t^2 + 64t + 80$. Factor this polynomial.

$-16t^2 + 64t + 80$

Factor each trinomial completely.

84. $x^2 + x + \dfrac{1}{4}$

85. $x^2 + \dfrac{1}{2}x + \dfrac{1}{16}$

86. $y^2(x + 1) - 2y(x + 1) - 15(x + 1)$

87. $z^2(x + 1) - 3z(x + 1) - 70(x + 1)$

Find a positive value of c so that each trinomial is factorable.

88. $y^2 - 4y + c$

89. $n^2 - 16n + c$

Find a positive value of b so that each trinomial is factorable.

90. $x^2 + bx + 15$

91. $y^2 + by + 20$

Factor each trinomial. (Hint: Notice that $x^{2n} + 4x^n + 3$ factors as $(x^n + 1)(x^n + 3)$. Remember: $x^n \cdot x^n = x^{n+n}$ or x^{2n}.)

92. $x^{2n} + 5x^n + 6$

93. $x^{2n} + 8x^n - 20$

A Factor Trinomials of the Form $ax^2 + bx + c$, where $a \neq 1$.

B Factor Out the GCF before Factoring a Trinomial of the Form $ax^2 + bx + c$.

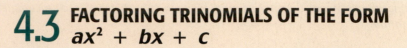

4.3 FACTORING TRINOMIALS OF THE FORM $ax^2 + bx + c$

Objective **A** Factoring Trinomials of the Form $ax^2 + bx + c$

In this section, we factor trinomials of the form $ax^2 + bx + c$, such as

$$3x^2 + 11x + 6, \qquad 8x^2 - 22x + 5, \quad \text{and} \quad 2x^2 + 13x - 7$$

Notice that the coefficient of the squared variable in these trinomials is a number other than 1. We will factor these trinomials using a trial-and-check method based on our work in the last section.

To begin, let's review the relationship between the numerical coefficients of the trinomial and the numerical coefficients of its factored form. For example, since $(2x + 1)(x + 6) = 2x^2 + 13x + 6$,

a factored form of $2x^2 + 13x + 6$ is $(2x + 1)(x + 6)$

Notice that $2x$ and x are factors of $2x^2$, the first term of the trinomial. Also, 6 and 1 are factors of 6, the last term of the trinomial, as shown:

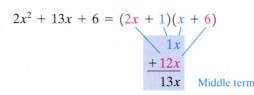

$$2x^2 + 13x + 6 = (2x + 1)(x + 6)$$

Also notice that $13x$, the middle term, is the sum of the following products:

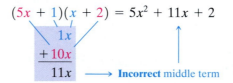

Let's use this pattern to factor $5x^2 + 7x + 2$. First, we find factors of $5x^2$. Since all numerical coefficients in this trinomial are positive, we will use factors with positive numerical coefficients only. Thus, the factors of $5x^2$ are $5x$ and x. Let's try these factors as first terms of the binomials. Thus far, we have

$$5x^2 + 7x + 2 = (5x + \square)(x + \square)$$

Next, we need to find positive factors of 2. Positive factors of 2 are 1 and 2. Now we try possible combinations of these factors as second terms of the binomials until we obtain a middle term of $7x$.

$(5x + 1)(x + 2) = 5x^2 + 11x + 2$

$\begin{array}{r} 1x \\ + 10x \\ \hline 11x \end{array}$ ⟶ **Incorrect** middle term

Let's try switching factors 2 and 1.

$(5x + 2)(x + 1) = 5x^2 + 7x + 2$

$\begin{array}{r} 2x \\ + 5x \\ \hline 7x \end{array}$ ⟶ **Correct** middle term

Thus a factored form of $5x^2 + 7x + 2$ is $(5x + 2)(x + 1)$. To check, we multiply $(5x + 2)$ and $(x + 1)$. The product is $5x^2 + 7x + 2$.

EXAMPLE 1 Factor: $3x^2 + 11x + 6$

Solution: Since all numerical coefficients are positive, we use factors with positive numerical coefficients. We first find factors of $3x^2$.

Factors of $3x^2$: $3x^2 = 3x \cdot x$

If factorable, the trinomial will be of the form

$3x^2 + 11x + 6 = (3x + \square)(x + \square)$

Next we factor 6.

Factors of 6: $6 = 1 \cdot 6$, $\quad 6 = 2 \cdot 3$

Now we try combinations of factors of 6 until a middle term of $11x$ is obtained. Let's try 1 and 6 first.

$(3x + 1)(x + 6) = 3x^2 + 19x + 6$

$\quad 1x$
$+18x$
$\overline{\quad 19x} \longrightarrow$ **Incorrect** middle term

Now let's next try 6 and 1.

$(3x + 6)(x + 1)$

Before multiplying, notice that the terms of the factor $3x + 6$ have a common factor of 3. The terms of the original trinomial $3x^2 + 11x + 6$ have no common factor other than 1, so the terms of its factors will also contain no common factor other than 1. This means that $(3x + 6)(x + 1)$ is not a factored form.

Next let's try 2 and 3 as last terms.

$(3x + 2)(x + 3) = 3x^2 + 11x + 6$

$\quad 2x$
$+9x$
$\overline{\quad 11x} \longrightarrow$ **Correct** middle term

Thus a factored form of $3x^2 + 11x + 6$ is $(3x + 2)(x + 3)$.

☐ **Work Practice Problem 1**

✔**Concept Check** Do the terms of $3x^2 + 29x + 18$ have a common factor? Without multiplying, decide which of the following factored forms could not be a factored form of $3x^2 + 29x + 18$.

a. $(3x + 18)(x + 1)$ **b.** $(3x + 2)(x + 9)$
c. $(3x + 6)(x + 3)$ **d.** $(3x + 9)(x + 2)$

EXAMPLE 2 Factor: $8x^2 - 22x + 5$

Solution: Factors of $8x^2$: $8x^2 = 8x \cdot x$, $\quad 8x^2 = 4x \cdot 2x$

We'll try $8x$ and x.

$8x^2 - 22x + 5 = (8x + \square)(x + \square)$

Since the middle term, $-22x$, has a negative numerical coefficient, we factor 5 into negative factors.

Factors of 5: $5 = -1 \cdot -5$

Continued on next page

PRACTICE PROBLEM 1

Factor each trinomial.
a. $5x^2 + 27x + 10$
b. $4x^2 + 12x + 5$

Helpful Hint

This is true in general: If the terms of a trinomial have no common factor (other than 1), then the terms of each of its binomial factors will contain no common factor (other than 1).

PRACTICE PROBLEM 2

Factor each trinomial.
a. $2x^2 - 11x + 12$
b. $6x^2 - 5x + 1$

Answers
1. a. $(5x + 2)(x + 5)$,
b. $(2x + 5)(2x + 1)$,
2. a. $(2x - 3)(x - 4)$,
b. $(3x - 1)(2x - 1)$

✔ **Concept Check Answer**
no; a, c, d

Let's try -1 and -5.

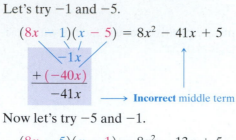

$(8x - 1)(x - 5) = 8x^2 - 41x + 5$

$\begin{array}{c} -1x \\ + (-40x) \\ \hline -41x \end{array}$ $\longrightarrow$ **Incorrect** middle term

Now let's try -5 and -1.

$(8x - 5)(x - 1) = 8x^2 - 13x + 5$

$\begin{array}{c} -5x \\ + (-8x) \\ \hline -13x \end{array}$ $\longrightarrow$ **Incorrect** middle term

Don't give up yet! We can still try other factors of $8x^2$. Let's try $4x$ and $2x$ with -1 and -5.

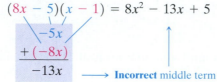

$(4x - 1)(2x - 5) = 8x^2 - 22x + 5$

$\begin{array}{c} -2x \\ + (-20x) \\ \hline -22x \end{array}$ $\longrightarrow$ **Correct** middle term

A factored form of $8x^2 - 22x + 5$ is $(4x - 1)(2x - 5)$.

🔲 **Work Practice Problem 2**

PRACTICE PROBLEM 3

Factor each trinomial.
a. $3x^2 + 14x - 5$
b. $35x^2 + 4x - 4$

EXAMPLE 3 Factor: $2x^2 + 13x - 7$

Solution: Factors of $2x^2$: $2x^2 = 2x \cdot x$

Factors of -7: $-7 = -1 \cdot 7$, $\quad -7 = 1 \cdot -7$

We try possible combinations of these factors:

$(2x + 1)(x - 7) = 2x^2 - 13x - 7$ **Incorrect** middle term
$(2x - 1)(x + 7) = 2x^2 + 13x - 7$ **Correct** middle term

A factored form of $2x^2 + 13x - 7$ is $(2x - 1)(x + 7)$.

🔲 **Work Practice Problem 3**

PRACTICE PROBLEM 4

Factor each trinomial.
a. $14x^2 - 3xy - 2y^2$
b. $12a^2 - 16ab - 3b^2$

EXAMPLE 4 Factor: $10x^2 - 13xy - 3y^2$

Solution: Factors of $10x^2$: $10x^2 = 10x \cdot x$, $\quad 10x^2 = 2x \cdot 5x$

Factors of $-3y^2$: $-3y^2 = -3y \cdot y$, $\quad -3y^2 = 3y \cdot -y$

We try some combinations of these factors:

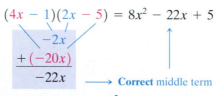

$\qquad\qquad\qquad$ Correct $\qquad\qquad$ Correct

$(10x - 3y)(x + y) = 10x^2 + 7xy - 3y^2$
$(x + 3y)(10x - y) = 10x^2 + 29xy - 3y^2$
$(5x + 3y)(2x - y) = 10x^2 + xy - 3y^2$
$(2x - 3y)(5x + y) = 10x^2 - 13xy - 3y^2$ **Correct** middle term

A factored form of $10x^2 - 13xy - 3y^2$ is $(2x - 3y)(5x + y)$.

🔲 **Work Practice Problem 4**

PRACTICE PROBLEM 5

Factor: $2x^4 - 5x^2 - 7$

EXAMPLE 5 Factor: $3x^4 - 5x^2 - 8$

Solution: Factors of $3x^4$: $3x^4 = 3x^2 \cdot x^2$

$\qquad\qquad$ Factors of -8: $-8 = -2 \cdot 4, 2 \cdot -4, -1 \cdot 8, 1 \cdot -8$

Answers

3. a. $(3x - 1)(x + 5)$,
b. $(5x + 2)(7x - 2)$,
4. a. $(7x + 2y)(2x - y)$,
b. $(6a + b)(2a - 3b)$,
5. $(2x^2 - 7)(x^2 + 1)$

Try combinations of these factors:

Correct   Correct

$$(3x^2 - 2)(x^2 + 4) = 3x^4 + 10x^2 - 8$$
$$(3x^2 + 4)(x^2 - 2) = 3x^4 - 2x^2 - 8$$
$$(3x^2 + 8)(x^2 - 1) = 3x^4 + 5x^2 - 8 \quad \text{Incorrect sign on middle term, so switch signs in binomial factors.}$$
$$(3x^2 - 8)(x^2 + 1) = 3x^4 - 5x^2 - 8 \quad \text{Correct middle term.}$$

🔲 **Work Practice Problem 5**

> **Helpful Hint**
>
> Study the last two lines of Example 5. If a factoring attempt gives you a middle term whose numerical coefficient is the opposite of the desired numerical coefficient, try switching the signs of the last terms in the binomials.
>
> Switched signs
> $$(3x^2 + 8)(x^2 - 1) = 3x^4 + 5x^2 - 8 \quad \text{Middle term: } +5x$$
> $$(3x^2 - 8)(x^2 + 1) = 3x^4 - 5x^2 - 8 \quad \text{Middle term: } -5x$$

Objective B Factoring Out the Greatest Common Factor

Don't forget that the first step in factoring any polynomial is to look for a common factor to factor out.

EXAMPLE 6   Factor: $24x^4 + 40x^3 + 6x^2$

Solution: Notice that all three terms have a common factor of $2x^2$. Thus we factor out $2x^2$ first.

$$24x^4 + 40x^3 + 6x^2 = 2x^2(12x^2 + 20x + 3)$$

Next we factor $12x^2 + 20x + 3$.

Factors of $12x^2$: $12x^2 = 4x \cdot 3x$, $12x^2 = 12x \cdot x$, $12x^2 = 6x \cdot 2x$

Since all terms in the trinomial have positive numerical coefficients, we factor 3 using positive factors only.

Factors of 3: $3 = 1 \cdot 3$

We try some combinations of the factors.

$$2x^2(4x + 3)(3x + 1) = 2x^2(12x^2 + 13x + 3)$$
$$2x^2(12x + 1)(x + 3) = 2x^2(12x^2 + 37x + 3)$$
$$2x^2(2x + 3)(6x + 1) = 2x^2(12x^2 + 20x + 3) \quad \text{Correct middle term}$$

A factored form of $24x^4 + 40x^3 + 6x^2$ is $2x^2(2x + 3)(6x + 1)$.

🔲 **Work Practice Problem 6**

When the term containing the squared variable has a negative coefficient, you may want to first factor out a common factor of -1.

EXAMPLE 7   Factor: $-6x^2 - 13x + 5$

Solution: We begin by factoring out a common factor of -1.

$$-6x^2 - 13x + 5 = -1(6x^2 + 13x - 5) \quad \text{Factor out } -1.$$
$$= -1(3x - 1)(2x + 5) \quad \text{Factor } 6x^2 + 13x - 5.$$

🔲 **Work Practice Problem 7**

PRACTICE PROBLEM 6

Factor each trinomial.
a. $3x^3 + 17x^2 + 10x$
b. $6xy^2 + 33xy - 18x$

> **Helpful Hint**
>
> Don't forget to include the common factor in the factored form.

PRACTICE PROBLEM 7

Factor: $-5x^2 - 19x + 4$

Answers
6. a. $x(3x + 2)(x + 5)$,
b. $3x(2y - 1)(y + 6)$,
7. $-1(x + 4)(5x - 1)$

Objective A *Complete each factored form. See Examples 1 through 5.*

1. $5x^2 + 22x + 8 = (5x + 2)(\quad)$

2. $2y^2 + 15y + 25 = (2y + 5)(\quad)$

3. $50x^2 + 15x - 2 = (5x + 2)(\quad)$

4. $6y^2 + 11y - 10 = (2y + 5)(\quad)$

5. $20x^2 - 7x - 6 = (5x + 2)(\quad)$

6. $8y^2 - 2y - 55 = (2y + 5)(\quad)$

Factor each trinomial completely. See Examples 1 through 5.

7. $2x^2 + 13x + 15$

8. $3x^2 + 8x + 4$

9. $8y^2 - 17y + 9$

10. $21x^2 - 41x + 10$

11. $2x^2 - 9x - 5$

12. $36r^2 - 5r - 24$

13. $20r^2 + 27r - 8$

14. $3x^2 + 20x - 63$

15. $10x^2 + 17x + 3$

16. $2x^2 + 7x + 5$

17. $x + 3x^2 - 2$

18. $y + 8y^2 - 9$

19. $6x^2 - 13xy + 5y^2$

20. $8x^2 - 14xy + 3y^2$

21. $15m^2 - 16m - 15$

22. $25n^2 - 5n - 6$

23. $-9x + 20 + x^2$

24. $-7x + 12 + x^2$

25. $2x^2 - 7x - 99$

26. $2x^2 + 7x - 72$

27. $-27t + 7t^2 - 4$

28. $-3t + 4t^2 - 7$

29. $3a^2 + 10ab + 3b^2$

30. $2a^2 + 11ab + 5b^2$

31. $49p^2 - 7p - 2$

32. $3r^2 + 10r - 8$

33. $18x^2 - 9x - 14$

34. $42a^2 - 43a + 6$

35. $2m^2 + 17m + 10$

36. $3n^2 + 20n + 5$

37. $24x^2 + 41x + 12$

38. $24x^2 - 49x + 15$

Objectives A B Mixed Practice *Factor each trinomial completely. See Examples 1 through 7.*

39. $12x^3 + 11x^2 + 2x$

40. $8a^3 + 14a^2 + 3a$

41. $21b^2 - 48b - 45$

42. $12x^2 - 14x - 10$

43. $7z + 12z^2 - 12$

44. $16t + 15t^2 - 15$

45. $6x^2y^2 - 2xy^2 - 60y^2$

46. $8x^2y + 34xy - 84y$

47. $4x^2 - 8x - 21$

48. $6x^2 - 11x - 10$

49. $3x^2 - 42x + 63$

50. $5x^2 - 75x + 60$

51. $8x^2 + 6xy - 27y^2$

52. $54a^2 + 39ab - 8b^2$

53. $-x^2 + 2x + 24$

54. $-x^2 + 4x + 21$

55. $4x^3 - 9x^2 - 9x$

56. $6x^3 - 31x^2 + 5x$

57. $24x^2 - 58x + 9$

58. $36x^2 + 55x - 14$

59. $40a^2b + 9ab - 9b$ **60.** $24y^2x + 7yx - 5x$ **61.** $30x^3 + 38x^2 + 12x$ **62.** $6x^3 - 28x^2 + 16x$

63. $6y^3 - 8y^2 - 30y$ **64.** $12x^3 - 34x^2 + 24x$ **65.** $10x^4 + 25x^3y - 15x^2y^2$ **66.** $42x^4 - 99x^3y - 15x^2y^2$

67. $-14x^2 + 39x - 10$ **68.** $-15x^2 + 26x - 8$ **69.** $16p^4 - 40p^3 + 25p^2$ **70.** $9q^4 - 42q^3 + 49q^2$

71. $-2x^2 + 9x + 5$ **72.** $-3x^2 + 8x + 16$ **73.** $-4 + 52x - 48x^2$ **74.** $-5 + 55x - 50x^2$

75. $2t^4 + 3t^2 - 27$ **76.** $4r^4 - 17r^2 - 15$ **77.** $5x^2y^2 + 20xy + 1$ **78.** $3a^2b^2 + 12ab + 1$

79. $6a^5 + 37a^3b^2 + 6ab^4$ **80.** $5m^5 + 26m^3h^2 + 5mh^4$

Review

Multiply. See Section 3.6.

81. $(x - 4)(x + 4)$ **82.** $(2x - 9)(2x + 9)$ **83.** $(x + 2)^2$

84. $(x + 3)^2$ **85.** $(2x - 1)^2$ **86.** $(3x - 5)^2$

Concept Extensions

See the Concept Check in this section.

87. Do the terms of $4x^2 + 19x + 12$ have a common factor (other than 1)?

88. Without multiplying, decide which of the following factored forms is not a factored form of $4x^2 + 19x + 12$.
    a. $(2x + 4)(2x + 3)$ **b.** $(4x + 4)(x + 3)$
    c. $(4x + 3)(x + 4)$ **d.** $(2x + 2)(2x + 6)$

Write the perimeter of each figure as a simplified polynomial. Then factor the polynomial.

89.

$3x^2 + 1$ $6x + 4$

$x^2 + 15x$

90.

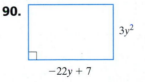

$3y^2$

$-22y + 7$

Factor each trinomial completely.

91. $4x^2 + 2x + \dfrac{1}{4}$ **92.** $27x^2 + 2x - \dfrac{1}{9}$

93. $4x^2(y - 1)^2 + 10x(y - 1)^2 + 25(y - 1)^2$ **94.** $3x^2(a + 3)^3 - 28x(a + 3)^3 + 25(a + 3)^3$

Find a positive value of b so that each trinomial is factorable.

95. $3x^2 + bx - 5$

96. $2z^2 + bz - 7$

Find a positive value of c so that each trinomial is factorable.

97. $5x^2 + 7x + c$

98. $3x^2 - 8x + c$

 99. In your own words, describe the steps you will use to factor a trinomial.

STUDY SKILLS BUILDER

Are You Satisfied with Your Performance on a Particular Quiz or Exam?

If not, don't forget to analyze your quiz or exam and look for common errors. Were most of your errors a result of:

- *Carelessness?* Did you turn in your quiz or exam before the allotted time expired? If so, resolve to use the entire time allotted next time. Any extra time can be spent checking your work.

- *Running out of time?* If so, make a point to better manage your time on your next quiz or exam. Try completing any questions that you are unsure of last and delay checking your work until all questions have been answered.

- *Not understanding a concept?* If so, review that concept and correct your work. Try to understand how this happened so that you make sure it doesn't happen before the next quiz or exam.

- *Test conditions?* When studying for a quiz or exam, make sure you place yourself in conditions similar to test conditions. For example, before your next quiz or exam, use a few sheets of blank paper and take a sample test without the aid of your notes or text.

  (See your instructor or use the Chapter Test at the end of each chapter.)

Exercises

1. Have you corrected all your previous quizzes and exams?

2. List any errors you have found common to two or more of your graded papers.

3. Is one of your common errors not understanding a concept? If so, are you making sure you understand all the concepts for the next quiz or exam?

4. Is one of your common errors making careless mistakes? If so, are you now taking all the time allotted to check over your work so that you can minimize the number of careless mistakes?

5. Are you satisfied with your grades thus far on quizzes and tests?

6. If your answer to Exercise 5 is no, are there any more suggestions you can make to your instructor or yourself to help? If so, list them here and share these with your instructor.

Objective A Using the Grouping Method

There is an alternative method that can be used to factor trinomials of the form $ax^2 + bx + c$, $a \neq 1$. This method is called the **grouping method** because it uses factoring by grouping as we learned in Section 4.1.

To see how this method works, recall from Section 4.1 that to factor a trinomial such as $x^2 + 11x + 30$, we find two numbers such that

Product is 30
↓
$x^2 + 11x + 30$
↓
Sum is 11.

To factor a trinomial such as $2x^2 + 11x + 12$ by grouping, we use an extension of the method in Section 4.1. Here we look for two numbers such that

Product is $2 \cdot 12 = 24$
↓
$2x^2 + 11x + 12$
↓
Sum is 11.

This time, we use the two numbers to write

$2x^2 + 11x + 12$ as
$= 2x^2 + \Box x + \Box x + 12$

Then we factor by grouping. Since we want a positive product, 24, and a positive sum, 11, we consider pairs of positive factors of 24 only.

Factors of 24	Sum of Factors	
1, 24	25	
2, 12	14	
3, 8	11	**Correct** sum

The factors are 3 and 8. Now we use these factors to write the middle term $11x$ as $3x + 8x$ (or $8x + 3x$). We replace $11x$ with $3x + 8x$ in the original trinomial and then we can factor by grouping.

$2x^2 + 11x + 12 = 2x^2 + 3x + 8x + 12$
$= (2x^2 + 3x) + (8x + 12)$ Group the terms.
$= x(2x + 3) + 4(2x + 3)$ Factor each group.
$= (2x + 3)(x + 4)$ Factor out $(2x + 3)$.

In general, we have the following procedure.

To Factor Trinomials by Grouping

Step 1: Factor out a greatest common factor, if there is one other than 1.

Step 2: For the resulting trinomial $ax^2 + bx + c$, find two numbers whose product is $a \cdot c$ and whose sum is b.

Step 3: Write the middle term, bx, using the factors found in Step 2.

Step 4: Factor by grouping.

PRACTICE PROBLEM 1

Factor each trinomial by grouping.

a. $3x^2 + 14x + 8$

b. $12x^2 + 19x + 5$

EXAMPLE 1 Factor $8x^2 - 14x + 5$ by grouping.

Solution:

Step 1: The terms of this trinomial contain no greatest common factor other than 1.

Step 2: This trinomial is of the form $ax^2 + bx + c$ with $a = 8$, $b = -14$, and $c = 5$. Find two numbers whose product is $a \cdot c$ or $8 \cdot 5 = 40$, and whose sum is b or -14.

The numbers are -4 and -10.

Factors of 40	Sum of Factors
$-40, -1$	-41
$-20, -2$	-22
$-10, -4$	-14

Step 3: Write $-14x$ as $-4x - 10x$ so that

$$8x^2 - 14x + 5 = 8x^2 - 4x - 10x + 5$$

Correct sum

Step 4: Factor by grouping.

$$8x^2 - 4x - 10x + 5 = 4x(2x - 1) - 5(2x - 1)$$
$$= (2x - 1)(4x - 5)$$

🔲 **Work Practice Problem 1**

PRACTICE PROBLEM 2

Factor each trinomial by grouping.

a. $30x^2 - 26x + 4$

b. $6x^2y - 7xy - 5y$

EXAMPLE 2 Factor $6x^2 - 2x - 20$ by grouping.

Solution:

Step 1: First factor out the greatest common factor, 2.

$$6x^2 - 2x - 20 = 2(3x^2 - x - 10)$$

Step 2: Next notice that $a = 3$, $b = -1$, and $c = -10$ in the resulting trinomial. Find two numbers whose product is $a \cdot c$ or $3(-10) = -30$ and whose sum is b, -1. The numbers are -6 and 5.

Step 3: $3x^2 - x - 10 = 3x^2 - 6x + 5x - 10$

Step 4: $3x^2 - 6x + 5x - 10 = 3x(x - 2) + 5(x - 2)$
$$= (x - 2)(3x + 5)$$

The factored form of $6x^2 - 2x - 20 = 2(x - 2)(3x + 5)$.

└─ Don't forget to include the common factor of 2.

🔲 **Work Practice Problem 2**

PRACTICE PROBLEM 3

Factor $12y^5 + 10y^4 - 42y^3$ by grouping.

EXAMPLE 3 Factor $18y^4 + 21y^3 - 60y^2$ by grouping.

Solution:

Step 1: First factor out the greatest common factor, $3y^2$.

$$18y^4 + 21y^3 - 60y^2 = 3y^2(6y^2 + 7y - 20)$$

Step 2: Notice that $a = 6$, $b = 7$, and $c = -20$ in the resulting trinomial. Find two numbers whose product is $a \cdot c$ or $6(-20) = -120$ and whose sum is 7. It may help to factor -120 as a product of primes and -1.

$$-120 = 2 \cdot 2 \cdot 2 \cdot 3 \cdot 5 \cdot (-1)$$

Then choose pairings of factors until you have two pairings whose sum is 7.

$$2 \cdot 2 \cdot \overset{-8}{\overbrace{2 \cdot 3 \cdot 5}} \cdot (-1)$$ The numbers are -8 and 15.

Step 3: $6y^2 + 7y - 20 = 6y^2 - 8y + 15y - 20$

Step 4: $6y^2 - 8y + 15y - 20 = 2y(3y - 4) + 5(3y - 4)$
$$= (3y - 4)(2y + 5)$$

The factored form of $18y^4 + 21y^3 - 60y^2$ is $3y^2(3y - 4)(2y + 5)$

└─ Don't forget to include the common factor of $3y^2$.

🔲 **Work Practice Problem 3**

Answers

1. a. $(x + 4)(3x + 2)$,
b. $(4x + 5)(3x + 1)$,
2. a. $2(5x - 1)(3x - 2)$,
b. $y(2x + 1)(3x - 5)$,
3. $2y^3(3y + 7)(2y - 3)$

Objective **A** *Factor each polynomial by grouping. Notice that Step 3 has already been done in these exercises. See Examples 1 through 3.*

1. $x^2 + 3x + 2x + 6$

2. $x^2 + 5x + 3x + 15$

3. $y^2 + 8y - 2y - 16$

4. $z^2 + 10z - 7z - 70$

5. $8x^2 - 5x - 24x + 15$

6. $4x^2 - 9x - 32x + 72$

7. $5x^4 - 3x^2 + 25x^2 - 15$

8. $2y^4 - 10y^2 + 7y^2 - 35$

Factor each trinomial by grouping. Exercises 9–12 are broken into parts to help you get started. See Examples 1 through 3.

9. $6x^2 + 11x + 3$

   a. Find two numbers whose product is $6 \cdot 3 = 18$ and whose sum is 11.

   b. Write $11x$ using the factors from part (a).

   c. Factor by grouping.

10. $8x^2 + 14x + 3$

   a. Find two numbers whose product is $8 \cdot 3 = 24$ and whose sum is 14.

   b. Write $14x$ using the factors from part (a).

   c. Factor by grouping.

11. $15x^2 - 23x + 4$

   a. Find two numbers whose product is $15 \cdot 4 = 60$ and whose sum is -23.

   b. Write $-23x$ using the factors from part (a).

   c. Factor by grouping.

12. $6x^2 - 13x + 5$

   a. Find two numbers whose product is $6 \cdot 5 = 30$ and whose sum is -13.

   b. Write $-13x$ using the factors from part (a).

   c. Factor by grouping.

13. $21y^2 + 17y + 2$

14. $15x^2 + 11x + 2$

15. $7x^2 - 4x - 11$

16. $8x^2 - x - 9$

17. $10x^2 - 9x + 2$

18. $30x^2 - 23x + 3$

19. $2x^2 - 7x + 5$

20. $2x^2 - 7x + 3$

21. $12x + 4x^2 + 9$

22. $20x + 25x^2 + 4$

23. $4x^2 - 8x - 21$

24. $6x^2 - 11x - 10$

25. $10x^2 - 23x + 12$

26. $21x^2 - 13x + 2$

27. $2x^3 + 13x^2 + 15x$

28. $3x^3 + 8x^2 + 4x$

29. $16y^2 - 34y + 18$

30. $4y^2 - 2y - 12$

31. $-13x + 6 + 6x^2$

32. $-25x + 12 + 12x^2$

33. $54a^2 - 9a - 30$

34. $30a^2 + 38a - 20$

35. $20a^3 + 37a^2 + 8a$

36. $10a^3 + 17a^2 + 3a$

37. $12x^3 - 27x^2 - 27x$

38. $30x^3 - 155x^2 + 25x$

39. $3x^2y + 4xy^2 + y^3$

40. $6r^2t + 7rt^2 + t^3$

41. $20z^2 + 7z + 1$

42. $36z^2 + 6z + 1$

43. $24a^2 - 6ab - 30b^2$

44. $30a^2 + 5ab - 25b^2$

45. $15p^4 + 31p^3q + 2p^2q^2$

46. $20s^4 + 61s^3t + 3s^2t^2$

47. $35 + 12x + x^2$

48. $33 + 14x + x^2$

49. $6 - 11x + 5x^2$

50. $5 - 12x + 7x^2$

Review

Multiply. See Section 3.6.

51. $(x - 2)(x + 2)$

52. $(y - 5)(y + 5)$

53. $(y + 4)(y + 4)$

54. $(x + 7)(x + 7)$

55. $(9z + 5)(9z - 5)$

56. $(8y + 9)(8y - 9)$

57. $(4x - 3)^2$

58. $(2z - 1)^2$

Concept Extensions

Write the perimeter of each figure as a simplified polynomial. Then factor the polynomial.

59.

Regular Pentagon

$2x^2 + 9x + 9$

60.

Equilateral Triangle

$7x^2 + 11xy + 4y^2$

Factor each polynomial by grouping.

61. $x^{2n} + 2x^n + 3x^n + 6$
(*Hint:* Don't forget that $x^{2n} = x^n \cdot x^n$.)

62. $x^{2n} + 6x^n + 10x^n + 60$

63. $3x^{2n} + 16x^n - 35$

64. $12x^{2n} - 40x^n + 25$

65. In your own words, explain how to factor a trinomial by grouping.

4.5 FACTORING PERFECT SQUARE TRINOMIALS AND THE DIFFERENCE OF TWO SQUARES

Objectives

A Recognize Perfect Square Trinomials.

B Factor Perfect Square Trinomials.

C Factor the Difference of Two Squares.

Objective A Recognizing Perfect Square Trinomials

A trinomial that is the square of a binomial is called a **perfect square trinomial.** For example,

$$(x + 3)^2 = (x + 3)(x + 3)$$
$$= x^2 + 6x + 9$$

Thus $x^2 + 6x + 9$ is a perfect square trinomial.

In Chapter 3, we discovered special product formulas for squaring binomials.

$$(a + b)^2 = a^2 + 2ab + b^2 \quad \text{and} \quad (a - b)^2 = a^2 - 2ab + b^2$$

Because multiplication and factoring are reverse processes, we can now use these special products to help us factor perfect square trinomials. If we reverse these equations, we have the following.

Factoring Perfect Square Trinomials

$$a^2 + 2ab + b^2 = (a + b)^2$$

$$a^2 - 2ab + b^2 = (a - b)^2$$

Helpful Hint

Notice that for both given forms of a perfect square trinomial, the last term is positive. This is because the last term is a square.

To use these equations to help us factor, we must first be able to recognize a perfect square trinomial. A trinomial is a perfect square when

1. two terms, a^2 and b^2, are squares and
2. another term is $2 \cdot a \cdot b$ or $-2 \cdot a \cdot b$. That is, this term is twice the product of a and b, or its opposite.

EXAMPLE 1 Decide whether $x^2 + 8x + 16$ is a perfect square trinomial.

Solution:

1. Two terms, x^2 and 16, are squares ($16 = 4^2$).
2. Twice the product of x and 4 is the other term of the trinomial.
   $2 \cdot x \cdot 4 = 8x$

Thus, $x^2 + 8x + 16$ is a perfect square trinomial.

■ **Work Practice Problem 1**

EXAMPLE 2 Decide whether $4x^2 + 10x + 9$ is a perfect square trinomial.

Solution:

1. Two terms, $4x^2$ and 9, are squares.
   $4x^2 = (2x)^2 \quad \text{and} \quad 9 = 3^2$
2. Twice the product of $2x$ and 3 is *not* the other term of the trinomial.
   $2 \cdot 2x \cdot 3 = 12x,\ not\ 10x$

The trinomial is *not* a perfect square trinomial.

■ **Work Practice Problem 2**

PRACTICE PROBLEM 1

Decide whether each trinomial is a perfect square trinomial.

a. $x^2 + 12x + 36$
b. $x^2 + 20x + 100$

PRACTICE PROBLEM 2

Decide whether each trinomial is a perfect square trinomial.

a. $9x^2 + 20x + 25$
b. $4x^2 + 8x + 11$

Answers

1. a. yes, b. yes, 2. a. no, b. no

283

PRACTICE PROBLEM 3

Decide whether each trinomial is a perfect square trinomial.
a. $25x^2 - 10x + 1$
b. $9x^2 - 42x + 49$

EXAMPLE 3 Decide whether $9x^2 - 12xy + 4y^2$ is a perfect square trinomial.

Solution:

1. Two terms, $9x^2$ and $4y^2$, are squares.

$$9x^2 = (3x)^2 \quad \text{and} \quad 4y^2 = (2y)^2$$

2. Twice the product of $3x$ and $2y$ is the opposite of the other term of the trinomial.

$$2 \cdot 3x \cdot 2y = 12xy, \text{ the opposite of } -12xy$$

Thus, $9x^2 - 12xy + 4y^2$ is a perfect square trinomial.

■ **Work Practice Problem 3**

Objective B Factoring Perfect Square Trinomials

Now that we can recognize perfect square trinomials, we are ready to factor them.

PRACTICE PROBLEM 4

Factor: $x^2 + 16x + 64$

EXAMPLE 4 Factor: $x^2 + 12x + 36$

Solution:

$$x^2 + 12x + 36 = x^2 + 2 \cdot x \cdot 6 + 6^2 \quad \text{\small{36 = 6^2 and 12x = 2·x·6}}$$
$$a^2 + 2 \cdot a \cdot b + b^2$$
$$= (x + 6)^2$$
$$(a + b)^2$$

■ **Work Practice Problem 4**

PRACTICE PROBLEM 5

Factor: $9r^2 + 24rs + 16s^2$

EXAMPLE 5 Factor: $25x^2 + 20xy + 4y^2$

Solution:

$$25x^2 + 20xy + 4y^2 = (5x)^2 + 2 \cdot 5x \cdot 2y + (2y)^2$$
$$= (5x + 2y)^2$$

■ **Work Practice Problem 5**

PRACTICE PROBLEM 6

Factor: $9n^4 - 6n^2 + 1$

EXAMPLE 6 Factor: $4m^4 - 4m^2 + 1$

Solution:

$$4m^4 - 4m^2 + 1 = (2m^2)^2 - 2 \cdot 2m^2 \cdot 1 + 1^2$$
$$a^2 \quad - 2 \cdot a \cdot b + b^2$$
$$= (2m^2 - 1)^2$$
$$(a - b)^2$$

■ **Work Practice Problem 6**

PRACTICE PROBLEM 7

Factor: $9x^2 + 15x + 4$

EXAMPLE 7 Factor: $25x^2 + 50x + 9$

Solution: Notice that this trinomial is not a perfect square trinomial.

$$25x^2 = (5x)^2, 9 = 3^2$$

but

$$2 \cdot 5x \cdot 3 = 30x$$

and $30x$ is not the middle term $50x$.

Answers
3. **a.** yes, **b.** yes, **4.** $(x + 8)^2$,
5. $(3r + 4s)^2$, **6.** $(3n^2 - 1)^2$,
7. $(3x + 1)(3x + 4)$

Although $25x^2 + 50x + 9$ is not a perfect square trinomial, it is factorable. Using techniques we learned in Sections 4.3 or 4.4, we find that

$$25x^2 + 50x + 9 = (5x + 9)(5x + 1)$$

Work Practice Problem 7

Helpful Hint
A perfect square trinomial can also be factored by the methods found in Sections 4.2 through 4.4.

EXAMPLE 8 Factor: $162x^3 - 144x^2 + 32x$

Solution: Don't forget to first look for a common factor. There is a greatest common factor of $2x$ in this trinomial.

$$162x^3 - 144x^2 + 32x = 2x(81x^2 - 72x + 16)$$
$$= 2x[(9x)^2 - 2 \cdot 9x \cdot 4 + 4^2]$$
$$= 2x(9x - 4)^2$$

Work Practice Problem 8

PRACTICE PROBLEM 8

Factor:

a. $8n^2 + 40n + 50$

b. $12x^3 - 84x^2 + 147x$

Objective C Factoring the Difference of Two Squares

In Chapter 3, we discovered another special product, the product of the sum and difference of two terms a and b:

$$(a + b)(a - b) = a^2 - b^2$$

Reversing this equation gives us another factoring pattern, which we use to factor the difference of two squares.

Factoring the Difference of Two Squares

$$a^2 - b^2 = (a + b)(a - b)$$

To use this equation to help us factor, we must first be able to recognize the difference of two squares. A binomial is a difference of two squares if

1. both terms are squares and
2. the signs of the terms are different.

Let's practice using this pattern.

EXAMPLES Factor each binomial.

9. $z^2 - 4 = z^2 - 2^2 = (z + 2)(z - 2)$

$$a^2 - b^2 = (a + b)(a - b)$$

10. $y^2 - 25 = y^2 - 5^2 = (y + 5)(y - 5)$

11. $y^2 - \dfrac{4}{9} = y^2 - \left(\dfrac{2}{3}\right)^2 = \left(y + \dfrac{2}{3}\right)\left(y - \dfrac{2}{3}\right)$

12. $x^2 + 4$

Note that the binomial $x^2 + 4$ is the *sum* of two squares since we can write $x^2 + 4$ as $x^2 + 2^2$. We might try to factor using $(x + 2)(x + 2)$ or $(x - 2)(x - 2)$. But when we multiply to check, we find that neither factoring is correct.

$$(x + 2)(x + 2) = x^2 + 4x + 4$$
$$(x - 2)(x - 2) = x^2 - 4x + 4$$

In both cases, the product is a trinomial, not the required binomial. In fact, $x^2 + 4$ is a prime polynomial.

Work Practice Problems 9–12

PRACTICE PROBLEMS 9–12

Factor each binomial.

9. $x^2 - 9$ **10.** $a^2 - 16$

11. $c^2 - \dfrac{9}{25}$ **12.** $s^2 + 9$

Helpful Hint
After the greatest common factor has been removed, the *sum* of two squares cannot be factored further using real numbers.

Answers

8. a. $2(2n + 5)^2$, **b.** $3x(2x - 7)^2$,

9. $(x - 3)(x + 3)$,

10. $(a - 4)(a + 4)$,

11. $\left(c - \dfrac{3}{5}\right)\left(c + \dfrac{3}{5}\right)$,

12. prime polynomial

PRACTICE PROBLEMS 13-15

Factor each difference of two squares.

13. $9s^2 - 1$

14. $16x^2 - 49y^2$

15. $p^4 - 81$

EXAMPLES Factor each difference of two squares.

13. $4x^2 - 1 = (2x)^2 - 1^2 = (2x + 1)(2x - 1)$

14. $25a^2 - 9b^2 = (5a)^2 - (3b)^2 = (5a + 3b)(5a - 3b)$

15. $y^4 - 16 = (y^2)^2 - 4^2$

$\qquad = (y^2 + 4)(y^2 - 4)$ Factor the difference of two squares.

$\qquad = (y^2 + 4)(y + 2)(y - 2)$ Factor the difference of two squares.

■ **Work Practice Problems 13-15**

Helpful Hint

1. Don't forget to first see whether there's a greatest common factor (other than 1) that can be factored out.
2. Factor completely. In other words, check to see whether any factors can be factored further (as in Example 15).

PRACTICE PROBLEMS 16-18

Factor each difference of two squares.

16. $9x^3 - 25x$

17. $48x^4 - 3$

18. $-9x^2 + 100$

EXAMPLES Factor each difference of two squares.

16. $4x^3 - 49x = x(4x^2 - 49)$ Factor out the common factor, x.

$\qquad = x[(2x)^2 - 7^2]$

$\qquad = x(2x + 7)(2x - 7)$ Factor the difference of two squares.

17. $162x^4 - 2 = 2(81x^4 - 1)$ Factor out the common factor, 2.

$\qquad = 2(9x^2 + 1)(9x^2 - 1)$ Factor the difference of two squares.

$\qquad = 2(9x^2 + 1)(3x + 1)(3x - 1)$ Factor the difference of two squares.

18. $-49x^2 + 16 = -1(49x^2 - 16)$ Factor out -1.

$\qquad = -1(7x + 4)(7x - 4)$ Factor the difference of two squares.

■ **Work Practice Problems 16-18**

PRACTICE PROBLEM 19

Factor: $121 - m^2$

EXAMPLE 19 Factor: $36 - x^2$.

Solution: This is the difference of two squares. Factor as is or if you like, first write the binomial with variable term first.

Factor as is: $36 - x^2 = 6^2 - x^2 = (6 + x)(6 - x)$.

Rewrite binomial: $36 - x^2 = -x^2 + 36 = -1(x^2 - 36)$

$\qquad\qquad\qquad\qquad\qquad\qquad = -1(x + 6)(x - 6)$.

Both factorizations are correct and are equal. To see this, factor -1 from $(6 - x)$ in the first factorization.

■ **Work Practice Problem 19**

Helpful Hint

When rearranging terms, keep in mind that the sign of a term is in front of the term.

Answers

13. $(3s - 1)(3s + 1)$,

14. $(4x - 7y)(4x + 7y)$,

15. $(p^2 + 9)(p + 3)(p - 3)$,

16. $x(3x - 5)(3x + 5)$,

17. $3(4x^2 + 1)(2x + 1)(2x - 1)$,

18. $-1(3x - 10)(3x + 10)$,

19. $(11 + m)(11 - m)$ or $-1(m + 11)(m - 11)$

A graphing calculator is a convenient tool for evaluating an expression at a given replacement value. For example, let's evaluate $x^2 - 6x$ when $x = 2$. To do so, store the value 2 in the variable x and then enter and evaluate the algebraic expression.

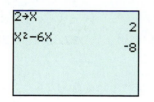

```
2→X
                    2
X²-6X
                    -8
```

The value of $x^2 - 6x$ when $x = 2$ is -8. You may want to use this method for evaluating expressions as you explore the following.

We can use a graphing calculator to explore factoring patterns numerically. Use your calculator to evaluate $x^2 - 2x + 1$, $x^2 - 2x - 1$, and $(x - 1)^2$ for each value of x given in the table. What do you observe?

	$x^2 - 2x + 1$	$x^2 - 2x - 1$	$(x - 1)^2$
$x = 5$			
$x = -3$			
$x = 2.7$			
$x = -12.1$			
$x = 0$			

Notice in each case that $x^2 - 2x - 1 \neq (x - 1)^2$. Because for each x in the table the value of $x^2 - 2x + 1$ and the value of $(x - 1)^2$ are the same, we might guess that $x^2 - 2x + 1 = (x - 1)^2$. We can verify our guess algebraically with multiplication:

$$(x - 1)(x - 1) = x^2 - x - x + 1 = x^2 - 2x + 1$$

Mental Math

Write each number as a square.

1. 1 **2.** 25 **3.** 81 **4.** 64 **5.** 9 **6.** 100

Write each term as a square.

7. $9x^2$ **8.** $16y^2$ **9.** $25a^2$ **10.** $81b^2$ **11.** $36p^4$ **12.** $4q^4$

4.5 EXERCISE SET

Objective **A** *Determine whether each trinomial is a perfect square trinomial. See Examples 1 through 3.*

1. $x^2 + 16x + 64$ **2.** $x^2 + 22x + 121$ **3.** $y^2 + 5y + 25$ **4.** $y^2 + 4y + 16$

5. $m^2 - 2m + 1$ **6.** $p^2 - 4p + 4$ **7.** $a^2 - 16a + 49$ **8.** $n^2 - 20n + 144$

9. $4x^2 + 12xy + 8y^2$ **10.** $25x^2 + 20xy + 2y^2$ **11.** $25a^2 - 40ab + 16b^2$ **12.** $36a^2 - 12ab + b^2$

Objective **B** *Factor each trinomial completely. See Examples 4 through 8.*

13. $x^2 + 22x + 121$ **14.** $x^2 + 18x + 81$ **15.** $x^2 - 16x + 64$ **16.** $x^2 - 12x + 36$

17. $16a^2 - 24a + 9$ **18.** $25x^2 - 20x + 4$ **19.** $x^4 + 4x^2 + 4$ **20.** $m^4 + 10m^2 + 25$

21. $2n^2 - 28n + 98$ **22.** $3y^2 - 6y + 3$ **23.** $16y^2 + 40y + 25$ **24.** $9y^2 + 48y + 64$

25. $x^2y^2 - 10xy + 25$ **26.** $4x^2y^2 - 28xy + 49$ **27.** $m^3 + 18m^2 + 81m$ **28.** $y^3 + 12y^2 + 36y$

29. $1 + 6x^2 + x^4$ **30.** $1 + 16x^2 + x^4$ **31.** $9x^2 - 24xy + 16y^2$ **32.** $25x^2 - 60xy + 36y^2$

Objective **C** *Factor each binomial completely. See Examples 9 through 19.*

33. $x^2 - 4$ **34.** $x^2 - 36$ **35.** $81 - p^2$ **36.** $100 - t^2$

37. $-4r^2 + 1$ **38.** $-9t^2 + 1$ **39.** $9x^2 - 16$ **40.** $36y^2 - 25$

41. $16r^2 + 1$ **42.** $49y^2 + 1$ **43.** $-36 + x^2$ **44.** $-1 + y^2$

45. $m^4 - 1$ **46.** $n^4 - 16$ **47.** $x^2 - 169y^2$ **48.** $x^2 - 225y^2$

49. $18r^2 - 8$ **50.** $32t^2 - 50$ **51.** $9xy^2 - 4x$ **52.** $36x^2y - 25y$

53. $16x^4 - 64x^2$ **54.** $25y^4 - 100y^2$ **55.** $xy^3 - 9xyz^2$ **56.** $x^3y - 4xy^3$

57. $36x^2 - 64y^2$ **58.** $225a^2 - 81b^2$ **59.** $144 - 81x^2$ **60.** $12x^2 - 27$

61. $25y^2 - 9$ **62.** $49a^2 - 16$ **63.** $121m^2 - 100n^2$ **64.** $169a^2 - 49b^2$

65. $x^2y^2 - 1$ **66.** $a^2b^2 - 16$ **67.** $x^2 - \dfrac{1}{4}$

68. $y^2 - \dfrac{1}{16}$ **69.** $49 - \dfrac{9}{25}m^2$ **70.** $100 - \dfrac{4}{81}n^2$

Objectives **B** **C** **Mixed Practice** *Factor each binomial or trinomial completely. See Examples 4 through 19.*

71. $81a^2 - 25b^2$ **72.** $49y^2 - 100z^2$ **73.** $x^2 + 14xy + 49y^2$ **74.** $x^2 + 10xy + 25y^2$

75. $32n^4 - 112n^2 + 98$

76. $162a^4 - 72a^2 + 8$

77. $x^6 - 81x^2$

78. $n^9 - n^5$

79. $64p^3q - 81pq^3$

80. $100x^3y - 49xy^3$

Review

Solve each equation. See Section 2.3.

81. $x - 6 = 0$

82. $y + 5 = 0$

83. $2m + 4 = 0$

84. $3x - 9 = 0$

85. $5z - 1 = 0$

86. $4a + 2 = 0$

Concept Extensions

Factor each expression completely.

87. $x^2 - \dfrac{2}{3}x + \dfrac{1}{9}$

88. $x^2 - \dfrac{1}{25}$

89. $(x + 2)^2 - y^2$

90. $(y - 6)^2 - z^2$

91. $a^2(b - 4) - 16(b - 4)$

92. $m^2(n + 8) - 9(n + 8)$

93. $(x^2 + 6x + 9) - 4y^2$ (*Hint:* Factor the trinomial in parentheses first.)

94. $(x^2 + 2x + 1) - 36y^2$

95. $x^{2n} - 100$

96. $x^{2n} - 81$

97. Fill in the blank so that $x^2 + $ _____ $x + 16$ is a perfect square trinomial.

98. Fill in the blank so that $9x^2 + $ _____ $x + 25$ is a perfect square trinomial.

99. Describe a perfect square trinomial.

100. Write a perfect square trinomial that factors as $(x + 3y)^2$.

101. What binomial multiplied by $(x - 6)$ gives the difference of two squares?

102. What binomial multiplied by $(5 + y)$ gives the difference of two squares?

The area of the largest square in the figure is $(a + b)^2$. Use this figure to answer Exercises 103 and 104.

103. Write the area of the largest square as the sum of the areas of the smaller squares and rectangles.

104. What factoring formula from this section is visually represented by this square?

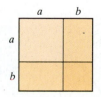

105. An object is dropped from the top of Pittsburgh's USX Towers, which is 841 feet tall. (*Source: World Almanac* research) The height of the object after t seconds is given by the expression $841 - 16t^2$.

    a. Find the height of the object after 2 seconds.

    b. Find the height of the object after 5 seconds.

    c. To the nearest whole second, estimate when the object hits the ground.

    d. Factor $841 - 16t^2$.

841 feet

106. A worker on the top of the Aetna Life Building in San Francisco accidentally drops a bolt. The Aetna Life Building is 529 feet tall. (*Source: World Almanac* research) The height of the bolt after t seconds is given by the expression $529 - 16t^2$.

    a. Find the height of the bolt after 1 second.

    b. Find the height of the bolt after 4 seconds.

    c. To the nearest whole second, estimate when the bolt hits the ground.

    d. Factor $529 - 16t^2$.

107. At this writing, the world's tallest building is the Taipei 101 in Taipei, Taiwan, at a height of 1671 feet. (*Source:* Council on Tall Buildings and Urban Habitat) Suppose a worker is suspended 71 feet below the top of the pinnacle atop the building, at a height of 1600 feet above the ground. If the worker accidentally drops a bolt, the height of the bolt after t seconds is given by the expression $1600 - 16t^2$.

    a. Find the height of the bolt after 3 seconds.

    b. Find the height of the bolt after 7 seconds.

    c. To the nearest whole second, estimate when the bolt hits the ground.

    d. Factor $1600 - 16t^2$.

108. A performer with the Moscow Circus is planning a stunt involving a free fall from the top of the Moscow State University building, which is 784 feet tall. (*Source:* Council on Tall Buildings and Urban Habitat) Neglecting air resistance, the performer's height above gigantic cushions positioned at ground level after t seconds is given by the expression $784 - 16t^2$.

    a. Find the performer's height after 2 seconds.

    b. Find the performer's height after 5 seconds.

    c. To the nearest whole second, estimate when the performer reaches the cushions positioned at ground level.

    d. Factor $784 - 16t^2$.

Choosing a Factoring Strategy

The following steps may be helpful when factoring polynomials.

To Factor a Polynomial

Step 1: Are there any common factors? If so, factor out the GCF.

Step 2: How many terms are in the polynomial?

    a. Two terms: Is it the difference of two squares? $a^2 - b^2 = (a - b)(a + b)$

    b. Three terms: Try one of the following.

      i. Perfect square trinomial: $a^2 + 2ab + b^2 = (a + b)^2$
$$a^2 - 2ab + b^2 = (a - b)^2$$

      ii. If not a perfect square trinomial, factor using the methods presented in Sections 4.2 through 4.4.

    c. Four terms: Try factoring by grouping.

Step 3: See if any factors in the factored polynomial can be factored further.

Step 4: Check by multiplying.

Factor each polynomial completely.

1. $x^2 + x - 12$

2. $x^2 - 10x + 16$

3. $x^2 - x - 6$

4. $x^2 + 2x + 1$

5. $x^2 - 6x + 9$

6. $x^2 + x - 2$

7. $x^2 + x - 6$

8. $x^2 + 7x + 12$

9. $x^2 - 7x + 10$

10. $x^2 - x - 30$

11. $2x^2 - 98$

12. $3x^2 - 75$

13. $x^2 + 3x + 5x + 15$

14. $3y - 21 + xy - 7x$

15. $x^2 + 6x - 16$

16. $x^2 - 3x - 28$

17. $4x^3 + 20x^2 - 56x$

18. $6x^3 - 6x^2 - 120x$

19. $12x^2 + 34x + 24$

20. $8a^2 + 6ab - 5b^2$

21. $4a^2 - b^2$

22. $x^2 - 25y^2$

23. $28 - 13x - 6x^2$

24. $20 - 3x - 2x^2$

25. $4 - 2x + x^2$

26. $a + a^2 - 3$

27. $6y^2 + y - 15$

28. $4x^2 - x - 5$

29. $18x^3 - 63x^2 + 9x$

30. $12a^3 - 24a^2 + 4a$

31. $16a^2 - 56a + 49$

32. $25p^2 - 70p + 49$

33. $14 + 5x - x^2$

34. $3 - 2x - x^2$

35. $3x^4y + 6x^3y - 72x^2y$

36. $2x^3y + 8x^2y^2 - 10xy^3$

Answers

1. _____
2. _____
3. _____
4. _____
5. _____
6. _____
7. _____
8. _____
9. _____
10. _____
11. _____
12. _____
13. _____
14. _____
15. _____
16. _____
17. _____
18. _____
19. _____
20. _____
21. _____
22. _____
23. _____
24. _____
25. _____
26. _____
27. _____
28. _____
29. _____
30. _____
31. _____
32. _____
33. _____
34. _____
35. _____
36. _____

37. _____

38. _____

39. _____

40. _____

41. _____

42. _____

43. _____

44. _____

45. _____

46. _____

47. _____

48. _____

49. _____

50. _____

51. _____

52. _____

53. _____

54. _____

55. _____

56. _____

57. _____

58. _____

59. _____

60. _____

61. _____

62. _____

63. _____

64. _____

65. _____

66. _____

67. _____

68. _____

69. _____

70. _____

71. _____

72. _____

73. _____

74. _____

75. _____

76. _____

37. $12x^3y + 243xy$ **38.** $6x^3y^2 + 8xy^2$ **39.** $2xy - 72x^3y$

40. $2x^3 - 18x$ **41.** $x^3 + 6x^2 - 4x - 24$ **42.** $x^3 - 2x^2 - 36x + 72$

43. $6a^3 + 10a^2$ **44.** $4n^2 - 6n$ **45.** $3x^3 - x^2 + 12x - 4$

46. $x^3 - 2x^2 + 3x - 6$ **47.** $6x^2 + 18xy + 12y^2$ **48.** $12x^2 + 46xy - 8y^2$

49. $5(x + y) + x(x + y)$ **50.** $7(x - y) + y(x - y)$ **51.** $14t^2 - 9t + 1$

52. $3t^2 - 5t + 1$ **53.** $-3x^2 - 2x + 5$ **54.** $-7x^2 - 19x + 6$

55. $1 - 8a - 20a^2$ **56.** $1 - 7a - 60a^2$ **57.** $x^4 - 10x^2 + 9$

58. $x^4 - 13x^2 + 36$ **59.** $x^2 - 23x + 120$ **60.** $y^2 + 22y + 96$

61. $x^2 - 14x - 48$ **62.** $16a^2 - 56ab + 49b^2$ **63.** $25p^2 - 70pq + 49q^2$

64. $7x^2 + 24xy + 9y^2$ **65.** $-x^2 - x + 30$ **66.** $-x^2 + 6x - 8$

67. $3rs - s + 12r - 4$ **68.** $x^3 - 2x^2 + x - 2$ **69.** $4x^2 - 8xy - 3x + 6y$

70. $4x^2 - 2xy - 7yz + 14xz$ **71.** $x^2 + 9xy - 36y^2$ **72.** $3x^2 + 10xy - 8y^2$

73. $x^4 - 14x^2 - 32$ **74.** $x^4 - 22x^2 - 75$

75. Explain why it makes good sense to factor out the GCF first, before using other methods of factoring.

76. The sum of two squares usually does not factor. Is the sum of two squares $9x^2 + 81y^2$ factorable?

4.6 SOLVING QUADRATIC EQUATIONS BY FACTORING

Objectives

A Solve Quadratic Equations by Factoring.

B Solve Equations with Degree Greater Than 2 by Factoring.

In this section, we introduce a new type of equation—the **quadratic equation.**

Quadratic Equation

A quadratic equation is one that can be written in the form

$$ax^2 + bx + c = 0$$

where a, b, and c are real numbers and $a \neq 0$.

Some examples of quadratic equations are shown below.

$$x^2 - 9x - 22 = 0 \qquad 4x^2 - 28 = -49 \qquad x(2x - 7) = 4$$

The form $ax^2 + bx + c = 0$ is called the **standard form** of a quadratic equation. The quadratic equation $x^2 - 9x - 22 = 0$ is the only equation above that is in standard form.

Quadratic equations model many real-life situations. For example, let's suppose we want to know how long before a person diving from a 144-foot cliff reaches the ocean. The answer to this question is found by solving the quadratic equation $-16t^2 + 144 = 0$. (See Example 1 in Section 4.7.)

144 feet

Objective **A** Solving Quadratic Equations by Factoring

Some quadratic equations can be solved by making use of factoring and the **zero factor property.**

Zero Factor Property

If a and b are real numbers and if $ab = 0$, then $a = 0$ or $b = 0$.

In other words, if the product of two numbers is 0, then at least one of the numbers must be 0.

EXAMPLE 1 Solve: $(x - 3)(x + 1) = 0$

Solution: If this equation is to be a true statement, then either the factor $x - 3$ must be 0 or the factor $x + 1$ must be 0. In other words, either

$$x - 3 = 0 \qquad \text{or} \qquad x + 1 = 0$$

If we solve these two linear equations, we have

$$x = 3 \qquad \text{or} \qquad x = -1$$

Continued on next page

PRACTICE PROBLEM 1

Solve: $(x - 7)(x + 2) = 0$

Answer

1. 7 and -2

Thus, 3 and -1 are both solutions of the equation $(x - 3)(x + 1) = 0$. To check, we replace x with 3 in the original equation. Then we replace x with -1 in the original equation.

Check:

$$(x - 3)(x + 1) = 0 \qquad\qquad (x - 3)(x + 1) = 0$$

$$(3 - 3)(3 + 1) \stackrel{?}{=} 0 \quad \text{Replace } x \text{ with 3.} \qquad (-1 - 3)(-1 + 1) \stackrel{?}{=} 0 \quad \text{Replace } x \text{ with } -1.$$

$$0(4) = 0 \quad \text{True} \qquad\qquad\qquad (-4)(0) = 0 \quad \text{True}$$

The solutions are 3 and -1.

■ **Work Practice Problem 1**

Helpful Hint

The zero factor property says that *if a product is 0, then a factor is 0.*

If $a \cdot b = 0$, then $a = 0$ or $b = 0$.

If $x(x + 5) = 0$, then $x = 0$ or $x + 5 = 0$.

If $(x + 7)(2x - 3) = 0$, then $x + 7 = 0$ or $2x - 3 = 0$.

Use this property only when the product is 0. For example, if $a \cdot b = 8$, we do not know the value of a or b. The values may be $a = 2$, $b = 4$ or $a = 8$, $b = 1$, or any other two numbers whose product is 8.

PRACTICE PROBLEM 2

Solve: $(x - 10)(3x + 1) = 0$

EXAMPLE 2 Solve: $(x - 5)(2x + 7) = 0$

Solution: The product is 0. By the zero factor property, this is true only when a factor is 0. To solve, we set each factor equal to 0 and solve the resulting linear equations.

$$(x - 5)(2x + 7) = 0$$

$$x - 5 = 0 \quad \text{or} \quad 2x + 7 = 0$$

$$x = 5 \qquad\qquad 2x = -7$$

$$x = -\frac{7}{2}$$

Check: Let $x = 5$.

$$(x - 5)(2x + 7) = 0$$

$$(5 - 5)(2 \cdot 5 + 7) \stackrel{?}{=} 0 \quad \text{Replace } x \text{ with 5.}$$

$$0 \cdot 17 \stackrel{?}{=} 0$$

$$0 = 0 \quad \text{True}$$

Let $x = -\frac{7}{2}$.

$$(x - 5)(2x + 7) = 0$$

$$\left(-\frac{7}{2} - 5\right)\left(2\left(-\frac{7}{2}\right) + 7\right) \stackrel{?}{=} 0 \quad \text{Replace } x \text{ with } -\frac{7}{2}.$$

$$\left(-\frac{17}{2}\right)(-7 + 7) \stackrel{?}{=} 0$$

$$\left(-\frac{17}{2}\right) \cdot 0 \stackrel{?}{=} 0$$

$$0 = 0 \quad \text{True}$$

The solutions are 5 and $-\frac{7}{2}$.

■ **Work Practice Problem 2**

Answer

2. 10 and $-\frac{1}{3}$

EXAMPLE 3 Solve: $x(5x - 2) = 0$

Solution: $x(5x - 2) = 0$

$x = 0$ or $5x - 2 = 0$ Use the zero factor property.

$5x = 2$

$x = \dfrac{2}{5}$

Check these solutions in the original equation. The solutions are 0 and $\dfrac{2}{5}$.

◻ **Work Practice Problem 3**

PRACTICE PROBLEM 3
Solve each equation.
a. $y(y + 3) = 0$
b. $x(4x - 3) = 0$

EXAMPLE 4 Solve: $x^2 - 9x - 22 = 0$

Solution: One side of the equation is 0. However, to use the zero factor property, one side of the equation must be 0 *and* the other side must be written as a product (must be factored). Thus, we must first factor this polynomial.

$x^2 - 9x - 22 = 0$

$(x - 11)(x + 2) = 0$ Factor.

Now we can apply the zero factor property.

$x - 11 = 0$ or $x + 2 = 0$

$x = 11$ $x = -2$

Check: Let $x = 11$.   Let $x = -2$.

$x^2 - 9x - 22 = 0$ $x^2 - 9x - 22 = 0$

$11^2 - 9 \cdot 11 - 22 \overset{?}{=} 0$ $(-2)^2 - 9(-2) - 22 \overset{?}{=} 0$

$121 - 99 - 22 \overset{?}{=} 0$ $4 + 18 - 22 \overset{?}{=} 0$

$22 - 22 \overset{?}{=} 0$ $22 - 22 \overset{?}{=} 0$

$0 = 0$ True $0 = 0$ True

The solutions are 11 and -2.

◻ **Work Practice Problem 4**

PRACTICE PROBLEM 4
Solve: $x^2 - 3x - 18 = 0$

EXAMPLE 5 Solve: $4x^2 - 28x = -49$

Solution: First we rewrite the equation in standard form so that one side is 0. Then we factor the polynomial.

$4x^2 - 28x = -49$

$4x^2 - 28x + 49 = 0$ Write in standard form by adding 49 to both sides.

$(2x - 7)(2x - 7) = 0$ Factor.

Next we use the zero factor property and set each factor equal to 0. Since the factors are the same, the related equations will give the same solution.

$2x - 7 = 0$ or $2x - 7 = 0$ Set each factor equal to 0.

$2x = 7$ $2x = 7$ Solve.

$x = \dfrac{7}{2}$ $x = \dfrac{7}{2}$

Check: Check this solution in the original equation. The solution is $\dfrac{7}{2}$.

◻ **Work Practice Problem 5**

PRACTICE PROBLEM 5
Solve: $9x^2 - 24x = -16$

Answers

3. a. 0 and -3,   **b.** 0 and $\dfrac{3}{4}$,

4. 6 and -3,   **5.** $\dfrac{4}{3}$

The following steps may be used to solve a quadratic equation by factoring.

> ### To Solve Quadratic Equations by Factoring
>
> **Step 1:** Write the equation in standard form so that one side of the equation is 0.
>
> **Step 2:** Factor the quadratic equation completely.
>
> **Step 3:** Set each factor containing a variable equal to 0.
>
> **Step 4:** Solve the resulting equations.
>
> **Step 5:** Check each solution in the original equation.

Since it is not always possible to factor a quadratic polynomial, not all quadratic equations can be solved by factoring. Other methods of solving quadratic equations are presented in Chapter 9.

PRACTICE PROBLEM 6

Solve each equation.

a. $x(x - 4) = 5$

b. $x(3x + 7) = 6$

EXAMPLE 6 Solve: $x(2x - 7) = 4$

Solution: First we write the equation in standard form; then we factor.

$$x(2x - 7) = 4$$
$$2x^2 - 7x = 4 \qquad \text{Multiply.}$$
$$2x^2 - 7x - 4 = 0 \qquad \text{Write in standard form.}$$
$$(2x + 1)(x - 4) = 0 \qquad \text{Factor.}$$
$$2x + 1 = 0 \quad \text{or} \quad x - 4 = 0 \qquad \text{Set each factor equal to zero.}$$
$$2x = -1 \qquad\qquad x = 4 \qquad \text{Solve.}$$
$$x = -\frac{1}{2}$$

Check the solutions in the original equation. The solutions are $-\frac{1}{2}$ and 4.

🟧 **Work Practice Problem 6**

☁ Helpful Hint

To solve the equation $x(2x - 7) = 4$, do **not** set each factor equal to 4. Remember that to apply the zero factor property, one side of the equation must be 0 and the other side of the equation must be in factored form.

✔**Concept Check** Explain the error and solve the equation correctly.

$$(x - 3)(x + 1) = 5$$
$$x - 3 = 0 \quad \text{or} \quad x + 1 = 0$$
$$x = 3 \quad \text{or} \quad x = -1$$

Answers

6. **a.** 5 and −1, **b.** $\frac{2}{3}$ and −3

✔ Concept Check Answer

To use the zero factor property, one side of the equation must be 0, not 5. Correctly, $(x - 3)(x + 1) = 5$, $x^2 - 2x - 3 = 5$, $x^2 - 2x - 8 = 0$, $(x - 4)(x + 2) = 0$, $x - 4 = 0$ or $x + 2 = 0$, $x = 4$ or $x = -2$.

Objective 🅱 Solving Equations with Degree Greater Than Two by Factoring

Some equations with degree greater than 2 can be solved by factoring and then using the zero factor property.

EXAMPLE 7 Solve: $3x^3 - 12x = 0$

Solution: To factor the left side of the equation, we begin by factoring out the greatest common factor, $3x$.

$$3x^3 - 12x = 0$$
$$3x(x^2 - 4) = 0 \quad \text{Factor out the GCF, } 3x.$$
$$3x(x + 2)(x - 2) = 0 \quad \text{Factor } x^2 - 4, \text{ a difference of two squares.}$$
$$3x = 0 \quad \text{or} \quad x + 2 = 0 \quad \text{or} \quad x - 2 = 0 \quad \text{Set each factor equal to 0.}$$
$$x = 0 \qquad\qquad x = -2 \qquad\qquad x = 2 \quad \text{Solve.}$$

Thus, the equation $3x^3 - 12x = 0$ has three solutions: $0, -2,$ and 2.

Check: Replace x with each solution in the original equation.

Let $x = 0$.

$$3(0)^3 - 12(0) \stackrel{?}{=} 0$$
$$0 = 0 \quad \text{True}$$

Let $x = -2$.

$$3(-2)^3 - 12(-2) \stackrel{?}{=} 0$$
$$3(-8) + 24 \stackrel{?}{=} 0$$
$$0 = 0 \quad \text{True}$$

Let $x = 2$.

$$3(2)^3 - 12(2) \stackrel{?}{=} 0$$
$$3(8) - 24 \stackrel{?}{=} 0$$
$$0 = 0 \quad \text{True}$$

The solutions are $0, -2,$ and 2.

🔲 **Work Practice Problem 7**

EXAMPLE 8 Solve: $(5x - 1)(2x^2 + 15x + 18) = 0$

Solution:

$$(5x - 1)(2x^2 + 15x + 18) = 0$$
$$(5x - 1)(2x + 3)(x + 6) = 0 \quad \text{Factor the trinomial.}$$
$$5x - 1 = 0 \quad \text{or} \quad 2x + 3 = 0 \quad \text{or} \quad x + 6 = 0 \quad \text{Set each factor equal to 0.}$$
$$5x = 1 \qquad\qquad 2x = -3 \qquad\qquad x = -6 \quad \text{Solve.}$$
$$x = \frac{1}{5} \qquad\qquad x = -\frac{3}{2}$$

Check each solution in the original equation. The solutions are $\frac{1}{5}, -\frac{3}{2},$ and -6.

🔲 **Work Practice Problem 8**

PRACTICE PROBLEM 7

Solve: $2x^3 - 18x = 0$

PRACTICE PROBLEM 8

Solve:
$(x + 3)(3x^2 - 20x - 7) = 0$

Answers

7. $0, 3,$ and -3, **8.** $-3, -\frac{1}{3},$ and 7

Mental Math

Solve each equation by inspection.

1. $(a - 3)(a - 7) = 0$

2. $(a - 5)(a - 2) = 0$

3. $(x + 8)(x + 6) = 0$

4. $(x + 2)(x + 3) = 0$

5. $(x + 1)(x - 3) = 0$

6. $(x - 1)(x + 2) = 0$

4.6 EXERCISE SET

Objective Ⓐ *Solve each equation. See Examples 1 through 3.*

1. $(x - 2)(x + 1) = 0$

2. $(x + 3)(x + 2) = 0$

3. $(x - 6)(x - 7) = 0$

4. $(x + 4)(x - 10) = 0$

5. $(x + 9)(x + 17) = 0$

6. $(x - 11)(x - 1) = 0$

7. $x(x + 6) = 0$

8. $x(x - 7) = 0$

9. $3x(x - 8) = 0$

10. $2x(x + 12) = 0$

11. $(2x + 3)(4x - 5) = 0$

12. $(3x - 2)(5x + 1) = 0$

13. $(2x - 7)(7x + 2) = 0$

14. $(9x + 1)(4x - 3) = 0$

15. $\left(x - \dfrac{1}{2}\right)\left(x + \dfrac{1}{3}\right) = 0$

16. $\left(x + \dfrac{2}{9}\right)\left(x - \dfrac{1}{4}\right) = 0$

17. $(x + 0.2)(x + 1.5) = 0$

18. $(x + 1.7)(x + 2.3) = 0$

Solve. See Examples 4 through 6.

19. $x^2 - 13x + 36 = 0$

20. $x^2 + 2x - 63 = 0$

21. $x^2 + 2x - 8 = 0$

22. $x^2 - 5x + 6 = 0$

23. $x^2 - 7x = 0$

24. $x^2 - 3x = 0$

25. $x^2 + 20x = 0$

26. $x^2 + 15x = 0$

27. $x^2 = 16$

28. $x^2 = 9$

29. $x^2 - 4x = 32$

30. $x^2 - 5x = 24$

31. $(x + 4)(x - 9) = 4x$

32. $(x + 3)(x + 8) = x$

33. $x(3x - 1) = 14$

34. $x(4x - 11) = 3$

35. $3x^2 + 19x - 72 = 0$ **36.** $36x^2 + x - 21 = 0$

Objectives Ⓐ Ⓑ **Mixed Practice** *Solve each equation. See Examples 1 through 8. (A few exercises are linear equations.)*

37. $4x^3 - x = 0$

38. $4y^3 - 36y = 0$

39. $4(x - 7) = 6$

40. $5(3 - 4x) = 9$

41. $(4x - 3)(16x^2 - 24x + 9) = 0$

42. $(2x + 5)(4x^2 + 20x + 25) = 0$

43. $4y^2 - 1 = 0$

44. $4y^2 - 81 = 0$

45. $(2x + 3)(2x^2 - 5x - 3) = 0$

46. $(2x - 9)(x^2 + 5x - 36) = 0$

47. $x^2 - 15 = -2x$

48. $x^2 - 26 = -11x$

49. $30x^2 - 11x = 30$

50. $12x^2 + 7x - 12 = 0$

51. $5x^2 - 6x - 8 = 0$

52. $9x^2 + 7x = -2$

53. $6y^2 - 22y - 40 = 0$

54. $3x^2 - 6x - 9 = 0$

55. $(y - 2)(y + 3) = 6$

56. $(y - 5)(y - 2) = 28$

57. $x^3 - 12x^2 + 32x = 0$

58. $x^3 - 14x^2 + 49x = 0$

59. $x^2 + 14x + 49 = 0$

60. $x^2 + 22x + 121 = 0$

61. $12y = 8y^2$

62. $9y = 6y^2$

63. $7x^3 - 7x = 0$

64. $3x^3 - 27x = 0$

65. $3x^2 + 8x - 11 = 13 - 6x$

66. $2x^2 + 12x - 1 = 4 + 3x$

67. $3x^2 - 20x = -4x^2 - 7x - 6$

68. $4x^2 - 20x = -5x^2 - 6x - 5$

Review

Perform each indicated operation. Write all results in lowest terms. See Section R.2.

69. $\dfrac{3}{5} + \dfrac{4}{9}$

70. $\dfrac{2}{3} + \dfrac{3}{7}$

71. $\dfrac{7}{10} - \dfrac{5}{12}$

72. $\dfrac{5}{9} - \dfrac{5}{12}$

73. $\dfrac{4}{5} \cdot \dfrac{7}{8}$

74. $\dfrac{3}{7} \cdot \dfrac{12}{17}$

Concept Extensions

For Exercises 75 and 76, see the Concept Check in this section.

75. Explain the error and solve correctly:

$$x(x - 2) = 8$$
$$x = 8 \quad \text{or} \quad x - 2 = 8$$
$$x = 10$$

76. Explain the error and solve correctly:

$$(x - 4)(x + 2) = 0$$
$$x = -4 \quad \text{or} \quad x = 2$$

77. Write a quadratic equation that has two solutions, 6 and -1. Leave the polynomial in the equation in factored form.

78. Write a quadratic equation that has two solutions, 0 and -2. Leave the polynomial in the equation in factored form.

79. Write a quadratic equation in standard form that has two solutions, 5 and 7.

80. Write an equation that has three solutions, 0, 1, and 2.

81. A compass is accidentally thrown upward and out of an air balloon at a height of 300 feet. The height, y, of the compass at time x is given by the equation $y = -16x^2 + 20x + 300$.

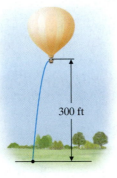

300 ft

a. Find the height of the compass at the given times by filling in the table below.

Time, x (in seconds)	0	1	2	3	4	5	6
Height, y (in feet)							

b. Use the table to determine when the compass strikes the ground.
c. Use the table to approximate the maximum height of the compass.

82. A rocket is fired upward from the ground with an initial velocity of 100 feet per second. The height, y, of the rocket at any time x is given by the equation $y = -16x^2 + 100x$.

y

a. Find the height of the rocket at the given times by filling in the table below.

Time, x (in seconds)	0	1	2	3	4	5	6	7
Height, y (in feet)								

b. Use the table to determine between what two whole-numbered seconds the rocket strikes the ground.
c. Use the table to approximate the maximum height of the rocket.

Solve each equation.

83. $(x - 3)(3x + 4) = (x + 2)(x - 6)$

84. $(2x - 3)(x + 6) = (x - 9)(x + 2)$

85. $(2x - 3)(x + 8) = (x - 6)(x + 4)$

86. $(x + 6)(x - 6) = (2x - 9)(x + 4)$

THE BIGGER PICTURE   Simplifying Expressions and Solving Equations

Now we continue our outline from Sections 1.6, 2.7, and 3.7. Although suggestions are given, this outline should be in your own words. Once you complete this new portion, try the exercises below.

I. Simplifying Expressions

    A. Real Numbers

        1. Add (Section 1.4)

        2. Subtract (Section 1.5)

        3. Multiply or Divide (Section 1.6)

    B. Exponents (Section 3.2)

    C. Polynomials

        1. Add (Section 3.4)

        2. Subtract (Section 3.4)

        3. Multiply (Section 3.5)

        4. Divide (Section 3.7)

    D. Factoring Polynomials—see the Chapter 4 Integrated Review for steps.

$$3x^4 - 78x^2 + 75$$

$$= 3(x^4 - 26x^2 + 25) \qquad \text{Factor out GCF—always first step.}$$

$$= 3(x^2 - 25)(x^2 - 1) \qquad \text{Factor trinomial.}$$

$$= 3(x + 5)(x - 5)(x + 1)(x - 1) \qquad \text{Factor further—each difference of squares.}$$

II. Solving Equations and Inequalities

    A. Linear Equations (Section 2.3)

    B. Linear Inequalities (Section 2.7)

C. Quadratic & Higher Degree Equations (Solving by Factoring)—highest power on variable is at least 2 when equation is written in standard form (set equal to 0).

$$x^2 + x = 6$$

$$x^2 + x - 6 = 0 \qquad \text{Write the equation in standard form (set it equal to 0).}$$

$$(x - 2)(x + 3) = 0 \qquad \text{Factor.}$$

$$x = 2 \quad \text{or} \quad x = -3 \qquad \text{Set each factor equal to 0 and solve.}$$

Simplify each expression.

1. $-7 + (-27)$

2. $\dfrac{(x^3)^4}{(x^{-2})^5}$

3. $(x^3 - 6x^2 + 2) - (5x^3 - 6)$

4. $\dfrac{3y^3 - 3y^2 + 9}{3y^2}$

Factor each expression.

5. $10x^3 - 250x$

6. $x^2 - 36x + 35$

7. $6xy + 15x - 6y - 15$

8. $5xy^2 - 2xy - 7x$

Solve each equation. Remember to use your outline to determine whether the equation is linear or quadratic and how to proceed with solving.

9. $(x - 5)(2x + 1) = 0$

10. $5x - 5 = 0$

11. $x(x - 12) = 28$

12. $7(x - 3) + 2(5x + 1) = 14$

A Solve Problems That Can Be Modeled by Quadratic Equations.

Objective A Solving Problems Modeled by Quadratic Equations

Some problems may be modeled by quadratic equations. To solve these problems, we use the same problem-solving steps that were introduced in Section 2.5. When solving these problems, keep in mind that a solution of an equation that models a problem may not be a solution to the problem. For example, a person's age or the length of a rectangle is always a positive number. Thus we discard solutions that do not make sense as solutions of the problem.

PRACTICE PROBLEM 1

Cliff divers also frequent the falls at Waimea Falls Park in Oahu, Hawaii. Here, a diver can jump from a ledge 64 feet up the waterfall into a rocky pool below. Neglecting air resistance, the height of a diver above the pool after t seconds is $h = -16t^2 + 64$. Find how long it takes the diver to reach the pool.

EXAMPLE 1 Finding Free-Fall Time

Since the 1940s, one of the top tourist attractions in Acapulco, Mexico is watching the cliff divers off the La Quebrada. The divers' platform is about 144 feet above the sea. These divers must time their descent just right, since they land in the crashing Pacific, in an inlet that is at most $9\frac{1}{2}$ feet deep. Neglecting air resistance, the height h in feet of a cliff diver above the ocean after t seconds is given by the quadratic equation $h = -16t^2 + 144$.

Find out how long it takes the diver to reach the ocean.

Solution:

1. UNDERSTAND. Read and reread the problem. Then draw a picture of the problem.

   The equation $h = -16t^2 + 144$ models the height of the falling diver at time t. Familiarize yourself with this equation by find the height of the diver at time $t = 1$ second and $t = 2$ seconds.

   When $t = 1$ second, the height of the diver is $h = -16(1)^2 + 144 = 128$ feet.
   When $t = 2$ seconds, the height of the diver is $h = -16(2)^2 + 144 = 80$ feet.

2. TRANSLATE. To find out how long it takes the diver to reach the ocean, we want to know the value of t for which $h = 0$.

   $$0 = -16t^2 + 144$$
   $$0 = -16(t^2 - 9) \qquad \text{Factor out } -16.$$
   $$0 = -16(t - 3)(t + 3) \qquad \text{Factor completely.}$$
   $$t - 3 = 0 \quad \text{or} \quad t + 3 = 0 \qquad \text{Set each factor containing a variable equal to 0.}$$
   $$t = 3 \quad \text{or} \qquad t = -3 \qquad \text{Solve.}$$

3. INTERPRET. Since the time t cannot be negative, the proposed solution is 3 seconds.

Check: Verify that the height of the diver when t is 3 seconds is 0.

When $t = 3$ seconds, $h = -16(3)^2 + 144 = -144 + 144 = 0$.

■ **Work Practice Problem 1**

Answer

1. 2 sec

EXAMPLE 2 **Finding a Number**

The square of a number plus three times the number is 70. Find the number.

Solution:

1. UNDERSTAND. Read and reread the problem. Suppose that the number is 5. The square of 5 is 5^2 or 25. Three times 5 is 15. Then $25 + 15 = 40$, not 70, so the number must be greater than 5. Remember, the purpose of proposing a number, such as 5, is to better understand the problem. Now that we do, we will let $x =$ the number.

2. TRANSLATE.

the square of a number	plus	three times the number	is	70
↓	↓	↓	↓	↓
x^2	$+$	$3x$	$=$	70

3. SOLVE.

$$x^2 + 3x = 70$$
$$x^2 + 3x - 70 = 0 \qquad \text{Subtract 70 from both sides.}$$
$$(x + 10)(x - 7) = 0 \qquad \text{Factor.}$$
$$x + 10 = 0 \quad \text{or} \quad x - 7 = 0 \qquad \text{Set each factor equal to 0.}$$
$$x = -10 \qquad \qquad x = 7 \qquad \text{Solve.}$$

4. INTERPRET.

Check: The square of -10 is $(-10)^2$, or 100. Three times -10 is $3(-10)$ or -30. Then $100 + (-30) = 70$, the correct sum, so -10 checks.
The square of 7 is 7^2 or 49. Three times 7 is $3(7)$, or 21. Then $49 + 21 = 70$, the correct sum, so 7 checks.

State: There are two numbers. They are -10 and 7.

🟧 **Work Practice Problem 2**

△ **EXAMPLE 3** **Finding the Dimensions of a Sail**

The height of a triangular sail is 2 meters less than twice the length of the base. If the sail has an area of 30 square meters, find the length of its base and the height.

Solution:

1. UNDERSTAND. Read and reread the problem. Since we are finding the length of the base and the height, we let

$x =$ the length of the base

Since the height is 2 meters less than twice the length of the base,

$2x - 2 =$ the height

An illustration is shown on the next page.

2. TRANSLATE. We are given that the area of the triangle is 30 square meters, so we use the formula for area of a triangle.

area of triangle	$=$	$\frac{1}{2}$	$\cdot$	base	$\cdot$	height
↓		↓		↓		↓
30	$=$	$\frac{1}{2}$	$\cdot$	x	$\cdot$	$(2x - 2)$

Continued on next page

PRACTICE PROBLEM 2

The square of a number minus twice the number is 63. Find the number.

PRACTICE PROBLEM 3

The length of a rectangular garden is 5 feet more than its width. The area of the garden is 176 square feet. Find the length and the width of the garden.

Answers
2. 9 and -7,
3. length: 16 ft; width: 11 ft

Height = $2x - 2$

Base = x

3. SOLVE. Now we solve the quadratic equation.

$$30 = \frac{1}{2}x(2x - 2)$$

$30 = x^2 - x$	Multiply.
$0 = x^2 - x - 30$	Write in standard form.
$0 = (x - 6)(x + 5)$	Factor.
$x - 6 = 0$ or $x + 5 = 0$	Set each factor equal to 0.
$x = 6$ $\qquad$ $x = -5$	

4. INTERPRET. Since x represents the length of the base, we discard the solution -5. The base of a triangle cannot be negative. The base is then 6 meters and the height is $2(6) - 2 = 10$ meters.

Check: To check this problem, we recall that

$$\text{area} = \frac{1}{2}\,\text{base} \cdot \text{height or}$$

$$30 \overset{?}{=} \frac{1}{2}(6)(10)$$

$30 = 30$	True

State: The base of the triangular sail is 6 meters and the height is 10 meters.

🔲 **Work Practice Problem 3**

The next examples make use of the **Pythagorean theorem** and consecutive integers. Before we review this theorem, recall that a **right triangle** is a triangle that contains a 90° or right angle. The **hypotenuse** of a right triangle is the side opposite the right angle and is the longest side of the triangle. The **legs** of a right triangle are the other sides of the triangle.

Pythagorean Theorem

In a right triangle, the sum of the squares of the lengths of the two legs is equal to the square of the length of the hypotenuse.

$$(\text{leg})^2 + (\text{leg})^2 = (\text{hypotenuse})^2 \quad \text{or} \quad a^2 + b^2 = c^2$$

Leg b — Hypotenuse c — Leg a

Study the following diagrams for a review of consecutive integers.

Examples

If x is the first integer, then consecutive integers are
$x, x + 1, x + 2, \ldots$

If x is the first even integer, then consecutive even integers are
$x, x + 2, x + 4, \ldots$

If x is the first odd integer, then consecutive odd integers are
$x, x + 2, x + 4, \ldots$

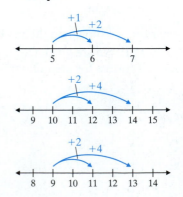

EXAMPLE 4 **Finding Consecutive Even Integers**

Find two consecutive even integers whose product is 34 more than their sum.

Solution:

1. UNDERSTAND. Read and reread the problem. Let's just choose two consecutive even integers to help us better understand the problem. Let's choose 10 and 12. Their product is $10(12) = 120$ and their sum is $10 + 12 = 22$. The product is $120 - 22$, or 98 greater than the sum. Thus our guess is incorrect, but we have a better understanding of this example.

   Let's let x and $x + 2$ be the consecutive even integers.

2. TRANSLATE.

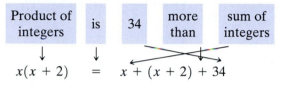

$$x(x + 2) \quad = \quad x + (x + 2) + 34$$

3. SOLVE. Now we solve the equation.

$x(x + 2) = x + (x + 2) + 34$	
$x^2 + 2x = x + x + 2 + 34$	Multiply.
$x^2 + 2x = 2x + 36$	Combine like terms.
$x^2 - 36 = 0$	Write in standard form.
$(x + 6)(x - 6) = 0$	Factor.
$x + 6 = 0 \quad \text{or} \quad x - 6 = 0$	Set each factor equal to 0.
$x = -6 \qquad\qquad x = 6$	Solve.

4. INTERPRET. If $x = -6$, then $x + 2 = -6 + 2$, or -4.

   If $x = 6$, then $x + 2 = 6 + 2$, or 8.

Check: $-6, -4$

$-6(-4) \stackrel{?}{=} -6 + (-4) + 34$

$24 \stackrel{?}{=} -10 + 34$

$24 = 24$ True

$6, 8$

$6(8) \stackrel{?}{=} 6 + 8 + 34$

$48 \stackrel{?}{=} 14 + 34$

$48 = 48$ True

State: The two consecutive even integers are -6 and -4 or 6 and 8.

🔲 **Work Practice Problem 4**

△ **EXAMPLE 5** **Finding the Dimensions of a Triangle**

Find the lengths of the sides of a right triangle if the lengths can be expressed as three consecutive even integers.

Solution:

1. UNDERSTAND. Read and reread the problem. Let's suppose that the length of one leg of the right triangle is 4 units. Then the other leg is the next even integer, or 6 units, and the hypotenuse of the triangle is the next even integer, or 8 units. Remember that the hypotenuse is the longest side. Let's see if a triangle with sides of these lengths forms a right triangle. To do this, we check to see whether the Pythagorean theorem holds true.

$4^2 + 6^2 \stackrel{?}{=} 8^2$

$16 + 36 \stackrel{?}{=} 64$

$52 = 64$ False

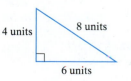

4 units

8 units

6 units

Continued on next page

PRACTICE PROBLEM 4

Find two consecutive odd integers whose product is 23 more than their sum.

PRACTICE PROBLEM 5

The length of one leg of a right triangle is 7 meters less than the length of the other leg. The length of the hypotenuse is 13 meters. Find the lengths of the legs.

Answers

4. 5 and 7 or -5 and -3,

5. 5 meters, 12 meters

Our proposed numbers do not check, but we now have a better understanding of the problem.

We let x, $x + 2$, and $x + 4$ be three consecutive even integers. Since these integers represent lengths of the sides of a right triangle, we have the following.

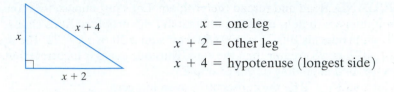

x = one leg

$x + 2$ = other leg

$x + 4$ = hypotenuse (longest side)

2. **TRANSLATE.** By the Pythagorean theorem, we have that

$$(\text{leg})^2 + (\text{leg})^2 = (\text{hypotenuse})^2$$
$$(x)^2 + (x + 2)^2 = (x + 4)^2$$

3. **SOLVE.** Now we solve the equation.

$$x^2 + (x + 2)^2 = (x + 4)^2$$

$x^2 + x^2 + 4x + 4 = x^2 + 8x + 16$	Multiply.
$2x^2 + 4x + 4 = x^2 + 8x + 16$	Combine like terms.
$x^2 - 4x - 12 = 0$	Write in standard form.
$(x - 6)(x + 2) = 0$	Factor.
$x - 6 = 0$ or $x + 2 = 0$	Set each factor equal to 0.
$x = 6$ $x = -2$	

4. **INTERPRET.** We discard $x = -2$ since length cannot be negative. If $x = 6$, then $x + 2 = 8$ and $x + 4 = 10$.

Check: Verify that

$$(\text{leg})^2 + (\text{leg})^2 = (\text{hypotenuse})^2$$
$$6^2 + 8^2 \overset{?}{=} 10^2$$
$$36 + 64 \overset{?}{=} 100$$
$$100 = 100 \qquad \text{True}$$

State: The sides of the right triangle have lengths 6 units, 8 units, and 10 units.

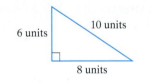

🟧 **Work Practice Problem 5**

Objective A *See Examples 1 through 5 for all exercises. For Exercises 1 through 6, represent each given condition using a single variable, x.*

△ **1.** The length and width of a rectangle whose length is 4 centimeters more than its width

△ **2.** The length and width of a rectangle whose length is twice its width

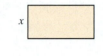

3. Two consecutive odd integers

4. Two consecutive even integers

△ **5.** The base and height of a triangle whose height is one more than four times its base

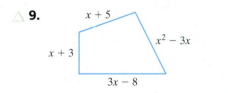

△ **6.** The base and height of a trapezoid whose base is three less than five times its height

base

Use the information given to find the dimensions of each figure.

△ **7.**

x

The *area* of the square is 121 square units. Find the length of its sides.

△ **8.**

x − 2

x + 3

The *area* of the rectangle is 84 square inches. Find its length and width.

△ **9.**

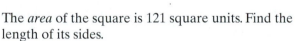

x + 5

x² − 3x

x + 3

3x − 8

The *perimeter* of the quadrilateral is 120 centimeters. Find the lengths of the sides.

△ **10.**

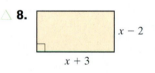

2x   2x + 5

x² + 3

The *perimeter* of the triangle is 85 feet. Find the lengths of its sides.

△ **11.**

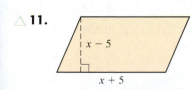

x − 5

x + 5

The *area* of the parallelogram is 96 square miles. Find its base and height.

△ **12.**

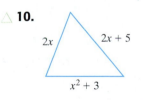

x

The *area* of the circle is 25π square kilometers. Find its radius.

Solve.

13. An object is thrown upward from the top of an 80-foot building with an initial velocity of 64 feet per second. The height h of the object after t seconds is given by the quadratic equation $h = -16t^2 + 64t + 80$. When will the object hit the ground?

14. A hang glider accidentally drops her compass from the top of a 400-foot cliff. The height h of the compass after t seconds is given by the quadratic equation $h = -16t^2 + 400$. When will the compass hit the ground?

15. The width of a rectangle is 7 centimeters less than twice its length. Its area is 30 square centimeters. Find the dimensions of the rectangle.

16. The length of a rectangle is 9 inches more than its width. Its area is 112 square inches. Find the dimensions of the rectangle.

△ *The equation* $D = \frac{1}{2}n(n-3)$ *gives the number of diagonals D for a polygon with n sides. For example, a polygon with 6 sides has* $D = \frac{1}{2} \cdot 6(6-3)$ *or D = 9 diagonals. (See if you can count all 9 diagonals. Some are shown in the figure.) Use this equation,* $D = \frac{1}{2}n(n-3)$, *for Exercises 17 through 20.*

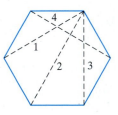

17. Find the number of diagonals for a polygon that has 12 sides.

18. Find the number of diagonals for a polygon that has 15 sides.

19. Find the number of sides n for a polygon that has 35 diagonals.

20. Find the number of sides n for a polygon that has 14 diagonals.

21. The sum of a number and its square is 132. Find the number.

22. The sum of a number and its square is 182. Find the number.

23. The product of two consecutive room numbers is 210. Find the room numbers.

24. The product of two consecutive page numbers is 420. Find the page numbers.

25. A ladder is leaning against a building so that the distance from the ground to the top of the ladder is one foot less than the length of the ladder. Find the length of the ladder if the distance from the bottom of the ladder to the building is 5 feet.

26. Use the given figure to find the length of the guy wire.

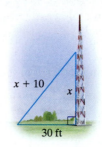

△ **27.** If the sides of a square are increased by 3 inches, the area becomes 64 square inches. Find the length of the sides of the original square.

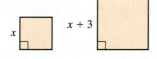

△ **28.** If the sides of a square are increased by 5 meters, the area becomes 100 square meters. Find the length of the sides of the original square.

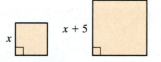

△ **29.** One leg of a right triangle is 4 millimeters longer than the smaller leg and the hypotenuse is 8 millimeters longer than the smaller leg. Find the lengths of the sides of the triangle.

△ **30.** One leg of a right triangle is 9 centimeters longer than the other leg and the hypotenuse is 45 centimeters. Find the lengths of the legs of the triangle.

△ **31.** The length of the base of a triangle is twice its height. If the area of the triangle is 100 square kilometers, find the height.

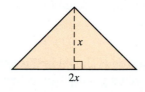

△ **32.** The height of a triangle is 2 millimeters less than the base. If the area is 60 square millimeters, find the base.

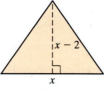

△ **33.** Find the length of the shorter leg of a right triangle if the longer leg is 12 feet more than the shorter leg and the hypotenuse is 12 feet less than twice the shorter leg.

△ **34.** Find the length of the shorter leg of a right triangle if the longer leg is 10 miles more than the shorter leg and the hypotenuse is 10 miles less than twice the shorter leg.

35. An object is dropped from 39 feet below the tip of the pinnacle atop one of the 1483-foot-tall Petronas Twin Towers in Kuala Lumpur, Malaysia. (*Source:* Council on Tall Buildings and Urban Habitat) The height h of the object after t seconds is given by the equation $h = -16t^2 + 1444$. Find how many seconds pass before the object reaches the ground.

36. An object is dropped from the top of 311 South Wacker Drive, a 961-foot-tall office building in Chicago. (*Source:* Council on Tall Buildings and Urban Habitat) The height h of the object after t seconds is given by the equation $h = -16t^2 + 961$. Find how many seconds pass before the object reaches the ground.

37. At the end of 2 years, P dollars invested at an interest rate r compounded annually increases to an amount, A dollars, given by

$$A = P(1 + r)^2$$

Find the interest rate if $100 increased to $144 in 2 years. Write your answer as a percent.

38. At the end of 2 years, P dollars invested at an interest rate r compounded annually increases to an amount, A dollars, given by

$$A = P(1 + r)^2$$

Find the interest rate if $2000 increased to $2420 in 2 years. Write your answer as a percent.

△ **39.** Find the dimensions of a rectangle whose width is 7 miles less than its length and whose area is 120 square miles.

△ **40.** Find the dimensions of a rectangle whose width is 2 inches less than half its length and whose area is 160 square inches.

41. If the cost, C, for manufacturing x units of a certain product is given by $C = x^2 - 15x + 50$, find the number of units manufactured at a cost of $9500.

42. If a switchboard handles n telephones, the number C of telephone connections it can make simultaneously is given by the equation $C = \dfrac{n(n-1)}{2}$. Find how many telephones are handled by a switchboard making 120 telephone connections simultaneously.

Review

The following double line graph shows a comparison of the amount of land (in thousand acres) occupied by farms in Florida during the years shown with the amount of land occupied by farms in Georgia. Use this graph to answer Exercises 43 through 49. See Section 2.4.

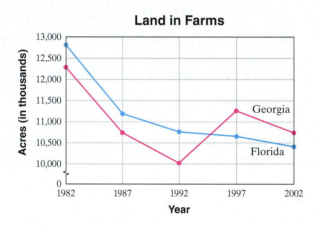

Land in Farms

43. Approximate the amount of land occupied by farms in Georgia in 1997.

44. Approximate the amount of land occupied by farms in Florida in 1997.

45. Approximate the amount of land occupied by farms in Georgia in 1987.

46. Approximate the amount of land occupied by farms in Florida in 1987.

47. Approximate the year that the colored lines in this graph intersect.

48. In your own words, explain the meaning of the point of intersection in the graph.

49. Describe the trends shown in this graph and speculate as to why these trends have occurred.

Concept Extensions

△ **50.** Two boats travel at right angles to each other after leaving the same dock at the same time. One hour later the boats are 17 miles apart. If one boat travels 7 miles per hour faster than the other boat, find the rate of each boat.

52. The sum of two numbers is 20, and the sum of their squares is 218. Find the numbers.

△ **51.** The side of a square equals the width of a rectangle. The length of the rectangle is 6 meters longer than its width. The sum of the areas of the square and the rectangle is 176 square meters. Find the side of the square.

53. The sum of two numbers is 25, and the sum of their squares is 325. Find the numbers.

△ **54.** According to the International America's Cup Class (IACC) rule, a sailboat competing in the America's Cup match must have a 110-foot-tall mast and a combined mainsail and jib sail area of 3000 square feet. (*Source:* America's Cup Organizing Committee) A design for an IACC-class sailboat calls for the mainsail to be 60% of the combined sail area. If the height of the triangular mainsail is 28 feet more than twice the length of the boom, find the length of the boom and the height of the mainsail.

△ **55.** A rectangular pool is surrounded by a walk 4 meters wide. The pool is 6 meters longer than its width. If the total area of the pool and walk is 576 square meters more than the area of the pool, find the dimensions of the pool.

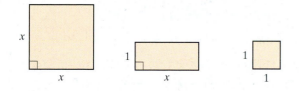

△ **56.** A rectangular garden is surrounded by a walk of uniform width. The area of the garden is 180 square yards. If the dimensions of the garden plus the walk are 16 yards by 24 yards, find the width of the walk.

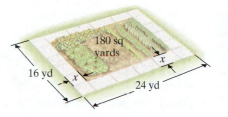

CHAPTER 4 Group Activity

Factoring polynomials can be visualized using areas of rectangles. To see this, let's first find the areas of the following squares and rectangles. (Recall that Area = Length · Width.)

To use these areas to visualize factoring the polynomial $x^2 + 3x + 2$, for example, use the shapes below to form a rectangle. The factored form is found by reading the length and the width of the rectangle as shown below.

Thus, $x^2 + 3x + 2 = (x + 2)(x + 1)$.

Try using this method to visualize the factored form of each polynomial below.

Work in a group and use tiles to find the factored form of the polynomials below. (Tiles can be hand made from index cards.)

1. $x^2 + 6x + 5$

2. $x^2 + 5x + 6$

3. $x^2 + 5x + 4$

4. $x^2 + 4x + 3$

5. $x^2 + 6x + 9$

6. $x^2 + 4x + 4$

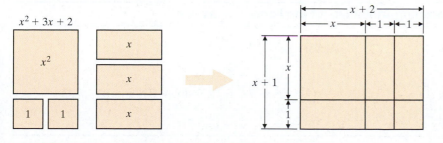

Chapter 4 Vocabulary Check

Fill in each blank with one of the words or phrases listed below.

factoring   quadratic equation

greatest common factor   perfect square trinomial

1. An equation that can be written in the form $ax^2 + bx + c = 0$ (with *a* not 0) is called a _____.
2. _____ is the process of writing an expression as a product.
3. The _____ of a list of terms is the product of all common factors.
4. A trinomial that is the square of some binomial is called a _____.

Helpful Hint

Are you preparing for your test? Don't forget to take the Chapter 4 Test on page 319. Then check your answers at the back of the text and use the Chapter Test Prep Video CD to see the fully worked-out solutions to any of the exercises you want to review.

4 Chapter Highlights

DEFINITIONS AND CONCEPTS	EXAMPLES
Section 4.1 The Greatest Common Factor	

Factoring is the process of writing an expression as a product.

The GCF of a list of variable terms contains the smallest exponent on each common variable.

The GCF of a list of terms is the product of all common factors.

Factor: $6 = 2 \cdot 3$
Factor: $x^2 + 5x + 6 = (x + 2)(x + 3)$
The GCF of z^5, z^3, and z^{10} is z^3.

Find the GCF of $8x^2y$, $10x^3y^2$, and $50x^2y^3$.

$$8x^2y = 2 \cdot 2 \cdot 2 \cdot x^2 \cdot y$$
$$10x^3y^2 = 2 \cdot 5 \cdot x^3 \cdot y^2$$
$$50x^2y^3 = 2 \cdot 5 \cdot 5 \cdot x^2 \cdot y^3$$
$$GCF = 2 \cdot x^2 \cdot y \quad \text{or} \quad 2x^2y$$

TO FACTOR BY GROUPING

Step 1. Group the terms in two groups so that each group has a common factor.

Step 2. Factor out the GCF from each group.

Step 3. If there is a common binomial factor, factor it out.

Step 4. If not, rearrange the terms and try these steps again.

Factor: $10ax + 15a - 6xy - 9y$

Step 1. $(10ax + 15a) + (-6xy - 9y)$
Step 2. $5a(2x + 3) - 3y(2x + 3)$
Step 3. $(2x + 3)(5a - 3y)$

Section 4.2 Factoring Trinomials of the Form $x^2 + bx + c$	

The product of these numbers is *c*.

$$x^2 + bx + c = (x + \square)(x + \square)$$

The sum of these numbers is *b*.

Factor: $x^2 + 7x + 12$

$3 + 4 = 7 \quad 3 \cdot 4 = 12$
$x^2 + 7x + 12 = (x + 3)(x + 4)$

DEFINITIONS AND CONCEPTS	**EXAMPLES**

Section 4.3   Factoring Trinomials of the Form $ax^2 + bx + c$

To factor $ax^2 + bx + c$, try various combinations of factors of ax^2 and c until a middle term of bx is obtained when checking.	Factor: $3x^2 + 14x - 5$ $\qquad$ Factors of $3x^2$: $3x, x$ $\qquad$ Factors of -5: $-1, 5$ and $1, -5$. $\qquad (3x - 1)(x + 5)$ $\qquad\qquad -1x$ $\qquad\qquad +15x$ $\qquad\qquad\ \ 14x\quad$ **Correct** middle term

Section 4.4   Factoring Trinomials of the Form $ax^2 + bx + c$ by Grouping

TO FACTOR $ax^2 + bx + c$ BY GROUPING **Step 1.** Find two numbers whose product is $a \cdot c$ and whose sum is b. **Step 2.** Rewrite bx, using the factors found in Step 1. **Step 3.** Factor by grouping.	Factor: $3x^2 + 14x - 5$ **Step 1.** Find two numbers whose product is $3 \cdot (-5)$ or -15 and whose sum is 14. They are 15 and -1. **Step 2.** $3x^2 + 14x - 5$ $\qquad = 3x^2 + 15x - 1x - 5$ **Step 3.** $\quad = 3x(x + 5) - 1(x + 5)$ $\qquad = (x + 5)(3x - 1)$

Section 4.5   Factoring Perfect Square Trinomials and the Difference of Two Squares

A **perfect square trinomial** is a trinomial that is the square of some binomial.	**PERFECT SQUARE TRINOMIAL = SQUARE OF BINOMIAL** $\qquad x^2 + 4x + 4 = (x + 2)^2$ $\qquad 25x^2 - 10x + 1 = (5x - 1)^2$
Factoring Perfect Square Trinomials $\qquad a^2 + 2ab + b^2 = (a + b)^2$ $\qquad a^2 - 2ab + b^2 = (a - b)^2$	Factor. $\qquad x^2 + 6x + 9 = x^2 + 2(x \cdot 3) + 3^2 = (x + 3)^2$ $\qquad 4x^2 - 12x + 9 = (2x)^2 - 2(2x \cdot 3) + 3^2$ $\qquad\qquad\qquad\qquad = (2x - 3)^2$
Difference of Two Squares $\qquad a^2 - b^2 = (a + b)(a - b)$	Factor. $\qquad x^2 - 9 = x^2 - 3^2 = (x + 3)(x - 3)$

Section 4.6   Solving Quadratic Equations by Factoring

A **quadratic equation** is an equation that can be written in the form $ax^2 + bx + c = 0$ with a not 0. The form $ax^2 + bx + c = 0$ is called the **standard form** of a quadratic equation.	**Quadratic Equation** $\qquad$ **Standard Form** $x^2 = 16 \qquad\qquad\qquad x^2 - 16 = 0$ $y = -2y^2 + 5 \qquad\qquad 2y^2 + y - 5 = 0$
Zero Factor Property If a and b are real numbers and if $ab = 0$, then $a = 0$ or $b = 0$.	If $(x + 3)(x - 1) = 0$, then $x + 3 = 0$ or $x - 1 = 0$.

continued

DEFINITIONS AND CONCEPTS	**EXAMPLES**

Section 4.6   Solving Quadratic Equations by Factoring (*continued*)

TO SOLVE QUADRATIC EQUATIONS BY FACTORING	Solve: $3x^2 = 13x - 4$
Step 1. Write the equation in standard form so that one side of the equation is 0.	**Step 1.** $3x^2 - 13x + 4 = 0$
Step 2. Factor completely.	**Step 2.** $(3x - 1)(x - 4) = 0$
Step 3. Set each factor containing a variable equal to 0.	**Step 3.** $3x - 1 = 0$ or $x - 4 = 0$
Step 4. Solve the resulting equations.	**Step 4.** $3x = 1$ $x = 4$ $$x = \frac{1}{3}$$
Step 5. Check solutions in the original equation.	**Step 5.** Check both $\frac{1}{3}$ and 4 in the original equation.

Section 4.7   Quadratic Equations and Problem Solving

PROBLEM-SOLVING STEPS	A garden is in the shape of a rectangle whose length is two feet more than its width. If the area of the garden is 35 square feet, find its dimensions.
1. UNDERSTAND the problem.	**1.** Read and reread the problem. Guess a solution and check your guess. Draw a diagram. Let x be the width of the rectangular garden. Then $x + 2$ is the length.
2. TRANSLATE.	**2.** length · width = area $(x + 2)$ · x = 35
3. SOLVE.	**3.** $(x + 2)x = 35$ $x^2 + 2x - 35 = 0$ $(x - 5)(x + 7) = 0$ $x - 5 = 0$ or $x + 7 = 0$ $x = 5$ $x = -7$
4. INTERPRET.	**4.** Discard the solution $x = -7$ since x represents width. *Check:* If x is 5 feet, then $x + 2 = 5 + 2 = 7$ feet. The area of a rectangle whose width is 5 feet and whose length is 7 feet is (5 feet)(7 feet) or 35 square feet. *State:* The garden is 5 feet by 7 feet.

Are You Prepared for a Test on Chapter 4?

Below is a list of some *common trouble areas* for students in Chapter 4. After studying for your test—but before taking your test—read these.

- The difference of two squares such as $x^2 - 25$ factors as $x^2 - 25 = (x + 5)(x - 5)$.

- The sum of two squares, for example, $x^2 + 25$, cannot be factored using real numbers.

- Don't forget that the first step to factor any polynomial is to first factor out any common factors.

  $$9x^2 - 36 = 9(x^2 - 4) = 9(x + 2)(x - 2)$$

- Can you completely factor $x^4 - 24x^2 - 25$?

  $$x^4 - 24x^2 - 25 = (x^2 - 25)(x^2 + 1)$$
  $$= (x + 5)(x - 5)(x^2 + 1)$$

- Remember that to use the zero factor property to solve a quadratic equation, one side of the equation must be 0 and the other side must be a factored polynomial.

  $$x(x - 2) = 3 \quad \text{Cannot use zero factor property.}$$
  $$x^2 - 2x - 3 = 0$$
  $$(x - 3)(x + 1) = 0 \quad \text{Now we can use zero factor property.}$$
  $$x - 3 = 0 \quad \text{or} \quad x + 1 = 0$$
  $$x = 3 \quad \text{or} \quad x = -1$$

Remember: This is simply a sampling of selected topics given to check your understanding. For a review of Chapter 4 in your text, see the material at the end of this chapter.

4 CHAPTER REVIEW

(4.1) *Complete each factoring.*

1. $6x^2 - 15x = 3x(\qquad)$

2. $4x^5 + 2x - 10x^4 = 2x(\qquad)$

Factor out the GCF from each polynomial.

3. $5m + 30$

4. $20x^3 + 12x^2 + 24x$

5. $3x(2x + 3) - 5(2x + 3)$

6. $5x(x + 1) - (x + 1)$

Factor each polynomial by grouping.

7. $3x^2 - 3x + 2x - 2$

8. $6x^2 + 10x - 3x - 5$

9. $3a^2 + 9ab + 3b^2 + ab$

(4.2) *Factor each trinomial.*

10. $x^2 + 6x + 8$

11. $x^2 - 11x + 24$

12. $x^2 + x + 2$

13. $x^2 - 5x - 6$

14. $x^2 + 2x - 8$

15. $x^2 + 4xy - 12y^2$

16. $x^2 + 8xy + 15y^2$

17. $72 - 18x - 2x^2$

18. $32 + 12x - 4x^2$

19. $5y^3 - 50y^2 + 120y$

20. To factor $x^2 + 2x - 48$, think of two numbers whose product is _____ and whose sum is _____.

21. What is the first step to factoring $3x^2 + 15x + 30$?

(4.3) or (4.4) *Factor each trinomial.*

22. $2x^2 + 13x + 6$

23. $4x^2 + 4x - 3$

24. $6x^2 + 5xy - 4y^2$

25. $x^2 - x + 2$

26. $2x^2 - 23x - 39$

27. $18x^2 - 9xy - 20y^2$

28. $10y^3 + 25y^2 - 60y$

Write the perimeter of each figure as a simplified polynomial. Then factor each polynomial.

△ **29.**

△ **30.**

(4.5) *Determine whether each polynomial is a perfect square trinomial.*

31. $x^2 + 6x + 9$

32. $x^2 + 8x + 64$

33. $9m^2 - 12m + 16$

34. $4y^2 - 28y + 49$

Determine whether each binomial is a difference of two squares.

35. $x^2 - 9$

36. $x^2 + 16$

37. $4x^2 - 25y^2$

38. $9a^3 - 1$

Factor each polynomial completely.

39. $x^2 - 81$

40. $x^2 + 12x + 36$

41. $4x^2 - 9$

42. $9t^2 - 25s^2$

43. $16x^2 + y^2$

44. $n^2 - 18n + 81$

45. $3r^2 + 36r + 108$

46. $9y^2 - 42y + 49$

47. $5m^8 - 5m^6$

48. $4x^2 - 28xy + 49y^2$

49. $3x^2y + 6xy^2 + 3y^3$

50. $16x^4 - 1$

(4.6) *Solve each equation.*

51. $(x + 6)(x - 2) = 0$

52. $3x(x + 1)(7x - 2) = 0$

53. $4(5x + 1)(x + 3) = 0$

54. $x^2 + 8x + 7 = 0$

55. $x^2 - 2x - 24 = 0$

56. $x^2 + 10x = -25$

57. $x(x - 10) = -16$

58. $(3x - 1)(9x^2 + 3x + 1) = 0$

59. $56x^2 - 5x - 6 = 0$

60. $m^2 = 6m$

61. $r^2 = 25$

62. Write a quadratic equation that has the two solutions 4 and 5.

(4.7) *Use the given information to choose the correct dimensions.*

△ **63.** The perimeter of a rectangle is 24 inches. The length is twice the width. Find the dimensions of the rectangle.
  a. 5 inches by 7 inches   **b.** 5 inches by 10 inches
  c. 4 inches by 8 inches   **d.** 2 inches by 10 inches

△ **64.** The area of a rectangle is 80 meters. The length is one more than three times the width. Find the dimensions of the rectangle.
  a. 8 meters by 10 meters   **b.** 4 meters by 13 meters
  c. 4 meters by 20 meters   **d.** 5 meters by 16 meters

Use the given information to find the dimensions of each figure.

△ **65.** The *area* of the square is 81 square units. Find the length of a side.

△ **66.** The *perimeter* of the quadrilateral is 47 units. Find the lengths of the sides.

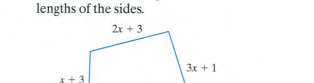

$2x + 3$

$3x + 1$

$x + 3$

$x^2 - 3x$

Solve.

△ **67.** A flag for a local organization is in the shape of a rectangle whose length is 15 inches less than twice its width. If the area of the flag is 500 square inches, find its dimensions.

PRAIRIEVIEW

GOLF CLUB

x

△ **68.** The base of a triangular sail is four times its height. If the area of the triangle is 162 square yards, find the base.

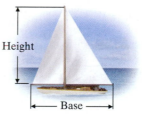

Height

Base

69. Find two consecutive positive integers whose product is 380.

70. A rocket is fired from the ground with an initial velocity of 440 feet per second. Its height h after t seconds is given by the equation $h = -16t^2 + 440t$.

  a. Find how many seconds pass before the rocket reaches a height of 2800 feet. Explain why two answers are obtained.

  b. Find how many seconds pass before the rocket reaches the ground again.

△ **71.** An architect's squaring instrument is in the shape of a right triangle. Find the length of the longer leg of the right triangle if the hypotenuse is 8 centimeters longer than the longer leg and the shorter leg is 8 centimeters shorter than the longer leg.

Mixed Review

Factor Completely.

72. $6x + 24$

73. $7x - 63$

74. $11x(4x - 3) - 6(4x - 3)$

75. $2x(x - 5) - (x - 5)$

76. $3x^3 - 4x^2 + 6x - 8$

77. $xy + 2x - y - 2$

78. $2x^2 + 2x - 24$

79. $3x^3 - 30x^2 + 27x$

80. $4x^2 - 81$

81. $2x^2 - 18$

82. $16x^2 - 24x + 9$

83. $5x^2 + 20x + 20$

Solve.

84. $2x^2 - x - 28 = 0$

85. $x^2 - 2x = 15$

86. $2x(x + 7)(x + 4) = 0$

87. $x(x - 5) = -6$

88. $x^2 = 16x$

Solve.

89. The perimeter of the following triangle is 48 inches. Find the lengths of its sides.

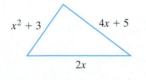

90. The width of a rectangle is 4 inches less than its length. Its area is 12 square inches. Find the dimensions of the rectangle.

4 CHAPTER TEST

 Remember to use the Chapter Test Prep Video CD to see the fully worked-out solutions to any of the exercises you want to review.

Factor each polynomial completely. If a polynomial cannot be factored, write "prime."

1. $9x^2 - 3x$

2. $x^2 + 11x + 28$

3. $49 - m^2$

4. $y^2 + 22y + 121$

5. $x^4 - 16$

6. $4(a + 3) - y(a + 3)$

7. $x^2 + 4$

8. $y^2 - 8y - 48$

9. $3a^2 + 3ab - 7a - 7b$

10. $3x^2 - 5x + 2$

11. $180 - 5x^2$

12. $3x^3 - 21x^2 + 30x$

13. $6t^2 - t - 5$

14. $xy^2 - 7y^2 - 4x + 28$

15. $x - x^5$

16. $x^2 + 14xy + 24y^2$

Solve each equation.

17. $(x - 3)(x + 9) = 0$

18. $x^2 + 5x = 14$

19. $x(x + 6) = 7$

20. $3x(2x - 3)(3x + 4) = 0$

21. $5t^3 - 45t = 0$

22. $t^2 - 2t - 15 = 0$

23. $6x^2 = 15x$

1. _____

2. _____

3. _____

4. _____

5. _____

6. _____

7. _____

8. _____

9. _____

10. _____

11. _____

12. _____

13. _____

14. _____

15. _____

16. _____

17. _____

18. _____

19. _____

20. _____

21. _____

22. _____

23. _____

24. _____

Solve.

△ **24.** A deck for a home is in the shape of a triangle. The length of the base of the triangle is 9 feet longer than its height. If the area of the triangle is 68 square feet find the length of the base.

Base

Altitude

△ **25.** The *area* of the rectangle is 54 square units. Find the dimensions of the rectangle.

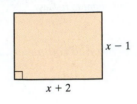

$x - 1$

$x + 2$

25. _____

26. _____

26. An object is dropped from the top of the Woolworth Building on Broadway in New York City. The height h of the object after t seconds is given by the equation

$$h = -16t^2 + 784$$

Find how many seconds pass before the object reaches the ground.

△ **27.** Find the lengths of the sides of a right triangle if the hypotenuse is 10 centimeters longer than the shorter leg and 5 centimeters longer than the longer leg.

27. _____

28. A window washer is suspended 38 feet below the roof of the 1127-foot-tall John Hancock Center in Chicago. (*Source:* Council on Tall Buildings and Urban Habitat) If the window washer drops an object from this height, the object's height h after t seconds is given by the equation $h = -16t^2 + 1089$. Find how many seconds pass before the object reaches the ground.

28. _____

1. Translate each sentence into a mathematical statement.
   a. Nine is less than or equal to eleven.
   b. Eight is greater than one.
   c. Three is not equal to four.

2. Insert $<$ or $>$ in the space to make each statement true.
   a. $|-5|$ ___ $|-3|$
   b. $|0|$ ___ $|-2|$

3. Decide whether 2 is a solution of $3x + 10 = 8x$.

4. Evaluate $\dfrac{x}{y} + 5x$ if $x = 20$ and $y = 10$.

5. Subtract 8 from -4.

6. Evaluate $\dfrac{x}{y} + 5x$ if $x = -20$ and $y = 10$.

7. If $x = -2$ and $y = -4$, evaluate each expression.
   a. $\dfrac{3x}{2y}$
   b. $x^3 - y^2$

8. Evaluate $\dfrac{x}{y} + 5x$ if $x = -20$ and $y = -10$.

Simplify each expression by combining like terms.

9. $2x + 3x + 5 + 2$

10. $5 - 2(3x - 7)$

11. $-5a - 3 + a + 2$

12. $5(x - 6) + 9(-2x + 1)$

13. $2.3x + 5x - 6$

Solve each equation.

14. $0.8y + 0.2(y - 1) = 1.8$

15. $-3x = 33$

16. $\dfrac{x}{-7} = -4$

17. $3(x - 4) = 3x - 12$

18. $-\dfrac{2}{3}x = -22$

19. Solve for *l*: $V = lwh$

20. Solve for *y*: $3x + 2y = -7$

1. a. _____

b. _____

c. _____

2. a. _____

b. _____

3. _____

4. _____

5. _____

6. _____

7. a. _____

b. _____

8. _____

9. _____

10. _____

11. _____

12. _____

13. _____

14. _____

15. _____

16. _____

17. _____

18. _____

19. _____

20. _____

Simplify each expression.

21. $(5^3)^6$

22. $5^2 + 5^1$

23. $(y^8)^2$

24. $y^8 \cdot y^2$

Simplify the following expressions. Write each result using positive exponents only.

25. $\dfrac{(x^3)^4 x}{x^7}$

26. 3^{-2}

27. $(y^{-3} z^6)^{-6}$

28. $\dfrac{x^{-3}}{x^{-7}}$

29. $\dfrac{x^{-7}}{(x^4)^3}$

30. $\dfrac{(5a^7)^2}{a^5}$

Simplify each polynomial by combining any like terms.

31. $-3x + 7x$

32. $\dfrac{2}{3}x + 23 + \dfrac{1}{6}x - 100$

33. $11x^2 + 5 + 2x^2 - 7$

34. $0.2x - 1.1 + 2.3 - 0.7x$

35. Multiply: $(2x - y)^2$

36. Multiply: $(3x - 7y)^2$

Use a special product to square each binomial.

37. $(t + 2)^2$

38. $(x - 13)^2$

39. $(x^2 - 7y)^2$

40. $(7x + y)^2$

41. Divide: $\dfrac{8x^2y^2 - 16xy + 2x}{4xy}$

21. _____

22. _____

23. _____

24. _____

25. _____

26. _____

27. _____

28. _____

29. _____

30. _____

31. _____

32. _____

33. _____

34. _____

35. _____

36. _____

37. _____

38. _____

39. _____

40. _____

41. _____

Factor each polynomial.

42. $z^3 + 7z + z^2 + 7$

43. $5(x + 3) + y(x + 3)$

44. $2x^3 + 2x^2 - 84x$

45. $x^4 + 5x^2 + 6$

46. $-4x^2 - 23x + 6$

47. $6x^2 - 2x - 20$

48. $9xy^2 - 16x$

49. The platform for the cliff divers in Acapulco, Mexico, is about 144 feet above the sea. Neglecting air resistance, the height h in feet of a cliff diver above the ocean after t seconds is given by the quadratic equation $h = -16t^2 + 144$. Find how long it takes the diver to reach the ocean.

50. Solve $x^2 - 13x = -36$.

42. _____

43. _____

44. _____

45. _____

46. _____

47. _____

48. _____

49. _____

50. _____

5

Rational Expressions

In this chapter, we expand our knowledge of algebraic expressions to include algebraic fractions, called *rational expressions*. We explore the operations of addition, subtraction, multiplication, and division using principles similar to the principles for numerical fractions.

American football is one of this nation's most followed sports. It has its roots in English rugby, a game played with the same shaped ball, but where the ball can only advance through running. American college players were the first to add advancing the ball by throwing or kicking it past the opponents. In 1867, both Rutgers and Princeton established a basic set of rules and played the first intercollegiate football game. In 1876, Walter Camp, the coach at Yale, helped establish that teams were to consist of 11 men, standardized the size of the field, and generally instituted the first cohesive set of rules.

In Exercise 87, Section 5.1, you will have the opportunity to calculate a quarterback's rating.

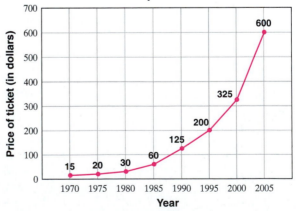

Price of Super Bowl Tickets*

Source: JS Online: Super Bowl Ticket Prices and NFL research
* For years with multiple ticket prices, highest price is shown

5.1 SIMPLIFYING RATIONAL EXPRESSIONS

Objectives

A Find the Value of a Rational Expression Given a Replacement Number.

B Identify Values for Which a Rational Expression Is Undefined.

C Simplify, or Write Rational Expressions in Lowest Terms.

D Write Equivalent Forms of Rational Expressions.

Objective A Evaluating Rational Expressions

A rational number is a number that can be written as a quotient of integers. A *rational expression* is also a quotient; it is a quotient of polynomials. Examples are

$$\frac{2}{3}, \quad \frac{3y^3}{8}, \quad \frac{-4p}{p^3 + 2p + 1}, \quad \text{and} \quad \frac{5x^2 - 3x + 2}{3x + 7}$$

Rational Expression

A **rational expression** is an expression that can be written in the form

$$\frac{P}{Q}$$

where P and Q are polynomials and $Q \neq 0$.

Rational expressions have different numerical values depending on what values replace the variables.

EXAMPLE 1 Find the numerical value of $\dfrac{x + 4}{2x - 3}$ for each replacement value.

a. $x = 5$ **b.** $x = -2$

Solution:

a. We replace each x in the expression with 5 and then simplify.

$$\frac{x + 4}{2x - 3} = \frac{5 + 4}{2(5) - 3} = \frac{9}{10 - 3} = \frac{9}{7}$$

b. We replace each x in the expression with -2 and then simplify.

$$\frac{x + 4}{2x - 3} = \frac{-2 + 4}{2(-2) - 3} = \frac{2}{-7} \quad \text{or} \quad -\frac{2}{7}$$

□ **Work Practice Problem 1**

In the example above, we wrote $\dfrac{2}{-7}$ as $-\dfrac{2}{7}$. For a negative fraction such as $\dfrac{2}{-7}$, recall from Section 1.7 that

$$\frac{2}{-7} = \frac{-2}{7} = -\frac{2}{7}$$

In general, for any fraction,

$$\frac{-a}{b} = \frac{a}{-b} = -\frac{a}{b}, \qquad b \neq 0$$

This is also true for rational expressions. For example,

$$\underbrace{\frac{-(x + 2)}{x}}_{\substack{\uparrow \\ \text{Notice the parentheses.}}} = \frac{x + 2}{-x} = -\frac{x + 2}{x}$$

PRACTICE PROBLEM 1

Find the value of $\dfrac{x - 3}{5x + 1}$ for each replacement value.

a. $x = 4$

b. $x = -3$

Answers

1. **a.** $\dfrac{1}{21}$, **b.** $\dfrac{3}{7}$

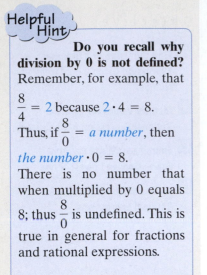

Objective B Identifying when a Rational Expression Is Undefined

In the definition of rational expression (first "box" in this section), notice that we wrote $Q \neq 0$ for the denominator Q. The denominator of a rational expression must not equal 0 since division by 0 is not defined. (See the Helpful Hint to the left.) This means we must be careful when replacing the variable in a rational expression by a number.

For example, suppose we replace x with 5 in the rational expression $\dfrac{3 + x}{x - 5}$. The expression becomes

$$\frac{3 + x}{x - 5} = \frac{3 + 5}{5 - 5} = \frac{8}{0}$$

But division by 0 is undefined. Therefore, in this expression we can allow x to be any real number *except* 5. **A rational expression is undefined for values that make the denominator 0.** Thus,

> To find values for which a rational expression is undefined, find values for which the denominator is 0.

PRACTICE PROBLEM 2

Are there any values for x for which each rational expression is undefined?

a. $\dfrac{x}{x + 8}$

b. $\dfrac{x - 3}{x^2 + 5x + 4}$

c. $\dfrac{x^2 - 3x + 2}{5}$

EXAMPLE 2 Are there any values for x for which each expression is undefined?

a. $\dfrac{x}{x - 3}$ b. $\dfrac{x^2 + 2}{x^2 - 3x + 2}$ c. $\dfrac{x^3 - 6x^2 - 10x}{3}$

Solution: To find values for which a rational expression is undefined, we find values that make the denominator 0.

a. The denominator of $\dfrac{x}{x - 3}$ is 0 when $x - 3 = 0$ or when $x = 3$. Thus, when $x = 3$, the expression $\dfrac{x}{x - 3}$ is undefined.

b. We set the denominator equal to 0.

$$x^2 - 3x + 2 = 0$$
$$(x - 2)(x - 1) = 0 \qquad \text{Factor.}$$
$$x - 2 = 0 \qquad \text{or} \qquad x - 1 = 0 \quad \text{Set each factor equal to 0.}$$
$$x = 2 \qquad\qquad x = 1 \quad \text{Solve.}$$

Thus, when $x = 2$ or $x = 1$, the denominator $x^2 - 3x + 2$ is 0. So the rational expression $\dfrac{x^2 + 2}{x^2 - 3x + 2}$ is undefined when $x = 2$ or when $x = 1$.

c. The denominator of $\dfrac{x^3 - 6x^2 - 10x}{3}$ is never 0, so there are no values of x for which this expression is undefined.

Note: Unless otherwise stated, we will now assume that variables in rational expressions are only replaced by values for which the expressions are defined.

☐ **Work Practice Problem 2**

Objective C Simplifying Rational Expressions

A fraction is said to be written in lowest terms or simplest form when the numerator and denominator have no common factors other than 1 (or -1). For example, the fraction $\dfrac{7}{10}$ is written in lowest terms since the numerator and denominator have no common factors other than 1 (or -1).

The process of writing a rational expression in lowest terms or simplest form is called **simplifying** a rational expression.

Simplifying a rational expression is similar to simplifying a fraction. Recall from Chapter R that to simplify a fraction, we essentially "remove factors of 1." Our ability to do this comes from these facts:

- Any nonzero number over itself simplifies to 1 $\left(\dfrac{5}{5} = 1, \dfrac{-7.26}{-7.26} = 1, \text{ or } \dfrac{c}{c} = 1 \text{ as long as } c \text{ is not } 0\right)$, and

- The product of any number and 1 is that number $\left(19 \cdot 1 = 19, -8.9 \cdot 1 = -8.9, \dfrac{a}{b} \cdot 1 = \dfrac{a}{b}\right)$.

In other words, we have the following:

$$\frac{a \cdot c}{b \cdot c} = \frac{a}{b} \cdot \frac{c}{c} = \frac{a}{b}$$

$$\text{Since } \frac{a}{b} \cdot 1 = \frac{a}{b}$$

Simplify: $\dfrac{15}{20}$

$$\frac{15}{20} = \frac{3 \cdot 5}{2 \cdot 2 \cdot 5} \quad \text{Factor the numerator and the denominator.}$$

$$= \frac{3 \cdot 5}{2 \cdot 2 \cdot 5} \quad \text{Look for common factors.}$$

$$= \frac{3}{2 \cdot 2} \cdot \frac{5}{5} \quad \text{Common factors in the numerator and denominator form factors of 1.}$$

$$= \frac{3}{2 \cdot 2} \cdot 1 \quad \text{Write } \frac{5}{5} \text{ as 1.}$$

$$= \frac{3}{2 \cdot 2} = \frac{3}{4} \quad \text{Multiply to remove a factor of 1.}$$

Before we use the same technique to simplify a rational expression, remember that as long as the denominator is not 0, $\dfrac{a^3 b}{a^3 b} = 1$, $\dfrac{x+3}{x+3} = 1$, and $\dfrac{7x^2 + 5x - 100}{7x^2 + 5x - 100} = 1$.

Simplify: $\dfrac{x^2 - 9}{x^2 + x - 6}$

$$\frac{x^2 - 9}{x^2 + x - 6} = \frac{(x-3)(x+3)}{(x-2)(x+3)} \quad \text{Factor the numerator and the denominator.}$$

$$= \frac{(x-3)(x+3)}{(x-2)(x+3)} \quad \text{Look for common factors.}$$

$$= \frac{x-3}{x-2} \cdot \frac{x+3}{x+3}$$

$$= \frac{x-3}{x-2} \cdot 1 \quad \text{Write } \frac{x+3}{x+3} \text{ as 1.}$$

$$= \frac{x-3}{x-2} \quad \text{Multiply to remove a factor of 1.}$$

Just as for numerical fractions, we can use a shortcut notation. Remember that as long as exact factors in both the numerator and denominator are divided out, we are "removing a factor of 1." We will use the following notation to show this:

$$\frac{x^2 - 9}{x^2 + x - 6} = \frac{(x-3)(x+3)}{(x-2)(x+3)} \quad \text{A factor of 1 is identified by the shading.}$$

$$= \frac{x-3}{x-2} \quad \text{Remove a factor of 1.}$$

Thus, the rational expression $\dfrac{x^2 - 9}{x^2 + x - 6}$ has the same value as the rational expression $\dfrac{x - 3}{x - 2}$ for all values of x except 2 and -3. (Remember that when x is 2, the denominator of both rational expressions is 0 and when x is -3, the original rational expression has a denominator of 0.)

As we simplify rational expressions, we will assume that the simplified rational expression is equal to the original rational expression for all real numbers except those for which either denominator is 0. The following steps may be used to simplify rational expressions.

> ### To Simplify a Rational Expression
>
> **Step 1:** Completely factor the numerator and denominator.
>
> **Step 2:** Divide out factors common to the numerator and denominator. (This is the same as "removing a factor of 1.")

PRACTICE PROBLEM 3

Simplify: $\dfrac{x^4 + x^3}{5x + 5}$

EXAMPLE 3 Simplify: $\dfrac{5x - 5}{x^3 - x^2}$

Solution: To begin, we factor the numerator and denominator if possible. Then we look for common factors.

$$\frac{5x - 5}{x^3 - x^2} = \frac{5\,(x - 1)}{x^2\,(x - 1)} = \frac{5}{x^2}$$

□ **Work Practice Problem 3**

PRACTICE PROBLEM 4

Simplify: $\dfrac{x^2 + 11x + 18}{x^2 + x - 2}$

EXAMPLE 4 Simplify: $\dfrac{x^2 + 8x + 7}{x^2 - 4x - 5}$

Solution: We factor the numerator and denominator and then look for common factors.

$$\frac{x^2 + 8x + 7}{x^2 - 4x - 5} = \frac{(x + 7)\,(x + 1)}{(x - 5)\,(x + 1)} = \frac{x + 7}{x - 5}$$

□ **Work Practice Problem 4**

PRACTICE PROBLEM 5

Simplify: $\dfrac{x^2 + 10x + 25}{x^2 + 5x}$

EXAMPLE 5 Simplify: $\dfrac{x^2 + 4x + 4}{x^2 + 2x}$

Solution: We factor the numerator and denominator and then look for common factors.

$$\frac{x^2 + 4x + 4}{x^2 + 2x} = \frac{(x + 2)\,(x + 2)}{x\,(x + 2)} = \frac{x + 2}{x}$$

□ **Work Practice Problem 5**

Helpful Hint

When simplifying a rational expression, we look for **common *factors*, not** common *terms*.

$$\frac{x \cdot (x + 2)}{x \cdot x} = \frac{x + 2}{x} \qquad \qquad \frac{x + 2}{x}$$

Common factors. These can be divided out.

Common terms. There is no factor of 1 that can be generated.

Answers

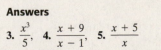
3. $\dfrac{x^3}{5}$, 4. $\dfrac{x + 9}{x - 1}$, 5. $\dfrac{x + 5}{x}$

✔**Concept Check** Recall that we can only remove *factors* of 1. Which of the following are *not* true? Explain why.

a. $\dfrac{3-1}{3+5}$ simplifies to $-\dfrac{1}{5}$

b. $\dfrac{2x+10}{2}$ simplifies to $x+5$

c. $\dfrac{37}{72}$ simplifies to $\dfrac{3}{2}$

d. $\dfrac{2x+3}{2}$ simplifies to $x+3$

EXAMPLE 6 Simplify: $\dfrac{x+9}{x^2-81}$

Solution: We factor and then apply the fundamental principle.

$$\dfrac{x+9}{x^2-81} = \dfrac{x+9}{(x+9)(x-9)} = \dfrac{1}{x-9}$$

🟧 **Work Practice Problem 6**

PRACTICE PROBLEM 6

Simplify: $\dfrac{x+5}{x^2-25}$

EXAMPLE 7 Simplify each rational expression.

a. $\dfrac{x+y}{y+x}$

b. $\dfrac{x-y}{y-x}$

Solution:

a. The expression $\dfrac{x+y}{y+x}$ can be simplified by using the commutative property of addition to rewrite the denominator $y+x$ as $x+y$.

$$\dfrac{x+y}{y+x} = \dfrac{x+y}{x+y} = 1$$

b. The expression $\dfrac{x-y}{y-x}$ can be simplified by recognizing that $y-x$ and $x-y$ are opposites. In other words, $y-x = -1(x-y)$. We proceed as follows:

$$\dfrac{x-y}{y-x} = \dfrac{1 \cdot (x-y)}{(-1)(x-y)} = \dfrac{1}{-1} = -1$$

🟧 **Work Practice Problem 7**

PRACTICE PROBLEM 7

Simplify each rational expression.

a. $\dfrac{x+4}{4+x}$

b. $\dfrac{x-4}{4-x}$

Objective **D** **Writing Equivalent Forms of Rational Expressions**

From Example 7a, we have $y+x = x+y$. $\qquad$ $y+x$ and $x+y$ are equivalent.

From Example 7b, we have $y-x = -1(x-y)$. $\quad$ $y-x$ and $x-y$ are opposites.

Thus, $\dfrac{x+y}{y+x} = \dfrac{x+y}{x+y} = 1$ $\quad$ and $\quad$ $\dfrac{x-y}{y-x} = \dfrac{x-y}{-1(x-y)} = \dfrac{1}{-1} = -1$.

When performing operations on rational expressions, equivalent forms of answers often result. For this reason, it is very important to be able to recognize equivalent answers.

PRACTICE PROBLEM 8

List 4 equivalent forms of

$$-\frac{3x + 7}{x - 6}.$$

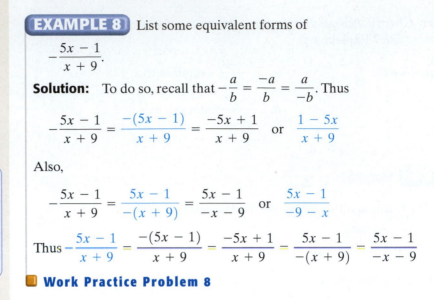

EXAMPLE 8 List some equivalent forms of

$$-\frac{5x - 1}{x + 9}.$$

Solution: To do so, recall that $-\dfrac{a}{b} = \dfrac{-a}{b} = \dfrac{a}{-b}$. Thus

$$-\frac{5x - 1}{x + 9} = \frac{-(5x - 1)}{x + 9} = \frac{-5x + 1}{x + 9} \quad \text{or} \quad \frac{1 - 5x}{x + 9}$$

Also,

$$-\frac{5x - 1}{x + 9} = \frac{5x - 1}{-(x + 9)} = \frac{5x - 1}{-x - 9} \quad \text{or} \quad \frac{5x - 1}{-9 - x}$$

Thus $-\dfrac{5x - 1}{x + 9} = \dfrac{-(5x - 1)}{x + 9} = \dfrac{-5x + 1}{x + 9} = \dfrac{5x - 1}{-(x + 9)} = \dfrac{5x - 1}{-x - 9}$

Helpful Hint

Remember, a negative sign in front of a fraction or rational expression may be moved to the numerator or the denominator, but *not* both.

◼ **Work Practice Problem 8**

Keep in mind that many rational expressions may look different, but in fact be equivalent.

Answer

8. $\dfrac{-(3x + 7)}{x - 6}; \dfrac{-3x - 7}{x - 6}; \dfrac{3x + 7}{-(x - 6)};$

$\dfrac{3x + 7}{-x + 6}$

Mental Math

Find any numbers for which each rational expression is undefined. See Example 2.

1. $\dfrac{x + 5}{x}$

2. $\dfrac{x^2 - 5x}{x - 3}$

3. $\dfrac{x^2 + 4x - 2}{x(x - 1)}$

4. $\dfrac{x + 2}{(x - 5)(x - 6)}$

Objective A *Find the value of the following expressions when $x = 2$, $y = -2$, and $z = -5$. See Example 1.*

1. $\dfrac{x + 5}{x + 2}$

2. $\dfrac{x + 8}{x + 1}$

3. $\dfrac{y^3}{y^2 - 1}$

4. $\dfrac{z}{z^2 - 5}$

5. $\dfrac{x^2 + 8x + 2}{x^2 - x - 6}$

6. $\dfrac{x + 5}{x^2 + 4x - 8}$

7. The average cost per DVD, in dollars, for a company to produce x DVDs on exercising is given by the formula: $A = \dfrac{3x + 400}{x}$, where A is the average cost per DVD, and x is the number of DVDs produced.

   a. Find the cost for producing 1 DVD.

   b. Find the average cost for producing 100 DVDs.

   c. Does the cost per DVD decrease or increase when more DVDs are produced? Explain your answer.

8. For a certain model of fax machine, the manufacturing cost C per machine is given by the equation

$$C = \dfrac{250x + 10,000}{x}$$

where x is the number of fax machines manufactured and cost C is in dollars per machine.

   a. Find the cost per fax machine when manufacturing 100 fax machines.

   b. Find the cost per fax machine when manufacturing 1000 fax machines.

   c. Does the cost per machine decrease or increase when more machines are manufactured? Explain why this is so.

Objective B *Find any numbers for which each rational expression is undefined. See Example 2.*

9. $\dfrac{7}{2x}$

10. $\dfrac{3}{5x}$

11. $\dfrac{x + 3}{x + 2}$

12. $\dfrac{5x + 1}{x - 9}$

13. $\dfrac{x - 4}{2x - 5}$

14. $\dfrac{x + 1}{5x - 2}$

15. $\dfrac{9x^3 + 4}{15x^2 + 30x}$

16. $\dfrac{19x^3 + 2}{x^2 - x}$

17. $\dfrac{x^2 - 5x - 2}{4}$

18. $\dfrac{9y^5 + y^3}{9}$

19. $\dfrac{3x^2 + 9}{x^2 - 5x - 6}$

20. $\dfrac{11x^2 + 1}{x^2 - 5x - 14}$

21. $\dfrac{x}{3x^2 + 13x + 14}$

22. $\dfrac{x}{2x^2 + 15x + 27}$

Objective **C** *Simplify each expression. See Examples 3 through 7.*

23. $\dfrac{x + 7}{7 + x}$

24. $\dfrac{y + 9}{9 + y}$

25. $\dfrac{x - 7}{7 - x}$

26. $\dfrac{y - 9}{9 - y}$

27. $\dfrac{2}{8x + 16}$

28. $\dfrac{3}{9x + 6}$

29. $\dfrac{x - 2}{x^2 - 4}$

30. $\dfrac{x + 5}{x^2 - 25}$

31. $\dfrac{2x - 10}{3x - 30}$

32. $\dfrac{3x - 9}{4x - 16}$

33. $\dfrac{-5a - 5b}{a + b}$

34. $\dfrac{-4x - 4y}{x + y}$

35. $\dfrac{7x + 35}{x^2 + 5x}$

36. $\dfrac{9x + 99}{x^2 + 11x}$

37. $\dfrac{x + 5}{x^2 - 4x - 45}$

38. $\dfrac{x - 3}{x^2 - 6x + 9}$

39. $\dfrac{5x^2 + 11x + 2}{x + 2}$

40. $\dfrac{12x^2 + 4x - 1}{2x + 1}$

41. $\dfrac{x^3 + 7x^2}{x^2 + 5x - 14}$

42. $\dfrac{x^4 - 10x^3}{x^2 - 17x + 70}$

43. $\dfrac{14x^2 - 21x}{2x - 3}$

44. $\dfrac{4x^2 + 24x}{x + 6}$

45. $\dfrac{x^2 + 7x + 10}{x^2 - 3x - 10}$

46. $\dfrac{2x^2 + 7x - 4}{x^2 + 3x - 4}$

47. $\dfrac{3x^2 + 7x + 2}{3x^2 + 13x + 4}$

48. $\dfrac{4x^2 - 4x + 1}{2x^2 + 9x - 5}$

49. $\dfrac{2x^2 - 8}{4x - 8}$

50. $\dfrac{5x^2 - 500}{35x + 350}$

51. $\dfrac{4 - x^2}{x - 2}$

52. $\dfrac{49 - y^2}{y - 7}$

53. $\dfrac{x^2 - 1}{x^2 - 2x + 1}$

54. $\dfrac{x^2 - 16}{x^2 - 8x + 16}$

Simplify each expression. Each exercise contains a four-term polynomial that should be factored by grouping.

55. $\dfrac{x^2 + xy + 2x + 2y}{x + 2}$

56. $\dfrac{ab + ac + b^2 + bc}{b + c}$

57. $\dfrac{5x + 15 - xy - 3y}{2x + 6}$

58. $\dfrac{xy - 6x + 2y - 12}{y^2 - 6y}$

59. $\dfrac{2xy + 5x - 2y - 5}{3xy + 4x - 3y - 4}$

60. $\dfrac{2xy + 2x - 3y - 3}{2xy + 4x - 3y - 6}$

Objective **D** *Study Example 8. Then list four equivalent forms for each rational expression.*

61. $-\dfrac{x - 10}{x + 8}$

62. $-\dfrac{x + 11}{x - 4}$

63. $-\dfrac{5y - 3}{y - 12}$

64. $-\dfrac{8y - 1}{y - 15}$

Objectives **C** **D** **Mixed Practice** *Simplify each expression. Then determine whether the given answer is correct. See Examples 3 through 8.*

65. $\dfrac{9 - x^2}{x - 3}$; Answer: $-3 - x$

66. $\dfrac{100 - x^2}{x - 10}$; Answer: $-10 - x$

67. $\dfrac{7 - 34x - 5x^2}{25x^2 - 1}$; Answer: $\dfrac{x + 7}{-5x - 1}$

68. $\dfrac{2 - 15x - 8x^2}{64x^2 - 1}$; Answer: $\dfrac{x + 2}{-8x - 1}$

Review

Perform each indicated operation. See Section R.2.

69. $\dfrac{1}{3} \cdot \dfrac{9}{11}$

70. $\dfrac{5}{27} \cdot \dfrac{2}{5}$

71. $\dfrac{1}{3} \div \dfrac{1}{4}$

72. $\dfrac{7}{8} \div \dfrac{1}{2}$

73. $\dfrac{13}{20} \div \dfrac{2}{9}$

74. $\dfrac{8}{15} \div \dfrac{5}{8}$

Concept Extensions

Which of the following are incorrect and why? See the Concept Check in this section.

75. $\dfrac{5a - 15}{5}$ simplifies to $a - 3$

76. $\dfrac{7m - 9}{7}$ simplifies to $m - 9$

77. $\dfrac{1 + 2}{1 + 3}$ simplifies to $\dfrac{2}{3}$

78. $\dfrac{46}{54}$ simplifies to $\dfrac{6}{5}$

79. Explain how to write a fraction in lowest terms.

80. Explain how to write a rational expression in lowest terms.

81. Explain why the denominator of a fraction or a rational expression must not equal 0.

82. Does $\dfrac{(x - 3)(x + 3)}{x - 3}$ have the same value as $x + 3$ for all real numbers? Explain why or why not.

83. The dose of medicine prescribed for a child depends on the child's age A in years and the adult dose D for the medication. Young's Rule is a formula used by pediatricians that gives a child's dose C as

$$C = \frac{DA}{A + 12}$$

Suppose that an 8-year-old child needs medication, and the normal adult dose is 1000 mg. What size dose should the child receive?

84. Calculating body-mass index is a way to gauge whether a person should lose weight. Doctors recommend that body-mass index values fall between 19 and 25. The formula for body-mass index B is

$$B = \frac{705w}{h^2}$$

where w is weight in pounds and h is height in inches. Should a 148-pound person who is 5 feet 6 inches tall lose weight?

85. A baseball player's slugging percentage S can be calculated with the following formula:

$$S = \frac{h + d + 2t + 3r}{b},$$ where h = number of hits,

d = number of doubles, t = number of triples, r = number of home runs, and b = number of at-bats. In 2004, Ichiro Suzuki of the Seattle Mariners led the American League in slugging percentage. During the 2004 season, Suzuki had 704 at-bats, 262 hits, 24 doubles, 5 triples, and 8 home runs. (*Source:* Major League Baseball) Calculate Suzuki's 2004 slugging percentage. Round to the nearest tenth of a percent.

86. A company's gross profit margin P can be computed with the formula $P = \dfrac{R - C}{R}$, where

R = the company's revenue and C = cost of goods sold. For fiscal year 2004, consumer electronics retailer Best Buy had revenues of $24.5 billion and cost of goods sold of $18.3 billion. (*Source:* Best Buy Company, Inc.) What was Best Buy's gross profit margin in 2004? Express the answer as a percent, rounded to the nearest tenth of a percent.

87. To calculate a quarterback's rating in football, you may use the formula

$$\left[\frac{20C + 0.5A + Y + 80T - 100I}{A}\right]\left(\frac{25}{6}\right),$$ where

C = the number of completed passes, A = the number of attempted passes, Y = total yards thrown for passes, T = the number of touchdown passes, and I = the number of interceptions. The New England Patriots were the winners of the Super Bowl in 2005. Their quarterback, Tom Brady, boasted the final season totals of 527 attempts, 317 completions, 3620 yards, 23 touchdown passes, and 12 interceptions. Calculate Brady's quarterback rating for the 2004–2005 season. Round the answer to the nearest tenth. (*Source:* The NFL)

88. Anthropologists and forensic scientists use a measure called the cephalic index to help classify skulls. The cephalic index of a skull with width W and length L from front to back is given by the formula

$$C = \frac{100W}{L}$$

A long skull has an index value less than 75, a medium skull has an index value between 75 and 85, and a broad skull has an index value over 85. Find the cephalic index of a skull that is 5 inches wide and 6.4 inches long. Classify the skull.

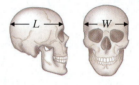

 STUDY SKILLS BUILDER

Is Your Notebook Still Organized?

It's never too late to organize your material in a course. Let's see how you are doing.

1. Are all your graded papers in one place in your math notebook or binder?

2. Flip through the pages of your notebook. Are your notes neat and readable?

3. Are your notes complete with no sections missing?

4. Are important notes marked in some way (like an exclamation point) so that you will know to review them before a quiz or task?

5. Are your assignments complete?

6. Do exercises that have given you trouble have a mark (like a question mark) so that you will remember to talk to your instructor or a tutor about them?

7. Describe your attitude toward this course.

8. List ways your attitude can improve and make a commitment to work on at least one of those during the next week.

5.2 MULTIPLYING AND DIVIDING RATIONAL EXPRESSIONS

Objectives

A Multiply Rational Expressions.

B Divide Rational Expressions.

C Multiply and Divide Rational Expressions.

D Convert between Units of Measure.

Objective **A** Multiplying Rational Expressions

Just as simplifying rational expressions is similar to simplifying number fractions, multiplying and dividing rational expressions is similar to multiplying and dividing number fractions.

Fractions	**Rational Expressions**
Multiply: $\dfrac{3}{5} \cdot \dfrac{10}{11}$	Multiply: $\dfrac{x-3}{x+5} \cdot \dfrac{2x+10}{x^2-9}$

Multiply numerators and then multiply denominators.

$$\frac{3}{5} \cdot \frac{10}{11} = \frac{3 \cdot 10}{5 \cdot 11} \qquad \frac{x-3}{x+5} \cdot \frac{2x+10}{x^2-9} = \frac{(x-3) \cdot (2x+10)}{(x+5) \cdot (x^2-9)}$$

Simplify by factoring numerators and denominators.

$$= \frac{3 \cdot 2 \cdot 5}{5 \cdot 11} \qquad = \frac{(x-3) \cdot 2 (x+5)}{(x+5)(x+3)(x-3)}$$

Apply the fundamental principle.

$$= \frac{3 \cdot 2}{11} \text{ or } \frac{6}{11} \qquad = \frac{2}{x+3}$$

Multiplying Rational Expressions

If $\dfrac{P}{Q}$ and $\dfrac{R}{S}$ are rational expressions, then

$$\frac{P}{Q} \cdot \frac{R}{S} = \frac{PR}{QS}$$

To multiply rational expressions, multiply the numerators and then multiply the denominators.

Note: Recall that for Sections 5.1 through 5.4, we assume variables in rational expressions have only those replacement values for which the expressions are defined.

EXAMPLE 1 Multiply.

a. $\dfrac{25x}{2} \cdot \dfrac{1}{y^3}$ **b.** $\dfrac{-7x^2}{5y} \cdot \dfrac{3y^5}{14x^2}$

Solution: To multiply rational expressions, we first multiply the numerators and then multiply the denominators of both expressions. Then we write the product in lowest terms.

a. $\dfrac{25x}{2} \cdot \dfrac{1}{y^3} = \dfrac{25x \cdot 1}{2 \cdot y^3} = \dfrac{25x}{2y^3}$

The expression $\dfrac{25x}{2y^3}$ is in lowest terms.

b. $\dfrac{-7x^2}{5y} \cdot \dfrac{3y^5}{14x^2} = \dfrac{-7x^2 \cdot 3y^5}{5y \cdot 14x^2}$ *Multiply.*

PRACTICE PROBLEM 1

Multiply.

a. $\dfrac{16y}{3} \cdot \dfrac{1}{x^2}$

b. $\dfrac{-5a^3}{3b^3} \cdot \dfrac{2b^2}{15a}$

Answers

1. **a.** $\dfrac{16y}{3x^2}$, **b.** $-\dfrac{2a^2}{9b}$

Continued on next page

The expression $\dfrac{-7x^2 \cdot 3y^5}{5y \cdot 14x^2}$ is not in lowest terms, so we factor the numerator and the denominator and apply the fundamental principle.

$$= \frac{-1 \cdot 7 \cdot 3 \cdot x^2 \cdot y \cdot y^4}{5 \cdot 2 \cdot 7 \cdot x^2 \cdot y}$$

$$= -\frac{3y^4}{10}$$

◾ **Work Practice Problem 1**

When multiplying rational expressions, it is usually best to factor each numerator and denominator first. This will help us when we apply the fundamental principle to write the product in lowest terms.

PRACTICE PROBLEM 2

Multiply: $\dfrac{3x + 6}{14} \cdot \dfrac{7x^2}{x^3 + 2x^2}$

EXAMPLE 2 Multiply: $\dfrac{x^2 + x}{3x} \cdot \dfrac{6}{5x + 5}$

Solution:

$$\frac{x^2 + x}{3x} \cdot \frac{6}{5x + 5} = \frac{x(x + 1)}{3x} \cdot \frac{2 \cdot 3}{5(x + 1)} \quad \text{Factor numerators and denominators.}$$

$$= \frac{x(x + 1) \cdot 2 \cdot 3}{3x \cdot 5 (x + 1)} \quad \text{Multiply.}$$

$$= \frac{2}{5} \quad \text{Divide out common factors.}$$

◾ **Work Practice Problem 2**

The following steps may be used to multiply rational expressions.

To Multiply Rational Expressions

Step 1: Completely factor numerators and denominators.

Step 2: Multiply numerators and multiply denominators.

Step 3: Simplify or write the product in lowest terms by dividing out common factors.

✔ **Concept Check** Which of the following is a true statement?

a. $\dfrac{1}{3} \cdot \dfrac{1}{2} = \dfrac{1}{5}$ **b.** $\dfrac{2}{x} \cdot \dfrac{5}{x} = \dfrac{10}{x}$ **c.** $\dfrac{3}{x} \cdot \dfrac{1}{2} = \dfrac{3}{2x}$ **d.** $\dfrac{x}{7} \cdot \dfrac{x + 5}{4} = \dfrac{2x + 5}{28}$

PRACTICE PROBLEM 3

Multiply:

$\dfrac{4x + 8}{7x^2 - 14x} \cdot \dfrac{3x^2 - 5x - 2}{9x^2 - 1}$

EXAMPLE 3 Multiply: $\dfrac{3x + 3}{5x^2 - 5x} \cdot \dfrac{2x^2 + x - 3}{4x^2 - 9}$

Solution:

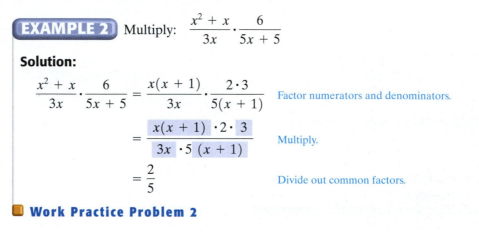

$$\frac{3x + 3}{5x^2 - 5x} \cdot \frac{2x^2 + x - 3}{4x^2 - 9} = \frac{3(x + 1)}{5x(x - 1)} \cdot \frac{(2x + 3)(x - 1)}{(2x - 3)(2x + 3)} \quad \text{Factor.}$$

$$= \frac{3(x + 1) (2x + 3)(x - 1)}{5x (x - 1) (2x - 3) (2x + 3)} \quad \text{Multiply.}$$

$$= \frac{3(x + 1)}{5x(2x - 3)} \quad \text{Simplify.}$$

◾ **Work Practice Problem 3**

Answers

2. $\dfrac{3}{2}$, 3. $\dfrac{4(x + 2)}{7x(3x - 1)}$

✔ **Concept Check Answer**

c

Objective B Dividing Rational Expressions

We can divide by a rational expression in the same way we divide by a number fraction. Recall that to divide by a fraction, we multiply by its reciprocal.

For example, to divide $\frac{3}{2}$ by $\frac{7}{8}$, we multiply $\frac{3}{2}$ by $\frac{8}{7}$.

$$\frac{3}{2} \div \frac{7}{8} = \frac{3}{2} \cdot \frac{8}{7} = \frac{3 \cdot 4 \cdot 2}{2 \cdot 7} = \frac{12}{7}$$

Helpful Hint

Don't forget how to find reciprocals. The reciprocal of $\frac{a}{b}$ is $\frac{b}{a}$, $a \neq 0$, $b \neq 0$.

Dividing Rational Expressions

If $\frac{P}{Q}$ and $\frac{R}{S}$ are rational expressions and $\frac{R}{S}$ is not 0, then

$$\frac{P}{Q} \div \frac{R}{S} = \frac{P}{Q} \cdot \frac{S}{R} = \frac{PS}{QR}$$

To divide two rational expressions, multiply the first rational expression by the reciprocal of the second rational expression.

EXAMPLE 4 Divide: $\dfrac{3x^3}{40} \div \dfrac{4x^3}{y^2}$

Solution:

$$\frac{3x^3}{40} \div \frac{4x^3}{y^2} = \frac{3x^3}{40} \cdot \frac{y^2}{4x^3} \qquad \text{Multiply by the reciprocal of } \frac{4x^3}{y^2}.$$

$$= \frac{3 \; x^3 \cdot y^2}{160 \; x^3}$$

$$= \frac{3y^2}{160} \qquad \text{Simplify.}$$

🔲 **Work Practice Problem 4**

EXAMPLE 5 Divide: $\dfrac{(x+2)^2}{10} \div \dfrac{2x+4}{5}$

Solution:

$$\frac{(x+2)^2}{10} \div \frac{2x+4}{5} = \frac{(x+2)^2}{10} \cdot \frac{5}{2x+4} \qquad \text{Multiply by the reciprocal of } \frac{2x+4}{5}.$$

$$= \frac{(x+2)(x+2) \cdot 5}{5 \cdot 2 \cdot 2 \cdot (x+2)} \qquad \text{Factor and multiply.}$$

$$= \frac{x+2}{4} \qquad \text{Simplify.}$$

🔲 **Work Practice Problem 5**

PRACTICE PROBLEM 4

Divide: $\dfrac{7x^2}{6} \div \dfrac{x}{2y}$

PRACTICE PROBLEM 5

Divide: $\dfrac{(x-4)^2}{6} \div \dfrac{3x-12}{2}$

Helpful Hint

Remember, **to Divide by a Rational Expression,** multiply by its reciprocal.

Answers

4. $\dfrac{7xy}{3}$,   5. $\dfrac{x-4}{9}$

Divide: $\dfrac{10x + 4}{x^2 - 4} \div \dfrac{5x^3 + 2x^2}{x + 2}$

EXAMPLE 6 Divide: $\dfrac{6x + 2}{x^2 - 1} \div \dfrac{3x^2 + x}{x - 1}$

Solution:

$$\dfrac{6x + 2}{x^2 - 1} \div \dfrac{3x^2 + x}{x - 1} = \dfrac{6x + 2}{x^2 - 1} \cdot \dfrac{x - 1}{3x^2 + x} \qquad \text{Multiply by the reciprocal.}$$

$$= \dfrac{2\,(3x + 1)(x - 1)}{(x + 1)\,(x - 1) \cdot x\,(3x + 1)} \qquad \text{Factor and multiply.}$$

$$= \dfrac{2}{x(x + 1)} \qquad \text{Simplify.}$$

■ **Work Practice Problem 6**

PRACTICE PROBLEM 7

Divide:

$\dfrac{3x^2 - 10x + 8}{7x - 14} \div \dfrac{9x - 12}{21}$

EXAMPLE 7 Divide: $\dfrac{2x^2 - 11x + 5}{5x - 25} \div \dfrac{4x - 2}{10}$

Solution:

$$\dfrac{2x^2 - 11x + 5}{5x - 25} \div \dfrac{4x - 2}{10} = \dfrac{2x^2 - 11x + 5}{5x - 25} \cdot \dfrac{10}{4x - 2} \qquad \text{Multiply by the reciprocal.}$$

$$= \dfrac{(2x - 1)(x - 5) \cdot 2 \cdot 5}{5(x - 5) \cdot 2(2x - 1)} \qquad \text{Factor and multiply.}$$

$$= \dfrac{1}{1} \quad \text{or} \quad 1 \qquad \text{Simplify.}$$

■ **Work Practice Problem 7**

Objective **C** Multiplying and Dividing Rational Expressions

Let's make sure that we understand the difference between multiplying and dividing rational expressions.

Rational Expressions	
Multiplication	Multiply the numerators and multiply the denominators.
Division	Multiply by the reciprocal of the divisor.

PRACTICE PROBLEM 8

Multiply or divide as indicated.

a. $\dfrac{x + 3}{x} \cdot \dfrac{7}{x + 3}$

b. $\dfrac{x + 3}{x} \div \dfrac{7}{x + 3}$

c. $\dfrac{3 - x}{x^2 + 6x + 5} \cdot \dfrac{2x + 10}{x^2 - 7x + 12}$

EXAMPLE 8 Multiply or divide as indicated.

a. $\dfrac{x - 4}{5} \cdot \dfrac{x}{x - 4}$

b. $\dfrac{x - 4}{5} \div \dfrac{x}{x - 4}$

c. $\dfrac{x^2 - 4}{2x + 6} \cdot \dfrac{x^2 + 4x + 3}{2 - x}$

Solution:

a. $\dfrac{x - 4}{5} \cdot \dfrac{x}{x - 4} = \dfrac{(x - 4) \cdot x}{5 \cdot (x - 4)} = \dfrac{x}{5}$

b. $\dfrac{x - 4}{5} \div \dfrac{x}{x - 4} = \dfrac{x - 4}{5} \cdot \dfrac{x - 4}{x} = \dfrac{(x - 4)^2}{5x}$

c. $\dfrac{x^2 - 4}{2x + 6} \cdot \dfrac{x^2 + 4x + 3}{2 - x} = \dfrac{(x - 2)(x + 2) \cdot (x + 1)(x + 3)}{2(x + 3) \cdot (2 - x)}$ Factor and multiply.

Answers

6. $\dfrac{2}{x^2(x - 2)}$,   7. 1,

8. a. $\dfrac{7}{x}$,   b. $\dfrac{(x + 3)^2}{7x}$,

c. $-\dfrac{2}{(x + 1)(x - 4)}$

Recall from Section 5.1 that $x - 2$ and $2 - x$ are opposites. This means that $\dfrac{x-2}{2-x} = -1$. Thus,

$$\frac{(x-2)\,(x+2)\cdot(x+1)\,(x+3)}{2\,(x+3)\cdot(2-x)} = \frac{-1(x+2)(x+1)}{2}$$

$$= -\frac{(x+2)(x+1)}{2}$$

◻ **Work Practice Problem 8**

Objective D Converting between Units of Measure

How many square inches are in 1 square foot?

How many cubic feet are in a cubic yard?

If you have trouble answering these questions, this section will be helpful to you.

Now that we know how to multiply fractions and rational expressions, we can use this knowledge to help us convert between units of measure. To do so, we will use **unit fractions.** A unit fraction is a fraction that equals 1. For example, since 12 in. = 1 ft, we have the unit fractions

$$\frac{12 \text{ in.}}{1 \text{ ft}} = 1 \quad \text{and} \quad \frac{1 \text{ ft}}{12 \text{ in.}} = 1$$

EXAMPLE 9 18 square feet = _____ square yards

Solution: Let's multiply 18 square feet by a unit fraction that has square feet in denominator and square yards in the numerator. From the diagram, you can see that

1 square yard = 9 square feet

Thus,

$$18 \text{ sq ft} = \frac{18 \text{ sq ft}}{1} \cdot 1 = \frac{\overset{2}{\cancel{18}} \text{ sq ft}}{1} \cdot \frac{1 \text{ sq yd}}{\underset{1}{\cancel{9}} \text{ sq ft}}$$

$$= \frac{2 \cdot 1}{1 \cdot 1} \text{ sq yd} = 2 \text{ sq yd}$$

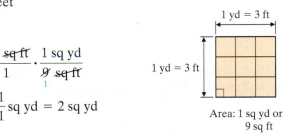

1 yd = 3 ft

Area: 1 sq yd or
9 sq ft

Thus, 18 sq ft = 2 sq yd.

Draw a diagram of 18 sq ft to help you see that this is reasonable.

◻ **Work Practice Problem 9**

EXAMPLE 10 5.2 square yards = _____ square feet

Solution:

$$5.2 \text{ sq yd} = \frac{5.2 \text{ sq yd}}{1} \cdot 1 = \frac{5.2 \,\cancel{\text{sq yd}}}{1} \cdot \frac{9 \text{ sq ft}}{1 \,\cancel{\text{sq yd}}} \quad \begin{array}{l}\leftarrow \text{ Units converting to} \\ \leftarrow \text{ Units given}\end{array}$$

$$= \frac{5.2 \cdot 9}{1 \cdot 1} \text{ sq ft}$$

$$= 46.8 \text{ sq ft}$$

Thus, 5.2 sq yd = 46.8 sq ft.

Draw a diagram to see that this is reasonable.

◻ **Work Practice Problem 10**

PRACTICE PROBLEM 9

288 square inches = _____ square feet

PRACTICE PROBLEM 10

3.5 square feet = _____ square inches

Answers

9. 2 sq ft,   **10.** 504 sq in.

PRACTICE PROBLEM 11

The largest casino in the world is the Foxwoods Resort Casino in Ledyard, CT. The gaming area for this casino is approximately 35,000 *square yards*. Find the size of the gaming area in *square feet*. (*Source:* Foxwoods Resort)

EXAMPLE 11　Converting from Cubic Feet to Cubic Yards

The largest building in the world by volume is The Boeing Company's Everett, Washington, factory complex where Boeing's wide-body jetliners, the 747, 767, and 777, are built. The volume of this factory complex is 472,370,319 cubic feet. Find the volume of this Boeing facility in cubic yards. (*Source:* The Boeing Company)

Solution:　There are 27 cubic feet in 1 cubic yard. (See the diagram.)

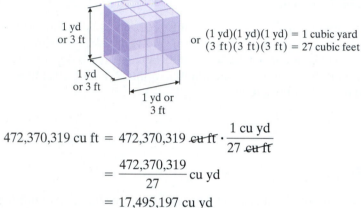

or (1 yd)(1 yd)(1 yd) = 1 cubic yard
(3 ft)(3 ft)(3 ft) = 27 cubic feet

$$472{,}370{,}319 \text{ cu ft} = 472{,}370{,}319 \text{ cu ft} \cdot \frac{1 \text{ cu yd}}{27 \text{ cu ft}}$$

$$= \frac{472{,}370{,}319}{27} \text{ cu yd}$$

$$= 17{,}495{,}197 \text{ cu yd}$$

▣ **Work Practice Problem 11**

Helpful Hint

When converting among units of measurement, if possible write the unit fraction so that **the numerator contains the units you are converting to** and **the denominator contains the original units.**

Unit fraction

$$48 \text{ in.} = \frac{48 \text{ in.}}{1} \cdot \overbrace{\frac{1 \text{ ft}}{12 \text{ in.}}}^{} \quad \leftarrow \text{Units converting to}$$
$$\leftarrow \text{Original units}$$

$$= \frac{48}{12} \text{ ft} = 4 \text{ ft}$$

PRACTICE PROBLEM 12

The cheetah is the fastest land animal, being clocked at about 102.7 feet per second. Convert this to miles per hour. Round to the nearest tenth. (*Source: World Almanac and Book of Facts*)

EXAMPLE 12

At the 2004 Summer Olympics, U.S. athlete Justin Gatlin won the gold medal in the men's 100-meter track event. He ran the distance at an average speed of 33.3 feet per second. Convert this speed to miles per hour. (*Source:* Athens Olympic Committee)

Solution:　Recall that 1 mile = 5280 feet and 1 hour = 3600 seconds (60 · 60).

Unit fractions

$$33.3 \text{ feet/second} = \frac{33.3 \text{ feet}}{1 \text{ second}} \cdot \frac{3600 \text{ seconds}}{1 \text{ hour}} \cdot \frac{1 \text{ mile}}{5280 \text{ feet}}$$

$$= \frac{33.3 \cdot 3600}{5280} \text{ miles/hour}$$

$$\approx 22.7 \text{ miles/hour (rounded to the nearest tenth)}$$

▣ **Work Practice Problem 12**

Answers

11. 315,000 sq ft,
12. 70.0 miles per hour

Mental Math

Find each product. See Example 1.

1. $\dfrac{2}{y} \cdot \dfrac{x}{3}$

2. $\dfrac{3x}{4} \cdot \dfrac{1}{y}$

3. $\dfrac{5}{7} \cdot \dfrac{y^2}{x^2}$

4. $\dfrac{x^5}{11} \cdot \dfrac{4}{z^3}$

5. $\dfrac{9}{x} \cdot \dfrac{x}{5}$

6. $\dfrac{y}{7} \cdot \dfrac{3}{y}$

5.2 EXERCISE SET

Objective A *Find each product and simplify if possible. See Examples 1 through 3.*

1. $\dfrac{3x}{y^2} \cdot \dfrac{7y}{4x}$

2. $\dfrac{9x^2}{y} \cdot \dfrac{4y}{3x^3}$

3. $\dfrac{8x}{2} \cdot \dfrac{x^5}{4x^2}$

4. $\dfrac{6x^2}{10x^3} \cdot \dfrac{5x}{12}$

5. $-\dfrac{5a^2b}{30a^2b^2} \cdot b^3$

6. $-\dfrac{9x^3y^2}{18xy^5} \cdot y^3$

7. $\dfrac{x}{2x - 14} \cdot \dfrac{x^2 - 7x}{5}$

8. $\dfrac{4x - 24}{20x} \cdot \dfrac{5}{x - 6}$

9. $\dfrac{6x + 6}{5} \cdot \dfrac{10}{36x + 36}$

10. $\dfrac{x^2 + x}{8} \cdot \dfrac{16}{x + 1}$

11. $\dfrac{(m + n)^2}{m - n} \cdot \dfrac{m}{m^2 + mn}$

12. $\dfrac{(m - n)^2}{m + n} \cdot \dfrac{m}{m^2 - mn}$

13. $\dfrac{x^2 - 25}{x^2 - 3x - 10} \cdot \dfrac{x + 2}{x}$

14. $\dfrac{a^2 - 4a + 4}{a^2 - 4} \cdot \dfrac{a + 3}{a - 2}$

15. $\dfrac{x^2 + 6x + 8}{x^2 + x - 20} \cdot \dfrac{x^2 + 2x - 15}{x^2 + 8x + 16}$

16. $\dfrac{x^2 + 9x + 20}{x^2 - 15x + 44} \cdot \dfrac{x^2 - 11x + 28}{x^2 + 12x + 35}$

Objective B *Find each quotient and simplify. See Examples 4 through 7.*

17. $\dfrac{5x^7}{2x^5} \div \dfrac{15x}{4x^3}$

18. $\dfrac{9y^4}{6y} \div \dfrac{y^2}{3}$

19. $\dfrac{8x^2}{y^3} \div \dfrac{4x^2y^3}{6}$

20. $\dfrac{7a^2b}{3ab^2} \div \dfrac{21a^2b^2}{14ab}$

21. $\dfrac{(x - 6)(x + 4)}{4x} \div \dfrac{2x - 12}{8x^2}$

22. $\dfrac{(x + 3)^2}{5} \div \dfrac{5x + 15}{25}$

23. $\dfrac{3x^2}{x^2 - 1} \div \dfrac{x^5}{(x + 1)^2}$

24. $\dfrac{9x^5}{a^2 - b^2} \div \dfrac{27x^2}{3b - 3a}$

25. $\dfrac{m^2 - n^2}{m + n} \div \dfrac{m}{m^2 + nm}$

26. $\dfrac{(m - n)^2}{m + n} \div \dfrac{m^2 - mn}{m}$

27. $\dfrac{x + 2}{7 - x} \div \dfrac{x^2 - 5x + 6}{x^2 - 9x + 14}$

28. $\dfrac{x - 3}{2 - x} \div \dfrac{x^2 + 3x - 18}{x^2 + 2x - 8}$

29. $\dfrac{x^2 + 7x + 10}{x - 1} \div \dfrac{x^2 + 2x - 15}{x - 1}$

30. $\dfrac{x + 1}{(x + 1)(2x + 3)} \div \dfrac{20x + 100}{2x + 3}$

Objective 🄒 **Mixed Practice** *Multiply or divide as indicated. See Example 8.*

31. $\dfrac{5x - 10}{12} \div \dfrac{4x - 8}{8}$

32. $\dfrac{6x + 6}{5} \div \dfrac{9x + 9}{10}$

33. $\dfrac{x^2 + 5x}{8} \cdot \dfrac{9}{3x + 15}$

34. $\dfrac{3x^2 + 12x}{6} \cdot \dfrac{9}{2x + 8}$

35. $\dfrac{7}{6p^2 + q} \div \dfrac{14}{18p^2 + 3q}$

36. $\dfrac{3x + 6}{20} \div \dfrac{4x + 8}{8}$

37. $\dfrac{3x + 4y}{x^2 + 4xy + 4y^2} \cdot \dfrac{x + 2y}{2}$

38. $\dfrac{x^2 - y^2}{3x^2 + 3xy} \cdot \dfrac{3x^2 + 6x}{3x^2 - 2xy - y^2}$

39. $\dfrac{(x + 2)^2}{x - 2} \div \dfrac{x^2 - 4}{2x - 4}$

40. $\dfrac{x + 3}{x^2 - 9} \div \dfrac{5x + 15}{(x - 3)^2}$

41. $\dfrac{x^2 - 4}{24x} \div \dfrac{2 - x}{6xy}$

42. $\dfrac{3y}{3 - x} \div \dfrac{12xy}{x^2 - 9}$

43. $\dfrac{a^2 + 7a + 12}{a^2 + 5a + 6} \cdot \dfrac{a^2 + 8a + 15}{a^2 + 5a + 4}$

44. $\dfrac{b^2 + 2b - 3}{b^2 + b - 2} \cdot \dfrac{b^2 - 4}{b^2 + 6b + 8}$

45. $\dfrac{5x - 20}{3x^2 + x} \cdot \dfrac{3x^2 + 13x + 4}{x^2 - 16}$

46. $\dfrac{9x + 18}{4x^2 - 3x} \cdot \dfrac{4x^2 - 11x + 6}{x^2 - 4}$

47. $\dfrac{8n^2 - 18}{2n^2 - 5n + 3} \div \dfrac{6n^2 + 7n - 3}{n^2 - 9n + 8}$

48. $\dfrac{36n^2 - 64}{3n^2 - 10n + 8} \div \dfrac{3n^2 - 13n + 12}{n^2 - 5n - 14}$

Objective 🄓 *Convert as indicated. See Examples 9 through 12.*

49. 10 square feet = _____ square inches.

50. 1008 square inches = _____ square feet.

51. 45 square feet = _____ square yards.

52. 2 square yards = _____ square inches.

53. 3 cubic yards = _____ cubic feet.

54. 2 cubic yards = _____ cubic inches.

55. 50 miles per hour = _____ feet per second (round to the nearest whole).

56. 10 feet per second = _____ miles per hour (round to the nearest tenth).

57. 6.3 square yards = _____ square feet.

58. 3.6 square yards = _____ square feet.

59. The Pentagon, headquarters for the Department of Defense, contains 3,705,793 square feet of office and storage space. Convert this to square yards. Round to the nearest square yard. (*Source:* U.S. Department of Defense)

60. The world's tallest building, Taipei 101 in Taipei, Taiwan, has 427,831 square yards of floor space. Convert this to square feet. (*Source:* Taipei 101)

61. On October 4, 2004, the rocket plane *SpaceShipOne* shot to an altitude of more than 100 km for the second time inside a week to claim the $10 million Ansari X-Prize. At one point in its flight, *SpaceShipOne* was traveling past Mach 1, about 930 miles per hour. Find this speed in feet per second. Round to the nearest whole. (*Source:* Space.com)

62. In 2002, Tim Montgomery of the United States held the current world record for the men's 100-meter track event. In that year, he covered the distance at an average speed of 33.55 feet per second. Convert this speed to miles per hour. Round to the nearest tenth. (*Source:* International Amateur Athletic Association)

Review

Perform each indicated operation. See Section R.2.

63. $\dfrac{1}{5} + \dfrac{4}{5}$

64. $\dfrac{3}{15} + \dfrac{6}{15}$

65. $\dfrac{9}{9} - \dfrac{19}{9}$

66. $\dfrac{4}{3} - \dfrac{8}{3}$

67. $\dfrac{6}{5} + \left(\dfrac{1}{5} - \dfrac{8}{5}\right)$

68. $-\dfrac{3}{2} + \left(\dfrac{1}{2} - \dfrac{3}{2}\right)$

Concept Extensions

Identify each statement as true or false. If false, correct the multiplication. See the Concept Check in this section.

69. $\dfrac{4}{a} \cdot \dfrac{1}{b} = \dfrac{4}{ab}$

70. $\dfrac{2}{3} \cdot \dfrac{2}{4} = \dfrac{2}{7}$

71. $\dfrac{x}{5} \cdot \dfrac{x+3}{4} = \dfrac{2x+3}{20}$

72. $\dfrac{7}{a} \cdot \dfrac{3}{a} = \dfrac{21}{a}$

73. Find the area of the rectangle.

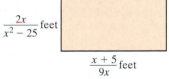

$\dfrac{2x}{x^2 - 25}$ feet

$\dfrac{x+5}{9x}$ feet

74. Find the area of the square.

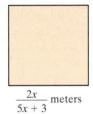

$\dfrac{2x}{5x+3}$ meters

Multiply or divide as indicated.

75. $\left(\dfrac{x^2 - y^2}{x^2 + y^2} \div \dfrac{x^2 - y^2}{3x}\right) \cdot \dfrac{x^2 + y^2}{6}$

76. $\left(\dfrac{x^2 - 9}{x^2 - 1} \cdot \dfrac{x^2 + 2x + 1}{2x^2 + 9x + 9}\right) \div \dfrac{2x + 3}{1 - x}$

77. $\left(\dfrac{2a + b}{b^2} \cdot \dfrac{3a^2 - 2ab}{ab + 2b^2}\right) \div \dfrac{a^2 - 3ab + 2b^2}{5ab - 10b^2}$

78. $\left(\dfrac{x^2 y^2 - xy}{4x - 4y} \div \dfrac{3y - 3x}{8x - 8y}\right) \cdot \dfrac{y - x}{8}$

79. In your own words, explain how you multiply rational expressions.

80. Explain how dividing rational expressions is similar to dividing rational numbers.

81. On November 14, 2004, 1 euro was equivalent to 1.2955 U.S. dollars. If you had wanted to exchange $2000 U.S. for euros on that day for a European vacation, how much would you have received? Round to the nearest hundredth. (*Source:* International Monetary Fund)

82. An environmental technician finds that warm water from an industrial process is being discharged into a nearby pond at a rate of 30 gallons per minute. Plant regulations state that the flow rate should be no more than 0.1 cubic feet per second. Is the flow rate of 30 gallons per minute in violation of the plant regulations? (*Hint:* 1 cubic foot is equivalent to 7.48 gallons.)

Objectives

A Add and Subtract Rational Expressions with Common Denominators.

B Find the Least Common Denominator of a List of Rational Expressions.

C Write a Rational Expression as an Equivalent Expression Whose Denominator Is Given.

5.3 ADDING AND SUBTRACTING RATIONAL EXPRESSIONS WITH THE SAME DENOMINATOR AND LEAST COMMON DENOMINATORS

Objective **A** Adding and Subtracting Rational Expressions with the Same Denominator

Like multiplication and division, addition and subtraction of rational expressions is similar to addition and subtraction of rational numbers. In this section, we add and subtract rational expressions with a common denominator.

Add: $\dfrac{6}{5} + \dfrac{2}{5}$ | Add: $\dfrac{9}{x+2} + \dfrac{3}{x+2}$

Add the numerators and place the sum over the common denominator.

$$\dfrac{6}{5} + \dfrac{2}{5} = \dfrac{6+2}{5}$$
$$= \dfrac{8}{5} \quad \text{Simplify.}$$

$$\dfrac{9}{x+2} + \dfrac{3}{x+2} = \dfrac{9+3}{x+2}$$
$$= \dfrac{12}{x+2} \quad \text{Simplify.}$$

Adding and Subtracting Rational Expressions with Common Denominators

If $\dfrac{P}{R}$ and $\dfrac{Q}{R}$ are rational expressions, then

$$\dfrac{P}{R} + \dfrac{Q}{R} = \dfrac{P+Q}{R} \qquad \text{and} \qquad \dfrac{P}{R} - \dfrac{Q}{R} = \dfrac{P-Q}{R}$$

To add or subtract rational expressions, add or subtract numerators and place the sum or difference over the common denominator.

PRACTICE PROBLEM 1

Add: $\dfrac{8x}{3y} + \dfrac{x}{3y}$

EXAMPLE 1 Add: $\dfrac{5m}{2n} + \dfrac{m}{2n}$

Solution:

$$\dfrac{5m}{2n} + \dfrac{m}{2n} = \dfrac{5m+m}{2n} \quad \text{Add the numerators.}$$

$$= \dfrac{6m}{2n} \quad \text{Simplify the numerator by combining like terms.}$$

$$= \dfrac{3m}{n} \quad \text{Simplify by applying the fundamental principle.}$$

🔲 **Work Practice Problem 1**

PRACTICE PROBLEM 2

Subtract: $\dfrac{3x}{3x-7} - \dfrac{7}{3x-7}$

EXAMPLE 2 Subtract: $\dfrac{2y}{2y-7} - \dfrac{7}{2y-7}$

Solution:

$$\dfrac{2y}{2y-7} - \dfrac{7}{2y-7} = \boxed{\dfrac{2y-7}{2y-7}} \quad \text{Subtract the numerators.}$$

$$= \dfrac{1}{1} \text{ or } 1 \quad \text{Simplify.}$$

🔲 **Work Practice Problem 2**

Answers

1. $\dfrac{3x}{y}$, 2. 1

EXAMPLE 3 Subtract: $\dfrac{3x^2 + 2x}{x - 1} - \dfrac{10x - 5}{x - 1}$

Solution:

$$\dfrac{3x^2 + 2x}{x - 1} - \dfrac{10x - 5}{x - 1} = \dfrac{3x^2 + 2x - (10x - 5)}{x - 1} \qquad \text{Subtract the numerators. Notice the parentheses.}$$

$$= \dfrac{3x^2 + 2x - 10x + 5}{x - 1} \qquad \text{Use the distributive property.}$$

$$= \dfrac{3x^2 - 8x + 5}{x - 1} \qquad \text{Combine like terms.}$$

$$= \dfrac{(x - 1)(3x - 5)}{x - 1} \qquad \text{Factor.}$$

$$= 3x - 5 \qquad \text{Simplify.}$$

◼ **Work Practice Problem 3**

PRACTICE PROBLEM 3

Subtract: $\dfrac{2x^2 + 5x}{x + 2} - \dfrac{4x + 6}{x + 2}$

Helpful Hint

Notice how the numerator $10x - 5$ was subtracted in Example 3.

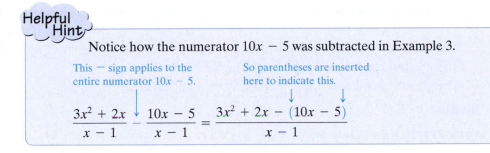

This − sign applies to the entire numerator $10x - 5$.

So parentheses are inserted here to indicate this.

$$\dfrac{3x^2 + 2x}{x - 1} - \dfrac{10x - 5}{x - 1} = \dfrac{3x^2 + 2x - (10x - 5)}{x - 1}$$

Objective B Finding the Least Common Denominator

Recall from Chapter R that to add and subtract fractions with different denominators, we first find a least common denominator (LCD). Then we write all fractions as equivalent fractions with the LCD.

For example, suppose we want to add $\dfrac{3}{8}$ and $\dfrac{1}{6}$. To find the LCD of the denominators, factor 8 and 6. Remember, the LCD is the same as the least common multiple LCM. It is the smallest number that is a multiple of 6 and also 8.

$$8 = 2 \cdot 2 \cdot 2$$
$$6 = 2 \cdot 3$$

The LCM is a multiple of 6.

$$\text{LCM} = 2 \cdot 2 \cdot 2 \cdot 3 = 24$$

The LCM is a multiple of 8.

In the next section, we will continue and find the sum: $\dfrac{3}{8} + \dfrac{1}{6}$, but for now, let's concentrate on the LCD.

To add or subtract rational expressions with different denominators, we also first find an LCD and then write all rational expressions as equivalent expressions with the LCD. The **least common denominator (LCD) of a list of rational expressions** is a polynomial of least degree whose factors include all the factors of the denominators in the list.

To Find the Least Common Denominator (LCD)

Step 1: Factor each denominator completely.

Step 2: The least common denominator (LCD) is the product of all unique factors found in Step 1, each raised to a power equal to the greatest number of times that the factor appears in any one factored denominator.

Answer

3. $2x - 3$

PRACTICE PROBLEM 4

Find the LCD for each pair.

a. $\dfrac{2}{9}, \dfrac{7}{15}$

b. $\dfrac{5}{6x^3}, \dfrac{11}{8x^5}$

EXAMPLE 4 Find the LCD for each pair.

a. $\dfrac{1}{8}, \dfrac{3}{22}$ **b.** $\dfrac{7}{5x}, \dfrac{6}{15x^2}$

Solution:

a. We start by finding the prime factorization of each denominator.

$$8 = 2^3 \quad \text{and}$$
$$22 = 2 \cdot 11$$

Next we write the product of all the unique factors, each raised to a power equal to the greatest number of times that the factor appears.

The greatest number of times that the factor 2 appears is 3.

The greatest number of times that the factor 11 appears is 1.

$$\text{LCD} = 2^3 \cdot 11^1 = 8 \cdot 11 = 88$$

b. We factor each denominator.

$$5x = 5 \cdot x \quad \text{and}$$
$$15x^2 = 3 \cdot 5 \cdot x^2$$

The greatest number of times that the factor 5 appears is 1.

The greatest number of times that the factor 3 appears is 1.

The greatest number of times that the factor x appears is 2.

$$\text{LCD} = 3^1 \cdot 5^1 \cdot x^2 = 15x^2$$

🔲 **Work Practice Problem 4**

PRACTICE PROBLEM 5

Find the LCD of $\dfrac{3a}{a + 5}$ and $\dfrac{7a}{a - 5}$.

EXAMPLE 5 Find the LCD of $\dfrac{7x}{x + 2}$ and $\dfrac{5x^2}{x - 2}$.

Solution: The denominators $x + 2$ and $x - 2$ are completely factored already. The factor $x + 2$ appears once and the factor $x - 2$ appears once.

$$\text{LCD} = (x + 2)(x - 2)$$

🔲 **Work Practice Problem 5**

PRACTICE PROBLEM 6

Find the LCD of $\dfrac{7x^2}{(x - 4)^2}$ and $\dfrac{5x}{3x - 12}$.

EXAMPLE 6 Find the LCD of $\dfrac{6m^2}{3m + 15}$ and $\dfrac{2}{(m + 5)^2}$.

Solution: We factor each denominator.

$$3m + 15 = 3(m + 5)$$
$$(m + 5)^2 = (m + 5)^2 \quad \text{This denominator is already factored.}$$

The greatest number of times that the factor 3 appears is 1.

The greatest number of times that the factor $m + 5$ appears *in any one denominator* is 2.

$$\text{LCD} = 3(m + 5)^2$$

🔲 **Work Practice Problem 6**

✔**Concept Check** Choose the correct LCD of $\dfrac{x}{(x + 1)^2}$ and $\dfrac{5}{x + 1}$.

a. $x + 1$ **b.** $(x + 1)^2$ **c.** $(x + 1)^3$ **d.** $5x(x + 1)^2$

EXAMPLE 7 Find the LCD of $\dfrac{t-10}{2t^2+t-6}$ and $\dfrac{t+5}{t^2+3t+2}$.

Solution:

$$2t^2 + t - 6 = (2t-3)(t+2)$$
$$t^2 + 3t + 2 = (t+1)(t+2)$$
$$\text{LCD} = (2t-3)(t+2)(t+1)$$

🔲 **Work Practice Problem 7**

PRACTICE PROBLEM 7

Find the LCD of $\dfrac{y+5}{y^2+2y-3}$

and $\dfrac{y+4}{y^2-3y+2}$.

EXAMPLE 8 Find the LCD of $\dfrac{2}{x-2}$ and $\dfrac{10}{2-x}$.

Solution: The denominators $x-2$ and $2-x$ are opposites. That is, $2-x = -1(x-2)$. We can use either $x-2$ or $2-x$ as the LCD.

$$\text{LCD} = x-2 \qquad \text{or} \qquad \text{LCD} = 2-x$$

🔲 **Work Practice Problem 8**

PRACTICE PROBLEM 8

Find the LCD of $\dfrac{6}{x-4}$ and

$\dfrac{9}{4-x}$.

Objective C Writing Equivalent Rational Expressions

Next we practice writing a rational expression as an equivalent rational expression with a given denominator. To do this, we multiply by a form of 1. Recall that multiplying an expression by 1 produces an equivalent expression. In other words,

$$\frac{P}{Q} = \frac{P}{Q} \cdot 1 = \frac{P}{Q} \cdot \frac{R}{R} = \frac{PR}{QR}$$

EXAMPLE 9 Write each rational expression as an equivalent rational expression with the given denominator.

a. $\dfrac{4b}{9a} = \dfrac{}{27a^2b}$ **b.** $\dfrac{7x}{2x+5} = \dfrac{}{6x+15}$

Solution:

a. We can ask ourselves: "What do we multiply $9a$ by to get $27a^2b$?" The answer is $3ab$, since $9a(3ab) = 27a^2b$. So we multiply by 1 in the form of $\dfrac{3ab}{3ab}$.

$$\frac{4b}{9a} = \frac{4b}{9a} \cdot 1 = \frac{4b}{9a} \cdot \frac{3ab}{3ab}$$

$$= \frac{4b(3ab)}{9a(3ab)} = \frac{12ab^2}{27a^2b}$$

b. First, factor the denominator on the right.

$$\frac{7x}{2x+5} = \frac{}{3(2x+5)}$$

To obtain the denominator on the right from the denominator on the left, we multiply by 1 in the form of $\dfrac{3}{3}$.

$$\frac{7x}{2x+5} = \frac{7x}{2x+5} \cdot \frac{3}{3} = \frac{7x \cdot 3}{(2x+5) \cdot 3} = \frac{21x}{3(2x+5)}$$

🔲 **Work Practice Problem 9**

PRACTICE PROBLEM 9

Write the rational expression as an equivalent rational expression with the given denominator.

$$\frac{2x}{5y} = \frac{}{20x^2y^2}$$

Answers

7. $(y+3)(y-2)(y-1)$,

8. $x-4$ or $4-x$,

9. $\dfrac{8x^3y}{20x^2y^2}$

PRACTICE PROBLEM 10

Write the rational expression as an equivalent rational expression with the given denominator.

$$\frac{3}{x^2 - 25} = \frac{}{(x + 5)(x - 5)(x - 3)}$$

EXAMPLE 10 Write the rational expression as an equivalent rational expression with the given denominator.

$$\frac{5}{x^2 - 4} = \frac{}{(x - 2)(x + 2)(x - 4)}$$

Solution: First we factor the denominator $x^2 - 4$ as $(x - 2)(x + 2)$. If we multiply the original denominator $(x - 2)(x + 2)$ by $x - 4$, the result is the new denominator $(x - 2)(x + 2)(x - 4)$. Thus, we multiply by 1 in the form of $\frac{x - 4}{x - 4}$.

$$\frac{5}{x^2 - 4} = \frac{5}{(x - 2)(x + 2)} = \frac{5}{(x - 2)(x + 2)} \cdot \frac{x - 4}{x - 4}$$

$$= \frac{5(x - 4)}{(x - 2)(x + 2)(x - 4)}$$

$$= \frac{5x - 20}{(x - 2)(x + 2)(x - 4)}$$

⬛ **Work Practice Problem 10**

Answer

10. $\dfrac{3x - 9}{(x + 5)(x - 5)(x - 3)}$

Mental Math

Perform each indicated operation.

1. $\dfrac{2}{3} + \dfrac{1}{3}$

2. $\dfrac{5}{11} + \dfrac{1}{11}$

3. $\dfrac{3x}{9} + \dfrac{4x}{9}$

4. $\dfrac{3y}{8} + \dfrac{2y}{8}$

5. $\dfrac{8}{9} - \dfrac{7}{9}$

6. $\dfrac{14}{12} - \dfrac{3}{12}$

7. $\dfrac{7y}{5} + \dfrac{10y}{5}$

8. $\dfrac{12x}{7} - \dfrac{4x}{7}$

5.3 EXERCISE SET

Objective **A** *Add or subtract as indicated. Simplify the result if possible. See Examples 1 through 3.*

1. $\dfrac{a}{13} + \dfrac{9}{13}$

2. $\dfrac{x+1}{7} + \dfrac{6}{7}$

3. $\dfrac{4m}{3n} + \dfrac{5m}{3n}$

4. $\dfrac{3p}{2q} + \dfrac{11p}{2q}$

5. $\dfrac{4m}{m-6} - \dfrac{24}{m-6}$

6. $\dfrac{8y}{y-2} - \dfrac{16}{y-2}$

7. $\dfrac{9}{3+y} + \dfrac{y+1}{3+y}$

8. $\dfrac{9}{y+9} + \dfrac{y-5}{y+9}$

9. $\dfrac{5x^2+4x}{x-1} - \dfrac{6x+3}{x-1}$

10. $\dfrac{x^2+9x}{x+7} - \dfrac{4x+14}{x+7}$

11. $\dfrac{4a}{a^2+2a-15} - \dfrac{12}{a^2+2a-15}$

12. $\dfrac{3y}{y^2+3y-10} - \dfrac{6}{y^2+3y-10}$

13. $\dfrac{2x+3}{x^2-x-30} - \dfrac{x-2}{x^2-x-30}$

14. $\dfrac{3x-1}{x^2+5x-6} - \dfrac{2x-7}{x^2+5x-6}$

Objective **B** *Find the LCD for each list of rational expressions. See Examples 4 through 8.*

15. $\dfrac{19}{2x}, \ \dfrac{5}{4x^3}$

16. $\dfrac{17x}{4y^5}, \ \dfrac{2}{8y}$

17. $\dfrac{9}{8x}, \ \dfrac{3}{2x+4}$

18. $\dfrac{1}{6y}, \ \dfrac{3x}{4y+12}$

19. $\dfrac{2}{x+3}, \ \dfrac{5}{x-2}$

20. $\dfrac{-6}{x-1}, \ \dfrac{4}{x+5}$

21. $\dfrac{x}{x+6}, \ \dfrac{10}{3x+18}$

22. $\dfrac{12}{x+5}, \ \dfrac{x}{4x+20}$

23. $\dfrac{8x^2}{(x-6)^2}, \ \dfrac{13x}{5x-30}$

24. $\dfrac{9x^2}{7x-14}, \ \dfrac{6x}{(x-2)^2}$

25. $\dfrac{1}{3x+3}, \dfrac{8}{2x^2+4x+2}$

26. $\dfrac{19x+5}{4x-12}, \ \dfrac{3}{2x^2-12x+18}$

27. $\dfrac{5}{x-8}, \ \dfrac{3}{8-x}$

28. $\dfrac{2x+5}{3x-7}, \dfrac{5}{7-3x}$

29. $\dfrac{5x+1}{x^2+3x-4}, \dfrac{3x}{x^2+2x-3}$

30. $\dfrac{4}{x^2+4x+3}, \dfrac{4x-2}{x^2+10x+21}$

31. $\dfrac{2x}{3x^2+4x+1}, \dfrac{7}{2x^2-x-1}$

32. $\dfrac{3x}{4x^2+5x+1}, \dfrac{5}{3x^2-2x-1}$

33. $\dfrac{1}{x^2-16}, \dfrac{x+6}{2x^3-8x^2}$

34. $\dfrac{5}{x^2-25}, \dfrac{x+9}{3x^3-15x^2}$

Objective **C** *Rewrite each rational expression as an equivalent rational expression with the given denominator. See Examples 9 and 10.*

35. $\dfrac{3}{2x} = \dfrac{}{4x^2}$

36. $\dfrac{3}{9y^5} = \dfrac{}{72y^9}$

37. $\dfrac{6}{3a} = \dfrac{}{12ab^2}$

38. $\dfrac{5}{4y^2x} = \dfrac{}{32y^3x^2}$

39. $\dfrac{9}{2x+6} = \dfrac{}{2y(x+3)}$

40. $\dfrac{4x+1}{3x+6} = \dfrac{}{3y(x+2)}$

41. $\dfrac{9a+2}{5a+10} = \dfrac{}{5b(a+2)}$

42. $\dfrac{5+y}{2x^2+10} = \dfrac{}{4(x^2+5)}$

43. $\dfrac{x}{x^3+6x^2+8x} = \dfrac{}{x(x+4)(x+2)(x+1)}$

44. $\dfrac{5x}{x^3+2x^2-3x} = \dfrac{}{x(x-1)(x-5)(x+3)}$

45. $\dfrac{9y-1}{15x^2-30} = \dfrac{}{30x^2-60}$

46. $\dfrac{6m-5}{3x^2-9} = \dfrac{}{12x^2-36}$

Mixed Practice (*Sections 5.2, 5.3*) *Perform the indicated operations.*

47. $\dfrac{5x}{7} + \dfrac{9x}{7}$

48. $\dfrac{5x}{7} \cdot \dfrac{9x}{7}$

49. $\dfrac{x+3}{4} \div \dfrac{2x-1}{4}$

50. $\dfrac{x+3}{4} - \dfrac{2x-1}{4}$

51. $\dfrac{x^2}{x-6} - \dfrac{5x+6}{x-6}$

52. $\dfrac{x^2+5x}{x^2-25} \cdot \dfrac{3x-15}{x^2}$

53. $\dfrac{-2x}{x^3-8x} + \dfrac{3x}{x^3-8x}$

54. $\dfrac{-2x}{x^3-8x} \div \dfrac{3x}{x^3-8x}$

55. $\dfrac{12x-6}{x^2+3x} \cdot \dfrac{4x^2+13x+3}{4x^2-1}$

56. $\dfrac{x^3+7x^2}{3x^3-x^2} \div \dfrac{5x^2+36x+7}{9x^2-1}$

Review

Perform each indicated operation. See Section R.2.

57. $\dfrac{2}{3} + \dfrac{5}{7}$ **58.** $\dfrac{9}{10} - \dfrac{3}{5}$ **59.** $\dfrac{2}{6} - \dfrac{3}{4}$ **60.** $\dfrac{11}{15} + \dfrac{5}{9}$ **61.** $\dfrac{1}{12} + \dfrac{3}{20}$ **62.** $\dfrac{7}{30} + \dfrac{3}{18}$

Concept Extensions

63. Choose the correct LCD of $\dfrac{11a^3}{4a - 20}$ and $\dfrac{15a^3}{(a - 5)^2}$.
See the Concept Check in this section.
  a. $4a(a - 5)(a + 5)$ **b.** $a - 5$
  c. $(a - 5)^2$ **d.** $4(a - 5)^2$
  e. $(4a - 20)(a - 5)^2$

64. An algebra student approaches you with a problem. He's tried to subtract two rational expressions, but his result does not match the book's. Check to see if the student has made an error. If so, correct his work shown below.

$$\dfrac{2x - 6}{x - 5} - \dfrac{x + 4}{x - 5}$$
$$= \dfrac{2x - 6 - x + 4}{x - 5}$$
$$= \dfrac{x - 2}{x - 5}$$

△ **65.** A square has a side of length $\dfrac{5}{x - 2}$ meters. Express its perimeter as a rational expression.

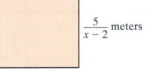

$\dfrac{5}{x - 2}$ meters

△ **66.** A trapezoid has sides of the indicated lengths. Find its perimeter.

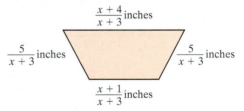

$\dfrac{x + 4}{x + 3}$ inches

$\dfrac{5}{x + 3}$ inches $\dfrac{5}{x + 3}$ inches

$\dfrac{x + 1}{x + 3}$ inches

67. Write two rational expressions with the same denominator whose sum is $\dfrac{5}{3x - 1}$.

68. Write two rational expressions with the same denominator whose difference is $\dfrac{x - 7}{x^2 + 1}$.

69. You are throwing a barbecue and you want to make sure that you purchase the same number of hot dogs as hot dog buns. Hot dogs come 8 to a package and hot dog buns come 12 to a package. What is the least number of each type of package you should buy?

70. The planet Mercury revolves around the sun in 88 Earth days. It takes Jupiter 4332 Earth days to make one revolution around the Sun. (*Source:* National Space Science Data Center) If the two planets are aligned as shown in the figure, how long will it take for them to align again?

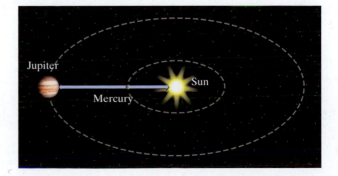

71. Write some instructions to help a friend who is having difficulty finding the LCD of two rational expressions.

72. Explain why the LCD of the rational expressions $\dfrac{7}{x + 1}$ and $\dfrac{9x}{(x + 1)^2}$ is $(x + 1)^2$ and not $(x + 1)^3$.

73. In your own words, describe how to add or subtract two rational expressions with the same denominators.

74. Explain the similarities between subtracting $\dfrac{3}{8}$ from $\dfrac{7}{8}$ and subtracting $\dfrac{6}{x + 3}$ from $\dfrac{9}{x + 3}$.

STUDY SKILLS BUILDER

How Are You Doing?

If you haven't done so yet, take a few moments and think about how you are doing in this course. Are you working toward your goal of successfully completing this course? Is your performance on homework, quizzes, and tests satisfactory? If not, you might want to see your instructor to see if he/she has any suggestions on how you can improve your performance. Reread Section 1.1 for ideas on places to get help with your mathematics course.

Answer the following

1. List any textbook supplements you are using to help you through this course.

2. List any campus resources you are using to help you through this course.

3. Write a short paragraph describing how you are doing in your mathematics course.

4. If improvement is needed, list ways that you can work toward improving your situation as described in Exercise 3.

5.4 ADDING AND SUBTRACTING RATIONAL EXPRESSIONS WITH DIFFERENT DENOMINATORS

Objective **A** Adding and Subtracting Rational Expressions with Different Denominators

Let's add $\frac{3}{8}$ and $\frac{1}{6}$. In the previous section, we found the LCD to be 24. Now let's write equivalent fractions with denominators of 24 by multiplying by different forms of 1.

$$\frac{3}{8} = \frac{3}{8} \cdot 1 = \frac{3}{8} \cdot \frac{3}{3} = \frac{3 \cdot 3}{8 \cdot 3} = \frac{9}{24}.$$

$$\frac{1}{6} = \frac{1}{6} \cdot 1 = \frac{1}{6} \cdot \frac{4}{4} = \frac{1 \cdot 4}{6 \cdot 4} = \frac{4}{24}.$$

Now that the denominators are the same, we may add.

$$\frac{3}{8} + \frac{1}{6} = \frac{9}{24} + \frac{4}{24} = \frac{9 + 4}{24} = \frac{13}{24}$$

We add or subtract rational expressions the same way. You may want to use the steps below.

To Add or Subtract Rational Expressions with Different Denominators

Step 1: Find the LCD of the rational expressions.

Step 2: Rewrite each rational expression as an equivalent expression whose denominator is the LCD found in Step 1.

Step 3: Add or subtract numerators and write the sum or difference over the common denominator.

Step 4: Simplify or write the rational expression in lowest terms.

EXAMPLE 1 Perform each indicated operation.

a. $\dfrac{a}{4} - \dfrac{2a}{8}$

b. $\dfrac{3}{10x^2} + \dfrac{7}{25x}$

Solution:

a. First, we must find the LCD. Since $4 = 2^2$ and $8 = 2^3$, the LCD $= 2^3 = 8$. Next we write each fraction as an equivalent fraction with the denominator 8, and then we subtract.

$$\frac{a}{4} = \frac{a}{4} \cdot 1 = \frac{a}{4} \cdot \frac{2}{2} = \frac{a \cdot 2}{4 \cdot 2} = \frac{2a}{8}$$

$$\frac{a}{4} - \frac{2a}{8} = \frac{a(2)}{4(2)} - \frac{2a}{8} = \frac{2a}{8} - \frac{2a}{8} = \frac{2a - 2a}{8} = \frac{0}{8} = 0$$

Notice that we wrote $\dfrac{a}{4}$ as the equivalent expression $\dfrac{2a}{8}$. Multiplying by a form of 1 means we multiply the numerator and the denominator by the same number. Since this is so, we will start using the shorthand notation on the next page.

Continued on next page

PRACTICE PROBLEM 1

Perform each indicated operation.

a. $\dfrac{y}{5} - \dfrac{3y}{15}$

b. $\dfrac{5}{8x} + \dfrac{11}{10x^2}$

Answers

1. a. 0, **b.** $\dfrac{25x + 44}{40x^2}$

$$\frac{a}{4} = \frac{a(2)}{4(2)} = \frac{2a}{8}$$

Multiplying the numerator and denominator by 2 is the same as multiplying by $\frac{2}{2}$ or 1.

b. Since $10x^2 = 2 \cdot 5 \cdot x \cdot x$ and $25x = 5 \cdot 5 \cdot x$, the LCD $= 2 \cdot 5^2 \cdot x^2 = 50x^2$. We write each fraction as an equivalent fraction with a denominator of $50x^2$.

$$\frac{3}{10x^2} + \frac{7}{25x} = \frac{3(5)}{10x^2(5)} + \frac{7(2x)}{25x(2x)}$$

$$= \frac{15}{50x^2} + \frac{14x}{50x^2}$$

$$= \frac{15 + 14x}{50x^2} \qquad \text{Add numerators. Write the sum over the common denominator.}$$

■ **Work Practice Problem 1**

PRACTICE PROBLEM 2

Subtract: $\dfrac{10x}{x^2 - 9} - \dfrac{5}{x + 3}$

EXAMPLE 2 Subtract: $\dfrac{6x}{x^2 - 4} - \dfrac{3}{x + 2}$

Solution: Since $x^2 - 4 = (x + 2)(x - 2)$, the LCD $= (x + 2)(x - 2)$. We write equivalent expressions with the LCD as denominators.

$$\frac{6x}{x^2 - 4} - \frac{3}{x + 2} = \frac{6x}{(x + 2)(x - 2)} - \frac{3(x - 2)}{(x + 2)(x - 2)}$$

$$= \frac{6x - 3(x - 2)}{(x + 2)(x - 2)} \qquad \text{Subtract numerators. Write the difference over the common denominator.}$$

$$= \frac{6x - 3x + 6}{(x + 2)(x - 2)} \qquad \text{Apply the distributive property in the numerator.}$$

$$= \frac{3x + 6}{(x + 2)(x - 2)} \qquad \text{Combine like terms in the numerator.}$$

Next we factor the numerator to see if this rational expression can be simplified.

$$\frac{3x + 6}{(x + 2)(x - 2)} = \frac{3(x + 2)}{(x + 2)(x - 2)} \qquad \text{Factor.}$$

$$= \frac{3}{x - 2} \qquad \text{Apply the fundamental principle to simplify.}$$

■ **Work Practice Problem 2**

PRACTICE PROBLEM 3

Add: $\dfrac{5}{7x} + \dfrac{2}{x + 1}$

EXAMPLE 3 Add: $\dfrac{2}{3t} + \dfrac{5}{t + 1}$

Solution: The LCD is $3t(t + 1)$. We write each rational expression as an equivalent rational expression with a denominator of $3t(t + 1)$.

$$\frac{2}{3t} + \frac{5}{t + 1} = \frac{2(t + 1)}{3t(t + 1)} + \frac{5(3t)}{(t + 1)(3t)}$$

$$= \frac{2(t + 1) + 5(3t)}{3t(t + 1)} \qquad \text{Add numerators. Write the sum over the common denominator.}$$

$$= \frac{2t + 2 + 15t}{3t(t + 1)} \qquad \text{Apply the distributive property in the numerator.}$$

$$= \frac{17t + 2}{3t(t + 1)} \qquad \text{Combine like terms in the numerator.}$$

Answers

2. $\dfrac{5}{x - 3}$, **3.** $\dfrac{19x + 5}{7x(x + 1)}$

■ **Work Practice Problem 3**

EXAMPLE 4 Subtract: $\dfrac{7}{x-3} - \dfrac{9}{3-x}$

Solution: To find a common denominator, we notice that $x - 3$ and $3 - x$ are opposites. That is, $3 - x = -(x - 3)$. We write the denominator $3 - x$ as $-(x - 3)$ and simplify.

$$\frac{7}{x-3} - \frac{9}{3-x} = \frac{7}{x-3} - \frac{9}{-(x-3)}$$

$$= \frac{7}{x-3} - \frac{-9}{x-3} \qquad \text{Apply } \frac{a}{-b} = \frac{-a}{b}.$$

$$= \frac{7 - (-9)}{x-3} \qquad \text{Subtract numerators. Write the difference over the common denominator.}$$

$$= \frac{16}{x-3}$$

🔲 **Work Practice Problem 4**

EXAMPLE 5 Add: $1 + \dfrac{m}{m+1}$

Solution: Recall that 1 is the same as $\dfrac{1}{1}$. The LCD of $\dfrac{1}{1}$ and $\dfrac{m}{m+1}$ is $m + 1$.

$$1 + \frac{m}{m+1} = \frac{1}{1} + \frac{m}{m+1} \qquad \text{Write 1 as } \frac{1}{1}.$$

$$= \frac{1(m+1)}{1(m+1)} + \frac{m}{m+1} \qquad \text{Multiply both the numerator and the denominator of } \frac{1}{1} \text{ by } m + 1.$$

$$= \frac{m+1+m}{m+1} \qquad \text{Add numerators. Write the sum over the common denominator.}$$

$$= \frac{2m+1}{m+1} \qquad \text{Combine like terms in the numerator.}$$

🔲 **Work Practice Problem 5**

EXAMPLE 6 Subtract: $\dfrac{3}{2x^2 + x} - \dfrac{2x}{6x + 3}$

Solution: First, we factor the denominators.

$$\frac{3}{2x^2 + x} - \frac{2x}{6x + 3} = \frac{3}{x(2x + 1)} - \frac{2x}{3(2x + 1)}$$

The LCD is $3x(2x + 1)$. We write equivalent expressions with denominators of $3x(2x + 1)$.

$$\frac{3}{x(2x + 1)} - \frac{2x}{3(2x + 1)} = \frac{3(3)}{x(2x + 1)(3)} - \frac{2x(x)}{3(2x + 1)(x)}$$

$$= \frac{9 - 2x^2}{3x(2x + 1)} \qquad \text{Subtract numerators. Write the difference over the common denominator.}$$

🔲 **Work Practice Problem 6**

PRACTICE PROBLEM 4

Subtract: $\dfrac{10}{x-6} - \dfrac{15}{6-x}$

PRACTICE PROBLEM 5

Add: $2 + \dfrac{x}{x+5}$

PRACTICE PROBLEM 6

Subtract: $\dfrac{4}{3x^2 + 2x} - \dfrac{3x}{12x + 8}$

Answers

4. $\dfrac{25}{x-6}$, **5.** $\dfrac{3x+10}{x+5}$, **6.** $\dfrac{16-3x^2}{4x(3x+2)}$

PRACTICE PROBLEM 7

Add: $\dfrac{6x}{x^2 + 4x + 4} + \dfrac{x}{x^2 - 4}$

EXAMPLE 7 Add: $\dfrac{2x}{x^2 + 2x + 1} + \dfrac{x}{x^2 - 1}$

Solution: First we factor the denominators.

$$\frac{2x}{x^2 + 2x + 1} + \frac{x}{x^2 - 1}$$

$$= \frac{2x}{(x + 1)(x + 1)} + \frac{x}{(x + 1)(x - 1)}$$

Rewrite each expression with LCD $(x + 1)(x + 1)(x - 1)$.

$$= \frac{2x(x - 1)}{(x + 1)(x + 1)(x - 1)} + \frac{x(x + 1)}{(x + 1)(x - 1)(x + 1)}$$

$$= \frac{2x(x - 1) + x(x + 1)}{(x + 1)^2(x - 1)} \qquad \text{Add numerators. Write the sum over the common denominator.}$$

$$= \frac{2x^2 - 2x + x^2 + x}{(x + 1)^2(x - 1)} \qquad \text{Apply the distributive property in the numerator.}$$

$$= \frac{3x^2 - x}{(x + 1)^2(x - 1)} \quad \text{or} \quad \frac{x(3x - 1)}{(x + 1)^2(x - 1)}$$

The numerator was factored as a last step to see if the rational expression could be simplified further. Since there are no factors common to the numerator and the denominator, we can't simplify further.

■ **Work Practice Problem 7**

Answer

7. $\dfrac{x(7x - 10)}{(x + 2)^2(x - 2)}$

Objective **A** *Perform each indicated operation. Simplify if possible. See Examples 1 through 7.*

1. $\dfrac{4}{2x} + \dfrac{9}{3x}$

2. $\dfrac{15}{7a} + \dfrac{8}{6a}$

3. $\dfrac{15a}{b} + \dfrac{6b}{5}$

4. $\dfrac{4c}{d} - \dfrac{8d}{5}$

5. $\dfrac{3}{x} + \dfrac{5}{2x^2}$

6. $\dfrac{14}{3x^2} + \dfrac{6}{x}$

7. $\dfrac{6}{x+1} + \dfrac{10}{2x+2}$

8. $\dfrac{8}{x+4} - \dfrac{3}{3x+12}$

9. $\dfrac{3}{x+2} - \dfrac{2x}{x^2-4}$

10. $\dfrac{5}{x-4} + \dfrac{4x}{x^2-16}$

11. $\dfrac{3}{4x} + \dfrac{8}{x-2}$

12. $\dfrac{5}{y^2} - \dfrac{y}{2y+1}$

13. $\dfrac{6}{x-3} + \dfrac{8}{3-x}$

14. $\dfrac{15}{y-4} + \dfrac{20}{4-y}$

15. $\dfrac{9}{x-3} + \dfrac{9}{3-x}$

16. $\dfrac{5}{a-7} + \dfrac{5}{7-a}$

17. $\dfrac{-8}{x^2-1} - \dfrac{7}{1-x^2}$

18. $\dfrac{-9}{25x^2-1} + \dfrac{7}{1-25x^2}$

19. $\dfrac{5}{x} + 2$

20. $\dfrac{7}{x^2} - 5x$

21. $\dfrac{5}{x-2} + 6$

22. $\dfrac{6y}{y+5} + 1$

23. $\dfrac{y+2}{y+3} - 2$

24. $\dfrac{7}{2x-3} - 3$

25. $\dfrac{-x+2}{x} - \dfrac{x-6}{4x}$

26. $\dfrac{-y+1}{y} - \dfrac{2y-5}{3y}$

27. $\dfrac{5x}{x+2} - \dfrac{3x-4}{x+2}$

28. $\dfrac{7x}{x-3} - \dfrac{4x+9}{x-3}$

29. $\dfrac{3x^4}{7} - \dfrac{4x^2}{21}$

30. $\dfrac{5x}{6} + \dfrac{11x^2}{2}$

31. $\dfrac{1}{x+3} - \dfrac{1}{(x+3)^2}$

32. $\dfrac{5x}{(x-2)^2} - \dfrac{3}{x-2}$

33. $\dfrac{4}{5b} + \dfrac{1}{b-1}$

34. $\dfrac{1}{y+5} + \dfrac{2}{3y}$

35. $\dfrac{2}{m} + 1$

36. $\dfrac{6}{x} - 1$

37. $\dfrac{2x}{x-7} - \dfrac{x}{x-2}$

38. $\dfrac{9x}{x-10} - \dfrac{x}{x-3}$

39. $\dfrac{6}{1-2x} - \dfrac{4}{2x-1}$

40. $\dfrac{10}{3n-4} - \dfrac{5}{4-3n}$

357

41. $\dfrac{7}{(x+1)(x-1)} + \dfrac{8}{(x+1)^2}$

42. $\dfrac{5}{(x+1)(x+5)} - \dfrac{2}{(x+5)^2}$

43. $\dfrac{x}{x^2-1} - \dfrac{2}{x^2-2x+1}$

44. $\dfrac{x}{x^2-4} - \dfrac{5}{x^2-4x+4}$

45. $\dfrac{3a}{2a+6} - \dfrac{a-1}{a+3}$

46. $\dfrac{1}{x+y} - \dfrac{y}{x^2-y^2}$

47. $\dfrac{y-1}{2y+3} + \dfrac{3}{(2y+3)^2}$

48. $\dfrac{x-6}{5x+1} + \dfrac{6}{(5x+1)^2}$

49. $\dfrac{5}{2-x} + \dfrac{x}{2x-4}$

50. $\dfrac{-1}{a-2} + \dfrac{4}{4-2a}$

51. $\dfrac{15}{x^2+6x+9} + \dfrac{2}{x+3}$

52. $\dfrac{2}{x^2+4x+4} + \dfrac{1}{x+2}$

53. $\dfrac{13}{x^2-5x+6} - \dfrac{5}{x-3}$

54. $\dfrac{-7}{y^2-3y+2} - \dfrac{2}{y-1}$

55. $\dfrac{70}{m^2-100} + \dfrac{7}{2(m+10)}$

56. $\dfrac{27}{y^2-81} + \dfrac{3}{2(y+9)}$

57. $\dfrac{x+8}{x^2-5x-6} + \dfrac{x+1}{x^2-4x-5}$

58. $\dfrac{x+4}{x^2+12x+20} + \dfrac{x+1}{x^2+8x-20}$

59. $\dfrac{5}{4n^2-12n+8} - \dfrac{3}{3n^2-6n}$

60. $\dfrac{6}{5y^2-25y+30} - \dfrac{2}{4y^2-8y}$

Mixed Practice (Sections 5.2, 5.3, 5.4)   *Perform the indicated operations. Addition, subtraction, multiplication, and division of rational expressions are included here.*

61. $\dfrac{15x}{x+8} \cdot \dfrac{2x+16}{3x}$

62. $\dfrac{9z+5}{15} \cdot \dfrac{5z}{81z^2-25}$

63. $\dfrac{8x+7}{3x+5} - \dfrac{2x-3}{3x+5}$

64. $\dfrac{2z^2}{4z-1} - \dfrac{z-2z^2}{4z-1}$

65. $\dfrac{5a+10}{18} \div \dfrac{a^2-4}{10a}$

66. $\dfrac{9}{x^2-1} \div \dfrac{12}{3x+3}$

67. $\dfrac{5}{x^2-3x+2} + \dfrac{1}{x-2}$

68. $\dfrac{4}{2x^2+5x-3} + \dfrac{2}{x+3}$

Review

Solve each linear or quadratic equation. See Sections 2.3 and 4.5.

69. $3x+5=7$

70. $5x-1=8$

71. $2x^2-x-1=0$

72. $4x^2 - 9 = 0$

73. $4(x + 6) + 3 = -3$

74. $2(3x + 1) + 15 = -7$

Concept Extensions

Perform each indicated operation.

75. $\dfrac{3}{x} - \dfrac{2x}{x^2 - 1} + \dfrac{5}{x + 1}$

76. $\dfrac{5}{x - 2} + \dfrac{7x}{x^2 - 4} - \dfrac{11}{x}$

77. $\dfrac{5}{x^2 - 4} + \dfrac{2}{x^2 - 4x + 4} - \dfrac{3}{x^2 - x - 6}$

78. $\dfrac{8}{x^2 + 6x + 5} - \dfrac{3x}{x^2 + 4x - 5} + \dfrac{2}{x^2 - 1}$

79. $\dfrac{9}{x^2 + 9x + 14} - \dfrac{3x}{x^2 + 10x + 21} + \dfrac{x + 4}{x^2 + 5x + 6}$

80. $\dfrac{x + 10}{x^2 - 3x - 4} - \dfrac{8}{x^2 + 6x + 5} - \dfrac{9}{x^2 + x - 20}$

81. A board of length $\dfrac{3}{x + 4}$ inches was cut into two pieces. If one piece is $\dfrac{1}{x - 4}$ inches, express the length of the other board as a rational expression.

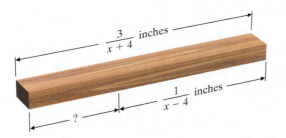

△ 82. The length of a rectangle is $\dfrac{3}{y - 5}$ feet, while its width is $\dfrac{2}{y}$ feet. Find its perimeter and then find its area.

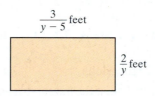

83. In ice hockey, penalty killing percentage is a statistic calculated as $1 - \dfrac{G}{P}$, where $G =$ opponent's power play goals and $P =$ opponent's power play opportunities. Simplify this expression.

84. The dose of medicine prescribed for a child depends on the child's age A in years and the adult dose D for the medication. Two expressions that give a child's dose are Young's Rule, $\dfrac{DA}{A + 12}$, and Cowling's Rule, $\dfrac{D(A + 1)}{24}$. Find an expression for the difference in the doses given by these expressions.

85. Explain when the LCD of the rational expressions in a sum is the product of the denominators.

86. Explain when the LCD is the same as one of the denominators of a rational expression to be added or subtracted.

△ **87.** Two angles are said to be complementary if the sum of their measures is 90°. If one angle measures $\dfrac{40}{x}$ degrees, find the measure of its complement.

△ **88.** Two angles are said to be supplementary if the sum of their measures is 180°. If one angle measures $\dfrac{x + 2}{x}$ degrees, find the measure of its supplement.

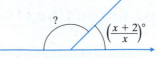

89. In your own words, explain how to add two rational expressions with different denominators.

90. In your own words, explain how to subtract two rational expressions with different denominators.

THE BIGGER PICTURE   Simplifying Expressions and Solving Equations

Now we continue our outline from Sections 1.6, 2.7, 3.7, and 4.6. Although suggestions are given, this outline should be in your own words. Once you complete this new portion, try the exercises below.

I. Simplifying Expressions

  A. Real Numbers

    1. Add (Section 1.4)

    2. Subtract (Section 1.5)

    3. Multiply or Divide (Section 1.6)

  B. Exponents (Section 3.2)

  C. Polynomials

    1. Add (Section 3.4)

    2. Subtract (Section 3.4)

    3. Multiply (Section 3.5)

    4. Divide (Section 3.7)

  D. Factoring Polynomials (Chapter 4 Integrated Review)

  E. Rational Expressions

    1. Simplify: Factor the numerator and denominator. Then divide out factors of 1 by dividing out common factors in the numerator and denominator.

$$\frac{x^2 - 9}{7x^2 - 21x} = \frac{(x + 3)(x - 3)}{7x(x - 3)} = \frac{x + 3}{7x}$$

    2. Multiply: Multiply numerators, then multiply denominators.

$$\frac{5z}{2z^2 - 9z - 18} \cdot \frac{22z + 33}{10z}$$
$$= \frac{5 \cdot z}{(2z + 3)(z - 6)} \cdot \frac{11(2z + 3)}{2 \cdot 5 \cdot z} = \frac{11}{2(z - 6)}$$

    3. Divide: First fraction times the reciprocal of the second fraction.

$$\frac{14}{x + 5} \div \frac{x + 1}{2} = \frac{14}{x + 5} \cdot \frac{2}{x + 1}$$
$$= \frac{28}{(x + 5)(x + 1)}$$

    4. Add or Subtract: Must have same denominator. If not find the LCD and write each fraction as an equivalent fraction with the LCD as denominator.

$$\frac{9}{10} - \frac{x + 1}{x + 5} = \frac{9(x + 5)}{10(x + 5)} - \frac{10(x + 1)}{10(x + 5)}$$
$$= \frac{9x + 45 - 10x - 10}{10(x + 5)}$$
$$= \frac{-x + 35}{10(x + 5)}$$

II. Solving Equations and Inequalities

  A. Linear Equations (Section 2.3)

  B. Linear Inequalities (Section 2.7)

  C. Quadratic & Higher Degree Equations (Section 4.6)

Perform indicated operations and simplify.

1. $-8.6 + (-9.1)$

2. $(-8.6)(-9.1)$

3. $14 - (-14)$

4. $3x^4 - 7 + x^4 - x^2 - 10$

5. $\dfrac{5x^2 - 5}{25x + 25}$

6. $\dfrac{7x}{x^2 + 4x + 3} \div \dfrac{x}{2x + 6}$

7. $\dfrac{2}{9} - \dfrac{5}{6}$

8. $\dfrac{x}{9} - \dfrac{x + 3}{5}$

Factor.

9. $9x^3 - 2x^2 - 11x$

10. $12xy - 21x + 4y - 7$

Solve.

11. $7x - 14 = 5x + 10$

12. $\dfrac{-x + 2}{5} < \dfrac{3}{10}$

13. $1 + 4(x + 4) = 3^2 + x$

14. $x(x - 2) = 24$

5.5 SOLVING EQUATIONS CONTAINING RATIONAL EXPRESSIONS

Objectives

A Solve Equations Containing Rational Expressions.

B Solve Equations Containing Rational Expressions for a Specified Variable.

Objective **A** Solving Equations Containing Rational Expressions

In Chapter 2, we solved equations containing fractions. In this section, we continue the work we began in Chapter 2 by solving equations containing rational expressions. For example,

$$\frac{x}{2} + \frac{8}{3} = \frac{1}{6} \quad \text{and} \quad \frac{4x}{x^2 + x - 30} + \frac{2}{x - 5} = \frac{1}{x + 6}$$

are equations containing rational expressions. To solve equations such as these, we use the multiplication property of equality to clear the equation of fractions by multiplying both sides of the equation by the LCD.

EXAMPLE 1 Solve: $\frac{x}{2} + \frac{8}{3} = \frac{1}{6}$

Solution: The LCD of denominators 2, 3, and 6 is 6, so we multiply both sides of the equation by 6.

$$6\left(\frac{x}{2} + \frac{8}{3}\right) = 6\left(\frac{1}{6}\right)$$

$$6\left(\frac{x}{2}\right) + 6\left(\frac{8}{3}\right) = 6\left(\frac{1}{6}\right) \quad \text{Apply the distributive property.}$$

$$3 \cdot x + 16 = 1 \qquad \text{Multiply and simplify.}$$

$$3x = -15 \qquad \text{Subtract 16 from both sides.}$$

$$x = -5 \qquad \text{Divide both sides by 3.}$$

Check: To check, we replace x with -5 in the original equation.

$$\frac{-5}{2} + \frac{8}{3} \stackrel{?}{=} \frac{1}{6} \qquad \text{Replace } x \text{ with } -5.$$

$$\frac{1}{6} = \frac{1}{6} \qquad \text{True}$$

This number checks, so the solution is -5.

▢ Work Practice Problem 1

EXAMPLE 2 Solve: $\frac{t - 4}{2} - \frac{t - 3}{9} = \frac{5}{18}$

Solution: The LCD of denominators 2, 9, and 18 is 18, so we multiply both sides of the equation by 18.

$$18\left(\frac{t - 4}{2} - \frac{t - 3}{9}\right) = 18\left(\frac{5}{18}\right)$$

$$18\left(\frac{t - 4}{2}\right) - 18\left(\frac{t - 3}{9}\right) = 18\left(\frac{5}{18}\right) \quad \text{Apply the distributive property.}$$

$$9(t - 4) - 2(t - 3) = 5 \qquad \text{Simplify.}$$

$$9t - 36 - 2t + 6 = 5 \qquad \text{Use the distributive property.}$$

$$7t - 30 = 5 \qquad \text{Combine like terms.}$$

$$7t = 35$$

$$t = 5 \qquad \text{Solve for } t.$$

PRACTICE PROBLEM 1

Solve: $\frac{x}{4} + \frac{4}{5} = \frac{1}{20}$

Helpful Hint

Make sure that *each* term is multiplied by the LCD.

PRACTICE PROBLEM 2

Solve: $\frac{x + 2}{3} - \frac{x - 1}{5} = \frac{1}{15}$

Helpful Hint

Multiply *each* term by 18.

Answers

1. $x = -3$, **2.** $x = -6$

Check: $\dfrac{t-4}{2} - \dfrac{t-3}{9} = \dfrac{5}{18}$

$$\dfrac{5-4}{2} - \dfrac{5-3}{9} \overset{?}{=} \dfrac{5}{18} \qquad \text{Replace } t \text{ with 5.}$$

$$\dfrac{1}{2} - \dfrac{2}{9} \overset{?}{=} \dfrac{5}{18} \qquad \text{Simplify.}$$

$$\dfrac{5}{18} = \dfrac{5}{18} \qquad \text{True}$$

The solution is 5.

🔲 **Work Practice Problem 2**

Recall from Section 5.1 that a rational expression is defined for all real numbers except those that make the denominator of the expression 0. This means that if an equation contains *rational expressions with variables in the denominator*, we must be certain that the proposed solution does not make the denominator 0. If replacing the variable with the proposed solution makes the denominator 0, the rational expression is undefined and this proposed solution must be rejected.

PRACTICE PROBLEM 3

Solve: $2 + \dfrac{6}{x} = x + 7$

EXAMPLE 3 Solve: $3 - \dfrac{6}{x} = x + 8$

Solution: In this equation, 0 cannot be a solution because if x is 0, the rational expression $\dfrac{6}{x}$ is undefined. The LCD is x, so we multiply both sides of the equation by x.

$$x\left(3 - \dfrac{6}{x}\right) = x(x + 8)$$

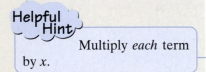

Helpful Hint

Multiply *each* term by x.

$$x(3) - x\left(\dfrac{6}{x}\right) = x \cdot x + x \cdot 8 \qquad \text{Apply the distributive property.}$$

$$3x - 6 = x^2 + 8x \qquad \text{Simplify.}$$

Now we write the quadratic equation in standard form and solve for x.

$$0 = x^2 + 5x + 6$$

$$0 = (x + 3)(x + 2) \qquad \text{Factor.}$$

$$x + 3 = 0 \quad \text{or} \quad x + 2 = 0 \qquad \text{Set each factor equal to 0 and solve.}$$

$$x = -3 \qquad\qquad x = -2$$

Notice that neither -3 nor -2 makes the denominator in the original equation equal to 0.

Check: To check these solutions, we replace x in the original equation by -3, and then by -2.

If $x = -3$:

$$3 - \dfrac{6}{x} = x + 8$$

$$3 - \dfrac{6}{-3} \overset{?}{=} -3 + 8$$

$$3 - (-2) \overset{?}{=} 5$$

$$5 = 5 \qquad \text{True}$$

If $x = -2$:

$$3 - \dfrac{6}{x} = x + 8$$

$$3 - \dfrac{6}{-2} \overset{?}{=} -2 + 8$$

$$3 - (-3) \overset{?}{=} 6$$

$$6 = 6 \qquad \text{True}$$

Both -3 and -2 are solutions.

🔲 **Work Practice Problem 3**

The following steps may be used to solve an equation containing rational expressions.

Answer

3. $x = -6, x = 1$

To Solve an Equation Containing Rational Expressions

Step 1: Multiply both sides of the equation by the LCD of all rational expressions in the equation.

Step 2: Remove any grouping symbols and solve the resulting equation.

Step 3: Check the solution in the original equation.

EXAMPLE 4 Solve: $\dfrac{4x}{x^2 + x - 30} + \dfrac{2}{x - 5} = \dfrac{1}{x + 6}$

Solution: The denominator $x^2 + x - 30$ factors as $(x + 6)(x - 5)$. The LCD is then $(x + 6)(x - 5)$, so we multiply both sides of the equation by this LCD.

$$(x + 6)(x - 5)\left(\dfrac{4x}{x^2 + x - 30} + \dfrac{2}{x - 5}\right) = (x + 6)(x - 5)\left(\dfrac{1}{x + 6}\right) \quad \text{Multiply by the LCD.}$$

$$(x + 6)(x - 5) \cdot \dfrac{4x}{x^2 + x - 30} + (x + 6)(x - 5) \cdot \dfrac{2}{x - 5} \quad \text{Apply the distributive property.}$$

$$= (x + 6)(x - 5) \cdot \dfrac{1}{x + 6}$$

$$4x + 2(x + 6) = x - 5 \quad \text{Simplify.}$$

$$4x + 2x + 12 = x - 5 \quad \text{Apply the distributive property.}$$

$$6x + 12 = x - 5 \quad \text{Combine like terms.}$$

$$5x = -17$$

$$x = -\dfrac{17}{5} \quad \text{Divide both sides by 5.}$$

Check: Check by replacing x with $-\dfrac{17}{5}$ in the original equation. The solution is $-\dfrac{17}{5}$.

🔲 **Work Practice Problem 4**

EXAMPLE 5 Solve: $\dfrac{2x}{x - 4} = \dfrac{8}{x - 4} + 1$

Solution: Multiply both sides by the LCD, $x - 4$.

$$(x - 4)\left(\dfrac{2x}{x - 4}\right) = (x - 4)\left(\dfrac{8}{x - 4} + 1\right) \quad \text{Multiply by the LCD.}$$

$$(x - 4) \cdot \dfrac{2x}{x - 4} = (x - 4) \cdot \dfrac{8}{x - 4} + (x - 4) \cdot 1 \quad \text{Use the distributive property.}$$

$$2x = 8 + (x - 4) \quad \text{Simplify.}$$

$$2x = 4 + x$$

$$x = 4$$

Notice that 4 makes the denominator 0 in the original equation. Therefore, 4 is *not* a solution and this equation has *no solution*.

🔲 **Work Practice Problem 5**

✔**Concept Check** When can we clear fractions by multiplying through by the LCD?

a. When adding or subtracting rational expressions

b. When solving an equation containing rational expressions

c. Both of these

d. Neither of these

PRACTICE PROBLEM 4

Solve:

$$\dfrac{2}{x + 3} + \dfrac{3}{x - 3} = \dfrac{-2}{x^2 - 9}$$

PRACTICE PROBLEM 5

Solve: $\dfrac{5x}{x - 1} = \dfrac{5}{x - 1} + 3$

Answers

4. $x = -1$, **5.** No solution

✔ **Concept Check Answer**

b

> **Helpful Hint**
>
> As we can see from Example 5, it is important to check the proposed solution(s) in the original equation.

PRACTICE PROBLEM 6

Solve:

$$x - \frac{6}{x + 3} = \frac{2x}{x + 3} + 2$$

EXAMPLE 6 Solve: $x + \dfrac{14}{x - 2} = \dfrac{7x}{x - 2} + 1$

Solution: Notice the denominators in this equation. We can see that 2 can't be a solution. The LCD is $x - 2$, so we multiply both sides of the equation by $x - 2$.

$$(x - 2)\left(x + \frac{14}{x - 2}\right) = (x - 2)\left(\frac{7x}{x - 2} + 1\right)$$

$$(x - 2)(x) + (x - 2)\left(\frac{14}{x - 2}\right) = (x - 2)\left(\frac{7x}{x - 2}\right) + (x - 2)(1)$$

$x^2 - 2x + 14 = 7x + x - 2$	Simplify.
$x^2 - 2x + 14 = 8x - 2$	Combine like terms.
$x^2 - 10x + 16 = 0$	Write the quadratic equation in standard form.
$(x - 8)(x - 2) = 0$	Factor.
$x - 8 = 0 \quad \text{or} \quad x - 2 = 0$	Set each factor equal to 0.
$x = 8 \qquad\qquad x = 2$	Solve.

As we have already noted, 2 can't be a solution of the original equation. So we need only replace x with 8 in the original equation. We find that 8 is a solution; the only solution is 8.

◻ **Work Practice Problem 6**

Objective B Solving Equations for a Specified Variable

The last example in this section is an equation containing several variables, and we are directed to solve for one of the variables. The steps used in the preceding examples can be applied to solve equations for a specified variable as well.

PRACTICE PROBLEM 7

Solve $\dfrac{1}{a} + \dfrac{1}{b} = \dfrac{1}{x}$ for a.

EXAMPLE 7 Solve $\dfrac{1}{a} + \dfrac{1}{b} = \dfrac{1}{x}$ for x.

Solution: (This type of equation often models a work problem, as we shall see in the next section.) The LCD is abx, so we multiply both sides by abx.

$$abx\left(\frac{1}{a} + \frac{1}{b}\right) = abx\left(\frac{1}{x}\right)$$

$$abx\left(\frac{1}{a}\right) + abx\left(\frac{1}{b}\right) = abx \cdot \frac{1}{x}$$

$bx + ax = ab$	Simplify.
$x(b + a) = ab$	Factor out x from each term on the left side.
$\dfrac{x(b + a)}{b + a} = \dfrac{ab}{b + a}$	Divide both sides by $b + a$.
$x = \dfrac{ab}{b + a}$	Simplify.

This equation is now solved for x.

◻ **Work Practice Problem 7**

Answers

6. $x = 4$, 7. $a = \dfrac{bx}{b - x}$

Mental Math

Solve each equation for the variable.

1. $\dfrac{x}{5} = 2$

2. $\dfrac{x}{8} = 4$

3. $\dfrac{z}{6} = 6$

4. $\dfrac{y}{7} = 8$

FOR EXTRA HELP

Student Solutions Manual   PH Math/Tutor Center   CD/Video for Review   Math XL MathXL®   MyMathLab

5.5 EXERCISE SET

Objective Ⓐ *Solve each equation and check each solution. See Examples 1 through 3.*

1. $\dfrac{x}{5} + 3 = 9$

2. $\dfrac{x}{5} - 2 = 9$

3. $\dfrac{x}{2} + \dfrac{5x}{4} = \dfrac{x}{12}$

4. $\dfrac{x}{6} + \dfrac{4x}{3} = \dfrac{x}{18}$

5. $2 - \dfrac{8}{x} = 6$

6. $5 + \dfrac{4}{x} = 1$

7. $2 + \dfrac{10}{x} = x + 5$

8. $6 + \dfrac{5}{y} = y - \dfrac{2}{y}$

9. $\dfrac{a}{5} = \dfrac{a - 3}{2}$

10. $\dfrac{b}{5} = \dfrac{b + 2}{6}$

⊙ 11. $\dfrac{x - 3}{5} + \dfrac{x - 2}{2} = \dfrac{1}{2}$

12. $\dfrac{a + 5}{4} + \dfrac{a + 5}{2} = \dfrac{a}{8}$

Solve each equation and check each proposed solution. See Examples 4 through 6.

13. $\dfrac{3}{2a - 5} = -1$

14. $\dfrac{6}{4 - 3x} = -3$

15. $\dfrac{4y}{y - 4} + 5 = \dfrac{5y}{y - 4}$

16. $\dfrac{2a}{a + 2} - 5 = \dfrac{7a}{a + 2}$

⊙ 17. $2 + \dfrac{3}{a - 3} = \dfrac{a}{a - 3}$

18. $\dfrac{2y}{y - 2} - \dfrac{4}{y - 2} = 4$

19. $\dfrac{1}{x + 3} + \dfrac{6}{x^2 - 9} = 1$

20. $\dfrac{1}{x + 2} + \dfrac{4}{x^2 - 4} = 1$

21. $\dfrac{2y}{y + 4} + \dfrac{4}{y + 4} = 3$

22. $\dfrac{5y}{y + 1} - \dfrac{3}{y + 1} = 4$

23. $\dfrac{2x}{x + 2} - 2 = \dfrac{x - 8}{x - 2}$

24. $\dfrac{4y}{y - 3} - 3 = \dfrac{3y - 1}{y + 3}$

Solve each equation. See Examples 1 through 6.

⊙ 25. $\dfrac{2}{y} + \dfrac{1}{2} = \dfrac{5}{2y}$

26. $\dfrac{6}{3y} + \dfrac{3}{y} = 1$

27. $\dfrac{a}{a - 6} = \dfrac{-2}{a - 1}$

28. $\dfrac{5}{x - 6} = \dfrac{x}{x - 2}$

29. $\dfrac{11}{2x} + \dfrac{2}{3} = \dfrac{7}{2x}$

30. $\dfrac{5}{3} - \dfrac{3}{2x} = \dfrac{3}{2}$

⊙ 31. $\dfrac{2}{x - 2} + 1 = \dfrac{x}{x + 2}$

32. $1 + \dfrac{3}{x + 1} = \dfrac{x}{x - 1}$

33. $\dfrac{x + 1}{3} - \dfrac{x - 1}{6} = \dfrac{1}{6}$

34. $\dfrac{3x}{5} - \dfrac{x - 6}{3} = -\dfrac{2}{5}$

⊙ 35. $\dfrac{t}{t - 4} = \dfrac{t + 4}{6}$

36. $\dfrac{15}{x + 4} = \dfrac{x - 4}{x}$

37. $\dfrac{y}{2y + 2} + \dfrac{2y - 16}{4y + 4} = \dfrac{2y - 3}{y + 1}$

38. $\dfrac{1}{x + 2} = \dfrac{4}{x^2 - 4} - \dfrac{1}{x - 2}$

39. $\dfrac{4r - 4}{r^2 + 5r - 14} + \dfrac{2}{r + 7} = \dfrac{1}{r - 2}$

40. $\dfrac{3}{x + 3} = \dfrac{12x + 19}{x^2 + 7x + 12} - \dfrac{5}{x + 4}$

41. $\dfrac{x + 1}{x + 3} = \dfrac{x^2 - 11x}{x^2 + x - 6} - \dfrac{x - 3}{x - 2}$

42. $\dfrac{2t + 3}{t - 1} - \dfrac{2}{t + 3} = \dfrac{5 - 6t}{t^2 + 2t - 3}$

Objective B *Solve each equation for the indicated variable. See Example 7.*

43. $R = \dfrac{E}{I}$ for I (Electronics: resistance of a circuit)

44. $T = \dfrac{V}{Q}$ for Q (Water purification: settling time)

45. $T = \dfrac{2U}{B + E}$ for B (Merchandising: stock turnover rate)

46. $i = \dfrac{A}{t + B}$ for t (Hydrology: rainfall intensity)

47. $B = \dfrac{705w}{h^2}$ for w (Health: body-mass index)

△ **48.** $\dfrac{A}{W} = L$ for W (Geometry: area of a rectangle)

49. $N = R + \dfrac{V}{G}$ for G (Urban forestry: tree plantings per year)

50. $C = \dfrac{D(A + 1)}{24}$ for A (Medicine: Cowling's Rule for child's dose)

△ **51.** $\dfrac{C}{\pi r} = 2$ for r (Geometry: circumference of a circle)

52. $W = \dfrac{CE^2}{2}$ for C (Electronics: energy stored in a capacitor)

53. $\dfrac{1}{y} + \dfrac{1}{3} = \dfrac{1}{x}$ for x

54. $\dfrac{1}{5} + \dfrac{2}{y} = \dfrac{1}{x}$ for x

Review

Write each phrase as an expression.

55. The reciprocal of x

56. The reciprocal of $x + 1$

57. The reciprocal of x, added to the reciprocal of 2

58. The reciprocal of x, subtracted from the reciprocal of 5

Answer each question.

59. If a tank is filled in 3 hours, what part of the tank is filled in 1 hour?

60. If a strip of beach is cleaned in 4 hours, what part of the beach is cleaned in 1 hour?

Concept Extensions

Solve each equation.

61. $\dfrac{4}{a^2 + 4a + 3} + \dfrac{2}{a^2 + a - 6} - \dfrac{3}{a^2 - a - 2} = 0$

62. $\dfrac{-4}{a^2 + 2a - 8} + \dfrac{1}{a^2 + 9a + 20} = \dfrac{-4}{a^2 + 3a - 10}$

Recall that two angles are supplementary if the sum of their measures is 180°. Find the measures of the supplementary angles.

△ **63.**

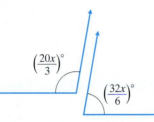

△ **64.**

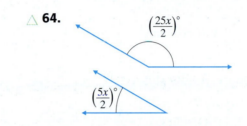

Recall that two angles are complementary if the sum of their measures is 90°. Find the measures of the complementary angles.

△ **65.**

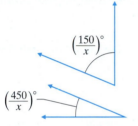

△ **66.**

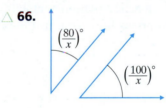

67. When adding the expressions in $\dfrac{3x}{2} + \dfrac{x}{4}$, can you multiply each term by 4? Why or why not?

68. When solving the equation $\dfrac{3x}{2} + \dfrac{x}{4} = 1$, can you multiply both sides of the equation by 4? Why or why not?

Summary on Rational Expressions

It is important to know the difference between performing operations with rational expressions and solving an equation containing rational expressions. Study the examples below.

Performing Operations with Rational Expressions

Adding: $\dfrac{1}{x} + \dfrac{1}{x+5} = \dfrac{1 \cdot (x+5)}{x(x+5)} + \dfrac{1 \cdot x}{x(x+5)} = \dfrac{x+5+x}{x(x+5)} = \dfrac{2x+5}{x(x+5)}$

Subtracting: $\dfrac{3}{x} - \dfrac{5}{x^2 y} = \dfrac{3 \cdot xy}{x \cdot xy} - \dfrac{5}{x^2 y} = \dfrac{3xy - 5}{x^2 y}$

Multiplying: $\dfrac{2}{x} \cdot \dfrac{5}{x-1} = \dfrac{2 \cdot 5}{x(x-1)} = \dfrac{10}{x(x-1)}$

Dividing: $\dfrac{4}{2x+1} \div \dfrac{x-3}{x} = \dfrac{4}{2x+1} \cdot \dfrac{x}{x-3} = \dfrac{4x}{(2x+1)(x-3)}$

Solving an Equation Containing Rational Expressions

To solve an equation containing rational expressions, we clear the equation of fractions by multiplying both sides by the LCD.

$$\dfrac{3}{x} - \dfrac{5}{x-1} = \dfrac{1}{x(x-1)} \qquad \text{Note that } x \text{ can't be 0 or 1.}$$

$$x(x-1)\left(\dfrac{3}{x}\right) - x(x-1)\left(\dfrac{5}{x-1}\right) = x(x-1) \cdot \dfrac{1}{x(x-1)} \qquad \text{Multiply both sides by the LCD.}$$

$$3(x-1) - 5x = 1 \qquad \text{Simplify.}$$

$$3x - 3 - 5x = 1 \qquad \text{Use the distributive property.}$$

$$-2x - 3 = 1 \qquad \text{Combine like terms.}$$

$$-2x = 4 \qquad \text{Add 3 to both sides.}$$

$$x = -2 \qquad \text{Divide both sides by } -2.$$

Don't forget to check to make sure our proposed solution of -2 does not make any denominators 0. If it does, this proposed solution is *not* a solution of the equation. -2 checks and is the solution.

Determine whether each of the following is an equation or an expression. If it is an equation, solve it for its variable. If it is an expression, perform the indicated operation.

1. $\dfrac{1}{x} + \dfrac{2}{3}$ **2.** $\dfrac{3}{a} + \dfrac{5}{6}$ **3.** $\dfrac{1}{x} + \dfrac{2}{3} = \dfrac{3}{x}$

4. $\dfrac{3}{a} + \dfrac{5}{6} = 1$ **5.** $\dfrac{2}{x+1} - \dfrac{1}{x}$ **6.** $\dfrac{4}{x-3} - \dfrac{1}{x}$

7. $\dfrac{2}{x+1} - \dfrac{1}{x} = 1$ **8.** $\dfrac{4}{x-3} - \dfrac{1}{x} = \dfrac{6}{x(x-3)}$

9. $\dfrac{15x}{x+8} \cdot \dfrac{2x+16}{3x}$ **10.** $\dfrac{9z+5}{15} \cdot \dfrac{5z}{81z^2 - 25}$

11. $\dfrac{2x + 1}{x - 3} + \dfrac{3x + 6}{x - 3}$

12. $\dfrac{4p - 3}{2p + 7} + \dfrac{3p + 8}{2p + 7}$

13. $\dfrac{x + 5}{7} = \dfrac{8}{2}$

14. $\dfrac{1}{2} = \dfrac{x + 1}{8}$

15. $\dfrac{5a + 10}{18} \div \dfrac{a^2 - 4}{10a}$

16. $\dfrac{9}{x^2 - 1} \div \dfrac{12}{3x + 3}$

17. $\dfrac{x + 2}{3x - 1} + \dfrac{5}{(3x - 1)^2}$

18. $\dfrac{4}{(2x - 5)^2} + \dfrac{x + 1}{2x - 5}$

19. $\dfrac{x - 7}{x} - \dfrac{x + 2}{5x}$

20. $\dfrac{9}{x^2 - 4} + \dfrac{2}{x + 2} = \dfrac{-1}{x - 2}$

21. $\dfrac{3}{x + 3} = \dfrac{5}{x^2 - 9} - \dfrac{2}{x - 3}$

22. $\dfrac{10x - 9}{x} - \dfrac{x - 4}{3x}$

23. Explain the difference between solving an equation such as $\dfrac{x}{2} + \dfrac{3}{4} = \dfrac{x}{4}$ for x and performing an operation such as adding $\dfrac{x}{2} + \dfrac{3}{4}$.

24. When solving an equation such as $\dfrac{y}{4} = \dfrac{y}{2} - \dfrac{1}{4}$, we may multiply all terms by 4. When subtracting two rational expressions such as $\dfrac{y}{2} - \dfrac{1}{4}$, we may not. Explain why.

11. _____

12. _____

13. _____

14. _____

15. _____

16. _____

17. _____

18. _____

19. _____

20. _____

21. _____

22. _____

23. _____

24. _____

A Solve Proportions.

B Use Proportions to Solve Problems, Including Similar Triangle Problems.

C Solve Problems about Numbers.

D Solve Problems about Work.

E Solve Problems about Distance.

5.6 PROPORTIONS AND PROBLEM SOLVING WITH RATIONAL EQUATIONS

Objective A Solving Proportions

A **ratio** is the quotient of two numbers or two quantities. For example, the ratio of 2 to 5 can be written in fraction form as $\frac{2}{5}$, the quotient of 2 and 5.

A **rate** is a special type of ratio with different kinds of measurement. For example, the ratio "110 miles in 2 hours" written as a fraction in simplest form is $\frac{110 \text{ miles}}{2 \text{ hours}} = \frac{55 \text{ mi}}{1 \text{ hr}}$ or 55 mph.

If two ratios are equal, we say the ratios are **in proportion** to each other. A **proportion** is a mathematical statement that two ratios are equal.

For example, the equation $\frac{1}{2} = \frac{4}{8}$ is a proportion, as is $\frac{x}{5} = \frac{8}{10}$, because both sides of the equations are ratios. When we want to emphasize the equation as a proportion, we

> **read the proportion $\frac{1}{2} = \frac{4}{8}$ as "one is to two as four is to eight"**

In a proportion, cross products are equal. To understand cross products, let's start with the proportion

$$\frac{a}{b} = \frac{c}{d}$$

and multiply both sides by the LCD, bd.

$$bd\left(\frac{a}{b}\right) = bd\left(\frac{c}{d}\right) \quad \text{Multiply both sides by the LCD, } bd.$$
$$\underbrace{ad} = \underbrace{bc} \quad \text{Simplify.}$$
$$\text{Cross product} \quad \text{Cross product}$$

Notice why ad and bc are called cross products.

$$ad \qquad\qquad\qquad bc$$
$$\frac{a}{b} = \frac{c}{d}$$

Cross Products

If $\frac{a}{b} = \frac{c}{d}$, then $ad = bc$.

For example, if

$$\frac{1}{2} = \frac{4}{8}, \quad \text{then} \quad \begin{array}{c} 1 \cdot 8 = 2 \cdot 4 \\ 8 = 8 \end{array} \quad \text{or}$$

Notice that a proportion contains four numbers (or expressions). If any three numbers are known, we can solve and find the fourth number.

EXAMPLE 1 Solve for x: $\dfrac{45}{x} = \dfrac{5}{7}$

Solution: This is an equation with rational expressions, and also a proportion. Below are two ways to solve.

Since this is a rational equation, we can use the methods of the previous section.	Since this is also a proportion, we may set cross products equal.

$$\frac{45}{x} = \frac{5}{7}$$

$7x \cdot \dfrac{45}{x} = 7x \cdot \dfrac{5}{7}$ Multiply both sides by LCD, $7x$.

$7 \cdot 45 = x \cdot 5$ Divide out common factors.

$315 = 5x$ Multiply.

$\dfrac{315}{5} = \dfrac{5x}{5}$ Divide both sides by 5.

$63 = x$ Simplify.

$$\frac{45}{x} \diagup\!\!\!\!\diagup \frac{5}{7}$$

$45 \cdot 7 = x \cdot 5$ Set cross products equal.

$315 = 5x$ Multiply.

$\dfrac{315}{5} = \dfrac{5x}{5}$ Divide both sides by 5.

$63 = x$ Simplify.

Check: Both methods give us a solution of 63. To check, substitute 63 for x in the original proportion. The solution is 63.

🔲 **Work Practice Problem 1**

In this section, if the rational equation is a proportion, we will use cross products to solve.

EXAMPLE 2 Solve for x: $\dfrac{x-5}{3} = \dfrac{x+2}{5}$

Solution:

$$\frac{x-5}{3} \diagup\!\!\!\!\diagup \frac{x+2}{5}$$

$5(x-5) = 3(x+2)$ Set cross products equal.

$5x - 25 = 3x + 6$ Multiply.

$5x = 3x + 31$ Add 25 to both sides.

$2x = 31$ Subtract $3x$ from both sides.

$\dfrac{2x}{2} = \dfrac{31}{2}$ Divide both sides by 2.

$x = \dfrac{31}{2}$

Check: Verify that $\dfrac{31}{2}$ is the solution.

🔲 **Work Practice Problem 2**

Objective B Using Proportions to Solve Problems

Proportions can be used to model and solve many real-life problems. When using proportions in this way, it is important to judge whether the solution is reasonable. Doing so helps us to decide if the proportion has been formed correctly. We use the same problem-solving steps that were introduced in Section 2.4.

PRACTICE PROBLEM 1

Solve for x: $\dfrac{3}{8} = \dfrac{63}{x}$

PRACTICE PROBLEM 2

Solve for x: $\dfrac{2x+1}{7} = \dfrac{x-3}{5}$

Answers

1. $x = 168$,   **2.** $x = -\dfrac{26}{3}$

PRACTICE PROBLEM 3

To estimate the number of people in Jackson, population 50,000, who have a flu shot, 250 people were polled. Of those polled, 26 had a flu shot. How many people in the city might we expect to have a flu shot?

EXAMPLE 3 **Calculating the Cost of Recordable Compact Discs**

Three boxes of CD-Rs (recordable compact discs) cost $37.47. How much should 5 boxes cost?

Solution:

1. UNDERSTAND. Read and reread the problem. We know that the cost of 5 boxes is more than the cost of 3 boxes, or $37.47, and less than the cost of 6 boxes, which is double the cost of 3 boxes, or 2($37.47) = $74.94. Let's suppose that 5 boxes cost $60.00. To check, we see if 3 boxes is to 5 boxes as the *price* of 3 boxes is to the *price* of 5 boxes. In other words, we see if

$$\frac{3 \text{ boxes}}{5 \text{ boxes}} = \frac{\text{price of 3 boxes}}{\text{price of 5 boxes}}$$

or

$$\frac{3}{5} = \frac{37.47}{60.00}$$

$$3(60.00) = 5(37.47) \quad \text{Set cross products equal.}$$

or

$$180.00 = 187.35 \quad \text{Not a true statement.}$$

Thus, $60 is not correct, but we now have a better understanding of the problem.

Let x = price of 5 boxes of CD-Rs.

2. TRANSLATE.

$$\frac{3 \text{ boxes}}{5 \text{ boxes}} = \frac{\text{price of 3 boxes}}{\text{price of 5 boxes}}$$

$$\frac{3}{5} = \frac{37.47}{x}$$

3. SOLVE.

$$\frac{3}{5} = \frac{37.47}{x}$$

$$3x = 5(37.47) \quad \text{Set cross products equal.}$$
$$3x = 187.35$$
$$x = 62.45 \quad \text{Divide both sides by 3.}$$

4. INTERPRET.

Check: Verify that 3 boxes is to 5 boxes as $37.47 is to $62.45. Also, notice that our solution is a reasonable one as discussed in Step 1.

State: Five boxes of CD-Rs cost $62.45.

◻ **Work Practice Problem 3**

Answer

3. 5200 people

Helpful Hint

The proportion $\dfrac{5 \text{ boxes}}{3 \text{ boxes}} = \dfrac{\text{price of 5 boxes}}{\text{price of 3 boxes}}$ could also have been used to solve Example 3. Notice that the cross products are the same.

Similar triangles have the same shape but not necessarily the same size. In similar triangles, the measures of corresponding angles are equal, and corresponding sides are in proportion.

If triangle *ABC* and triangle *XYZ* shown are similar, then we know that the measure of angle *A* = the measure of angle *X*, the measure of angle *B* = the measure of angle *Y*, and the measure of angle *C* = the measure of angle *Z*. We also know that corresponding sides are in proportion: $\dfrac{a}{x} = \dfrac{b}{y} = \dfrac{c}{z}$.

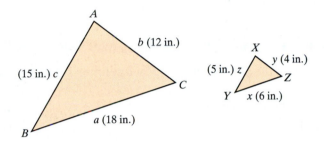

In this section, we will position similar triangles so that they have the same orientation.

To show that corresponding sides are in proportion for the triangles above, we write the ratios of the corresponding sides.

$$\frac{a}{x} = \frac{18}{6} = 3 \qquad \frac{b}{y} = \frac{12}{4} = 3 \qquad \frac{c}{z} = \frac{15}{5} = 3$$

⚠️ **EXAMPLE 4** **Finding the Length of a Side of a Triangle**

If the following two triangles are similar, find the missing length *x*.

Solution:

1. **UNDERSTAND.** Read the problem and study the figure.
2. **TRANSLATE.** Since the triangles are similar, their corresponding sides are in proportion and we have

$$\frac{2}{3} = \frac{10}{x}$$

3. **SOLVE.** To solve, we multiply both sides by the LCD, $3x$, or cross multiply.

   $2x = 30$

   $x = 15$ Divide both sides by 2.

4. **INTERPRET.**

Check: To check, replace *x* with 15 in the original proportion and see that a true statement results.

State: The missing length is 15 yards.

 Work Practice Problem 4

PRACTICE PROBLEM 4

For the similar triangles, find *x*.

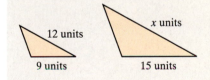

Answer

4. 20

Objective **C** Solving Problems about Numbers

Let's continue to solve problems. The remaining problems are all modeled by rational equations.

EXAMPLE 5 **Finding an Unknown Number**

The quotient of a number and 6, minus $\frac{5}{3}$, is the quotient of the number and 2. Find the number.

Solution:

1. UNDERSTAND. Read and reread the problem. Suppose that the unknown number is 2, then we see if the quotient of 2 and 6, or $\frac{2}{6}$, minus $\frac{5}{3}$ is equal to the quotient of 2 and 2, or $\frac{2}{2}$.

$$\frac{2}{6} - \frac{5}{3} = \frac{1}{3} - \frac{5}{3} = -\frac{4}{3}, \text{ not } \frac{2}{2}$$

Don't forget that the purpose of a proposed solution is to better understand the problem.

Let x = the unknown number.

2. TRANSLATE.

In words:	the quotient of x and 6	minus	$\frac{5}{3}$	is	the quotient of x and 2
	↓	↓	↓	↓	↓
Translate:	$\frac{x}{6}$	$-$	$\frac{5}{3}$	$=$	$\frac{x}{2}$

3. SOLVE. Here, we solve the equation $\frac{x}{6} - \frac{5}{3} = \frac{x}{2}$. We begin by multiplying both sides of the equation by the LCD, 6.

$$6\left(\frac{x}{6} - \frac{5}{3}\right) = 6\left(\frac{x}{2}\right)$$

$$6\left(\frac{x}{6}\right) - 6\left(\frac{5}{3}\right) = 6\left(\frac{x}{2}\right) \quad \text{Apply the distributive property.}$$

$$x - 10 = 3x \quad \text{Simplify.}$$

$$-10 = 2x \quad \text{Subtract } x \text{ from both sides.}$$

$$\frac{-10}{2} = \frac{2x}{2} \quad \text{Divide both sides by 2.}$$

$$-5 = x \quad \text{Simplify.}$$

4. INTERPRET.

Check: To check, we verify that "the quotient of -5 and 6 minus $\frac{5}{3}$ is the quotient of -5 and 2," or $-\frac{5}{6} - \frac{5}{3} = -\frac{5}{2}$.

State: The unknown number is -5.

◻ **Work Practice Problem 5**

Objective **D** Solving Problems about Work

The next example is often called a work problem. Work problems usually involve people or machines doing a certain task.

EXAMPLE 6 **Finding Work Rates**

Sam Waterton and Frank Schaffer work in a plant that manufactures automobiles. Sam can complete a quality control tour of the plant in 3 hours while his assistant, Frank, needs 7 hours to complete the same job. The regional manager is coming to inspect the plant facilities, so both Sam and Frank are directed to complete a quality control tour together. How long will this take?

Solution:

1. UNDERSTAND. Read and reread the problem. The key idea here is the relationship between the **time** (hours) it takes to complete the job and the **part of the job** completed in 1 unit of time (hour). For example, if the **time** it takes Sam to complete the job is 3 hours, the **part of the job** he can complete in 1 hour is $\frac{1}{3}$. Similarly, Frank can complete $\frac{1}{7}$ of the job in 1 hour.

   Let x = the **time** in hours it takes Sam and Frank to complete the job together. Then $\frac{1}{x}$ = the **part of the job** they complete in 1 hour.

	Hours to Complete Total Job	Part of Job Completed in 1 Hour
Sam	3	$\frac{1}{3}$
Frank	7	$\frac{1}{7}$
Together	x	$\frac{1}{x}$

2. TRANSLATE.

In words:	part of job Sam completed in 1 hour	added to	part of job Frank completed in 1 hour	is equal to	part of job they completed together in 1 hour
	↓	↓	↓	↓	↓
Translate:	$\frac{1}{3}$	$+$	$\frac{1}{7}$	$=$	$\frac{1}{x}$

3. SOLVE. Here, we solve the equation $\frac{1}{3} + \frac{1}{7} = \frac{1}{x}$. We begin by multiplying both sides of the equation by the LCD, $21x$.

$$21x\left(\frac{1}{3}\right) + 21x\left(\frac{1}{7}\right) = 21x\left(\frac{1}{x}\right)$$

$$7x + 3x = 21 \qquad \text{Simplify.}$$

$$10x = 21$$

$$x = \frac{21}{10} \quad \text{or} \quad 2\frac{1}{10} \text{ hours}$$

4. INTERPRET.

Check: Our proposed solution is $2\frac{1}{10}$ hours. This proposed solution is reasonable since $2\frac{1}{10}$ hours is more than half of Sam's time and less than half of Frank's time. Check this solution in the originally *stated* problem.

State: Sam and Frank can complete the quality control tour in $2\frac{1}{10}$ hours.

■ **Work Practice Problem 6**

PRACTICE PROBLEM 6

Andrew and Timothy Larson volunteer at a local recycling plant. Andrew can sort a batch of recyclables in 2 hours alone while his brother Timothy needs 3 hours to complete the same job. If they work together, how long will it take them to sort one batch?

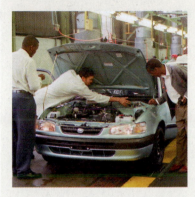

Answer

6. $1\frac{1}{5}$ hours

✔**Concept Check** Solve $E = mc^2$

a. for m **b.** for c^2.

Objective E Solving Problems about Distance

Next we look at a problem solved by the distance formula,

$$d = r \cdot t$$

PRACTICE PROBLEM 7

A car travels 600 miles in the same time that a motorcycle travels 450 miles. If the car's speed is 15 miles per hour more than the motorcycle's, find the speed of the car and the speed of the motorcycle.

EXAMPLE 7 **Finding Speeds of Vehicles**

A car travels 180 miles in the same time that a truck travels 120 miles. If the car's speed is 20 miles per hour faster than the truck's, find the car's speed and the truck's speed.

Solution:

1. UNDERSTAND. Read and reread the problem. Suppose that the truck's speed is 45 miles per hour. Then the car's speed is 20 miles per hour more, or 65 miles per hour.

   We are given that the car travels 180 miles in the same time that the truck travels 120 miles. To find the time it takes the car to travel 180 miles, remember that since $d = rt$, we know that $\dfrac{d}{r} = t$.

Car's Time	*Truck's Time*
$t = \dfrac{d}{r} = \dfrac{180}{65} = 2\dfrac{50}{65} = 2\dfrac{10}{13}$ hours	$t = \dfrac{d}{r} = \dfrac{120}{45} = 2\dfrac{30}{45} = 2\dfrac{2}{3}$ hours

   Since the times are not the same, our proposed solution is not correct. But we have a better understanding of the problem.

   Let x = the speed of the truck.

   Since the car's speed is 20 miles per hour faster than the truck's, then

   $x + 20$ = the speed of the car

   Use the formula $d = r \cdot t$ or **d**istance = **r**ate $\cdot$ **t**ime. Prepare a chart to organize the information in the problem.

Helpful Hint

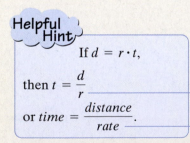

If $d = r \cdot t$,

then $t = \dfrac{d}{r}$

or $time = \dfrac{distance}{rate}$.

	Distance	**=**	**Rate**	**·**	**Time**
Truck	120		x		$\dfrac{120}{x}$ ← distance ← rate
Car	180		$x + 20$		$\dfrac{180}{x + 20}$ ← distance ← rate

2. TRANSLATE. Since the car and the truck traveled the same amount of time, we have that

   In words: car's time = truck's time

   Translate: $\dfrac{180}{x + 20} = \dfrac{120}{x}$

Answer

7. car: 60 mph; motorcycle: 45 mph

✔ **Concept Check Answer**

a. $m = \dfrac{E}{c^2}$, **b.** $c^2 = \dfrac{E}{m}$

3. Solve. We begin by multiplying both sides of the equation by the LCD, $x(x + 20)$, or cross multiplying.

$$\frac{180}{x + 20} = \frac{120}{x}$$

$$180x = 120(x + 20)$$

$$180x = 120x + 2400 \quad \text{Use the distributive property.}$$

$$60x = 2400 \qquad\qquad \text{Subtract } 120x \text{ from both sides.}$$

$$x = 40 \qquad\qquad\;\; \text{Divide both sides by 60.}$$

4. INTERPRET. The speed of the truck is 40 miles per hour. The speed of the car must then be $x + 20$ or 60 miles per hour.

Check: Find the time it takes the car to travel 180 miles and the time it takes the truck to travel 120 miles.

Car's Time   **Truck's Time**

$$t = \frac{d}{r} = \frac{180}{60} = 3 \text{ hours} \qquad t = \frac{d}{r} = \frac{120}{40} = 3 \text{ hours}$$

Since both travel the same amount of time, the proposed solution is correct.

State: The car's speed is 60 miles per hour and the truck's speed is 40 miles per hour.

■ **Work Practice Problem 7**

Mental Math

Without solving algebraically, select the best choice for each exercise.

1. One person can complete a job in 7 hours. A second person can complete the same job in 5 hours. How long will it take them to complete the job if they work together?
   a. more than 7 hours
   b. between 5 and 7 hours
   c. less than 5 hours

2. One inlet pipe can fill a pond in 30 hours. A second inlet pipe can fill the same pond in 25 hours. How long before the pond is filled if both inlet pipes are on?
   a. less than 25 hours
   b. between 25 and 30 hours
   c. more than 30 hours

5.6 EXERCISE SET

FOR EXTRA HELP

Student Solutions Manual   PH Math/Tutor Center   CD/Video for Review   MathXL®   MyMathLab

Objective **A** *Solve each proportion. See Examples 1 and 2.*

1. $\dfrac{2}{3} = \dfrac{x}{6}$

2. $\dfrac{x}{2} = \dfrac{16}{6}$

3. $\dfrac{x}{10} = \dfrac{5}{9}$

4. $\dfrac{9}{4x} = \dfrac{6}{2}$

5. $\dfrac{x+1}{2x+3} = \dfrac{2}{3}$

6. $\dfrac{x+1}{x+2} = \dfrac{5}{3}$

7. $\dfrac{9}{5} = \dfrac{12}{3x+2}$

8. $\dfrac{6}{11} = \dfrac{27}{3x-2}$

Objective **B** *Solve. See Example 3.*

9. The ratio of the weight of an object on Earth to the weight of the same object on Pluto is 100 to 3. If an elephant weighs 4100 pounds on Earth, find the elephant's weight on Pluto.

10. If a 170-pound person weighs approximately 65 pounds on Mars, about how much does a 9000-pound satellite weigh? Round your answer to the nearest pound.

11. There are 110 calories per 28.4 grams of Crispy Rice cereal. Find how many calories are in 42.6 grams of this cereal.

12. On an architect's blueprint, 1 inch corresponds to 4 feet. Find the length of a wall represented by a line that is $3\dfrac{7}{8}$ inches long on the blueprint.

Find the unknown length x or y in the following pairs of similar triangles. See Example 4.

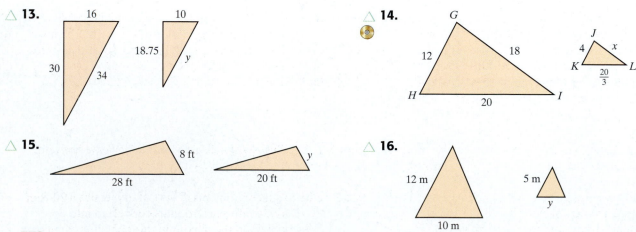

△ 13.

△ 14.

△ 15.

△ 16.

378

Objective C *Solve the following. See Example 5.*

17. Three times the reciprocal of a number equals 9 times the reciprocal of 6. Find the number.

18. Twelve divided by the sum of x and 2 equals the quotient of 4 and the difference of x and 2. Find x.

19. If twice a number added to 3 is divided by the number plus 1, the result is three halves. Find the number.

20. A number added to the product of 6 and the reciprocal of the number equals -5. Find the number.

Objective D *See Example 6.*

21. Smith Engineering found that an experienced surveyor surveys a roadbed in 4 hours. An apprentice surveyor needs 5 hours to survey the same stretch of road. If the two work together, find how long it takes them to complete the job.

22. An experienced bricklayer constructs a small wall in 3 hours. The apprentice completes the job in 6 hours. Find how long it takes if they work together.

23. In 2 minutes, a conveyor belt moves 300 pounds of recyclable aluminum from the delivery truck to a storage area. A smaller belt moves the same quantity of cans the same distance in 6 minutes. If both belts are used, find how long it takes to move the cans to the storage area.

24. Find how long it takes the conveyor belts described in Exercise 23 to move 1200 pounds of cans. (*Hint:* Think of 1200 pounds as four 300-pound jobs.)

Objective E *See Example 7.*

25. A jogger begins her workout by jogging to the park, a distance of 12 miles. She then jogs home at the same speed but along a different route. This return trip is 18 miles and her time is one hour longer. Find her jogging speed. Complete the accompanying chart and use it to find her jogging speed.

	Distance	=	Rate	·	Time
Trip to Park	12				
Return Trip	18				

26. A boat can travel 9 miles upstream in the same amount of time it takes to travel 11 miles downstream. If the current of the river is 3 miles per hour, complete the chart below and use it to find the speed of the boat in still water.

	Distance	=	Rate	·	Time
Upstream	9		$r - 3$		
Downstream	11		$r + 3$		

27. A cyclist rode the first 20-mile portion of his workout at a constant speed. For the 16-mile cooldown portion of his workout, he reduced his speed by 2 miles per hour. Each portion of the workout took the same time. Find the cyclist's speed during the first portion and find his speed during the cooldown portion.

28. A semi-truck travels 300 miles through the flatland in the same amount of time that it travels 180 miles through mountains. The rate of the truck is 20 miles per hour slower in the mountains than in the flatland. Find both the flatland rate and mountain rate.

Objectives A B C D E **Mixed Practice** *Solve the following. See Examples 1 through 7.* (Note: *Some exercises can be modeled by equations without rational expressions.*)

29. A human factors expert recommends that there be at least 9 square feet of floor space in a college classroom for every student in the class. Find the minimum floor space that 40 students need.

30. Due to space problems at a local university, a 20-foot by 12-foot conference room is converted into a classroom. Find the maximum number of students the room can accommodate. (See Exercise 29.)

31. One-fourth equals the quotient of a number and 8. Find the number.

32. Four times a number added to 5 is divided by 6. The result is $\frac{7}{2}$. Find the number.

33. Marcus and Tony work for Lombardo's Pipe and Concrete. Mr. Lombardo is preparing an estimate for a customer. He knows that Marcus lays a slab of concrete in 6 hours. Tony lays the same size slab in 4 hours. If both work on the job and the cost of labor is $45.00 per hour, decide what the labor estimate should be.

34. Mr. Dodson can paint his house by himself in 4 days. His son needs an additional day to complete the job if he works by himself. If they work together, find how long it takes to paint the house.

35. A pilot can travel 400 miles with the wind in the same amount of time as 336 miles against the wind. Find the speed of the wind if the pilot's speed in still air is 230 miles per hour.

36. A fisherman on Pearl River rows 9 miles downstream in the same amount of time he rows 3 miles upstream. If the current is 6 miles per hour, find how long it takes him to cover the 12 miles.

37. Find the unknown length y.

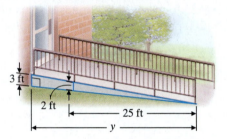

38. Find the unknown length y.

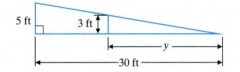

39. Ken Hall, a tailback, holds the high school sports record for total yards rushed in a season. In 1953, he rushed for 4045 total yards in 12 games. Find his average rushing yards per game. Round your answer to the nearest whole yard.

40. To estimate the number of people in Jackson, population 50,000, who have no health insurance, 250 people were polled. Of those polled, 39 had no insurance. How many people in the city might we expect to be uninsured?

41. Two divided by the difference of a number and 3 minus 4 divided by a number plus 3, equals 8 times the reciprocal of the difference of the number squared and 9. What is the number?

42. If 15 times the reciprocal of a number is added to the ratio of 9 times a number minus 7 and the number plus 2, the result is 9. What is the number?

43. A pilot flies 630 miles with a tail wind of 35 miles per hour. Against the wind, he flies only 455 miles in the same amount of time. Find the rate of the plane in still air.

44. A marketing manager travels 1080 miles in a corporate jet and then an additional 240 miles by car. If the car ride takes one hour longer than the jet ride takes, and if the rate of the jet is 6 times the rate of the car, find the time the manager travels by jet and find the time the manager travels by car.

45. To mix weed killer with water correctly, it is necessary to mix 8 teaspoons of weed killer with 2 gallons of water. Find how many gallons of water are needed to mix with the entire box if it contains 36 teaspoons of weed killer.

46. The directions for a certain bug spray concentrate is to mix 3 ounces of concentrate with 2 gallons of water. How many ounces of concentrate are needed to mix with 5 gallons of water?

47. A boater travels 16 miles per hour on the water on a still day. During one particular windy day, he finds that he travels 48 miles with the wind behind him in the same amount of time that he travels 16 miles into the wind. Find the rate of the wind.

48. The current on a portion of the Mississippi River is 3 miles per hour. A barge can go 6 miles upstream in the same amount of time it takes to go 10 miles downstream. Find the speed of the boat in still water.

49. Two hikers are 11 miles apart and walking toward each other. They meet in 2 hours. Find the rate of each hiker if one hiker walks 1.1 mph faster than the other.

50. On a 255-mile trip, Gary Alessandrini traveled at an average speed of 70 mph, got a speeding ticket, and then traveled at 60 mph for the remainder of the trip. If the entire trip took 4.5 hours and the speeding ticket stop took 30 minutes, how long did Gary speed before getting stopped?

51. One custodian cleans a suite of offices in 3 hours. When a second worker is asked to join the regular custodian, the job takes only $1\frac{1}{2}$ hours. How long does it take the second worker to do the same job alone?

52. One person proofreads a copy for a small newspaper in 4 hours. If a second proofreader is also employed, the job can be done in $2\frac{1}{2}$ hours. How long does it take for the second proofreader to do the same job alone?

△ **53.** An architect is completing the plans for a triangular deck. Use the diagram below to find the missing dimension.

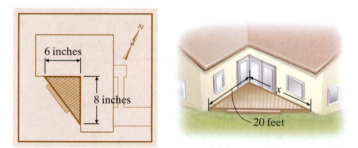

△ **54.** A student wishes to make a small model of a triangular mainsail in order to study the effects of wind on the sail. The smaller model will be the same shape as a regular-size sailboat's mainsail. Use the following diagram to find the missing dimensions.

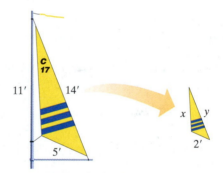

55. The manufacturers of cans of salted mixed nuts state that the ratio of peanuts to other nuts is 3 to 2. If 324 peanuts are in a can, find how many other nuts should also be in the can.

56. There are 1280 calories in a 14-ounce portion of Eagle Brand Milk. Find how many calories are in 2 ounces of Eagle Brand Milk.

57. A jet plane traveling at 500 mph overtakes a propeller plane traveling at 200 mph that had a 2-hour head start. How far from the starting point are the planes?

58. How long will it take a bus traveling at 60 miles per hour to overtake a car traveling at 40 miles per hour if the car had a 1.5-hour head start?

59. One pipe fills a storage pool in 20 hours. A second pipe fills the same pool in 15 hours. When a third pipe is added and all three are used to fill the pool, it takes only 6 hours. Find how long it takes the third pipe to do the job.

60. One pump fills a tank 2 times as fast as another pump. If the pumps work together, they fill the tank in 18 minutes. How long does it take for each pump to fill the tank?

61. A car travels 280 miles in the same time that a motorcycle travels 240 miles. If the car's speed is 10 miles per hour more than the motorcycle's, find the speed of the car and the speed of the motorcycle.

62. A bus traveled on a level road for 3 hours at an average speed 20 miles per hour faster than it traveled on a winding road. The time spent on the winding road was 4 hours. Find the average speed on the level road if the entire trip was 305 miles.

63. In 6 hours, an experienced cook prepares enough pies to supply a local restaurant's daily order. Another cook prepares the same number of pies in 7 hours. Together with a third cook, they prepare the pies in 2 hours. Find how long it takes the third cook to prepare the pies alone.

64. Mrs. Smith balances the company books in 8 hours. It takes her assistant 12 hours to do the same job. If they work together, find how long it takes them to balance the books.

65. One pump fills a tank 3 times as fast as another pump. If the pumps work together, they fill the tank in 21 minutes. How long does it take for each pump to fill the tank?

Given that the following pairs of triangles are similar, find each missing length.

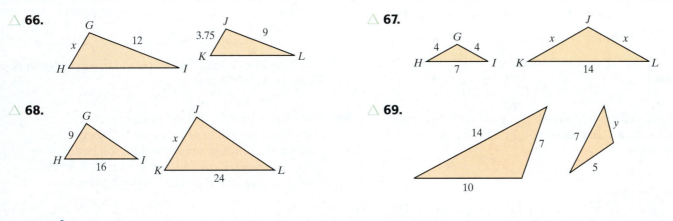

△ **66.**

△ **67.**

△ **68.**

△ **69.**

Review

Simplify. Follow the circled steps in the order shown. See Section R.2.

70. $\dfrac{\dfrac{3}{4} + \dfrac{1}{4}}{\dfrac{3}{8} + \dfrac{13}{8}}$ ← ① Add. ← ② Add.

71. $\dfrac{\dfrac{9}{5} + \dfrac{6}{5}}{\dfrac{17}{6} + \dfrac{7}{6}}$ ← ① Add. ← ② Add.

72. $\dfrac{\dfrac{2}{5} + \dfrac{1}{5}}{\dfrac{7}{10} + \dfrac{7}{10}}$ ← ① Add ← ③ Divide. ← ② Add.

73. $\dfrac{\dfrac{1}{4} + \dfrac{5}{4}}{\dfrac{3}{8} + \dfrac{7}{8}}$ ← ① Add ← ③ Divide. ← ② Add.

74. Person A can complete a job in 5 hours, and person B can complete the same job in 3 hours. Without solving algebraically, discuss reasonable and unreasonable answers for how long it would take them to complete the job together.

75. For which of the following equations can we immediately use cross products to solve for x?

  a. $\dfrac{2 - x}{5} = \dfrac{1 + x}{3}$

  b. $\dfrac{2}{5} - x = \dfrac{1 + x}{3}$

76. For what value of x is $\dfrac{x}{x - 1}$ in proportion to $\dfrac{x + 1}{x}$? Explain your result.

Concept Extensions

Solve. See the Concept Check in this section.

Solve $D = RT$

77. for R

78. for T

79. A hyena spots a giraffe 0.5 mile away and begins running toward it. The giraffe starts running away from the hyena just as the hyena begins running toward it. A hyena can run at a speed of 40 mph and a giraffe can run at 32 mph. How long will it take for the hyena to overtake the giraffe? (*Source: The World Almanac and Book of Facts*)

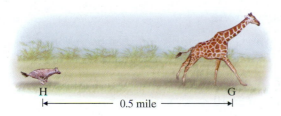

80. The Andretti Green Racing team boasts the proud name of one of the best known Indy car drivers, Mario Andretti. Two of its drivers, Tony Kanaan and Bryan Herta, placed second and fourth, respectively, in the 2004 Indianapolis 500. The track is 2.5 miles long. When traveling at their fastest lap speeds, Herta drove 2.479 miles in the same time that Kanaan completed an entire 2.5-mile lap. Kanaan's fastest lap speed was 1.822 mph faster than Herta's fastest lap speed. Find each driver's fastest lap speed. Round each speed to the nearest tenth. (*Source:* Indy Racing League)

THE BIGGER PICTURE   Simplifying Expressions and Solving Equations

Now we continue our outline from Sections 1.6, 2.7, 3.7, 4.6, and 5.4. Although suggestions are given, this outline should be in your own words. Once you complete this new portion, try the exercises below.

I. Simplifying Expressions

   A. Real Numbers

      1. Add (Section 1.4)
      2. Subtract (Section 1.5)
      3. Multiply or Divide (Section 1.6)

   B. Exponents (Section 3.2)

   C. Polynomials

      1. Add (Section 3.4)
      2. Subtract (Section 3.4)
      3. Multiply (Section 3.5)
      4. Divide (Section 3.7)

   D. Factoring Polynomials (Chapter 4 Integrated Review)

   E. Rational Expressions

      1. Simplify (Section 5.1)
      2. Multiply (Section 5.2)
      3. Divide (Section 5.2)
      4. Add or Subtract (Section 5.4)

II. Solving Equations and Inequalities

   A. Linear Equations (Section 2.3)

   B. Linear Inequalities (Section 2.7)

   C. Quadratic and Higher Degree Equations (Section 4.6)

   D. Equations with Rational Expressions—solving equations with rational expressions

$$\frac{3}{x} - \frac{1}{x-1} = \frac{4}{x-1} \qquad \text{Equation with rational expressions.}$$

$$x(x-1)\cdot\frac{3}{x} - x(x-1)\frac{1}{x-1} \qquad \text{Multiply through by } x(x-1).$$

$$= x(x-1)\frac{4}{x-1}$$

$$3(x-1) - x\cdot 1 = x\cdot 4 \qquad \text{Simplify.}$$

$$3x - 3 - x = 4x \qquad \text{Use the distributive property.}$$

$$-3 = 2x \qquad \text{Simplify and move variable terms to right side.}$$

$$-\frac{3}{2} = x \qquad \text{Divide both sides by 2.}$$

E. Proportions—an equation with two ratios equal. Set cross products equal, then solve.

$$\frac{5}{x} = \frac{9}{2x-3}, \text{ or } 5(2x-3) = 9\cdot x$$

or $10x - 15 = 9x$ or $x = 15$

Multiply.

   1. $(3x-2)(4x^2 - x - 5)$
   2. $(2x - y)^2$

Factor.

   3. $8y^3 - 20y^5$
   4. $9m^2 - 11mn + 2n^2$

Simplify or solve.

If an expression, perform indicated operations and simplify. If an equation or inequality, solve it.

   5. $\dfrac{7}{x} = \dfrac{9}{x-10}$

   6. $\dfrac{7}{x} + \dfrac{9}{x-10}$

   7. $(-3x^5)\left(\dfrac{1}{2}x^7\right)(8x)$

   8. $5x - 1 = |-4| + |-5|$

   9. $\dfrac{8 - 12}{12 \div 3\cdot 2}$

   10. $-2(3y - 4) \le 5y - 7 - 7y - 1$

   11. $\dfrac{7}{x} + \dfrac{5}{2x+3} = \dfrac{-2}{x}$

   12. $\dfrac{(a^{-3}b^2)^{-5}}{ab^4}$

5.7 SIMPLIFYING COMPLEX FRACTIONS

Objectives

A Simplify Complex Fractions Using Method 1.

B Simplify Complex Fractions Using Method 2.

A rational expression whose numerator or denominator or both numerator and denominator contain fractions is called a **complex rational expression** or a **complex fraction.** Some examples are

$$\dfrac{4}{2-\dfrac{1}{2}} \qquad \dfrac{\dfrac{3}{2}}{\dfrac{4}{7}-x} \qquad \dfrac{\dfrac{1}{x+2}}{x+2-\dfrac{1}{x}}\Bigg\}$$

⟵ Numerator of complex fraction
⟵ Main fraction bar
⟵ Denominator of complex fraction

Our goal in this section is to write complex fractions in simplest form. A complex fraction is in simplest form when it is in the form $\dfrac{P}{Q}$, where P and Q are polynomials that have no common factors.

Objective **A** Simplifying Complex Fractions—Method 1

In this section, two methods of simplifying complex fractions are represented. The first method presented uses the fact that the main fraction bar indicates division.

Method 1: To Simplify a Complex Fraction

Step 1: Add or subtract fractions in the numerator or denominator so that the numerator is a single fraction and the denominator is a single fraction.

Step 2: Perform the indicated division by multiplying the numerator of the complex fraction by the reciprocal of the denominator of the complex fraction.

Step 3: Write the rational expression in lowest terms.

EXAMPLE 1 Simplify the complex fraction $\dfrac{\dfrac{5}{8}}{\dfrac{2}{3}}$.

Solution: Since the numerator and denominator of the complex fraction are already single fractions, we proceed to Step 2: perform the indicated division by multiplying the numerator $\dfrac{5}{8}$ by the reciprocal of the denominator $\dfrac{2}{3}$.

$$\dfrac{\dfrac{5}{8}}{\dfrac{2}{3}} = \dfrac{5}{8} \div \dfrac{2}{3} = \dfrac{5}{8} \cdot \dfrac{3}{2} = \dfrac{15}{16}$$

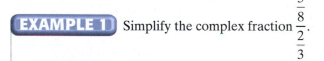

The reciprocal of $\dfrac{2}{3}$ is $\dfrac{3}{2}$.

🟧 **Work Practice Problem 1**

PRACTICE PROBLEM 1

Simplify the complex fraction $\dfrac{\dfrac{3}{7}}{\dfrac{5}{9}}$.

Answer

1. $\dfrac{27}{35}$

385

PRACTICE PROBLEM 2

Simplify: $\dfrac{\dfrac{3}{4} - \dfrac{2}{3}}{\dfrac{1}{2} + \dfrac{3}{8}}$

EXAMPLE 2 Simplify: $\dfrac{\dfrac{2}{3} + \dfrac{1}{5}}{\dfrac{2}{3} - \dfrac{2}{9}}$

Solution: We simplify the numerator and denominator of the complex fraction separately. First we add $\dfrac{2}{3}$ and $\dfrac{1}{5}$ to obtain a single fraction in the numerator. Then we subtract $\dfrac{2}{9}$ from $\dfrac{2}{3}$ to obtain a single fraction in the denominator.

$$\dfrac{\dfrac{2}{3} + \dfrac{1}{5}}{\dfrac{2}{3} - \dfrac{2}{9}} = \dfrac{\dfrac{2(5)}{3(5)} + \dfrac{1(3)}{5(3)}}{\dfrac{2(3)}{3(3)} - \dfrac{2}{9}}$$ The LCD of the numerator's fractions is 15.

The LCD of the denominator's fractions is 9.

$$= \dfrac{\dfrac{10}{15} + \dfrac{3}{15}}{\dfrac{6}{9} - \dfrac{2}{9}}$$ Simplify.

$$= \dfrac{\dfrac{13}{15}}{\dfrac{4}{9}}$$ Add the numerator's fractions.

Subtract the denominator's fractions.

Next we perform the indicated division by multiplying the numerator of the complex fraction by the reciprocal of the denominator of the complex fraction.

$$\dfrac{\dfrac{13}{15}}{\dfrac{4}{9}} = \dfrac{13}{15} \cdot \dfrac{9}{4}$$ The reciprocal of $\dfrac{4}{9}$ is $\dfrac{9}{4}$.

$$= \dfrac{13 \cdot 3 \cdot \boxed{3}}{\boxed{3} \cdot 5 \cdot 4} = \dfrac{39}{20}$$

🟧 **Work Practice Problem 2**

PRACTICE PROBLEM 3

Simplify: $\dfrac{\dfrac{2}{5} - \dfrac{1}{x}}{\dfrac{2x}{15} - \dfrac{1}{3}}$

EXAMPLE 3 Simplify: $\dfrac{\dfrac{1}{z} - \dfrac{1}{2}}{\dfrac{1}{3} - \dfrac{z}{6}}$

Solution: Subtract to get a single fraction in the numerator and a single fraction in the denominator of the complex fraction.

$$\dfrac{\dfrac{1}{z} - \dfrac{1}{2}}{\dfrac{1}{3} - \dfrac{z}{6}} = \dfrac{\dfrac{2}{2z} - \dfrac{z}{2z}}{\dfrac{2}{6} - \dfrac{z}{6}}$$ The LCD of the numerator's fractions is $2z$.

The LCD of the denominator's fractions is 6.

$$= \dfrac{\dfrac{2 - z}{2z}}{\dfrac{2 - z}{6}}$$

$$= \dfrac{2 - z}{2z} \cdot \dfrac{6}{2 - z}$$ Multiply by the reciprocal of $\dfrac{2 - z}{6}$.

$$= \dfrac{2 \cdot 3 \cdot \boxed{(2 - z)}}{2 \cdot z \cdot \boxed{(2 - z)}}$$ Factor.

$$= \dfrac{3}{z}$$ Write in lowest terms.

🟧 **Work Practice Problem 3**

Answers

2. $\dfrac{2}{21}$, **3.** $\dfrac{3}{x}$

Objective B Simplifying Complex Fractions—Method 2

Next we study a second method for simplifying complex fractions. In this method, we multiply the numerator and the denominator of the complex fraction by the LCD of all fractions in the complex fraction.

Method 2: To Simplify a Complex Fraction

Step 1: Find the LCD of all the fractions in the complex fraction.

Step 2: Multiply both the numerator and the denominator of the complex fraction by the LCD from Step 1.

Step 3: Perform the indicated operations and write the result in lowest terms.

We use method 2 to rework Example 2.

EXAMPLE 4 Simplify: $\dfrac{\dfrac{2}{3} + \dfrac{1}{5}}{\dfrac{2}{3} - \dfrac{2}{9}}$

Solution: The LCD of $\dfrac{2}{3}, \dfrac{1}{5}, \dfrac{2}{3}$, and $\dfrac{2}{9}$ is 45, so we multiply the numerator and the denominator of the complex fraction by 45. Then we perform the indicated operations, and write in lowest terms.

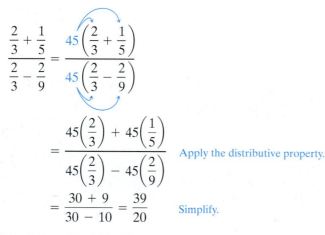

$$\frac{\dfrac{2}{3} + \dfrac{1}{5}}{\dfrac{2}{3} - \dfrac{2}{9}} = \frac{45\left(\dfrac{2}{3} + \dfrac{1}{5}\right)}{45\left(\dfrac{2}{3} - \dfrac{2}{9}\right)}$$

$$= \frac{45\left(\dfrac{2}{3}\right) + 45\left(\dfrac{1}{5}\right)}{45\left(\dfrac{2}{3}\right) - 45\left(\dfrac{2}{9}\right)} \quad \text{Apply the distributive property.}$$

$$= \frac{30 + 9}{30 - 10} = \frac{39}{20} \quad \text{Simplify.}$$

■ **Work Practice Problem 4**

Helpful Hint

The same complex fraction was simplified using two different methods in Examples 2 and 4. Notice that the simplified results are the same.

PRACTICE PROBLEM 4

Use method 2 to simplify the complex fraction in Practice Problem 2:

$$\frac{\dfrac{3}{4} - \dfrac{2}{3}}{\dfrac{1}{2} + \dfrac{3}{8}}$$

Answer

4. $\dfrac{2}{21}$

PRACTICE PROBLEM 5

Simplify: $\dfrac{1 + \dfrac{x}{y}}{\dfrac{2x + 1}{y}}$

 **EXAMPLE 5** Simplify: $\dfrac{\dfrac{x + 1}{y}}{\dfrac{x}{y} + 2}$

Solution: The LCD of $\dfrac{x + 1}{y}$ and $\dfrac{x}{y}$ is y, so we multiply the numerator and the denominator of the complex fraction by y.

$$\dfrac{\dfrac{x + 1}{y}}{\dfrac{x}{y} + 2} = \dfrac{y\left(\dfrac{x + 1}{y}\right)}{y\left(\dfrac{x}{y} + 2\right)}$$

$$= \dfrac{y\left(\dfrac{x + 1}{y}\right)}{y\left(\dfrac{x}{y}\right) + y \cdot 2} \qquad \text{Apply the distributive property in the denominator.}$$

$$= \dfrac{x + 1}{x + 2y} \qquad \text{Simplify.}$$

🔲 **Work Practice Problem 5**

PRACTICE PROBLEM 6

Simplify: $\dfrac{\dfrac{5}{6y} + \dfrac{y}{x}}{\dfrac{y}{3} - x}$

EXAMPLE 6 Simplify: $\dfrac{\dfrac{x}{y} + \dfrac{3}{2x}}{\dfrac{x}{2} + y}$

Solution: The LCD of $\dfrac{x}{y}, \dfrac{3}{2x}, \dfrac{x}{2}$, and $\dfrac{y}{1}$ is $2xy$, so we multiply both the numerator and the denominator of the complex fraction by $2xy$.

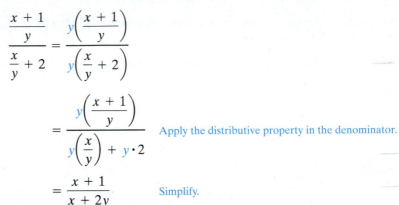

$$\dfrac{\dfrac{x}{y} + \dfrac{3}{2x}}{\dfrac{x}{2} + y} = \dfrac{2xy\left(\dfrac{x}{y} + \dfrac{3}{2x}\right)}{2xy\left(\dfrac{x}{2} + y\right)}$$

$$= \dfrac{2xy\left(\dfrac{x}{y}\right) + 2xy\left(\dfrac{3}{2x}\right)}{2xy\left(\dfrac{x}{2}\right) + 2xy(y)} \qquad \text{Apply the distributive property.}$$

$$= \dfrac{2x^2 + 3y}{x^2 y + 2xy^2}$$

$$\text{or } \dfrac{2x^2 + 3y}{xy(x + 2y)}$$

🔲 **Work Practice Problem 6**

Answers

5. $\dfrac{y + x}{2x + 1}$,

6. $\dfrac{5x + 6y^2}{2xy^2 - 6x^2 y} \text{ or } \dfrac{5x + 6y^2}{2xy(y - 3x)}$

Objectives A B **Mixed Practice** *Simplify each complex fraction. See Examples 1 through 6.*

1. $\dfrac{\dfrac{1}{2}}{\dfrac{3}{4}}$

2. $\dfrac{\dfrac{1}{8}}{-\dfrac{5}{12}}$

3. $\dfrac{-\dfrac{4x}{9}}{-\dfrac{2x}{3}}$

4. $\dfrac{-\dfrac{6y}{11}}{\dfrac{4y}{9}}$

5. $\dfrac{\dfrac{1+x}{6}}{\dfrac{1+x}{3}}$

6. $\dfrac{\dfrac{6x-3}{5x^2}}{\dfrac{2x-1}{10x}}$

7. $\dfrac{\dfrac{1}{2}+\dfrac{2}{3}}{\dfrac{5}{9}-\dfrac{5}{6}}$

8. $\dfrac{\dfrac{3}{4}-\dfrac{1}{2}}{\dfrac{3}{8}+\dfrac{1}{6}}$

9. $\dfrac{2+\dfrac{7}{10}}{1+\dfrac{3}{5}}$

10. $\dfrac{4-\dfrac{11}{12}}{5+\dfrac{1}{4}}$

11. $\dfrac{\dfrac{1}{3}}{\dfrac{1}{2}-\dfrac{1}{4}}$

12. $\dfrac{\dfrac{7}{10}-\dfrac{3}{5}}{\dfrac{1}{2}}$

13. $\dfrac{-\dfrac{2}{9}}{-\dfrac{14}{3}}$

14. $\dfrac{\dfrac{3}{8}}{\dfrac{4}{15}}$

15. $\dfrac{-\dfrac{5}{12x^2}}{\dfrac{25}{16x^3}}$

16. $\dfrac{-\dfrac{7}{8y}}{\dfrac{21}{4y}}$

17. $\dfrac{\dfrac{m}{n}-1}{\dfrac{m}{n}+1}$

18. $\dfrac{\dfrac{x}{2}+2}{\dfrac{x}{2}-2}$

19. $\dfrac{\dfrac{1}{5}-\dfrac{1}{x}}{\dfrac{7}{10}+\dfrac{1}{x^2}}$

20. $\dfrac{\dfrac{1}{y^2}+\dfrac{2}{3}}{\dfrac{1}{y}-\dfrac{5}{6}}$

21. $\dfrac{1+\dfrac{1}{y-2}}{y+\dfrac{1}{y-2}}$

22. $\dfrac{x-\dfrac{1}{2x+1}}{1-\dfrac{x}{2x+1}}$

23. $\dfrac{\dfrac{4y-8}{16}}{\dfrac{6y-12}{4}}$

24. $\dfrac{\dfrac{7y+21}{3}}{\dfrac{3y+9}{8}}$

25. $\dfrac{\dfrac{x}{y}+1}{\dfrac{x}{y}-1}$

26. $\dfrac{\dfrac{3}{5y}+8}{\dfrac{3}{5y}-8}$

27. $\dfrac{1}{2+\dfrac{1}{3}}$

28. $\dfrac{3}{1-\dfrac{4}{3}}$

29. $\dfrac{\dfrac{ax+ab}{x^2-b^2}}{\dfrac{x+b}{x-b}}$

30. $\dfrac{\dfrac{m+2}{m-2}}{\dfrac{2m+4}{m^2-4}}$

31. $\dfrac{\dfrac{-3+y}{4}}{\dfrac{8+y}{28}}$

32. $\dfrac{\dfrac{-x+2}{18}}{\dfrac{8}{9}}$

33. $\dfrac{3+\dfrac{12}{x}}{1-\dfrac{16}{x^2}}$

34. $\dfrac{2+\dfrac{6}{x}}{1-\dfrac{9}{x^2}}$

35. $\dfrac{\dfrac{8}{x+4}+2}{\dfrac{12}{x+4}-2}$

36. $\dfrac{\dfrac{25}{x+5}+5}{\dfrac{3}{x+5}-5}$

37. $\dfrac{\dfrac{s}{r} + \dfrac{r}{s}}{\dfrac{s}{r} - \dfrac{r}{s}}$

38. $\dfrac{\dfrac{2}{x} + \dfrac{x}{2}}{\dfrac{2}{x} - \dfrac{x}{2}}$

39. $\dfrac{\dfrac{6}{x - 5} + \dfrac{x}{x - 2}}{\dfrac{3}{x - 6} - \dfrac{2}{x - 5}}$

40. $\dfrac{\dfrac{4}{x} + \dfrac{x}{x + 1}}{\dfrac{1}{2x} + \dfrac{1}{x + 6}}$

Review

Use the bar graph below to answer Exercises 41 through 44. See Section 1.2.

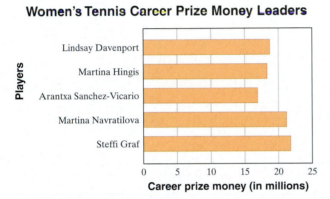

Women's Tennis Career Prize Money Leaders

Players: Lindsay Davenport, Martina Hingis, Arantxa Sanchez-Vicario, Martina Navratilova, Steffi Graf

Career prize money (in millions)

Source: Sanex WTA Tour Media Information System

41. Which women's tennis player has earned the most prize money in her career?

42. Estimate how much more prize money Lindsay Davenport has earned over her career than Arantxa Sanchez-Vicario.

43. Which of the players shown have earned over $20 million in prize money over their careers?

44. During her career, through July 4, 2004, Martina Navratilova has won 347 singles and doubles tournaments. Assuming her prize money was earned only for tournament titles, how much prize money did she earn per tournament title, on average?

Concept Extensions

45. Explain how to simplify a complex fraction using method 1.

46. Explain how to simplify a complex fraction using method 2.

To find the average of two numbers, we find their sum and divide by 2. For example, the average of 65 and 81 is found by simplifying $\dfrac{65 + 81}{2}$. *This simplifies to* $\dfrac{146}{2} = 73$.

47. Find the average of $\dfrac{1}{3}$ and $\dfrac{3}{4}$.

48. Write the average of $\dfrac{3}{n}$ and $\dfrac{5}{n^2}$ as a simplified rational expression.

49. In electronics, when two resistors R_1 (read R sub 1) and R_2 (read R sub 2) are connected in parallel, the total resistance is given by the complex fraction

$$\dfrac{1}{\dfrac{1}{R_1} + \dfrac{1}{R_2}}.$$

Simplify this expression.

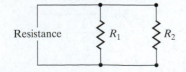

Resistance R_1 R_2

50. Astronomers occasionally need to know the day of the week a particular date fell on. The complex fraction

$$\dfrac{J + \dfrac{3}{2}}{7}$$

where J is the *Julian day number*, is used to make this calculation. Simplify this expression.

Simplify each of the following. First, write each expression with positive exponents. Then simplify the complex fraction. The first step has been completed for Exercise 51.

51. $\dfrac{x^{-1} + 2^{-1}}{x^{-2} - 4^{-1}} = \dfrac{\dfrac{1}{x} + \dfrac{1}{2}}{\dfrac{1}{x^2} - \dfrac{1}{4}}$

52. $\dfrac{3^{-1} - x^{-1}}{9^{-1} - x^{-2}}$

53. $\dfrac{y^{-2}}{1 - y^{-2}}$

54. $\dfrac{4 + x^{-1}}{3 + x^{-1}}$

55. If the distance formula $d = r \cdot t$ is solved for t, then $t = \dfrac{d}{r}$. Use this formula to find t if distance d is $\dfrac{20x}{3}$ miles and rate r is $\dfrac{5x}{9}$ miles per hour. Write t in simplified form.

△ **56.** If the formula for area of a rectangle, $A = l \cdot w$, is solved for w, then $w = \dfrac{A}{l}$. Use this formula to find w if area A is $\dfrac{4x - 2}{3}$ square meters and length l is $\dfrac{6x - 3}{5}$ meters. Write w in simplified form.

CHAPTER 5 Group Activity

Fast-Growing Careers

According to U.S. Bureau of Labor Statistics projections, the careers listed below will have the largest job growth in the next decade.

Occupation	Employment (number in thousands)		
	2002	2012	Change
1 Registered nurses	2284	2908	+623
2 Postsecondary teachers	1581	2184	+603
3 Retail salespersons	4076	4672	+596
4 Customer service representatives	1894	2354	+460
5 Combined food preparation and serving workers, including fast food	1990	2444	+454
6 Cashiers, except gaming	3432	3886	+454
7 Janitors and cleaners, except maids and housekeeping cleaners	2267	2681	+414
8 General and operations managers	2049	2425	+376
9 Waiters and waitresses	2097	2464	+367
10 Nursing aides, orderlies, and attendants	1375	1718	+343

What do all of these in-demand occupations have in common? They all require a knowledge of math! For some careers, like nurses, postsecondary teachers, and salespersons, the ways math is used on the job may be obvious. For other occupations, the use of math may not be quite as obvious. However, tasks common to many jobs, such as filling in a time sheet or a medication log, writing up an expense report, planning a budget, figuring a bill, ordering supplies, and even making a work schedule, all require math.

Activity

Suppose that your college placement office is planning to publish an occupational handbook on math in popular occupations. Choose one of the occupations from the given list that interests you. Research the occupation. Then write a brief entry for the occupational handbook that describes how a person in that career would use math in his or her job. Include an example if possible.

Chapter 5 Vocabulary Check

Fill in each blank with one of the words or phrases listed below.

rational expression   complex fraction   proportion

cross products   ratio   rate

1. A _____ is the quotient of two numbers.

2. $\dfrac{x}{2} = \dfrac{7}{16}$ is an example of a _____.

3. If $\dfrac{a}{b} = \dfrac{c}{d}$, the *ad* and *bc* are called _____.

4. A _____ is an expression that can be written in the form $\dfrac{P}{Q}$, where P and Q are polynomials and Q is not 0.

5. In a _____, the numerator or denominator or both may contain fractions.

6. A _____ is a special type of ratio where different measurements are used.

> **Helpful Hint**
>
> Are you preparing for your test? Don't forget to take the Chapter 5 Test on page 401. Then check your answers at the back of the text and use the Chapter Test Prep Video CD to see the fully worked-out solutions to any of the exercises you want to review.

5 Chapter Highlights

DEFINITIONS AND CONCEPTS	EXAMPLES
Section 5.1 Simplifying Rational Expressions	

A **rational expression** is an expression that can be written in the form $\dfrac{P}{Q}$, where P and Q are polynomials and Q does not equal 0.	$\dfrac{7y^3}{4}, \; \dfrac{x^2 + 6x + 1}{x - 3}, \; \dfrac{-5}{s^3 + 8}$
To find values for which a rational expression is undefined, find values for which the denominator is 0.	Find any values for which the expression $\dfrac{5y}{y^2 - 4y + 3}$ is undefined. $$y^2 - 4y + 3 = 0 \quad \text{Set the denominator equal to 0.}$$ $$(y - 3)(y - 1) = 0 \quad \text{Factor.}$$ $$y - 3 = 0 \quad \text{or} \quad y - 1 = 0 \quad \text{Set each factor equal to 0.}$$ $$y = 3 \qquad\qquad y = 1 \quad \text{Solve.}$$ The expression is undefined when y is 3 and when y is 1.
TO SIMPLIFY A RATIONAL EXPRESSION **Step 1.** Factor the numerator and denominator. **Step 2.** Divide out factors common to the numerator and denominator. (This is the same as removing a factor of 1.)	Simplify: $\dfrac{4x + 20}{x^2 - 25}$ $$\dfrac{4x + 20}{x^2 - 25} = \dfrac{4\,(x + 5)}{(x + 5)\,(x - 5)} = \dfrac{4}{x - 5}$$

DEFINITIONS AND CONCEPTS	**EXAMPLES**

Section 5.2 Multiplying and Dividing Rational Expressions

TO MULTIPLY RATIONAL EXPRESSIONS

Step 1. Factor numerators and denominators.

Step 2. Multiply numerators and multiply denominators.

Step 3. Write the product in lowest terms.

$$\frac{P}{Q} \cdot \frac{R}{S} = \frac{PR}{QS}$$

Multiply: $\dfrac{4x + 4}{2x - 3} \cdot \dfrac{2x^2 + x - 6}{x^2 - 1}$

$$\frac{4x + 4}{2x - 3} \cdot \frac{2x^2 + x - 6}{x^2 - 1}$$

$$= \frac{4(x + 1)}{2x - 3} \cdot \frac{(2x - 3)(x + 2)}{(x + 1)(x - 1)}$$

$$= \frac{4\,(x + 1)(2x - 3)\,(x + 2)}{(2x - 3)(x + 1)\,(x - 1)}$$

$$= \frac{4(x + 2)}{x - 1}$$

To divide by a rational expression, multiply by the reciprocal.

$$\frac{P}{Q} \div \frac{R}{S} = \frac{P}{Q} \cdot \frac{S}{R} = \frac{PS}{QR}$$

Divide: $\dfrac{15x + 5}{3x^2 - 14x - 5} \div \dfrac{15}{3x - 12}$

$$\frac{15x + 5}{3x^2 - 14x - 5} \div \frac{15}{3x - 12}$$

$$= \frac{5(3x + 1)}{(3x + 1)\,(x - 5)} \cdot \frac{3\,(x - 4)}{3 \cdot 5}$$

$$= \frac{x - 4}{x - 5}$$

Section 5.3 Adding and Subtracting Rational Expressions with the Same Denominator and Least Common Denominators

To add or subtract rational expressions with the same denominator, add or subtract numerators, and place the sum or difference over the common denominator.

$$\frac{P}{R} + \frac{Q}{R} = \frac{P + Q}{R}$$

$$\frac{P}{R} - \frac{Q}{R} = \frac{P - Q}{R}$$

Perform each indicated operation.

$$\frac{5}{x + 1} + \frac{x}{x + 1} = \frac{5 + x}{x + 1}$$

$$\frac{2y + 7}{y^2 - 9} - \frac{y + 4}{y^2 - 9}$$

$$= \frac{(2y + 7) - (y + 4)}{y^2 - 9}$$

$$= \frac{2y + 7 - y - 4}{y^2 - 9}$$

$$= \frac{y + 3}{(y + 3)\,(y - 3)}$$

$$= \frac{1}{y - 3}$$

TO FIND THE LEAST COMMON DENOMINATOR (LCD)

Step 1. Factor the denominators.

Step 2. The LCD is the product of all unique factors, each raised to a power equal to the greatest number of times that it appears in any one factored denominator.

Find the LCD for

$$\frac{7x}{x^2 + 10x + 25} \quad \text{and} \quad \frac{11}{3x^2 + 15x}$$

$$x^2 + 10x + 25 = (x + 5)(x + 5)$$

$$3x^2 + 15x = 3x(x + 5)$$

$$\text{LCD} = 3x(x + 5)(x + 5) \text{ or}$$

$$3x(x + 5)^2$$

DEFINITIONS AND CONCEPTS	EXAMPLES

Section 5.4   Adding and Subtracting Rational Expressions with Different Denominators

To Add or Subtract Rational Expressions with Different Denominators

Step 1. Find the LCD.

Step 2. Rewrite each rational expression as an equivalent expression whose denominator is the LCD.

Step 3. Add or subtract numerators and place the sum or difference over the common denominator.

Step 4. Write the result in lowest terms.

Perform the indicated operation.

$$\frac{9x + 3}{x^2 - 9} - \frac{5}{x - 3}$$

$$= \frac{9x + 3}{(x + 3)(x - 3)} - \frac{5}{x - 3}$$

LCD is $(x + 3)(x - 3)$.

$$= \frac{9x + 3}{(x + 3)(x - 3)} - \frac{5(x + 3)}{(x - 3)(x + 3)}$$

$$= \frac{9x + 3 - 5(x + 3)}{(x + 3)(x - 3)}$$

$$= \frac{9x + 3 - 5x - 15}{(x + 3)(x - 3)}$$

$$= \frac{4x - 12}{(x + 3)(x - 3)}$$

$$= \frac{4(x - 3)}{(x + 3)(x - 3)} = \frac{4}{x + 3}$$

Section 5.5   Solving Equations Containing Rational Expressions

To Solve an Equation Containing Rational Expressions

Step 1. Multiply both sides of the equation by the LCD of all rational expressions in the equation.

Step 2. Remove any grouping symbols and solve the resulting equation.

Step 3. Check the solution in the original equation.

Solve: $\dfrac{5x}{x + 2} + 3 = \dfrac{4x - 6}{x + 2}$ The LCD is $x + 2$.

$$(x + 2)\left(\frac{5x}{x + 2} + 3\right) = (x + 2)\left(\frac{4x - 6}{x + 2}\right)$$

$$(x + 2)\left(\frac{5x}{x + 2}\right) + (x + 2)(3) = (x + 2)\left(\frac{4x - 6}{x + 2}\right)$$

$$5x + 3x + 6 = 4x - 6$$

$$4x = -12$$

$$x = -3$$

The solution checks; the solution is -3.

Section 5.6   Proportions and Problem Solving with Rational Equations

A **ratio** is the quotient of two numbers or two quantities.
A **proportion** is a mathematical statement that two ratios are equal.

Cross products:

$$\text{If } \frac{a}{b} = \frac{c}{d}, \text{ then } ad = bc.$$

Proportions

$$\frac{2}{3} = \frac{8}{12} \qquad \frac{x}{7} = \frac{15}{35}$$

Cross Products

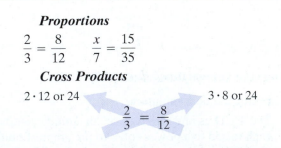

$2 \cdot 12$ or 24 $3 \cdot 8$ or 24

$$\frac{2}{3} = \frac{8}{12}$$

DEFINITIONS AND CONCEPTS	**EXAMPLES**
Section 5.6 Proportions and Problem Solving with Rational Equations (continued)	

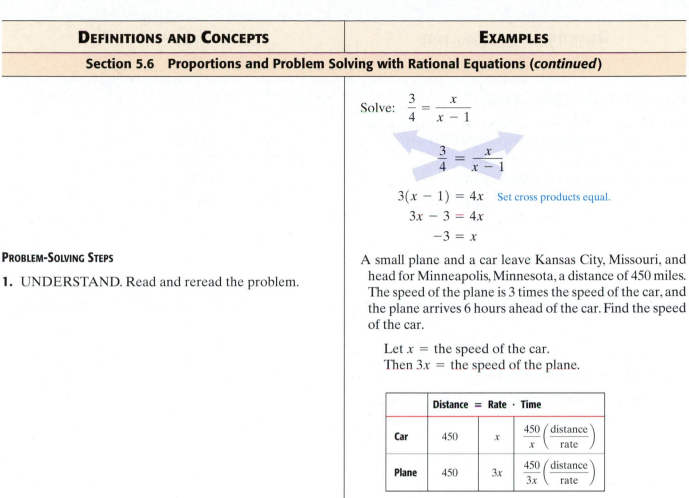

Solve: $\dfrac{3}{4} = \dfrac{x}{x-1}$

$\dfrac{3}{4} = \dfrac{x}{x-1}$

$3(x-1) = 4x$ Set cross products equal.

$3x - 3 = 4x$

$-3 = x$

PROBLEM-SOLVING STEPS

1. UNDERSTAND. Read and reread the problem.

A small plane and a car leave Kansas City, Missouri, and head for Minneapolis, Minnesota, a distance of 450 miles. The speed of the plane is 3 times the speed of the car, and the plane arrives 6 hours ahead of the car. Find the speed of the car.

Let x = the speed of the car.
Then $3x$ = the speed of the plane.

	Distance	**=** **Rate**	**· Time**
Car	450	x	$\dfrac{450}{x}\left(\dfrac{\text{distance}}{\text{rate}}\right)$
Plane	450	$3x$	$\dfrac{450}{3x}\left(\dfrac{\text{distance}}{\text{rate}}\right)$

2. TRANSLATE.

In words: $\boxed{\text{plane's time}}$ + $\boxed{\text{6 hours}}$ = $\boxed{\text{car's time}}$

Translate: $\dfrac{450}{3x}$ $+$ 6 $=$ $\dfrac{450}{x}$

3. SOLVE.

$\dfrac{450}{3x} + 6 = \dfrac{450}{x}$

$3x\left(\dfrac{450}{3x}\right) + 3x(6) = 3x\left(\dfrac{450}{x}\right)$

$450 + 18x = 1350$

$18x = 900$

$x = 50$

4. INTERPRET.

Check this solution in the originally stated problem. **State** the conclusion: The speed of the car is 50 miles per hour.

DEFINITIONS AND CONCEPTS	EXAMPLES
Section 5.7 Simplifying Complex Fractions	

METHOD 1: TO SIMPLIFY A COMPLEX FRACTION

Step 1. Add or subtract fractions in the numerator and the denominator of the complex fraction.

Step 2. Perform the indicated division.

Step 3. Write the result in lowest terms.

Simplify:

$$\frac{\dfrac{1}{x} + 2}{\dfrac{1}{x} - \dfrac{1}{y}} = \frac{\dfrac{1}{x} + \dfrac{2x}{x}}{\dfrac{y}{xy} - \dfrac{x}{xy}}$$

$$= \frac{\dfrac{1 + 2x}{x}}{\dfrac{y - x}{xy}}$$

$$= \frac{1 + 2x}{x} \cdot \frac{x\,y}{y - x}$$

$$= \frac{y(1 + 2x)}{y - x}$$

METHOD 2: TO SIMPLIFY A COMPLEX FRACTION

Step 1. Find the LCD of all fractions in the complex fraction.

Step 2. Multiply the numerator and the denominator of the complex fraction by the LCD.

Step 3. Perform the indicated operations and write the result in lowest terms.

$$\frac{\dfrac{1}{x} + 2}{\dfrac{1}{x} - \dfrac{1}{y}} = \frac{xy\left(\dfrac{1}{x} + 2\right)}{xy\left(\dfrac{1}{x} - \dfrac{1}{y}\right)}$$

$$= \frac{xy\left(\dfrac{1}{x}\right) + xy(2)}{xy\left(\dfrac{1}{x}\right) - xy\left(\dfrac{1}{y}\right)}$$

$$= \frac{y + 2xy}{y - x} \quad \text{or} \quad \frac{y(1 + 2x)}{y - x}$$

STUDY SKILLS BUILDER

Are You Prepared for a Test on Chapter 5?

Below I have listed *a common trouble* area for students in Chapter 5. After studying for your test, but before taking your test, read this.

Do you know the differences between how to perform operations such as $\dfrac{4}{x} + \dfrac{2}{3}$ or $\dfrac{4}{x} \div \dfrac{2}{x}$ and how to solve an equation such as $\dfrac{4}{x} + \dfrac{2}{3} = 1$?

$$\frac{4}{x} + \frac{2}{3} = \frac{4 \cdot 3}{x \cdot 3} + \frac{2 \cdot x}{3 \cdot x}$$

Addition—write each expression as an equivalent expression with the same LCD denominator.

$$= \frac{12}{3x} + \frac{2x}{3x} = \frac{12 + 2x}{3x} \quad \text{or} \quad \frac{2(6 + x)}{3x}, \text{ the sum.}$$

$$\frac{4}{x} \div \frac{2}{x} = \frac{4}{x} \cdot \frac{x}{2} = \frac{4 \cdot x}{x \cdot 2} = \frac{4}{2} = 2, \text{ the quotient.}$$

Division—multiply the first rational expression by the reciprocal of the second.

$$\frac{4}{x} + \frac{2}{3} = 1$$ Equation to be solved.

$$3x\left(\frac{4}{x} + \frac{2}{3}\right) = 3x \cdot 1$$ Multiply both sides of the equation by the LCD, $3x$.

$$3x\left(\frac{4}{x}\right) + 3x\left(\frac{2}{3}\right) = 3x \cdot 1$$ Use the distributive property.

$$12 + 2x = 3x$$ Multiply and simplify.

$$12 = x$$ Subtract $2x$ from both sides.

The solution is 12.

For more examples and exercises, see the Chapter 5 Integrated Review.

5 CHAPTER REVIEW

(5.1) *Find any real number for which each rational expression is undefined.*

1. $\dfrac{x + 5}{x^2 - 4}$

2. $\dfrac{5x + 9}{4x^2 - 4x - 15}$

Find the value of each rational expression when $x = 5$, $y = 7$, and $z = -2$.

3. $\dfrac{2 - z}{z + 5}$

4. $\dfrac{x^2 + xy - y^2}{x + y}$

Simplify each rational expression.

5. $\dfrac{2x + 6}{x^2 + 3x}$

6. $\dfrac{3x - 12}{x^2 - 4x}$

7. $\dfrac{x + 2}{x^2 - 3x - 10}$

8. $\dfrac{x + 4}{x^2 + 5x + 4}$

9. $\dfrac{x^3 - 4x}{x^2 + 3x + 2}$

10. $\dfrac{5x^2 - 125}{x^2 + 2x - 15}$

11. $\dfrac{x^2 - x - 6}{x^2 - 3x - 10}$

12. $\dfrac{x^2 - 2x}{x^2 + 2x - 8}$

Simplify each expression. First, factor the four-term polynomials by grouping.

13. $\dfrac{x^2 + xa + xb + ab}{x^2 - xc + bx - bc}$

14. $\dfrac{x^2 + 5x - 2x - 10}{x^2 - 3x - 2x + 6}$

(5.2) *Perform each indicated operation and simplify.*

15. $\dfrac{15x^3 y^2}{z} \cdot \dfrac{z}{5xy^3}$

16. $\dfrac{-y^3}{8} \cdot \dfrac{9x^2}{y^3}$

17. $\dfrac{x^2 - 9}{x^2 - 4} \cdot \dfrac{x - 2}{x + 3}$

18. $\dfrac{2x + 5}{x - 6} \cdot \dfrac{2x}{-x + 6}$

19. $\dfrac{x^2 - 5x - 24}{x^2 - x - 12} \div \dfrac{x^2 - 10x + 16}{x^2 + x - 6}$

20. $\dfrac{4x + 4y}{xy^2} \div \dfrac{3x + 3y}{x^2 y}$

21. $\dfrac{x^2 + x - 42}{x - 3} \cdot \dfrac{(x - 3)^2}{x + 7}$

22. $\dfrac{2a + 2b}{3} \cdot \dfrac{a - b}{a^2 - b^2}$

23. $\dfrac{2x^2 - 9x + 9}{8x - 12} \div \dfrac{x^2 - 3x}{2x}$

24. $\dfrac{x^2 - y^2}{x^2 + xy} \div \dfrac{3x^2 - 2xy - y^2}{3x^2 + 6x}$

(5.3) *Perform each indicated operation and simplify.*

25. $\dfrac{x}{x^2 + 9x + 14} + \dfrac{7}{x^2 + 9x + 14}$

26. $\dfrac{x}{x^2 + 2x - 15} + \dfrac{5}{x^2 + 2x - 15}$

27. $\dfrac{4x - 5}{3x^2} - \dfrac{2x + 5}{3x^2}$

28. $\dfrac{9x + 7}{6x^2} - \dfrac{3x + 4}{6x^2}$

Find the LCD of each pair of rational expressions.

29. $\dfrac{x + 4}{2x}, \dfrac{3}{7x}$

30. $\dfrac{x - 2}{x^2 - 5x - 24}, \dfrac{3}{x^2 + 11x + 24}$

Rewrite each rational expression as an equivalent expression whose denominator is the given polynomial.

31. $\dfrac{5}{7x} = \dfrac{}{14x^3y}$

32. $\dfrac{9}{4y} = \dfrac{}{16y^3x}$

33. $\dfrac{x + 2}{x^2 + 11x + 18} = \dfrac{}{(x + 2)(x - 5)(x + 9)}$

34. $\dfrac{3x - 5}{x^2 + 4x + 4} = \dfrac{}{(x + 2)^2(x + 3)}$

(5.4) *Perform each indicated operation and simplify.*

35. $\dfrac{4}{5x^2} - \dfrac{6}{y}$

36. $\dfrac{2}{x - 3} - \dfrac{4}{x - 1}$

37. $\dfrac{4}{x + 3} - 2$

38. $\dfrac{3}{x^2 + 2x - 8} + \dfrac{2}{x^2 - 3x + 2}$

39. $\dfrac{2x - 5}{6x + 9} - \dfrac{4}{2x^2 + 3x}$

40. $\dfrac{x - 1}{x^2 - 2x + 1} - \dfrac{x + 1}{x - 1}$

(5.5) *Solve each equation.*

41. $\dfrac{n}{10} = 9 - \dfrac{n}{5}$

42. $\dfrac{2}{x + 1} - \dfrac{1}{x - 2} = -\dfrac{1}{2}$

43. $\dfrac{y}{2y + 2} + \dfrac{2y - 16}{4y + 4} = \dfrac{y - 3}{y + 1}$

44. $\dfrac{2}{x - 3} - \dfrac{4}{x + 3} = \dfrac{8}{x^2 - 9}$

45. $\dfrac{x - 3}{x + 1} - \dfrac{x - 6}{x + 5} = 0$

46. $x + 5 = \dfrac{6}{x}$

(5.6) *Solve each proportion.*

47. $\dfrac{2}{x - 1} = \dfrac{3}{x + 3}$

48. $\dfrac{4}{y - 3} = \dfrac{2}{y - 3}$

Solve.

49. A machine can process 300 parts in 20 minutes. Find how many parts can be processed in 45 minutes.

50. As his consulting fee, Mr. Visconti charges $90.00 per day. Find how much he charges for 3 hours of consulting. Assume an 8-hour work day.

51. Five times the reciprocal of a number equals the sum of $\frac{3}{2}$ the reciprocal of the number and $\frac{7}{6}$. What is the number?

52. The reciprocal of a number equals the reciprocal of the difference of 4 and the number. Find the number.

53. A car travels 90 miles in the same time that a car traveling 10 miles per hour slower travels 60 miles. Find the speed of each car.

54. The current in a bayou near Lafayette, Louisiana, is 4 miles per hour. A paddle boat travels 48 miles upstream in the same amount of time it takes to travel 72 miles downstream. Find the speed of the boat in still water.

55. When Mark and Maria manicure Mr. Stergeon's lawn, it takes them 5 hours. If Mark works alone, it takes 7 hours. Find how long it takes Maria alone.

56. It takes pipe A 20 days to fill a fish pond. Pipe B takes 15 days. Find how long it takes both pipes together to fill the pond.

Given that the pairs of triangles are similar, find each missing length x.

△ **57.**

△ **58.**

(5.7) *Simplify each complex fraction.*

59. $\dfrac{\dfrac{5x}{27}}{-\dfrac{10xy}{21}}$

60. $\dfrac{\dfrac{3}{5} + \dfrac{2}{7}}{\dfrac{1}{5} + \dfrac{5}{6}}$

61. $\dfrac{3 - \dfrac{1}{y}}{2 - \dfrac{1}{y}}$

62. $\dfrac{\dfrac{6}{x+2} + 4}{\dfrac{8}{x+2} - 4}$

Mixed Review

Simplify each rational expression.

63. $\dfrac{4x + 12}{8x^2 + 24x}$

64. $\dfrac{x^3 - 6x^2 + 9x}{x^2 + 4x - 21}$

Perform the indicated operations and simplify.

65. $\dfrac{x^2 + 9x + 20}{x^2 - 25} \cdot \dfrac{x^2 - 9x + 20}{x^2 + 8x + 16}$

66. $\dfrac{x^2 - x - 72}{x^2 - x - 30} \div \dfrac{x^2 + 6x - 27}{x^2 - 9x + 18}$

67. $\dfrac{x}{x^2 - 36} + \dfrac{6}{x^2 - 36}$

68. $\dfrac{5x - 1}{4x} - \dfrac{3x - 2}{4x}$

69. $\dfrac{4}{3x^2 + 8x - 3} + \dfrac{2}{3x^2 - 7x + 2}$

70. $\dfrac{3x}{x^2 + 9x + 14} - \dfrac{6x}{x^2 + 4x - 21}$

Solve.

71. $\dfrac{4}{a-1} + 2 = \dfrac{3}{a-1}$

72. $\dfrac{x}{x+3} + 4 = \dfrac{x}{x+3}$

Solve.

73. The quotient of twice a number and three, minus one-sixth is the quotient of the number and two. Find the number.

74. Mr. Crocker can paint his shed by himself in three days. His son will need an additional day to complete the job if he works alone. If they work together, find how long it takes to paint the house.

Given that the following pairs of triangles are similar, find each missing length.

75.

76.

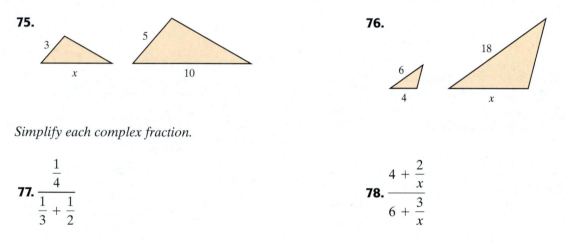

Simplify each complex fraction.

77. $\dfrac{\dfrac{1}{4}}{\dfrac{1}{3} + \dfrac{1}{2}}$

78. $\dfrac{4 + \dfrac{2}{x}}{6 + \dfrac{3}{x}}$

5 CHAPTER TEST

Remember to use the Chapter Test Prep Video CD to see the fully worked-out solutions to any of the exercises you want to review.

1. Find any real numbers for which the following expression is undefined.

$$\frac{x + 5}{x^2 + 4x + 3}$$

2. For a certain computer desk, the average manufacturing cost C per desk (in dollars) is

$$C = \frac{100x + 3000}{x}$$

   where x is the number of desks manufactured.
   a. Find the average cost per desk when manufacturing 200 computer desks.
   b. Find the average cost per desk when manufacturing 1000 computer desks.

Simplify each rational expression.

3. $\dfrac{3x - 6}{5x - 10}$

4. $\dfrac{x + 6}{x^2 + 12x + 36}$

5. $\dfrac{7 - x}{x - 7}$

6. $\dfrac{y - x}{x^2 - y^2}$

7. $\dfrac{2m^3 - 2m^2 - 12m}{m^2 - 5m + 6}$

8. $\dfrac{ay + 3a + 2y + 6}{ay + 3a + 5y + 15}$

Perform each indicated operation and simplify if possible.

9. $\dfrac{x^2 - 13x + 42}{x^2 + 10x + 21} \div \dfrac{x^2 - 4}{x^2 + x - 6}$

10. $\dfrac{3}{x - 1} \cdot (5x - 5)$

11. $\dfrac{y^2 - 5y + 6}{2y + 4} \cdot \dfrac{y + 2}{2y - 6}$

12. $\dfrac{5}{2x + 5} - \dfrac{6}{2x + 5}$

13. $\dfrac{5a}{a^2 - a - 6} - \dfrac{2}{a - 3}$

14. $\dfrac{6}{x^2 - 1} + \dfrac{3}{x + 1}$

15. $\dfrac{x^2 - 9}{x^2 - 3x} \div \dfrac{x^2 + 4x + 1}{2x + 10}$

16. $\dfrac{x + 2}{x^2 + 11x + 18} + \dfrac{5}{x^2 - 3x - 10}$

17. $\dfrac{4y}{y^2 + 6y + 5} - \dfrac{3}{y^2 + 5y + 4}$

1. _____

2. a. _____

   b. _____

3. _____

4. _____

5. _____

6. _____

7. _____

8. _____

9. _____

10. _____

11. _____

12. _____

13. _____

14. _____

15. _____

16. _____

17. _____

18. _____

19. _____

20. _____

21. _____

22. _____

23. _____

24. _____

25. _____

26. _____

27. _____

28. _____

29. _____

30. _____

Solve each equation.

18. $\dfrac{4}{y} - \dfrac{5}{3} = \dfrac{-1}{5}$

19. $\dfrac{5}{y+1} = \dfrac{4}{y+2}$

20. $\dfrac{a}{a-3} = \dfrac{3}{a-3} - \dfrac{3}{2}$

21. $\dfrac{10}{x^2 - 25} = \dfrac{3}{x+5} + \dfrac{1}{x-5}$

22. $x - \dfrac{14}{x-1} = 4 - \dfrac{2x}{x-1}$

Simplify each complex fraction.

23. $\dfrac{\dfrac{5x^2}{yz^2}}{\dfrac{10x}{z^3}}$

24. $\dfrac{\dfrac{b}{a} - \dfrac{a}{b}}{\dfrac{1}{b} + \dfrac{1}{a}}$

25. $\dfrac{5 - \dfrac{1}{y^2}}{\dfrac{1}{y} + \dfrac{2}{y^2}}$

△ **26.** Given that the two triangles are similar, find x.

27. One number plus five times its reciprocal is equal to six. Find the number.

28. A pleasure boat traveling down the Red River takes the same time to go 14 miles upstream as it takes to go 16 miles downstream. If the current of the river is 2 miles per hour, find the speed of the boat in still water.

29. An inlet pipe can fill a tank in 12 hours. A second pipe can fill the tank in 15 hours. If both pipes are used, find how long it takes to fill the tank.

30. In a sample of 85 fluorescent bulbs, 3 were found to be defective. At this rate, how many defective bulbs should be found in 510 bulbs?

1. Write each sentence as an equation. Let x represent the unknown number.

   a. The quotient of 15 and a number is 4.

   b. Three subtracted from 12 is a number.

   c. 17 added to four times a number is 21.

2. Write each sentence as an equation. Let x represent the unknown number.

   a. The difference of 12 and a number is -45.

   b. The product of 12 and a number is -45.

   c. A number less 10 is twice the number.

3. Find the sums.

   a. $3 + (-7) + (-8)$

   b. $[7 + (-10)] + [-2 + (-4)]$

4. Find the differences.

   a. $28 - 6 - 30$

   b. $7 - 2 - 22$

For Exercises 5 through 8, name the property illustrated by each true statement.

5. $3(x + y) = 3 \cdot x + 3 \cdot y$

6. $3 + y = y + 3$

7. $(x + 7) + 9 = x + (7 + 9)$

8. $(x \cdot 7) \cdot 9 = x \cdot (7 \cdot 9)$

9. Solve: $3 - x = 7$

10. Solve: $7x - 6 = 6x - 6$

11. A 10-foot board is to be cut into two pieces so that the longer piece is 4 times the shorter. Find the length of each piece.

12. Find two consecutive even integers whose sum is 382.

13. Solve: $y = mx + b$ for x.

14. Solve: $3x - 2y = 6$ for y.

Answers

1. a. _____
  b. _____
  c. _____

2. a. _____
  b. _____
  c. _____

3. a. _____
  b. _____

4. a. _____
  b. _____

5. _____

6. _____

7. _____

8. _____

9. _____

10. _____

11. _____

12. _____

13. _____

14. _____

15. _____

16. _____

17. _____

18. _____

19. _____

20. _____

21. _____

22. _____

23. _____

24. _____

25. _____

26. _____

27. _____

28. _____

29. _____

30. _____

31. _____

32. _____

33. _____

34. _____

35. _____

36. _____

15. Solve $x + 4 \leq -6$. Graph the solutions.

16. Solve: $-3x + 7 > -x + 9$

Simplify.

17. $\dfrac{x^5}{x^2}$

18. $\dfrac{y^{14}}{y^{14}}$

19. $\dfrac{4^7}{4^3}$

20. $(x^5 y^2)^3$

21. $\dfrac{(-3)^5}{(-3)^2}$

22. $\dfrac{x^{19} y^5}{xy}$

23. $\dfrac{2x^5 y^2}{xy}$

24. $(-3a^2 b)(5a^3 b)$

Simplify by writing each expression with positive exponents only.

25. $2x^{-3}$

26. 7^{-2}

27. $(-2)^{-4}$

28. $5z^{-7}$

Multiply.

29. $5x(2x^3 + 6)$

30. $(x + 9)^2$

31. $-3x^2(5x^2 + 6x - 1)$

32. $(2x + 1)(2x - 1)$

Perform the indicated operations.

33. Divide: $\dfrac{4x^2 + 7 + 8x^3}{2x + 3}$

34. Divide: $(4x^3 - 9x + 2)$ by $(x - 4)$

35. Factor: $x^2 + 7x + 12$

36. Factor: $-2a^2 + 10a + 12$

37. Factor: $25x^2 + 20xy + 4y^2$

38. Factor: $x^2 - 4$

39. Solve: $x^2 - 9x - 22 = 0$

40. Solve: $3x^2 + 5x = 2$

41. Multiply: $\dfrac{x^2 + x}{3x} \cdot \dfrac{6}{5x + 5}$

42. Simplify: $\dfrac{2x^2 - 50}{4x^4 - 20x^3}$

43. Subtract: $\dfrac{3x^2 + 2x}{x - 1} - \dfrac{10x - 5}{x - 1}$

44. Factor: $7x^6 - 7x^5 + 7x^4$

45. Subtract: $\dfrac{6x}{x^2 - 4} - \dfrac{3}{x + 2}$

46. Factor: $4x^2 + 12x + 9$

47. Solve: $\dfrac{t - 4}{2} - \dfrac{t - 3}{9} = \dfrac{5}{18}$

48. Multiply: $\dfrac{6x^2 - 18x}{3x^2 - 2x} \cdot \dfrac{15x - 10}{x^2 - 9}$

49. Sam Waterton and Frank Schaffer work in a plant that manufactures automobiles. Sam can complete a quality control tour of the plant in 3 hours while his assistant, Frank, needs 7 hours to complete the same job. The regional manager is coming to inspect the plant facilities, so both Sam and Frank are directed to complete a quality control tour together. How long will this take?

50. Simplify: $\dfrac{\dfrac{m}{3} + \dfrac{n}{6}}{\dfrac{m + n}{12}}$

37. _____

38. _____

39. _____

40. _____

41. _____

42. _____

43. _____

44. _____

45. _____

46. _____

47. _____

48. _____

49. _____

50. _____

6

Graphing Equations and Inequalities

In Chapter 2 we learned to solve and graph the solutions of linear equations and inequalities in one variable on number lines. Now we define and present techniques for solving and graphing linear equations and inequalities in two variables on grids. Two-variable equations lead directly to the concept of *function*, perhaps the most important concept in all mathematics. Functions are introduced in Section 6.6.

Americans enjoy pets more than ever before. Currently 62% of all U.S. households, or about 64.2 million households, have at least one pet. According to an American Pet Products Manufacturing Association survey, companionship, love, company, and affection eclipse all other benefits of pet ownership and are cited as the primary benefits of sharing their lives with their pet.

In Exercise 33 on page 417, we will examine the growth of these pet-related expenditures, such as food, veterinary care, supplies, and pet care and grooming.

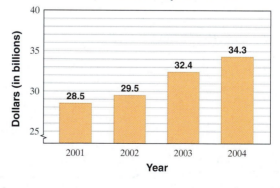

Pet-Related Expenditures

6.1 READING GRAPHS AND THE RECTANGULAR COORDINATE SYSTEM

In today's world, where the exchange of information must be fast and entertaining, graphs are becoming increasingly popular. They provide a quick way of making comparisons, drawing conclusions, and approximating quantities.

Objective A Reading Bar and Line Graphs

A **bar graph** consists of a series of bars arranged vertically or horizontally. The bar graph in Example 1 shows a comparison of worldwide Internet users by region. The names of the regions are listed vertically and a bar is shown for each region. Corresponding to the length of the bar for each region is a number along a horizontal axis. These horizontal numbers are number of Internet users in millions.

Objectives

A Read Bar and Line Graphs.

B Plot Ordered Pairs of Numbers on the Rectangular Coordinate System.

C Graph Paired Data to Create a Scatter Diagram.

D Find the Missing Coordinate of an Ordered Pair Solution, Given One Coordinate of the Pair.

EXAMPLE 1

The following bar graph shows the estimated number of Internet users worldwide by region, as of a recent year.

a. Find the region that has the most Internet users and approximate the number of users.

b. How many more users are in the U.S./Canada region than the Latin America region?

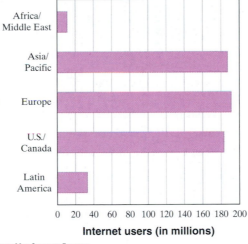

Worldwide Internet Users

Region: Africa/Middle East, Asia/Pacific, Europe, U.S./Canada, Latin America

Internet users (in millions): 0 20 40 60 80 100 120 140 160 180 200

Source: Nua Internet Surveys

PRACTICE PROBLEM 1

Use the graph from Example 1 to answer the following.

a. Find the region with the fewest Internet users and approximate the number of users.

b. How many more users are in the Asia/Pacific region than in the Latin America region?

Solution:

a. Since these bars are arranged horizontally, we look for the longest bar, which is the bar representing Europe. To approximate the number associated with this region, we move from the right edge of this bar vertically downward to the Internet user axis. This region has approximately 191 million Internet users.

b. The U.S./Canada region has approximately 183 million Internet users. The Latin America region has approximately 33 million Internet users. To find how many more users are in the U.S./Canada region, we subtract 183 − 33 = 150 or 150 million more Internet users.

Worldwide Internet Users

Region: Africa/Middle East, Asia/Pacific, Europe, U.S./Canada, Latin America

33, 183, 191

Internet users (in millions): 0 20 40 60 80 100 120 140 160 180 200

Source: Nua Internet Surveys

▢ Work Practice Problem 1

Answers

1. a. Africa/Middle East region, 11 million Internet users,
b. 154 million Internet users

407

A **line graph** consists of a series of points connected by a line. The next graph is an example of a line graph. It is also sometimes called a **broken line graph.**

PRACTICE PROBLEM 2

Use the graph from Example 2 to answer the following.

a. What is the pulse rate 40 minutes after lighting a cigarette?

b. What is the pulse rate when the cigarette is being lit?

c. When is the pulse rate the highest?

EXAMPLE 2

The line graph shows the relationship between time spent smoking a cigarette and pulse rate. Time is recorded along the horizontal axis in minutes, with 0 minutes being the moment a smoker lights a cigarette. Pulse is recorded along the vertical axis in heartbeats per minute.

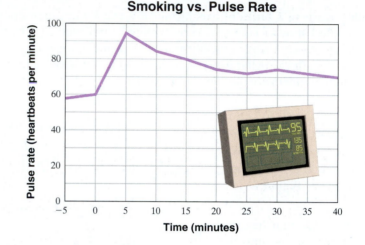

a. What is the pulse rate 15 minutes after a cigarette is lit?

b. When is the pulse rate the lowest?

c. When does the pulse rate show the greatest change?

Solution:

a. We locate the number 15 along the time axis and move vertically upward until the line is reached. From this point on the line, we move horizontally to the left until the pulse rate axis is reached. Reading the number of beats per minute, we find that the pulse rate is 80 beats per minute 15 minutes after a cigarette is lit.

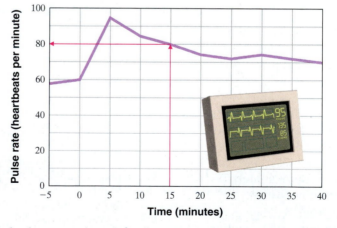

b. We find the lowest point of the line graph, which represents the lowest pulse rate. From this point, we move vertically downward to the time axis. We find that the pulse rate is the lowest at −5 minutes, which means 5 minutes *before* lighting a cigarette.

c. The pulse rate shows the greatest change during the 5 minutes between 0 and 5. Notice that the line graph is *steepest* between 0 and 5 minutes.

Answers

2. a. 70 beats per minute,

b. 60 beats per minute,

c. 5 minutes after lighting

■ **Work Practice Problem 2**

Notice in the graph on the previous page that there are two numbers associated with each point of the graph. For example, we discussed earlier that 15 minutes after "lighting up," the pulse rate is 80 beats per minute. If we agree to write the time first and the pulse rate second, we can say there is a point on the graph corresponding to the **ordered pair** of numbers (15, 80). A few more ordered pairs are shown alongside their corresponding points.

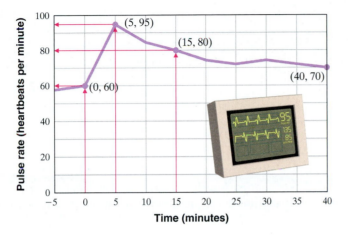

Objective B Plotting Ordered Pairs of Numbers

In general, we use the idea of ordered pairs to describe the location of a point in a plane (such as a piece of paper). We start with a horizontal and a vertical axis. Each axis is a number line, and for the sake of consistency we construct our axes to intersect at the 0 coordinate of both. This point of intersection is called the **origin.** Notice that these two number lines or axes divide the plane into four regions called **quadrants.** The quadrants are usually numbered with Roman numerals as shown. The axes are not considered to be in any quadrant.

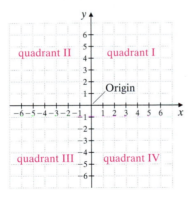

It is helpful to label axes, so we label the horizontal axis the **x-axis** and the vertical axis the **y-axis.** We call the system described above the **rectangular coordinate system,** or the **coordinate plane.** Just as with other graphs shown, we can then describe the locations of points by ordered pairs of numbers. We list the horizontal **x-axis** measurement first and the vertical **y-axis** measurement second.

To plot or graph the point corresponding to the ordered pair

(a, b)

we start at the origin. We then move a units left or right (right if a is positive, left if a is negative). From there, we move b units up or down (up if b is positive, down if b is negative). For example, to plot the point corresponding to the ordered pair (3, 2), we start at the origin, move 3 units right, and from there move 2 units up. (See the figure on next page.) The x-value, 3, is also called the **x-coordinate** and the y-value, 2, is also called the **y-coordinate.** From now on, we will call the point with coordinates (3, 2) simply the point (3, 2). The point (−2, 5) is also graphed below.

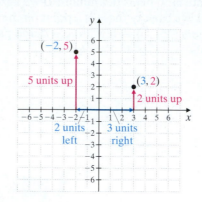

PRACTICE PROBLEM 3

On a single coordinate system, plot each ordered pair. State in which quadrant, or on which axis each point lies.

a. (4, 2)   **b.** (−1, −3)

c. (2, −2)   **d.** (−5, 1)

e. (0, 3)   **f.** (3, 0)

g. (0, −4)   **h.** $\left(-2\frac{1}{2}, 0\right)$

i. $\left(1, -3\frac{3}{4}\right)$

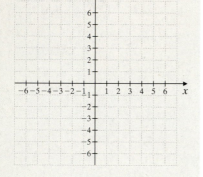

Helpful Hint

Don't forget that **each ordered pair corresponds to exactly one point in the plane and that each point in the plane corresponds to exactly one ordered pair.**

✔ **Concept Check**   Is the graph of the point (−5, 1) in the same location as the graph of the point (1, −5)? Explain.

EXAMPLE 3   On a single coordinate system, plot each ordered pair. State in which quadrant, or on which axis each point lies.

a. (5, 3)   **b.** (−2, −4)   **c.** (1, −2)   **d.** (−5, 3)   **e.** (0, 0)

f. (0, 2)   **g.** (−5, 0)   **h.** $\left(0, -5\frac{1}{2}\right)$ **i.** $\left(4\frac{2}{3}, -3\right)$

Solution:

a. Point (5, 3) lies in quadrant I.

b. Point (−2, −4) lies in quadrant III.

c. Point (1, −2) lies in quadrant IV.

d. Point (−5, 3) lies in quadrant II.

e.–h. Points (0, 0), (0, 2), and $\left(0, -5\frac{1}{2}\right)$ lie on the y-axis. Points (0, 0) and (−5, 0) lie on the x-axis.

i. Point $\left(4\frac{2}{3}, -3\right)$ lies in quadrant IV.

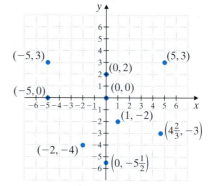

■ **Work Practice Problem 3**

Answers

3.

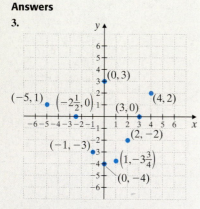

a. Point (4, 2) lies in quadrant I.
b. Point (−1, −3) lies in quadrant III.
c. Point (2, −2) lies in quadrant IV.
d. Point (−5, 1) lies in quadrant II.

e.–h. Points (3, 0) and $\left(-2\frac{1}{2}, 0\right)$ lie on the x-axis. Points (0, 3) and (0, −4) lie on the y-axis.

i. Point $\left(1, -3\frac{3}{4}\right)$ lies in quadrant IV.

✔ **Concept Check Answer**

The graph of point (−5, 1) lies in quadrant II and the graph of point (1, −5) lies in quadrant IV. They are *not* in the same location.

Helpful Hint

In Example 3, notice that the point (0, 0) lies on both the x-axis and the y-axis. It is the only point in the entire rectangular coordinate system that has this feature. Why? It is the only point of intersection of the x-axis and the y-axis.

✔ **Concept Check** For each description of a point in the rectangular coordinate system, write an ordered pair that represents it.

a. Point A is located three units to the left of the *y*-axis and five units above the *x*-axis.

b. Point B is located six units below the origin.

From Example 3, notice that the *y*-coordinate of any point on the *x*-axis is 0. For example, the point $(-5, 0)$ lies on the *x*-axis. Also, the *x*-coordinate of any point on the *y*-axis is 0. For example, the point $(0, 2)$ lies on the *y*-axis.

Objective **C** Creating Scatter Diagrams

Data that can be represented as ordered pairs are called **paired data.** Many types of data collected from the real world are paired data. For instance, the annual measurements of a child's height can be written as ordered pairs of the form (year, height in inches) and are paired data. The graph of paired data as points in the rectangular coordinate system is called a **scatter diagram.** Scatter diagrams can be used to look for patterns and trends in paired data.

EXAMPLE 4 The table gives the annual net sales for Target Stores for the years shown. (*Source:* TargetCorp.com)

a. Write this paired data as a set of ordered pairs of the form (year, net sales in billions of dollars).

b. Create a scatter diagram of the paired data.

c. What trend in the paired data does the scatter diagram show?

Year	Target Net Sales (in billions of dollars)
1999	34
2000	37
2001	40
2002	44
2003	48

Solution:

a. The ordered pairs are $(1999, 34)$, $(2000, 37)$, $(2001, 40)$, $(2002, 44)$, and $(2003, 48)$.

b. We begin by plotting the ordered pairs. Because the *x*-coordinate in each ordered pair is a year, we label the *x*-axis "Year" and mark the horizontal axis with the years given. Then we label the *y*-axis or vertical axis "Target Net Sales (in billions of dollars)." In this case, it is convenient to mark the vertical axis in multiples of 5, starting with 0. In Practice Problem 4, since there are no years when the number of tornadoes is less than 900, we use the notation ⸝ to skip to 900, then proceed by multiples of 100.

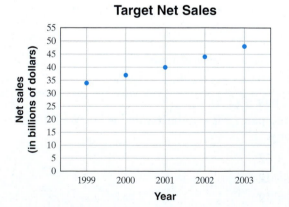

c. The scatter diagram shows that Target net sales steadily increased over the years 1999–2003.

⬛ **Work Practice Problem 4**

PRACTICE PROBLEM 4

The table gives the number of tornadoes that have occurred in the United States for the years shown. (*Source:* Storm Prediction Center, National Weather Service)

Year	Tornadoes
1998	1424
1999	1343
2000	997
2001	1216
2002	941
2003	1376

a. Write this paired data as a set of ordered pairs of the form (year, number of tornadoes).

b. Create a scatter diagram of the paired data.

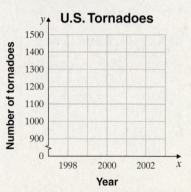

c. What trend in the paired data, if any, does the scatter diagram show?

Answers

4. a. (1998, 1424), (1999, 1343), (2000, 997), (2001, 1216), (2002, 941), (2003, 1376)

b.

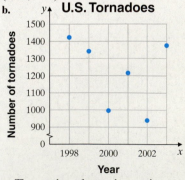

c. The number of tornadoes varies greatly from year to year.

✔ **Concept Check Answers**

a. $(-3, 5)$, **b.** $(0, -6)$

Objective D Completing Ordered Pair Solutions

Let's see how we can use ordered pairs to record solutions of equations containing two variables. An equation in one variable such as $x + 1 = 5$ has one solution, 4: the number 4 is the value of the variable x that makes the equation true.

An equation in two variables, such as $2x + y = 8$, has solutions consisting of two values, one for x and one for y. For example, $x = 3$ and $y = 2$ is a solution of $2x + y = 8$ because, if x is replaced with 3 and y with 2, we get a true statement.

$$2x + y = 8$$
$$2(3) + 2 \overset{?}{=} 8 \quad \text{Replace } x \text{ with 3 and } y \text{ with 2.}$$
$$8 = 8 \quad \text{True}$$

The solution $x = 3$ and $y = 2$ can be written as $(3, 2)$, an ordered pair of numbers.

In general, an ordered pair is a **solution** of an equation in two variables if replacing the variables by the values of the ordered pair results in a *true statement*.

For example, another ordered pair solution of $2x + y = 8$ is $(5, -2)$. Replacing x with 5 and y with -2 results in a true statement.

$$2x + y = 8$$
$$2(5) + (-2) \overset{?}{=} 8 \quad \text{Replace } x \text{ with 5 and } y \text{ with } -2.$$
$$10 - 2 \overset{?}{=} 8$$
$$8 = 8 \quad \text{True}$$

EXAMPLE 5 Complete each ordered pair so that it is a solution to the equation $3x + y = 12$.

a. $(0, \ )$ **b.** $(\ , 6)$ **c.** $(-1, \ )$

Solution:

a. In the ordered pair $(0, \ )$, the x-value is 0. We let $x = 0$ in the equation and solve for y.

$$3x + y = 12$$
$$3(0) + y = 12 \quad \text{Replace } x \text{ with 0.}$$
$$0 + y = 12$$
$$y = 12$$

The completed ordered pair is $(0, 12)$.

b. In the ordered pair $(\ , 6)$, the y-value is 6. We let $y = 6$ in the equation and solve for x.

$$3x + y = 12$$
$$3x + 6 = 12 \quad \text{Replace } y \text{ with 6.}$$
$$3x = 6 \quad \text{Subtract 6 from both sides.}$$
$$x = 2 \quad \text{Divide both sides by 3.}$$

The ordered pair is $(2, 6)$.

c. In the ordered pair $(-1, \ )$, the x-value is -1. We let $x = -1$ in the equation and solve for y.

$$3x + y = 12$$
$$3(-1) + y = 12 \quad \text{Replace } x \text{ with } -1.$$
$$-3 + y = 12$$
$$y = 15 \quad \text{Add 3 to both sides.}$$

The ordered pair is $(-1, 15)$.

Work Practice Problem 5

PRACTICE PROBLEM 5

Complete each ordered pair so that it is a solution to the equation $x + 2y = 8$.

a. $(0, \ )$
b. $(\ , 3)$
c. $(-4, \ )$

Answers

5. a. $(0, 4)$,   **b.** $(2, 3)$,   **c.** $(-4, 6)$

Solutions of equations in two variables can also be recorded in a **table of paired values,** as shown in the next example.

EXAMPLE 6 Complete the table for the equation $y = 3x$.

	x	y
a.	−1	
b.		0
c.		−9

Solution:

a. We replace x with −1 in the equation and solve for y.

$y = 3x$

$y = 3(-1)$ Let $x = -1$.

$y = -3$

The ordered pair is $(-1, -3)$.

b. We replace y with 0 in the equation and solve for x.

$y = 3x$

$0 = 3x$ Let $y = 0$.

$0 = x$ Divide both sides by 3.

The ordered pair is $(0, 0)$.

c. We replace y with −9 in the equation and solve for x.

$y = 3x$

$-9 = 3x$ Let $y = -9$.

$-3 = x$ Divide both sides by 3.

The ordered pair is $(-3, -9)$. The completed table is shown to the right.

x	y
−1	−3
0	0
−3	−9

🟧 **Work Practice Problem 6**

EXAMPLE 7 Complete the table for the equation

$$y = \frac{1}{2}x - 5.$$

	x	y
a.	−2	
b.	0	
c.		0

Solution:

a. Let $x = -2$.

$y = \frac{1}{2}x - 5$

$y = \frac{1}{2}(-2) - 5$

$y = -1 - 5$

$y = -6$

b. Let $x = 0$.

$y = \frac{1}{2}x - 5$

$y = \frac{1}{2}(0) - 5$

$y = 0 - 5$

$y = -5$

c. Let $y = 0$.

$y = \frac{1}{2}x - 5$

$0 = \frac{1}{2}x - 5$ Now, solve for x.

$5 = \frac{1}{2}x$ Add 5.

$10 = x$ Multiply by 2.

Ordered Pairs: $(-2, -6)$ $(0, -5)$ $(10, 0)$

The completed table is

x	−2	0	10
y	−6	−5	0

🟧 **Work Practice Problem 7**

PRACTICE PROBLEM 6

Complete the table for the equation $y = -2x$.

	x	y
a.	−3	
b.		0
c.		10

PRACTICE PROBLEM 7

Complete the table for the equation $y = \frac{1}{3}x - 1$.

	x	y
a.	−3	
b.	0	
c.		0

Answers

6.

	x	y
a.	−3	6
b.	0	0
c.		−5	10

7.

	x	y
a.	−3	−2
b.	0	−1
c.	3	0

By now, you have noticed that equations in two variables often have more than one solution. We discuss this more in the next section.

A table showing ordered pair solutions may be written vertically, or horizontally as shown in the next example.

EXAMPLE 8 A small business purchased a computer for $2000. The business predicts that the computer will be used for 5 years and the value in dollars y of the computer in x years is $y = -300x + 2000$. Complete the table.

x	0	1	2	3	4	5
y						

Solution:

To find the value of y when x is 0, we replace x with 0 in the equation. We use this same procedure to find y when x is 1 and when x is 2.

When x = 0,

$y = -300x + 2000$
$y = -300 \cdot 0 + 2000$
$y = 0 + 2000$
$y = 2000$

When x = 1,

$y = -300x + 2000$
$y = -300 \cdot 1 + 2000$
$y = -300 + 2000$
$y = 1700$

When x = 2,

$y = -300x + 2000$
$y = -300 \cdot 2 + 2000$
$y = -600 + 2000$
$y = 1400$

We have the ordered pairs (0, 2000), (1, 1700), and (2, 1400). This means that in 0 years the value of the computer is $2000, in 1 year the value of the computer is $1700, and in 2 years the value is $1400. To complete the table of values, we continue the procedure for $x = 3$, $x = 4$, and $x = 5$.

When x = 3,

$y = -300x + 2000$
$y = -300 \cdot 3 + 2000$
$y = -900 + 2000$
$y = 1100$

When x = 4,

$y = -300x + 2000$
$y = -300 \cdot 4 + 2000$
$y = -1200 + 2000$
$y = 800$

When x = 5,

$y = -300x + 2000$
$y = -300 \cdot 5 + 2000$
$y = -1500 + 2000$
$y = 500$

The completed table is shown below.

x	0	1	2	3	4	5
y	2000	1700	1400	1100	800	500

Work Practice Problem 8

The ordered pair solutions recorded in the completed table for Example 6 are another set of paired data. They are graphed next. Notice that this scatter diagram gives a visual picture of the decrease in value of the computer.

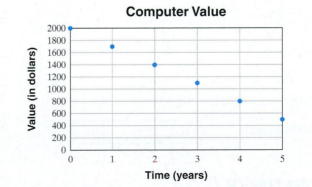

Computer Value

Mental Math

Give two ordered pair solutions for each linear equation.

1. $x + y = 10$

2. $x + y = 6$

6.1 EXERCISE SET

FOR EXTRA HELP

Student Solutions Manual | PH Math/Tutor Center | CD/Video for Review | MathXL® | MyMathLab

Objective **A** *The following bar graph shows the top 10 tourist destinations and the number of tourists that visit each country per year. Use this graph to answer Exercises 1 through 6. See Example 1.*

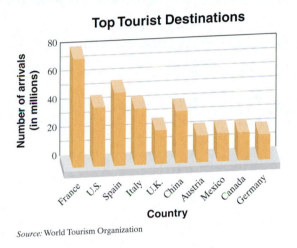

Top Tourist Destinations

Number of arrivals (in millions)

France, U.S., Spain, Italy, U.K., China, Austria, Mexico, Canada, Germany

Country

Source: World Tourism Organization

1. Which country shown is the most popular tourist destination?

2. Which country shown is the least popular tourist destination?

3. Which countries shown have more than 40 million tourists per year?

4. Which countries shown have fewer than 20 million tourists per year?

5. Estimate the number of tourists per year whose destination is Italy.

6. Estimate the number of tourists per year whose destination is Canada.

The following line graph shows the attendance at each Super Bowl game from 1998 through 2004. Use this graph to answer Exercises 7 through 10. See Example 2.

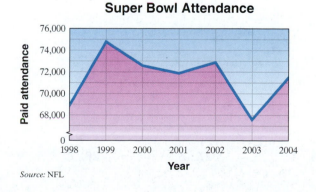

Super Bowl Attendance

Paid attendance

76,000
74,000
72,000
70,000
68,000
0

1998 1999 2000 2001 2002 2003 2004

Year

Source: NFL

7. Estimate the Super Bowl attendance in 2000.

8. Estimate the Super Bowl attendance in 2004.

9. Find the year on the graph with the greatest Super Bowl attendance and approximate that attendance.

10. Find the year on the graph with the least Super Bowl attendance and approximate that attendance.

The line graph below shows the number of students per computer in U.S. public schools. Use this graph for Exercises 11 through 16. See Example 2.

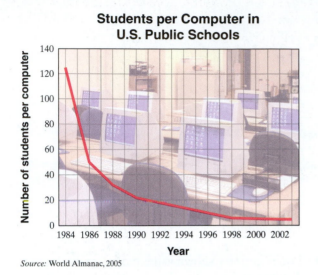

Students per Computer in U.S. Public Schools

Source: World Almanac, 2005

11. Approximate the number of students per computer in 1986.

12. Approximate the number of students per computer in 2002.

13. Between what years did the greatest decrease in number of students per computer occur?

14. What was the first year that the number of students per computer fell below 20?

15. What was the first year that the number of students per computer fell below 15?

16. Discuss any trends shown by this line graph.

Objective **B** *Plot each ordered pair. State in which quadrant or on which axis each point lies. See Example 3.*

17. a. $(1, 5)$ **b.** $(-5, -2)$ **c.** $(-3, 0)$ **d.** $(0, -1)$

     e. $(2, -4)$ **f.** $\left(-1, 4\frac{1}{2}\right)$ **g.** $(3.7, 2.2)$ **h.** $\left(\frac{1}{2}, -3\right)$

18. a. $(2, 4)$ **b.** $(0, 2)$ **c.** $(-2, 1)$ **d.** $(-3, -3)$

     e. $\left(3\frac{3}{4}, 0\right)$ **f.** $(5, -4)$ **g.** $(-3.4, 4.8)$ **h.** $\left(\frac{1}{3}, -5\right)$

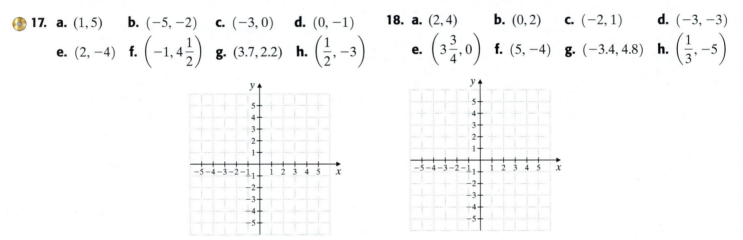

Find the x- and y-coordinates of each labeled point. See Example 3.

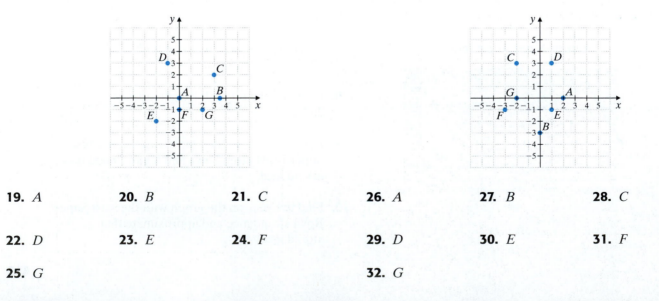

19. *A*   **20.** *B*   **21.** *C*

22. *D*   **23.** *E*   **24.** *F*

25. *G*

26. *A*   **27.** *B*   **28.** *C*

29. *D*   **30.** *E*   **31.** *F*

32. *G*

Objective **C** *Solve. See Example 4.*

33. The table shows the amount of money (in billions of dollars) that Americans spent on their pets for the years shown. (*Source:* American Pet Products Manufacturers Association)

Year	Pet-Related Expenditures (in billions of dollars)
2001	28.5
2002	29.5
2003	32.4
2004	34.3

a. Write this paired data as a set of ordered pairs of the form (year, pet-related expenditures).

b. In your own words, write the meaning of the ordered pair (2004, 34.3).

c. Create a scatter diagram of the paired data. Be sure to label the axes appropriately.

Pet-Related Expenditures

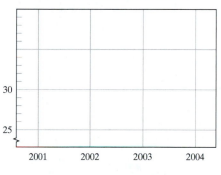

d. What trend in the paired data does the scatter diagram show?

34. The table shows the average farm size (in acres) in the United States during the years shown. (*Source:* National Agricultural Statistics Service)

Year	Average Farm Size (in acres)
1998	435
1999	432
2000	434
2001	438
2002	440
2003	441

a. Write this paired data as a set of ordered pairs of the form (year, average farm size).

b. In your own words, write the meaning of the ordered pair (2003, 441).

c. Create a scatter diagram of the paired data. Be sure to label the axes appropriately.

U.S. Average Farm Size

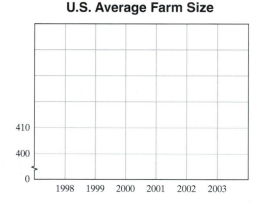

35. Minh, a psychology student, kept a record of how much time she spent studying for each of her 20-point psychology quizzes and her score on each quiz.

Hours Spent Studying	0.50	0.75	1.00	1.25	1.50	1.50	1.75	2.00
Quiz Score	10	12	15	16	18	19	19	20

a. Write each paired data as an ordered pair of the form (hours spent studying, quiz score).

b. In your own words, write the meaning of the ordered pair (1.25, 16).

c. Create a scatter diagram of the paired data. Be sure to label the axes appropriately.

d. What might Minh conclude from the scatter diagram?

Minh's Chart for Psychology

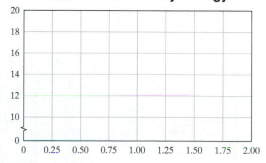

36. A local lumberyard uses quantity pricing. The table shows the price per board for different amounts of lumber purchased.

Price per Board (in dollars)	Number of Boards Purchased
8.00	1
7.50	10
6.50	25
5.00	50
2.00	100

a. Write each paired data as an ordered pair of the form (price per board, number of boards purchased).

✏ **b.** In your own words, write the meaning of the ordered pair $(2.00, 100)$.

c. Create a scatter diagram of the paired data. Be sure to label the axes appropriately.

Lumberyard Board Pricing

✏ **d.** What trend in the paired data does the scatter diagram show?

Objective **D** *Complete each ordered pair so that it is a solution of the given linear equation. See Example 5.*

💿 **37.** $x - 4y = 4; (\quad, -2), (4, \quad)$

38. $x - 5y = -1; (\quad, -2), (4, \quad)$

39. $y = \dfrac{1}{4}x - 3; (-8, \quad), (\quad, 1)$

40. $y = \dfrac{1}{5}x - 2; (-10, \quad), (\quad, 1)$

Complete the table of ordered pairs for each linear equation. See Examples 6 and 7.

41. $y = -7x$

x	y
0	
−1	
	2

42. $y = -9x$

x	y
	0
−3	
	2

43. $y = -x + 2$

x	y
0	
	0
−3	

44. $x = -y + 4$

x	y
	0
0	
	−3

45. $y = \dfrac{1}{2}x$

x	y
0	
−6	
	1

46. $y = \dfrac{1}{3}x$

x	y
0	
−6	
	1

47. $x + 3y = 6$

x	y
0	
	0
	1

48. $2x + y = 4$

x	y
0	
	0
	2

49. $y = 2x - 12$

x	y
0	
	−2
3	

50. $y = 5x + 10$

x	y
	0
	5
0	

51. $2x + 7y = 5$

x	y
0	
	0
	1

52. $x - 6y = 3$

x	y
0	
1	
	-1

Objectives B C D **Mixed Practice** *Complete the table of ordered pairs for each equation. Then plot the ordered pair solutions. See Examples 1 through 7.*

53. $x = -5y$

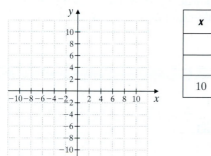

x	y
	0
	1
10	

54. $y = -3x$

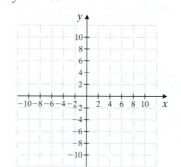

x	y
0	
-2	
	9

55. $y = \dfrac{1}{3}x + 2$

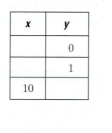

x	y
0	
-3	
	0

56. $y = \dfrac{1}{2}x + 3$

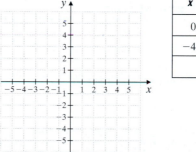

x	y
0	
-4	
	0

Solve. See Example 8.

57. The cost in dollars y of producing x computer desks is given by $y = 80x + 5000$.

   a. Complete the table.

x	100	200	300
y			

   b. Find the number of computer desks that can be produced for $8600. (*Hint:* Find x when $y = 8600$.)

58. The hourly wage y of an employee at a certain production company is given by $y = 0.25x + 9$ where x is the number of units produced by the employee in an hour.

   a. Complete the table.

x	0	1	5	10
y				

   b. Find the number of units that an employee must produce each hour to earn an hourly wage of $12.25. (*Hint:* Find x when $y = 12.25$.)

59. The percent y of recorded music sales that were in cassette format from 1998 through 2003 is given by $y = -2.4x + 13$. In the equation, x represents the number of years after 1998. (*Source:* Recording Industry Association of America)

a. Complete the table.

x	1	3	5
y			

b. Find the year in which approximately 3% of recorded music sales were cassettes. (*Hint:* Find x when $y = 3$ and round to the nearest whole number.)

60. The amount y of land occupied by farms in the United States (in million acres) from 1993 through 2003 is given by $y = -3.2x + 968$. In the equation, x represents the number of years after 1993. (*Source:* National Agricultural Statistics Service)

a. Complete the table.

x	4	7	10
y			

b. Find the year in which there were approximately 943 million acres of land occupied by farms. (*Hint:* Find x when $y = 943$ and round to the nearest whole number.)

Review

Solve each equation for y. See Section 2.5.

61. $x + y = 5$

62. $x - y = 3$

63. $2x + 4y = 5$

64. $5x + 2y = 7$

65. $10x = -5y$

66. $4y = -8x$

Concept Extensions

Answer each exercise with true or false.

67. Point $(-1, 5)$ lies in quadrant IV.

68. Point $(3, 0)$ lies on the y-axis.

69. For the point $\left(-\frac{1}{2}, 1.5\right)$, the first value, $-\frac{1}{2}$, is the x-coordinate and the second value, 1.5, is the y-coordinate.

70. The ordered pair $\left(2, \frac{2}{3}\right)$ is a solution of $2x - 3y = 6$.

For Exercises 71 through 75, fill in each blank with "0," "positive," or "negative." For Exercises 76 and 77, fill in each blank with "x" or "y."

	Point	Location
71.	(,)	quadrant III
72.	(,)	quadrant I
73.	(,)	quadrant IV
74.	(,)	quadrant II
75.	(,)	origin
76.	(number, 0)	-axis
77.	(0, number)	-axis

78. Give an example of an ordered pair whose location is in (or on)

a. quadrant I   **b.** quadrant II   **c.** quadrant III
d. quadrant IV   **e.** x-axis   **f.** y-axis

Solve. See the Concept Check in this section.

79. Is the graph of $(3, 0)$ in the same location as the graph of $(0, 3)$? Explain why or why not.

80. Give the coordinates of a point such that if the coordinates are reversed, their location is the same.

81. In general, what points can have coordinates reversed and still have the same location?

82. In your own words, describe how to plot or graph an ordered pair of numbers.

Write an ordered pair for each point described.

83. Point C is four units to the right of the y-axis and seven units below the x-axis.

84. Point D is three units to the left of the origin.

85. Find the perimeter of the rectangle whose vertices are the points with coordinates $(-1, 5)$, $(3, 5)$, $(3, -4)$, and $(-1, -4)$.

86. Find the area of the rectangle whose vertices are the points with coordinates $(5, 2)$, $(5, -6)$, $(0, -6)$, and $(0, 2)$.

The scatter diagram below shows Walt Disney Company's annual revenues. The horizontal axis represents the number of years after 1999.

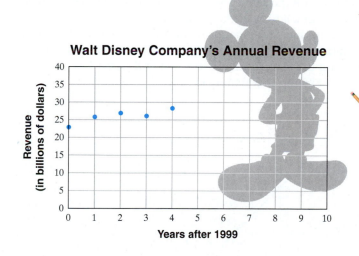

Walt Disney Company's Annual Revenue

87. Estimate the annual revenues for years 1, 2, 3, and 4.

88. Use a straight edge or ruler and this scatter diagram to predict Disney's revenue in the year 2008.

89. Discuss any similarities in the graphs of the ordered pair solutions for Exercises 53–56.

📖 **STUDY SKILLS BUILDER**

Are You Satisfied with Your Performance in This Course Thus Far?

To see if there is room for improvement, answer these questions:

1. Am I attending all classes and arriving on time?

2. Am I working and checking my homework assignments on time?

3. Am I getting help (from my instructor or a campus learning resource lab) when I need it?

4. In addition to my instructor, am I using the text supplements that might help me?

5. Am I satisfied with my performance on quizzes and exams?

If you answered no to any of these questions, read or reread Section 1.1 for suggestions in these areas. Also, you might want to contact your instructor for additional feedback.

6.2 GRAPHING LINEAR EQUATIONS

In the previous section, we found that equations in two variables may have more than one solution. For example, both $(2, 2)$ and $(0, 4)$ are solutions of the equation $x + y = 4$. In fact, this equation has an infinite number of solutions. Other solutions include $(-2, 6)$, $(4, 0)$, and $(6, -2)$. Notice the pattern that appears in the graph of these solutions.

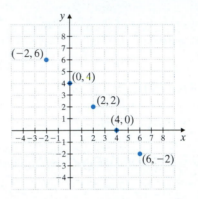

These solutions all appear to lie on the same line, as seen in the second graph. It can be shown that every ordered pair solution of the equation corresponds to a point on this line, and every point on this line corresponds to an ordered pair solution. Thus, we say that this line is the **graph of the equation** $x + y = 4$. Notice that we can only show a part of a line on a graph. The arrowheads on each end of the line below remind us that the line actually extends indefinitely in both directions.

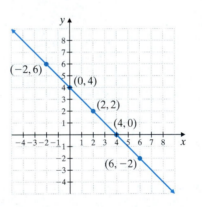

The equation $x + y = 4$ is called a *linear equation in two variables* and *the graph of every linear equation in two variables is a straight line.*

Linear Equation in Two Variables

A **linear equation in two variables** is an equation that can be written in the form

$$Ax + By = C$$

where A, B, and C are real numbers and A and B are not both 0. This form is called **standard form.** **The graph of a linear equation in two variables is a straight line.**

A linear equation in two variables may be written in many forms. Standard form, $Ax + By = C$, is just one of many of these forms.

Following are examples of linear equations in two variables.

$2x + y = 8$ $-2x = 7y$ $y = \dfrac{1}{3}x + 2$ $y = 7$
(Standard Form)

Objective Ⓐ Graphing Linear Equations

From geometry, we know that a straight line is determined by just two points. Thus, to graph a linear equation in two variables we need to find just two of its infinitely many solutions. Once we do so, we plot the solution points and draw the line connecting the points. Usually, we find a third solution as well, as a check.

EXAMPLE 1 Graph the linear equation $2x + y = 5$.

Solution: To graph this equation, we find three ordered pair solutions of $2x + y = 5$. To do this, we choose a value for one variable, x or y, and solve for the other variable. For example, if we let $x = 1$, then $2x + y = 5$ becomes

$$2x + y = 5$$
$$2(1) + y = 5 \quad \text{Replace } x \text{ with 1.}$$
$$2 + y = 5 \quad \text{Multiply.}$$
$$y = 3 \quad \text{Subtract 2 from both sides.}$$

Since $y = 3$ when $x = 1$, the ordered pair $(1, 3)$ is a solution of $2x + y = 5$. Next, we let $x = 0$.

$$2x + y = 5$$
$$2(0) + y = 5 \quad \text{Replace } x \text{ with 0.}$$
$$0 + y = 5$$
$$y = 5$$

The ordered pair $(0, 5)$ is a second solution.

The two solutions found so far allow us to draw the straight line that is the graph of all solutions of $2x + y = 5$. However, we will find a third ordered pair as a check. Let $y = -1$.

$$2x + y = 5$$
$$2x + (-1) = 5 \quad \text{Replace } y \text{ with } -1.$$
$$2x - 1 = 5$$
$$2x = 6 \quad \text{Add 1 to both sides.}$$
$$x = 3 \quad \text{Divide both sides by 2.}$$

The third solution is $(3, -1)$. These ordered pair solutions are listed in the table and plotted on the coordinate plane. The graph of $2x + y = 5$ is the line through the three points.

x	y
1	3
0	5
3	-1

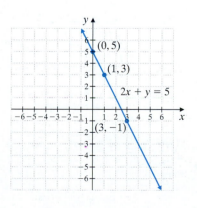

 Work Practice Problem 1

PRACTICE PROBLEM 1

Graph the linear equation $x + 3y = 6$.

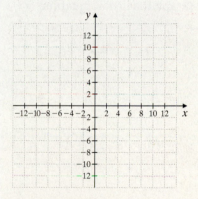

Helpful Hint

All three points should fall on the same straight line. If not, check your ordered pair solutions for a mistake.

Answer

1.

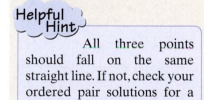

PRACTICE PROBLEM 2

Graph the linear equation
$-2x + 4y = 8$.

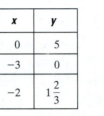

PRACTICE PROBLEM 3

Graph the linear equation
$y = 2x$.

Answers

2.

3.

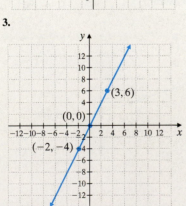

EXAMPLE 2 Graph the linear equation $-5x + 3y = 15$.

Solution: We find three ordered pair solutions of $-5x + 3y = 15$.

Let $x = 0$.	Let $y = 0$.	Let $x = -2$.
$-5x + 3y = 15$	$-5x + 3y = 15$	$-5x + 3y = 15$
$-5 \cdot 0 + 3y = 15$	$-5x + 3 \cdot 0 = 15$	$-5 \cdot -2 + 3y = 15$
$0 + 3y = 15$	$-5x + 0 = 15$	$10 + 3y = 15$
$3y = 15$	$-5x = 15$	$3y = 5$
$y = 5$	$x = -3$	$y = \dfrac{5}{3}$ or $1\dfrac{2}{3}$

The ordered pairs are $(0, 5)$, $(-3, 0)$, and $\left(-2, 1\dfrac{2}{3}\right)$. The graph of $-5x + 3y = 15$ is the line through the three points.

x	y
0	5
-3	0
-2	$1\dfrac{2}{3}$

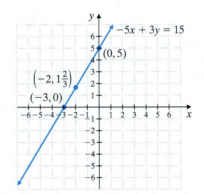

🔲 **Work Practice Problem 2**

EXAMPLE 3 Graph the linear equation $y = 3x$.

Solution: We find three ordered pair solutions. Since this equation is solved for y, we'll choose three x values.

If $x = 2$, $y = 3 \cdot 2 = 6$.
If $x = 0$, $y = 3 \cdot 0 = 0$.
If $x = -1$, $y = 3 \cdot -1 = -3$.

Next, we plot the ordered pair solutions and draw a line through the plotted points. The line is the graph of $y = 3x$.

Think about the following for a moment: A line is made up of an infinite number of points. Every point on the line defined by $y = 3x$ represents an ordered pair solution of the equation and every ordered pair solution is a point on this line.

x	y
2	6
0	0
-1	-3

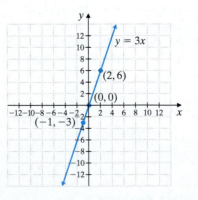

🔲 **Work Practice Problem 3**

Helpful Hint

When graphing a linear equation in two variables, if it is

- solved for y, it may be easier to find ordered-pair solutions by choosing x-values. If it is
- solved for x, it may be easier to find ordered-pair solutions by choosing y-values.

EXAMPLE 4 Graph the linear equation $y = -\frac{1}{3}x + 2$.

Solution: We find three ordered pair solutions, plot the solutions, and draw a line through the plotted solutions. To avoid fractions, we'll choose x values that are multiples of 3 to substitute into the equation.

If $x = 6$, then $y = -\frac{1}{3} \cdot 6 + 2 = -2 + 2 = 0$

If $x = 0$, then $y = -\frac{1}{3} \cdot 0 + 2 = 0 + 2 = 2$

If $x = -3$, then $y = -\frac{1}{3} \cdot -3 + 2 = 1 + 2 = 3$

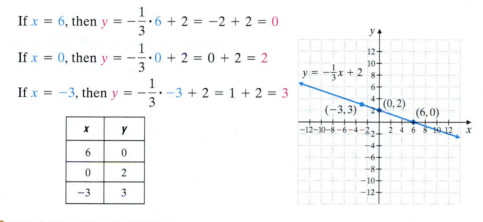

x	y
6	0
0	2
-3	3

Work Practice Problem 4

Let's take a moment and compare the graphs in Examples 3 and 4. The graph of $y = 3x$ tilts upward (as we follow the line from left to right) and the graph of $y = -\frac{1}{3}x + 2$ tilts downward (as we follow the line from left to right). We will learn more about the tilt, or slope, of a line in Section 6.4.

EXAMPLE 5 Graph the linear equation $y = -2$.

Solution: The equation $y = -2$ can be written in standard form as $0x + y = -2$. No matter what value we replace x with, y is always -2.

x	y
0	-2
3	-2
-2	-2

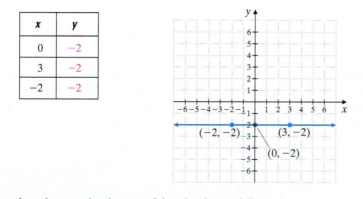

Notice that the graph of $y = -2$ is a horizontal line.

Work Practice Problem 5

Linear equations are often used to model real data, as seen in the next example.

PRACTICE PROBLEM 4

Graph the linear equation $y = -\frac{1}{2}x + 4$.

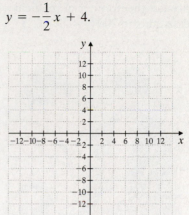

PRACTICE PROBLEM 5

Graph the linear equation $x = 3$.

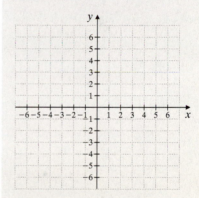

Answers

4.

5.

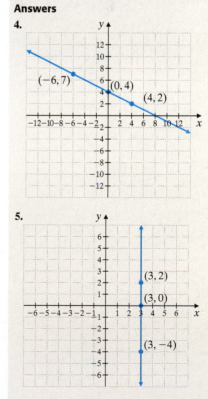

PRACTICE PROBLEM 6

Use the graph in Example 6 to predict the number of medical assistants in 2004.

EXAMPLE 6 **Estimating the Number of Medical Assistants**

One of the occupations expected to have the most growth in the next few years is medical assistant. The number of people y (in thousands) employed as medical assistants in the United States can be estimated by the linear equation $y = 31.8x + 180$, where x is the number of years after the year 1995. (*Source:* Based on data from the Bureau of Labor Statistics)

a. Graph the equation.

b. Use the graph to predict the number of medical assistants in the year 2010.

Solution:

a. To graph $y = 31.8x + 180$, choose x-values and substitute in the equation.

If $x = 0$, then $y = 31.8(0) + 180 = 180$.
If $x = 2$, then $y = 31.8(2) + 180 = 243.6$.
If $x = 7$, then $y = 31.8(7) + 180 = 402.6$.

x	y
0	180
2	243.6
7	402.6

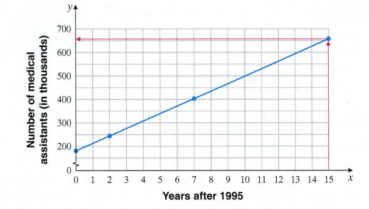

Years after 1995

b. To use the graph to *predict* the number of medical assistants in the year 2010, we need to find the y-coordinate that corresponds to $x = 15$. (15 years after 1995 is the year 2010.) To do so, find 15 on the x-axis. Move vertically upward to the graphed line and then horizontally to the left. We approximate the number on the y-axis to be 655. Thus in the year 2010, we predict that there will be 655 thousand medical assistants. (The actual value, using 15 for x, is 657.)

■ **Work Practice Problem 6**

 Helpful Hint

Make sure you understand that models are mathematical approximations of the data for the known years. (For example, see the model in Example 6.) Any number of unknown factors can affect future years, so be cautious when using models to predict.

Answer

6. 465 thousand

CALCULATOR EXPLORATIONS Graphing

In this section, we begin an optional study of graphing calculators and graphing software packages for computers. These graphers use the same point plotting technique that was introduced in this section. The advantage of this graphing technology is, of course, that graphing calculators and computers can find and plot ordered pair solutions much faster than we can. Note, however, that the features described in these boxes may not be available on all graphing calculators.

The rectangular screen where a portion of the rectangular coordinate system is displayed is called a **window.** We call it a **standard window** for graphing when both the x- and y-axes show coordinates between -10 and 10. This information is often displayed in the window menu on a graphing calculator as follows.

Xmin = -10
Xmax = 10
Xscl = 1 The scale on the x-axis is one unit per tick mark.
Ymin = -10
Ymax = 10
Yscl = 1 The scale on the y-axis is one unit per tick mark.

To use a graphing calculator to graph the equation $y = 2x + 3$, press the $\boxed{Y=}$ key and enter the keystrokes $\boxed{2}\ \boxed{x}\ \boxed{+}\ \boxed{3}$. The top row should now read $Y_1 = 2x + 3$. Next press the $\boxed{GRAPH}$ key, and the display should look like this:

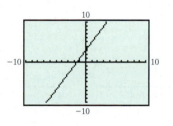

Graph the following linear equations. (Unless otherwise stated, use a standard window when graphing.)

1. $y = -3x + 7$

2. $y = -x + 5$

3. $y = 2.5x - 7.9$

4. $y = -1.3x + 5.2$

5. $y = -\dfrac{3}{10}x + \dfrac{32}{5}$

6. $y = \dfrac{2}{9}x - \dfrac{22}{3}$

Objective A *For each equation, find three ordered pair solutions by completing the table. Then use the ordered pairs to graph the equation. See Examples 1 through 5.*

1. $x - y = 6$

x	y
	0
4	
	-1

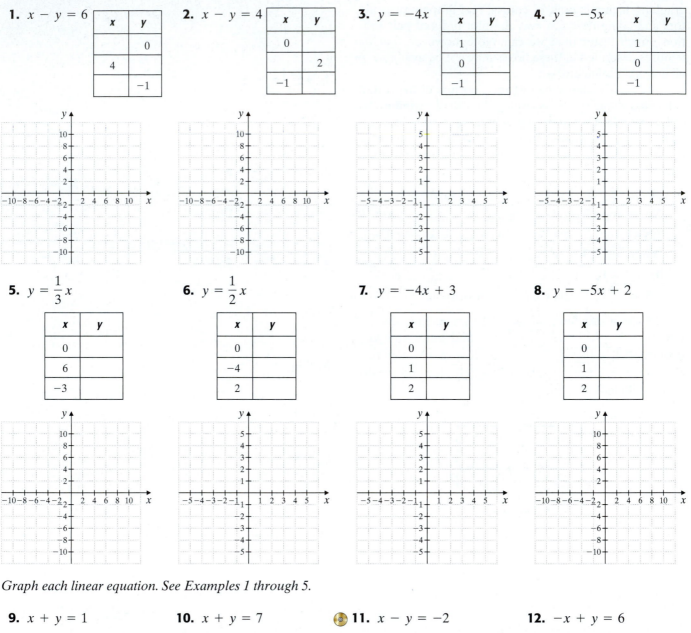

2. $x - y = 4$

x	y
0	
	2
-1	

3. $y = -4x$

x	y
1	
0	
-1	

4. $y = -5x$

x	y
1	
0	
-1	

5. $y = \dfrac{1}{3}x$

x	y
0	
6	
-3	

6. $y = \dfrac{1}{2}x$

x	y
0	
-4	
2	

7. $y = -4x + 3$

x	y
0	
1	
2	

8. $y = -5x + 2$

x	y
0	
1	
2	

Graph each linear equation. See Examples 1 through 5.

9. $x + y = 1$

10. $x + y = 7$

11. $x - y = -2$

12. $-x + y = 6$

13. $x - 2y = 6$

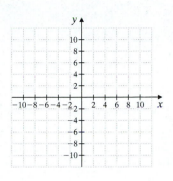

14. $-x + 5y = 5$

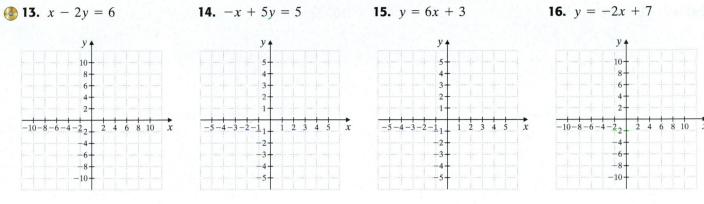

15. $y = 6x + 3$

16. $y = -2x + 7$

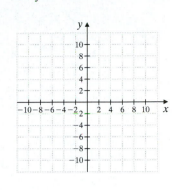

17. $x = -4$

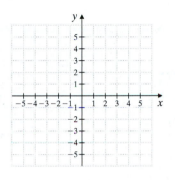

18. $y = 5$

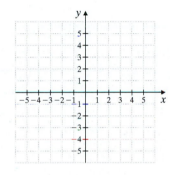

19. $y = 3$

20. $x = -1$

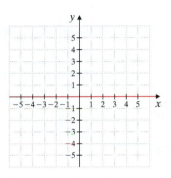

21. $y = x$

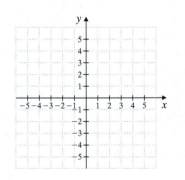

22. $y = -x$

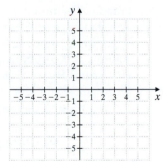

23. $y = 5x$

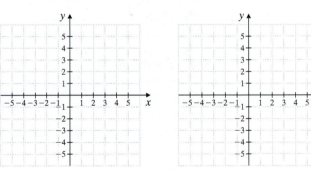

24. $y = 4x$

25. $x + 3y = 9$

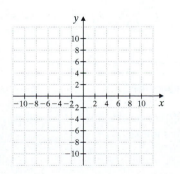

26. $2x + y = 2$

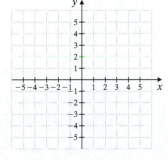

27. $y = \dfrac{1}{2}x - 1$

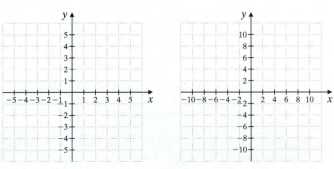

28. $y = \dfrac{1}{4}x + 3$

29. $3x - 2y = 12$

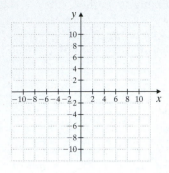

30. $2x - 7y = 14$

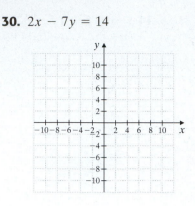

31. $y = -3.5x + 4$

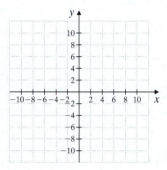

32. $y = -1.5x - 3$

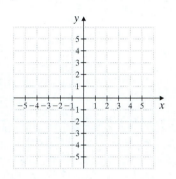

Solve. See Example 6.

33. One of the top five occupations in terms of growth in the next few years is expected to be physician's assistants. The number of people y (in hundreds) employed as physician's assistants in the United States can be estimated by the linear equation $y = 31x + 630$ where x is the number of years after 2002. (*Source:* Based on data from the Bureau of Labor Statistics)

a. Graph the linear equation. The break in the vertical axis means that the numbers between 0 and 600 have been skipped.

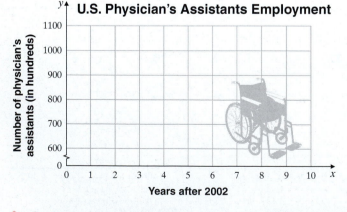

b. Does the point $(6, 816)$ lie on the line? If so, what does this ordered pair mean?

34. Head Start is a comprehensive child development program serving young children in low-income families. The number of children y (in thousands) enrolled in Head Start from 1998 to 2003 can be approximated by the linear equation $y = 21x + 822$, where x is the number of years after 1998. (*Source:* Head Start Bureau, the Administration on Children, Youth and Families)

a. Graph the linear equation.

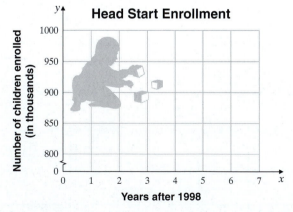

b. Does the point $(3, 885)$ lie on the line? If so, what does this ordered pair mean?

35. The number of U.S. households y in millions that have at least one television set can be estimated by the linear equation $y = 1.5x + 99$ where x is the number of years after 1999. (*Source:* Nielsen Media Research)

a. Graph the linear equation.

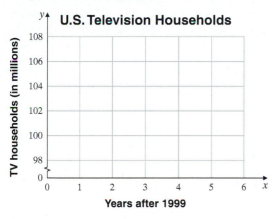

U.S. Television Households

b. Complete the ordered pair (5,).

c. Write a sentence explaining the meaning of the ordered pair found in part b.

36. The restaurant industry is busier than ever. The yearly revenue for restaurants in the United States can be estimated by $y = 11.9x + 284$ where x is the number of years after 2001 and y is the revenue in billions of dollars. (*Source:* National Restaurant Assn.)

a. Graph the linear equation.

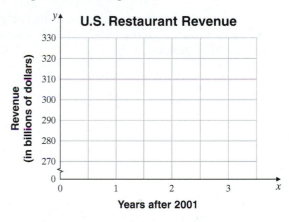

U.S. Restaurant Revenue

b. Complete the ordered pair (3,).

c. Write a sentence explaining the meaning of the ordered pair found in part b.

Review

△ **37.** The coordinates of three vertices of a rectangle are $(-2, 5)$, $(4, 5)$, and $(-2, -1)$. Find the coordinates of the fourth vertex. See Section 6.1.

△ **38.** The coordinates of two vertices of a square are $(-3, -1)$ and $(2, -1)$. Find the coordinates of two pairs of points possible for the third and fourth vertices. See Section 6.1.

Complete each table. See Section 6.1.

39. $x - y = -3$

x	y
0	
	0

40. $y - x = 5$

x	y
0	
	0

41. $y = 2x$

x	y
0	
	0

42. $x = -3y$

x	y
0	
	0

Concept Extensions

Graph each pair of linear equations on the same set of axes. Discuss how the graphs are similar and how they are different.

43. $y = 5x$
$y = 5x + 4$

44. $y = 2x$
$y = 2x + 5$

45. $y = -2x$
$y = -2x - 3$

46. $y = x$
$y = x - 7$

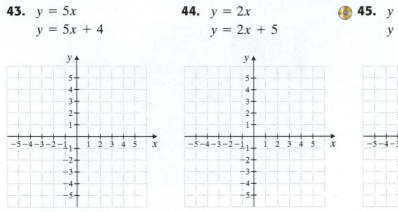
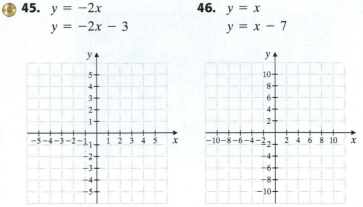

47. Graph the nonlinear equation $y = x^2$ by completing the table shown. Plot the ordered pairs and connect them with a smooth curve.

x	y
0	
1	
−1	
2	
−2	

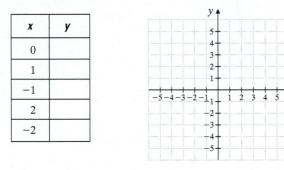

48. Graph the nonlinear equation $y = |x|$ by completing the table shown. Plot the ordered pairs and connect them. This curve is "V" shaped.

x	y
0	
1	
−1	
2	
−2	

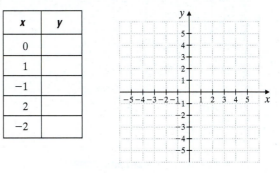

△ **49.** The perimeter of the trapezoid below is 22 centimeters. Write a linear equation in two variables for the perimeter. Find y if x is 3 centimeters.

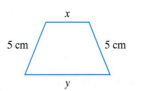

△ **50.** The perimeter of the rectangle below is 50 miles. Write a linear equation in two variables for the perimeter. Use this equation to find x when y is 20.

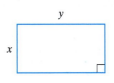

51. If (a, b) is an ordered pair solution of $x + y = 5$, is (b, a) also a solution? Explain why or why not.

6.3 INTERCEPTS

Objectives

A Identify Intercepts of a Graph.

B Graph a Linear Equation by Finding and Plotting Intercept Points.

C Identify and Graph Vertical and Horizontal Lines.

Objective A Identifying Intercepts

The graph of $y = 4x - 8$ is shown below. Notice that this graph crosses the y-axis at the point $(0, -8)$. This point is called the **y-intercept.** Likewise the graph crosses the x-axis at $(2, 0)$. This point is called the **x-intercept.**

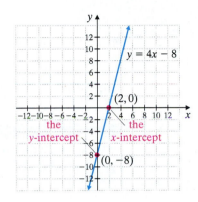

The intercepts are $(2, 0)$ and $(0, -8)$.

> **Helpful Hint**
>
> If a graph crosses the x-axis at $(2, 0)$ and the y-axis at $(0, -8)$, then
>
> $$\underset{\substack{\uparrow \\ x\text{-intercept}}}{(2, 0)} \qquad \underset{\substack{\uparrow \\ y\text{-intercept}}}{(0, -8)}$$
>
> Notice that for the x-intercept, the y-value is 0 and for the y-intercept, the x-value is 0.
>
> *Note:* Sometimes in mathematics, you may see just the number -8 stated as the y-intercept, and 2 stated as the x-intercept.

EXAMPLES Identify the x- and y-intercepts.

1.

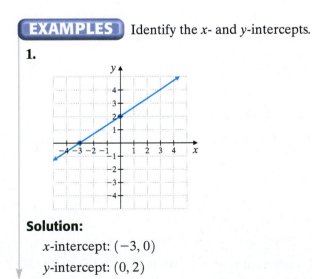

Solution:

x-intercept: $(-3, 0)$

y-intercept: $(0, 2)$

Continued on next page

PRACTICE PROBLEM 1

Identify the x- and y-intercepts.

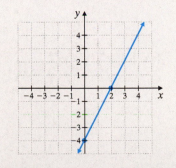

Answer

1. x-intercept: $(2, 0)$; y-intercept: $(0, -4)$

433

PRACTICE PROBLEMS 2–3

Identify the *x*- and *y*-intercepts.

2.

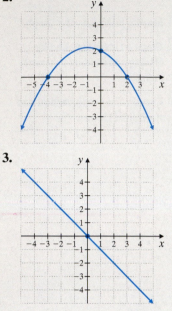

3.

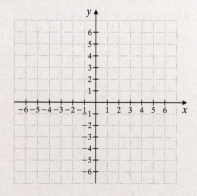

2.

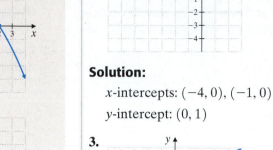

Solution:

x-intercepts: $(-4, 0)$, $(-1, 0)$

y-intercept: $(0, 1)$

3.

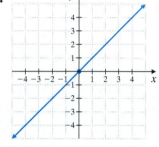

Solution:

x-intercept: $(0, 0)$

y-intercept: $(0, 0)$

Here, the *x*- and *y*-intercept happen to be the same point.

Helpful Hint

Notice that any time $(0, 0)$ is a point of a graph, then it is an *x*-intercept and a *y*-intercept. Why? It is the *only* point that lies on both axes.

🔶 **Work Practice Problems 1–3**

PRACTICE PROBLEM 4

Graph $2x - y = 4$ by finding and plotting its intercepts.

Objective B Finding and Plotting Intercepts

Given an equation of a line, we can usually find intercepts easily since one coordinate is 0.

To find the *x*-intercept of a line from its equation, let $y = 0$, since a point on the *x*-axis has a *y*-coordinate of 0. To find the *y*-intercept of a line from its equation, let $x = 0$, since a point on the *y*-axis has an *x*-coordinate of 0.

Finding x- and y-Intercepts

To find the *x*-intercept, let $y = 0$ and solve for *x*.
To find the *y*-intercept, let $x = 0$ and solve for *y*.

EXAMPLE 4 Graph $x - 3y = 6$ by finding and plotting its intercepts.

Solution: We let $y = 0$ to find the *x*-intercept and $x = 0$ to find the *y*-intercept.

$$\text{Let } y = 0. \qquad \text{Let } x = 0.$$
$$x - 3y = 6 \qquad x - 3y = 6$$
$$x - 3(0) = 6 \qquad 0 - 3y = 6$$
$$x - 0 = 6 \qquad -3y = 6$$
$$x = 6 \qquad y = -2$$

The *x*-intercept is $(6, 0)$ and the *y*-intercept is $(0, -2)$. We find a third ordered pair solution to check our work. If we let $y = -1$, then $x = 3$. We plot the points $(6, 0)$,

Answers

2. *x*-intercepts: $(-4, 0)$, $(2, 0)$; *y*-intercept: $(0, 2)$,

3. *x*-intercept and *y*-intercept: $(0, 0)$,

4. See page 435.

(0, −2), and (3, −1). The graph of $x - 3y = 6$ is the line drawn through these points as shown.

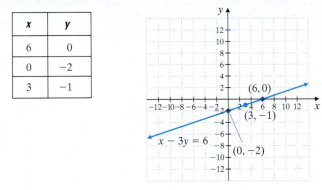

x	y
6	0
0	−2
3	−1

🔲 **Work Practice Problem 4**

EXAMPLE 5 Graph $x = -2y$ by finding and plotting its intercepts.

Solution: We let $y = 0$ to find the x-intercept and $x = 0$ to find the y-intercept.

Let $y = 0$.	Let $x = 0$.
$x = -2y$	$x = -2y$
$x = -2(0)$	$0 = -2y$
$x = 0$	$0 = y$

Both the x-intercept and y-intercept are (0, 0). In other words, when $x = 0$, then $y = 0$, which gives the ordered pair (0, 0). Also, when $y = 0$, then $x = 0$, which gives the same ordered pair (0, 0). This happens when the graph passes through the origin. Since two points are needed to determine a line, we must find at least one more ordered pair that satisfies $x = -2y$. Since the equation is solved for x, we choose y-values so that there is no need to solve to find the corresponding x-value. We let $y = -1$ to find a second ordered pair solution and let $y = 1$ as a check point.

Let $y = -1$.
$x = -2(-1)$
$x = 2$ Multiply.

Let $y = 1$.
$x = -2(1)$
$x = -2$ Multiply.

The ordered pairs are (0, 0), (2, −1), and (−2, 1). We plot these points to graph $x = -2y$.

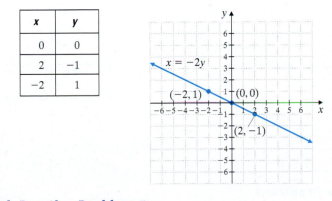

x	y
0	0
2	−1
−2	1

🔲 **Work Practice Problem 5**

PRACTICE PROBLEM 5

Graph $y = 3x$ by finding and plotting its intercepts.

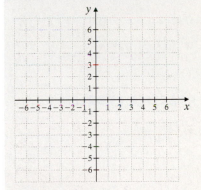

Answers

4.

5.

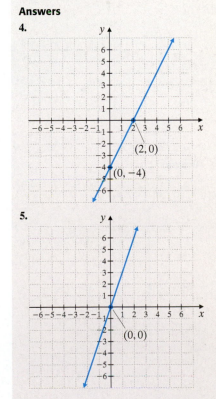

Objective C Graphing Vertical and Horizontal Lines

From Section 6.2, recall that the equation $x = 2$, for example, is a linear equation in two variables because it can be written in the form $x + 0y = 2$. The graph of this equation is a vertical line, as reviewed in the next example.

PRACTICE PROBLEM 6

Graph: $x = -3$

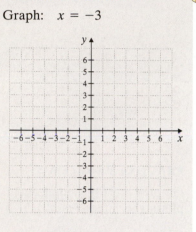

EXAMPLE 6 Graph: $x = 2$

Solution: The equation $x = 2$ can be written as $x + 0y = 2$. For any y-value chosen, notice that x is 2. No other value for x satisfies $x + 0y = 2$. Any ordered pair whose x-coordinate is 2 is a solution of $x + 0y = 2$. We will use the ordered pair solutions $(2, 3)$, $(2, 0)$, and $(2, -3)$ to graph $x = 2$.

x	y
2	3
2	0
2	-3

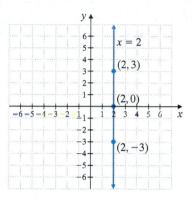

The graph is a vertical line with x-intercept 2. Note that this graph has no y-intercept because x is never 0.

Work Practice Problem 6

PRACTICE PROBLEM 7

Graph: $y = 4$

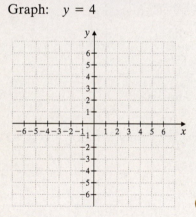

In general, we have the following.

Vertical Lines

The graph of $x = c$, where c is a real number, is a **vertical line** with x-intercept $(c, 0)$.

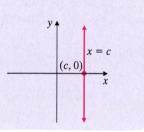

EXAMPLE 7 Graph: $y = -3$

Solution: The equation $y = -3$ can be written as $0x + y = -3$. For any x-value chosen, y is -3. If we choose 4, 1, and -2 as x-values, the ordered pair solutions are $(4, -3)$, $(1, -3)$, and $(-2, -3)$. We use these ordered pairs to graph $y = -3$. The graph is a horizontal line with y-intercept -3 and no x-intercept.

x	y
4	-3
1	-3
-2	-3

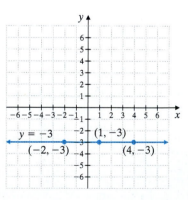

Work Practice Problem 7

Answers
6.

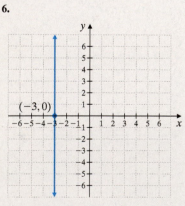

7. See page 437.

In general, we have the following.

Horizontal Lines

The graph of $y = c$, where c is a real number, is a **horizontal line** with y-intercept $(0, c)$.

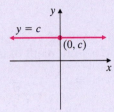

Answer

7.
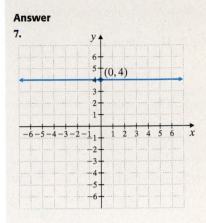

📟 CALCULATOR EXPLORATIONS Graphing

You may have noticed that to use the $\boxed{Y=}$ key on a graphing calculator to graph an equation, the equation must be solved for y. For example, to graph $2x + 3y = 7$, we solve this equation for y.

$$2x + 3y = 7$$

$$3y = -2x + 7 \quad \text{Subtract } 2x \text{ from both sides.}$$

$$\frac{3y}{3} = -\frac{2x}{3} + \frac{7}{3} \quad \text{Divide both sides by 3.}$$

$$y = -\frac{2}{3}x + \frac{7}{3} \quad \text{Simplify.}$$

To graph $2x + 3y = 7$ or $y = -\dfrac{2}{3}x + \dfrac{7}{3}$, press the $\boxed{Y=}$ key and enter

$$Y_1 = -\frac{2}{3}x + \frac{7}{3}$$

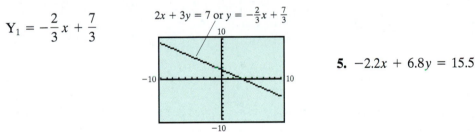

Graph each linear equation.

1. $x = 3.78y$

2. $-2.61y = x$

3. $3x + 7y = 21$

4. $-4x + 6y = 12$

5. $-2.2x + 6.8y = 15.5$

6. $5.9x - 0.8y = -10.4$

Mental Math

Answer the following true or false.

1. The graph of $x = 2$ is a horizontal line.

2. All lines have an x-intercept *and* a y-intercept.

3. The graph of $y = 4x$ contains the point $(0, 0)$.

4. The graph of $x + y = 5$ has an x-intercept of $(5, 0)$ and a y-intercept of $(0, 5)$.

5. The graph of $y = 5x$ contains the point $(5, 1)$.

6. The graph of $y = 5$ is a horizontal line.

6.3 EXERCISE SET

Objective **A** *Identify the intercepts. See Examples 1 through 3.*

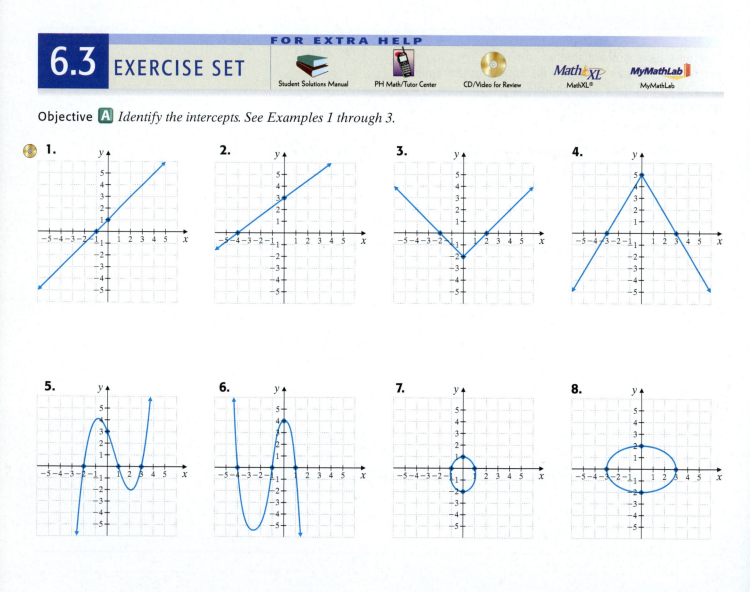

1.
2.
3.
4.

5.
6.
7.
8.

Objective B *Graph each linear equation by finding and plotting its intercepts. See Examples 4 and 5.*

9. $x - y = 3$

10. $x - y = -4$

11. $x = 5y$

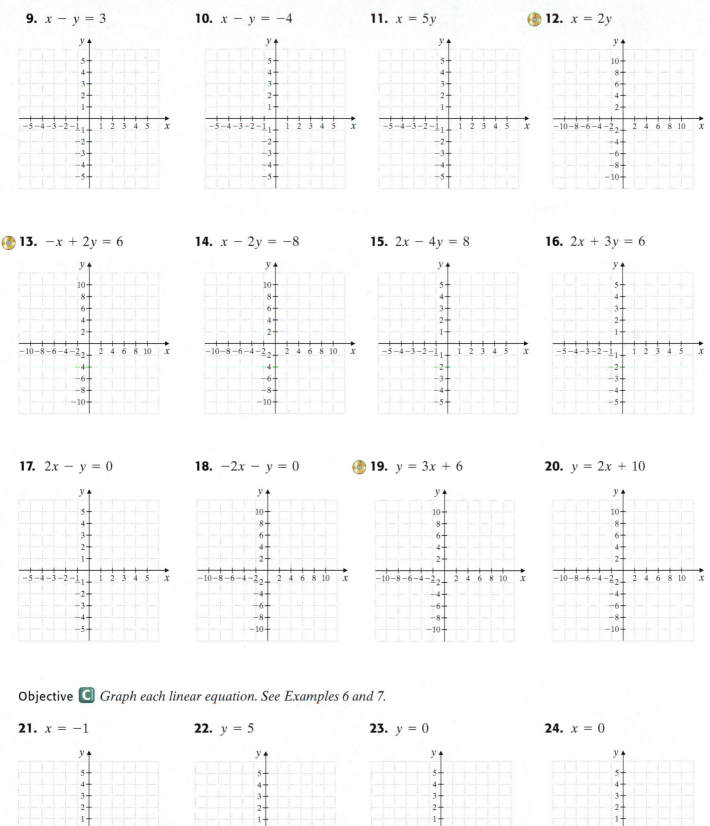

12. $x = 2y$

13. $-x + 2y = 6$

14. $x - 2y = -8$

15. $2x - 4y = 8$

16. $2x + 3y = 6$

17. $2x - y = 0$

18. $-2x - y = 0$

19. $y = 3x + 6$

20. $y = 2x + 10$

Objective C *Graph each linear equation. See Examples 6 and 7.*

21. $x = -1$

22. $y = 5$

23. $y = 0$

24. $x = 0$

25. $y + 7 = 0$ **26.** $x - 2 = 0$ **27.** $x + 3 = 0$ **28.** $y - 6 = 0$

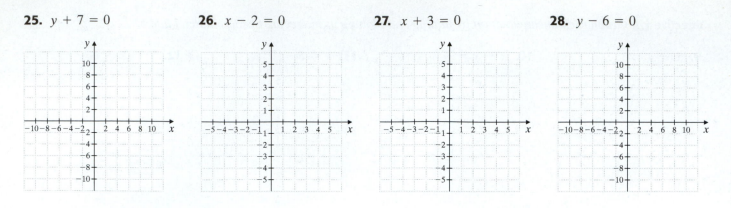

Objectives **B** **C** **Mixed Practice** *Graph each linear equation. See Examples 4 through 7.*

29. $x = y$ **30.** $x = -y$ **31.** $x + 8y = 8$ **32.** $x + 3y = 9$

33. $5 = 6x - y$ **34.** $4 = x - 3y$ **35.** $-x + 10y = 11$ **36.** $-x + 9y = 10$

37. $x = -4\frac{1}{2}$ **38.** $x = -1\frac{3}{4}$ **39.** $y = 3\frac{1}{4}$ **40.** $y = 2\frac{1}{2}$

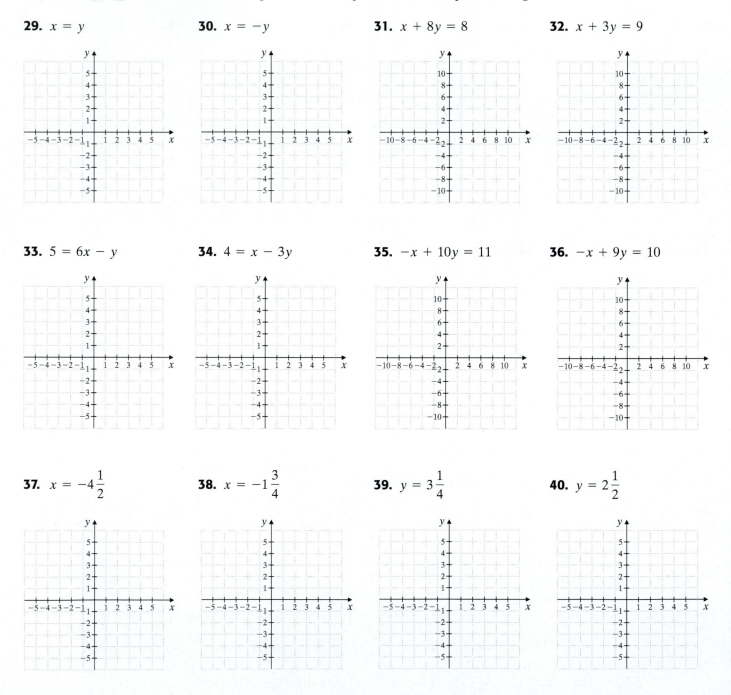

41. $y = -\dfrac{2}{3}x + 1$

42. $y = -\dfrac{3}{5}x + 3$

43. $4x - 6y + 2 = 0$

44. $9x - 6y + 3 = 0$

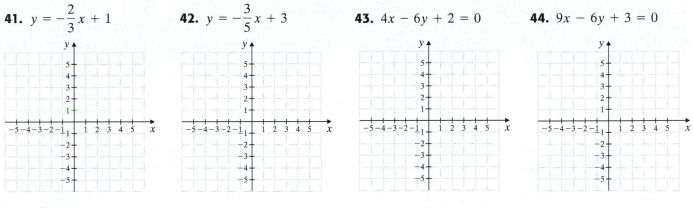

Review

Simplify. See Sections 1.5 and 1.6.

45. $\dfrac{-6 - 3}{2 - 8}$

46. $\dfrac{4 - 5}{-1 - 0}$

47. $\dfrac{-8 - (-2)}{-3 - (-2)}$

48. $\dfrac{12 - 3}{10 - 9}$

49. $\dfrac{0 - 6}{5 - 0}$

50. $\dfrac{2 - 2}{3 - 5}$

Concept Extensions

Match each equation with its graph.

51. $y = 3$

52. $y = 2x + 2$

53. $x = 3$

54. $y = 2x + 3$

a.

b.

c.

d.

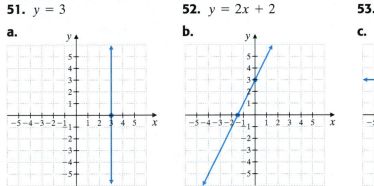
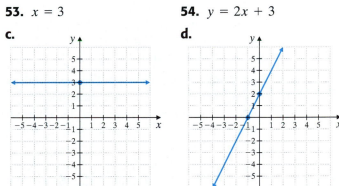

55. What is the greatest number of *x*- and *y*-intercepts that a line can have?

56. What is the smallest number of *x*- and *y*-intercepts that a line can have?

57. What is the smallest number of *x*- and *y*-intercepts that a circle can have?

58. What is the greatest number of *x*- and *y*-intercepts that a circle can have?

59. Discuss whether a vertical line ever has a *y*-intercept.

60. Discuss whether a horizontal line ever has an *x*-intercept.

61. The production supervisor at Alexandra's Office Products finds that it takes 3 hours to manufacture a particular office chair and 6 hours to manufacture an office desk. A total of 1200 hours is available to produce office chairs and desks of this style. The linear equation that models this situation is $3x + 6y = 1200$, where *x* represents the number of chairs produced and *y* the number of desks manufactured.

　a. Complete the ordered pair solution (0,) of this equation. Describe the manufacturing situation that corresponds to this solution.

　b. Complete the ordered pair solution (, 0) of this equation. Describe the manufacturing situation that corresponds to this solution.

　c. If 50 desks are manufactured, find the greatest number of chairs that can be made.

*Two lines in the same plane that do not intersect are called **parallel lines.***

62. Use your own graph paper to draw a line parallel to the line $x = 5$ that intersects the *x*-axis at 1. What is the equation of this line?

63. Use your own graph paper to draw a line parallel to the line $y = -1$ that intersects the *y*-axis at -4. What is the equation of this line?

Solve.

64. The number of music videos *y*, in millions, shipped to retailers in the United States from 1998 to 2003 can be modeled by the equation $y = -2.71x + 25$, where *x* represents the number of years after 1998. (*Source:* Recording Industry Association of America)

   a. Find the *x*-intercept of this equation (round to the nearest tenth).

   b. What does this *x*-intercept mean?

65. The number of a certain chain of stores *y* for the years 1999–2003 can be modeled by the equation $y = 29.2x + 919$, where *x* represents the number of years after 1999. (*Source:* Limited Brands)

   a. Find the *y*-intercept of this equation.

   b. What does this *y*-intercept mean?

 STUDY SKILLS BUILDER

Are You Familiar with Your Textbook Supplements?

Below is a review of some of the student supplements available for additional study. Check to see if you are using the ones most helpful to you.

- Chapter Test Prep Videos on CD. This material is found with your textbook and is fully explained there. The CD contains video clip solutions to the Chapter Test exercises in this text and are excellent help when studying for chapter tests.

- Lecture Videos on CD-ROM. These video segments are keyed to each section of the text. The material is presented by me, Elayn Martin-Gay, and I have placed a 💿 by the exercises in the text that I have worked on the video.

- The *Student Solutions Manual*. This contains worked out solutions to odd-numbered exercises as well as every exercise in the Integrated Reviews. Chapter Reviews, Chapter Tests, and Cumulative Reviews.

- Prentice Hall Tutor Center. Mathematics questions may be phoned, faxed, or emailed to this center.

- MyMathLab, MathXL, and Internet Math. These are computer and Internet tutorials. This supplement may already be available to you somewhere on campus, for example at your local learning resource lab. Take a moment and find the name and location of any such lab on campus.

   As usual, your instructor is your best source of information.

Let's see how you are doing with textbook supplements.

1. Name one way the Lecture Videos can be helpful to you.

2. Name one way the Chapter Test Prep Video can help you prepare for a chapter test.

3. List any textbook supplements that you have found useful.

4. Have you located and visited a learning resource lab located on your campus?

5. List the textbook supplements that are currently housed in your campus' learning resource lab.

6.4 SLOPE AND RATE OF CHANGE

Objectives

A Find the Slope of a Line Given Two Points of the Line.

B Find the Slope of a Line Given Its Equation.

C Find the Slopes of Horizontal and Vertical Lines.

D Compare the Slopes of Parallel and Perpendicular Lines.

E Slope as a Rate of Change.

Objective **A** Finding the Slope of a Line Given Two Points

Thus far, much of this chapter has been devoted to graphing lines. You have probably noticed by now that a key feature of a line is its slant or steepness. In mathematics, the slant or steepness of a line is formally known as its **slope.** We measure the slope of a line by the ratio of vertical change (rise) to the corresponding horizontal change (run) as we move along the line.

On the line below, for example, suppose that we begin at the point $(1, 2)$ and move to the point $(4, 6)$. The vertical change is the change in y-coordinates: $6 - 2$ or 4 units. The corresponding horizontal change is the change in x-coordinates: $4 - 1 = 3$ units. The ratio of these changes is

$$\text{slope} = \frac{\text{change in } y \text{ (vertical change or rise)}}{\text{change in } x \text{ (horizontal change or run)}} = \frac{4}{3}$$

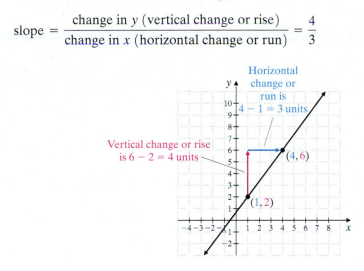

The slope of this line, then, is $\frac{4}{3}$. This means that for every 4 units of change in y-coordinates, there is a corresponding change of 3 units in x-coordinates.

> ### Helpful Hint
>
> It makes no difference what two points of a line are chosen to find its slope. The slope of a line is the same everywhere on the line.
>
>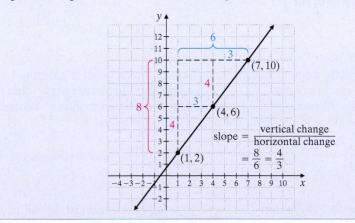

To find the slope of a line, then, choose two points of the line. Label the two x-coordinates of two points x_1 and x_2 (read "x sub one" and "x sub two"), and label the corresponding y-coordinates y_1 and y_2.

443

The vertical change or **rise** between these points is the difference in the y-coordinates: $y_2 - y_1$. The horizontal change or **run** between the points is the difference of the x-coordinates: $x_2 - x_1$. The slope of the line is the ratio of $y_2 - y_1$ to $x_2 - x_1$, and we traditionally use the letter m to denote slope $m = \dfrac{y_2 - y_1}{x_2 - x_1}$.

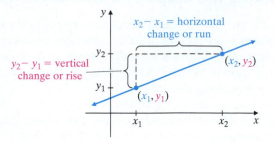

Slope of a Line

The slope m of the line containing the points (x_1, y_1) and (x_2, y_2) is given by

$$m = \frac{\text{rise}}{\text{run}} = \frac{\text{change in } y}{\text{change in } x} = \frac{y_2 - y_1}{x_2 - x_1}, \qquad \text{as long as } x_2 \neq x_1$$

PRACTICE PROBLEM 1

Find the slope of the line through $(-2, 3)$ and $(4, -1)$. Graph the line.

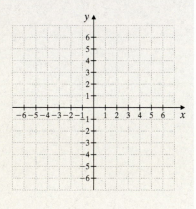

Answer

1. $-\dfrac{2}{3}$

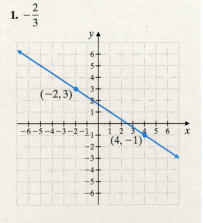

✔ **Concept Check Answer**

$m = \dfrac{3}{2}$

EXAMPLE 1 Find the slope of the line through $(-1, 5)$ and $(2, -3)$. Graph the line.

Solution: Let (x_1, y_1) be $(-1, 5)$ and (x_2, y_2) be $(2, -3)$. Then, by the definition of slope, we have the following.

$$m = \frac{y_2 - y_1}{x_2 - x_1}$$

$$= \frac{-3 - 5}{2 - (-1)}$$

$$= \frac{-8}{3} = -\frac{8}{3}$$

The slope of the line is $-\dfrac{8}{3}$.

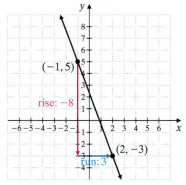

⬛ Work Practice Problem 1

Helpful Hint

When finding slope, it makes no difference which point is identified as (x_1, y_1) and which is identified as (x_2, y_2). Just remember that whatever y-value is first in the numerator, its corresponding x-value is first in the denominator. Another way to calculate the slope in Example 1 is

$$m = \frac{y_2 - y_1}{x_2 - x_1} = \frac{5 - (-3)}{-1 - 2} = \frac{8}{-3} \quad \text{or} \quad -\frac{8}{3} \quad \leftarrow \text{Same slope as found in Example 1.}$$

✔ **Concept Check** The points $(-2, -5)$, $(0, -2)$, $(4, 4)$, and $(10, 13)$ all lie on the same line. Work with a partner and verify that the slope is the same no matter which points are used to find slope.

EXAMPLE 2 Find the slope of the line through $(-1, -2)$ and $(2, 4)$. Graph the line.

Solution: Let (x_1, y_1) be $(2, 4)$ and (x_2, y_2) be $(-1, -2)$.

$$m = \frac{y_2 - y_1}{x_2 - x_1}$$

$$= \frac{-2 - 4}{-1 - 2} \quad \begin{array}{l} \text{\textit{y}-value} \\ \text{corresponding \textit{x}-value} \end{array}$$

$$= \frac{-6}{-3} = 2$$

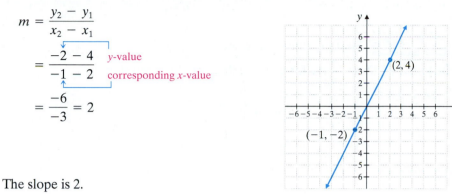

The slope is 2.

■ **Work Practice Problem 2**

✔**Concept Check** What is wrong with the following slope calculation for the points $(3, 5)$ and $(-2, 6)$?

$$m = \frac{5 - 6}{-2 - 3} = \frac{-1}{-5} = \frac{1}{5}$$

Notice that the slope of the line in Example 1 is negative, and the slope of the line in Example 2 is positive. Let your eye follow the line with negative slope from left to right and notice that the line "goes down." If you follow the line with positive slope from left to right, you will notice that the line "goes up." This is true in general.

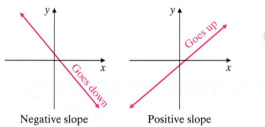

Negative slope   Positive slope

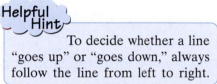

Helpful Hint To decide whether a line "goes up" or "goes down," always follow the line from left to right.

Objective B Finding the Slope of a Line Given Its Equation

As we have seen, the slope of a line is defined by two points on the line. Thus, if we know the equation of a line, we can find its slope by finding two of its points. For example, let's find the slope of the line

$$y = 3x - 2$$

To find two points, we can choose two values for x and substitute to find corresponding y-values. If $x = 0$, for example, $y = 3 \cdot 0 - 2$ or $y = -2$. If $x = 1$, $y = 3 \cdot 1 - 2$ or $y = 1$. This gives the ordered pairs $(0, -2)$ and $(1, 1)$. Using the definition for slope, we have

$$m = \frac{1 - (-2)}{1 - 0} = \frac{3}{1} = 3 \quad \text{The slope is 3.}$$

Notice that the slope, 3, is the same as the coefficient of x in the equation $y = 3x - 2$. This is true in general.

PRACTICE PROBLEM 2

Find the slope of the line through $(-2, 1)$ and $(3, 5)$. Graph the line.

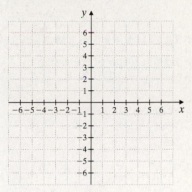

Answer

2. $\dfrac{4}{5}$

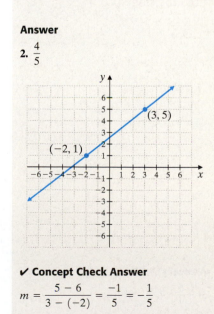

✔ **Concept Check Answer**

$$m = \frac{5 - 6}{3 - (-2)} = \frac{-1}{5} = -\frac{1}{5}$$

If a linear equation is solved for y, the coefficient of x is the line's slope. In other words, the slope of the line given by $y = mx + b$ is m, the coefficient of x.

$$y = mx + b$$
$$\underset{\text{slope}}{\uparrow}$$

PRACTICE PROBLEM 3

Find the slope of the line $5x + 4y = 10$.

EXAMPLE 3 Find the slope of the line $-2x + 3y = 11$.

Solution: When we solve for y, the coefficient of x is the slope.

$$-2x + 3y = 11$$
$$3y = 2x + 11 \qquad \text{Add } 2x \text{ to both sides.}$$
$$y = \frac{2}{3}x + \frac{11}{3} \qquad \text{Divide both sides by 3.}$$

The slope is $\frac{2}{3}$.

▣ **Work Practice Problem 3**

PRACTICE PROBLEM 4

Find the slope of the line $-y = -2x + 7$.

EXAMPLE 4 Find the slope of the line $-y = 5x - 2$.

Solution: Remember, the equation must be solved for y (not $-y$) in order for the coefficient of x to be the slope.

To solve for y, let's divide both sides of the equation by -1.

$$-y = 5x - 2$$
$$\frac{-y}{-1} = \frac{5x}{-1} - \frac{2}{-1} \qquad \text{Divide both sides by } -1.$$
$$y = -5x + 2 \qquad \text{Simplify.}$$

The slope is -5.

▣ **Work Practice Problem 4**

Objective ⓒ Finding Slopes of Horizontal and Vertical Lines

PRACTICE PROBLEM 5

Find the slope of $y = 3$.

EXAMPLE 5 Find the slope of the line $y = -1$.

Solution: Recall that $y = -1$ is a horizontal line with y-intercept -1. To find the slope, we find two ordered pair solutions of $y = -1$, knowing that solutions of $y = -1$ must have a y-value of -1. We will use $(2, -1)$ and $(-3, -1)$. We let (x_1, y_1) be $(2, -1)$ and (x_2, y_2) be $(-3, -1)$.

$$m = \frac{y_2 - y_1}{x_2 - x_1} = \frac{-1 - (-1)}{-3 - 2} = \frac{0}{-5} = 0$$

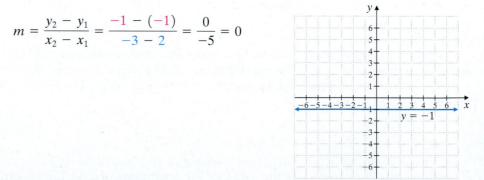

The slope of the line $y = -1$ is 0. Since the y-values will have a difference of 0 for every horizontal line, we can say that all **horizontal lines have a slope of 0.**

▣ **Work Practice Problem 5**

EXAMPLE 6 Find the slope of the line $x = 5$.

Solution: Recall that the graph of $x = 5$ is a vertical line with x-intercept 5. To find the slope, we find two ordered pair solutions of $x = 5$. Ordered pair solutions of $x = 5$ must have an x-value of 5. We will use $(5, 0)$ and $(5, 4)$. We let $(x_1, y_1) = (5, 0)$ and $(x_2, y_2) = (5, 4)$.

$$m = \frac{y_2 - y_1}{x_2 - x_1} = \frac{4 - 0}{5 - 5} = \frac{4}{0}$$

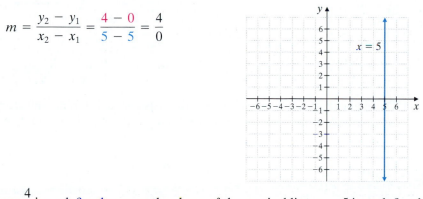

Since $\frac{4}{0}$ is undefined, we say the slope of the vertical line $x = 5$ is undefined.

Since the x-values will have a difference of 0 for every vertical line, we can say that all **vertical lines have undefined slope.**

■ **Work Practice Problem 6**

Here is a general review of slope.

PRACTICE PROBLEM 6

Find the slope of the line $x = -2$.

Helpful Hint
Slope of 0 and undefined slope are not the same. Vertical lines have undefined slope, while horizontal lines have a slope of 0.

Summary of Slope

Slope m of the line through (x_1, y_1) and (x_2, y_2) is given by the equation

$$m = \frac{y_2 - y_1}{x_2 - x_1}.$$

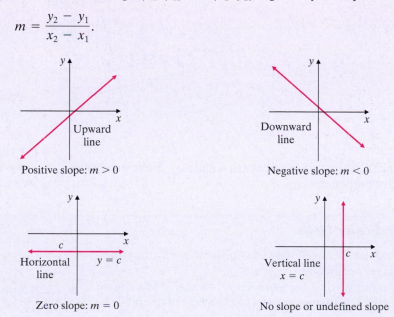

Objective D Slopes of Parallel and Perpendicular Lines

Two lines in the same plane are **parallel** if they do not intersect. Slopes of lines can help us determine whether lines are parallel. Since parallel lines have the same steepness, it follows that they have the same slope.

Answer
6. undefined slope

For example, the graphs of

$$y = -2x + 4$$

and

$$y = -2x - 3$$

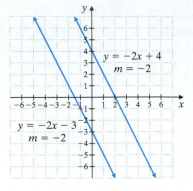

are shown. These lines have the same slope, -2. They also have different y-intercepts, so the lines are parallel. (If the y-intercepts were the same also, the lines would be the same.)

Parallel Lines

Nonvertical parallel lines have the same slope and different y-intercepts.

Two lines are **perpendicular** if they lie in the same plane and meet at a $90°$ (right) angle. How do the slopes of perpendicular lines compare? The product of the slopes of two perpendicular lines is -1.

For example, the graphs of

$$y = 4x + 1$$

and

$$y = -\frac{1}{4}x - 3$$

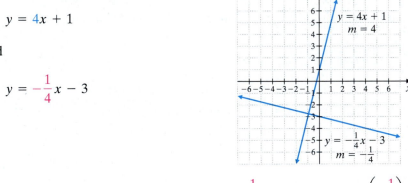

are shown. The slopes of the lines are 4 and $-\frac{1}{4}$. Their product is $4\left(-\frac{1}{4}\right) = -1$, so the lines are perpendicular.

Perpendicular Lines

If the product of the slopes of two lines is -1, then the lines are perpendicular.

(Two nonvertical lines are perpendicular if the slope of one is the negative reciprocal of the slope of the other.)

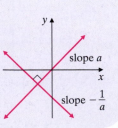

Helpful Hint

Here are examples of numbers that are negative (opposite) reciprocals.

Number	Negative Reciprocal	Their product is −1.
$\dfrac{2}{3}$	$-\dfrac{3}{2}$	$\dfrac{2}{3} \cdot -\dfrac{3}{2} = -\dfrac{6}{6} = -1$
-5 or $-\dfrac{5}{1}$	$\dfrac{1}{5}$	$-5 \cdot \dfrac{1}{5} = -\dfrac{5}{5} = -1$

Here are a few important points about vertical and horizontal lines.

- Two distinct vertical lines are parallel.
- Two distinct horizontal lines are parallel.
- A horizontal line and a vertical line are always perpendicular.

EXAMPLE 7 Determine whether each pair of lines is parallel, perpendicular, or neither.

a. $y = -\dfrac{1}{5}x + 1$ **b.** $x + y = 3$ **c.** $3x + y = 5$
$2x + 10y = 3$ $-x + y = 4$ $2x + 3y = 6$

Solution:

a. The slope of the line $y = -\dfrac{1}{5}x + 1$ is $-\dfrac{1}{5}$. We find the slope of the second line by solving its equation for y.

$$2x + 10y = 3$$
$$10y = -2x + 3 \qquad \text{Subtract } 2x \text{ from both sides.}$$
$$y = \dfrac{-2}{10}x + \dfrac{3}{10} \qquad \text{Divide both sides by 10.}$$
$$y = -\dfrac{1}{5}x + \dfrac{3}{10} \qquad \text{Simplify.}$$

The slope of this line is $-\dfrac{1}{5}$ also. Since the lines have the same slope and different y-intercepts, they are parallel, as shown in the figure below.

b. To find each slope, we solve each equation for y.

$x + y = 3$ $-x + y = 4$
$y = -x + 3$ $y = x + 4$
The slope is −1.   The slope is 1.

The slopes are not the same, so the lines are not parallel. Next we check the product of the slopes: $(-1)(1) = -1$. Since the product is −1, the lines are perpendicular, as shown in the figure.

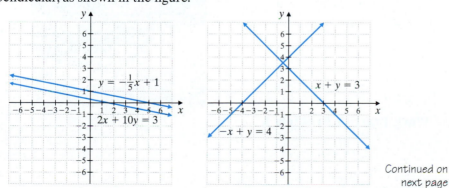

$y = -\dfrac{1}{5}x + 1$
$2x + 10y = 3$

$x + y = 3$
$-x + y = 4$

Continued on next page

PRACTICE PROBLEM 7

Determine whether each pair of lines is parallel, perpendicular, or neither.

a. $x + y = 5$
$2x + y = 5$

b. $5y = 2x - 3$
$5x + 2y = 1$

c. $y = 2x + 1$
$4x - 2y = 8$

Answers

7. a. neither, **b.** perpendicular,
c. parallel

c. We solve each equation for y to find each slope. The slopes are -3 and $-\dfrac{2}{3}$. The slopes are not the same and their product is not -1. Thus, the lines are neither parallel nor perpendicular.

☐ **Work Practice Problem 7**

✔ **Concept Check**   Consider the line $-6x + 2y = 1$.

a. Write the equations of two lines parallel to this line.
b. Write the equations of two lines perpendicular to this line.

Objective **E** Slope as a Rate of Change

Slope can also be interpreted as a rate of change. In other words, slope tells us how fast y is changing with respect to x. To see this, let's look at a few of the many real-world applications of slope. For example, the pitch of a roof, used by builders and architects, is its slope. The pitch of the roof on the left is $\dfrac{7}{10}\left(\dfrac{\text{rise}}{\text{run}}\right)$. This means that the roof rises vertically 7 feet for every horizontal 10 feet. The rate of change for the roof is 7 vertical feet (y) per 10 horizontal feet (x).

The grade of a road is its slope written as a percent. A 7% grade, as shown below, means that the road rises (or falls) 7 feet for every horizontal 100 feet. $\Big($ Recall that $7\% = \dfrac{7}{100}.\Big)$ Here, the slope of $\dfrac{7}{100}$ gives us the rate of change. The road rises (in our diagram) 7 vertical feet (y) for every 100 horizontal feet (x).

$$\frac{7}{100} = 7\%\text{ grade}$$

7 feet

100 feet

PRACTICE PROBLEM 8

Find the grade of the road shown.

3 feet

20 feet

Answer
8. 15%

✔ **Concept Check Answers**
Answers may vary; for example,
a. $y = 3x - 3$, $y = 3x - 1$,
b. $y = -\dfrac{1}{3}x$, $y = -\dfrac{1}{3}x + 1$

EXAMPLE 8   **Finding the Grade of a Road**

At one part of the road to the summit of Pike's Peak, the road rises 15 feet for a horizontal distance of 250 feet. Find the grade of the road.

Solution: Recall that the grade of a road is its slope written as a percent.

$$\text{grade} = \frac{\text{rise}}{\text{run}} = \frac{15}{250} = 0.06 = 6\%$$

15 feet

250 feet

The grade is 6%.

☐ **Work Practice Problem 8**

Slope can also be interpreted as a rate of change. In other words, slope tells us how fast y is changing with respect to x.

EXAMPLE 9 **Finding the Slope of a Line**

The following graph shows the cost y (in cents) of a nationwide long-distance telephone call from Texas with a certain telephone-calling plan, where x is the length of the call in minutes. Find the slope of the line and attach the proper units for the rate of change. Then write a sentence explaining the meaning of slope in this application.

Solution: Use (2, 34) and (6, 62) to calculate slope.

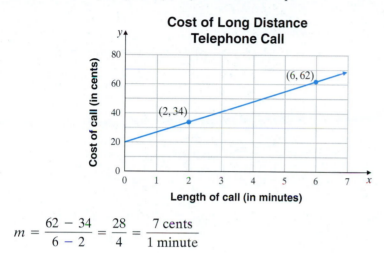

$$m = \frac{62 - 34}{6 - 2} = \frac{28}{4} = \frac{7 \text{ cents}}{1 \text{ minute}}$$

This means that the rate of change of a phone call is 7 cents per 1 minute or the cost of the phone call is 7 cents per minute.

☐ **Work Practice Problem 9**

PRACTICE PROBLEM 9

Find the slope of the line and write the slope as a rate of change. This graph represents annual food and drink sales y (in billions of dollars) for year x. Write a sentence explaining the meaning of slope in this application.

Source: National Restaurant Assn.

Answer

9. $m = 12$; Each year the sales of food and drink from restaurants increases by $12 billion dollars per year.

▦ CALCULATOR EXPLORATIONS   Graphing

It is possible to use a graphing calculator and sketch the graph of more than one equation on the same set of axes. This feature can be used to see parallel lines with the same slope. For example, graph the equations $y = \frac{2}{5}x$, $y = \frac{2}{5}x + 7$, and $y = \frac{2}{5}x - 4$ on the same set of axes. To do so, press the $\boxed{Y=}$ key and enter the equations on the first three lines.

$$Y_1 = \left(\frac{2}{5}\right)x$$

$$Y_2 = \left(\frac{2}{5}\right)x + 7$$

$$Y_3 = \left(\frac{2}{5}\right)x - 4$$

The displayed equations should look like this:

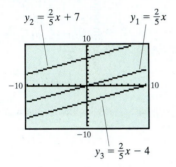

These lines are parallel as expected since they all have a slope of $\frac{2}{5}$. The graph of $y = \frac{2}{5}x + 7$ is the graph of $y = \frac{2}{5}x$ moved 7 units upward with a y-intercept of 7. Also, the graph of $y = \frac{2}{5}x - 4$ is the graph of $y = \frac{2}{5}x$ moved 4 units downward with a y-intercept of -4.

Graph the parallel lines on the same set of axes. Describe the similarities and differences in their graphs.

1. $y = 3.8x$, $y = 3.8x - 3$, $y = 3.8x + 9$

2. $y = -4.9x$, $y = -4.9x + 1$, $y = -4.9x + 8$

3. $y = \frac{1}{4}x$, $y = \frac{1}{4}x + 5$, $y = \frac{1}{4}x - 8$

4. $y = -\frac{3}{4}x$, $y = -\frac{3}{4}x - 5$, $y = -\frac{3}{4}x + 6$

Mental Math

State whether the slope of the line is positive, negative, 0, or is undefined.

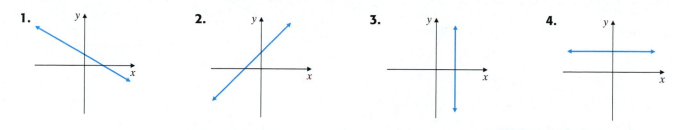

1.   2.   3.   4.

Decide whether a line with the given slope is upward sloping, downward sloping, horizontal, or vertical.

5. $m = \dfrac{7}{6}$

6. $m = -3$

7. $m = 0$

8. m is undefined

6.4 EXERCISE SET

Objective A Find the slope of the line that passes through the given points. See Examples 1 and 2.

1. $(-1, 5)$ and $(6, -2)$

2. $(-1, 16)$ and $(3, 4)$

3. $(1, 4)$ and $(5, 3)$

4. $(3, 1)$ and $(2, 6)$

5. $(5, 1)$ and $(-2, 1)$

6. $(-8, 3)$ and $(-2, 3)$

7. $(5, 4)$ and $(5, 0)$

8. $(-2, -3)$ and $(-2, 5)$

Use the points shown on each graph to find the slope of each line. See Examples 1 and 2.

9.   **10.**   **11.**   **12.**

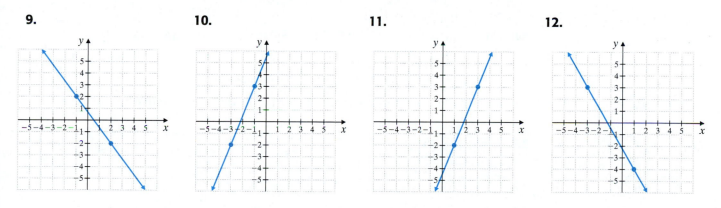

For each graph, determine which line has the greater slope.

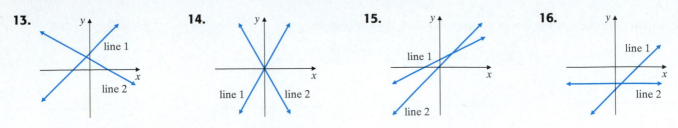

13. line 1, line 2

14. line 1, line 2

15. line 1, line 2

16. line 1, line 2

Objectives **B** **C** **Mixed Practice** *Find the slope of each line. See Examples 3 through 6.*

17. $y = 5x - 2$

18. $y = -2x + 6$

19. $y = -0.3x + 2.5$

20. $y = -7.6x - 0.1$

21. $2x + y = 7$

22. $-5x + y = 10$

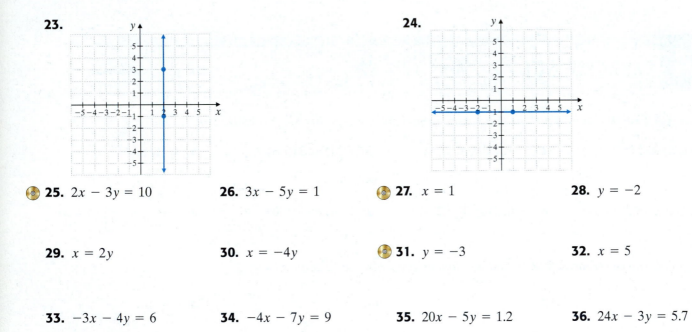

23.

24.

25. $2x - 3y = 10$

26. $3x - 5y = 1$

27. $x = 1$

28. $y = -2$

29. $x = 2y$

30. $x = -4y$

31. $y = -3$

32. $x = 5$

33. $-3x - 4y = 6$

34. $-4x - 7y = 9$

35. $20x - 5y = 1.2$

36. $24x - 3y = 5.7$

△ Objective **D** *Determine whether each pair of lines is parallel, perpendicular, or neither. See Example 7.*

37. $y = \dfrac{2}{9}x + 3$
$y = -\dfrac{2}{9}x$

38. $y = \dfrac{1}{5}x + 20$
$y = -\dfrac{1}{5}x$

39. $x - 3y = -6$
$y = 3x - 9$

40. $y = 4x - 2$
$4x + y = 5$

41. $6x = 5y + 1$
$-12x + 10y = 1$

42. $-x + 2y = -2$
$2x = 4y + 3$

43. $6 + 4x = 3y$
$3x + 4y = 8$

44. $10 + 3x = 5y$
$5x + 3y = 1$

△ *Find the slope of the line that is (a) parallel and (b) perpendicular to the line through each pair of points. See Example 7.*

45. $(-3, -3)$ and $(0, 0)$

46. $(6, -2)$ and $(1, 4)$

47. $(-8, -4)$ and $(3, 5)$

48. $(6, -1)$ and $(-4, -10)$

Objective **E** *The pitch of a roof is its slope. Find the pitch of each roof shown. See Example 8.*

49.

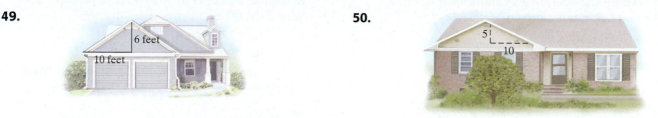

6 feet

10 feet

50.

5

10

The grade of a road is its slope written as a percent. Find the grade of each road shown. See Example 8.

51.

2 meters

16 meters

52.

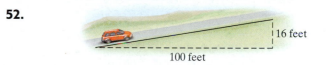

16 feet

100 feet

53. One of Japan's superconducting "bullet" trains is researched and tested at the Yamanashi Maglev Test Line near Otsuki City. The steepest section of the track has a rise of 2580 meters for a horizontal distance of 6450 meters. What is the grade of this section of track? (*Source:* Japan Railways Central Co.)

2580 meters

6450 meters

54. Professional plumbers suggest that a sewer pipe should rise 0.25 inch for every horizontal foot. Find the recommended slope for a sewer pipe. Round to the nearest hundredth.

0.25 inch

12 inches

55. The steepest street is Baldwin Street in Dunedin, New Zealand. It has a maximum rise of 10 meters for a horizontal distance of 12.66 meters. Find the grade of this section of road. Round to the nearest whole percent. (*Source: The Guinness Book of Records*)

56. According to federal regulations, a wheelchair ramp should rise no more than 1 foot for a horizontal distance of 12 feet. Write the slope as a grade. Round to the nearest tenth of a percent.

Find the slope of each line and write the slope as a rate of change. Don't forget to attach the proper units. See Example 9.

57. This graph approximates the number of U.S. households that have personal computers *y* (in millions) for year *x*.

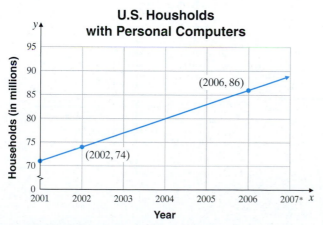

U.S. Housholds with Personal Computers

Households (in millions)

95
90
85
80
75
70

(2006, 86)

(2002, 74)

2001 2002 2003 2004 2005 2006 2007* *x*

Year

Source: Statistical Abstract of the United States, *projected numbers

58. This graph approximates the total number of cosmetic surgeons for year *x*.

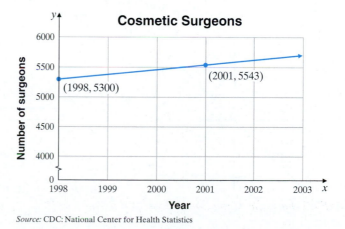

Cosmetic Surgeons

Number of surgeons

6000
5500
5000
4500
4000

(2001, 5543)

(1998, 5300)

1998 1999 2000 2001 2002 2003 *x*

Year

Source: CDC: National Center for Health Statistics

59. The graph below shows the total cost y (in dollars) of owning and operating a compact car where x is the number of miles driven.

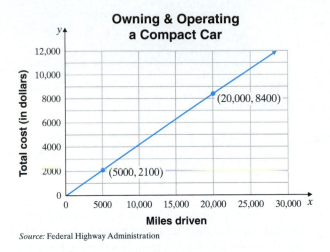

Owning & Operating a Compact Car

$(20,000, 8400)$

$(5000, 2100)$

Total cost (in dollars)

Miles driven

Source: Federal Highway Administration

60. The graph below shows the total cost y (in dollars) of owning and operating a standard pickup truck, where x is the number of miles driven.

Owning & Operating a Standard Truck

$(40,000, 19,200)$

$(10,000, 4800)$

Total cost (in dollars)

Miles driven

Source: Federal Highway Administration

Review

Solve each equation for y. See Section 2.5.

61. $y - (-6) = 2(x - 4)$

62. $y - 7 = -9(x - 6)$

63. $y - 1 = -6(x - (-2))$

64. $y - (-3) = 4(x - (-5))$

Concept Extensions

Match each line with its slope.

a. $m = 0$

b. undefined slope

c. $m = 3$

d. $m = 1$

e. $m = -\dfrac{1}{2}$

f. $m = -\dfrac{3}{4}$

65.

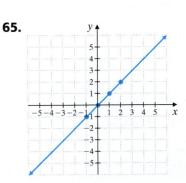

66.

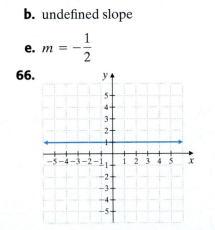

67.

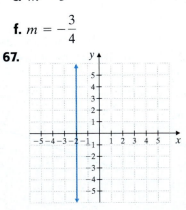

68.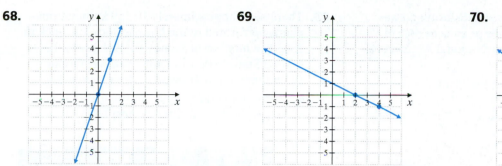

69.

70.

Solve. See a Concept Check in this section.

71. Verify that the points $(2, 1)$, $(0, 0)$, $(-2, -1)$ and $(-4, -2)$ are all on the same line by computing the slope between each pair of points. (See the first Concept Check.)

72. Given the points $(2, 3)$ and $(-5, 1)$, can the slope of the line through these points be calculated by $\dfrac{1 - 3}{2 - (-5)}$? Why or why not? (See the second Concept Check.)

73. Write the equations of three lines parallel to $10x - 5y = -7$. (See the third Concept Check.)

74. Write the equations of two lines perpendicular to $10x - 5y = -7$. (See the third Concept Check.)

The following line graph shows the average fuel economy (in miles per gallon) by passenger automobiles produced during each of the model years shown. Use this graph to answer Exercises 75 through 79.

75. What was the average fuel economy (in miles per gallon) for automobiles produced during 2001?

76. Find the decrease in average fuel economy for automobiles between the years 1998 to 2000.

77. During which of the model years shown was average fuel economy the lowest?
What was the average fuel economy for that year?

Average Fuel Economy for Autos

Source: U.S. Environmental Protection Agency, Office of Transportation and Air Quality

78. During which of the model years shown was average fuel economy the highest?
What was the average fuel economy for that year?

79. What line segment has the greatest slope?

80. Find x so that the pitch of the roof is $\dfrac{1}{3}$.

18 feet

81. Find x so that the pitch of the roof is $\dfrac{2}{5}$.

4 feet

x

82. The average price of an acre of U.S. farmland was $974 in 1998. In 2003, the price of an acre rose to approximately $1275. (*Source:* National Agricultural Statistics Services)

   a. Write two ordered pairs of the form (year, price of acre)

   b. Find the slope of the line through the two points.

   c. Write a sentence explaining the meaning of the slope as a rate of change.

83. There were approximately 10,359 kidney transplants performed in the United States in 1993. In 2003, the number of kidney transplants in the United States rose to 15,138. (*Source:* Organ Procurement and Transplantation Network)

   a. Write two ordered pairs of the form (year, number of kidney transplants).

   b. Find the slope of the line between the two points.

   c. Write a sentence explaining the meaning of the slope as a rate of change.

84. Show that a triangle with vertices at the points $(1, 1), (-4, 4)$, and $(-3, 0)$ is a right triangle.

85. Show that the quadrilateral with vertices $(1, 3), (2, 1), (-4, 0)$, and $(-3, -2)$ is a parallelogram.

Find the slope of the line through the given points.

86. $(2.1, 6.7)$ and $(-8.3, 9.3)$

87. $(-3.8, 1.2)$ and $(-2.2, 4.5)$

88. $(2.3, 0.2)$ and $(7.9, 5.1)$

89. $(14.3, -10.1)$ and $(9.8, -2.9)$

90. The graph of $y = -\frac{1}{3}x + 2$ has a slope of $-\frac{1}{3}$. The graph of $y = -2x + 2$ has a slope of -2. The graph of $y = -4x + 2$ has a slope of -4. Graph all three equations on a single coordinate system. As the absolute value of the slope becomes larger, how does the steepness of the line change?

91. The graph of $y = \frac{1}{2}x$ has a slope of $\frac{1}{2}$. The graph of $y = 3x$ has a slope of 3. The graph of $y = 5x$ has a slope of 5. Graph all three equations on a single coordinate system. As slope becomes larger, how does the steepness of the line change?

6.5 EQUATIONS OF LINES

Objectives

A Use the Slope-Intercept Form to Write an Equation of a Line.

B Use the Slope-Intercept Form to Graph a Linear Equation.

C Use the Point-Slope Form to Find an Equation of a Line Given Its Slope and a Point of the Line.

D Use the Point-Slope Form to Find an Equation of a Line Given Two Points of the Line.

E Use the Point-Slope Form to Solve Problems.

We know that when a linear equation is solved for y, the coefficient of x is the slope of the line. For example, the slope of the line whose equation is $y = 3x + 1$ is 3. In this equation, $y = 3x + 1$, what does 1 represent? To find out, let $x = 0$ and watch what happens.

$$y = 3x + 1$$
$$y = 3 \cdot 0 + 1 \quad \text{Let } x = 0.$$
$$y = 1$$

We now have the ordered pair $(0, 1)$, which means that 1 is the y-intercept.

This is true in general. To see this, let $x = 0$ and solve for y in $y = mx + b$.

$$y = m \cdot 0 + b \quad \text{Let } x = 0.$$
$$y = b$$

We obtain the ordered pair $(0, b)$, which means that point is the y-intercept.

The form $y = mx + b$ is appropriately called the *slope-intercept form* of a linear equation.

$$y = mx + b$$
↑ slope   ↑ y-intercept is $(0, b)$

Slope-Intercept Form

When a linear equation in two variables is written in **slope-intercept form,**

$$y = mx + b$$
↑ slope   ↑ $(0, b)$, y-intercept

then m is the slope of the line and $(0, b)$ is the y-intercept of the line.

Objective **A** Using the Slope-Intercept Form to Write an Equation

The slope-intercept form can be used to write the equation of a line when we know its slope and y-intercept.

EXAMPLE 1 Find an equation of the line with y-intercept $(0, -3)$ and slope of $\frac{1}{4}$.

Solution: We are given the slope and the y-intercept. We let $m = \frac{1}{4}$ and $b = -3$ and write the equation in slope-intercept form, $y = mx + b$.

$$y = mx + b$$
$$y = \frac{1}{4}x + (-3) \quad \text{Let } m = \frac{1}{4} \text{ and } b = -3.$$
$$y = \frac{1}{4}x - 3 \quad \text{Simplify.}$$

Work Practice Problem 1

Objective **B** Using the Slope-Intercept Form to Graph an Equation

We also can use the slope-intercept form of the equation of a line to graph a linear equation.

PRACTICE PROBLEM 1

Find an equation of the line with y-intercept $(0, -2)$ and slope of $\frac{3}{5}$.

Answer

1. $y = \frac{3}{5}x - 2$

459

PRACTICE PROBLEM 2

Graph the equation
$y = \dfrac{2}{3}x - 4$.

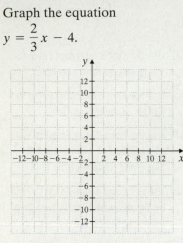

PRACTICE PROBLEM 3

Use the slope-intercept form to graph $3x + y = 2$.

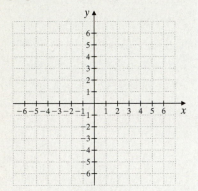

EXAMPLE 2 Use the slope-intercept form to graph the equation

$$y = \frac{3}{5}x - 2$$

Solution: Since the equation $y = \dfrac{3}{5}x - 2$ is written in slope-intercept form $y = mx + b$, the slope of its graph is $\dfrac{3}{5}$ and the y-intercept is $(0, -2)$. To graph this equation, we begin by plotting the point $(0, -2)$.

From this point, we can find another point of the graph by using the slope $\dfrac{3}{5}$ and recalling that slope is $\dfrac{\text{rise}}{\text{run}}$. We start at the y-intercept and move 3 units up since the numerator of the slope is 3; then we move 5 units to the right since the denominator of the slope is 5. We stop at the point $(5, 1)$. The line through $(0, -2)$ and $(5, 1)$ is the graph of $y = \dfrac{3}{5}x - 2$.

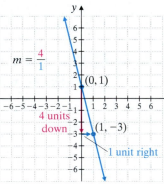

🔲 **Work Practice Problem 2**

EXAMPLE 3 Use the slope-intercept form to graph the equation $4x + y = 1$.

Solution: First we write the given equation in slope-intercept form.

$$4x + y = 1$$
$$y = -4x + 1$$

The graph of this equation will have slope -4 and y-intercept $(0, 1)$. To graph this line, we first plot the point $(0, 1)$. To find another point of the graph, we use the slope -4, which can be written as $\dfrac{-4}{1}\left(\dfrac{4}{-1}\text{ could also be used}\right)$. We start at the point $(0, 1)$ and move 4 units down (since the numerator of the slope is -4), and then 1 unit to the right (since the denominator of the slope is 1).

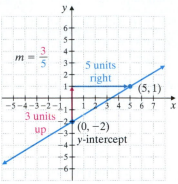

We arrive at the point $(1, -3)$. The line through $(0, 1)$ and $(1, -3)$ is the graph of $4x + y = 1$.

🔲 **Work Practice Problem 3**

Answers

2.

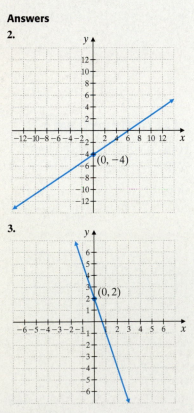

Helpful Hint

In Example 3, if we interpret the slope of -4 as $\dfrac{4}{-1}$, we arrive at $(-1, 5)$ for a second point. Notice that this point is also on the line.

Objective C Writing an Equation Given Its Slope and a Point

Thus far, we have seen that we can write an equation of a line if we know its slope and y-intercept. We can also write an equation of a line if we know its slope and any point on the line. To see how we do this, let m represent slope and (x_1, y_1) represent the point on the line. Then if (x, y) is any other point of the line, we have that

$$\frac{y - y_1}{x - x_1} = m$$

$$y - y_1 = m(x - x_1) \quad \text{Multiply both sides by } (x - x_1).$$

$$\uparrow$$
slope

This is the *point-slope form* of the equation of a line.

Point-Slope Form of the Equation of a Line

The **point-slope form** of the equation of a line is $y - y_1 = m(x - x_1)$, where m is the slope of the line and (x_1, y_1) is a point on the line.

EXAMPLE 4   Find an equation of the line with slope -2 that passes through $(-1, 5)$. Write the equation in slope-intercept form, $y = mx + b$, and in standard form, $Ax + By = C$.

Solution: Since the slope and a point on the line are given, we use point-slope form $y - y_1 = m(x - x_1)$ to write the equation. Let $m = -2$ and $(-1, 5) = (x_1, y_1)$.

$$y - y_1 = m(x - x_1)$$

$$y - 5 = -2[x - (-1)] \quad \text{Let } m = -2 \text{ and } (x_1, y_1) = (-1, 5).$$

$$y - 5 = -2(x + 1) \quad \text{Simplify.}$$

$$y - 5 = -2x - 2 \quad \text{Use the distributive property.}$$

To write the equation in slope-intercept form, $y = mx + b$, we simply solve the equation for y. To do this, we add 5 to both sides.

$$y - 5 = -2x - 2$$

$$y = -2x + 3 \quad \text{Slope-intercept form.}$$

$$2x + y = 3 \quad \text{Add } 2x \text{ to both sides and we have standard form.}$$

◻ **Work Practice Problem 4**

PRACTICE PROBLEM 4

Find an equation of the line with slope -3 that passes through $(2, -4)$. Write the equation in slope-intercept form $y = mx + b$.

Objective D Writing an Equation Given Two Points

We can also find the equation of a line when we are given any two points of the line.

EXAMPLE 5   Find an equation of the line through $(2, 5)$ and $(-3, 4)$. Write the equation in the form $Ax + By = C$.

Solution: First, use the two given points to find the slope of the line.

$$m = \frac{4 - 5}{-3 - 2} = \frac{-1}{-5} = \frac{1}{5}$$

Next we use the slope $\frac{1}{5}$ and either one of the given points to write the equation in point-slope form. We use $(2, 5)$. Let $x_1 = 2$, $y_1 = 5$, and $m = \frac{1}{5}$.

$$y - y_1 = m(x - x_1) \quad \text{Use point-slope form.}$$

$$y - 5 = \frac{1}{5}(x - 2) \quad \text{Let } x_1 = 2, y_1 = 5, \text{ and } m = \frac{1}{5}.$$

$$5(y - 5) = 5 \cdot \frac{1}{5}(x - 2) \quad \text{Multiply both sides by 5 to clear fractions.}$$

$$5y - 25 = x - 2 \quad \text{Use the distributive property and simplify.}$$

$$-x + 5y - 25 = -2 \quad \text{Subtract } x \text{ from both sides.}$$

$$-x + 5y = 23 \quad \text{Add 25 to both sides.}$$

◻ **Work Practice Problem 5**

PRACTICE PROBLEM 5

Find an equation of the line through $(1, 3)$ and $(5, -2)$. Write the equation in the form $Ax + By = C$.

Answers
4. $y = -3x + 2$,   **5.** $5x + 4y = 17$

Copyright 2007 Pearson Education, Inc.

> **Helpful Hint**
>
> When you multiply both sides of the equation from Example 5, $-x + 5y = 23$ by -1, it becomes $x - 5y = -23$.
>
> Both $-x + 5y = 23$ and $x - 5y = -23$ are in the form $Ax + By = C$ and both are equations of the same line.

Objective E Using the Point-Slope Form to Solve Problems

Problems occurring in many fields can be modeled by linear equations in two variables. The next example is from the field of marketing and shows how consumer demand of a product depends on the price of the product.

PRACTICE PROBLEM 6

The Pool Entertainment Company learned that by pricing a new pool toy at $10, local sales will reach 200 a week. Lowering the price to $9 will cause sales to rise to 250 a week.

a. Assume that the relationship between sales price and number of toys sold is linear, and write an equation describing this relationship. Write the equation in slope-intercept form. Use ordered pairs of the form (sales price, number sold).

b. Predict the weekly sales of the toy if the price is $7.50.

EXAMPLE 6 The Whammo Company has learned that by pricing a newly released Frisbee at $6, sales will reach 2000 Frisbees per day. Raising the price to $8 will cause the sales to fall to 1500 Frisbees per day.

a. Assume that the relationship between sales price and number of Frisbees sold is linear and write an equation describing this relationship. Write the equation in slope-intercept form. Use ordered pairs of the form (sales price, number sold).

b. Predict the daily sales of Frisbees if the price is $7.50.

Solution:

a. We use the given information and write two ordered pairs. Our ordered pairs are $(6, 2000)$ and $(8, 1500)$. To use the point-slope form to write an equation, we find the slope of the line that contains these points.

$$m = \frac{2000 - 1500}{6 - 8} = \frac{500}{-2} = -250$$

Next we use the slope and either one of the points to write the equation in point-slope form. We use $(6, 2000)$.

$$y - y_1 = m(x - x_1) \qquad \text{Use point-slope form.}$$
$$y - 2000 = -250(x - 6) \qquad \text{Let } x_1 = 6, y_1 = 2000, \text{ and } m = -250.$$
$$y - 2000 = -250x + 1500 \qquad \text{Use the distributive property.}$$
$$y = -250x + 3500 \qquad \text{Write in slope-intercept form.}$$

b. To predict the sales if the price is $7.50, we find y when $x = 7.50$.

$$y = -250x + 3500$$
$$y = -250(7.50) + 3500 \qquad \text{Let } x = 7.50.$$
$$y = -1875 + 3500$$
$$y = 1625$$

If the price is $7.50, sales will reach 1625 Frisbees per day.

Answers

6. **a.** $y = -50x + 700$,   **b.** 325

■ **Work Practice Problem 6**

We could have solved Example 6 by using ordered pairs of the form (number sold, sales price).

Here is a summary of our discussion on linear equations thus far.

Forms of Linear Equations

$Ax + By = C$	**Standard form** of a linear equation. A and B are not both 0.
$y = mx + b$	**Slope-intercept form** of a linear equation. The slope is m and the y-intercept is $(0, b)$.
$y - y_1 = m(x - x_1)$	**Point-slope form** of a linear equation. The slope is m and (x_1, y_1) is a point on the line.
$y = c$	**Horizontal line** The slope is 0 and the y-intercept is $(0, c)$.
$x = c$	**Vertical line** The slope is undefined and the x-intercept is $(c, 0)$.

Parallel and Perpendicular Lines

Nonvertical parallel lines have the same slope.
The product of the slopes of two nonvertical perpendicular lines is -1.

CALCULATOR EXPLORATIONS Graphing

A graphing calculator is a very useful tool for discovering patterns. To discover the change in the graph of a linear equation caused by a change in slope, try the following. Use a standard window and graph a linear equation in the form $y = mx + b$. Recall that the graph of such an equation will have slope m and y-intercept $(0, b)$.

First graph $y = x + 3$. To do so, press the $\boxed{Y=}$ key and enter $Y_1 = x + 3$. Notice that this graph has slope 1 and that the y-intercept is 3. Next, on the same set of axes, graph $y = 2x + 3$ and $y = 3x + 3$ by pressing $\boxed{Y=}$ and entering $Y_2 = 2x + 3$ and $Y_3 = 3x + 3$.

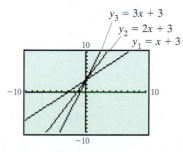

Notice the difference in the graph of each equation as the slope changes from 1 to 2 to 3. How would the graph of $y = 5x + 3$ appear? To see the change in the graph caused by a change in negative slope, try graphing $y = -x + 3$, $y = -2x + 3$, and $y = -3x + 3$ on the same set of axes.

Use a graphing calculator to graph the following equations. For each exercise, graph the first equation and use its graph to predict the appearance of the other equations. Then graph the other equations on the same set of axes and check your prediction.

1. $y = x$; $y = 6x$, $y = -6x$

2. $y = -x$; $y = -5x$, $y = -10x$

3. $y = \dfrac{1}{2}x + 2$; $y = \dfrac{3}{4}x + 2$, $y = x + 2$

4. $y = x + 1$; $y = \dfrac{5}{4}x + 1$, $y = \dfrac{5}{2}x + 1$

Mental Math

Use the equation to identify the slope and the y-intercept of the graph of each equation.

1. $y = 2x - 1$

2. $y = -7x + 3$

3. $y = x + \dfrac{1}{3}$

4. $y = -x - \dfrac{2}{9}$

5. $y = \dfrac{5}{7}x - 4$

6. $y = -\dfrac{1}{4}x + \dfrac{3}{5}$

Use the equation to identify the slope and a point on the line.

7. $y - 8 = 3(x - 4)$

8. $y - 1 = 5(x - 2)$

9. $y + 3 = -2(x - 10)$

10. $y + 6 = -7(x - 2)$

11. $y = \dfrac{2}{5}(x + 1)$

12. $y = \dfrac{3}{7}(x + 4)$

6.5 EXERCISE SET

FOR EXTRA HELP

Student Solutions Manual · PH Math/Tutor Center · CD/Video for Review · MathXL® · MyMathLab

Objective **A** *Write an equation of the line with each given slope, m, and y-intercept, (0, b). See Example 1.*

1. $m = 5, b = 3$

2. $m = -3, b = -3$

3. $m = -4, b = -\dfrac{1}{6}$

4. $m = 2, b = \dfrac{3}{4}$

5. $m = \dfrac{2}{3}, b = 0$

6. $m = -\dfrac{4}{5}, b = 0$

7. $m = 0, b = -8$

8. $m = 0, b = -2$

9. $m = -\dfrac{1}{5}, b = \dfrac{1}{9}$

10. $m = \dfrac{1}{2}, b = -\dfrac{1}{3}$

Objective **B** *Use the slope-intercept form to graph each equation. See Examples 2 and 3.*

11. $y = 2x + 1$

12. $y = -4x - 1$

13. $y = \dfrac{2}{3}x + 5$

14. $y = \dfrac{1}{4}x - 3$

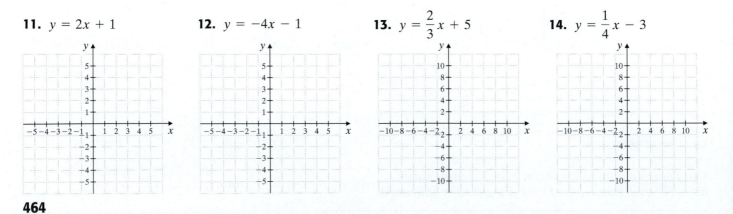

15. $y = -5x$ **16.** $y = -6x$ **17.** $4x + y = 6$ **18.** $-3x + y = 2$

19. $4x - 7y = -14$ **20.** $3x - 4y = 4$ **21.** $x = \dfrac{5}{4}y$ **22.** $x = \dfrac{3}{2}y$

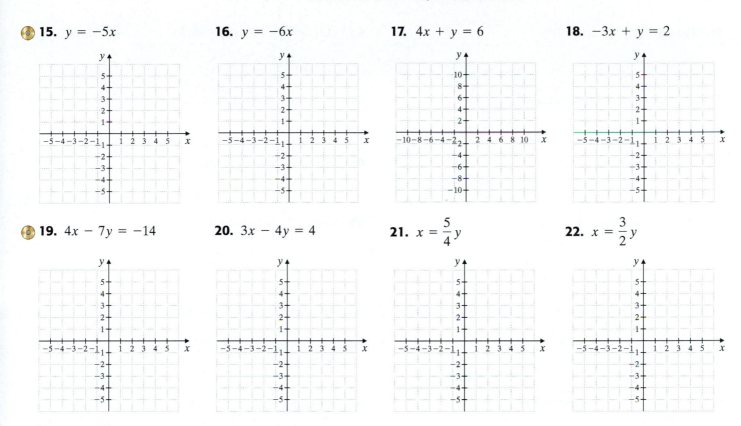

Objective **C** *Find an equation of each line with the given slope that passes through the given point. Write the equation in the form $Ax + By = C$. See Example 4.*

23. $m = 6$; $(2, 2)$ **24.** $m = 4$; $(1, 3)$

25. $m = -8$; $(-1, -5)$ **26.** $m = -2$; $(-11, -12)$

27. $m = \dfrac{3}{2}$; $(5, -6)$ **28.** $m = \dfrac{2}{3}$; $(-8, 9)$

29. $m = -\dfrac{1}{2}$; $(-3, 0)$ **30.** $m = -\dfrac{1}{5}$; $(4, 0)$

Objective **D** *Find an equation of the line passing through each pair of points. Write the equation in the form $Ax + By = C$. See Example 5.*

31. $(3, 2)$ and $(5, 6)$ **32.** $(6, 2)$ and $(8, 8)$

33. $(-1, 3)$ and $(-2, -5)$ **34.** $(-4, 0)$ and $(6, -1)$

35. $(2, 3)$ and $(-1, -1)$ **36.** $(7, 10)$ and $(-1, -1)$

37. $(0, 0)$ and $\left(-\dfrac{1}{8}, \dfrac{1}{13}\right)$ **38.** $(0, 0)$ and $\left(-\dfrac{1}{2}, \dfrac{1}{3}\right)$

Objectives **A** **C** **D** **Mixed Practice** *See Examples 1, 4, and 5. Find an equation of each line described. Write each equation in slope-intercept form when possible.*

39. With slope $-\dfrac{1}{2}$, through $\left(0, \dfrac{5}{3}\right)$ **40.** With slope $\dfrac{5}{7}$, through $(0, -3)$

41. Through $(10, 7)$ and $(7, 10)$

42. Through $(5, -6)$ and $(-6, 5)$

43. With undefined slope, through $\left(-\dfrac{3}{4}, 1\right)$

44. With slope 0, through $(6.7, 12.1)$

45. Slope 1, through $(-7, 9)$

46. Slope 5, through $(6, -8)$

47. Slope -5, y-intercept $(0, 7)$

48. Slope -2; y-intercept $(0, -4)$

49. Through $(6, 7)$, parallel to the x-axis

50. Through $(1, -5)$, parallel to the y-axis

51. Through $(2, 3)$ and $(0, 0)$

52. Through $(4, 7)$ and $(0, 0)$

53. Through $(-2, -3)$, perpendicular to the y-axis

54. Through $(0, 12)$, perpendicular to the x-axis

55. Slope $-\dfrac{4}{7}$, through $(-1, -2)$

56. Slope $-\dfrac{3}{5}$, through $(4, 4)$

Objective E *Solve. Assume each exercise describes a linear relationship. Write the equations in slope-intercept form. See Example 6.*

57. A rock is dropped from the top of a 400-foot cliff. After 1 second, the rock is traveling 32 feet per second. After 3 seconds, the rock is traveling 96 feet per second.

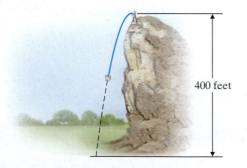

400 feet

a. Assume that the relationship between time and speed is linear and write an equation describing this relationship. Use ordered pairs of the form (time, speed).
b. Use this equation to determine the speed of the rock 4 seconds after it was dropped.

58. A Hawaiian fruit company is studying the sales of a pineapple sauce to see if this product is to be continued. At the end of its first year, profits on this product amounted to $30,000. At the end of the fourth year, profits were $66,000.

a. Assume that the relationship between years on the market and profit is linear and write an equation describing this relationship. Use ordered pairs of the form (years on the market, profit).
b. Use this equation to predict the profit at the end of 7 years.

59. In 2003 there were approximately 54,000 gas-electric hybrid vehicles sold in the United States. In 2001, there were only 22,000 such vehicles sold. (*Source: Energy Information Administration, Department of Energy*)

a. Write an equation describing the relationship between time and the number of gas-electric hybrid vehicles sold. Use ordered pairs of the form (years past 2001, number of vehicles sold).

b. Use this equation to predict the number of gas-electric hybrid sales in 2006.

60. In 2004, there were approximately 875 thousand restaurants in the United States. In 1972, there were 491 thousand restaurants. (*Source: National Restaurant Association*)

a. Write an equation describing the relationship between time and the number of restaurants. Use ordered pairs of the form (years past 1972, numbers of restaurants in thousands).

b. Use this equation to predict the number of eating establishments in 2012.

61. In 2003 there were approximately 5700 cinema sites in the United States. In 1999 there were 7032 cinema sites. (*Source: National Association of Theater Owners*)

a. Write an equation describing this relationship. Use ordered pairs of the form (years past 1999, number of cinema sites).

b. Use this equation to predict the number of cinema sites in 2007.

62. In 2000, the U.S. population per square mile of land area was 79.6. In 1990, this person per square mile population was 70.3.

a. Write an equation describing the relationship between year and person per square mile. Use ordered pairs of the form (years past 1990, person per square mile).

b. Use this equation to predict the person per square mile population in 2007.

63. In 1997 there were 1509 daily newspapers in the United States. By 2003, there were only 1456 daily newspapers. (*Source: Statistical Abstract of the United States*)

a. Write two ordered pairs of the form (years after 1997, number of daily newspapers) for this situation.

b. The relationship between years after 1997 and numbers of daily newspapers is linear over this period. Use the ordered pairs from part (a) to write an equation for the line relating year after 1997 to numbers of daily newspapers. (Round the slope to one decimal place.)

c. Use the linear equation in part (b) to estimate numbers of daily newspapers in 1999. (Round to the nearest whole.)

64. In 1999, crude oil production by the European Union countries was 3803 thousand barrels per day. In 2002, European Union oil production had decreased to 3482 thousand barrels per day. (*Source: Energy Information Administration*)

a. Write two ordered pairs of the form (years after 1999, crude oil production).

b. Assume that the relationship between years after 1999 and crude oil production is linear over this period. Use the ordered pairs from part (a) to write an equation of the line relating year to crude oil production.

c. Use the linear equation from part (b) to estimate the crude oil production by European Union countries in 2005, if this trend were to continue.

65. The Pool Fun Company has learned that, by pricing a newly released Fun Noodle at $3, sales will reach 10,000 Fun Noodles per day during the summer. Raising the price to $5 will cause the sales to fall to 8000 Fun Noodles per day.

   a. Assume that the relationship between sales price and number of Fun Noodles sold is linear and write an equation describing this relationship. Use ordered pairs of the form (sales price, number sold).

   b. Predict the daily sales of Fun Noodles if the price is $3.50.

66. The value of a building bought in 1990 may be depreciated (or decreased) as time passes for income tax purposes. Seven years after the building was bought, this value was $225,000 and 12 years after it was bought, this value was $195,000.

   a. If the relationship between number of years past 1990 and the depreciated value of the building is linear, write an equation describing this relationship. Use ordered pairs of the form (years past 1990, value of building).

   b. Use this equation to estimate the depreciated value of the building in 2008.

Review

Find the value of $x^2 - 3x + 1$ for each given value of x. See Section 1.6.

67. 2   **68.** 5   **69.** -1 **70.** -3

Concept Extensions

Match each linear equation with its graph.

71. $y = 2x + 1$ **72.** $y = -x + 1$ **73.** $y = -3x - 2$ **74.** $y = \dfrac{5}{3}x - 2$

a.   **b.**   **c.**   **d.**

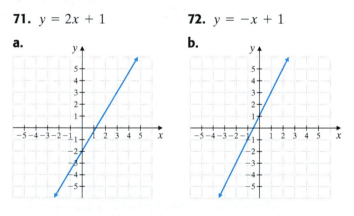
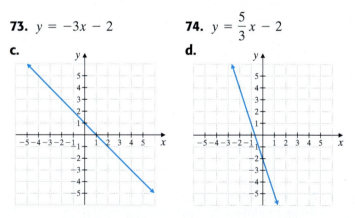

75. Write an equation of the line that contains the point $(-1, 2)$ and has the same slope as the line $y = 3x - 1$.

76. Write an equation of the line that contains the point $(4, 0)$ and has the same slope as the line $y = -2x + 3$.

△ **77.** Write an equation in standard form of the line that contains the point $(-1, 2)$ and is

   a. parallel to the line $y = 3x - 1$.

   b. perpendicular to the line $y = 3x - 1$.

△ **78.** Write an equation in standard form of the line that contains the point $(4, 0)$ and is

   a. parallel to the line $y = -2x + 3$.

   b. perpendicular to the line $y = -2x + 3$.

Summary on Linear Equations

Find the slope of each line.

1.

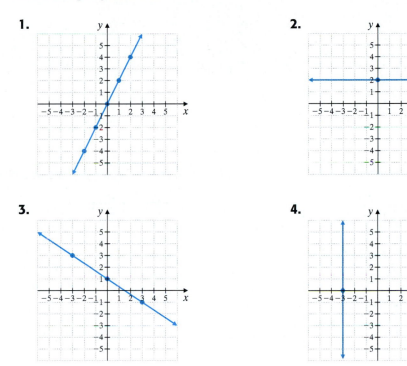

2.

3.

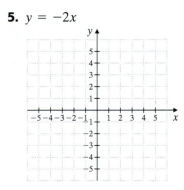

4.

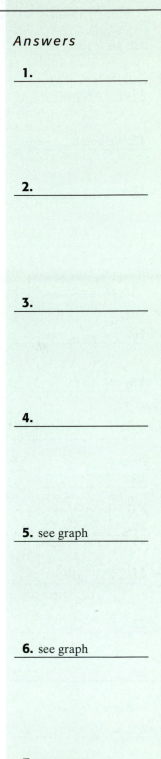

Graph each linear equation. For Exercises 11 and 12, label the intercepts.

5. $y = -2x$

6. $x + y = 3$

7. $x = -1$

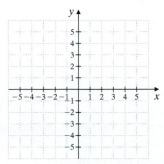

8. $y = 4$

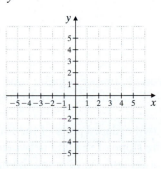

9. see graph

10. see graph

11. see graph

12. see graph

13. _____

14. _____

15. _____

16. _____

17. _____

18. _____

19. _____

20. _____

21. _____

22. _____

23. _____

24. a. _____

b. _____

c. _____

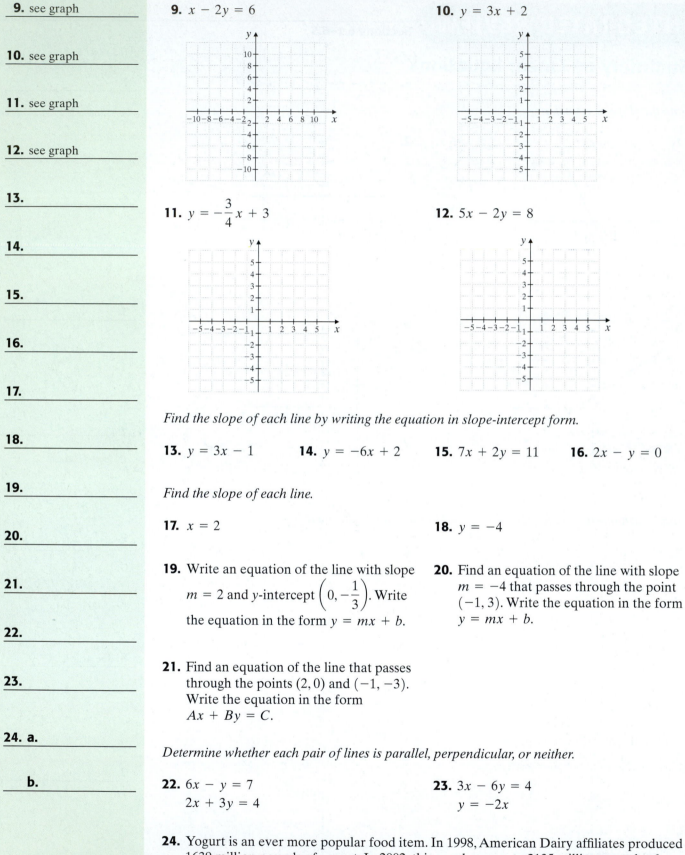

9. $x - 2y = 6$

10. $y = 3x + 2$

11. $y = -\dfrac{3}{4}x + 3$

12. $5x - 2y = 8$

Find the slope of each line by writing the equation in slope-intercept form.

13. $y = 3x - 1$
14. $y = -6x + 2$
15. $7x + 2y = 11$
16. $2x - y = 0$

Find the slope of each line.

17. $x = 2$

18. $y = -4$

19. Write an equation of the line with slope $m = 2$ and y-intercept $\left(0, -\dfrac{1}{3}\right)$. Write the equation in the form $y = mx + b$.

20. Find an equation of the line with slope $m = -4$ that passes through the point $(-1, 3)$. Write the equation in the form $y = mx + b$.

21. Find an equation of the line that passes through the points $(2, 0)$ and $(-1, -3)$. Write the equation in the form $Ax + By = C$.

Determine whether each pair of lines is parallel, perpendicular, or neither.

22. $6x - y = 7$
    $2x + 3y = 4$

23. $3x - 6y = 4$
    $y = -2x$

24. Yogurt is an ever more popular food item. In 1998, American Dairy affiliates produced 1639 million pounds of yogurt. In 2002, this number rose to 2135 million pounds of yogurt.
    a. Write two ordered pairs of the form (year, millions of pounds of yogurt produced).
    b. Find the slope of the line between these two points.
    c. Write a sentence explaining the meaning of the slope as a rate of change.

6.6 INTRODUCTION TO FUNCTIONS

Objectives

A Identify Relations, Domains, and Ranges.

B Identify Functions.

C Use the Vertical Line Test.

D Use Function Notation.

Objective A Identifying Relations, Domains, and Ranges

In this chapter, we have studied paired data in the form of ordered pairs. For example, when we list an ordered pair such as $(3, 1)$, we are saying that when x is 3, then y is 1. In other words $x = 3$ and $y = 1$ are related to each other.

For this reason, we call a set of ordered pairs a **relation.** The set of all x-coordinates is called the **domain** of a relation, and the set of all y-coordinates is called the **range** of a relation.

EXAMPLE 1 Find the domain and the range of the relation $\{(0, 2), (3, 3), (-1, 0), (3, -2)\}$.

Solution: The domain is the set of all x-coordinates, or $\{-1, 0, 3\}$, and the range is the set of all y-coordinates, or $\{-2, 0, 2, 3\}$.

◻ **Work Practice Problem 1**

PRACTICE PROBLEM 1

Find the domain and range of the relation $\{(-3, 5), (-3, 1), (4, 6), (7, 0)\}$.

Objective B Identifying Functions

Paired data occur often in real-life applications. Some special sets of paired data, or ordered pairs, are called *functions*.

Function

A **function** is a set of ordered pairs in which each x-coordinate has exactly one y-coordinate.

In other words, a function cannot have two ordered pairs with the same x-coordinate but different y-coordinates.

EXAMPLE 2 Which of the following relations are also functions?

a. $\{(-1, 1), (2, 3), (7, 3), (8, 6)\}$
b. $\{(0, -2), (1, 5), (0, 3), (7, 7)\}$

Solution:

a. Although the ordered pairs $(2, 3)$ and $(7, 3)$ have the same y-value, each x-value is assigned to only one y-value, so this set of ordered pairs is a function.
b. The x-value 0 is paired with two y-values, -2 and 3, so this set of ordered pairs is not a function.

◻ **Work Practice Problem 2**

Relations and functions can be described by graphs of their ordered pairs.

PRACTICE PROBLEM 2

Are the following relations also functions?

a. $\{(2, 5), (-3, 7), (4, 5), (0, -1)\}$
b. $\{(1, 4), (6, 6), (1, -3), (7, 5)\}$

Answers
1. domain: $\{-3, 4, 7\}$; range: $\{0, 1, 5, 6\}$,
2. a. a function, **b.** not a function

PRACTICE PROBLEM 3

Is each graph the graph of a function?

a.

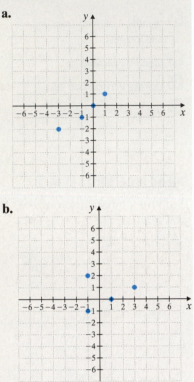

b.

EXAMPLE 3 Which graph is the graph of a function?

a.

b.

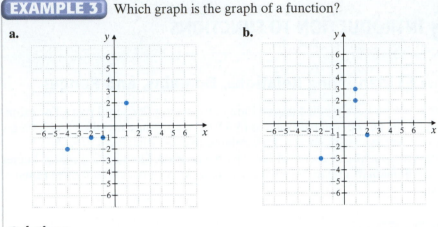

Solution:

a. This is the graph of the relation $\{(-4, -2), (-2, -1), (-1, -1), (1, 2)\}$. Each x-coordinate has exactly one y-coordinate, so this is the graph of a function.

b. This is the graph of the relation $\{(-2, -3), (1, 2), (1, 3), (2, -1)\}$. The x-coordinate 1 is paired with two y-coordinates, 2 and 3, so this is not the graph of a function.

■ **Work Practice Problem 3**

Objective **C** Using the Vertical Line Test

The graph in Example 3(b) was not the graph of a function because the x-coordinate 1 was paired with two y-coordinates, 2 and 3. Notice that when an x-coordinate is paired with more than one y-coordinate, a vertical line can be drawn that will intersect the graph at more than one point. We can use this fact to determine whether a relation is also a function. We call this the vertical line test.

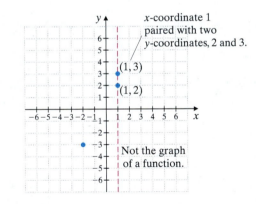

Vertical Line Test

If a vertical line can be drawn so that it intersects a graph more than once, the graph is not the graph of a function. (If no such vertical line can be drawn, the graph is that of a function.)

This vertical line test works for all types of graphs on the rectangular coordinate system.

Answers

3. **a.** a function, **b.** not a function

EXAMPLE 4 Use the vertical line test to determine whether each graph is the graph of a function.

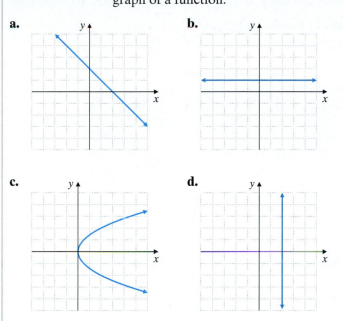

a.

b.

c.

d.

Solution:

a. This graph is the graph of a function since no vertical line will intersect this graph more than once.

b. This graph is also the graph of a function; no vertical line will intersect it more than once.

c. This graph is not the graph of a function. Vertical lines can be drawn that intersect the graph in two points. An example of one is shown.

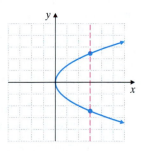

d. This graph is not the graph of a function. A vertical line can be drawn that intersects this line at every point.

☐ **Work Practice Problem 4**

Examples of functions can often be found in magazines, newspapers, books, and other printed material in the form of tables or graphs such as that in Example 5.

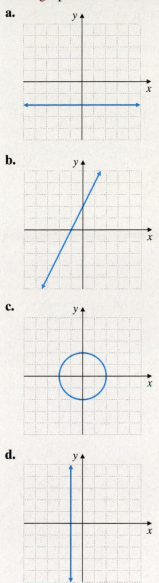

PRACTICE PROBLEM 4

Determine whether each graph is the graph of a function.

a.

b.

c.

d.

Answers

4. a. a function, **b.** a function,
c. not a function, **d.** not a function

PRACTICE PROBLEM 5

Use the graph in Example 5 to answer the questions.

a. Approximate the time of sunrise on March 1.

b. Approximate the date(s) when the sun rises at 6 A.M.

EXAMPLE 5 The graph shows the sunrise time for Indianapolis, Indiana, for the year. Use this graph to answer the questions.

a. Approximate the time of sunrise on February 1.

b. Approximate the date(s) when the sun rises at 5 A.M.

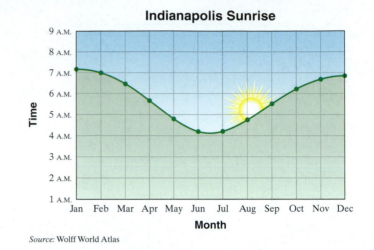

Indianapolis Sunrise

Source: Wolff World Atlas

c. Is this the graph of a function?

Solution:

a. To approximate the time of sunrise on February 1, we find the mark on the horizontal axis that corresponds to February 1. From this mark, we move vertically upward (shown in blue) until the graph is reached. From that point on the graph, we move horizontally to the left until the vertical axis is reached. The vertical axis there reads 7 A.M. as shown below.

b. To approximate the date(s) when the sun rises at 5 A.M., we find 5 A.M. on the time axis and move horizontally to the right (shown in red). Notice that we will hit the graph at two points, corresponding to two dates for which the sun rises at 5 A.M. We follow both points on the graph vertically downward until the horizontal axis is reached. The sun rises at 5 A.M. at approximately the end of the month of April and the middle of the month of August.

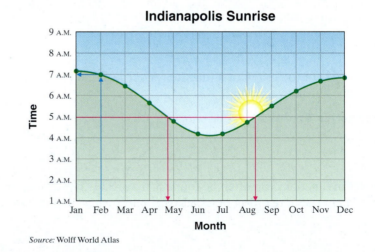

Indianapolis Sunrise

Source: Wolff World Atlas

c. The graph is the graph of a function since it passes the vertical line test. In other words, for every day of the year in Indianapolis, there is exactly one sunrise time.

Answers

5. a. 6:30 A.M., **b.** middle of March and middle of September

🟧 **Work Practice Problem 5**

Objective D Using Function Notation

The graph of the linear equation $y = 2x + 1$ passes the vertical line test, so we say that $y = 2x + 1$ is a function. In other words, $y = 2x + 1$ gives us a rule for writing ordered pairs where every x-coordinate is paired with at most one y-coordinate.

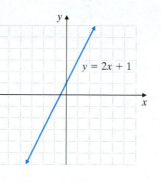

We often use letters such as f, g, and h to name functions. For example, the symbol $f(x)$ means *function of x* and is read "f of x." This notation is called **function notation.** The equation $y = 2x + 1$ can be written as $f(x) = 2x + 1$ using function notation, and these equations mean the same thing. In other words $y = f(x)$.

The notation $f(1)$ means to replace x with 1 and find the resulting y or function value. Since

$$f(x) = 2x + 1$$

then

$$f(1) = 2(1) + 1 = 3$$

This means that, when $x = 1$, y or $f(x) = 3$, and we have the ordered pair $(1, 3)$. Now let's find $f(2)$, $f(0)$, and $f(-1)$.

$f(x) = 2x + 1$	$f(x) = 2x + 1$	$f(x) = 2x + 1$
$f(2) = 2(2) + 1$	$f(0) = 2(0) + 1$	$f(-1) = 2(-1) + 1$
$= 4 + 1$	$= 0 + 1$	$= -2 + 1$
$= 5$	$= 1$	$= -1$

Ordered Pair: $(2, 5)$ $(0, 1)$ $(-1, -1)$

> **Helpful Hint**
>
> Note that $f(x)$ is a special symbol in mathematics used to denote a function. The symbol $f(x)$ is read "f of x." It does **not** mean $f \cdot x$ (f times x).

EXAMPLE 6 Given $g(x) = x^2 - 3$, find the following and list the corresponding ordered pair.

a. $g(2)$ **b.** $g(-2)$ **c.** $g(0)$

Solution:

a.	**b.**	**c.**
$g(x) = x^2 - 3$	$g(x) = x^2 - 3$	$g(x) = x^2 - 3$
$g(2) = 2^2 - 3$	$g(-2) = (-2)^2 - 3$	$g(0) = 0^2 - 3$
$= 4 - 3$	$= 4 - 3$	$= 0 - 3$
$= 1$	$= 1$	$= -3$

Ordered Pair: $(2, 1)$ $(-2, 1)$ $(0, -3)$

📖 **Work Practice Problem 6**

✔ **Concept Check** Suppose that the value of a function f is -7 when the function is evaluated at 2. Write this situation in function notation.

PRACTICE PROBLEM 6

Given $f(x) = x^2 + 1$, find the following and list the corresponding ordered pair.

a. $f(1)$

b. $f(-3)$

c. $f(0)$

Answers

6. a. $2; (1, 2)$,   **b.** $10; (-3, 10)$,
c. $1; (0, 1)$

✔ **Concept Check Answer**

$f(2) = -7$

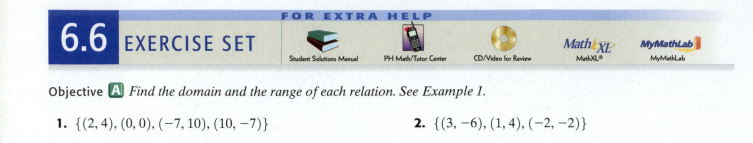
Objective **A** *Find the domain and the range of each relation. See Example 1.*

1. $\{(2, 4), (0, 0), (-7, 10), (10, -7)\}$

2. $\{(3, -6), (1, 4), (-2, -2)\}$

3. $\{(0, -2), (1, -2), (5, -2)\}$

4. $\{(5, 0), (5, -3), (5, 4), (5, 3)\}$

Objective **B** *Determine whether each relation is also a function. See Example 2.*

5. $\{(1, 1), (2, 2), (-3, -3), (0, 0)\}$

6. $\{(11, 6), (-1, -2), (0, 0), (3, -2)\}$

7. $\{(-1, 0), (-1, 6), (-1, 8)\}$

8. $\{(1, 2), (3, 2), (1, 4)\}$

Objectives **B** **C** **Mixed Practice** *Determine whether each graph is the graph of a function. For Exercises 9 through 12, either write down the ordered pairs or use the vertical line test. See Examples 3 and 4.*

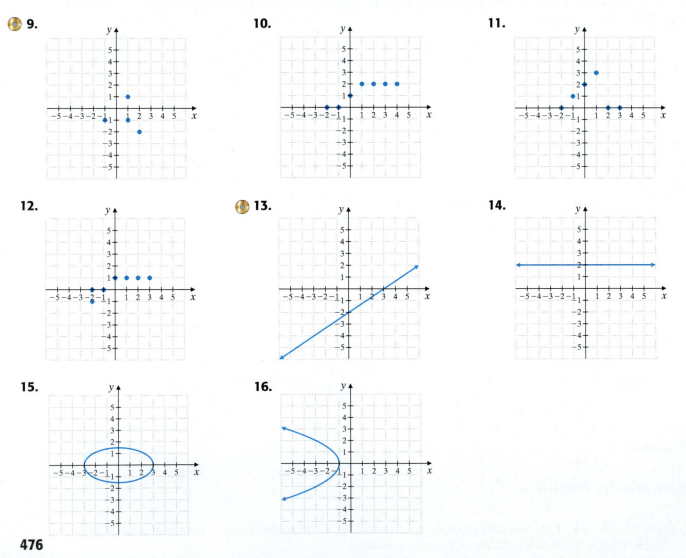

The graph shows the sunset times for Seward, Alaska. Use this graph to answer Exercises 17 through 22.

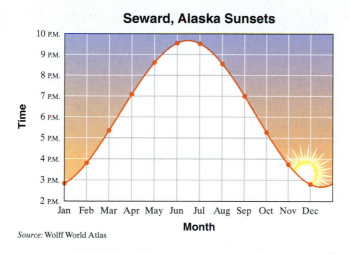

Seward, Alaska Sunsets

Source: Wolff World Atlas

17. Approximate the time of sunset on June 1.

18. Approximate the time of sunset on November 1.

19. Approximate the date(s) when the sunset is at 3 P.M.

20. Approximate the date(s) when the sunset is at 9 P.M.

21. Is this graph the graph of a function? Why or why not?

22. Do you think a graph of sunset times for any location will always be a function? Why or why not?

This graph shows the U.S. hourly minimum wage for each year shown. Use this graph to answer Exercises 23 through 28.

U.S. Hourly Minimum Wage

* Already passed by Congress

23. Approximate the minimum wage at the beginning of 1997.

24. Approximate the minimum wage at the beginning of 1999.

25. Approximate the year when the minimum wage will increase to over $5.75 per hour.

26. Approximate the year when the minimum wage increased to over $5.00 per hour.

27. Is this graph the graph of a function? Why or why not?

28. Do you think that a similar graph of your hourly wage on January 1 of every year (whether you are working or not) will be the graph of a function? Why or why not?

Objective **D** *Find $f(-2)$, $f(0)$, and $f(3)$ for each function. See Example 6.*

29. $f(x) = 2x - 5$

30. $f(x) = 3 - 7x$

31. $f(x) = x^2 + 2$

32. $f(x) = x^2 - 4$

33. $f(x) = 3x$

34. $f(x) = -3x$

35. $f(x) = |x|$

36. $f(x) = |2 - x|$

Find $h(-1)$, $h(0)$, and $h(4)$ for each function. See Example 6.

37. $h(x) = -5x$

38. $h(x) = -3x$

39. $h(x) = 2x^2 + 3$

40. $h(x) = 3x^2$

41. If $f(3) = 6$, write a corresponding ordered-pair solution.

42. If $f(7) = -2$, write a corresponding ordered-pair solution.

Use the graph of f below to answer Exercises 43 through 48.

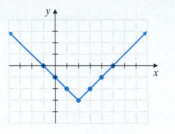

43. Complete the ordered-pair solution for f. (0,   )

44. Complete the ordered-pair solution for f. (3,   )

45. $f(0) =$ _____?

46. $f(3) =$ _____?

47. If $f(x) = 0$, find the value(s) of x.

48. If $f(x) = -1$, find the value(s) of x.

Review

Solve each inequality. See Section 2.7.

49. $2x + 5 < 7$

50. $3x - 1 \geq 11$

51. $-x + 6 \leq 9$

52. $-2x + 3 > 3$

Find the perimeter of each figure. See Section 5.4.

△ **53.**

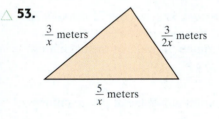

$\dfrac{3}{x}$ meters

$\dfrac{3}{2x}$ meters

$\dfrac{5}{x}$ meters

△ **54.**

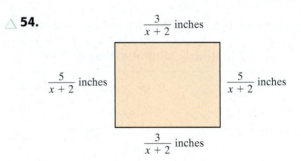

$\dfrac{3}{x+2}$ inches

$\dfrac{5}{x+2}$ inches

$\dfrac{5}{x+2}$ inches

$\dfrac{3}{x+2}$ inches

Concept Extensions

Solve. See the Concept Check in this section.

55. If a function f is evaluated at -5, the value of the function is 12. Write this situation using function notation.

56. Suppose $(9, 20)$ is an ordered-pair solution for the function g. Write this situation using function notation.

57. In your own words define (a) function; (b) domain; (c) range.

58. Explain the vertical line test and how it is used.

59. Since $y = x + 7$ is a function, rewrite the equation using function notation.

60. Forensic scientists use the function

$$f(x) = 2.59x + 47.24$$

to estimate the height of a woman, in centimeters, given the length x of her femur bone in centimeters.

   a. Estimate the height of a woman whose femur measures 46 centimeters.
   b. Estimate the height of a woman whose femur measures 39 centimeters.

61. The dosage in milligrams of Ivermectin, a heartworm preventive for a dog who weighs x pounds, is given by the function

$$f(x) = \frac{136}{25}x$$

   a. Find the proper dosage for a dog that weighs 35 pounds.
   b. Find the proper dosage for a dog that weighs 70 pounds.

6.7 GRAPHING LINEAR INEQUALITIES IN TWO VARIABLES

Objectives

A Determine Whether an Ordered Pair is a Solution of a Linear Inequality in Two Variables.

B Graph a Linear Inequality in Two Variables.

Recall that a linear equation in two variables is an equation that can be written in the form $Ax + By = C$, where A, B, and C are real numbers and A and B are not both 0. A **linear inequality in two variables** is an inequality that can be written in one of the forms

$$Ax + By < C \qquad Ax + By \le C$$
$$Ax + By > C \qquad Ax + By \ge C$$

where A, B, and C are real numbers and A and B are not both 0.

Objective **A** Determining Solutions of Linear Inequalities in Two Variables

Just as for linear equations in x and y, an ordered pair is a **solution** of an inequality in x and y if replacing the variables with the coordinates of the ordered pair results in a true statement.

EXAMPLE 1 Determine whether each ordered pair is a solution of the inequality $2x - y < 6$.

a. $(5, -1)$ **b.** $(2, 7)$

Solution:

a. We replace x with 5 and y with -1 and see if a true statement results.

$$2x - y < 6$$
$$2(5) - (-1) < 6 \quad \text{Replace } x \text{ with 5 and } y \text{ with } -1.$$
$$10 + 1 < 6$$
$$11 < 6 \quad \text{False}$$

The ordered pair $(5, -1)$ is not a solution since $11 < 6$ is a false statement.

b. We replace x with 2 and y with 7 and see if a true statement results.

$$2x - y < 6$$
$$2(2) - (7) < 6 \quad \text{Replace } x \text{ with 2 and } y \text{ with 7.}$$
$$4 - 7 < 6$$
$$-3 < 6 \quad \text{True}$$

The ordered pair $(2, 7)$ is a solution since $-3 < 6$ is a true statement.

Work Practice Problem 1

Objective **B** Graphing Linear Inequalities in Two Variables

The linear equation $x - y = 1$ is graphed next. Recall that all points on the line correspond to ordered pairs that satisfy the equation $x - y = 1$.

Notice the line defined by $x - y = 1$ divides the rectangular coordinate system plane into 2 sides. All points on one side of the line satisfy the inequality $x - y < 1$ and all points on the other side satisfy the inequality $x - y > 1$. The graph on the next page shows a few examples of this.

PRACTICE PROBLEM 1

Determine whether each ordered pair is a solution of $x - 4y > 8$.

a. $(-3, 2)$

b. $(9, 0)$

Answers

1. a. no, **b.** yes

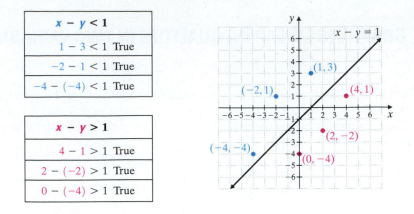

$x - y < 1$
$1 - 3 < 1$ True
$-2 - 1 < 1$ True
$-4 - (-4) < 1$ True

$x - y > 1$
$4 - 1 > 1$ True
$2 - (-2) > 1$ True
$0 - (-4) > 1$ True

The graph of $x - y < 1$ is the region shaded blue and the graph of $x - y > 1$ is the region shaded red below.

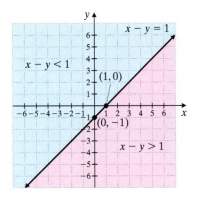

The region to the left of the line and the region to the right of the line are called **half-planes.** Every line divides the plane (similar to a sheet of paper extending indefinitely in all directions) into two half-planes; the line is called the **boundary.**

Recall that the inequality $x - y \leq 1$ means

$$x - y = 1 \quad \text{or} \quad x - y < 1$$

Thus, the graph of $x - y \leq 1$ is the half-plane $x - y < 1$ along with the boundary line $x - y = 1$.

To Graph a Linear Inequality in Two Variables

Step 1: Graph the boundary line found by replacing the inequality sign with an equal sign. If the inequality sign is $>$ or $<$, graph a dashed boundary line (indicating that the points on the line are not solutions of the inequality). If the inequality sign is $\geq$ or $\leq$, graph a solid boundary line (indicating that the points on the line are solutions of the inequality).

Step 2: Choose a point, *not* on the boundary line, as a test point. Substitute the coordinates of this test point into the *original* inequality.

Step 3: If a true statement is obtained in Step 2, shade the half-plane that contains the test point. If a false statement is obtained, shade the half-plane that does not contain the test point.

EXAMPLE 2 Graph: $x + y < 7$

Solution:

Step 1: First we graph the boundary line by graphing the equation $x + y = 7$. We graph this boundary as a *dashed line* because the inequality sign is $<$, and thus the points on the line are not solutions of the inequality $x + y < 7$.

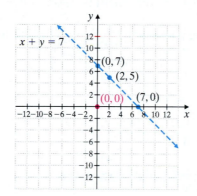

Step 2: Next we choose a test point, being careful *not* to choose a point on the boundary line. We choose $(0, 0)$, and substitute the coordinates of $(0, 0)$ into $x + y < 7$.

$x + y < 7$ Original inequality

$0 + 0 < 7$ Replace x with 0 and y with 0.

$\quad\ 0 < 7$ True

Step 3: Since the result is a true statement, $(0, 0)$ is a solution of $x + y < 7$, and every point in the same half-plane as $(0, 0)$ is also a solution. To indicate this, we shade the entire half-plane containing $(0, 0)$, as shown.

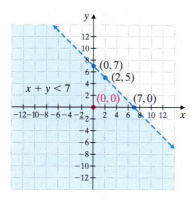

▢ **Work Practice Problem 2**

✔ **Concept Check** Determine whether $(0, 0)$ is included in the graph of

a. $y \geq 2x + 3$

b. $x < 7$

c. $2x - 3y < 6$

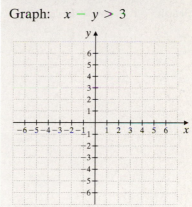
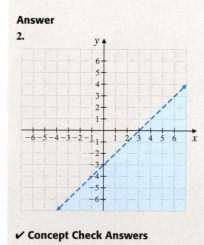

PRACTICE PROBLEM 3

Graph: $x - 4y \leq 4$

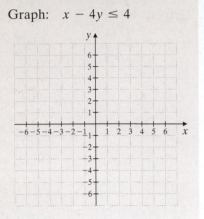

PRACTICE PROBLEM 4

Graph: $y < 3x$

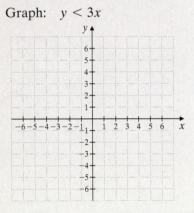

EXAMPLE 3 Graph: $2x - y \geq 3$

Solution:

Step 1: We graph the boundary line by graphing $2x - y = 3$. We draw this line as a solid line because the inequality sign is $\geq$, and thus the points on the line are solutions of $2x - y \geq 3$.

Step 2: Once again, $(0, 0)$ is a convenient test point since it is not on the boundary line.

We substitute 0 for x and 0 for y into the original inequality.

$2x - y \geq 3$

$2(0) - 0 \geq 3$ Let $x = 0$ and $y = 0$.

$0 \geq 3$ False

Step 3: Since the statement is false, no point in the half-plane containing $(0, 0)$ is a solution. Therefore, we shade the half-plane that does not contain $(0, 0)$. Every point in the shaded half-plane and every point on the boundary line is a solution of $2x - y \geq 3$.

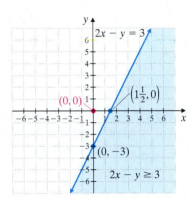

🔲 **Work Practice Problem 3**

Helpful Hint

When graphing an inequality, make sure the test point is substituted into the **original inequality.** For Example 3, we substituted the test point $(0, 0)$ into the **original inequality** $2x - y \geq 3$, *not* $2x - y = 3$.

Answers

3.

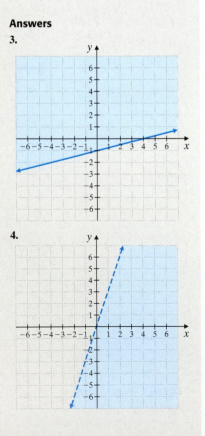

EXAMPLE 4 Graph: $x > 2y$

Solution:

Step 1: We find the boundary line by graphing $x = 2y$. The boundary line is a dashed line since the inequality symbol is $>$.

Step 2: We cannot use $(0, 0)$ as a test point because it is a point on the boundary line. We choose instead $(0, 2)$.

$x > 2y$

$0 > 2(2)$ Let $x = 0$ and $y = 2$.

$0 > 4$ False

Step 3: Since the statement is false, we shade the half-plane that does not contain the test point $(0, 2)$, as shown.

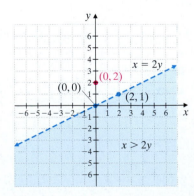

🔲 **Work Practice Problem 4**

EXAMPLE 5 Graph: $5x + 4y \leq 20$

Solution: We graph the solid boundary line $5x + 4y = 20$ and choose $(0, 0)$ as the test point.

$$5x + 4y \leq 20$$
$$5(0) + 4(0) \leq 20 \quad \text{Let } x = 0 \text{ and } y = 0.$$
$$0 \leq 20 \quad \text{True}$$

We shade the half-plane that contains $(0, 0)$, as shown.

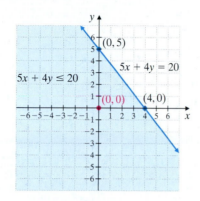

Work Practice Problem 5

EXAMPLE 6 Graph: $y > 3$

Solution: We graph the dashed boundary line $y = 3$ and choose $(0, 0)$ as the test point. (Recall that the graph of $y = 3$ is a horizontal line with y-intercept 3.)

$$y > 3$$
$$0 > 3 \quad \text{Let } y = 0.$$
$$0 > 3 \quad \text{False}$$

We shade the half-plane that does not contain $(0, 0)$, as shown.

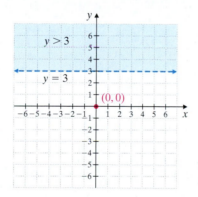

Work Practice Problem 6

PRACTICE PROBLEM 5

Graph: $3x + 2y \geq 12$

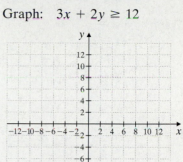

PRACTICE PROBLEM 6

Graph: $x < 2$

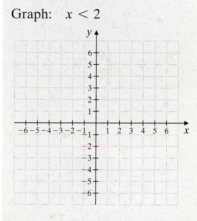

Answers

5.

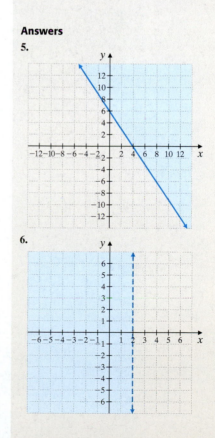

6.

PRACTICE PROBLEM 7

Graph: $y \geq \dfrac{1}{4}x + 3$

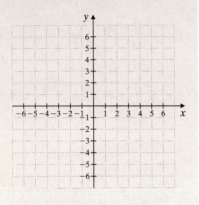

EXAMPLE 7 Graph: $y \leq \dfrac{2}{3}x - 4$

Solution: Graph the solid boundary line $y = \dfrac{2}{3}x - 4$. This equation is in slope-intercept form with slope $\dfrac{2}{3}$ and y-intercept -4.

We use this information to graph the line. Then we choose $(0, 0)$ as our test point.

$$y \leq \frac{2}{3}x - 4$$

$$0 \leq \frac{2}{3} \cdot 0 - 4$$

$$0 \leq -4 \quad \text{False}$$

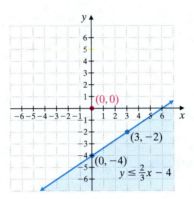

We shade the half-plane that does not contain $(0, 0)$, as shown.

🔲 **Work Practice Problem 7**

Answer

7.

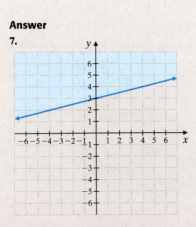

Mental Math

State whether the graph of each inequality includes its corresponding boundary line.

1. $y \geq x + 4$ **2.** $x - y > -7$ **3.** $y \geq x$ **4.** $x > 0$

Decide whether $(0, 0)$ is a solution of each given inequality.

5. $x + y > -5$ **6.** $2x + 3y < 10$ **7.** $x - y \leq -1$ **8.** $\dfrac{2}{3}x + \dfrac{5}{6}y > 4$

6.7 EXERCISE SET

FOR EXTRA HELP

Student Solutions Manual   PH Math/Tutor Center   CD/Video for Review   MathXL®   MyMathLab

Objective A *Determine whether the ordered pairs given are solutions of the linear inequality in two variables. See Example 1.*

1. $x - y > 3; (0, 3), (2, -1)$

2. $y - x < -2; (2, 1), (5, -1)$

3. $3x - 5y \leq -4; (2, 3), (-1, -1)$

4. $2x + y \geq 10; (0, 11), (5, 0)$

5. $x < -y; (0, 2), (-5, 1)$

6. $y > 3x; (0, 0), (1, 4)$

Objective B *Graph each inequality. See Examples 2 through 7.*

7. $x + y \leq 1$

8. $x + y \geq -2$

9. $2x - y > -4$

10. $x - 3y < 3$

11. $y > 2x$

12. $y < 3x$

13. $x \leq -3y$

14. $x \geq -2y$

15. $y \geq x + 5$

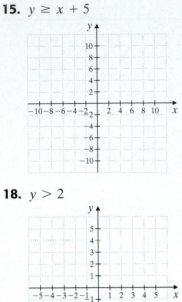

16. $y \leq x + 1$

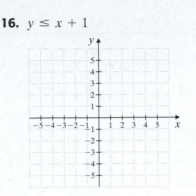

17. $y < 4$

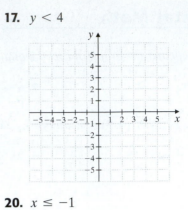

18. $y > 2$

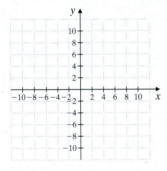

19. $x \geq -3$

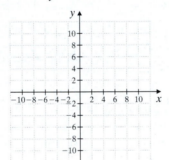

20. $x \leq -1$

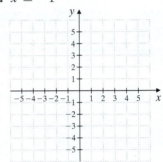

21. $5x + 2y \leq 10$

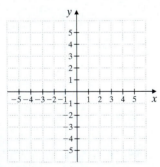

22. $4x + 3y \geq 12$

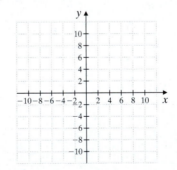

23. $x > y$

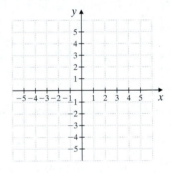

24. $x \leq -y$

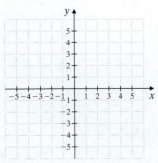

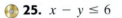 **25.** $x - y \leq 6$

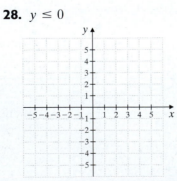

26. $x - y > 10$

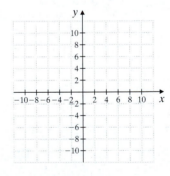

27. $x \geq 0$

28. $y \leq 0$

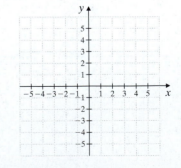 **29.** $2x + 7y > 5$

30. $3x + 5y \leq -2$ **31.** $y \geq \dfrac{1}{2}x - 4$ **32.** $y < \dfrac{2}{5}x - 3$

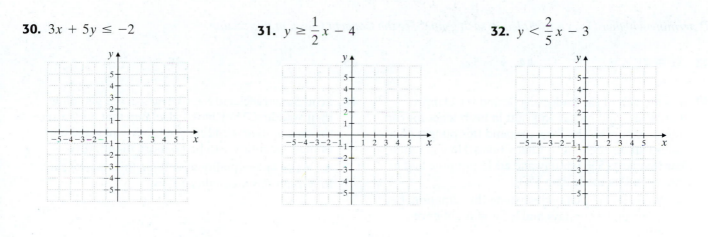

Review

Approximate the coordinates of each point of intersection. See Section 6.1.

33.   **34.**   **35.**   **36.**

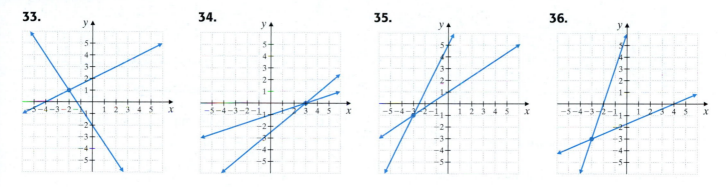

Concept Extensions

Match each inequality with its graph.

a. $x > 2$ **b.** $y < 2$ **c.** $y \leq 2x$ **d.** $y \leq -3x$

37.   **38.**   **39.**   **40.**

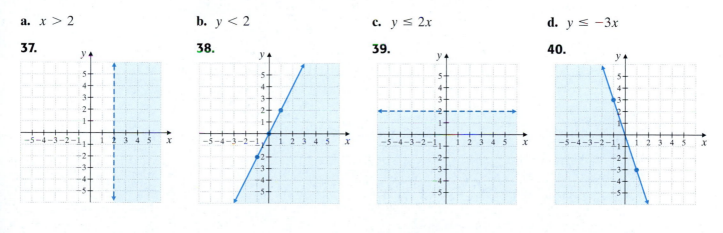

41. Explain why a point on the boundary line should not be chosen as the test point.

42. Write an inequality whose solutions are all points of numbers whose sum is at least 13.

Determine whether $(1, 1)$ *is included in each graph. See the Concept Check in this section.*

43. $3x + 4y < 8$ **44.** $y > 5x$ **45.** $y \geq -\dfrac{1}{2}x$ **46.** $x > 3$

47. It's the end of the budgeting period for Dennis Fernandes and he has $500 left in his budget for car rental expenses. He plans to spend this budget on a sales trip throughout southern Texas. He will rent a car that costs $30 per day and $0.15 per mile and he can spend no more than $500.

a. Write an inequality describing this situation. Let x = number of days and let y = number of miles.

b. Graph this inequality below.

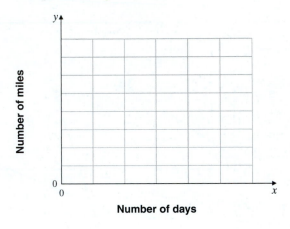

c. Why is the grid showing quadrant I only?

48. Scott Sambracci and Sara Thygeson are planning their wedding. They have calculated that they want the cost of their wedding ceremony x plus the cost of their reception y to be no more than $5000.

a. Write an inequality describing this relationship.

b. Graph this inequality below.

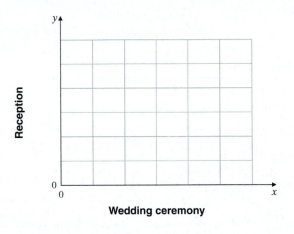

c. Why is the grid showing quadrant I only?

6.8 DIRECT AND INVERSE VARIATION

Objectives

A Solve Problems Involving Direct Variation.

B Solve Problems Involving Inverse Variation.

C Solve Problems Involving Other Types of Direct and Inverse Variation.

D Solve Applications of Variation.

In Chapter 2, we studied linear equations in two variables. Recall that such an equation can be written in the form $Ax + By = C$, where A and B are not both 0. Also recall that the graph of a linear equation in two variables is a line. In this section, we begin by looking at a particular family of linear equations—those that can be written in the form

$$y = kx$$

where k is a constant. This family of equations is called *direct variation*.

Objective **A** Solving Direct Variation Problems

Let's suppose that you are earning \$7.25 per hour at a part-time job. The amount of money you earn depends on the number of hours you work. This is illustrated by the following table:

Hours Worked	0	1	2	3	4
Money Earned (before deductions)	0	7.25	14.50	21.75	29.00

and so on

In general, to calculate your earnings (before deductions), multiply the constant \$7.25 by the number of hours you work. If we let y represent the amount of money earned and x represent the number of hours worked, we get the direct variation equation

$$y = 7.25 \cdot x$$

earnings = \$7.25 · hours worked

Notice that in this direct variation equation, as the number of hours increases, the pay increases as well.

> ### Direct Variation
>
> **y varies directly as x, or y is directly proportional to x,** if there is a nonzero constant k such that
>
> $$y = kx$$
>
> The number k is called the **constant of variation** or the **constant of proportionality.**

In our direct variation example, $y = 7.25x$, the constant of variation is 7.25.

Let's use the previous table to graph $y = 7.25x$. We begin our graph at the ordered-pair solution $(0, 0)$. Why? We assume that the least amount of hours worked is 0. If 0 hours are worked, then the pay is \$0.

As illustrated in this graph, a direct variation equation $y = kx$ is linear. Also notice that $y = 7.25x$ is a function since its graph passes the vertical line test.

PRACTICE PROBLEM 1

Write a direct variation equation that satisfies:

x	4	$\frac{1}{2}$	1.5	6
y	8	1	3	12

EXAMPLE 1 Write a direct variation equation of the form $y = kx$ that satisfies the ordered pairs in the table below.

x	2	9	1.5	−1
y	6	27	4.5	−3

Solution: We are given that there is a direct variation relationship between x and y. This means that

$$y = kx$$

By studying the given values, you may be able to mentally calculate k. If not, to find k, we simply substitute one given ordered pair into this equation and solve for k. We'll use the given pair $(2, 6)$.

$$y = kx$$
$$6 = k \cdot 2$$
$$\frac{6}{2} = \frac{k \cdot 2}{2}$$
$$3 = k \qquad \text{Solve for } k.$$

Since $k = 3$, we have the equation $y = 3x$.
To check, see that each given y is 3 times the given x.

☐ **Work Practice Problem 1**

Let's try another type of direct variation example.

PRACTICE PROBLEM 2

Suppose that y varies directly as x. If y is 15 when x is 45, find the constant of variation and the direct variation equation. Then find y when x is 3.

EXAMPLE 2 Suppose that y varies directly as x. If y is 17 when x is 34, find the constant of variation and the direct variation equation. Then find y when x is 12.

Solution: Let's use the same method as in Example 1 to find k. Since we are told that y varies directly as x, we know the relationship is of the form

$$y = kx$$

Let $y = 17$ and $x = 34$ and solve for k.

$$17 = k \cdot 34$$
$$\frac{17}{34} = \frac{k \cdot 34}{34}$$
$$\frac{1}{2} = k \qquad \text{Solve for } k.$$

Thus, the constant of variation is $\frac{1}{2}$ and the equation is $y = \frac{1}{2}x$.

To find y when $x = 12$, use $y = \frac{1}{2}x$ and replace x with 12.

$$y = \frac{1}{2}x$$
$$y = \frac{1}{2} \cdot 12 \qquad \text{Replace } x \text{ with 12.}$$
$$y = 6$$

Thus, when x is 12, y is 6.

☐ **Work Practice Problem 2**

Let's review a few facts about linear equations of the form $y = kx$.

Answers
1. $y = 2x$, **2.** $k = \frac{1}{3}$; $y = \frac{1}{3}x$; $y = 1$

Direct Variation: y = kx

- There is a direct variation relationship between x and y.
- The graph is a line.
- The line will always go through the origin $(0, 0)$. Why?

    Let $x = 0$. Then $y = k \cdot 0$ or $y = 0$.

- The slope of the graph of $y = kx$ is k, the constant of variation. Why? Remember that the slope of an equation of the form $y = mx + b$ is m, the coefficient of x.
- The equation $y = kx$ describes a function. Each x has a unique y and its graph passes the vertical line test.

EXAMPLE 3 The line is the graph of a direct variation equation. Find the constant of variation and the direct variation equation.

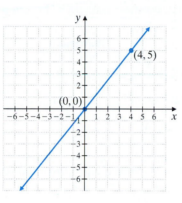

PRACTICAL PROBLEM 3

Find the constant of variation and the direct variation equation for the line below.

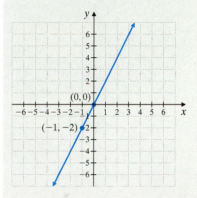

Solution: Recall that k, the constant of variation, is the same as the slope of the line. Thus, to find k, we use the slope formula and find slope.

Using the given points $(0, 0)$, and $(4, 5)$, we have

$$\text{slope} = \frac{5 - 0}{4 - 0} = \frac{5}{4}$$

Thus, $k = \dfrac{5}{4}$ and the variation equation is $y = \dfrac{5}{4}x$.

Work Practice Problem 3

Objective B Solving Inverse Variation Problems

In this section, we introduce another type of variation called inverse variation.

Let's suppose you need to drive a distance of 40 miles. You know that the faster you drive the distance, the sooner you arrive at your destination. Recall that there is a mathematical relationship between distance, rate, and time. It is $d = r \cdot t$. In our example, distance is a constant 40 miles, so we have $40 = r \cdot t$ or $t = \dfrac{40}{r}$.

For example, if you drive 10 mph, the time to drive the 40 miles is

$$t = \frac{40}{r} = \frac{40}{10} = 4 \text{ hours}$$

If you drive 20 mph, the time is

$$t = \frac{40}{r} = \frac{40}{20} = 2 \text{ hours}$$

Answer

3. $k = 2$; $y = 2x$

Again, notice that as speed increases, time decreases. Below are some ordered-pair solutions of $t = \dfrac{40}{r}$ and its graph.

Rate (mph)	r	5	10	20	40	60	80
Time (hr)	t	8	4	2	1	$\frac{2}{3}$	$\frac{1}{2}$

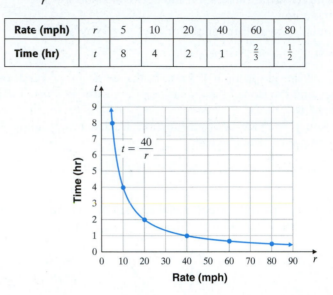

Notice that the graph of this variation is not a line, but it passes the vertical line test so $t = \dfrac{40}{r}$ does describe a function. This is an example of inverse variation.

Inverse Variation

y varies inversely as x, or **y is inversely proportional to x,** if there is a nonzero constant k such that

$$y = \frac{k}{x}$$

The number k is called the **constant of variation** or the **constant of proportionality.**

In our inverse variation example, $t = \dfrac{40}{r}$ or $y = \dfrac{40}{x}$, the constant of variation is 40.

We can immediately see differences and similarities in direct variation and inverse variation.

Direct Variation	$y = kx$	linear equation	both functions
Inverse Variation	$y = \frac{k}{x}$	rational equation	

Remember from Chapter 5 that $y = \dfrac{k}{x}$ is a rational equation and not a linear equation. Also notice that because x is in the denominator, x can be any value except 0.

We can still derive an inverse variation equation from a table of values.

EXAMPLE 4 Write an inverse variation equation of the form $y = \dfrac{k}{x}$ that satisfies the ordered pairs in the table below.

x	2	4	$\frac{1}{2}$
y	6	3	24

Solution: Since there is an inverse variation relationship between x and y, we know that $y = \dfrac{k}{x}$.

PRACTICE PROBLEM 4

Write an inverse variation equation of the form $y = \dfrac{k}{x}$ that satisfies:

x	4	10	40	-2
y	5	2	$\frac{1}{2}$	-10

Answer

4. $y = \dfrac{20}{x}$

To find k, choose one given ordered pair and substitute the values into the equation. We'll use $(2, 6)$.

$$y = \frac{k}{x}$$

$$6 = \frac{k}{2}$$

$$2 \cdot 6 = 2 \cdot \frac{k}{2} \quad \text{Multiply both sides by 2.}$$

$$12 = k \quad \text{Solve.}$$

Since $k = 12$, we have the equation $y = \frac{12}{x}$.

📖 **Work Practice Problem 4**

Helpful Hint

Multiply both sides of the inverse variation relationship equation $y = \frac{k}{x}$ by x (as long as x is not 0), and we have $xy = k$. This means that if y varies inversely as x, their product is always the constant of variation k. For an example of this, check the table from Example 4:

x	2	4	$\frac{1}{2}$
y	6	3	24

$2 \cdot 6 = 12 \qquad 4 \cdot 3 = 12 \qquad \frac{1}{2} \cdot 24 = 12$

EXAMPLE 5 Suppose that y varies inversely as x. If $y = 0.02$ when $x = 75$, find the constant of variation and the inverse variation equation. Then find y when x is 30.

Solution: Since y varies inversely as x, the constant of variation may be found by simply finding the product of the given x and y.

$$k = xy = 75(0.02) = 1.5$$

To check, we will use the inverse variation equation

$$y = \frac{k}{x}$$

Let $y = 0.02$ and $x = 75$ and solve for k.

$$0.02 = \frac{k}{75}$$

$$75(0.02) = 75 \cdot \frac{k}{75} \quad \text{Multiply both sides by 75.}$$

$$1.5 = k \quad \text{Solve for } k.$$

Thus, the constant of variation is 1.5 and the equation is $y = \frac{1.5}{x}$. To find y when $x = 30$, use $y = \frac{1.5}{x}$ and replace x with 30.

$$y = \frac{1.5}{x}$$

$$y = \frac{1.5}{30} \quad \text{Replace } x \text{ with 30.}$$

$$y = 0.05$$

Thus, when x is 30, y is 0.05.

📖 **Work Practice Problem 5**

PRACTICE PROBLEM 5

Suppose that y varies inversely as x. If y is 4 when x is 0.8, find the constant of variation and the direct variation equation. Then find y when x is 20.

Answer

5. $k = 3.2$; $y = \frac{3.2}{x}$; $y = 0.16$

Objective C Solving Other Types of Direct and Inverse Variation Problems

It is possible for y to vary directly or inversely as powers of x.

Direct and Inverse Variation as nth Powers of x

y varies directly as a power of x if there is a nonzero constant k and a natural number n such that

$$y = kx^n$$

y varies inversely as a power of x if there is a nonzero constant k and a natural number n such that

$$y = \frac{k}{x^n}$$

PRACTICE PROBLEM 6

The area of a circle varies directly as the square of its radius. A circle with radius 7 inches has an area of 49π square inches. Find the area of a circle whose radius is 4 feet.

EXAMPLE 6 The surface area of a cube A varies directly as the square of a length of its sides. If A is 54 when s is 3, find A when $s = 4.2$.

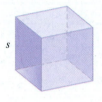

Solution: Since the surface area A varies directly as the square of side s, we have

$$A = ks^2$$

To find k, let $A = 54$ and $s = 3$.

$A = k \cdot s^2$

$54 = k \cdot 3^2$ ⠀⠀Let $A = 54$ and $s = 3$.

$54 = 9k$ ⠀⠀⠀$3^2 = 9$.

$6 = k$ ⠀⠀⠀⠀Divide by 9.

The formula for surface area of a cube is then

$A = 6s^2$ where s is the length of a side.

To find the surface area when $s = 4.2$, substitute.

$A = 6s^2$

$A = 6 \cdot (4.2)^2$

$A = 105.84$

The surface area of a cube whose side measures 4.2 units is 105.84 sq units.

◻ **Work Practice Problem 6**

Answer

6. 16π sq ft

Objective D Solving Applications of Variation

There are many real-life applications of direct and inverse variation.

EXAMPLE 7 The weight of a body w varies inversely with the square of its distance from the center of Earth, d. If a person weighs 160 pounds on the surface of Earth, what is the person's weight 200 miles above the surface? (Assume that the radius of Earth is 4000 miles.)

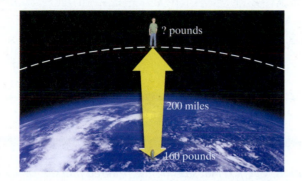

Solution:

1. UNDERSTAND. Make sure you read and reread the problem.

2. TRANSLATE. Since we are told that weight w varies inversely with the square of its distance from the center of Earth, d, we have

   $$w = \frac{k}{d^2}$$

3. SOLVE. To solve the problem, we first find k. To do so, use the fact that the person weighs 160 pounds on Earth's surface, which is a distance of 4000 miles from the Earth's center.

   $$w = \frac{k}{d^2}$$

   $$160 = \frac{k}{(4000)^2}$$

   $$2{,}560{,}000{,}000 = k$$

   Thus, we have $w = \dfrac{2{,}560{,}000{,}000}{d^2}$.

   Since we want to know the person's weight 200 miles above the Earth's surface, we let $d = 4200$ and find w.

   $$w = \frac{2{,}560{,}000{,}000}{d^2}$$

   $$w = \frac{2{,}560{,}000{,}000}{(4200)^2} \quad \text{A person 200 miles above the Earth's surface is 4200 miles from the Earth's center.}$$

   $$w \approx 145 \quad \text{Simplify.}$$

4. INTERPRET. *Check:* Your answer is reasonable since the farther a person is from Earth, the less the person weighs. *State:* Thus, 200 miles above the surface of the Earth, a 160-pound person weighs approximately 145 pounds.

Work Practice Problem 7

Answer

7. 400 feet

Mental Math

State whether each equation represents direct or indirect variation.

1. $y = 5x$

2. $y = \dfrac{5}{x}$

3. $y = \dfrac{7}{x^2}$

4. $y = 6.5x^4$

5. $y = \dfrac{11}{x}$

6. $y = 18x$

7. $y = 12x^2$

8. $y = \dfrac{20}{x^3}$

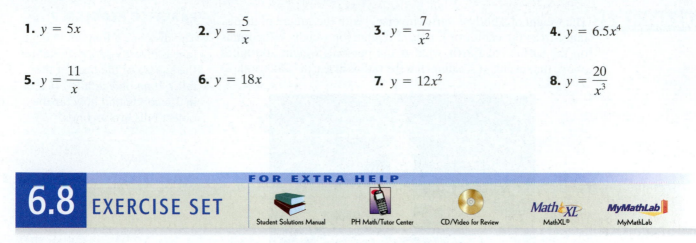

6.8 EXERCISE SET

FOR EXTRA HELP
Student Solutions Manual   PH Math/Tutor Center   CD/Video for Review   MathXL   MathXL®   MyMathLab   MyMathLab

Objective A *Write a direct variation equation, $y = kx$, that satisfies the ordered pairs in each table. See Example 1.*

1.

x	0	6	10
y	0	3	5

2.

x	0	2	−1	3
y	0	14	−7	21

3.

x	−2	2	4	5
y	−12	12	24	30

4.

x	3	9	−2	12
y	1	3	$-\dfrac{2}{3}$	4

Write a direct variation equation, $y = kx$, that describes each graph. See Example 3.

5.

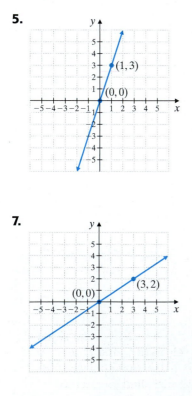

(1, 3)
(0, 0)

6.

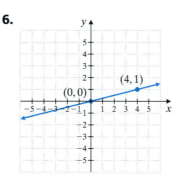

(4, 1)
(0, 0)

7.

(0, 0)
(3, 2)

8.

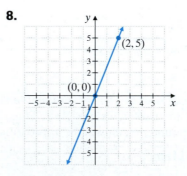

(2, 5)
(0, 0)

Objective **B** *Write an inverse variation equation, $y = \dfrac{k}{x}$, that satisfies the ordered pairs in each table. See Example 4.*

9.

x	1	−7	3.5	−2
y	7	−1	2	−3.5

10.

x	2	−11	4	−4
y	11	−2	5.5	−5.5

11.

x	10	$\frac{1}{2}$	$-\frac{1}{4}$
y	0.05	1	−2

12.

x	4	$\frac{1}{5}$	−8
y	0.1	2	−0.05

Objectives **A** **B** *Write an equation to describe each variation. Use k for the constant of proportionality. See Examples 1 through 5.*

13. y varies directly as x

14. a varies directly as b

15. h varies inversely as t

16. s varies inversely as t

17. z varies directly as x^2

18. p varies inversely as x^2

19. y varies inversely as z^3

20. x varies directly as y^4

21. x varies inversely as $\sqrt{y}$

22. y varies directly as d^2

Objectives **A** **B** **C** *Solve. See Examples 2, 5, and 6.*

23. y varies directly as x. If $y = 20$ when $x = 5$, find y when x is 10.

24. y varies directly as x. If $y = 27$ when $x = 3$, find y when x is 2.

25. y varies inversely as x. If $y = 5$ when $x = 60$, find y when x is 100.

26. y varies inversely as x. If $y = 200$ when $x = 5$, find y when x is 4.

27. z varies directly as x^2. If $z = 96$ when $x = 4$, find z when $x = 3$.

28. s varies directly as t^3. If $s = 270$ when $t = 3$, find s when $x = 1$.

29. a varies inversely as b^3. If $a = \dfrac{3}{2}$ when $b = 2$, find a when b is 3.

30. p varies inversely as q^2. If $p = \dfrac{5}{16}$ when $q = 8$, find p when $q = \dfrac{1}{2}$.

Objectives **C** **D** *Solve. See Examples 1 through 7.*

31. Your paycheck (before deductions) varies directly as the number of hours you work. If your paycheck is $112.50 for 18 hours, find your pay for 10 hours.

32. If your paycheck (before deductions) is $244.50 for 30 hours, find your pay for 34 hours. (See Exercise 31.)

33. The cost of manufacturing a certain type of headphone varies inversely as the number of headphones increases. If 5000 headphones can be manufactured for $9.00 each, find the cost to manufacture 7500 headphones.

34. The cost of manufacturing a certain composition notebook varies inversely as the number of notebooks increases. If 10,000 notebooks can be manufactured for $0.50 each, find the cost to manufacture 18,000 notebooks. Round your answer to the nearest cent.

35. The distance a spring stretches varies directly with the weight attached to the spring. If a 60-pound weight stretches the spring 4 inches, find the distance that an 80-pound weight stretches the spring.

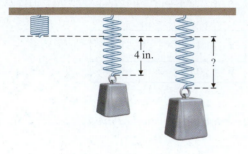

4 in.

?

36. If a 30-pound weight stretches a spring 10 inches, find the distance a 20-pound weight stretches the spring. (See Exercise 35.)

37. The weight of an object varies inversely as the square of its distance from the *center* of the Earth. If a person weighs 180 pounds on Earth's surface, what is his weight 10 miles above the surface of the Earth? (Assume that the Earth's radius is 4000 miles and round your answer to one decimal place.)

38. For a constant distance, the rate of travel varies inversely as the time traveled. If a family travels 55 mph and arrives at a destination in 4 hours, how long will the return trip take traveling at 60 mph?

39. The distance d that an object falls is directly proportional to the square of the time of the fall, t. A person who is parachuting for the first time is told to wait ten seconds before opening the parachute. If the person falls 64 feet in 2 seconds, find how far he falls in 10 seconds.

40. The distance needed for a car to stop, d, is directly proportional to the square of its rate of travel, r. Under certain driving conditions, a car traveling 60 mph needs 300 feet to stop. With these same driving conditions, how long does it take a car to stop if the car is traveling 30 mph when the brakes are applied?

Review

Add the equations. See Section 3.4.

41. $-3x + 4y = 7$
$\underline{3x - 2y = 9}$

42. $x - y = -9$
$\underline{-x - y = -14}$

43. $5x - 0.4y = 0.7$
$\underline{-9x + 0.4y = -0.2}$

44. $1.9x - 2y = 5.7$
$\underline{-1.9x - 0.1y = 2.3}$

Concept Extensions

45. Suppose that y varies directly as x. If x is tripled, what is the effect on y?

46. Suppose that y varies directly as x^2. If x is tripled, what is the effect on y?

47. The period of a pendulum p (the time of one complete back-and-forth swing) varies directly with the square root of its length, ℓ. If the length of the pendulum is quadrupled, what is the effect on the period, p?

48. For a constant distance, the rate of travel r varies inversely with the time traveled, t. If a car traveling 100 mph completes a test track in 6 minutes, find the rate needed to complete the same test track in 4 minutes. (*Hint:* Convert minutes to hours.)

CHAPTER 6   Group Activity

Finding a Linear Model

This activity may be completed by working in groups or individually.

The following table shows the actual number of foreign visitors (in millions) to the United States for the years 2000 through 2003.

Year	Foreign Visitors to the United States (in millions)
2000	50.9
2001	44.8
2002	41.9
2003	40.4

(*Source:* Tourism Industries/International Trade Administration, U.S. Department of Commerce)

1. Make a scatter diagram of the paired data in the table.

2. Use what you have learned in this chapter to write an equation of the line representing the paired data in the table. Explain how you found the equation, and what each variable represents.

3. What is the slope of your line? What does the slope mean in this context?

4. Use your linear equation to predict the number of foreign visitors to the United States in 2010.

5. Compare your linear equation to that found by other students or groups. Is it the same, similar, or different? How?

6. Compare your prediction from question 3 to that of other students or groups. Describe what you find.

7. The number of visitors to the United States for 2004 was estimated to be 45.7 million. If this data point is added to the chart above, how does it affect your results?

Chapter 6 Vocabulary Check

Fill in each blank with one of the words listed below.

y-axis	*x*-axis	solution	linear	standard	point-slope
x-intercept	*y*-intercept	*y*	*x*	slope	relation
domain	range	direct	inverse	slope-intercept	function

1. An ordered pair is a _____ of an equation in two variables if replacing the variables by the coordinates of the ordered pair results in a true statement.
2. The vertical number line in the rectangular coordinate system is called the _____.
3. A _____ equation can be written in the form $Ax + By = C$.
4. A(n) _____ is a point of the graph where the graph crosses the *x*-axis.
5. The form $Ax + By = C$ is called _____ form.
6. A(n) _____ is a point of the graph where the graph crosses the *y*-axis.
7. A set of ordered pairs that assigns to each *x*-value exactly one *y*-value is called a _____.
8. The equation $y = 7x - 5$ is written in _____ form.
9. The set of all *x*-coordinates of a relation is called the _____ of the relation.
10. The set of all *y*-coordinates of a relation is called the _____ of the relation.
11. A set of ordered pairs is called a _____.
12. The equation $y + 1 = 7(x - 2)$ is written in _____ form.
13. To find an *x*-intercept of a graph, let _____ = 0.
14. The horizontal number line in the rectangular coordinate system is called the _____.
15. To find a *y*-intercept of a graph, let _____ = 0.
16. The _____ of a line measures the steepness or tilt of a line.
17. The equation $y = kx$ is an example of _____ variation.
18. The equation $y = \dfrac{k}{x}$ is an example of _____ variation.

> **Helpful Hint**
>
> Are you preparing for your test? Don't forget to take the Chapter 6 Test on page 511. Then check your answers at the back of the text and use the Chapter Test Prep Video CD to see the fully worked-out solutions to any of the exercises you want to review.

6 Chapter Highlights

DEFINITIONS AND CONCEPTS	**EXAMPLES**
Section 6.1 Reading Graphs and the Rectangular Coordinate System	

The **rectangular coordinate system** consists of a plane and a vertical and a horizontal number line intersecting at their 0 coordinate. The vertical number line is called the **y-axis** and the horizontal number line is called the **x-axis.** The point of intersection of the axes is called the **origin.**

To **plot** or **graph** an ordered pair means to find its corresponding point on a rectangular coordinate system.

To plot or graph an ordered pair such as $(3, -2)$, start at the origin. Move 3 units to the right and from there, 2 units down.

To plot or graph $(-3, 4)$; start at the origin. Move 3 units to the left and from there, 4 units up.

An ordered pair is a **solution** of an equation in two variables if replacing the variables with the coordinates of the ordered pair results in a true statement.

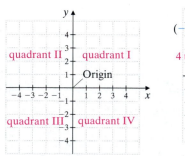

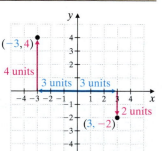

DEFINITIONS AND CONCEPTS	**EXAMPLES**

Section 6.1   Reading Graphs and The Rectangular Coordinate System (*continued*)

If one coordinate of an ordered pair solution is known, the other value can be determined by substitution.	Complete the ordered pair $(0,\ )$ for the equation $x - 6y = 12$. $$x - 6y = 12$$ $$0 - 6y = 12 \quad \text{Let } x = 0.$$ $$\frac{-6y}{-6} = \frac{12}{-6} \quad \text{Divide by } -6.$$ $$y = -2$$ The ordered pair solution is $(0, -2)$.

Section 6.2   Graphing Linear Equations

A **linear equation in two variables** is an equation that can be written in the form $Ax + By = C$, where A and B are not both 0. The form $Ax + By = C$ is called **standard form.**

$$3x + 2y = -6 \qquad x = -5$$
$$y = 3 \qquad y = -x + 10$$

$x + y = 10$ is in standard form.

To graph a linear equation in two variables, find three ordered pair solutions. Plot the solution points and draw the line connecting the points.

Graph: $x - 2y = 5$

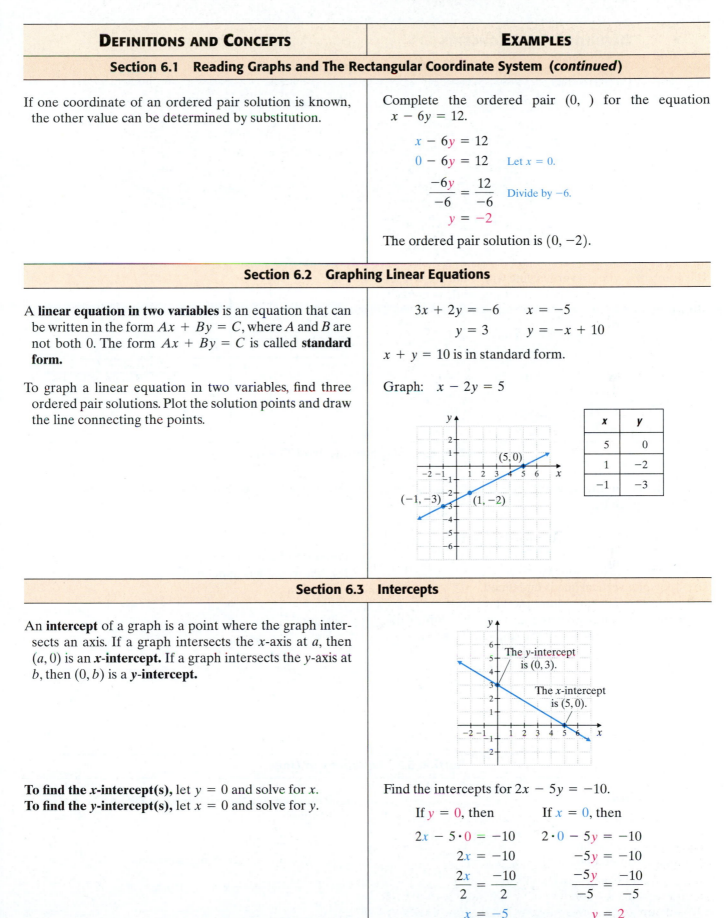

x	y
5	0
1	-2
-1	-3

Section 6.3   Intercepts

An **intercept** of a graph is a point where the graph intersects an axis. If a graph intersects the x-axis at a, then $(a, 0)$ is an x**-intercept.** If a graph intersects the y-axis at b, then $(0, b)$ is a y**-intercept.**

The y-intercept is $(0, 3)$.

The x-intercept is $(5, 0)$.

To find the x-intercept(s), let $y = 0$ and solve for x.
To find the y-intercept(s), let $x = 0$ and solve for y.

Find the intercepts for $2x - 5y = -10$.

If $y = 0$, then

$$2x - 5 \cdot 0 = -10$$
$$2x = -10$$
$$\frac{2x}{2} = \frac{-10}{2}$$
$$x = -5$$

If $x = 0$, then

$$2 \cdot 0 - 5y = -10$$
$$-5y = -10$$
$$\frac{-5y}{-5} = \frac{-10}{-5}$$
$$y = 2$$

continued

DEFINITIONS AND CONCEPTS	EXAMPLES

Section 6.3   Intercepts (*continued*)

The x-intercept is $(-5, 0)$. The y-intercept is $(0, 2)$.

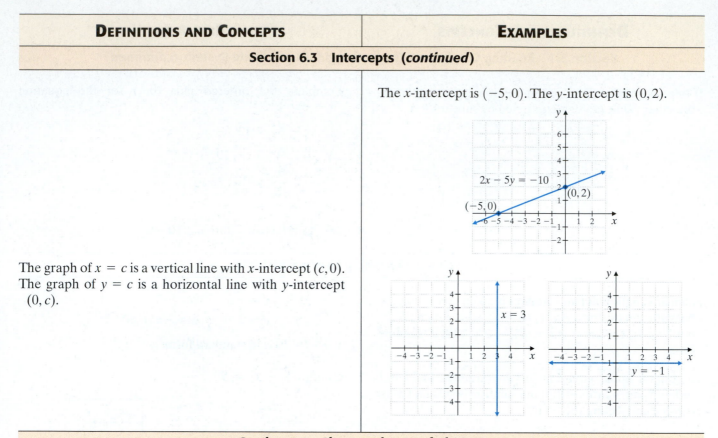

The graph of $x = c$ is a vertical line with x-intercept $(c, 0)$. The graph of $y = c$ is a horizontal line with y-intercept $(0, c)$.

Section 6.4   Slope and Rate of Change

The **slope** m of the line through points (x_1, y_1) and (x_2, y_2) is given by

$$m = \frac{y_2 - y_1}{x_2 - x_1} \qquad \text{as long as } x_2 \neq x_1$$

A horizontal line has slope 0.
The slope of a vertical line is undefined.
Nonvertical parallel lines have the same slope.
Two nonvertical lines are perpendicular if the slope of one is the negative reciprocal of the slope of the other.

The slope of the line through points $(-1, 6)$ and $(-5, 8)$ is

$$m = \frac{y_2 - y_1}{x_2 - x_1} = \frac{8 - 6}{-5 - (-1)} = \frac{2}{-4} = -\frac{1}{2}$$

The slope of the line $y = -5$ is 0.
The line $x = 3$ has undefined slope.

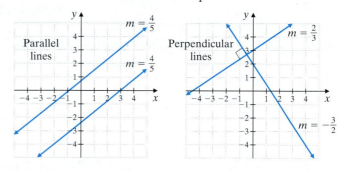

Section 6.5   Equations of Lines

SLOPE-INTERCEPT FORM

$$y = mx + b$$

m is the slope of the line.
$(0, b)$ is the y-intercept.

Find the slope and the y-intercept of the line $2x + 3y = 6$.
Solve for y:

$$2x + 3y = 6$$
$$3y = -2x + 6 \qquad \text{Subtract } 2x.$$
$$y = -\frac{2}{3}x + 2 \qquad \text{Divide by 3.}$$

The slope of the line is $-\dfrac{2}{3}$ and the y-intercept is $(0, 2)$.

DEFINITIONS AND CONCEPTS	**EXAMPLES**

Section 6.5   Equations of Lines (*continued*)

POINT-SLOPE FORM

$$y - y_1 = m(x - x_1)$$

m is the slope.
(x_1, y_1) is a point of the line.

Find an equation of the line with slope $\dfrac{3}{4}$ that contains the point $(-1, 5)$.

$$y - 5 = \frac{3}{4}[x - (-1)]$$

$4(y - 5) = 3(x + 1)$	Multiply by 4.
$4y - 20 = 3x + 3$	Distribute.
$-3x + 4y = 23$	Subtract $3x$ and add 20.

Section 6.6   Introduction to Functions

A set of ordered pairs is a **relation.** The set of all x-coordinates is called the **domain** of the relation and the set of all y-coordinates is called the **range** of the relation.

The domain of the relation

$$\{(0, 5), (2, 5), (4, 5), (5, -2)\}$$

is $\{0, 2, 4, 5\}$. The range is $\{-2, 5\}$.

A **function** is a set of ordered pairs that assigns to each x-value exactly one y-value.

Which are graphs of functions?

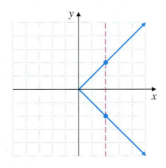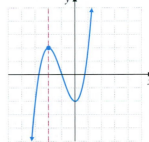

VERTICAL LINE TEST

If a vertical line can be drawn so that it intersects a graph more than once, the graph is not the graph of a function. (If no such line can be drawn, the graph is that of a function.)

This graph is not the graph of a function.   This graph is the graph of a function.

The symbol $f(x)$ means **function of x.** This notation is called **function notation.**

If $f(x) = 3x - 7$, then

$$f(-1) = 3(-1) - 7$$
$$= -3 - 7$$
$$= -10$$

Section 6.7   Graphing Linear Inequalities in Two Variables

A **linear inequality in two variables** is an inequality that can be written in one of these forms:

$Ax + By < C$	$Ax + By \le C$
$Ax + By > C$	$Ax + By \ge C$

where A and B are not both 0.

$2x - 5y < 6$	$x \ge -5$
$y > -8x$	$y \le 2$

continued

DEFINITIONS AND CONCEPTS	**EXAMPLES**

Section 6.7   Graphing Linear Inequalities in Two Variables (*continued*)

TO GRAPH A LINEAR INEQUALITY

1. Graph the boundary line by graphing the related equation. Draw the line solid if the inequality symbol is $\leq$ or $\geq$. Draw the line dashed if the inequality symbol is $<$ or $>$.

2. Choose a test point not on the line. Substitute its coordinates into the original inequality.

3. If the resulting inequality is true, shade the half-plane that contains the test point. If the inequality is not true, shade the half-plane that does not contain the test point.

Graph: $2x - y \leq 4$

1. Graph $2x - y = 4$. Draw a solid line because the inequality symbol is $\leq$.

2. Check the test point $(0, 0)$ in the original inequality, $2x - y \leq 4$.

$$2 \cdot 0 - 0 \leq 4 \quad \text{Let } x = 0 \text{ and } y = 0.$$
$$0 \leq 4 \quad \text{True}$$

3. The inequality is true, so shade the half-plane containing $(0, 0)$ as shown.

Section 6.8   Direct and Inverse Variation

y **varies directly as** *x*, or *y* is **directly proportional to** *x*, if there is a nonzero constant *k* such that

$$y = kx$$

y **varies inversely as** *x*, or *y* is **inversely proportional to** *x*, if there is a nonzero constant *k* such that

$$y = \frac{k}{x}$$

The circumference of a circle C varies directly as its radius r.

$$C = \underbrace{2\pi}_{k} r$$

Pressure P varies inversely with volume V.

$$P = \frac{k}{V}$$

📖 STUDY SKILLS BUILDER

Are You Prepared for a Test on Chapter 6?

Below I have listed some common trouble areas for students in Chapter 6. After studying for your test—but before taking your test—read these.

- If you are having trouble with graphing, you might want to ask your instructor if you can use graph paper on your test. This will save you time and keep your graphs neat.

- Don't forget that the graph of an ordered pair is a *single* point in the rectangular coordinate system.

- Make sure you remember that to find the slope of a linear equation using its equation, *first* solve the equation for *y*. *Then* the coefficient of *x* is its slope.

$$2x + 3y = 7$$
$$3y = -2x + 7 \quad \text{Subtract } 2x \text{ from both sides.}$$
$$\frac{3y}{3} = -\frac{2}{3}x + \frac{7}{3} \quad \text{Divide both sides by 3.}$$
$$y = -\frac{2}{3}x + \frac{7}{3} \leftarrow y\text{-intercept}$$
$$\underset{\text{slope}}{}$$

- Remember that a point that is an x-intercept will have a y-value of 0 and a point that is a y-intercept will have an x-value of 0. Also—the point $(0, 0)$ will be both an x- and y-intercept.

- If you studied functions, remember that $f(x)$ *does not* mean $f \cdot x$. It is a special function notation. If $f(x) = x^2 - 6$, then $f(-3) = (-3)^2 - 6 = 9 - 6 = 3$.

(6.1) *Plot each pair on the same rectangular coordinate system.*

1. $(-7, 0)$

2. $\left(0, 4\frac{4}{5}\right)$

3. $(-2, -5)$

4. $(1, -3)$

5. $(0.7, 0.7)$

6. $(-6, 4)$

Complete each ordered pair so that it is a solution of the given equation.

7. $-2 + y = 6x;\ (7,\ \ )$

8. $y = 3x + 5;\ (\ \ , -8)$

Complete the table of values for each given equation.

9. $9 = -3x + 4y$

x	y
	0
	3
9	

10. $y = 5$

x	y
7	
-7	
0	

11. $x = 2y$

x	y
	0
	5
	-5

12. The cost in dollars of producing x compact disc holders is given by $y = 5x + 2000$.

   a. Complete the table.

x	1	100	1000
y			

   b. Find the number of compact disc holders that can be produced for $6430.

(6.2) *Graph each linear equation.*

13. $x - y = 1$

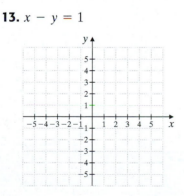

14. $x + y = 6$

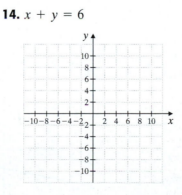

15. $x - 3y = 12$

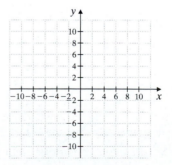

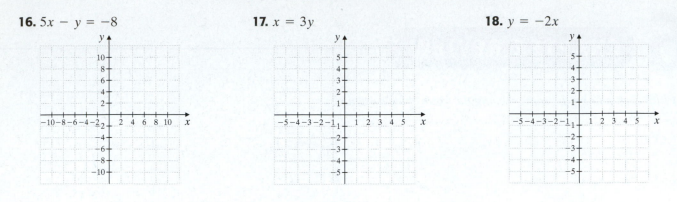

16. $5x - y = -8$

17. $x = 3y$

18. $y = -2x$

(6.3) *Identify the intercepts in each graph.*

19.

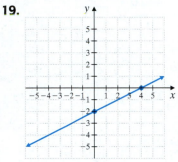

20.

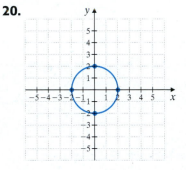

Graph each linear equation.

21. $y = -3$

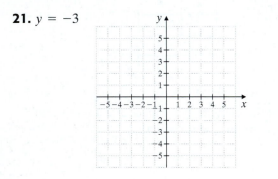

22. $x = 5$

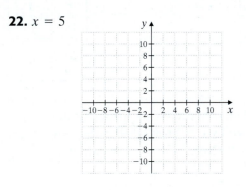

Find the intercepts of each equation.

23. $x - 3y = 12$

24. $-4x + y = 8$

(6.4) *Find the slope of each line.*

25.

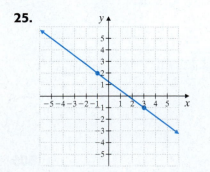

26.

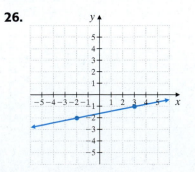

Match each line with its slope.

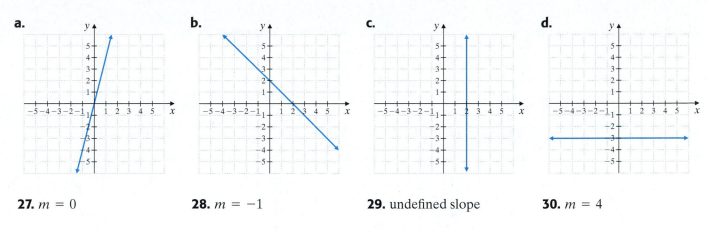

a.

b.

c.

d.

27. $m = 0$

28. $m = -1$

29. undefined slope

30. $m = 4$

Find the slope of the line that passes through each pair of points.

31. $(2, 5)$ and $(6, 8)$

32. $(4, 7)$ and $(1, 2)$

33. $(1, 3)$ and $(-2, -9)$

34. $(-4, 1)$ and $(3, -6)$

Find the slope of each line.

35. $y = 3x + 7$

36. $x - 2y = 4$

37. $y = -2$

38. $x = 0$

△ *Determine whether each pair of lines is parallel, perpendicular, or neither.*

39. $x - y = -6$
$\quad x + y = 3$

40. $3x + y = 7$
$\quad -3x - y = 10$

41. $y = 4x + \dfrac{1}{2}$
$\quad 4x + 2y = 1$

Find the slope of each line and write the slope as a rate of change. Don't forget to attach the proper units.

42. The graph below approximates the number of U.S. persons y (in millions) who have a bachelor's degree or higher per year x.

43. The graph below approximates the number of U.S. travelers y (in millions) that are vacationing per year x.

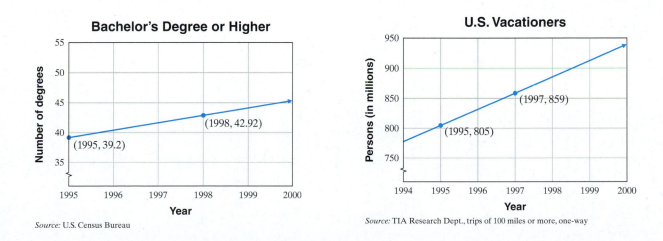

Bachelor's Degree or Higher

Number of degrees

$(1998, 42.92)$

$(1995, 39.2)$

Year

Source: U.S. Census Bureau

U.S. Vacationers

Persons (in millions)

$(1997, 859)$

$(1995, 805)$

Year

Source: TIA Research Dept., trips of 100 miles or more, one-way

(6.5) *Determine the slope and the y-intercept of the graph of each equation.*

44. $3x + y = 7$

45. $x - 6y = -1$

Write an equation of each line.

46. slope -5; y-intercept $\left(0, \dfrac{1}{2}\right)$

47. slope $\dfrac{2}{3}$; y-intercept $(0, 6)$

Match each equation with its graph.

48. $y = 2x + 1$ **49.** $y = -4x$ **50.** $y = 2x$ **51.** $y = 2x - 1$

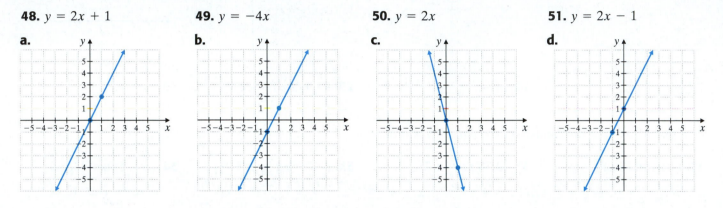

a.   **b.**   **c.**   **d.**

Write an equation of the line with the given slope that passes through the given point. Write the equation in the form $Ax + By = C$.

52. $m = 4$; $(2, 0)$ **53.** $m = -3$; $(0, -5)$ **54.** $m = \dfrac{3}{5}$; $(1, 4)$ **55.** $m = -\dfrac{1}{3}$; $(-3, 3)$

Write an equation of the line passing through each pair of points. Write the equation in the form $y = mx + b$.

56. $(1, 7)$ and $(2, -7)$

57. $(-2, 5)$ and $(-4, 6)$

(6.6) *Determine whether each relation or graph is a function.*

58. $\{(7, 1), (7, 5), (2, 6)\}$

59. $\{(0, -1), (5, -1), (2, 2)\}$

60.   **61.**   **62.**   **63.**

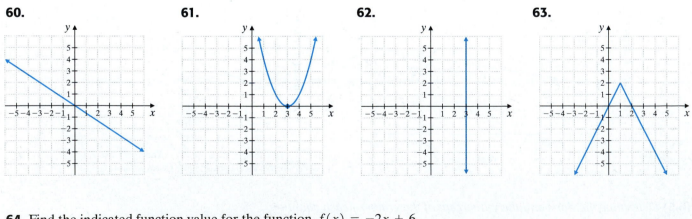

64. Find the indicated function value for the function, $f(x) = -2x + 6$.

a. $f(0)$ **b.** $f(-2)$ **c.** $f\left(\dfrac{1}{2}\right)$

(6.7) *Graph each inequality.*

65. $x + 6y < 6$

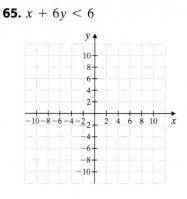

66. $x + y > -2$

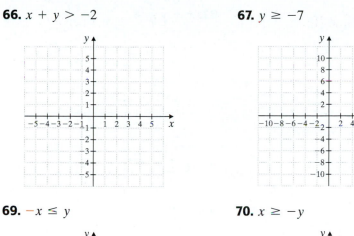

67. $y \geq -7$

68. $y \leq -4$

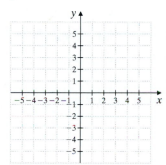

69. $-x \leq y$

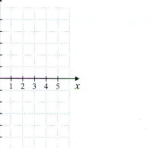

70. $x \geq -y$

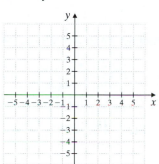

(6.8) *Solve.*

71. y varies directly as x. If $y = 40$ when $x = 4$, find y when x is 11.

72. y varies inversely as x. If $y = 4$ when $x = 6$, find y when x is 48.

73. y varies inversely as x^3. If $y = 12.5$ when $x = 2$, find y when x is 3.

74. y varies directly as x^2. If $y = 175$ when $x = 5$, find y when $x = 10$.

75. The cost of manufacturing a certain medicine varies inversely as the amount of medicine manufactured increases. If 3000 milliliters can be manufactured for \$6600, find the cost to manufacture 5000 milliliters.

76. The distance a spring stretches varies directly with the weight attached to the spring. If a 150-pound weight stretches the spring 8 inches, find the distance that a 90-pound weight stretches the spring.

Mixed Review

Complete the table of values for each given equation.

77. $2x - 5y = 9$

x	y
	1
2	
	-3

78. $x = -3y$

x	y
0	
	1
6	

Find the intercepts for each equation.

79. $2x - 3y = 6$

80. $-5x + y = 10$

Graph each linear equation.

81. $x - 5y = 10$

82. $x + y = 4$

83. $y = -4x$

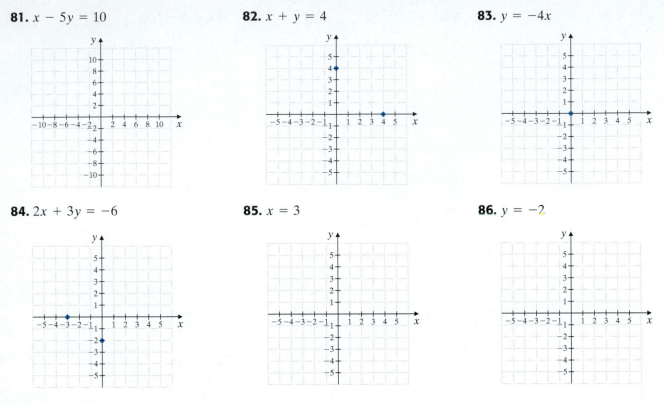

84. $2x + 3y = -6$

85. $x = 3$

86. $y = -2$

Find the slope of the line that passes through each pair of points.

87. $(3, -5)$ and $(-4, 2)$

88. $(1, 3)$ and $(-6, -8)$

Find the slope of each line.

89.

90.

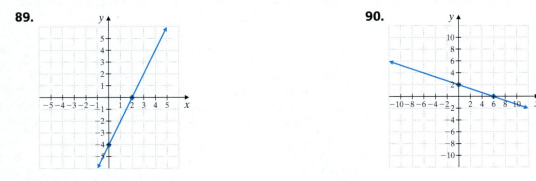

Determine the slope and y-intercept of the graph of each equation.

91. $-2x + 3y = -15$

92. $6x + y - 2 = 0$

Write an equation of the line with the given slope that passes through the given point. Write the equation in the form $Ax + By = C$.

93. $m = -5; (3, -7)$

94. $m = 3; (0, 6)$

Write an equation of the line passing through each pair of points. Write the equation in the form $Ax + By = C$.

95. $(-3, 9)$ and $(-2, 5)$

96. $(3, 1)$ and $(5, -9)$

6 CHAPTER TEST

Remember to use the Chapter Test Prep Video CD to see the fully worked-out solutions to any of the exercises you want to review.

Complete each ordered pair so that it is a solution of the given equation.

1. $12y - 7x = 5; (1, \quad)$

2. $y = 17; (-4, \quad)$

Find the slope of each line.

3.

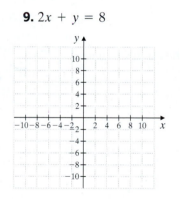

4.

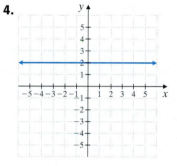

5. Passes through $(6, -5)$ and $(-1, 2)$

6. Passes through $(0, -8)$ and $(-1, -1)$

7. $-3x + y = 5$

8. $x = 6$

Graph.

9. $2x + y = 8$

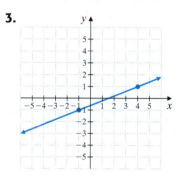

10. $-x + 4y = 5$

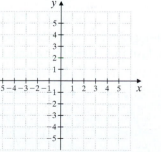

11. $x - y \geq -2$

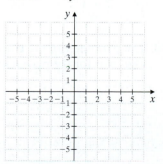

12. $y \geq -4x$

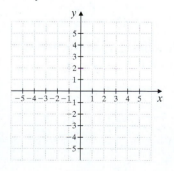

13. $5x - 7y = 10$

14. $2x - 3y > -6$

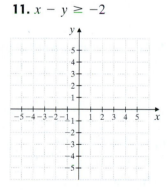

1. _____

2. _____

3. _____

4. _____

5. _____

6. _____

7. _____

8. _____

9. see graph

10. see graph

11. see graph

12. see graph

13. see graph

14. see graph

15. see graph _____

16. see graph _____

17. _____

18. _____

19. _____

20. _____

21. _____

22. _____

23. _____

24. _____

25. _____

26. a. _____

b. _____

c. _____

27. a. _____

b. _____

c. _____

28. _____

15. $6x + y > -1$

16. $y = -1$

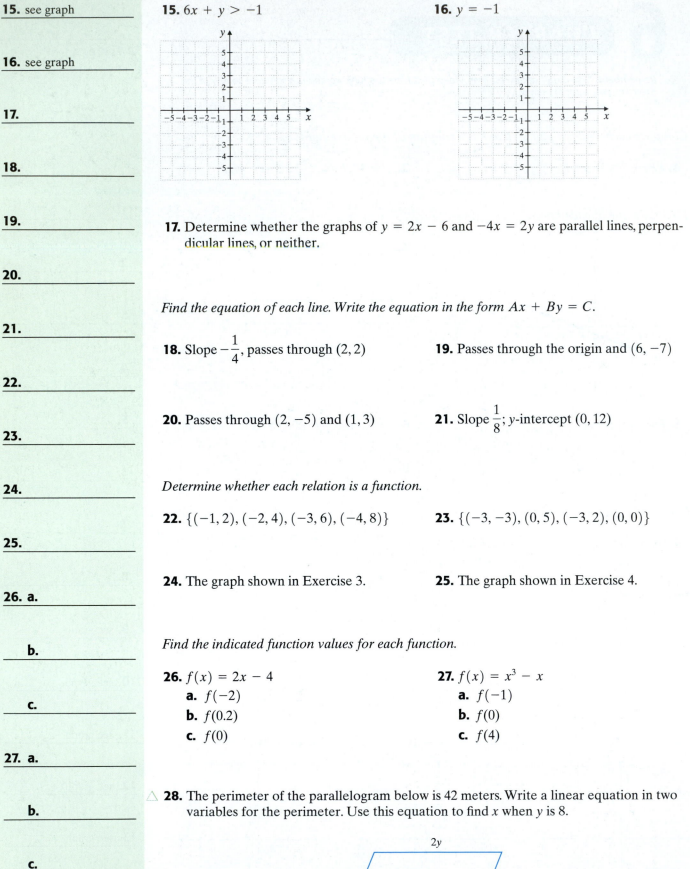

17. Determine whether the graphs of $y = 2x - 6$ and $-4x = 2y$ are parallel lines, perpendicular lines, or neither.

Find the equation of each line. Write the equation in the form $Ax + By = C$.

18. Slope $-\dfrac{1}{4}$, passes through $(2, 2)$

19. Passes through the origin and $(6, -7)$

20. Passes through $(2, -5)$ and $(1, 3)$

21. Slope $\dfrac{1}{8}$; y-intercept $(0, 12)$

Determine whether each relation is a function.

22. $\{(-1, 2), (-2, 4), (-3, 6), (-4, 8)\}$

23. $\{(-3, -3), (0, 5), (-3, 2), (0, 0)\}$

24. The graph shown in Exercise 3.

25. The graph shown in Exercise 4.

Find the indicated function values for each function.

26. $f(x) = 2x - 4$
   a. $f(-2)$
   b. $f(0.2)$
   c. $f(0)$

27. $f(x) = x^3 - x$
   a. $f(-1)$
   b. $f(0)$
   c. $f(4)$

△ **28.** The perimeter of the parallelogram below is 42 meters. Write a linear equation in two variables for the perimeter. Use this equation to find x when y is 8.

29. The table gives the number of basic cable TV subscribers (in millions) for the years shown. (*Source:* Cisco Systems)

Year	Basic Cable TV Subscribers (in millions)
2000	69.3
2001	70.0
2002	69.9
2003	70.1
2004	70.3
2005	70.5 (estimated)

a. Write this data as a set of ordered pairs of the form (year, number of basic cable TV subscribers in millions).

b. Create a scatter diagram of the data. Be sure to label the axes properly.

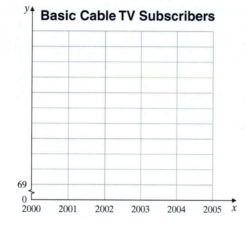

30. This graph approximates the movie ticket sales *y* (in millions) for the year *x*. Find the slope of the line and write the slope as a rate of change. Don't forget to attach the proper units.

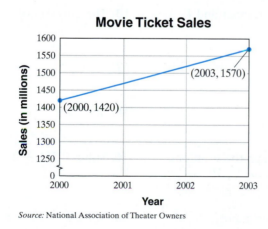

Source: National Association of Theater Owners

31. *y* varies directly as *x*. If $y = 10$ when $x = 15$, find *y* when *x* is 42.

32. *y* varies inversely as x^2. If $y = 8$ when $x = 5$, find *y* when *x* is 15.

29. a. _____

b. see diagram _____

30. _____

31. _____

32. _____

Simplify each expression.

Answers

1. $6 \div 3 + 5^2$

2. $\dfrac{10}{3} + \dfrac{5}{21}$

3. $1 + 2[5(2 \cdot 3 + 1) - 10]$

4. $16 - 3 \cdot 3 + 2^4$

5. The lowest point on the surface of the Earth is the Dead Sea, at an elevation of 1349 feet below sea level. The highest point is Mt. Everest, at an elevation of 29,035 feet. What is the difference in elevation between these two world extremes? (*Source:* National Geographic Society)

29,035 feet — Mt. Everest -----------
above sea level
(+29,035)

Sea level (0)
1349 feet — Dead Sea
below sea level
(−1349)

6. Simplify: $1.7x - 11 - 0.9x - 25$

Write each phrase as an algebraic expression and simplify if possible. Let x represent the unknown number.

7. Twice a number, plus 6.

8. The product of −15 and the sum of a number and $\dfrac{2}{3}$.

9. The difference of a number and 4, divided by 7.

10. The quotient of −9 and twice a number.

11. Five plus the sum of a number and 1.

12. A number subtracted from −86.

13. Solve for x: $\dfrac{5}{2}x = 15$

14. Solve for x: $\dfrac{x}{4} - 1 = -7$

15. Solve $2x < -4$. Graph the solutions.

$\begin{array}{ccccccccccc} & & & & & & & & & & \\ \hline -5 & -4 & -3 & -2 & -1 & 0 & 1 & 2 & 3 & 4 & 5 \end{array}$

16. Solve: $5(x + 4) \geq 4(2x + 3)$

17. Find the degree of each polynomial and tell whether the polynomial is a monomial, binomial, trinomial, or none of these.
 a. $-2t^2 + 3t + 6$
 b. $15x - 10$
 c. $7x + 3x^3 + 2x^2 - 1$

18. Solve $x + 2y = 6$ for y.

Answers

1. _____

2. _____

3. _____

4. _____

5. _____

6. _____

7. _____

8. _____

9. _____

10. _____

11. _____

12. _____

13. _____

14. _____

15. _____

16. _____

17. a. _____

 b. _____

 c. _____

18. _____

19. Add: $(-2x^2 + 5x - 1) + (-2x^2 + x + 3)$

20. Subtract: $(-2x^2 + 5x - 1) - (-2x^2 + x + 3)$

21. Multiply: $(3y + 1)^2$

22. Multiply: $(x - 12)^2$

23. Factor: $-9a^5 + 18a^2 - 3a$

24. Factor: $4x^2 - 36$

25. Factor: $x^2 + 4x - 12$

26. Factor: $3x^2 - 20xy - 7y^2$

27. Factor: $8x^2 - 22x + 5$

28. Factor: $18x^2 + 35x - 2$

29. Solve: $x^2 - 9x - 22 = 0$

30. Solve: $x^2 = x$

31. Divide: $\dfrac{2x^2 - 11x + 5}{5x - 25} \div \dfrac{4x - 2}{10}$

32. Simplify: $\dfrac{2x^2 - 50}{4x^4 - 20x^3}$

Write the rational expression as an equivalent rational expression with the given denominator.

33. $\dfrac{4b}{9a} = \dfrac{}{27a^2b}$

34. $\dfrac{1}{2x} = \dfrac{}{14x^3}$

35. Add: $1 + \dfrac{m}{m + 1}$

36. Subtract: $\dfrac{2x + 1}{x - 6} - \dfrac{x - 4}{x - 6}$

37. Solve: $3 - \dfrac{6}{x} = x + 8$

38. Solve: $3x^2 + 5x = 2$

19. _____

20. _____

21. _____

22. _____

23. _____

24. _____

25. _____

26. _____

27. _____

28. _____

29. _____

30. _____

31. _____

32. _____

33. _____

34. _____

35. _____

36. _____

37. _____

38. _____

39. _____

40. _____

41. a. _____

b. _____

c. _____

42. see table _____

43. see graph _____

44. _____

45. _____

46. _____

47. _____

48. _____

49. a. _____

b. _____

c. _____

50. _____

39. Simplify: $\dfrac{\dfrac{x+1}{y}}{\dfrac{x}{y}+2}$

40. Simplify: $\dfrac{\dfrac{x}{2}-\dfrac{y}{6}}{\dfrac{x}{12}-\dfrac{y}{3}}$

41. Complete each ordered pair solution so that it is a solution to the equation $3x + y = 12$.

  a. $(0, \ )$

  b. $(\ , 6)$

  c. $(-1, \ )$

42. Complete the table for $y = -5x$.

x	y
0	
-1	
	-10

43. Graph: $2x + y = 5$

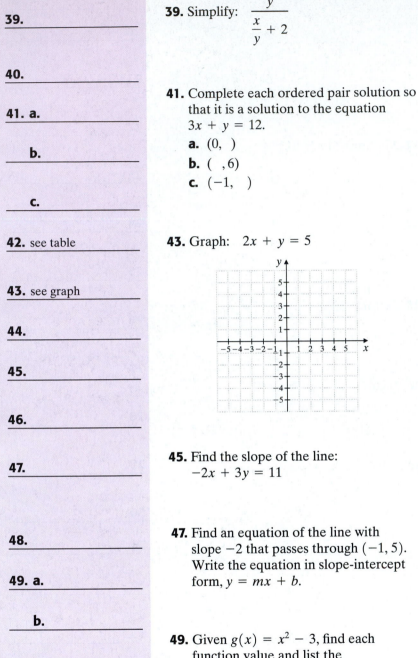

44. Find the slope of the line through $(0, 5)$ and $(-5, 4)$.

45. Find the slope of the line:
$-2x + 3y = 11$

46. Find the slope of the line $x = -10$.

47. Find an equation of the line with slope -2 that passes through $(-1, 5)$. Write the equation in slope-intercept form, $y = mx + b$.

48. Find the slope and y-intercept of the line whose equation is $2x - 5y = 10$.

49. Given $g(x) = x^2 - 3$, find each function value and list the corresponding ordered pair.

  a. $g(2)$ **b.** $g(-2)$ **c.** $g(0)$

50. Write an equation of the line through $(2, 3)$ and $(0, 0)$. Write the equation in standard form.

7

Systems of Equations

In Chapter 6, we graphed equations containing two variables. As we have seen, equations like these are often needed to represent relationships between two different quantities. There are also many opportunities to compare and contrast two such equations, called a *system of equations*. This chapter presents *linear systems* and ways we solve these systems and apply them to real-life situations.

Many of the occupations predicted to have the largest increase in number of jobs are in the fields of medicine and computer science. For example, from 2002 to 2012 the job growth predicted for medical assistants is 59% and for computer software engineers is 46%. Although both jobs are growing, they are growing at different rates. On page 542, Exercise 59, we will predict when these occupations will have the same number of jobs.

Job Title	Job Description	Employment (in thousands)	
		2002	2012
Medical Assistant	Perform administrative and clinical tasks to keep the offices of physicians running smoothly.	365	579
Computer Software Engineers	Apply the principles of computer science, engineering, and mathematics to design, test, and evaluate new software and systems for computers.	394	573

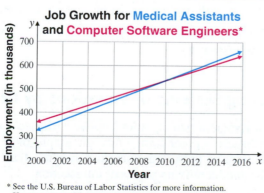

Job Growth for Medical Assistants and Computer Software Engineers*

* See the U.S. Bureau of Labor Statistics for more information. Here, we assumed linear growth.

A Decide Whether an Ordered Pair Is a Solution of a System of Linear Equations.

B Solve a System of Linear Equations by Graphing.

C Identify Special Systems: Those with no Solution and Those with an Infinite Number of Solutions.

7.1 SOLVING SYSTEMS OF LINEAR EQUATIONS BY GRAPHING

A **system of linear equations** consists of two or more linear equations. In this section, we focus on solving systems of linear equations containing two equations in two variables. Examples of such linear systems are

$$\begin{cases} 3x - 3y = 0 \\ x = 2y \end{cases} \qquad \begin{cases} x - y = 0 \\ 2x + y = 10 \end{cases} \qquad \begin{cases} y = 7x - 1 \\ y = 4 \end{cases}$$

Objective **A** Deciding Whether an Ordered Pair Is a Solution

A **solution** of a system of two equations in two variables is an ordered pair of numbers that is a solution of both equations in the system.

PRACTICE PROBLEM 1

Determine whether $(3, 9)$ is a solution of the system

$$\begin{cases} 5x - 2y = -3 \\ y = 3x \end{cases}$$

EXAMPLE 1 Determine whether $(12, 6)$ is a solution of the system

$$\begin{cases} 2x - 3y = 6 \\ x = 2y \end{cases}$$

Solution: To determine whether $(12, 6)$ is a solution of the system, we replace x with 12 and y with 6 in both equations.

$2x - 3y = 6$ First equation	$x = 2y$ Second equation
$2(12) - 3(6) \stackrel{?}{=} 6$ Let $x = 12$ and $y = 6$.	$12 \stackrel{?}{=} 2(6)$ Let $x = 12$ and $y = 6$.
$24 - 18 \stackrel{?}{=} 6$ Simplify.	$12 = 12$ True
$6 = 6$ True	

Since $(12, 6)$ is a solution of both equations, it is a solution of the system.

🔲 **Work Practice Problem 1**

PRACTICE PROBLEM 2

Determine whether $(3, -2)$ is a solution of the system

$$\begin{cases} 2x - y = 8 \\ x + 3y = 4 \end{cases}$$

EXAMPLE 2 Determine whether $(-1, 2)$ is a solution of the system

$$\begin{cases} x + 2y = 3 \\ 4x - y = 6 \end{cases}$$

Solution: We replace x with -1 and y with 2 in both equations.

$x + 2y = 3$ First equation	$4x - y = 6$ Second equation
$-1 + 2(2) \stackrel{?}{=} 3$ Let $x = -1$ and $y = 2$.	$4(-1) - 2 \stackrel{?}{=} 6$ Let $x = -1$ and $y = 2$.
$-1 + 4 \stackrel{?}{=} 3$ Simplify.	$-4 - 2 \stackrel{?}{=} 6$ Simplify.
$3 = 3$ True	$-6 = 6$ False

$(-1, 2)$ is not a solution of the second equation, $4x - y = 6$, so it is not a solution of the system.

🔲 **Work Practice Problem 2**

Objective **B** Solving Systems of Equations by Graphing

Since a solution of a system of two equations in two variables is a solution common to both equations, it is also a point common to the graphs of both equations. Let's practice finding solutions of both equations in a system—that is, solutions of the system—by graphing and identifying points of intersection.

Answers

1. $(3, 9)$ is a solution of the system,
2. $(3, -2)$ is not a solution of the system

EXAMPLE 3 Solve the system of equations by graphing.

$$\begin{cases} -x + 3y = 10 \\ x + y = 2 \end{cases}$$

Solution: On a single set of axes, graph each linear equation.

$-x + 3y = 10$

x	y
0	$\frac{10}{3}$
−4	2
2	4

$x + y = 2$

x	y
0	2
2	0
1	1

Helpful Hint The point of intersection gives the solution of the system.

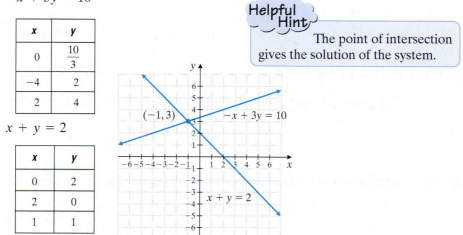

The two lines appear to intersect at the point $(-1, 3)$. To check, we replace x with -1 and y with 3 in both equations.

$-x + 3y = 10$	First equation	$x + y = 2$	Second equation
$-(-1) + 3(3) \stackrel{?}{=} 10$	Let $x = -1$ and $y = 3$.	$-1 + 3 \stackrel{?}{=} 2$	Let $x = -1$ and $y = 3$.
$1 + 9 \stackrel{?}{=} 10$	Simplify.	$2 = 2$	True
$10 = 10$	True		

$(-1, 3)$ checks, so it is the solution of the system.

■ **Work Practice Problem 3**

Helpful Hint Neatly drawn graphs can help when "guessing" the solution of a system of linear equations by graphing.

EXAMPLE 4 Solve the system of equations by graphing.

$$\begin{cases} 2x + 3y = -2 \\ x = 2 \end{cases}$$

Solution: We graph each linear equation on a single set of axes.

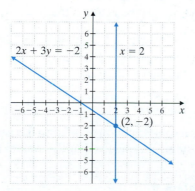

Continued on next page

PRACTICE PROBLEM 3

Solve the system of equations by graphing.

$$\begin{cases} -3x + y = -10 \\ x - y = 6 \end{cases}$$

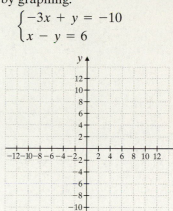

PRACTICE PROBLEM 4

Solve the system of equations by graphing.

$$\begin{cases} x + 3y = -1 \\ y = 1 \end{cases}$$

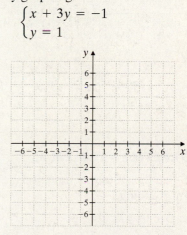

Answers

3. $(2, -4)$,

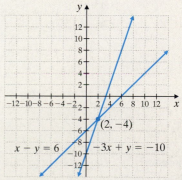

4. See page 521.

The two lines appear to intersect at the point $(2, -2)$. To determine whether $(2, -2)$ is the solution, we replace x with 2 and y with -2 in both equations.

$2x + 3y = -2$ First equation	$x = 2$ Second equation
$2(2) + 3(-2) \overset{?}{=} -2$ Let $x = 2$ and $y = -2$.	$2 \overset{?}{=} 2$ Let $x = 2$.
$4 + (-6) \overset{?}{=} -2$ Simplify.	$2 = 2$ True
$-2 = -2$ True	

Since a true statement results in both equations, $(2, -2)$ is the solution of the system.

☐ **Work Practice Problem 4**

Objective **C** Identifying Special Systems of Linear Equations

Not all systems of linear equations have a single solution. Some systems have no solution and some have an infinite number of solutions.

EXAMPLE 5 Solve the system of equations by graphing.

$$\begin{cases} 2x + y = 7 \\ 2y = -4x \end{cases}$$

Solution: We graph the two equations in the system. The equations in slope-intercept form are $y = -2x + 7$ and $y = -2x$. Notice from the equations that the lines have the same slope, -2, and different y-intercepts. This means that the lines are parallel.

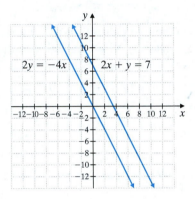

Since the lines are parallel, they do not intersect. This means that the system has *no solution*.

☐ **Work Practice Problem 5**

EXAMPLE 6 Solve the system of equations by graphing.

$$\begin{cases} x - y = 3 \\ -x + y = -3 \end{cases}$$

Solution: We graph each equation. The graphs of the equations are the same line. To see this, notice that if both sides of the first equation in the system are multiplied by -1, the result is the second equation.

$x - y = 3$	First equation
$-1(x - y) = -1(3)$	Multiply both sides by -1.
$-x + y = -3$	Simplify. This is the second equation.

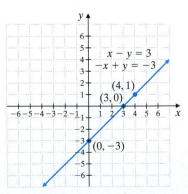

Any ordered pair that is a solution of one equation is a solution of the other and is then a solution of the system. This means that the system has an infinite number of solutions.

☐ **Work Practice Problem 6**

PRACTICE PROBLEM 5

Solve the system of equations by graphing.

$$\begin{cases} 3x - y = 6 \\ 6x = 2y \end{cases}$$

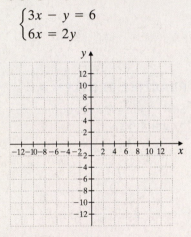

PRACTICE PROBLEM 6

Solve the system of equations by graphing.

$$\begin{cases} x + y = -4 \\ -2x - 2y = 8 \end{cases}$$

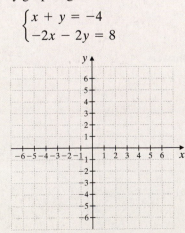

Answers

5. See page 521.
6. See page 521.

Examples 5 and 6 are special cases of systems of linear equations. A system that has no solution is said to be an **inconsistent system.** If the graphs of the two equations of a system are identical, we call the equations **dependent equations.** Thus, the system in Example 5 is an inconsistent system and the equations in the system in Example 6 are dependent equations.

As we have seen, three different situations can occur when graphing the two lines associated with the equations in a linear system. These situations are shown in the figures.

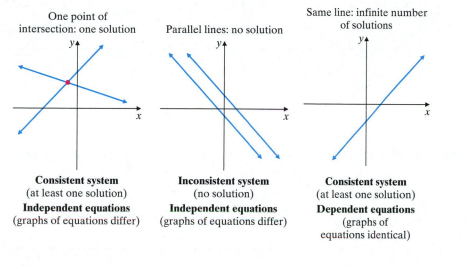

One point of
intersection: one solution

Parallel lines: no solution

Same line: infinite number
of solutions

Consistent system
(at least one solution)
Independent equations
(graphs of equations differ)

Inconsistent system
(no solution)
Independent equations
(graphs of equations differ)

Consistent system
(at least one solution)
Dependent equations
(graphs of
equations identical)

🖩 **CALCULATOR EXPLORATIONS** Graphing

A graphing calculator may be used to approximate solutions of systems of equations. For example, to approximate the solution of the system

$$\begin{cases} y = -3.14x - 1.35 \\ y = 4.88x + 5.25, \end{cases}$$

first graph each equation on the same set of axes. Then use the intersect feature of your calculator to approximate the point of intersection.

The approximate point of intersection is $(-0.82, 1.23)$.

Solve each system of equations. Approximate the solutions to two decimal places.

1. $\begin{cases} y = -2.68x + 1.21 \\ y = 5.22x - 1.68 \end{cases}$ **2.** $\begin{cases} y = 4.25x + 3.89 \\ y = -1.88x + 3.21 \end{cases}$

3. $\begin{cases} 4.3x - 2.9y = 5.6 \\ 8.1x + 7.6y = -14.1 \end{cases}$ **4.** $\begin{cases} -3.6x - 8.6y = 10 \\ -4.5x + 9.6y = -7.7 \end{cases}$

Answers

4. $(-4, 1)$,

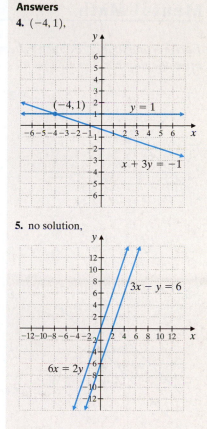

5. no solution,

6. infinite number of solutions

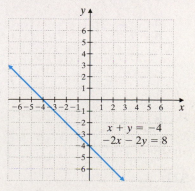

Mental Math

Each rectangular coordinate system shows the graph of the equations in a system of equations. Use each graph to determine the number of solutions for each associated system. If the system has only one solution, give its coordinates.

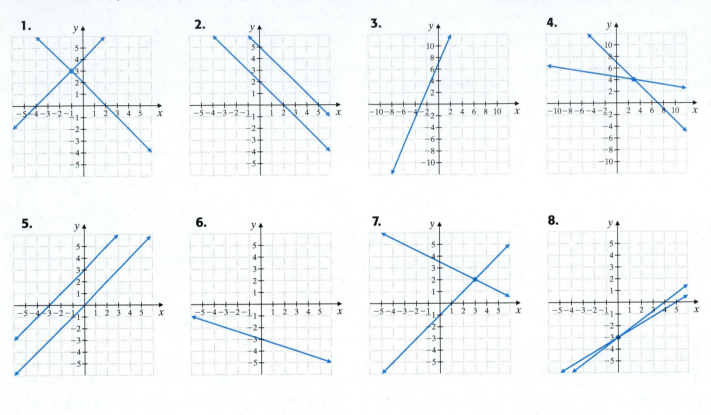

1. **2.** **3.** **4.**

5. **6.** **7.** **8.**

7.1 EXERCISE SET

FOR EXTRA HELP

Student Solutions Manual   PH Math/Tutor Center   CD/Video for Review   MathXL®   MyMathLab

Objective A *Determine whether each ordered pair is a solution of the system of linear equations. See Examples 1 and 2.*

1. $\begin{cases} x + y = 8 \\ 3x + 2y = 21 \end{cases}$
 a. $(2, 4)$
 b. $(5, 3)$

2. $\begin{cases} 2x + y = 5 \\ x + 3y = 5 \end{cases}$
 a. $(5, 0)$
 b. $(2, 1)$

3. $\begin{cases} 3x - y = 5 \\ x + 2y = 11 \end{cases}$
 a. $(3, 4)$
 b. $(0, -5)$

4. $\begin{cases} 2x - 3y = 8 \\ x - 2y = 6 \end{cases}$
 a. $(-2, -4)$
 b. $(7, 2)$

5. $\begin{cases} 2y = 4x + 6 \\ 2x - y = -3 \end{cases}$
 a. $(-3, -3)$
 b. $(0, 3)$

6. $\begin{cases} x + 5y = -4 \\ -2x = 10y + 8 \end{cases}$
 a. $(-4, 0)$
 b. $(6, -2)$

7. $\begin{cases} -2 = x - 7y \\ 6x - y = 13 \end{cases}$
 a. $(-2, 0)$
 b. $\left(\dfrac{1}{2}, \dfrac{5}{14}\right)$

8. $\begin{cases} 4x = 1 - y \\ x - 3y = -8 \end{cases}$
 a. $(0, 1)$
 b. $\left(\dfrac{1}{6}, \dfrac{1}{3}\right)$

522

Objectives **B** **C** **Mixed Practice** *Solve each system of linear equations by graphing. See Examples 3 through 6.*

9. $\begin{cases} x + y = 4 \\ x - y = 2 \end{cases}$

10. $\begin{cases} x + y = 3 \\ x - y = 5 \end{cases}$

11. $\begin{cases} x + y = 6 \\ -x + y = -6 \end{cases}$

12. $\begin{cases} x + y = 1 \\ -x + y = -3 \end{cases}$

13. $\begin{cases} y = 2x \\ 3x - y = -2 \end{cases}$

14. $\begin{cases} y = -3x \\ 2x - y = -5 \end{cases}$

15. $\begin{cases} y = x + 1 \\ y = 2x - 1 \end{cases}$

16. $\begin{cases} y = 3x - 4 \\ y = x + 2 \end{cases}$

17. $\begin{cases} 2x + y = 0 \\ 3x + y = 1 \end{cases}$

18. $\begin{cases} 2x + y = 1 \\ 3x + y = 0 \end{cases}$

19. $\begin{cases} y = -x - 1 \\ y = 2x + 5 \end{cases}$

20. $\begin{cases} y = x - 1 \\ y = -3x - 5 \end{cases}$

21. $\begin{cases} x + y = 5 \\ x + y = 6 \end{cases}$

22. $\begin{cases} x - y = 4 \\ x - y = 1 \end{cases}$

23. $\begin{cases} 2x - y = 6 \\ y = 2 \end{cases}$

24. $\begin{cases} x + y = 5 \\ x = 4 \end{cases}$

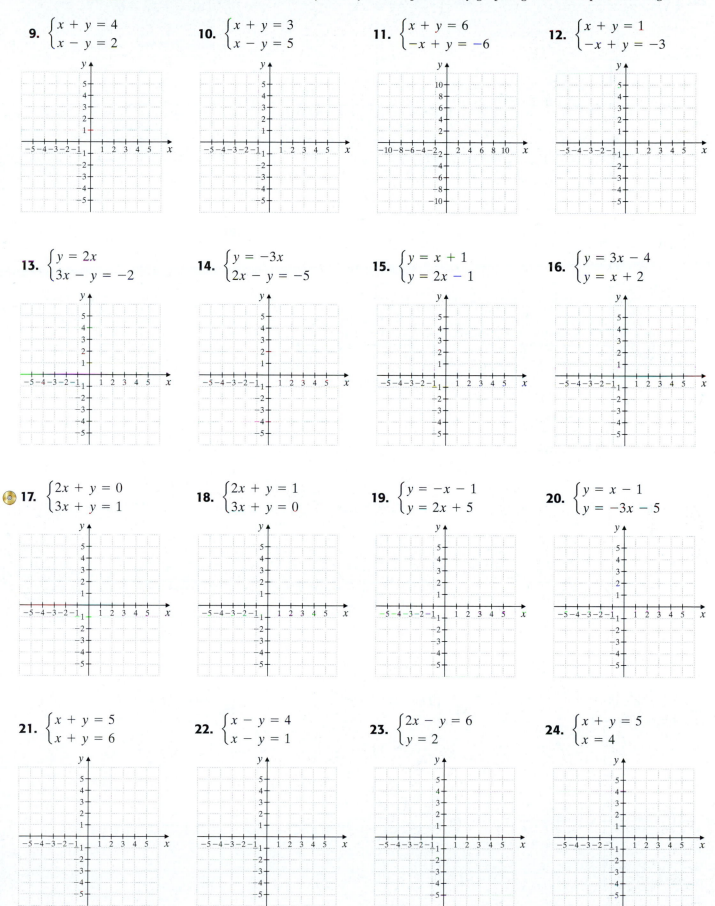

25. $\begin{cases} x - 2y = 2 \\ 3x + 2y = -2 \end{cases}$

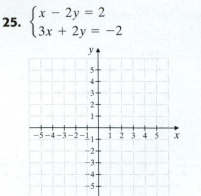

26. $\begin{cases} x + 3y = 7 \\ 2x - 3y = -4 \end{cases}$

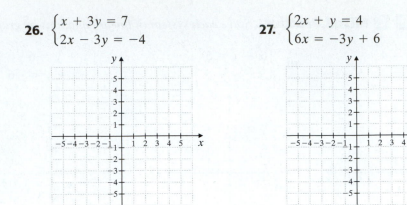

27. $\begin{cases} 2x + y = 4 \\ 6x = -3y + 6 \end{cases}$

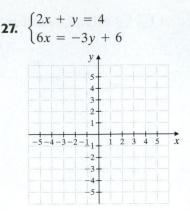

28. $\begin{cases} y + 2x = 3 \\ 4x = 2 - 2y \end{cases}$

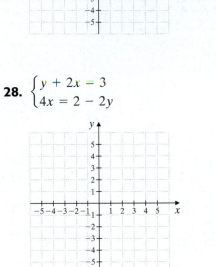

29. $\begin{cases} y - 3x = -2 \\ 6x - 2y = 4 \end{cases}$

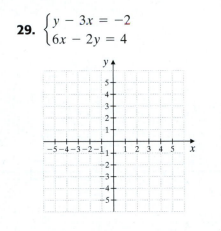

30. $\begin{cases} x - 2y = -6 \\ -2x + 4y = 12 \end{cases}$

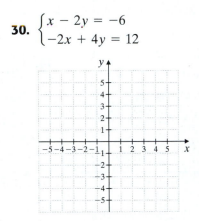

31. $\begin{cases} x = 3 \\ y = -1 \end{cases}$

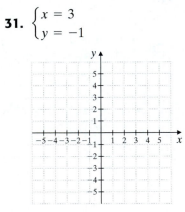

32. $\begin{cases} x = -5 \\ y = 3 \end{cases}$

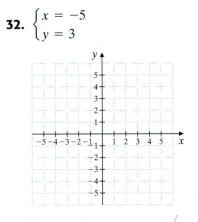

33. $\begin{cases} y = x - 2 \\ y = 2x + 3 \end{cases}$

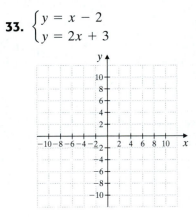

34. $\begin{cases} y = x + 5 \\ y = -2x - 4 \end{cases}$

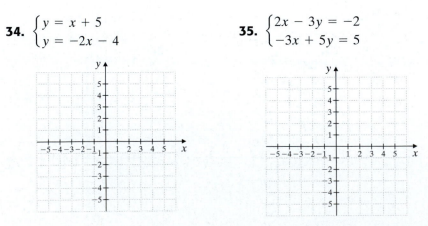

35. $\begin{cases} 2x - 3y = -2 \\ -3x + 5y = 5 \end{cases}$

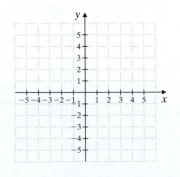

36. $\begin{cases} 4x - y = 7 \\ 2x - 3y = -9 \end{cases}$

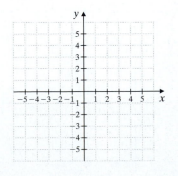

37. $\begin{cases} 6x - y = 4 \\ \dfrac{1}{2}y = -2 + 3x \end{cases}$

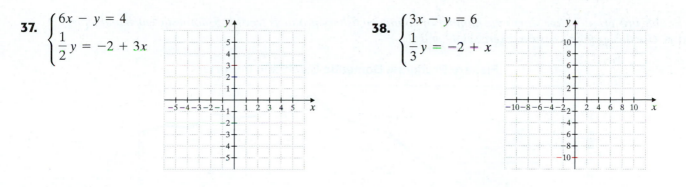

38. $\begin{cases} 3x - y = 6 \\ \dfrac{1}{3}y = -2 + x \end{cases}$

Review

Solve each equation. See Section 2.3.

39. $5(x - 3) + 3x = 1$

40. $-2x + 3(x + 6) = 17$

41. $4\left(\dfrac{y + 1}{2}\right) + 3y = 0$

42. $-y + 12\left(\dfrac{y - 1}{4}\right) = 3$

43. $8a - 2(3a - 1) = 6$

44. $3z - (4z - 2) = 9$

Concept Extensions

45. Draw a graph of two linear equations whose associated system has the solution $(-1, 4)$.

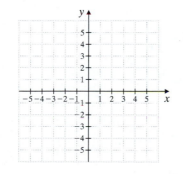

46. Draw a graph of two linear equations whose associated system has the solution $(3, -2)$.

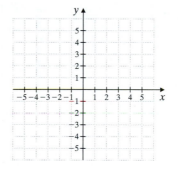

47. Draw a graph of two linear equations whose associated system has no solution.

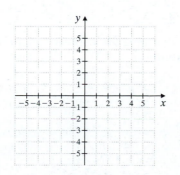

48. Draw a graph of two linear equations whose associated system has an infinite number of solutions.

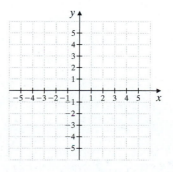

The double line graph below shows the number of pounds of fishery products from U.S. domestic catch and from imports. Use this graph to answer Exercises 49 and 50.

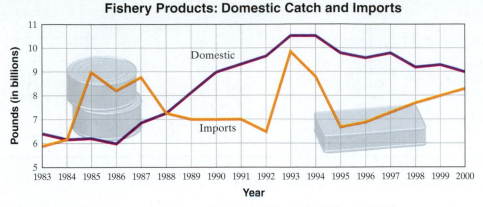

Fishery Products: Domestic Catch and Imports

Source: U.S. Bureau of the Census, *Statistical Abstract of the United States:* 2003, 115th ed., Washington, DC, 1995.

49. In what year(s) was the number of pounds of imported fishery products equal to the number of pounds of domestic catch?

50. In what year(s) was the number of pounds of imported fishery products less than or equal to the number of pounds of domestic catch?

The double line graph below shows the number of Target Stores versus the number of Wal-Mart discount stores. Use this for Exercises 51 and 52. (Note: This does not include Wal-Mart Supercenters or Sam's Club) (Sources: Target.com and Walmart.com)

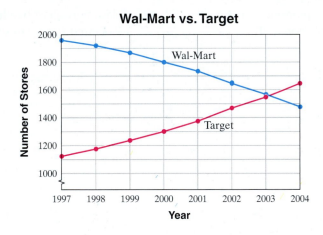

Wal-Mart vs. Target

51. In what year(s) was the number of Target stores approximately equal to the number of Wal-Mart discount stores?

52. In what year(s) was the number of Wal-Mart discount stores greater than the number of Target stores?

53. Construct a system of two linear equations that has (2, 5) as a solution.

54. Construct a system of two linear equations that has (0, 1) as a solution.

55. The ordered pair $(-2, 3)$ is a solution of the three linear equations below:

$$x + y = 1$$
$$2x - y = -7$$
$$x + 3y = 7$$

If each equation has a distinct graph, describe the graph of all three equations on the same axes.

56. Below are tables of values for two linear equations.
   a. Find a solution of the corresponding system.

   b. Graph several ordered pairs from each table and sketch the two lines.

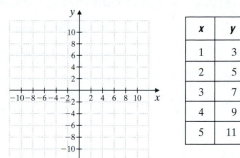

x	y
1	3
2	5
3	7
4	9
5	11

x	y
1	6
2	7
3	8
4	9
5	10

   c. Does your graph confirm the solution from part (a)?

57. Explain how to use a graph to determine the number of solutions of a system.

STUDY SKILLS BUILDER

Have You Decided to Successfully Complete This Course?

Hopefully by now, one of your current goals is to successfully complete this course.

   If it is not a goal of yours, ask yourself why? One common reason is fear of failure. Amazingly enough, fear of failure alone can be strong enough to keep many of us from doing our best in any endeavor. Another common reason is that you simply haven't taken the time to make successfully completing this course one of your goals.

   Anytime you are registered for a course, successfully completing the course should probably be a goal. How do you do this? Start by writing this goal in your mathematics notebook. Then list steps you will take to ensure success. A great first step is to read or reread Section 1.1 and make a commitment to try the suggestions in that section.

   Good luck and don't forget that a positive attitude will make a big difference.

Let's see how you are doing.

1. Have you made the decision to make "successfully completing this course" a goal of yours? If no, please list reasons that this has not happened. Study your list and talk to your instructor about this.

2. If your answer to Exercise 1 is yes, take a moment and list, in your notebook, further specific goals that will help you achieve this goal of successfully completing this course. (For example, my goal this semester is not to miss any of my mathematics classes.)

3. Rate your commitment to this course with a number between 1 and 5. Use the diagram below to help.

High Commitment		Average Commitment		Not committed at all
5	4	3	2	1

4. If you have rated your personal commitment level (from the exercise above) as a 1, 2, or 3, list the reasons why this is so. Then determine whether it is possible to increase your commitment level to a 4 or 5.

Objective

A Use the Substitution Method to Solve a System of Linear Equations.

Objective **A** Using the Substitution Method

You may have suspected by now that graphing alone is not an accurate way to solve a system of linear equations. For example, a solution of $\left(\frac{1}{2}, \frac{2}{9}\right)$ is unlikely to be read correctly from a graph. In this section, we discuss a second, more accurate method for solving systems of equations. This method is called the **substitution method** and is introduced in the next example.

PRACTICE PROBLEM 1

Use the substitution method to solve the system:

$$\begin{cases} 2x + 3y = 13 \\ x = y + 4 \end{cases}$$

EXAMPLE 1 Solve the system:

$$\begin{cases} 2x + y = 10 & \text{First equation} \\ x = y + 2 & \text{Second equation} \end{cases}$$

Solution: The second equation in this system is $x = y + 2$. This tells us that x and $y + 2$ have the same value. This means that we may substitute $y + 2$ for x in the first equation.

$$2x + y = 10 \quad \text{First equation}$$

$$2(y + 2) + y = 10 \quad \text{Substitute } y + 2 \text{ for } x \text{ since } x = y + 2.$$

Notice that this equation now has one variable, y. Let's now solve this equation for y.

> **Helpful Hint**
> Don't forget the distributive property.

$$2(y + 2) + y = 10$$
$$2y + 4 + y = 10 \quad \text{Apply the distributive property.}$$
$$3y + 4 = 10 \quad \text{Combine like terms.}$$
$$3y = 6 \quad \text{Subtract 4 from both sides.}$$
$$y = 2 \quad \text{Divide both sides by 3.}$$

Now we know that the y-value of the ordered pair solution of the system is 2. To find the corresponding x-value, we replace y with 2 in the second equation, $x = y + 2$, and solve for x.

$$x = y + 2 \quad \text{Second equation.}$$
$$x = 2 + 2 \quad \text{Let } y = 2.$$
$$x = 4$$

The solution of the system is the ordered pair $(4, 2)$. Since an ordered pair solution must satisfy both linear equations in the system, we could have chosen the equation $2x + y = 10$ to find the corresponding x-value. The resulting x-value is the same.

Check: We check to see that $(4, 2)$ satisfies both equations of the original system.

First Equation	**Second Equation**
$2x + y = 10$	$x = y + 2$
$2(4) + 2 \stackrel{?}{=} 10$	$4 \stackrel{?}{=} 2 + 2$ Let $x = 4$ and $y = 2$.
$10 = 10$ True	$4 = 4$ True

Answer

1. $(5, 1)$

The solution of the system is $(4, 2)$.
A graph of the two equations shows the two lines intersecting at the point $(4, 2)$.

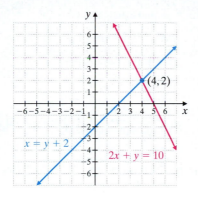

🟧 Work Practice Problem 1

EXAMPLE 2 Solve the system:

$$\begin{cases} 5x - y = -2 \\ y = 3x \end{cases}$$

Solution: The second equation is solved for y in terms of x. We substitute $3x$ for y in the first equation.

$5x - y = -2$ First equation

$5x - (3x) = -2$ Substitute $3x$ for y.

Now we solve for x.

$5x - 3x = -2$

$2x = -2$ Combine like terms.

$x = -1$ Divide both sides by 2.

The x-value of the ordered pair solution is -1. To find the corresponding y-value, we replace x with -1 in the second equation $y = 3x$.

$y = 3x$ Second equation

$y = 3(-1)$ Let $x = -1$.

$y = -3$

Check to see that the solution of the system is $(-1, -3)$.

🟧 Work Practice Problem 2

To solve a system of equations by substitution, we first need an equation solved for one of its variables, as in Examples 1 and 2. If neither equation in a system is solved for x or y, this will be our first step.

EXAMPLE 3 Solve the system:

$$\begin{cases} x + 2y = 7 \\ 2x + 2y = 13 \end{cases}$$

Solution: Notice that neither equation is solved for x or y. Thus, we choose one of the equations and solve for x or y. We will solve the first equation for x so that we will not introduce tedious fractions when solving. To solve the first equation for x, we subtract $2y$ from both sides.

$x + 2y = 7$ First equation

$x = 7 - 2y$ Subtract $2y$ from both sides.

Continued on next page

PRACTICE PROBLEM 2

Use the substitution method to solve the system:

$$\begin{cases} 4x - y = 2 \\ y = 5x \end{cases}$$

PRACTICE PROBLEM 3

Solve the system:

$$\begin{cases} 3x + y = 5 \\ 3x - 2y = -7 \end{cases}$$

Answers

2. $(-2, -10)$, **3.** $\left(\dfrac{1}{3}, 4\right)$

Since $x = 7 - 2y$, we now substitute $7 - 2y$ for x in the second equation and solve for y.

$$2x + 2y = 13 \quad \text{Second equation}$$
$$2(7 - 2y) + 2y = 13 \quad \text{Let } x = 7 - 2y.$$
$$14 - 4y + 2y = 13 \quad \text{Apply the distributive property.}$$
$$14 - 2y = 13 \quad \text{Simplify.}$$
$$-2y = -1 \quad \text{Subtract 14 from both sides.}$$
$$y = \frac{1}{2} \quad \text{Divide both sides by } -2.$$

Helpful Hint Don't forget to insert parentheses when substituting $7 - 2y$ for x.

To find x, we let $y = \frac{1}{2}$ in the equation $x = 7 - 2y$.

$$x = 7 - 2y$$
$$x = 7 - 2\left(\frac{1}{2}\right) \quad \text{Let } y = \frac{1}{2}.$$
$$x = 7 - 1$$
$$x = 6$$

Helpful Hint To find x, any equation in two variables equivalent to the original equations of the system may be used. We used this equation since it is solved for x.

Check the solution in both equations of the original system. The solution is $\left(6, \frac{1}{2}\right)$.

■ **Work Practice Problem 3**

The following steps summarize how to solve a system of equations by the substitution method.

To Solve a System of Two Linear Equations by the Substitution Method

Step 1: Solve one of the equations for one of its variables.

Step 2: Substitute the expression for the variable found in Step 1 into the other equation.

Step 3: Solve the equation from Step 2 to find the value of one variable.

Step 4: Substitute the value found in Step 3 in any equation containing both variables to find the value of the other variable.

Step 5: Check the proposed solution in the original system.

✔ **Concept Check** As you solve the system

$$\begin{cases} 2x + y = -5 \\ x - y = 5 \end{cases}$$

you find that $y = -5$. Is this the solution of the system?

PRACTICE PROBLEM 4

Solve the system:
$$\begin{cases} 5x - 2y = 6 \\ -3x + y = -3 \end{cases}$$

Answer
4. $(0, -3)$

✔ **Concept Check Answer**
no, the solution will be an ordered pair

EXAMPLE 4 Solve the system:
$$\begin{cases} 7x - 3y = -14 \\ -3x + y = 6 \end{cases}$$

Solution: Since the coefficient of y is 1 in the second equation, we will solve the second equation for y. This way, we avoid introducing tedious fractions.

$$-3x + y = 6 \quad \text{Second equation}$$
$$y = 3x + 6$$

Next, we substitute $3x + 6$ for y in the first equation.

$$7x - 3y = -14 \quad \text{First equation}$$
$$7x - 3(3x + 6) = -14 \quad \text{Let } y = 3x + 6.$$
$$7x - 9x - 18 = -14 \quad \text{Use the distributive property.}$$
$$-2x - 18 = -14 \quad \text{Simplify.}$$
$$-2x = 4 \quad \text{Add 18 to both sides.}$$
$$x = -2 \quad \text{Divide both sides by } -2.$$

To find the corresponding y-value, we substitute -2 for x in the equation $y = 3x + 6$. Then $y = 3(-2) + 6$ or $y = 0$. The solution of the system is $(-2, 0)$. Check this solution in both equations of the system.

🔲 **Work Practice Problem 4**

✔**Concept Check** To avoid fractions, which of the equations below would you use to solve for x?

a. $3x - 4y = 15$ **b.** $14 - 3y = 8x$ **c.** $7y + x = 12$

> **Helpful Hint**
>
> When solving a system of equations by the substitution method, begin by solving an equation for one of its variables. If possible, solve for a variable that has a coefficient of 1 or -1 to avoid working with time-consuming fractions.

EXAMPLE 5 Solve the system:
$$\begin{cases} \dfrac{1}{2}x - y = 3 \\ x = 6 + 2y \end{cases}$$

PRACTICE PROBLEM 5

Solve the system:
$$\begin{cases} -x + 3y = 6 \\ y = \dfrac{1}{3}x + 2 \end{cases}$$

Solution: The second equation is already solved for x in terms of y. Thus we substitute $6 + 2y$ for x in the first equation and solve for y.

$$\frac{1}{2}x - y = 3 \quad \text{First equation}$$

$$\frac{1}{2}(6 + 2y) - y = 3 \quad \text{Let } x = 6 + 2y.$$

$$3 + y - y = 3 \quad \text{Apply the distributive property.}$$
$$3 = 3 \quad \text{Simplify.}$$

Arriving at a true statement such as $3 = 3$ indicates that the two linear equations in the original system are equivalent. This means that their graphs are identical, as shown in the figure. There is an infinite number of solutions to the system, and any solution of one equation is also a solution of the other.

$$\frac{1}{2}x - y = 3$$
$$x = 6 + 2y$$

🔲 **Work Practice Problem 5**

Answer

5. infinite number of solutions

✔ **Concept Check Answer**

c

PRACTICE PROBLEM 6

Solve the system:

$$\begin{cases} 2x - 3y = 6 \\ -4x + 6y = -12 \end{cases}$$

EXAMPLE 6 Solve the system:

$$\begin{cases} 6x + 12y = 5 \\ -4x - 8y = 0 \end{cases}$$

Solution: We choose the second equation and solve for y. (*Note:* Although you might not see this beforehand, if you solve the second equation for x, the result is $x = -2y$ and no fractions are introduced. Either way will lead to the correct solution.)

$$-4x - 8y = 0 \qquad \text{Second equation}$$

$$-8y = 4x \qquad \text{Add } 4x \text{ to both sides.}$$

$$\frac{-8y}{-8} = \frac{4x}{-8} \qquad \text{Divide both sides by } -8.$$

$$y = -\frac{1}{2}x \qquad \text{Simplify.}$$

Now we replace y with $-\frac{1}{2}x$ in the first equation.

$$6x + 12y = 5 \qquad \text{First equation}$$

$$6x + 12\left(-\frac{1}{2}x\right) = 5 \qquad \text{Let } y = -\frac{1}{2}x.$$

$$6x + (-6x) = 5 \qquad \text{Simplify.}$$

$$0 = 5 \qquad \text{Combine like terms.}$$

The false statement $0 = 5$ indicates that this system has no solution. The graph of the linear equations in the system is a pair of parallel lines, as shown in the figure.

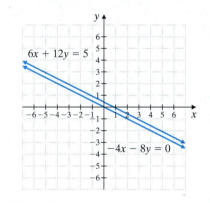

■ **Work Practice Problem 6**

✔**Concept Check** Describe how the graphs of the equations in a system appear if the system has

a. no solution

b. one solution

c. an infinite number of solutions

Answer

6. infinite number of solutions

✔ **Concept Check Answers**

a. parallel lines, **b.** intersect at one point, **c.** identical graphs

Objective **A** *Solve each system of equations by the substitution method. See Examples 1 and 2.*

1. $\begin{cases} x + y = 3 \\ x = 2y \end{cases}$

2. $\begin{cases} x + y = 20 \\ x = 3y \end{cases}$

3. $\begin{cases} x + y = 6 \\ y = -3x \end{cases}$

4. $\begin{cases} x + y = 6 \\ y = -4x \end{cases}$

5. $\begin{cases} y = 3x + 1 \\ 4y - 8x = 12 \end{cases}$

6. $\begin{cases} y = 2x + 3 \\ 5y - 7x = 18 \end{cases}$

7. $\begin{cases} y = 2x + 9 \\ y = 7x + 10 \end{cases}$

8. $\begin{cases} y = 5x - 3 \\ y = 8x + 4 \end{cases}$

Solve each system of equations by the substitution method. See Examples 1 through 6.

9. $\begin{cases} 3x - 4y = 10 \\ y = x - 3 \end{cases}$

10. $\begin{cases} 4x - 3y = 10 \\ y = x - 5 \end{cases}$

11. $\begin{cases} x + 2y = 6 \\ 2x + 3y = 8 \end{cases}$

12. $\begin{cases} x + 3y = -5 \\ 2x + 2y = 6 \end{cases}$

13. $\begin{cases} 3x + 2y = 16 \\ x = 3y - 2 \end{cases}$

14. $\begin{cases} 2x + 3y = 18 \\ x = 2y - 5 \end{cases}$

15. $\begin{cases} 2x - 5y = 1 \\ 3x + y = -7 \end{cases}$

16. $\begin{cases} 3y - x = 6 \\ 4x + 12y = 0 \end{cases}$

17. $\begin{cases} 4x + 2y = 5 \\ -2x = y + 4 \end{cases}$

18. $\begin{cases} 2y = x + 2 \\ 6x - 12y = 0 \end{cases}$

19. $\begin{cases} 4x + y = 11 \\ 2x + 5y = 1 \end{cases}$

20. $\begin{cases} 3x + y = -14 \\ 4x + 3y = -22 \end{cases}$

21. $\begin{cases} x + 2y + 5 = -4 + 5y - x \\ 2x + x = y + 4 \end{cases}$
(*Hint:* First simplify each equation.)

22. $\begin{cases} 5x + 4y - 2 = -6 + 7y - 3x \\ 3x + 4x = y + 3 \end{cases}$
(*Hint:* See Exercise 21.)

23. $\begin{cases} 6x - 3y = 5 \\ x + 2y = 0 \end{cases}$

24. $\begin{cases} 10x - 5y = -21 \\ x + 3y = 0 \end{cases}$

25. $\begin{cases} 3x - y = 1 \\ 2x - 3y = 10 \end{cases}$

26. $\begin{cases} 2x - y = -7 \\ 4x - 3y = -11 \end{cases}$

27. $\begin{cases} -x + 2y = 10 \\ -2x + 3y = 18 \end{cases}$

28. $\begin{cases} -x + 3y = 18 \\ -3x + 2y = 19 \end{cases}$

29. $\begin{cases} 5x + 10y = 20 \\ 2x + 6y = 10 \end{cases}$

30. $\begin{cases} 6x + 3y = 12 \\ 9x + 6y = 15 \end{cases}$

31. $\begin{cases} 3x + 6y = 9 \\ 4x + 8y = 16 \end{cases}$

32. $\begin{cases} 2x + 4y = 6 \\ 5x + 10y = 16 \end{cases}$

33. $\begin{cases} \dfrac{1}{3}x - y = 2 \\ x - 3y = 6 \end{cases}$

34. $\begin{cases} \dfrac{1}{4}x - 2y = 1 \\ x - 8y = 4 \end{cases}$

35. $\begin{cases} x = \dfrac{3}{4}y - 1 \\ 8x - 5y = -6 \end{cases}$

36. $\begin{cases} x = \dfrac{5}{6}y - 2 \\ 12x - 5y = -9 \end{cases}$

533

Review

Write equivalent equations by multiplying both sides of each given equation by the given nonzero number. See Section 2.2.

37. $3x + 2y = 6$ by -2 **38.** $-x + y = 10$ by 5 **39.** $-4x + y = 3$ by 3 **40.** $5a - 7b = -4$ by -4

Add the binomials. See Section 3.4.

41.
$$\begin{array}{r} 3n + 6m \\ 2n - 6m \\ \hline \end{array}$$

42.
$$\begin{array}{r} -2x + 5y \\ 2x + 11y \\ \hline \end{array}$$

43.
$$\begin{array}{r} -5a - 7b \\ 5a - 8b \\ \hline \end{array}$$

44.
$$\begin{array}{r} 9q + p \\ -9q - p \\ \hline \end{array}$$

Concept Extensions

Solve each system by the substitution method. First simplify each equation by combining like terms.

45. $\begin{cases} -5y + 6y = 3x + 2(x - 5) - 3x + 5 \\ 4(x + y) - x + y = -12 \end{cases}$

46. $\begin{cases} 5x + 2y - 4x - 2y = 2(2y + 6) - 7 \\ 3(2x - y) - 4x = 1 + 9 \end{cases}$

47. Explain how to identify a system with no solution when using the substitution method.

48. Occasionally, when using the substitution method, we obtain the equation $0 = 0$. Explain how this result indicates that the graphs of the equations in the system are identical.

Solve. See a Concept Check in this section.

49. As you solve the system $\begin{cases} 3x - y = -6 \\ -3x + 2y = 7 \end{cases}$, you find that $y = 1$. Is this the solution to the system.

50. As you solve the system $\begin{cases} x = 5y \\ y = 2x \end{cases}$, you find that $x = 0$ and $y = 0$. What is the solution to this system?

51. To avoid fractions, which of the equations below would you use if solving for y? Explain why.

   a. $\dfrac{1}{2}x - 4y = \dfrac{3}{4}$

   b. $8x - 5y = 13$

   c. $7x - y = 19$

52. Give the number of solutions for a system if the graphs of the equations in the system are

   a. lines intersecting in one point

   b. parallel lines

   c. same line

Use a graphing calculator to solve each system.

53. $\begin{cases} y = 5.1x + 14.56 \\ y = -2x - 3.9 \end{cases}$

54. $\begin{cases} y = 3.1x - 16.35 \\ y = -9.7x + 28.45 \end{cases}$

55. $\begin{cases} 3x + 2y = 14.04 \\ 5x + y = 18.5 \end{cases}$

56. $\begin{cases} x + y = -15.2 \\ -2x + 5y = -19.3 \end{cases}$

57. For the years 1973 through 2003, the annual percent y of U.S. households that used fuel oil to heat their homes is given by the equation $y = -0.50x + 21.92$, where x is the number of years after 1973. For the same period the annual percent y of U.S. households that used electricity to heat their homes is given by the equation $y = 0.71x + 11.03$, where x is the number of years after 1973. (*Source:* U.S. Census Bureau, American Housing Survey Branch)

a. Use the substitution method to solve this system of equations.

$$\begin{cases} y = -0.50x + 21.92 \\ y = 0.71x + 11.03 \end{cases}$$

(Round your final results to the nearest whole numbers.)

b. Explain the meaning of your answer to part (a).

c. Sketch a graph of the system of equations. Write a sentence describing the use of fuel oil and electricity for heating homes between 1973 and 2003.

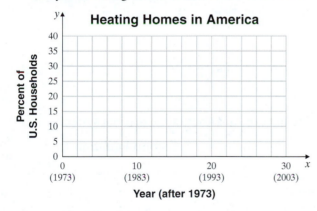

58. The number y of VHS movie format units (in billions) shipped to retailers in the United States from 1998 to 2003 is given by $y = -0.5x + 8.7$, where x is the number of years after 1998. The number y of DVD movie format units (in billions) shipped to retailers in the United States from 1998 to 2003 is given by $y = 1.67x - 0.6$, where x is the number of years after 1998. (*Source:* Business 2.0 e-magazine)

a. Use the substitution method to solve this system of equations:

$$\begin{cases} y = -0.5x + 8.7 \\ y = 1.67x - 0.6 \end{cases}$$

(Round x to the nearest tenth and y to the nearest whole.)

b. Explain the meaning of your answer to part (a).

c. Sketch a graph of the system of equations. Write a sentence describing the trends in the popularity of these two types of movie formats.

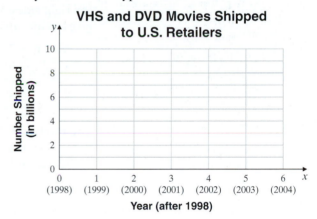

How Are Your Homework Assignments Going?

Remember that it is important to keep up with homework. Why? Many concepts in mathematics build on each other. Often, your understanding of a day's lecture depends on an understanding of the previous day's material.

To complete a homework assignment, remember these 4 things:

- Attempt all of it.
- Check it.
- Correct it.
- If needed, ask questions about it.

Take a moment and review your completed homework assignments. Answer the exercises below based on this review.

1. Approximate the fraction of your homework you have attempted.

2. Approximate the fraction of your homework you have checked (if possible).

3. If you are able to check your homework, have you corrected it when errors have been found?

4. When working homework, if you do not understand a concept, what do you personally do?

7.3 SOLVING SYSTEMS OF LINEAR EQUATIONS BY ADDITION

Objective **A** Using the Addition Method

We have seen that substitution is an accurate method for solving a system of linear equations. Another accurate method is the **addition** or **elimination method.** The addition method is based on the addition property of equality: Adding equal quantities to both sides of an equation does not change the solution of the equation. In symbols,

if $A = B$ and $C = D$, then $A + C = B + D$

To see how we use this to solve a system of equations, study Example 1.

PRACTICE PROBLEM 1

Use the addition method to solve the system:
$$\begin{cases} x + y = 13 \\ x - y = 5 \end{cases}$$

EXAMPLE 1 Solve the system: $\begin{cases} x + y = 7 \\ x - y = 5 \end{cases}$

Solution: Since the left side of each equation is equal to its right side, we are adding equal quantities when we add the left sides of the equations together and the right sides of the equations together. This adding eliminates the variable y and gives us an equation in one variable, x. We can then solve for x.

$$\begin{array}{ll} x + y = 7 & \text{First equation} \\ \underline{x - y = 5} & \text{Second equation} \\ 2x \quad = 12 & \text{Add the equations to eliminate } y. \\ \quad x = 6 & \text{Divide both sides by 2.} \end{array}$$

The x-value of the solution is 6. To find the corresponding y-value, we let $x = 6$ in either equation of the system. We will use the first equation.

> **Helpful Hint**
> Notice in Example 1 that our goal when solving a system of equations by the addition method is to eliminate a variable when adding the equations.

$$\begin{array}{ll} x + y = 7 & \text{First equation} \\ 6 + y = 7 & \text{Let } x = 6. \\ \quad y = 1 & \text{Solve for } y. \end{array}$$

The solution is $(6, 1)$.

Check: Check the solution in both equations of the original system.

First Equation	**Second Equation**
$x + y = 7$	$x - y = 5$
$6 + 1 \overset{?}{=} 7$ Let $x = 6$ and $y = 1$.	$6 - 1 \overset{?}{=} 5$ Let $x = 6$ and $y = 1$.
$7 = 7$ True	$5 = 5$ True

Thus, the solution of the system is $(6, 1)$.

If we graph the two equations in the system, we have two lines that intersect at the point $(6, 1)$ as shown.

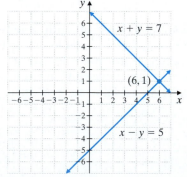

■ **Work Practice Problem 1**

Answer

1. $(9, 4)$

EXAMPLE 2 Solve the system: $\begin{cases} -2x + y = 2 \\ -x + 3y = -4 \end{cases}$

Solution: If we simply add these two equations, the result is still an equation in two variables. However, from Example 1, remember that our goal is to eliminate one of the variables so that we have an equation in the other variable. To do this, notice what happens if we multiply *both sides* of the first equation by -3. We are allowed to do this by the multiplication property of equality. Then the system

$\begin{cases} -3(-2x + y) = -3(2) \\ -x + 3y = -4 \end{cases}$ simplifies to $\begin{cases} 6x - 3y = -6 \\ -x + 3y = -4 \end{cases}$

When we add the resulting equations, the y variable is eliminated.

$$\begin{array}{r} 6x - 3y = -6 \\ \underline{-x + 3y = -4} \\ 5x \quad\quad = -10 \end{array} \quad \text{Add.}$$

$$x = -2 \quad \text{Divide both sides by 5.}$$

To find the corresponding y-value, we let $x = -2$ in either of the original equations. We use the first equation of the original system.

$$\begin{aligned} -2x + y &= 2 \quad \text{First equation} \\ -2(-2) + y &= 2 \quad \text{Let } x = -2. \\ 4 + y &= 2 \\ y &= -2 \end{aligned}$$

Check the ordered pair $(-2, -2)$ in both equations of the *original* system. The solution is $(-2, -2)$.

🔲 **Work Practice Problem 2**

> **Helpful Hint**
>
> When finding the second value of an ordered pair solution, any equation equivalent to one of the original equations in the system may be used.

In Example 2, the decision to multiply the first equation by -3 was no accident. **To eliminate a variable** when adding two equations, **the coefficient of the variable in one equation must be the opposite of its coefficient in the other equation.**

> **Helpful Hint**
>
> Be sure to multiply *both sides* of an equation by a chosen number when solving by the addition method. A common mistake is to multiply only the side containing the variables.

EXAMPLE 3 Solve the system: $\begin{cases} 2x - y = 7 \\ 8x - 4y = 1 \end{cases}$

Solution: When we multiply both sides of the first equation by -4, the resulting coefficient of x is -8. This is the opposite of 8, the coefficient of x in the second equation. Then the system

$\begin{cases} -4(2x - y) = -4(7) \\ 8x - 4y = 1 \end{cases}$ simplifies to

> **Helpful Hint**
>
> Don't forget to multiply both sides by -4.

$$\begin{cases} -8x + 4y = -28 \\ \underline{8x - 4y = 1} \end{cases}$$

$$0 = -27 \quad \text{Add the equations.}$$

Continued on next page

PRACTICE PROBLEM 2

Solve the system:

$$\begin{cases} 2x - y = -6 \\ -x + 4y = 17 \end{cases}$$

PRACTICE PROBLEM 3

Solve the system:

$$\begin{cases} x - 3y = -2 \\ -3x + 9y = 5 \end{cases}$$

Answers

2. $(-1, 4)$,   **3.** no solution

When we add the equations, both variables are eliminated and we have $0 = -27$, a false statement. This means that the system has no solution. The equations, if graphed, represent parallel lines.

▮ **Work Practice Problem 3**

EXAMPLE 4 Solve the system: $\begin{cases} 3x - 2y = 2 \\ -9x + 6y = -6 \end{cases}$

Solution: First we multiply both sides of the first equation by 3 and then we add the resulting equations.

$$\begin{cases} 3(3x - 2y) = 3(2) \\ -9x + 6y = -6 \end{cases} \quad \text{simplifies to} \quad \begin{cases} 9x - 6y = 6 \\ \underline{-9x + 6y = -6} \quad \text{Add the equations.} \\ 0 = 0 \end{cases}$$

Both variables are eliminated and we have $0 = 0$, a true statement. This means that the system has an infinite number of solutions. The equations, if graphed, are the same line.

▮ **Work Practice Problem 4**

✔ **Concept Check** Suppose you are solving the system

$$\begin{cases} 3x + 8y = -5 \\ 2x - 4y = 3 \end{cases}$$

You decide to use the addition method by multiplying both sides of the second equation by 2. In which of the following was the multiplication performed correctly? Explain.

a. $4x - 8y = 3$ **b.** $4x - 8y = 6$

In the next example, we multiply both equations by numbers so that coefficients of a variable are opposites.

EXAMPLE 5 Solve the system: $\begin{cases} 3x + 4y = 13 \\ 5x - 9y = 6 \end{cases}$

Solution: We can eliminate the variable y by multiplying the first equation by 9 and the second equation by 4. Then we add the resulting equations.

$$\begin{cases} 9(3x + 4y) = 9(13) \\ 4(5x - 9y) = 4(6) \end{cases} \quad \text{simplifies to} \quad \begin{cases} 27x + 36y = 117 \\ \underline{20x - 36y = 24} \\ 47x = 141 \quad \text{Add the equations.} \\ x = 3 \quad \text{Solve for } x. \end{cases}$$

To find the corresponding y-value, we let $x = 3$ in one of the original equations of the system. Doing so in any of these equations will give $y = 1$. Check to see that $(3, 1)$ satisfies each equation in the original system. The solution is $(3, 1)$.

▮ **Work Practice Problem 5**

If we had decided to eliminate x instead of y in Example 5, the first equation could have been multiplied by 5 and the second by -3. Try solving the original system this way to check that the solution is $(3, 1)$.

The following steps summarize how to solve a system of linear equations by the addition method.

To Solve a System of Two Linear Equations by the Addition Method

Step 1: Rewrite each equation in standard form $Ax + By = C$.

Step 2: If necessary, multiply one or both equations by a nonzero number so that the coefficients of a chosen variable in the system are opposites.

Step 3: Add the equations.

Step 4: Find the value of one variable by solving the resulting equation from Step 3.

Step 5: Find the value of the second variable by substituting the value found in Step 4 into either of the original equations.

Step 6: Check the proposed solution in the original system.

✔**Concept Check** Suppose you are solving the system

$$\begin{cases} -4x + 7y = 6 \\ x + 2y = 5 \end{cases}$$

by the addition method.

a. What step(s) should you take if you wish to eliminate x when adding the equations?

b. What step(s) should you take if you wish to eliminate y when adding the equations?

EXAMPLE 6 Solve the system: $\begin{cases} -x - \dfrac{y}{2} = \dfrac{5}{2} \\ \dfrac{x}{6} - \dfrac{y}{2} = 0 \end{cases}$

Solution: We begin by clearing each equation of fractions. To do so, we multiply both sides of the first equation by the LCD 2 and both sides of the second equation by the LCD 6. Then the system

simplifies to $\begin{cases} -2x - y = 5 \\ x - 3y = 0 \end{cases}$

We can now eliminate the variable x by multiplying the second equation by 2.

$$\begin{cases} -2x - y = 5 \\ 2(x - 3y) = 2(0) \end{cases} \quad \text{simplifies to} \quad \begin{cases} -2x - y = 5 \\ \underline{2x - 6y = 0} \\ -7y = 5 \quad \text{Add the equations.} \\ y = -\dfrac{5}{7} \quad \text{Solve for } y. \end{cases}$$

To find x, we could replace y with $-\dfrac{5}{7}$ in one of the equations with two variables.

Instead, let's go back to the simplified system and multiply by appropriate factors to eliminate the variable y and solve for x. To do this, we multiply the first equation by -3. Then the system

$$\begin{cases} -3(-2x - y) = -3(5) \\ x - 3y = 0 \end{cases} \quad \text{simplifies to} \quad \begin{cases} 6x + 3y = -15 \\ \underline{x - 3y = 0} \\ 7x = -15 \quad \text{Add the equations.} \\ x = -\dfrac{15}{7} \quad \text{Solve for } x. \end{cases}$$

Check the ordered pair $\left(-\dfrac{15}{7}, -\dfrac{5}{7}\right)$ in both equations of the original system. The solution is $\left(-\dfrac{15}{7}, -\dfrac{5}{7}\right)$.

■ **Work Practice Problem 6**

PRACTICE PROBLEM 6

Solve the system:

$$\begin{cases} -\dfrac{x}{3} + y = \dfrac{4}{3} \\ \dfrac{x}{2} - \dfrac{5}{2}y = -\dfrac{1}{2} \end{cases}$$

Answer

6. $\left(-\dfrac{17}{2}, -\dfrac{3}{2}\right)$

✔ **Concept Check Answer**

a. multiply the second equation by 4,

b. possible answer: multiply the first equation by -2 and the second equation by 7

Objective Ⓐ *Solve each system of equations by the addition method. See Example 1.*

1. $\begin{cases} 3x + y = 5 \\ 6x - y = 4 \end{cases}$

2. $\begin{cases} 4x + y = 13 \\ 2x - y = 5 \end{cases}$

3. $\begin{cases} x - 2y = 8 \\ -x + 5y = -17 \end{cases}$

4. $\begin{cases} x - 2y = -11 \\ -x + 5y = 23 \end{cases}$

Solve each system of equations by the addition method. If a system contains fractions or decimals, you may want to first clear each equation of fractions or decimals. See Examples 2 through 6.

5. $\begin{cases} 3x + y = -11 \\ 6x - 2y = -2 \end{cases}$

6. $\begin{cases} 4x + y = -13 \\ 6x - 3y = -15 \end{cases}$

7. $\begin{cases} 3x + 2y = 11 \\ 5x - 2y = 29 \end{cases}$

8. $\begin{cases} 4x + 2y = 2 \\ 3x - 2y = 12 \end{cases}$

9. $\begin{cases} x + 5y = 18 \\ 3x + 2y = -11 \end{cases}$

10. $\begin{cases} x + 4y = 14 \\ 5x + 3y = 2 \end{cases}$

11. $\begin{cases} x + y = 6 \\ x - y = 6 \end{cases}$

12. $\begin{cases} x - y = 1 \\ -x + 2y = 0 \end{cases}$

13. $\begin{cases} 2x + 3y = 0 \\ 4x + 6y = 3 \end{cases}$

14. $\begin{cases} 3x + y = 4 \\ 9x + 3y = 6 \end{cases}$

15. $\begin{cases} -x + 5y = -1 \\ 3x - 15y = 3 \end{cases}$

16. $\begin{cases} 2x + y = 6 \\ 4x + 2y = 12 \end{cases}$

17. $\begin{cases} 3x - 2y = 7 \\ 5x + 4y = 8 \end{cases}$

18. $\begin{cases} 6x - 5y = 25 \\ 4x + 15y = 13 \end{cases}$

19. $\begin{cases} 8x = -11y - 16 \\ 2x + 3y = -4 \end{cases}$

20. $\begin{cases} 10x + 3y = -12 \\ 5x = -4y - 16 \end{cases}$

21. $\begin{cases} 4x - 3y = 7 \\ 7x + 5y = 2 \end{cases}$

22. $\begin{cases} -2x + 3y = 10 \\ 3x + 4y = 2 \end{cases}$

23. $\begin{cases} 4x - 6y = 8 \\ 6x - 9y = 12 \end{cases}$

24. $\begin{cases} 9x - 3y = 12 \\ 12x - 4y = 18 \end{cases}$

25. $\begin{cases} 2x - 5y = 4 \\ 3x - 2y = 4 \end{cases}$

26. $\begin{cases} 6x - 5y = 7 \\ 4x - 6y = 7 \end{cases}$

27. $\begin{cases} \dfrac{x}{3} + \dfrac{y}{6} = 1 \\[2mm] \dfrac{x}{2} - \dfrac{y}{4} = 0 \end{cases}$

28. $\begin{cases} \dfrac{x}{2} + \dfrac{y}{8} = 3 \\[2mm] x - \dfrac{y}{4} = 0 \end{cases}$

29. $\begin{cases} \dfrac{10}{3}x + 4y = -4 \\[2mm] 5x + 6y = -6 \end{cases}$

30. $\begin{cases} \dfrac{3}{2}x + 4y = 1 \\[2mm] 9x + 24y = 5 \end{cases}$

31. $\begin{cases} x - \dfrac{y}{3} = -1 \\[2mm] -\dfrac{x}{2} + \dfrac{y}{8} = \dfrac{1}{4} \end{cases}$

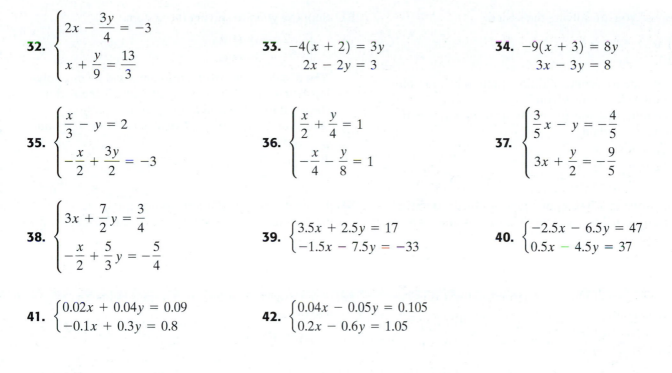

32. $\begin{cases} 2x - \dfrac{3y}{4} = -3 \\ x + \dfrac{y}{9} = \dfrac{13}{3} \end{cases}$

33. $-4(x + 2) = 3y$
$\ 2x - 2y = 3$

34. $-9(x + 3) = 8y$
$\ 3x - 3y = 8$

35. $\begin{cases} \dfrac{x}{3} - y = 2 \\ -\dfrac{x}{2} + \dfrac{3y}{2} = -3 \end{cases}$

36. $\begin{cases} \dfrac{x}{2} + \dfrac{y}{4} = 1 \\ -\dfrac{x}{4} - \dfrac{y}{8} = 1 \end{cases}$

37. $\begin{cases} \dfrac{3}{5}x - y = -\dfrac{4}{5} \\ 3x + \dfrac{y}{2} = -\dfrac{9}{5} \end{cases}$

38. $\begin{cases} 3x + \dfrac{7}{2}y = \dfrac{3}{4} \\ -\dfrac{x}{2} + \dfrac{5}{3}y = -\dfrac{5}{4} \end{cases}$

39. $\begin{cases} 3.5x + 2.5y = 17 \\ -1.5x - 7.5y = -33 \end{cases}$

40. $\begin{cases} -2.5x - 6.5y = 47 \\ 0.5x - 4.5y = 37 \end{cases}$

41. $\begin{cases} 0.02x + 0.04y = 0.09 \\ -0.1x + 0.3y = 0.8 \end{cases}$

42. $\begin{cases} 0.04x - 0.05y = 0.105 \\ 0.2x - 0.6y = 1.05 \end{cases}$

Review

Rewrite each sentence using mathematical symbols. Do not solve the equations. See Sections 2.4 and 2.5.

43. Twice a number, added to 6, is 3 less than the number.

44. The sum of three consecutive integers is 66.

45. Three times a number, subtracted from 20, is 2.

46. Twice the sum of 8 and a number is the difference of the number and 20.

47. The product of 4 and the sum of a number and 6 is twice the number.

48. If the quotient of twice a number and 7 is subtracted from the reciprocal of the number, the result is 2.

Concept Extensions

Solve. See a Concept Check in this section.

49. To solve this system by the addition method and eliminate the variable y,

$$\begin{cases} 4x + 2y = -7 \\ 3x - y = -12 \end{cases}$$

by what value would you multiply the second equation? What do you get when you complete the multiplication?

Given the system of linear equations $\begin{cases} 3x - y = -8 \\ 5x + 3y = 2 \end{cases}$

50. Use the addition method and
   a. Solve the system by eliminating x.
   b. Solve the system by eliminating y.

51. Suppose you are solving the system

$$\begin{cases} 3x + 8y = -5 \\ 2x - 4y = 3. \end{cases}$$

You decide to use the addition method by multiplying both sides of the second equation by 2. In which of the following was the multiplication performed correctly? Explain.

a. $4x - 8y = 3$
b. $4x - 8y = 6$

52. Suppose you are solving the system

$$\begin{cases} -2x - y = 0 \\ -2x + 3y = 6. \end{cases}$$

You decide to use the addition method by multiplying both sides of the first equation by 3, then adding the resulting equation to the second equation. Which of the following is the correct sum? Explain.

a. $-8x = 6$
b. $-8x = 9$

53. When solving a system of equations by the addition method, how do we know when the system has no solution?

54. Explain why the addition method might be preferred over the substitution method for solving the system $\begin{cases} 2x - 3y = 5 \\ 5x + 2y = 6. \end{cases}$

55. Use the system of linear equations below to answer the questions.

$$\begin{cases} x + y = 5 \\ 3x + 3y = b \end{cases}$$

a. Find the value of b so that the system has an infinite number of solutions.
b. Find a value of b so that there are no solutions to the system.

56. Use the system of linear equations below to answer the questions.

$$\begin{cases} x + y = 4 \\ 2x + by = 8 \end{cases}$$

a. Find the value of b so that the system has an infinite number of solutions.
b. Find a value of b so that the system has a single solution.

Solve each system by the addition method.

57. $\begin{cases} 2x + 3y = 14 \\ 3x - 4y = -69.1 \end{cases}$

58. $\begin{cases} 5x - 2y = -19.8 \\ -3x + 5y = -3.7 \end{cases}$

59. Two occupations predicted to greatly increase in number of jobs are medical assistants and computer software engineers. The number of medical assistant jobs predicted for 2002 through 2012 can be approximated by $21.4x - y = -365$. The number of computer software engineer jobs for the same years can be approximated by $17.9x - y = -394$. For both equations, x is the number of years since 2002 and y is the number of jobs in thousands.

a. Use the addition method to solve this system of equations:

$$\begin{cases} 21.4x - y = -365 \\ 17.9x - y = -394 \end{cases}$$

(Round answer to the nearest whole number.)
b. Interpret your solution from part (a).
c. Use the year in your answer to part (b) and estimate the number of medical assistant jobs and computer software engineer jobs in that year.

60. In recent years, the number of daily newspapers printed as morning editions has been increasing and the number of daily newspapers printed as evening editions has been decreasing. The number y of daily morning newspapers in existence from 1993 through 2003 is given by the equation $88x - 5y = -3498$, where x is the number of years since 1993. The number y of daily evening newspapers in existence from 1993 through 2003 is given by the equation $291x + 10y = 9940$, where x is the number of years since 1993. (*Source:* Newspaper Association of America)

a. Use the addition method to solve this system of equations:

$$\begin{cases} 88x - 5y = -3498 \\ 291x + 10y = 9940. \end{cases}$$

(Round to the nearest whole number. Because of rounding, the y-value of your ordered-pair solution may vary.)
b. Interpret your solution from part (a).
c. How many of each type of newspaper were in existence in that year?

Summary on Solving Systems of Equations

Solve each system by either the addition method or the substitution method.

1. $\begin{cases} 2x - 3y = -11 \\ y = 4x - 3 \end{cases}$

2. $\begin{cases} 4x - 5y = 6 \\ y = 3x - 10 \end{cases}$

3. $\begin{cases} x + y = 3 \\ x - y = 7 \end{cases}$

4. $\begin{cases} x - y = 20 \\ x + y = -8 \end{cases}$

5. $\begin{cases} x + 2y = 1 \\ 3x + 4y = -1 \end{cases}$

6. $\begin{cases} x + 3y = 5 \\ 5x + 6y = -2 \end{cases}$

7. $\begin{cases} y = x + 3 \\ 3x = 2y - 6 \end{cases}$

8. $\begin{cases} y = -2x \\ 2x - 3y = -16 \end{cases}$

9. $\begin{cases} y = 2x - 3 \\ y = 5x - 18 \end{cases}$

10. $\begin{cases} y = 6x - 5 \\ y = 4x - 11 \end{cases}$

11. $\begin{cases} x + \dfrac{1}{6}y = \dfrac{1}{2} \\ 3x + 2y = 3 \end{cases}$

12. $\begin{cases} x + \dfrac{1}{3}y = \dfrac{5}{12} \\ 8x + 3y = 4 \end{cases}$

13. $\begin{cases} x - 5y = 1 \\ -2x + 10y = 3 \end{cases}$

14. $\begin{cases} -x + 2y = 3 \\ 3x - 6y = -9 \end{cases}$

15. $\begin{cases} 0.2x - 0.3y = -0.95 \\ 0.4x + 0.1y = 0.55 \end{cases}$

16. $\begin{cases} 0.08x - 0.04y = -0.11 \\ 0.02x - 0.06y = -0.09 \end{cases}$

17. $\begin{cases} x = 3y - 7 \\ 2x - 6y = -14 \end{cases}$

18. $\begin{cases} y = \dfrac{x}{2} - 3 \\ 2x - 4y = 0 \end{cases}$

19. $\begin{cases} 2x + 5y = -1 \\ 3x - 4y = 33 \end{cases}$

20. $\begin{cases} 7x - 3y = 2 \\ 6x + 5y = -21 \end{cases}$

21. Which method, substitution or addition, would you prefer to use to solve the system below? Explain your reasoning.

$$\begin{cases} 3x + 2y = -2 \\ y = -2x \end{cases}$$

22. Which method, substitution or addition, would you prefer to use to solve the system below? Explain your reasoning.

$$\begin{cases} 3x - 2y = -3 \\ 6x + 2y = 12 \end{cases}$$

Answers
1.
2.
3.
4.
5.
6.
7.
8.
9.
10.
11.
12.
13.
14.
15.
16.
17.
18.
19.
20.
21.
22.

7.4 SYSTEMS OF LINEAR EQUATIONS AND PROBLEM SOLVING

Objective **A** Using a System of Equations for Problem Solving

Many of the word problems solved earlier with one-variable equations can also be solved with two equations in two variables. We use the same problem-solving steps that have been used throughout this text. The only difference is that two variables are assigned to represent the two unknown quantities and that the problem is translated into two equations.

> **Problem-Solving Steps**
>
> **1.** UNDERSTAND the problem. During this step, become comfortable with the problem. Some ways of doing this are to
>
> > Read and reread the problem.
> >
> > Choose two variables to represent the two unknowns.
> >
> > Construct a drawing.
> >
> > Propose a solution and check. Pay careful attention to how you check your proposed solution. This will help when writing equations to model the problem.
>
> **2.** TRANSLATE the problem into two equations.
>
> **3.** SOLVE the system of equations.
>
> **4.** INTERPRET the results: *Check* the proposed solution in the stated problem and *state* your conclusion.

PRACTICE PROBLEM 1

Find two numbers whose sum is 50 and whose difference is 22.

EXAMPLE 1 Finding Unknown Numbers

Find two numbers whose sum is 37 and whose difference is 21.

Solution:

1. UNDERSTAND. Read and reread the problem. Suppose that one number is 20. If their sum is 37, the other number is 17 because $20 + 17 = 37$. Is their difference 21? No; $20 - 17 = 3$. Our proposed solution is incorrect, but we now have a better understanding of the problem.

Since we are looking for two numbers, we let

x = first number and

y = second number

2. TRANSLATE. Since we have assigned two variables to this problem, we translate our problem into two equations.

In words:	two numbers whose sum	is	37
Translate:	$x + y$	$=$	37

In words:	two numbers whose difference	is	21
Translate:	$x - y$	$=$	21

Answer

1. 36 and 14

544

3. SOLVE. Now we solve the system.

$$\begin{cases} x + y = 37 \\ x - y = 21 \end{cases}$$

Notice that the coefficients of the variable y are opposites. Let's then solve by the addition method and begin by adding the equations.

$$\begin{array}{ll} x + y = 37 & \\ \underline{x - y = 21} & \\ 2x \quad = 58 & \text{Add the equations.} \\ \quad\; x = 29 & \text{Divide both sides by 2.} \end{array}$$

Now we let $x = 29$ in the first equation to find y.

$$\begin{array}{ll} x + y = 37 & \text{First equation} \\ 29 + y = 37 & \\ \quad\quad y = 8 & \text{Subtract 29 from both sides.} \end{array}$$

4. INTERPRET. The solution of the system is $(29, 8)$.

Check: Notice that the sum of 29 and 8 is $29 + 8 = 37$, the required sum. Their difference is $29 - 8 = 21$, the required difference.

State: The numbers are 29 and 8.

◻ **Work Practice Problem 1**

EXAMPLE 2 **Solving a Problem about Prices**

The Cirque du Soleil show Varekai is performing locally. Matinee admission for 4 adults and 2 children is $374, while admission for 2 adults and 3 children is $285.

a. What is the price of an adult's ticket?

b. What is the price of a child's ticket?

c. Suppose that a special rate of $1000 is offered for groups of 20 persons. Should a group of 4 adults and 16 children use the group rate? Why or why not?

Solution:

1. UNDERSTAND. Read and reread the problem and guess a solution. Let's suppose that the price of an adult's ticket is $50 and the price of a child's ticket is $40. To check our proposed solution, let's see if admission for 4 adults and 2 children is $374. Admission for 4 adults is 4($50) or $200 and admission for 2 children is 2($40) or $80. This gives a total admission of $200 + $80 = $280, not the required $374. Again though, we have accomplished the purpose of this process: We have a better understanding of the problem. To continue, we let

A = the price of an adult's ticket and

C = the price of a child's ticket

Continued on next page

PRACTICE PROBLEM 2

Admission prices at a local weekend fair were $5 for children and $7 for adults. The total money collected was $3379, and 587 people attended the fair. How many children and how many adults attended the fair?

Answer

2. 365 children and 222 adults

2. TRANSLATE. We translate the problem into two equations using both variables.

In words:

admission for 4 adults	and	admission for 2 children	is	$374
↓	↓	↓	↓	↓

Translate: $4A$ $+$ $2C$ $=$ 374

In words:

admission for 2 adults	and	admission for 3 children	is	$285
↓	↓	↓	↓	↓

Translate: $2A$ $+$ $3C$ $=$ 285

3. SOLVE. We solve the system.

$$\begin{cases} 4A + 2C = 374 \\ 2A + 3C = 285 \end{cases}$$

Since both equations are written in standard form, we solve by the addition method. First we multiply the second equation by -2 so that when we add the equations we eliminate the variable A. Then the system

$$\begin{cases} 4A + 2C = 374 \\ -2(2A + 3C) = -2(285) \end{cases}$$

simplifies to

$$\begin{cases} 4A + 2C = 374 \\ \underline{-4A - 6C = -570} \\ -4C = -196 \\ C = 49 \text{ or } \$49, \text{ the} \\ \text{children's} \\ \text{ticket price.} \end{cases}$$

Add the equations.

To find A, we replace C with 49 in the first equation.

$$4A + 2C = 374 \qquad \text{First equation}$$
$$4A + 2(49) = 374 \qquad \text{Let } C = 49.$$
$$4A + 98 = 374$$
$$4A = 276$$
$$A = 69 \text{ or } \$69, \text{ the adult's ticket price.}$$

4. INTERPRET.

Check: Notice that 4 adults and 2 children will pay

$4(\$69) + 2(\$49) = \$276 + \$98 = \$374$, the required amount. Also, the price for 2 adults and 3 children is $2(\$69) + 3(\$49) = \$138 + \$147 = \$285$, the required amount.

State: Answer the three original questions.

a. Since $A = 69$, the price of an adult's ticket is $69.

b. Since $C = 49$, the price of a child's ticket is $49.

c. The regular admission price for 4 adults and 16 children is

$$4(\$69) + 16(\$49) = \$276 + \$784$$
$$= \$1060$$

This is $60 more than the special group rate of $1000, so they should request the group rate.

🟧 **Work Practice Problem 2**

EXAMPLE 3 **Finding Rates**

As part of an exercise program, Louisa and Alfredo start walking each morning. They live 15 miles away from each other. They decide to meet one day by walking toward one another. After 2 hours they meet. If Louisa walks one mile per hour faster than Alfredo, find both walking speeds.

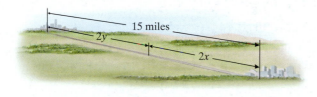

Solution:

1. **UNDERSTAND.** Read and reread the problem. Let's propose a solution and use the formula $d = r \cdot t$ to check. Suppose that Louisa's rate is 4 miles per hour. Since Louisa's rate is 1 mile per hour faster, Alfredo's rate is 3 miles per hour. To check, see if they can walk a total of 15 miles in 2 hours. Louisa's distance is rate · time = 4(2) = 8 miles and Alfredo's distance is rate · time = 3(2) = 6 miles. Their total distance is 8 miles + 6 miles = 14 miles, not the required 15 miles. Now that we have a better understanding of the problem, let's model it with a system of equations.

First, we let

$x =$ Alfredo's rate in miles per hour and

$y =$ Louisa's rate in miles per hour

Now we use the facts stated in the problem and the formula $d = rt$ to fill in the following chart.

	r	$\cdot$ t	$=$ d
Alfredo	x	2	$2x$
Louisa	y	2	$2y$

2. **TRANSLATE.** We translate the problem into two equations using both variables.

In words: | Alfredo's distance | + | Louisa's distance | = | 15 miles |

Translate: $2x$ + $2y$ = 15

In words: | Louisa's rate | is | 1 mile per hour faster than Alfredo's |

Translate: y = $x + 1$

3. **SOLVE.** The system of equations we are solving is

$$\begin{cases} 2x + 2y = 15 \\ y = x + 1 \end{cases}$$

Continued on next page

PRACTICE PROBLEM 3

Two cars are 440 miles apart and traveling toward each other. They meet in 3 hours. If one car's speed is 10 miles per hour faster than the other car's speed, find the speed of each car.

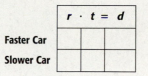

	r	$\cdot$ t	$=$ d
Faster Car			
Slower Car			

Answer

3. One car's speed is $68\frac{1}{3}$ mph and the other car's speed is $78\frac{1}{3}$ mph.

Let's use substitution to solve the system since the second equation is solved for y.

$$2x + 2y = 15 \qquad\qquad \text{First equation}$$

$$2x + 2(x + 1) = 15 \qquad\qquad \text{Replace } y \text{ with } x + 1.$$

$$2x + 2x + 2 = 15$$

$$4x = 13$$

$$x = \frac{13}{4} = 3\frac{1}{4} \text{ or } 3.25$$

$$y = x + 1 = 3\frac{1}{4} + 1 = 4\frac{1}{4} \text{ or } 4.25$$

4. INTERPRET. Alfredo's proposed rate is $3\frac{1}{4}$ miles per hour and Louisa's proposed rate is $4\frac{1}{4}$ miles per hour.

Check: Use the formula $d = rt$ and find that in 2 hours, Alfredo's distance is $(3.25)(2)$ miles or 6.5 miles. In 2 hours, Louisa's distance is $(4.25)(2)$ miles or 8.5 miles. The total distance walked is 6.5 miles + 8.5 miles or 15 miles, the given distance.

State: Alfredo walks at a rate of 3.25 miles per hour and Louisa walks at a rate of 4.25 miles per hour.

■ **Work Practice Problem 3**

EXAMPLE 4   **Finding Amounts of Solutions**

Eric Daly, a chemistry teaching assistant, needs 10 liters of a 20% saline solution (salt water) for his 2 p.m. laboratory class. Unfortunately, the only mixtures on hand are a 5% saline solution and a 25% saline solution. How much of each solution should he mix to produce the 20% solution?

Solution:

1. UNDERSTAND. Read and reread the problem. Suppose that we need 4 liters of the 5% solution. Then we need $10 - 4 = 6$ liters of the 25% solution. To see if this gives us 10 liters of a 20% saline solution, let's find the amount of pure salt in each solution.

	concentration rate	×	amount of solution	=	amount of pure salt
	↓		↓		↓
5% solution:	0.05	×	4 liters	=	0.2 liters
25% solution:	0.25	×	6 liters	=	1.5 liters
20% solution:	0.20	×	10 liters	=	2 liters

Since 0.2 liters + 1.5 liters = 1.7 liters, not 2 liters, our proposed solution is incorrect. But we have gained some insight into how to model and check this problem.

   We let

x = number of liters of 5% solution and

y = number of liters of 25% solution

5% saline   25% saline   20% saline
solution   solution   solution

Now we use a table to organize the given data.

	Concentration Rate	Liters of Solution	Liters of Pure Salt
First Solution	5%	x	$0.05x$
Second Solution	25%	y	$0.25y$
Mixture Needed	20%	10	$(0.20)(10)$

2. TRANSLATE. We translate into two equations using both variables.

In words: liters of 5% solution $+$ liters of 25% solution $=$ 10 liters

$\qquad\qquad\qquad\downarrow\qquad\qquad\qquad\downarrow\qquad\qquad\quad\downarrow$

Translate: x $+$ y $=$ 10

In words: salt in 5% solution $+$ salt in 25% solution $=$ salt in mixture

$\qquad\qquad\qquad\downarrow\qquad\qquad\qquad\downarrow\qquad\qquad\quad\downarrow$

Translate: $0.05x$ $+$ $0.25y$ $= (0.20)(10)$

3. SOLVE. Here we solve the system

$$\begin{cases} x + y = 10 \\ 0.05x + 0.25y = 2 \end{cases}$$

To solve by the addition method, we first multiply the first equation by -25 and the second equation by 100. Then the system

$$\begin{cases} -25(x + y) = -25(10) \\ 100(0.05x + 0.25y) = 100(2) \end{cases} \quad \text{simplifies to} \quad \begin{cases} \begin{aligned} -25x - 25y &= -250 \\ \underline{5x + 25y} &= \underline{200} \\ -20x &= -50 \quad \text{Add.} \\ x &= 2.5 \end{aligned} \end{cases}$$

To find y, we let $x = 2.5$ in the first equation of the original system.

$x + y = 10$

$2.5 + y = 10$ Let $x = 2.5$.

$\qquad y = 7.5$

4. INTERPRET. Thus, we propose that Eric needs to mix 2.5 liters of 5% saline solution with 7.5 liters of 25% saline solution.

Check: Notice that $2.5 + 7.5 = 10$, the required number of liters. Also, the sum of the liters of salt in the two solutions equals the liters of salt in the required mixture:

$0.05(2.5) + 0.25(7.5) = 0.20(10)$

$\qquad 0.125 + 1.875 = 2$

State: Eric needs 2.5 liters of the 5% saline solution and 7.5 liters of the 25% saline solution.

📘 **Work Practice Problem 4**

✔**Concept Check** Suppose you mix an amount of a 30% acid solution with an amount of a 50% acid solution. Which of the following acid strengths would be possible for the resulting acid mixture?

a. 22%   **b.** 44%   **c.** 63%

Without actually solving each problem, choose the correct solution by deciding which choice satisfies the given conditions.

△ **1.** The length of a rectangle is 3 feet longer than the width. The perimeter is 30 feet. Find the dimensions of the rectangle.
   a. length = 8 feet; width = 5 feet
   b. length = 8 feet; width = 7 feet
   c. length = 9 feet; width = 6 feet

△ **2.** An isosceles triangle, a triangle with two sides of equal length, has a perimeter of 20 inches. Each of the equal sides is one inch longer than the third side. Find the lengths of the three sides.
   a. 6 inches, 6 inches, and 7 inches
   b. 7 inches, 7 inches, and 6 inches
   c. 6 inches, 7 inches, and 8 inches

3. Two computer disks and three notebooks cost $17. However, five computer disks and four notebooks cost $32. Find the price of each.
   a. notebook = $4; computer disk = $3
   b. notebook = $3; computer disk = $4
   c. notebook = $5; computer disk = $2

4. Two music CDs and four DVDs cost a total of $40. However, three music CDs and five DVDs cost $55. Find the price of each.
   a. CD = $12; DVD = $4
   b. CD = $15; DVD = $2
   c. CD = $10; DVD = $5

5. Kesha has a total of 100 coins, all of which are either dimes or quarters. The total value of the coins is $13.00. Find the number of each type of coin.
   a. 80 dimes; 20 quarters
   b. 20 dimes; 44 quarters
   c. 60 dimes; 40 quarters

6. Samuel has 28 gallons of saline solution available in two large containers at his pharmacy. One container holds three times as much as the other container. Find the capacity of each container.
   a. 15 gallons; 5 gallons
   b. 20 gallons; 8 gallons
   c. 21 gallons; 7 gallons

Objective Ⓐ *Write a system of equations describing each situation. Do not solve the system. See Example 1.*

7. Two numbers add up to 15 and have a difference of 7.

8. The total of two numbers is 16. The first number plus 2 more than 3 times the second equals 18.

9. Keiko has a total of $6500, which she has invested in two accounts. The larger account is $800 greater than the smaller account.

10. Dominique has four times as much money in his savings account as in his checking account. The total amount is $2300.

Solve. See Examples 1 through 4.

11. Two numbers total 83 and have a difference of 17. Find the two numbers.

12. The sum of two numbers is 76 and their difference is 52. Find the two numbers.

13. A first number plus twice a second number is 8. Twice the first number plus the second totals 25. Find the numbers.

14. One number is 4 more than twice the second number. Their total is 25. Find the numbers.

15. The highest scorer during the WNBA 2004 regular season was Lauren Jackson of the Seattle Storm. Over the season, Jackson scored 36 more points than the second-highest scorer, Lisa Leslie of the Los Angeles Sparks. Together, Jackson and Leslie scored 1232 points during the 2004 regular season. How many points did each player score over the course of the season? (*Source:* Women's National Basketball Association)

16. Ilya Kovalchuk of the Atlanta Thrashers was tied for the title NHL's leading goal scorer during the 2003–2004 regular season. Bill Guerin of the Dallas Stars, who was ranked ninth for goals, scored 7 fewer goals than Kovalchuk. Together, these two players made a total of 75 goals during the 2003–2004 regular season. How many goals each did Kovalchuk and Guerin make? (*Source:* National Hockey League)

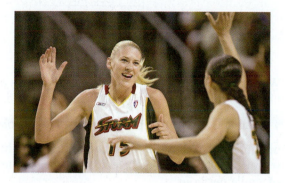

17. Ann Marie Jones has been pricing Amtrak train fares for a group trip to New York. Three adults and four children must pay $159. Two adults and three children must pay $112. Find the price of an adult's ticket, and find the price of a child's ticket.

18. Last month, Jerry Papa purchased five DVDs and two CDs at Wall-to-Wall Sound for $65. This month he bought three DVDs and four CDs for $81. Find the price of each DVD, and find the price of each CD.

19. Johnston and Betsy Waring have a jar containing 80 coins, all of which are either quarters or nickels. The total value of the coins is $14.60. How many of each type of coin do they have?

20. Sarah and Keith Robinson purchased 40 stamps, a mixture of 37¢ and 23¢ stamps. Find the number of each type of stamp if they spent $14.10.

21. Norman and Suzanne Scarpulla own 35 shares of McDonald's stock and 69 shares of The Ohio Art Company stock (makers of Etch A Sketch and other toys). At the close of the markets on a particular day in 2004, their stock portfolio consisting of these two stocks was worth $1551.00. The closing price of the McDonald's stock was $25 more per share than the closing price of The Ohio Art Company stock on that day. What was the closing price of each stock on that day? (*Source:* Yahoo finance)

22. Saralee Rose has an investment in Google and Nintendo stock. On a particular day in 2004, Google stock closed at $169.98 per share, and Nintendo stock closed at $115.40 per share. Saralee's portfolio made up of these two stocks was worth $8712.66 at the end of the day. If Saralee owns 16 more shares of Google stock than she owns of Nintendo stock, how many shares of each type of stock does she own?

23. Twice last month, Judy Carter rented a car from Enterprise in Fresno, California, and traveled around the Southwest on business. Enterprise rents its cars for a daily fee, plus an additional charge per mile driven. Judy recalls that her first trip lasted 4 days, she drove 450 miles, and the rental cost her $240.50. On her second business trip she drove 200 miles in 3 days, and paid $146.00 for the rental. Find the daily fee and the mileage charge.

24. Joan Gundersen rented a car from Hertz, which rents its cars for a daily fee plus an additional charge per mile driven. Joan recalls that a car rented for 5 days and driven for 300 miles cost her $178, while a car rented for 4 days and driven for 500 miles cost $197. Find the daily fee, and find the mileage charge.

25. Pratap Puri rowed 18 miles down the Delaware River in 2 hours, but the return trip took him $4\frac{1}{2}$ hours. Find the rate Pratap can row in still water, and find the rate of the current.

Let x = rate Pratap can row in still water and
     y = rate of the current

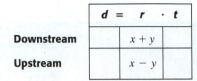

	d =	r ·	t
Downstream		$x + y$	
Upstream		$x - y$	

26. The Jonathan Schultz family took a canoe 10 miles down the Allegheny River in $1\frac{1}{4}$ hours. After lunch it took them 4 hours to return. Find the rate of the current.

Let x = rate the family can row in still water and
     y = rate of the current

	d =	r ·	t
Downstream		$x + y$	
Upstream		$x - y$	

27. Dave and Sandy Hartranft are frequent flyers with Delta Airlines. They often fly from Philadelphia to Chicago, a distance of 780 miles. On one particular trip they fly into the wind, and the flight takes 2 hours. The return trip, with the wind behind them, only takes $1\frac{1}{2}$ hours. Find the speed of the wind and find the speed of the plane in still air.

28. With a strong wind behind it, a United Airlines jet flies 2400 miles from Los Angeles to Orlando in $4\frac{3}{4}$ hours. The return trip takes 6 hours, as the plane flies into the wind. Find the speed of the plane in still air, and find the wind speed to the nearest tenth of a mile per hour.

29. Jim Williamson began a 186-mile bicycle trip to build up stamina for a triathlete competition. Unfortunately, his bicycle chain broke, so he finished the trip walking. The whole trip took 6 hours. If Jim walks at a rate of 4 miles per hour and rides at 40 miles per hour, find the amount of time he spent on the bicycle.

30. In Canada, eastbound and westbound trains travel along the same track, with sidings to pull onto to avoid accidents. Two trains are now 150 miles apart, with the westbound train traveling twice as fast as the eastbound train. A warning must be issued to pull one train onto a siding or else the trains will crash in $1\frac{1}{4}$ hours. Find the speed of the eastbound train and the speed of the westbound train.

31. Dorren Schmidt is a chemist with Gemco Pharmaceutical. She needs to prepare 12 ounces of a 9% hydrochloric acid solution. Find the amount of a 4% solution and the amount of a 12% solution she should mix to get this solution.

Concentration Rate	Liters of Solution	Liters of Pure Acid
0.04	x	0.04x
0.12	y	?
0.09	12	?

32. Elise Everly is preparing 15 liters of a 25% saline solution. Elise has two other saline solutions with strengths of 40% and 10%. Find the amount of 40% solution and the amount of 10% solution she should mix to get 15 liters of a 25% solution.

Concentration Rate	Liters of Solution	Liters of Pure Salt
0.40	x	0.40x
0.10	y	?
0.25	15	?

33. Wayne Osby blends coffee for a local coffee café. He needs to prepare 200 pounds of blended coffee beans selling for $3.95 per pound. He intends to do this by blending together a high-quality bean costing $4.95 per pound and a cheaper bean costing $2.65 per pound. To the nearest pound, find how much high-quality coffee bean and how much cheaper coffee bean he should blend.

34. Macadamia nuts cost an astounding $16.50 per pound, but research by an independent firm says that mixed nuts sell better if macadamias are included. The standard mix costs $9.25 per pound. Find how many pounds of macadamias and how many pounds of the standard mix should be combined to produce 40 pounds that will cost $10 per pound. Find the amounts to the nearest tenth of a pound.

35. Recall that two angles are complementary if the sum of their measures is 90°. Find the measures of two complementary angles if one angle is twice the other.

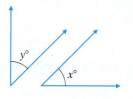

36. Recall that two angles are supplementary if the sum of their measures is 180°. Find the measures of two supplementary angles if one angle is 20° more than four times the other.

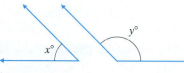

37. Find the measures of two complementary angles if one angle is 10° more than three times the other.

38. Find the measures of two supplementary angles if one angle is 18° more than twice the other.

39. Kathi and Robert Hawn had a pottery stand at the annual Skippack Craft Fair. They sold some of their pottery at the original price of $9.50 each, but later decreased the price of each by $2. If they sold all 90 pieces and took in $721, find how many they sold at the original price and how many they sold at the reduced price.

40. A charity fund-raiser consisted of a spaghetti supper where a total of 387 people were fed. They charged $6.80 for adults and half-price for children. If they took in $2444.60, find how many adults and how many children attended the supper.

41. The Santa Fe National Historic Trail is approximately 1200 miles between Old Franklin, Missouri, and Santa Fe, New Mexico. Suppose that a group of hikers start from each town and walk the trail toward each other. They meet after a total hiking time of 240 hours. If one group travels $\frac{1}{2}$ mile per hour slower than the other group, find the rate of each group. (*Source:* National Park Service)

42. California 1 South is a historic highway that stretches 123 miles along the coast from Monterey to Morro Bay. Suppose that two cars start driving this highway, one from each town. They meet after 3 hours. Find the rate of each car if one car travels 1 mile per hour faster than the other car. (*Source:* National Geographic)

43. A 30% solution of fertilizer is to be mixed with a 60% solution of fertilizer in order to get 150 gallons of a 50% solution. How many gallons of the 30% solution and 60% solution should be mixed?

44. A 10% acid solution is to be mixed with a 50% acid solution in order to get 120 ounces of a 20% acid solution. How many ounces of the 10% solution and 50% solution should be mixed?

45. Traffic signs are regulated by the *Manual on Uniform Traffic Control Devices* (MUTCD). According to this manual, if the sign below is placed on a freeway, its perimeter must be 144 inches. Also, its length is 12 inches longer than its width. Find the dimensions of this sign.

46. According to the MUTCD (see Exercise 45), this sign must have a perimeter of 60 inches. Also, its length must be 6 inches longer than its width. Find the perimeter of this sign.

Review

Find the square of each expression. For example, the square of 7 is 7^2 or 49. The square of 5x is $(5x)^2$ or $25x^2$. See Section 3.1.

47. 4 **48.** 3 **49.** $6x$ **50.** $11y$ **51.** $10y^3$ **52.** $8x^5$

Concept Extensions

Solve. See the Concept Check in the section.

53. Suppose you mix an amount of candy costing $0.49 a pound with candy costing $0.65 a pound. Which of the following costs per pound could result?

   a. $0.58   **b.** $0.72   **c.** $0.29

54. Suppose you mix a 50% acid solution with pure acid (100%). Which of the following acid strengths are possible for the resulting acid mixture?

   a. 25%   **b.** 150%   **c.** 62%   **d.** 90%

△ **55.** Dale and Sharon Mahnke have decided to fence off a garden plot behind their house, using their house as the "fence" along one side of the garden. The length (which runs parallel to the house) is 3 feet less than twice the width. Find the dimensions if 33 feet of fencing is used along the three sides requiring it.

△ **56.** Judy McElroy plans to erect 152 feet of fencing around her rectangular horse pasture. A river bank serves as one side length of the rectangle. If each width is 4 feet longer than half the length, find the dimensions.

Break-Even Point

Sections 7.1, 7.2, 7.3, 7.4

When a business sells a new product, it generally does not start making a profit right away. There are usually many expenses associated with creating a new product. These expenses might include an advertising blitz to introduce the product to the public. These start-up expenses might also include the cost of market research and product development or any brand-new equipment needed to manufacture the product. Start-up costs like these are generally called *fixed costs* because they don't depend on the number of items manufactured. Expenses that depend on the number of items manufactured, such as the cost of materials and shipping, are called *variable costs*. The total cost of manufacturing the new product is given by the cost equation: Total cost = Fixed costs + Variable costs.

For instance, suppose a greeting card company is launching a new line of greeting cards. The company spent $7000 doing product research and development for the new line and spent $15,000 on advertising the new line. The company does not need to buy any new equipment to manufacture the cards, but the paper and ink needed to make each card will cost $0.20 per card. The total cost y in dollars for manufacturing x cards is $y = 22{,}000 + 0.20x$.

Once a business sets a price for the new product, the company can find the product's expected *revenue*. Revenue is the amount of money the company takes in from the sales of its product. The revenue from selling a product is given by the revenue equation: Revenue = Price per item × Number of items sold.

For instance, suppose that the card company plans to sell its new cards for $1.50 each. The revenue y, in dollars, that the company can expect to receive from the sales of x cards is $y = 1.50x$.

If the total cost and revenue equations are graphed on the same coordinate system, the graphs should intersect. The point of intersection is where total cost equals revenue and is called the *break-even point*. The break-even point gives the number of items x that must be manufactured and sold for the company to recover its expenses. If fewer than this number of items are produced and sold, the company loses money. If more than this number of items are produced and sold, the company makes a profit. In the case of the greeting card company, approximately 16,923 cards must be manufactured and sold for the company to break even on this new card line. The total cost and revenue of producing and selling 16,923 cards is the same. It is approximately $25,385.

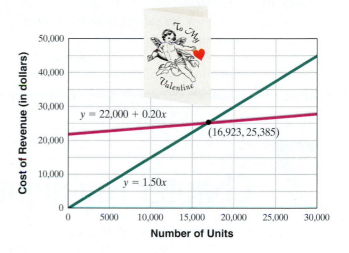

Group Activity

Suppose your group is starting a small business near your campus.

a. Choose a business and decide what campus-related product or service you will provide.

b. Research the fixed costs of starting up such a business.

c. Research the variable costs of producing such a product or providing such a service.

d. Decide how much you would charge per unit of your product or service.

e. Find a system of equations for the total cost and revenue of your product or service.

f. How many units of your product or service must be sold before your business will break even?

Chapter 7 Vocabulary Check

Fill in each blank with one of the words or phrases listed below.

system of linear equations   solution   consistent   independent

dependent   inconsistent   substitution   addition

1. In a system of linear equations in two variables, if the graphs of the equations are the same, the equations are _____ equations.

2. Two or more linear equations are called a _____.

3. A system of equations that has at least one solution is called a(n) _____ system.

4. A _____ of a system of two equations in two variables is an ordered pair of numbers that is a solution of both equations in the system.

5. Two algebraic methods for solving systems of equations are _____ and _____.

6. A system of equations that has no solution is called a(n) _____ system.

7. In a system of linear equations in two variables, if the graphs of the equations are different, the equations are _____ equations.

> **Helpful Hint**
>
> Are you preparing for your test? Don't forget to take the Chapter 7 Test on page 562. Then check your answers at the back of the text and use the Chapter Test Prep Video CD to see the fully worked-out solutions to any of the exercises you want to review.

7 Chapter Highlights

DEFINITIONS AND CONCEPTS	EXAMPLES
Section 7.1 Solving Systems of Linear Equations by Graphing	

A **system of linear equations** consists of two or more linear equations.	$\begin{cases} 2x + y = 6 \\ x = -3y \end{cases}$ $\begin{cases} -3x + 5y = 10 \\ x - 4y = -2 \end{cases}$
A **solution** of a system of two equations in two variables is an ordered pair of numbers that is a solution of both equations in the system.	Determine whether $(-1, 3)$ is a solution of the system. $$\begin{cases} 2x - y = -5 \\ x = 3y - 10 \end{cases}$$ Replace x with -1 and y with 3 in both equations. $$2x - y = -5$$ $$2(-1) - 3 \stackrel{?}{=} -5$$ $$-5 = -5 \quad \text{True}$$ $$x = 3y - 10$$ $$-1 \stackrel{?}{=} 3(3) - 10$$ $$-1 = -1 \quad \text{True}$$ $(-1, 3)$ is a solution of the system.
Graphically, a solution of a system is a point common to the graphs of both equations.	Solve by graphing: $\begin{cases} 3x - 2y = -3 \\ x + y = 4 \end{cases}$

DEFINITIONS AND CONCEPTS	**EXAMPLES**

Section 7.1 Solving Systems of Linear Equations by Graphing (*continued*)

Three different situations can occur when graphing the two lines associated with the equations in a linear system.

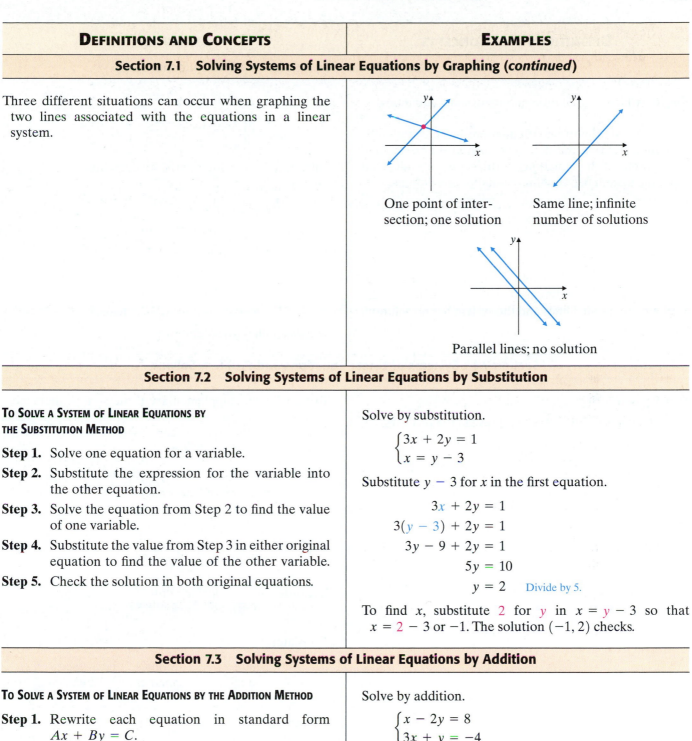

One point of intersection; one solution

Same line; infinite number of solutions

Parallel lines; no solution

Section 7.2 Solving Systems of Linear Equations by Substitution

TO SOLVE A SYSTEM OF LINEAR EQUATIONS BY THE SUBSTITUTION METHOD

Step 1. Solve one equation for a variable.

Step 2. Substitute the expression for the variable into the other equation.

Step 3. Solve the equation from Step 2 to find the value of one variable.

Step 4. Substitute the value from Step 3 in either original equation to find the value of the other variable.

Step 5. Check the solution in both original equations.

Solve by substitution.

$$\begin{cases} 3x + 2y = 1 \\ x = y - 3 \end{cases}$$

Substitute $y - 3$ for x in the first equation.

$$3x + 2y = 1$$
$$3(y - 3) + 2y = 1$$
$$3y - 9 + 2y = 1$$
$$5y = 10$$
$$y = 2 \quad \text{Divide by 5.}$$

To find x, substitute 2 for y in $x = y - 3$ so that $x = 2 - 3$ or -1. The solution $(-1, 2)$ checks.

Section 7.3 Solving Systems of Linear Equations by Addition

TO SOLVE A SYSTEM OF LINEAR EQUATIONS BY THE ADDITION METHOD

Step 1. Rewrite each equation in standard form $Ax + By = C$.

Step 2. Multiply one or both equations by a nonzero number so that the coefficients of a variable are opposites.

Step 3. Add the equations.

Step 4. Find the value of one variable by solving the resulting equation.

Step 5. Substitute the value from Step 4 into either original equation to find the value of the other variable.

Solve by addition.

$$\begin{cases} x - 2y = 8 \\ 3x + y = -4 \end{cases}$$

Multiply both sides of the first equation by -3.

$$\begin{cases} -3x + 6y = -24 \\ \underline{3x + y = -4} \end{cases}$$
$$7y = -28 \quad \text{Add.}$$
$$y = -4 \quad \text{Divide by 7.}$$

To find x, let $y = -4$ in an original equation.

$$x - 2(-4) = 8 \quad \text{First equation}$$
$$x + 8 = 8$$
$$x = 0$$

continued

DEFINITIONS AND CONCEPTS	**EXAMPLES**

Section 7.3   Solving Systems of Linear Equations by Addition (*continued*)

Step 6. Check the solution in both original equations.

If solving a system of linear equations by substitution or addition yields a true statement such as $-2 = -2$, then the graphs of the equations in the system are identical and the system has an infinite number of solutions.

The solution $(0, -4)$ checks.

Solve: $\begin{cases} 2x - 6y = -2 \\ x = 3y - 1 \end{cases}$

Substitute $3y - 1$ for x in the first equation.

$$2(3y - 1) - 6y = -2$$
$$6y - 2 - 6y = -2$$
$$-2 = -2 \quad \text{True}$$

The system has an infinite number of solutions.

If solving a system of linear equations yields a false statement such as $0 = 3$, the graphs of the equations in the system are parallel lines and the system has no solution.

Solve: $\begin{cases} 5x - 2y = 6 \\ -5x + 2y = -3 \end{cases}$

$$0 = 3 \quad \text{False}$$

The system has no solution.

Section 7.4   Systems of Linear Equations and Problem Solving

PROBLEM-SOLVING STEPS

1. UNDERSTAND. Read and reread the problem.

Two angles are supplementary if the sum of their measures is $180°$. The larger of two supplementary angles is three times the smaller, decreased by twelve. Find the measure of each angle. Let

x = measure of smaller angle and

y = measure of larger angle

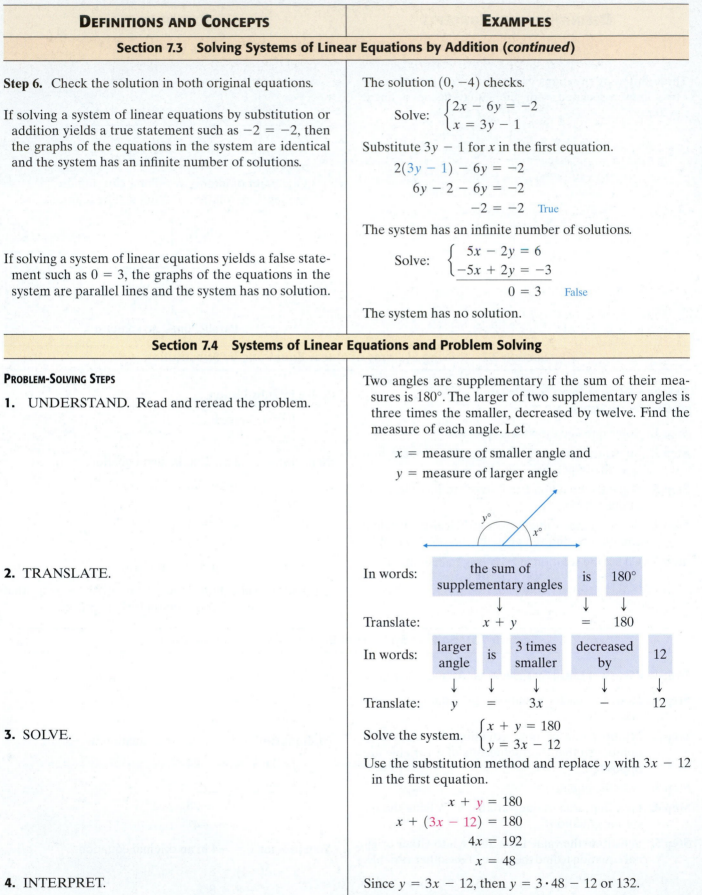

2. TRANSLATE.

In words: the sum of supplementary angles   is $180°$

Translate: $x + y$ $=$ 180

In words: larger angle   is   3 times smaller   decreased by   12

Translate: y $=$ $3x$ $-$ 12

3. SOLVE.

Solve the system. $\begin{cases} x + y = 180 \\ y = 3x - 12 \end{cases}$

Use the substitution method and replace y with $3x - 12$ in the first equation.

$$x + y = 180$$
$$x + (3x - 12) = 180$$
$$4x = 192$$
$$x = 48$$

4. INTERPRET.

Since $y = 3x - 12$, then $y = 3 \cdot 48 - 12$ or 132.

The solution checks. The smaller angle measures $48°$ and the larger angle measures $132°$.

Are You Prepared for a Test on Chapter 7?

Below I have listed some common trouble areas for students in Chapter 7. After studying for your test—but before taking your test—read these.

- If you are having trouble drawing a neat graph, remember to ask your instructor if you can use graph paper on your test. This will save you time and keep your graphs neat.
- Do you remember how to check solutions of systems of equations? If $(-1, 5)$ is a solution of the system

$$\begin{cases} 3x - y = -8 \\ -x + y = 6 \end{cases}$$

then the ordered pair will make *both* equations a true statement.

$$3x - y = -8$$
$$3(-1) - 5 = -8 \quad \text{Let } x = -1 \text{ and } y = 5.$$
$$-8 = -8 \quad \text{True}$$

$$-x + y = 6$$
$$-(-1) + 5 = 6 \quad \text{Let } x = -1 \text{ and } y = 5.$$
$$6 = 6 \quad \text{True}$$

Remember: This is simply a list of a few common trouble areas. For a review of Chapter 7, see the Highlights and Chapter Review at the end of this chapter.

7 CHAPTER REVIEW

(7.1) *Determine whether each ordered pair is a solution of the system of linear equations.*

1. $\begin{cases} 2x - 3y = 12 \\ 3x + 4y = 1 \end{cases}$

   a. $(12, 4)$
   b. $(3, -2)$

2. $\begin{cases} 4x + y = 0 \\ -8x - 5y = 9 \end{cases}$

   a. $\left(\dfrac{3}{4}, -3\right)$
   b. $(-2, 8)$

3. $\begin{cases} 5x - 6y = 18 \\ 2y - x = -4 \end{cases}$

   a. $(-6, -8)$
   b. $\left(3, \dfrac{5}{2}\right)$

4. $\begin{cases} 2x + 3y = 1 \\ 3y - x = 4 \end{cases}$

   a. $(2, 2)$
   b. $(-1, 1)$

Solve each system of equations by graphing.

5. $\begin{cases} x + y = 5 \\ x - y = 1 \end{cases}$

6. $\begin{cases} x + y = 3 \\ x - y = -1 \end{cases}$

7. $\begin{cases} x = 5 \\ y = -1 \end{cases}$

8. $\begin{cases} x = -3 \\ y = 2 \end{cases}$

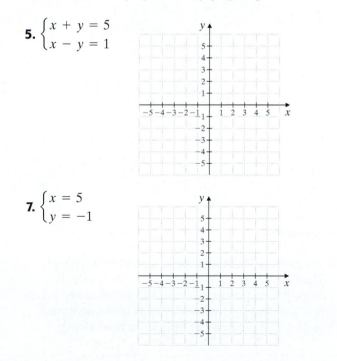
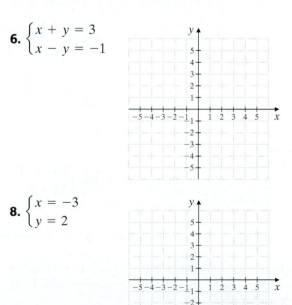

559

9. $\begin{cases} 2x + y = 5 \\ x = -3y \end{cases}$

10. $\begin{cases} 3x + y = -2 \\ y = -5x \end{cases}$

11. $\begin{cases} y = 3x \\ -6x + 2y = 6 \end{cases}$

12. $\begin{cases} x - 2y = 2 \\ -2x + 4y = -4 \end{cases}$

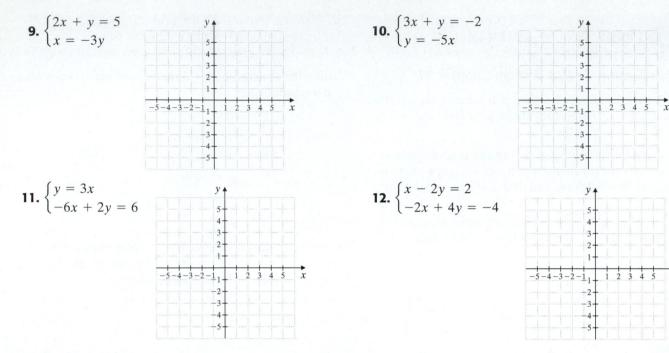

(7.2) *Solve each system of equations by the substitution method.*

13. $\begin{cases} y = 2x + 6 \\ 3x - 2y = -11 \end{cases}$

14. $\begin{cases} y = 3x - 7 \\ 2x - 3y = 7 \end{cases}$

15. $\begin{cases} x + 3y = -3 \\ 2x + y = 4 \end{cases}$

16. $\begin{cases} 3x + y = 11 \\ x + 2y = 12 \end{cases}$

17. $\begin{cases} 4y = 2x + 6 \\ x - 2y = -3 \end{cases}$

18. $\begin{cases} 9x = 6y + 3 \\ 6x - 4y = 2 \end{cases}$

19. $\begin{cases} x + y = 6 \\ y = -x - 4 \end{cases}$

20. $\begin{cases} -3x + y = 6 \\ y = 3x + 2 \end{cases}$

(7.3) *Solve each system of equations by the addition method.*

21. $\begin{cases} 2x + 3y = -6 \\ x - 3y = -12 \end{cases}$

22. $\begin{cases} 4x + y = 15 \\ -4x + 3y = -19 \end{cases}$

23. $\begin{cases} 2x - 3y = -15 \\ x + 4y = 31 \end{cases}$

24. $\begin{cases} x - 5y = -22 \\ 4x + 3y = 4 \end{cases}$

25. $\begin{cases} 2x - 6y = -1 \\ -x + 3y = \dfrac{1}{2} \end{cases}$

26. $\begin{cases} 0.6x - 0.3y = -1.5 \\ 0.04x - 0.02y = -0.1 \end{cases}$

27. $\begin{cases} \dfrac{3}{4}x + \dfrac{2}{3}y = 2 \\ x + \dfrac{y}{3} = 6 \end{cases}$

28. $\begin{cases} 10x + 2y = 0 \\ 3x + 5y = 33 \end{cases}$

(7.4) *Solve each problem by writing and solving a system of linear equations.*

29. The sum of two numbers is 16. Three times the larger number decreased by the smaller number is 72. Find the two numbers.

30. The Forrest Theater can seat a total of 360 people. They take in $15,150 when every seat is sold. If orchestra section tickets cost $45 and balcony tickets cost $35, find the number of seats in the orchestra section and the number of seats in the balcony.

31. A riverboat can head 340 miles upriver in 19 hours, but the return trip takes only 14 hours. Find the current of the river and find the speed of the riverboat in still water to the nearest tenth of a mile.

	d =	r ·	t
Upriver		$x - y$	
Downriver		$x + y$	

32. Find the amount of a 6% acid solution and the amount of a 14% acid solution Pat Mayfield should combine to prepare 50 cc (cubic centimeters) of a 12% solution.

33. A deli charges $3.80 for a breakfast of three eggs and four strips of bacon. The charge is $2.75 for two eggs and three strips of bacon. Find the cost of each egg and the cost of each strip of bacon.

34. An exercise enthusiast alternates between jogging and walking. He traveled 15 miles during the past 3 hours. He jogs at a rate of 7.5 miles per hour and walks at a rate of 4 miles per hour. Find how much time, to the nearest hundredth of an hour, he actually spent jogging and how much time he spent walking.

Mixed Review

Solve each system of equations by graphing.

35. $\begin{cases} x - 2y = 1 \\ 2x + 3y = -12 \end{cases}$

36. $\begin{cases} 3x - y = -4 \\ 6x - 2y = -8 \end{cases}$

Solve each system of equations.

37. $\begin{cases} x + 4y = 11 \\ 5x - 9y = -3 \end{cases}$

38. $\begin{cases} x + 9y = 16 \\ 3x - 8y = 13 \end{cases}$

39. $\begin{cases} y = -2x \\ 4x + 7y = -15 \end{cases}$

40. $\begin{cases} 3y = 2x + 15 \\ -2x + 3y = 21 \end{cases}$

41. $\begin{cases} 3x - y = 4 \\ 4y = 12x - 16 \end{cases}$

42. $\begin{cases} x + y = 19 \\ x - y = -3 \end{cases}$

43. $\begin{cases} x - 3y = -11 \\ 4x + 5y = -10 \end{cases}$

44. $\begin{cases} -x - 15y = 44 \\ 2x + 3y = 20 \end{cases}$

45. $\begin{cases} 2x + y = 3 \\ 6x + 3y = 9 \end{cases}$

46. $\begin{cases} -3x + y = 5 \\ -3x + y = -2 \end{cases}$

Solve each problem by writing and solving a system of linear equations.

47. The sum of two numbers is 12. Three times the smaller number increased by the larger number is 20. Find the numbers.

48. The difference of two numbers is −18. Twice the smaller decreased by the larger is −23. Find the two numbers.

49. Emma Hodges has a jar containing 65 coins, all of which are either nickels or dimes. The total value of the coins is $5.30. How many of each type does she have?

50. Sarah and Owen Hebert purchased 25 stamps, a mixture of 13¢ and 22¢ stamps. Find the number of each type of stamp if they spent $4.19.

Remember to use the Chapter Test Prep Video CD to see the fully worked-out solutions to any of the exercises you want to review.

Answers

Answer each question true or false.

1. A system of two linear equations in two variables can have exactly two solutions.

2. Although $(1, 4)$ is not a solution of $x + 2y = 6$, it can still be a solution of the system $\begin{cases} x + 2y = 6 \\ x + y = 5 \end{cases}$

3. If the two equations in a system of linear equations are added and the result is $3 = 0$, the system has no solution.

4. If the two equations in a system of linear equations are added and the result is $3x = 0$, the system has no solution.

Is the ordered pair a solution of the given linear system?

5. $\begin{cases} 2x - 3y = 5 \\ 6x + y = 1 \end{cases}; (1, -1)$

6. $\begin{cases} 4x - 3y = 24 \\ 4x + 5y = -8 \end{cases}; (3, -4)$

Solve each system by graphing.

7. $\begin{cases} x - y = 2 \\ 3x - y = -2 \end{cases}$

8. $\begin{cases} y = -3x \\ 3x + y = 6 \end{cases}$

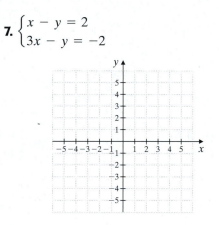

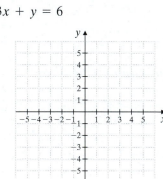

Solve each system by the substitution method.

9. $\begin{cases} 3x - 2y = -14 \\ y = x + 5 \end{cases}$

10. $\begin{cases} \dfrac{1}{2}x + 2y = -\dfrac{15}{4} \\ 4x = -y \end{cases}$

Solve each system by the addition method.

11. $\begin{cases} x + y = 28 \\ x - y = 12 \end{cases}$

12. $\begin{cases} 4x - 6y = 7 \\ -2x + 3y = 0 \end{cases}$

Solve each system using the substitution method or the addition method.

13. $\begin{cases} 3x + y = 7 \\ 4x + 3y = 1 \end{cases}$

14. $\begin{cases} 3(2x + y) = 4x + 20 \\ x - 2y = 3 \end{cases}$

1. _____

2. _____

3. _____

4. _____

5. _____

6. _____

7. see graph _____

8. see graph _____

9. _____

10. _____

11. _____

12. _____

13. _____

14. _____

15.
$$\begin{cases} \dfrac{x-3}{2} = \dfrac{2-y}{4} \\ \dfrac{7-2x}{3} = \dfrac{y}{2} \end{cases}$$

16.
$$\begin{cases} 8x - 4y = 12 \\ y = 2x - 3 \end{cases}$$

17.
$$\begin{cases} 0.01x - 0.06y = -0.23 \\ 0.2x + 0.4y = 0.2 \end{cases}$$

18.
$$\begin{cases} x - \dfrac{2}{3}y = 3 \\ -2x + 3y = 10 \end{cases}$$

Solve each problem by writing and using a system of linear equations.

19. Two numbers have a sum of 124 and a difference of 32. Find the numbers.

20. Find the amount of a 12% saline solution a lab assistant should add to 80 cc (cubic centimeters) of a 22% saline solution in order to have a 16% solution.

21. Although the number of farms in the United States is still decreasing, small farms are making a comeback. Texas and Missouri are the states with the most number of farms. Texas has 116 thousand more farms than Missouri and the total number of farms for these two states is 336 thousand. Find the number of farms for each state.

22. Two hikers start at opposite ends of the St. Tammany Trails and walk toward each other. The trail is 36 miles long and they meet in 4 hours. If one hiker is twice as fast as the other, find both hiking speeds.

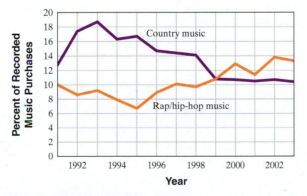

The graph below shows the percent of recorded music purchases that fell within the rap/hip-hop or country music genres for the years shown. Use this graph to answer Exercises 23 and 24.

Source: Recording Industry Association of America

23. In what year were purchases of country music equal to purchases of rap/hip-hop music?

24. In what year(s) were there more purchases of country music than rap/hip-hop music?

15. _____

16. _____

17. _____

18. _____

19. _____

20. _____

21. _____

22. _____

23. _____

24. _____

Answers

1. a. _____

  b. _____

2. a. _____

  b. _____

3. _____

4. _____

5. _____

6. _____

7. _____

8. _____

9. _____

10. _____

11. _____

12. _____

13. a. _____

  b. _____

14. _____

15. _____

16. _____

17. _____

18. _____

1. Simplify each expression.
   a. $-14 - 8 + 10 - (-6)$
   b. $1.6 - (-10.3) + (-5.6)$

2. Evaluate:
   a. 5^2
   b. 2^5

Find the reciprocal or opposite of each number.

3. reciprocal of 22

4. opposite of 22

5. reciprocal of $\dfrac{3}{16}$

6. opposite of $\dfrac{3}{16}$

7. reciprocal of -10

8. opposite of -10

9. reciprocal of $-\dfrac{9}{13}$

10. opposite of $-\dfrac{9}{13}$

11. reciprocal of 1.7

12. opposite of 1.7

13. a. The sum of two numbers is 8. If one number is 3, find the other number.
   b. The sum of two numbers is 8. If one number is x, write an expression representing the other number.

14. Five times the sum of a number and -1 is the same as 6 times the number. Find the number.

15. Solve:
   $-2(x - 5) + 10 = -3(x + 2) + x$

16. Solve: $5(y - 5) = 5y + 10$

17. Solve: $\dfrac{x}{2} - 1 = \dfrac{2}{3}x - 3$

18. Solve: $7(x - 2) - 6(x + 1) = 20$

564

19. Solve $-5x + 7 < 2(x - 3)$. Graph the solution set.

20. Solve $P = a + b + c$ for b.

Simplify each expression.

21. $\left(\dfrac{m}{n}\right)^7$

22. $\dfrac{a^7 b^{10}}{ab^{15}}$

23. $\left(\dfrac{2x^4}{3y^5}\right)^4$

24. $(7a^2 b^{-3})^2$

25. Subtract: $(2x^3 + 8x^2 - 6x) - (2x^3 - x^2 + 1)$

26. Add: $\left(5x^2 + 6x + \dfrac{1}{2}\right) + \left(x^2 - \dfrac{4}{3}x - \dfrac{10}{21}\right)$

27. Divide $6x^2 + 10x - 5$ by $3x - 1$.

28. Find the GCF of $9x^2$, $6x^3$, and $21x^5$.

29. Solve: $x(2x - 7) = 4$

30. Solve: $x(2x - 7) = 0$

△ **31.** Find the lengths of the sides of a right triangle if the lengths can be expressed by three consecutive even integers.

△ **32.** The height of a parallelogram is 5 feet more than three times its base. If the area of the parallelogram is 182 square feet, find the length of its base and height.

33. Subtract: $\dfrac{2y}{2y - 7} - \dfrac{7}{2y - 7}$

34. Add: $\dfrac{2}{x - 6} + \dfrac{3}{x + 1}$

35. Simplify: $\dfrac{\dfrac{x}{y} + \dfrac{3}{2x}}{\dfrac{x}{2} + y}$

36. Find the slope of the line parallel to the line passing through $(-1, 3)$ and $(2, -8)$.

37. Find the slope of the line $y = -1$.

19. _____

20. _____

21. _____

22. _____

23. _____

24. _____

25. _____

26. _____

27. _____

28. _____

29. _____

30. _____

31. _____

32. _____

33. _____

34. _____

35. _____

36. _____

37. _____

38. _____

39. _____

40. _____

41. _____

42. _____

43. _____

44. a. _____

   b. _____

   c. _____

45. _____

46. _____

47. _____

48. _____

49. _____

50. a. _____

   b. _____

   c. _____

38. Find the slope of the line $x = 2$.

39. Find an equation of the line through $(2, 5)$ and $(-3, 4)$. Write the equation in the form $Ax + By = C$.

40. Write an equation of the line with slope -5, through $(-2, 3)$.

41. Find the domain and the range of the relation
$\{(0, 2), (3, 3), (-1, 0), (3, -2)\}$.

42. If $f(x) = 5x^2 - 6$, find $f(0)$ and $f(-2)$.

43. Determine whether $(12, 6)$ is a solution of the system $\begin{cases} 2x - 3y = 6 \\ x = 2y \end{cases}$

44. Determine whether each ordered pair is a solution of the given system.
$\begin{cases} 2x - y = 6 \\ 3x + 2y = -5 \end{cases}$
a. $(1, -4)$
b. $(0, 6)$
c. $(3, 0)$

45. Solve the system: $\begin{cases} x + 2y = 7 \\ 2x + 2y = 13 \end{cases}$

Solve each system.

46. $\begin{cases} 3x - 4y = 10 \\ \quad\;\; y = 2x \end{cases}$

47. Solve the system: $\begin{cases} -x - \dfrac{y}{2} = \dfrac{5}{2} \\ \dfrac{x}{6} - \dfrac{y}{2} = 0 \end{cases}$

48. $\begin{cases} x = 5y - 3 \\ x = 8y + 4 \end{cases}$

49. Find two numbers whose sum is 37 and whose difference is 21.

50. Determine which graph(s) are graphs of functions.

a.   **b.**   **c.**

8

Roots and Radicals

Having spent the last chapter studying equations, we return now to algebraic expressions. We expand on our skills of operating on expressions—adding, subtracting, multiplying, dividing, and raising to powers—to include finding roots. Just as subtraction is defined by addition and division by multiplication, finding roots is defined by raising to powers. As we master finding roots, we will work with equations that contain roots and solve problems that can be modeled by such equations.

When we think of pendulums, we often think of grandfather clocks. In fact, pendulums can be used to provide accurate timekeeping. But, did you know that pendulums can also be used to demonstrate that the earth rotates on its axis?

In 1851, French physicist Léon Foucault developed a special pendulum in an experiment to demonstrate that the Earth rotated on its axis. He connected his tall pendulum, capable of running for many hours, to the roof of the Paris Observatory. The pendulum's bob was able to swing back and forth in one plane, but not to twist in other directions. So, when the pendulum bob appeared to move in a circle over time, he demonstrated that it was not the pendulum but the building that moved. And since the building was firmly attached to the earth, it must be the earth rotating which created the apparent circular motion of the bob. In Section 8.1, Exercise 85 on page 574, roots are used to explore the time it takes Foucault's pendulum to complete one swing of its bob.

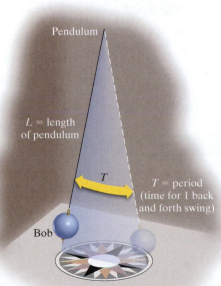

Pendulum

L = length of pendulum

T

T = period (time for 1 back and forth swing)

Bob

567

8.1 INTRODUCTION TO RADICALS

Objective **A** Finding Square Roots

In this section, we define finding the **root** of a number by its reverse operation, raising a number to a power. We begin with squares and square roots.

The *square* of 5 is $5^2 = 25$.

The *square* of -5 is $(-5)^2 = 25$

The *square* of $\frac{1}{2}$ is $\left(\frac{1}{2}\right)^2 = \frac{1}{4}$.

The reverse operation of squaring a number is finding a **square root** of a number. For example,

A *square root* of 25 is 5, because $5^2 = 25$.

A *square root* of 25 is also -5, because $(-5)^2 = 25$.

A *square root* of $\frac{1}{4}$ is $\frac{1}{2}$, because $\left(\frac{1}{2}\right)^2 = \frac{1}{4}$.

> In general, the number b is a square root of a number a if $b^2 = a$.

The symbol $\sqrt{}$ is used to denote the **positive** or **principal square root** of a number. For example,

$\sqrt{25} = 5$ only, since $5^2 = 25$ and 5 is positive.

The symbol $-\sqrt{}$ is used to denote the **negative square root**. For example,

$-\sqrt{25} = -5$

The symbol $\sqrt{}$ is called a **radical** or **radical sign.** The expression within or under a radical sign is called the **radicand.** An expression containing a radical is called a **radical expression.**

radical sign

$\sqrt{a}$

radicand

Square Root

If a is a positive number, then

$\sqrt{a}$ is the **positive square root** of a and

$-\sqrt{a}$ is the **negative square root** of a.

Also, $\sqrt{0} = 0$.

PRACTICE PROBLEMS 1–5

Find each square root.

1. $\sqrt{100}$ **2.** $-\sqrt{36}$

3. $\sqrt{\dfrac{25}{81}}$ **4.** $\sqrt{1}$

5. $\sqrt{0.81}$

Answers

1. 10,   **2.** -6,   **3.** $\dfrac{5}{9}$,   **4.** 1,   **5.** 0.9

EXAMPLES   Find each square root.

1. $\sqrt{36} = 6$, because $6^2 = 36$ and 6 is positive.

2. $-\sqrt{64} = -8$. The negative sign in front of the radical indicates the negative square root of 64.

3. $\sqrt{\dfrac{9}{100}} = \dfrac{3}{10}$ because $\left(\dfrac{3}{10}\right)^2 = \dfrac{9}{100}$ and $\dfrac{3}{10}$ is positive.

4. $\sqrt{0} = 0$ because $0^2 = 0$.

5. $\sqrt{0.36} = 0.6$ because $(0.6)^2 = 0.36$ and 0.6 is positive.

■ **Work Practice Problems 1–5**

Is the square root of a negative number a real number? For example, is $\sqrt{-4}$ a real number? To answer this question, we ask ourselves, is there a real number whose square is -4? Since there is no real number whose square is -4, we say that $\sqrt{-4}$ is not a real number. In general,

---

A square root of a negative number is not a real number.

Study the following table to make sure you understand the differences discussed earlier.

Number	Square Roots of Number	$\sqrt{\text{number}}$	$-\sqrt{\text{number}}$
25	$-5, 5$	$\sqrt{25} = 5$ only	$-\sqrt{25} = -5$
$\dfrac{1}{4}$	$-\dfrac{1}{2}, \dfrac{1}{2}$	$\sqrt{\dfrac{1}{4}} = \dfrac{1}{2}$ only	$-\sqrt{\dfrac{1}{4}} = -\dfrac{1}{2}$
-9	No real square roots.	$\sqrt{-9}$ is not a real number.	

Objective B Finding Cube Roots

We can find roots other than square roots. For example, since $2^3 = 8$, we call 2 the **cube root** of 8. In symbols, we write

$\sqrt[3]{8} = 2$ The number 3 is called the **index.**

Also,

$\sqrt[3]{-64} = -4$ Since $(-4)^3 = -64$

Notice that unlike the square root of a negative number, the cube root of a negative number is a real number. This is so because while we cannot find a real number whose *square* is negative, we *can* find a real number whose *cube* is negative. In fact, the cube of a negative number is a negative number. Therefore, the cube root of a negative number is a negative number.

EXAMPLES   Find each cube root.

6. $\sqrt[3]{1} = 1$ because $1^3 = 1$.

7. $\sqrt[3]{-27} = -3$ because $(-3)^3 = -27$.

8. $\sqrt[3]{\dfrac{1}{125}} = \dfrac{1}{5}$ because $\left(\dfrac{1}{5}\right)^3 = \dfrac{1}{125}$.

🔲 **Work Practice Problems 6–8**

Objective C Finding *n*th Roots

Just as we can raise a real number to powers other than 2 or 3, we can find roots other than square roots and cube roots. In fact, we can take the *n*th root of a number where n is any natural number. An ***n*th root** of a number a is a number whose *n*th power is a.

In symbols, the *n*th root of a is written as $\sqrt[n]{a}$. Recall that n is called the **index.** The index 2 is usually omitted for square roots.

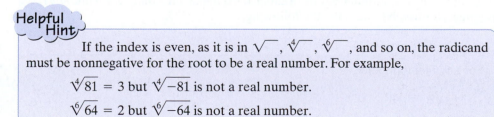

Helpful Hint

If the index is even, as it is in $\sqrt{}$, $\sqrt[4]{}$, $\sqrt[6]{}$, and so on, the radicand must be nonnegative for the root to be a real number. For example,

$\sqrt[4]{81} = 3$ but $\sqrt[4]{-81}$ is not a real number.

$\sqrt[6]{64} = 2$ but $\sqrt[6]{-64}$ is not a real number.

✔ **Concept Check**   Which of the following is a real number?

a. $\sqrt{-64}$ **b.** $\sqrt[4]{-64}$ **c.** $\sqrt[5]{-64}$ **d.** $\sqrt[6]{-64}$

EXAMPLES   Find each root.

9. $\sqrt[4]{16} = 2$ because $2^4 = 16$ and 2 is positive.

10. $\sqrt[5]{-32} = -2$ because $(-2)^5 = -32$.

11. $-\sqrt[6]{1} = -1$ because $\sqrt[6]{1} = 1$.

12. $\sqrt[4]{-81}$ is not a real number since the index 4 is even and the radicand -81 is negative. In other words, there is no real number that when raised to the 4th power gives -81.

☐ **Work Practice Problems 9–12**

PRACTICE PROBLEMS 9–12

Find each root.

9. $\sqrt[4]{-16}$

10. $\sqrt[5]{-1}$

11. $\sqrt[4]{81}$

12. $\sqrt[6]{-1}$

Objective D Approximating Square Roots

Recall that numbers such as 1, 4, 9, 25, and $\dfrac{4}{25}$ are called **perfect squares,** since

$1^2 = 1, 2^2 = 4, 3^2 = 9, 5^2 = 25,$ and $\left(\dfrac{2}{5}\right)^2 = \dfrac{4}{25}$. Square roots of perfect square radicands simplify to rational numbers.

What happens when we try to simplify a root such as $\sqrt{3}$? Since 3 is not a perfect square, $\sqrt{3}$ is not a rational number. It cannot be written as a quotient of integers. It is called an **irrational number** and we can find a decimal **approximation** of it. To find decimal approximations, use a calculator or Appendix F.1. (For calculator help, see the next example or the box at the end of this section.)

PRACTICE PROBLEM 13

Use a calculator or Appendix F.1 to approximate $\sqrt{22}$ to three decimal places.

EXAMPLE 13   Use a calculator or Appendix F.1 to approximate $\sqrt{3}$ to three decimal places.

Solution: We may use Appendix F.1 or a calculator to approximate $\sqrt{3}$. To use a calculator, find the square root key $\boxed{\sqrt{}}$.

$$\sqrt{3} \approx 1.732050808$$

To three decimal places, $\sqrt{3} \approx 1.732$.

☐ **Work Practice Problem 13**

Objective E Simplifying Radicals Containing Variables

Radicals can also contain variables. To simplify radicals containing variables, special care must be taken. To see how we simplify $\sqrt{x^2}$, let's look at a few examples in this form.

If $x = 3$, we have $\sqrt{3^2} = \sqrt{9} = 3$, or x.

If x is 5, we have $\sqrt{5^2} = \sqrt{25} = 5$, or x.

From these two examples, you may think that $\sqrt{x^2}$ simplifies to x. Let's now look at an example where x is a negative number. If $x = -3$, we have $\sqrt{(-3)^2} = \sqrt{9} = 3$, not -3, our original x. To make sure that $\sqrt{x^2}$ simplifies to a nonnegative number, we have the following.

Answers

9. not a real number,   **10.** -1,   **11.** 3,
12. not a real number,   **13.** 4.690

✔ **Concept Check Answer**

c

For any real number a,

$$\sqrt{a^2} = |a|.$$

Thus,

$$\sqrt{x^2} = |x|,$$
$$\sqrt{(-8)^2} = |-8| = 8$$
$$\sqrt{(7y)^2} = |7y|, \quad \text{and so on.}$$

To avoid this, for the rest of the chapter we assume that **if a variable appears in the radicand of a radical expression, it represents positive numbers only.** Then

$\sqrt{x^2} = |x| = x$ since x is a positive number.

$\sqrt{y^2} = y$ Because $(y)^2 = y^2$

$\sqrt{x^8} = x^4$ Because $(x^4)^2 = x^8$

$\sqrt{9x^2} = 3x$ Because $(3x)^2 = 9x^2$

EXAMPLES   Simplify each expression. Assume that all variables represent positive numbers.

14. $\sqrt{z^2} = z$ because $(z)^2 = z^2$.

15. $\sqrt{x^6} = x^3$ because $(x^3)^2 = x^6$.

16. $\sqrt[3]{27y^6} = 3y^2$ because $(3y^2)^3 = 27y^6$.

17. $\sqrt{16x^{16}} = 4x^8$ because $(4x^8)^2 = 16x^{16}$.

18. $\sqrt[3]{-125a^{12}b^{15}} = -5a^4b^5$ because $(-5a^4b^5)^3 = -125a^{12}b^{15}$.

🟧 **Work Practice Problems 14–18**

PRACTICE PROBLEMS 14–18

Simplify each expression. Assume that all variables represent positive numbers.

14. $\sqrt{z^8}$ **15.** $\sqrt{x^{20}}$

16. $\sqrt{4x^6}$ **17.** $\sqrt[3]{8y^{12}}$

18. $\sqrt[3]{-64x^9y^{24}}$

Answers

14. z^4, **15.** x^{10}, **16.** $2x^3$, **17.** $2y^4$, **18.** $-4x^3y^8$

📇 CALCULATOR EXPLORATIONS

To simplify or approximate square roots using a calculator, locate the key marked [√]. To simplify $\sqrt{25}$ using a scientific calculator, press [25] [√]. The display should read [5]. To simplify $\sqrt{25}$ using a graphing calculator, press [√] [25] [ENTER].

To approximate $\sqrt{30}$, press [30] [√] (or [√] [30]). The display should read [5.477225575]. This is an approximation for $\sqrt{30}$. A three-decimal-place approximation is

$$\sqrt{30} \approx 5.477$$

Is this answer reasonable? Since 30 is between perfect squares 25 and 36, $\sqrt{30}$ is between $\sqrt{25} = 5$ and $\sqrt{36} = 6$. The calculator result is then reasonable since 5.477225575 is between 5 and 6.

Use a calculator to approximate each expression to three decimal places. Decide whether each result is reasonable.

1. $\sqrt{6}$ **2.** $\sqrt{14}$

3. $\sqrt{11}$ **4.** $\sqrt{200}$

5. $\sqrt{82}$ **6.** $\sqrt{46}$

Many scientific calculators have a key, such as [ⁿ√y], that can be used to approximate roots other than square roots. To approximate these roots using a graphing calculator, look under the [MATH] menu or consult your manual. To use a [ⁿ√y] key to find $\sqrt[3]{8}$, press [3] [ⁿ√y] [8] (press [ENTER] if needed.) The display should read [2].

Use a calculator to approximate each expression to three decimal places. Decide whether each result is reasonable.

7. $\sqrt[3]{40}$ **8.** $\sqrt[3]{71}$

9. $\sqrt[4]{20}$ **10.** $\sqrt[4]{15}$

11. $\sqrt[5]{18}$ **12.** $\sqrt[6]{2}$

Objective **A** *Find each square root. See Examples 1 through 5.*

1. $\sqrt{16}$

2. $\sqrt{64}$

3. $\sqrt{\dfrac{1}{25}}$

4. $\sqrt{\dfrac{1}{64}}$

5. $-\sqrt{100}$

6. $-\sqrt{36}$

7. $\sqrt{-4}$

8. $\sqrt{-25}$

9. $-\sqrt{121}$

10. $-\sqrt{49}$

11. $\sqrt{\dfrac{9}{25}}$

12. $\sqrt{\dfrac{4}{81}}$

13. $\sqrt{900}$

14. $\sqrt{400}$

15. $\sqrt{144}$

16. $\sqrt{169}$

17. $\sqrt{\dfrac{1}{100}}$

18. $\sqrt{\dfrac{1}{121}}$

19. $\sqrt{0.25}$

20. $\sqrt{0.49}$

Objective **B** *Find each cube root. See Examples 6 through 8.*

21. $\sqrt[3]{125}$

22. $\sqrt[3]{64}$

23. $\sqrt[3]{-64}$

24. $\sqrt[3]{-27}$

25. $-\sqrt[3]{8}$

26. $-\sqrt[3]{27}$

27. $\sqrt[3]{\dfrac{1}{8}}$

28. $\sqrt[3]{\dfrac{1}{64}}$

29. $\sqrt[3]{-125}$

30. $\sqrt[3]{-1}$

Objectives **A** **B** **C** **Mixed Practice** *Find each root. See Examples 1 through 12.*

31. $\sqrt[5]{32}$

32. $\sqrt[4]{81}$

33. $\sqrt{81}$

34. $\sqrt{49}$

35. $\sqrt[4]{-16}$

36. $\sqrt{-9}$

37. $\sqrt[3]{-\dfrac{27}{64}}$

38. $\sqrt[3]{-\dfrac{8}{27}}$

39. $-\sqrt[4]{625}$

40. $-\sqrt[5]{32}$

41. $\sqrt[6]{1}$

42. $\sqrt[5]{1}$

Objective **D** *Approximate each square root to three decimal places. See Example 13.*

43. $\sqrt{7}$

44. $\sqrt{10}$

45. $\sqrt{37}$

46. $\sqrt{27}$

47. $\sqrt{136}$

48. $\sqrt{8}$

49. A standard baseball diamond is a square with 90-foot sides connecting the bases. The distance from home plate to second base is $90 \cdot \sqrt{2}$ feet. Approximate $\sqrt{2}$ to two decimal places and use your result to approximate the distance $90 \cdot \sqrt{2}$ feet.

50. The roof of the warehouse shown needs to be shingled. The total area of the roof is exactly $240 \cdot \sqrt{41}$ square feet. Approximate $\sqrt{41}$ to two decimal places and use your result to approximate the area $240 \cdot \sqrt{41}$ square feet. Approximate this area to the nearest whole number.

8 feet
20 feet
60 feet

90 ft
$90\sqrt{2}$ ft
90 ft

Objective **E** *Find each root. Assume that all variables represent positive numbers. See Examples 14 through 18.*

51. $\sqrt{m^2}$ **52.** $\sqrt{y^{10}}$ **53.** $\sqrt{x^4}$ **54.** $\sqrt{x^6}$

55. $\sqrt{9x^8}$ **56.** $\sqrt{36x^{12}}$ **57.** $\sqrt{81x^2}$ **58.** $\sqrt{100z^4}$

59. $\sqrt{a^2b^4}$ **60.** $\sqrt{x^{12}y^{20}}$ **61.** $\sqrt{16a^6b^4}$ **62.** $\sqrt{4m^{14}n^2}$

63. $\sqrt[3]{a^6b^{18}}$ **64.** $\sqrt[3]{x^{12}y^{18}}$ **65.** $\sqrt[3]{-8x^3y^{27}}$ **66.** $\sqrt[3]{-27a^6b^{30}}$

Review

Write each integer as a product of two integers such that one of the factors is a perfect square. For example, we can write $18 = 9 \cdot 2$, *where 9 is a perfect square. See Section R.1.*

67. 50   **68.** 8   **69.** 32   **70.** 75

71. 28   **72.** 44   **73.** 27   **74.** 90

Concept Extensions

Solve. See the Concept Check in this section.

75. Which of the following is a real number?
     a. $\sqrt[7]{-1}$ **b.** $\sqrt[3]{-125}$
     c. $\sqrt[6]{-128}$ **d.** $\sqrt[8]{-1}$

76. a. $\sqrt{-1}$ **b.** $\sqrt[3]{-1}$
    c. $\sqrt[4]{-1}$ **d.** $\sqrt[5]{-1}$

The length of a side of a square in given by the expression $\sqrt{A}$, *where A is the square's area. Use this expression for Exercises 77 through 80. Be sure to attach the appropriate units.*

77. The area of a square is 49 square miles. Find the length of a side of the square.

78. The area of a square is $\dfrac{1}{81}$ square meters. Find the length of a side of the square.

Square

$\sqrt{A}$

79. Sony makes the current smallest mini disc player. It is in the shape of a square with area of 9.0601 square inches. Find the length of a side. (*Source:* SONY)

80. A parking lot is in the shape of a square with area 2500 square yards. Find the length of a side.

81. Simplify $\sqrt{\sqrt{81}}$.

82. Simplify $\sqrt[3]{\sqrt[3]{1}}$.

83. Simplify $\sqrt{\sqrt{10{,}000}}$.

84. Simplify $\sqrt{\sqrt{1{,}600{,}000{,}000}}$.

85. The formula for calculating the period (one back and forth swing) of a pendulum is $T = 2\pi\sqrt{\frac{L}{g}}$, where T is time of the period of the swing, L is the length of the pendulum, and g is the acceleration of gravity. At the California Academy of Sciences, one can see a Foucault's pendulum with a length = 30 ft, and $g = 32$ ft/sec^2. Using $\pi = 3.14$, find the period of this pendulum. (Round to the nearest tenth of a second.)

86. If the amount of gold discovered by humankind could be assembled in one place, it would make a cube with a volume of 195,112 cubic feet. Each side of the cube would be $\sqrt[3]{195{,}112}$ feet long. How long would one side of the cube be? (*Source: Reader's Digest*)

87. Explain why the square root of a negative number is not a real number.

88. Explain why the cube root of a negative number is a real number.

89. Graph $y = \sqrt{x}$. (Complete the table below, plot the ordered pair solutions, and draw a smooth curve through the points. Remember that since the radicand cannot be negative, this particular graph begins at the point with coordinates (0, 0).)

x	y	
0	0	
1		
3		(approximate)
4		
9		

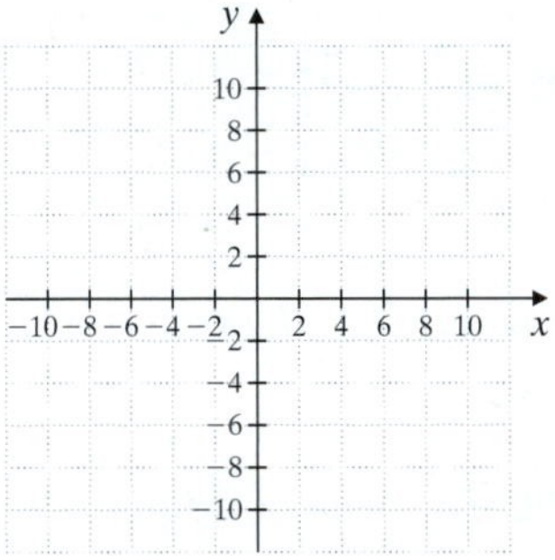

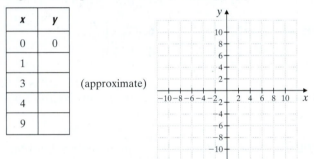

90. Graph $y = \sqrt[3]{x}$. (Complete the table below, plot the ordered pair solutions, and draw a smooth curve through the points.)

x	y	
−8		
−2		(approximate)
−1		
0		
1		
2		(approximate)
8		

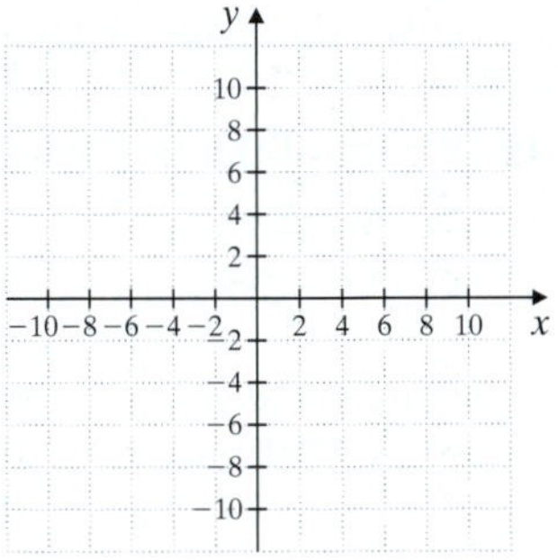

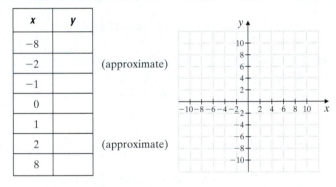

Use a graphing calculator and graph each function. Observe the graph from left to right and give the ordered pair that corresponds to the "beginning" of the graph. Then tell why the graph starts at that point.

91. $y = \sqrt{x - 2}$

92. $y = \sqrt{x + 3}$

93. $y = \sqrt{x + 4}$

94. $y = \sqrt{x - 5}$

STUDY SKILLS BUILDER

How Well Do You Know Your Textbook?

Let's check to see whether you are familiar with your textbook yet. Remember, for help, see Section 1.1 in this text.

1. What does the icon mean?
2. What does the icon mean?
3. What does the △ icon mean?
4. Where can you find a review for each chapter? What answers to this review can be found in the back of your text?
5. Each chapter contains an overview of the chapter along with examples. What is this feature called?
6. Each chapter contains a review of vocabulary. What is this feature called?
7. There are free CDs in your text. What content is contained on these CDs?
8. What is the location of the section that is entirely devoted to study skills?
9. There are Practice Problems that are contained in the margin of the text. What are they and how can they be used?

8.2 SIMPLIFYING RADICALS

Objectives

A Use the Product Rule to Simplify Radicals.

B Use the Quotient Rule to Simplify Radicals.

C Use Both Rules to Simplify Radicals Containing Variables.

D Simplify Cube Roots.

Objective **A** Simplifying Radicals Using the Product Rule

A square root is simplified when the radicand contains no perfect square factors (other than 1). For example, $\sqrt{20}$ is not simplified because $\sqrt{20} = \sqrt{4 \cdot 5}$ and 4 is a perfect square.

To begin simplifying square roots, we notice the following pattern.

$$\sqrt{9 \cdot 16} = \sqrt{144} = 12$$
$$\sqrt{9} \cdot \sqrt{16} = 3 \cdot 4 = 12$$

Since both expressions simplify to 12, we can write

$$\sqrt{9 \cdot 16} = \sqrt{9} \cdot \sqrt{16}$$

This suggests the following product rule for square roots.

Product Rule for Square Roots

If $\sqrt{a}$ and $\sqrt{b}$ are real numbers, then

$$\sqrt{a \cdot b} = \sqrt{a} \cdot \sqrt{b}$$

In other words, the square root of a product is equal to the product of the square roots.

To simplify $\sqrt{45}$, for example, we factor 45 so that one of its factors is a perfect square factor.

$$
\begin{aligned}
\sqrt{45} &= \sqrt{9 \cdot 5} && \text{Factor 45.} \\
&= \sqrt{9} \cdot \sqrt{5} && \text{Use the product rule.} \\
&= 3\sqrt{5} && \text{Write } \sqrt{9} \text{ as 3.}
\end{aligned}
$$

The notation $3\sqrt{5}$ means $3 \cdot \sqrt{5}$. Since the radicand 5 has no perfect square factor other than 1, the expression $3\sqrt{5}$ is in simplest form.

Helpful Hint

A radical expression in simplest form *does not mean* a decimal approximation. The simplest form of a radical expression is an exact form and may still contain a radical.

$$\underbrace{\sqrt{45} = 3\sqrt{5}}_{\text{exact}} \qquad \underbrace{\sqrt{45} \approx 6.71}_{\text{decimal approximation}}$$

EXAMPLES   Simplify.

1.
$$
\begin{aligned}
\sqrt{54} &= \sqrt{9 \cdot 6} && \text{Factor 54 so that one factor is a perfect square.} \\
& && \text{9 is a perfect square.} \\
&= \sqrt{9} \cdot \sqrt{6} && \text{Use the product rule.} \\
&= 3\sqrt{6} && \text{Write } \sqrt{9} \text{ as 3.}
\end{aligned}
$$

2.
$$
\begin{aligned}
\sqrt{12} &= \sqrt{4 \cdot 3} && \text{Factor 12 so that one factor is a perfect square.} \\
& && \text{4 is a perfect square.} \\
&= \sqrt{4} \cdot \sqrt{3} && \text{Use the product rule.} \\
&= 2\sqrt{3} && \text{Write } \sqrt{4} \text{ as 2.}
\end{aligned}
$$

PRACTICE PROBLEMS 1–4

Simplify.

1. $\sqrt{40}$ **2.** $\sqrt{18}$

3. $\sqrt{700}$ **4.** $\sqrt{15}$

Answers

1. $2\sqrt{10}$,   **2.** $3\sqrt{2}$,   **3.** $10\sqrt{7}$,   **4.** $\sqrt{15}$

Continued on next page

3. $\sqrt{200} = \sqrt{100 \cdot 2}$ Factor 200 so that one factor is a perfect square.
               100 is a perfect square.

         $= \sqrt{100} \cdot \sqrt{2}$ Use the product rule.

         $= 10\sqrt{2}$ Write $\sqrt{100}$ as 10.

4. $\sqrt{35}$ The radicand 35 contains no perfect square factors other than 1. Thus $\sqrt{35}$ is in simplest form.

🔲 **Work Practice Problems 1–4**

In Example 3, 100 is the largest perfect square factor of 200. What happens if we don't use the largest perfect square factor? Although using the largest perfect square factor saves time, the result is the same no matter what perfect square factor is used. For example, it is also true that $200 = 4 \cdot 50$. Then

$$\sqrt{200} = \sqrt{4} \cdot \sqrt{50}$$
$$= 2 \cdot \sqrt{50}$$

Since $\sqrt{50}$ is not in simplest form, we continue.

$$\sqrt{200} = 2 \cdot \sqrt{50}$$
$$= 2 \cdot \sqrt{25 \cdot 2}$$
$$= 2 \cdot \sqrt{25} \cdot \sqrt{2}$$
$$= 2 \cdot 5 \cdot \sqrt{2}$$
$$= 10\sqrt{2}$$

PRACTICE PROBLEM 5

Simplify $7\sqrt{75}$.

EXAMPLE 5 Simplify $3\sqrt{8}$.

Solution: Remember that $3\sqrt{8}$ means $3 \cdot \sqrt{8}$.

$3 \cdot \sqrt{8} = 3 \cdot \sqrt{4 \cdot 2}$ Factor 8 so that one factor is a perfect square.

         $= 3 \cdot \sqrt{4} \cdot \sqrt{2}$ Use the product rule.

         $= 3 \cdot 2 \cdot \sqrt{2}$ Write $\sqrt{4}$ as 2.

         $= 6 \cdot \sqrt{2}$ or $6\sqrt{2}$ Write $3 \cdot 2$ as 6.

🔲 **Work Practice Problem 5**

Objective **B** **Simplifying Radicals Using the Quotient Rule**

Next, let's examine the square root of a quotient.

$$\sqrt{\frac{16}{4}} = \sqrt{4} = 2$$

Also,

$$\frac{\sqrt{16}}{\sqrt{4}} = \frac{4}{2} = 2$$

Since both expressions equal 2, we can write

$$\sqrt{\frac{16}{4}} = \frac{\sqrt{16}}{\sqrt{4}}$$

This suggests the following quotient rule.

Answer

5. $35\sqrt{3}$

Quotient Rule for Square Roots

If $\sqrt{a}$ and $\sqrt{b}$ are real numbers and $b \neq 0$, then

$$\sqrt{\frac{a}{b}} = \frac{\sqrt{a}}{\sqrt{b}}$$

In other words, the square root of a quotient is equal to the quotient of the square roots.

EXAMPLES Use the quotient rule to simplify.

6. $\sqrt{\dfrac{25}{36}} = \dfrac{\sqrt{25}}{\sqrt{36}} = \dfrac{5}{6}$

7. $\sqrt{\dfrac{3}{64}} = \dfrac{\sqrt{3}}{\sqrt{64}} = \dfrac{\sqrt{3}}{8}$

8. $\sqrt{\dfrac{40}{81}} = \dfrac{\sqrt{40}}{\sqrt{81}}$ Use the quotient rule.

$\qquad = \dfrac{\sqrt{4} \cdot \sqrt{10}}{9}$ Use the product rule and write $\sqrt{81}$ as 9.

$\qquad = \dfrac{2\sqrt{10}}{9}$ Write $\sqrt{4}$ as 2.

▢ **Work Practice Problems 6–8**

PRACTICE PROBLEMS 6–8

Use the quotient rule to simplify.

6. $\sqrt{\dfrac{16}{81}}$

7. $\sqrt{\dfrac{2}{25}}$

8. $\sqrt{\dfrac{45}{49}}$

Objective **C** **Simplifying Radicals Containing Variables**

Recall that $\sqrt{x^6} = x^3$ because $(x^3)^2 = x^6$. If a variable radicand has an odd exponent, we write the exponential expression so that one factor is the greatest even power contained in the expression. Then we use the product rule to simplify.

EXAMPLES Simplify each radical. Assume that all variables represent positive numbers.

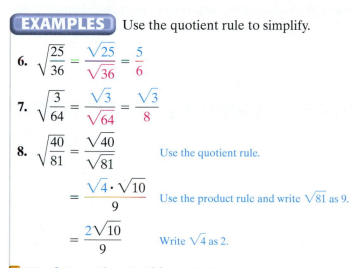

9. $\sqrt{x^5} = \sqrt{x^4 \cdot x} = \sqrt{x^4} \cdot \sqrt{x} = x^2\sqrt{x}$

10. $\sqrt{8y^2} = \sqrt{4 \cdot 2 \cdot y^2} = \sqrt{4y^2 \cdot 2} = \sqrt{4y^2} \cdot \sqrt{2} = 2y\sqrt{2}$ 4 and y^2 are both perfect square factors so we grouped them under one radical.

11. $\sqrt{\dfrac{45}{x^6}} = \dfrac{\sqrt{45}}{\sqrt{x^6}} = \dfrac{\sqrt{9 \cdot 5}}{x^3} = \dfrac{\sqrt{9} \cdot \sqrt{5}}{x^3} = \dfrac{3\sqrt{5}}{x^3}$

12. $\sqrt{\dfrac{5p^3}{9}} = \dfrac{\sqrt{5p^3}}{\sqrt{9}} = \dfrac{\sqrt{p^2 \cdot 5p}}{3} = \dfrac{\sqrt{p^2} \cdot \sqrt{5p}}{3} = \dfrac{p\sqrt{5p}}{3}$

▢ **Work Practice Problems 9–12**

PRACTICE PROBLEMS 9–12

Simplify each radical. Assume that all variables represent positive numbers.

9. $\sqrt{x^{11}}$ **10.** $\sqrt{18x^4}$

11. $\sqrt{\dfrac{27}{x^8}}$ **12.** $\sqrt{\dfrac{7y^7}{25}}$

Answers

6. $\dfrac{4}{9}$, **7.** $\dfrac{\sqrt{2}}{5}$, **8.** $\dfrac{3\sqrt{5}}{7}$, **9.** $x^5\sqrt{x}$,

10. $3x^2\sqrt{2}$, **11.** $\dfrac{3\sqrt{3}}{x^4}$, **12.** $\dfrac{y^3\sqrt{7y}}{5}$

Objective D Simplifying Cube Roots

The product and quotient rules also apply to roots other than square roots. For example, to simplify cube roots, we look for perfect cube factors of the radicand. Recall that 8 is a perfect cube since $2^3 = 8$. Therefore, to simplify $\sqrt[3]{48}$, we factor 48 as $8 \cdot 6$.

$$\sqrt[3]{48} = \sqrt[3]{8 \cdot 6} \qquad \text{Factor 48.}$$
$$= \sqrt[3]{8} \cdot \sqrt[3]{6} \qquad \text{Use the product rule.}$$
$$= 2\sqrt[3]{6} \qquad \text{Write } \sqrt[3]{8} \text{ as 2.}$$

$2\sqrt[3]{6}$ is in simplest form since the radicand 6 contains no perfect cube factors other than 1.

PRACTICE PROBLEMS 13–16

Simplify each radical.

13. $\sqrt[3]{88}$ **14.** $\sqrt[3]{50}$

15. $\sqrt[3]{\dfrac{10}{27}}$ **16.** $\sqrt[3]{\dfrac{81}{8}}$

EXAMPLES Simplify each radical.

13. $\sqrt[3]{54} = \sqrt[3]{27 \cdot 2} = \sqrt[3]{27} \cdot \sqrt[3]{2} = 3\sqrt[3]{2}$

14. $\sqrt[3]{18}$ The number 18 contains no perfect cube factors, so $\sqrt[3]{18}$ cannot be simplified further.

15. $\sqrt[3]{\dfrac{7}{8}} = \dfrac{\sqrt[3]{7}}{\sqrt[3]{8}} = \dfrac{\sqrt[3]{7}}{2}$

16. $\sqrt[3]{\dfrac{40}{27}} = \dfrac{\sqrt[3]{40}}{\sqrt[3]{27}} = \dfrac{\sqrt[3]{8 \cdot 5}}{3} = \dfrac{\sqrt[3]{8} \cdot \sqrt[3]{5}}{3} = \dfrac{2\sqrt[3]{5}}{3}$

Work Practice Problems 13–16

Answers

13. $2\sqrt[3]{11}$,   **14.** $\sqrt[3]{50}$,   **15.** $\dfrac{\sqrt[3]{10}}{3}$,

16. $\dfrac{3\sqrt[3]{3}}{2}$

Mental Math

Simplify each radical. Assume that all variables represent positive numbers.

1. $\sqrt{4 \cdot 9}$

2. $\sqrt{9 \cdot 36}$

3. $\sqrt{x^2}$

4. $\sqrt{y^4}$

5. $\sqrt{0}$

6. $\sqrt{1}$

7. $\sqrt{25x^4}$

8. $\sqrt{49x^2}$

8.2 EXERCISE SET

Objective A *Use the product rule to simplify each radical. See Examples 1 through 4.*

1. $\sqrt{20}$

2. $\sqrt{44}$

3. $\sqrt{50}$

4. $\sqrt{28}$

5. $\sqrt{33}$

6. $\sqrt{21}$

7. $\sqrt{98}$

8. $\sqrt{125}$

9. $\sqrt{60}$

10. $\sqrt{90}$

11. $\sqrt{180}$

12. $\sqrt{150}$

13. $\sqrt{52}$

14. $\sqrt{75}$

Use the product rule to simplify each radical. See Example 5.

15. $3\sqrt{25}$

16. $9\sqrt{36}$

17. $7\sqrt{63}$

18. $11\sqrt{99}$

19. $-5\sqrt{27}$

20. $-6\sqrt{75}$

Objective B *Use the quotient rule and the product rule to simplify each radical. See Examples 6 through 8.*

21. $\sqrt{\dfrac{8}{25}}$

22. $\sqrt{\dfrac{63}{16}}$

23. $\sqrt{\dfrac{27}{121}}$

24. $\sqrt{\dfrac{24}{169}}$

25. $\sqrt{\dfrac{9}{4}}$

26. $\sqrt{\dfrac{100}{49}}$

27. $\sqrt{\dfrac{125}{9}}$

28. $\sqrt{\dfrac{27}{100}}$

29. $\sqrt{\dfrac{11}{36}}$

30. $\sqrt{\dfrac{30}{49}}$

31. $-\sqrt{\dfrac{27}{144}}$

32. $-\sqrt{\dfrac{84}{121}}$

Objective C *Simplify each radical. Assume that all variables represent positive numbers. See Examples 9 through 12.*

33. $\sqrt{x^7}$

34. $\sqrt{y^3}$

35. $\sqrt{x^{13}}$

36. $\sqrt{y^{17}}$

37. $\sqrt{36a^3}$

38. $\sqrt{81b^5}$

39. $\sqrt{96x^4}$

40. $\sqrt{40y^{10}}$

41. $\sqrt{\dfrac{12}{m^2}}$

42. $\sqrt{\dfrac{63}{p^2}}$

43. $\sqrt{\dfrac{9x}{y^{10}}}$

44. $\sqrt{\dfrac{6y^2}{z^{16}}}$

45. $\sqrt{\dfrac{88}{x^{12}}}$

46. $\sqrt{\dfrac{500}{y^{22}}}$

Objectives **A** **B** **C** **Mixed Practice** *Simplify each radical. See Examples 1 through 10.*

47. $8\sqrt{4}$

48. $6\sqrt{49}$

49. $\sqrt{\dfrac{36}{121}}$

50. $\sqrt{\dfrac{25}{144}}$

51. $\sqrt{175}$

52. $\sqrt{700}$

53. $\sqrt{\dfrac{20}{9}}$

54. $\sqrt{\dfrac{45}{64}}$

55. $\sqrt{24m^7}$

56. $\sqrt{50n^{13}}$

57. $\sqrt{\dfrac{23y^3}{4x^6}}$

58. $\sqrt{\dfrac{41x^5}{9y^8}}$

Objective **D** *Simplify each radical. See Examples 13 through 16.*

59. $\sqrt[3]{24}$

60. $\sqrt[3]{81}$

61. $\sqrt[3]{250}$

62. $\sqrt[3]{56}$

63. $\sqrt[3]{\dfrac{5}{64}}$

64. $\sqrt[3]{\dfrac{32}{125}}$

65. $\sqrt[3]{\dfrac{23}{8}}$

66. $\sqrt[3]{\dfrac{37}{27}}$

67. $\sqrt[3]{\dfrac{15}{64}}$

68. $\sqrt[3]{\dfrac{4}{27}}$

69. $\sqrt[3]{80}$

70. $\sqrt[3]{108}$

Review

Perform each indicated operation. See Sections 3.4 and 3.5.

71. $6x + 8x$

72. $(6x)(8x)$

73. $(2x + 3)(x - 5)$

74. $(2x + 3) + (x - 5)$

75. $9y^2 - 9y^2$

76. $(9y^2)(-8y^2)$

Concept Extensions

Simplify each radical. Assume that all variables represent positive numbers.

77. $\sqrt{x^6y^3}$

78. $\sqrt{a^{13}b^{14}}$

79. $\sqrt{98x^5y^4}$

80. $\sqrt{27x^8y^{11}}$

81. $\sqrt[3]{-8x^6}$

82. $\sqrt[3]{27x^{12}}$

83. If a cube is to have a volume of 80 cubic inches, then each side must be $\sqrt[3]{80}$ inches long. Simplify the radical representing the side length.

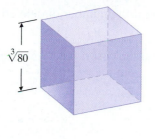

84. Jeannie Boswell is swimming across a 40-foot-wide river, trying to head straight across to the opposite shore. However, the current is strong enough to move her downstream 100 feet by the time she reaches land. (See the figure.) Because of the current, the actual distance she swam is $\sqrt{11,600}$ feet. Simplify this radical.

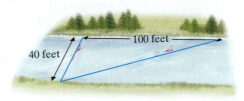

100 feet

40 feet

85. By using replacement values for a and b, show that $\sqrt{a^2 + b^2}$ does not equal $a + b$.

86. By using replacement values for a and b, show that $\sqrt{a + b}$ does not equal $\sqrt{a} + \sqrt{b}$.

The length of a side of a cube is given by the expression $\dfrac{\sqrt{6A}}{6}$*, where A is the cube's surface area. Use this expression for Exercises 87 through 90. Be sure to attach the appropriate units.*

△ **87.** The surface area of a cube is 120 square inches. Find the exact length of a side of the cube.

△ **88.** The surface area of a cube is 594 square feet. Find the exact length of a side of the cube.

$\sqrt{A/6}$

89. A Guinness World record was set in December 2004, when an electrical engineering student from Johannesburg, South Africa, solved 42 Rubik's cubes in one hour, the most ever in that time. Rubik's cube, named after its inventor, Erno Rubik, was first imagined by him in 1974, and by 1980 was a world-wide phenomenon. These cubes have remained un-changed in size, and a standard Rubik's cube has a surface area of 30.375 square inches. Find the length of one side of a Rubik's cube. (*Source: Guinness Book of World Records*)

△ **90.** The Borg spaceship on *Star Trek: The Next Generation* is in the shape of a cube. Suppose a model of this ship has a surface area of 121 square inches. Find the length of a side of the ship.

The cost C in dollars per day to operate a small delivery service is given by $C = 100\sqrt[3]{n} + 700$, where n is the number of deliveries per day.

91. Find the cost if the number of deliveries is 1000.

92. Approximate the cost if the number of deliveries is 500.

The Mosteller formula for calculating body surface area is $B = \sqrt{\dfrac{hw}{3600}}$, where B is an individual's body surface area in square meters, h is the individual's height in centimeters, and w is the individual's weight in kilograms. Use this formula in Exercises 93 and 94. Round answers to the nearest tenth.

93. Find the body surface area of a person who is 169 cm tall and weighs 64 kilograms.

94. Approximate the body surface area of a person who is 183 cm tall and weighs 85 kilograms.

Copyright 2007 Pearson Education, Inc.

8.3 ADDING AND SUBTRACTING RADICALS

A Add or Subtract Like Radicals.

B Simplify Square Root Radical Expressions, and Then Add or Subtract Any Like Radicals.

C Simplify Cube Root Radical Expressions, and Then Add or Subtract Any Like Radicals.

Objective **A** Adding and Subtracting Radicals

Recall that to combine like terms, we use the distributive property.

$$5x + 3x = (5 + 3)x = 8x$$

The distributive property can also be applied to expressions containing the same radicals. For example,

$$5\sqrt{2} + 3\sqrt{2} = (5 + 3)\sqrt{2} = 8\sqrt{2}$$

Also,

$$9\sqrt{5} - 6\sqrt{5} = (9 - 6)\sqrt{5} = 3\sqrt{5}$$

Radical terms such as $5\sqrt{2}$ and $3\sqrt{2}$ are **like radicals,** as are $9\sqrt{5}$ and $6\sqrt{5}$. Like radicals have the same index and the same radicand.

EXAMPLES Add or subtract as indicated.

1. $4\sqrt{5} + 3\sqrt{5} = (4 + 3)\sqrt{5} = 7\sqrt{5}$

2. $\sqrt{10} - 6\sqrt{10} = 1\sqrt{10} - 6\sqrt{10} = (1 - 6)\sqrt{10} = -5\sqrt{10}$

3. $2\sqrt{6} + 2\sqrt{5}$ cannot be simplified further since the radicands are not the same.

4. $\sqrt{15} + \sqrt{15} - \sqrt{2} = 1\sqrt{15} + 1\sqrt{15} - \sqrt{2}$

$$= (1 + 1)\sqrt{15} - \sqrt{2}$$

$$= 2\sqrt{15} - \sqrt{2}$$

This expression cannot be simplified further since the radicands are not the same.

Work Practice Problems 1–4

PRACTICE PROBLEMS 1–4

Add or subtract as indicated.

1. $6\sqrt{11} + 9\sqrt{11}$

2. $\sqrt{7} - 3\sqrt{7}$

3. $\sqrt{2} + \sqrt{2}$

4. $3\sqrt{3} - 3\sqrt{2}$

✔**Concept Check** Which is true?

a. $2 + 3\sqrt{5} = 5\sqrt{5}$

b. $2\sqrt{3} + 2\sqrt{7} = 2\sqrt{10}$

c. $\sqrt{3} + \sqrt{5} = \sqrt{8}$

d. $\sqrt{3} + \sqrt{3} = 3$

e. None of the above is true. In each case, the left-hand side cannot be simplified further.

Objective **B** Simplifying Square Root Radicals before Adding or Subtracting

At first glance, it appears that the expression $\sqrt{50} + \sqrt{8}$ cannot be simplified further because the radicands are different. However, the product rule can be used to simplify each radical, and then further simplification might be possible.

Answers

1. $15\sqrt{11}$, **2.** $-2\sqrt{7}$, **3.** $2\sqrt{2}$,
4. $3\sqrt{3} - 3\sqrt{2}$

✔ **Concept Check Answer**

e

PRACTICE PROBLEMS 5–8

Simplify each radical expression.

5. $\sqrt{27} + \sqrt{75}$

6. $3\sqrt{20} - 7\sqrt{45}$

7. $\sqrt{36} - \sqrt{48} - 4\sqrt{3} - \sqrt{9}$

8. $\sqrt{9x^4} - \sqrt{36x^3} + \sqrt{x^3}$

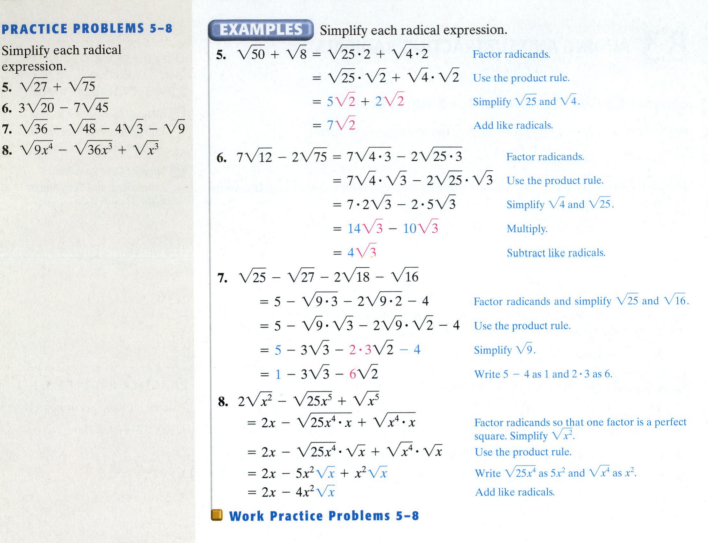

EXAMPLES Simplify each radical expression.

5. $\sqrt{50} + \sqrt{8} = \sqrt{25 \cdot 2} + \sqrt{4 \cdot 2}$ Factor radicands.

$\qquad\qquad\quad = \sqrt{25} \cdot \sqrt{2} + \sqrt{4} \cdot \sqrt{2}$ Use the product rule.

$\qquad\qquad\quad = 5\sqrt{2} + 2\sqrt{2}$ Simplify $\sqrt{25}$ and $\sqrt{4}$.

$\qquad\qquad\quad = 7\sqrt{2}$ Add like radicals.

6. $7\sqrt{12} - 2\sqrt{75} = 7\sqrt{4 \cdot 3} - 2\sqrt{25 \cdot 3}$ Factor radicands.

$\qquad\qquad\qquad\quad = 7\sqrt{4} \cdot \sqrt{3} - 2\sqrt{25} \cdot \sqrt{3}$ Use the product rule.

$\qquad\qquad\qquad\quad = 7 \cdot 2\sqrt{3} - 2 \cdot 5\sqrt{3}$ Simplify $\sqrt{4}$ and $\sqrt{25}$.

$\qquad\qquad\qquad\quad = 14\sqrt{3} - 10\sqrt{3}$ Multiply.

$\qquad\qquad\qquad\quad = 4\sqrt{3}$ Subtract like radicals.

7. $\sqrt{25} - \sqrt{27} - 2\sqrt{18} - \sqrt{16}$

$\qquad = 5 - \sqrt{9 \cdot 3} - 2\sqrt{9 \cdot 2} - 4$ Factor radicands and simplify $\sqrt{25}$ and $\sqrt{16}$.

$\qquad = 5 - \sqrt{9} \cdot \sqrt{3} - 2\sqrt{9} \cdot \sqrt{2} - 4$ Use the product rule.

$\qquad = 5 - 3\sqrt{3} - 2 \cdot 3\sqrt{2} - 4$ Simplify $\sqrt{9}$.

$\qquad = 1 - 3\sqrt{3} - 6\sqrt{2}$ Write $5 - 4$ as 1 and $2 \cdot 3$ as 6.

8. $2\sqrt{x^2} - \sqrt{25x^5} + \sqrt{x^5}$

$\qquad = 2x - \sqrt{25x^4 \cdot x} + \sqrt{x^4 \cdot x}$ Factor radicands so that one factor is a perfect square. Simplify $\sqrt{x^2}$.

$\qquad = 2x - \sqrt{25x^4} \cdot \sqrt{x} + \sqrt{x^4} \cdot \sqrt{x}$ Use the product rule.

$\qquad = 2x - 5x^2\sqrt{x} + x^2\sqrt{x}$ Write $\sqrt{25x^4}$ as $5x^2$ and $\sqrt{x^4}$ as x^2.

$\qquad = 2x - 4x^2\sqrt{x}$ Add like radicals.

🟧 **Work Practice Problems 5–8**

Objective C Simplifying Cube Root Radicals before Adding or Subtracting

PRACTICE PROBLEM 9

Simplify the radical expression.
$10\sqrt[3]{81p^6} - \sqrt[3]{24p^6}$

EXAMPLE 9 Simplify the radical expression.

$5\sqrt[3]{16x^3} - \sqrt[3]{54x^3}$

$= 5\sqrt[3]{8x^3 \cdot 2} - \sqrt[3]{27x^3 \cdot 2}$ Factor radicands so that one factor is a perfect cube.

$= 5 \cdot \sqrt[3]{8x^3} \cdot \sqrt[3]{2} - \sqrt[3]{27x^3} \cdot \sqrt[3]{2}$ Use the product rule.

$= 5 \cdot 2x \cdot \sqrt[3]{2} - 3x \cdot \sqrt[3]{2}$ Write $\sqrt[3]{8x^3}$ as $2x$ and $\sqrt[3]{27x^3}$ as $3x$.

$= 10x\sqrt[3]{2} - 3x\sqrt[3]{2}$ Write $5 \cdot 2x$ as $10x$.

$= 7x\sqrt[3]{2}$ Subtract like radicands.

🟧 **Work Practice Problem 9**

Answers

5. $8\sqrt{3}$,   **6.** $-15\sqrt{5}$,   **7.** $3 - 8\sqrt{3}$,

8. $3x^2 - 5x\sqrt{x}$,   **9.** $28p^2\sqrt[3]{3}$

Mental Math

Simplify each expression by combining like radicals.

1. $3\sqrt{2} + 5\sqrt{2}$

2. $3\sqrt{5} + 7\sqrt{5}$

3. $5\sqrt{x} + 2\sqrt{x}$

4. $8\sqrt{x} + 3\sqrt{x}$

5. $5\sqrt{7} - 2\sqrt{7}$

6. $8\sqrt{6} - 5\sqrt{6}$

8.3 EXERCISE SET

Objective A *Add or subtract as indicated. See Examples 1 through 4.*

1. $4\sqrt{3} - 8\sqrt{3}$

2. $\sqrt{5} - 9\sqrt{5}$

3. $3\sqrt{6} + 8\sqrt{6} - 2\sqrt{6} - 5$

4. $12\sqrt{2} - 3\sqrt{2} + 8\sqrt{2} + 10$

5. $6\sqrt{5} - 5\sqrt{5} + \sqrt{2}$

6. $4\sqrt{3} + \sqrt{5} - 3\sqrt{3}$

7. $2\sqrt{3} + 5\sqrt{3} - \sqrt{2}$

8. $8\sqrt{14} + 2\sqrt{14} + \sqrt{5}$

9. $2\sqrt{2} - 7\sqrt{2} - 6$

10. $5\sqrt{7} + 2 - 11\sqrt{7}$

Objective B *Add or subtract by first simplifying each radical and then combining any like radicals. Assume that all variables represent positive numbers. See Examples 5 through 8.*

11. $\sqrt{12} + \sqrt{27}$

12. $\sqrt{50} + \sqrt{18}$

13. $\sqrt{45} + 3\sqrt{20}$

14. $5\sqrt{32} - \sqrt{72}$

15. $2\sqrt{54} - \sqrt{20} + \sqrt{45} - \sqrt{24}$

16. $2\sqrt{8} - \sqrt{128} + \sqrt{48} + \sqrt{18}$

17. $4x - 3\sqrt{x^2} + \sqrt{x}$

18. $x - 6\sqrt{x^2} + 2\sqrt{x}$

19. $\sqrt{25x} + \sqrt{36x} - 11\sqrt{x}$

20. $\sqrt{9x} - \sqrt{16x} + 2\sqrt{x}$

21. $\sqrt{\dfrac{5}{9}} + \sqrt{\dfrac{5}{81}}$

22. $\sqrt{\dfrac{3}{64}} + \sqrt{\dfrac{3}{16}}$

23. $\sqrt{\dfrac{3}{4}} - \sqrt{\dfrac{3}{64}}$

24. $\sqrt{\dfrac{2}{25}} + \sqrt{\dfrac{2}{9}}$

Objectives A B Mixed Practice *See Examples 1 through 8.*

25. $12\sqrt{5} - \sqrt{5} - 4\sqrt{5}$

26. $\sqrt{6} + 3\sqrt{6} + \sqrt{6}$

27. $\sqrt{75} + \sqrt{48}$

28. $2\sqrt{80} - \sqrt{45}$

29. $\sqrt{5} + \sqrt{15}$

30. $\sqrt{5} + \sqrt{5}$

31. $3\sqrt{x^3} - x\sqrt{4x}$

32. $\sqrt{16x} - \sqrt{x^3}$

33. $\sqrt{8} + \sqrt{9} + \sqrt{18} + \sqrt{81}$

34. $\sqrt{6} + \sqrt{16} + \sqrt{24} + \sqrt{25}$

35. $4 + 8\sqrt{2} - 9$

36. $11 - 5\sqrt{7} - 8$

37. $2\sqrt{45} - 2\sqrt{20}$ **38.** $5\sqrt{18} + 2\sqrt{32}$ **39.** $\sqrt{35} - \sqrt{140}$ **40.** $\sqrt{15} - \sqrt{135}$

41. $6 - 2\sqrt{3} - \sqrt{3}$ **42.** $8 - \sqrt{2} - 5\sqrt{2}$ **43.** $3\sqrt{9x} + 2\sqrt{x}$ **44.** $5\sqrt{x} + 4\sqrt{4x}$

45. $\sqrt{9x^2} + \sqrt{81x^2} - 11\sqrt{x}$ **46.** $\sqrt{100x^2} + 3\sqrt{x} - \sqrt{36x^2}$ **47.** $\sqrt{3x^3} + 3x\sqrt{x}$

48. $x\sqrt{4x} + \sqrt{9x^3}$ **49.** $\sqrt{32x^2} + \sqrt{32x^2} + \sqrt{4x^2}$ **50.** $\sqrt{18x^2} + \sqrt{24x^3} + \sqrt{2x^2}$

51. $\sqrt{40x} + \sqrt{40x^4} - 2\sqrt{10x} - \sqrt{5x^4}$ **52.** $\sqrt{72x^2} + \sqrt{54x} - x\sqrt{50} - 3\sqrt{2x}$

Objective C *Simplify each radical expression.*

53. $2\sqrt[3]{9} + 5\sqrt[3]{9} - \sqrt[3]{25}$ **54.** $8\sqrt[3]{4} + 2\sqrt[3]{4} - \sqrt[3]{49}$ **55.** $2\sqrt[3]{2} - 7\sqrt[3]{2} - 6$ **56.** $5\sqrt[3]{11} - 9\sqrt[3]{11} - 5$

57. $\sqrt[3]{81} + \sqrt[3]{24}$ **58.** $\sqrt[3]{32} + \sqrt[3]{4}$ **59.** $2\sqrt[3]{8x^3} + 2\sqrt[3]{16x^3}$ **60.** $3\sqrt[3]{27z^3} + 3\sqrt[3]{81z^3}$

61. $\sqrt{40x} + x\sqrt[3]{40} - 2\sqrt{10x} - x\sqrt[3]{5}$ **62.** $\sqrt{72x^2} + \sqrt[3]{54} - x\sqrt{50} - 3\sqrt[3]{2}$

63. $12\sqrt[3]{y^7} - y^2\sqrt[3]{8y}$ **64.** $19\sqrt[3]{z^{11}} - z^3\sqrt[3]{125z^2}$

Review

Square each binomial. See Section 3.6.

65. $(x + 6)^2$ **66.** $(3x + 2)^2$ **67.** $(2x - 1)^2$ **68.** $(x - 5)^2$

Concept Extensions

69. In your own words, describe like radicals.

70. In the expression $\sqrt{5} + 2 - 3\sqrt{5}$, explain why 2 and −3 cannot be combined.

△ **71.** Find the perimeter of the rectangular picture frame.   △ **72.** Find the perimeter of the plot of land.

$\sqrt{5}$ inches

$3\sqrt{5}$ inches

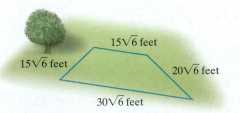

$15\sqrt{6}$ feet

$15\sqrt{6}$ feet

$20\sqrt{6}$ feet

$30\sqrt{6}$ feet

△ **73.** A water trough is to be made of wood. Each of the two triangular end pieces has an area of $\dfrac{3\sqrt{27}}{4}$ square feet. The two side panels are both rectangular. In simplest radical form, find the total area of the wood needed.

74. Eight wooden braces are to be attached along the diagonals of the vertical sides of a storage bin. Each of four of these diagonals has a length of $\sqrt{52}$ feet, while each of the other four has a length of $\sqrt{80}$ feet. In simplest radical form, find the total length of the wood needed for these braces.

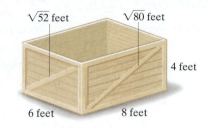

Determine whether each expression can be simplified. If yes, then simplify. See the Concept Check in this section.

75. $4\sqrt{2} + 3\sqrt{2}$

76. $3\sqrt{7} + 3\sqrt{6}$

77. $6 + 7\sqrt{6}$

78. $5x\sqrt{2} + 8x\sqrt{2}$

79. $\sqrt{7} + \sqrt{7} + \sqrt{7}$

80. $6\sqrt{5} - \sqrt{5}$

Simplify.

81. $\sqrt{\dfrac{x^3}{16}} - x\sqrt{\dfrac{9x}{25}} + \dfrac{\sqrt{81x^3}}{2}$

82. $7\sqrt{x^{11}y^7} - x^2y\sqrt{25x^7y^5} + \sqrt{8x^8y^2}$

STUDY SKILLS BUILDER

Learning New Terms?

By now, you have encountered many new terms. It's never too late to make a list of new terms and review them frequently. Remember that placing these new terms (including page references) on 3 × 5 index cards might help you later when you're preparing for a quiz.

Answer the following.

1. How do new terms stand out in this text so that they can be found?

2. Name one way placing a word and its definition on a 3 × 5 card might be helpful.

8.4 MULTIPLYING AND DIVIDING RADICALS

Objective **A** Multiplying Radicals

In Section 8.2, we used the product and quotient rules for radicals to help us simplify radicals. In this section, we use these rules to simplify products and quotients of radicals.

> ### Product Rule for Radicals
>
> If $\sqrt{a}$ and $\sqrt{b}$ are real numbers, then
> $$\sqrt{a} \cdot \sqrt{b} = \sqrt{a \cdot b}$$

In other words, the product of the square roots of two numbers is the square root of the product of the two numbers. For example,

$$\sqrt{3} \cdot \sqrt{2} = \sqrt{3 \cdot 2} = \sqrt{6}$$

PRACTICE PROBLEMS 1–4

Multiply. Then simplify each product if possible.

1. $\sqrt{5} \cdot \sqrt{2}$
2. $\sqrt{7} \cdot \sqrt{7}$
3. $\sqrt{6} \cdot \sqrt{3}$
4. $\sqrt{10x} \cdot \sqrt{2x}$

EXAMPLES Multiply. Then simplify each product if possible.

1. $\sqrt{7} \cdot \sqrt{3} = \sqrt{7 \cdot 3}$
   $= \sqrt{21}$

2. $\sqrt{3} \cdot \sqrt{3} = \sqrt{3 \cdot 3} = \sqrt{9} = 3$

3. $\sqrt{3} \cdot \sqrt{15} = \sqrt{45}$ Use the product rule.
   $= \sqrt{9 \cdot 5}$ Factor the radicand.
   $= \sqrt{9} \cdot \sqrt{5}$ Use the product rule.
   $= 3\sqrt{5}$ Simplify $\sqrt{9}$.

4. $\sqrt{2x^3} \cdot \sqrt{6x} = \sqrt{2x^3 \cdot 6x}$ Use the product rule.
   $= \sqrt{12x^4}$ Multiply.
   $= \sqrt{4x^4 \cdot 3}$ Write $12x^4$ so that one factor is a perfect square.
   $= \sqrt{4x^4} \cdot \sqrt{3}$ Use the product rule.
   $= 2x^2\sqrt{3}$ Simplify.

■ **Work Practice Problems 1–4**

From Example 2, we found that

$$\sqrt{3} \cdot \sqrt{3} = 3 \quad \text{or} \quad (\sqrt{3})^2 = 3$$

This is true in general.

> If a is a positive number,
> $$\sqrt{a} \cdot \sqrt{a} = a \quad \text{or} \quad (\sqrt{a})^2 = a$$

✔ Concept Check Identify the true statement(s).

a. $\sqrt{7} \cdot \sqrt{7} = 7$ c. $(\sqrt{131})^2 = 131$

b. $\sqrt{2} \cdot \sqrt{3} = 6$ d. $\sqrt{5x} \cdot \sqrt{5x} = 5x$ (Here x is a positive number.)

When multiplying radical expressions containing more than one term, we use the same techniques we use to multiply other algebraic expressions with more than one term.

Answers

1. $\sqrt{10}$,   2. 7,   3. $3\sqrt{2}$,   4. $2x\sqrt{5}$

✔ Concept Check Answers

a, c, d

EXAMPLE 5 Multiply.

a. $\sqrt{5}(\sqrt{5} - \sqrt{2})$ **c.** $(\sqrt{x} + \sqrt{2})(\sqrt{x} - \sqrt{7})$

b. $\sqrt{3x}(\sqrt{x} - 5\sqrt{3})$

Solution:

a. Using the distributive property, we have

$$\sqrt{5}(\sqrt{5} - \sqrt{2}) = \sqrt{5}\cdot\sqrt{5} - \sqrt{5}\cdot\sqrt{2}$$

$$= 5 - \sqrt{10} \qquad \text{Since } \sqrt{5}\cdot\sqrt{5} = 5 \text{ and } \sqrt{5}\cdot\sqrt{2} = \sqrt{10}$$

b. $\sqrt{3x}(\sqrt{x} - 5\sqrt{3}) = \sqrt{3x}\cdot\sqrt{x} - \sqrt{3x}\cdot5\sqrt{3}$ Use the distributive property.

$$= \sqrt{3x\cdot x} - 5\sqrt{3x\cdot 3} \qquad \text{Use the product rule.}$$

$$= \sqrt{3\cdot x^2} - 5\sqrt{9\cdot x} \qquad \text{Factor each radicand so that one factor is a perfect square.}$$

$$= \sqrt{3}\cdot\sqrt{x^2} - 5\cdot\sqrt{9}\cdot\sqrt{x} \qquad \text{Use the product rule.}$$

$$= x\sqrt{3} - 5\cdot3\cdot\sqrt{x} \qquad \text{Simplify.}$$

$$= x\sqrt{3} - 15\sqrt{x} \qquad \text{Simplify.}$$

c. Using the FOIL method of multiplication, we have

$$(\sqrt{x} + \sqrt{2})(\sqrt{x} - \sqrt{7})$$

$$\overset{\text{F}}{=} \sqrt{x}\cdot\sqrt{x} \overset{\text{O}}{-} \sqrt{x}\cdot\sqrt{7} \overset{\text{I}}{+} \sqrt{2}\cdot\sqrt{x} \overset{\text{L}}{-} \sqrt{2}\cdot\sqrt{7}$$

$$= x - \sqrt{7x} + \sqrt{2x} - \sqrt{14} \qquad \text{Use the product rule.}$$

◻ **Work Practice Problem 5**

The special product formulas also can be used to multiply expressions containing radicals.

EXAMPLE 6 Multiply.

a. $(\sqrt{5} - 7)(\sqrt{5} + 7)$ **b.** $(\sqrt{7x} + 2)^2$

Solution:

a. $(\sqrt{5} - 7)(\sqrt{5} + 7) = (\sqrt{5})^2 - 7^2$ Recall that $(a - b)(a + b) = a^2 - b^2$.

$$= 5 - 49$$

$$= -44$$

b. $(\sqrt{7x} + 2)^2$

$$= (\sqrt{7x})^2 + 2(\sqrt{7x})(2) + (2)^2 \qquad \text{Recall that } (a + b)^2 = a^2 + 2ab + b^2.$$

$$= 7x + 4\sqrt{7x} + 4$$

◻ **Work Practice Problem 6**

Objective B Dividing Radicals

To simplify quotients of rational expressions, we use the quotient rule.

Quotient Rule for Radicals

If $\sqrt{a}$ and $\sqrt{b}$ are real numbers and $b \neq 0$, then

$$\frac{\sqrt{a}}{\sqrt{b}} = \sqrt{\frac{a}{b}}$$

PRACTICE PROBLEM 5

Multiply.

a. $\sqrt{7}(\sqrt{7} - \sqrt{3})$

b. $\sqrt{5x}(\sqrt{x} - 3\sqrt{5})$

c. $(\sqrt{x} + \sqrt{5})(\sqrt{x} - \sqrt{3})$

PRACTICE PROBLEM 6

Multiply.

a. $(\sqrt{3} + 6)(\sqrt{3} - 6)$

b. $(\sqrt{5x} + 4)^2$

Answers

5. a. $7 - \sqrt{21}$, **b.** $x\sqrt{5} - 15\sqrt{x}$,

c. $x - \sqrt{3x} + \sqrt{5x} - \sqrt{15}$,

6. a. -33, **b.** $5x + 8\sqrt{5x} + 16$

PRACTICE PROBLEMS 7–9

Divide. Then simplify the quotient if possible.

7. $\dfrac{\sqrt{15}}{\sqrt{3}}$

8. $\dfrac{\sqrt{90}}{\sqrt{2}}$

9. $\dfrac{\sqrt{75x^3}}{\sqrt{5x}}$

EXAMPLES Divide. Then simplify the quotient if possible.

7. $\dfrac{\sqrt{14}}{\sqrt{2}} = \sqrt{\dfrac{14}{2}} = \sqrt{7}$

8. $\dfrac{\sqrt{100}}{\sqrt{5}} = \sqrt{\dfrac{100}{5}} = \sqrt{20} = \sqrt{4 \cdot 5} = \sqrt{4} \cdot \sqrt{5} = 2\sqrt{5}$

9. $\dfrac{\sqrt{12x^3}}{\sqrt{3x}} = \sqrt{\dfrac{12x^3}{3x}} = \sqrt{4x^2} = 2x$

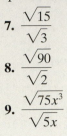 **Work Practice Problems 7–9**

Objective C Rationalizing Denominators

It is sometimes easier to work with radical expressions if the denominator does not contain a radical. To rewrite the expression so that the denominator does not contain a radical expression, we use the fact that we can multiply the numerator and the denominator of a fraction by the same nonzero number without changing the value of the expression. This is the same as multiplying the fraction by 1. For example, to get rid of the radical in the denominator of $\dfrac{\sqrt{5}}{\sqrt{2}}$, we multiply by 1 in the form of $\dfrac{\sqrt{2}}{\sqrt{2}}$. Then

$$\frac{\sqrt{5}}{\sqrt{2}} = \frac{\sqrt{5}}{\sqrt{2}} \cdot 1 = \frac{\sqrt{5}}{\sqrt{2}} \cdot \frac{\sqrt{2}}{\sqrt{2}} = \frac{\sqrt{5} \cdot \sqrt{2}}{\sqrt{2} \cdot \sqrt{2}} = \frac{\sqrt{10}}{2}$$

This process is called **rationalizing** the denominator.

PRACTICE PROBLEM 10

Rationalize the denominator of $\dfrac{5}{\sqrt{3}}$.

EXAMPLE 10 Rationalize the denominator of $\dfrac{2}{\sqrt{7}}$.

Solution: To rewrite $\dfrac{2}{\sqrt{7}}$ so that there is no radical in the denominator, we multiply by 1 in the form of $\dfrac{\sqrt{7}}{\sqrt{7}}$.

$$\frac{2}{\sqrt{7}} = \frac{2}{\sqrt{7}} \cdot \frac{\sqrt{7}}{\sqrt{7}} = \frac{2 \cdot \sqrt{7}}{\sqrt{7} \cdot \sqrt{7}} = \frac{2\sqrt{7}}{7}$$

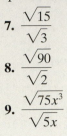 **Work Practice Problem 10**

PRACTICE PROBLEM 11

Rationalize the denominator of $\dfrac{\sqrt{7}}{\sqrt{20}}$.

EXAMPLE 11 Rationalize the denominator of $\dfrac{\sqrt{5}}{\sqrt{12}}$.

Solution: We can multiply by $\dfrac{\sqrt{12}}{\sqrt{12}}$, but see what happens if we simplify first.

$$\frac{\sqrt{5}}{\sqrt{12}} = \frac{\sqrt{5}}{\sqrt{4 \cdot 3}} = \frac{\sqrt{5}}{2\sqrt{3}}$$

To rationalize the denominator now, we multiply by $\dfrac{\sqrt{3}}{\sqrt{3}}$.

$$\frac{\sqrt{5}}{2\sqrt{3}} = \frac{\sqrt{5}}{2\sqrt{3}} \cdot \frac{\sqrt{3}}{\sqrt{3}} = \frac{\sqrt{5} \cdot \sqrt{3}}{2\sqrt{3} \cdot \sqrt{3}} = \frac{\sqrt{15}}{2 \cdot 3} = \frac{\sqrt{15}}{6}$$

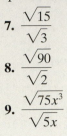 **Work Practice Problem 11**

Answers

7. $\sqrt{5}$,   **8.** $3\sqrt{5}$,   **9.** $x\sqrt{15}$,

10. $\dfrac{5\sqrt{3}}{3}$,   **11.** $\dfrac{\sqrt{35}}{10}$

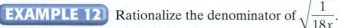

EXAMPLE 12 Rationalize the denominator of $\sqrt{\dfrac{1}{18x}}$.

Solution: First we simplify.

$$\sqrt{\dfrac{1}{18x}} = \dfrac{\sqrt{1}}{\sqrt{18x}} = \dfrac{1}{\sqrt{9} \cdot \sqrt{2x}} = \dfrac{1}{3\sqrt{2x}}$$

Now to rationalize the denominator, we multiply by $\dfrac{\sqrt{2x}}{\sqrt{2x}}$.

$$\dfrac{1}{3\sqrt{2x}} = \dfrac{1}{3\sqrt{2x}} \cdot \dfrac{\sqrt{2x}}{\sqrt{2x}} = \dfrac{1 \cdot \sqrt{2x}}{3\sqrt{2x} \cdot \sqrt{2x}} = \dfrac{\sqrt{2x}}{3 \cdot 2x} = \dfrac{\sqrt{2x}}{6x}$$

🟧 **Work Practice Problem 12**

Objective D Rationalizing Denominators Using Conjugates

To rationalize a denominator that is a sum or a difference, such as the denominator in

$$\dfrac{2}{4 + \sqrt{3}}$$

we multiply the numerator and the denominator by $4 - \sqrt{3}$. The expressions $4 + \sqrt{3}$ and $4 - \sqrt{3}$ are called conjugates of each other. When a radical expression such as $4 + \sqrt{3}$ is multiplied by its conjugate $4 - \sqrt{3}$, the product simplifies to an expression that contains no radicals.

In general, the expressions $a + b$ and $a - b$ are **conjugates** of each other.

$$(a + b)(a - b) = a^2 - b^2$$
$$\left(4 + \sqrt{3}\right)\left(4 - \sqrt{3}\right) = 4^2 - \left(\sqrt{3}\right)^2 = 16 - 3 = 13$$

Then

$$\dfrac{2}{4 + \sqrt{3}} = \dfrac{2\left(4 - \sqrt{3}\right)}{\left(4 + \sqrt{3}\right)\left(4 - \sqrt{3}\right)} = \dfrac{2\left(4 - \sqrt{3}\right)}{13}$$

EXAMPLE 13 Rationalize the denominator of $\dfrac{2}{1 + \sqrt{3}}$.

Solution: We multiply the numerator and the denominator of this fraction by the conjugate of $1 + \sqrt{3}$, that is, by $1 - \sqrt{3}$.

$$\dfrac{2}{1 + \sqrt{3}} = \dfrac{2\left(1 - \sqrt{3}\right)}{\left(1 + \sqrt{3}\right)\left(1 - \sqrt{3}\right)}$$

$$= \dfrac{2\left(1 - \sqrt{3}\right)}{1^2 - \left(\sqrt{3}\right)^2}$$

Helpful Hint
Don't forget that $\left(\sqrt{3}\right)^2 = 3$.

$$= \dfrac{2\left(1 - \sqrt{3}\right)}{1 - 3}$$

$$= \dfrac{2\left(1 - \sqrt{3}\right)}{-2}$$

$$= -\dfrac{2\left(1 - \sqrt{3}\right)}{2} \qquad \dfrac{a}{-b} = -\dfrac{a}{b}$$

$$= -1\left(1 - \sqrt{3}\right) \qquad \text{Simplify.}$$

$$= -1 + \sqrt{3} \qquad \text{Multiply.}$$

🟧 **Work Practice Problem 13**

PRACTICE PROBLEM 12

Rationalize the denominator of $\sqrt{\dfrac{2}{45x}}$.

PRACTICE PROBLEM 13

Rationalize the denominator of $\dfrac{3}{2 + \sqrt{7}}$.

Answers

12. $\dfrac{\sqrt{10x}}{15x}$, 13. $-2 + \sqrt{7}$

Copyright 2007 Pearson Education, Inc.

PRACTICE PROBLEM 14

Rationalize the denominator

of $\dfrac{\sqrt{2} + 5}{\sqrt{2} - 1}$.

EXAMPLE 14 Rationalize the denominator of $\dfrac{\sqrt{5} + 4}{\sqrt{5} - 1}$.

Solution:

$$\frac{\sqrt{5} + 4}{\sqrt{5} - 1} = \frac{\left(\sqrt{5} + 4\right)\left(\sqrt{5} + 1\right)}{\left(\sqrt{5} - 1\right)\left(\sqrt{5} + 1\right)} \quad \text{Multiply the numerator and denominator by } \sqrt{5} + 1, \text{ the conjugate of } \sqrt{5} - 1.$$

$$= \frac{5 + \sqrt{5} + 4\sqrt{5} + 4}{5 - 1} \quad \text{Multiply.}$$

$$= \frac{9 + 5\sqrt{5}}{4} \quad \text{Simplify.}$$

🔶 **Work Practice Problem 14**

PRACTICE PROBLEM 15

Rationalize the denominator

of $\dfrac{7}{2 - \sqrt{x}}$.

EXAMPLE 15 Rationalize the denominator of $\dfrac{3}{1 + \sqrt{x}}$.

Solution:

$$\frac{3}{1 + \sqrt{x}} = \frac{3(1 - \sqrt{x})}{(1 + \sqrt{x})(1 - \sqrt{x})} \quad \text{Multiply the numerator and denominator by } 1 - \sqrt{x}, \text{ the conjugate of } 1 + \sqrt{x}.$$

$$= \frac{3(1 - \sqrt{x})}{1 - x}$$

🔶 **Work Practice Problem 15**

Answers

14. $7 + 6\sqrt{2}$, **15.** $\dfrac{7(2 + \sqrt{x})}{4 - x}$

Mental Math

Multiply. Assume that all variables represent positive numbers.

1. $\sqrt{2} \cdot \sqrt{11}$ **2.** $\sqrt{5} \cdot \sqrt{7}$ **3.** $\sqrt{1} \cdot \sqrt{6}$ **4.** $\sqrt{7} \cdot \sqrt{x}$ **5.** $\sqrt{10} \cdot \sqrt{y}$ **6.** $\sqrt{x} \cdot \sqrt{y}$

8.4 EXERCISE SET

FOR EXTRA HELP

Student Solutions Manual · PH Math/Tutor Center · CD/Video for Review · MathXL · MyMathLab

Objective A *Multiply and simplify. Assume that all variables represent positive real numbers. See Examples 1 through 6.*

1. $\sqrt{8} \cdot \sqrt{2}$ **2.** $\sqrt{3} \cdot \sqrt{12}$ **3.** $\sqrt{10} \cdot \sqrt{5}$ **4.** $\sqrt{2} \cdot \sqrt{14}$

5. $(\sqrt{6})^2$ **6.** $(\sqrt{10})^2$ **7.** $\sqrt{2x} \cdot \sqrt{2x}$ **8.** $\sqrt{5y} \cdot \sqrt{5y}$

9. $(2\sqrt{5})^2$ **10.** $(3\sqrt{10})^2$ **11.** $(6\sqrt{x})^2$ **12.** $(8\sqrt{y})^2$

13. $\sqrt{3x^5} \cdot \sqrt{6x}$ **14.** $\sqrt{21y^7} \cdot \sqrt{3y}$ **15.** $\sqrt{2xy^2} \cdot \sqrt{8xy}$ **16.** $\sqrt{18x^2y^2} \cdot \sqrt{2x^2y}$

17. $\sqrt{6}(\sqrt{5} + \sqrt{7})$ **18.** $\sqrt{10}(\sqrt{3} - \sqrt{7})$ **19.** $\sqrt{10}(\sqrt{2} + \sqrt{5})$

20. $\sqrt{6}(\sqrt{3} + \sqrt{2})$ **21.** $\sqrt{7y}(\sqrt{y} - 2\sqrt{7})$ **22.** $\sqrt{5b}(2\sqrt{b} + \sqrt{5})$

23. $(\sqrt{3} + 6)(\sqrt{3} - 6)$ **24.** $(\sqrt{5} + 2)(\sqrt{5} - 2)$ **25.** $(\sqrt{3} + \sqrt{5})(\sqrt{2} - \sqrt{5})$

26. $(\sqrt{7} + \sqrt{5})(\sqrt{2} - \sqrt{5})$ **27.** $(2\sqrt{11} + 1)(\sqrt{11} - 6)$ **28.** $(5\sqrt{3} + 2)(\sqrt{3} - 1)$

29. $(\sqrt{x} + 6)(\sqrt{x} - 6)$ **30.** $(\sqrt{y} + 5)(\sqrt{y} - 5)$ **31.** $(\sqrt{x} - 7)^2$

32. $(\sqrt{x} + 4)^2$ **33.** $(\sqrt{6y} + 1)^2$ **34.** $(\sqrt{3y} - 2)^2$

Objective B *Divide and simplify. Assume that all variables represent positive real numbers. See Examples 7 through 9.*

35. $\dfrac{\sqrt{32}}{\sqrt{2}}$ **36.** $\dfrac{\sqrt{40}}{\sqrt{10}}$ **37.** $\dfrac{\sqrt{21}}{\sqrt{3}}$ **38.** $\dfrac{\sqrt{55}}{\sqrt{5}}$ **39.** $\dfrac{\sqrt{90}}{\sqrt{5}}$

40. $\dfrac{\sqrt{96}}{\sqrt{8}}$ **41.** $\dfrac{\sqrt{75y^5}}{\sqrt{3y}}$ **42.** $\dfrac{\sqrt{24x^7}}{\sqrt{6x}}$ **43.** $\dfrac{\sqrt{150}}{\sqrt{2}}$ **44.** $\dfrac{\sqrt{120}}{\sqrt{3}}$

45. $\dfrac{\sqrt{72y^5}}{\sqrt{3y^3}}$ **46.** $\dfrac{\sqrt{54x^3}}{\sqrt{2x}}$ **47.** $\dfrac{\sqrt{24x^3y^4}}{\sqrt{2xy}}$ **48.** $\dfrac{\sqrt{96x^5y^3}}{\sqrt{3x^2y}}$

Objective **C** *Rationalize each denominator and simplify. Assume that all variables represent positive real numbers. See Examples 10 through 12.*

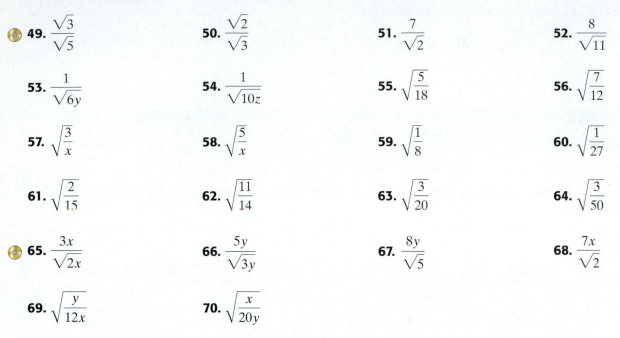

49. $\dfrac{\sqrt{3}}{\sqrt{5}}$

50. $\dfrac{\sqrt{2}}{\sqrt{3}}$

51. $\dfrac{7}{\sqrt{2}}$

52. $\dfrac{8}{\sqrt{11}}$

53. $\dfrac{1}{\sqrt{6y}}$

54. $\dfrac{1}{\sqrt{10z}}$

55. $\sqrt{\dfrac{5}{18}}$

56. $\sqrt{\dfrac{7}{12}}$

57. $\sqrt{\dfrac{3}{x}}$

58. $\sqrt{\dfrac{5}{x}}$

59. $\sqrt{\dfrac{1}{8}}$

60. $\sqrt{\dfrac{1}{27}}$

61. $\sqrt{\dfrac{2}{15}}$

62. $\sqrt{\dfrac{11}{14}}$

63. $\sqrt{\dfrac{3}{20}}$

64. $\sqrt{\dfrac{3}{50}}$

65. $\dfrac{3x}{\sqrt{2x}}$

66. $\dfrac{5y}{\sqrt{3y}}$

67. $\dfrac{8y}{\sqrt{5}}$

68. $\dfrac{7x}{\sqrt{2}}$

69. $\sqrt{\dfrac{y}{12x}}$

70. $\sqrt{\dfrac{x}{20y}}$

Objective **D** *Rationalize each denominator and simplify. Assume that all variables represent positive real numbers. See Examples 13 through 15.*

71. $\dfrac{3}{\sqrt{2}+1}$

72. $\dfrac{6}{\sqrt{5}+2}$

73. $\dfrac{4}{2-\sqrt{5}}$

74. $\dfrac{2}{\sqrt{10}-3}$

75. $\dfrac{\sqrt{5}+1}{\sqrt{6}-\sqrt{5}}$

76. $\dfrac{\sqrt{3}+1}{\sqrt{3}-\sqrt{2}}$

77. $\dfrac{\sqrt{3}+1}{\sqrt{2}-1}$

78. $\dfrac{\sqrt{2}-2}{2-\sqrt{3}}$

79. $\dfrac{5}{2+\sqrt{x}}$

80. $\dfrac{9}{3+\sqrt{x}}$

81. $\dfrac{3}{\sqrt{x}-4}$

82. $\dfrac{4}{\sqrt{x}-1}$

Review

Solve each equation. See Sections 2.3 and 4.6.

83. $x+5=7^2$

84. $2y-1=3^2$

85. $4z^2+6z-12=(2z)^2$

86. $16x^2+x+9=(4x)^2$

87. $9x^2+5x+4=(3x+1)^2$

88. $x^2+3x+4=(x+2)^2$

Concept Extensions

△ **89.** Find the area of a rectangular room whose length is $13\sqrt{2}$ meters and width is $5\sqrt{6}$ meters.

5$\sqrt{6}$ meters

13$\sqrt{2}$ meters

△ **90.** Find the volume of a microwave oven whose length is $\sqrt{3}$ feet, width is $\sqrt{2}$ feet, and height is $\sqrt{2}$ feet.

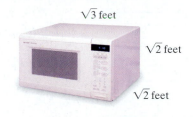

$\sqrt{3}$ feet

$\sqrt{2}$ feet

$\sqrt{2}$ feet

△ **91.** If a circle has area A, then the formula for the radius r of the circle is

$$r = \sqrt{\frac{A}{\pi}}$$

Rationalize the denominator of this expression.

△ **92.** If a round ball has volume V, then the formula for the radius r of the ball is

$$r = \sqrt[3]{\frac{3V}{4\pi}}$$

Simplify this expression by rationalizing the denominator.

Identify each statement as true or false. See the Concept Check in this section.

93. $\sqrt{5} \cdot \sqrt{5} = 5$

94. $\sqrt{5} \cdot \sqrt{3} = 15$

95. $\sqrt{3x} \cdot \sqrt{3x} = 2\sqrt{3x}$

96. $\sqrt{3x} + \sqrt{3x} = 2\sqrt{3x}$

97. $\sqrt{11} + \sqrt{2} = \sqrt{13}$

98. $\sqrt{11} \cdot \sqrt{2} = \sqrt{22}$

99. When rationalizing the denominator of $\dfrac{\sqrt{2}}{\sqrt{3}}$, explain why both the numerator and the denominator must be multiplied by $\sqrt{3}$.

100. In your own words, explain why $\sqrt{6} + \sqrt{2}$ cannot be simplified further, but $\sqrt{6} \cdot \sqrt{2}$ can be.

101. When rationalizing the denominator of $\dfrac{\sqrt[3]{2}}{\sqrt[3]{3}}$, explain why both the numerator and the denominator must be multiplied by $\sqrt[3]{9}$.

102. When rationalizing the denominator of $\dfrac{5}{1 + \sqrt{2}}$, explain why multiplying by $\dfrac{\sqrt{2}}{\sqrt{2}}$ will not accomplish this, but multiplying by $\dfrac{1 - \sqrt{2}}{1 - \sqrt{2}}$ will.

It is often more convenient to work with a radical expression whose numerator is rationalized. Rationalize the numerator of each expression by multiplying the numerator and denominator by the conjugate of the numerator.

103. $\dfrac{\sqrt{3} + 1}{\sqrt{2} - 1}$

104. $\dfrac{\sqrt{2} - 2}{2 - \sqrt{3}}$

Simplifying Radicals

Copyright 2007 Pearson Education, Inc.

Answers

1. _____

2. _____

3. _____

4. _____

5. _____

6. _____

7. _____

8. _____

9. _____

10. _____

11. _____

12. _____

13. _____

14. _____

15. _____

16. _____

17. _____

18. _____

19. _____

20. _____

21. _____

22. _____

23. _____

24. _____

Simplify. Assume that all variables represent positive numbers.

1. $\sqrt{36}$

2. $\sqrt{48}$

3. $\sqrt{x^4}$

4. $\sqrt{y^7}$

5. $\sqrt{16x^2}$

6. $\sqrt{18x^{11}}$

7. $\sqrt[3]{8}$

8. $\sqrt[4]{81}$

9. $\sqrt[3]{-27}$

10. $\sqrt{-4}$

11. $\sqrt{\dfrac{11}{9}}$

12. $\sqrt[3]{\dfrac{7}{64}}$

13. $-\sqrt{16}$

14. $-\sqrt{25}$

15. $\sqrt{\dfrac{9}{49}}$

16. $\sqrt{\dfrac{1}{64}}$

17. $\sqrt{a^8 b^2}$

18. $\sqrt{x^{10} y^{20}}$

19. $\sqrt{25m^6}$

20. $\sqrt{9n^{16}}$

Add or subtract as indicated.

21. $5\sqrt{7} + \sqrt{7}$

22. $\sqrt{50} - \sqrt{8}$

23. $5\sqrt{2} - 5\sqrt{3}$

24. $2\sqrt{x} + \sqrt{25x} - \sqrt{36x} + 3x$

596

Multiply and simplify if possible.

25. $\sqrt{2} \cdot \sqrt{15}$ **26.** $\sqrt{3} \cdot \sqrt{3}$ **27.** $(2\sqrt{7})^2$ **28.** $(3\sqrt{5})^2$

29. $\sqrt{3}(\sqrt{11} + 1)$ **30.** $\sqrt{6}(\sqrt{3} - 2)$ **31.** $\sqrt{8y} \cdot \sqrt{2y}$

32. $\sqrt{15x^2} \cdot \sqrt{3x^2}$ **33.** $(\sqrt{x} - 5)(\sqrt{x} + 2)$ **34.** $(3 + \sqrt{2})^2$

Divide and simplify if possible.

35. $\dfrac{\sqrt{8}}{\sqrt{2}}$ **36.** $\dfrac{\sqrt{45}}{\sqrt{15}}$ **37.** $\dfrac{\sqrt{24x^5}}{\sqrt{2x}}$ **38.** $\dfrac{\sqrt{75a^4b^5}}{\sqrt{5ab}}$

Rationalize each denominator.

39. $\sqrt{\dfrac{1}{6}}$ **40.** $\dfrac{x}{\sqrt{20}}$ **41.** $\dfrac{4}{\sqrt{6} + 1}$ **42.** $\dfrac{\sqrt{2} + 1}{\sqrt{x} - 5}$

25. ____ 26. ____ 27. ____ 28. ____ 29. ____ 30. ____ 31. ____ 32. ____ 33. ____ 34. ____ 35. ____ 36. ____ 37. ____ 38. ____ 39. ____ 40. ____ 41. ____ 42. ____

A Solve Radical Equations by Using the Squaring Property of Equality Once.

B Solve Radical Equations by Using the Squaring Property of Equality Twice.

8.5 SOLVING EQUATIONS CONTAINING RADICALS

Objective **A** Using the Squaring Property of Equality Once

In this section, we solve **radical equations** such as

$$\sqrt{x + 3} = 5 \quad \text{and} \quad \sqrt{2x + 1} = \sqrt{3x}$$

Radical equations contain variables in the radicand. To solve these equations, we rely on the following squaring property.

The Squaring Property of Equality

If $a = b$, then $a^2 = b^2$.

Unfortunately, this squaring property does not guarantee that all solutions of the new equation are solutions of the original equation. For example, if we square both sides of the equation

$$x = 2$$

we have

$$x^2 = 4$$

This new equation has two solutions, 2 and -2, while the original equation $x = 2$ has only one solution. For this reason, we must **always check proposed solutions of radical equations in the original equation.**

PRACTICE PROBLEM 1

Solve: $\sqrt{x - 2} = 7$

EXAMPLE 1 Solve: $\sqrt{x + 3} = 5$

Solution: To solve this radical equation, we use the squaring property of equality and square both sides of the equation.

$$\sqrt{x + 3} = 5$$
$$\left(\sqrt{x + 3}\right)^2 = 5^2 \quad \text{Square both sides.}$$
$$x + 3 = 25 \quad \text{Simplify.}$$
$$x = 22 \quad \text{Subtract 3 from both sides.}$$

Check: We replace x with 22 in the original equation.

$$\sqrt{x + 3} = 5 \quad \text{Original equation}$$
$$\sqrt{22 + 3} \stackrel{?}{=} 5 \quad \text{Let } x = 22.$$
$$\sqrt{25} \stackrel{?}{=} 5$$
$$5 = 5 \quad \text{True}$$

Since a true statement results, 22 is the solution.

Helpful Hint

Don't forget to check the proposed solutions of radical equations in the original equation.

■ **Work Practice Problem 1**

PRACTICE PROBLEM 2

Solve: $\sqrt{6x - 1} = \sqrt{x}$

EXAMPLE 2 Solve: $\sqrt{x} = \sqrt{5x - 2}$

Solution: Each radical is by itself on one side of the equation. Let's begin solving by squaring both sides.

$$\sqrt{x} = \sqrt{5x - 2} \quad \text{Original equation}$$
$$(\sqrt{x})^2 = \left(\sqrt{5x - 2}\right)^2 \quad \text{Square both sides.}$$
$$x = 5x - 2 \quad \text{Simplify.}$$
$$-4x = -2 \quad \text{Subtract 5x from both sides.}$$
$$x = \frac{-2}{-4} = \frac{1}{2} \quad \text{Divide both sides by } -4 \text{ and simplify.}$$

Answers

1. $x = 51$, 2. $x = \frac{1}{5}$

Check: We replace x with $\frac{1}{2}$ in the original equation.

$$\sqrt{x} = \sqrt{5x - 2} \quad \text{Original equation}$$

$$\sqrt{\frac{1}{2}} \stackrel{?}{=} \sqrt{5 \cdot \frac{1}{2} - 2} \quad \text{Let } x = \frac{1}{2}.$$

$$\sqrt{\frac{1}{2}} \stackrel{?}{=} \sqrt{\frac{5}{2} - 2} \quad \text{Multiply.}$$

$$\sqrt{\frac{1}{2}} \stackrel{?}{=} \sqrt{\frac{5}{2} - \frac{4}{2}} \quad \text{Write 2 as } \frac{4}{2}.$$

$$\sqrt{\frac{1}{2}} = \sqrt{\frac{1}{2}} \quad \text{True}$$

This statement is true, so the solution is $\frac{1}{2}$.

🔲 **Work Practice Problem 2**

EXAMPLE 3 Solve: $\sqrt{x} + 6 = 4$

Solution: First we write the equation so that the radical is by itself on one side of the equation.

$$\sqrt{x} + 6 = 4$$

$$\sqrt{x} = -2 \quad \text{Subtract 6 from both sides to get the radical by itself.}$$

Normally we would now square both sides. Recall, however, that $\sqrt{x}$ is the principal or nonnegative square root of x so that $\sqrt{x}$ cannot equal -2 and thus this equation has no solution. We arrive at the same conclusion if we continue by applying the squaring property.

$$\sqrt{x} = -2$$

$$(\sqrt{x})^2 = (-2)^2 \quad \text{Square both sides.}$$

$$x = 4 \quad \text{Simplify.}$$

Check: We replace x with 4 in the original equation.

$$\sqrt{x} + 6 = 4 \quad \text{Original equation}$$

$$\sqrt{4} + 6 \stackrel{?}{=} 4 \quad \text{Let } x = 4.$$

$$2 + 6 = 4 \quad \text{False}$$

Since 4 *does not* satisfy the original equation, this equation has no solution.

🔲 **Work Practice Problem 3**

Example 3 makes it very clear that we *must* check proposed solutions in the original equation to determine if they are truly solutions. If a proposed solution does not work, we say that the value is an **extraneous solution.**

The following steps can be used to solve radical equations containing square roots.

To Solve a Radical Equation Containing Square Roots

Step 1: Arrange terms so that one radical is by itself on one side of the equation. That is, isolate a radical.

Step 2: Square both sides of the equation.

Step 3: Simplify both sides of the equation.

Step 4: If the equation still contains a radical term, repeat Steps 1 through 3.

Step 5: Solve the equation.

Step 6: Check all solutions in the original equation for extraneous solutions.

PRACTICE PROBLEM 3
Solve: $\sqrt{x} + 9 = 2$

Answer
3. no solution

PRACTICE PROBLEM 4

Solve: $\sqrt{9y^2 + 2y - 10} = 3y$

EXAMPLE 4 Solve: $\sqrt{4y^2 + 5y - 15} = 2y$

Solution: The radical is already isolated, so we start by squaring both sides.

$$\sqrt{4y^2 + 5y - 15} = 2y$$

$$\left(\sqrt{4y^2 + 5y - 15}\right)^2 = (2y)^2 \qquad \text{Square both sides.}$$

$$4y^2 + 5y - 15 = 4y^2 \qquad \text{Simplify.}$$

$$5y - 15 = 0 \qquad \text{Subtract } 4y^2 \text{ from both sides.}$$

$$5y = 15 \qquad \text{Add 15 to both sides.}$$

$$y = 3 \qquad \text{Divide both sides by 5.}$$

Check: We replace y with 3 in the original equation.

$$\sqrt{4y^2 + 5y - 15} = 2y \qquad \text{Original equation}$$

$$\sqrt{4 \cdot 3^2 + 5 \cdot 3 - 15} \stackrel{?}{=} 2 \cdot 3 \qquad \text{Let } y = 3.$$

$$\sqrt{4 \cdot 9 + 15 - 15} \stackrel{?}{=} 6 \qquad \text{Simplify.}$$

$$\sqrt{36} \stackrel{?}{=} 6$$

$$6 = 6 \qquad \text{True}$$

This statement is true, so the solution is 3.

🔲 **Work Practice Problem 4**

PRACTICE PROBLEM 5

Solve: $\sqrt{x + 1} - x = -5$

EXAMPLE 5 Solve: $\sqrt{x + 3} - x = -3$

Solution: First we isolate the radical by adding x to both sides. Then we square both sides.

$$\sqrt{x + 3} - x = -3$$

$$\sqrt{x + 3} = x - 3 \qquad \text{Add } x \text{ to both sides.}$$

$$\left(\sqrt{x + 3}\right)^2 = (x - 3)^2 \qquad \text{Square both sides.}$$

$$x + 3 = \underbrace{x^2 - 6x + 9} \qquad \text{Simplify.}$$

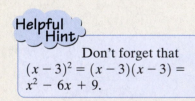

Helpful Hint

Don't forget that $(x - 3)^2 = (x - 3)(x - 3) = x^2 - 6x + 9.$

To solve the resulting quadratic equation, we write the equation in standard form by subtracting x and 3 from both sides.

$$x + 3 = x^2 - 6x + 9$$

$$3 = x^2 - 7x + 9 \qquad \text{Subtract } x \text{ from both sides.}$$

$$0 = x^2 - 7x + 6 \qquad \text{Subtract 3 from both sides.}$$

$$0 = (x - 6)(x - 1) \qquad \text{Factor.}$$

$$0 = x - 6 \quad \text{or} \quad 0 = x - 1 \qquad \text{Set each factor equal to zero.}$$

$$6 = x \qquad\qquad 1 = x \qquad \text{Solve for } x.$$

Check: We replace x with 6 and then x with 1 in the original equation.

Let $x = 6$.

$$\sqrt{x + 3} - x = -3$$

$$\sqrt{6 + 3} - 6 \stackrel{?}{=} -3$$

$$\sqrt{9} - 6 \stackrel{?}{=} -3$$

$$3 - 6 \stackrel{?}{=} -3$$

$$-3 = -3 \qquad \text{True}$$

Let $x = 1$.

$$\sqrt{x + 3} - x = -3$$

$$\sqrt{1 + 3} - 1 \stackrel{?}{=} -3$$

$$\sqrt{4} - 1 \stackrel{?}{=} -3$$

$$2 - 1 \stackrel{?}{=} -3$$

$$1 = -3 \qquad \text{False}$$

Since replacing x with 1 resulted in a false statement, 1 is an extraneous solution. The only solution is 6.

🔲 **Work Practice Problem 5**

Answers

4. $y = 5$, **5.** $x = 8$

Objective B Using the Squaring Property of Equality Twice

If a radical equation contains two radicals, we may need to use the squaring property twice.

EXAMPLE 6 Solve: $\sqrt{x-4} = \sqrt{x} - 2$

Solution:

$$\sqrt{x-4} = \sqrt{x} - 2$$

$$\left(\sqrt{x-4}\right)^2 = \left(\sqrt{x} - 2\right)^2 \quad \text{Square both sides.}$$

$$x - 4 = \underbrace{x - 4\sqrt{x} + 4}$$

$$-8 = -4\sqrt{x} \quad \text{To get the radical term alone, subtract } x \text{ and}$$
$$\qquad\qquad\qquad 4 \text{ from both sides.}$$

$$2 = \sqrt{x} \quad \text{Divide both sides by } -4.$$

$$4 = x \quad \text{Square both sides again.}$$

Check the proposed solution in the original equation. The solution is 4.

■ **Work Practice Problem 6**

Helpful Hint

Don't forget:

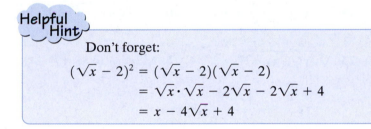

$$(\sqrt{x} - 2)^2 = (\sqrt{x} - 2)(\sqrt{x} - 2)$$
$$= \sqrt{x} \cdot \sqrt{x} - 2\sqrt{x} - 2\sqrt{x} + 4$$
$$= x - 4\sqrt{x} + 4$$

PRACTICE PROBLEM 6

Solve: $\sqrt{x+3} = \sqrt{x+15}$

Answer

6. $x = 1$

Objective A *Solve each equation. See Examples 1 through 3.*

1. $\sqrt{x} = 9$

2. $\sqrt{x} = 4$

3. $\sqrt{x + 5} = 2$

4. $\sqrt{x + 12} = 3$

5. $\sqrt{x} - 2 = 5$

6. $4\sqrt{x} - 7 = 5$

7. $3\sqrt{x} + 5 = 2$

8. $3\sqrt{x} + 8 = 5$

9. $\sqrt{x} = \sqrt{3x - 8}$

10. $\sqrt{x} = \sqrt{4x - 3}$

11. $\sqrt{4x - 3} = \sqrt{x + 3}$

12. $\sqrt{5x - 4} = \sqrt{x + 8}$

Solve each equation. See Examples 4 and 5.

13. $\sqrt{9x^2 + 2x - 4} = 3x$

14. $\sqrt{4x^2 + 3x - 9} = 2x$

15. $\sqrt{x} = x - 6$

16. $\sqrt{x} = x - 2$

17. $\sqrt{x + 7} = x + 5$

18. $\sqrt{x + 5} = x - 1$

19. $\sqrt{3x + 7} - x = 3$

20. $x = \sqrt{4x - 7} + 1$

21. $\sqrt{16x^2 + 2x + 2} = 4x$

22. $\sqrt{4x^2 + 3x + 2} = 2x$

23. $\sqrt{2x^2 + 6x + 9} = 3$

24. $\sqrt{3x^2 + 6x + 4} = 2$

Objective B *Solve each equation. See Example 6.*

25. $\sqrt{x - 7} = \sqrt{x} - 1$

26. $\sqrt{x - 8} = \sqrt{x} - 2$

27. $\sqrt{x + 2} = \sqrt{x + 24}$

28. $\sqrt{x} + 5 = \sqrt{x + 55}$

29. $\sqrt{x + 8} = \sqrt{x} + 2$

30. $\sqrt{x} + 1 = \sqrt{x + 15}$

Objectives A B Mixed Practice *Solve each equation. See Examples 1 through 6.*

31. $\sqrt{2x + 6} = 4$

32. $\sqrt{3x + 7} = 5$

33. $\sqrt{x + 6} + 1 = 3$

34. $\sqrt{x + 5} + 2 = 5$

35. $\sqrt{x + 6} + 5 = 3$

36. $\sqrt{2x - 1} + 7 = 1$

37. $\sqrt{16x^2 - 3x + 6} = 4x$

38. $\sqrt{9x^2 - 2x + 8} = 3x$

39. $-\sqrt{x} = -6$

40. $-\sqrt{y} = -8$

41. $\sqrt{x + 9} = \sqrt{x} - 3$

42. $\sqrt{x} - 6 = \sqrt{x + 36}$

43. $\sqrt{2x + 1} + 3 = 5$

44. $\sqrt{3x - 1} + 1 = 4$

45. $\sqrt{x} + 3 = 7$

46. $\sqrt{x} + 5 = 10$

47. $\sqrt{4x} = \sqrt{2x + 6}$

48. $\sqrt{5x + 6} = \sqrt{8x}$

49. $\sqrt{2x + 1} = x - 7$

50. $\sqrt{2x + 5} = x - 5$

51. $x = \sqrt{2x - 2} + 1$

52. $\sqrt{2x - 4} + 2 = x$

53. $\sqrt{1 - 8x} - x = 4$

54. $\sqrt{2x + 5} - 1 = x$

Review

Translate each sentence into an equation and then solve. See Section 2.4.

55. If 8 is subtracted from the product of 3 and x, the result is 19. Find x.

56. If 3 more than x is subtracted from twice x, the result is 11. Find x.

57. The length of a rectangle is twice the width. The perimeter is 24 inches. Find the length.

58. The length of a rectangle is 2 inches longer than the width. The perimeter is 24 inches. Find the length.

Concept Extensions

Solve each equation.

59. $\sqrt{x - 3} + 3 = \sqrt{3x + 4}$

60. $\sqrt{2x + 3} = \sqrt{x - 2} + 2$

61. Explain why proposed solutions of radical equations must be checked in the original equation.

62. Is 8 a solution of the equation $\sqrt{x - 4} - 5 = \sqrt{x + 1}$? Explain why or why not.

63. The formula $b = \sqrt{\dfrac{V}{2}}$ can be used to determine the length b of a side of the base of a square-based pyramid with height 6 units and volume V cubic units.

    a. Find the length of the side of the base that produces a pyramid with each volume. (Round to the nearest tenth of a unit.)

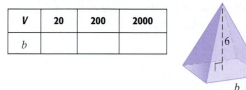

V	20	200	2000
b			

    b. Notice in the table that volume V has been increased by a factor of 10 each time. Does the corresponding length b of a side increase by a factor of 10 each time also?

64. The formula $r = \sqrt{\dfrac{V}{2\pi}}$ can be used to determine the radius r of a cylinder with height 2 units and volume V cubic units.

    a. Find the radius needed to manufacture a cylinder with each volume. (Round to the nearest tenth of a unit.)

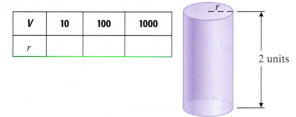

V	10	100	1000
r			

2 units

    b. Notice in the table that volume V has been increased by a factor of 10 each time. Does the corresponding radius increase by a factor of 10 each time also?

Graphing calculators can be used to solve equations. To solve $\sqrt{x - 2} = x - 5$, for example, graph $y_1 = \sqrt{x - 2}$ and $y_2 = x - 5$ on the same set of axes. Use the Trace and Zoom features or an Intersect feature to find the point of intersection of the graphs. The x-value of the point is the solution of the equation. Use a graphing calculator to solve the equations below. Approximate solutions to the nearest hundredth.

65. $\sqrt{x - 2} = x - 5$

66. $\sqrt{x + 1} = 2x - 3$

67. $-\sqrt{x + 4} = 5x - 6$

68. $-\sqrt{x + 5} = -7x + 1$

THE BIGGER PICTURE Simplifying Expressions and Solving Equations

Now we continue our outline from Sections 1.6, 2.7, 3.7, 4.6, 5.4, and 5.6. Although suggestions are given, this outline should be in your own words. Once you complete this new portion, try the exercises below.

I. Simplifying Expressions

  A. Real Numbers

   1. Add (Section 1.4)

   2. Subtract (Section 1.5)

   3. Multiply or Divide (Section 1.6)

  B. Exponents (Section 3.2)

  C. Polynomials

   1. Add (Section 3.4)

   2. Subtract (Section 3.4)

   3. Multiply (Section 3.5)

   4. Divide (Section 3.7)

  D. Factoring Polynomials (Chapter 4 Integrated Review)

  E. Rational Expressions

   1. Simplify (Section 5.1)

   2. Multiply (Section 5.2)

   3. Divide (Section 5.2)

   4. Add or Subtract (Section 5.4)

  F. Radicals

   1. Simplify square roots: If possible, factor the radicand so that one factor is a perfect square. Then use the product rule, and simplify.

   $$\sqrt{75} = \sqrt{25 \cdot 3} = \sqrt{25} \cdot \sqrt{3} = 5\sqrt{3}$$

   2. Add or subtract: Only like radicals (same index and radicand) can be added or subtracted.

   $$8\sqrt{10} - \sqrt{40} + \sqrt{5}$$
   $$= 8\sqrt{10} - 2\sqrt{10} + \sqrt{5}$$
   $$= 6\sqrt{10} + \sqrt{5}$$

   3. Multiply or divide:

   $$\sqrt{a} \cdot \sqrt{b} = \sqrt{ab}; \frac{\sqrt{a}}{\sqrt{b}} = \sqrt{\frac{a}{b}}.$$
   $$\sqrt{11} \cdot \sqrt{3} = \sqrt{33};$$
   $$\frac{\sqrt{140}}{\sqrt{7}} = \sqrt{\frac{140}{7}} = \sqrt{20} = \sqrt{4 \cdot 5} = 2\sqrt{5}$$

   4. Rationalizing the denominator:

   a. If the denominator is one term,

   $$\frac{5}{\sqrt{11}} = \frac{5 \cdot \sqrt{11}}{\sqrt{11} \cdot \sqrt{11}} = \frac{5\sqrt{11}}{11}$$

   b. If the denominator has two terms, multiply by 1 in the form of $\dfrac{\text{conjugate of denominator}}{\text{conjugate of denominator}}$.

   $$\frac{13}{3 + \sqrt{2}} = \frac{13}{3 + \sqrt{2}} \cdot \frac{3 - \sqrt{2}}{3 - \sqrt{2}}$$
   $$= \frac{13(3 - \sqrt{2})}{9 - 2} = \frac{13(3 - \sqrt{2})}{7}$$

II. Solving Equations and Inequalities

  A. Linear Equations (Section 2.3)

  B. Linear Inequalities (Section 2.7)

  C. Quadratic and Higher Degree Equations (Section 4.6)

  D. Equations with Rational Expressions (Section 5.5)

  E. Proportions (Section 5.6)

  F. Equations with Radicals To solve, isolate a radical, then square both sides. You may have to repeat this. Check possible solution in the original equation.

   $$\sqrt{x + 49} + 7 = x$$
   $$\sqrt{x + 49} = x - 7 \qquad \text{Subtract 7 from both sides.}$$
   $$x + 49 = x^2 - 14x + 49 \qquad \text{Square both sides.}$$
   $$0 = x^2 - 15x \qquad \text{Set terms equal to 0.}$$
   $$0 = x(x - 15) \qquad \text{Factor.}$$
   $$\cancel{x = 0} \text{ or } x = 15 \qquad \text{Set each factor equal to 0 and solve.}$$

Perform indicated operations and simplify. If necessary, rationalize the denominator.

1. $\sqrt{56}$

2. $\sqrt{\dfrac{20x^5}{49}}$

3. $(-5x^{12}y^{-3})(3x^{-7}y^{14})$

4. $\sqrt{\dfrac{10}{11}}$

5. $\dfrac{8}{\sqrt{5} - 1}$

6. $\dfrac{1}{2}(6x^2 - 4) + \dfrac{1}{3}(6x^2 - 9) - 14$

Solve each equation or inequality.

7. $9x - 7 = 7x - 9$

8. $\dfrac{x}{5} = \dfrac{x - 3}{11}$

9. $-5(2y + 1) \leq 3y - 2 - 2y + 1$

10. $x(x + 1) = 42$

11. $\dfrac{-6}{x - 7} + \dfrac{8}{x} = \dfrac{-4}{x - 7}$

12. $1 + 4(x - 2) = x(x - 6) - x^2 + 13$

8.6 RADICAL EQUATIONS AND PROBLEM SOLVING

Objectives

A Use the Pythagorean Theorem to Solve Problems.

B Solve Problems Using Formulas Containing Radicals.

Objective **A** Using the Pythagorean Theorem

Applications of radicals can be found in geometry, finance, science, and other areas of technology. Our first application involves the Pythagorean theorem, which gives a formula that relates the lengths of the three sides of a right triangle. We first studied the Pythagorean theorem in Chapter 4 and we review it here.

The Pythagorean Theorem

If a and b are lengths of the legs of a right triangle and c is the length of the hypotenuse, then $a^2 + b^2 = c^2$.

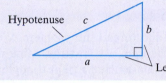

PRACTICE PROBLEM 1

Find the length of the hypotenuse of the right triangle shown.

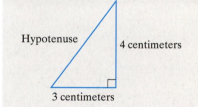

EXAMPLE 1 Find the length of the hypotenuse of a right triangle whose legs are 6 inches and 8 inches long.

Solution: Because this is a right triangle, we use the Pythagorean theorem. We let $a = 6$ inches and $b = 8$ inches. Length c must be the length of the hypotenuse.

$$a^2 + b^2 = c^2 \quad \text{Use the Pythagorean theorem.}$$
$$6^2 + 8^2 = c^2 \quad \text{Substitute the lengths of the legs.}$$
$$36 + 64 = c^2 \quad \text{Simplify.}$$
$$100 = c^2$$

Since c represents a length, we know that c is positive and is the principal square root of 100.

$$100 = c^2$$
$$\sqrt{100} = c \quad \text{Use the definition of principal square root.}$$
$$10 = c \quad \text{Simplify.}$$

The hypotenuse has a length of 10 inches.

☐ **Work Practice Problem 1**

EXAMPLE 2 Find the length of the leg of the right triangle shown. Give the exact length and a two-decimal-place approximation.

Solution: We let $a = 2$ meters and b be the unknown length of the other leg. The hypotenuse is $c = 5$ meters.

$$a^2 + b^2 = c^2 \quad \text{Use the Pythagorean theorem.}$$
$$2^2 + b^2 = 5^2 \quad \text{Let } a = 2 \text{ and } c = 5.$$
$$4 + b^2 = 25$$
$$b^2 = 21$$
$$b = \sqrt{21} \approx 4.58 \text{ meters}$$

The length of the leg is exactly $\sqrt{21}$ meters or approximately 4.58 meters.

☐ **Work Practice Problem 2**

PRACTICE PROBLEM 2

Find the length of the leg of the right triangle shown. Give the exact length and a two-decimal-place approximation.

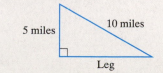

Answers

1. 5 cm, **2.** $5\sqrt{3}$ mi; 8.66 mi

PRACTICE PROBLEM 3

Evan Saacks wants to determine the distance at certain points across a pond on his property. He is able to measure the distances shown on the following diagram. Find how wide the pond is to the nearest tenth of a foot.

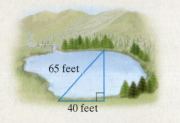

65 feet

40 feet

⚠️ **EXAMPLE 3** **Finding a Distance**

A surveyor must determine the distance across a lake at points P and Q as shown in the figure. To do this, she finds a third point R perpendicular to line PQ. If the length of $\overline{PR}$ is 320 feet and the length of $\overline{QR}$ is 240 feet, what is the distance across the lake? Approximate this distance to the nearest whole foot.

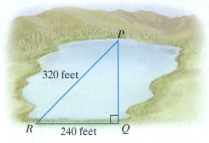

320 feet

R 240 feet Q

Solution:

1. UNDERSTAND. Read and reread the problem. We will set up the problem using the Pythagorean theorem. By creating a line perpendicular to line PQ, the surveyor deliberately constructed a right triangle. The hypotenuse, $\overline{PR}$, has a length of 320 feet, so we let $c = 320$ in the Pythagorean theorem. The side $\overline{QR}$ is one of the legs, so we let $a = 240$ and $b =$ the unknown length.

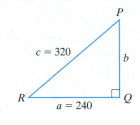

$c = 320$ b

R $a = 240$ Q

2. TRANSLATE.

$$a^2 + b^2 = c^2 \qquad \text{Use the Pythagorean theorem.}$$
$$240^2 + b^2 = 320^2 \qquad \text{Let } a = 240 \text{ and } c = 320.$$

3. SOLVE.

$$57{,}600 + b^2 = 102{,}400$$
$$b^2 = 44{,}800 \qquad \text{Subtract 57,600 from both sides.}$$
$$b = \sqrt{44{,}800} \qquad \text{Use the definition of principal square root.}$$
$$= 80\sqrt{7} \qquad \text{Simplify.}$$

4. INTERPRET.

Check: See that $240^2 + \left(\sqrt{44{,}800}\right)^2 = 320^2$.

State: The distance across the lake is *exactly* $\sqrt{44{,}800}$ or $80\sqrt{7}$ feet. The surveyor can now use a calculator to find that $80\sqrt{7}$ feet is *approximately* 211.6601 feet, so the distance across the lake is roughly 212 feet.

🟧 **Work Practice Problem 3**

Objective B Using Formulas Containing Radicals

The Pythagorean theorem is an extremely important result in mathematics and should be memorized. But there are other applications involving formulas containing radicals that are not quite as well known, such as the velocity formula used in the next example.

Answer

3. 51.2 feet

EXAMPLE 4 **Finding the Velocity of an Object**

A formula used to determine the velocity v, in feet per second, of an object after it has fallen a certain height (neglecting air resistance) is $v = \sqrt{2gh}$, where g is the acceleration due to gravity and h is the height the object has fallen. On Earth, the acceleration g due to gravity is approximately 32 feet per second per second. Find the velocity of a person after falling 5 feet.

Solution: We are told that $g = 32$ feet per second per second. To find the velocity v when $h = 5$ feet, we use the velocity formula.

$$v = \sqrt{2gh} \qquad \text{Use the velocity formula.}$$
$$= \sqrt{2 \cdot 32 \cdot 5} \qquad \text{Substitute known values.}$$
$$= \sqrt{320}$$
$$= 8\sqrt{5} \qquad \text{Simplify the radicand.}$$

The velocity of the person after falling 5 feet is *exactly* $8\sqrt{5}$ feet per second, or *approximately* 17.9 feet per second.

 Work Practice Problem 4

STUDY SKILLS BUILDER

Are You Prepared for a Test on Chapter 8?

Below I have listed some *common trouble areas* for students in Chapter 8. After studying for your test—but before taking your test—read these.

- Do you understand the difference between $\sqrt{3} \cdot \sqrt{2}$ and $\sqrt{3} + \sqrt{2}$?

  $$\sqrt{3} \cdot \sqrt{2} = \sqrt{3 \cdot 2} = \sqrt{6}$$

  $\sqrt{3} + \sqrt{2}$ cannot be simplified further. The terms are unlike terms.

- Do you understand the difference between rationalizing the denominator of $\dfrac{\sqrt{3}}{\sqrt{7}}$ and rationalizing the denominator of $\dfrac{\sqrt{3}}{\sqrt{7} + 1}$?

  $$\frac{\sqrt{3}}{\sqrt{7}} = \frac{\sqrt{3} \cdot \sqrt{7}}{\sqrt{7} \cdot \sqrt{7}} = \frac{\sqrt{21}}{7}$$

$$\frac{\sqrt{3}}{\sqrt{7} + 1} = \frac{\sqrt{3}(\sqrt{7} - 1)}{(\sqrt{7} + 1)(\sqrt{7} - 1)}$$
$$= \frac{\sqrt{3}(\sqrt{7} - 1)}{7 - 1} = \frac{\sqrt{3}(\sqrt{7} - 1)}{6}$$

- To solve an equation containing a radical, don't forget to first isolate the radical.

  $$\sqrt{x} - 10 = -4$$
  $$\sqrt{x} = 6 \qquad \text{Isolate the radical.}$$
  $$(\sqrt{x})^2 = 6^2 \qquad \text{Square both sides.}$$
  $$x = 36 \qquad \text{Simplify.}$$

Make sure you check the proposed solution in the original equation.

Remember: This is simply a listing of a few common trouble areas. For a review of Chapter 8, see the Highlights and Chapter Review at the end of the chapter.

Objective Ⓐ *Use the Pythagorean theorem to find the length of the unknown side of each right triangle. Give an exact answer and a two-decimal-place approximation. See Examples 1 and 2.*

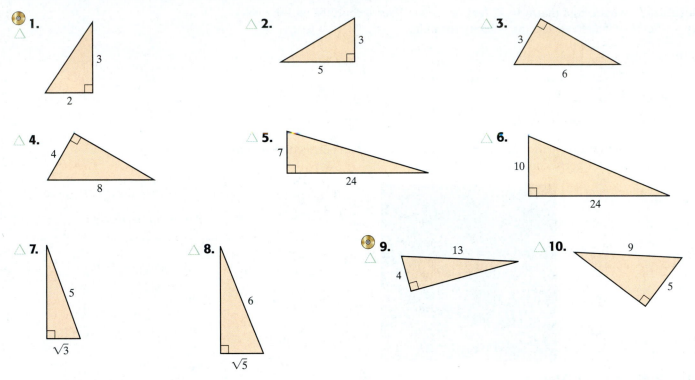

1.
3
2

2.
3
5

3.
3
6

4.
4
8

5.
7
24

6.
10
24

7.
5
$\sqrt{3}$

8.
6
$\sqrt{5}$

9.
13
4

10.
9
5

Find the length of the unknown side of each right triangle with sides a, b, and c, where c is the hypotenuse. See Examples 1 and 2. Give an exact answer and a two-decimal-place approximation.

11. $a = 4, b = 5$

12. $a = 2, b = 7$

13. $b = 2, c = 6$

14. $b = 1, c = 5$

15. $a = \sqrt{10}, c = 10$

16. $a = \sqrt{7}, c = \sqrt{35}$

Solve each problem. See Example 3.

17. A wire is used to anchor a 20-foot-tall pole. One end of the wire is attached to the top of the pole. The other end is fastened to a stake five feet away from the bottom of the pole. Find the length of the wire, to the nearest tenth of a foot.

20 feet

5 feet

18. Jim Spivey needs to connect two underground pipelines, which are offset by 3 feet, as pictured in the diagram. Neglecting the joints needed to join the pipes, find the length of the shortest possible connecting pipe rounded to the nearest hundredth of a foot.

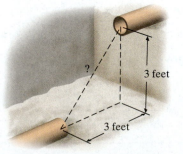

?

3 feet

3 feet

△ **19.** Robert Weisman needs to attach a diagonal brace to a rectangular frame in order to make it structurally sound. If the framework is 6 feet by 10 feet, find how long the brace needs to be to the nearest tenth of a foot.

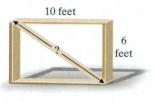

10 feet

6 feet

?

△ **20.** Elizabeth Kaster is flying a kite. She let out 80 feet of string and attached the string to a stake in the ground. The kite is now directly above her brother Mike, who is 32 feet away from the stake. Find the height of the kite to the nearest foot.

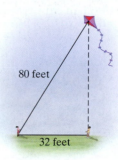

80 feet

32 feet

Objective B *Solve each problem. See Example 4.*

△ **21.** For a square-based pyramid, the formula $b = \sqrt{\dfrac{3V}{h}}$ describes the relationship between the length b of one side of the base, the volume V, and the height h. Find the volume if each side of the base is 6 feet long, and the pyramid is 2 feet high.

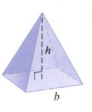

h

b

22. The formula $t = \dfrac{\sqrt{d}}{4}$ relates the distance d, in feet, that an object falls in t seconds, assuming that air resistance does not slow down the object. Find how long, to the nearest hundredth of a second, it takes an object to reach the ground from the top of the Sears Tower in Chicago, a distance of 1730 feet. (*Source:* Council on Tall Buildings and Urban Habitat)

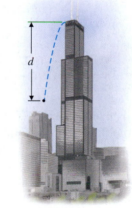

d

23. Police use the formula $s = \sqrt{30fd}$ to estimate the speed s of a car just before it skidded. In this formula, the speed s is measured in miles per hour, d represents the distance the car skidded in feet and f represents the coefficient of friction. The value of f depends on the type of road surface, and for wet concrete f is 0.35. Find how fast a car was moving if it skidded 280 feet on wet concrete. Round your result to the nearest mile per hour.

d

24. The coefficient of friction of a certain dry road is 0.95. Use the formula in Exercise 23 to find how far a car will skid on this dry road if it is traveling at a rate of 60 mph. Round the length to the nearest foot.

25. The formula $v = \sqrt{2.5r}$ can be used to estimate the maximum safe velocity, v, in miles per hour, at which a car can travel if it is driven along a curved road with a **radius of curvature** r in feet. Find the maximum safe speed to the nearest whole number if a cloverleaf exit on an expressway has a radius of curvature of 300 feet.

26. Use the formula from Exercise 25 to find the radius of curvature if the safe velocity is 30 mph.

27. The maximum distance d in kilometers that you can see from a height of h meters is given by $d = 3.5\sqrt{h}$. Find how far you can see from the top of the Bank One Tower in Indianapolis, a height of 285.4 meters. Round to the nearest tenth of a kilometer. (*Source: World Almanac and Book of Facts*, 2001)

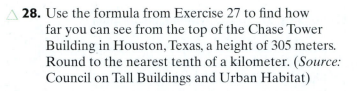

28. Use the formula from Exercise 27 to find how far you can see from the top of the Chase Tower Building in Houston, Texas, a height of 305 meters. Round to the nearest tenth of a kilometer. (*Source:* Council on Tall Buildings and Urban Habitat)

29. Use the formula from Exercise 27 to find how far you can see from the top of the First Interstate Tower in Houston, Texas, a height of 295.7 meters. Round to the nearest tenth of a kilometer. (*Source:* Council on Tall Buildings and Urban Habitat)

30. Use the formula from Exercise 27 to find how far you can see from the top of the Gas Company Tower in Los Angeles, California, a height of 228.3 m. Round to the nearest tenth of a kilometer. (*Source:* Council on Tall Buildings and Urban Habitat)

Review

Find two numbers whose square is the given number. See Section 8.1.

31. 9

32. 25

33. 100

34. 49

35. 64

36. 121

Concept Extensions

For each triangle, find the length of y, then x.

37.

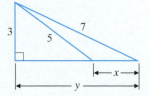

38.

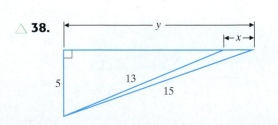

△ 39. Mike and Sandra Hallahan leave the seashore at the same time. Mike drives northward at a rate of 30 miles per hour, while Sandra drives west at 60 mph. Find how far apart they are after 3 hours to the nearest mile.

Distance apart

30 mph for 3 hours

60 mph for 3 hours

△ 40. Railroad tracks are invariably made up of relatively short sections of rail connected by expansion joints. To see why this construction is necessary, consider a single rail 100 feet long (or 1200 inches). On an extremely hot day, suppose it expands 1 inch in the hot sun to a new length of 1201 inches. Theoretically, the track would bow upward as pictured.

100 feet = 1200 inches

1201 inches

Let us approximate the bulge in the railroad this way.

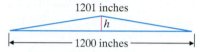

1201 inches

h

1200 inches

Calculate the height h of the bulge to the nearest tenth of an inch.

41. Based on the results of Exercise 40, explain why railroads use short sections of rail connected by expansion joints.

CHAPTER 8 Group Activity

Graphing and the Distance Formula

One application of radicals is finding the distance between two points in the coordinate plane. This can be very useful in graphing.

The distance d between two points with coordinates (x_1, y_1) and (x_2, y_2) is given by the **distance formula** $d = \sqrt{(x_2 - x_1)^2 + (y_2 - y_1)^2}$.

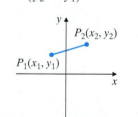

$P_2(x_2, y_2)$

$P_1(x_1, y_1)$

Suppose we want to find the distance between the two points $(-1, 9)$ and $(3, 5)$. We can use the distance formula with $(x_1, y_1) = (-1, 9)$ and $(x_2, y_2) = (3, 5)$. Then we have

$$d = \sqrt{(x_2 - x_1)^2 + (y_2 - y_1)^2}$$
$$= \sqrt{[3 - (-1)]^2 + (5 - 9)^2}$$
$$= \sqrt{(4)^2 + (-4)^2}$$
$$= \sqrt{16 + 16}$$
$$= \sqrt{32} = 4\sqrt{2}$$

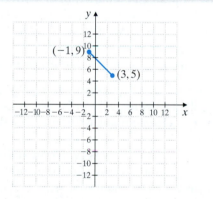

$(-1, 9)$

$(3, 5)$

The distance between the two points is exactly $4\sqrt{2}$ units or approximately 5.66 units.

Group Activity

Brainstorm to come up with several disciplines or activities in which the distance formula might be useful. Make up an example that shows how the distance formula would be used in one of the activities on your list. Then present your example to the rest of the class.

Chapter 8 Vocabulary Check

Fill in each blank with one of the words or phrases listed below.

index	radicand	like radicals
rationalizing the denominator	conjugate	
principal square root	radical	

1. The expressions $5\sqrt{x}$ and $7\sqrt{x}$ are examples of _____.
2. In the expression $\sqrt[3]{45}$ the number 3 is the _____, the number 45 is the _____, and $\sqrt{}$ is called the _____ sign.
3. The _____ of $a + b$ is $a - b$.
4. The _____ of 25 is 5.
5. The process of eliminating the radical in the denominator of a radical expression is called

   _____.

Helpful Hint

Are you preparing for your test? Don't forget to take the Chapter 8 Test on page 618. Then check your answers at the back of the text and use the Chapter Test Prep Video CD to see the fully worked-out solutions to any of the exercises you want to review.

8 Chapter Highlights

DEFINITIONS AND CONCEPTS	**EXAMPLES**
Section 8.1 Introduction to Radicals	

The **positive or principal square root** of a positive number a is written as $\sqrt{a}$. The **negative square root** of a is written as $-\sqrt{a}$. $\sqrt{a} = b$ only if $b^2 = a$ and $b > 0$.	$\sqrt{25} = 5$ $\sqrt{100} = 10$ $-\sqrt{9} = -3$ $\sqrt{\dfrac{4}{49}} = \dfrac{2}{7}$
A square root of a negative number is not a real number.	$\sqrt{-4}$ is not a real number.
The **cube root** of a real number a is written as $\sqrt[3]{a}$ and $\sqrt[3]{a} = b$ only if $b^3 = a$.	$\sqrt[3]{64} = 4$ $\sqrt[3]{-8} = -2$
The **nth root** of a number a is written as $\sqrt[n]{a}$ and $\sqrt[n]{a} = b$ only if $b^n = a$.	$\sqrt[4]{81} = 3$
In $\sqrt[n]{a}$, the natural number n is called the **index,** the symbol $\sqrt{}$ is called a **radical,** and the expression within the radical is called the **radicand.**	$\sqrt[5]{-32} = -2$
(*Note:* If the index is even, the radicand must be non-negative for the root to be a real number.)	index $\downarrow$ $\sqrt[n]{a}$ $\uparrow$ radicand

Section 8.2 Simplifying Radicals	

PRODUCT RULE FOR RADICALS	
If $\sqrt{a}$ and $\sqrt{b}$ are real numbers, then $$\sqrt{a \cdot b} = \sqrt{a} \cdot \sqrt{b}$$	
A square root is in **simplified form** if the radicand contains no perfect square factors other than 1. To simplify a square root, factor the radicand so that one of its factors is a perfect square factor.	$\begin{aligned} \sqrt{45} &= \sqrt{9 \cdot 5} \\ &= \sqrt{9} \cdot \sqrt{5} \\ &= 3\sqrt{5} \end{aligned}$

DEFINITIONS AND CONCEPTS	**EXAMPLES**

Section 8.2   Simplifying Radicals (*continued*)

QUOTIENT RULE FOR RADICALS

If $\sqrt{a}$ and $\sqrt{b}$ are real numbers and $b \neq 0$, then

$$\sqrt{\frac{a}{b}} = \frac{\sqrt{a}}{\sqrt{b}}$$

$$\sqrt{\frac{18}{x^6}} = \frac{\sqrt{9 \cdot 2}}{\sqrt{x^6}} = \frac{\sqrt{9} \cdot \sqrt{2}}{x^3} = \frac{3\sqrt{2}}{x^3}$$

Section 8.3   Adding and Subtracting Radicals

Like radicals are radical expressions that have the same index and the same radicand.

To **combine like radicals** use the distributive property.

$$5\sqrt{2}, -7\sqrt{2}, \sqrt{2}$$

$$2\sqrt{7} - 13\sqrt{7} = (2 - 13)\sqrt{7} = -11\sqrt{7}$$
$$\sqrt{8} + \sqrt{50} = 2\sqrt{2} + 5\sqrt{2} = 7\sqrt{2}$$

Section 8.4   Multiplying and Dividing Radicals

The product and quotient rules for radicals may be used to simplify products and quotients of radicals.

Perform each indicated operation and simplify.
Multiply.

$$\sqrt{2} \cdot \sqrt{8} = \sqrt{16} = 4$$
$$(\sqrt{3x} + 1)(\sqrt{5} - \sqrt{3})$$
$$= \sqrt{15x} - \sqrt{9x} + \sqrt{5} - \sqrt{3}$$
$$= \sqrt{15x} - 3\sqrt{x} + \sqrt{5} - \sqrt{3}$$

Divide.

$$\frac{\sqrt{20}}{\sqrt{2}} = \sqrt{\frac{20}{2}} = \sqrt{10}$$

The process of eliminating the radical in the denominator of a radical expression is called **rationalizing the denominator.**

Rationalize the denominator.

$$\frac{5}{\sqrt{11}} = \frac{5 \cdot \sqrt{11}}{\sqrt{11} \cdot \sqrt{11}} = \frac{5\sqrt{11}}{11}$$

The **conjugate** of $a + b$ is $a - b$.

The conjugate of $2 + \sqrt{3}$ is $2 - \sqrt{3}$.

To rationalize a denominator that is a sum or difference of radicals, multiply the numerator and the denominator by the conjugate of the denominator.

Rationalize the denominator.

$$\frac{5}{6 - \sqrt{5}} = \frac{5(6 + \sqrt{5})}{(6 - \sqrt{5})(6 + \sqrt{5})}$$
$$= \frac{5(6 + \sqrt{5})}{36 - 5}$$
$$= \frac{5(6 + \sqrt{5})}{31}$$

DEFINITIONS AND CONCEPTS	**EXAMPLES**
Section 8.5 Solving Equations Containing Radicals	

TO SOLVE A RADICAL EQUATION CONTAINING SQUARE ROOTS

Step 1. Get one radical by itself on one side of the equation.

Step 2. Square both sides of the equation.

Step 3. Simplify both sides of the equation.

Step 4. If the equation still contains a radical term, repeat Steps 1 through 3.

Step 5. Solve the equation.

Step 6. Check solutions in the original equation.

Solve:

$$\sqrt{2x - 1} - x = -2$$
$$\sqrt{2x - 1} = x - 2$$
$$\left(\sqrt{2x - 1}\right)^2 = (x - 2)^2 \qquad \text{Square both sides.}$$
$$2x - 1 = x^2 - 4x + 4$$
$$0 = x^2 - 6x + 5$$
$$0 = (x - 1)(x - 5) \qquad \text{Factor.}$$
$$x - 1 = 0 \quad \text{or} \quad x - 5 = 0$$
$$x = 1 \qquad\qquad x = 5 \quad \text{Solve.}$$

Check both proposed solutions in the original equation. Here, 5 checks but 1 does not. The only solution is 5.

Section 8.6 Radical Equations and Problem Solving	

PROBLEM-SOLVING STEPS

1. UNDERSTAND. Read and reread the problem.

A rain gutter is to be mounted on the eaves of a house 15 feet above the ground. A garden is adjacent to the house so that the closest a ladder can be placed to the house is 6 feet. How long a ladder is needed for installing the gutter?

Let x = the length of the ladder.

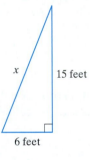

2. TRANSLATE.

Here, we use the Pythagorean theorem. The unknown length x is the hypotenuse.

In words:

$$(\text{leg})^2 \; + \; (\text{leg})^2 \; = \; (\text{hypotenuse})^2$$

3. SOLVE.

Translate:

$$6^2 + 15^2 = x^2$$
$$36 + 225 = x^2$$
$$261 = x^2$$
$$\sqrt{261} = x \quad \text{or} \quad x = 3\sqrt{29}$$

4. INTERPRET.

Check and state. The ladder needs to be $3\sqrt{29}$ feet or approximately 16.2 feet long.

8 CHAPTER REVIEW

(8.1) *Find each root.*

1. $\sqrt{81}$

2. $-\sqrt{49}$

3. $\sqrt[3]{27}$

4. $\sqrt[4]{81}$

5. $-\sqrt{\dfrac{9}{64}}$

6. $\sqrt{\dfrac{36}{81}}$

7. $\sqrt[4]{16}$

8. $\sqrt[3]{-8}$

9. Which radical(s) is not a real number?

   a. $\sqrt{4}$ **b.** $-\sqrt{4}$ **c.** $\sqrt{-4}$ **d.** $\sqrt[3]{-4}$

10. Which radical(s) is not a real number?

   a. $\sqrt{-5}$ **b.** $\sqrt[3]{-5}$ **c.** $\sqrt[4]{-5}$ **d.** $\sqrt[5]{-5}$

Find each root. Assume that all variables represent positive numbers.

11. $\sqrt{x^{12}}$

12. $\sqrt{x^8}$

13. $\sqrt{9y^2}$

14. $\sqrt{25x^4}$

(8.2) *Simplify each expression using the product rule. Assume that all variables represent positive numbers.*

15. $\sqrt{40}$

16. $\sqrt{24}$

17. $\sqrt{54}$

18. $\sqrt{88}$

19. $\sqrt{x^5}$

20. $\sqrt{y^7}$

21. $\sqrt{20x^2}$

22. $\sqrt{50y^4}$

23. $\sqrt[3]{54}$

24. $\sqrt[3]{88}$

Simplify each expression using the quotient rule. Assume that all variables represent positive numbers.

25. $\sqrt{\dfrac{18}{25}}$

26. $\sqrt{\dfrac{75}{64}}$

27. $-\sqrt{\dfrac{50}{9}}$

28. $-\sqrt{\dfrac{12}{49}}$

29. $\sqrt{\dfrac{11}{x^2}}$

30. $\sqrt{\dfrac{7}{y^4}}$

31. $\sqrt{\dfrac{y^5}{100}}$

32. $\sqrt{\dfrac{x^3}{81}}$

(8.3) *Add or subtract by combining like radicals.*

33. $5\sqrt{2} - 8\sqrt{2}$

34. $\sqrt{3} - 6\sqrt{3}$

35. $6\sqrt{5} + 3\sqrt{6} - 2\sqrt{5} + \sqrt{6}$

36. $-\sqrt{7} + 8\sqrt{2} - \sqrt{7} - 6\sqrt{2}$

Add or subtract by simplifying each radical and then combining like terms. Assume that all variables represent positive numbers.

37. $\sqrt{28} + \sqrt{63} + \sqrt{56}$

38. $\sqrt{75} + \sqrt{48} - \sqrt{16}$

39. $\sqrt{\dfrac{5}{9}} - \sqrt{\dfrac{5}{36}}$

40. $\sqrt{\dfrac{11}{25}} + \sqrt{\dfrac{11}{16}}$

41. $\sqrt{45x^2} + 3\sqrt{5x^2} - 7x\sqrt{5} + 10$

42. $\sqrt{50x} - 9\sqrt{2x} + \sqrt{72x} - \sqrt{3x}$

(8.4) *Multiply and simplify if possible. Assume that all variables represent positive numbers.*

43. $\sqrt{3} \cdot \sqrt{6}$

44. $\sqrt{5} \cdot \sqrt{15}$

45. $\sqrt{2}(\sqrt{5} - \sqrt{7})$ **46.** $\sqrt{5}(\sqrt{11} + \sqrt{3})$

47. $(\sqrt{3} + 2)(\sqrt{6} - 5)$ **48.** $(\sqrt{5} + 1)(\sqrt{5} - 3)$

49. $(\sqrt{x} - 2)^2$ **50.** $(\sqrt{y} + 4)^2$

Divide and simplify if possible. Assume that all variables represent positive numbers.

51. $\dfrac{\sqrt{27}}{\sqrt{3}}$ **52.** $\dfrac{\sqrt{20}}{\sqrt{5}}$ **53.** $\dfrac{\sqrt{160}}{\sqrt{8}}$

54. $\dfrac{\sqrt{96}}{\sqrt{3}}$ **55.** $\dfrac{\sqrt{30x^6}}{\sqrt{2x^3}}$ **56.** $\dfrac{\sqrt{54x^5y^2}}{\sqrt{3xy^2}}$

Rationalize each denominator and simplify.

57. $\dfrac{\sqrt{2}}{\sqrt{11}}$ **58.** $\dfrac{\sqrt{3}}{\sqrt{13}}$ **59.** $\sqrt{\dfrac{5}{6}}$ **60.** $\sqrt{\dfrac{7}{10}}$

61. $\dfrac{1}{\sqrt{5x}}$ **62.** $\dfrac{5}{\sqrt{3y}}$ **63.** $\sqrt{\dfrac{3}{x}}$ **64.** $\sqrt{\dfrac{6}{y}}$

65. $\dfrac{3}{\sqrt{5} - 2}$ **66.** $\dfrac{8}{\sqrt{10} - 3}$

67. $\dfrac{\sqrt{2} + 1}{\sqrt{3} - 1}$ **68.** $\dfrac{\sqrt{3} - 2}{\sqrt{5} + 2}$

69. $\dfrac{10}{\sqrt{x} + 5}$ **70.** $\dfrac{8}{\sqrt{x} - 1}$

(8.5) *Solve each radical equation.*

71. $\sqrt{2x} = 6$ **72.** $\sqrt{x + 3} = 4$ **73.** $\sqrt{x} + 3 = 8$ **74.** $\sqrt{x} + 8 = 3$

75. $\sqrt{2x + 1} = x - 7$ **76.** $\sqrt{3x + 1} = x - 1$ **77.** $\sqrt{x + 3} = \sqrt{x + 15}$ **78.** $\sqrt{x - 5} = \sqrt{x} - 1$

(8.6) *Use the Pythagorean theorem to find the length of each unknown side. Give an exact answer and a two-decimal-place approximation.*

△ **79.**

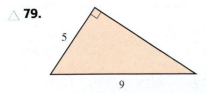

△ **80.**

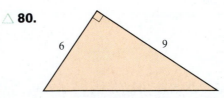

△ **81.** Romeo is standing 20 feet away from the wall below Juliet's balcony during a school play. Juliet is on the balcony, 12 feet above the ground. Find how far apart Romeo and Juliet are.

△ **82.** The diagonal of a rectangle is 10 inches long. If the width of the rectangle is 5 inches, find the length of the rectangle.

Use the formula $r = \sqrt{\dfrac{S}{4\pi}}$, where $r =$ the radius of a sphere and $S =$ the surface area of the sphere, for Exercises 83 and 84.

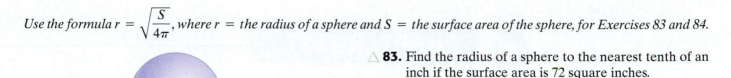

△ **83.** Find the radius of a sphere to the nearest tenth of an inch if the surface area is 72 square inches.

△ **84.** Find the exact surface area of a sphere if its radius is 6 inches. (Do not approximate π.)

Mixed Review

Find each root. Assume all variables represent positive numbers.

85. $\sqrt{144}$ **86.** $-\sqrt[3]{64}$ **87.** $\sqrt{16x^{16}}$ **88.** $\sqrt{4x^{24}}$

Simplify each expression. Assume all variables represent positive numbers.

89. $\sqrt{18x^7}$ **90.** $\sqrt{48y^6}$ **91.** $\sqrt{\dfrac{y^4}{81}}$ **92.** $\sqrt{\dfrac{x^9}{9}}$

Add or subtract by simplifying and then combining like terms. Assume all variables represent positive numbers.

93. $\sqrt{12} + \sqrt{75}$ **94.** $\sqrt{63} + \sqrt{28} - \sqrt{9}$

95. $\sqrt{\dfrac{3}{16}} - \sqrt{\dfrac{3}{4}}$ **96.** $\sqrt{45x^3} + x\sqrt{20x} - \sqrt{5x^3}$

Multiply and simplify if possible. Assume all variables represent positive numbers.

97. $\sqrt{7} \cdot \sqrt{14}$ **98.** $\sqrt{3}\left(\sqrt{9} - \sqrt{2}\right)$ **99.** $\left(\sqrt{2} + 4\right)\left(\sqrt{5} - 1\right)$ **100.** $\left(\sqrt{x} + 3\right)^2$

Divide and simplify if possible. Assume all variables represent positive numbers.

101. $\dfrac{\sqrt{120}}{\sqrt{5}}$ **102.** $\dfrac{\sqrt{60x^9}}{\sqrt{15x^7}}$

Rationalize each denominator and simplify.

103. $\sqrt{\dfrac{2}{7}}$ **104.** $\dfrac{3}{\sqrt{2x}}$

105. $\dfrac{3}{\sqrt{x} - 6}$ **106.** $\dfrac{\sqrt{7} - 5}{\sqrt{5} + 3}$

Solve each radical equation.

107. $\sqrt{4x} = 2$ **108.** $\sqrt{x - 4} = 3$ **109.** $\sqrt{4x + 8} + 6 = x$ **110.** $\sqrt{x - 8} = \sqrt{x} - 2$

111. Use the Pythagorean theorem to find the length of the unknown side. Give an exact answer and a two-decimal-place approximation.

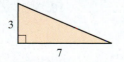

112. The diagonal of a rectangle is 6 inches long. If the width of the rectangle is 2 inches, find the length of the rectangle.

Remember to use the Chapter Test Prep Video CD to see the fully worked-out solutions to any of the exercises you want to review.

Answers

Simplify each radical. Indicate if the radical is not a real number. Assume that x represents a positive number.

1. $\sqrt{16}$　　　　**2.** $\sqrt[3]{125}$　　　　**3.** $\sqrt[4]{81}$

1. _____

2. _____

3. _____

4. $\sqrt{\dfrac{9}{16}}$　　　**5.** $\sqrt[4]{-81}$　　　**6.** $\sqrt{x^{10}}$

4. _____

5. _____

6. _____

Simplify each radical. Assume that all variables represent positive numbers.

7. _____

7. $\sqrt{54}$　　**8.** $\sqrt{92}$　　**9.** $\sqrt{y^7}$　　**10.** $\sqrt{24x^8}$

8. _____

9. _____

10. _____

11. _____

11. $\sqrt[3]{27}$　　**12.** $\sqrt[3]{16}$　　**13.** $\sqrt{\dfrac{5}{16}}$　　**14.** $\sqrt{\dfrac{y^3}{25}}$

12. _____

13. _____

14. _____

Perform each indicated operation. Assume that all variables represent positive numbers.

15. _____

15. $\sqrt{13} + \sqrt{13} - 4\sqrt{13}$　　　　**16.** $\sqrt{18} - \sqrt{75} + 7\sqrt{3} - \sqrt{8}$

16. _____

17. _____

18. _____

17. $\sqrt{\dfrac{3}{4}} + \sqrt{\dfrac{3}{25}}$　　　**18.** $\sqrt{7} \cdot \sqrt{14}$　　　**19.** $\sqrt{2}\left(\sqrt{6} - \sqrt{5}\right)$

19. _____

20. $(\sqrt{x} + 2)(\sqrt{x} - 3)$ **21.** $\dfrac{\sqrt{50}}{\sqrt{10}}$ **22.** $\dfrac{\sqrt{40x^4}}{\sqrt{2x}}$

Rationalize each denominator. Assume that all variables represent positive numbers.

23. $\sqrt{\dfrac{2}{3}}$ **24.** $\dfrac{8}{\sqrt{5y}}$ **25.** $\dfrac{8}{\sqrt{6} + 2}$ **26.** $\dfrac{1}{3 - \sqrt{x}}$

Solve each radical equation.

27. $\sqrt{x} + 8 = 11$ **28.** $\sqrt{3x - 6} = \sqrt{x + 4}$ **29.** $\sqrt{2x - 2} = x - 5$

△ **30.** Find the length of the unknown leg of the right triangle shown. Give an exact answer.

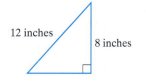

12 inches

8 inches

△ **31.** The formula $r = \sqrt{\dfrac{A}{\pi}}$ can be used to find the radius r of a circle given its area A. Use this formula to approximate the radius of the given circle. Round to two decimal places.

Area is
15 square
meters.

20. _____

21. _____

22. _____

23. _____

24. _____

25. _____

26. _____

27. _____

28. _____

29. _____

30. _____

31. _____

Multiply.

Answers

1. $-2(-14)$

2. $9(-5.2)$

3. $-\dfrac{2}{3} \cdot \dfrac{4}{7}$

4. $-3\dfrac{3}{8} \cdot 5\dfrac{1}{3}$

5. Solve: $4(2x - 3) + 7 = 3x + 5$

6. Solve: $6y - 11 + 4 + 2y = 8 + 15y - 8y$

7. The circle graph below shows the purpose of trips made by American travelers. Use this graph to answer the questions below.

  a. What percent of trips made by American travelers is solely for the purpose of business?

  b. What percent of trips made by American travelers is for the purpose of business or combined business/pleasure?

  c. On an airplane flight of 253 Americans, how many of these people might we expect to be traveling solely for business?

Purpose of Trip

Personal/Other 13%

Combined Business/Pleasure 4%

Business 17%

Pleasure 66%

Source: Travel Industry Association of America

8. Simplify each expression.

  a. $\dfrac{4(-3) - (-6)}{-8 + 4}$

  b. $\dfrac{3 + (-3)(-2)^3}{-1 - (-4)}$

9. Write the following numbers in standard notation, without exponents.

  a. 1.02×10^5

  b. 7.358×10^{-3}

  c. 8.4×10^7

  d. 3.007×10^{-5}

10. Write the following in scientific notation:

  a. 7,200,000

  b. 0.000308

11. Find the product: $(3x + 2)(2x - 5)$

Answers

1. _____

2. _____

3. _____

4. _____

5. _____

6. _____

7. a. _____

 b. _____

 c. _____

8. a. _____

 b. _____

9. a. _____

 b. _____

 c. _____

 d. _____

10. a. _____

 b. _____

11. _____

12. Multiply: $(7x + 1)^2$

13. Factor $xy + 2x + 3y + 6$ by grouping.

14. Factor $xy^2 + 5x - y^2 - 5$ by grouping.

15. Factor: $3x^2 + 11x + 6$

16. Factor: $3x^2 + 15x + 18$

17. Are there any values for x for which each expression is undefined?

    a. $\dfrac{x}{x - 3}$

    b. $\dfrac{x^2 + 2}{x^2 - 3x + 2}$

    c. $\dfrac{x^3 - 6x^2 - 10x}{3}$

18. Simplify: $\dfrac{2x^2 + 7x + 3}{x^2 - 9}$

19. Simplify: $\dfrac{x^2 + 4x + 4}{x^2 + 2x}$

20. Divide: $\dfrac{12x^2y^3}{5} \div \dfrac{3y^3}{x}$

21. Perform each indicated operation.

    a. $\dfrac{a}{4} - \dfrac{2a}{8}$ **b.** $\dfrac{3}{10x^2} + \dfrac{7}{25x}$

22. Find an equation of a line with y-intercept $(0, 4)$ and slope of -2.

23. Solve:

$$\dfrac{4x}{x^2 + x - 30} + \dfrac{2}{x - 5} = \dfrac{1}{x + 6}$$

24. Combine like terms to simplify.

$$4a^2 + 3a - 2a^2 + 7a - 5$$

25. Graph $y = -3$.

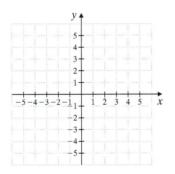

26. Complete the table for the equation $2x + y = 6$.

x	y
0	
	-2
3	

27. Find an equation of the line with y-intercept $(0, -3)$ and slope of $\dfrac{1}{4}$.

28. Find an equation of the line perpendicular to $y = 2x + 4$ and passing through $(1, 5)$.

29. Solve the system:

$$\begin{cases} 3x + 4y = 13 \\ 5x - 9y = 6 \end{cases}$$

30. Solve the system:

$$\begin{cases} \dfrac{x}{2} + y = \dfrac{5}{6} \\ 2x - y = \dfrac{5}{6} \end{cases}$$

12. _____

13. _____

14. _____

15. _____

16. _____

17. a. _____

    b. _____

    c. _____

18. _____

19. _____

20. _____

21. a. _____

    b. _____

22. _____

23. _____

24. _____

25. _____

26. see table

27. _____

28. _____

29. _____

30. _____

31. As part of an exercise program, Louisa and Alfredo start walking each morning. They live 15 miles away from each other. They decide to meet one day by walking toward one another. After 2 hours they meet. If Louisa walks one mile per hour faster than Alfredo, find both walking speeds.

32. Two streetcars are 11 miles apart and traveling toward each other on parallel tracks. They meet in 12 minutes. Find the speed of each streetcar if one travels 15 miles per hour faster than the other.

Find each root.

33. $\sqrt[3]{1}$

34. $\sqrt{121}$

35. $\sqrt[3]{-27}$

36. $\sqrt{\dfrac{1}{4}}$

37. $\sqrt[3]{\dfrac{1}{125}}$

38. $\sqrt{\dfrac{25}{144}}$

Simplify.

39. $\sqrt{54}$

40. $\sqrt{63}$

41. $\sqrt{200}$

42. $\sqrt{500}$

Perform indicated operations. If possible, first simplify each radical.

43. $7\sqrt{12} - 2\sqrt{75}$

44. $(\sqrt{x} + 5)(\sqrt{x} - 5)$

45. $2\sqrt{x^2} - \sqrt{25x^5} + \sqrt{x^5}$

46. $\left(\sqrt{6} + 2\right)^2$

47. Rationalize the denominator of $\dfrac{2}{\sqrt{7}}$.

48. Simplify: $\dfrac{x+3}{\dfrac{1}{x} + \dfrac{1}{3}}$

49. Solve: $\sqrt{x} = \sqrt{5x - 2}$

50. Solve: $\sqrt{x + 4} = \sqrt{3x - 1}$

31. _____

32. _____

33. _____

34. _____

35. _____

36. _____

37. _____

38. _____

39. _____

40. _____

41. _____

42. _____

43. _____

44. _____

45. _____

46. _____

47. _____

48. _____

49. _____

50. _____

9

Quadratic Equations

An important part of the study of algebra is learning to use methods for solving equations. In Chapter 2, we presented techniques for solving linear equations in one variable. In Chapter 4, we solved quadratic equations in one variable by factoring the quadratic expressions. We now present other methods for solving quadratic equations in one variable.

Your heart is a muscular pump that continuously circulates blood throughout your body. Coronary heart disease and congestive heart failure are two examples of cardiovascular diseases that affect more than 64 million Americans and accounted for about 900,000 deaths in a recent year. Unfortunately, many deaths occur while severe heart failure patients wait for donor hearts.

In 2001, history was made when the AbioCor® artificial heart, developed by ABIOMED, was successfully implanted in a 58-year-old telephone company employee and teacher. This artificial heart is completely self-contained and is the first artificial heart to be used in nearly 20 years. In Section 9.3, Exercise 70 on page 643, we will use a method called the quadratic formula to predict when the net income of the ABIOMED Corporation will reach a certain level.

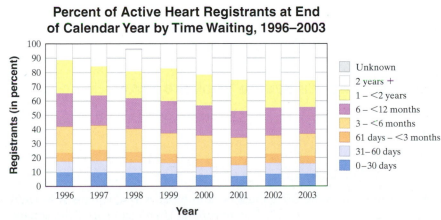

Percent of Active Heart Registrants at End of Calendar Year by Time Waiting, 1996–2003

Source: 2004 OPTN/SRTR Annual Report, Additional Analyses.

623

A Review Factoring to Solve Quadratic Equations.

B Use the Square Root Property to Solve Quadratic Equations.

C Use the Square Root Property to Solve Applications.

9.1 SOLVING QUADRATIC EQUATIONS BY THE SQUARE ROOT PROPERTY

Recall that a quadratic equation is an equation that can be written in the form

$$ax^2 + bx + c = 0$$

where a, b, and c are real numbers and $a \neq 0$.

Objective **A** Solving Quadratic Equations by Factoring

Recall from Section 4.6 that to solve quadratic equations by factoring, we use the **zero factor property:** If the product of two numbers is zero, then at least one of the two numbers is zero. Examples 1 and 2 review the process of solving quadratic equations by factoring.

PRACTICE PROBLEM 1

Solve: $x^2 - 25 = 0$

EXAMPLE 1 Solve: $x^2 - 4 = 0$

Solution:

$$x^2 - 4 = 0$$
$$(x + 2)(x - 2) = 0 \quad \text{Factor.}$$
$$x + 2 = 0 \quad \text{or} \quad x - 2 = 0 \quad \text{Use the zero factor property.}$$
$$x = -2 \qquad\qquad x = 2 \quad \text{Solve each equation.}$$

The solutions are -2 and 2.

▪ Work Practice Problem 1

PRACTICE PROBLEM 2

Solve: $2x^2 - 3x = 9$

EXAMPLE 2 Solve: $3y^2 + 13y = 10$

Solution: Recall that to use the zero factor property, one side of the equation must be 0 and the other side must be factored.

$$3y^2 + 13y = 10$$
$$3y^2 + 13y - 10 = 0 \quad \text{Subtract 10 from both sides.}$$
$$(3y - 2)(y + 5) = 0 \quad \text{Factor.}$$
$$3y - 2 = 0 \quad \text{or} \quad y + 5 = 0 \quad \text{Use the zero factor property.}$$
$$3y = 2 \qquad\qquad y = -5 \quad \text{Solve each equation.}$$
$$y = \frac{2}{3}$$

The solutions are $\dfrac{2}{3}$ and -5.

▪ Work Practice Problem 2

Objective **B** Using the Square Root Property

Consider solving Example 1, $x^2 - 4 = 0$, another way. First, add 4 to both sides of the equation.

$$x^2 - 4 = 0$$
$$x^2 = 4 \quad \text{Add 4 to both sides.}$$

Now we see that the value for x must be a number whose square is 4. Therefore $x = \sqrt{4} = 2$ or $x = -\sqrt{4} = -2$. This reasoning is an example of the square root property.

Answers

1. 5 and -5,   2. $-\dfrac{3}{2}$ and 3

Square Root Property

If $x^2 = a$ for $a \geq 0$, then

$$x = \sqrt{a} \quad \text{or} \quad x = -\sqrt{a}$$

EXAMPLE 3 Use the square root property to solve $x^2 - 9 = 0$.

Solution: First we solve for x^2 by adding 9 to both sides.

$$x^2 - 9 = 0$$
$$x^2 = 9 \quad \text{Add 9 to both sides.}$$

Next we use the square root property.

$$x = \sqrt{9} \quad \text{or} \quad x = -\sqrt{9}$$
$$x = 3 \qquad\qquad x = -3$$

Check:

$x^2 - 9 = 0$ Original equation	$x^2 - 9 = 0$ Original equation
$3^2 - 9 \stackrel{?}{=} 0$ Let $x = 3$.	$(-3)^2 - 9 \stackrel{?}{=} 0$ Let $x = -3$.
$0 = 0$ True	$0 = 0$ True

The solutions are 3 and −3.

■ **Work Practice Problem 3**

EXAMPLE 4 Use the square root property to solve $2x^2 = 7$.

Solution: First we solve for x^2 by dividing both sides by 2. Then we use the square root property.

$$2x^2 = 7$$
$$x^2 = \frac{7}{2} \qquad \text{Divide both sides by 2.}$$
$$x = \sqrt{\frac{7}{2}} \quad \text{or} \quad x = -\sqrt{\frac{7}{2}} \qquad \text{Use the square root property.}$$
$$x = \frac{\sqrt{7} \cdot \sqrt{2}}{\sqrt{2} \cdot \sqrt{2}} \qquad x = -\frac{\sqrt{7} \cdot \sqrt{2}}{\sqrt{2} \cdot \sqrt{2}} \qquad \text{Rationalize the denominator.}$$
$$x = \frac{\sqrt{14}}{2} \qquad x = -\frac{\sqrt{14}}{2} \qquad \text{Simplify.}$$

Remember to check both solutions in the original equation. The solutions are $\frac{\sqrt{14}}{2}$ and $-\frac{\sqrt{14}}{2}$.

■ **Work Practice Problem 4**

EXAMPLE 5 Use the square root property to solve $(x - 3)^2 = 16$.

Solution: Instead of x^2, here we have $(x - 3)^2$. But the square root property can still be used.

$$(x - 3)^2 = 16$$
$$x - 3 = \sqrt{16} \quad \text{or} \quad x - 3 = -\sqrt{16} \qquad \text{Use the square root property.}$$
$$x - 3 = 4 \qquad\qquad x - 3 = -4 \qquad \text{Write } \sqrt{16} \text{ as 4 and } -\sqrt{16} \text{ as } -4.$$
$$x = 7 \qquad\qquad\quad x = -1 \qquad \text{Solve.}$$

Continued on next page

PRACTICE PROBLEM 3
Use the square root property to solve $x^2 - 16 = 0$.

PRACTICE PROBLEM 4
Use the square root property to solve $3x^2 = 11$.

PRACTICE PROBLEM 5
Use the square root property to solve $(x - 4)^2 = 49$.

Answers
3. 4 and −4, **4.** $\frac{\sqrt{33}}{3}$ and $-\frac{\sqrt{33}}{3}$,
5. 11 and −3

Check:

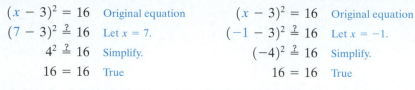

$(x - 3)^2 = 16$	Original equation	
$(7 - 3)^2 \stackrel{?}{=} 16$	Let $x = 7$.	
$4^2 \stackrel{?}{=} 16$	Simplify.	
$16 = 16$	True	

$(x - 3)^2 = 16$	Original equation	
$(-1 - 3)^2 \stackrel{?}{=} 16$	Let $x = -1$.	
$(-4)^2 \stackrel{?}{=} 16$	Simplify.	
$16 = 16$	True	

Both 7 and -1 are solutions.

🔲 **Work Practice Problem 5**

PRACTICE PROBLEM 6

Use the square root property to solve $(x - 5)^2 = 18$.

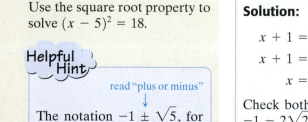

Helpful Hint

read "plus or minus"
↓

The notation $-1 \pm \sqrt{5}$, for example, is just a shorthand notation for both $-1 + \sqrt{5}$ and $-1 - \sqrt{5}$.

EXAMPLE 6 Use the square root property to solve $(x + 1)^2 = 8$.

Solution: $(x + 1)^2 = 8$

$x + 1 = \sqrt{8}$	or	$x + 1 = -\sqrt{8}$	Use the square root property.
$x + 1 = 2\sqrt{2}$		$x + 1 = -2\sqrt{2}$	Simplify the radical.
$x = -1 + 2\sqrt{2}$		$x = -1 - 2\sqrt{2}$	Solve for x.

Check both solutions in the original equation. The solutions are $-1 + 2\sqrt{2}$ and $-1 - 2\sqrt{2}$. This can be written compactly as $-1 \pm 2\sqrt{2}$. The notation $\pm$ is read as "plus or minus."

🔲 **Work Practice Problem 6**

PRACTICE PROBLEM 7

Use the square root property to solve $(x + 3)^2 = -5$.

EXAMPLE 7 Use the square root property to solve $(x - 1)^2 = -2$.

Solution: This equation has no real solution because the square root of -2 is not a real number.

🔲 **Work Practice Problem 7**

PRACTICE PROBLEM 8

Use the square root property to solve $(4x + 1)^2 = 15$.

EXAMPLE 8 Use the square root property to solve $(5x - 2)^2 = 10$.

Solution: $(5x - 2)^2 = 10$

$5x - 2 = \sqrt{10}$	or	$5x - 2 = -\sqrt{10}$	Use the square root property.
$5x = 2 + \sqrt{10}$		$5x = 2 - \sqrt{10}$	Add 2 to both sides.
$x = \dfrac{2 + \sqrt{10}}{5}$		$x = \dfrac{2 - \sqrt{10}}{5}$	Divide both sides by 5.

Check both solutions in the original equation. The solutions are $\dfrac{2 + \sqrt{10}}{5}$ and $\dfrac{2 - \sqrt{10}}{5}$, which can be written as $\dfrac{2 \pm \sqrt{10}}{5}$.

🔲 **Work Practice Problem 8**

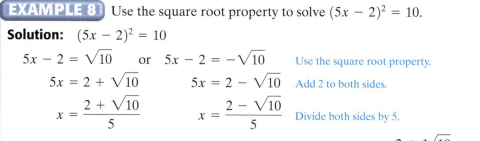

Helpful Hint

For some applications and graphing purposes, decimal approximations of exact solutions to quadratic equations may be desired.

Exact solutions from Example 8		Decimal approximations
$\dfrac{2 + \sqrt{10}}{5}$	$\approx$	1.032
$\dfrac{2 - \sqrt{10}}{5}$	$\approx$	-0.232

Answers

6. $5 \pm 3\sqrt{2}$,   **7.** no real solution,

8. $\dfrac{-1 \pm \sqrt{15}}{4}$

Objective C Using the Square Root Property to Solve Applications

Many real-world applications are modeled by quadratic equations.

In the next example, we use the quadratic formula $h = 16t^2$. This formula gives the distance h traveled by a free-falling object in time t. One important note is that this formula does not take into account any air resistance.

EXAMPLE 9 **Finding the Length of Time of a Dive**

The record for the highest dive into a lake was made by Harry Froboess of Switzerland. In 1936 he dove 394 feet from the airship Hindenburg into Lake Constance. To the nearest tenth of a second, how long did his dive take? (*Source: The Guinness Book of Records*)

Solution:

1. UNDERSTAND. To approximate the time of the dive, we use the formula*
   $h = 16t^2$ where t is time in seconds and h is the distance in feet traveled by a free-falling body or object. For example, to find the distance traveled in 1 second, or 3 seconds, we let $t = 1$ and then $t = 3$.

   If $t = 1, h = 16(1)^2 = 16 \cdot 1 = 16$ feet

   If $t = 3, h = 16(3)^2 = 16 \cdot 9 = 144$ feet

   Since a body travels 144 feet in 3 seconds, we now know the dive of 394 feet lasted longer than 3 seconds.

2. TRANSLATE. Use the formula $h = 16t^2$, let the distance $h = 394$, and we have the equation $394 = 16t^2$.

3. SOLVE. To solve $394 = 16t^2$ for t, we will use the square root property.

$$394 = 16t^2$$

$$\frac{394}{16} = t^2 \qquad \text{Divide both sides by 16.}$$

$$24.625 = t^2 \qquad \text{Simplify.}$$

$$\sqrt{24.625} = t \quad \text{ or } \quad -\sqrt{24.625} = t \quad \text{Use the square root property.}$$

$$5.0 \approx t \quad \text{ or } \quad -5.0 \approx t \quad \text{Approximate.}$$

4. INTERPRET.

Check: We reject the solution -5.0 since the length of the dive is not a negative number.

State: The dive lasted approximately 5 seconds.

☐ **Work Practice Problem 9**

---

*The formula $h = 16t^2$ does not take into account air resistance.

PRACTICE PROBLEM 9

Use the formula $h = 16t^2$ (see Example 9) to find how long, to the nearest tenth of a second, it takes a free-falling body to fall 650 feet.

Answer

9. 6.4 sec

Objective **A** *Solve each equation by factoring. See Examples 1 and 2.*

1. $k^2 - 49 = 0$ **2.** $k^2 - 9 = 0$ **3.** $m^2 + 2m = 15$ **4.** $m^2 + 6m = 7$ **5.** $2x^2 - 32 = 0$

6. $2x^2 - 98 = 0$ **7.** $4a^2 - 36 = 0$ **8.** $7a^2 - 175 = 0$ **9.** $x^2 + 7x = -10$ **10.** $x^2 + 10x = -24$

Objective **B** *Use the square root property to solve each quadratic equation. See Examples 3 and 4.*

11. $x^2 = 64$ **12.** $x^2 = 121$ **13.** $x^2 = 21$ **14.** $x^2 = 22$ **15.** $x^2 = \dfrac{1}{25}$

16. $x^2 = \dfrac{1}{16}$ **17.** $x^2 = -4$ **18.** $x^2 = -25$ **19.** $3x^2 = 13$

20. $5x^2 = 2$ **21.** $7x^2 = 4$ **22.** $2x^2 = 9$ **23.** $x^2 - 2 = 0$ **24.** $x^2 - 15 = 0$

Use the square root property to solve each quadratic equation. See Examples 5 through 8.

25. $(x - 5)^2 = 49$ **26.** $(x + 2)^2 = 25$ **27.** $(x + 2)^2 = 7$ **28.** $(x - 7)^2 = 2$

29. $\left(m - \dfrac{1}{2}\right)^2 = \dfrac{1}{4}$ **30.** $\left(m + \dfrac{1}{3}\right)^2 = \dfrac{1}{9}$ **31.** $(p + 2)^2 = 10$ **32.** $(p - 7)^2 = 13$

33. $(3y + 2)^2 = 100$ **34.** $(4y - 3)^2 = 81$ **35.** $(z - 4)^2 = -9$ **36.** $(z + 7)^2 = -20$

37. $(2x - 11)^2 = 50$ **38.** $(3x - 17)^2 = 28$ **39.** $(3x - 7)^2 = 32$ **40.** $(5x - 11)^2 = 54$

Use the square root property to solve. See Examples 3 through 8.

41. $x^2 - 2 = 0$ **42.** $x^2 - 15 = 0$ **43.** $(x + 6)^2 = 24$

44. $(x + 5)^2 = 20$ **45.** $\dfrac{1}{2}n^2 = 5$ **46.** $\dfrac{1}{5}y^2 = 2$

47. $(4x - 1)^2 = 5$ **48.** $(7x - 2)^2 = 11$ **49.** $3z^2 = 36$

50. $3z^2 = 24$ **51.** $(8 - 3x)^2 - 45 = 0$ **52.** $(10 - 9x)^2 - 75 = 0$

Objective **C** *Solve. For Exercises 53 through 56, use the formula* $h = 16t^2$. *See Example 9. Round each answer to the nearest tenth of a second.*

53. The highest regularly performed dives are made by professional divers from La Quebrada. If this cliff in Acapulco has a height of 87.6 feet, determine the time of a dive. (*Source: The Guinness Book of Records*)

54. In 1988, Eddie Turner saved Frank Fanan, who became unconscious after an injury while jumping out of an airplane. Fanan fell 11,136 feet before Turner pulled his ripcord. Determine the time of Fanan's unconscious free-fall.

55. In 1997, stuntman Stig Gunther of Denmark jumped from a height of 343 feet off a crane onto an airbag. Determine the time of Gunther's stunt fall. (*Source: Guinness Book of World Records*, 2005)

56. Eugene Andreev holds the official Federation Aeronautique Internationale (FAI) world's record for the longest free-fall jump. On November 1, 1962, he fell 80,380 feet before opening his parachute. How long did Andreev free-fall? (*Source: Guinness Book of World Records*, 2005)

The formula for area of a square is $A = s^2$ *where s is the length of a side. Use this formula for Exercises 57 through 60. For each exercise, give an exact answer and a two-decimal-place approximation.*

57. If the area of a square is 20 square inches, find the length of a side.

58. If the area of a square is 32 square meters, find the length of a side.

59. The "Water Cube" National Swimming Center is being constructed in Beijing for the 2008 Summer Olympics. Its square base has an area of 31,329 sq meters. Find the length of a side of this building. (*Source:* ARUP East Asia)

60. The Washington Monument has a square base whose area is approximately 3039 square feet. Find the length of a side. (*Source: The World Almanac*)

Review

Factor each perfect square trinomial. See Section 4.5.

61. $x^2 + 6x + 9$ **62.** $y^2 + 10y + 25$ **63.** $x^2 - 4x + 4$ **64.** $x^2 - 20x + 100$

Concept Extensions

65. Explain why the equation $x^2 = -9$ has no real solution.

66. Explain why the equation $x^2 = 9$ has two solutions.

Solve each quadratic equation by first factoring the perfect square trinomial on the left side. Then apply the square root property.

67. $x^2 + 4x + 4 = 16$

68. $y^2 - 10y + 25 = 11$

△ **69.** The area of a circle is found by the equation $A = \pi r^2$. If the area A of a certain circle is 36π square inches, find its radius r.

70. Neglecting air resistance, the distance d in feet that an object falls in t seconds is given by the equation $d = 16t^2$. If a sandblaster drops his goggles from a bridge 400 feet from the water below, find how long it takes for the goggles to hit the water.

400 feet

Solve each quadratic equation by using the square root property. Use a calculator and round each solution to the nearest hundredth.

 71. $x^2 = 1.78$

 72. $(x - 1.37)^2 = 5.71$

73. The number y of Barnes & Noble Booksellers open for business from 1999 through 2003 is given by the equation $y = -0.07(x - 192.5)^2 + 3135$, where $x = 0$ represents the year 1999. Assume that this trend continues, and find the first year in which there will be 727 stores open for business. (*Hint:* Replace y with 727 in the equation and solve for x. Round to the nearest year.) (*Source:* Based on data from Barnes & Noble Corporation)

74. World cotton production y (in millions of bales) from 2002 to 2004 can be represented by the equation $y = 6.4(x + 0.0065)^2 + 88.31$, where $x = 0$ represents the year 2002. Assume that this trend continues and find the year in which there will be 249 millions of bales produced. (*Hint:* Replace y with 249 in the equation and solve for x. Round to the nearest year.) (*Source:* Based on data from U.S. Department of Agriculture, Foreign Agricultural Service)

STUDY SKILLS BUILDER

Let's review the tips for preparing for your final exam.

Are You Preparing for Your Final Exam?

To prepare for your final exam, try the following study techniques:

- Review the material that you will be responsible for on your exam. This includes material from your textbook, your notebook, and any handouts from your instructor.

- Review any formulas that you may need to memorize.

- Check to see if your instructor or mathematics department will be conducting a final exam review.

- Check with your instructor to see whether final exams from previous semesters/quarters are available to students for review.

- Use your previously taken exams as a practice final exam. To do so, rewrite the test questions in mixed order on blank sheets of paper. This will help you prepare for exam conditions.

- If you are unsure of a few concepts, see your instructor or visit a learning lab for assistance. Also, view the video segment of any troublesome sections.

- If you need further exercises to work, try the Cumulative Reviews at the end of the chapters.

Once again, good luck! I hope you are enjoying this textbook and your mathematics course.

9.2 SOLVING QUADRATIC EQUATIONS BY COMPLETING THE SQUARE

Objectives

A Solve Quadratic Equations of the Form $x^2 + bx + c = 0$ by Completing the Square.

B Solve Quadratic Equations of the Form $ax^2 + bx + c = 0$ by Completing the Square.

Objective **A** Completing the Square to Solve $x^2 + bx + c = 0$

In the last section, we used the square root property to solve equations such as

$$(x + 1)^2 = 8 \quad \text{and} \quad (5x - 2)^2 = 3$$

Notice that one side of each equation is a quantity squared and that the other side is a constant. To solve

$$x^2 + 2x = 4$$

notice that if we add 1 to both sides of the equation, the left side is a perfect square trinomial that can be factored.

$$x^2 + 2x + 1 = 4 + 1 \quad \text{Add 1 to both sides.}$$
$$(x + 1)^2 = 5 \quad \text{Factor.}$$

Now we can solve this equation as we did in the previous section by using the square root property.

$$x + 1 = \sqrt{5} \quad \text{or} \quad x + 1 = -\sqrt{5} \quad \text{Use the square root property.}$$
$$x = -1 + \sqrt{5} \qquad x = -1 - \sqrt{5} \quad \text{Solve.}$$

The solutions are $-1 \pm \sqrt{5}$.

Adding a number to $x^2 + 2x$ to form a perfect square trinomial is called **completing the square** on $x^2 + 2x$.

In general, we have the following:

Completing the Square

To complete the square on $x^2 + bx$, add $\left(\dfrac{b}{2}\right)^2$. To find $\left(\dfrac{b}{2}\right)^2$, **find half the coefficient of x, and then square the result.**

EXAMPLE 1 Solve $x^2 + 6x + 3 = 0$ by completing the square.

Solution: First we get the variable terms alone by subtracting 3 from both sides of the equation.

$$x^2 + 6x + 3 = 0$$
$$x^2 + 6x = -3 \quad \text{Subtract 3 from both sides.}$$

Next we find half the coefficient of the x-term, and then square it. We add this result to *both sides* of the equation. This will make the left side a perfect square trinomial. The coefficient of x is 6, and half of 6 is 3. So we add 3^2 or 9 to both sides.

$$x^2 + 6x + 9 = -3 + 9 \quad \text{Complete the square.}$$
$$(x + 3)^2 = 6 \quad \text{Factor the trinomial } x^2 + 6x + 9.$$
$$x + 3 = \sqrt{6} \quad \text{or} \quad x + 3 = -\sqrt{6} \quad \text{Use the square root property.}$$
$$x = -3 + \sqrt{6} \qquad x = -3 - \sqrt{6} \quad \text{Subtract 3 from both sides.}$$

Check by substituting $-3 + \sqrt{6}$ and $-3 - \sqrt{6}$ in the original equation. The solutions are $-3 \pm \sqrt{6}$.

🔲 **Work Practice Problem 1**

PRACTICE PROBLEM 1

Solve $x^2 + 8x + 1 = 0$ by completing the square.

Answer

1. $-4 \pm \sqrt{15}$

Helpful Hint

Remember, when completing the square, add the number that completes the square to **both sides of the equation.**

PRACTICE PROBLEM 2

Solve $x^2 - 14x = -32$ by completing the square.

EXAMPLE 2 Solve $x^2 - 10x = -14$ by completing the square.

Solution: The variable terms are already alone on one side of the equation. The coefficient of x is -10. Half of -10 is -5, and $(-5)^2 = 25$. So we add 25 to both sides.

$$x^2 - 10x = -14$$
$$x^2 - 10x + 25 = -14 + 25$$

Helpful Hint

Add 25 to *both* sides of the equation.

$$(x - 5)^2 = 11$$ Factor the trinomial and simplify $-14 + 25$.
$$x - 5 = \sqrt{11} \quad \text{or} \quad x - 5 = -\sqrt{11}$$ Use the square root property.
$$x = 5 + \sqrt{11} \qquad x = 5 - \sqrt{11}$$ Add 5 to both sides.

The solutions are $5 \pm \sqrt{11}$.

■ **Work Practice Problem 2**

Objective B **Completing the Square to Solve** $ax^2 + bx + c = 0$

The method of completing the square can be used to solve *any* quadratic equation whether the coefficient of the squared variable is 1 or not. When the coefficient of the squared variable is not 1, we first divide both sides of the equation by the coefficient of the squared variable so that the new coefficient is 1. Then we complete the square.

PRACTICE PROBLEM 3

Solve $4x^2 - 16x - 9 = 0$ by completing the square.

EXAMPLE 3 Solve $4x^2 - 8x - 5 = 0$ by completing the square.

Solution: Since the coefficient of x^2 is 4, not 1, we first divide both sides of the equation by 4 so that the coefficient of x^2 is 1.

$$4x^2 - 8x - 5 = 0$$

$$x^2 - 2x - \frac{5}{4} = 0 \quad \text{Divide both sides by 4.}$$

$$x^2 - 2x = \frac{5}{4} \quad \text{Get the variable terms alone on one side of the equation.}$$

The coefficient of x is -2. Half of -2 is -1, and $(-1)^2 = 1$. So we add 1 to both sides.

$$x^2 - 2x + 1 = \frac{5}{4} + 1$$

$$(x - 1)^2 = \frac{9}{4} \quad \text{Factor } x^2 - 2x + 1 \text{ and simplify } \frac{5}{4} + 1.$$

$$x - 1 = \sqrt{\frac{9}{4}} \quad \text{or} \quad x - 1 = -\sqrt{\frac{9}{4}} \quad \text{Use the square root property.}$$

$$x = 1 + \frac{3}{2} \qquad x = 1 - \frac{3}{2} \quad \text{Add 1 to both sides and simplify the radical.}$$

$$x = \frac{5}{2} \qquad x = -\frac{1}{2} \quad \text{Simplify.}$$

Both $\frac{5}{2}$ and $-\frac{1}{2}$ are solutions.

■ **Work Practice Problem 3**

Answers

2. $7 \pm \sqrt{17}$, **3.** $\frac{9}{2}$ and $-\frac{1}{2}$

The following steps may be used to solve a quadratic equation in x by completing the square.

To Solve a Quadratic Equation in x by Completing the Square

Step 1: If the coefficient of x^2 is 1, go to Step 2. If not, divide both sides of the equation by the coefficient of x^2.

Step 2: Get all terms with variables on one side of the equation and constants on the other side.

Step 3: Find half the coefficient of x and then square the result. Add this number to both sides of the equation.

Step 4: Factor the resulting perfect square trinomial.

Step 5: Use the square root property to solve the equation.

EXAMPLE 4 Solve $2x^2 + 6x = -7$ by completing the square.

Solution: The coefficient of x^2 is not 1. We divide both sides by 2, the coefficient of x^2.

$$2x^2 + 6x = -7$$

$$x^2 + 3x = -\frac{7}{2} \qquad \text{Divide both sides by 2.}$$

$$x^2 + 3x + \frac{9}{4} = -\frac{7}{2} + \frac{9}{4} \qquad \text{Add } \left(\frac{3}{2}\right)^2 \text{ or } \frac{9}{4} \text{ to both sides.}$$

$$\left(x + \frac{3}{2}\right)^2 = -\frac{5}{4} \qquad \text{Factor the left side and simplify the right.}$$

There is no real solution to this equation since the square root of a negative number is not a real number.

Work Practice Problem 4

EXAMPLE 5 Solve $2x^2 = 10x + 1$ by completing the square.

Solution: First we divide both sides of the equation by 2, the coefficient of x^2.

$$2x^2 = 10x + 1$$

$$x^2 = 5x + \frac{1}{2} \qquad \text{Divide both sides by 2.}$$

Next we get the variable terms alone by subtracting $5x$ from both sides.

$$x^2 - 5x = \frac{1}{2}$$

$$x^2 - 5x + \frac{25}{4} = \frac{1}{2} + \frac{25}{4} \qquad \text{Add } \left(-\frac{5}{2}\right)^2 \text{ or } \frac{25}{4} \text{ to both sides.}$$

$$\left(x - \frac{5}{2}\right)^2 = \frac{27}{4} \qquad \text{Factor the left side and simplify the right side.}$$

$$x - \frac{5}{2} = \sqrt{\frac{27}{4}} \quad \text{or} \quad x - \frac{5}{2} = -\sqrt{\frac{27}{4}} \qquad \text{Use the square root property.}$$

$$x - \frac{5}{2} = \frac{3\sqrt{3}}{2} \qquad\qquad x - \frac{5}{2} = -\frac{3\sqrt{3}}{2} \qquad \text{Simplify.}$$

$$x = \frac{5}{2} + \frac{3\sqrt{3}}{2} \qquad\qquad x = \frac{5}{2} - \frac{3\sqrt{3}}{2}$$

The solutions are $\dfrac{5 \pm 3\sqrt{3}}{2}$.

Work Practice Problem 5

PRACTICE PROBLEM 4

Solve $2x^2 + 10x = -13$ by completing the square.

PRACTICE PROBLEM 5

Solve $2x^2 = -6x + 5$ by completing the square.

Answers
4. no real solution, 5. $\dfrac{-3 \pm \sqrt{19}}{2}$ and -2

Mental Math

Determine the number to add to make each expression a perfect square trinomial.

1. $p^2 + 8p$ **2.** $p^2 + 6p$ **3.** $x^2 + 20x$ **4.** $x^2 + 18x$ **5.** $y^2 + 14y$ **6.** $y^2 + 2y$

9.2 EXERCISE SET

FOR EXTRA HELP

Student Solutions Manual   PH Math/Tutor Center   CD/Video for Review   MathXL MathXL®   MyMathLab MyMathLab

Objective A *Solve each quadratic equation by completing the square. See Examples 1 and 2.*

1. $x^2 + 8x = -12$ **2.** $x^2 - 10x = -24$ **3.** $x^2 + 2x - 7 = 0$ **4.** $z^2 + 6z - 9 = 0$

5. $x^2 - 6x = 0$ **6.** $y^2 + 4y = 0$ **7.** $z^2 + 5z = 7$ **8.** $x^2 - 7x = 5$

9. $x^2 - 2x - 1 = 0$ **10.** $x^2 - 4x + 2 = 0$ **11.** $y^2 + 5y + 4 = 0$ **12.** $y^2 - 5y + 6 = 0$

Objective B *Solve each quadratic equation by completing the square. See Examples 3 through 5.*

13. $3x^2 - 6x = 24$ **14.** $2x^2 + 18x = -40$ **15.** $5x^2 + 10x + 6 = 0$ **16.** $3x^2 - 12x + 14 = 0$

17. $2x^2 = 6x + 5$ **18.** $4x^2 = -20x + 3$ **19.** $2y^2 + 8y + 5 = 0$ **20.** $4z^2 - 8z + 1 = 0$

Objectives A B Mixed Practice *Solve each quadratic equation by completing the square. See Examples 1 through 5.*

21. $x^2 + 6x - 25 = 0$ **22.** $x^2 - 6x + 7 = 0$ **23.** $x^2 - 3x - 3 = 0$

24. $x^2 - 9x + 3 = 0$ **25.** $2y^2 - 3y + 1 = 0$ **26.** $2y^2 - y - 1 = 0$

27. $x(x + 3) = 18$ **28.** $x(x - 3) = 18$ **29.** $3z^2 + 6z + 4 = 0$

30. $2y^2 + 8y + 9 = 0$ **31.** $4x^2 + 16x = 48$ **32.** $6x^2 - 30x = -36$

Review

Simplify each expression. See Section 8.2.

33. $\dfrac{3}{4} - \sqrt{\dfrac{25}{16}}$ **34.** $\dfrac{3}{5} + \sqrt{\dfrac{16}{25}}$ **35.** $\dfrac{1}{2} - \sqrt{\dfrac{9}{4}}$ **36.** $\dfrac{9}{10} - \sqrt{\dfrac{49}{100}}$

634

Simplify each expression. See Section 8.4.

37. $\dfrac{6 + 4\sqrt{5}}{2}$ **38.** $\dfrac{10 - 20\sqrt{3}}{2}$ **39.** $\dfrac{3 - 9\sqrt{2}}{6}$ **40.** $\dfrac{12 - 8\sqrt{7}}{16}$

Concept Extensions

41. In your own words, describe a perfect square trinomial.

42. Describe how to find the number to add to $x^2 - 7x$ to make a perfect square trinomial.

43. Write your own quadratic equation to be solved by completing the square. Write it in the form

     perfect square trinomial = a number that is not a perfect square

$$x^2 + 6x + 9 = 11$$

For example,

a. Solve $x^2 + 6x + 9 = 11$.

b. Solve your quadratic equation by completing the square.

44. Follow the directions of Exercise 43, except write your equation in the form

     perfect square trinomial = negative number

Solve your quadratic equation by completing the square.

45. Find a value of k that will make $x^2 + kx + 16$ a perfect square trinomial.

46. Find a value of k that will make $x^2 + kx + 25$ a perfect square trinomial.

47. Retail sales y (in millions of dollars) for bookstores in the United States from 2001 through 2003 can be represented by the equation $y = 10x^2 + 513x + 15{,}743$. In this equation x is the number of years after 2001. Assume that this trend continues and predict the years after 2001 in which the retail sales for U.S. bookstores will be $20,487 million. (*Source:* Based on data from the U.S. Bureau of the Census, Monthly Retail Surveys Branch)

48. The average price of gold y (in dollars per ounce) from 2000 through 2003 is given by the equation $y = 12x^2 - 10x + 278$. Assume that this trend continues and find the year after 2000 in which the price of gold will be $1620 per ounce. (*Source:* Based on data from U.S. Geological survey, Minerals Information)

Recall that a graphing calculator may be used to solve an equation. For example, to solve $x^2 + 8x = -12$ (Exercise 1), graph

$y_1 = x^2 + 8x$ *(left side of equation) and*

$y_2 = -12$ *(right side of equation)*

The x-coordinate of the point of intersection of the graphs is the solution. Use a graphing calculator and solve each equation. Round solutions to the nearest hundredth.

49. Exercise 1   **50.** Exercise 2   **51.** Exercise 17   **52.** Exercise 8

9.3 SOLVING QUADRATIC EQUATIONS BY THE QUADRATIC FORMULA

Objective **A** Using the Quadratic Formula

We can use the technique of completing the square to develop a formula to find solutions of any quadratic equation. We develop and use the **quadratic formula** in this section.

Recall that a quadratic equation in **standard form** is

$$ax^2 + bx + c = 0, \quad a \neq 0$$

To develop the quadratic formula, let's complete the square for this quadratic equation in standard form.

First we divide both sides of the equation by the coefficient of x^2 and then get the variable terms alone on one side of the equation.

$$x^2 + \frac{b}{a}x + \frac{c}{a} = 0 \qquad \text{Divide by } a; \text{ recall that } a \text{ cannot be 0.}$$

$$x^2 + \frac{b}{a}x = -\frac{c}{a} \qquad \text{Get the variable terms alone on one side of the equation.}$$

The coefficient of x is $\frac{b}{a}$. Half of $\frac{b}{a}$ is $\frac{b}{2a}$ and $\left(\frac{b}{2a}\right)^2 = \frac{b^2}{4a^2}$. So we add $\frac{b^2}{4a^2}$ to both sides of the equation.

$$x^2 + \frac{b}{a}x + \frac{b^2}{4a^2} = -\frac{c}{a} + \frac{b^2}{4a^2} \qquad \text{Add } \frac{b^2}{4a^2} \text{ to both sides.}$$

$$\left(x + \frac{b}{2a}\right)^2 = -\frac{c}{a} + \frac{b^2}{4a^2} \qquad \text{Factor the left side.}$$

$$\left(x + \frac{b}{2a}\right)^2 = -\frac{4ac}{4a^2} + \frac{b^2}{4a^2} \qquad \text{Multiply } -\frac{c}{a} \text{ by } \frac{4a}{4a} \text{ so that the terms on the right side have a common denominator.}$$

$$\left(x + \frac{b}{2a}\right)^2 = \frac{b^2 - 4ac}{4a^2} \qquad \text{Simplify the right side.}$$

Now we use the square root property.

$$x + \frac{b}{2a} = \sqrt{\frac{b^2 - 4ac}{4a^2}} \quad \text{or} \quad x + \frac{b}{2a} = -\sqrt{\frac{b^2 - 4ac}{4a^2}} \qquad \text{Use the square root property.}$$

$$x + \frac{b}{2a} = \frac{\sqrt{b^2 - 4ac}}{2a} \qquad x + \frac{b}{2a} = -\frac{\sqrt{b^2 - 4ac}}{2a} \qquad \text{Simplify the radical.}$$

$$x = -\frac{b}{2a} + \frac{\sqrt{b^2 - 4ac}}{2a} \qquad x = -\frac{b}{2a} - \frac{\sqrt{b^2 - 4ac}}{2a} \qquad \text{Subtract } \frac{b}{2a} \text{ from both sides.}$$

$$x = \frac{-b + \sqrt{b^2 - 4ac}}{2a} \qquad x = \frac{-b - \sqrt{b^2 - 4ac}}{2a} \qquad \text{Simplify.}$$

The solutions are $\dfrac{-b \pm \sqrt{b^2 - 4ac}}{2a}$. This final equation is called the **quadratic formula** and gives the solutions of any quadratic equation.

Quadratic Formula

If a, b, and c are real numbers and $a \neq 0$, a quadratic equation written in the standard form $ax^2 + bx + c = 0$ has solutions

$$x = \frac{-b \pm \sqrt{b^2 - 4ac}}{2a}$$

Helpful Hint

Don't forget that to correctly identify a, b, and c in the quadratic formula, you should write the equation in standard form.

Quadratic Equations in Standard Form

$$5x^2 - 6x + 2 = 0 \qquad a = 5, b = -6, c = 2$$
$$4y^2 - 9 = 0 \qquad a = 4, b = 0, c = -9$$
$$x^2 + x = 0 \qquad a = 1, b = 1, c = 0$$
$$\sqrt{2}x^2 + \sqrt{5}x + \sqrt{3} = 0 \qquad a = \sqrt{2}, b = \sqrt{5}, c = \sqrt{3}$$

EXAMPLE 1 Solve $3x^2 + x - 3 = 0$ using the quadratic formula.

Solution: This equation is in standard form with $a = 3$, $b = 1$, and $c = -3$. By the quadratic formula, we have

$$x = \frac{-b \pm \sqrt{b^2 - 4ac}}{2a}$$

$$x = \frac{-1 \pm \sqrt{1^2 - 4 \cdot 3 \cdot (-3)}}{2 \cdot 3} \qquad \text{Let } a = 3, b = 1, \text{ and } c = -3.$$

$$= \frac{-1 \pm \sqrt{1 + 36}}{6} \qquad \text{Simplify.}$$

$$= \frac{-1 \pm \sqrt{37}}{6}$$

Check both solutions in the original equation. The solutions are $\dfrac{-1 + \sqrt{37}}{6}$ and $\dfrac{-1 - \sqrt{37}}{6}$.

□ **Work Practice Problem 1**

EXAMPLE 2 Solve $2x^2 - 9x = 5$ using the quadratic formula.

Solution: First we write the equation in standard form by subtracting 5 from both sides.

$$2x^2 - 9x = 5$$
$$2x^2 - 9x - 5 = 0$$

Next we note that $a = 2$, $b = -9$, and $c = -5$. We substitute these values into the quadratic formula.

$$x = \frac{-b \pm \sqrt{b^2 - 4ac}}{2a}$$

$$x = \frac{-(-9) \pm \sqrt{(-9)^2 - 4 \cdot 2 \cdot (-5)}}{2 \cdot 2} \qquad \text{Substitute in the formula.}$$

$$= \frac{9 \pm \sqrt{81 + 40}}{4} \qquad \text{Simplify.}$$

$$= \frac{9 \pm \sqrt{121}}{4} = \frac{9 \pm 11}{4}$$

Continued on next page

PRACTICE PROBLEM 1

Solve $2x^2 - x - 5 = 0$ using the quadratic formula.

PRACTICE PROBLEM 2

Solve $3x^2 + 8x = 3$ using the quadratic formula.

Helpful Hint

Notice that the fraction bar is under the entire numerator $-b \pm \sqrt{b^2 - 4ac}$.

Answers

1. $\dfrac{1 + \sqrt{41}}{4}$ and $\dfrac{1 - \sqrt{41}}{4}$,

2. $\dfrac{1}{3}$ and -3

Then,

$$x = \frac{9 - 11}{4} = -\frac{1}{2} \quad \text{or} \quad x = \frac{9 + 11}{4} = 5$$

Check $-\frac{1}{2}$ and 5 in the original equation. Both $-\frac{1}{2}$ and 5 are solutions.

🔲 **Work Practice Problem 2**

The following steps may be useful when solving a quadratic equation by the quadratic formula.

To Solve a Quadratic Equation by the Quadratic Formula

Step 1: Write the quadratic equation in standard form: $ax^2 + bx + c = 0$.

Step 2: If necessary, clear the equation of fractions to simplify calculations.

Step 3: Identify a, b, and c.

Step 4: Replace a, b, and c in the quadratic formula with the identified values, and simplify.

✔**Concept Check**   For the quadratic equation $2x^2 - 5 = 7x$, if $a = 2$ and $c = -5$ in the quadratic formula, the value of b is which of the following?

a. $\dfrac{7}{2}$ **b.** 7   **c.** -5 **d.** -7

PRACTICE PROBLEM 3

Solve $5x^2 = 2$ using the quadratic formula.

EXAMPLE 3   Solve $7x^2 = 1$ using the quadratic formula.

Solution: First we write the equation in standard form by subtracting 1 from both sides.

$$7x^2 = 1$$

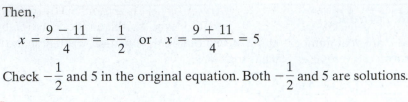

$$7x^2 - 1 = 0$$

> **Helpful Hint**
> $7x^2 - 1 = 0$ can be written as $7x^2 + 0x - 1 = 0$. This form helps you see that $b = 0$.

Next we replace a, b, and c with the identified values: $a = 7, b = 0, c = -1$.

$$x = \frac{0 \pm \sqrt{0^2 - 4 \cdot 7 \cdot (-1)}}{2 \cdot 7} \qquad \text{\color{blue}Substitute in the formula.}$$

$$= \frac{\pm \sqrt{28}}{14} \qquad\qquad\qquad \text{\color{blue}Simplify.}$$

$$= \frac{\pm 2\sqrt{7}}{14}$$

$$= \pm\frac{2 \; \sqrt{7}}{2 \; \cdot 7}$$

$$= \pm\frac{\sqrt{7}}{7}$$

The solutions are $\dfrac{\sqrt{7}}{7}$ and $-\dfrac{\sqrt{7}}{7}$.

🔲 **Work Practice Problem 3**

Answer

3. $\dfrac{\sqrt{10}}{5}$ and $-\dfrac{\sqrt{10}}{5}$

✔ **Concept Check Answer**

d

Notice that we could have solved the equation $7x^2 = 1$ in Example 3 by dividing both sides by 7 and then using the square root property. We solved the equation by the quadratic formula to show that this formula can be used to solve any quadratic equation.

EXAMPLE 4 Solve $x^2 = -x - 1$ using the quadratic formula.

Solution: First we write the equation in standard form.

$$x^2 + x + 1 = 0$$

Next we replace a, b, and c in the quadratic formula with $a = 1$, $b = 1$, and $c = 1$.

$$x = \frac{-1 \pm \sqrt{1^2 - 4 \cdot 1 \cdot 1}}{2 \cdot 1} \qquad \text{Substitute in the formula.}$$

$$= \frac{-1 \pm \sqrt{-3}}{2} \qquad \text{Simplify.}$$

There is no real number solution because $\sqrt{-3}$ is not a real number.

☐ **Work Practice Problem 4**

PRACTICE PROBLEM 4

Solve $x^2 = -2x - 3$ using the quadratic formula.

EXAMPLE 5 Solve $\frac{1}{2}x^2 - x = 2$ using the quadratic formula.

Solution: We write the equation in standard form and then clear the equation of fractions by multiplying both sides by the LCD, 2.

$$\frac{1}{2}x^2 - x = 2$$

$$\frac{1}{2}x^2 - x - 2 = 0 \qquad \text{Write in standard form.}$$

$$x^2 - 2x - 4 = 0 \qquad \text{Multiply both sides by 2.}$$

Here, $a = 1$, $b = -2$, and $c = -4$, so we substitute these values into the quadratic formula.

$$x = \frac{-(-2) \pm \sqrt{(-2)^2 - 4 \cdot 1 \cdot (-4)}}{2 \cdot 1}$$

$$= \frac{2 \pm \sqrt{20}}{2} = \frac{2 \pm 2\sqrt{5}}{2} \qquad \text{Simplify.}$$

$$= \frac{2\left(1 \pm \sqrt{5}\right)}{2} = 1 \pm \sqrt{5} \qquad \text{Factor and simplify.}$$

The solutions are $1 - \sqrt{5}$ and $1 + \sqrt{5}$.

☐ **Work Practice Problem 5**

PRACTICE PROBLEM 5

Solve $\frac{1}{3}x^2 - x = 1$ using the quadratic formula.

Notice that in Example 5, although we cleared the equation of fractions, the coefficients $a = \frac{1}{2}$, $b = -1$, and $c = -2$ will give the same results.

Answers
4. no real solution,

5. $\dfrac{3 + \sqrt{21}}{2}$ and $\dfrac{3 - \sqrt{21}}{2}$

Helpful Hint

When simplifying an expression such as

$$\frac{3 \pm 6\sqrt{2}}{6}$$

first factor out a common factor from the terms of the numerator and then simplify.

$$\frac{3 \pm 6\sqrt{2}}{6} = \frac{3\left(1 \pm 2\sqrt{2}\right)}{2 \cdot 3} = \frac{1 \pm 2\sqrt{2}}{2}$$

Objective B Approximate Solutions to Quadratic Equations

Sometimes approximate solutions for quadratic equations are appropriate.

PRACTICE PROBLEM 6

Approximate the exact solutions of the quadratic equation in Practice Problem 1. Round the approximations to the nearest tenth.

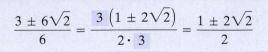 **EXAMPLE 6** Approximate the exact solutions of the quadratic equation in Example 1. Round the approximations to the nearest tenth.

Solution: From Example 1, we have exact solutions $\dfrac{-1 \pm \sqrt{37}}{6}$. Thus,

$$\frac{-1 + \sqrt{37}}{6} \approx 0.847127088 \approx 0.8 \text{ to the nearest tenth.}$$

$$\frac{-1 - \sqrt{37}}{6} \approx -1.180460422 \approx -1.2 \text{ to the nearest tenth.}$$

Thus approximate solutions to the quadratic equation in Example 1 are 0.8 and −1.2.

▢ **Work Practice Problem 6**

Mental Math

Identify the values of a, b, and c in each quadratic equation.

1. $2x^2 + 5x + 3 = 0$

2. $5x^2 - 7x + 1 = 0$

3. $10x^2 - 13x - 2 = 0$

4. $x^2 + 3x - 7 = 0$

5. $x^2 - 6 = 0$

6. $9x^2 - 4 = 0$

9.3 EXERCISE SET

FOR EXTRA HELP

Student Solutions Manual   PH Math/Tutor Center   CD/Video for Review   MathXL   MathXL®   MyMathLab   MyMathLab

Objective **A** *Use the quadratic formula to solve each quadratic equation. See Examples 1 through 4.*

1. $x^2 - 3x + 2 = 0$

2. $x^2 - 5x - 6 = 0$

3. $3k^2 + 7k + 1 = 0$

4. $7k^2 + 3k - 1 = 0$

5. $4x^2 - 3 = 0$

6. $25x^2 - 15 = 0$

7. $5z^2 - 4z + 3 = 0$

8. $3x^2 + 2x + 1 = 0$

9. $y^2 = 7y + 30$

10. $y^2 = 5y + 36$

11. $2x^2 = 10$

12. $5x^2 = 15$

13. $m^2 - 12 = m$

14. $m^2 - 14 = 5m$

15. $3 - x^2 = 4x$

16. $10 - x^2 = 2x$

17. $6x^2 + 9x = 2$

18. $3x^2 - 9x = 8$

19. $7p^2 + 2 = 8p$

20. $11p^2 + 2 = 10p$

21. $x^2 - 6x + 2 = 0$

22. $x^2 - 10x + 19 = 0$

23. $2x^2 - 6x + 3 = 0$

24. $5x^2 - 8x + 2 = 0$

25. $3x^2 = 1 - 2x$

26. $5y^2 = 4 - y$

27. $4y^2 = 6y + 1$

28. $6z^2 = 2 - 3z$

29. $20y^2 = 3 - 11y$

30. $2z^2 = z + 3$

31. $x^2 + x + 2 = 0$

32. $k^2 + 2k + 5 = 0$

Use the quadratic formula to solve each quadratic equation. See Example 5.

33. $\dfrac{m^2}{2} = m + \dfrac{1}{2}$

34. $\dfrac{m^2}{2} = 3m - 1$

35. $3p^2 - \dfrac{2}{3}p + 1 = 0$

36. $\dfrac{5}{2}p^2 - p + \dfrac{1}{2} = 0$

37. $4p^2 + \dfrac{3}{2} = -5p$

38. $4p^2 + \dfrac{3}{2} = 5p$

39. $5x^2 = \dfrac{7}{2}x + 1$

40. $2x^2 = \dfrac{5}{2}x + \dfrac{7}{2}$

41. $x^2 - \dfrac{11}{2}x - \dfrac{1}{2} = 0$

42. $\dfrac{2}{3}x^2 - 2x - \dfrac{2}{3} = 0$

43. $5z^2 - 2z = \dfrac{1}{5}$

44. $9z^2 + 12z = -1$

Objectives A B Mixed Practice *Use the quadratic formula to solve each quadratic equation. Find the exact solutions; then approximate these solutions to the nearest tenth.*

45. $3x^2 = 21$

46. $2x^2 = 26$

47. $x^2 + 6x + 1 = 0$

48. $x^2 + 4x + 2 = 0$

49. $x^2 = 9x + 4$

50. $x^2 = 7x + 5$

51. $3x^2 - 2x - 2 = 0$

52. $5x^2 - 3x - 1 = 0$

Review

Graph the following linear equations in two variables. See Section 6.2.

53. $y = -3$ **54.** $x = 4$ **55.** $y = 3x - 2$ **56.** $y = 2x + 3$

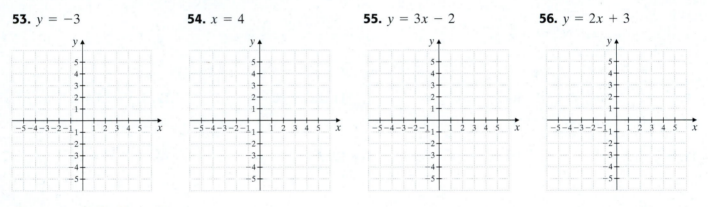

Concept Extensions

Solve. See the Concept Check in this section. For the quadratic equation $5x^2 + 2 = x$, if $a = 5$,

57. What is the value of b?

   a. $\dfrac{1}{5}$ **b.** 0 **c.** -1 **d.** 1

58. What is the value of c?

   a. 5 **b.** x **c.** -2 **d.** 2

For the quadratic equation $7y^2 = 3y$, if $b = 3$,

59. What is the value of a?

   a. 7 **b.** -7 **c.** 0 **d.** 1

60. What is the value of c?

   a. 7 **b.** 3 **c.** 0 **d.** 1

△ **61.** The largest chocolate bar was a 5026-lb scaled-up model of a Novi chocolate bar, made by the Elah-Dufour United Food Company in 2000. The bar had a base area of 50.8 square feet and its length was 0.5 feet longer than twice its width. Find the length and the width of the bar, rounded to one decimal place. (*Source: Guinness Book of World Records*, 2005)

△ **62.** The area of a rectangular conference room table is 95 square feet. If its length is six feet longer than its width, find the dimensions of the table. Round each dimension to the nearest tenth.

Solve each equation using the quadratic formula.

63. $x^2 + 3\sqrt{2}x - 5 = 0$

64. $y^2 - 2\sqrt{5}y - 1 = 0$

65. Explain how the quadratic formula is developed and why it is useful.

Use the quadratic formula and a calculator to solve each equation. Round solutions to the nearest tenth.

66. $1.2x^2 - 5.2x - 3.9 = 0$

67. $7.3z^2 + 5.4z - 1.1 = 0$

A rocket is launched from the top of an 80-foot cliff with an initial velocity of 120 feet per second. The height, h, of the rocket after t seconds is given by the equation $h = -16t^2 + 120t + 80$.

68. How long after the rocket is launched will it be 30 feet from the ground? Round to the nearest tenth of a second.

69. How long after the rocket is launched will it strike the ground? Round to the nearest tenth of a second. (*Hint:* The rocket will strike the ground when its height $h = 0$.)

80 feet

70. The net revenues y (in thousands of dollars) of ABIOMED, Inc., maker of the AbioCor® artificial heart, from 2002 through 2004, can be modeled by the equation $y = 1351.5x^2 + 4670.5x - 24{,}193$, where $x = 0$ represents 2002. Assume that this trend continues and predict the year in which ABIOMED's net revenues will be $16,113 thousand. (*Source:* Based on data from ABIOMED Corporation)

71. The average annual salary y (in dollars) for NFL players for the years 2000 through 2003 is approximated by $y = 21{,}400x^2 - 16{,}100x + 1{,}111{,}000$, where $x = 0$ represents the year 2000. Assume that this trend continues and predict the year in which the average NFL salary will be approximately $2,351,800. (*Source:* Based on data from NFL Players Association and *USA Today*)

THE BIGGER PICTURE   Simplifying Expressions and Solving Equations

Now we continue our outline from Sections 1.6, 2.7, 3.7, 4.6, 5.4, 5.6, and 8.5. Although suggestions are given, this outline should be in your own words. Once you complete this new portion, try the exercises below.

I. Simplifying Expressions

A. Real Numbers

  1. Add (Section 1.4)
  2. Subtract (Section 1.5)
  3. Multiply or Divide (Section 1.6)

B. Exponents (Section 3.2)

C. Polynomials

  1. Add (Section 3.4)
  2. Subtract (Section 3.4)
  3. Multiply (Section 3.5)
  4. Divide (Section 3.7)

D. Factoring Polynomials (Chapter 4 Integrated Review)

E. Rational Expressions

  1. Simplify (Section 5.1)
  2. Multiply (Section 5.2)
  3. Divide (Section 5.2)
  4. Add or Subtract (Section 5.4)

F. Radicals

  1. Simplify (Section 8.2)
  2. Add or Subtract (Section 8.3)
  3. Multiply or Divide (Section 8.4)
  4. Rationalize the denominator (Section 8.4)

II. Solving Equations and Inequalities

A. Linear Equations (Section 2.3)

B. Linear Inequalities (Section 2.7)

C. Quadratic and Higher Degree Equations (Section 4.6)

  1. If in the form $x^2 = a$, solve by the Square Root Property. If not, write the equation in standard form (one side is 0).
  2. If the polynomial on one side factors, solve by factoring.
  3. If the polynomial does not factor, solve by the quadratic formula.

D. Equations with Rational Expressions (Section 5.5)

E. Proportions (Section 5.6)

F. Equations with Radicals (Section 8.5)

Perform indicated operations and simplify.

1. $7.9 - 9.7$

2. $5 + (-3) + (-7)$

3. $(-4)^2 - 5^2$

4. $7x - 2 + \dfrac{1}{3}(9x - 3) + 5$

5. $\left(\dfrac{1}{2}x + 5\right)\left(\dfrac{1}{2}x - 5\right)$

6. $\dfrac{9x^2y + 3xy - 12y}{3xy}$

7. $\dfrac{x^2}{(x-5)(x-4)} - \dfrac{3x+10}{(x-5)(x-4)}$

8. $\dfrac{x}{x-10} + \dfrac{5}{x+3}$

9. $\sqrt{50}$

10. $\dfrac{\sqrt{30a^2b^3}}{\sqrt{3ab}}$

11. $\sqrt{\dfrac{2}{3}}$

12. $\dfrac{7x - 14}{x^2 - 4} \cdot \dfrac{x^2 + 5x + 6}{49}$

Solve.

13. $x^2 + 3x - 5 = 0$

14. $x^2 + x = x^2 + 6$

15. $-2x \le 5.6$

16. $2x^2 + 15x = 8$

17. $\sqrt{x + 2} + 4 = x$

18. $\dfrac{5}{x} - \dfrac{3}{x - 4} = \dfrac{7 + x}{x(x - 4)}$

By factoring:

$$x^2 + x = 6$$
$$x^2 + x - 6 = 0$$
$$(x - 2)(x + 3) = 0$$
$$x - 2 = 0 \quad \text{or} \quad x + 3 = 0$$
$$x = 2 \quad \text{or} \qquad x = -3$$

By Square Root Property:

$$9x^2 = 2$$
$$x^2 = \dfrac{2}{9}$$
$$x = \pm\sqrt{\dfrac{2}{9}} = \dfrac{\pm 2}{3}$$

By quadratic formula:

$$x^2 + x = 5$$
$$x^2 + x - 5 = 0$$
$$a = 1, b = 1, c = -5$$
$$x = \dfrac{-1 \pm \sqrt{1^2 - 4(1)(-5)}}{2 \cdot 1}$$
$$x = \dfrac{-1 \pm \sqrt{21}}{2}$$

Summary on Quadratic Equations

An important skill in mathematics is learning when to use one technique in favor of another. We now practice this by deciding which method to use when solving quadratic equations. Although both the quadratic formula and completing the square can be used to solve any quadratic equation, the quadratic formula is usually less tedious and thus preferred. The following steps may be used to solve a quadratic equation.

To Solve a Quadratic Equation

Step 1: If the equation is in the form $ax^2 = c$ or $(ax + b)^2 = c$, use the square root property and solve. If not, go to Step 2.

Step 2: Write the equation in standard form: $ax^2 + bx + c = 0$.

Step 3: Try to solve the equation by the factoring method. If not possible, go to Step 4.

Step 4: Solve the equation by the quadratic formula.

Choose and use a method to solve each equation.

1. $5x^2 - 11x + 2 = 0$

2. $5x^2 + 13x - 6 = 0$

3. $x^2 - 1 = 2x$

4. $x^2 + 7 = 6x$

5. $a^2 = 20$

6. $a^2 = 72$

7. $x^2 - x + 4 = 0$

8. $x^2 - 2x + 7 = 0$

9. $3x^2 - 12x + 12 = 0$

10. $5x^2 - 30x + 45 = 0$

11. $9 - 6p + p^2 = 0$

12. $49 - 28p + 4p^2 = 0$

13. $4y^2 - 16 = 0$

14. $3y^2 - 27 = 0$

15. $x^2 - 3x + 2 = 0$

16. $x^2 + 7x + 12 = 0$

17. $(2z + 5)^2 = 25$

18. $(3z - 4)^2 = 16$

19. $30x = 25x^2 + 2$

20. $12x = 4x^2 + 4$

Answers

1. _____

2. _____

3. _____

4. _____

5. _____

6. _____

7. _____

8. _____

9. _____

10. _____

11. _____

12. _____

13. _____

14. _____

15. _____

16. _____

17. _____

18. _____

19. _____

20. _____

645

21. $\frac{2}{3}m^2 - \frac{1}{3}m - 1 = 0$

22. $\frac{5}{8}m^2 + m - \frac{1}{2} = 0$

23. $x^2 - \frac{1}{2}x - \frac{1}{5} = 0$

24. $x^2 + \frac{1}{2}x - \frac{1}{8} = 0$

25. $4x^2 - 27x + 35 = 0$

26. $9x^2 - 16x + 7 = 0$

27. $(7 - 5x)^2 = 18$

28. $(5 - 4x)^2 = 75$

29. $3z^2 - 7z = 12$

30. $6z^2 + 7z = 6$

31. $x = x^2 - 110$

32. $x = 56 - x^2$

33. $\frac{3}{4}x^2 - \frac{5}{2}x - 2 = 0$

34. $x^2 - \frac{6}{5}x - \frac{8}{5} = 0$

35. $x^2 - 0.6x + 0.05 = 0$

36. $x^2 - 0.1x - 0.06 = 0$

37. $10x^2 - 11x + 2 = 0$

38. $20x^2 - 11x + 1 = 0$

39. $\frac{1}{2}z^2 - 2z + \frac{3}{4} = 0$

40. $\frac{1}{5}z^2 - \frac{1}{2}z - 2 = 0$

41. Explain how you will decide what method to use when solving quadratic equations.

---

21. _____

22. _____

23. _____

24. _____

25. _____

26. _____

27. _____

28. _____

29. _____

30. _____

31. _____

32. _____

33. _____

34. _____

35. _____

36. _____

37. _____

38. _____

39. _____

40. _____

41. _____

9.4 GRAPHING QUADRATIC EQUATIONS IN TWO VARIABLES

Objectives

A. Graph Quadratic Equations of the Form $y = ax^2$.

B. Graph Quadratic Equations of the Form $y = ax^2 + bx + c$.

Recall from Section 6.2 that the graph of a linear equation in two variables $Ax + By = C$ is a straight line. In this section, we will find that the graph of a quadratic equation in the form $y = ax^2 + bx + c$ is a parabola.

Objective A Graphing $y = ax^2$

We begin our work by graphing $y = x^2$. To do so, we will find and plot ordered pair solutions of this equation. Let's select a few values for x, find the corresponding y-values, and record them in a table of values to keep track. Then we can plot the points corresponding to these solutions on a coordinate plane.

If $x = -3$, then $y = (-3)^2$, or 9.
If $x = -2$, then $y = (-2)^2$, or 4.
If $x = -1$, then $y = (-1)^2$, or 1.
If $x = 0$, then $y = 0^2$, or 0.
If $x = 1$, then $y = 1^2$, or 1.
If $x = 2$, then $y = 2^2$, or 4.
If $x = 3$, then $y = 3^2$, or 9.

x	y
-3	9
-2	4
-1	1
0	0
1	1
2	4
3	9

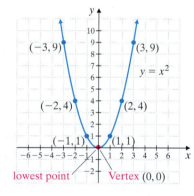

The graph of $y = x^2$ is a smooth curve through the plotted points. This curve is called a **parabola.** The lowest point on a parabola opening upward is called the **vertex.** The vertex is $(0, 0)$ for the parabola $y = x^2$. If we fold the graph paper along the y-axis, the two pieces of the parabola match perfectly. For this reason, we say the graph is **symmetric about the y-axis,** and we call the y-axis the **line of symmetry.**

Notice that the parabola that corresponds to the equation $y = x^2$ opens upward. This happens when the coefficient of x^2 is positive. In the equation $y = x^2$, the coefficient of x^2 is 1. Example 1 shows the graph of a quadratic equation where the coefficient of x^2 is negative.

EXAMPLE 1 Graph: $y = -2x^2$

Solution: We begin by selecting x-values and calculating the corresponding y-values. Then we plot the ordered pairs found and draw a smooth curve through those points. Notice that when the coefficient of x^2 is negative, the corresponding

Continued on next page

PRACTICE PROBLEM 1

Graph: $y = -3x^2$

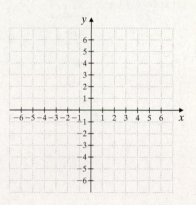

Answer

1.

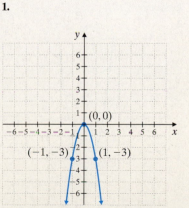

647

parabola opens downward. When a parabola opens downward, the vertex is the highest point of the parabola. The vertex of this parabola is $(0, 0)$.

$y = -2x^2$

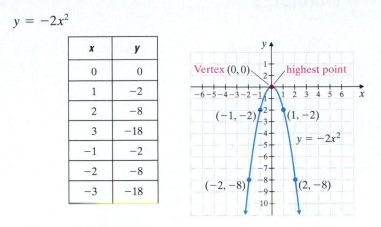

x	y
0	0
1	-2
2	-8
3	-18
-1	-2
-2	-8
-3	-18

🟧 **Work Practice Problem 1**

Objective B Graphing $y = ax^2 + bx + c$

Just as for linear equations, we can use x- and y-intercepts to help graph quadratic equations. Recall from Chapter 6 that an x-intercept is the point where the graph crosses the x-axis. A y-intercept is the point where the graph crosses the y-axis. We find intercepts just as we did in Chapter 6.

> **Helpful Hint**
>
> Recall that:
>
> To find x-intercepts, let $y = 0$ and solve for x.
> To find y-intercepts, let $x = 0$ and solve for y.

EXAMPLE 2 Graph: $y = x^2 - 4$

Solution: If we write this equation as $y = x^2 + 0x + (-4)$, we can see that it is in the form $y = ax^2 + bx + c$. To graph it, we first find the intercepts. To find the y-intercept, we let $x = 0$. Then

$$y = 0^2 - 4 = -4$$

To find x-intercepts, we let $y = 0$.

$$0 = x^2 - 4$$
$$0 = (x - 2)(x + 2)$$
$$x - 2 = 0 \quad \text{or} \quad x + 2 = 0$$
$$x = 2 \qquad\qquad x = -2$$

Thus far, we have the y-intercept $(0, -4)$ and the x-intercepts $(2, 0)$ and $(-2, 0)$. Now we can select additional x-values, find the corresponding y-values, plot the points, and draw a smooth curve through the points.

PRACTICE PROBLEM 2

Graph: $y = x^2 - 9$

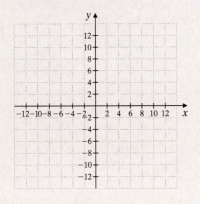

Answer

2.

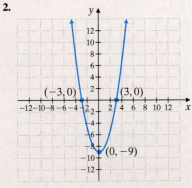

$y = x^2 - 4$

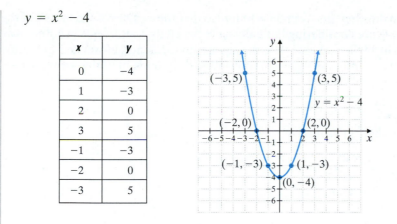

x	y
0	−4
1	−3
2	0
3	5
−1	−3
−2	0
−3	5

Notice that the vertex of this parabola is $(0, -4)$.

◼ **Work Practice Problem 2**

✔ **Concept Check** Tell whether the graph of each equation opens upward or downward.

a. $y = 2x^2$ **b.** $y = 3x^2 + 4x - 5$ **c.** $y = -5x^2 + 2$

Helpful Hint

For the graph of $y = ax^2 + bx + c$,

If a is positive, the parabola opens upward.

If a is negative, the parabola opens downward.

✔ **Concept Check** For which of the following graphs of $y = ax^2 + bx + c$ would the value of a be negative?

a.

b.

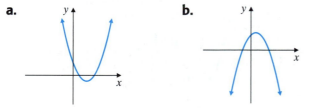

Thus far, we have accidentally stumbled upon the vertex of each parabola that we have graphed. However, our choice of values for x may not yield an ordered pair for the vertex of the parabola. It would be helpful if we could first find the vertex of a parabola. Next we would determine whether the parabola opens upward or downward. Finally we would calculate additional points such as x- and y-intercepts as needed. In fact, there is a formula that may be used to find the vertex of a parabola.

Vertex Formula

The vertex of the parabola $y = ax^2 + bx + c$ has x-coordinate

$$\frac{-b}{2a}$$

The corresponding y-coordinate of the vertex is obtained by substituting the x-coordinate into the equation and finding y.

✔ **Concept Check Answer**

a. upward, b. upward, c. downward; b

One way to develop this formula is to notice that the x-value of the vertex of the parabolas that we are considering lies halfway between its x-intercepts. Another way to develop this formula is to complete the square on the general form of a quadratic equation: $y = ax^2 + bx + c$. We will not show the development of this formula here.

PRACTICE PROBLEM 3

Graph: $y = x^2 - 2x - 3$

EXAMPLE 3 Graph: $y = x^2 - 6x + 8$

Solution: In the equation $y = x^2 - 6x + 8$, $a = 1$ and $b = -6$.

Vertex: The x-coordinate of the vertex is

$$\frac{-b}{2a} = \frac{-(-6)}{2 \cdot 1} = 3 \quad \text{Use the vertex formula, } \frac{-b}{2a}.$$

To find the corresponding y-coordinate, we let $x = 3$ in the original equation.

$$y = x^2 - 6x + 8 = 3^2 - 6 \cdot 3 + 8 = -1$$

The vertex is $(3, -1)$ and the parabola opens upward since a is positive. We now find and plot the intercepts.

Intercepts: To find the x-intercepts, we let $y = 0$.

$$0 = x^2 - 6x + 8$$

We factor the expression $x^2 - 6x + 8$ to find $(x - 4)(x - 2) = 0$. The x-intercepts are $(4, 0)$ and $(2, 0)$.

If we let $x = 0$ in the original equation, then $y = 8$ gives us the y-intercept $(0, 8)$. Now we plot the vertex $(3, -1)$ and the intercepts $(4, 0)$, $(2, 0)$, and $(0, 8)$. Then we can sketch the parabola.

These and two additional points are shown in the table.

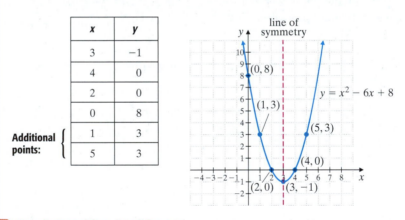

x	y
3	-1
4	0
2	0
0	8
1	3
5	3

Additional points: { 1 → 3, 5 → 3 }

Work Practice Problem 3

Answer

3.

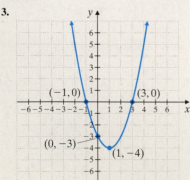

Study Example 3 and let's use it to write down a general procedure for graphing quadratic equations.

Graphing Parabolas Defined by $y = ax^2 + bx + c$

1. **Find the vertex by using the formula $x = -\dfrac{b}{2a}$.** Don't forget to find the y-value of the vertex.

2. **Find the intercepts.**
   - Let $x = 0$ and solve for y to find the y-intercept. There will be only one.
   - Let $y = 0$ and solve for x to find any x-intercepts. There may be 0, 1, or 2.

3. **Plot the vertex and the intercepts.**

4. **Find and plot additional points on the graph.** Then draw a smooth curve through the plotted points. Keep in mind if $a > 0$, the parabola opens up and if $a < 0$, the parabola opens down.

EXAMPLE 4 Graph: $y = x^2 + 2x - 5$

Solution: In the equation $y = x^2 + 2x - 5$, $a = 1$ and $b = 2$. Using the vertex formula, we find that the x-coordinate of the vertex is

$$x = \frac{-b}{2a} = \frac{-2}{2 \cdot 1} = -1$$

The y-coordinate is

$$y = (-1)^2 + 2(-1) - 5 = -6$$

Thus the vertex is $(-1, -6)$.
To find the x-intercepts, we let $y = 0$.

$$0 = x^2 + 2x - 5$$

This cannot be solved by factoring, so we use the quadratic formula.

$$x = \frac{-2 \pm \sqrt{2^2 - 4(1)(-5)}}{2 \cdot 1} \quad \text{Let } a = 1, b = 2, \text{ and } c = -5.$$

$$x = \frac{-2 \pm \sqrt{24}}{2}$$

$$x = \frac{-2 \pm 2\sqrt{6}}{2} \quad \text{Simplify the radical.}$$

$$x = \frac{\boxed{2}\left(-1 \pm \sqrt{6}\right)}{\boxed{2}} = -1 \pm \sqrt{6}$$

The x-intercepts are $\left(-1 + \sqrt{6}, 0\right)$ and $\left(-1 - \sqrt{6}, 0\right)$. We use a calculator to approximate these so that we can easily graph these intercepts.

$$-1 + \sqrt{6} \approx 1.4 \quad \text{and} \quad -1 - \sqrt{6} \approx -3.4$$

To find the y-intercept, we let $x = 0$ in the original equation and find that $y = -5$. Thus the y-intercept is $(0, -5)$. You will find because of symmetry, that $(-2, -5)$ is also an ordered-pair solution.

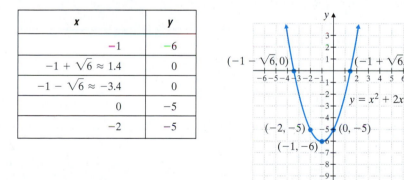

x	y
-1	-6
$-1 + \sqrt{6} \approx 1.4$	0
$-1 - \sqrt{6} \approx -3.4$	0
0	-5
-2	-5

■ **Work Practice Problem 4**

PRACTICE PROBLEM 4

Graph: $y = x^2 - 4x + 1$

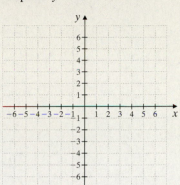

Answer

4.

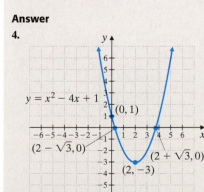

Helpful Hint

Notice that the number of x-intercepts of the graph of the parabola $y = ax^2 + bx + c$ is the same as the number of real solutions of $0 = ax^2 + bx + c$.

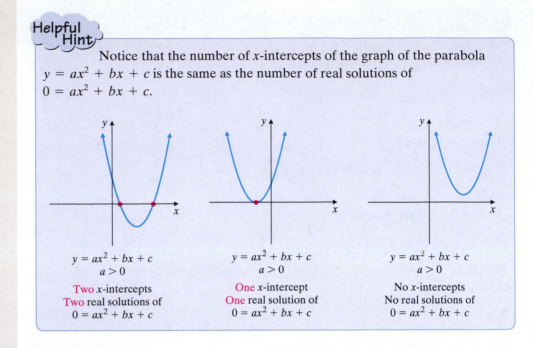

$y = ax^2 + bx + c$
$a > 0$

Two x-intercepts
Two real solutions of
$0 = ax^2 + bx + c$

$y = ax^2 + bx + c$
$a > 0$

One x-intercept
One real solution of
$0 = ax^2 + bx + c$

$y = ax^2 + bx + c$
$a > 0$

No x-intercepts
No real solutions of
$0 = ax^2 + bx + c$

▥ CALCULATOR EXPLORATIONS Graphing

Recall that a graphing calculator may be used to solve quadratic equations. The x-intercepts of the graph of $y = ax^2 + bx + c$ are solutions of $0 = ax^2 + bx + c$. To solve $x^2 - 7x - 3 = 0$, for example, graph $y_1 = x^2 - 7x - 3$. The x-intercepts of the graph are the solutions of the equation.

Use a graphing calculator to solve each quadratic equation. Round solutions to two decimal places.

1. $x^2 - 7x - 3 = 0$

2. $2x^2 - 11x - 1 = 0$

3. $-1.7x^2 + 5.6x - 3.7 = 0$

4. $-5.8x^2 + 2.3x - 3.9 = 0$

5. $5.8x^2 - 2.6x - 1.9 = 0$

6. $7.5x^2 - 3.7x - 1.1 = 0$

9.4 EXERCISE SET

Objective Ⓐ *Graph each quadratic equation by finding and plotting ordered pair solutions. See Example 1.*

1. $y = 2x^2$ **2.** $y = 3x^2$ **3.** $y = -x^2$ **4.** $y = -4x^2$

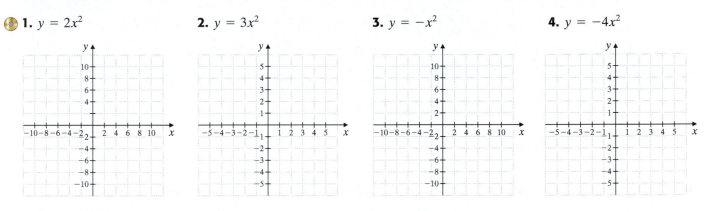

Objective Ⓑ *Sketch the graph of each equation. Label the vertex and the intercepts. See Examples 2 through 4.*

5. $y = x^2 - 1$ **6.** $y = x^2 - 16$ **7.** $y = x^2 + 4$ **8.** $y = x^2 + 9$

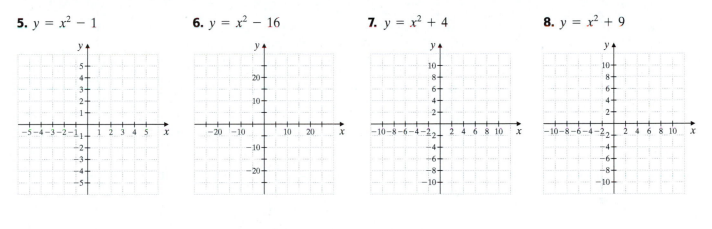

9. $y = -x^2 + 4x - 4$ **10.** $y = -x^2 - 2x - 1$ **11.** $y = x^2 + 5x + 4$ **12.** $y = x^2 + 7x + 10$

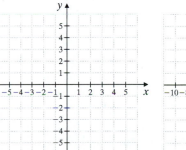

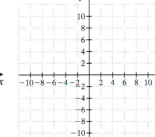

13. $y = x^2 - 4x + 5$

14. $y = x^2 - 6x + 10$

15. $y = 2 - x^2$

16. $y = 3 - x^2$

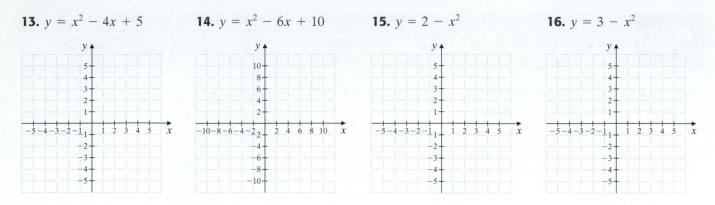

Objectives **A** **B** **Mixed Practice** *Sketch the graph of each equation. Label the vertex and the intercepts. See Examples 1 through 4.*

17. $y = \dfrac{1}{3}x^2$

18. $y = \dfrac{1}{2}x^2$

19. $y = x^2 + 6x$

20. $y = x^2 - 4x$

21. $y = x^2 + 2x - 8$

22. $y = x^2 - 2x - 3$

23. $y = -\dfrac{1}{2}x^2$

24. $y = -\dfrac{1}{3}x^2$

25. $y = 2x^2 - 11x + 5$

26. $y = 2x^2 + x - 3$

27. $y = -x^2 + 4x - 3$

28. $y = -x^2 + 6x - 8$

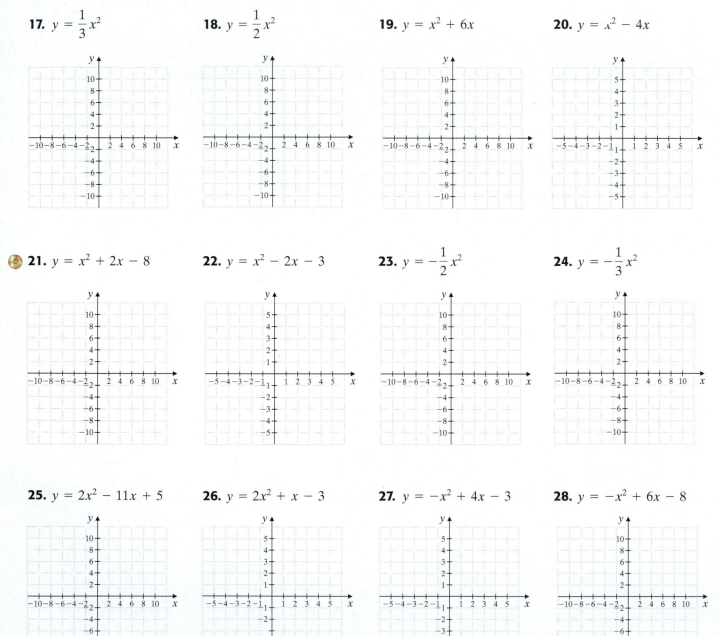

Review

Simplify each complex fraction. See Section 5.7.

29. $\dfrac{\frac{1}{7}}{\frac{2}{5}}$

30. $\dfrac{\frac{3}{8}}{\frac{1}{7}}$

31. $\dfrac{\frac{1}{x}}{\frac{2}{x^2}}$

32. $\dfrac{\frac{x}{5}}{\frac{2}{x}}$

33. $\dfrac{2x}{1 - \frac{1}{x}}$

34. $\dfrac{x}{x - \frac{1}{x}}$

35. $\dfrac{\frac{a-b}{2b}}{\frac{b-a}{8b^2}}$

36. $\dfrac{\frac{2a^2}{a-3}}{\frac{a}{3-a}}$

Concept Extensions

37. The height h of a fireball launched from a Roman candle with an initial velocity of 128 feet per second is given by the equation $h = -16t^2 + 128t$, where t is time in seconds after launch.

   Use the graph of this equation to answer each question.

   a. Estimate the maximum height of the fireball.
   b. Estimate the time when the fireball is at its maximum height.
   c. Estimate the time when the fireball returns to the ground.

38. Determine the maximum number and the minimum number of x-intercepts for a parabola. Explain your answers.

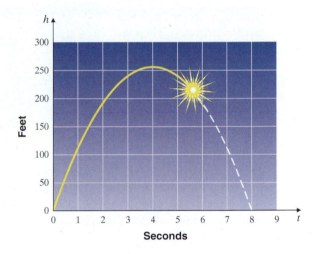

Match the graph of each equation of the form $y = ax^2 + bx + c$ with the given description.

39. $a > 0$, two x-intercepts

40. $a < 0$, one x-intercept

41. $a < 0$, no x-intercept

42. $a > 0$, no x-intercept

43. $a > 0$, one x-intercept

44. $a < 0$, two x-intercepts

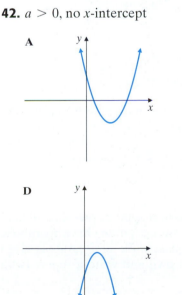

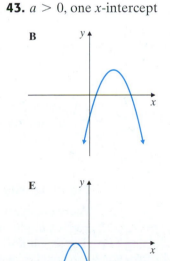

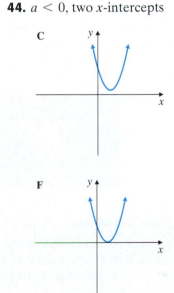

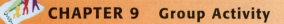

CHAPTER 9   Group Activity

Uses of Parabolas

In this chapter, we learned that the graph of a quadratic equation in two variables of the form $y = ax^2 + bx + c$ is a shape called a **parabola.** The figure to the right shows the general shape of a parabola.

The shape of a parabola shows up in many situations, both natural and human-made, in the world around us.

Natural Situations

- **Hurricanes** The paths of many hurricanes are roughly shaped like a parabola. In the northern hemisphere, hurricanes generally begin moving to the northwest. Then, as they move farther from the equator, they swing around to head in a northeastern direction.

- **Projectiles** The force of the earth's gravity acts on a projectile launched into the air. The resulting path of the projectile, anything from a bullet to a football, is generally shaped like a parabola.

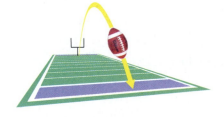

- **Orbits** There are several different possible shapes of orbits of satellites, planets, moons, and comets in outer space. One of the possible types of orbits is in the shape of a parabola. A parabolic orbit is most often seen with comets.

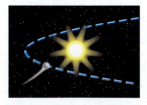

Human-Made Situations

- **Telescopes** Because a parabola has nice reflecting properties, its shape is used in many kinds of telescopes. The largest non-steerable radio telescope is

Arecibo Observatory in Puerto Rico. This telescope consists of a huge parabolic dish built into a valley. The dish is about 1000 feet across.

- **Training Astronauts** Astronauts must be able to work in zero-gravity conditions on missions in space. However, it's nearly impossible to escape the force of gravity on earth. To help astronauts train to work in weightlessness, a specially modified jet can be flown in a parabolic path. At the top of the parabola, weightlessness can be simulated for up to 30 seconds at a time.

- **Architecture** The reinforced concrete arches used in many modern buildings are based on the shape of a parabola.

- **Music** The design of the modern flute incorporates a parabolic head joint.

Group Activity

There are many other physical applications of parabolas. For example, satellite dishes often have parabolic shapes. Choose a physical example of a parabola given here or use one of your own and write a report (with diagrams).

Chapter 9 Vocabulary Check

Fill in each blank with one of the words or phrases listed below.

> square root vertex
>
> completing the square quadratic

1. If $x^2 = a$, then $x = \sqrt{a}$ or $x = -\sqrt{a}$. This property is called the _____ property.

2. The formula $\dfrac{-b}{2a}$ where $y = ax^2 + bx + c$ is called the _____ formula.

3. The process of solving a quadratic equation by writing it in the form $(x + a)^2 = c$ is called _____.

4. The formula $x = \dfrac{-b \pm \sqrt{b^2 - 4ac}}{2a}$ is called the _____ formula.

Helpful Hint

Are you preparing for your test? Don't forget to take the Chapter 9 Test on page 663. Then check your answers at the back of the text and use the Chapter Test Prep Video CD to see the fully worked-out solutions to any of the exercises you want to review.

9 Chapter Highlights

DEFINITIONS AND CONCEPTS	EXAMPLES

Section 9.1   Solving Quadratic Equations by the Square Root Property

SQUARE ROOT PROPERTY

If $x^2 = a$ for $a \geq 0$, then $x = \sqrt{a}$ or $x = -\sqrt{a}$.

Solve the equation.

$$(x - 1)^2 = 15$$

$$x - 1 = \sqrt{15} \quad \text{or} \quad x - 1 = -\sqrt{15}$$

$$x = 1 + \sqrt{15} \qquad x = 1 - \sqrt{15}$$

Section 9.2   Solving Quadratic Equations by Completing the Square

TO SOLVE A QUADRATIC EQUATION BY COMPLETING THE SQUARE

Step 1. If the coefficient of x^2 is not 1, divide both sides of the equation by the coefficient.

Step 2. Get all terms with variables alone on one side.

Step 3. Complete the square by adding the square of half of the coefficient of x to both sides.

Step 4. Factor the perfect square trinomial.

Step 5. Use the square root property to solve.

Solve $2x^2 + 12x - 10 = 0$ by completing the square.

$$\dfrac{2x^2}{2} + \dfrac{12x}{2} - \dfrac{10}{2} = \dfrac{0}{2} \quad \text{Divide by 2.}$$

$$x^2 + 6x - 5 = 0 \quad \text{Simplify.}$$

$$x^2 + 6x = 5 \quad \text{Add 5.}$$

The coefficient of x is 6. Half of 6 is 3 and $3^2 = 9$. Add 9 to both sides.

$$x^2 + 6x + 9 = 5 + 9$$

$$(x + 3)^2 = 14 \quad \text{Factor.}$$

$$x + 3 = \sqrt{14} \quad \text{or} \quad x + 3 = -\sqrt{14}$$

$$x = -3 + \sqrt{14} \quad x = -3 - \sqrt{14}$$

DEFINITIONS AND CONCEPTS	**EXAMPLES**

Section 9.3   Solving Quadratic Equations by the Quadratic Formula

QUADRATIC FORMULA

If a, b, and c are real numbers and $a \neq 0$, the quadratic equation $ax^2 + bx + c = 0$ has solutions

$$x = \frac{-b \pm \sqrt{b^2 - 4ac}}{2a}$$

TO SOLVE A QUADRATIC EQUATION BY THE QUADRATIC FORMULA

Step 1. Write the equation in standard form: $ax^2 + bx + c = 0$.

Step 2. If necessary, clear the equation of fractions.

Step 3. Identify a, b, and c.

Step 4. Replace a, b, and c in the quadratic formula with the identified values, and simplify.

Identify a, b, and c in the quadratic equation

$$4x^2 - 6x = 5$$

First, subtract 5 from both sides.

$$4x^2 - 6x - 5 = 0$$

$a = 4$, $b = -6$, and $c = -5$.

Solve $3x^2 - 2x - 2 = 0$.

In this equation, $a = 3$, $b = -2$, and $c = -2$.

$$x = \frac{-(-2) \pm \sqrt{(-2)^2 - 4(3)(-2)}}{2 \cdot 3}$$

$$= \frac{2 \pm \sqrt{4 - (-24)}}{6}$$

$$= \frac{2 \pm \sqrt{28}}{6} = \frac{2 \pm \sqrt{4 \cdot 7}}{6} = \frac{2 \pm 2\sqrt{7}}{6}$$

$$= \frac{2\left(1 \pm \sqrt{7}\right)}{2 \cdot 3} = \frac{1 \pm \sqrt{7}}{3}$$

Section 9.4   Graphing Quadratic Equations in Two Variables

The graph of a quadratic equation $y = ax^2 + bx + c$, $a \neq 0$, is called a **parabola.** The lowest point on a parabola opening upward or the highest point on a parabola opening downward is called the **vertex.** The vertical line through the vertex is the **line of symmetry.**

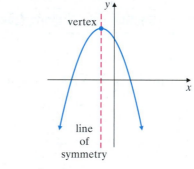

VERTEX FORMULA

The vertex of the parabola $y = ax^2 + bx + c$ has x-coordinate $\dfrac{-b}{2a}$. To find the corresponding y-coordinate, substitute the x-coordinate into the original equation and solve for y.

Graph: $y = 2x^2 - 6x + 4$

The x-coordinate of the vertex is

$$x = \frac{-b}{2a} = \frac{-(-6)}{2(2)} = \frac{6}{4} = \frac{3}{2}$$

The y-coordinate is

$$y = 2\left(\frac{3}{2}\right)^2 - 6\left(\frac{3}{2}\right) + 4 = 2\left(\frac{9}{4}\right) - 9 + 4 = -\frac{1}{2}$$

The vertex is $\left(\dfrac{3}{2}, -\dfrac{1}{2}\right)$.

The y-intercept is

$$y = 2 \cdot 0^2 - 6 \cdot 0 + 4 = 4$$

The x-intercepts are

$$0 = 2x^2 - 6x + 4$$
$$0 = 2(x - 2)(x - 1)$$
$$x = 2 \quad \text{or} \quad x = 1$$

9 CHAPTER REVIEW

(9.1) *Solve each quadradic equation by factoring.*

1. $x^2 - 121 = 0$　　　**2.** $y^2 - 100 = 0$　　　**3.** $3m^2 - 5m = 2$　　　**4.** $7m^2 + 2m = 5$

Use the square root property to solve each quadratic equation.

5. $x^2 = 36$　　　**6.** $x^2 = 81$　　　**7.** $k^2 = 50$　　　**8.** $k^2 = 45$

9. $(x - 11)^2 = 49$　　　**10.** $(x + 3)^2 = 100$　　　**11.** $(4p + 5)^2 = 41$　　　**12.** $(3p + 7)^2 = 37$

Solve. For Exercises 13 and 14, use the formula $h = 16t^2$, where h is the height in feet at time t seconds.

13. If Kara Washington dives from a height of 100 feet, how long before she hits the water?

14. How long does a 5-mile free-fall take? Round your result to the nearest tenth of a second. (*Hint:* 1 mi = 5280 ft)

(9.2) *Solve each quadratic equation by completing the square.*

15. $x^2 - 9x = -8$　　　**16.** $x^2 + 8x = 20$　　　**17.** $x^2 + 4x = 1$　　　**18.** $x^2 - 8x = 3$

19. $x^2 - 6x + 7 = 0$　　　**20.** $x^2 + 6x + 7 = 0$　　　**21.** $2y^2 + y - 1 = 0$　　　**22.** $y^2 + 3y - 1 = 0$

(9.3) *Use the quadratic formula to solve each quadratic equation.*

23. $9x^2 + 30x + 25 = 0$　　　**24.** $16x^2 - 72x + 81 = 0$　　　**25.** $7x^2 = 35$　　　**26.** $11x^2 = 33$

27. $x^2 - 10x + 7 = 0$　　　**28.** $x^2 + 4x - 7 = 0$　　　**29.** $3x^2 + x - 1 = 0$　　　**30.** $x^2 + 3x - 1 = 0$

31. $2x^2 + x + 5 = 0$　　　**32.** $7x^2 - 3x + 1 = 0$

For the Exercise numbers given, approximate the exact solutions to the nearest tenth.

33. Exercise 29

34. Exercise 30

35. The average price of silver (in cents per ounce) from 2001 to 2003 is modeled by the equation $y = 38x^2 - 43x + 446$. In this equation, x is the number of years since 2001. Assume that this trend continues and find the year after 2001 in which the price of silver will be 1556 cents per ounce. (*Source:* U.S. Geological Survey, Minerals Information)

36. The average price of platinum (in dollars per ounce) from 2001 to 2004 is modeled by the equation $y = 64x^2 - 87x + 545$. In this equation, x is the number of years since 2001. Assume that this trend continues and find the year after 2001 in which the price of platinum will be 2327 dollars per ounce. (*Source:* U.S. Geological Survey, Minerals Information)

(9.4) *Graph each quadratic equation and find and plot any intercept points.*

37. $y = 5x^2$

38. $y = -\dfrac{1}{2}x^2$

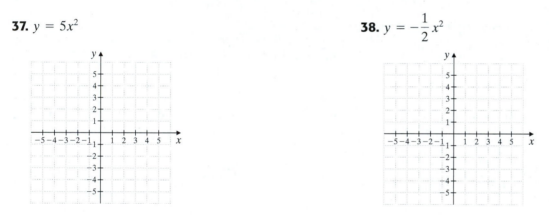

Graph each quadratic equation. Label the vertex and the intercept points with their coordinates.

39. $y = x^2 - 25$ **40.** $y = x^2 - 36$ **41.** $y = x^2 + 3$ **42.** $y = x^2 + 8$

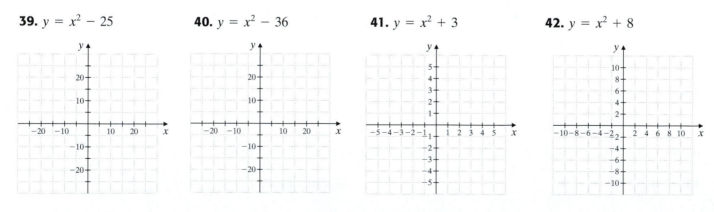

43. $y = -4x^2 + 8$ **44.** $y = -3x^2 + 9$ **45.** $y = x^2 + 3x - 10$ **46.** $y = x^2 + 3x - 4$

47. $y = -x^2 - 5x - 6$ **48.** $y = 3x^2 - x - 2$ **49.** $y = 2x^2 - 11x - 6$ **50.** $y = -x^2 + 4x + 8$

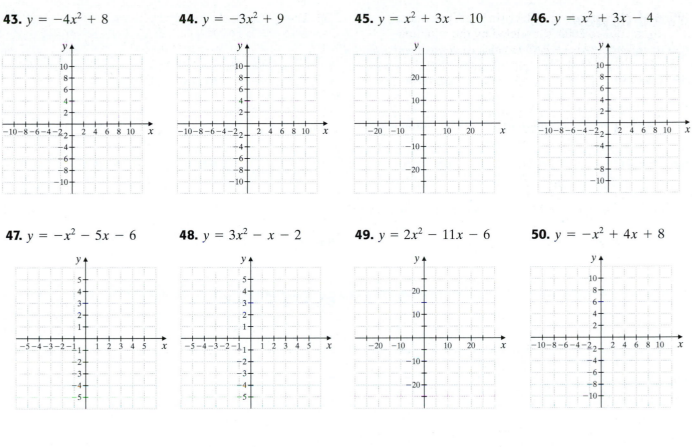

Match each quadratic equation with its graph.

51. $y = 2x^2$ **52.** $y = -x^2$ **53.** $y = x^2 + 4x + 4$ **54.** $y = x^2 + 5x + 4$

A   B   C   D

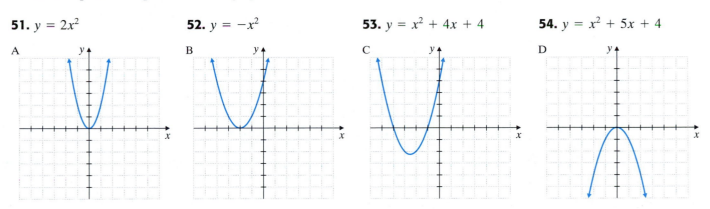

Quadratic equations in the form $y = ax^2 + bx + c$ are graphed below. Determine the number of real solutions for the related equation $0 = ax^2 + bx + c$ from each graph.

55.   **56.**   **57.**   **58.**

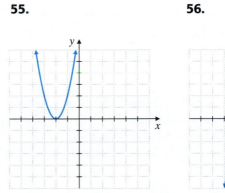
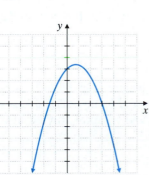
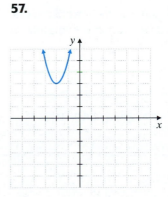
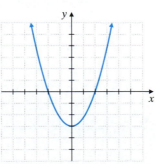

Mixed Review

Use the square root property to solve each quadratic equation.

59. $x^2 = 49$

60. $y^2 = 75$

61. $(x - 7)^2 = 64$

Solve each quadratic equation by completing the square.

62. $x^2 + 4x = 6$

63. $3x^2 + x = 2$

64. $4x^2 - x - 2 = 0$

Use the quadratic formula to solve each quadratic equation.

65. $4x^2 - 3x - 2 = 0$

66. $5x^2 + x - 2 = 0$

67. $4x^2 + 12x + 9 = 0$

68. $2x^2 + x + 4 = 0$

Graph each quadratic equation. Label the vertex and the intercept points with their coordinates.

69. $y = 4 - x^2$

70. $y = x^2 + 4$

71. $y = x^2 + 6x + 8$

72. $y = x^2 - 2x - 4$

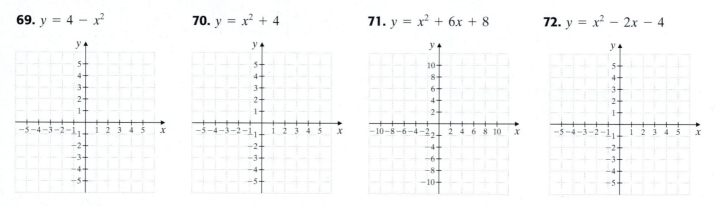

![book icon] **STUDY SKILLS BUILDER**

Are You Prepared for a Test on Chapter 9?

Below I have listed some common trouble areas for students in Chapter 9. After studying for your test—but before taking your test—read these.

- Don't forget that to use the square root property, one side of your equation should be a squared variable or variable expression.

  Solve: $3x^2 = 15$

  $x^2 = 5$ Divide both sides by 3 to isolate x^2.

  $x = \sqrt{5}$ or $x = -\sqrt{5}$ Use the square root property.

- Remember that to identify a, b, and c for the quadratic formula, write the quadratic equation in standard form: $ax^2 + bx + c = 0$

  Solve: $x^2 = -x + 1$

  $x^2 + x - 1 = 0$ Write in standard form.

  Here, $a = 1$, $b = 1$, and $c = -1$.

  $$x = \frac{-1 \pm \sqrt{1^2 - 4(1)(-1)}}{2(1)} = \frac{-1 \pm \sqrt{5}}{2}$$

  Remember: This is simply a listing of a few common trouble areas. For a review of Chapter 9, see the Highlights and Chapter Review at the end of this chapter.

9 CHAPTER TEST

Remember to use the Chapter Test Prep Video CD to see the fully worked-out solutions to any of the exercises you want to review.

Solve by factoring.

1. $x^2 - 400 = 0$

2. $2x^2 - 11x = 21$

Solve using the square root property.

3. $5k^2 = 80$

4. $(3m - 5)^2 = 8$

Solve by completing the square.

5. $x^2 - 26x + 160 = 0$

6. $3x^2 + 12x - 4 = 0$

Solve using the quadratic formula.

7. $x^2 - 3x - 10 = 0$

8. $p^2 - \dfrac{5}{3}p - \dfrac{1}{3} = 0$

Solve by the most appropriate method.

9. $(3x - 5)(x + 2) = -6$ **10.** $(3x - 1)^2 = 16$ **11.** $3x^2 - 7x - 2 = 0$

12. $x^2 - 4x - 5 = 0$ **13.** $3x^2 - 7x + 2 = 0$ **14.** $2x^2 - 6x + 1 = 0$

△ **15.** The height of a triangle is 4 times the length of the base. The area of the triangle is 18 square feet. Find the height and base of the triangle.

4x

x

1. _____

2. _____

3. _____

4. _____

5. _____

6. _____

7. _____

8. _____

9. _____

10. _____

11. _____

12. _____

13. _____

14. _____

15. _____

Graph each quadratic equation. Label the vertex and the intercept points with their coordinates.

16. see graph

16. $y = -5x^2$

17. see graph

17. $y = x^2 - 4$

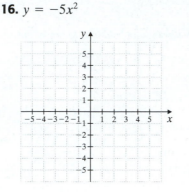

18. see graph

18. $y = x^2 - 7x + 10$

19. $y = 2x^2 + 4x - 1$

19. see graph

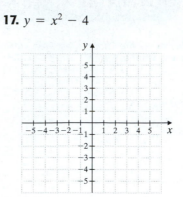

20. _____

△ **20.** The number of diagonals d that a polygon with n sides has is given by the formula

$$d = \frac{n^2 - 3n}{2}$$

Find the number of sides of a polygon if it has 9 diagonals.

Polygon

A diagonal

21. _____

Solve.

22. _____

▦ **21.** The highest dive from a diving board by a woman was made by Lucy Wardle of the United States. She dove from a height of 120.75 feet at Ocean Park, Hong Kong, in 1985. To the nearest tenth of a second, how long did the dive take? Use the formula $h = 16t^2$.

22. The value of Washington State's mineral production y (in millions of dollars) from 2000 through 2003 is modeled by the equation $y = 26x^2 - 136x + 607$. In this equation, $x = 0$ represents the year 2000. Assume that this trend continues and find the year after 2000 in which the value of mineral production is $727 million. (*Source:* U.S. Geological Survey, Minerals Information)

Solve each equation.

1. $y + 0.6 = -1.0$

2. $8x - 14 = 6x - 20$

3. $8(2 - t) = -5t$

4. $2(x + 7) = 5(2x - 3)$

5. In a recent year, the U.S. House of Representatives had a total of 431 Democrats and Republicans. There were 15 more Republican representatives than Democratic. Find the number of representatives from each party.

6. The sum of three consecutive integers is 438. Find the integers.

Simplify the following expressions.

7. 3^0

8. $\left(\dfrac{-6x}{y^3}\right)^3$

9. $(5x^3y^2)^0$

10. $\dfrac{a^2b^7}{(2b^2)^5}$

11. -4^0

12. $\dfrac{(3y)^2}{y^2}$

13. Multiply: $(3y + 2)^2$

14. Multiply: $(x^2 + 5)(y - 1)$

15. Divide $x^2 + 7x + 12$ by $x + 3$ using long division.

16. Simplify by combining like terms: $2 + 8.1a + a - 6$

17. Factor: $r^2 - r - 42$

18. Find the value of each expression when $x = -4$ and $y = 7$.

    a. $\dfrac{x - y}{7 - x}$

    b. $x^2 + 2y$

19. Factor: $10x^2 - 13xy - 3y^2$

20. Add: $\dfrac{1}{x + 2} + \dfrac{7}{x - 1}$

21. Factor $8x^2 - 14x + 5$ by grouping.

Answers

1. _____

2. _____

3. _____

4. _____

5. _____

6. _____

7. _____

8. _____

9. _____

10. _____

11. _____

12. _____

13. _____

14. _____

15. _____

16. _____

17. _____

18. a. _____

    b. _____

19. _____

20. _____

21. _____

665

22. _____

23. a. _____

b. _____

24. _____

25. _____

26. a. _____

b. _____

c. _____

d. _____

27. _____

28. _____

29. _____

30. _____

31. _____

32. a. _____

b. _____

33. a. _____

b. _____

c. _____

34. _____

35. a. _____

b. _____

22. Multiply: $\dfrac{x^2 + 7x}{5x} \cdot \dfrac{10x + 25}{x^2 - 49}$

23. Factor the difference of squares.
  a. $4x^3 - 49x$ **b.** $162x^4 - 2$

24. Solve $\dfrac{2x + 7}{3} = \dfrac{x - 6}{2}$.

25. Solve: $(5x - 1)(2x^2 + 15x + 18) = 0$

26. Simplify each expression by combining like terms.
  a. $4x - 3 + 7 - 5x$ **c.** $2 + 8.1a + a - 6$
  b. $-6y + 3y - 8 + 8y$ **d.** $2x^2 - 2x$

27. Simplify: $\dfrac{x^2 + 8x + 7}{x^2 - 4x - 5}$

28. Solve $2x^2 + 5x = 7$.

29. The quotient of a number and 6, minus $\dfrac{5}{3}$ is the quotient of the number and 2. Find the number.

30. Find the distance between $(-7, 4)$ and $(2, 5)$.

31. Complete the table for the equation $y = 3x$.

	x	**y**
a.	-1	
b.		0
c.		-9

32. Identify the x- and y-intercepts.
  a.   **b.**

33. Determine whether each pair of lines is parallel, perpendicular, or neither.
  a. $y = -\dfrac{1}{5}x + 1$ **b.** $x + y = 3$ **c.** $3x + y = 5$
       $2x + 10y = 3$ $-x + y = 4$ $2x + 3y = 6$

34. Determine whether the graphs of $y = 3x + 7$ and $x + 3y = -15$ are parallel lines, perpendicular lines, or neither.

35. Which of the following relations are also functions?
  a. $\{(-1, 1), (2, 3), (7, 3), (8, 6)\}$
  b. $\{(0, -2), (1, 5), (0, 3), (7, 7)\}$

36. Add or subtract by first simplifying each radical.

   a. $\sqrt{80} + \sqrt{20}$

   b. $2\sqrt{98} - 2\sqrt{18}$

   c. $\sqrt{32} + \sqrt{121} - \sqrt{12}$

37. Solve the system: $\begin{cases} 2x + y = 10 \\ x = y + 2 \end{cases}$

38. Solve the system. $\begin{cases} 5x + y = 3 \\ y = -5x \end{cases}$

39. Solve the system: $\begin{cases} 2x - y = 7 \\ 8x - 4y = 1 \end{cases}$

40. Solve the system. $\begin{cases} -2x + y = 7 \\ 6x - 3y = -21 \end{cases}$

Find each square root.

41. $\sqrt{36}$

42. $\sqrt{\dfrac{4}{25}}$

43. $\sqrt{\dfrac{9}{100}}$

44. $\sqrt{\dfrac{16}{121}}$

45. Rationalize the denominator of $\dfrac{2}{1 + \sqrt{3}}$.

46. Rationalize the denominator of $\dfrac{5}{\sqrt{8}}$

47. Use the square root property to solve $(x - 3)^2 = 16$.

48. Use the square root property to solve $3(x - 4)^2 = 9$.

49. Solve $\dfrac{1}{2}x^2 - x = 2$ by using the quadratic formula.

50. Solve $x^2 + 4x = 8$ by using the quadratic formula.

36. a. _____ b. _____ c. _____
37. _____
38. _____
39. _____
40. _____
41. _____
42. _____
43. _____
44. _____
45. _____
46. _____
47. _____
48. _____
49. _____
50. _____

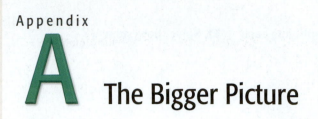

Appendix

A The Bigger Picture

A.1 SIMPLIFYING EXPRESSIONS AND SOLVING EQUATIONS

I. Simplifying Expressions

A. Real Numbers

1. Add: (Sec. 1.4)

$$-1.7 + (-0.21) = -1.91$$ Adding like signs.
Add absolute values. Attach common sign.

$$-7 + 3 = -4$$ Adding different signs.
Subtract absolute values. Attach the sign of the number with the larger absolute value.

2. Subtract: Add the first number to the opposite of the second number. (Sec. 1.5)

$$17 - 25 = 17 + (-25) = -8$$

3. Multiply or divide: Multiply or divide the two numbers as usual. If the signs are the same, the answer is positive. If the signs are different, the answer is negative. (Sec. 1.6)

$$-10 \cdot 3 = -30, \qquad -81 \div (-3) = 27$$

B. Exponents (Sec. 3.2)

$$x^7 \cdot x^5 = x^{12}; \ (x^7)^5 = x^{35}; \ \frac{x^7}{x^5} = x^2; \ x^0 = 1; \ 8^{-2} = \frac{1}{8^2} = \frac{1}{64}$$

C. Polynomials

1. Add: Combine like terms. (Sec. 3.4)

$$(3y^2 + 6y + 7) + (9y^2 - 11y - 15) = 3y^2 + 6y + 7 + 9y^2 - 11y - 15$$
$$= 12y^2 - 5y - 8$$

2. Subtract: Change the sign of the terms of the polynomial being subtracted, then add. (Sec. 3.4)

$$(3y^2 + 6y + 7) - (9y^2 - 11y - 15) = 3y^2 + 6y + 7 - 9y^2 + 11y + 15$$
$$= -6y^2 + 17y + 22$$

3. Multiply: Multiply each term of one polynomial by each term of the other polynomial. (Sec. 3.5)

$$(x + 5)(2x^2 - 3x + 4) = x(2x^2 - 3x + 4) + 5(2x^2 - 3x + 4)$$
$$= 2x^3 - 3x^2 + 4x + 10x^2 - 15x + 20$$
$$= 2x^3 + 7x^2 - 11x + 20$$

4. **Divide:** (Sec. 3.7)

   a. To divide by a monomial, divide each term of the polynomial by the monomial.

   $$\frac{8x^2 + 2x - 6}{2x} = \frac{8x^2}{2x} + \frac{2x}{2x} - \frac{6}{2x} = 4x + 1 - \frac{3}{x}$$

   b. To divide by a polynomial other than a monomial, use long division.

   $$\begin{array}{r} x - 6 + \dfrac{40}{2x + 5} \\[2pt] 2x + 5 \overline{\smash{)}\, 2x^2 - 7x + 10} \\ \underline{2x^2 + 5x} \\ -12x + 10 \\ \underline{-12x - 30} \\ 40 \end{array}$$

D. Factoring Polynomials

See the Chapter 4 Integrated Review for steps. (Sec. 4.5)

$$\begin{aligned} 3x^4 - 78x^2 + 75 &= 3(x^4 - 26x^2 + 25) \quad \text{Factor out GCF—always first step.} \\ &= 3(x^2 - 25)(x^2 - 1) \quad \text{Factor trinomial.} \\ &= 3(x + 5)(x - 5)(x + 1)(x - 1) \quad \text{Factor further—each} \\ &\qquad\qquad\qquad\qquad\qquad\qquad\quad \text{difference of squares.} \end{aligned}$$

E. Rational Expressions

1. **Simplify:** Factor the numerator and denominator. Then divide out factors of 1 by dividing out common factors in the numerator and denominator. (Sec. 5.1)

   $$\frac{x^2 - 9}{7x^2 - 21x} = \frac{(x + 3)(x - 3)}{7x(x - 3)} = \frac{x + 3}{7x}$$

2. **Multiply:** Multiply numerators, then multiply denominators. (Sec. 5.2)

   $$\frac{5z}{2z^2 - 9z - 18} \cdot \frac{22z + 33}{10z} = \frac{5 \cdot z}{(2z + 3)(z - 6)} \cdot \frac{11(2z + 3)}{2 \cdot 5 \cdot z} = \frac{11}{2(z - 6)}$$

3. **Divide:** First fraction times the reciprocal of the second fraction. (Sec. 5.2)

   $$\frac{14}{x + 5} \div \frac{x + 1}{2} = \frac{14}{x + 5} \cdot \frac{2}{x + 1} = \frac{28}{(x + 5)(x + 1)}$$

4. **Add or subtract:** Must have same denominator. If not, find the LCD and write each fraction as an equivalent fraction with the LCD as denominator. (Sec. 5.4)

   $$\begin{aligned} \frac{9}{10} - \frac{x + 1}{x + 5} &= \frac{9(x + 5)}{10(x + 5)} - \frac{10(x + 1)}{10(x + 5)} \\ &= \frac{9x + 45 - 10x - 10}{10(x + 5)} = \frac{-x + 35}{10(x + 5)} \end{aligned}$$

F. Radicals

1. **Simplify square roots:** If possible, factor the radicand so that one factor is a perfect square. Then use the product rule and simplify. (Sec. 8.2)

   $$\sqrt{75} = \sqrt{25 \cdot 3} = \sqrt{25} \cdot \sqrt{3} = 5\sqrt{3}$$

2. **Add or subtract:** Only like radicals (same index and radicand) can be added or subtracted. (Sec. 8.3)

$$8\sqrt{10} - \sqrt{40} + \sqrt{5} = 8\sqrt{10} - 2\sqrt{10} + \sqrt{5} = 6\sqrt{10} + \sqrt{5}$$

3. **Multiply or divide:** $\sqrt{a} \cdot \sqrt{b} = \sqrt{ab}; \dfrac{\sqrt{a}}{\sqrt{b}} = \sqrt{\dfrac{a}{b}}$. (Sec. 8.4)

$$\sqrt{11} \cdot \sqrt{3} = \sqrt{33}; \dfrac{\sqrt{140}}{\sqrt{7}} = \sqrt{\dfrac{140}{7}} = \sqrt{20} = \sqrt{4 \cdot 5} = 2\sqrt{5}$$

4. **Rationalizing the denominator:** (Sec. 8.4)
   a. If denominator is one term,

$$\frac{5}{\sqrt{11}} = \frac{5 \cdot \sqrt{11}}{\sqrt{11} \cdot \sqrt{11}} = \frac{5\sqrt{11}}{11}$$

   b. If denominator is two terms, multiply by 1 in the form of $\dfrac{\text{conjugate of denominator}}{\text{conjugate of denominator}}$.

$$\frac{13}{3 + \sqrt{2}} = \frac{13}{3 + \sqrt{2}} \cdot \frac{3 - \sqrt{2}}{3 - \sqrt{2}} = \frac{13(3 - \sqrt{2})}{9 - 2} = \frac{13(3 - \sqrt{2})}{7}$$

II. Solving Equations and Inequalities

A. **Linear Equations:** Power on variable is 1 and there are no variables in denominator. (Sec. 2.3)

$7(x - 3) = 4x + 6$ Linear equation. (If fractions, multiply by LCD.)

$7x - 21 = 4x + 6$ Use the distributive property.

$7x = 4x + 27$ Add 21 to both sides.

$3x = 27$ Subtract $4x$ from both sides.

$x = 9$ Divide both sides by 3.

B. **Linear Inequalities:** Same as linear equation except if you multiply or divide by a negative number, then reverse direction of inequality. (Sec. 2.7)

$-4x + 11 \le -1$ Linear inequality.

$-4x \le -12$ Subtract 11 from both sides.

$\dfrac{-4x}{-4} \ge \dfrac{-12}{-4}$ Divide both sides by -4 and reverse the direction of the inequality symbol.

$x \ge 3$ Simplify.

C. **Quadratic and Higher Degree Equations:** Solve: first write the equation in standard form (one side is 0.)

1. If the polynomial on one side factors, solve by factoring. (Sec. 4.6)
2. If the polynomial does not factor, solve by the quadratic formula. (Sec. 9.3)

By factoring:	**By quadratic formula:**
$x^2 + x = 6$	$x^2 + x = 5$
$x^2 + x - 6 = 0$	$x^2 + x - 5 = 0$
$(x - 2)(x + 3) = 0$	$a = 1, b = 1, c = -5$
$x - 2 = 0 \text{ or } x + 3 = 0$	$x = \dfrac{-1 \pm \sqrt{1^2 - 4(1)(-5)}}{2 \cdot 1}$
$x = 2 \text{ or } x = -3$	$x = \dfrac{-1 \pm \sqrt{21}}{2}$

D. Equations with Rational Expressions: Make sure the proposed solution does not make the denominator 0. (Sec. 5.5)

$$\frac{3}{x} - \frac{1}{x-1} = \frac{4}{x-1} \qquad \text{Equation with rational expressions.}$$

$$x(x-1) \cdot \frac{3}{x} - x(x-1) \cdot \frac{1}{x-1} = x(x-1) \cdot \frac{4}{x-1} \qquad \text{Multiply through by } x(x-1).$$

$$3(x-1) - x \cdot 1 = x \cdot 4 \qquad \text{Simplify.}$$

$$3x - 3 - x = 4x \qquad \text{Use the distributive property.}$$

$$-3 = 2x \qquad \text{Simplify and move variable terms to right side.}$$

$$-\frac{3}{2} = x \qquad \text{Divide both sides by 2.}$$

E. Proportions: An equation with two ratios equal. Set cross products equal, then solve. Make sure the proposed solution does not make the denominator 0. (Sec. 5.6)

$$\frac{5}{x} = \frac{9}{2x-3}$$

$$5(2x-3) = 9 \cdot x \qquad \text{Set cross products equal.}$$

$$10x - 15 = 9x \qquad \text{Multiply.}$$

$$x = 15 \qquad \text{Write equation with variable terms on one side and constants on the other.}$$

F. Equations with Radicals: To solve, isolate a radical, then square both sides. You may have to repeat this. Check possible solution in the original equation. (Sec. 8.5)

$$\sqrt{x + 49} + 7 = x$$

$$\sqrt{x + 49} = x - 7 \qquad \text{Subtract 7 from both sides.}$$

$$x + 49 = x^2 - 14x + 49 \qquad \text{Square both sides.}$$

$$0 = x^2 - 15x \qquad \text{Set terms equal to 0.}$$

$$0 = x(x - 15) \qquad \text{Factor.}$$

$$\cancel{x = 0} \text{ or } x = 15 \qquad \text{Set each factor equal to 0 and solve.}$$

B Factoring Sums and Differences of Cubes

Although the sum of two squares usually does not factor, the sum or difference of two cubes can be factored and reveal factoring patterns. The pattern for the sum of cubes can be checked by multiplying the binomial $x + y$ and the trinomial $x^2 - xy + y^2$. The pattern for the difference of two cubes can be checked by multiplying the binomial $x - y$ by the trinomial $x^2 + xy + y^2$.

Sum or Difference of Two Cubes

$$a^3 + b^3 = (a + b)(a^2 - ab + b^2)$$
$$a^3 - b^3 = (a - b)(a^2 + ab + b^2)$$

EXAMPLE 1 Factor $x^3 + 8$.

Solution: First, write the binomial in the form $a^3 + b^3$.

$x^3 + 8 = x^3 + 2^3$ Write in the form $a^3 + b^3$.

If we replace a with x and b with 2 in the formula above, we have

$$x^3 + 2^3 = (x + 2)[x^2 - (x)(2) + 2^2]$$
$$= (x + 2)(x^2 - 2x + 4)$$

Helpful Hint

When factoring sums or differences of cubes, notice the sign patterns.

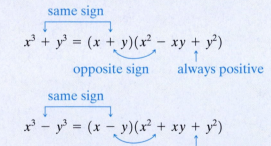

$$x^3 + y^3 = (x + y)(x^2 - xy + y^2)$$

same sign   opposite sign   always positive

$$x^3 - y^3 = (x - y)(x^2 + xy + y^2)$$

same sign   opposite sign   always positive

EXAMPLE 2 Factor $y^3 - 27$.

Solution:

$$
\begin{aligned}
y^3 - 27 &= y^3 - 3^3 && \text{Write in the form } a^3 - b^3. \\
&= (y - 3)[y^2 + (y)(3) + 3^2] \\
&= (y - 3)(y^2 + 3y + 9)
\end{aligned}
$$

EXAMPLE 3 Factor $64x^3 + 1$.

Solution:

$$
\begin{aligned}
64x^3 + 1 &= (4x)^3 + 1^3 \\
&= (4x + 1)[(4x)^2 - (4x)(1) + 1^2] \\
&= (4x + 1)(16x^2 - 4x + 1)
\end{aligned}
$$

EXAMPLE 4 Factor $54a^3 - 16b^3$.

Solution: Remember to factor out common factors first before using other factoring methods.

$$
\begin{aligned}
54a^3 - 16b^3 &= 2(27a^3 - 8b^3) && \text{Factor out the GCF 2.} \\
&= 2[(3a)^3 - (2b)^3] && \text{Difference of two cubes.} \\
&= 2(3a - 2b)[(3a)^2 + (3a)(2b) + (2b)^2] \\
&= 2(3a - 2b)(9a^2 + 6ab + 4b^2)
\end{aligned}
$$

B EXERCISE SET

FOR EXTRA HELP

Student Solutions Manual | PH Math/Tutor Center | CD/Video for Review | MathXL® | MyMathLab

Factor the sum or difference of two cubes. See Examples 1 through 4.

1. $a^3 + 27$

2. $b^3 - 8$

3. $8a^3 + 1$

4. $64x^3 - 1$

5. $5k^3 + 40$

6. $6r^3 - 162$

7. $x^3y^3 - 64$

8. $8x^3 - y^3$

9. $x^3 + 125$

10. $a^3 - 216$

11. $24x^4 - 81xy^3$

12. $375y^6 - 24y^3$

Factor the binomials completely.

13. $27 - t^3$

14. $125 + r^3$

15. $8r^3 - 64$

16. $54r^3 + 2$

17. $t^3 - 343$

18. $s^3 + 216$

19. $s^3 - 64t^3$

20. $8t^3 + s^3$

C Mean, Median, and Mode

It is sometimes desirable to be able to describe a set of data, or a set of numbers, by a single "middle" number. Three such **measures of central tendency** are the mean, the median, and the mode.

The most common measure of central tendency is the mean (sometimes called the arithmetic mean or the average). The **mean** of a set of data items, denoted by $\bar{x}$, is the sum of the items divided by the number of items.

EXAMPLE 1

Seven students in a psychology class conducted an experiment on mazes. Each student was given a pencil and asked to successfully complete the same maze. The timed results are below.

Student	Ann	Thanh	Carlos	Jesse	Melinda	Ramzi	Dayni
Time (seconds)	13.2	11.8	10.7	16.2	15.9	13.8	18.5

a. Who completed the maze in the shortest time? Who completed the maze in the longest time?

b. Find the mean.

c. How many students took longer than the mean time? How many students took shorter than the mean time?

Solution:

a. Carlos completed the maze in 10.7 seconds, the shortest time. Dayni completed the maze in 18.5 seconds, the longest time.

b. To find the mean, $\bar{x}$, find the sum of the data items and divide by 7, the number of items.

$$\bar{x} = \frac{13.2 + 11.8 + 10.7 + 16.2 + 15.9 + 13.8 + 18.5}{7} = \frac{100.1}{7} = 14.3$$

c. Three students, Jesse, Melinda, and Dayni, had times longer than the mean time. Four students, Ann, Thanh, Carlos, and Ramzi, had times shorter than the mean time.

Two other measures of central tendency are the median and the mode.

The **median** of an ordered set of numbers is the middle number. If the number of items is even, the median is the mean of the two middle numbers. The **mode** of a set of numbers is the number that occurs most often. It is possible for a data set to have no mode or more than one mode.

EXAMPLE 2

Find the median and the mode of the following set of numbers. These numbers were high temperatures for fourteen consecutive days in a city in Montana.

76, 80, 85, 86, 89, 87, 82, 77, 76, 79, 82, 89, 89, 92

Solution:

First, write the numbers in order.

76, 76, 77, 79, 80, 82, $\underbrace{82, 85}_{\substack{\text{two} \\ \text{middle numbers}}}$, 86, 87, $\underbrace{89, 89, 89}_{\text{mode}}$, 92

Since there are an even number of items, the median is the mean of the two middle numbers.

$$\text{median} = \frac{82 + 85}{2} = 83.5$$

The mode is 89, since 89 occurs most often.

For each of the following data sets, find the mean, the median, and the mode. If necessary, round the mean to one decimal place.

1. 21, 28, 16, 42, 38

2. 42, 35, 36, 40, 50

3. 7.6, 8.2, 8.2, 9.6, 5.7, 9.1

4. 4.9, 7.1, 6.8, 6.8, 5.3, 4.9

5. 0.2, 0.3, 0.5, 0.6, 0.6, 0.9, 0.2, 0.7, 1.1

6. 0.6, 0.6, 0.8, 0.4, 0.5, 0.3, 0.7, 0.8, 0.1

7. 231, 543, 601, 293, 588, 109, 334, 268

8. 451, 356, 478, 776, 892, 500, 467, 780

The eight tallest buildings in the United States are listed below. Use this table for Exercises 9 through 12.

Building	Height (Feet)
Sears Tower, Chicago, IL	1454
Empire State, New York, NY	1250
Amoco, Chicago, IL	1136
John Hancock Center, Chicago, IL	1127
First Interstate World Center, Los Angeles, CA	1107
Chrysler, New York, NY	1046
NationsBank Tower, Atlanta, GA	1023
Texas Commerce Tower, Houston, TX	1002

9. Find the mean height for the five tallest buildings.

10. Find the median height for the five tallest buildings.

11. Find the median height for the eight tallest buildings.

12. Find the mean height for the eight tallest buildings. Round to the nearest tenth.

During an experiment, the following times (in seconds) were recorded: 7.8, 6.9, 7.5, 4.7, 6.9, 7.0.

13. Find the mean.

14. Find the median.

15. Find the mode.

In a mathematics class, the following test scores were recorded for a student: 86, 95, 91, 74, 77, 85.

16. Find the mean. Round to the nearest hundredth.

17. Find the median.

18. Find the mode.

The following pulse rates were recorded for a group of fifteen students: 78, 80, 66, 68, 71, 64, 82, 71, 70, 65, 70, 75, 77, 86, 72.

19. Find the mean.

20. Find the median.

21. Find the mode.

22. How many rates were higher than the mean?

23. How many rates were lower than the mean?

24. Have each student in your algebra class take his/her pulse rate. Record the data and find the mean, the median, and the mode.

Find the missing numbers in each set of numbers. (These numbers are not necessarily in numerical order.)

25. _____, _____ , 16, 18, _____
The mode is 21.
The median is 20.

26. _____, _____ , _____ , _____ , 40
The mode is 35.
The median is 37.
The mean is 38.

Appendix

Sets

Objectives

A Determine Whether an Object Is an Element of a Set.

B Determine Whether a Set Is a Subset of Another Set.

C Find Unions and Intersection of Sets.

Objective A Determining Whether an Object Is an Element of a Set

A **set** is a collection of objects, called **elements** or **members.** The elements of a set are listed or described between a pair of braces, { }. Two common ways of representing a set are by **roster form** or by **set builder** notation. Examples of each are as follows:

$A = \{1, 2, 3, 4, 5\}$ Roster form—elements are listed.

$B = \{x \mid x \leq 3\}$ Set builder notation

Set B is read as "the set of all x such that x is less than or equal to 3."
The symbol $\in$ means "is an element of." For set A above, we can write

$3 \in A$

means the number 3 is an element of set A.
  Also,

$6 \notin A$

means the number 6 is not an element of set A.
  A set that has no elements is called the **empty set** or the **null set.** The empty set (or null set) is symbolized by { } or $\varnothing$. For example, if set C is the set of all positive numbers less than 0, then

$$C = \{ \ \} \quad \text{or} \quad C = \varnothing.$$

Helpful Hint

The set $\{\varnothing\}$ is *not* the empty set. It is a set with one element, $\varnothing$.

EXAMPLE 1 Given the following sets, determine whether each statement is true or false.

$N = \{0, 1, 2, 3, 4, 5\}$ $E = \{0, 2, 4, 6, 8, 10\}$ $O = \{x \mid x \text{ is an odd number}\}$

a. $2 \in N$
b. $5 \in E$
c. $5 \in O$
d. $10 \in O$

Solution:

a. True, since 2 is a listed element of N
b. False, since 5 is not listed as an element of E
c. True, since 5 is an odd number
d. False, since 10 is not an odd number

679

Objective B Determining Whether a Set Is a Subset of Another Set

Set A is a subset of set B if every element of A is also an element of B. In symbols, we write $A \subseteq B$, and this is illustrated to the right.

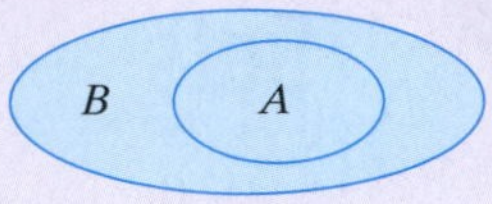

$A \subseteq B$ means set A is a subset of set B. The symbol $\not\subseteq$ means "is not a subset." Thus,

$A \not\subseteq B$ means set A is not a subset of set B.

> **Helpful Hint**
> If $A \not\subseteq B$ is true, then there must be at least one element of set A that is not in set B.

EXAMPLE 2   Determine whether $A \subseteq B$.

a. $A = \{0, 7\}, B = \{0, 1, 2, 3, 4, 5, 6, 7\}$
b. $A = \{2, 4, 6\}, B = \{2, 4, 6\}$
c. $A = \{x \mid x < 5\}, B = \{x \mid x < 1\}$
d. $A = \{\bigcirc, \triangle, \square\}, B = \{\bigcirc, \triangle\}$

Solution:

a. $A \subseteq B$ since every element of A is also an element of B.

b. $A \subseteq B$ since every element of A is also an element of B.

c. $A \not\subseteq B$ because there are elements of A that are not in B. For example, 3 is an element of A but is not an element of B.

d. $A \not\subseteq B$ because $\square \in A$, but $\square \notin B$.

> **Helpful Hint**
> From Example 2b, we see that every set is a subset of itself. For example, $A \subseteq A$. Why? Because every element of A is always an element of A.

Objective C Finding Unions and Intersections of Sets

The **union** of two sets A and B, written as $A \cup B$, is the set of all elements that are in set A or in set B (or in both sets).

$A \cup B$

The **intersection** of two sets A and B, written as $A \cap B$, is the set of all elements that are common to both set A and set B.

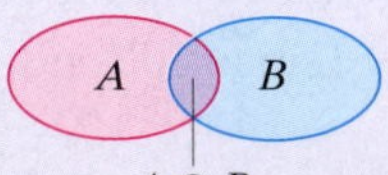

$A \cap B$

EXAMPLE 3 Let $R = \{0, 2, 5\}$ and $S = \{0, 1, 2, 3, 4\}$. Find

a. $R \cup S$

b. $R \cap S$

Solution:

a. $R \cup S = \{0, 1, 2, 3, 4, 5\}$, all the numbers in set R or set S (or both)

b. $R \cap S = \{0, 2\}$, the only numbers in both sets

EXAMPLE 4 Let $M = \{10, 20\}$ and $N = \{5, 15, 25\}$. Find

a. $M \cup N$

b. $M \cap N$

Solution:

a. $M \cup N = \{5, 10, 15, 20, 25\}$ Any order of the elements is fine.

b. $M \cap N = \{\ \}$ or $\varnothing$ There are no elements common to both sets. The intersection is the empty set.

Helpful Hint

- You may list the elements of a set in any order.
- There is no need to list an element of a set more than once.

For each set described, write the set in roster form.

1. The set of negative integers from -10 to -5.

2. The set of integers between -1 and 1.

3. The set of the days of the week starting with the letter *T*.

4. The set of the first five letters of the alphabet

5. The set of whole numbers

6. The set of natural numbers

7. $\{x \mid x \text{ is an integer between 1 and 2}\}$

8. $\{x \mid x \text{ is a number that is both even and odd}\}$

Objectives A B Mixed Practice *Determine whether each statement is true or false. See Examples 1 and 2.*

9. $3 \in \{1, 3, 5, 7, 9\}$

10. $6 \in \{1, 3, 5, 7, 9\}$

11. $\{3\} \subseteq \{1, 3, 5, 7, 9\}$

12. $\{6\} \subseteq \{1, 3, 5, 7, 9\}$

13. $\{a, e, i, o, u\} \subseteq \{a, e, i, o, u\}$

14. $\{\triangle, \square, \bigcirc\} \subseteq \{\triangle, \square, \bigcirc\}$

15. $\{\text{May}\} \subseteq$ the set of days of the week

16. $\{\text{Sunday}\} \subseteq$ the set of months of the year

17. $9 \notin \{x \mid x \text{ is an even number}\}$

18. $10 \notin \{x \mid x \text{ is an odd number}\}$

19. $\{a\} \not\subseteq$ the set of vowels

20. $\{\triangle\} \not\subseteq$ the set of polygons

Objective C *Given the sets, find each union or intersection. See Examples 3 and 4.*

$A = \{1, 2, 3, 4, 5, 6\} \quad B = \{2, 4, 6\} \quad C = \{1, 3, 5\} \quad D = \{7\}$

21. $A \cup B$

22. $A \cup C$

23. $A \cap B$

24. $A \cap C$

25. $C \cup D$

26. $C \cap D$

27. $B \cap D$

28. $B \cup D$

Objectives B C Mixed Practice *Use sets A, B, C, and D above to determine whether each statement is true or false. See Examples 2 through 4.*

29. $B \subseteq A$

30. $A \subseteq B$

31. $D \subseteq C$

32. $C \subseteq D$

33. $A \subseteq C$

34. $C \subseteq A$

35. $B \not\subseteq C$

36. $C \not\subseteq B$

37. $\varnothing \subseteq A$

38. $\varnothing \subseteq D$

39. $A \cup D$ is $\{1, 2, 3, 4, 5, 6, 7\}$

40. $A \cap D$ is $\varnothing$

41. $\{a, b, c\} \cup \{\ \}$ is $\{a, b, c\}$

42. $\{a, b, c\} \cap \{\ \}$ is $\{\ \}$

Appendix

E Review of Angles, Lines, and Special Triangles

The word **geometry** is formed from the Greek words, **geo,** meaning earth, and **metron,** meaning measure. Geometry literally means to measure the earth.

This appendix contains a review of some basic geometric ideas. It will be assumed that fundamental ideas of geometry such as point, line, ray, and angle are known. In this appendix, the notation $\angle 1$ is read "angle 1" and the notation $m\angle 1$ is read "the measure of angle 1."

We first review types of angles.

Angles

An angle whose measure is greater than $0°$ but less than $90°$ is called an **acute angle.**

A **right angle** is an angle whose measure is $90°$. A right angle can be indicated by a square drawn at the vertex of the angle, as shown below.

An angle whose measure is greater than $90°$ but less than $180°$ is called an **obtuse angle.**

An angle whose measure is $180°$ is called a **straight angle.**

Two angles are said to be **complementary** if the sum of their measures is $90°$. Each angle is called the **complement** of the other.

Two angles are said to be **supplementary** if the sum of their measures is $180°$. Each angle is called the **supplement** of the other.

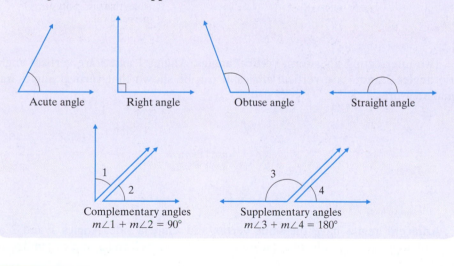

| Acute angle | Right angle | Obtuse angle | Straight angle |

Complementary angles
$m\angle 1 + m\angle 2 = 90°$

Supplementary angles
$m\angle 3 + m\angle 4 = 180°$

EXAMPLE 1 If an angle measures $28°$, find its complement.

Solution: Two angles are complementary if the sum of their measures is $90°$. The complement of a $28°$ angle is an angle whose measure is $90° - 28° = 62°$. To check, notice that $28° + 62° = 90°$.

Plane is an undefined term that we will describe. A plane can be thought of as a flat surface with infinite length and width, but no thickness. A plane is two dimensional. The arrows in the following diagram indicate that a plane extends indefinitely and has no boundaries.

Figures that lie on a plane are called **plane figures.** Lines that lie in the same plane are called **coplanar.**

Lines

Two lines are **parallel** if they lie in the same plane but never meet. **Intersecting lines** meet or cross in one point.

Two lines that form right angles when they intersect are said to be **perpendicular.**

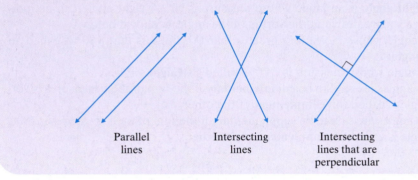

| Parallel lines | Intersecting lines | Intersecting lines that are perpendicular |

Two intersecting lines form **vertical angles.** Angles 1 and 3 are vertical angles. Also angles 2 and 4 are vertical angles. It can be shown that **vertical angles have equal measures.**

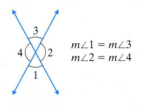

$$m\angle 1 = m\angle 3$$
$$m\angle 2 = m\angle 4$$

Adjacent angles have the same vertex and share a side. Angles 1 and 2 are adjacent angles. Other pairs of adjacent angles are angles 2 and 3, angles 3 and 4, and angles 4 and 1.

A **transversal** is a line that intersects two or more lines in the same plane. Line l is a transversal that intersects lines m and n. The eight angles formed are numbered and certain pairs of these angles are given special names.

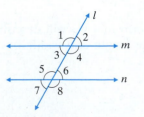

Corresponding angles: ∠1 and ∠5, ∠3 and ∠7, ∠2 and ∠6, and ∠4 and ∠8.
Exterior angles: ∠1, ∠2, ∠7, and ∠8.
Interior angles: ∠3, ∠4, ∠5, and ∠6.
Alternate interior angles: ∠3 and ∠6, ∠4 and ∠5.

These angles and parallel lines are related in the following manner.

Parallel Lines Cut by a Transversal

1. If two parallel lines are cut by a transversal, then

    a. corresponding angles are equal and
    b. alternate interior angles are equal.

2. If corresponding angles formed by two lines and a transversal are equal, then the lines are parallel.

3. If alternate interior angles formed by two lines and a transversal are equal, then the lines are parallel.

EXAMPLE 2 Given that lines m and n are parallel and that the measure of angle 1 is 100°, find the measures of angles 2, 3, and 4.

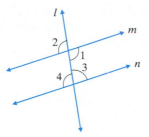

Solution:

$m\angle 2 = 100°$ since angles 1 and 2 are vertical angles.

$m\angle 4 = 100°$ since angles 1 and 4 are alternate interior angles.

$m\angle 3 = 180° - 100° = 80°$ since angles 4 and 3 are supplementary angles.

A **polygon** is the union of three or more coplanar line segments that intersect each other only at each end point, with each end point shared by exactly two segments.

A **triangle** is a polygon with three sides. The sum of the measures of the three angles of a triangle is 180°. In the following figure, $m\angle 1 + m\angle 2 + m\angle 3 = 180°$.

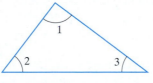

EXAMPLE 3 Find the measure of the third angle of the triangle shown.

Solution: The sum of the measures of the angles of a triangle is 180°. Since one angle measures 45° and the other angle measures 95°, the third angle measures $180° - 45° - 95° = 40°$.

Two triangles are **congruent** if they have the same size and the same shape. In congruent triangles, the measures of corresponding angles are equal and the lengths of corresponding sides are equal. The following triangles are congruent.

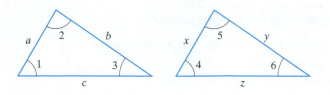

Corresponding angles are equal: $m\angle 1 = m\angle 4, m\angle 2 = m\angle 5$, and $m\angle 3 = m\angle 6$. Also, lengths of corresponding sides are equal: $a = x, b = y$, and $c = z$.

Any one of the following may be used to determine whether two triangles are congruent.

Congruent Triangles

1. If the measures of two angles of a triangle equal the measures of two angles of another triangle and the lengths of the sides between each pair of angles are equal, the triangles are congruent.

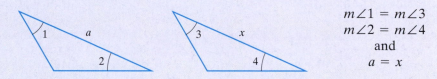

$$m\angle 1 = m\angle 3$$
$$m\angle 2 = m\angle 4$$
$$\text{and}$$
$$a = x$$

2. If the lengths of the three sides of a triangle equal the lengths of corresponding sides of another triangle, the triangles are congruent.

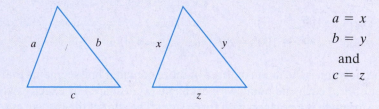

$$a = x$$
$$b = y$$
$$\text{and}$$
$$c = z$$

3. If the lengths of two sides of a triangle equal the lengths of corresponding sides of another triangle, and the measures of the angles between each pair of sides are equal, the triangles are congruent.

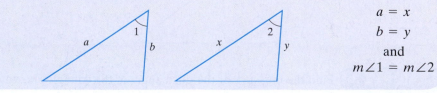

$$a = x$$
$$b = y$$
$$\text{and}$$
$$m\angle 1 = m\angle 2$$

Two triangles are **similar** if they have the same shape but not necessarily the same size. In similar triangles, the measures of corresponding angles are equal and corresponding sides are in proportion. The following triangles are similar. (All similar triangles drawn in this appendix will be oriented the same.)

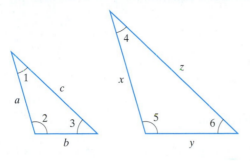

Corresponding angles are equal: $m\angle 1 = m\angle 4$, $m\angle 2 = m\angle 5$, and $m\angle 3 = m\angle 6$.

Also, corresponding sides are proportional: $\dfrac{a}{x} = \dfrac{b}{y} = \dfrac{c}{z}$.

Any one of the following may be used to determine whether two triangles are similar.

Similar Triangles

1. If the measures of two angles of a triangle equal the measures of two angles of another triangle, the triangles are similar.

$$m\angle 1 = m\angle 2$$
$$\text{and}$$
$$m\angle 3 = m\angle 4$$

2. If three sides of one triangle are proportional to three sides of another triangle, the triangles are similar.

$$\frac{a}{x} = \frac{b}{y} = \frac{c}{z}$$

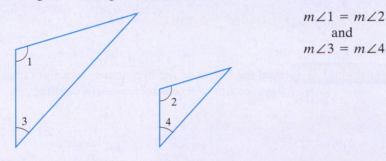

3. If two sides of a triangle are proportional to two sides of another triangle and the measures of the included angles are equal, the triangles are similar.

$$m\angle 1 = m\angle 2$$
$$\text{and}$$
$$\frac{a}{x} = \frac{b}{y}$$

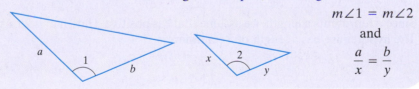

EXAMPLE 4 Given that the following triangles are similar, find the missing length x.

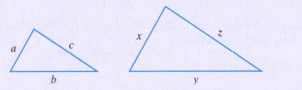

Solution: Since the triangles are similar, corresponding sides are in proportion. Thus, $\frac{2}{3} = \frac{10}{x}$. To solve this equation for x, we cross multiply.

$$\frac{2}{3} = \frac{10}{x}$$

$$2x = 30$$

$$x = 15$$

The missing length is 15 units.

A **right triangle** contains a right angle. The side opposite the right angle is called the **hypotenuse,** and the other two sides are called the **legs.** The **Pythagorean theorem** gives a formula that relates the lengths of the three sides of a right triangle.

The Pythagorean Theorem

If a and b are the lengths of the legs of a right triangle, and c is the length of the hypotenuse, then $a^2 + b^2 = c^2$.

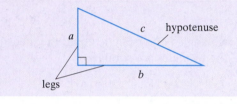

EXAMPLE 5 Find the length of the hypotenuse of a right triangle whose legs have lengths of 3 centimeters and 4 centimeters.

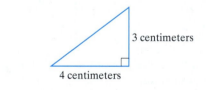

Solution: Because we have a right triangle, we use the Pythagorean theorem. The legs are 3 centimeters and 4 centimeters, so let $a = 3$ and $b = 4$ in the formula.

$$a^2 + b^2 = c^2$$

$$3^2 + 4^2 = c^2$$

$$9 + 16 = c^2$$

$$25 = c^2$$

Since c represents a length, we assume that c is positive. Thus, if c^2 is 25, c must be 5. The hypotenuse has a length of 5 centimeters.

FOR EXTRA HELP

| E | **EXERCISE SET** | Student Solutions Manual | PH Math/Tutor Center | CD/Video for Review | Math XL MathXL® | MyMathLab MyMathLab |

Find the complement of each angle. See Example 1.

1. 19°

2. 65°

3. 70.8°

4. $45\frac{2}{3}°$

5. $11\frac{1}{4}°$

6. 19.6°

Find the supplement of each angle.

7. 150°

8. 90°

9. 30.2°

10. 81.9°

11. $79\frac{1}{2}°$

12. $165\frac{8}{9}°$

13. If lines *m* and *n* are parallel, find the measures of angles 1 through 7. See Example 2.

14. If lines *m* and *n* are parallel, find the measures of angles 1 through 5. See Example 2.

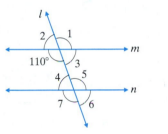

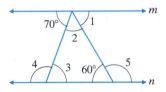

In each of the following, the measures of two angles of a triangle are given. Find the measure of the third angle. See Example 3.

15. 11°, 79°

16. 8°, 102°

17. 25°, 65°

18. 44°, 19°

19. 30°, 60°

20. 67°, 23°

In each of the following, the measure of one angle of a right triangle is given. Find the measures of the other two angles.

21. 45°

22. 60°

23. 17°

24. 30°

25. $39\frac{3}{4}°$

26. 72.6°

Given that each of the following pairs of triangles is similar, find the missing length x. See Example 4.

27.

28.

29.

30.

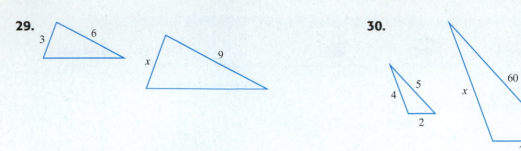

Use the Pythagorean theorem to find the missing lengths in the right triangles. See Example 5.

31.

32.

33.

34.

Appendix

F Tables

n	n^2	$\sqrt{n}$	n	n^2	$\sqrt{n}$
1	1	1.000	51	2601	7.141
2	4	1.414	52	2704	7.211
3	9	1.732	53	2809	7.280
4	16	2.000	54	2916	7.348
5	25	2.236	55	3025	7.416
6	36	2.449	56	3136	7.483
7	49	2.646	57	3249	7.550
8	64	2.828	58	3364	7.616
9	81	3.000	59	3481	7.681
10	100	3.162	60	3600	7.746
11	121	3.317	61	3721	7.810
12	144	3.464	62	3844	7.874
13	169	3.606	63	3969	7.937
14	196	3.742	64	4096	8.000
15	225	3.873	65	4225	8.062
16	256	4.000	66	4356	8.124
17	289	4.123	67	4489	8.185
18	324	4.243	68	4624	8.246
19	361	4.359	69	4761	8.307
20	400	4.472	70	4900	8.367
21	441	4.583	71	5041	8.426
22	484	4.690	72	5184	8.485
23	529	4.796	73	5329	8.544
24	576	4.899	74	5476	8.602
25	625	5.000	75	5625	8.660
26	676	5.099	76	5776	8.718
27	729	5.196	77	5929	8.775
28	784	5.292	78	6084	8.832
29	841	5.385	79	6241	8.888
30	900	5.477	80	6400	8.944
31	961	5.568	81	6561	9.000
32	1024	5.657	82	6724	9.055
33	1089	5.745	83	6889	9.110
34	1156	5.831	84	7056	9.165
35	1225	5.916	85	7225	9.220
36	1296	6.000	86	7396	9.274
37	1369	6.083	87	7569	9.327
38	1444	6.164	88	7744	9.381
39	1521	6.245	89	7921	9.434
40	1600	6.325	90	8100	9.487
41	1681	6.403	91	8281	9.539
42	1764	6.481	92	8464	9.592
43	1849	6.557	93	8649	9.644
44	1936	6.633	94	8836	9.695
45	2025	6.708	95	9025	9.747
46	2116	6.782	96	9216	9.798
47	2209	6.856	97	9409	9.849
48	2304	6.928	98	9604	9.899
49	2401	7.000	99	9801	9.950
50	2500	7.071	100	10,000	10.000

Percent, Decimal, and Fraction Equivalents		
Percent	**Decimal**	**Fraction**
1%	0.01	$\frac{1}{100}$
5%	0.05	$\frac{1}{20}$
10%	0.1	$\frac{1}{10}$
12.5% or $12\frac{1}{2}$%	0.125	$\frac{1}{8}$
$16.\overline{6}$% or $16\frac{2}{3}$%	$0.1\overline{6}$	$\frac{1}{6}$
20%	0.2	$\frac{1}{5}$
25%	0.25	$\frac{1}{4}$
30%	0.3	$\frac{3}{10}$
$33.\overline{3}$% or $33\frac{1}{3}$%	$0.\overline{3}$	$\frac{1}{3}$
37.5% or $37\frac{1}{2}$%	0.375	$\frac{3}{8}$
40%	0.4	$\frac{2}{5}$
50%	0.5	$\frac{1}{2}$
60%	0.6	$\frac{3}{5}$
62.5% or $62\frac{1}{2}$%	0.625	$\frac{5}{8}$
$66.\overline{6}$% or $66\frac{2}{3}$%	$0.\overline{6}$	$\frac{2}{3}$
70%	0.7	$\frac{7}{10}$
75%	0.75	$\frac{3}{4}$
80%	0.8	$\frac{4}{5}$
$83.\overline{3}$% or $83\frac{1}{3}$%	$08.\overline{3}$	$\frac{5}{6}$
87.5% or $87\frac{1}{2}$%	0.875	$\frac{7}{8}$
90%	0.9	$\frac{9}{10}$
100%	1.0	1
110%	1.1	$1\frac{1}{10}$
125%	1.25	$1\frac{1}{4}$
$133.\overline{3}$% or $133\frac{1}{3}$%	$1.\overline{3}$	$1\frac{1}{3}$
150%	1.5	$1\frac{1}{2}$
$166.\overline{6}$% or $166\frac{2}{3}$%	$1.\overline{6}$	$1\frac{2}{3}$
175%	1.75	$1\frac{3}{4}$
200%	2.0	2

ANSWERS TO SELECTED EXERCISES

CHAPTER R   Prealgebra Review

Exercise Set R.1   1. 1, 3, 9   **3.** 1, 2, 3, 4, 6, 8, 12, 24   **5.** 1, 2, 3, 6, 7, 14, 21, 42   **7.** 1, 2, 4, 5, 8, 10, 16, 20, 40, 80   **9.** 1, 19   **11.** prime
13. composite   **15.** prime   **17.** composite   **19.** composite   **21.** $2 \cdot 3 \cdot 3$ **23.** $2 \cdot 2 \cdot 5$ **25.** $2 \cdot 2 \cdot 2 \cdot 7$ **27.** $3 \cdot 3 \cdot 3 \cdot 3$
29. $2 \cdot 2 \cdot 3 \cdot 5 \cdot 5$ **31.** $2 \cdot 2 \cdot 3 \cdot 7 \cdot 7$ **33.** d   **35.** 12   **37.** 42   **39.** 60   **41.** 35   **43.** 36   **45.** 80   **47.** 360   **49.** 72   **51.** 120
53. 42   **55.** 48   **57.** 360   **59. a.** $2 \cdot 2 \cdot 2 \cdot 5$ **b.** $2 \cdot 2 \cdot 2 \cdot 5$ **c.** answers may vary   **61.** true   **63.** every 140 days   **65.** 9000

Exercise Set R.2   1. 1   **3.** 10   **5.** 13   **7.** 0   **9.** undefined   **11.** $\frac{21}{30}$ **13.** $\frac{4}{18}$ **15.** $\frac{16}{20}$ **17.** $\frac{1}{2}$ **19.** $\frac{2}{3}$ **21.** $\frac{3}{7}$ **23.** $\frac{3}{5}$

25. $\frac{4}{5}$ **27.** $\frac{11}{8}$ **29.** $\frac{30}{61}$ **31.** $\frac{8}{11}$ **33.** $\frac{3}{8}$ **35.** $\frac{1}{2}$ **37.** $\frac{6}{7}$ **39.** 15   **41.** $18\frac{20}{27}$ **43.** $2\frac{28}{29}$ **45.** 1   **47.** $\frac{11}{60}$ **49.** $\frac{23}{21}$

51. $\frac{65}{21}$ **53.** $1\frac{3}{4}$ **55.** $12\frac{7}{40}$ **57.** $\frac{9}{35}$ **59.** $\frac{1}{3}$ **61.** $\frac{1}{6}$ **63.** $\frac{3}{80}$ **65.** $\frac{5}{66}$ **67.** $48\frac{1}{15}$ **69.** 37   **71.** $10\frac{5}{11}$ **73.** $\frac{7}{5}$ **75.** $7\frac{1}{12}$

77. $\frac{17}{18}$ **79.** b   **81.** answers may vary   **83.** $\frac{6}{11}$ **85.** $\frac{1}{12}$ **87.** $\frac{9}{50}$ **89. a.** $\frac{36}{459} = \frac{4}{51}$ **b.** $\frac{239}{459}$ **91.** $\frac{6}{55}$ sq m

Exercise Set R.3   1. $\frac{6}{10}$ **3.** $\frac{186}{100}$ **5.** $\frac{114}{1000}$ **7.** $\frac{1231}{10}$ **9.** 6.83   **11.** 27.0578   **13.** 6.5   **15.** 15.22   **17.** 0.12   **19.** 0.2646
21. 1.68   **23.** 5.8   **25.** 56.431   **27.** 67.5   **29.** 70   **31.** 598.23   **33.** 43.274   **35.** 840   **37.** 84.97593   **39.** 0.6   **41.** 0.23
43. 0.595   **45.** 98,207.2   **47.** 12.35   **49.** 0.75   **51.** $0.\overline{3} \approx 0.33$ **53.** 0.4375   **55.** $0.\overline{54} \approx 0.55$ **57.** $4.8\overline{3} \approx 4.83$ **59.** 0.28
61. 0.031   **63.** 1.35   **65.** 2   **67.** 0.9655   **69.** 0.001   **71.** 0.51   **73.** 68%   **75.** 87.6%   **77.** 100%   **79.** 50%   **81.** 192%
83. 0.4%   **85. a.** tenths   **b.** thousandths   **c.** ones   **87.** 78.7879   **89.** answers may vary   **91. a.** 182.6 lb   **b.** 248.6 lb

Chapter R Vocabulary Check   1. factor   **2.** multiple   **3.** composite number   **4.** percent   **5.** equivalent   **6.** improper fraction
7. prime number   **8.** simplified   **9.** proper fraction   **10.** mixed number

Chapter R Review   1. $2 \cdot 3 \cdot 7$ **2.** $2 \cdot 2 \cdot 2 \cdot 2 \cdot 2 \cdot 5 \cdot 5$ **3.** 60   **4.** 42   **5.** 60   **6.** 70   **7.** $\frac{15}{24}$ **8.** $\frac{40}{60}$ **9.** $\frac{2}{5}$ **10.** $\frac{3}{20}$ **11.** 2

12. 1   **13.** $\frac{8}{77}$ **14.** $\frac{11}{20}$ **15.** $\frac{1}{20}$ **16.** $\frac{11}{18}$ **17.** $14\frac{11}{32}$ **18.** $\frac{1}{2}$ **19.** $20\frac{17}{30}$ **20.** $2\frac{6}{7}$ **21.** $\frac{11}{20}$ sq mi   **22.** $\frac{5}{16}$ sq m   **23.** $\frac{181}{100}$

24. $\frac{35}{1000}$ **25.** 95.118   **26.** 36.785   **27.** 13.38   **28.** 691.573   **29.** 91.2   **30.** 46.816   **31.** 28.6   **32.** 230   **33.** 0.77   **34.** 25.6
35. 0.5   **36.** 0.375   **37.** $0.\overline{36} \approx 0.364$ **38.** $0.8\overline{3} \approx 0.833$ **39.** 0.29   **40.** 0.014   **41.** 39%   **42.** 120%   **43.** 0.683   **44.** b

Chapter R Test   1. $2 \cdot 2 \cdot 2 \cdot 3 \cdot 3$ **2.** 180   **3.** $\frac{25}{60}$ **4.** $\frac{3}{4}$ **5.** $\frac{12}{25}$ **6.** $\frac{13}{10}$ **7.** $\frac{53}{40}$ **8.** $\frac{18}{49}$ **9.** $\frac{1}{20}$ **10.** $\frac{29}{36}$ **11.** $4\frac{8}{9}$

12. $2\frac{5}{22}$ **13.** 55   **14.** $13\frac{13}{20}$ **15.** 45.11   **16.** 65.88   **17.** 12.688   **18.** 320   **19.** 23.73   **20.** 0.875   **21.** $0.1\overline{6} \approx 0.167$

22. 0.632   **23.** 9%   **24.** 75%   **25.** $\frac{3}{4}$ **26.** $\frac{1}{200}$ **27.** $\frac{49}{200}$ **28.** $\frac{199}{200}$ **29.** $\frac{1}{8}$ sq ft   **30.** $\frac{63}{64}$ sq cm

CHAPTER 1   Real Numbers and Introduction to Algebra

Exercise Set 1.2   1. $<$ **3.** $>$ **5.** $=$ **7.** $<$ **9.** $32 < 212$ **11.** $30 \le 45$ **13.** true   **15.** false   **17.** true   **19.** false
21. $20 \le 25$ **23.** $6 > 0$ **25.** $-12 < -10$ **27.** $7 < 11$ **29.** $5 \ge 4$ **31.** $15 \ne -2$ **33.** $14,494; -282$ **35.** $-43,413$
37. $475; -195$ **39.** ◄━●━●━●━━●━━●━► $-4\,-3\,-2\,-1\ 0\ 1\ 2\ 3\ 4\ 5$ **41.** $-\frac{1}{4}\,\frac{1}{3}$ ◄━━●━●━●●━━━► $-4\,-3\,-2\,-1\ 0\ 1\ 2\ 3\ 4$ **43.** -4.5 $-\frac{3}{2}$ $\frac{7}{4}$ 3.25 ◄━●━━━●━●━━●━► $-5\,-4\,-3\,-2\,-1\ 0\ 1\ 2\ 3\ 4$

45. whole, integers, rational, real   **47.** integers, rational, real   **49.** natural, whole, integers, rational, real   **51.** rational, real   **53.** false
55. true   **57.** false   **59.** false   **61.** 8.9   **63.** 20   **65.** $\frac{9}{2}$ **67.** $\frac{12}{13}$ **69.** $>$ **71.** $=$ **73.** $<$ **75.** $<$ **77.** $762 < 1548$
79. went down by 261 or -261 **81.** $-0.04 > -26.7$ **83.** sun   **85.** sun   **87.** answers may vary

Calculator Explorations   1. 125   **3.** 59,049   **5.** 30   **7.** 9857   **9.** 2376

Exercise Set 1.3   1. 243   **3.** 27   **5.** 1   **7.** 5   **9.** 49   **11.** $\frac{16}{81}$ **13.** $\frac{1}{125}$ **15.** 1.44   **17.** 0.343   **19.** 5^2 sq m   **21.** 17

23. 20   **25.** 12   **27.** 21   **29.** 45   **31.** 0   **33.** $\frac{2}{7}$ **35.** 30   **37.** 2   **39.** $\frac{7}{18}$ **41.** $\frac{27}{10}$ **43.** $\frac{7}{5}$ **45.** 32   **47.** $\frac{23}{27}$ **49.** 9

51. 1   **53.** 1   **55.** 11   **57.** 8   **59.** 45   **61.** 15   **63.** $\frac{1}{4}$ **65.** 6   **67.** solution   **69.** not a solution   **71.** not a solution

73. solution **75.** not a solution **77.** solution **79.** $x + 15$ **81.** $x - 5$ **83.** $\dfrac{x}{4}$ **85.** $3x + 22$ **87.** $1 + 2 = 9 \div 3$

89. $3 \neq 4 \div 2$ **91.** $5 + x = 20$ **93.** $7.6x = 17$ **95.** $13 - 3x = 13$ **97.** no **99. a.** 64 **b.** 43 **c.** 19 **d.** 22
101. 14 in., 12 sq in.; 14 in., 6 sq in.; 14 in., 10 sq in. **103.** $(20 - 4) \cdot 4 \div 2$ **105.** answers may vary **107. a.** expression **b.** equation
c. equation **d.** expression **e.** expression **109.** answers may vary

Exercise Set 1.4 **1.** 3 **3.** -14 **5.** 1 **7.** -12 **9.** -5 **11.** -12 **13.** -4 **15.** 7 **17.** -2 **19.** 0 **21.** -19 **23.** 31

25. -47 **27.** -2.1 **29.** 38 **31.** -13.1 **33.** $\dfrac{1}{4}$ **35.** $-\dfrac{3}{16}$ **37.** $-\dfrac{13}{10}$ **39.** -8 **41.** -8 **43.** -59 **45.** -9 **47.** 5

49. 11 **51.** -18 **53.** 19 **55.** -7 **57.** -26 **59.** -6 **61.** 2 **63.** 0 **65.** -6 **67.** -2 **69.** 7 **71.** 7.9 **73.** $5z$

75. $\dfrac{2}{3}$ **77.** 107°F **79.** -95 m **81.** -3.36 points **83.** -8 **85.** $-\$356$ million **87.** answers may vary **89.** July **91.** October

93. 4.7°F **95.** negative **97.** positive **99.** answers may vary **101.** answers may vary

Exercise Set 1.5 **1.** -10 **3.** -5 **5.** 19 **7.** 11 **9.** -8 **11.** -11 **13.** 37 **15.** 5 **17.** -71 **19.** 0 **21.** $\dfrac{2}{11}$

23. -6.4 **25.** 4.1 **27.** $-\dfrac{1}{6}$ **29.** $-\dfrac{11}{12}$ **31.** 8.92 **33.** -8.92 **35.** 13 **37.** -5 **39.** -1 **41.** -23 **43.** -26 **45.** -24

47. 3 **49.** -45 **51.** -4 **53.** 13 **55.** 6 **57.** 9 **59.** -9 **61.** $\dfrac{7}{5}$ **63.** -7 **65.** 21 **67.** $\dfrac{1}{4}$ **69.** not a solution

71. not a solution **73.** solution **75.** 100° **77.** lost 23 yd **79.** -569 or 569 B.C. **81.** 30° **83.** -308 ft **85.** 19,852 ft **87.** 130°
89. $-4.4°, 2.6°, 12°, 23.5°, 15.3°, 3.9°, -0.3°, -6.3°, -18.2°, -15.7°, -10.3°$ **91.** October **93.** answers may vary **95.** true; answers may vary
97. true; answers may vary **99.** negative, -2.6466 **101.** sometimes positive and sometimes negative

Integrated Review—Operations on Real Numbers **1.** negative **2.** negative **3.** positive **4.** 0 **5.** positive **6.** 0 **7.** positive

8. positive **9.** $-\dfrac{1}{7}; \dfrac{1}{7}$ **10.** $\dfrac{12}{5}; \dfrac{12}{5}$ **11.** 3; 3 **12.** $-\dfrac{9}{11}; \dfrac{9}{11}$ **13.** -42 **14.** 10 **15.** 2 **16.** -18 **17.** -7

18. -39 **19.** -2 **20.** -9 **21.** -3.4 **22.** -9.8 **23.** $-\dfrac{25}{28}$ **24.** $-\dfrac{5}{24}$ **25.** -4 **26.** -24 **27.** 6 **28.** 20 **29.** 6

30. 61 **31.** -6 **32.** -16 **33.** -19 **34.** -13 **35.** -4 **36.** -1 **37.** $\dfrac{13}{20}$ **38.** $-\dfrac{29}{40}$ **39.** 4 **40.** 9 **41.** -1 **42.** -3

43. 8 **44.** 10 **45.** 47 **46.** $\dfrac{2}{3}$

Calculator Explorations **1.** 38 **3.** -441 **5.** 490 **7.** 54,499 **9.** 15,625

Mental Math **1.** positive **3.** negative **5.** positive

Exercise Set 1.6 **1.** -24 **3.** -2 **5.** 50 **7.** -45 **9.** $\dfrac{3}{10}$ **11.** -7 **13.** -15 **15.** 0 **17.** 16 **19.** -16 **21.** $\dfrac{9}{16}$

23. -0.49 **25.** $\dfrac{3}{2}$ **27.** $-\dfrac{1}{14}$ **29.** $-\dfrac{11}{3}$ **31.** $\dfrac{1}{0.2}$ **33.** -9 **35.** -4 **37.** 0 **39.** undefined **41.** $-\dfrac{18}{7}$ **43.** 160 **45.** 64

47. $-\dfrac{8}{27}$ **49.** 3 **51.** -15 **53.** -125 **55.** -0.008 **57.** $\dfrac{2}{3}$ **59.** $\dfrac{20}{27}$ **61.** 0.84 **63.** -40 **65.** 81 **67.** -1 **69.** -121

71. -1 **73.** -19 **75.** 90 **77.** -84 **79.** -5 **81.** $-\dfrac{9}{2}$ **83.** 18 **85.** 17 **87.** -20 **89.** 16 **91.** 2 **93.** $-\dfrac{34}{7}$ **95.** 0

97. $\dfrac{6}{5}$ **99.** $\dfrac{3}{2}$ **101.** $-\dfrac{5}{38}$ **103.** 3 **105.** -1 **107.** undefined **109.** $-\dfrac{22}{9}$ **111.** solution **113.** not a solution

115. solution **117.** true **119.** false **121.** $-162°$F **123.** answers may vary **125.** 1, -1 **127.** $\dfrac{0}{5} - 7 = -7$
129. $-8(-5) + (-1) = 39$

The Bigger Picture **1.** -5 **2.** -14 **3.** $-\dfrac{26}{35}$ **4.** 8 **5.** 49 **6.** -49 **7.** -21 **8.** undefined **9.** 0 **10.** -12 **11.** -16.6

12. $\dfrac{1}{6}$ **13.** -79 **14.** $\dfrac{10}{13}$ **15.** 50 **16.** -12

Exercise Set 1.7 **1.** $16 + x$ **3.** $y \cdot (-4)$ **5.** yx **7.** $13 + 2x$ **9.** $x \cdot (yz)$ **11.** $(2 + a) + b$ **13.** $(4a) \cdot b$ **15.** $a + (b + c)$

17. $17 + b$ **19.** $24y$ **21.** y **23.** $26 + a$ **25.** $-72x$ **27.** s **29.** $-\dfrac{5}{2}x$ **31.** $4x + 4y$ **33.** $9x - 54$ **35.** $6x + 10$

37. $28x - 21$ **39.** $18 + 3x$ **41.** $-2y + 2z$ **43.** $-y - \dfrac{5}{3}$ **45.** $5x + 20m + 10$ **47.** $8m - 4n$ **49.** $-5x - 2$ **51.** $-r + 6 + 7p$

53. $3x + 4$ **55.** $-x + 3y$ **57.** $6r + 8$ **59.** $-36x - 70$ **61.** $-1.6x - 2.5$ **63.** $4(1 + y)$ **65.** $11(x + y)$ **67.** $-1(5 + x)$
69. $30(a + b)$ **71.** commutative property of multiplication **73.** associative property of addition **75.** commutative property of addition
77. associative property of multiplication **79.** identity element for addition **81.** distributive property **83.** multiplicative inverse property

85. identity element for multiplication **87.** $-8; \dfrac{1}{8}$ **89.** $-x; \dfrac{1}{x}$ **91.** $2x; -2x$ **93.** false **95.** no **97.** yes **99.** yes **101.** yes

103. a. commutative property of addition **b.** commutative property of addition **c.** associative property of addition **105.** answers may vary
107. answers may vary

Mental Math   1. -7 **3.** 1   **5.** 17   **7.** like   **9.** unlike   **11.** like

Exercise Set 1.8   1. $15y$ **3.** $13w$ **5.** $-7b - 9$ **7.** $-m - 6$ **9.** -8 **11.** $7.2x - 5.2$ **13.** $k - 6$ **15.** $-15x + 18$ **17.** $4x - 3$
19. $5x^2$ **21.** -11 **23.** $1.3x + 3.5$ **25.** $5y + 20$ **27.** $-2x - 4$ **29.** $-10x + 15y - 30$ **31.** $-3x + 2y - 1$ **33.** $7d - 11$

35. 16   **37.** $x + 5$ **39.** $x + 2$ **41.** $2k + 10$ **43.** $-3x + 5$ **45.** $2x + 14$ **47.** $3y + \dfrac{5}{6}$ **49.** $-22 + 24x$ **51.** $0.9m + 1$

53. $10 - 6x - 9y$ **55.** $-x - 38$ **57.** $5x - 7$ **59.** $10x - 3$ **61.** $-4x - 9$ **63.** $-4m - 3$ **65.** $2x - 4$ **67.** $\dfrac{3}{4}x + 12$

69. $12x - 2$ **71.** $8x + 48$ **73.** $x - 10$ **75.** 2   **77.** -23 **79.** -25 **81.** balanced   **83.** balanced   **85.** answers may vary
87. $(18x - 2)$ ft   **89.** $(15x + 23)$ in.   **91.** answers may vary

Chapter 1 Vocabulary Check   1. inequality symbols   **2.** equation   **3.** absolute value   **4.** variable   **5.** opposites   **6.** numerator
7. solution   **8.** reciprocals   **9.** base; exponent   **10.** numerical coefficient   **11.** denominator   **12.** grouping symbols   **13.** term
14. like terms   **15.** unlike terms

Chapter 1 Review   1. $<$ **2.** $>$ **3.** $>$ **4.** $>$ **5.** $<$ **6.** $>$ **7.** $=$ **8.** $=$ **9.** $>$ **10.** $<$ **11.** $4 \geq -3$ **12.** $6 \neq 5$

13. $0.03 < 0.3$ **14.** $400 < 155$ **15. a.** $1, 3$ **b.** $0, 1, 3$ **c.** $-6, 0, 1, 3$ **d.** $-6, 0, 1, 1\dfrac{1}{2}, 3, 9.62$ **e.** π **f.** all numbers in set

16. a. $2, 5$ **b.** $2, 5$ **c.** $-3, 2, 5$ **d.** $-3, -1.6, 2, 5, \dfrac{11}{2}, 15.1$ **e.** $\sqrt{5}, 2\pi$ **f.** all numbers in set   **17.** Friday   **18.** Wednesday   **19.** c

20. b   **21.** 37   **22.** 41   **23.** $\dfrac{18}{7}$ **24.** 80   **25.** $20 - 12 = 2 \cdot 4$ **26.** $\dfrac{9}{2} > -5$ **27.** 18   **28.** 108   **29.** 5   **30.** 24

31. $63°$ **32.** solution   **33.** not a solution   **34.** 9   **35.** $-\dfrac{2}{3}$ **36.** -2 **37.** 7   **38.** -11 **39.** -17 **40.** $-\dfrac{3}{16}$ **41.** -5

42. -13.9 **43.** 3.9   **44.** -14 **45.** -11.5 **46.** 5   **47.** -11 **48.** -19 **49.** 4   **50.** a   **51.** a   **52.** 51 **53.** $-\dfrac{1}{6}$ **54.** $\dfrac{5}{3}$

55. -48 **56.** 28   **57.** 3   **58.** -14 **59.** -36 **60.** 0   **61.** undefined   **62.** $-\dfrac{1}{2}$ **63.** commutative property of addition

64. multiplicative identity property   **65.** distributive property   **66.** additive inverse property   **67.** associative property of addition
68. commutative property of multiplication   **69.** distributive property   **70.** associative property of multiplication
71. multiplicative inverse property   **72.** additive identity property   **73.** commutative property of addition   **74.** distributive property
75. $6x$ **76.** $-11.8z$ **77.** $4x - 2$ **78.** $2y + 3$ **79.** $3n - 18$ **80.** $4w - 6$ **81.** $-6x + 7$ **82.** $-0.4y + 2.3$ **83.** $3x - 7$

84. $5x + 5.6$ **85.** $<$ **86.** $>$ **87.** -15.3 **88.** -6 **89.** -80 **90.** -5 **91.** $-\dfrac{1}{4}$ **92.** 0.15   **93.** 16   **94.** 16   **95.** -5

96. 9   **97.** $-\dfrac{5}{6}$ **98.** undefined   **99.** $16x - 41$ **100.** $18x - 12$

Chapter 1 Test   1. $|-7| > 5$ **2.** $(9 + 5) \geq 4$ **3.** -5 **4.** -11 **5.** -14 **6.** -39 **7.** 12   **8.** -2 **9.** undefined   **10.** -8

11. $-\dfrac{1}{3}$ **12.** $4\dfrac{5}{8}$ **13.** $\dfrac{51}{40}$ **14.** -32 **15.** -48 **16.** 3   **17.** 0   **18.** $>$ **19.** $>$ **20.** $>$ **21.** $=$

22. a. $1, 7$ **b.** $0, 1, 7$ **c.** $-5, -1, 0, 1, 7$ **d.** $-5, -1, \dfrac{1}{4}, 0, 1, 7, 11.6$ **e.** $\sqrt{7}, 3\pi$ **f.** $-5, -1, \dfrac{1}{4}, 0, 1, 7, 11.6, \sqrt{7}, 3\pi$ **23.** 40   **24.** 12   **25.** 22

26. -1 **27.** associative property of addition   **28.** commutative property of multiplication   **29.** distributive property
30. multiplicative inverse   **31.** 9   **32.** -3 **33.** second down   **34.** yes   **35.** $17°$ **36.** loss of $420   **37.** $y - 10$ **38.** $5.9x + 1.2$
39. $-2x + 10$ **40.** $-15y + 1$

CHAPTER 2   Equations, Inequalities, and Problem Solving

Mental Math   1. 2   **3.** 12   **5.** 17

Exercise Set 2.1   1. 3   **3.** -2 **5.** -14 **7.** 0.5   **9.** $\dfrac{1}{4}$ **11.** $\dfrac{5}{12}$ **13.** -3 **15.** -9 **17.** -10 **19.** 2   **21.** -7 **23.** -1

25. -9 **27.** -12 **29.** $-\dfrac{1}{2}$ **31.** 11   **33.** 21   **35.** 25   **37.** -3 **39.** -0.7 **41.** 11   **43.** 13   **45.** -30 **47.** -0.4 **49.** -7

51. $-\dfrac{1}{3}$ **53.** -17.9 **55.** $(10 - x)$ ft   **57.** $(180 - x)°$ **59.** $n - 28,000$ **61.** $7x$ sq mi   **63.** $\dfrac{8}{5}$ **65.** $\dfrac{1}{2}$ **67.** -9 **69.** x

71. y **73.** x **75.** answers may vary   **77.** 4   **79.** answers may vary   **81.** $(173 - 3x)°$ **83.** answers may vary   **85.** $x = -145.478$

Mental Math   1. 9   **3.** 2   **5.** -5

Exercise Set 2.2   1. 4   **3.** 0   **5.** 12   **7.** -12 **9.** 3   **11.** 2   **13.** 0   **15.** 6.3   **17.** 10   **19.** -20 **21.** 0   **23.** -9 **25.** 1

27. -30 **29.** 3   **31.** $\dfrac{10}{9}$ **33.** -1 **35.** -4 **37.** $-\dfrac{1}{2}$ **39.** 0   **41.** 4   **43.** $-\dfrac{1}{14}$ **45.** 0.21   **47.** 5   **49.** 6   **51.** -5.5

53. -5 **55.** 0   **57.** -3 **59.** $-\dfrac{9}{28}$ **61.** $\dfrac{14}{3}$ **63.** -9 **65.** -2 **67.** $\dfrac{11}{2}$ **69.** $-\dfrac{1}{4}$ **71.** $\dfrac{9}{10}$ **73.** $-\dfrac{17}{20}$ **75.** -16

77. $2x + 2$ **79.** $2x + 2$ **81.** $5x + 20$ **83.** $7x - 12$ **85.** $12z + 44$ **87.** 1   **89.** -48 **91.** answers may vary
93. answers may vary   **95.** 2

Calculator Explorations   1. solution   **3.** not a solution   **5.** solution

Exercise Set 2.3   1. -6 **3.** 3   **5.** 1   **7.** $\frac{3}{2}$ **9.** 0   **11.** 1   **13.** 4   **15.** -4 **17.** -3 **19.** 2   **21.** 50   **23.** 1   **25.** $\frac{7}{3}$
27. 0.2   **29.** all real numbers   **31.** no solution   **33.** no solution   **35.** all real numbers   **37.** 18   **39.** $\frac{19}{9}$ **41.** $\frac{14}{3}$ **43.** 13
45. 4   **47.** all real numbers   **49.** $-\frac{3}{5}$ **51.** -5 **53.** 10   **55.** no solution   **57.** 3   **59.** -17 **61.** $(6x - 8)$ m   **63.** $-8 - x$
65. $-3 + 2x$ **67.** $9(x + 20)$ **69. a.** all real numbers   **b.** answers may vary   **c.** answers may vary   **71.** a   **73.** b   **75.** c
77. answers may vary   **79. a.** $x + x + x + 2x + 2x = 28$ **b.** $x = 4$ **c.** $x = 4$ cm; $2x = 8$ cm   **81.** answers may vary   **83.** 15.3
85. -0.2

Integrated Review   1. 6   **2.** -17 **3.** 12   **4.** -26 **5.** -3 **6.** -1 **7.** $\frac{27}{2}$ **8.** $\frac{25}{2}$ **9.** 8   **10.** -64 **11.** 2
12. -3 **13.** 5   **14.** -1 **15.** 2   **16.** 2   **17.** -2 **18.** -2 **19.** $-\frac{5}{6}$ **20.** $\frac{1}{6}$ **21.** 1   **22.** 6   **23.** 4   **24.** 1   **25.** $\frac{9}{5}$
26. $-\frac{6}{5}$ **27.** all real numbers   **28.** all real numbers   **29.** 0   **30.** -1.6 **31.** $\frac{4}{19}$ **32.** $-\frac{5}{19}$ **33.** $\frac{7}{2}$ **34.** $-\frac{1}{4}$ **35.** no solution
36. no solution   **37.** $\frac{7}{6}$ **38.** $\frac{1}{15}$

Exercise Set 2.4   1. -25 **3.** $-\frac{3}{4}$ **5.** 234, 235   **7.** Belgium: 32; France: 33; Spain: 34   **9.** 3 in.; 6 in.; 16 in.   **11.** 1st piece: 5 in.;
2nd piece: 10 in.; 3rd piece: 25 in.   **13.** Governor of California: $175,000;   Governor of Florida: $124,575   **15.** 172 mi   **17.** 25 mi
19. 1st angle: 37.5°; 2nd angle: 37.5°; 3rd angle: 105°   **21.** A: 60°; B: 120°; C: 120°; D: 60°   **23.** 5 ft, 12 ft   **25.** 1997: 15.1 million prescriptions;
2001: 20.6 million prescriptions   **27.** 45°, 135°   **29.** 58°, 60°, 62°   **31.** 1   **33.** 280 mi   **35.** Johnson: 4932; Kenseth: 5022
37. Montana: 56 counties; California: 58 counties   **39.** Neptune: 8 moons; Uranus: 21 moons; Saturn: 18 moons   **41.** -16
43. Sahara: 3,500,000 sq mi; Gobi: 500,000 sq mi   **45.** Korea: 9, Italy: 10, France: 11   **47.** Brown: 66,362; Randall: 53,074   **49.** Illinois
51. Texas: $29.4 million; Florida: $27.2 million   **53.** answers may vary   **55.** 34   **57.** 225π **59.** 15 ft by 24 ft
61. 720 blinks per hour; 11,520 blinks per day; 4,204,800 blinks per year   **63.** answers may vary   **65.** answers may vary

Exercise Set 2.5   1. $h = 3$ **3.** $h = 3$ **5.** $h = 20$ **7.** $c = 12$ **9.** $r = 2.5$ **11.** $h = \frac{f}{5g}$ **13.** $w = \frac{V}{lh}$ **15.** $y = 7 - 3x$
17. $R = \frac{A - P}{PT}$ **19.** $A = \frac{3V}{h}$ **21.** $a = P - b - c$ **23.** $h = \frac{S - 2\pi r^2}{2\pi r}$ **25. a.** area: 103.5 sq ft; perimeter: 41 ft   **b.** baseboard: perimeter;
carpet: area   **27. a.** area: 480 sq in.; perimeter: 120 in.   **b.** frame: perimeter; glass: area   **29.** 70 ft   **31.** $-10°C$ **33.** 6.25 hr
35. length: 78 ft; width: 52 ft   **37.** 18 ft, 36 ft, 48 ft   **39.** 137.5 mi   **41.** 96 piranhas   **43.** 2 bags   **45.** one 16-in. pizza   **47.** 4.65 min
49. 13 in.   **51.** 2.25 hr   **53.** 12,090 ft   **55.** 50°C   **57.** 515,509.5 cu in.   **59.** 449 cu in.   **61.** 333°F   **63.** 0.32   **65.** 2.00 or 2
67. 17%   **69.** 720%   **71.** $V = G(N - R)$ **73.** multiplies the volume by 8   **75.** $-40°$ **77.** $\frac{\triangle - \square}{\blacksquare}$ **79.** 44.3 sec
81. $P = 3,200,000$ **83.** $V = 113.1$

Mental Math   1. no   **3.** yes

Exercise Set 2.6   1. 11.2   **3.** 55%   **5.** 180   **7.** 4%   **9.** 9990   **11.** discount: $1480; new price: $17,020   **13.** $46.58   **15.** 73%
17. 30%   **19.** $104   **21.** $42,500   **23.** 2 gal   **25.** 7 lb   **27.** 4.6   **29.** 50   **31.** 30%   **33.** 71%   **35.** 176,118
37. 5462; 60%; 27%; 6%   **39.** 75% increase   **41.** $3900   **43.** 300%   **45.** mark-up: $0.11; new price: $2.31   **47.** 400 oz   **49.** 51.7%
51. 120 employees   **53.** decrease: $64; sale price: $192   **55.** 854 thousand Scoville units   **57.** 361   **59.** 400 oz   **61.** $>$ **63.** $=$
65. $>$ **67.** no; answers may vary   **69.** 9.6%   **71.** 26.9%; yes   **73.** 17.1%

Mental Math   1. $x > 2$ **3.** $x \geq 8$ **5.** -5 **7.** 4.1

Exercise Set 2.7   1. [number line, closed dot at -1, arrow left]   **3.** [number line, open dot at $\frac{1}{2}$, arrow left]   **5.** [number line, open dot at 4, arrow right]   **7.** [number line, closed dot at -2, arrow right]   **9.** [number line, open dots at -1 and 3, segment between]
11. [number line, closed dot at 0, open dot at 2]   **13.** $\{x \mid x \geq -5\}$ [closed dot at -5, arrow right]   **15.** $\{y \mid y < 9\}$ [open dot at 9, arrow left]   **17.** $\{x \mid x > -3\}$ [open dot at -3, arrow right]
19. $\{x \mid x \leq 1\}$ [closed dot at 1, arrow left]   **21.** $\{x \mid x < -3\}$ [open dot at -3, arrow left]   **23.** $\{x \mid x \geq -2\}$ [closed dot at -2, arrow left]   **25.** $\{x \mid x < 0\}$ [open dot at 0, arrow right]
27. $\left\{y \mid y \geq -\frac{8}{3}\right\}$ [closed dot at $-\frac{8}{3}$, arrow left]   **29.** $\{y \mid y > 3\}$ [open dot at 3, arrow right]   **31.** $\{x \mid x > -15\}$ **33.** $\{x \mid x \geq -11\}$ **35.** $\left\{x \mid x > \frac{1}{4}\right\}$
37. $\{y \mid y \geq -12\}$ **39.** $\{z \mid z < 0\}$ **41.** $\{x \mid x > -3\}$ **43.** $\left\{x \mid x \geq -\frac{2}{3}\right\}$ **45.** $\{x \mid x \leq -2\}$ **47.** $\{x \mid x > -13\}$ **49.** $\{x \mid x \leq -8\}$
51. $\{x \mid x > 4\}$ **53.** $\left\{x \mid x \leq \frac{5}{4}\right\}$ **55.** $\left\{x \mid x > \frac{8}{3}\right\}$ **57.** $\{x \mid x \geq 0\}$ **59.** all numbers greater than -10 **61.** 35 cm   **63.** 193
65. 86 people   **67.** 35 min   **69.** 81   **71.** 1   **73.** $\frac{49}{64}$ **75.** about 120   **77.** 2003 and 2004   **79.** 2001   **81.** $>$ **83.** $\geq$
85. when multiplying or dividing by a negative number   **87.** final exam score ≥ 78.5

The Bigger Picture   1. -3 **2.** $\{x \mid x < -3\}$ **3.** $\frac{2}{9}$ **4.** $-\frac{1}{4}$ **5.** $\{x \mid x \geq -15\}$ **6.** no solution   **7.** 7   **8.** $\{x \mid x < 37\}$
9. all real numbers   **10.** $\frac{41}{29}$

Chapter 2 Vocabulary Check   1. linear equation in one variable   **2.** equivalent equations   **3.** formula   **4.** linear inequality in one variable
5. all real numbers   **6.** no solution   **7.** the same   **8.** reversed

Chapter 2 Review 1. 4 **2.** −3 **3.** 6 **4.** −6 **5.** 0 **6.** −9 **7.** −23 **8.** 28 **9.** b **10.** a **11.** b **12.** c **13.** −12
14. 4 **15.** 0 **16.** −7 **17.** 0.75 **18.** −3 **19.** −6 **20.** −1 **21.** −1 **22.** $\frac{3}{2}$ **23.** $-\frac{1}{5}$ **24.** 7 **25.** $3x + 3$
26. $2x + 6$ **27.** −4 **28.** −4 **29.** 2 **30.** −3 **31.** no solution **32.** no solution **33.** $\frac{3}{4}$ **34.** $-\frac{8}{9}$ **35.** 20 **36.** $-\frac{6}{23}$
37. $\frac{23}{7}$ **38.** $-\frac{2}{5}$ **39.** 102 **40.** 0.25 **41.** 6665.5 in. **42.** short piece: 4 ft; long piece: 8 ft **43.** Kellogg: 35 plants; Keebler: 18 plants
44. −39, −38, −37 **45.** 3 **46.** −4 **47.** $w = 9$ **48.** $h = 4$ **49.** $m = \frac{y - b}{x}$ **50.** $s = \frac{r + 5}{vt}$ **51.** $x = \frac{2y - 7}{5}$
52. $y = \frac{2 + 3x}{6}$ **53.** $\pi = \frac{C}{D}$ **54.** $\pi = \frac{C}{2r}$ **55.** 15 m **56.** 18 ft by 12 ft **57.** 1 hr and 20 min **58.** 40°C **59.** 20%
60. 70% **61.** 110 **62.** 1280 **63.** mark-up: $209; new price: $2109 **64.** 50,844 **65.** 40% solution: 10 gal; 10% solution: 20 gal
66. 1.45% increase **67.** 18% **68.** swerving into another lane **69.** 966 customers **70.** no; answers may vary **71.** ←———————●——→ −2
72. ←———○————●——→ 0 5 **73.** $\{x \mid x \le 1\}$ **74.** $\{x \mid x > -5\}$ **75.** $\{x \mid x \le 10\}$ **76.** $\{x \mid x < -4\}$ **77.** $\{x \mid x < -4\}$
78. $\{x \mid x \le 4\}$ **79.** $\{y \mid y > 9\}$ **80.** $\{y \mid y \ge -15\}$ **81.** $\left\{x \mid x < \frac{7}{4}\right\}$ **82.** $\left\{x \mid x \le \frac{19}{3}\right\}$ **83.** at least $2500
84. score must be less than 83 **85.** $x = 4$ **86.** $y = -14$ **87.** $a = -\frac{3}{2}$ **88.** $x = 21$ **89.** all real numbers **90.** no solution
91. −13 **92.** shorter piece: 4 in.; longer piece: 19 in. **93.** $h = \frac{3v}{A}$ **94.** 22.1 **95.** 160 **96.** 20% **97.** $\{x \mid x > 9\}$ ←————○→ 9
98. $\{x \mid x > -4\}$ ←——○———●→ −4 **99.** $\{x \mid x \le 0\}$ ←———●→ 0

Chapter 2 Test 1. −5 **2.** 8 **3.** $\frac{7}{10}$ **4.** 0 **5.** 27 **6.** $-\frac{19}{6}$ **7.** 3 **8.** $\frac{3}{11}$ **9.** 0.25 **10.** $\frac{25}{7}$ **11.** no solution **12.** 21
13. 7 gal **14.** $x = 6$ **15.** $h = \frac{V}{\pi r^2}$ **16.** $y = \frac{3x - 10}{4}$ **17.** $\{x \mid x < -2\}$ **18.** $\{x \mid x < 4\}$ **19.** $\{x \mid x \le -8\}$ **20.** $\{x \mid x \ge 11\}$
21. $\left\{x \mid x > \frac{2}{5}\right\}$ **22.** 29% **23.** 552 **24.** 40% **25.** New York: 1077; Indiana: 427

Cumulative Review 1. True; Sec. 1.2, Ex. 3 **2.** False; Sec. 1.2 **3.** True; Sec. 1.2, Ex. 4 **4.** True; Sec. 1.2 **5.** False; Sec. 1.2, Ex. 5
6. True; Sec. 1.2 **7.** True; Sec. 1.2, Ex. 6 **8.** True; Sec. 1.2 **9. a.** < **b.** = **c.** > **d.** < **e.** >; Sec. 1.2, Ex. 12 **10. a.** 5 **b.** 8 **c.** $\frac{2}{3}$; Sec. 1.2
11. $\frac{8}{3}$; Sec. 1.3, Ex. 6 **12.** 33; Sec. 1.3 **13.** −19; Sec. 1.4, Ex. 7 **14.** −10; Sec. 1.4 **15.** 8; Sec. 1.4, Ex. 8 **16.** 10; Sec. 1.4
17. −0.3; Sec. 1.4, Ex. 9 **18.** 0; Sec. 1.4 **19. a.** −12 **b.** −3; Sec. 1.5, Ex. 7 **20. a.** 5 **b.** $\frac{2}{3}$ **c.** a **d.** −3; Sec. 1.5 **21. a.** 0 **b.** −24 **c.** 45
d. 54; Sec. 1.6, Ex. 7 **22. a.** −11.1 **b.** $-\frac{1}{5}$ **c.** $\frac{3}{4}$; Sec. 1.5 **23. a.** −6 **b.** 7 **c.** −5; Sec. 1.6, Ex. 6 **24. a.** −0.36 **b.** $\frac{6}{17}$; Sec. 1.6
25. $15 - 10z$; Sec. 1.7, Ex. 8 **26.** $2x^3 - 6x^2 + 8x$; Sec. 1.7 **27.** $12x + 38$; Sec. 1.7, Ex. 12 **28.** $2x + 8$; Sec. 1.7 **29. a.** unlike **b.** like
c. like **d.** like; Sec. 1.8, Ex. 2 **30. a.** −4 **b.** 9 **c.** $\frac{10}{63}$; Sec. 1.6 **31.** $-2x - 1$; Sec. 1.8, Ex. 15 **32.** $-15x - 2$; Sec. 1.8 **33.** 17; Sec. 2.1, Ex. 1
34. $-\frac{1}{6}$; Sec. 2.1 **35.** −10; Sec. 2.2, Ex. 7 **36.** 3; Sec. 2.3 **37.** 0; Sec. 3, Ex. 4 **38.** 72; Sec. 2.2 **39.** Republicans: 223; Democrats: 208;
Sec. 2.4, Ex. 4 **40.** 5; Sec. 2.3 **41.** 79.2 yr; Sec. 2.5, Ex. 1 **42.** 6; Sec. 2.4 **43.** 87.5%; Sec. 2.6, Ex. 1 **44.** $\frac{C}{2\pi} = r$; Sec. 2.5
45. $-\frac{9}{10}$; Sec. 2.2, Ex. 10 **46.** $\{x \mid x > 5\}$; Sec. 2.7 **47.** ←————○———→; Sec. 2.7, Ex. 2 −1 **48.** $\{x \mid x \le -10\}$; Sec. 2.7
49. $\{x \mid x \ge 1\}$; Sec. 2.7, Ex. 8 **50.** $\{x \mid x \le -3\}$; Sec. 2.7

CHAPTER 3 Exponents and Polynomials

Mental Math 1. base: 3; exponent: 2 **3.** base: −3; exponent: 6 **5.** base: 4; exponent: 2 **7.** base: 5; exponent: 1; base: 3; exponent: 4
9. base: 5; exponent: 1; base: x; exponent: 2

Exercise Set 3.1 1. 49 **3.** −5 **5.** −16 **7.** 16 **9.** $\frac{1}{27}$ **11.** 112 **13.** 4 **15.** 135 **17.** 150 **19.** $\frac{80}{7}$ **21.** x^7
23. $(-3)^{12}$ **25.** $15y^5$ **27.** $x^{19}y^6$ **29.** $-72m^3n^8$ **31.** $-24z^{20}$ **33.** $20x^5$ sq ft **35.** x^{36} **37.** p^8q^8 **39.** $8a^{15}$ **41.** $x^{10}y^{15}$
43. $49a^4b^{10}c^2$ **45.** $\frac{r^9}{s^9}$ **47.** $\frac{m^9p^9}{n^9}$ **49.** $\frac{4x^2z^2}{y^{10}}$ **51.** $64z^{10}$ sq dm **53.** $27y^{12}$ cu ft **55.** x^2 **57.** −64 **59.** p^6q^5 **61.** $\frac{y^3}{2}$
63. 1 **65.** 1 **67.** −7 **69.** 2 **71.** −81 **73.** $\frac{1}{64}$ **75.** b^6 **77.** a^9 **79.** $-16x^7$ **81.** $a^{11}b^{20}$ **83.** $26m^9n^7$ **85.** z^{40}
87. $64a^3b^3$ **89.** $36x^2y^2z^6$ **91.** z^8 **93.** $3x$ **95.** 1 **97.** $81x^2y^2$ **99.** 40 **101.** $\frac{y^{15}}{8x^{12}}$ **103.** $2x^2y$ **105.** −2 **107.** 5
109. −7 **111.** c **113.** e **115.** answers may vary **117.** answers may vary **119.** 343 cu m **121.** volume
123. answers may vary **125.** answers may vary **127.** x^{9a} **129.** a^{5b} **131.** x^{5a}

Calculator Explorations 1. 5.31 EE 3 **3.** 6.6 EE −9 **5.** 1.5×10^{13} **7.** 8.15×10^{19}

Mental Math 1. $\dfrac{5}{x^2}$ **3.** y^6 **5.** $4y^3$

Exercise Set 3.2 1. $\dfrac{1}{64}$ **3.** $\dfrac{7}{x^3}$ **5.** -64 **7.** $\dfrac{5}{6}$ **9.** p^3 **11.** $\dfrac{q^4}{p^5}$ **13.** $\dfrac{1}{x^3}$ **15.** z^3 **17.** $\dfrac{4}{9}$ **19.** $\dfrac{1}{9}$ **21.** $-p^4$ **23.** -2

25. x^4 **27.** p^4 **29.** m^{11} **31.** r^6 **33.** $\dfrac{1}{x^{15}y^9}$ **35.** $\dfrac{1}{x^4}$ **37.** $\dfrac{1}{a^2}$ **39.** $4k^3$ **41.** $3m$ **43.** $-\dfrac{4a^5}{b}$ **45.** $-\dfrac{6}{7y^2}$ **47.** $\dfrac{27a^6}{b^{12}}$

49. $\dfrac{a^{30}}{b^{12}}$ **51.** $\dfrac{1}{x^{10}y^6}$ **53.** $\dfrac{z^2}{4}$ **55.** $\dfrac{x^{11}}{81}$ **57.** $\dfrac{49a^4}{b^6}$ **59.** $-\dfrac{3m^7}{n^4}$ **61.** $a^{24}b^8$ **63.** 200 **65.** x^9y^{19} **67.** $-\dfrac{y^8}{8x^2}$ **69.** $\dfrac{25b^{33}}{a^{16}}$

71. $\dfrac{27}{z^3x^6}$ cu in.   **73.** 7.8×10^4 **75.** 1.67×10^{-6} **77.** 6.35×10^{-3} **79.** 1.16×10^6 **81.** 1.36×10^4 **83.** 0.0000000008673

85. 0.033 **87.** $20,320$ **89.** $700,000,000$ **91.** 9.4×10^8 **93.** $1,230,000,000,000$
95. Yahoo!: $115,000,000$: 1.15×10^8; eBay: $58,000,000$; 5.8×10^7 **97.** 0.000036 **99.** 0.0000000000000000028 **101.** 0.0000005
103. $200,000$ **105.** 2.7×10^9 gal   **107.** $-2x + 7$ **109.** $2y - 10$ **111.** $-x - 4$ **113.** $9a^{13}$ **115.** -5 **117.** answers may vary

119. a. 1.3×10^1 **b.** 4.4×10^7 **c.** 6.1×10^{-2} **121. a.** false **b.** true **c.** false   **123.** $\dfrac{1}{x^{9s}}$ **125.** a^{4m+5}

Exercise Set 3.3 1. $1, -3x, 5$ **3.** $-5; 3.2; 1; -5$ **5.** 1; binomial   **7.** 3; none of these   **9.** 6; trinomial   **11.** 4; binomial
13. a. -6 **b.** -11 **15. a.** -2 **b.** 4 **17. a.** -15 **b.** -10 **19.** 184 ft   **21.** 595.84 ft   **23.** 1044 thousand, or $1,044,000$ visitors

25. $-11x$ **27.** $23x^3$ **29.** $16x^2 - 7$ **31.** $12x^2 - 13$ **33.** $7s$ **35.** $-1.1y^2 + 4.8$ **37.** $\dfrac{5}{6}x^4 - 7x^3 - 19$

39. $\dfrac{3}{20}x^3 + 6x^2 - \dfrac{13}{20}x - \dfrac{1}{10}$ **41.** $2, 1, 1, 0; 2$ **43.** $4, 0, 4, 3; 4$ **45.** $9ab - 11a$ **47.** $4x^2 - 7xy + 3y^2$ **49.** $-3xy^2 + 4$

51. $14y^3 - 19 - 16a^2b^2$ **53.** $7x^2 + 0x + 3$ **55.** $x^3 + 0x^2 + 0x - 64$ **57.** $5y^3 + 0y^2 + 2y - 10$
59. $2y^4 + 0y^3 + 0y^2 + 8y + 0y^0$ or $2y^4 + 0y^3 + 0y^2 + 8y + 0$ **61.** $6x^5 + 0x^4 + x^3 + 0x^2 - 3x + 15$ **63.** $4x^2 + 7x + x^2 + 5x; 5x^2 + 12x$
65. $5x + 3 + 4x + 3 + 2x + 6 + 3x + 7x; 21x + 12$ **67.** $10x + 19$ **69.** $-x + 5$ **71.** answers may vary   **73.** answers may vary
75. x^{13} **77.** a^3b^{10} **79.** $2y^{20}$ **81.** answers may vary   **83.** answers may vary   **85.** $11.1x^2 - 7.97x + 10.76$

Exercise Set 3.4 1. $12x + 12$ **3.** $-3x^2 + 10$ **5.** $-3x^2 + 4$ **7.** $-y^2 - 3y - 1$ **9.** $7.9x^3 + 4.4x^2 - 3.4x - 3$ **11.** $\dfrac{1}{2}m^2 - \dfrac{7}{10}m + \dfrac{13}{16}$

13. $8t^2 - 4$ **15.** $15a^3 + a^2 - 3a + 16$ **17.** $-x + 14$ **19.** $5x^2 + 2y^2$ **21.** $-2x + 9$ **23.** $2x^2 + 7x - 16$ **25.** $2x^2 + 11x$

27. $-0.2x^2 + 0.2x - 2.2$ **29.** $\dfrac{2}{5}z^2 - \dfrac{3}{10}z + \dfrac{7}{20}$ **31.** $-2z^2 - 16z + 6$ **33.** $2u^5 - 10u^2 + 11u - 9$ **35.** $5x - 9$ **37.** $4x - 3$

39. $11y + 7$ **41.** $-2x^2 + 8x - 1$ **43.** $14x + 18$ **45.** $3a^2 - 6a + 11$ **47.** $3x - 3$ **49.** $7x^2 - 4x + 2$ **51.** $7x^2 - 2x + 2$
53. $4y^2 + 12y + 19$ **55.** $-15x + 7$ **57.** $-2a - b + 1$ **59.** $3x^2 + 5$ **61.** $6x^2 - 2xy + 19y^2$ **63.** $8r^2s + 16rs - 8 + 7r^2s^2$
65. $(x^2 + 7x + 4)$ ft   **67.** $\left(\dfrac{19}{2}x + 3\right)$ units   **69.** $(3y^2 + 4y + 11)$ m   **71.** $-6.6x^2 - 1.8x - 1.8$ **73.** $6x^2$ **75.** $-12x^8$
77. $200x^3y^2$ **79.** $2; 2$ **81.** $4; 3; 3; 4$ **83.** b   **85.** e   **87. a.** $4z$ **b.** $3z^2$ **c.** $-4z$ **d.** $3z^2$; answers may vary
89. a. m^3 **b.** $3m$ **c.** $-m^3$ **d.** $-3m$; answers may vary   **91.** $-0.325x^2 + 10.14x + 83.58$

Mental Math 1. x^8 **3.** cannot simplify   **5.** y^5 **7.** x^{14} **9.** $2x^7$

Exercise Set 3.5 1. $24x^3$ **3.** x^4 **5.** $-28n^{10}$ **7.** $-12.4x^{12}$ **9.** $-\dfrac{2}{15}y^3$ **11.** $-24x^8$ **13.** $6x^2 + 15x$ **15.** $7x^3 + 14x^2 - 7x$

17. $-2a^2 - 8a$ **19.** $6x^3 - 9x^2 + 12x$ **21.** $12a^5 + 45a^2$ **23.** $-6a^4 + 4a^3 - 6a^2$ **25.** $6x^5y - 3x^4y^3 + 24x^2y^4$

27. $-4x^3y + 7x^2y^2 - xy^3 - 3y^4$ **29.** $4x^4 - 3x^3 + \dfrac{1}{2}x^2$ **31.** $x^2 + 7x + 12$ **33.** $a^2 + 5a - 14$ **35.** $x^2 + \dfrac{1}{3}x - \dfrac{2}{9}$

37. $12x^4 + 25x^2 + 7$ **39.** $12x^2 - 29x + 15$ **41.** $1 - 7a + 12a^2$ **43.** $4y^2 - 16y + 16$ **45.** $x^3 - 5x^2 + 13x - 14$
47. $x^4 + 5x^3 - 3x^2 - 11x + 20$ **49.** $10a^3 - 27a^2 + 26a - 12$ **51.** $49x^2y^2 - 14xy + y^2$ **53.** $12x^2 - 64x - 11$
55. $2x^3 + 10x^2 + 11x - 3$ **57.** $2x^4 + 3x^3 - 58x^2 + 4x + 63$ **59.** $8.4y^7$ **61.** $-3x^3 - 6x^2 + 24x$ **63.** $2x^2 + 39x + 19$

65. $x^2 - \dfrac{2}{7}x - \dfrac{3}{49}$ **67.** $9y^2 + 30y + 25$ **69.** $a^3 - 2a^2 - 18a + 24$ **71.** $(4x^2 - 25)$ sq yd   **73.** $(6x^2 - 4x)$ sq in.   **75.** $25x^2$

77. $9y^6$ **79. a.** $6x + 12$ **b.** $9x^2 + 36x + 35$; answers may vary   **81.** $13x - 7$ **83.** $30x^2 - 28x + 6$ **85.** $-7x + 5$
87. $x^2 + 3x$ **89.** $x^2 + 5x + 6$ **91.** $11a$ **93.** $25x^2 + 4y^2$ **95. a.** $a^2 - b^2$ **b.** $4x^2 - 9y^2$ **c.** $16x^2 - 49$ **d.** answers may vary

Exercise Set 3.6 1. $x^2 + 7x + 12$ **3.** $x^2 + 5x - 50$ **5.** $5x^2 + 4x - 12$ **7.** $4y^2 - 25y + 6$ **9.** $6x^2 + 13x - 5$

11. $6y^3 + 4y^2 + 42y + 28$ **13.** $x^2 + \dfrac{1}{3}x - \dfrac{2}{9}$ **15.** $0.08 - 2.6a + 15a^2$ **17.** $2x^2 + 9xy - 5y^2$ **19.** $x^2 + 4x + 4$ **21.** $4x^2 - 4x + 1$

23. $9a^2 - 30a + 25$ **25.** $x^4 + x^2 + 0.25$ **27.** $y^2 - \dfrac{4}{7}y + \dfrac{4}{49}$ **29.** $4a^2 - 12a + 9$ **31.** $25x^2 + 90x + 81$ **33.** $9x^2 - 42xy + 49y^2$

35. $16m^2 + 40mn + 25n^2$ **37.** $25x^8 - 15x^4 + 9$ **39.** $a^2 - 49$ **41.** $x^2 - 36$ **43.** $9x^2 - 1$ **45.** $x^4 - 25$ **47.** $4y^4 - 1$

49. $16 - 49x^2$ **51.** $9x^2 - \dfrac{1}{4}$ **53.** $81x^2 - y^2$ **55.** $4m^2 - 25n^2$ **57.** $a^2 + 9a + 20$ **59.** $a^2 - 14a + 49$ **61.** $12a^2 - a - 1$

63. $x^2 - 4$ **65.** $9a^2 + 6a + 1$ **67.** $4x^2 + 3xy - y^2$ **69.** $a^2 - \dfrac{1}{4}y^2$ **71.** $6b^2 - b - 35$ **73.** $x^4 - 100$ **75.** $16x^2 - 25$

77. $25x^2 - 60xy + 36y^2$ **79.** $4r^2 - 9s^2$ **81.** $(4x^2 + 4x + 1)$ sq ft   **83.** $\dfrac{5b^5}{7}$ **85.** $-\dfrac{2a^{10}}{b^5}$ **87.** $\dfrac{2y^8}{3}$ **89.** c   **91.** d   **93.** 2

95. $(24x^2 - 32x + 8)$ sq m   **97.** answers may vary

Integrated Review   **1.** $35x^5$ **2.** $-32y^9$ **3.** -16 **4.** 16 **5.** $2x^2 - 9x - 5$ **6.** $3x^2 + 13x - 10$ **7.** $3x - 4$ **8.** $4x + 3$

9. $7x^6y^2$ **10.** $\dfrac{10b^6}{7}$ **11.** $144m^{14}n^{12}$ **12.** $64y^{27}z^{30}$ **13.** $16y^2 - 9$ **14.** $49x^2 - 1$ **15.** $\dfrac{y^{45}}{x^{63}}$ **16.** $\dfrac{1}{64}$ **17.** $\dfrac{x^{27}}{27}$ **18.** $\dfrac{r^{58}}{16s^{14}}$

19. $2x^2 - 2x - 6$ **20.** $6x^2 + 13x - 11$ **21.** $2.5y^2 - 6y - 0.2$ **22.** $8.4x^2 - 6.8x - 4.2$ **23.** $2y^2 - 6y - 1$ **24.** $6z^2 + 2z + \dfrac{11}{2}$

25. $x^2 + 8x + 16$ **26.** $y^2 - 18y + 81$ **27.** $2x + 8$ **28.** $2y - 18$ **29.** $7x^2 - 10xy + 4y^2$ **30.** $-a^2 - 3ab + 6b^2$

31. $x^3 + 2x^2 - 16x + 3$ **32.** $x^3 - 2x^2 - 5x - 2$ **33.** $6x^2 - x - 70$ **34.** $20x^2 + 21x - 5$ **35.** $2x^3 - 19x^2 + 44x - 7$

36. $5x^3 + 9x^2 - 17x + 3$ **37.** $4x^2 - \dfrac{25}{81}$ **38.** $144y^2 - \dfrac{9}{49}$

Mental Math   **1.** a^2 **3.** a^2 **5.** k^3

Exercise Set 3.7   **1.** $12x^3 + 3x$ **3.** $4x^3 - 6x^2 + x + 1$ **5.** $5p^2 + 6p$ **7.** $-\dfrac{3}{2x} + 3$ **9.** $-3x^2 + x - \dfrac{4}{x^3}$ **11.** $-1 + \dfrac{3}{2x} - \dfrac{7}{4x^4}$

13. $x + 1$ **15.** $2x + 3$ **17.** $2x + 1 + \dfrac{7}{x - 4}$ **19.** $3a^2 - 3a + 1 + \dfrac{2}{3a + 2}$ **21.** $4x + 3 - \dfrac{2}{2x + 1}$ **23.** $2x^2 + 6x - 5 - \dfrac{2}{x - 2}$

25. $x + 6$ **27.** $x^2 + 3x + 9$ **29.** $-3x + 6 - \dfrac{11}{x + 2}$ **31.** $2b - 1 - \dfrac{6}{2b - 1}$ **33.** $ab - b^2$ **35.** $4x + 9$ **37.** $x + 4xy - \dfrac{y}{2}$

39. $2b^2 + b + 2 - \dfrac{12}{b + 4}$ **41.** $y^2 + 5y + 10 + \dfrac{24}{y - 2}$ **43.** $-6x - 12 - \dfrac{19}{x - 2}$ **45.** $x^3 - x^2 + x$ **47.** 3 **49.** -4 **51.** $3x$

53. $9x$ **55.** $(3x^3 + x - 4)$ ft   **57.** $(2x + 5)$ m   **59.** answers may vary   **61.** c

The Bigger Picture   **1.** -5.93 **2.** $-\dfrac{2}{5}$ **3.** $5x^9y^4$ **4.** $\dfrac{1}{8a^4}$ **5.** $6y^3 - 2y^2 - 6y$ **6.** $8y^2 - 3y - 7$ **7.** $4x^3 - 13x^2 + 10x - 21$

8. $36m^2 - 60m + 25$ **9.** $4n - 1 + \dfrac{2}{n}$ **10.** $2x - 6 + \dfrac{14}{3x - 1}$ **11.** -0.6 **12.** $\{x \mid x > 0.6\}$ **13.** $\{x \mid x \le 2\}$ **14.** $\dfrac{2}{3}$

Vocabulary Check   **1.** term   **2.** FOIL   **3.** trinomial   **4.** degree of polynomial   **5.** binomial   **6.** coefficient   **7.** degree of a term
8. monomial   **9.** polynomials

Chapter 3 Review   **1.** base: 3; exponent: 2   **2.** base: -5; exponent: 4   **3.** base: 5; exponent: 4   **4.** base: x; exponent: 6   **5.** 512
6. 36   **7.** -36 **8.** -65 **9.** 1   **10.** 1   **11.** y^9 **12.** x^{14} **13.** $-6x^{11}$ **14.** $-20y^7$ **15.** x^8 **16.** y^{15} **17.** $81y^{24}$ **18.** $8x^9$

19. x^5 **20.** z^7 **21.** a^4b^3 **22.** x^3y^5 **23.** $\dfrac{4}{x^3y^4}$ **24.** $\dfrac{x^6y^6}{4}$ **25.** $40a^{19}$ **26.** $36x^3$ **27.** 3   **28.** 9   **29.** b   **30.** c   **31.** $\dfrac{1}{49}$

32. $-\dfrac{1}{49}$ **33.** $\dfrac{2}{x^4}$ **34.** $\dfrac{1}{16x^4}$ **35.** 125   **36.** $\dfrac{9}{4}$ **37.** $\dfrac{17}{16}$ **38.** $\dfrac{1}{42}$ **39.** x^8 **40.** z^8 **41.** r **42.** y^3 **43.** c^4 **44.** $\dfrac{x^3}{y^3}$

45. $\dfrac{1}{x^6y^{13}}$ **46.** $\dfrac{a^{10}}{b^{10}}$ **47.** 2.7×10^{-4} **48.** 8.868×10^{-1} **49.** 8.08×10^7 **50.** -8.68×10^5 **51.** 1.124×10^8 **52.** 1.5×10^5

53. 867,000   **54.** 0.00386   **55.** 0.00086   **56.** 893,600   **57.** 1,431,280,000,000,000 cu km   **58.** 0.0000000001 m   **59.** 0.016
60. 400,000,000,000   **61.** 5   **62.** 2   **63.** 5   **64.** 6   **65.** 22; 78; 154.02; 400   **66.** $2a^2$ **67.** $-4y$ **68.** $15a^2 + 4a$
69. $22x^2 + 3x + 6$ **70.** $-6a^2b - 3b^2 - q^2$ **71.** cannot be combined   **72.** $8x^2 + 3x + 6$ **73.** $2x^5 + 3x^4 + 4x^3 + 9x^2 + 7x + 6$
74. $-7y^2 - 1$ **75.** $-6m^7 - 3x^4 + 7m^6 - 4m^2$ **76.** $-x^2 - 6xy - 2y^2$ **77.** $-5x^2 + 5x + 1$ **78.** $-2x^2 - x + 20$ **79.** $6x + 30$
80. $9x - 63$ **81.** $8a + 28$ **82.** $54a - 27$ **83.** $-7x^3 - 35x$ **84.** $-32y^3 + 48y$ **85.** $-2x^3 + 18x^2 - 2x$ **86.** $-3a^3b - 3a^2b - 3ab^2$
87. $-6a^4 + 8a^2 - 2a$ **88.** $42b^4 - 28b^2 + 14b$ **89.** $2x^2 - 12x - 14$ **90.** $6x^2 - 11x - 10$ **91.** $4a^2 + 27a - 7$ **92.** $42a^2 + 11a - 3$
93. $x^4 + 7x^3 + 4x^2 + 23x - 35$ **94.** $x^6 + 2x^5 + x^2 + 3x + 2$ **95.** $x^4 + 4x^3 + 4x^2 - 16$ **96.** $x^6 + 8x^4 + 16x^2 - 16$
97. $x^3 + 21x^2 + 147x + 343$ **98.** $8x^3 - 60x^2 + 150x - 125$ **99.** $x^2 + 14x + 49$ **100.** $x^2 - 10x + 25$ **101.** $9x^2 - 42x + 49$
102. $16x^2 + 16x + 4$ **103.** $25x^2 - 90x + 81$ **104.** $25x^2 - 1$ **105.** $49x^2 - 16$ **106.** $a^2 - 4b^2$ **107.** $4x^2 - 36$ **108.** $16a^4 - 4b^2$

109. $(9x^2 - 6x + 1)$ sq m   **110.** $(5x^2 - 3x - 2)$ sq mi   **111.** $\dfrac{1}{7} + \dfrac{3}{x} + \dfrac{7}{x^2}$ **112.** $-a^2 + 3b - 4$ **113.** $a + 1 + \dfrac{6}{a - 2}$

114. $4x + \dfrac{7}{x + 5}$ **115.** $a^2 + 3a + 8 + \dfrac{22}{a - 2}$ **116.** $3b^2 - 4b - \dfrac{1}{3b - 2}$ **117.** $2x^3 - x^2 + 2 - \dfrac{1}{2x - 1}$

118. $-x^2 - 16x - 117 - \dfrac{684}{x - 6}$ **119.** $\left(5x - 1 + \dfrac{20}{x^2}\right)$ ft   **120.** $(7a^3b^6 + a - 1)$ units   **121.** $-\dfrac{1}{8}$ **122.** $4x^4y^7$ **123.** $\dfrac{2x^6}{3}$

124. $\dfrac{27a^{12}}{b^6}$ **125.** $\dfrac{x^{16}}{16y^{12}}$ **126.** $9a^2b^8$ **127.** $11x - 5$ **128.** $2y^2 - 10$ **129.** $5y^2 - 3y - 1$ **130.** $5x^2 + 3x - 2$ **131.** $28x^3 + 12x$

132. $6x^2 + 11x - 10$ **133.** $x^3 + x^2 - 18x + 18$ **134.** $28x^2 - 71x + 18$ **135.** $25x^2 + 40x + 16$ **136.** $36x^2 - 9$

137. $4a - 1 + \dfrac{2}{a^2} - \dfrac{5}{2a^3}$ **138.** $x - 3 + \dfrac{25}{x + 5}$ **139.** $2x^2 + 7x + 5 + \dfrac{19}{2x - 3}$

Chapter 3 Test   **1.** 32   **2.** 81   **3.** -81 **4.** $\dfrac{1}{64}$ **5.** $-15x^{11}$ **6.** y^5 **7.** $\dfrac{1}{r^5}$ **8.** $\dfrac{y^{14}}{x^2}$ **9.** $\dfrac{1}{6xy^8}$ **10.** 5.63×10^5

11. 8.63×10^{-5} **12.** 0.0015   **13.** 62,300   **14.** 0.036   **15. a.** $4, 3; 7, 3; 1, 4; -2, 0$ **b.** 4   **16.** $-2x^2 + 12x + 11$

17. $16x^3 + 7x^2 - 3x - 13$ **18.** $-3x^3 + 5x^2 + 4x + 5$ **19.** $x^3 + 8x^2 + 3x - 5$ **20.** $3x^3 + 22x^2 + 41x + 14$ **21.** $6x^4 - 9x^3 + 21x^2$

22. $3x^2 + 16x - 35$ **23.** $9x^2 - \dfrac{1}{25}$ **24.** $16x^2 - 16x + 4$ **25.** $64x^2 + 48x + 9$ **26.** $x^4 - 81b^2$ **27.** 1001 ft; 985 ft; 857 ft; 601 ft

28. $(4x^2 - 9)$ sq in.   **29.** $\dfrac{x}{2y} + \dfrac{1}{4} - \dfrac{7}{8y}$ **30.** $x + 2$ **31.** $9x^2 - 6x + 4 - \dfrac{16}{3x + 2}$

Cumulative Review   1. a. $11, 112$ **b.** $0, 11, 112$ **c.** $-3, -2, 0, 11, 112$ **d.** $-3, -2, 0, \dfrac{1}{4}, 11, 112$ **e.** $\sqrt{2}$

f. $-2, 0, \dfrac{1}{4}, 112, -3, 11, \sqrt{2}$; Sec. 1.2, Ex. 11   **2. a.** 7.2 **b.** 0 **c.** $\dfrac{1}{2}$; Sec. 1.2   **3. a.** 9 **b.** 125 **c.** 16 **d.** 7 **e.** $\dfrac{9}{49}$; Sec. 1.3, Ex. 1

4. a. $\dfrac{1}{4}$ **b.** $2\dfrac{5}{12}$; Sec. R.2   **5.** $\dfrac{1}{4}$; Sec. 1.3, Ex. 4   **6.** $\dfrac{3}{25}$; Sec. 1.3   **7. a.** $x + 3$ **b.** $3x$ **c.** $7.3 \div x$ or $\dfrac{7.3}{x}$ **d.** $10 - x$ **e.** $5x + 7$; Sec. 1.3, Ex. 8

8. 41; Sec. 1.3   **9.** 6.7; Sec. 1.4, Ex. 10   **10.** no; Sec. 1.5   **11. a.** $\dfrac{1}{2}$ **b.** 9; Sec. 1.5, Ex. 8   **12. a.** -33 **b.** 5; Sec. 1.5   **13.** 3; Sec. 1.6, Ex. 11a

14. -8; Sec. 1.6   **15.** -70; Sec. 1.6, Ex. 11d   **16.** 150; Sec. 1.6   **17.** $15x + 10$; Sec. 1.8, Ex. 8   **18.** $-6x + 9$; Sec. 1.8

19. $-2y - 0.6z + 2$; Sec. 1.8, Ex. 9   **20.** $-4x^3 + 24x - 4x$; Sec. 1.8   **21.** $-9x - y + 2z - 6$; Sec. 1.8, Ex. 10   **22.** $4xy - 6y + 2$; Sec. 1.8

23. $a = 19$; Sec. 2.1, Ex. 6   **24.** $x = -\dfrac{1}{2}$; Sec. 2.1   **25.** $y = 140$; Sec. 2.2, Ex. 4   **26.** $x = \dfrac{12}{5}$; Sec. 2.2   **27.** $x = 4$; Sec. 2.3, Ex. 5

28. $x = 1$; Sec. 2.3   **29.** 10; Sec. 2.4, Ex. 1   **30.** $(x + 7) - 2x$ or $-x - 7$; Sec. 2.1   **31.** 40 feet; Sec. 2.5, Ex. 2   **32.** undefined; Sec. 1.6

33. 800; Sec. 2.6, Ex. 2   **34.** ⟵─○─⟶ ; Sec. 2.7   **35.** ⟵─●─⟶ $\{x \mid x \le 4\}$; Sec. 2.7, Ex. 7   **36. a.** 25   **b.** -25 **c.** 50; Sec. 3.1
                         5   4

37. a. x^{11} **b.** $\dfrac{t^4}{16}$ **c.** $81y^{10}$; Sec. 3.1, Ex. 33   **38.** z^4; Sec. 3.1   **39.** $\dfrac{b^3}{27a^6}$; Sec. 3.2, Ex. 10   **40.** $-15x^{16}$; Sec. 3.2   **41.** $\dfrac{1}{25y^6}$; Sec. 3.2, Ex. 14

42. $\dfrac{1}{9}$; Sec. 3.2   **43.** $10x^3$; Sec. 3.3, Ex. 8   **44.** $4y^2 - 8$; Sec. 3.3   **45.** $5x^2 - 3x - 3$; Sec. 3.3, Ex. 9   **46.** $100x^4 - 9$; Sec. 3.5

47. $7x^3 + 14x^2 + 35x$; Sec. 3.5, Ex. 4   **48.** $100x^4 + 60x^2 + 9$; Sec. 3.5   **49.** $3x^3 - 4 + \dfrac{1}{x}$; Sec. 3.7, Ex. 2

CHAPTER 4   Factoring Polynomials

Mental Math   1. 2   **3.** 1   **5.** 7

Exercise Set 4.1   1. 4   **3.** 6   **5.** 1   **7.** y^2 **9.** z^7 **11.** xy^2 **13.** 7   **15.** $4y^3$ **17.** $5x^2$ **19.** $3x^3$ **21.** $9x^2y$ **23.** $10a^6b$

25. $3(a + 2)$ **27.** $15(2x - 1)$ **29.** $x^2(x + 5)$ **31.** $2y^3(3y + 1)$ **33.** $2x(16y - 9x)$ **35.** $4(x - 2y + 1)$ **37.** $3x(2x^2 - 3x + 4)$

39. $a^2b^2(a^5b^4 - a + b^3 - 1)$ **41.** $5xy(x^2 - 3x + 2)$ **43.** $4(2x^5 + 4x^4 - 5x^3 + 3)$ **45.** $\dfrac{1}{3}x(x^3 + 2x^2 - 4x^4 + 1)$ **47.** $(x^2 + 2)(y + 3)$

49. $(y + 4)(z + 3)$ **51.** $(z^2 - 6)(r + 1)$ **53.** $-1(x + 7)$ **55.** $-1(2 - z)$ **57.** $-1(-3a + b - 2)$ **59.** $(x + 2)(x^2 + 5)$

61. $(x + 3)(5 + y)$ **63.** $(3x - 2)(2x^2 + 5)$ **65.** $(5m^2 + 6n)(m + 1)$ **67.** $(y - 4)(2 + x)$ **69.** $(2x + 1)(x^2 + 4)$

71. $(x - 2y)(4x - 3)$ **73.** $(5q - 4p)(q - 1)$ **75.** $2(2y - 7)(3x^2 - 1)$ **77.** $x^2 + 7x + 10$ **79.** $b^2 - 3b - 4$ **81.** 2, 6

83. $-1, -8$ **85.** $-2, 5$ **87.** $-8, 3$ **89.** b   **91.** factored   **93.** not factored   **95. a.** 8684 thousand barrels per day
b. 9022 thousand barrels per day   **c.** $-13(x^2 - 17x - 652)$ or $13(-x^2 + 17x + 652)$ **97.** $4x^2 - \pi x^2$; $x^2(4 - \pi)$ **99.** $(x^3 - 1)$ units
101. answers may vary   **103.** answers may vary

Mental Math   1. $+5$ **3.** -3 **5.** $+2$

Exercise Set 4.2   1. $(x + 6)(x + 1)$ **3.** $(y - 9)(y - 1)$ **5.** $(x - 3)(x - 3)$ or $(x - 3)^2$ **7.** $(x - 6)(x + 3)$ **9.** $(x + 10)(x - 7)$
11. prime   **13.** $(x + 5y)(x + 3y)$ **15.** $(a^2 - 5)(a^2 + 3)$ **17.** $(m + 13)(m + 1)$ **19.** $(t - 2)(t + 12)$ **21.** $(a - 2b)(a - 8b)$
23. $2(z + 8)(z + 2)$ **25.** $2x(x - 5)(x - 4)$ **27.** $(x - 4y)(x + y)$ **29.** $(x + 12)(x + 3)$ **31.** $(x - 2)(x + 1)$ **33.** $(r - 12)(r - 4)$
35. $(x + 2y)(x - y)$ **37.** $3(x + 5)(x - 2)$ **39.** $3(x - 18)(x - 2)$ **41.** $(x - 24)(x + 6)$ **43.** prime   **45.** $(x - 5)(x - 3)$
47. $6x(x + 4)(x + 5)$ **49.** $4y(x^2 + x - 3)$ **51.** $(x - 7)(x + 3)$ **53.** $(x + 5y)(x + 2y)$ **55.** $2(t + 8)(t + 4)$ **57.** $x(x - 6)(x + 4)$

59. $2t^3(t - 4)(t - 3)$ **61.** $5xy(x - 8y)(x + 3y)$ **63.** $3(m - 9)(m - 6)$ **65.** $-1(x - 11)(x - 1)$ **67.** $\dfrac{1}{2}(y - 11)(y + 2)$

69. $x(xy - 4)(xy + 5)$ **71.** $2x^2 + 11x + 5$ **73.** $15y^2 - 17y + 4$ **75.** $9a^2 + 23ab - 12b^2$ **77.** $x^2 + 5x - 24$ **79.** answers may vary

81. $2x^2 + 28x + 66$; $2(x + 3)(x + 11)$ **83.** $-16(t - 5)(t + 1)$ **85.** $\left(x + \dfrac{1}{4}\right)\left(x + \dfrac{1}{4}\right)$ or $\left(x + \dfrac{1}{4}\right)^2$ **87.** $(x + 1)(z - 10)(z + 7)$

89. 15; 28; 39; 48; 55; 60; 63; 64   **91.** 9; 12; 21   **93.** $(x^n + 10)(x^n - 2)$

Exercise Set 4.3   1. $x + 4$ **3.** $10x - 1$ **5.** $4x - 3$ **7.** $(2x + 3)(x + 5)$ **9.** $(y - 1)(8y - 9)$ **11.** $(2x + 1)(x - 5)$
13. $(4r - 1)(5r + 8)$ **15.** $(5x + 1)(2x + 3)$ **17.** $(3x - 2)(x + 1)$ **19.** $(3x - 5y)(2x - y)$ **21.** $(3m - 5)(5m + 3)$
23. $(x - 4)(x - 5)$ **25.** $(2x + 11)(x - 9)$ **27.** $(7t + 1)(t - 4)$ **29.** $(3a + b)(a + 3b)$ **31.** $(7p + 1)(7p - 2)$
33. $(6x - 7)(3x + 2)$ **35.** prime   **37.** $(8x + 3)(3x + 4)$ **39.** $x(3x + 2)(4x + 1)$ **41.** $3(7b + 5)(b - 3)$ **43.** $(3z + 4)(4z - 3)$
45. $2y^2(3x - 10)(x + 3)$ **47.** $(2x - 7)(2x + 3)$ **49.** $3(x^2 - 14x + 21)$ **51.** $(4x + 9y)(2x - 3y)$ **53.** $-1(x - 6)(x + 4)$
55. $x(4x + 3)(x - 3)$ **57.** $(4x - 9)(6x - 1)$ **59.** $b(8a - 3)(5a + 3)$ **61.** $2x(3x + 2)(5x + 3)$ **63.** $2y(3y + 5)(y - 3)$
65. $5x^2(2x - y)(x + 3y)$ **67.** $-1(2x - 5)(7x - 2)$ **69.** $p^2(4p - 5)(4p - 5)$ or $p^2(4p - 5)^2$ **71.** $-1(2x + 1)(x - 5)$
73. $-4(12x - 1)(x - 1)$ **75.** $(2t^2 + 9)(t^2 - 3)$ **77.** prime   **79.** $a(6a^2 + b^2)(a^2 + 6b^2)$ **81.** $x^2 - 16$ **83.** $x^2 + 4x + 4$

85. $4x^2 + 4x + 1$ **87.** no   **89.** $4x^2 + 21x + 5$; $(4x + 1)(x + 5)$ **91.** $\left(2x + \dfrac{1}{2}\right)\left(2x + \dfrac{1}{2}\right)$ or $\left(2x + \dfrac{1}{2}\right)^2$

93. $(y - 1)^2(4x^2 + 10x + 25)$ **95.** 2; 14   **97.** 2   **99.** answers may vary

Exercise Set 4.4 **1.** $(x + 3)(x + 2)$ **3.** $(y + 8)(y - 2)$ **5.** $(8x - 5)(x - 3)$ **7.** $(5x^2 - 3)(x^2 + 5)$ **9. a.** 9.2 **b.** $9x + 2x$
c. $(2x + 3)(3x + 1)$ **11. a.** $-20, -3$ **b.** $-20x - 3x$ **c.** $(3x - 4)(5x - 1)$ **13.** $(3y + 2)(7y + 1)$ **15.** $(7x - 11)(x + 1)$
17. $(5x - 2)(2x - 1)$ **19.** $(2x - 5)(x - 1)$ **21.** $(2x + 3)(2x + 3)$ or $(2x + 3)^2$ **23.** $(2x + 3)(2x - 7)$ **25.** $(5x - 4)(2x - 3)$
27. $x(2x + 3)(x + 5)$ **29.** $2(8y - 9)(y - 1)$ **31.** $(2x - 3)(3x - 2)$ **33.** $3(3a + 2)(6a - 5)$ **35.** $a(4a + 1)(5a + 8)$
37. $3x(4x + 3)(x - 3)$ **39.** $y(3x + y)(x + y)$ **41.** prime **43.** $6(a + b)(4a - 5b)$ **45.** $p^2(15p + q)(p + 2q)$
47. $(7 + x)(5 + x)$ or $(x + 7)(x + 5)$ **49.** $(6 - 5x)(1 - x)$ or $(5x - 6)(x - 1)$ **51.** $x^2 - 4$ **53.** $y^2 + 8y + 16$ **55.** $81z^2 - 25$
57. $16x^2 - 24x + 9$ **59.** $10x^2 + 45x + 45; 5(2x + 3)(x + 3)$ **61.** $(x^n + 2)(x^n + 3)$ **63.** $(3x^n - 5)(x^n + 7)$ **65.** answers may vary

Calculator Explorations

	$x^2 - 2x + 1$	$x^2 - 2x - 1$	$(x - 1)^2$
$x = 5$	16	14	16
$x = -3$	16	14	16
$x = 2.7$	2.89	0.89	2.89
$x = -12.1$	171.61	169.61	171.61
$x = 0$	1	-1	1

Mental Math **1.** 1^2 **3.** 9^2 **5.** 3^2 **7.** $(3x)^2$ **9.** $(5a)^2$ **11.** $(6p^2)^2$

Exercise Set 4.5 **1.** yes **3.** no **5.** yes **7.** no **9.** no **11.** yes **13.** $(x + 11)^2$ **15.** $(x - 8)^2$ **17.** $(4a - 3)^2$
19. $(x^2 + 2)^2$ **21.** $2(n - 7)^2$ **23.** $(4y + 5)^2$ **25.** $(xy - 5)^2$ **27.** $m(m + 9)^2$ **29.** prime **31.** $(3x - 4y)^2$
33. $(x + 2)(x - 2)$ **35.** $(9 + p)(9 - p)$ or $-1(p + 9)(p - 9)$ **37.** $-1(2r + 1)(2r - 1)$ **39.** $(3x + 4)(3x - 4)$ **41.** prime
43. $(-6 + x)(6 + x)$ or $-1(6 + x)(6 - x)$ **45.** $(m^2 + 1)(m + 1)(m - 1)$ **47.** $(x + 13y)(x - 13y)$ **49.** $2(3r + 2)(3r - 2)$
51. $x(3y + 2)(3y - 2)$ **53.** $16x^2(x + 2)(x - 2)$ **55.** $xy(y - 3z)(y + 3z)$ **57.** $4(3x - 4y)(3x + 4y)$ **59.** $9(4 - 3x)(4 + 3x)$

61. $(5y - 3)(5y + 3)$ **63.** $(11m + 10n)(11m - 10n)$ **65.** $(xy - 1)(xy + 1)$ **67.** $\left(x - \frac{1}{2}\right)\left(x + \frac{1}{2}\right)$ **69.** $\left(7 - \frac{3}{5}m\right)\left(7 + \frac{3}{5}m\right)$

71. $(9a + 5b)(9a - 5b)$ **73.** $(x + 7y)^2$ **75.** $2(4n^2 - 7)^2$ **77.** $x^2(x^2 + 9)(x + 3)(x - 3)$ **79.** $pq(8p + 9q)(8p - 9q)$

81. $x = 6$ **83.** $m = -2$ **85.** $z = \frac{1}{5}$ **87.** $\left(x - \frac{1}{3}\right)^2$ **89.** $(x + 2 + y)(x + 2 - y)$ **91.** $(b - 4)(a + 4)(a - 4)$

93. $(x + 3 + 2y)(x + 3 - 2y)$ **95.** $(x^n + 10)(x^n - 10)$ **97.** 8 **99.** answers may vary **101.** $(x + 6)$ **103.** $a^2 + 2ab + b^2$
105. a. 777 ft **b.** 441 ft **c.** 7 sec **d.** $(29 + 4t)(29 - 4t)$ **107. a.** 1456 feet **b.** 816 feet **c.** 10 seconds **d.** $16(10 + t)(10 - t)$

Integrated Review **1.** $(x - 3)(x + 4)$ **2.** $(x - 8)(x - 2)$ **3.** $(x + 2)(x - 3)$ **4.** $(x + 1)^2$ **5.** $(x - 3)^2$ **6.** $(x + 2)(x - 1)$
7. $(x + 3)(x - 2)$ **8.** $(x + 3)(x + 4)$ **9.** $(x - 5)(x - 2)$ **10.** $(x - 6)(x + 5)$ **11.** $2(x - 7)(x + 7)$ **12.** $3(x - 5)(x + 5)$
13. $(x + 3)(x + 5)$ **14.** $(y - 7)(3 + x)$ **15.** $(x + 8)(x - 2)$ **16.** $(x - 7)(x + 4)$ **17.** $4x(x + 7)(x - 2)$ **18.** $6x(x - 5)(x + 4)$
19. $2(3x + 4)(2x + 3)$ **20.** $(2a - b)(4a + 5b)$ **21.** $(2a + b)(2a - b)$ **22.** $(x + 5y)(x - 5y)$ **23.** $(4 - 3x)(7 + 2x)$
24. $(5 - 2x)(4 + x)$ **25.** prime **26.** prime **27.** $(3y + 5)(2y - 3)$ **28.** $(4x - 5)(x + 1)$ **29.** $9x(2x^2 - 7x + 1)$
30. $4a(3a^2 - 6a + 1)$ **31.** $(4a - 7)^2$ **32.** $(5p - 7)^2$ **33.** $(7 - x)(2 + x)$ **34.** $(3 + x)(1 - x)$ **35.** $3x^2y(x + 6)(x - 4)$
36. $2xy(x + 5y)(x - y)$ **37.** $3xy(4x^2 + 81)$ **38.** $2xy^2(3x^2 + 4)$ **39.** $2xy(1 + 6x)(1 - 6x)$ **40.** $2x(x - 3)(x + 3)$
41. $(x + 6)(x + 2)(x - 2)$ **42.** $(x - 2)(x + 6)(x - 6)$ **43.** $2a^2(3a + 5)$ **44.** $2n(2n - 3)$ **45.** $(3x - 1)(x^2 + 4)$
46. $(x - 2)(x^2 + 3)$ **47.** $6(x + 2y)(x + y)$ **48.** $2(x + 4y)(6x - y)$ **49.** $(x + y)(5 + x)$ **50.** $(x - y)(7 + y)$
51. $(7t - 1)(2t - 1)$ **52.** prime **53.** $-1(3x + 5)(x - 1)$ **54.** $-1(7x - 2)(x + 3)$ **55.** $(1 - 10a)(1 + 2a)$ **56.** $(1 + 5a)(1 - 12a)$
57. $(x + 3)(x - 3)(x - 1)(x + 1)$ **58.** $(x + 3)(x - 3)(x + 2)(x - 2)$ **59.** $(x - 15)(x - 8)$ **60.** $(y + 16)(y + 6)$ **61.** prime
62. $(4a - 7b)^2$ **63.** $(5p - 7q)^2$ **64.** $(7x + 3y)(x + 3y)$ **65.** $-1(x - 5)(x + 6)$ **66.** $-1(x - 2)(x - 4)$ **67.** $(3r - 1)(s + 4)$
68. $(x - 2)(x^2 + 1)$ **69.** $(x - 2y)(4x - 3)$ **70.** $(2x - y)(2x + 7z)$ **71.** $(x + 12y)(x - 3y)$ **72.** $(3x - 2y)(x + 4y)$
73. $(x^2 + 2)(x + 4)(x - 4)$ **74.** $(x^2 + 3)(x + 5)(x - 5)$ **75.** answers may vary **76.** yes; $9(x^2 + 9y^2)$

Mental Math **1.** $3, 7$ **3.** $-8, -6$ **5.** $-1, 3$

Exercise Set 4.6 **1.** $2, -1$ **3.** $6, 7$ **5.** $-9, -17$ **7.** $0, -6$ **9.** $0, 8$ **11.** $-\frac{3}{2}, \frac{5}{4}$ **13.** $\frac{7}{2}, -\frac{2}{7}$ **15.** $\frac{1}{2}, -\frac{1}{3}$ **17.** $-0.2, -1.5$

19. $9, 4$ **21.** $-4, 2$ **23.** $0, 7$ **25.** $0, -20$ **27.** $4, -4$ **29.** $8, -4$ **31.** $-3, 12$ **33.** $\frac{7}{3}, -2$ **35.** $\frac{8}{3}, -9$ **37.** $0, -\frac{1}{2}, \frac{1}{2}$ **39.** $\frac{17}{2}$

41. $\frac{3}{4}$ **43.** $-\frac{1}{2}, \frac{1}{2}$ **45.** $-\frac{3}{2}, -\frac{1}{2}, 3$ **47.** $-5, 3$ **49.** $-\frac{5}{6}, \frac{6}{5}$ **51.** $2, -\frac{4}{5}$ **53.** $-\frac{4}{3}, 5$ **55.** $-4, 3$ **57.** $0, 8, 4$

59. -7 **61.** $0, \frac{3}{2}$ **63.** $0, 1, -1$ **65.** $-6, \frac{4}{3}$ **67.** $\frac{6}{7}, 1$ **69.** $\frac{47}{45}$ **71.** $\frac{17}{60}$ **73.** $\frac{7}{10}$

75. didn't write equation in standard form; should be $x = 4$ or $x = -2$ **77.** answers may vary, for example, $(x - 6)(x + 1) = 0$

79. answers may vary, for example, $x^2 - 12x + 35 = 0$ **81. a.** $300; 304; 276; 216; 124; 0; -156$ **b.** 5 sec **c.** 304 ft **83.** $0, \frac{1}{2}$ **85.** $0, -15$

The Bigger Picture **1.** -34 **2.** x^{22} **3.** $-4x^3 - 6x^2 + 8$ **4.** $y - 1 + \frac{3}{y^2}$ **5.** $10x(x + 5)(x - 5)$ **6.** $(x - 1)(x - 35)$

7. $3(2y + 5)(x - 1)$ **8.** $x(5y - 7)(y + 1)$ **9.** $5, -\frac{1}{2}$ **10.** 1 **11.** $-2, 14$ **12.** $\frac{33}{17}$

Exercise Set 4.7   **1.** width: x; length: $x + 4$ **3.** x and $x + 2$ if x is an odd integer   **5.** base: x; height: $4x + 1$ **7.** 11 units
9. 15 cm, 13 cm, 22 cm, 70 cm   **11.** base: 16 mi; height: 6 mi   **13.** 5 sec   **15.** width: 5 cm; length: 6 cm   **17.** 54 diagonals
19. 10 sides   **21.** -12 or 11   **23.** 14, 15   **25.** 13 feet   **27.** 5 in.   **29.** 12 mm, 16 mm, 20 mm   **31.** 10 km   **33.** 36 ft   **35.** 9.5 sec
37. 20%   **39.** length: 15 mi; width: 8 mi   **41.** 105 units   **43.** 11,250 thousand acres   **45.** 10,750 thousand acres   **47.** 1995
49. answers may vary   **51.** 8 m   **53.** 10 and 15   **55.** width of pool: 29 m; length of pool: 35 m

Chapter 4 Vocabulary Check   **1.** quadratic equation   **2.** factoring   **3.** greatest common factor   **4.** perfect square trinomial

Chapter 4 Review   **1.** $2x - 5$ **2.** $2x^4 + 1 - 5x^3$ **3.** $5(m + 6)$ **4.** $4x(5x^2 + 3x + 6)$ **5.** $(2x + 3)(3x - 5)$ **6.** $(x + 1)(5x - 1)$
7. $(x - 1)(3x + 2)$ **8.** $(3x + 5)(2x - 1)$ **9.** $(a + 3b)(3a + b)$ **10.** $(x + 4)(x + 2)$ **11.** $(x - 8)(x - 3)$ **12.** prime
13. $(x - 6)(x + 1)$ **14.** $(x + 4)(x - 2)$ **15.** $(x + 6y)(x - 2y)$ **16.** $(x + 5y)(x + 3y)$ **17.** $2(3 - x)(12 + x)$
18. $4(8 + 3x - x^2)$ **19.** $5y(y - 6)(y - 4)$ **20.** $-48, 2$ **21.** factor out the GCF, 3   **22.** $(2x + 1)(x + 6)$ **23.** $(2x + 3)(2x - 1)$
24. $(3x + 4y)(2x - y)$ **25.** prime   **26.** $(2x + 3)(x - 13)$ **27.** $(6x + 5y)(3x - 4y)$ **28.** $5y(2y - 3)(y + 4)$
29. $5x^2 - 9x - 2; (5x + 1)(x - 2)$ **30.** $16x^2 - 28x + 6; 2(4x - 1)(2x - 3)$ **31.** yes   **32.** no   **33.** no   **34.** yes   **35.** yes
36. no   **37.** yes   **38.** no   **39.** $(x + 9)(x - 9)$ **40.** $(x + 6)^2$ **41.** $(2x + 3)(2x - 3)$ **42.** $(3t + 5s)(3t - 5s)$ **43.** prime
44. $(n - 9)^2$ **45.** $3(r + 6)^2$ **46.** $(3y - 7)^2$ **47.** $5m^6(m + 1)(m - 1)$ **48.** $(2x - 7y)^2$ **49.** $3y(x + y)^2$

50. $(4x^2 + 1)(2x + 1)(2x - 1)$ **51.** $-6, 2$ **52.** $0, -1, \dfrac{2}{7}$ **53.** $-\dfrac{1}{5}, -3$ **54.** $-7, -1$ **55.** $-4, 6$ **56.** -5 **57.** $2, 8$ **58.** $\dfrac{1}{3}$

59. $-\dfrac{2}{7}, \dfrac{3}{8}$ **60.** $0, 6$ **61.** $5, -5$ **62.** $x^2 - 9x + 20 = 0$ **63.** c   **64.** d   **65.** 9 units   **66.** 8 units, 13 units, 16 units, 10 units

67. width: 20 in.; length: 25 in.   **68.** 36 yd   **69.** 19 and 20   **70. a.** 17.5 sec and 10 sec; answers may vary   **b.** 27.5 sec   **71.** 32 cm
72. $6(x + 4)$ **73.** $7(x - 9)$ **74.** $(4x - 3)(11x - 6)$ **75.** $(x - 5)(2x - 1)$ **76.** $(3x - 4)(x^2 + 2)$ **77.** $(y + 2)(x - 1)$
78. $2(x + 4)(x - 3)$ **79.** $3x(x - 9)(x - 1)$ **80.** $(2x + 9)(2x - 9)$ **81.** $2(x + 3)(x - 3)$ **82.** $(4x - 3)^2$ **83.** $5(x + 2)^2$
84. $-\dfrac{7}{2}, 4$ **85.** $-3, 5$ **86.** $0, -7, -4$ **87.** $3, 2$ **88.** $0, 16$ **89.** 19 in.; 8 in.; 21 in.   **90.** length: 6 in.; width: 2 in.

Chapter 4 Test   **1.** $3x(3x - 1)$ **2.** $(x + 7)(x + 4)$ **3.** $(7 + m)(7 - m)$ **4.** $(y + 11)^2$ **5.** $(x^2 + 4)(x + 2)(x - 2)$
6. $(a + 3)(4 - y)$ **7.** prime   **8.** $(y - 12)(y + 4)$ **9.** $(a + b)(3a - 7)$ **10.** $(3x - 2)(x - 1)$ **11.** $5(6 + x)(6 - x)$
12. $3x(x - 5)(x - 2)$ **13.** $(6t + 5)(t - 1)$ **14.** $(x - 7)(y - 2)(y + 2)$ **15.** $x(1 + x^2)(1 + x)(1 - x)$ **16.** $(x + 12y)(x + 2y)$
17. $3, -9$ **18.** $-7, 2$ **19.** $-7, 1$ **20.** $0, \dfrac{3}{2}, -\dfrac{4}{3}$ **21.** $0, 3, -3$ **22.** $-3, 5$ **23.** $0, \dfrac{5}{2}$ **24.** 17 ft   **25.** width: 6 units; length: 9 units
26. 7 sec   **27.** hypotenuse: 25 cm; legs: 15 cm, 20 cm   **28.** 8.25 Sec

Cumulative Review   **1. a.** $9 \leq 11$ **b.** $8 > 1$ **c.** $3 \neq 4$; Sec. 1.2, Ex. 7   **2. a.** $>$ **b.** $<$; Sec. 1.2   **3.** solution; Sec. 1.3, Ex. 8   **4.** 102; Sec. 1.3

5. -12; Sec. 1.5, Ex. 5   **6.** -102; Sec. 1.5   **7. a.** $\dfrac{3}{4}$ **b.** -24; Sec. 1.6, Ex. 16   **8.** -98; Sec. 1.6   **9.** $5x + 7$; Sec. 1.8, Ex. 4

10. $19 - 6x$; Sec. 1.8   **11.** $-4a - 1$; Sec. 1.8, Ex. 5   **12.** $-13x - 21$; Sec. 1.8   **13.** $7.3x - 6$; Sec. 1.8, Ex. 7   **14.** 2; Sec. 2.2

15. -11; Sec. 2.3, Ex. 3   **16.** 28; Sec. 2.2   **17.** every real number; Sec. 2.3, Ex. 7   **18.** 33; Sec. 2.2   **19.** $l = \dfrac{V}{wh}$; Sec. 2.5, Ex. 5

20. $y = \dfrac{-3x - 7}{2}$ or $y = -\dfrac{3}{2}x - \dfrac{7}{2}$; Sec. 2.5   **21.** 5^{18}; Sec. 3.1, Ex. 14   **22.** 30; Sec. 3.1   **23.** y^{16}; Sec. 3.1, Ex. 15   **24.** y^{10}; Sec. 3.1

25. x^6; Sec. 3.2, Ex. 9   **26.** $\dfrac{1}{9}$; Sec. 3.2   **27.** $\dfrac{y^{18}}{z^{36}}$; Sec. 3.2, Ex. 11   **28.** x^4; Sec. 3.2   **29.** $\dfrac{1}{x^{19}}$; Sec. 3.2, Ex. 13   **30.** $25a^9$; Sec. 3.2

31. $4x$; Sec. 3.3, Ex. 6   **32.** $\dfrac{5}{6}x - 77$; Sec. 3.3   **33.** $13x^2 - 2$; Sec. 3.3, Ex. 7   **34.** $-0.5x + 1.2$; Sec. 3.3   **35.** $4x^2 - 4xy + y^2$; Sec. 3.5, Ex. 8

36. $9x^2 - 42xy + 49y^2$; Sec. 3.5   **37.** $t^2 + 4t + 4$; Sec. 3.6, Ex. 5   **38.** $x^2 - 26x + 169$; Sec. 3.6   **39.** $x^4 - 14x^2y + 49y^2$; Sec. 3.6, Ex. 8

40. $49x^2 + 14xy + y^2$; Sec. 3.6   **41.** $2xy - 4 + \dfrac{1}{2y}$; Sec. 3.7, Ex. 3   **42.** $(z^2 + 7)(z + 1)$; Sec. 4.1   **43.** $(x + 3)(5 + y)$; Sec. 4.1, Ex. 7

44. $2x(x + 7)(x - 6)$; Sec. 4.1   **45.** $(x^2 + 2)(x^2 + 3)$; Sec. 4.2, Ex. 7   **46.** $(-4x + 1)(x + 6)$ or $-1(4x - 1)(x + 6)$; Sec. 4.3

47. $2(x - 2)(3x + 5)$; Sec. 4.4, Ex. 2   **48.** $x(3y + 4)(3y - 4)$; Sec. 4.5   **49.** 3 sec; Sec. 4.7, Ex. 1   **50.** 9, 4; Sec. 4.6

CHAPTER 5   Rational Expressions

Mental Math   **1.** $x = 0$ **3.** $x = 0, x = 1$

Exercise Set 5.1   **1.** $\dfrac{7}{4}$ **3.** $-\dfrac{8}{3}$ **5.** $-\dfrac{11}{2}$ **7. a.** \$403   **b.** \$7   **c.** decrease; answers may vary   **9.** $x = 0$ **11.** $x = -2$ **13.** $x = \dfrac{5}{2}$

15. $x = 0, x = -2$ **17.** none   **19.** $x = 6, x = -1$ **21.** $x = -2, x = -\dfrac{7}{3}$ **23.** 1   **25.** -1 **27.** $\dfrac{1}{4(x + 2)}$ **29.** $\dfrac{1}{x + 2}$

31. can't simplify   **33.** -5 **35.** $\dfrac{7}{x}$ **37.** $\dfrac{1}{x - 9}$ **39.** $5x + 1$ **41.** $\dfrac{x^2}{x - 2}$ **43.** $7x$ **45.** $\dfrac{x + 5}{x - 5}$ **47.** $\dfrac{x + 2}{x + 4}$ **49.** $\dfrac{x + 2}{2}$

51. $-(x + 2)$ **53.** $\dfrac{x + 1}{x - 1}$ **55.** $x + y$ **57.** $\dfrac{5 - y}{2}$ **59.** $\dfrac{2y + 5}{3y + 4}$ **61.** $\dfrac{-(x - 10)}{x + 8}; \dfrac{-x + 10}{x + 8}; \dfrac{x - 10}{-(x + 8)}; \dfrac{x - 10}{-x - 8}$

63. $\dfrac{-(5y - 3)}{y - 12}; \dfrac{-5y + 3}{y - 12}; \dfrac{5y - 3}{-(y - 12)}; \dfrac{5y - 3}{-y + 12}$ **65.** correct   **67.** correct   **69.** $\dfrac{3}{11}$ **71.** $\dfrac{4}{3}$ **73.** $\dfrac{117}{40}$ **75.** correct

77. incorrect; $\dfrac{1 + 2}{1 + 3} = \dfrac{3}{4}$ **79.** answers may vary   **81.** answers may vary   **83.** 400 mg   **85.** 45.5%   **87.** 85.9

Mental Math   **1.** $\dfrac{2x}{3y}$ **3.** $\dfrac{5y^2}{7x^2}$ **5.** $\dfrac{9}{5}$

Exercise Set 5.2 **1.** $\dfrac{21}{4y}$ **3.** x^4 **5.** $-\dfrac{b^2}{6}$ **7.** $\dfrac{x^2}{10}$ **9.** $\dfrac{1}{3}$ **11.** $\dfrac{m+n}{m-n}$ **13.** $\dfrac{x+5}{x}$ **15.** $\dfrac{(x+2)(x-3)}{(x-4)(x+4)}$ **17.** $\dfrac{2x^4}{3}$ **19.** $\dfrac{12}{y^6}$

21. $x(x+4)$ **23.** $\dfrac{3(x+1)}{x^3(x-1)}$ **25.** m^2-n^2 **27.** $-\dfrac{x+2}{x-3}$ **29.** $\dfrac{x+2}{x-3}$ **31.** $\dfrac{5}{6}$ **33.** $\dfrac{3x}{8}$ **35.** $\dfrac{3}{2}$ **37.** $\dfrac{3x+4y}{2(x+2y)}$

39. $\dfrac{2(x+2)}{x-2}$ **41.** $-\dfrac{y(x+2)}{4}$ **43.** $\dfrac{(a+5)(a+3)}{(a+2)(a+1)}$ **45.** $\dfrac{5}{x}$ **47.** $\dfrac{2(n-8)}{3n-1}$ **49.** 1440 **51.** 5 **53.** 81 **55.** 73 **57.** 56.7

59. 411,755 sq yd **61.** 1364 feet per second **63.** 1 **65.** $-\dfrac{10}{9}$ **67.** $-\dfrac{1}{5}$ **69.** true **71.** false; $\dfrac{x^2+3x}{20}$ **73.** $\dfrac{2}{9(x-5)}$ sq ft

75. $\dfrac{x}{2}$ **77.** $\dfrac{5a(2a+b)(3a-2b)}{b^2(a-b)(a+2b)}$ **79.** answers may vary **81.** 1543.81 euros

Mental Math **1.** 1 **3.** $\dfrac{7x}{9}$ **5.** $\dfrac{1}{9}$ **7.** $\dfrac{17y}{5}$

Exercise Set 5.3 **1.** $\dfrac{a+9}{13}$ **3.** $\dfrac{3m}{n}$ **5.** 4 **7.** $\dfrac{y+10}{3+y}$ **9.** $5x+3$ **11.** $\dfrac{4}{a+5}$ **13.** $\dfrac{1}{x-6}$ **15.** $4x^3$ **17.** $8x(x+2)$

19. $(x+3)(x-2)$ **21.** $3(x+6)$ **23.** $5(x-6)^2$ **25.** $6(x+1)^2$ **27.** $x-8$ or $8-x$ **29.** $(x-1)(x+4)(x+3)$

31. $(3x+1)(x+1)(x-1)(2x+1)$ **33.** $2x^2(x+4)(x-4)$ **35.** $\dfrac{6x}{4x^2}$ **37.** $\dfrac{24b^2}{12ab^2}$ **39.** $\dfrac{9y}{2y(x+3)}$ **41.** $\dfrac{9ab+2b}{5b(a+2)}$

43. $\dfrac{x^2+x}{x(x+4)(x+2)(x+1)}$ **45.** $\dfrac{18y-2}{30x^2-60}$ **47.** $2x$ **49.** $\dfrac{x+3}{2x-1}$ **51.** $x+1$ **53.** $\dfrac{1}{x^2-8}$ **55.** $\dfrac{6(4x+1)}{x(2x+1)}$ **57.** $\dfrac{29}{21}$

59. $-\dfrac{5}{12}$ **61.** $\dfrac{7}{30}$ **63.** d **65.** $\dfrac{20}{x-2}$ m **67.** answers may vary **69.** 3 packages hot dogs and 2 packages buns

71. answers may vary **73.** answers may vary

Exercise Set 5.4 **1.** $\dfrac{5}{x}$ **3.** $\dfrac{75a+6b^2}{5b}$ **5.** $\dfrac{6x+5}{2x^2}$ **7.** $\dfrac{11}{x+1}$ **9.** $\dfrac{x-6}{(x-2)(x+2)}$ **11.** $\dfrac{35x-6}{4x(x-2)}$ **13.** $-\dfrac{2}{x-3}$ **15.** 0

17. $-\dfrac{1}{x^2-1}$ **19.** $\dfrac{5+2x}{x}$ **21.** $\dfrac{6x-7}{x-2}$ **23.** $-\dfrac{y+4}{y+3}$ **25.** $\dfrac{-5x+14}{4x}$ or $-\dfrac{5x-14}{4x}$ **27.** 2 **29.** $\dfrac{9x^4-4x^2}{21}$ **31.** $\dfrac{x+2}{(x+3)^2}$

33. $\dfrac{9b-4}{5b(b-1)}$ **35.** $\dfrac{2+m}{m}$ **37.** $\dfrac{x^2+3x}{(x-7)(x-2)}$ or $\dfrac{x(x+3)}{(x-7)(x-2)}$ **39.** $\dfrac{10}{1-2x}$ **41.** $\dfrac{15x-1}{(x+1)^2(x-1)}$ **43.** $\dfrac{x^2-3x-2}{(x-1)^2(x+1)}$

45. $\dfrac{a+2}{2(a+3)}$ **47.** $\dfrac{y(2y+1)}{(2y+3)^2}$ **49.** $\dfrac{x-10}{2(x-2)}$ **51.** $\dfrac{2x+21}{(x+3)^2}$ **53.** $\dfrac{-5x+23}{(x-2)(x-3)}$ **55.** $\dfrac{7}{2(m-10)}$

57. $\dfrac{2x^2-2x-46}{(x+1)(x-6)(x-5)}$ **59.** $\dfrac{n+4}{4n(n-1)(n-2)}$ **61.** 10 **63.** 2 **65.** $\dfrac{25a}{9(a-2)}$ **67.** $\dfrac{x+4}{(x-2)(x-1)}$ **69.** $x=\dfrac{2}{3}$

71. $x=-\dfrac{1}{2},x=1$ **73.** $x=-\dfrac{15}{2}$ **75.** $\dfrac{6x^2-5x-3}{x(x+1)(x-1)}$ **77.** $\dfrac{4x^2-15x+6}{(x-2)^2(x+2)(x-3)}$ **79.** $\dfrac{-2x^2+14x+55}{(x+2)(x+7)(x+3)}$

81. $\dfrac{2x-16}{(x+4)(x-4)}$ in. **83.** $\dfrac{P-G}{P}$ **85.** answers may vary **87.** $\left(\dfrac{90x-40}{x}\right)^\circ$ **89.** answers may vary

The Bigger Picture **1.** -17.7 **2.** 78.26 **3.** 28 **4.** $4x^4-x^2-17$ **5.** $\dfrac{x-1}{5}$ **6.** $\dfrac{14}{x+1}$ **7.** $-\dfrac{11}{18}$ **8.** $\dfrac{-4x-27}{45}$ or $-\dfrac{4x+27}{45}$

9. $x(9x-11)(x+1)$ **10.** $(4y-7)(3x+1)$ **11.** 12 **12.** $\left\{x\mid x>\dfrac{1}{2}\right\}$ **13.** $-\dfrac{8}{3}$ **14.** $-4,6$

Mental Math **1.** 10 **3.** 36

Exercise Set 5.5 **1.** 30 **3.** 0 **5.** -2 **7.** $-5,2$ **9.** 5 **11.** 3 **13.** 1 **15.** 5 **17.** no solution **19.** 4 **21.** -8

23. $6,-4$ **25.** 1 **27.** $3,-4$ **29.** -3 **31.** 0 **33.** -2 **35.** $8,-2$ **37.** no solution **39.** 3 **41.** $-11,1$ **43.** $I=\dfrac{E}{R}$

45. $B=\dfrac{2U-TE}{T}$ **47.** $W=\dfrac{Bh^2}{705}$ **49.** $G=\dfrac{V}{N-R}$ **51.** $r=\dfrac{C}{2\pi}$ **53.** $x=\dfrac{3y}{3+y}$ **55.** $\dfrac{1}{x}$ **57.** $\dfrac{1}{x}+\dfrac{1}{2}$ **59.** $\dfrac{1}{3}$ **61.** 5

63. $100^\circ,80^\circ$ **65.** $22.5^\circ,67.5^\circ$ **67.** no; multiplying both terms in the expression by 4 changes the value of the original expression.

Integrated Review **1.** expression; $\dfrac{3+2x}{3x}$ **2.** expression; $\dfrac{18+5a}{6a}$ **3.** equation; 3 **4.** equation; 18 **5.** expression; $\dfrac{x-1}{x(x+1)}$

6. expression; $\dfrac{3(x+1)}{x(x-3)}$ **7.** equation; no solution **8.** equation; 1 **9.** expression; 10 **10.** expression; $\dfrac{z}{3(9z-5)}$

11. expression; $\dfrac{5x+7}{x-3}$ **12.** expression; $\dfrac{7p+5}{2p+7}$ **13.** equation; 23 **14.** equation; 3 **15.** expression; $\dfrac{25a}{9(a-2)}$

16. expression; $\dfrac{9}{4(x-1)}$ **17.** expression; $\dfrac{3x^2+5x+3}{(3x-1)^2}$ **18.** expression; $\dfrac{2x^2-3x-1}{(2x-5)^2}$ **19.** expression; $\dfrac{4x-37}{5x}$ **20.** equation; $-\dfrac{7}{3}$

21. equation; $\dfrac{8}{5}$ **22.** expression; $\dfrac{29x-23}{3x}$ **23.** answers may vary **24.** answers may vary

Mental Math   1. c

Exercise Set 5.6   1. 4   **3.** $\dfrac{50}{9}$ **5.** -3 **7.** $\dfrac{14}{9}$ **9.** 123 lb   **11.** 165 cal   **13.** $y = 21.25$ **15.** $y = 5\dfrac{5}{7}$ ft   **17.** 2   **19.** -3

21. $2\dfrac{2}{9}$ hr   **23.** $1\dfrac{1}{2}$ min   **25.** trip to park rate: r; to park time: $\dfrac{12}{r}$; return trip rate: r; return time: $\dfrac{18}{r} = \dfrac{12}{r} + 1$; $r = 6$ mph

27. 1st portion: 10 mph; cooldown: 8 mph   **29.** 360 sq ft   **31.** 2   **33.** \$108.00   **35.** 20 mph   **37.** $y = 37\dfrac{1}{2}$ ft   **39.** 337 yd/game

41. 5   **43.** 217 mph   **45.** 9 gal   **47.** 8 mph   **49.** 2.2 mph; 3.3 mph   **51.** 3 hr   **53.** $26\dfrac{2}{3}$ ft   **55.** 216 nuts   **57.** $666\dfrac{2}{3}$ mi

59. 20 hr   **61.** car: 70 mph; motorcycle: 60 mph   **63.** $5\dfrac{1}{4}$ hr   **65.** first pump: 28 min; second pump: 84 min   **67.** $x = 8$ **69.** $y = 3.5$

71. $\dfrac{3}{4}$ **73.** $\dfrac{6}{5}$ **75.** a   **77.** $R = \dfrac{D}{T}$ **79.** 3.75 min

The Bigger Picture   1. $12x^3 - 11x^2 - 13x + 10$ **2.** $4x^2 - 4xy + y^2$ **3.** $4y^3(2 - 5y^2)$ **4.** $(9m - 2n)(m - n)$ **5.** -35

6. $\dfrac{16x - 70}{x(x - 10)}$ or $\dfrac{2(8x - 35)}{x(x - 10)}$ **7.** $-12x^{13}$ **8.** 2   **9.** $-\dfrac{1}{2}$ **10.** $\{y \mid y \geq 4\}$ **11.** $-\dfrac{27}{23}$ **12.** $\dfrac{a^{14}}{b^{14}}$

Exercise Set 5.7   1. $\dfrac{2}{3}$ **3.** $\dfrac{2}{3}$ **5.** $\dfrac{1}{2}$ **7.** $-\dfrac{21}{5}$ **9.** $\dfrac{27}{16}$ **11.** $\dfrac{4}{3}$ **13.** $\dfrac{1}{21}$ **15.** $-\dfrac{4x}{15}$ **17.** $\dfrac{m - n}{m + n}$ **19.** $\dfrac{2x(x - 5)}{7x^2 + 10}$ **21.** $\dfrac{1}{y - 1}$

23. $\dfrac{1}{6}$ **25.** $\dfrac{x + y}{x - y}$ **27.** $\dfrac{3}{7}$ **29.** $\dfrac{a}{x + b}$ **31.** $\dfrac{7(y - 3)}{8 + y}$ **33.** $\dfrac{3x}{x - 4}$ **35.** $-\dfrac{x + 8}{x - 2}$ **37.** $\dfrac{s^2 + r^2}{s^2 - r^2}$ **39.** $\dfrac{(x - 6)(x + 4)}{x - 2}$ **41.** Steffi Graf

43. Martina Navratilova and Steffi Graf   **45.** answers may vary   **47.** $\dfrac{13}{24}$ **49.** $\dfrac{R_1 R_2}{R_2 + R_1}$ **51.** $\dfrac{2x}{2 - x}$ **53.** $\dfrac{1}{y^2 - 1}$ **55.** 12 hr

Chapter 5 Vocabulary Check   1. ratio   **2.** proportion   **3.** cross products   **4.** rational expression   **5.** complex fraction   **6.** rate

Chapter 5 Review   1. $x = 2, x = -2$ **2.** $x = \dfrac{5}{2}, x = -\dfrac{3}{2}$ **3.** $\dfrac{4}{3}$ **4.** $\dfrac{11}{12}$ **5.** $\dfrac{2}{x}$ **6.** $\dfrac{3}{x}$ **7.** $\dfrac{1}{x - 5}$ **8.** $\dfrac{1}{x + 1}$ **9.** $\dfrac{x(x - 2)}{x + 1}$

10. $\dfrac{5(x - 5)}{x - 3}$ **11.** $\dfrac{x - 3}{x - 5}$ **12.** $\dfrac{x}{x + 4}$ **13.** $\dfrac{x + a}{x - c}$ **14.** $\dfrac{x + 5}{x - 3}$ **15.** $\dfrac{3x^2}{y}$ **16.** $-\dfrac{9x^2}{8}$ **17.** $\dfrac{x - 3}{x + 2}$ **18.** $\dfrac{-2x(2x + 5)}{(x - 6)^2}$

19. $\dfrac{x + 3}{x - 4}$ **20.** $\dfrac{4x}{3y}$ **21.** $(x - 6)(x - 3)$ **22.** $\dfrac{2}{3}$ **23.** $\dfrac{1}{2}$ **24.** $\dfrac{3(x + 2)}{3x + y}$ **25.** $\dfrac{1}{x + 2}$ **26.** $\dfrac{1}{x - 3}$ **27.** $\dfrac{2x - 10}{3x^2}$ **28.** $\dfrac{2x + 1}{2x^2}$

29. $14x$ **30.** $(x - 8)(x + 8)(x + 3)$ **31.** $\dfrac{10x^2 y}{14x^3 y}$ **32.** $\dfrac{36y^2 x}{16y^3 x}$ **33.** $\dfrac{x^2 - 3x - 10}{(x + 2)(x - 5)(x + 9)}$ **34.** $\dfrac{3x^2 + 4x - 15}{(x + 2)^2(x + 3)}$ **35.** $\dfrac{4y - 30x^2}{5x^2 y}$

36. $\dfrac{-2x + 10}{(x - 3)(x - 1)}$ **37.** $\dfrac{-2x - 2}{x + 3}$ **38.** $\dfrac{5x + 5}{(x + 4)(x - 2)(x - 1)}$ **39.** $\dfrac{x - 4}{3x}$ **40.** $-\dfrac{x}{x - 1}$ **41.** 30   **42.** $3, -4$ **43.** no solution

44. 5   **45.** $\dfrac{9}{7}$ **46.** $-6, 1$ **47.** $x = 9$ **48.** no solution   **49.** 675 parts   **50.** \$33.75   **51.** 3   **52.** 2

53. fast car speed: 30 mph; slow car speed: 20 mph   **54.** 20 mph   **55.** $17\dfrac{1}{2}$ hr   **56.** $8\dfrac{4}{7}$ days   **57.** $x = 15$ **58.** $x = 6$ **59.** $-\dfrac{7}{18y}$

60. $\dfrac{6}{7}$ **61.** $\dfrac{3y - 1}{2y - 1}$ **62.** $-\dfrac{7 + 2x}{2x}$ **63.** $\dfrac{1}{2x}$ **64.** $\dfrac{x(x - 3)}{x + 7}$ **65.** $\dfrac{x - 4}{x + 4}$ **66.** $\dfrac{(x - 9)(x + 8)}{(x + 5)(x + 9)}$ **67.** $\dfrac{1}{x - 6}$ **68.** $\dfrac{2x + 1}{4x}$

69. $\dfrac{2}{(x + 3)(x - 2)}$ **70.** $-\dfrac{3x}{(x + 2)(x - 3)}$ **71.** $\dfrac{1}{2}$ **72.** no solution   **73.** 1   **74.** $1\dfrac{5}{7}$ days   **75.** $x = 6$ **76.** $x = 12$ **77.** $\dfrac{3}{10}$ **78.** $\dfrac{2}{3}$

Chapter 5 Test   1. $x = -1, x = -3$ **2. a.** \$115   **b.** \$103   **3.** $\dfrac{3}{5}$ **4.** $\dfrac{1}{x + 6}$ **5.** -1 **6.** $-\dfrac{1}{x + y}$ **7.** $\dfrac{2m(m + 2)}{m - 2}$ **8.** $\dfrac{a + 2}{a + 5}$

9. $\dfrac{(x - 6)(x - 7)}{(x + 7)(x + 2)}$ **10.** 15   **11.** $\dfrac{y - 2}{4}$ **12.** $-\dfrac{1}{2x + 5}$ **13.** $\dfrac{3a - 4}{(a - 3)(a + 2)}$ **14.** $\dfrac{3}{x - 1}$ **15.** $\dfrac{2(x + 3)(x + 5)}{x(x^2 + 4x + 1)}$

16. $\dfrac{x^2 + 2x + 35}{(x + 9)(x + 2)(x - 5)}$ **17.** $\dfrac{4y^2 + 13y - 15}{(y + 5)(y + 1)(y + 4)}$ **18.** $\dfrac{30}{11}$ **19.** -6 **20.** no solution   **21.** no solution   **22.** $-2, 5$

23. $\dfrac{xz}{2y}$ **24.** $b - a$ **25.** $\dfrac{5y^2 - 1}{y + 2}$ **26.** $x = 12$ **27.** $x = 1$ and $x = 5$ **28.** 30 mph   **29.** $6\dfrac{2}{3}$ hr   **30.** 18 bulbs

Cumulative Review   1. a. $\dfrac{15}{x} = 4$ **b.** $12 - 3 = x$ **c.** $4x + 17 = 21$; Sec. 1.3, Ex. 10   **2. a.** $12 - x = -45$; Sec. 1.3   **b.** $12x = -45$

c. $x - 10 = 2x$ **3. a.** -12 **b.** -9; Sec. 1.4, Ex. 12   **4. a.** -8 **b.** -17; Sec. 1.4   **5.** distributive property; Sec. 1.7, Ex. 15

6. commutative property of addition; Sec. 1.7   **7.** associative property of addition; Sec. 1.7, Ex. 16   **8.** associative property of multiplication; Sec. 1.7

9. $x = -4$; Sec. 2.1, Ex. 7   **10.** 0; Sec. 2.1   **11.** shorter piece, 2 ft; longer piece, 8 ft; Sec. 2.4, Ex. 3   **12.** 190, 192; Sec. 2.4

13. $\dfrac{y - b}{m} = x$; Sec. 2.5, Ex. 6   **14.** $x = \dfrac{2y + 6}{3}$; Sec. 2.5   **15.** $x \leq -10$;   ; Sec. 2.7, Ex. 4

16. $\{x \mid x < -1\}$; Sec. 2.7   **17.** x^3; Sec. 3.1, Ex. 24   **18.** 1; Sec. 3.1   **19.** 256; Sec. 3.1, Ex. 25   **20.** $x^{15}y^6$; Sec. 3.1

21. -27; Sec. 3.1, Ex. 26   **22.** $x^{18}y^4$; Sec. 3.1   **23.** $2x^4 y$; Sec. 3.1, Ex. 27   **24.** $-15a^5 b^2$; Sec. 3.1

25. $\dfrac{2}{x^3}$; Sec. 3.2, Ex. 2 **26.** $\dfrac{1}{49}$; Sec. 3.2 **27.** $\dfrac{1}{16}$; Sec. 3.2, Ex. 4 **28.** $\dfrac{5}{z^7}$; Sec. 3.2 **29.** $10x^4 + 30x$; Sec. 3.5, Ex. 5 **30.** $x^2 + 18x + 81$; Sec. 3.5

31. $-15x^4 - 18x^3 + 3x^2$; Sec. 3.5, Ex. 6 **32.** $4x^2 - 1$; Sec. 3.5 **33.** $4x^2 - 4x + 6 + \dfrac{-11}{2x + 3}$; Sec. 3.7, Ex. 7

34. $4x^2 + 16x + 55 + \dfrac{222}{x - 4}$; Sec. 3.7 **35.** $(x + 3)(x + 4)$; Sec. 4.2, Ex. 1 **36.** $-2(a + 1)(a - 6)$; Sec. 4.2 **37.** $(5x + 2y)^2$; Sec. 4.5, Ex. 5

38. $(x + 2)(x - 2)$; Sec. 4.5 **39.** $x = 11, x = -2$; Sec. 4.6, Ex. 4 **40.** $-2, \dfrac{1}{3}$; Sec. 4.6 **41.** $\dfrac{2}{5}$; Sec. 5.2, Ex. 2 **42.** $\dfrac{x + 5}{2x^3}$; Sec. 5.1

43. $3x - 5$; Sec. 5.3, Ex. 3 **44.** $7x^4(x^2 - x + 1)$; Sec. 4.1 **45.** $\dfrac{3}{x - 2}$; Sec. 5.4, Ex. 2 **46.** $(2x + 3)^2$; Sec. 4.5 **47.** $t = 5$; Sec. 5.5, Ex. 2

48. $\dfrac{30}{x + 3}$; Sec. 5.2 **49.** $2\dfrac{1}{10}$ hr; Sec. 5.6, Ex. 6 **50.** $\dfrac{4m + 2n}{m + n}$ or $\dfrac{2(2m + n)}{m + n}$; Sec. 5.7

CHAPTER 6 Graphing Equations and Inequalities

Mental Math 1. answers may vary; Ex. $(5, 5), (7, 3)$

Exercise Set 6.1 1. France **3.** France, U.S., Spain **5.** 40 million **7.** 72,600 **9.** 1999; 74,800 **11.** 50 **13.** from 1984 to 1986 **15.** 1994

17.

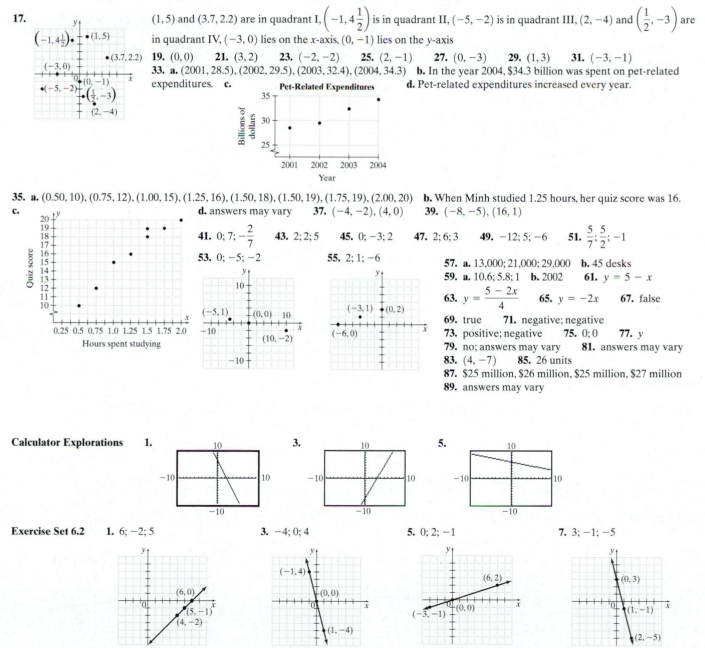

$(1, 5)$ and $(3.7, 2.2)$ are in quadrant I, $\left(-1, 4\frac{1}{2}\right)$ is in quadrant II, $(-5, -2)$ is in quadrant III, $(2, -4)$ and $\left(\frac{1}{2}, -3\right)$ are in quadrant IV, $(-3, 0)$ lies on the x-axis, $(0, -1)$ lies on the y-axis

19. $(0, 0)$ **21.** $(3, 2)$ **23.** $(-2, -2)$ **25.** $(2, -1)$ **27.** $(0, -3)$ **29.** $(1, 3)$ **31.** $(-3, -1)$

33. a. $(2001, 28.5), (2002, 29.5), (2003, 32.4), (2004, 34.3)$ **b.** In the year 2004, \$34.3 billion was spent on pet-related expenditures. **c.** **d.** Pet-related expenditures increased every year.

Pet-Related Expenditures

35. a. $(0.50, 10), (0.75, 12), (1.00, 15), (1.25, 16), (1.50, 18), (1.50, 19), (1.75, 19), (2.00, 20)$ **b.** When Minh studied 1.25 hours, her quiz score was 16.
c. **d.** answers may vary **37.** $(-4, -2), (4, 0)$ **39.** $(-8, -5), (16, 1)$

41. $0; 7; -\dfrac{2}{7}$ **43.** $2; 2; 5$ **45.** $0; -3; 2$ **47.** $2; 6; 3$ **49.** $-12; 5; -6$ **51.** $\dfrac{5}{7}; \dfrac{5}{2}; -1$

53. $0; -5; -2$ **55.** $2; 1; -6$

57. a. $13,000; 21,000; 29,000$ **b.** 45 desks
59. a. $10.6; 5.8; 1$ **b.** 2002 **61.** $y = 5 - x$
63. $y = \dfrac{5 - 2x}{4}$ **65.** $y = -2x$ **67.** false
69. true **71.** negative; negative
73. positive; negative **75.** $0; 0$ **77.** y
79. no; answers may vary **81.** answers may vary
83. $(4, -7)$ **85.** 26 units
87. \$25 million, \$26 million, \$25 million, \$27 million
89. answers may vary

Calculator Explorations 1. **3.** **5.**

Exercise Set 6.2 1. $6; -2; 5$ **3.** $-4; 0; 4$ **5.** $0; 2; -1$ **7.** $3; -1; -5$

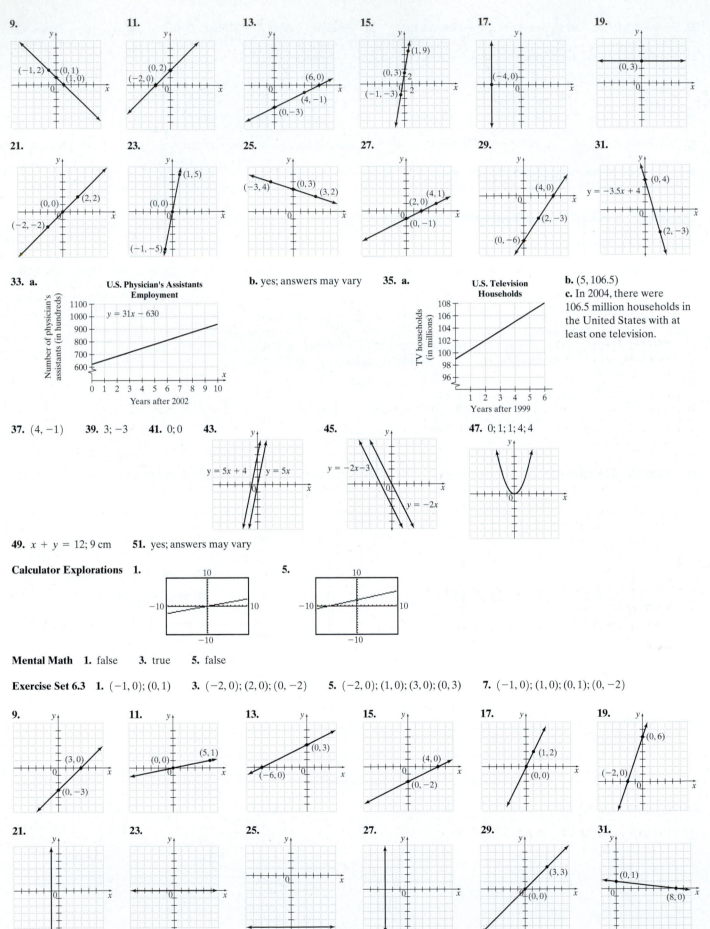

9.

11.

13.

15.

17.

19.

21.

23.

25.

27.

29.

31.

33. a. U.S. Physician's Assistants Employment; $y = 31x - 630$; Years after 2002 **b.** yes; answers may vary

35. a. U.S. Television Households; Years after 1999 **b.** $(5, 106.5)$ **c.** In 2004, there were 106.5 million households in the United States with at least one television.

37. $(4, -1)$ **39.** $3; -3$ **41.** $0; 0$ **43.** $y = 5x + 4$; $y = 5x$

45. $y = -2x - 3$; $y = -2x$

47. $0; 1; 1; 4; 4$

49. $x + y = 12; 9$ cm **51.** yes; answers may vary

Calculator Explorations **1.**

5.

Mental Math **1.** false **3.** true **5.** false

Exercise Set 6.3 **1.** $(-1, 0); (0, 1)$ **3.** $(-2, 0); (2, 0); (0, -2)$ **5.** $(-2, 0); (1, 0); (3, 0); (0, 3)$ **7.** $(-1, 0); (1, 0); (0, 1); (0, -2)$

9.

11.

13.

15.

17.

19.

21.

23.

25.

27.

29.

31.

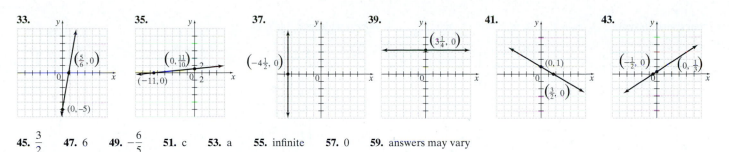

33. **35.** **37.** **39.** **41.** **43.**

45. $\dfrac{3}{2}$ **47.** 6   **49.** $-\dfrac{6}{5}$ **51.** c   **53.** a   **55.** infinite   **57.** 0   **59.** answers may vary

61. a. $(0, 200)$; no chairs and 200 desks are manufactured.   **b.** $(400, 0)$; 400 chairs and no desks are manufactured.   **c.** 300 chairs   **63.** $y = -4$
65. a. $(0, 919)$ **b.** In 1999, the number of stores was 919.

Calculator Explorations   1. **3.**

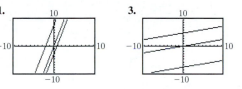

Mental Math   1. negative   **3.** undefined   **5.** upward   **7.** horizontal

Exercise Set 6.4   1. $m = -1$ **3.** $m = -\dfrac{1}{4}$ **5.** $m = 0$ **7.** undefined slope   **9.** $m = -\dfrac{4}{3}$ **11.** $m = \dfrac{5}{2}$ **13.** line 1   **15.** line 2

17. $m = 5$ **19.** $m = -0.3$ **21.** $m = -2$ **23.** undefined slope   **25.** $m = \dfrac{2}{3}$ **27.** undefined slope   **29.** $m = \dfrac{1}{2}$ **31.** $m = 0$

33. $m = -\dfrac{3}{4}$ **35.** $m = 4$ **37.** neither   **39.** neither   **41.** parallel   **43.** perpendicular   **45. a.** 1   **b.** -1 **47. a.** $\dfrac{9}{11}$ **b.** $-\dfrac{11}{9}$

49. $\dfrac{3}{5}$ **51.** 12.5%   **53.** 40%   **55.** 79%   **57.** $m = 3$; Every 1 year, there are/should be 3 million more U.S. households with personal computers.

59. $m = 0.42$; It costs \$0.42 per 1 mile to own and operate a compact car.   **61.** $y = 2x - 14$ **63.** $y = -6x - 11$ **65.** d   **67.** b   **69.** e

71. $m = \dfrac{1}{2}$ **73.** answers may vary   **75.** 28.5   **77.** 1994 and 2000; 28.1 miles per gallon   **79.** from 2000 to 2001   **81.** $x = 20$

83. a. $(1993, 10{,}359)$; $(2003, 15{,}139)$ **b.** 478   **c.** For the years 1993 through 2003, the number of kidney transplants increased at a rate of 478 per year.
85. Opposite sides are parallel since their slopes are equal, so the figure is a parallelogram.   **87.** 2.0625   **89.** -1.6 **91.** the line becomes steeper

Calculator Explorations   1. **3.**

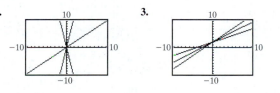

Mental Math   1. $m = 2$; $(0, -1)$ **3.** $m = 1$; $\left(0, \dfrac{1}{3}\right)$ **5.** $m = \dfrac{5}{7}$; $(0, -4)$ **7.** $m = 3$; answers may vary, Ex. $(4, 8)$ **9.** $m = -2$; answers

may vary, Ex. $(10, -3)$ **11.** $m = \dfrac{2}{5}$; answers may vary, Ex. $(-1, 0)$

Exercise Set 6.5   1. $y = 5x + 3$ **3.** $y = -4x - \dfrac{1}{6}$ **5.** $y = \dfrac{2}{3}x$ **7.** $y = -8$ **9.** $y = -\dfrac{1}{5}x + \dfrac{1}{9}$

11. **13.** **15.** **17.** **19.** **21.**

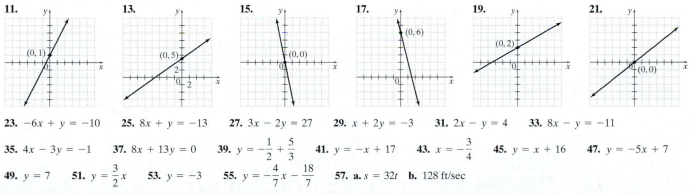

23. $-6x + y = -10$ **25.** $8x + y = -13$ **27.** $3x - 2y = 27$ **29.** $x + 2y = -3$ **31.** $2x - y = 4$ **33.** $8x - y = -11$

35. $4x - 3y = -1$ **37.** $8x + 13y = 0$ **39.** $y = -\dfrac{1}{2} + \dfrac{5}{3}$ **41.** $y = -x + 17$ **43.** $x = -\dfrac{3}{4}$ **45.** $y = x + 16$ **47.** $y = -5x + 7$

49. $y = 7$ **51.** $y = \dfrac{3}{2}x$ **53.** $y = -3$ **55.** $y = -\dfrac{4}{7}x - \dfrac{18}{7}$ **57. a.** $s = 32t$ **b.** 128 ft/sec

59. a. $y = 16{,}000x + 22{,}000$ **b.** 102,000 vehicles   **61. a.** $y = -333x + 7032$ **b.** 4368 cinema sites
63. a. $(0, 1509)$; $(6, 1456)$ **b.** $y = -8.8x + 1509$ **c.** 1491 daily newspapers   **65. a.** $S = -1000p + 13{,}000$ **b.** 9500 Fun Noodles   **67.** -1
69. 5   **71.** b   **73.** d   **75.** $3x - y = -5$ **77. a.** $3x - y = -5$ **b.** $x + 3y = 5$

Integrated Review   **1.** $m = 2$ **2.** $m = 0$ **3.** $m = -\dfrac{2}{3}$ **4.** slope is undefined

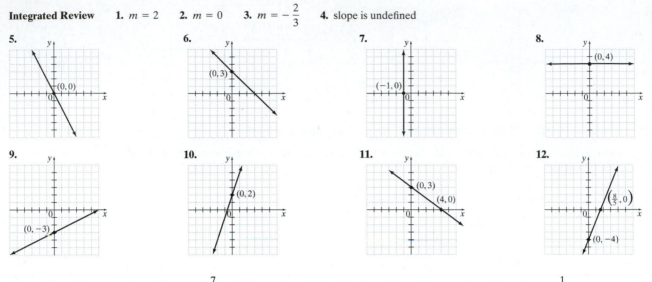

13. $m = 3$ **14.** $m = -6$ **15.** $m = -\dfrac{7}{2}$ **16.** $m = 2$ **17.** undefined slope   **18.** $m = 0$ **19.** $y = 2x - \dfrac{1}{3}$ **20.** $y = -4x - 1$

21. $-x + y = -2$ **22.** neither   **23.** perpendicular   **24. a.** $(1998, 1639); (2002, 2135)$ **b.** 124 **c.** For the years 1998 through 2002, the amount of yogurt produced increased at a rate of 124 million pounds per year.

Exercise Set 6.6 **1.** $\{-7, 0, 2, 10\}; \{-7, 0, 4, 10\}$ **3.** $\{0, 1, 5\}; \{-2\}$ **5.** yes **7.** no **9.** no **11.** yes **13.** yes **15.** no
17. 9:30 P.M. **19.** January 1 and December 1 **21.** yes; it passes the vertical line test **23.** $4.75 per hour **25.** 2005
27. yes; answers may vary **29.** $-9, -5, 1$ **31.** $6, 2, 11$ **33.** $-6, 0, 9$ **35.** $2, 0, 3$ **37.** $5, 0, -20$ **39.** $5, 3, 35$ **41.** $(3, 6)$ **43.** -1

45. -1 **47.** $-1, 5$ **49.** $x < 1$ **51.** $x \geq -3$ **53.** $\dfrac{19}{2x}$ m **55.** $f(-5) = 12$ **57.** answers may vary **59.** $f(x) = x + 7$
61. a. 190.4 mg **b.** 380.8 mg

Mental Math **1.** yes **3.** yes **5.** yes **7.** no

Exercise Set 6.7 **1.** no; no **3.** yes; no **5.** no; yes

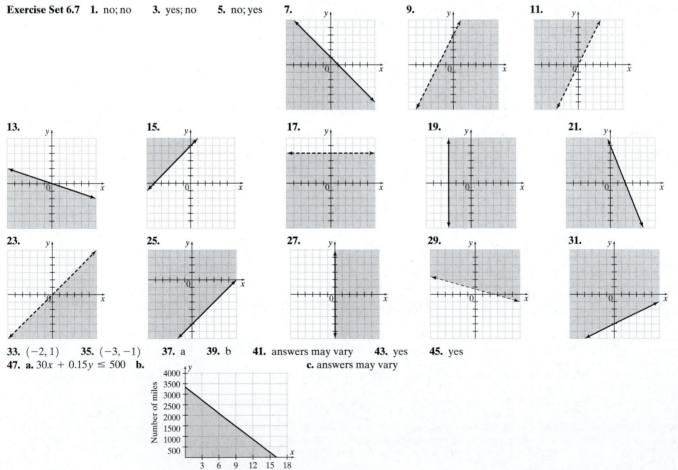

33. $(-2, 1)$ **35.** $(-3, -1)$ **37.** a **39.** b **41.** answers may vary **43.** yes **45.** yes
47. a. $30x + 0.15y \leq 500$ **b.**   **c.** answers may vary

Mental Math   **1.** direct   **3.** inverse   **5.** inverse   **7.** direct

Exercise Set 6.8   **1.** $y = \frac{1}{2}x$ **3.** $y = 6x$ **5.** $y = 3x$ **7.** $y = \frac{2}{3}x$ **9.** $y = \frac{7}{x}$ **11.** $y = \frac{0.5}{x}$ **13.** $y = kx$ **15.** $h = \frac{k}{t}$ **17.** $z = kx^2$ **19.** $y = \frac{k}{z^3}$ **21.** $x = \frac{k}{\sqrt{y}}$ **23.** $y = 40$ **25.** $y = 3$ **27.** $z = 54$ **29.** $a = \frac{4}{9}$ **31.** \$62.50   **33.** \$6   **35.** $5\frac{1}{3}$ in.   **37.** 179.1 lb   **39.** 1600 feet   **41.** $2y = 16$ **43.** $-4x = 0.5$ **45.** multiplied by 3   **47.** it is doubled

Chapter 6 Vocabulary Check   **1.** solution   **2.** y-axis   **3.** linear   **4.** x-intercept   **5.** standard   **6.** y-intercept   **7.** function   **8.** slope-intercept   **9.** domain   **10.** range   **11.** relation   **12.** point-slope   **13.** y **14.** x-axis   **15.** x **16.** slope   **17.** direct   **18.** inverse

Chapter 6 Review   **1–6.**   **7.** $(7, 44)$ **8.** $\left(-\frac{13}{3}, -8\right)$ **9.** $-3; 1; 9$ **10.** $5; 5; 5$ **11.** $0; 10; -10$ **12. a.** $2005; 2500; 7000$ **b.** 886 compact disc holders

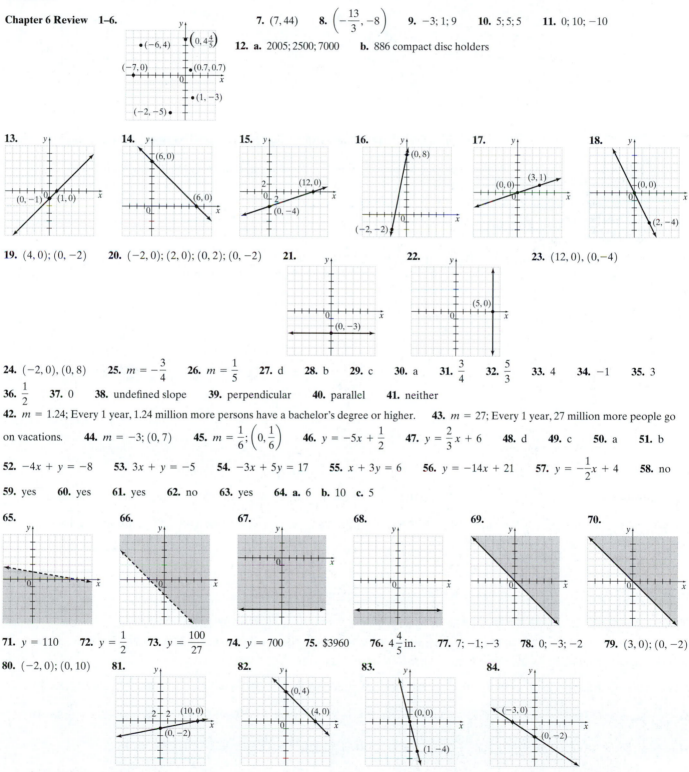

19. $(4, 0); (0, -2)$ **20.** $(-2, 0); (2, 0); (0, 2); (0, -2)$ **21.**   **22.**   **23.** $(12, 0), (0, -4)$

24. $(-2, 0), (0, 8)$ **25.** $m = -\frac{3}{4}$ **26.** $m = \frac{1}{5}$ **27.** d   **28.** b   **29.** c   **30.** a   **31.** $\frac{3}{4}$ **32.** $\frac{5}{3}$ **33.** 4   **34.** -1 **35.** 3   **36.** $\frac{1}{2}$ **37.** 0   **38.** undefined slope   **39.** perpendicular   **40.** parallel   **41.** neither   **42.** $m = 1.24$; Every 1 year, 1.24 million more persons have a bachelor's degree or higher.   **43.** $m = 27$; Every 1 year, 27 million more people go on vacations.   **44.** $m = -3; (0, 7)$ **45.** $m = \frac{1}{6}; \left(0, \frac{1}{6}\right)$ **46.** $y = -5x + \frac{1}{2}$ **47.** $y = \frac{2}{3}x + 6$ **48.** d   **49.** c   **50.** a   **51.** b   **52.** $-4x + y = -8$ **53.** $3x + y = -5$ **54.** $-3x + 5y = 17$ **55.** $x + 3y = 6$ **56.** $y = -14x + 21$ **57.** $y = -\frac{1}{2}x + 4$ **58.** no   **59.** yes   **60.** yes   **61.** yes   **62.** no   **63.** yes   **64. a.** 6   **b.** 10   **c.** 5

71. $y = 110$ **72.** $y = \frac{1}{2}$ **73.** $y = \frac{100}{27}$ **74.** $y = 700$ **75.** \$3960   **76.** $4\frac{4}{5}$ in.   **77.** $7; -1; -3$ **78.** $0; -3; -2$ **79.** $(3, 0); (0, -2)$ **80.** $(-2, 0); (0, 10)$ **81.**   **82.**   **83.**   **84.**

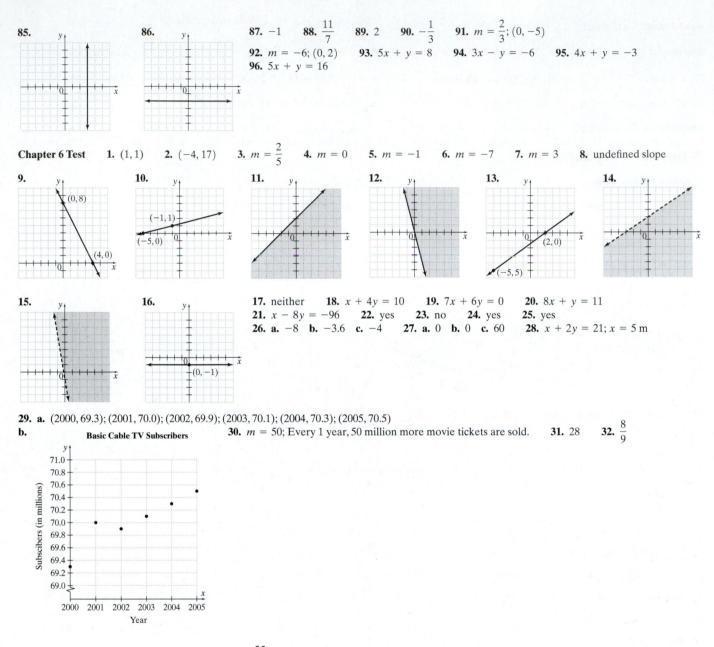

85.

86.

87. -1 **88.** $\dfrac{11}{7}$ **89.** 2 **90.** $-\dfrac{1}{3}$ **91.** $m = \dfrac{2}{3}; (0, -5)$

92. $m = -6; (0, 2)$ **93.** $5x + y = 8$ **94.** $3x - y = -6$ **95.** $4x + y = -3$

96. $5x + y = 16$

Chapter 6 Test   **1.** $(1, 1)$ **2.** $(-4, 17)$ **3.** $m = \dfrac{2}{5}$ **4.** $m = 0$ **5.** $m = -1$ **6.** $m = -7$ **7.** $m = 3$ **8.** undefined slope

9.   **10.**   **11.**   **12.**   **13.**   **14.**

15.   **16.**

17. neither   **18.** $x + 4y = 10$ **19.** $7x + 6y = 0$ **20.** $8x + y = 11$

21. $x - 8y = -96$ **22.** yes   **23.** no   **24.** yes   **25.** yes

26. a. -8 **b.** -3.6 **c.** -4 **27. a.** 0 **b.** 0 **c.** 60 **28.** $x + 2y = 21; x = 5$ m

29. a. $(2000, 69.3); (2001, 70.0); (2002, 69.9); (2003, 70.1); (2004, 70.3); (2005, 70.5)$

b.

Basic Cable TV Subscribers

30. $m = 50$; Every 1 year, 50 million more movie tickets are sold.   **31.** 28 **32.** $\dfrac{8}{9}$

Cumulative Review   **1.** 27; Sec. 1.3, Ex. 2   **2.** $\dfrac{25}{7}$; Sec. R.2   **3.** 51; Sec. 1.3, Ex. 5   **4.** 23; Sec. 1.3   **5.** $30,384$ feet; Sec. 1.5, Ex. 10

6. $0.8x - 36$; Sec. 1.8   **7.** $2x + 6$; Sec. 1.8, Ex. 16   **8.** $-15\left(x + \dfrac{2}{3}\right)$; Sec. 1.8   **9.** $(x - 4) \div 7$; Sec. 1.8, Ex. 17   **10.** $\dfrac{-9}{2x}$; Sec. 1.8

11. $5 + (x + 1) = 6 + x$; Sec. 1.8, Ex. 18   **12.** $-86 - x$; Sec. 1.8   **13.** 6; Sec. 2.2, Ex. 1   **14.** -24; Sec. 2.2   **15.**

$\{x \mid x < -2\}$; Sec. 2.7, Ex. 6

16. $\left\{x \mid x \le \dfrac{8}{3}\right\}$; Sec. 2.7   **17. a.** 2; trinomial   **b.** 1; binomial   **c.** 3; none of these; Sec. 3.3, Ex. 3

18. $y = \dfrac{6 - x}{2}$; Sec. 2.5   **19.** $-4x^2 + 6x + 2$; Sec. 3.4, Ex. 2   **20.** $4x - 4$; Sec. 3.4   **21.** $9y^2 + 6y + 1$; Sec. 3.6, Ex. 4

22. $x^2 - 24x + 144$; Sec. 3.6   **23.** $3a(-3a^4 + 6a - 1)$; Sec. 4.1, Ex. 5   **24.** $4(x + 3)(x - 3)$; Sec. 4.5   **25.** $(x - 2)(x + 6)$; Sec. 4.2, Ex. 3

26. $(3x + y)(x - 7y)$; Sec. 4.3   **27.** $(4x - 1)(2x - 5)$; Sec. 4.3, Ex. 2   **28.** $(18x - 1)(x + 2)$; Sec. 4.3

29. $x = 11, x = -2$; Sec. 4.6, Ex. 4   **30.** $x = 0, x = 1$; Sec. 4.6   **31.** 1; Sec. 5.2, Ex. 7   **32.** $\dfrac{x + 5}{2x^3}$; Sec. 5.1   **33.** $\dfrac{12ab^2}{27a^2b}$; Sec. 5.3, Ex. 9

34. $\dfrac{7x^2}{14x^3}$; Sec. 5.3   **35.** $\dfrac{2m + 1}{m + 1}$; Sec. 5.4, Ex. 5   **36.** $\dfrac{x + 5}{x - 6}$; Sec. 5.4   **37.** $x = -3, x = -2$; Sec. 5.5, Ex. 3   **38.** $x = -2, x = \dfrac{1}{3}$; Sec. 4.6

39. $\dfrac{x + 1}{x + 2y}$; Sec. 5.7, Ex. 5   **40.** $\dfrac{6x - 2y}{x - 4y}$ or $\dfrac{2(3x - y)}{x - 4y}$; Sec. 5.7   **41. a.** $(0, 12)$ **b.** $(2, 6)$ **c.** $(-1, 15)$; Sec. 6.1, Ex. 5

42. $0; 5; -2;$ Sec. 6.1 **43.** Sec. 6.2, Ex. 1 **44.** $\frac{1}{5}$; Sec. 6.4 **45.** $\frac{2}{3}$; Sec. 6.4, Ex. 3 **46.** undefined slope; Sec. 6.4

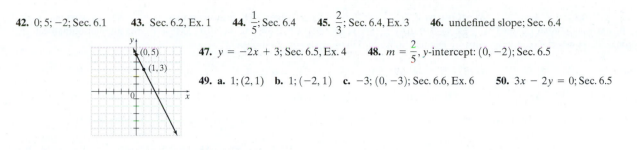

47. $y = -2x + 3$; Sec. 6.5, Ex. 4 **48.** $m = \frac{2}{5}$, y-intercept: $(0, -2)$; Sec. 6.5

49. a. $1; (2, 1)$ **b.** $1; (-2, 1)$ **c.** $-3; (0, -3)$; Sec. 6.6, Ex. 6 **50.** $3x - 2y = 0$; Sec. 6.5

CHAPTER 7 Systems of Equations

Calculator Explorations **1.** $(0.37, 0.23)$ **3.** $(0.03, -1.89)$

Mental Math **1.** 1 solution, $(-1, 3)$ **3.** infinite number of solutions **5.** no solution **7.** 1 solution, $(3, 2)$

Exercise Set 7.1 **1. a.** no **b.** yes **3. a.** yes **b.** no **5. a.** yes **b.** yes **7. a.** no **b.** no

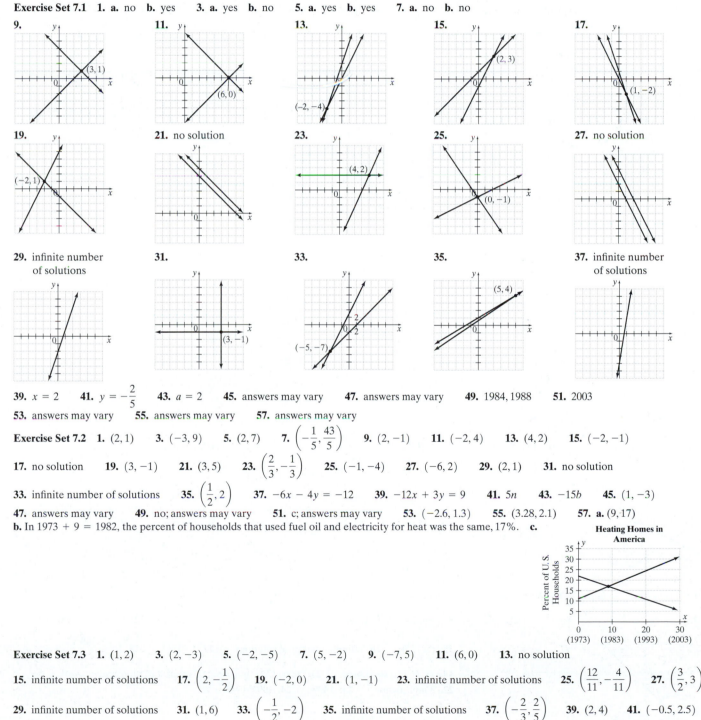

9. **11.** **13.** **15.** **17.**

19. **21.** no solution **23.** **25.** **27.** no solution

29. infinite number of solutions **31.** **33.** **35.** **37.** infinite number of solutions

39. $x = 2$ **41.** $y = -\frac{2}{5}$ **43.** $a = 2$ **45.** answers may vary **47.** answers may vary **49.** 1984, 1988 **51.** 2003
53. answers may vary **55.** answers may vary **57.** answers may vary

Exercise Set 7.2 **1.** $(2, 1)$ **3.** $(-3, 9)$ **5.** $(2, 7)$ **7.** $\left(-\frac{1}{5}, \frac{43}{5}\right)$ **9.** $(2, -1)$ **11.** $(-2, 4)$ **13.** $(4, 2)$ **15.** $(-2, -1)$

17. no solution **19.** $(3, -1)$ **21.** $(3, 5)$ **23.** $\left(\frac{2}{3}, -\frac{1}{3}\right)$ **25.** $(-1, -4)$ **27.** $(-6, 2)$ **29.** $(2, 1)$ **31.** no solution

33. infinite number of solutions **35.** $\left(\frac{1}{2}, 2\right)$ **37.** $-6x - 4y = -12$ **39.** $-12x + 3y = 9$ **41.** $5n$ **43.** $-15b$ **45.** $(1, -3)$
47. answers may vary **49.** no; answers may vary **51.** c; answers may vary **53.** $(-2.6, 1.3)$ **55.** $(3.28, 2.1)$ **57. a.** $(9, 17)$
b. In $1973 + 9 = 1982$, the percent of households that used fuel oil and electricity for heat was the same, 17%. **c.**

Heating Homes in America

Exercise Set 7.3 **1.** $(1, 2)$ **3.** $(2, -3)$ **5.** $(-2, -5)$ **7.** $(5, -2)$ **9.** $(-7, 5)$ **11.** $(6, 0)$ **13.** no solution

15. infinite number of solutions **17.** $\left(2, -\frac{1}{2}\right)$ **19.** $(-2, 0)$ **21.** $(1, -1)$ **23.** infinite number of solutions **25.** $\left(\frac{12}{11}, -\frac{4}{11}\right)$ **27.** $\left(\frac{3}{2}, 3\right)$

29. infinite number of solutions **31.** $(1, 6)$ **33.** $\left(-\frac{1}{2}, -2\right)$ **35.** infinite number of solutions **37.** $\left(-\frac{2}{3}, \frac{2}{5}\right)$ **39.** $(2, 4)$ **41.** $(-0.5, 2.5)$

43. $2x + 6 = x - 3$ **45.** $20 - 3x = 2$ **47.** $4(x + 6) = 2x$ **49.** $2; 6x - 2y = -24$ **51.** b; answers may vary   **53.** answers may vary
55. a. $b = 15$ **b.** any real number except 15   **57.** $(-8.9, 10.6)$ **59. a.** $(8, 536)$ or $(8, 537)$ **b.** In 2010 $(2002 + 8)$, the number of medical assistant jobs equals the number of computer software engineer jobs.   **c.** 536 thousand or 537 thousand

Integrated Review   1. $(2, 5)$ **2.** $(4, 2)$ **3.** $(5, -2)$ **4.** $(6, -14)$ **5.** $(-3, 2)$ **6.** $(-4, 3)$ **7.** $(0, 3)$ **8.** $(-2, 4)$ **9.** $(5, 7)$

10. $(-3, -23)$ **11.** $\left(\frac{1}{3}, 1\right)$ **12.** $\left(-\frac{1}{4}, 2\right)$ **13.** no solution   **14.** infinite number of solutions   **15.** $(0.5, 3.5)$ **16.** $(-0.75, 1.25)$

17. infinite number of solutions   **18.** no solution   **19.** $(7, -3)$ **20.** $(-1, -3)$ **21.** answers may vary   **22.** answers may vary

Exercise Set 7.4   1. c   **3.** b   **5.** a   **7.** $\begin{cases} x + y = 15 \\ x - y = 7 \end{cases}$ **9.** $\begin{cases} x + y = 6500 \\ x = y + 800 \end{cases}$ **11.** 33 and 50   **13.** 14 and -3

15. Jackson: 634 points; Leslie: 598 points   **17.** child's ticket: $18; adult's ticket: $29   **19.** quarters: 53; nickels: 27
21. McDonald's: $31.50; The Ohio Art Company: $6.50   **23.** daily fee: $32; mileage charge: $0.25 per mi
25. distance downstream = distance upstream = 18 mi; time downstream: 2 hr; time upstream: $4\frac{1}{2}$ hr; still water: 6.5 mph; current: 2.5 mph

27. still air: 455 mph; wind: 65 mph   **29.** $4\frac{1}{2}$ hr   **31.** 12% solution: $7\frac{1}{2}$ oz; 4% solution: $4\frac{1}{2}$ oz   **33.** $4.95 beans: 113 lb; $2.65 beans: 87 lb

35. $60°, 30°$ **37.** $20°, 70°$ **39.** number sold at $9.50: 23; number sold at $7.50: 67   **41.** $2\frac{1}{4}$ mph and $2\frac{3}{4}$ mph   **43.** 30%: 50 gal; 60%: 100 gal

45. length: 42 in.; width: 30 in.   **47.** 16   **49.** $36x^2$ **51.** $100y^6$ **53.** a   **55.** width: 9 ft; length: 15 ft

Chapter 7 Vocabulary Check   1. dependent   **2.** system of linear equations   **3.** consistent   **4.** solution   **5.** addition; substitution
6. inconsistent   **7.** independent

Chapter 7 Review   1. a. no   **b.** yes   **2. a.** yes   **b.** no   **3. a.** no   **b.** no   **4. a.** no   **b.** yes

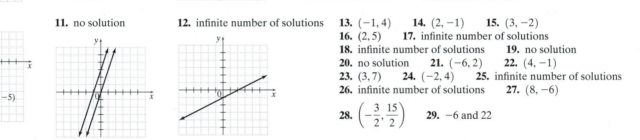

5.   **6.**   **7.**   **8.**   **9.**

10.   **11.** no solution   **12.** infinite number of solutions   **13.** $(-1, 4)$ **14.** $(2, -1)$ **15.** $(3, -2)$
16. $(2, 5)$ **17.** infinite number of solutions
18. infinite number of solutions   **19.** no solution
20. no solution   **21.** $(-6, 2)$ **22.** $(4, -1)$
23. $(3, 7)$ **24.** $(-2, 4)$ **25.** infinite number of solutions
26. infinite number of solutions   **27.** $(8, -6)$

28. $\left(-\frac{3}{2}, \frac{15}{2}\right)$ **29.** -6 and 22

30. orchestra: 255 seats; balcony: 105 seats   **31.** current of river: 3.2 mph; speed in still water: 21.1 mph

32. 6% solution: $12\frac{1}{2}$ cc; 14% solution: $37\frac{1}{2}$ cc   **33.** egg: $0.40; strip of bacon: $0.65   **34.** jogging: 0.86 hr; walking: 2.14 hr

35.   **36.** infinite number of solutions   **37.** $(3, 2)$ **38.** $(7, 1)$ **39.** $\left(1\frac{1}{2}, -3\right)$

40. no solution   **41.** infinite number of solutions   **42.** $(8, 11)$ **43.** $(-5, 2)$
44. $(16, -4)$ **45.** infinite number of solutions   **46.** no solution   **47.** 4 and 8
48. -5 and 13   **49.** 24 nickels and 41 dimes   **50.** 13¢ stamps: 17; 22¢ stamps: 9

Chapter 7 Test   1. false   **2.** false   **3.** true   **4.** false   **5.** no   **6.** yes

7.   **8.**   **9.** $(-4, 1)$ **10.** $\left(\frac{1}{2}, -2\right)$ **11.** $(20, 8)$ **12.** no solution   **13.** $(4, -5)$ **14.** $(7, 2)$

15. $(5, -2)$ **16.** infinite number of solutions   **17.** $(-5, 3)$ **18.** $\left(\frac{47}{5}, \frac{48}{5}\right)$

19. 78, 46   **20.** 120 cc   **21.** Texas: 226 thousand; Missouri: 110 thousand
22. 3 mph; 6 mph   **23.** 1999   **24.** 1991–1999

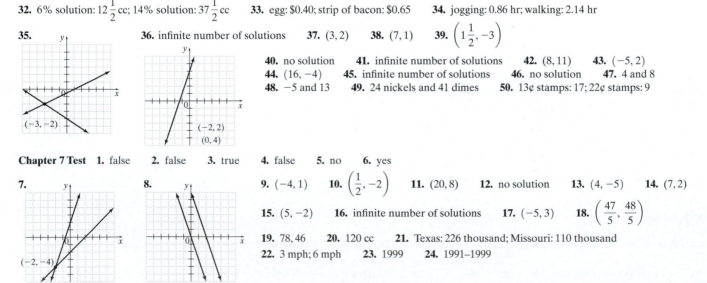

Cumulative Review **1. a.** -6 **b.** 6.3; Sec. 1.5, Ex. 6 **2. a.** 25; Sec. 1.3 **b.** 32 **3.** $\frac{1}{22}$; Sec. 1.6, Ex. 9a **4.** -22; Sec. 1.4 **5.** $\frac{16}{3}$; Sec. 1.6, Ex. 9b

6. $-\frac{3}{16}$; Sec. 1.4 **7.** $-\frac{1}{10}$; Sec. 1.6, Ex. 9c **8.** 10; Sec. 1.4 **9.** $-\frac{13}{9}$; Sec. 1.6, Ex. 9d **10.** $\frac{9}{13}$; Sec. 1.4 **11.** $\frac{1}{1.7}$; Sec. 1.6, Ex. 9e **12.** -1.7; Sec. 1.4

13. a. 5 **b.** $8 - x$; Sec. 2.1, Ex. 8 **14.** -5; Sec. 2.1 **15.** no solution; Sec. 2.3, Ex. 6 **16.** no solution; Sec. 2.3 **17.** 12; Sec. 2.3, Ex. 4

18. 40; Sec. 2.3 **19.** $\left\{x \mid x > \frac{13}{7}\right\}$ 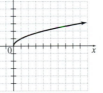 ; Sec. 2.7, Ex. 8 **20.** $b = P - a - c$; Sec. 2.5 **21.** $\frac{m^7}{n^7}$, $n \neq 0$; Sec. 3.1, Ex. 22

22. $\frac{a^6}{b^5}$, $b \neq 0$; Sec. 3.1 **23.** $\frac{16x^{16}}{81y^{20}}$, $y \neq 0$; Sec. 3.1, Ex. 23 **24.** $\frac{49a^4}{b^6}$, $b \neq 0$; Sec. 3.2 **25.** $9x^2 - 6x - 1$; Sec. 3.4, Ex. 5 **26.** $6x^2 - \frac{14}{3}x + \frac{1}{42}$; Sec. 3.4

27. $2x + 4 - \frac{1}{3x - 1}$; Sec. 3.7, Ex. 5 **28.** $3x^2$; Sec. 4.1 **29.** $-\frac{1}{2}, 4$; Sec. 4.6, Ex. 6 **30.** $0, \frac{7}{2}$; Sec. 4.6 **31.** 6 units, 8 units, 10 units; Sec. 4.7, Ex. 5

32. base: 7 ft; height: 26 ft; Sec. 4.7 **33.** 1; Sec. 5.3, Ex. 2 **34.** $\frac{5x - 16}{(x - 6)(x + 1)}$; Sec. 5.3 **35.** $\frac{2x^2 + 3y}{x^2y + 2xy^2}$; Sec. 5.7, Ex. 6 **36.** $-\frac{11}{3}$; Sec. 6.4

37. $m = 0$; Sec. 6.4, Ex. 5 **38.** undefined slope; Sec. 6.4 **39.** $-x + 5y = 23$; Sec. 6.5, Ex. 5 **40.** $y = -5x - 7$; Sec. 6.5

41. domain: $\{-1, 0, 3\}$; range: $\{-2, 0, 2, 3\}$; Sec. 6.6, Ex. 1 **42.** $-6; 14$; Sec. 6.6 **43.** It is a solution; Sec. 7.1, Ex. 1

44. a. yes **b.** no **c.** no; Sec. 7.1 **45.** $\left(6, \frac{1}{2}\right)$; Sec. 7.2, Ex. 3 **46.** $(-2, -4)$; Sec. 7.2 **47.** $\left(-\frac{15}{7}, -\frac{5}{7}\right)$; Sec. 7.3, Ex. 6

48. $\left(-\frac{44}{3}, -\frac{7}{3}\right)$; Sec. 7.3 **49.** 29 and 8; Sec. 7.4, Ex. 1 **50. a.** no **b.** yes **c.** no; Sec. 6.6

CHAPTER 8 Roots and Radicals

Calculator Explorations **1.** 2.449 **3.** 3.317 **5.** 9.055 **7.** 3.420 **9.** 2.115 **11.** 1.783

Exercise Set 8.1 **1.** 4 **3.** $\frac{1}{5}$ **5.** -10 **7.** not a real number **9.** -11 **11.** $\frac{3}{5}$ **13.** 30 **15.** 12 **17.** $\frac{1}{10}$ **19.** 0.5 **21.** 5

23. -4 **25.** -2 **27.** $\frac{1}{2}$ **29.** -5 **31.** 2 **33.** 9 **35.** not a real number **37.** $-\frac{3}{4}$ **39.** -5 **41.** 1 **43.** 2.646 **45.** 6.083

47. 11.662 **49.** $\sqrt{2} \approx 1.41$; 126.90 ft **51.** m **53.** x^2 **55.** $3x^4$ **57.** $9x$ **59.** ab^2 **61.** $4a^3b^2$ **63.** a^2b^6 **65.** $-2xy^9$

67. $25 \cdot 2$ **69.** $16 \cdot 2$ or $4 \cdot 8$ **71.** $4 \cdot 7$ **73.** $9 \cdot 3$ **75.** a, b **77.** 7 mi **79.** 3.01 in. **81.** 3 **83.** 10 **85.** $T = 6.1$ seconds
87. answers may vary **89.** $1; 1.7; 2; 3$ **91.** $(2, 0)$ **93.** $(-4, 0)$

Mental Math **1.** 6 **3.** x **5.** 0 **7.** $5x^2$

Exercise Set 8.2 **1.** $2\sqrt{5}$ **3.** $5\sqrt{2}$ **5.** $\sqrt{33}$ **7.** $7\sqrt{2}$ **9.** $2\sqrt{15}$ **11.** $6\sqrt{5}$ **13.** $2\sqrt{13}$ **15.** 15 **17.** $21\sqrt{7}$ **19.** $-15\sqrt{3}$

21. $\frac{2\sqrt{2}}{5}$ **23.** $\frac{3\sqrt{3}}{11}$ **25.** $\frac{3}{2}$ **27.** $\frac{5\sqrt{5}}{3}$ **29.** $\frac{\sqrt{11}}{6}$ **31.** $-\frac{\sqrt{3}}{4}$ **33.** $x^3\sqrt{x}$ **35.** $x^6\sqrt{x}$ **37.** $6a\sqrt{a}$ **39.** $4x^2\sqrt{6}$ **41.** $\frac{2\sqrt{3}}{m}$

43. $\frac{3\sqrt{x}}{y^5}$ **45.** $\frac{2\sqrt{22}}{x^6}$ **47.** 16 **49.** $\frac{6}{11}$ **51.** $5\sqrt{7}$ **53.** $\frac{2\sqrt{5}}{3}$ **55.** $2m^3\sqrt{6m}$ **57.** $\frac{y\sqrt{23y}}{2x^3}$ **59.** $2\sqrt[3]{3}$ **61.** $5\sqrt[3]{2}$ **63.** $\frac{\sqrt[3]{5}}{4}$

65. $\frac{\sqrt[3]{23}}{2}$ **67.** $\frac{\sqrt[3]{15}}{4}$ **69.** $2\sqrt[3]{10}$ **71.** $14x$ **73.** $2x^2 - 7x - 15$ **75.** 0 **77.** $x^3y\sqrt{y}$ **79.** $7x^2y^2\sqrt{2x}$ **81.** $-2x^2$ **83.** $2\sqrt[3]{10}$ in.

85. answers may vary **87.** $2\sqrt{5}$ in. **89.** 2.25 in. **91.** $\$1700$ **93.** 1.7 sq m

Mental Math **1.** $8\sqrt{2}$ **3.** $7\sqrt{x}$ **5.** $3\sqrt{7}$

Exercise Set 8.3 **1.** $-4\sqrt{3}$ **3.** $9\sqrt{6} - 5$ **5.** $\sqrt{5} + \sqrt{2}$ **7.** $7\sqrt{3} - \sqrt{2}$ **9.** $-5\sqrt{2} - 6$ **11.** $5\sqrt{3}$ **13.** $9\sqrt{5}$

15. $4\sqrt{6} + \sqrt{5}$ **17.** $x + \sqrt{x}$ **19.** 0 **21.** $\frac{4\sqrt{5}}{9}$ **23.** $\frac{3\sqrt{3}}{8}$ **25.** $7\sqrt{5}$ **27.** $9\sqrt{3}$ **29.** $\sqrt{5} + \sqrt{15}$ **31.** $x\sqrt{x}$

33. $5\sqrt{2} + 12$ **35.** $8\sqrt{2} - 5$ **37.** $2\sqrt{5}$ **39.** $-\sqrt{35}$ **41.** $6 - 3\sqrt{3}$ **43.** $11\sqrt{x}$ **45.** $12x - 11\sqrt{x}$ **47.** $x\sqrt{3x} + 3x\sqrt{x}$

49. $8x\sqrt{2} + 2x$ **51.** $2x^2\sqrt{10} - x^2\sqrt{5}$ **53.** $7\sqrt[3]{9} - \sqrt[3]{25}$ **55.** $-5\sqrt[3]{2} - 6$ **57.** $5\sqrt[3]{3}$ **59.** $4x + 4x\sqrt[3]{2}$ **61.** $x\sqrt[3]{5}$ **63.** $10y^2\sqrt[3]{y}$

65. $x^2 + 12x + 36$ **67.** $4x^2 - 4x + 1$ **69.** answers may vary   **71.** $8\sqrt{5}$ in.   **73.** $\left(48 + \dfrac{9\sqrt{3}}{2}\right)$ sq ft   **75.** yes; $7\sqrt{2}$

77. no   **79.** yes; $3\sqrt{7}$ **81.** $\dfrac{83x\sqrt{x}}{20}$

Mental Math   **1.** $\sqrt{22}$ **3.** $\sqrt{6}$ **5.** $\sqrt{10y}$

Exercise Set 8.4   **1.** 4   **3.** $5\sqrt{2}$ **5.** 6   **7.** $2x$ **9.** 20   **11.** $36x$ **13.** $3x^3\sqrt{2}$ **15.** $4xy\sqrt{y}$ **17.** $\sqrt{30} + \sqrt{42}$
19. $2\sqrt{5} + 5\sqrt{2}$ **21.** $y\sqrt{7} - 14\sqrt{y}$ **23.** -33 **25.** $\sqrt{6} - \sqrt{15} + \sqrt{10} - 5$ **27.** $16 - 11\sqrt{11}$ **29.** $x - 36$
31. $x - 14\sqrt{x} + 49$ **33.** $6y + 2\sqrt{6y} + 1$ **35.** 4   **37.** $\sqrt{7}$ **39.** $3\sqrt{2}$ **41.** $5y^2$ **43.** $5\sqrt{3}$ **45.** $2y\sqrt{6}$ **47.** $2xy\sqrt{3y}$
49. $\dfrac{\sqrt{15}}{5}$ **51.** $\dfrac{7\sqrt{2}}{2}$ **53.** $\dfrac{\sqrt{6y}}{6y}$ **55.** $\dfrac{\sqrt{10}}{6}$ **57.** $\dfrac{\sqrt{3x}}{x}$ **59.** $\dfrac{\sqrt{2}}{4}$ **61.** $\dfrac{\sqrt{30}}{15}$ **63.** $\dfrac{\sqrt{15}}{10}$ **65.** $\dfrac{3\sqrt{2x}}{2}$ **67.** $\dfrac{8y\sqrt{5}}{5}$
69. $\dfrac{\sqrt{3xy}}{6x}$ **71.** $3\sqrt{2} - 3$ **73.** $-8 - 4\sqrt{5}$ **75.** $5 + \sqrt{30} + \sqrt{6} + \sqrt{5}$ **77.** $\sqrt{6} + \sqrt{3} + \sqrt{2} + 1$ **79.** $\dfrac{10 - 5\sqrt{x}}{4 - x}$
81. $\dfrac{3\sqrt{x} + 12}{x - 16}$ **83.** $x = 44$ **85.** $z = 2$ **87.** $x = 3$ **89.** $130\sqrt{3}$ sq m   **91.** $\dfrac{\sqrt{A\pi}}{\pi}$ **93.** true   **95.** false   **97.** false
99. answers may vary   **101.** answers may vary   **103.** $\dfrac{2}{\sqrt{6} - \sqrt{2} - \sqrt{3} + 1}$

Integrated Review   **1.** 6   **2.** $4\sqrt{3}$ **3.** x^2 **4.** $y^3\sqrt{y}$ **5.** $4x$ **6.** $3x^5\sqrt{2x}$ **7.** 2   **8.** 3   **9.** -3 **10.** not a real number
11. $\dfrac{\sqrt{11}}{3}$ **12.** $\dfrac{\sqrt[3]{7}}{4}$ **13.** -4 **14.** -5 **15.** $\dfrac{3}{7}$ **16.** $\dfrac{1}{8}$ **17.** a^4b **18.** x^5y^{10} **19.** $5m^3$ **20.** $3n^8$ **21.** $6\sqrt{7}$ **22.** $3\sqrt{2}$
23. cannot be simplified   **24.** $\sqrt{x} + 3x$ **25.** $\sqrt{30}$ **26.** 3   **27.** 28   **28.** 45   **29.** $\sqrt{33} + \sqrt{3}$ **30.** $3\sqrt{2} - 2\sqrt{6}$ **31.** $4y$
32. $3x^2\sqrt{5}$ **33.** $x - 3\sqrt{x} - 10$ **34.** $11 + 6\sqrt{2}$ **35.** 2   **36.** $\sqrt{3}$ **37.** $2x^2\sqrt{3}$ **38.** $ab^2\sqrt{15a}$ **39.** $\dfrac{\sqrt{6}}{6}$ **40.** $\dfrac{x\sqrt{5}}{10}$
41. $\dfrac{4\sqrt{6} - 4}{5}$ **42.** $\dfrac{\sqrt{2x} + 5\sqrt{2} + \sqrt{x} + 5}{x - 25}$

Exercise Set 8.5   **1.** 81   **3.** -1 **5.** 49   **7.** no solution   **9.** 4   **11.** 2   **13.** 2   **15.** 9   **17.** -3 **19.** $-1, -2$
21. no solution   **23.** $0, -3$ **25.** 16   **27.** 25   **29.** 1   **31.** 5   **33.** -2 **35.** no solution   **37.** 2   **39.** 36   **41.** no solution
43. $\dfrac{3}{2}$ **45.** 16   **47.** 3   **49.** 12   **51.** 3, 1   **53.** -1 **55.** $3x - 8 = 19; x = 9$ **57.** $2(2x + x) = 24$; length $= 8$ in.   **59.** 4, 7
61. answers may vary   **63. a.** 3.2, 10, 31.6   **b.** no   **65.** 7.30   **67.** 0.76

The Bigger Picture
1. $2\sqrt{14}$ **2.** $\dfrac{2x^2\sqrt{5x}}{7}$ **3.** $-15x^5y^{11}$ **4.** $\dfrac{\sqrt{110}}{11}$ **5.** $2(\sqrt{5} + 1)$ or $2\sqrt{5} + 2$ **6.** $5x^2 - 19$ **7.** -1 **8.** $-\dfrac{5}{2}$
9. $\left\{y \mid y \geq -\dfrac{4}{11}\right\}$ **10.** 6, -7 **11.** $\dfrac{28}{3}$ **12.** 2

Exercise Set 8.6   **1.** $\sqrt{13}$; 3.61   **3.** $3\sqrt{3}$; 5.20   **5.** 25   **7.** $\sqrt{22}$; 4.69   **9.** $3\sqrt{17}$; 12.37   **11.** $\sqrt{41}$; 6.40   **13.** $4\sqrt{2}$; 5.66
15. $3\sqrt{10}$; 9.49   **17.** 20.6 ft   **19.** 11.7 ft   **21.** 24 cu ft   **23.** 54 mph   **25.** 27 mph   **27.** 59.1 km   **29.** 60.2 km   **31.** 3, -3
33. 10, -10 **35.** 8, -8 **37.** $y = 2\sqrt{10}; x = 2\sqrt{10} - 4$ **39.** 201 miles   **41.** answers may vary

Chapter 8 Vocabulary Check   **1.** like radicals   **2.** index; radicand; radical   **3.** conjugate   **4.** principal square root
5. rationalizing the denominator

Chapter 8 Review   **1.** 9   **2.** -7 **3.** 3   **4.** 3   **5.** $-\dfrac{3}{8}$ **6.** $\dfrac{2}{3}$ **7.** 2   **8.** -2 **9.** c   **10.** a, c   **11.** x^6 **12.** x^4 **13.** $3y$
14. $5x^2$ **15.** $2\sqrt{10}$ **16.** $2\sqrt{6}$ **17.** $3\sqrt{6}$ **18.** $2\sqrt{22}$ **19.** $x^2\sqrt{x}$ **20.** $y^3\sqrt{y}$ **21.** $2x\sqrt{5}$ **22.** $5y^2\sqrt{2}$ **23.** $3\sqrt[3]{2}$
24. $2\sqrt[3]{11}$ **25.** $\dfrac{3\sqrt{2}}{5}$ **26.** $\dfrac{5\sqrt{3}}{8}$ **27.** $-\dfrac{5\sqrt{2}}{3}$ **28.** $-\dfrac{2\sqrt{3}}{7}$ **29.** $\dfrac{\sqrt{11}}{x}$ **30.** $\dfrac{\sqrt{7}}{y^2}$ **31.** $\dfrac{y^2\sqrt{y}}{10}$ **32.** $\dfrac{x\sqrt{x}}{9}$ **33.** $-3\sqrt{2}$
34. $-5\sqrt{3}$ **35.** $4\sqrt{5} + 4\sqrt{6}$ **36.** $-2\sqrt{7} + 2\sqrt{2}$ **37.** $5\sqrt{7} + 2\sqrt{14}$ **38.** $9\sqrt{3} - 4$ **39.** $\dfrac{\sqrt{5}}{6}$ **40.** $\dfrac{9\sqrt{11}}{20}$ **41.** $10 - x\sqrt{5}$
42. $2\sqrt{2x} - \sqrt{3x}$ **43.** $3\sqrt{2}$ **44.** $5\sqrt{3}$ **45.** $\sqrt{10} - \sqrt{14}$ **46.** $\sqrt{55} + \sqrt{15}$ **47.** $3\sqrt{2} - 5\sqrt{3} + 2\sqrt{6} - 10$ **48.** $2 - 2\sqrt{5}$
49. $x - 4\sqrt{x} + 4$ **50.** $y + 8\sqrt{y} + 16$ **51.** 3   **52.** 2   **53.** $2\sqrt{5}$ **54.** $4\sqrt{2}$ **55.** $x\sqrt{15x}$ **56.** $3x^2\sqrt{2}$ **57.** $\dfrac{\sqrt{22}}{11}$
58. $\dfrac{\sqrt{39}}{13}$ **59.** $\dfrac{\sqrt{30}}{6}$ **60.** $\dfrac{\sqrt{70}}{10}$ **61.** $\dfrac{\sqrt{5x}}{5x}$ **62.** $\dfrac{5\sqrt{3y}}{3y}$ **63.** $\dfrac{\sqrt{3x}}{x}$ **64.** $\dfrac{\sqrt{6y}}{y}$ **65.** $3\sqrt{5} + 6$ **66.** $8\sqrt{10} + 24$
67. $\dfrac{\sqrt{6} + \sqrt{2} + \sqrt{3} + 1}{2}$ **68.** $\sqrt{15} - 2\sqrt{3} - 2\sqrt{5} + 4$ **69.** $\dfrac{10\sqrt{x} - 50}{x - 25}$ **70.** $\dfrac{8\sqrt{x} + 8}{x - 1}$ **71.** 18   **72.** 13   **73.** 25

74. no solution　**75.** 12　**76.** 5　**77.** 1　**78.** 9　**79.** $2\sqrt{14}$; 7.48　**80.** $\sqrt{117}$; 10.82　**81.** $4\sqrt{34}$ ft; 23.33 ft

82. $5\sqrt{3}$ in.; 8.66 in.　**83.** 2.4 in.　**84.** 144π sq in.　**85.** 12　**86.** -4　**87.** $4x^8$　**88.** $2x^{12}$　**89.** $3x^3\sqrt{2x}$　**90.** $4y^3\sqrt{3}$

91. $\dfrac{y^2}{9}$　**92.** $\dfrac{x^4\sqrt{x}}{3}$　**93.** $7\sqrt{3}$　**94.** $5\sqrt{7}-3$　**95.** $-\dfrac{\sqrt{3}}{4}$　**96.** $4x\sqrt{5x}$　**97.** $7\sqrt{2}$　**98.** $3\sqrt{3}-\sqrt{6}$

99. $\sqrt{10}-\sqrt{2}+4\sqrt{5}-4$　**100.** $x+6\sqrt{x}+9$　**101.** $2\sqrt{6}$　**102.** $2x$　**103.** $\dfrac{\sqrt{14}}{7}$　**104.** $\dfrac{3\sqrt{2x}}{2x}$　**105.** $\dfrac{3\sqrt{x}+18}{x-36}$

106. $\dfrac{\sqrt{35}-3\sqrt{7}-5\sqrt{5}+15}{-4}$　**107.** 1　**108.** 13　**109.** 14　**110.** 9　**111.** $\sqrt{58}$; 7.62　**112.** $4\sqrt{2}$ in.; 5.66 in.

Chapter 8 Test　**1.** 4　**2.** 5　**3.** 3　**4.** $\dfrac{3}{4}$　**5.** not a real number　**6.** x^5　**7.** $3\sqrt{6}$　**8.** $2\sqrt{23}$　**9.** $y^3\sqrt{y}$　**10.** $2x^4\sqrt{6}$　**11.** 3

12. $2\sqrt[3]{2}$　**13.** $\dfrac{\sqrt{5}}{4}$　**14.** $\dfrac{y\sqrt{y}}{5}$　**15.** $-2\sqrt{13}$　**16.** $\sqrt{2}+2\sqrt{3}$　**17.** $\dfrac{7\sqrt{3}}{10}$　**18.** $7\sqrt{2}$　**19.** $2\sqrt{3}-\sqrt{10}$　**20.** $x-\sqrt{x}-6$

21. $\sqrt{5}$　**22.** $2x\sqrt{5x}$　**23.** $\dfrac{\sqrt{6}}{3}$　**24.** $\dfrac{8\sqrt{5y}}{5y}$　**25.** $4\sqrt{6}-8$　**26.** $\dfrac{3+\sqrt{x}}{9-x}$　**27.** 9　**28.** 5　**29.** 9　**30.** $4\sqrt{5}$ in.　**31.** 2.19 m

Cumulative Review　**1.** 28; Sec. 1.6, Ex. 3　**2.** -46.8; Sec. 1.6　**3.** $-\dfrac{8}{21}$; Sec. 1.6, Ex. 4　**4.** -18; Sec. 1.6　**5.** 2; Sec. 2.3, Ex. 1　**6.** 15; Sec. 2.3

7. a. 17%　**b.** 21%　**c.** 43 American travelers; Sec. 2.6, Ex. 3　**8. a.** $\dfrac{3}{2}$　**b.** 9; Sec. 1.3　**9. a.** 102,000　**b.** 0.007358　**c.** 84,000,000

d. 0.00003007; Sec. 3.2, Ex. 18　**10. a.** 7.2×10^6　**b.** 3.08×10^{-4}; Sec. 3.2　**11.** $6x^2-11x-10$; Sec. 3.5; Ex. 7b

12. $49x^2+14x+1$; Sec. 3.5　**13.** $(y+2)(x+3)$; Sec. 4.1; Ex. 10　**14.** $(y^2+5)(x-1)$; Sec. 4.1　**15.** $(3x+2)(x+3)$; Sec. 4.3, Ex. 1

16. $3(x+2)(x+3)$; Sec. 4.2　**17. a.** $x=3$　**b.** $x=2,x=1$　**c.** none; Sec. 5.1, Ex. 2　**18.** $\dfrac{2x+1}{x-3}$; Sec. 5.1　**19.** $\dfrac{x+2}{x}$; Sec. 5.1, Ex. 5

20. $\dfrac{4x^3}{5}$; Sec. 5.2　**21. a.** 0　**b.** $\dfrac{15+14x}{50x^2}$; Sec. 5.4, Ex. 1　**22.** $y=-2x+4$; Sec. 6.5　**23.** $-\dfrac{17}{5}$; Sec. 5.5, Ex. 4　**24.** $2a^2+10a-5$; Sec. 3.3

25. ; Sec. 6.3, Ex. 7　**26.**

x	y
0	6
4	-2
3	0

; Sec. 6.1　**27.** $y=\dfrac{1}{4}x-3$; Sec. 6.5, Ex. 1

28. $y=-\dfrac{1}{2}x+\dfrac{11}{2}$; Sec. 6.5　**29.** $(3,1)$; Sec. 7.3, Ex. 5　**30.** $\left(\dfrac{2}{3},\dfrac{1}{2}\right)$; Sec. 7.3　**31.** Alfredo: 3.25 mph; Louisa: 4.25 mph; Sec. 7.4, Ex. 3

32. 20 mph, 35 mph; Sec. 7.4　**33.** 1; Sec. 8.1, Ex. 6　**34.** 11; Sec. 8.1　**35.** -3; Sec. 8.1, Ex. 7　**36.** $\dfrac{1}{2}$; Sec. 8.1　**37.** $\dfrac{1}{5}$; Sec. 8.1, Ex. 8

38. $\dfrac{5}{12}$; Sec. 8.1　**39.** $3\sqrt{6}$; Sec. 8.2, Ex. 1　**40.** $3\sqrt{7}$; Sec. 8.2　**41.** $10\sqrt{2}$; Sec. 8.2, Ex. 3　**42.** $10\sqrt{5}$; Sec. 8.2

43. $4\sqrt{3}$; Sec. 8.3, Ex. 6　**44.** $x-25$; Sec. 8.4　**45.** $2x-4x^2\sqrt{x}$; Sec. 8.3, Ex. 8　**46.** $10+4\sqrt{6}$; Sec. 8.4　**47.** $\dfrac{2\sqrt{7}}{7}$; Sec. 8.4, Ex. 10

48. $3x$; Sec. 5.7　**49.** $\dfrac{1}{2}$; Sec. 8.5, Ex. 2　**50.** $\dfrac{5}{2}$; Sec. 8.5

CHAPTER 9 Quadratic Equations

Exercise Set 9.1　**1.** ±7　**3.** $-5,3$　**5.** ±4　**7.** ±3　**9.** $-5,-2$　**11.** ±8　**13.** $\pm\sqrt{21}$　**15.** $\pm\dfrac{1}{5}$　**17.** no real solution

19. $\pm\dfrac{\sqrt{39}}{3}$　**21.** $\pm\dfrac{2\sqrt{7}}{7}$　**23.** $\pm\sqrt{2}$　**25.** $12,-2$　**27.** $-2\pm\sqrt{7}$　**29.** $1,0$　**31.** $-2\pm\sqrt{10}$　**33.** $\dfrac{8}{3},-4$　**35.** no real solution

37. $\dfrac{11\pm5\sqrt{2}}{2}$　**39.** $\dfrac{7\pm4\sqrt{2}}{3}$　**41.** $\pm\sqrt{2}$　**43.** $-6\pm2\sqrt{6}$　**45.** $\pm\sqrt{10}$　**47.** $\dfrac{1\pm\sqrt{5}}{4}$　**49.** $\pm2\sqrt{3}$

51. $\dfrac{-8\pm3\sqrt{5}}{-3}$ or $\dfrac{8\pm3\sqrt{5}}{3}$　**53.** 2.3 sec　**55.** 4.6 seconds　**57.** $2\sqrt{5}$ in. ≈4.47 in.　**59.** 177 meters　**61.** $(x+3)^2$　**63.** $(x-2)^2$

65. answers may vary　**67.** $2,-6$　**69.** $r=6$ in.　**71.** ±1.33　**73.** $x=7$, which is 2006

Mental Math　**1.** 16　**3.** 100　**5.** 49

Exercise Set 9.2　**1.** $-6,-2$　**3.** $-1\pm2\sqrt{2}$　**5.** $0,6$　**7.** $\dfrac{-5\pm\sqrt{53}}{2}$　**9.** $1\pm\sqrt{2}$　**11.** $-1,-4$　**13.** $-2,4$　**15.** no real solution

17. $\dfrac{3\pm\sqrt{19}}{2}$　**19.** $-2\pm\dfrac{\sqrt{6}}{2}$　**21.** $-3\pm\sqrt{34}$　**23.** $\dfrac{3\pm\sqrt{21}}{2}$　**25.** $\dfrac{1}{2},1$　**27.** $-6,3$　**29.** no real solution　**31.** $2,-6$　**33.** $-\dfrac{1}{2}$

35. -1　**37.** $3+2\sqrt{5}$　**39.** $\dfrac{1-3\sqrt{2}}{2}$　**41.** answers may vary　**43. a.** $-3\pm\sqrt{11}$　**b.** answers may vary　**45.** $k=8$ or $k=-8$

47. 8 years, or 2009　**49.** $-6,-2$　**51.** $\approx-0.68,3.68$

Mental Math   **1.** $a = 2, b = 5, c = 3$ **3.** $a = 10, b = -13, c = -2$ **5.** $a = 1, b = 0, c = -6$

Exercise Set 9.3   **1.** $x = 2, 1$ **3.** $k = \dfrac{-7 \pm \sqrt{37}}{6}$ **5.** $x = \pm\dfrac{\sqrt{3}}{2}$ **7.** no real solution   **9.** $y = 10, -3$ **11.** $x = \pm\sqrt{5}$

13. $m = -3, 4$ **15.** $x = -2 \pm \sqrt{7}$ **17.** $x = \dfrac{-9 \pm \sqrt{129}}{12}$ **19.** $p = \dfrac{4 \pm \sqrt{2}}{7}$ **21.** $x = 3 \pm \sqrt{7}$ **23.** $x = \dfrac{3 \pm \sqrt{3}}{2}$

25. $x = \dfrac{1}{3}, -1$ **27.** $y = \dfrac{3 \pm \sqrt{13}}{4}$ **29.** $y = \dfrac{1}{5}, -\dfrac{3}{4}$ **31.** no real solution   **33.** $m = 1 \pm \sqrt{2}$ **35.** no real solution

37. $p = -\dfrac{1}{2}, -\dfrac{3}{4}$ **39.** $x = \dfrac{7 \pm \sqrt{129}}{20}$ **41.** $x = \dfrac{11 \pm \sqrt{129}}{4}$ **43.** $z = \dfrac{1 \pm \sqrt{2}}{5}$ **45.** $\pm\sqrt{7}; -2.6, 2.6$ **47.** $-3 \pm 2\sqrt{2}; -5.8, -0.2$

49. $\dfrac{9 \pm \sqrt{97}}{2}; 9.4, -0.4$ **51.** $\dfrac{1 \pm \sqrt{7}}{3}; 1.2, -0.5$ **53.**   **55.**   **57.** c   **59.** b

61. 10.3 ft by 4.9 ft   **63.** $x = \dfrac{-3\sqrt{2} \pm \sqrt{38}}{2}$ **65.** answers may vary   **67.** $-0.9, 0.2$ **69.** 8.1 sec   **71.** 2008

The Bigger Picture   **1.** -1.8 **2.** -5 **3.** -9 **4.** $10x + 2$ **5.** $\dfrac{1}{4}x^2 - 25$ **6.** $3x + 1 - \dfrac{4}{x}$ **7.** $\dfrac{x + 2}{x - 4}$ **8.** $\dfrac{x^2 + 8x - 50}{(x - 10)(x + 3)}$

9. $5\sqrt{2}$ **10.** $b\sqrt{10a}$ **11.** $\dfrac{\sqrt{6}}{3}$ **12.** $\dfrac{x + 3}{7}$ **13.** $\dfrac{-3 \pm \sqrt{29}}{2}$ **14.** 6   **15.** $\{x \mid x \ge -2.8\}$ **16.** $\dfrac{1}{2}, -8$ **17.** 7   **18.** 27

Integrated Review   **1.** $x = 2, \dfrac{1}{5}$ **2.** $x = \dfrac{2}{5}, -3$ **3.** $x = 1 \pm \sqrt{2}$ **4.** $x = 3 \pm \sqrt{2}$ **5.** $a = \pm 2\sqrt{5}$ **6.** $a = \pm 6\sqrt{2}$

7. no real solution   **8.** no real solution   **9.** $x = 2$ **10.** $x = 3$ **11.** $p = 3$ **12.** $p = \dfrac{7}{2}$ **13.** $y = \pm 2$ **14.** $y = \pm 3$

15. $x = 1, 2$ **16.** $x = -3, -4$ **17.** $z = 0, -5$ **18.** $z = \dfrac{8}{3}, 0$ **19.** $x = \dfrac{3 \pm \sqrt{7}}{5}$ **20.** $x = \dfrac{3 \pm \sqrt{5}}{2}$ **21.** $m = \dfrac{3}{2}, -1$

22. $m = \dfrac{2}{5}, -2$ **23.** $x = \dfrac{5 \pm \sqrt{105}}{20}$ **24.** $x = \dfrac{-1 \pm \sqrt{3}}{4}$ **25.** $x = 5, \dfrac{7}{4}$ **26.** $x = 1, \dfrac{7}{9}$ **27.** $x = \dfrac{7 \pm 3\sqrt{2}}{5}$ **28.** $x = \dfrac{5 \pm 5\sqrt{3}}{4}$

29. $z = \dfrac{7 \pm \sqrt{193}}{6}$ **30.** $z = \dfrac{-7 \pm \sqrt{193}}{12}$ **31.** $x = 11, -10$ **32.** $x = 7, -8$ **33.** $x = 4, -\dfrac{2}{3}$ **34.** $x = 2, -\dfrac{4}{5}$ **35.** $x = 0.5, 0.1$

36. $x = 0.3, -0.2$ **37.** $x = \dfrac{11 \pm \sqrt{41}}{20}$ **38.** $x = \dfrac{11 \pm \sqrt{41}}{40}$ **39.** $z = \dfrac{4 \pm \sqrt{10}}{2}$ **40.** $z = \dfrac{5 \pm \sqrt{185}}{4}$ **41.** answers may vary

Calculator Explorations   **1.** $x = -0.41, 7.41$ **3.** $x = 0.91, 2.38$ **5.** $x = -0.39, 0.84$

Exercise Set 9.4   **1.**   **3.**   **5.**   **7.**

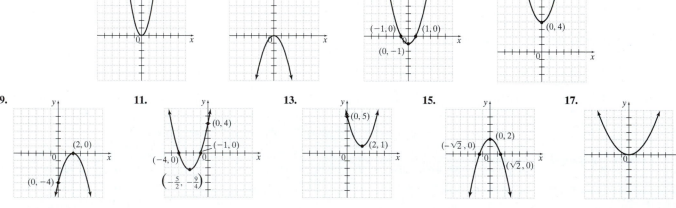

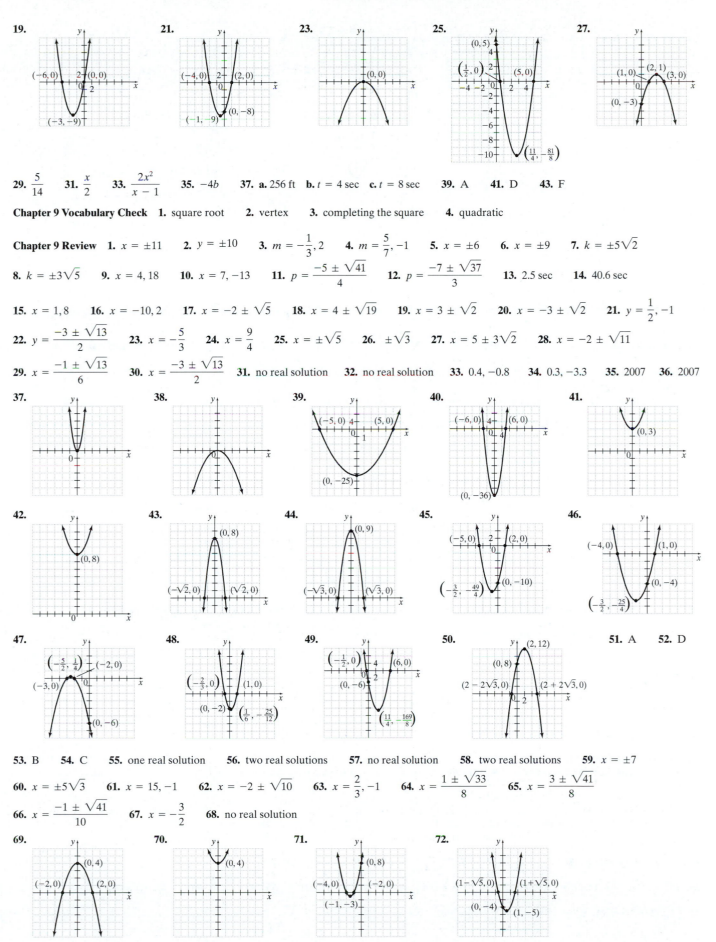

19. **21.** **23.** **25.** **27.**

29. $\dfrac{5}{14}$ **31.** $\dfrac{x}{2}$ **33.** $\dfrac{2x^2}{x-1}$ **35.** $-4b$ **37. a.** 256 ft   **b.** $t = 4$ sec   **c.** $t = 8$ sec   **39.** A   **41.** D   **43.** F

Chapter 9 Vocabulary Check   **1.** square root   **2.** vertex   **3.** completing the square   **4.** quadratic

Chapter 9 Review   **1.** $x = \pm 11$ **2.** $y = \pm 10$ **3.** $m = -\dfrac{1}{3}, 2$ **4.** $m = \dfrac{5}{7}, -1$ **5.** $x = \pm 6$ **6.** $x = \pm 9$ **7.** $k = \pm 5\sqrt{2}$

8. $k = \pm 3\sqrt{5}$ **9.** $x = 4, 18$ **10.** $x = 7, -13$ **11.** $p = \dfrac{-5 \pm \sqrt{41}}{4}$ **12.** $p = \dfrac{-7 \pm \sqrt{37}}{3}$ **13.** 2.5 sec   **14.** 40.6 sec

15. $x = 1, 8$ **16.** $x = -10, 2$ **17.** $x = -2 \pm \sqrt{5}$ **18.** $x = 4 \pm \sqrt{19}$ **19.** $x = 3 \pm \sqrt{2}$ **20.** $x = -3 \pm \sqrt{2}$ **21.** $y = \dfrac{1}{2}, -1$

22. $y = \dfrac{-3 \pm \sqrt{13}}{2}$ **23.** $x = -\dfrac{5}{3}$ **24.** $x = \dfrac{9}{4}$ **25.** $x = \pm\sqrt{5}$ **26.** $\pm\sqrt{3}$ **27.** $x = 5 \pm 3\sqrt{2}$ **28.** $x = -2 \pm \sqrt{11}$

29. $x = \dfrac{-1 \pm \sqrt{13}}{6}$ **30.** $x = \dfrac{-3 \pm \sqrt{13}}{2}$ **31.** no real solution   **32.** no real solution   **33.** $0.4, -0.8$ **34.** $0.3, -3.3$ **35.** 2007   **36.** 2007

37. **38.** **39.** **40.** **41.**

42. **43.** **44.** **45.** **46.**

47. **48.** **49.** **50.** **51.** A   **52.** D

53. B   **54.** C   **55.** one real solution   **56.** two real solutions   **57.** no real solution   **58.** two real solutions   **59.** $x = \pm 7$

60. $x = \pm 5\sqrt{3}$ **61.** $x = 15, -1$ **62.** $x = -2 \pm \sqrt{10}$ **63.** $x = \dfrac{2}{3}, -1$ **64.** $x = \dfrac{1 \pm \sqrt{33}}{8}$ **65.** $x = \dfrac{3 \pm \sqrt{41}}{8}$

66. $x = \dfrac{-1 \pm \sqrt{41}}{10}$ **67.** $x = -\dfrac{3}{2}$ **68.** no real solution

69. **70.** **71.** **72.**

Chapter 9 Test   **1.** $x = \pm 20$ **2.** $x = -\dfrac{3}{2}, 7$ **3.** $k = \pm 4$ **4.** $m = \dfrac{5 \pm 2\sqrt{2}}{3}$ **5.** $x = 10, 16$ **6.** $x = \dfrac{-6 \pm 4\sqrt{3}}{3}$ **7.** $x = -2, 5$

8. $p = \dfrac{5 \pm \sqrt{37}}{6}$ **9.** $x = 1, -\dfrac{4}{3}$ **10.** $x = -1, \dfrac{5}{3}$ **11.** $x = \dfrac{7 \pm \sqrt{73}}{6}$ **12.** $x = -1, 5$ **13.** $x = 2, \dfrac{1}{3}$ **14.** $x = \dfrac{3 \pm \sqrt{7}}{2}$

15. base: 3 ft; height: 12 ft   **16.**   **17.**   **18.**

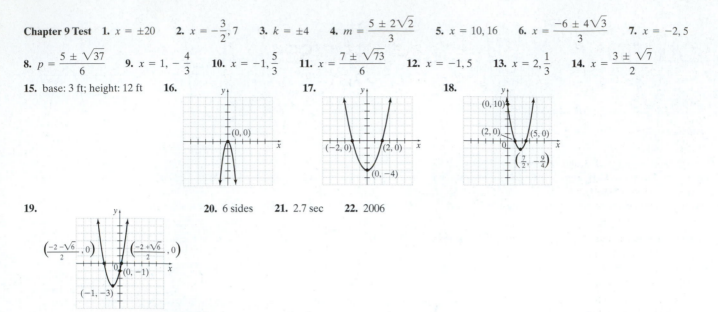

19.   **20.** 6 sides   **21.** 2.7 sec   **22.** 2006

Cumulative Review   **1.** $y = -1.6$; Sec. 2.1, Ex. 2   **2.** $x = -3$; Sec. 2.3   **3.** $t = \dfrac{16}{3}$; Sec. 2.3, Ex. 2   **4.** $\dfrac{29}{8}$; Sec. 2.3   **5.** Democratic: 208;

Republican: 223; Sec. 2.4, Ex. 4   **6.** 145, 146, 147; Sec. 2.4   **7.** 1; Sec. 3.1, Ex. 28   **8.** $-\dfrac{216x^3}{y^9}$; Sec. 3.1   **9.** 1; Sec. 3.1, Ex. 29

10. $\dfrac{a^2}{32b^3}$; Sec. 3.1   **11.** -1; Sec. 3.1, Ex. 30   **12.** 9; Sec. 3.1   **13.** $9y^2 + 12y + 4$; Sec. 3.6, Ex. 15   **14.** $x^2y - x^2 + 5y - 5$; Sec. 3.5

15. $x + 4$; Sec. 3.7, Ex. 4   **16.** $9.1a - 4$; Sec. 3.3   **17.** $(r + 6)(r - 7)$; Sec. 4.2, Ex. 4   **18. a.** -1 **b.** 30; Sec. 1.6   **19.** $(2x - 3y)(5x + 4)$;

Sec. 4.3, Ex. 4   **20.** $\dfrac{8x + 13}{(x + 2)(x - 1)}$; Sec. 5.4   **21.** $(2x - 1)(4x - 5)$; Sec. 4.4, Ex. 1   **22.** $\dfrac{2x + 5}{x - 7}$; Sec. 5.2   **23. a.** $x(2x + 7)(2x - 7)$;

Sec. 4.5, Ex. 16   **b.** $2(9x^2 + 1)(3x + 1)(3x - 1)$; Sec. 4.5, Ex. 17   **24.** -32; Sec. 5.5   **25.** $\dfrac{1}{5}, -\dfrac{3}{2}, -6$; Sec. 4.6, Ex. 8   **26. a.** $-x + 4$

b. $5y - 8$ **c.** $9.1a - 4$ **d.** $2x^2 - 2x$; Sec. 3.3   **27.** $\dfrac{x + 7}{x - 5}$; Sec. 5.1, Ex. 4   **28.** $x = -\dfrac{7}{2}, 1$; Sec. 4.6   **29.** -5; Sec. 5.6, Ex. 5

30. $\sqrt{82}$ units; Ch. 8 Group Activity   **31. a.** -3 **b.** $0; 0$ **c.** -3; Sec. 6.1, Ex. 4   **32. a.** x-int: $(4, 0)$; y-int: $(0, 1)$ **b.** x-int: $(-2, 0)$, $(0, 0)$, $(3, 0)$;

y-int: $(0, 0)$; Sec. 6.3   **33. a.** parallel   **b.** perpendicular   **c.** neither; Sec. 6.4, Ex. 7   **34.** perpendicular; Sec. 6.4   **35. a.** function

b. not a function; Sec. 6.6, Ex. 2   **36. a.** $6\sqrt{5}$ **b.** $8\sqrt{2}$ **c.** $11 - 2\sqrt{3} + 4\sqrt{2}$; Sec. 8.3   **37.** $(4, 2)$; Sec. 7.2, Ex. 1   **38.** no solution; Sec. 7.2

39. no solution; Sec. 7.3, Ex. 3   **40.** infinite number of solutions; Sec. 7.3   **41.** 6; Sec. 8.1, Ex. 1   **42.** $\dfrac{2}{5}$; Sec. 8.1   **43.** $\dfrac{3}{10}$; Sec. 8.1, Ex. 3

44. $\dfrac{4}{11}$; Sec. 8.1   **45.** $-1 + \sqrt{3}$; Sec. 8.4, Ex. 13   **46.** $\dfrac{5\sqrt{2}}{4}$; Sec. 8.4   **47.** $x = 7, -1$; Sec. 9.1, Ex. 5   **48.** $x = 4 \pm \sqrt{3}$; Sec. 9.1

49. $x = 1 \pm \sqrt{5}$; Sec. 9.3, Ex. 5   **50.** $x = -2 \pm 2\sqrt{3}$; Sec. 9.3

APPENDIX

Exercise Set Appendix B   **1.** $(a + 3)(a^2 - 3a + 9)$ **3.** $(2a + 1)(4a^2 - 2a + 1)$ **5.** $5(k + 2)(k^2 - 2k + 4)$
7. $(xy - 4)(x^2y^2 + 4xy + 16)$ **9.** $(x + 5)(x^2 - 5x + 25)$ **11.** $3x(2x - 3y)(4x^2 + 6xy + 9y^2)$ **13.** $(3 - t)(9 + 3t + t^2)$
15. $8(r - 2)(r^2 + 2r + 4)$ **17.** $(t - 7)(t^2 + 7t + 49)$ **19.** $(s - 4t)(s^2 + 4st + 16t^2)$

Exercise Set Appendix C   **1.** mean: 29, median: 28, no mode   **3.** mean: 8.1, median: 8.2, mode: 8.2   **5.** mean: 0.6, median: 0.6, mode: 0.2 and 0.6
7. mean: 370.9, median: 313.5, no mode   **9.** 1214.8 ft   **11.** 1117 ft   **13.** 6.8 sec   **15.** 6.9 sec   **17.** 85.5   **19.** 73   **21.** 70 and 71
23. 9   **25.** 21, 21, 20

Exercise Set Appendix D   **1.** $\{-9, -8, -7, -6\}$ **3.** {Tuesday, Thursday}   **5.** $\{0, 1, 2, 3, 4, \ldots\}$ **7.** $\{\ \}$ or $\varnothing$ **9.** true   **11.** true
13. true   **15.** false   **17.** true   **19.** false   **21.** $\{1, 2, 3, 4, 5, 6\}$ **23.** $\{2, 4, 6\}$ **25.** $\{1, 3, 5, 7\}$ **27.** $\varnothing$ **29.** true   **31.** false
33. false   **35.** true   **37.** true   **39.** true   **41.** true

Exercise Set Appendix E   **1.** $71°$ **3.** $19.2°$ **5.** $78\dfrac{3}{4}°$ **7.** $30°$ **9.** $149.8°$ **11.** $100\dfrac{1}{2}°$

13. $m\angle 1 = m\angle 5 = m\angle 7 = 110°$, $m\angle 2 = m\angle 3 = m\angle 4 = m\angle 6 = 70°$ **15.** $90°$ **17.** $90°$ **19.** $90°$ **21.** $45°, 90°$ **23.** $73°, 90°$

25. $50\dfrac{1}{4}°, 90°$ **27.** $x = 6$ **29.** $x = 4.5$ **31.** 10   **33.** 12

SOLUTIONS TO SELECTED EXERCISES

CHAPTER R

Exercise Set R.1

1. $9 = 1 \cdot 9, 9 = 3 \cdot 3$
The factors of 9 are 1, 3, and 9.

5. $42 = 1 \cdot 42, 42 = 2 \cdot 21, 42 = 3 \cdot 14, 42 = 6 \cdot 7$
The factors of 42 are 1, 2, 3, 6, 7, 14, 21, and 42.

9. $19 = 1 \cdot 19$
The factors of 19 are 1 and 19.

13. 39 is a composite number. Its factors are 1, 3, 13, and 39.

17. 201 is composite. Its factors are 1, 3, 67, and 201.

21. $18 = 2 \cdot 9 = 2 \cdot 3 \cdot 3$
The prime factorization of 18 is $2 \cdot 3 \cdot 3$.

25. $56 = 2 \cdot 28 = 2 \cdot 2 \cdot 14 = 2 \cdot 2 \cdot 2 \cdot 7$
The prime factorization of 56 is $2 \cdot 2 \cdot 2 \cdot 7$.

29.
$$\begin{array}{r} 5 \\ 5)\overline{25} \\ 3)\overline{75} \\ 2)\overline{150} \\ 2)\overline{300} \end{array}$$

The prime factorization of 300 is $2 \cdot 2 \cdot 3 \cdot 5 \cdot 5$.

33. $48 = 1 \cdot 48, 48 = 2 \cdot 24, 48 = 3 \cdot 16, 48 = 4 \cdot 12, 48 = 6 \cdot 8$
The factors of 48 are 1, 2, 3, 4, 6, 8, 12, 16, 24, and 48, which is choice d.

37. $6 = 2 \cdot 3$
$14 = 2 \cdot 7$
$\text{LCM} = 2 \cdot 3 \cdot 7 = 42$

41. $5 = 5$
$7 = 7$
$\text{LCM} = 5 \cdot 7 = 35$

45. $16 = 2 \cdot 2 \cdot 2 \cdot 2$
$20 = 2 \cdot 2 \cdot 5$
$\text{LCM} = 2 \cdot 2 \cdot 2 \cdot 2 \cdot 5 = 80$

49. $24 = 2 \cdot 2 \cdot 2 \cdot 3$
$36 = 2 \cdot 2 \cdot 3 \cdot 3$
$\text{LCM} = 2 \cdot 2 \cdot 2 \cdot 3 \cdot 3 = 72$

53. $2 = 2$
$3 = 3$
$7 = 7$
$\text{LCM} = 2 \cdot 3 \cdot 7 = 42$

57. $8 = 2 \cdot 2 \cdot 2$
$18 = 2 \cdot 3 \cdot 3$
$30 = 2 \cdot 3 \cdot 5$
$\text{LCM} = 2 \cdot 2 \cdot 2 \cdot 3 \cdot 3 \cdot 5 = 360$

61. True, the number 311 is a prime number.

65. $1000 = 25 \cdot 40 = 25 \cdot 5 \cdot 8$
$1125 = 25 \cdot 45 = 25 \cdot 5 \cdot 9$
$\text{LCM} = 25 \cdot 5 \cdot 8 \cdot 9 = 9000$

Exercise Set R.2

1. $\dfrac{14}{14} = 1$ since $14 \cdot 1 = 14$.

5. $\dfrac{13}{1} = 13$ since $1 \cdot 13 = 13$.

9. $\dfrac{9}{0}$ is undefined.

13. $\dfrac{2}{9} = \dfrac{2 \cdot 2}{9 \cdot 2} = \dfrac{4}{18}$

17. $\dfrac{2}{4} = \dfrac{2 \cdot 1}{2 \cdot 2} = \dfrac{1}{2}$

21. $\dfrac{3}{7}$ cannot be simplified further.

25. $\dfrac{16}{20} = \dfrac{4 \cdot 4}{4 \cdot 5} = \dfrac{4}{5}$

29. $\dfrac{120}{244} = \dfrac{4 \cdot 30}{4 \cdot 61} = \dfrac{30}{61}$

33. $\dfrac{1}{2} \cdot \dfrac{3}{4} = \dfrac{1 \cdot 3}{2 \cdot 4} = \dfrac{3}{8}$

37. $\dfrac{1}{2} \div \dfrac{7}{12} = \dfrac{1}{2} \cdot \dfrac{12}{7} = \dfrac{1 \cdot 12}{2 \cdot 7} = \dfrac{2 \cdot 6}{2 \cdot 7} = \dfrac{6}{7}$

41. $5\dfrac{1}{9} \cdot 3\dfrac{2}{3} = \dfrac{46}{9} \cdot \dfrac{11}{3} = \dfrac{506}{27} = 18\dfrac{20}{27}$

45. $\dfrac{4}{5} + \dfrac{1}{5} = \dfrac{4 + 1}{5} = \dfrac{5}{5} = 1$

49. $\dfrac{2}{3} + \dfrac{3}{7} = \dfrac{2 \cdot 7}{3 \cdot 7} + \dfrac{3 \cdot 3}{7 \cdot 3} = \dfrac{14}{21} + \dfrac{9}{21} = \dfrac{14 + 9}{21} = \dfrac{23}{21}$

53.
$$\begin{array}{r} 8\dfrac{1}{8} \quad 7\dfrac{9}{8} \\ -6\dfrac{3}{8} \quad -6\dfrac{3}{8} \\ \hline 1\dfrac{6}{8} = 1\dfrac{3}{4} \end{array}$$

57. $\dfrac{23}{105} + \dfrac{4}{105} = \dfrac{23 + 4}{105} = \dfrac{27}{105} = \dfrac{3 \cdot 9}{3 \cdot 35} = \dfrac{9}{35}$

61. $\dfrac{7}{10} \cdot \dfrac{5}{21} = \dfrac{7 \cdot 5}{10 \cdot 21} = \dfrac{7 \cdot 5}{5 \cdot 2 \cdot 7 \cdot 3} = \dfrac{1}{2 \cdot 3} = \dfrac{1}{6}$

65. $\dfrac{5}{22} - \dfrac{5}{33} = \dfrac{5 \cdot 3}{22 \cdot 3} - \dfrac{5 \cdot 2}{33 \cdot 2}$
$\qquad = \dfrac{15}{66} - \dfrac{10}{66}$
$\qquad = \dfrac{15 - 10}{66}$
$\qquad = \dfrac{5}{66}$

69. $7\dfrac{2}{5} \div \dfrac{1}{5} = \dfrac{37}{5} \div \dfrac{1}{5} = \dfrac{37}{5} \cdot \dfrac{5}{1} = \dfrac{37 \cdot 5}{5 \cdot 1} = \dfrac{37}{1} = 37$

73. $\dfrac{12}{5} - 1 = \dfrac{12}{5} - \dfrac{5}{5} = \dfrac{12 - 5}{5} = \dfrac{7}{5}$

77. $\dfrac{2}{3} - \dfrac{5}{9} + \dfrac{5}{6} = \dfrac{2 \cdot 6}{3 \cdot 6} - \dfrac{5 \cdot 2}{9 \cdot 2} + \dfrac{5 \cdot 3}{6 \cdot 3}$
$\qquad = \dfrac{12}{18} - \dfrac{10}{18} + \dfrac{15}{18}$
$\qquad = \dfrac{12 - 10 + 15}{18}$
$\qquad = \dfrac{17}{18}$

81. answers may vary

85. $1 - \dfrac{1}{3} - \dfrac{5}{12} - \dfrac{1}{6} = \dfrac{12}{12} - \dfrac{4}{12} - \dfrac{5}{12} - \dfrac{2}{12}$

$= \dfrac{12 - 4 - 5 - 2}{12}$

$= \dfrac{1}{12}$

The unknown part is $\dfrac{1}{12}$.

89. a. 36 of the 459 Federated stores were Bloomingdale's.

$\dfrac{36}{459} = \dfrac{9 \cdot 4}{9 \cdot 51} = \dfrac{4}{51}$

b. $95 + 144 = 239$ of the 459 Federated stores were either Macy's East or Macy's West. $\dfrac{239}{459}$ does not simplify further.

Exercise Set R.3

1. $0.6 = \dfrac{6}{10}$

5. $0.114 = \dfrac{114}{1000}$

9. $\begin{array}{r} 5.7 \\ + 1.13 \\ \hline 6.83 \end{array}$

13. $\begin{array}{r} 8.8 \\ - 2.3 \\ \hline 6.5 \end{array}$

17. $\begin{array}{r} 0.2 \\ \times 0.6 \\ \hline 0.12 \end{array}$

21. $\begin{array}{r} 1.68 \\ 5\overline{)\ 8.40} \\ -5 \\ \hline 3\,4 \\ -3\,0 \\ \hline 40 \\ -40 \\ \hline 0 \end{array}$

25. $\begin{array}{r} 45.02 \\ 3.006 \\ + 8.405 \\ \hline 56.431 \end{array}$

29. $0.6\overline{)42}$ $\begin{array}{r} 70 \\ 6\overline{)\ 420.} \\ -42 \\ \hline 00 \end{array}$

33. $\begin{array}{r} 5.62 \\ \times\ 7.7 \\ \hline 3\,934 \\ 39\,34 \\ \hline 43.274 \end{array}$

37. $\begin{array}{r} 16.003 \\ \times\ 5.31 \\ \hline 16\,003 \\ 480\,09 \\ 8001\,5 \\ \hline 84.97593 \end{array}$

41. 0.234 rounded to the nearest hundredth is 0.23.

45. 98,207.23 rounded to the nearest tenth is 98,207.2.

49. $\dfrac{3}{4} = \dfrac{3 \cdot 25}{4 \cdot 25} = \dfrac{75}{100} = 0.75$

53. $\dfrac{7}{16} = \dfrac{7 \cdot 625}{16 \cdot 625} = \dfrac{4375}{10,000} = 0.4375$

57. $\begin{array}{r} 4.833 \\ 6\overline{)\ 29.000} \\ -24 \\ \hline 5\,0 \\ -4\,8 \\ \hline 20 \\ -18 \\ \hline 20 \\ -18 \\ \hline 2 \end{array}$

$\dfrac{29}{6} = 4.8\overline{3} \approx 4.83$

61. $3.1\% = 0.031$

65. $200\% = 2.00$ or 2

69. $0.1\% = 0.001$

73. $0.68 = 68\%$

77. $1 = 1.00 = 100\%$

81. $1.92 = 192\%$

85. a. In the number 3.659, the place value of the 6 is tenths.
b. In the number 3.659, the place value of the 9 is thousandths.
c. In the number 3.659, the place value of the 3 is ones.

89. answers may vary

Chapter R Test

1. $\begin{array}{r} 3 \\ 3\overline{)\ 9} \\ 2\overline{)18} \\ 2\overline{)36} \\ 2\overline{)72} \end{array}$

The prime factorization of 72 is $2 \cdot 2 \cdot 2 \cdot 3 \cdot 3$.

5. $\dfrac{48}{100} = \dfrac{4 \cdot 12}{4 \cdot 25} = \dfrac{12}{25}$

9. $\dfrac{9}{10} \div 18 = \dfrac{9}{10} \div \dfrac{18}{1}$

$= \dfrac{9}{10} \cdot \dfrac{1}{18}$

$= \dfrac{9 \cdot 1}{10 \cdot 18}$

$= \dfrac{9 \cdot 1}{10 \cdot 9 \cdot 2}$

$= \dfrac{1}{20}$

13. $6\dfrac{7}{8} \div \dfrac{1}{8} = \dfrac{55}{8} \div \dfrac{1}{8} = \dfrac{55}{8} \cdot \dfrac{8}{1} = \dfrac{55 \cdot 8}{8 \cdot 1} = \dfrac{55}{1} = 55$

17. $\begin{array}{r} 7.93 \\ \times\ 1.6 \\ \hline 4758 \\ 793 \\ \hline 12.688 \end{array}$

21. $\begin{array}{r} 0.166 \\ 6\overline{)1.000} \\ -6 \\ \hline 40 \\ -36 \\ \hline 40 \\ -36 \\ \hline 4 \end{array}$

$\dfrac{1}{6} = 0.1\overline{6} \approx 0.167$

25. $\dfrac{3}{4}$ of the fresh water is ice caps and glaciers.

29. Area $= \dfrac{1}{2}$ (base)(height)

$$= \dfrac{1}{2} \cdot \dfrac{3}{4} \cdot \dfrac{1}{3}$$

$$= \dfrac{1 \cdot 3 \cdot 1}{2 \cdot 4 \cdot 3}$$

$$= \dfrac{1}{8}$$

The area is $\dfrac{1}{8}$ square foot.

CHAPTER 1

Exercise Set 1.2

1. Since 4 is to the left of 10 on the number line, $4 < 10$.

5. $6.26 = 6.26$

9. Since 32 is to the left of 212 on the number line, $32 < 212$.

13. Since $11 = 11$, the statement $11 \leq 11$ is true.

17. Comparing digits with the same place value, we have $0.0 < 0.9$. Thus the statement $5.092 < 5.902$ is true.

21. $25 \geq 20$ has the same meaning as $20 \leq 25$.

25. $-10 > -12$ has the same meaning as $-12 < -10$.

29. Five is greater than or equal to four is written as $5 \geq 4$.

33. The integer 14,494 represents 14,494 feet above sea level. The integer -282 represents 282 feet below sea level.

37. The integer 475 represents a \$475 deposit. The integer -195 represents a \$195 withdrawal.

41.

$$\underset{-4\,-3\,-2\,-1\ \ 0\ \ 1\ \ 2\ \ 3\ \ 4}{\overset{-\frac{1}{4}\,\frac{1}{3}}{\longleftrightarrow}}$$

45. 0 is a whole number, an integer, a rational number, and a real number.

49. 265 is a natural number, a whole number, an integer, a rational number, and a real number.

53. False; the rational number $\dfrac{2}{3}$ is not an integer.

57. False; the negative number $-\sqrt{2}$ is not a rational number.

61. $|8.9| = |8.9|$ since 8.9 is 8.9 units from 0 on the number line.

65. $\left|\dfrac{9}{2}\right| = \dfrac{9}{2}$ since $\dfrac{9}{2}$ is $\dfrac{9}{2}$ units from 0 on the number line.

69. $|-5| = 5$

$-4 = -4$

Since 5 is to the right of -4 on the number line, $|-5| > -4$.

73. $|-2| = 2$

$|-2.7| = 2.7$

Since 2 is to the left of 2.7 on the number line, $|-2| < |-2.7|$.

77. The apple production in 1998 was 762 thousand bushels, while the apple production in 1999 was 1548 thousand bushels. $762 < 1548$

81. Since -0.04 is to the right of -26.7 on the number line, $-0.04 > -26.7$.

85. Since the brightest star corresponds to the smallest apparent magnitude, which is -26.7, the brightest star is the sun.

Exercise Set 1.3

1. $3^5 = 3 \cdot 3 \cdot 3 \cdot 3 \cdot 3 = 243$

5. $1^5 = 1 \cdot 1 \cdot 1 \cdot 1 \cdot 1 = 1$

9. $7^2 = 7 \cdot 7 = 49$

13. $\left(\dfrac{1}{5}\right)^3 = \left(\dfrac{1}{5}\right)\left(\dfrac{1}{5}\right)\left(\dfrac{1}{5}\right) = \dfrac{1 \cdot 1 \cdot 1}{5 \cdot 5 \cdot 5} = \dfrac{1}{125}$

17. $(0.7)^3 = 0.7 \cdot 0.7 \cdot 0.7 = 0.343$

21. $5 + 6 \cdot 2 = 5 + 12 = 17$

25. $18 \div 3 \cdot 2 = 6 \cdot 2 = 12$

29. $5 \cdot 3^2 = 5 \cdot 9 = 45$

33. $\dfrac{6 - 4}{9 - 2} = \dfrac{2}{7}$

37. $\dfrac{19 - 3 \cdot 5}{6 - 4} = \dfrac{19 - 15}{2} = \dfrac{4}{2} = 2$

41. $\dfrac{3 + 3(5 + 3)}{3^2 + 1} = \dfrac{3 + 3(8)}{9 + 1} = \dfrac{3 + 24}{10} = \dfrac{27}{10}$

45. $2 + 3[10(4 \cdot 5 - 16) - 30] = 2 + 3[10(20 - 16) - 30]$

$$= 2 + 3[10(4) - 30]$$
$$= 2 + 3[40 - 30]$$
$$= 2 + 3[10]$$
$$= 2 + 30$$
$$= 32$$

49. Replace y with 3.

$3y = 3(3) = 9$

53. Replace x with 1.

$3x - 2 = 3(1) - 2 = 3 - 2 = 1$

57. Replace x with 1, y with 3, and z with 5.

$xy + z = 1(3) + 5 = 3 + 5 = 8$

61. Replace z with 3.

$5z = 5(3) = 15$

65. Replace x with 2 and y with 6.

$\dfrac{y}{x} + \dfrac{y}{x} = \dfrac{6}{2} + \dfrac{6}{2} = 3 + 3 = 6$

69. $2x + 6 = 5x - 1$

$2(0) + 6 \overset{?}{=} 5(0) - 1$

$0 + 6 \overset{?}{=} 0 - 1$

$6 = -1$ False

Since the result is false, 0 is not a solution of the given equation.

73. $x + 6 = x + 6$

$2 + 6 \overset{?}{=} 2 + 6$

$8 = 8$ True

Since the result is true, 2 is a solution of the given equation.

77. $\dfrac{1}{3}x = 9$

$\dfrac{1}{3}(27) \overset{?}{=} 9$

$9 = 9$ True

Since the result is true, 27 is a solution of the given equation.

81. Five subtracted from a number is written as $x - 5$.

85. Three times a number, increased by 22 is written as $3x + 22$.

89. Three is not equal to four divided by two is written as $3 \neq 4 \div 2$.

93. The product 7.6 and a number is 17 is written as $7.6x = 17$.

97. No, the parentheses are not necessary.

101.

Length, l	Width, w	Perimeter of Rectangle: $2l + 2w$	Area of Rectangle: lw
4 in.	3 in.	$2l + 2w$ $= 2(4 \text{ in.}) + 2(3 \text{ in.})$ $= 8 \text{ in.} + 6 \text{ in.}$ $= 14 \text{ in.}$	lw $= (4 \text{ in.})(3 \text{ in.})$ $= 12 \text{ sq in.}$
6 in.	1 in.	$2l + 2w$ $= 2(6 \text{ in.}) + 2(1 \text{ in.})$ $= 12 \text{ in.} + 2 \text{ in.}$ $= 14 \text{ in.}$	lw $= (6 \text{ in.})(1 \text{ in.})$ $= 6 \text{ sq in.}$
5 in.	2 in.	$2l + 2w$ $= 2(5 \text{ in.}) + 2(2 \text{ in.})$ $= 10 \text{ in.} + 4 \text{ in.}$ $= 14 \text{ in.}$	lw $= (5 \text{ in.})(2 \text{ in.})$ $= 10 \text{ sq in.}$

105. answers may vary

109. answers may vary

Exercise Set 1.4

1. $6 + (-3) = 3$

5. $8 + (-7) = 1$

9. $-2 + (-3) = -5$

13. $-7 + 3 = -4$

17. $5 + (-7) = -2$

21. $27 + (-46) = -19$

25. $-33 + (-14) = -47$

29. $117 + (-79) = 38$

33. $-\dfrac{3}{8} + \dfrac{5}{8} = \dfrac{2}{8} = \dfrac{2 \cdot 1}{2 \cdot 4} = \dfrac{1}{4}$

37. $-\dfrac{7}{10} + \left(-\dfrac{3}{5}\right) = -\dfrac{7}{10} + \left(-\dfrac{6}{10}\right) = -\dfrac{13}{10}$

41. $-15 + 9 + (-2) = -6 + (-2) = -8$

45. $-23 + 16 + (-2) = -7 + (-2) = -9$

49. $6 + (-4) + 9 = 2 + 9 = 11$

53. $|9 + (-12)| + |-16| = |-3| + |-16| = 3 + 16 = 19$

57. $-38 + 12 = -26$
The sum of -38 and 12 is -26.

61. The additive inverse of -2 is 2.

65. Since $|-6| = 6$, the additive inverse of $|-6|$ is -6.

69. $-(-7) = 7$

73. $-(-5z) = 5z$

77. $-35 + 142 = 107$
The highest recorded temperature in Massachusetts was $107°F$.

81. $-2.50 + (-0.86) = -3.36$
The combined change is -3.36 points.

85. $23 + (-581) + 93 + 155 = -604 + 93 + 155$
$\qquad\qquad\qquad\qquad\qquad = -511 + 155$
$\qquad\qquad\qquad\qquad\qquad = -356$
The total net income for fiscal year 2004 was $-\$356$ million.

89. The highest bar corresponds to July, so the highest low temperature was in July.

93. $\dfrac{-9.1 + 14.4 + 8.8}{3} = \dfrac{5.3 + 8.8}{3} = \dfrac{14.1}{3} = 4.7$

The average for the months of April, May, and October is $4.7°F$.

97. If p is a positive number, then $p + p$ is a positive number.

101. answers may vary

Exercise Set 1.5

1. $-6 - 4 = -6 + (-4) = -10$

5. $16 - (-3) = 16 + 3 = 19$

9. $-26 - (-18) = -26 + 18 = -8$

13. $16 - (-21) = 16 + 21 = 37$

17. $-44 - 27 = -44 + (-27) = -71$

21. $-\dfrac{3}{11} - \left(-\dfrac{5}{11}\right) = -\dfrac{3}{11} + \dfrac{5}{11} = \dfrac{2}{11}$

25. $-2.6 - (-6.7) = -2.6 + 6.7 = 4.1$

29. $-\dfrac{1}{6} - \dfrac{3}{4} = -\dfrac{1}{6} + \left(-\dfrac{3}{4}\right)$
$\qquad\qquad = -\dfrac{2}{12} + \left(-\dfrac{9}{12}\right)$
$\qquad\qquad = -\dfrac{11}{12}$

33. $0 - 8.92 = 0 + (-8.92) = -8.92$

37. $-6 - (-1) = -6 + 1 = -5$
The difference between -6 and -1 is -5.

41. $-8 - 15 = -8 + (-15) = -23$
-8 decreased by 15 is -23.

45. $5 - 9 + (-4) - 8 - 8 = 5 + (-9) + (-4) + (-8) + (-8)$
$\qquad\qquad\qquad\qquad\qquad = -4 + (-4) + (-8) + (-8)$
$\qquad\qquad\qquad\qquad\qquad = -8 + (-8) + (-8)$
$\qquad\qquad\qquad\qquad\qquad = -16 + (-8)$
$\qquad\qquad\qquad\qquad\qquad = -24$

49. $3^3 - 8 \cdot 9 = 27 - 8 \cdot 9 = 27 - 72 = 27 + (-72) = -45$

53. $(3 - 6) + 4^2 = [3 + (-6)] + 16 = [-3] + 16 = 13$

57. $|-3| + 2^2 + [-4 - (-6)] = |-3| + 4 + [-4 + 6]$
$\qquad\qquad\qquad\qquad\qquad = 3 + 4 + 2$
$\qquad\qquad\qquad\qquad\qquad = 7 + 2$
$\qquad\qquad\qquad\qquad\qquad = 9$

61. Replace x with -5 and y with 4.
$\dfrac{9 - x}{y + 6} = \dfrac{9 - (-5)}{4 + 6} = \dfrac{9 + 5}{10} = \dfrac{14}{10} = \dfrac{2 \cdot 7}{2 \cdot 5} = \dfrac{7}{5}$

65. Replace x with -5 and y with 4.
$y^2 - x = 4^2 - (-5) = 16 + 5 = 21$

69. $\quad x - 9 = 5$
$\quad -4 - 9 \overset{?}{=} 5$
$-4 + (-9) \overset{?}{=} 5$
$\qquad\quad -13 = 5 \quad$ False
Since the result is false, -4 is not a solution of the given equation.

73. $\quad -x - 13 = -15$
$\quad -2 - 13 \overset{?}{=} -15$
$-2 + (-13) \overset{?}{=} -15$
$\qquad\quad -15 = -15 \quad$ True
Since the result is true, 2 is a solution of the given equation.

77. $2 - 5 - 20 = 2 + (-5) + (-20) = -3 + (-20) = -23$
The total change in yardage is a loss of 23 yards.

81. Complementary angles sum to $90°$.
$\qquad 90° - 60° = x$
$\quad 90° + (-60°) = x$
$\qquad\qquad 30° = x$

85. $19,340 - (-512) = 19,340 + 512 = 19,852$
Mt. Kilimanjaro is $19,852$ feet higher than Lake Assal.

89.

Month	Monthly Increase or Decrease
February	$-23.7 - (-19.3) = -23.7 + 19.3 = -4.4°$
March	$-21.1 - (-23.7) = -21.1 + 23.7 = 2.6°$
April	$-9.1 - (-21.1) = -9.1 + 21.1 = 12°$
May	$14.4 - (-9.1) = 14.4 + 9.1 = 23.5°$
June	$29.7 - 14.4 = 29.7 + (-14.4) = 15.3°$
July	$33.6 - 29.7 = 33.6 + (-29.7) = 3.9°$
August	$33.3 - 33.6 = 33.3 + (-33.6) = -0.3°$
September	$27.0 - 33.3 = 27.0 + (-33.3) = -6.3°$
October	$8.8 - 27.0 = 8.8 + (-27.0) = -18.2°$
November	$-6.9 - 8.8 = -6.9 + (-8.8) = -15.7°$
December	$-17.2 - (-6.9) = -17.2 + 6.9 = -10.3°$

93. answers may vary

97. true; answers may vary

101. sometimes positive and sometimes negative

Exercise Set 1.6

1. $-6(4) = -24$

5. $-5(-10) = 50$

9. $-\dfrac{1}{2}\left(-\dfrac{3}{5}\right) = \dfrac{3}{10}$

13. $(-1)(-3)(-5) = 3(-5) = -15$

17. $(-4)^2 = (-4)(-4) = 16$

21. $\left(-\dfrac{3}{4}\right)^2 = \left(-\dfrac{3}{4}\right)\left(-\dfrac{3}{4}\right) = \dfrac{9}{16}$

25. The reciprocal of $\dfrac{2}{3}$ is $\dfrac{3}{2}$.

29. The reciprocal of $-\dfrac{3}{11}$ is $-\dfrac{11}{3}$.

33. $\dfrac{18}{-2} = -9$

37. $\dfrac{0}{-4} = 0$

41. $\dfrac{6}{7} \div \left(-\dfrac{1}{3}\right) = \dfrac{6}{7} \cdot \left(-\dfrac{3}{1}\right) = -\dfrac{18}{7}$

45. $(-8)(-8) = 64$

49. $\dfrac{-12}{-4} = 3$

53. $(-5)^3 = (-5)(-5)(-5) = 25(-5) = -125$

57. $\left(-\dfrac{3}{4}\right)\left(-\dfrac{8}{9}\right) = \dfrac{24}{36} = \dfrac{2 \cdot 12}{3 \cdot 12} = \dfrac{2}{3}$

61. $-2.1(-0.4) = 0.84$

65. $(-3)^4 = (-3)(-3)(-3)(-3)$
$= 9(-3)(-3)$
$= -27(-3)$
$= 81$

69. $-11 \cdot 11 = -121$

73. $-9 - 10 = -9 + (-10) = -19$

77. $7(-12) = -84$

81. $\dfrac{-9(-3)}{-6} = \dfrac{27}{-6} = -\dfrac{9 \cdot 3}{2 \cdot 3} = -\dfrac{9}{2}$

85. $-7(-2) - 3(-1) = 14 - (-3) = 14 + 3 = 17$

89. $\dfrac{-6^2 + 4}{-2} = \dfrac{-36 + 4}{-2} = \dfrac{-32}{-2} = 16$

93. $\dfrac{22 + (3)(-2)^2}{-5 - 2} = \dfrac{22 + 3(4)}{-5 + (-2)}$
$= \dfrac{22 + 12}{-7}$
$= \dfrac{34}{-7}$
$= -\dfrac{34}{7}$

97. $\dfrac{6 - 2(-3)}{4 - 3(-2)} = \dfrac{6 - (-6)}{4 - (-6)} = \dfrac{6 + 6}{4 + 6} = \dfrac{12}{10} = \dfrac{2 \cdot 6}{2 \cdot 5} = \dfrac{6}{5}$

101. $\dfrac{-7(-1) + (-3)4}{(-2)(5) + (-6)(-8)} = \dfrac{7 + (-12)}{-10 + 48} = \dfrac{-5}{38} = -\dfrac{5}{38}$

105. Replace x with -5 and y with -3.
$\dfrac{6 - y}{x - 4} = \dfrac{6 - (-3)}{-5 - 4} = \dfrac{6 + 3}{-5 + (-4)} = \dfrac{9}{-9} = -1$

109. Replace x with -5 and y with -3.
$\dfrac{x^2 + y}{3y} = \dfrac{(-5)^2 + (-3)}{3(-3)} = \dfrac{25 + (-3)}{-9} = \dfrac{22}{-9} = -\dfrac{22}{9}$

113. $\dfrac{x}{5} + 2 = -1$
$\dfrac{15}{5} + 2 \overset{?}{=} -1$
$3 + 2 \overset{?}{=} -1$
$\quad 5 = -1$ False
Since the result is false, 15 is not a solution of the given equation.

117. True since the product of an odd number of negative numbers is negative.

121. $2(-81) = -162$
The surface temperature of Jupiter is $-162°F$.

125. Since $\dfrac{1}{1} = 1$ and $\dfrac{1}{-1} = -1$, both 1 and -1 are their own reciprocals.

129. $-8(-5) + (-1) = 40 + (-1) = 39$

Exercise Set 1.7

1. $x + 16 = 16 + x$ by the commutative property of addition.

5. $xy = yx$ by the commutative property of multiplication.

9. $(xy) \cdot z = x \cdot (yz)$ by the associative property of multiplication.

13. $4 \cdot (ab) = (4a) \cdot b$ by the associative property of multiplication.

17. $8 + (9 + b) = (8 + 9) + b = 17 + b$

21. $\dfrac{1}{5}(5y) = \left(\dfrac{1}{5} \cdot 5\right)y = 1y = y$

25. $-9(8x) = (-9 \cdot 8)x = -72x$

29. $-\dfrac{1}{2}(5x) = \left(-\dfrac{1}{2} \cdot 5\right)x = -\dfrac{5}{2}x$

33. $9(x - 6) = 9(x) - 9(6) = 9x - 54$

37. $7(4x - 3) = 7(4x) - 7(3)$
$= (7 \cdot 4)x - 21$
$= 28x - 21$

41. $-2(y - z) = -2(y) - (-2)z = -2y + 2z$

45. $5(x + 4m + 2) = 5(x) + 5(4m) + 5(2)$
$= 5x + (5 \cdot 4)m + 10$
$= 5x + 20m + 10$

49. $-(5x + 2) = -1(5x + 2)$
$= -1(5x) + (-1)(2)$
$= (-1 \cdot 5)x - 2$
$= -5x - 2$

53. $\dfrac{1}{2}(6x + 7) + \dfrac{1}{2} = \dfrac{1}{2}(6x) + \dfrac{1}{2}(7) + \dfrac{1}{2}$
$= \left(\dfrac{1}{2} \cdot 6\right)x + \dfrac{7}{2} + \dfrac{1}{2}$
$= 3x + \dfrac{8}{2}$
$= 3x + 4$

57. $3(2r + 5) - 7 = 3(2r) + 3(5) - 7$
$= (3 \cdot 2)r + 15 - 7$
$= 6r + 8$

61. $-0.4(4x + 5) - 0.5 = -0.4(4x) + (-0.4)(5) - 0.5$
$= (-0.4 \cdot 4)x - 2 - 0.5$
$= -1.6x - 2.5$

65. $11x + 11y = 11 \cdot x + 11 \cdot y = 11(x + y)$

69. $30a + 30b = 30 \cdot a + 30 \cdot b = 30(a + b)$

73. $2 + (x + 5) = (2 + x) + 5$ illustrates the associative property of addition.

77. $(4 \cdot y) \cdot 9 = 4 \cdot (y \cdot 9)$ illustrates the associative property of multiplication.

81. $-4(y + 7) = -4 \cdot y + (-4) \cdot 7$ illustrates the distributive property.

85. $-6 \cdot 1 = -6$ illustrates the identity element for multiplication.

89. The opposite of x is $-x$.
The reciprocal of x is $\dfrac{1}{x}$.

93. False; the opposite of $-\dfrac{a}{2}$ is $\dfrac{a}{2}$. $-\dfrac{2}{a}$ is the reciprocal of $-\dfrac{a}{2}$.

97. "Putting on your left shoe" and "putting on your right shoe" are commutative, since the order in which they are performed does not affect the outcome.

101. "Feeding the dog" and "feeding the cat" are commutative, since the order in which they are performed does not affect the outcome.

105. answers may vary

Exercise Set 1.8

1. $7y + 8y = (7 + 8)y = 15y$

5. $3b - 5 - 10b - 4 = 3b - 10b - 5 - 4$
$$= (3 - 10)b + (-5 - 4)$$
$$= -7b - 9$$

9. $5g - 3 - 5 - 5g = 5g - 5g - 3 - 5$
$$= (5 - 5)g + (-3 - 5)$$
$$= 0g - 8$$
$$= -8$$

13. $2k - k - 6 = 2k - 1k - 6$
$$= (2 - 1)k - 6$$
$$= 1k - 6$$
$$= k - 6$$

17. $6x - 5x + x - 3 + 2x = 6x - 5x + x + 2x - 3$
$$= (6 - 5 + 1 + 2)x - 3$$
$$= 4x - 3$$

21. $3.4m - 4 - 3.4m - 7 = 3.4m - 3.4m - 4 - 7$
$$= (3.4 - 3.4)m + (-4 - 7)$$
$$= 0m - 11$$
$$= -11$$

25. $5(y + 4) = 5(y) + 5(4) = 5y + 20$

29. $-5(2x - 3y + 6) = -5(2x) - (-5)(3y) + (-5)(6)$
$$= -10x + 15y - 30$$

33. $7(d - 3) + 10 = 7(d) - 7(3) + 10$
$$= 7d - 21 + 10$$
$$= 7d - 11$$

37. $3(2x - 5) - 5(x - 4) = 3(2x) - 3(5) - 5(x) - 5(-4)$
$$= 6x - 15 - 5x + 20$$
$$= 6x - 5x - 15 + 20$$
$$= x + 5$$

41. $5k - (3k - 10) = 5k - 3k + 10 = 2k + 10$

45. $5(x + 2) - (3x - 4) = 5(x) + 5(2) - 3x + 4$
$$= 5x + 10 - 3x + 4$$
$$= 5x - 3x + 10 + 4$$
$$= 2x + 14$$

49. $2 + 4(6x - 6) = 2 + 4(6x) - 4(6)$
$$= 2 + 24x - 24$$
$$= 2 - 24 + 24x$$
$$= -22 + 24x$$

53. $10 - 3(2x + 3y) = 10 - 3(2x) - 3(3y)$
$$= 10 - 6x - 9y$$

57. $\frac{1}{2}(12x - 4) - (x + 5) = \frac{1}{2}(12x) - \frac{1}{2}(4) - x - 5$
$$= 6x - 2 - x - 5$$
$$= 6x - x - 2 - 5$$
$$= 5x - 7$$

61. $(3x - 8) - (7x + 1) = 3x - 8 - 7x - 1$
$$= 3x - 7x - 8 - 1$$
$$= -4x - 9$$

65. Twice a number, decreased by four is written as $2x - 4$.

69. The sum of 5 times a number and -2, added to 7 times the number is written as
$5x + (-2) + 7x = 5x + 7x - 2 = 12x - 2$.

73. Double a number minus the sum of the number and ten is written as
$2x - (x + 10) = 2x - x - 10 = x - 10$.

77. Replace a with 2 and b with -5.
$a - b^2 = 2 - (-5)^2 = 2 - 25 = -23$

81. Since 1 cone balances 1 cube and 1 cylinder balances 2 cubes, 1 cone and 1 cylinder balances $1 + 2 = 3$ cubes. The scale shown is balanced.

85. answers may vary

89. The length of the first board in inches is $12(x + 2)$.
$12(x + 2) + (3x - 1) = 12x + 24 + 3x - 1$
$$= 12x + 3x + 24 - 1$$
$$= 15x + 23$$
The total length is $(15x + 23)$ inches.

Chapter 1 Test

1. The absolute value of negative seven is greater than five is written as $|-7| > 5$.

5. $6 \cdot 3 - 8 \cdot 4 = 18 - 32 = 18 + (-32) = 14$

9. $\frac{-8}{0}$ is undefined.

13. $-\frac{3}{5} + \frac{15}{8} = -\frac{24}{40} + \frac{75}{40} = \frac{51}{40}$

17. $\frac{(-2)(0)(-3)}{-6} = \frac{0(-3)}{-6} = \frac{0}{-6} = 0$

21. $|-2| = 2$
$-1 - (-3) = -1 + 3 = 2$
Since $2 = 2$, $|-2| = -1 - (-3)$.

25. Replace x with 6 and y with -2.
$2 + 3x - y = 2 + 3(6) - (-2)$
$$= 2 + 18 + 2$$
$$= 20 + 2$$
$$= 22$$

29. $-6(2 + 4) = -6 \cdot 2 + (-6) \cdot 4$ illustrates the distributive property.

33. Losses of yardage occurred on the second and third downs. -10 indicates a loss of 10 yards while -2 indicates a loss of 2 yards, so the greatest loss of yardage occurred on the second down.

37. $2y - 6 - y - 4 = 2y - y - 6 - 4$
$$= 1y - 10$$
$$= y - 10$$

CHAPTER 2

Exercise Set 2.1

1. $x + 7 = 10$
$x + 7 - 7 = 10 - 7$
$x = 3$
Check: $x + 7 = 10$
$3 + 7 \stackrel{?}{=} 10$
$10 = 10$ True
The solution is 3.

5. $-11 = 3 + x$
$-11 - 3 = 3 + x - 3$
$-14 = x$
Check: $-11 = 3 + x$
$-11 \stackrel{?}{=} 3 + (-14)$
$-11 = -11$ True
The solution is -14.

9. $x - \frac{2}{5} = -\frac{3}{20}$
$x - \frac{2}{5} + \frac{2}{5} = -\frac{3}{20} + \frac{2}{5}$
$x = -\frac{3}{20} + \frac{8}{20}$
$x = \frac{5}{20}$
$x = \frac{1}{4}$
The solution is $\frac{1}{4}$.

Check: $x - \frac{2}{5} = -\frac{3}{20}$
$\frac{1}{4} - \frac{2}{5} \stackrel{?}{=} -\frac{3}{20}$
$\frac{5}{20} - \frac{8}{20} \stackrel{?}{=} -\frac{3}{20}$
$-\frac{3}{20} = -\frac{3}{20}$ True

13. $7x + 2x = 8x - 3$
$9x = 8x - 3$
$9x - 8x = 8x - 3 - 8x$
$x = -3$
Check: $7x + 2x = 8x - 3$
$7(-3) + 2(-3) \stackrel{?}{=} 8(-3) - 3$
$-21 - 6 \stackrel{?}{=} -24 - 3$
$-27 = -27$ True
The solution is -3.

17.
$$2y + 10 = 5y - 4y$$
$$2y + 10 = y$$
$$2y + 10 - 2y = y - 2y$$
$$10 = -y$$
$$-10 = y$$
Check:
$$2y + 10 = 5y - 4y$$
$$2(-10) + 10 \stackrel{?}{=} 5(-10) - 4(-10)$$
$$-20 + 10 \stackrel{?}{=} -50 + 40$$
$$-10 = -10 \quad \text{True}$$
The solution is -10.

21.
$$\frac{3}{7}x + 2 = -\frac{4}{7}x - 5$$
$$\frac{3}{7}x + 2 + \frac{4}{7}x = -\frac{4}{7}x - 5 + \frac{4}{7}x$$
$$x + 2 = -5$$
$$x + 2 - 2 = -5 - 2$$
$$x = -7$$
Check:
$$\frac{3}{7}x + 2 = -\frac{4}{7}x - 5$$
$$\frac{3}{7}(-7) + 2 \stackrel{?}{=} -\frac{4}{7}(-7) - 5$$
$$-3 + 2 \stackrel{?}{=} 4 - 5$$
$$-1 = -1 \quad \text{True}$$
The solution is -7.

25. $8y + 2 - 6y = 3 + y - 10$
$$2y + 2 = y - 7$$
$$2y + 2 - y = y - 7 - y$$
$$y + 2 = -7$$
$$y + 2 - 2 = -7 - 2$$
$$y = -9$$
Check:
$$8y + 2 - 6y = 3 + y - 10$$
$$8(-9) + 2 - 6(-9) \stackrel{?}{=} 3 + (-9) - 10$$
$$-72 + 2 + 54 \stackrel{?}{=} 3 - 9 - 10$$
$$-16 = -16 \quad \text{True}$$
The solution is -9.

29.
$$\frac{3}{8}x - \frac{1}{6} = -\frac{5}{8}x - \frac{2}{3}$$
$$\frac{3}{8}x - \frac{1}{6} + \frac{5}{8}x = -\frac{5}{8}x - \frac{2}{3} + \frac{5}{8}x$$
$$x - \frac{1}{6} = -\frac{2}{3}$$
$$x - \frac{1}{6} + \frac{1}{6} = -\frac{2}{3} + \frac{1}{6}$$
$$x = -\frac{4}{6} + \frac{1}{6}$$
$$x = -\frac{3}{6}$$
$$x = -\frac{1}{2}$$
Check:
$$\frac{3}{8}x - \frac{1}{6} = -\frac{5}{8}x - \frac{2}{3}$$
$$\frac{3}{8}\left(-\frac{1}{2}\right) - \frac{1}{6} \stackrel{?}{=} -\frac{5}{8}\left(-\frac{1}{2}\right) - \frac{2}{3}$$
$$-\frac{3}{16} - \frac{1}{6} \stackrel{?}{=} \frac{5}{16} - \frac{2}{3}$$
$$-\frac{9}{48} - \frac{8}{48} \stackrel{?}{=} \frac{15}{48} - \frac{32}{48}$$
$$-\frac{17}{48} = -\frac{17}{48} \quad \text{True}$$
The solution is $-\frac{1}{2}$.

33. $3(n - 5) - (6 - 2n) = 4n$
$$3n - 15 - 6 + 2n = 4n$$
$$5n - 21 = 4n$$
$$5n - 21 - 5n = 4n - 5n$$
$$-21 = -n$$
$$21 = n$$

Check:
$$3(n - 5) - (6 - 2n) = 4n$$
$$3(21 - 5) - (6 - 2 \cdot 21) \stackrel{?}{=} 4(21)$$
$$3(21 - 5) - (6 - 42) \stackrel{?}{=} 84$$
$$3(16) - (-36) \stackrel{?}{=} 84$$
$$48 + 36 \stackrel{?}{=} 84$$
$$84 = 84 \quad \text{True}$$
The solution is 21.

37.
$$13x - 3 = 14x$$
$$13x - 3 - 13x = 14x - 13x$$
$$-3 = x$$

41.
$$3x - 6 = 2x + 5$$
$$3x - 6 + 6 = 2x + 5 + 6$$
$$3x = 2x + 11$$
$$3x - 2x = 2x + 11 - 2x$$
$$x = 11$$

45.
$$7(6 + w) = 6(2 + w)$$
$$42 + 7w = 12 + 6w$$
$$42 + 7w - 6w = 12 + 6w - 6w$$
$$42 + w = 12$$
$$42 + w - 42 = 12 - 42$$
$$w = -30$$

49. $10 - (2x - 4) = 7 - 3x$
$$10 - 2x + 4 = 7 - 3x$$
$$14 - 2x = 7 - 3x$$
$$14 - 2x + 3x = 7 - 3x + 3x$$
$$14 + x = 7$$
$$14 + x - 14 = 7 - 14$$
$$x = -7$$

53. $-6.5 - 4x - 1.6 - 3x = -6x + 9.8$
$$-8.1 - 7x = -6x + 9.8$$
$$-8.1 - 7x + 7x = -6x + 9.8 + 7x$$
$$-8.1 = x + 9.8$$
$$-8.1 - 9.8 = x + 9.8 - 9.8$$
$$-17.9 = x$$

57. If the sum of the measures of two angles is $180°$ and one angle measures $x°$, then the other angle measures $(180 - x)°$.

61. If the area of the Gobi Desert is x square miles and the area of the Sahara Desert is 7 times the area of the Gobi Desert, then the area of the Sahara Desert is $7x$ square miles.

65. The multiplicative inverse of 2 is $\frac{1}{2}$, since $2 \cdot \frac{1}{2} = 1$.

69. $\frac{3x}{3} = \frac{3 \cdot x}{3 \cdot 1} = \frac{x}{1} = x$

73. $\frac{3}{5}\left(\frac{5}{3}x\right) = \left(\frac{3}{5} \cdot \frac{5}{3}\right)x = 1x = x$

77.
$$x - 4 = -9$$
$$x - 4 + 4 = -9 + 4$$
$$x = -5$$

81. $180 - x - (2x + 7) = 180 - x - 2x - 7$
$$= 173 - 3x$$
The measure of the third angle is $(173 - 3x)°$.

85.
$$36.766 + x = -108.712$$
$$36.766 + x - 36.766 = -108.712 - 36.766$$
$$x = -145.478$$

Exercise Set 2.2

1. $-5x = -20$
$$\frac{-5x}{-5} = \frac{-20}{-5}$$
$$x = 4$$
Check: $-5x = -20$
$$-5(4) \stackrel{?}{=} -20$$
$$-20 = -20 \quad \text{True}$$
The solution is 4.

5. $-x = -12$
$$\frac{-x}{-1} = \frac{-12}{-1}$$
$$x = 12$$
Check: $-x = -12$
$$-12 = -12 \quad \text{True}$$
The solution is 12.

9. $\dfrac{1}{6}d = \dfrac{1}{2}$

$6 \cdot \dfrac{1}{6}d = 6 \cdot \dfrac{1}{2}$

$d = 3$

Check: $\dfrac{1}{6}d = \dfrac{1}{2}$

$\dfrac{1}{6}(3) \stackrel{?}{=} \dfrac{1}{2}$

$\dfrac{3}{6} = \dfrac{1}{2}$ True

The solution is 3.

13. $\dfrac{k}{-7} = 0$

$-7\left(\dfrac{k}{-7}\right) = -7(0)$

$k = 0$

Check: $\dfrac{k}{-7} = 0$

$\dfrac{0}{-7} \stackrel{?}{=} 0$

$0 = 0$ True

The solution is 0.

17. $2x - 4 = 16$

$2x - 4 + 4 = 16 + 4$

$2x = 20$

$\dfrac{2x}{2} = \dfrac{20}{2}$

$x = 10$

Check: $2x - 4 = 16$

$2(10) - 4 \stackrel{?}{=} 16$

$20 - 4 \stackrel{?}{=} 16$

$16 = 16$ True

The solution is 10.

21. $6a + 3 = 3$

$6a + 3 - 3 = 3 - 3$

$6a = 0$

$\dfrac{6a}{6} = \dfrac{0}{6}$

$a = 0$

Check: $6a + 3 = 3$

$6(0) + 3 \stackrel{?}{=} 3$

$0 + 3 \stackrel{?}{=} 3$

$3 = 3$ True

The solution is 0.

25. $6z - 8 - z + 3 = 0$

$5z - 5 = 0$

$5z - 5 + 5 = 0 + 5$

$5z = 5$

$\dfrac{5z}{5} = \dfrac{5}{5}$

$z = 1$

Check: $6z - 8 - z + 3 = 0$

$6(1) - 8 - 1 + 3 \stackrel{?}{=} 0$

$6 - 8 - 1 + 3 \stackrel{?}{=} 0$

$0 = 0$ True

The solution is 1.

29. $\dfrac{2}{3}y - 11 = -9$

$\dfrac{2}{3}y - 11 + 11 = -9 + 11$

$\dfrac{2}{3}y = 2$

$\dfrac{3}{2}\left(\dfrac{2}{3}y\right) = \dfrac{3}{2}(2)$

$y = 3$

Check: $\dfrac{2}{3}y - 11 = -9$

$\dfrac{2}{3}(3) - 11 \stackrel{?}{=} -9$

$2 - 11 \stackrel{?}{=} -9$

$-9 = -9$ True

The solution is 3.

33. $8x + 20 = 6x + 18$

$8x + 20 - 6x = 6x + 18 - 6x$

$2x + 20 = 18$

$2x + 20 - 20 = 18 - 20$

$2x = -2$

$\dfrac{2x}{2} = \dfrac{-2}{2}$

$x = -1$

37. $2x - 5 = 20x + 4$

$2x - 5 - 20x = 20x + 4 - 20x$

$-18x - 5 = 4$

$-18x - 5 + 5 = 4 + 5$

$-18x = 9$

$\dfrac{-18x}{-18} = \dfrac{9}{-18}$

$x = -\dfrac{1}{2}$

41. $-6y - 3 = -5y - 7$

$-6y - 3 + 6y = -5y - 7 + 6y$

$-3 = y - 7$

$-3 + 7 = y - 7 + 7$

$4 = y$

45. $-10z - 0.5 = -20z + 1.6$

$-10z - 0.5 + 20z = -20z + 1.6 + 20z$

$10z - 0.5 = 1.6$

$10z - 0.5 + 0.5 = 1.6 + 0.5$

$10z = 2.1$

$\dfrac{10z}{10} = \dfrac{2.1}{10}$

$z = 0.21$

49. $42 = 7x$

$\dfrac{42}{7} = \dfrac{7x}{7}$

$6 = x$

53. $6x + 10 = -20$

$6x + 10 - 10 = -20 - 10$

$6x = -30$

$\dfrac{6x}{6} = \dfrac{-30}{6}$

$x = -5$

57. $13x - 5 = 11x - 11$

$13x - 5 + 5 = 11x - 11 + 5$

$13x = 11x - 6$

$13x - 11x = 11x - 11x - 6$

$2x = -6$

$\dfrac{2x}{2} = \dfrac{-6}{2}$

$x = -3$

61. $-\dfrac{3}{7}p = -2$

$-\dfrac{7}{3}\left(-\dfrac{3}{7}p\right) = -\dfrac{7}{3}(-2)$

$p = \dfrac{14}{3}$

65.
$$-2x - \frac{1}{2} = \frac{7}{2}$$
$$-2x - \frac{1}{2} + \frac{1}{2} = \frac{7}{2} + \frac{1}{2}$$
$$-2x = \frac{8}{2}$$
$$-2x = 4$$
$$\frac{-2x}{-2} = \frac{4}{-2}$$
$$x = -2$$

69.
$$10 - 3x - 6 - 9x = 7$$
$$4 - 12x = 7$$
$$4 - 12x - 4 = 7 - 4$$
$$12x = 3$$
$$\frac{12x}{12} = \frac{3}{12}$$
$$x = \frac{1}{4}$$

73.
$$-x - \frac{4}{5} = x + \frac{1}{2} + \frac{2}{5}$$
$$-x - \frac{4}{5} = x + \frac{5}{10} + \frac{4}{10}$$
$$-x - \frac{4}{5} = x + \frac{9}{10}$$
$$-x - \frac{4}{5} + x = x + \frac{9}{10} + x$$
$$-\frac{4}{5} = 2x + \frac{9}{10}$$
$$-\frac{4}{5} - \frac{9}{10} = 2x$$
$$-\frac{8}{10} - \frac{9}{10} = 2x$$
$$-\frac{17}{10} = 2x$$
$$\frac{1}{2}\left(-\frac{17}{10}\right) = \frac{1}{2}(2x)$$
$$-\frac{17}{20} = x$$

77. If x represents the first of two consecutive odd integers, then $x + 2$ represents the second. Thus, the sum is represented by $x + x + 2 = 2x + 2$.

81. If x represents the number on the first door, then the next four door numbers are represented by $x + 2$, $x + 4$, $x + 6$, and $x + 8$. The sum of the numbers is
$$x + x + 2 + x + 4 + x + 6 + x + 8 = x + 20.$$

85.
$$6(2z + 4) + 20 = 6 \cdot 2z + 6 \cdot 4 + 20$$
$$= 12z + 24 + 20$$
$$= 12z + 44$$

89. If the solution is -8, then replacing x by -8 results in a true statement.
$$6x = 6(-8) = -48$$
The missing number is -48.

93. answers may vary

Exercise Set 2.3

1.
$$-4y + 10 = -2(3y + 1)$$
$$-4y + 10 = -6y - 2$$
$$-4y + 10 - 10 = -6y - 2 - 10$$
$$-4y = -6y - 12$$
$$-4y + 6y = -6y - 12 + 6y$$
$$2y = -12$$
$$\frac{2y}{2} = \frac{-12}{2}$$
$$y = -6$$

5.
$$-2(3x - 4) = 2x$$
$$-6x + 8 = 2x$$
$$-6x + 8 + 6x = 2x + 6x$$
$$8 = 8x$$
$$\frac{8}{8} = \frac{8x}{8}$$
$$1 = x$$

9.
$$-6(x - 3) - 26 = -8$$
$$-6x + 18 - 26 = -8$$
$$-6x - 8 = -8$$
$$-6x - 8 + 8 = -8 + 8$$
$$-6x = 0$$
$$\frac{-6x}{-6} = \frac{0}{-6}$$
$$x = 0$$

13.
$$4x + 3 = -3 + 2x + 14$$
$$4x + 3 = 11 + 2x$$
$$4x + 3 - 2x = 11 + 2x - 2x$$
$$2x + 3 = 11$$
$$2x + 3 - 3 = 11 - 3$$
$$2x = 8$$
$$\frac{2x}{2} = \frac{8}{2}$$
$$x = 4$$

17.
$$\frac{2}{3}x + \frac{4}{3} = -\frac{2}{3}$$
$$3\left(\frac{2}{3}x + \frac{4}{3}\right) = 3\left(-\frac{2}{3}\right)$$
$$2x + 4 = -2$$
$$2x + 4 - 4 = -2 - 4$$
$$2x = -6$$
$$\frac{2x}{2} = \frac{-6}{2}$$
$$x = -3$$

21.
$$0.50x + 0.15(70) = 35.5$$
$$50x + 15(70) = 3550$$
$$50x + 1050 = 3550$$
$$50x + 1050 - 1050 = 3550 - 1050$$
$$50x = 2500$$
$$\frac{50x}{50} = \frac{2500}{50}$$
$$x = 50$$

25.
$$x + \frac{7}{6} = 2x - \frac{7}{6}$$
$$6\left(x + \frac{7}{6}\right) = 6\left(2x - \frac{7}{6}\right)$$
$$6x + 7 = 12x - 7$$
$$6x + 7 + 7 = 12x - 7 + 7$$
$$6x + 14 = 12x$$
$$6x + 14 - 6x = 12x - 6x$$
$$14 = 6x$$
$$\frac{14}{6} = \frac{6x}{6}$$
$$\frac{7}{3} = x$$

29. $4(3x + 2) = 12x + 8$
$$12x + 8 = 12x + 8$$
Since both sides of the equation are identical, the equation is an identity and every real number is a solution.

33.
$$3x - 7 = 3(x + 1)$$
$$3x - 7 = 3x + 3$$
$$3x - 7 - 3x = 3x + 3 - 3x$$
$$-7 = 3$$
Since the statement $-7 = 3$ is false, the equation has no solution.

37. $\dfrac{6(3-z)}{5} = -z$

$5 \cdot \dfrac{6(3-z)}{5} = 5(-z)$

$6(3-z) = -5z$

$18 - 6z = -5z$

$18 - 6z + 6z = -5z + 6z$

$18 = z$

41. $5y + 2(y-6) = 4(y+1) - 2$

$5y + 2y - 12 = 4y + 4 - 2$

$7y - 12 = 4y + 2$

$7y - 12 + 12 = 4y + 2 + 12$

$7y = 4y + 14$

$7y - 4y = 4y + 14 - 4y$

$3y = 14$

$\dfrac{3y}{3} = \dfrac{14}{3}$

$y = \dfrac{14}{3}$

45. $0.7x - 2.3 = 0.5$

$7x - 23 = 5$

$7x - 23 + 23 = 5 + 23$

$7x = 28$

$\dfrac{7x}{7} = \dfrac{28}{7}$

$x = 4$

49. $4(2n+1) = 3(6n+3) + 1$

$8n + 4 = 18n + 9 + 1$

$8n + 4 = 18n + 10$

$8n + 4 - 10 = 18n + 10 - 10$

$8n - 6 = 18n$

$8n - 6 - 8n = 18n - 8n$

$-6 = 10n$

$\dfrac{-6}{10} = \dfrac{10n}{10}$

$-\dfrac{3}{5} = n$

53. $\dfrac{x}{2} - 1 = \dfrac{x}{5} + 2$

$10\left(\dfrac{x}{2} - 1\right) = 10\left(\dfrac{x}{5} + 2\right)$

$5x - 10 = 2x + 20$

$5x - 10 + 10 = 2x + 20 + 10$

$5x = 2x + 30$

$5x - 2x = 2x + 30 - 2x$

$3x = 30$

$\dfrac{3x}{3} = \dfrac{30}{3}$

$x = 10$

57. $0.06 - 0.01(x+1) = -0.02(2-x)$

$6 - 1(x+1) = -2(2-x)$

$6 - x - 1 = -4 + 2x$

$5 - x = -4 + 2x$

$5 - x + x = -4 + 2x + x$

$5 = -4 + 3x$

$5 + 4 = -4 + 3x + 4$

$9 = 3x$

$\dfrac{9}{3} = \dfrac{3x}{3}$

$3 = x$

61. The perimeter is the sum of the lengths of the sides.

$x + (2x-3) + (3x-5) = x + 2x - 3 + 3x - 5$

$\qquad\qquad\qquad\qquad = 6x - 8$

The perimeter is $(6x - 8)$ meters.

65. The sum of -3 and twice a number is $-3 + 2x$.

69. a. Since both sides of the equation are identical, the equation is an identity and every real number is a solution.

b. answers may vary

c. answers may vary

73. $2x - 6x - 10 = -4x + 3 - 10$

$-4x - 10 = -4x - 7$

$-4x - 10 + 4x = -4x - 7 + 4x$

$-10 = -7$

Since the statement $-10 = -7$ is false, the equation has no solution.

77. answers may vary

81. answers may vary

85. $0.035x + 5.112 = 0.010x + 5.107$

$35x + 5112 = 10x + 5107$

$35x + 5112 - 10x = 10x + 5107 - 10x$

$25x + 5112 = 5107$

$25x + 5112 - 5112 = 5107 - 5112$

$25x = -5$

$\dfrac{25x}{25} = \dfrac{-5}{25}$

$x = -\dfrac{1}{5}$

$x = -0.2$

Exercise Set 2.4

1. $2(x-8) = 3(x+3)$

$2x - 16 = 3x + 9$

$2x - 16 - 2x = 3x + 9 - 2x$

$-16 = x + 9$

$-16 - 9 = x + 9 - 9$

$-25 = x$

The number is -25.

5. If x is the first integer, the next consecutive integer is $x + 1$.

$x + x + 1 = 469$

$2x + 1 = 469$

$2x + 1 - 1 = 469 - 1$

$2x = 468$

$\dfrac{2x}{2} = \dfrac{468}{2}$

$x = 234$

The page numbers are 234 and $234 + 1 = 235$.

9. The sum of the three lengths is 25 inches.

$x + 2x + 1 + 5x = 25$

$1 + 8x = 25$

$1 + 8x - 1 = 25 - 1$

$8x = 24$

$\dfrac{8x}{8} = \dfrac{24}{8}$

$x = 3$

$2x = 2(3) = 6$

$1 + 5x = 1 + 5(3) = 1 + 15 = 16$

The lengths are 3 inches, 6 inches, and 16 inches.

13. Let x represent the salary of the governor of Florida. Then $x + 50{,}425$ represents the salary of the governor of California.

$x + x + 50{,}425 = 299{,}575$

$2x + 50{,}425 = 299{,}575$

$2x + 50{,}425 - 50{,}425 = 299{,}575 - 50{,}425$

$2x = 249{,}150$

$\dfrac{2x}{2} = \dfrac{249{,}150}{2}$

$x = 124{,}575$

$x + 50{,}425 = 124{,}575 + 50{,}425 = 175{,}000$

The salary of the governor of Florida is \$124,575, while that of the governor of California is \$175,000.

17. Let x be the number of miles. Then the total fare is
$3 + 0.8x + 4.5.$

$$3 + 0.8x + 4.5 = 27.5$$
$$30 + 8x + 45 = 275$$
$$8x + 75 = 275$$
$$8x + 75 - 75 = 275 - 75$$
$$8x = 200$$
$$\frac{8x}{8} = \frac{200}{8}$$
$$x = 25$$

You can travel 25 miles from the airport by taxi for $27.50.

21. Angles A and D both measure $x°$, while angles C and B both measure $(2x)°$. The sum of the angle measures is $360°$.

$$x + 2x + x + 2x = 360$$
$$6x = 360$$
$$\frac{6x}{6} = \frac{360}{6}$$
$$x = 60$$
$$2x = 2(60) = 120$$

Angles A and D measure $60°$; angles B and C measure $120°$.

25. Let x represent the number of prescriptions written in 1997, in millions, then the number written in 2001 was $(x + 5.5)$ million.

$$x + x + 5.5 = 35.7$$
$$2x + 5.5 = 35.7$$
$$2x + 5.5 - 5.5 = 35.7 - 5.5$$
$$2x = 30.2$$
$$\frac{2x}{2} = \frac{30.2}{2}$$
$$x = 15.1$$
$$x + 5.5 = 15.1 + 5.5 = 20.6$$

There were 15.1 million prescriptions for ADHD drugs written in 1997, and 20.6 million prescriptions written in 2001.

29. Let x be the first even integer. Then the next two consecutive even integers are $x + 2$ and $x + 4$. The sum of the measures of the angles of a triangle is $180°$.

$$x + x + 2 + x + 4 = 180$$
$$3x + 6 = 180$$
$$3x + 6 - 6 = 180 - 6$$
$$3x = 174$$
$$\frac{3x}{3} = \frac{174}{3}$$
$$x = 58$$
$$x + 2 = 58 + 2 = 60$$
$$x + 4 = 58 + 4 = 62$$

The angles measure $58°$, $60°$, and $62°$.

33. Let x be the number of miles. Then the charge for driving x miles in one day is $39 + 0.2x.$

$$39 + 0.2x = 95$$
$$390 + 2x = 950$$
$$390 + 2x - 390 = 950 - 390$$
$$2x = 560$$
$$\frac{2x}{2} = \frac{560}{2}$$
$$x = 280$$

You drove 280 miles.

37. Let x represent the number of counties in Montana. Then $x + 2$ represents the number of counties in California.

$$x + x + 2 = 114$$
$$2x + 2 = 114$$
$$2x + 2 - 2 = 114 - 2$$
$$2x = 112$$
$$\frac{2x}{2} = \frac{112}{2}$$
$$x = 56$$
$$x + 2 = 56 + 2 = 58$$

Montana has 56 counties and California has 58 counties.

41.
$$3(x + 5) = 2x - 1$$
$$3x + 15 = 2x - 1$$

$$3x + 15 - 2x = 2x - 1 - 2x$$
$$x + 15 = -1$$
$$x + 15 - 15 = -1 - 15$$
$$x = -16$$

The number is -16.

45. Let x represent the number of gold medals won by Korea. Then Italy won $x + 1$ gold medals and France won $x + 2$ gold medals.

$$x + x + 1 + x + 2 = 30$$
$$3x + 3 = 30$$
$$3x + 3 - 3 = 30 - 3$$
$$3x = 27$$
$$\frac{3x}{3} = \frac{27}{3}$$
$$x = 9$$
$$x + 1 = 9 + 1 = 10$$
$$x + 2 = 9 + 2 = 11$$

Korea won 9 gold medals, Italy won 10, and France won 11.

49. The tallest bar represents the amount spent by Illinois, so Illinois spends the most on tourism.

53. answers may vary

57. Replace r by 15.
$$\pi r^2 = \pi(15)^2 = \pi(225) = 225\pi$$

61. One blink every 5 seconds is $\dfrac{1 \text{ blink}}{5 \text{ sec}}$.

There are $60 \cdot 60 = 3600$ seconds in one hour.

$$\frac{1 \text{ blink}}{5 \text{ sec}} \cdot 3600 \text{ sec} = 720 \text{ blinks}$$

The average eye blinks 720 times each hour.
$16 \cdot 720 = 11{,}520$
The average eye blinks 11,520 times while awake for a 16-hour day.
$11{,}520 \cdot 365 = 4{,}204{,}800$
The average eye blinks 4,204,800 times in one year.

65. answers may vary

Exercise Set 2.5

1. Use $A = bh$ when $A = 45$ and $b = 15$.
$$A = bh$$
$$45 = 15 \cdot h$$
$$\frac{45}{15} = \frac{15h}{15}$$
$$3 = h$$

5. Use $A = \dfrac{1}{2}h(B + b)$ when $A = 180$, $B = 11$, and $b = 7$.

$$A = \frac{1}{2}h(B + b)$$
$$180 = \frac{1}{2}h(11 + 7)$$
$$180 = \frac{1}{2}h(18)$$
$$180 = 9h$$
$$\frac{180}{9} = \frac{9h}{9}$$
$$20 = h$$

9. Use $C = 2\pi r$ when $C = 15.7$ and 3.14 is used as an approximation for π.
$$C = 2\pi r$$
$$15.7 = 2(3.14)r$$
$$15.7 = 6.28r$$
$$\frac{15.7}{6.28} = \frac{6.28r}{6.28}$$
$$2.5 = r$$

13. $V = lwh$
$$\frac{V}{lh} = \frac{lwh}{lh}$$
$$\frac{V}{lh} = w$$

17.
$$A = P + PRT$$
$$A - P = P + PRT - P$$
$$A - P = PRT$$
$$\frac{A - P}{PT} = \frac{PRT}{PT}$$
$$\frac{A - P}{PT} = R$$

21.
$$P = a + b + c$$
$$P - b - c = a + b + c - b - c$$
$$P - b - c = a$$

25. a. Area $= l \cdot w = (11.5)(9) = 103.5$
Perimeter $= 2l + 2w$
$$= 2(11.5) + 2(9)$$
$$= 23 + 18$$
$$= 41$$
The area is 103.5 square feet and the perimeter is 41 feet.

b. The baseboard goes around the edges of the room, so it involves the perimeter. The carpet covers the floor of the room, so it involves area.

29. Use $A = lw$ when $A = 3990$ and $w = 57$.
$$A = lw$$
$$3990 = l(57)$$
$$\frac{3990}{57} = \frac{57l}{57}$$
$$70 = l$$
The length (height) of the billboard was 70 feet.

33. Use $d = rt$ when $d = 25,000$ and $r = 4000$.
$$d = rt$$
$$25,000 = 4000t$$
$$\frac{25,000}{4000} = \frac{4000t}{4000}$$
$$6.25 = t$$
It will take the X-30 6.25 hours to travel around the Earth.

37. Let x represent the length of the shortest side. Then the second side has length $2x$ and the third side has length $30 + x$. The perimeter is the sum of the lengths of the sides.
$$x + 2x + 30 + x = 102$$
$$4x + 30 = 102$$
$$4x + 30 - 30 = 102 - 30$$
$$4x = 72$$
$$\frac{4x}{4} = \frac{72}{4}$$
$$x = 18$$
$2x = 2(18) = 36$
$30 + x = 30 + 18 = 48$
The flower bed has sides of length 18 feet, 36 feet, and 48 feet.

41. To find the amount of water in the tank, use $V = lwh$ with $l = 8, w = 3,$ and $h = 6$.
$$V = lwh = 8 \cdot 3 \cdot 6 = 144$$
The tank holds 144 cubic feet of water. Let x represent the number of piranhas the tank could hold. Then $1.5x = 144$.
$$1.5x = 144$$
$$\frac{1.5x}{1.5} = \frac{144}{1.5}$$
$$x = 96$$
The tank could hold 96 piranhas.

45. Use $A = \pi r^2$ to find the area of a pizza.
For the 16-inch pizza, $r = \frac{16}{2} = 8$.
$$A = \pi r^2 = \pi(8)^2 = 64\pi$$
For a 10-inch pizza, $r = \frac{10}{2} = 5$.
$$A = \pi r^2 = \pi(5)^2 = 25\pi$$
Two 10-inch pizzas have an area of $2 \cdot 25\pi = 50\pi$ square inches. Since $50\pi < 64\pi$, you get more pizza by buying the 16-inch pizza.

49. Let s represent the length of one side of the square, then the perimeter of the square is $4s$. A side of the triangle is $s + 5$ and the triangle's perimeter is $3(s + 5)$.
$$3(s + 5) = 4s + 7$$
$$3s + 15 = 4s + 7$$
$$3s + 15 - 3s = 4s + 7 - 3s$$
$$15 = s + 7$$
$$15 - 7 = s + 7 - 7$$
$$8 = s$$
$s + 5 = 8 + 5 = 13$
Each side of the triangle has length 13 inches.

53. Use $A = lw$ when $A = 1,813,500$ and $w = 150$.
$$A = lw$$
$$1,813,500 = l(150)$$
$$\frac{1,813,500}{150} = \frac{150l}{150}$$
$$12,090 = l$$
The length of the runway is 12,090 feet (more than 2 miles!).

57. Use $V = lwh$ when $l = 199, w = 78.5,$ and $h = 33$.
$V = lwh = 199(78.5)(33) = 515,509.5$
The smallest possible shipping crate has a volume of 515,509.5 cubic inches.

61. Use $F = \frac{9}{5}C + 32$ when $C = 167$.
$$F = \frac{9}{5}C + 32$$
$$= \frac{9}{5}(167) + 32$$
$$= 300.6 + 32$$
$$= 332.6$$
$$\approx 333$$
The average temperature on the planet Mercury is 333°F.

65. $200\% = 2.00$ or 2

69. $7.2 = 7.2(100\%) = 720\%$

73. Use $V = lwh$. If the length is doubled, the new length is $2l$. If the width and height are doubled, the new width and height are $2w$ and $2h$, respectively.
$$V = (2l)(2w)(2h) = 2 \cdot 2 \cdot 2lwh = 8lwh$$
The volume of the box is multiplied by 8.

77. △ − □

81. Use $I = PRT$ when $I = 1,056,000$ and $R = 0.055,$ and $T = 6$.
$$I = PRT$$
$$1,056,000 = P(0.055)(6)$$
$$1,056,000 = 0.33P$$
$$\frac{1,056,000}{0.33} = \frac{0.33P}{0.33}$$
$$3,200,000 = P$$

Exercise Set 2.6

1. Let x be the unknown number.
$$x = 16\% \cdot 70$$
$$x = 0.16 \cdot 70$$
$$x = 11.2$$
11.2 is 16% of 70.

5. Let x be the unknown number.
$$45 = 25\% \cdot x$$
$$45 = 0.25 \cdot x$$
$$\frac{45}{0.25} = \frac{0.25x}{0.25}$$
$$180 = x$$
45 is 25% of 180.

9. 37% of adults talk 16–60 minutes on the phone each day.
$37\% \cdot 27,000 = 0.37 \cdot 27,000 = 9990$
You would expect 9990 of the adults in Florence to talk 16–60 minutes each day.

13. $15\% \cdot 40.50 = 0.15 \cdot 40.5 = 6.075$
The tip is $6.08.
$40.5 + 6.08 = 46.58$
The total cost of the meal is $46.58.

17. percent decrease $= \dfrac{\text{amount of decrease}}{\text{original amount}} = \dfrac{40 - 28}{40} = \dfrac{12}{40} = 0.3$
The area decreased by 30%.

21. Let x represent last year's salary.
$x + 4\% \cdot x = 44{,}200$
$x + 0.04x = 44{,}200$
$1.04x = 44{,}200$
$\dfrac{1.04x}{1.04} = \dfrac{44{,}200}{1.04}$
$x = 42{,}500$
Last year's salary was $42,500.

25. Let x represent the number of pounds of coffee worth $7 a pound.

	Number of Pounds ·	Cost per Pound =	Value
$7/lb Coffee	x	7	$7x$
$4/lb Coffee	14	4	$4 \cdot 14 = 56$
$5/lb Coffee Wanted	$x + 14$	5	$5(x + 14)$

The value of the coffee being combined must be the same as the value of the mixture.
$7x + 56 = 5(x + 14)$
$7x + 56 = 5x + 70$
$7x + 56 - 5x = 5x + 70 - 5x$
$2x + 56 = 70$
$2x + 56 - 56 = 70 - 56$
$2x = 14$
$\dfrac{2x}{2} = \dfrac{14}{2}$
$x = 7$
7 pounds of the $4 a pound coffee should be used.

29. Let x represent the unknown number.
$40 = 80\% \cdot x$
$40 = 0.80 \cdot x$
$\dfrac{40}{0.8} = \dfrac{0.8x}{0.8}$
$50 = x$
40 is 80% of 50.

33. From the graph, it appears that 71% of the population of Fairbanks, Alaska shops by catalog.

37.

Ford Motor Company Model Year 2004 Vehicle Sales Worldwide		
	Thousands of Vehicles	Percent of Total (Rounded to Nearest Percent)
North America	3277	$\dfrac{3277}{5462} \approx 0.59996 \approx 60\%$
Europe	1474	$\dfrac{1474}{5462} \approx 0.26986 \approx 27\%$
Asia-Pacific	328	$\dfrac{328}{5462} \approx 0.06005 \approx 6\%$
Rest of the World	383	$\dfrac{383}{5462} \approx 0.07012 \approx 7\%$
Total	5462	

41. Let x represent the amount Charles paid for the car.
$x + 20\% \cdot x = 4680$
$x + 0.20x = 4680$
$1.2x = 4680$
$\dfrac{1.2x}{1.2} = \dfrac{4680}{1.2}$
$x = 3900$
Charles paid $3900 for the car.

45. Mark-up $= 5\% \cdot 2.20 = 0.05 \cdot 2.2 = 0.11$
New price $= 2.20 + 0.11 = 2.31$
The markup is $0.11 and the new price is $2.31.

49. percent decrease $= \dfrac{\text{amount of decrease}}{\text{original amount}}$
$= \dfrac{151 - 73}{151}$
$= \dfrac{78}{151}$
≈ 0.517
The number of decisions by the Supreme Court decreased by 51.7%.

53. decrease $= 25\% \cdot 256 = 0.25 \cdot 256 = 64$
$256 - 64 = 192$
The price of the cost decreased by $64. The sale price was $192.

57. $42\% \cdot 860 = 0.42 \cdot 860 = 361.2$
You would expect 361 students to rank flexible hours as their top priority.

61. $-5 > -7$ since -5 is to the right of -7 on a number line.

65. $(-3)^2 = (-3)(-3) = 9$
$-3^2 = -(3 \cdot 3) = -9$
Since $9 > -9$, $(-3)^2 > -3^2$.

69. 230 mg is what percent of 2400 mg?
Let x represent the unknown percent.
$x \cdot 2400 = 230$
$\dfrac{2400x}{2400} = \dfrac{230}{2400}$
$x = 0.0958\overline{3}$
This food contains 9.6% of the daily value of sodium in one serving.

73. 12 g $\cdot$ 4 calories/gram $= 48$ calories
48 of the 280 calories come from protein.
$\dfrac{48}{280} \approx 0.171$
17.1% of the calories in this food come from protein.

Exercise Set 2.7

1.
-1

5.
4

9.
-1 3

13. $x - 2 \geq -7$
$x - 2 + 2 \geq -7 + 2$
$x \geq -5$
$\{x \mid x \geq -5\}$
-5

17. $3x - 5 > 2x - 8$
$3x - 5 - 2x > 2x - 8 - 2x$
$x - 5 > -8$
$x - 5 + 5 > -8 + 5$
$x > -3$
$\{x \mid x > -3\}$
-3

21. $2x < -6$
$\dfrac{2x}{2} < \dfrac{-6}{2}$
$x < -3$
$\{x \mid x < -3\}$
-3

25.
$$-x > 0$$
$$(-1)(-x) < (-1)(0)$$
$$x < 0$$
$$\{x \mid x < 0\}$$

29.
$$-0.6y < -1.8$$
$$\frac{-0.6y}{-0.6} > \frac{-1.8}{-0.6}$$
$$y > 3$$
$$\{y \mid y > 3\}$$

33.
$$7(x + 1) - 6x \geq -4$$
$$7x + 7 - 6x \geq -4$$
$$x + 7 \geq -4$$
$$x + 7 - 7 \geq -4 - 7$$
$$x \geq -11$$
$$\{x \mid x \geq -11\}$$

37.
$$-\frac{2}{3}y \leq 8$$
$$-\frac{3}{2}\left(-\frac{2}{3}y\right) \geq -\frac{3}{2}(8)$$
$$y \geq -12$$
$$\{y \mid y \geq -12\}$$

41.
$$3x - 7 < 6x + 2$$
$$3x - 7 - 3x < 6x + 2 - 3x$$
$$-7 < 3x + 2$$
$$-7 - 2 < 3x + 2 - 2$$
$$-9 < 3x$$
$$\frac{-9}{3} < \frac{3x}{3}$$
$$-3 < x$$
$$\{x \mid x > -3\}$$

45.
$$-6x + 2 \geq 2(5 - x)$$
$$-6x + 2 \geq 10 - 2x$$
$$-6x + 2 + 6x \geq 10 - 2x + 6x$$
$$2 \geq 10 + 4x$$
$$2 - 10 \geq 10 + 4x - 10$$
$$-8 \geq 4x$$
$$\frac{-8}{4} \geq \frac{4x}{4}$$
$$-2 \geq x$$
$$\{x \mid x \leq -2\}$$

49.
$$4(3x - 1) \leq 5(2x - 4)$$
$$12x - 4 \leq 10x - 20$$
$$12x - 4 - 10x \leq 10x - 20 - 10x$$
$$2x - 4 \leq -20$$
$$2x - 4 + 4 \leq -20 + 4$$
$$2x \leq -16$$
$$\frac{2x}{2} \leq \frac{-16}{2}$$
$$x \leq -8$$
$$\{x \mid x \leq -8\}$$

53.
$$-5(1 - x) + x \leq -(6 - 2x) + 6$$
$$-5 + 5x + x \leq -6 + 2x + 6$$
$$-5 + 6x \leq 2x$$
$$-5 + 6x - 6x \leq 2x - 6x$$
$$-5 \leq -4x$$
$$\frac{-5}{-4} \geq \frac{-4x}{-4}$$
$$\frac{5}{4} \geq x$$
$$\left\{x \mid x \leq \frac{5}{4}\right\}$$

57.
$$-5x + 4 \leq -4(x - 1)$$
$$-5x + 4 \leq -4x + 4$$
$$-5x + 4 + 4x \leq -4x + 4 + 4x$$
$$-x + 4 \leq 4$$

$$-x + 4 - 4 \leq 4 - 4$$
$$-x \leq 0$$
$$-1(-x) \geq -1(0)$$
$$x \geq 0$$
$$\{x \mid x \geq 0\}$$

61. Use $P = 2l + 2w$ when $w = 15$ and $P \leq 100$.
$$2l + 2(15) \leq 100$$
$$2l + 30 \leq 100$$
$$2l + 30 - 30 \leq 100 - 30$$
$$2l \leq 70$$
$$\frac{2l}{2} \leq \frac{70}{2}$$
$$l \leq 35$$
The maximum length of the rectangle is 35 centimeters.

65. Let x represent the number of people. Then the cost is $50 + 34x$.
$$50 + 34x \leq 3000$$
$$50 + 34x - 50 \leq 3000 - 50$$
$$34x \leq 2950$$
$$\frac{34x}{34} \leq \frac{2950}{34}$$
$$x \leq \frac{2950}{34} \approx 86.76$$
They can invite at most 86 people.

69. $3^4 = 3 \cdot 3 \cdot 3 \cdot 3 = 81$

73. $\left(\frac{7}{8}\right)^2 = \left(\frac{7}{8}\right)\left(\frac{7}{8}\right) = \frac{49}{64}$

77. The greatest increase occurred between 2003 and 2004.

81. Since $3 > 5, 3(-4) < 5(-4)$.

85. When multiplying or dividing by a negative number.

Chapter 2 Test

1.
$$-\frac{4}{5}x = 4$$
$$-\frac{5}{4}\left(-\frac{4}{5}x\right) = -\frac{5}{4}(4)$$
$$x = -5$$

5.
$$\frac{2(x + 6)}{3} = x - 5$$
$$3\left(\frac{2(x + 6)}{3}\right) = 3(x - 5)$$
$$2(x + 6) = 3(x - 5)$$
$$2x + 12 = 3x - 15$$
$$2x + 12 - 2x = 3x - 15 - 2x$$
$$12 = x - 15$$
$$12 + 15 = x - 15 + 15$$
$$27 = x$$

9.
$$-0.3(x - 4) + x = 0.5(3 - x)$$
$$-0.3(x - 4) + 1.0x = 0.5(3 - x)$$
$$-3(x - 4) + 10x = 5(3 - x)$$
$$-3x + 12 + 10x = 15 - 5x$$
$$7x + 12 = 15 - 5x$$
$$7x + 12 + 5x = 15 - 5x + 5x$$
$$12x + 12 = 15$$
$$12x + 12 - 12 = 15 - 12$$
$$12x = 3$$
$$\frac{12x}{12} = \frac{3}{12}$$
$$x = \frac{1}{4} = 0.25$$

13. $A = lw = (35)(20) = 700$
The area of the deck is 700 square feet. To paint two coats of water seal means covering $2 \cdot 700 = 1400$ square feet.
$$1400 \text{ sq ft} \cdot \frac{1 \text{ gal}}{200 \text{ sq ft}} = 7 \text{ gal}$$
7 gallons of water seal are needed.

17.
$$3x - 5 > 7x + 3$$
$$3x - 5 - 3x > 7x + 3 - 3x$$
$$-5 > 4x + 3$$
$$-5 - 3 > 4x + 3 - 3$$
$$-8 > 4x$$
$$\frac{-8}{4} > \frac{4x}{4}$$
$$-2 > x$$
$$\{x \mid x < -2\}$$

21.
$$\frac{2(5x + 1)}{3} > 2$$
$$3 \cdot \frac{2(5x + 1)}{3} > 3(2)$$
$$2(5x + 1) > 6$$
$$10x + 2 > 6$$
$$10x + 2 - 2 > 6 - 2$$
$$10x > 4$$
$$\frac{10x}{10} > \frac{4}{10}$$
$$x > \frac{2}{5}$$
$$\left\{x \mid x > \frac{2}{5}\right\}$$

25. Let x represent the number of public libraries in Indiana, then there are $x + 650$ public libraries in New York.
$$x + x + 650 = 1504$$
$$2x + 650 = 1504$$
$$2x + 650 - 650 = 1504 - 650$$
$$2x = 854$$
$$\frac{2x}{2} = \frac{854}{2}$$
$$x = 427$$
$$x + 650 = 427 + 650 = 1077$$

Indiana has 427 public libraries and New York has 1077.

CHAPTER 3

Exercise Set 3.1

1. $7^2 = 7 \cdot 7 = 49$

5. $-2^4 = -(2 \cdot 2 \cdot 2 \cdot 2) = -16$

9. $\left(\dfrac{1}{3}\right)^3 = \dfrac{1}{3} \cdot \dfrac{1}{3} \cdot \dfrac{1}{3} = \dfrac{1}{27}$

13. When x is -2, $x^2 = (-2)^2 = (-2)(-2) = 4$.

17. When $x = 3$ and $y = -5$,
$$2xy^2 = 2(3)(-5)^2$$
$$= 2(3)(-5)(-5)$$
$$= 2(3)(25)$$
$$= 150$$

21. $x^2 \cdot x^5 = x^{2+5} = x^7$

25. $(5y^4)(3y) = (5 \cdot 3)(y^4 \cdot y) = 15y^{4+1} = 15y^5$

29. $(-8mn^6)(9m^2n^2) = (-8 \cdot 9)(m \cdot m^2)(n^6 \cdot n^2)$
$$= -72m^{1+2}n^{6+2}$$
$$= -72m^3n^8$$

33. Area $=$ (length)(width)
$$= (5x^3 \text{ feet})(4x^2 \text{ feet})$$
$$= (5 \cdot 4)(x^3 \cdot x^2) \text{ square feet}$$
$$= 20x^{3+2} \text{ square feet}$$
$$= 20x^5 \text{ square feet}$$

37. $(pq)^8 = p^8 \cdot q^8 = p^8q^8$

41. $(x^2y^3)^5 = (x^2)^5(y^3)^5 = x^{2 \cdot 5}y^{3 \cdot 5} = x^{10}y^{15}$

45. $\left(\dfrac{r}{s}\right)^9 = \dfrac{r^9}{s^9}$

49. $\left(\dfrac{-2xz}{y^5}\right)^2 = \dfrac{(-2)^2(x)^2(z)^2}{(y^5)^2}$
$$= \dfrac{4x^2z^2}{y^{5 \cdot 2}}$$
$$= \dfrac{4x^2z^2}{y^{10}}$$

53. Volume $=$ (length)(width)(height)
$$= (3y^4 \text{ feet})(3y^4 \text{ feet})(3y^4 \text{ feet})$$
$$= (3)^3(y^4)^3 \text{ cubic feet}$$
$$= 27y^{4 \cdot 3} \text{ cubic feet}$$
$$= 27y^{12} \text{ cubic feet}$$

57. $\dfrac{(-4)^6}{(-4)^3} = (-4)^{6-3} = (-4)^3 = -64$

61. $\dfrac{7x^2y^6}{14x^2y^3} = \dfrac{7}{14} \cdot \dfrac{x^2}{x^2} \cdot \dfrac{y^6}{y^3}$
$$= \dfrac{1}{2} \cdot x^{2-2} \cdot y^{6-3}$$
$$= \dfrac{1}{2}x^0y^3$$
$$= \dfrac{y^3}{2}$$

65. $(2x)^0 = 1$

69. $5^0 + y^0 = 1 + 1 = 2$

73. $\left(\dfrac{1}{4}\right)^3 = \dfrac{1}{4} \cdot \dfrac{1}{4} \cdot \dfrac{1}{4} = \dfrac{1}{64}$

77. $a^2a^3a^4 = a^{2+3+4} = a^9$

81. $(a^7b^{12})(a^4b^8) = a^7a^4 \cdot b^{12}b^8$
$$= a^{7+4}b^{12+8}$$
$$= a^{11}b^{20}$$

85. $(z^4)^{10} = z^{4 \cdot 10} = z^{40}$

89. $(-6xyz^3)^2 = (-6)^2x^2y^2(z^3)^2$
$$= 36x^2y^2z^{3 \cdot 2}$$
$$= 36x^2y^2z^6$$

93. $\dfrac{3x^5}{x^4} = 3 \cdot \dfrac{x^5}{x^4} = 3x^{5-4} = 3x^1 = 3x$

97. $(9xy)^2 = 9^2x^2y^2 = 81x^2y^2$

101. $\left(\dfrac{3y^5}{6x^4}\right)^3 = \left(\dfrac{y^5}{2x^4}\right)^3 = \dfrac{(y^5)^3}{2^3(x^4)^3} = \dfrac{y^{5 \cdot 3}}{8x^{4 \cdot 3}} = \dfrac{y^{15}}{8x^{12}}$

105. $5 - 7 = 5 + (-7) = -2$

109. $-11 - (-4) = -11 + 4 = -7$

113. The expression $x^{14} + x^{23}$ cannot be simplified by adding subtracting, multiplying, or dividing the exponents; e.

117. answers may vary

121. The volume of a cube measures the amount of material that the cube can hold, so to find the amount of water that a swimming pool can hold, the formula for volume should be used.

125. answers may vary

129. $(a^b)^5 = a^{b \cdot 5} = a^{5b}$

Exercise Set 3.2

1. $4^{-3} = \dfrac{1}{4^3} = \dfrac{1}{64}$

5. $\left(-\dfrac{1}{4}\right)^{-3} = \dfrac{(-1)^{-3}}{4^{-3}}$
$$= (-1)^{-3} \cdot \dfrac{1}{4^{-3}}$$
$$= \dfrac{1}{(-1)^3} \cdot 4^3$$
$$= \dfrac{1}{-1} \cdot 64$$
$$= -64$$

9. $\dfrac{1}{p^{-3}} = p^3$

13. $\dfrac{x^{-2}}{x} = \dfrac{x^{-2}}{x^1} = x^{-2-1} = x^{-3} = \dfrac{1}{x^3}$

17. $3^{-2} + 3^{-1} = \dfrac{1}{3^2} + \dfrac{1}{3} = \dfrac{1}{9} + \dfrac{1}{3} = \dfrac{4}{9}$

21. $\dfrac{-1}{p^{-4}} = -1 \cdot \dfrac{1}{p^{-4}} = -1 \cdot p^4 = -p^4$

25. $\dfrac{x^2 x^5}{x^3} = \dfrac{x^{2+5}}{x^3} = \dfrac{x^7}{x^3} = x^{7-3} = x^4$

29. $\dfrac{(m^5)^4 m}{m^{10}} = \dfrac{m^{5 \cdot 4} m^1}{m^{10}}$

$\qquad = \dfrac{m^{20} m^1}{m^{10}}$

$\qquad = \dfrac{m^{20+1}}{m^{10}}$

$\qquad = \dfrac{m^{21}}{m^{10}}$

$\qquad = m^{21-10}$

$\qquad = m^{11}$

33. $(x^5 y^3)^{-3} = (x^5)^{-3}(y^3)^{-3} = x^{-15} y^{-9} = \dfrac{1}{x^{15} y^9}$

37. $\dfrac{(a^5)^2}{(a^3)^4} = \dfrac{a^{10}}{a^{12}} = a^{10-12} = a^{-2} = \dfrac{1}{a^2}$

41. $\dfrac{-6m^4}{-2m^3} = \dfrac{-6}{-2} \cdot \dfrac{m^4}{m^3} = 3 \cdot m^{4-3} = 3m^1 = 3m$

45. $\dfrac{6x^2 y^3}{-7x^2 y^5} = \dfrac{6}{-7} \cdot \dfrac{x^2}{x^2} \cdot \dfrac{y^3}{y^5}$

$\qquad = -\dfrac{6}{7} \cdot x^{2-2} y^{3-5}$

$\qquad = -\dfrac{6}{7} \cdot x^0 y^{-2}$

$\qquad = -\dfrac{6}{7} \cdot 1 \cdot \dfrac{1}{y^2}$

$\qquad = -\dfrac{6}{7y^2}$

49. $(a^{-5} b^2)^{-6} = (a^{-5})^{-6}(b^2)^{-6} = a^{30} b^{-12} = \dfrac{a^{30}}{b^{12}}$

53. $\dfrac{4^2 z^{-3}}{4^3 z^{-5}} = \dfrac{4^2}{4^3} \cdot \dfrac{z^{-3}}{z^{-5}}$

$\qquad = 4^{2-3} z^{-3-(-5)}$

$\qquad = 4^{-1} z^{-3+5}$

$\qquad = \dfrac{1}{4} \cdot z^2$

$\qquad = \dfrac{z^2}{4}$

57. $\dfrac{7ab^{-4}}{7^{-1} a^{-3} b^2} = \dfrac{7}{7^{-1}} \cdot \dfrac{a^1}{a^{-3}} \cdot \dfrac{b^{-4}}{b^2}$

$\qquad = 7^{1-(-1)} a^{1-(-3)} b^{-4-2}$

$\qquad = 7^{1+1} a^{1+3} b^{-2}$

$\qquad = 7^2 a^4 b^{-2}$

$\qquad = \dfrac{49a^4}{b^2}$

61. $\left(\dfrac{a^{-5} b}{ab^3}\right)^{-4} = \dfrac{(a^{-5})^{-4} b^{-4}}{a^{-4}(b^3)^{-4}}$

$\qquad = \dfrac{a^{20} b^{-4}}{a^{-4} b^{-12}}$

$\qquad = a^{20-(-4)} b^{-4-(-12)}$

$\qquad = a^{20+4} b^{-4+12}$

$\qquad = a^{24} b^8$

65. $\dfrac{(xy^3)^5}{(xy)^{-4}} = \dfrac{x^5 (y^3)^5}{x^{-4} y^{-4}}$

$\qquad = \dfrac{x^5 y^{15}}{x^{-4} y^{-4}}$

$\qquad = x^{5-(-4)} y^{15-(-4)}$

$\qquad = x^{5+4} y^{15+4}$

$\qquad = x^9 y^{19}$

69. $\dfrac{(a^4 b^{-7})^{-5}}{(5a^2 b^{-1})^{-2}} = \dfrac{(a^4)^{-5}(b^{-7})^{-5}}{5^{-2}(a^2)^{-2}(b^{-1})^{-2}}$

$\qquad = \dfrac{a^{-20} b^{35}}{5^{-2} a^{-4} b^2}$

$\qquad = 5^2 a^{-20-(-4)} b^{35-2}$

$\qquad = 25 a^{-20+4} b^{33}$

$\qquad = 25 a^{-16} b^{33}$

$\qquad = \dfrac{25 b^{33}}{a^{16}}$

73. $78{,}000 = 7.8 \times 10^4$

77. $0.00635 = 6.35 \times 10^{-3}$

81. $13{,}600 = 1.36 \times 10^4$

85. $3.3 \times 10^{-2} = 0.033$

89. $7.0 \times 10^8 = 700{,}000{,}000$

93. $1.23 \times 10^{12} = 1{,}230{,}000{,}000{,}000$

97. $(1.2 \times 10^{-3})(3 \times 10^{-2}) = 1.2 \cdot 3 \cdot 10^{-3} \cdot 10^{-2}$

$\qquad = 3.6 \times 10^{-5}$

$\qquad = 0.000036$

101. $\dfrac{8 \times 10^{-1}}{16 \times 10^5} = \dfrac{8}{16} \times 10^{-1-5}$

$\qquad = 0.5 \times 10^{-6}$

$\qquad = 5.0 \times 10^{-7}$

$\qquad = 0.0000005$

105. $7.5 \times 10^5 \cdot 3600 = 7.5 \times 10^5 \cdot 3.6 \times 10^3$

$\qquad = 7.5 \cdot 3.6 \cdot 10^5 \cdot 10^3$

$\qquad = 27 \times 10^8$

$\qquad = 2.7 \times 10^9$

109. $y - 10 + y = y + y - 10 = 2y - 10$

113. $(2a^3)^3 a^4 + a^5 a^8 = 2^3(a^3)^3 a^4 + a^5 a^8$

$\qquad = 8a^9 a^4 + a^5 a^8$

$\qquad = 8a^{13} + a^{13}$

$\qquad = 9a^{13}$

117. answers may vary

121. a. $5^{-1} = \dfrac{1}{5}$

$\qquad 5^{-2} = \dfrac{1}{25}$

$\qquad$ Since $\dfrac{1}{5} > \dfrac{1}{25}$, the statement "$5^{-1} < 5^{-2}$" is false.

b. $\left(\dfrac{1}{5}\right)^{-1} = \dfrac{1}{5^{-1}} = 5$

$\qquad \left(\dfrac{1}{5}\right)^{-2} = \dfrac{1}{5^{-2}} = 5^2 = 25$

$\qquad$ Since $5 < 25$, the statement "$\left(\dfrac{1}{5}\right)^{-1} < \left(\dfrac{1}{5}\right)^{-2}$" is true.

c. From part a, the statement "$a^{-1} < a^{-2}$ for all nonzero numbers" is false.

125. $a^{4m+1} \cdot a^4 = a^{4m+1+4} = a^{4m+5}$

Exercise Set 3.3

1.

Term	Coefficient
x^2	1
$-3x$	-3
5	5

5. $x + 2 = x^1 + 2$
This is a binomial of degree 1.

9. $12x^4 - x^6 - 12x^2 = -x^6 + 12x^4 - 12x^2$
This is a trinomial of degree 6.

13. a. $5x - 6 = 5(0) - 6 = 0 - 6 = -6$
b. $5x - 6 = 5(-1) - 6 = -5 - 6 = -11$

17. a. $-x^3 + 4x^2 - 15 = -(0)^3 + 4(0)^2 - 15$
$= 0 + 0 - 15$
$= -15$
b. $-x^3 + 4x^2 - 15 = -(-1)^3 + 4(-1)^2 - 15$
$= -(-1) + 4(1) - 15$
$= 1 + 4 - 15$
$= -10$

21. $-16t^2 + 200t = -16(7.6)^2 + 200(7.6)$
$= -16(57.76) + 1520$
$= -924.16 + 1520$
$= 595.84$
After 7.6 seconds, the height of the rocket is 595.84 feet.

25. $9x - 20x = (9 - 20)x = -11x$

29. $7x^2 + 3 + 9x^2 - 10 = 7x^2 + 9x^2 + 3 - 10$
$= (7 + 9)x^2 + 3 - 10$
$= 16x^2 - 7$

33. $8s - 5s + 4s = (8 - 5 + 4)s = 7s$

37. $\dfrac{2}{3}x^4 + 12x^3 + \dfrac{1}{6}x^4 - 19x^3 - 19$

$= \dfrac{2}{3}x^4 + \dfrac{1}{6}x^4 + 12x^3 - 19x^3 - 19$

$= \left(\dfrac{4}{6} + \dfrac{1}{6}\right)x^4 + (12 - 19)x^3 - 19$

$= \dfrac{5}{6}x^4 - 7x^3 - 19$

41. $9ab = 9a^1b^1$ has degree $1 + 1 = 2$.
$-6a = -6a^1$ has degree 1.
$5b = 5b^1$ has degree 1.
$-3 = -3a^0b^0$ has degree 0.
$9ab - 6a + 5b - 3$ is a polynomial of degree 2.

45. $3ab - 4a + 6ab - 7a = 3ab + 6ab - 4a - 7a$
$= (3 + 6)ab - (4 + 7)a$
$= 9ab - 11a$

49. $5x^2y + 6xy^2 - 5yx^2 + 4 - 9y^2x$
$= 5x^2y - 5x^2y + 6xy^2 - 9xy^2 + 4$
$= (5 - 5)x^2y + (6 - 9)xy^2 + 4$
$= 0x^2y - 3xy^2 + 4$
$= -3xy^2 + 4$

53. $7x^2 + 3 = 7x^2 + 0x + 3$

57. $5y^3 + 2y - 10 = 5y^3 + 0y^2 + 2y - 10$

61. $6x^5 + x^3 - 3x + 15$
$= 6x^5 + 0x^4 + x^3 + 0x^2 - 3x + 15$

65. $5x + 3 + 4x + 3 + 2x + 6 + 3x + 7x$
$= 5x + 4x + 2x + 3x + 7x + 3 + 3 + 6$
$= (5 + 4 + 2 + 3 + 7)x + 12$
$= 12x + 12$

69. $2(x - 5) + 3(5 - x) = 2x - 10 + 15 - 3x$
$= 2x - 3x - 10 + 15$
$= (2 - 3)x + 5$
$= -x + 5$

73. answers may vary

77. $a \cdot b^3 \cdot a^2 \cdot b^7 = a^1 \cdot a^2 \cdot b^3 \cdot b^7$
$= a^{1+2}b^{3+7}$
$= a^3b^{10}$

81. answers may vary

85. $1.85x^2 - 3.76x + 9.25x^2 + 10.76 - 4.21x$
$= 1.85x^2 + 9.25x^2 - 3.76x - 4.21x + 10.76$
$= (1.85 + 9.25)x^2 - (3.76 + 4.21)x + 10.76$
$= 11.1x^2 - 7.97x + 10.76$

Exercise Set 3.4

1. $(3x + 7) + (9x + 5) = 3x + 7 + 9x + 5$
$= 3x + 9x + 7 + 5$
$= 12x + 12$

5. $(-5x^2 + 3) + (2x^2 + 1) = -5x^2 + 3 + 2x^2 + 1$
$= -5x^2 + 2x^2 + 3 + 1$
$= -3x^2 + 4$

9. $(1.2x^3 - 3.4x + 7.9) + (6.7x^3 + 4.4x^2 - 10.9)$
$= 1.2x^3 - 3.4x + 7.9 + 6.7x^3 + 4.4x^2 - 10.9$
$= 1.2x^3 + 6.7x^3 + 4.4x^2 - 3.4x + 7.9 - 10.9$
$= 7.9x^3 + 4.4x^2 - 3.4x - 3$

13. $\begin{array}{r} 3t^2 + 4 \\ 5t^2 - 8 \\ \hline 8t^2 - 4 \end{array}$

17. $(2x + 5) - (3x - 9) = (2x + 5) + (-3x + 9)$
$= 2x + 5 - 3x + 9$
$= 2x - 3x + 5 + 9$
$= -x + 14$

21. $3x - (5x - 9) = 3x + (-5x + 9)$
$= 3x - 5x + 9$
$= -2x + 9$

25. $(5x + 8) - (-2x^2 - 6x + 8)$
$= (5x + 8) + (2x^2 + 6x - 8)$
$= 5x + 8 + 2x^2 + 6x - 8$
$= 2x^2 + 5x + 6x + 8 - 8$
$= 2x^2 + 11x$

29. $\left(\dfrac{1}{4}z^2 - \dfrac{1}{5}z\right) - \left(-\dfrac{3}{20}z^2 + \dfrac{1}{10}z - \dfrac{7}{20}\right)$

$= \left(\dfrac{1}{4}z^2 - \dfrac{1}{5}z\right) + \left(\dfrac{3}{20}z^2 - \dfrac{1}{10}z + \dfrac{7}{20}\right)$

$= \dfrac{1}{4}z^2 - \dfrac{1}{5}z + \dfrac{3}{20}z^2 - \dfrac{1}{10}z + \dfrac{7}{20}$

$= \dfrac{5}{20}z^2 + \dfrac{3}{20}z^2 - \dfrac{2}{10}z - \dfrac{1}{10}z + \dfrac{7}{20}$

$= \dfrac{8}{20}z^2 - \dfrac{3}{10}z + \dfrac{7}{20}$

$= \dfrac{2}{5}z^2 - \dfrac{3}{10}z + \dfrac{7}{20}$

33. $\begin{array}{r} 5u^5 - 4u^2 + 3u - 7 \\ -(3u^5 + 6u^2 - 8u + 2) \end{array}$ $\qquad$ $\begin{array}{r} 5u^5 - 4u^2 + 3u - 7 \\ -3u^5 - 6u^2 + 8u - 2 \\ \hline 2u^5 - 10u^2 + 11u - 9 \end{array}$

37. $(9x - 1) - (5x + 2) = (9x - 1) + (-5x - 2)$
$= 9x - 1 - 5x - 2$
$= 4x - 3$

41. $(x^2 + 2x + 1) - (3x^2 - 6x + 2)$
$= (x^2 + 2x + 1) + (-3x^2 + 6x - 2)$
$= x^2 + 2x + 1 - 3x^2 + 6x - 2$
$= -2x^2 + 8x - 1$

45. $(-a^2 + 1) - (a^2 - 3) + (5a^2 - 6a + 7)$
$= (-a^2 + 1) + (-a^2 + 3) + (5a^2 - 6a + 7)$
$= -a^2 + 1 - a^2 + 3 + 5a^2 - 6a + 7$
$= 3a^2 - 6a + 11$

49. $(4x^2 - 6x + 1) + (3x^2 + 2x + 1)$
$= 4x^2 - 6x + 1 + 3x^2 + 2x + 1$
$= 7x^2 - 4x + 2$

53. $(8y^2 + 7) + (6y + 9) - (4y^2 - 6y - 3)$
$= (8y^2 + 7) + (6y + 9) + (-4y^2 + 6y + 3)$
$= 8y^2 + 7 + 6y + 9 - 4y^2 + 6y + 3$
$= 4y^2 + 12y + 19$

57. $(9a + 6b - 5) + (-11a - 7b + 6)$
$= 9a + 6b - 5 - 11a - 7b + 6$
$= -2a - b + 1$

61. $(x^2 + 2xy - y^2) + (5x^2 - 4xy + 20y^2)$
$= x^2 + 2xy - y^2 + 5x^2 - 4xy + 20y^2$
$= 6x^2 - 2xy + 19y^2$

65. $(2x^2 + 5) + (4x - 1) + (-x^2 + 3x)$
$= 2x^2 + 5 + 4x - 1 - x^2 + 3x$
$= x^2 + 7x + 4$
The perimeter is $(x^2 + 7x + 4)$ feet.

69. $(4y^2 + 4y + 1) - (y^2 - 10)$
$= (4y^2 + 4y + 1) + (-y^2 + 10)$
$= 4y^2 + 4y + 1 - y^2 + 10$
$= 3y^2 + 4y + 11$
The remaining piece is $(3y^2 + 4y + 11)$ meters long.

73. $3x(2x) = (3 \cdot 2)(x \cdot x) = 6x^2$

77. $10x^2(20xy^2) = 10 \cdot 20 \cdot (x^2 \cdot x)(y^2) = 200x^3y^2$

81. Since $2 + 4 = 6$ and $3 - 5 = -2$,
$2x^4 + 3x^3 - 5x^3 + 4x^4 = 6x^4 - 2x^3$ is a true statement.

85. $(5x - 3) + (5x - 3) = 5x - 3 + 5x - 3 = 10x - 6$; e

89. a. $m \cdot m \cdot m = m^1 \cdot m^1 \cdot m^1 = m^{1+1+1} = m^3$

    b. $m + m + m = 1m + 1m + 1m = (1 + 1 + 1)m = 3m$

    c. $(-m)(-m)(-m) = (-1 \cdot m^1)(-1 \cdot m^1)(-1 \cdot m^1)$
                    $= (-1)(-1)(-1)(m \cdot m \cdot m)$
                    $= -1m^3$
                    $= -m^3$

    d. $-m - m - m = -1m - 1m - 1m$
                   $= (-1 - 1 - 1)m$
                   $= -3m$

Exercise Set 3.5

1. $8x^2 \cdot 3x = (8 \cdot 3)(x^2 \cdot x) = 24x^3$

5. $-4n^3 \cdot 7n^7 = (-4 \cdot 7)(n^3 \cdot n^7) = -28n^{10}$

9. $\left(-\dfrac{1}{3}y^2\right)\left(\dfrac{2}{5}y\right) = \left(-\dfrac{1}{3}\right)\left(\dfrac{2}{5}\right)(y^2 \cdot y) = -\dfrac{2}{15}y^3$

13. $3x(2x + 5) = 3x(2x) + 3x(5) = 6x^2 + 15x$

17. $-2a(a + 4) = -2a(a) + (-2a)(4) = -2a^2 - 8a$

21. $3a^2(4a^3 + 15) = 3a^2(4a^3) + 3a^2(15)$
                  $= 12a^5 + 45a^2$

25. $3x^2y(2x^3 - x^2y^2 + 8y^3)$
$= 3x^2y(2x^3) + 3x^2y(-x^2y^2) + 3x^2y(8y^3)$
$= 6x^5y - 3x^4y^3 + 24x^2y^4$

29. $\dfrac{1}{2}x^2(8x^2 - 6x + 1)$
$= \dfrac{1}{2}x^2(8x^2) + \dfrac{1}{2}x^2(-6x) + \dfrac{1}{2}x^2(1)$
$= 4x^4 - 3x^3 + \dfrac{1}{2}x^2$

33. $(a + 7)(a - 2) = a(a - 2) + 7(a - 2)$
                  $= a(a) + a(-2) + 7(a) + 7(-2)$
                  $= a^2 - 2a + 7a - 14$
                  $= a^2 + 5a - 14$

37. $(3x^2 + 1)(4x^2 + 7)$
$= 3x^2(4x^2 + 7) + 1(4x^2 + 7)$
$= 3x^2(4x^2) + 3x^2(7) + 1(4x^2) + 1(7)$
$= 12x^4 + 21x^2 + 4x^2 + 7$
$= 12x^4 + 25x^2 + 7$

41. $(1 - 3a)(1 - 4a)$
$= 1(1 - 4a) + (-3a)(1 - 4a)$
$= 1(1) + 1(-4a) + (-3a)(1) + (-3a)(-4a)$
$= 1 - 4a - 3a + 12a^2$
$= 1 - 7a + 12a^2$

45. $(x - 2)(x^2 - 3x + 7)$
$= x(x^2 - 3x + 7) + (-2)(x^2 - 3x + 7)$
$= x(x^2) + x(-3x) + x(7) + (-2)(x^2) + (-2)(-3x) + (-2)(7)$
$= x^3 - 3x^2 + 7x - 2x^2 + 6x - 14$
$= x^3 - 5x^2 + 13x - 14$

49. $(2a - 3)(5a^2 - 6a + 4)$
$= 2a(5a^2 - 6a + 4) + (-3)(5a^2 - 6a + 4)$
$= 2a(5a^2) + 2a(-6a) + 2a(4) + (-3)(5a^2)$
    $+ (-3)(-6a) + (-3)(4)$
$= 10a^3 - 12a^2 + 8a - 15a^2 + 18a - 12$
$= 10a^3 - 27a^2 + 26a - 12$

53.
$$\begin{array}{r} 2x - 11 \\ 6x + 1 \\ \hline 2x - 11 \\ 12x^2 - 66x \\ \hline 12x^2 - 64x - 11 \end{array}$$

57.
$$\begin{array}{r} x^2 + 5x - 7 \\ 2x^2 - 7x - 9 \\ \hline -9x^2 - 45x + 63 \\ -7x^3 - 35x^2 + 49x \\ 2x^4 + 10x^3 - 14x^2 \\ \hline 2x^4 + 3x^3 - 58x^2 + 4x + 63 \end{array}$$

61. $-3x(x^2 + 2x - 8) = -3x(x^2) + (-3x)(2x) + (-3x)(-8)$
                          $= -3x^3 - 6x^2 + 24x$

65. $\left(x + \dfrac{1}{7}\right)\left(x - \dfrac{3}{7}\right) = x\left(x - \dfrac{3}{7}\right) + \dfrac{1}{7}\left(x - \dfrac{3}{7}\right)$
                 $= x(x) + x\left(-\dfrac{3}{7}\right) + \dfrac{1}{7}(x) + \dfrac{1}{7}\left(-\dfrac{3}{7}\right)$
                 $= x^2 - \dfrac{3}{7}x + \dfrac{1}{7}x - \dfrac{3}{49}$
                 $= x^2 - \dfrac{2}{7}x - \dfrac{3}{49}$

69. $(a + 4)(a^2 - 6a + 6)$
$= a(a^2 - 6a + 6) + 4(a^2 - 6a + 6)$
$= a(a^2) + a(-6a) + a(6) + 4(a^2) + 4(-6a) + 4(6)$
$= a^3 - 6a^2 + 6a + 4a^2 - 24a + 24$
$= a^3 - 2a^2 - 18a + 24$

73. Area $= \dfrac{1}{2}(\text{base})(\text{height})$
        $= \dfrac{1}{2}(3x - 2)(4x)$
        $= 2x(3x - 2)$
        $= 2x(3x) + 2x(-2)$
        $= 6x^2 - 4x$
The area is $(6x^2 - 4x)$ square inches.

77. $(-3y^3)^2 = (-3)^2(y^3)^2 = 9y^6$

81. $(3x - 1) + (10x - 6) = 3x - 1 + 10x - 6 = 13x - 7$

85. $(3x - 1) - (10x - 6) = (3x - 1) + (-10x + 6)$
                          $= 3x - 1 - 10x + 6$
                          $= -7x + 5$

89. The areas of the smaller rectangles are
$x \cdot x = x^2$
$x \cdot 3 = 3x$
$2 \cdot x = 2x$
$2 \cdot 3 = 6$
The area of the figure is
$x^2 + 3x + 2x + 6 = x^2 + 5x + 6$.

93. $(5x)^2 + (2y)^2 = 5^2x^2 + 2^2y^2 = 25x^2 + 4y^2$

Exercise Set 3.6

1. $(x + 3)(x + 4) = x^2 + 4x + 3x + 12 = x^2 + 7x + 12$

5. $(5x - 6)(x + 2) = 5x^2 + 10x - 6x - 12$
                          $= 5x^2 + 4x - 12$

9. $(2x + 5)(3x - 1) = 6x^2 - 2x + 15x - 5$
                          $= 6x^2 + 13x - 5$

13. $\left(x - \dfrac{1}{3}\right)\left(x + \dfrac{2}{3}\right) = x^2 + \dfrac{2}{3}x - \dfrac{1}{3}x - \dfrac{2}{9}$

$= x^2 + \dfrac{1}{3}x - \dfrac{2}{9}$

17. $(x + 5y)(2x - y) = 2x^2 - xy + 10xy - 5y^2$

$= 2x^2 + 9xy - 5y^2$

21. $(2x - 1)^2 = (2x)^2 - 2(2x)(1) + (1)^2 = 4x^2 - 4x + 1$

25. $(x^2 + 0.5)^2 = (x^2)^2 + 2(x^2)(0.5) + (0.5)^2$

$= x^4 + x^2 + 0.25$

29. $(2a - 3)^2 = (2a)^2 - 2(2a)(3) + (3)^2$

$= 4a^2 - 12a + 9$

33. $(3x - 7y)^2 = (3x)^2 - 2(3x)(7y) + (7y)^2$

$= 9x^2 - 42xy + 49y^2$

37. $(5x^4 - 3)^2 = (5x^4)^2 - 2(5x^4)(3) + (3)^2$

$= 25x^8 - 30x^4 + 9$

41. $(x + 6)(x - 6) = (x)^2 - (6)^2 = x^2 - 36$

45. $(x^2 + 5)(x^2 - 5) = (x^2)^2 - (5)^2 = x^4 - 25$

49. $(4 - 7x)(4 + 7x) = (4)^2 - (7x)^2 = 16 - 49x^2$

53. $(9x + y)(9x - y) = (9x)^2 - (y)^2 = 81x^2 - y^2$

57. $(a + 5)(a + 4) = a^2 + 4a + 5a + 20 = a^2 + 9a + 20$

61. $(4a + 1)(3a - 1) = 12a^2 - 4a + 3a - 1$

$= 12a^2 - a - 1$

65. $(3a + 1)^2 = (3a)^2 + 2(3a)(1) + (1)^2 = 9a^2 + 6a + 1$

69. $\left(a - \dfrac{1}{2}y\right)\left(a + \dfrac{1}{2}y\right) = (a)^2 - \left(\dfrac{1}{2}y\right)^2 = a^2 - \dfrac{1}{4}y^2$

73. $(x^2 + 10)(x^2 - 10) = (x^2)^2 - (10)^2 = x^4 - 100$

77. $(5x - 6y)^2 = (5x)^2 - 2(5x)(6y) + (6y)^2$

$= 25x^2 - 60xy + 36y^2$

81. $(2x + 1)^2 = (2x)^2 + 2(2x)(1) + (1)^2 = 4x^2 + 4x + 1$

The area of the rug is $(4x^2 + 4x + 1)$ square feet.

85. $\dfrac{8a^{17}b^5}{-4a^7b^{10}} = \dfrac{8}{-4} \cdot \dfrac{a^{17}}{a^7} \cdot \dfrac{b^5}{b^{10}}$

$= -2a^{17-7}b^{5-10}$

$= -2a^{10}b^{-5}$

$= -\dfrac{2a^{10}}{b^5}$

89. $(a - b)^2 = (a)^2 - 2(a)(b) + (b)^2 = a^2 - 2ab + b^2$, which is choice c.

93. $(x^2 + 7)(x^2 + 3) = x^4 + 3x^2 + 7x^2 + 21$

$= x^4 + 10x^2 + 21$

97. answers may vary

Exercise Set 3.7

1. $\dfrac{12x^4 + 3x^2}{x} = \dfrac{12x^4}{x} + \dfrac{3x^2}{x} = 12x^3 + 3x$

5. $\dfrac{15p^3 + 18p^2}{3p} = \dfrac{15p^3}{3p} + \dfrac{18p^2}{3p} = 5p^2 + 6p$

9. $\dfrac{-9x^5 + 3x^4 - 12}{3x^3} = \dfrac{-9x^5}{3x^3} + \dfrac{3x^4}{3x^3} - \dfrac{12}{3x^3}$

$= -3x^2 + x - \dfrac{4}{x^3}$

13.

$$\begin{array}{r} x + 1 \\ x + 3\overline{)x^2 + 4x + 3} \\ \underline{{}^-x^2 \not{+} 3x} \\ x + 3 \\ \underline{{}^-x \not{+} 3} \\ 0 \end{array}$$

$\dfrac{x^2 + 4x + 3}{x + 3} = x + 1$

17.

$$\begin{array}{r} 2x + 1 \\ x - 4\overline{)2x^2 - 7x + 3} \\ \underline{{}^-2x^2 \overset{+}{\not{}} 8x} \\ x + 3 \\ \underline{{}^-x \overset{+}{\not{}} 4} \\ 7 \end{array}$$

$\dfrac{2x^2 - 7x + 3}{x - 4} = 2x + 1 + \dfrac{7}{x - 4}$

21.

$$\begin{array}{r} 4x + 3 \\ 2x + 1\overline{)8x^2 + 10x + 1} \\ \underline{{}^-8x^2 \not{+} 4x} \\ 6x + 1 \\ \underline{{}^-6x \not{+} 3} \\ -2 \end{array}$$

$\dfrac{8x^2 + 10x + 1}{2x + 1} = 4x + 3 - \dfrac{2}{2x + 1}$

25.

$$\begin{array}{r} x + 6 \\ x - 6\overline{)x^2 + 0x - 36} \\ \underline{{}^-x^2 \overset{+}{\not{}} 6x} \\ 6x - 36 \\ \underline{{}^-6x \overset{+}{\not{}} 36} \\ 0 \end{array}$$

$\dfrac{x^2 - 36}{x - 6} = x + 6$

29. $1 - 3x^2 = -3x^2 + 0x + 1$

$$\begin{array}{r} -3x + 6 \\ x + 2\overline{)-3x^2 + 0x + 1} \\ \underline{\overset{+}{\not{}}3x^2 \overset{+}{\not{}} 6x} \\ 6x + 1 \\ \underline{{}^-6x \overset{+}{\not{}} 12} \\ -11 \end{array}$$

$\dfrac{1 - 3x^2}{x + 2} = -3x + 6 - \dfrac{11}{x + 2}$

33. $\dfrac{a^2b^2 - ab^3}{ab} = \dfrac{a^2b^2}{ab} - \dfrac{ab^3}{ab} = ab - b^2$

37. $\dfrac{2x^2y + 8x^2y^2 - xy^2}{2xy} = \dfrac{2x^2y}{2xy} + \dfrac{8x^2y^2}{2xy} - \dfrac{xy^2}{2xy}$

$= x + 4xy - \dfrac{y}{2}$

41.

$$\begin{array}{r} y^2 + 5y + 10 \\ y - 2\overline{)y^3 + 3y^2 + 0y + 4} \\ \underline{{}^-y^3 \overset{+}{\not{}} 2y^2} \\ 5y^2 + 0y \\ \underline{{}^-5y^2 \overset{+}{\not{}} 10y} \\ 10y + 4 \\ \underline{{}^-10y \overset{+}{\not{}} 20} \\ 24 \end{array}$$

$\dfrac{y^3 + 3y^2 - 4}{y - 2} = y^2 + 5y + 10 + \dfrac{24}{y - 2}$

45.

$$\begin{array}{r} x^3 - x^2 + x \\ x^2 + x\overline{)x^5 + 0x^4 + 0x^3 + x^2} \\ \underline{{}^-x^5 \overset{+}{\not{}} x^4} \\ -x^4 + 0x^3 \\ \underline{\overset{+}{\not{}}x^4 \overset{+}{\not{}} x^3} \\ x^3 + x^2 \\ \underline{{}^-x^3 \overset{+}{\not{}} x^2} \\ 0 \end{array}$$

$\dfrac{x^5 + x^2}{x^2 + x} = x^3 - x^2 + x$

49. $\dfrac{20}{-5} = -4$, so $20 = -5 \cdot -4$

53. $\dfrac{36x^2}{4x} = 9x$, so $36x^2 = 4x \cdot 9x$

57.
$$
\begin{array}{r}
2x + 5 \\
5x + 3 \overline{)10x^2 + 31x + 15} \\
\underline{10x^2 \mp 6x} \\
25x + 15 \\
\underline{25x \mp 15} \\
0
\end{array}
$$

The height of the parallelogram is $(2x + 5)$ meters.

61. $\dfrac{a + 7}{7} = \dfrac{a}{7} + \dfrac{7}{7} = \dfrac{a}{7} + 1$, which is choice c.

Chapter 3 Test

1. $2^5 = 2 \cdot 2 \cdot 2 \cdot 2 \cdot 2 = 32$

5. $(3x^2)(-5x^9) = 3(-5)(x^2 \cdot x^9) = -15x^{2+9} = -15x^{11}$

9.
$$
\begin{aligned}
\frac{6^2 x^{-4} y^{-1}}{6^3 x^{-3} y^7} &= \frac{6^2}{6^3} \cdot \frac{x^{-4}}{x^{-3}} \cdot \frac{y^{-1}}{y^7} \\
&= 6^{2-3} x^{-4-(-3)} y^{-1-7} \\
&= 6^{-1} x^{-4+3} y^{-1-7} \\
&= 6^{-1} x^{-1} y^{-8} \\
&= \frac{1}{6} \cdot \frac{1}{x} \cdot \frac{1}{y^8} \\
&= \frac{1}{6xy^8}
\end{aligned}
$$

13. $6.23 \times 10^4 = 62{,}300$

17.
$$
\begin{aligned}
&(8x^3 + 7x^2 + 4x - 7) + (8x^3 - 7x - 6) \\
&= 8x^3 + 7x^2 + 4x - 7 + 8x^3 - 7x - 6 \\
&= 8x^3 + 8x^3 + 7x^2 + 4x - 7x - 7 - 6 \\
&= 16x^3 + 7x^2 - 3x - 13
\end{aligned}
$$

21.
$$
\begin{aligned}
&3x^2(2x^2 - 3x + 7) \\
&= 3x^2(2x^2) + 3x^2(-3x) + 3x^2(7) \\
&= 6x^4 - 9x^3 + 21x^2
\end{aligned}
$$

25.
$$
\begin{aligned}
(8x + 3)^2 &= (8x)^2 + 2(8x)(3) + (3)^2 \\
&= 64x^2 + 48x + 9
\end{aligned}
$$

29.
$$
\begin{aligned}
\frac{4x^2 + 2xy - 7x}{8xy} &= \frac{4x^2}{8xy} + \frac{2xy}{8xy} - \frac{7x}{8xy} \\
&= \frac{x}{2y} + \frac{1}{4} - \frac{7}{8y}
\end{aligned}
$$

CHAPTER 4

Exercise Set 4.1

1. $32 = 2 \cdot 2 \cdot 2 \cdot 2 \cdot 2$
$36 = 2 \cdot 2 \cdot 3 \cdot 3$
$GCF = 2 \cdot 2 = 4$

5. $24 = 2 \cdot 2 \cdot 2 \cdot 3$
$14 = 2 \cdot 7$
$21 = 3 \cdot 7$
$GCF = 1$ since there are no common prime factors.

9. The GCF of z^7, z^9, and z^{11} is z^7.

13. $14x = 2 \cdot 7 \cdot x$
$21 = 3 \cdot 7$
$GCF = 7$

17. $-10x^2 = -1 \cdot 2 \cdot 5 \cdot x^2$
$15x^3 = 3 \cdot 5 \cdot x^3$
$GCF = 5 \cdot x^2 = 5x^2$

21. $-18x^2y = -1 \cdot 2 \cdot 3 \cdot 3 \cdot x^2 \cdot y$
$9x^3y^3 = 3 \cdot 3 \cdot x^3 \cdot y^3$
$36x^3y = 2 \cdot 2 \cdot 3 \cdot 3 \cdot x^3 \cdot y$
$GCF = 3 \cdot 3 \cdot x^2 \cdot y = 9x^2y$

25. $3a + 6 = 3(a + 2)$

29. $x^3 + 5x^2 = x^2(x + 5)$

33. $32xy - 18x^2 = 2x(16y - 9x)$

37. $6x^3 - 9x^2 + 12x = 3x(2x^2 - 3x + 4)$

41. $5x^3y - 15x^2y + 10xy = 5xy(x^2 - 3x + 2)$

45. $\dfrac{1}{3}x^4 + \dfrac{2}{3}x^3 - \dfrac{4}{3}x^5 + \dfrac{1}{3}x = \dfrac{1}{3}x(x^3 + 2x^2 - 4x^4 + 1)$

49. $z(y + 4) + 3(y + 4) = (y + 4)(z + 3)$

53. $-x - 7 = (-1)x + (-1)(7) = -1(x + 7)$

57. $3a - b + 2 = (-1)(-3a) + (-1)(b) - (-1)(2)$
$\qquad = -1(-3a + b - 2)$

61. $5x + 15 + xy + 3y = 5(x + 3) + y(x + 3)$
$\qquad = (x + 3)(5 + y)$

65. $5m^3 + 6mn + 5m^2 + 6n$
$= m(5m^2 + 6n) + 1(5m^2 + 6n)$
$= (5m^2 + 6n)(m + 1)$

69. $2x^3 + x^2 + 8x + 4 = x^2(2x + 1) + 4(2x + 1)$
$\qquad = (2x + 1)(x^2 + 4)$

73. $5q^2 - 4pq - 5q + 4p = q(5q - 4p) - 1(5q - 4p)$
$\qquad = (5q - 4p)(q - 1)$

77. $(x + 2)(x + 5) = x^2 + 5x + 2x + 10 = x^2 + 7x + 10$

81. $2 \cdot 6 = 12$
$2 + 6 = 8$
2 and 6 have a product of 12 and a sum of 8.

85. $-2 \cdot 5 = -10$
$-2 + 5 = 3$
-2 and 5 have a product of -10 and a sum of 3.

89. $8a - 24 = 8(a - 3)$, which is choice b.

93. Since $3x(a + 2b) + 2(a + 2b) = (a + 2b)(3x + 2)$, the given expression is not factored.

97. The area of the circle is πx^2. Since the sides of the square have length $2x$, the area of the square is $(2x)^2 = 4x^2$. The shaded region is the region inside the square but outside the circle. The area of the shaded region is $4x^2 - \pi x^2 = x^2(4 - \pi)$.

101. answers may vary

Exercise Set 4.2

1. Two factors of 6 whose sum is 7 are 6 and 1.
$x^2 + 7x + 6 = (x + 6)(x + 1)$

5. Two factors of 9 whose sum is -6 are -3 and -3.
$x^2 - 6x + 9 = (x - 3)(x - 3)$ or $(x - 3)^2$

9. Two factors of -70 whose sum is 3 and 10 and -7.
$x^2 + 3x - 70 = (x + 10)(x - 7)$

13. Two factors of $15y^2$ whose sum is $8y$ are $5y$ and $3y$.
$x^2 + 8xy + 15y^2 = (x + 5y)(x + 3y)$

17. $13 + 14m + m^2 = m^2 + 14m + 13$
Two factors of 13 whose sum is 14 are 13 and 1.
$13 + 14m + m^2 = m^2 + 14m + 13 = (m + 13)(m + 1)$

21. Two factors of $16b^2$ whose sum is $-10b$ are $-2b$ and $-8b$.
$a^2 - 10ab + 16b^2 = (a - 2b)(a - 8b)$

25. $2x^3 - 18x^2 + 40x = 2x(x^2 - 9x + 20)$
$\qquad = 2x(x - 5)(x - 4)$

29. $x^2 + 15x + 36 = (x + 12)(x + 3)$

33. $r^2 - 16r + 48 = (r - 12)(r - 4)$

37. $3x^2 + 9x - 30 = 3(x^2 + 3x - 10) = 3(x + 5)(x - 2)$

41. $x^2 - 18x - 144 = (x - 24)(x + 6)$

45. $x^2 - 8x + 15 = (x - 5)(x - 3)$

49. $4x^2y + 4xy - 12y = 4y(x^2 + x - 3)$

53. $x^2 + 7xy + 10y^2 = (x + 5y)(x + 2y)$

57. $x^3 - 2x^2 - 24x = x(x^2 - 2x - 24)$
$$= x(x - 6)(x + 4)$$

61. $5x^3y - 25x^2y^2 - 120xy^3$
$$= 5xy(x^2 - 5xy - 24y^2)$$
$$= 5xy(x - 8y)(x + 3y)$$

65. $-x^2 + 12x - 11 = -1(x^2 - 12x + 11)$
$$= -1(x - 11)(x - 1)$$

69. $x^3y^2 + x^2y - 20x = x(x^2y^2 + xy - 20)$
$$= x(xy - 4)(xy + 5)$$

73. $(5y - 4)(3y - 1) = 15y^2 - 5y - 12y + 4$
$$= 15y^2 - 17y + 4$$

77. $(x - 3)(x + 8) = x^2 + 8x - 3x - 24 = x^2 + 5x - 24$

81. $P = 2l + 2w$
$$= 2(x^2 + 10x) + 2(4x + 33)$$
$$= 2x^2 + 20x + 8x + 66$$
$$= 2x^2 + 28x + 66$$
$$= 2(x^2 + 14x + 33)$$
$$= 2(x + 3)(x + 11)$$

85. $x^2 + x + \dfrac{1}{4} = \left(x + \dfrac{1}{2}\right)\left(x + \dfrac{1}{2}\right)$ or $\left(x + \dfrac{1}{2}\right)^2$

89. The factors of c must sum to -16. Since c is positive, both factors must have the same sign.

$-1 + (-15) = -16; \ (-1)(-15) = 15$
$-2 + (-14) = -16; \ (-2)(-14) = 28$
$-3 + (-13) = -16; \ (-3)(-13) = 39$
$-4 + (-12) = -16; \ (-4)(-12) = 48$
$-5 + (-11) = -16; \ (-5)(-11) = 55$
$-6 + (-10) = -16; \ (-6)(-10) = 60$
$-7 + (-9) = -16; \ (-7)(-9) = 63$
$-8 + (-8) = -16; \ (-8)(-8) = 64$

The possible values of c are 15, 28, 39, 48, 55, 60, 63, and 64.

93. $x^{2n} + 8x^n - 20 = (x^n + 10)(x^n - 2)$

Exercise Set 4.3

1. $5x^2 = 5x \cdot x$
$8 = 2 \cdot 4$
$5x^2 + 22x + 8 = (5x + 2)(x + 4)$

5. $20x^2 = 5x \cdot 4x$
$-6 = 2 \cdot -3$
$20x^2 - 7x - 6 = (5x + 2)(4x - 3)$

9. Factors of $8y^2$: $8y^2 = 8y \cdot y$, $8y^2 = 4y \cdot 2y$.
Factors of 9: $9 = -1 \cdot -9, \ -3 \cdot -3$.
$8y^2 - 17y + 9 = (y - 1)(8y - 9)$

13. Factors of $20r^2$: $20r^2 = 20r \cdot r$, $20r^2 = 10r \cdot 2r$, $20r^2 = 5r \cdot 4r$.
Factors of -8: $-8 = -1 \cdot 8, \ -8 = -2 \cdot 4, \ -8 = -4 \cdot 2, \ -8 = -8 \cdot 1$
$20r^2 + 27r - 8 = (4r - 1)(5r + 8)$

17. $x + 3x^2 - 2 = 3x^2 + x - 2$
Factors of $3x^2$: $3x^2 = 3x \cdot x$
Factors of -2: $-2 = -1 \cdot 2, \ -2 = 2 \cdot -1$
$3x^2 + x - 2 = (3x - 2)(x + 1)$

21. Factors of $15m^2$: $15m^2 = 15m \cdot m$, $15m^2 = 5m \cdot 3m$.
Factors of -15: $-15 = -1 \cdot 15, \ -15 = -3 \cdot 5,$
$-15 = -5 \cdot 3, \ -15 = -15 \cdot 1$
$15m^2 - 16m - 15 = (3m - 5)(5m + 3)$

25. Factors of $2x^2$: $2x^2 = 2x \cdot x$
Factors of -99: $-99 = -1 \cdot 99, \ -99 = -3 \cdot 33,$
$-99 = -9 \cdot 11, \ -99 = -11 \cdot 9, \ -99 = -33 \cdot 3, \ -99 = -99 \cdot 1$
$2x^2 - 7x - 99 = (2x + 11)(x - 9)$

29. Factors of $3a^2$: $3a^2 = 3a \cdot a$
Factors of $3b^2$: $3b^2 = b \cdot 3b$
$3a^2 + 10ab + 3b^2 = (3a + b)(a + 3b)$

33. Factors of $18x^2$: $18x^2 = 18x \cdot x$, $18x^2 = 9x \cdot 2x$, $18x^2 = 6x \cdot 3x$
Factors of -14: $-14 = -1 \cdot 14, \ -14 = -2 \cdot 7,$
$-14 = -7 \cdot 2, \ -14 = -14 \cdot 1$
$18x^2 - 9x - 14 = (6x - 7)(3x + 2)$

37. Factors of $24x^2$: $24x^2 = 24x \cdot x$,
$24x^2 = 12x \cdot 2x, \ 24x^2 = 8x \cdot 3x, \ 24x^2 = 6x \cdot 4x$
Factors of 12: $12 = 1 \cdot 12, \ 12 = 2 \cdot 6, \ 12 = 3 \cdot 4$
$24x^2 + 41x + 12 = (3x + 4)(8x + 3)$

41. $21b^2 - 48b - 45 = 3(7b^2 - 16b - 15)$
Factors of $7b^2$: $7b^2 = 7b \cdot b$
Factors of -15: $-15 = -1 \cdot 15, \ -15 = -3 \cdot 5,$
$-15 = -5 \cdot 3, \ -15 = -15 \cdot 1$
$21b^2 - 48b - 45 = 3(7b + 5)(b - 3)$

45. $6x^2y^2 - 2xy^2 - 60y^2 = 2y^2(3x^2 - x - 30)$
Factors of $3x^2$: $3x^2 = 3x \cdot x$
Factors of -30: $-30 = -1 \cdot 30, \ -30 = -2 \cdot 15,$
$-30 = -3 \cdot 10, \ -30 = -5 \cdot 6, \ -30 = -6 \cdot 5,$
$-30 = -10 \cdot 3, \ -30 = -15 \cdot 2, \ -30 = -30 \cdot 1$
$6x^2y^2 - 2xy^2 - 60y^2 = 2y^2(3x - 10)(x + 3)$

49. $3x^2 - 42x + 63 = 3(x^2 - 14x + 21)$
$x^2 - 14x + 21$ is prime, so $3(x^2 - 14x + 21)$ is a factored form of $3x^2 - 42x + 63$.

53. $-x^2 + 2x + 24 = -1(x^2 - 2x - 24)$
$$= -1(x - 6)(x + 4)$$

57. Factors of $24x^2$: $24x^2 = 24x \cdot x$,
$24x^2 = 12x \cdot 2x, \ 24x^2 = 8x \cdot 3x, \ 24x^2 = 6x \cdot 4x$
Factors of 9: $9 = -1 \cdot -9, \ 9 = -3 \cdot -3$
$24x^2 - 58x + 9 = (4x - 9)(6x - 1)$

61. $30x^3 + 38x^2 + 12x = 2x(15x^2 + 19x + 6)$
Factors of $15x^2$: $15x^2 = 15x \cdot x$, $15x^2 = 5x \cdot 3x$
Factors of 6: $6 = 1 \cdot 6, \ 6 = 2 \cdot 3$
$30x^3 + 38x^2 + 12x = 2x(3x + 2)(5x + 3)$

65. $10x^4 + 25x^3y - 15x^2y^2 = 5x^2(2x^2 + 5xy - 3y^2)$
Factors of $2x^2$: $2x^2 = 2x \cdot x$
Factors of $-3y^2$: $-3y^2 = -y \cdot 3y, \ -3y^2 = -3y \cdot y$
$10x^4 + 25x^3y - 15x^2y^2 = 5x^2(2x - y)(x + 3y)$

69. $16p^4 - 40p^3 + 25p^2 = p^2(16p^2 - 40p + 25)$
Factors of $16p^2$: $16p^2 = 16p \cdot p$, $16p^2 = 8p \cdot 2p$, $16p^2 = 4p \cdot 4p$
Factors of 25: $25 = -1 \cdot -25, \ 25 = -5 \cdot -5$
$16p^4 - 40p^3 + 25p^2 = p^2(4p - 5)(4p - 5)$ or $p^2(4p - 5)^2$

73. $-4 + 52x - 48x^2 = -48x^2 + 52x - 4$
$$= -4(12x^2 - 13x + 1)$$
Factors of $12x^2$: $12x^2 = 12x \cdot x$, $12x^2 = 6x \cdot 2x$, $12x^2 = 4x \cdot 3x$
Factors of 1: $1 = -1 \cdot -1$
$-4 + 52x - 48x^2 = -4(12x - 1)(x - 1)$

77. Factors of $5x^2y^2$: $5x^2y^2 = 5xy \cdot xy$
Factors of 1: $1 = 1 \cdot 1$
There is no combination that gives the correct middle term, so $5x^2y^2 + 20xy + 1$ is prime.

81. $(x - 4)(x + 4) = (x)^2 - (4)^2 = x^2 - 16$

85. $(2x - 1)^2 = (2x)^2 - 2(2x)(1) + (1)^2 = 4x^2 - 4x + 1$

89. $(3x^2 + 1) + (6x + 4) + (x^2 + 15x) = 4x^2 + 21x + 5$
$$= (4x + 1)(x + 5)$$

93. $4x^2(y - 1)^2 + 10x(y - 1) + 25(y - 1)^2$
$$= (y - 1)^2(4x^2 + 10x + 25)$$

97. Note that $5 + 2 = 7$.
$(5x + 2)(x + 1) = 5x(x + 1) + 2(x + 1)$
$$= 5x^2 + 5x + 2x + 2$$
$$= 5x^2 + 7x + 2$$
If $c = 2$, then $5x^2 + 7x + c$ is factorable.

Exercise Set 4.4

1. $x^2 + 3x + 2x + 6 = x(x + 3) + 2(x + 3)$
$$= (x + 3)(x + 2)$$

5. $8x^2 - 5x - 24x + 15 = x(8x - 5) - 3(8x - 5)$
$$= (8x - 5)(x - 3)$$

9. a. $9 \cdot 2 = 18$
$9 + 2 = 11$
9 and 2 are numbers whose product is 18 and whose sum is 11.
b. $11x = 9x + 2x$
c. $6x^2 + 11x + 3 = 6x^2 + 9x + 2x + 3$
$= 3x(2x + 3) + 1(2x + 3)$
$= (2x + 3)(3x + 1)$

13. $21 \cdot 2 = 42$
$14 \cdot 3 = 42$
$14 + 3 = 17$
$21y^2 + 17y + 2 = 21y^2 + 14y + 3y + 2$
$= 7y(3y + 2) + 1(3y + 2)$
$= (3y + 2)(7y + 1)$

17. $10 \cdot 2 = 20$
$-4 \cdot -5 = 20$
$-4 + (-5) = -9$
$10x^2 - 9x + 2 = 10x^2 - 4x - 5x + 2$
$= 2x(5x - 2) - 1(5x - 2)$
$= (5x - 2)(2x - 1)$

21. $12x + 4x^2 + 9 = 4x^2 + 12x + 9$
$4 \cdot 9 = 36$
$6 \cdot 6 = 36$
$6 + 6 = 12$
$4x^2 + 12x + 9 = 4x^2 + 6x + 6x + 9$
$= 2x(2x + 3) + 3(2x + 3)$
$= (2x + 3)(2x + 3) \text{ or } (2x + 3)^2$

25. $10 \cdot 12 = 120$
$-8 \cdot -15 = 120$
$-8 + (-15) = -23$
$10x^2 - 23x + 12 = 10x^2 - 8x - 15x + 12$
$= 2x(5x - 4) - 3(5x - 4)$
$= (5x - 4)(2x - 3)$

29. $16y^2 - 34y + 18 = 2(8y^2 - 17y + 9)$
$8 \cdot 9 = 72$
$-9 \cdot -8 = 72$
$-9 + (-8) = -17$
$16y^2 - 34y + 18 = 2(8y^2 - 17y + 9)$
$= 2(8y^2 - 9y - 8y + 9)$
$= 2[y(8y - 9) - 1(8y - 9)]$
$= 2(8y - 9)(y - 1)$

33. $54a^2 - 9a - 30 = 3(18a^2 - 3a - 10)$
$18 \cdot -10 = -180$
$12 \cdot -15 = -180$
$12 + (-15) = -3$
$54a^2 - 9a - 30 = 3(18a^2 - 3a - 10)$
$= 3(18a^2 + 12a - 15a - 10)$
$= 3[6a(3a + 2) - 5(3a + 2)]$
$= 3(3a + 2)(6a - 5)$

37. $12x^3 - 27x^2 - 27x = 3x(4x^2 - 9x - 9)$
$4 \cdot -9 = -36$
$3 \cdot -12 = -36$
$3 + (-12) = -9$
$12x^3 - 27x^2 - 27x = 3x(4x^2 - 9x - 9)$
$= 3x(4x^2 + 3x - 12x - 9)$
$= 3x[x(4x + 3) - 3(4x + 3)]$
$= 3x(4x + 3)(x - 3)$

41. $20 \cdot 1 = 20$
There are no factors of 20 which sum to 7, so $20z^2 + 7z + 1$ is prime.

45. $15p^4 + 31p^3q + 2p^2q^2 = p^2(15p^2 + 31pq + 2q^2)$
$15 \cdot 2q^2 = 30q^2$
$q \cdot 30q = 30q^2$
$q + 30q = 31q$
$15p^4 + 31p^3q + 2p^2q^2 = p^2(15p^2 + 31pq + 2q^2)$
$= p^2(15p^2 + pq + 30pq + 2q^2)$
$= p^2[p(15p + q) + 2q(15p + q)]$
$= p^2(15p + q)(p + 2q)$

49. $6 - 11x + 5x^2 = 5x^2 - 11x + 6$
$5 \cdot 6 = 30$
$-6 \cdot -5 = 30$
$-6 + (-5) = -11$
$5x^2 - 11x + 6 = 5x^2 - 6x - 5x + 6$
$= x(5x - 6) - 1(5x - 6)$
$= (5x - 6)(x - 1)$

53. $(y + 4)(y + 4) = (y + 4)^2$
$= (y)^2 + 2(y)(4) + (4)^2$
$= y^2 + 8y + 16$

57. $(4x - 3)^2 = (4x)^2 - 2(4x)(3) + (3)^2$
$= 16x^2 - 24x + 9$

61. $x^{2n} + 2x^n + 3x^n + 6 = x^n(x^n + 2) + 3(x^n + 2)$
$= (x^n + 2)(x^n + 3)$

65. answers may vary

Exercise Set 4.5

1. Since $64 = 8^2$ and $16x = 2 \cdot 8 \cdot x$, $x^2 + 16x + 64$ is a perfect square trinomial.

5. Since $1 = 1^2$ and $-2m = -2 \cdot 1 \cdot m$, $m^2 - 2m + 1$ is a perfect square trinomial.

9. $4x^2 = (2x)^2$ but $8y^2$ is not a perfect square, so $4x^2 + 12xy + 8y^2$ is not a perfect square trinomial.

13. $x^2 + 22x + 121 = (x)^2 + 2 \cdot x \cdot 11 + (11)^2$
$= (x + 11)^2$

17. $16a^2 - 24a + 9 = (4a)^2 - 2 \cdot 4a \cdot 3 + (3)^2$
$= (4a - 3)^2$

21. $2n^2 - 28n + 98 = 2(n^2 - 14n + 49)$
$= 2[(n)^2 - 2 \cdot n \cdot 7 + (7)^2]$
$= 2(n - 7)^2$

25. $x^2y^2 - 10xy + 25 = (xy)^2 - 2 \cdot xy \cdot 5 + (5)^2$
$= (xy - 5)^2$

29. Since $1 = 1^2$ and $x^4 = (x^2)^2$, but $6x^2 \neq 2 \cdot 1 \cdot x^2$, the polynomial is not a perfect square trinomial. Since there are no factors of $1 \cdot 1 = 1$ that sum to 6, the trinomial is prime.

33. $x^2 - 4 = (x)^2 - (2)^2 = (x + 2)(x - 2)$

37. $-4r^2 + 1 = -1(4r^2 - 1)$
$= -1[(2r)^2 - (1)^2]$
$= -1(2r + 1)(2r - 1)$

41. $16r^2 + 1$ is the sum of two squares, which is prime.

45. $m^4 - 1 = (m^2)^2 - (1)^2$
$= (m^2 + 1)(m^2 - 1)$
$= (m^2 + 1)[(m)^2 - (1)^2]$
$= (m^2 + 1)(m + 1)(m - 1)$

49. $18r^2 - 8 = 2(9r^2 - 4)$
$= 2[(3r)^2 - (2)^2]$
$= 2(3r + 2)(3r - 2)$

53. $16x^4 - 64x^2 = 16x^2(x^2 - 4)$
$= 16x^2[(x)^2 - (2)^2]$
$= 16x^2(x + 2)(x - 2)$

57. $36x^2 - 64y^2 = 4(9x^2 - 16y^2)$
$= 4[(3x)^2 - (4y)^2]$
$= 4(3x + 4y)(3x - 4y)$

61. $25y^2 - 9 = (5y)^2 - (3)^2 = (5y + 3)(5y - 3)$

65. $x^2y^2 - 1 = (xy)^2 - (1)^2 = (xy + 1)(xy - 1)$

69. $49 - \dfrac{9}{25}m^2 = (7)^2 - \left(\dfrac{3}{5}m\right)^2$
$= \left(7 + \dfrac{3}{5}m\right)\left(7 - \dfrac{3}{5}m\right)$

73. $x^2 + 14xy + 49y^2 = (x)^2 + 2 \cdot x \cdot 7y + (7y)^2$
$= (x + 7y)^2$

77. $x^6 - 81x^2 = x^2(x^4 - 81)$
$= x^2[(x^2)^2 - (9)^2]$
$= x^2(x^2 + 9)(x^2 - 9)$
$= x^2(x^2 + 9)[(x)^2 - (3)^2]$
$= x^2(x^2 + 9)(x + 3)(x - 3)$

81. $x - 6 = 0$
$x - 6 + 6 = 0 + 6$
$x = 6$

85. $5z - 1 = 0$
$5z - 1 + 1 = 0 + 1$
$5z = 1$
$\dfrac{5z}{5} = \dfrac{1}{5}$
$z = \dfrac{1}{5}$

89. $(x + 2)^2 - y^2 = (x + 2)^2 - (y)^2$
$= [(x + 2) + y][(x + 2) - y]$
$= (x + 2 + y)(x + 2 - y)$

93. $(x^2 + 6x + 9) - 4y^2$
$= [(x)^2 + 2 \cdot x \cdot 3 + (3)^2] - 4y^2$
$= (x + 3)^2 - 4y^2$
$= (x + 3)^2 - (2y)^2$
$= [(x + 3) + 2y][(x + 3) - 2y]$
$= (x + 3 + 2y)(x + 3 - 2y)$

97. $x^2 = (x)^2$ and $16 = (4)^2$.
$(x + 4)^2 = (x)^2 + 2 \cdot x \cdot 4 + (4)^2 = x^2 + 8x + 16$, so the number 8 makes the given expression a perfect square trinomial.

101. The difference of two squares is the result of multiplying binomials of the form $(x + a)$ and $(x - a)$, so multiplying $(x - 6)$ by $(x + 6)$ results in the difference of two squares.

105. a. $841 - 16t^2 = 841 - 16(2)^2$
$= 841 - 16 \cdot 4$
$= 841 - 64$
$= 777$
After 2 seconds, the height of the object is 777 feet.

b. $841 - 16t^2 = 841 - 16(5)^2$
$= 841 - 16 \cdot 25$
$= 841 - 400$
$= 441$
After 5 seconds, the height of the object is 441 feet.

c. The object hits the ground when its height is 0 feet.
$841 - 16t^2 = 0$
$(29 + 4t)(29 - 4t) = 0$
$29 + 4t = 0$ or $29 - 4t = 0$
$4t = 29$ $-4t = 29$
$t = \dfrac{29}{4}$ $t = -\dfrac{29}{4}$

Discard $t = -\dfrac{29}{4}$ since time cannot be negative. The object hits the ground after $\dfrac{29}{4} \approx 7$ seconds.

d. $841 - 16t^2 = (29)^2 - (4t)^2$
$= (29 + 4t)(29 - 4t)$

Exercise Set 4.6

1. $(x - 2)(x + 1) = 0$
$x - 2 = 0$ or $x + 1 = 0$
$x = 2$ $x = -1$
The solutions are 2 and -1.

5. $(x + 9)(x + 17) = 0$
$x + 9 = 0$ or $x + 17 = 0$
$x = -9$ $x = -17$
The solutions are -9 and -17.

9. $3x(x - 8) = 0$
$3x = 0$ or $x - 8 = 0$
$x = 0$ $x = 8$
The solutions are 0 and 8.

13. $(2x - 7)(7x + 2) = 0$
$2x - 7 = 0$ or $7x + 2 = 0$
$2x = 7$ $7x = -2$
$x = \dfrac{7}{2}$ $x = -\dfrac{2}{7}$
The solutions are $\dfrac{7}{2}$ and $-\dfrac{2}{7}$.

17. $(x + 0.2)(x + 1.5) = 0$
$x + 0.2 = 0$ or $x + 1.5 = 0$
$x = -0.2$ $x = -1.5$
The solutions are -0.2 and -1.5.

21. $x^2 + 2x - 8 = 0$
$(x + 4)(x - 2) = 0$
$x + 4 = 0$ or $x - 2 = 0$
$x = -4$ $x = 2$
The solutions are -4 and 2.

25. $x^2 + 20x = 0$
$x(x + 20) = 0$
$x = 0$ or $x + 20 = 0$
$x = -20$
The solutions are 0 and -20.

29. $x^2 - 4x = 32$
$x^2 - 4x - 32 = 0$
$(x - 8)(x + 4) = 0$
$x - 8 = 0$ or $x + 4 = 0$
$x = 8$ $x = -4$
The solutions are 8 and -4.

33. $x(3x - 1) = 14$
$3x^2 - x = 14$
$3x^2 - x - 14 = 0$
$3x^2 - 7x + 6x - 14 = 0$
$x(3x - 7) + 2(3x - 7) = 0$
$(3x - 7)(x + 2) = 0$
$3x - 7 = 0$ or $x + 2 = 0$
$3x = 7$ $x = -2$
$x = \dfrac{7}{3}$
The solutions are $\dfrac{7}{3}$ and -2.

37. $4x^3 - x = 0$
$x(4x^2 - 1) = 0$
$x[(2x)^2 - (1)^2] = 0$
$x(2x + 1)(2x - 1) = 0$
$x = 0$ or $2x + 1 = 0$ or $2x - 1 = 0$
$2x = -1$ $2x = 1$
$x = -\dfrac{1}{2}$ $x = \dfrac{1}{2}$
The solutions are $0, -\dfrac{1}{2}$, and $\dfrac{1}{2}$.

41. $(4x - 3)(16x^2 - 24x + 9) = 0$
$(4x - 3)[(4x)^2 - 2 \cdot 4x \cdot 3 + (3)^2] = 0$
$(4x - 3)(4x - 3)^2 = 0$
$(4x - 3)(4x - 3)(4x - 3) = 0$
$4x - 3 = 0$
$4x = 3$
$x = \dfrac{3}{4}$
The solution is $\dfrac{3}{4}$.

45. $(2x + 3)(2x^2 - 5x - 3) = 0$
$(2x + 3)(2x^2 + x - 6x - 3) = 0$
$(2x + 3)[x(2x + 1) - 3(2x + 1)] = 0$
$(2x + 3)(2x + 1)(x - 3) = 0$
$2x + 3 = 0$ or $2x + 1 = 0$ or $x - 3 = 0$
$2x = -3$ $2x = -1$ $x = 3$
$x = -\dfrac{3}{2}$ $x = -\dfrac{1}{2}$
The solutions are $-\dfrac{3}{2}, -\dfrac{1}{2}$, and 3.

49.
$$30x^2 - 11x = 30$$
$$30x^2 - 11x - 30 = 0$$
$$30x^2 + 25x - 36x - 30 = 0$$
$$5x(6x + 5) - 6(6x + 5) = 0$$
$$(6x + 5)(5x - 6) = 0$$
$$6x + 5 = 0 \quad \text{or} \quad 5x - 6 = 0$$
$$6x = -5 \qquad\qquad 5x = 6$$
$$x = -\frac{5}{6} \qquad\qquad x = \frac{6}{5}$$

The solutions are $-\frac{5}{6}$ and $\frac{6}{5}$.

53.
$$6y^2 - 22y - 40 = 0$$
$$2(3y^2 - 11y - 20) = 0$$
$$2(3y^2 + 4y - 15y - 20) = 0$$
$$2[y(3y + 4) - 5(3y + 4)] = 0$$
$$2(3y + 4)(y - 5) = 0$$
$$3y + 4 = 0 \quad \text{or} \quad y - 5 = 0$$
$$3y = -4 \qquad\qquad y = 5$$
$$y = -\frac{4}{3}$$

The solutions are $-\frac{4}{3}$ and 5.

57.
$$x^3 - 12x^2 + 32x = 0$$
$$x(x^2 - 12x + 32) = 0$$
$$x(x - 8)(x - 4) = 0$$
$$x = 0 \quad \text{or} \quad x - 8 = 0 \quad \text{or} \quad x - 4 = 0$$
$$x = 8 \qquad\qquad x = 4$$

The solutions are $0, 8,$ and 4.

61.
$$12y = 8y^2$$
$$0 = 8y^2 - 12y$$
$$0 = 4y(2y - 3)$$
$$4y = 0 \quad \text{or} \quad 2y - 3 = 0$$
$$y = 0 \qquad\qquad 2y = 3$$
$$y = \frac{3}{2}$$

The solutions are 0 and $\frac{3}{2}$.

65.
$$3x^2 + 8x - 11 = 13 - 6x$$
$$3x^2 + 14x - 11 = 13$$
$$3x^2 + 14x - 24 = 0$$
$$3x^2 + 18x - 4x - 24 = 0$$
$$3x(x + 6) - 4(x + 6) = 0$$
$$(x + 6)(3x - 4) = 0$$
$$x + 6 = 0 \quad \text{or} \quad 3x - 4 = 0$$
$$x = -6 \qquad\qquad 3x = 4$$
$$x = \frac{4}{3}$$

The solutions are -6 and $\frac{4}{3}$.

69. $\dfrac{3}{5} + \dfrac{4}{9} = \dfrac{3 \cdot 9}{5 \cdot 9} + \dfrac{4 \cdot 5}{9 \cdot 5} = \dfrac{27}{45} + \dfrac{20}{25} = \dfrac{47}{45}$

73. $\dfrac{4}{5} \cdot \dfrac{7}{8} = \dfrac{4 \cdot 7}{5 \cdot 8} = \dfrac{4 \cdot 7}{5 \cdot 4 \cdot 2} = \dfrac{7}{5 \cdot 2} = \dfrac{7}{10}$

77. answers may vary, for example $(x - 6)(x + 1) = 0$

81. a.

Time, x (seconds)	Height, y (feet)
0	$-16(0)^2 + 20(0) + 300 = 300$
1	$-16(1)^2 + 20(1) + 300 = 304$
2	$-16(2)^2 + 20(2) + 300 = 276$
3	$-16(3)^2 + 20(3) + 300 = 216$
4	$-16(4)^2 + 20(4) + 300 = 124$
5	$-16(5)^2 + 20(5) + 300 = 0$
6	$-16(6)^2 + 20(6) + 300 = -156$

b. When the compass strikes the ground, its height is 0 feet, so it strikes the ground after 5 seconds.

c. The maximum height listed in the table is 304 feet, after 1 second.

85.
$$(2x - 3)(x + 8) = (x - 6)(x + 4)$$
$$2x^2 + 16x - 3x - 24 = x^2 + 4x - 6x - 24$$
$$2x^2 + 13x - 24 = x^2 - 2x - 24$$
$$x^2 + 15x = 0$$
$$x(x + 15) = 0$$
$$x = 0 \quad \text{or} \quad x + 15 = 0$$
$$x = -15$$

The solutions are 0 and -15.

Exercise Set 4.7

1. Let x be the width of the rectangle, then the length is $x + 4$.

5. Let x be the base of the triangle, then the height is $4x + 1$.

9. The perimeter is the sum of the lengths of the sides.
$$(x + 5) + (x^2 - 3x) + (3x - 8) + (x + 3) = 120$$
$$x^2 + 2x = 120$$
$$x^2 + 2x - 120 = 0$$
$$(x + 12)(x - 10) = 0$$
$$x + 12 = 0 \quad \text{or} \quad x - 10 = 0$$
$$x = -12 \qquad\qquad x = 10$$

For $x = -12$, $x + 5$, $x + 3$, and $3x - 8$ are negative. Since lengths cannot be negative, $x = 10$ is the only solution that works in this context.
$$x + 3 = 10 + 3 = 13$$
$$x + 5 = 10 + 5 = 15$$
$$x^2 - 3x = (10)^2 - 3(10) = 100 - 30 = 70$$
$$3x - 8 = 3(10) - 8 = 30 - 8 = 22$$

The sides have lengths 13 cm, 15 cm, 70 cm, and 22 cm.

13. The object will hit the ground when its height is 0.
$$0 = -16t^2 + 64t + 80$$
$$0 = -16(t^2 - 4t - 5)$$
$$0 = -16(t - 5)(t + 1)$$
$$0 = t - 5 \quad \text{or} \quad 0 = t + 1$$
$$5 = t \qquad\qquad -1 = t$$

Since the time t cannot be negative, we discard $t = -1$. The object hits the ground after 5 seconds.

17. $D = \dfrac{1}{2}n(n - 3)$
$$= \dfrac{1}{2}(12)(12 - 3)$$
$$= \dfrac{1}{2}(12)(9)$$
$$= 54$$

A polygon with 12 sides has 54 diagonals.

21. Let x be the number.
$$x + x^2 = 132$$
$$x^2 + x - 132 = 0$$
$$(x + 12)(x - 11) = 0$$
$$x + 12 = 0 \quad \text{or} \quad x - 11 = 0$$
$$x = -12 \qquad\qquad x = 11$$

The number is -12 or 11.

25. Use the Pythagorean theorem where $x =$ hypotenuse, $(x - 1) =$ one leg, and $5 =$ other leg.
$$x^2 = (x - 1)^2 + 5^2$$
$$x^2 = x^2 - 2x + 1 + 25$$
$$x^2 = x^2 - 2x + 26$$
$$0 = -2x + 26$$
$$2x = 26$$
$$x = 13$$

The length of the ladder is 13 feet.

29. Let x be the length of the shorter leg. Then the length of the other leg is $x + 4$ and the length of the hypotenuse is $x + 8$.

$$(x + 8)^2 = x^2 + (x + 4)^2$$
$$x^2 + 16x + 64 = x^2 + x^2 + 8x + 16$$
$$0 = x^2 - 8x - 48$$
$$0 = (x - 12)(x + 4)$$
$$0 = x - 12 \quad \text{or} \quad 0 = x + 4$$
$$12 = x \qquad\qquad -4 = x$$

Discard $x = -4$ since length cannot be negative.

$$x + 4 = 12 + 4 = 16$$
$$x + 8 = 12 + 8 = 20$$

The lengths of the sides are 12 mm, 16 mm, and 20 mm.

33. Let x be the length of the shorter leg. Then $x + 12$ is the length of the longer leg and $2x - 12$ is the length of the hypotenuse.

$$(2x - 12)^2 = x^2 + (x + 12)^2$$
$$4x^2 - 48x + 144 = x^2 + x^2 + 24x + 144$$
$$2x^2 - 72x = 0$$
$$2x(x - 36) = 0$$
$$2x = 0 \quad \text{or} \quad x - 36 = 0$$
$$x = 0 \qquad\qquad x = 36$$

Discard $x = 0$ since the length must be positive. The shorter leg of the triangle has length 36 feet.

37. Use $A = P(1 + r)^2$ when $P = 100$ and $A = 144$.

$$144 = 100(1 + r)^2$$
$$144 = 100(1 + 2r + r^2)$$
$$144 = 100 + 200r + 100r^2$$
$$0 = 100r^2 + 200r - 44$$
$$0 = 4(25r^2 + 50r - 11)$$
$$0 = 4(5r - 1)(5r + 11)$$
$$0 = 5r - 1 \quad \text{or} \quad 0 = 5r + 11$$
$$1 = 5r \qquad\qquad -11 = 5r$$
$$\frac{1}{5} = r \qquad\qquad -\frac{11}{5} = r$$

Discard $r = -\dfrac{11}{5}$ since the interest rate must be positive.

The interest rate is $\dfrac{1}{5} = 0.20 = 20\%$.

41. Let $C = 9500$ in $C = x^2 - 15x + 50$.

$$9500 = x^2 - 15x + 50$$
$$0 = x^2 - 15x - 9450$$
$$0 = (x + 90)(x - 105)$$
$$0 = x + 90 \quad \text{or} \quad 0 = x - 105$$
$$-90 = x \qquad\qquad 105 = x$$

Discard $x = -90$ since the number of units manufactured cannot be negative. 105 units are manufactured at a cost of $9500.

45. From the graph, there were approximately 10,750 thousand acres of farmland in Georgia in 1987.

49. answers may vary

53. Let x be one of the numbers. Since the numbers sum to 25, the other number is $25 - x$.

$$x^2 + (25 - x)^2 = 325$$
$$x^2 + 625 - 50x + x^2 = 325$$
$$2x^2 - 50x + 300 = 0$$
$$2(x^2 - 25x + 150) = 0$$
$$2(x - 10)(x - 15) = 0$$
$$x - 10 = 0 \quad \text{or} \quad x - 15 = 0$$
$$x = 10 \qquad\qquad x = 15$$

If $x = 10$, then $25 - x = 25 - 10 = 15$.
If $x = 15$, then $25 - x = 25 - 15 = 10$.
The numbers are 10 and 15.

Chapter 4 Test

1. $9x^2 - 3x = 3x(3x - 1)$

5. $x^4 - 16 = (x^2)^2 - (4)^2$
$$= (x^2 + 4)(x^2 - 4)$$
$$= (x^2 + 4)[(x)^2 - (2)^2]$$
$$= (x^2 + 4)(x + 2)(x - 2)$$

9. $3a^2 + 3ab - 7a - 7b = 3a(a + b) - 7(a + b)$
$$= (a + b)(3a - 7)$$

13. $6t^2 - t - 5 = 6t^2 + 5t - 6t - 5$
$$= t(6t + 5) - 1(6t + 5)$$
$$= (6t + 5)(t - 1)$$

17. $(x - 3)(x + 9) = 0$
$$x - 3 = 0 \quad \text{or} \quad x + 9 = 0$$
$$x = 3 \qquad\qquad x = -9$$
The solutions are 3 and -9.

21.
$$5t^3 - 45t = 0$$
$$5t(t^2 - 9) = 0$$
$$5t[(t)^2 - (3)^2] = 0$$
$$5t(t + 3)(t - 3) = 0$$
$$5t = 0 \quad \text{or} \quad t + 3 = 0 \quad \text{or} \quad t - 3 = 0$$
$$t = 0 \qquad\qquad t = -3 \qquad\qquad t = 3$$
The solutions are 0, -3, and 3.

25. $(x - 1)(x + 2) = 54$
$$x^2 + 2x - x - 2 = 54$$
$$x^2 + x - 56 = 0$$
$$(x + 8)(x - 7) = 0$$
$$x + 8 = 0 \quad \text{or} \quad x - 7 = 0$$
$$x = -8 \qquad\qquad x = 7$$
Discard $x = -8$ since length cannot be negative.
$$x - 1 = 7 - 1 = 6$$
$$x + 2 = 7 + 2 = 9$$
The width of the rectangle is 6 units and the length is 9 units.

CHAPTER 5

Exercise Set 5.1

1. $\dfrac{x + 5}{x + 2} = \dfrac{2 + 5}{2 + 2} = \dfrac{7}{4}$

5. $\dfrac{x^2 + 8x + 2}{x^2 - x - 6} = \dfrac{2^2 + 8(2) + 2}{2^2 - 2 - 6}$
$$= \dfrac{4 + 16 + 2}{4 - 2 - 6}$$
$$= \dfrac{22}{-4}$$
$$= -\dfrac{11}{2}$$

9. $2x = 0$
$$x = 0$$
$\dfrac{7}{2x}$ is undefined for $x = 0$.

13. $2x - 5 = 0$
$$2x = 5$$
$$x = \dfrac{5}{2}$$
$\dfrac{x - 4}{2x - 5}$ is undefined for $x = \dfrac{5}{2}$.

17. Since $4 \neq 0$, there are no values of x for which $\dfrac{x^2 - 5x - 2}{4}$ is undefined.

21. $3x^2 + 13x + 14 = 0$

$(x + 2)(3x + 7) = 0$

$x + 2 = 0 \quad$ or $\quad 3x + 7 = 0$

$x = -2 \qquad\qquad 3x = -7$

$$x = -\frac{7}{3}$$

$\dfrac{x}{3x^2 + 13x + 14}$ is undefined for $x = -2$ and $x = -\dfrac{7}{3}$.

25. $\dfrac{x - 7}{7 - x} = \dfrac{-1(-x + 7)}{7 - x} = \dfrac{-1(7 - x)}{7 - x} = -1$

29. $\dfrac{x - 2}{x^2 - 4} = \dfrac{x - 2}{(x + 2)(x - 2)} = \dfrac{1}{x + 2}$

33. $\dfrac{-5a - 5b}{a + b} = \dfrac{-5(a + b)}{a + b} = -5$

37. $\dfrac{x + 5}{x^2 - 4x - 45} = \dfrac{x + 5}{(x - 9)(x + 5)} = \dfrac{1}{x - 9}$

41. $\dfrac{x^3 + 7x^2}{x^2 + 5x - 14} = \dfrac{x^2(x + 7)}{(x - 2)(x + 7)} = \dfrac{x^2}{x - 2}$

45. $\dfrac{x^2 + 7x + 10}{x^2 - 3x - 10} = \dfrac{(x + 5)(x + 2)}{(x + 2)(x - 5)} = \dfrac{x + 5}{x - 5}$

49. $\dfrac{2x^2 - 8}{4x - 8} = \dfrac{2(x^2 - 4)}{4(x - 2)} = \dfrac{2(x + 2)(x - 2)}{2 \cdot 2(x - 2)} = \dfrac{x + 2}{2}$

53. $\dfrac{x^2 - 1}{x^2 - 2x + 1} = \dfrac{(x + 1)(x - 1)}{(x - 1)^2} = \dfrac{x + 1}{x - 1}$

57. $\dfrac{5x + 15 - xy - 3y}{2x + 6} = \dfrac{5(x + 3) - y(x + 3)}{2(x + 3)}$

$\qquad\qquad = \dfrac{(x + 3)(5 - y)}{2(x + 3)}$

$\qquad\qquad = \dfrac{5 - y}{2}$

61. $-\dfrac{x - 10}{x + 8} = \dfrac{-(x - 10)}{x + 8}$

$\qquad\quad = \dfrac{-x + 10}{x + 8}$

$\qquad\quad = \dfrac{x - 10}{-(x + 8)}$

$\qquad\quad = \dfrac{x - 10}{-x - 8}$

65. $\dfrac{9 - x^2}{x - 3} = \dfrac{9 - x^2}{-(3 - x)}$

$\qquad\quad = \dfrac{(3 + x)(3 - x)}{-(3 - x)}$

$\qquad\quad = \dfrac{3 + x}{-1}$

$\qquad\quad = -3 - x$

The given answer is correct.

69. $\dfrac{1}{3} \cdot \dfrac{9}{11} = \dfrac{1 \cdot 9}{3 \cdot 11} = \dfrac{1 \cdot 3 \cdot 3}{3 \cdot 11} = \dfrac{3}{11}$

73. $\dfrac{13}{20} \div \dfrac{2}{9} = \dfrac{13}{20} \cdot \dfrac{9}{2} = \dfrac{13 \cdot 9}{20 \cdot 2} = \dfrac{117}{40}$

77. $\dfrac{1 + 2}{1 + 3} = \dfrac{3}{4} \neq \dfrac{2}{3}$

The given answer is incorrect.

81. answers may vary

85. Use $S = \dfrac{h + d + 2t + 3r}{b}$ with $h = 262$, $d = 24$, $t = 5$, $r = 8$, and

$b = 704$.

$$S = \dfrac{262 + 24 + 2(5) + 3(8)}{704} = \dfrac{320}{704} \approx 0.4545$$

Suzuki's slugging percentage was about 45.5%.

Exercise Set 5.2

1. $\dfrac{3x}{y^2} \cdot \dfrac{7y}{4x} = \dfrac{3 \cdot 7 \cdot x \cdot y}{4 \cdot x \cdot y \cdot y} = \dfrac{21}{4y}$

5. $-\dfrac{5a^2b}{30a^2b^2} \cdot b^3 = -\dfrac{5a^2b}{30a^2b^2} \cdot \dfrac{b^3}{1}$

$\qquad\qquad\qquad = -\dfrac{5 \cdot a^2 b \cdot b \cdot b^2}{5 \cdot 6 \cdot a^2 \cdot b^2}$

$\qquad\qquad\qquad = -\dfrac{b \cdot b}{6}$

$\qquad\qquad\qquad = -\dfrac{b^2}{6}$

9. $\dfrac{6x + 6}{5} \cdot \dfrac{10}{36x + 36} = \dfrac{6(x + 1)}{5} \cdot \dfrac{10}{36(x + 1)}$

$\qquad\qquad\qquad = \dfrac{6 \cdot 10}{5 \cdot 36}$

$\qquad\qquad\qquad = \dfrac{6 \cdot 2 \cdot 5}{5 \cdot 6 \cdot 2 \cdot 3}$

$\qquad\qquad\qquad = \dfrac{1}{3}$

13. $\dfrac{x^2 - 25}{x^2 - 3x - 10} \cdot \dfrac{x + 2}{x} = \dfrac{(x + 5)(x - 5)}{(x + 2)(x - 5)} \cdot \dfrac{x + 2}{x}$

$\qquad\qquad\qquad = \dfrac{x + 5}{x}$

17. $\dfrac{5x^7}{2x^5} \div \dfrac{15x}{4x^3} = \dfrac{5x^7}{2x^5} \cdot \dfrac{4x^3}{15x}$

$\qquad\qquad\qquad = \dfrac{5 \cdot x^5 \cdot x \cdot x}{2 \cdot x^5} \cdot \dfrac{2 \cdot 2 \cdot x^3}{5 \cdot 3 \cdot x}$

$\qquad\qquad\qquad = \dfrac{2 \cdot x \cdot x^3}{3}$

$\qquad\qquad\qquad = \dfrac{2x^4}{3}$

21. $\dfrac{(x - 6)(x + 4)}{4x} \div \dfrac{2x - 12}{8x^2} = \dfrac{(x - 6)(x + 4)}{4x} \cdot \dfrac{8x^2}{2x - 12}$

$\qquad\qquad\qquad = \dfrac{(x - 6)(x + 4)}{4x} \cdot \dfrac{4x \cdot 2 \cdot x}{2(x - 6)}$

$\qquad\qquad\qquad = \dfrac{x(x + 4)}{1}$

$\qquad\qquad\qquad = x(x + 4)$

25. $\dfrac{m^2 - n^2}{m + n} \div \dfrac{m}{m^2 + nm} = \dfrac{m^2 - n^2}{m + n} \cdot \dfrac{m^2 + nm}{m}$

$\qquad\qquad\qquad = \dfrac{(m + n)(m - n)}{m + n} \cdot \dfrac{m(m + n)}{m}$

$\qquad\qquad\qquad = (m - n)(m + n)$

$\qquad\qquad\qquad = m^2 - n^2$

29. $\dfrac{x^2 + 7x + 10}{x - 1} \div \dfrac{x^2 + 2x - 15}{x - 1}$

$\qquad\quad = \dfrac{x^2 + 7x + 10}{x - 1} \cdot \dfrac{x - 1}{x^2 + 2x - 15}$

$\qquad\quad = \dfrac{(x + 5)(x + 2)}{x - 1} \cdot \dfrac{x - 1}{(x + 5)(x - 3)}$

$\qquad\quad = \dfrac{x + 2}{x - 3}$

33. $\dfrac{x^2 + 5x}{8} \cdot \dfrac{9}{3x + 15} = \dfrac{x(x + 5)}{8} \cdot \dfrac{3 \cdot 3}{3(x + 5)} = \dfrac{3x}{8}$

37. $\dfrac{3x + 4y}{x^2 + 4xy + 4y^2} \cdot \dfrac{x + 2y}{2} = \dfrac{3x + 4y}{(x + 2y)^2} \cdot \dfrac{x + 2y}{2}$

$\qquad\qquad\qquad = \dfrac{3x + 4y}{2(x + 2y)}$

41. $\dfrac{x^2 - 4}{24x} \div \dfrac{2 - x}{6xy} = \dfrac{x^2 - 4}{24x} \cdot \dfrac{6xy}{2 - x}$

$\qquad = \dfrac{(x + 2)(x - 2)}{6 \cdot 4 \cdot x} \cdot \dfrac{6 \cdot x \cdot y}{-1(x - 2)}$

$\qquad = \dfrac{y(x + 2)}{-4}$

$\qquad = -\dfrac{y(x + 2)}{4}$

45. $\dfrac{5x - 20}{3x^2 + x} \cdot \dfrac{3x^2 + 13x + 4}{x^2 - 16}$

$\qquad = \dfrac{5(x - 4)}{x(3x + 1)} \cdot \dfrac{(x + 4)(3x + 1)}{(x + 4)(x - 4)}$

$\qquad = \dfrac{5}{x}$

49. 10 square feet

$\qquad = \dfrac{10 \text{ square feet}}{1} \cdot \dfrac{144 \text{ square inches}}{1 \text{ square foot}}$

$\qquad = 1440 \text{ square inches}$

53. 3 cubic yards $= \dfrac{3 \text{ cubic yards}}{1} \cdot \dfrac{27 \text{ cubic feet}}{1 \text{ cubic yard}}$

$\qquad = 81 \text{ cubic feet}$

57. 6.3 square yards

$\qquad = \dfrac{6.3 \text{ square yards}}{1} \cdot \dfrac{9 \text{ square feet}}{1 \text{ square yard}}$

$\qquad = 56.7 \text{ square feet}$

61. 930 miles per hour

$\qquad = \dfrac{930 \text{ miles}}{1 \text{ hour}} \cdot \dfrac{5280 \text{ feet}}{1 \text{ mile}} \cdot \dfrac{1 \text{ hour}}{3600 \text{ seconds}}$

$\qquad = \dfrac{930 \cdot 5280 \text{ feet}}{3600 \text{ seconds}}$

$\qquad = 1364 \text{ feet per second}$

65. $\dfrac{9}{9} - \dfrac{19}{9} = \dfrac{9 - 19}{9} = \dfrac{-10}{9} = -\dfrac{10}{9}$

69. $\dfrac{4}{a} \cdot \dfrac{1}{b} = \dfrac{4 \cdot 1}{a \cdot b} = \dfrac{4}{ab}$

The statement is true.

73. $\dfrac{2x}{x^2 - 25} \cdot \dfrac{x + 5}{9x} = \dfrac{2 \cdot x}{(x + 5)(x - 5)} \cdot \dfrac{x + 5}{9 \cdot x}$

$\qquad = \dfrac{2}{9(x - 5)}$

The area is $\dfrac{2}{9(x - 5)}$ square feet.

77. $\left(\dfrac{2a + b}{b^2} \cdot \dfrac{3a^2 - 2ab}{ab + 2b^2} \right) \div \dfrac{a^2 - 3ab + 2b^2}{5ab - 10b^2}$

$\qquad = \dfrac{2a + b}{b^2} \cdot \dfrac{3a^2 - 2ab}{ab + 2b^2} \cdot \dfrac{5ab - 10b^2}{a^2 - 3ab + 2b^2}$

$\qquad = \dfrac{2a + b}{b^2} \cdot \dfrac{a(3a - 2b)}{b(a + 2b)} \cdot \dfrac{5 \cdot b(a - 2b)}{(a - b)(a - 2b)}$

$\qquad = \dfrac{5a(2a + b)(3a - 2b)}{b^2(a - b)(a + 2b)}$

81. $\$2000 \text{ US} = \dfrac{\$2000 \text{ US}}{1} \cdot \dfrac{1 \text{ euro}}{\$1.2955 \text{ US}}$

$\qquad = \dfrac{2000}{1.2955} \text{ euros}$

$\qquad \approx 1543.805 \text{ euros}$

On that day, $2000 US was worth 1543.81 euros.

Exercise Set 5.3

1. $\dfrac{a}{13} + \dfrac{9}{13} = \dfrac{a + 9}{13}$

5. $\dfrac{4m}{m - 6} - \dfrac{24}{m - 6} = \dfrac{4m - 24}{m - 6} = \dfrac{4(m - 6)}{m - 6} = 4$

9. $\dfrac{5x^2 + 4x}{x - 1} - \dfrac{6x + 3}{x - 1} = \dfrac{5x^2 + 4x - (6x + 3)}{x - 1}$

$\qquad = \dfrac{5x^2 - 2x - 3}{x - 1}$

$\qquad = \dfrac{(5x + 3)(x - 1)}{x - 1}$

$\qquad = 5x + 3$

13. $\dfrac{2x + 3}{x^2 - x - 30} - \dfrac{x - 2}{x^2 - x - 30} = \dfrac{2x + 3 - (x - 2)}{x^2 - x - 30}$

$\qquad = \dfrac{x + 5}{(x - 6)(x + 5)}$

$\qquad = \dfrac{1}{x - 6}$

17. $8x = 2 \cdot 2 \cdot 2 \cdot x$

$\qquad 2x + 4 = 2(x + 2)$

$\qquad \text{LCD} = 2 \cdot 2 \cdot 2 \cdot x(x + 2) = 8x(x + 2)$

21. $x + 6 = x + 6$

$\qquad 3x + 18 = 3(x + 6)$

$\qquad \text{LCD} = 3(x + 6)$

25. $3x + 3 = 3(x + 1)$

$\qquad 2x^2 + 4x + 2 = 2(x^2 + 2x + 1) = 2(x + 1)^2$

$\qquad \text{LCD} = 2 \cdot 3 \cdot (x + 1)^2 = 6(x + 1)^2$

29. $x^2 + 3x - 4 = (x - 1)(x + 4)$

$\qquad x^2 + 2x - 3 = (x - 1)(x + 3)$

$\qquad \text{LCD} = (x - 1)(x + 4)(x + 3)$

33. $x^2 - 16 = (x + 4)(x - 4)$

$\qquad 2x^3 - 8x^2 = 2x^2(x - 4)$

$\qquad \text{LCD} = 2x^2(x + 4)(x - 4)$

37. $\dfrac{6}{3a} = \dfrac{6 \cdot 4b^2}{3a \cdot 4b^2} = \dfrac{24b^2}{12ab^2}$

41. $\dfrac{9a + 2}{5a + 10} = \dfrac{9a + 2}{5(a + 2)} = \dfrac{(9a + 2) \cdot b}{5(a + 2) \cdot b} = \dfrac{9ab + 2b}{5b(a + 2)}$

45. $\dfrac{9y - 1}{15x^2 - 30} = \dfrac{(9y - 1) \cdot 2}{(15x^2 - 30) \cdot 2} = \dfrac{18y - 2}{30x^2 - 60}$

49. $\dfrac{x + 3}{4} \div \dfrac{2x - 1}{4} = \dfrac{x + 3}{4} \cdot \dfrac{4}{2x - 1} = \dfrac{x + 3}{2x - 1}$

53. $\dfrac{-2x}{x^3 - 8x} + \dfrac{3x}{x^3 - 8x} = \dfrac{-2x + 3x}{x^3 - 8x}$

$\qquad = \dfrac{x}{x(x^2 - 8)}$

$\qquad = \dfrac{1}{x^2 - 8}$

57. $\dfrac{2}{3} + \dfrac{5}{7} = \dfrac{2 \cdot 7}{3 \cdot 7} + \dfrac{5 \cdot 3}{7 \cdot 3} = \dfrac{14}{21} + \dfrac{15}{21} = \dfrac{29}{21}$

61. $\dfrac{1}{12} + \dfrac{3}{20} = \dfrac{1 \cdot 5}{12 \cdot 5} + \dfrac{3 \cdot 3}{20 \cdot 3} = \dfrac{5}{60} + \dfrac{9}{60} = \dfrac{14}{60} = \dfrac{7}{30}$

65. The perimeter of a square is 4 times the side length.

$\qquad 4 \cdot \dfrac{5}{x - 2} = \dfrac{4}{1} \cdot \dfrac{5}{x - 2} = \dfrac{20}{x - 2}$

The perimeter is $\dfrac{20}{x - 2}$ meters.

69. $8 = 2 \cdot 2 \cdot 2$
$12 = 2 \cdot 2 \cdot 3$
$LCD = 2 \cdot 2 \cdot 2 \cdot 3 = 24$
$24 = 8 \cdot 3$
$24 = 12 \cdot 2$

You should buy 3 packages of hot dogs and 2 packages of buns.

73. answers may vary

Exercise Set 5.4

1. $\dfrac{4}{2x} + \dfrac{9}{3x} = \dfrac{4 \cdot 3}{2x \cdot 3} + \dfrac{9 \cdot 2}{3x \cdot 2} = \dfrac{12}{6x} + \dfrac{18}{6x} = \dfrac{30}{6x} = \dfrac{5}{x}$

5. $\dfrac{3}{x} + \dfrac{5}{2x^2} = \dfrac{3 \cdot 2x}{x \cdot 2x} + \dfrac{5}{2x^2} = \dfrac{6x}{2x^2} + \dfrac{5}{2x^2} = \dfrac{6x + 5}{2x^2}$

9. $\dfrac{3}{x + 2} - \dfrac{2x}{x^2 - 4} = \dfrac{3(x - 2)}{(x + 2)(x - 2)} - \dfrac{2x}{(x + 2)(x - 2)}$

$= \dfrac{3x - 6 - 2x}{(x + 2)(x - 2)}$

$= \dfrac{x - 6}{(x + 2)(x - 2)}$

13. $\dfrac{6}{x - 3} + \dfrac{8}{3 - x} = \dfrac{6}{x - 3} + \dfrac{8}{-(x - 3)}$

$= \dfrac{6}{x - 3} + \dfrac{-8}{x - 3}$

$= \dfrac{6 + (-8)}{x - 3}$

$= \dfrac{-2}{x - 3}$

$= -\dfrac{2}{x - 3}$

17. $\dfrac{-8}{x^2 - 1} - \dfrac{7}{1 - x^2} = \dfrac{-8}{x^2 - 1} - \dfrac{7}{-(x^2 - 1)}$

$= \dfrac{-8}{x^2 - 1} - \dfrac{-7}{x^2 - 1}$

$= \dfrac{-8 - (-7)}{x^2 - 1}$

$= \dfrac{-1}{x^2 - 1}$

$= -\dfrac{1}{x^2 - 1}$

21. $\dfrac{5}{x - 2} + 6 = \dfrac{5}{x - 2} + \dfrac{6}{1}$

$= \dfrac{5}{x - 2} + \dfrac{6(x - 2)}{1(x - 2)}$

$= \dfrac{5 + 6x - 12}{x - 2}$

$= \dfrac{6x - 7}{x - 2}$

25. $\dfrac{-x + 2}{x} - \dfrac{x - 6}{4x} = \dfrac{4(-x + 2)}{4x} - \dfrac{x - 6}{4x}$

$= \dfrac{-4x + 8 - (x - 6)}{4x}$

$= \dfrac{-4x + 8 - x + 6}{4x}$

$= \dfrac{-5x + 14}{4x}$ or $-\dfrac{5x - 14}{4x}$

29. $\dfrac{3x^4}{7} - \dfrac{4x^2}{21} = \dfrac{3x^4 \cdot 3}{7 \cdot 3} - \dfrac{4x^2}{21}$

$= \dfrac{9x^4}{21} - \dfrac{4x^2}{21}$

$= \dfrac{9x^4 - 4x^2}{21}$

33. $\dfrac{4}{5b} + \dfrac{1}{b - 1} = \dfrac{4(b - 1)}{5b(b - 1)} + \dfrac{1 \cdot 5b}{(b - 1)(5b)}$

$= \dfrac{4b - 4}{5b(b - 1)} + \dfrac{5b}{5b(b - 1)}$

$= \dfrac{4b - 4 + 5b}{5b(b - 1)}$

$= \dfrac{9b - 4}{5b(b - 1)}$

37. $\dfrac{2x}{x - 7} - \dfrac{x}{x - 2} = \dfrac{2x(x - 2)}{(x - 7)(x - 2)} - \dfrac{x(x - 7)}{(x - 2)(x - 7)}$

$= \dfrac{2x^2 - 4x - (x^2 - 7x)}{(x - 7)(x - 2)}$

$= \dfrac{2x^2 - 4x - x^2 + 7x}{(x - 7)(x - 2)}$

$= \dfrac{x^2 + 3x}{(x - 7)(x - 2)}$ or $\dfrac{x(x + 3)}{(x - 7)(x - 2)}$

41. $\dfrac{7}{(x + 1)(x - 1)} + \dfrac{8}{(x + 1)^2}$

$= \dfrac{7(x + 1)}{(x + 1)(x - 1)(x + 1)} + \dfrac{8(x - 1)}{(x + 1)^2(x - 1)}$

$= \dfrac{7x + 7 + 8x - 8}{(x + 1)^2(x - 1)}$

$= \dfrac{15x - 1}{(x + 1)^2(x - 1)}$

45. $\dfrac{3a}{2a + 6} - \dfrac{a - 1}{a + 3} = \dfrac{3a}{2(a + 3)} - \dfrac{(a - 1)(2)}{(a + 3)(2)}$

$= \dfrac{3a - (2a - 2)}{2(a + 3)}$

$= \dfrac{3a - 2a + 2}{2(a + 3)}$

$= \dfrac{a + 2}{2(a + 3)}$

49. $\dfrac{5}{2 - x} + \dfrac{x}{2x - 4} = \dfrac{5}{-(x - 2)} + \dfrac{x}{2(x - 2)}$

$= \dfrac{-5(2)}{(x - 2)(2)} + \dfrac{x}{2(x - 2)}$

$= \dfrac{-10 + x}{2(x - 2)}$

$= \dfrac{x - 10}{2(x - 2)}$

53. $\dfrac{13}{x^2 - 5x + 6} - \dfrac{5}{x - 3}$

$= \dfrac{13}{(x - 3)(x - 2)} - \dfrac{5(x - 2)}{(x - 3)(x - 2)}$

$= \dfrac{13 - (5x - 10)}{(x - 3)(x - 2)}$

$= \dfrac{13 - 5x + 10}{(x - 3)(x - 2)}$

$= \dfrac{-5x + 23}{(x - 3)(x - 2)}$

57. $\dfrac{x+8}{x^2-5x-6}+\dfrac{x+1}{x^2-4x-5}$

$=\dfrac{x+8}{(x+1)(x-6)}+\dfrac{x+1}{(x+1)(x-5)}$

$=\dfrac{(x+8)(x-5)+(x+1)(x-6)}{(x+1)(x-6)(x-5)}$

$=\dfrac{x^2+3x-40+x^2-5x-6}{(x+1)(x-6)(x-5)}$

$=\dfrac{2x^2-2x-46}{(x+1)(x-6)(x-5)}$

61. $\dfrac{15x}{x+8}\cdot\dfrac{2x+16}{3x}=\dfrac{5\cdot3x}{x+8}\cdot\dfrac{2(x+8)}{3x}=\dfrac{5\cdot2}{1}=10$

65. $\dfrac{5a+10}{18}\div\dfrac{a^2-4}{10a}=\dfrac{5a+10}{18}\cdot\dfrac{10a}{a^2-4}$

$=\dfrac{5(a+2)}{9\cdot2}\cdot\dfrac{2\cdot5a}{(a+2)(a-2)}$

$=\dfrac{5\cdot5a}{9(a-2)}$

$=\dfrac{25a}{9(a-2)}$

69. $3x+5=7$

$3x=2$

$x=\dfrac{2}{3}$

73. $4(x+6)+3=-3$

$4x+24+3=-3$

$4x+27=-3$

$4x=-30$

$x=\dfrac{-30}{4}$

$x=-\dfrac{15}{2}$

77. $\dfrac{5}{x^2-4}+\dfrac{2}{x^2-4x+4}-\dfrac{3}{x^2-x-6}$

$=\dfrac{5}{(x+2)(x-2)}+\dfrac{2}{(x-2)^2}-\dfrac{3}{(x+2)(x-3)}$

$=\dfrac{5(x-2)(x-3)+2(x+2)(x-3)-3(x-2)^2}{(x+2)(x-2)^2(x-3)}$

$=\dfrac{5(x^2-5x+6)+2(x^2-x-6)\div3(x^2-4x+4)}{(x-2)^2(x+2)(x-3)}$

$=\dfrac{5x^2-25x+30+2x^2-2x-12-3x^2+12x-12}{(x-2)^2(x+2)(x-3)}$

$=\dfrac{4x^2-15x+6}{(x-2)^2(x+2)(x-3)}$

81. $\dfrac{3}{x+4}-\dfrac{1}{x-4}=\dfrac{3(x-4)-1(x+4)}{(x+4)(x-4)}$

$=\dfrac{3x-12-x-4}{(x+4)(x-4)}$

$=\dfrac{2x-16}{(x+4)(x-4)}$

The other piece of the board measures $\dfrac{2x-16}{(x+4)(x-4)}$ inches.

85. answers may vary

89. answers may vary

Exercise Set 5.5

1. The LCD is 5.

$\dfrac{x}{5}+3=9$

$5\left(\dfrac{x}{5}+3\right)=5(9)$

$x+15=45$

$x=30$

Check: $\dfrac{x}{5}+3=9$

$\dfrac{30}{5}+3\overset{?}{=}9$

$6+3\overset{?}{=}9$

$9=9$ True

The solution is 30.

5. The LCD is x.

$2-\dfrac{8}{x}=6$

$x\left(2-\dfrac{8}{x}\right)=x(6)$

$2x-8=6x$

$-8=4x$

$-2=x$

Check: $2-\dfrac{8}{x}=6$

$2-\dfrac{8}{-2}\overset{?}{=}6$

$2-(-4)\overset{?}{=}6$

$2+4=6$

$6=6$ True

The solution is -2.

9. The LCD is $5\cdot2=10$.

$\dfrac{a}{5}=\dfrac{a-3}{2}$

$10\left(\dfrac{a}{5}\right)=10\left(\dfrac{a-3}{2}\right)$

$2a=5(a-3)$

$2a=5a-15$

$-3a=-15$

$a=5$

Check: $\dfrac{a}{5}=\dfrac{a-3}{2}$

$\dfrac{5}{5}\overset{?}{=}\dfrac{5-3}{2}$

$1\overset{?}{=}\dfrac{2}{2}$

$1=1$ True

The solution is 5.

13. The LCD is $2a-5$.

$\dfrac{3}{2a-5}=-1$

$(2a-5)\left(\dfrac{3}{2a-5}\right)=(2a-5)(-1)$

$3=-2a+5$

$-2=-2a$

$1=a$

Check: $\dfrac{3}{2a - 5} = -1$

$\dfrac{3}{2(1) - 5} \overset{?}{=} -1$

$\dfrac{3}{2 - 5} \overset{?}{=} -1$

$\dfrac{3}{-3} \overset{?}{=} -1$

$-1 = -1$ True

The solution is 1.

17. The LCD is $a - 3$.

$2 + \dfrac{3}{a - 3} = \dfrac{a}{a - 3}$

$(a - 3)\left(2 + \dfrac{3}{a - 3}\right) = (a - 3)\dfrac{a}{a - 3}$

$2(a - 3) + 3 = a$

$2a - 6 + 3 = a$

$2a - 3 = a$

$-3 = -a$

$3 = a$

Check: $2 + \dfrac{3}{a - 3} = \dfrac{a}{a - 3}$

$2 + \dfrac{3}{3 - 3} \overset{?}{=} \dfrac{3}{3 - 3}$

$2 + \dfrac{3}{0} \overset{?}{=} \dfrac{3}{0}$

Since $\dfrac{3}{0}$ is undefined, $a = 3$ does not check and the equation has no solution.

21. The LCD is $y + 4$.

$\dfrac{2y}{y + 4} + \dfrac{4}{y + 4} = 3$

$(y + 4)\left(\dfrac{2y}{y + 4} + \dfrac{4}{y + 4}\right) = (y + 4)(3)$

$2y + 4 = 3y + 12$

$4 = y + 12$

$-8 = y$

Check: $\dfrac{2y}{y + 4} + \dfrac{4}{y + 4} = 3$

$\dfrac{2(-8)}{-8 + 4} + \dfrac{4}{-8 + 4} \overset{?}{=} 3$

$\dfrac{-16}{-4} + \dfrac{4}{-4} \overset{?}{=} 3$

$4 - 1 \overset{?}{=} 3$

$3 = 3$ True

The solution is -8.

25. The LCD is $2y$.

$\dfrac{2}{y} + \dfrac{1}{2} = \dfrac{5}{2y}$

$2y\left(\dfrac{2}{y} + \dfrac{1}{2}\right) = 2y\left(\dfrac{5}{2y}\right)$

$2(2) + y(1) = 5$

$4 + y = 5$

$y = 1$

The solution $y = 1$ checks.

29. The LCD is $2x \cdot 3 = 6x$.

$\dfrac{11}{2x} + \dfrac{2}{3} = \dfrac{7}{2x}$

$6x\left(\dfrac{11}{2x} + \dfrac{2}{3}\right) = 6x\left(\dfrac{7}{2x}\right)$

$3(11) + 2x(2) = 3(7)$

$33 + 4x = 21$

$4x = -12$

$x = -3$

The solution $x = -3$ checks.

33. The LCD is 6.

$\dfrac{x + 1}{3} - \dfrac{x - 1}{6} = \dfrac{1}{6}$

$6\left(\dfrac{x + 1}{3} - \dfrac{x - 1}{6}\right) = 6\left(\dfrac{1}{6}\right)$

$2(x + 1) - (x - 1) = 1$

$2x + 2 - x + 1 = 1$

$x + 3 = 1$

$x = -2$

The solution $x = -2$ checks.

37. $2y + 2 = 2(y + 1)$

$4y + 4 = 4(y + 1) = 2 \cdot 2(y + 1)$

The LCD is $4(y + 1)$.

$\dfrac{y}{2y + 2} + \dfrac{2y - 16}{4y + 4} = \dfrac{2y - 3}{y + 1}$

$4(y + 1)\left(\dfrac{y}{2(y + 1)} + \dfrac{2y - 16}{4(y + 1)}\right) = 4(y + 1)\left(\dfrac{2y - 3}{y + 1}\right)$

$2y + 2y - 16 = 4(2y - 3)$

$4y - 16 = 8y - 12$

$-16 = 4y - 12$

$-4 = 4y$

$-1 = y$

The solution $y = -1$ makes the denominators $2y + 2, 4y + 4$, and $y + 1$ zero, so the equation has no solution.

41. $x^2 + x - 6 = (x + 3)(x - 2)$

The LCD is $(x + 3)(x - 2)$.

$\dfrac{x + 1}{x + 3} = \dfrac{x^2 - 11x}{x^2 + x - 6} - \dfrac{x - 3}{x - 2}$

$(x + 3)(x - 2)\left(\dfrac{x + 1}{x + 3}\right) = (x + 3)(x - 2)\left(\dfrac{x^2 - 11x}{x^2 + x - 6} - \dfrac{x - 3}{x - 2}\right)$

$(x - 2)(x + 1) = x^2 - 11x - (x + 3)(x - 3)$

$x^2 - x - 2 = x^2 - 11x - (x^2 - 9)$

$x^2 - x - 2 = x^2 - 11x - x^2 + 9$

$x^2 - x - 2 = -11x + 9$

$x^2 + 10x - 11 = 0$

$(x + 11)(x - 1) = 0$

$x + 11 = 0$ or $x - 1 = 0$

$x = -11$ $x = 1$

The solutions $x = -11$ and $x = 1$ check.

45. The LCD is $B + E$.

$T = \dfrac{2U}{B + E}$

$(B + E)T = (B + E)\left(\dfrac{2U}{B + E}\right)$

$BT + ET = 2U$

$BT = 2U - ET$

$B = \dfrac{2U - ET}{T}$

49. The LCD is G.

$N = R + \dfrac{V}{G}$

$G(N) = G\left(R + \dfrac{V}{G}\right)$

$GN = GR + V$

$GN - GR = V$

$G(N - R) = V$

$G = \dfrac{V}{N - R}$

53. The LCD is $3xy$.

$$\frac{1}{y} + \frac{1}{3} = \frac{1}{x}$$

$$3xy\left(\frac{1}{y} + \frac{1}{3}\right) = 3xy\left(\frac{1}{x}\right)$$

$$3x + xy = 3y$$

$$x(3 + y) = 3y$$

$$x = \frac{3y}{3 + y}$$

57. The reciprocal of x, added to the reciprocal of 2 is $\frac{1}{x} + \frac{1}{2}$.

61. $a^2 + 4a + 3 = (a + 3)(a + 1)$

$a^2 + a - 6 = (a + 3)(a - 2)$

$a^2 - a - 2 = (a + 1)(a - 2)$

The LCD is $(a + 3)(a + 1)(a - 2)$.

$$\frac{4}{a^2 + 4a + 3} + \frac{2}{a^2 + a - 6} - \frac{3}{a^2 - a - 2} = 0$$

$$(a + 3)(a + 1)(a - 2)\left(\frac{4}{a^2 + 4a + 3} + \frac{2}{a^2 + a - 6} - \frac{3}{a^2 - a - 2}\right)$$
$$= (a + 3)(a + 1)(a - 2)(0)$$

$$4(a - 2) + 2(a + 1) - 3(a + 3) = 0$$

$$4a - 8 + 2a + 2 - 3a - 9 = 0$$

$$3a - 15 = 0$$

$$3a = 15$$

$$a = 5$$

The solution $a = 5$ checks.

65. $\dfrac{450}{x} + \dfrac{150}{x} = 90$

The LCD is x.

$$x\left(\frac{450}{x} + \frac{150}{x}\right) = x(90)$$

$$450 + 150 = 90x$$

$$600 = 90x$$

$$\frac{600}{90} = x$$

$$\frac{20}{3} = x$$

$$\frac{450}{x} = 450 \div \frac{20}{3} = \frac{450}{1} \cdot \frac{3}{20} = \frac{1350}{20} = 67.5$$

$$\frac{150}{x} = 150 \div \frac{20}{3} = \frac{150}{1} \cdot \frac{3}{20} = \frac{450}{20} = 22.5$$

The angles measure 22.5° and 67.5°.

Exercise Set 5.6

1. The LCD is 6.

$$\frac{2}{3} = \frac{x}{6}$$

$$6\left(\frac{2}{3}\right) = 6\left(\frac{x}{6}\right)$$

$$4 = x$$

5. $\dfrac{x + 1}{2x + 3} = \dfrac{2}{3}$

$$3(x + 1) = 2(2x + 3)$$

$$3x + 3 = 4x + 6$$

$$3 = x + 6$$

$$-3 = x$$

9. Let x be the elephant's weight on Pluto.

$$\frac{100}{3} = \frac{4100}{x}$$

$$100x = 3(4100)$$

$$100x = 12{,}300$$

$$x = 123$$

The elephant weighs 123 pounds on Pluto.

13. $\dfrac{16}{10} = \dfrac{34}{y}$

$$16y = 34(10)$$

$$16y = 340$$

$$y = \frac{340}{16}$$

$$y = 21.25$$

17. Let x be the number.

$$3\left(\frac{1}{x}\right) = 9\left(\frac{1}{6}\right)$$

$$\frac{3}{x} = \frac{9}{6}$$

$$3(6) = 9x$$

$$18 = 9x$$

$$2 = x$$

The number is 2.

21. Let x be the time in hours that it takes them to complete the job working together.

The experienced surveyor completes $\frac{1}{4}$ of the job in 1 hour. The apprentice surveyor completes $\frac{1}{5}$ of the job in 1 hour. Together, they complete $\frac{1}{x}$ of the job in 1 hour.

$$\frac{1}{4} + \frac{1}{5} = \frac{1}{x}$$

The LCD is $4 \cdot 5 \cdot x = 20x$.

$$20x\left(\frac{1}{4} + \frac{1}{5}\right) = 20x\left(\frac{1}{x}\right)$$

$$5x + 4x = 20$$

$$9x = 20$$

$$x = \frac{20}{9}$$

It takes them $\dfrac{20}{9} = 2\dfrac{2}{9}$ hours to survey the roadbed together.

25. Let r be her jogging rate.

	Distance	=	Rate	·	Time
Trip to Park	12		r		$\dfrac{12}{r}$
Return Trip	18		r		$\dfrac{18}{r}$

Since the return trip took 1 hour longer,

$\dfrac{18}{r} = 1 + \dfrac{12}{r}$. The LCD is r.

$$r\left(\frac{18}{r}\right) = r\left(1 + \frac{12}{r}\right)$$

$$18 = r + 12$$

$$6 = r$$

Her jogging rate is 6 miles per hour.

29. $40 \text{ students} \cdot \dfrac{9 \text{ square feet}}{1 \text{ student}} = 360 \text{ square feet}$

40 students need a minimum of 360 square feet.

33. Let x be the time in hours that it takes Marcus and Tony to do the job working together.

Marcus lays $\frac{1}{6}$ of a slab in 1 hour. Tony lays $\frac{1}{4}$ of a slab in 1 hour. Together, they lay $\frac{1}{x}$ of the slab in 1 hour.

$$\frac{1}{6} + \frac{1}{4} = \frac{1}{x}$$

The LCD is $12x$.

$$12x\left(\frac{1}{6} + \frac{1}{4}\right) = 12x\left(\frac{1}{x}\right)$$
$$2x + 3x = 12$$
$$5x = 12$$
$$x = \frac{12}{5}$$

It will take Tony and Marcus $\frac{12}{5}$ hours to lay the slab, so the labor estimate should be $\frac{12}{5}(\$45) = \$108$.

37. $\dfrac{2}{3} = \dfrac{25}{y}$

$$2y = 3(25)$$
$$2y = 75$$
$$y = \frac{75}{2}$$

The unknown length is $y = \dfrac{75}{2} = 37.5$ feet.

41. Let x be the number.

$$\frac{2}{x - 3} - \frac{4}{x + 3} = 8 \cdot \frac{1}{x^2 - 9}$$

The LCD is $x^2 - 9 = (x + 3)(x - 3)$.

$$(x + 3)(x - 3)\left(\frac{2}{x - 3} - \frac{4}{x + 3}\right) = (x^2 - 9)\left(\frac{8}{x^2 - 9}\right)$$
$$2(x + 3) - 4(x - 3) = 8$$
$$2x + 6 - 4x + 12 = 8$$
$$-2x + 18 = 8$$
$$-2x = -10$$
$$x = 5$$

The solution $x = 5$ checks, so the number is 5.

45. Let x be the number of gallons of water needed.

$$\frac{8 \text{ tsp}}{2 \text{ gal}} = \frac{36 \text{ tsp}}{x \text{ gal}}$$
$$\frac{8}{2} = \frac{36}{x}$$
$$8x = 36(2)$$
$$8x = 72$$
$$x = 9$$

9 gallons of water are needed to mix with a box of weed killer.

49. Let x be the rate of the slower hiker. Then the rate of the faster hiker is $x + 1.1$. In 2 hours, the slower hiker walks $2x$ miles, while the faster hiker walks $2(x + 1.1)$ miles.

$$2x + 2(x + 1.1) = 11$$
$$2x + 2x + 2.2 = 11$$
$$4x + 2.2 = 11$$
$$4x = 8.8$$
$$x = 2.2$$
$$x + 1.1 = 2.2 + 1.1 = 3.3$$

The hikers walk 2.2 miles per hour and 3.3 miles per hour.

53. $\dfrac{20 \text{ feet}}{6 \text{ inches}} = \dfrac{x \text{ feet}}{8 \text{ inches}}$

$$\frac{20}{6} = \frac{x}{8}$$
$$8(20) = 6x$$
$$160 = 6x$$
$$\frac{160}{6} = x$$
$$\frac{80}{3} = x$$

The missing dimension is $\dfrac{80}{3} = 26\dfrac{2}{3}$ feet.

57. Let t be the time in hours that the jet plane travels.

	Distance	=	Rate	·	Time
Jet Plane	$500t$		500		t
Propeller Plane	$200(t + 2)$		200		$t + 2$

$$500t = 200(t + 2)$$
$$500t = 200t + 400$$
$$300t = 400$$
$$t = \frac{400}{300}$$
$$t = \frac{4}{3}$$

$$\text{distance} = 500t = 500\left(\frac{4}{3}\right) = 666\frac{2}{3}$$

The planes are $666\frac{2}{3}$ miles from the starting point.

61. Let r be the motorcycle's speed.

	Distance	=	Rate	·	Time
Car	280		$r + 10$		$\dfrac{280}{r + 10}$
Motorcycle	240		r		$\dfrac{240}{r}$

$$\frac{280}{r + 10} = \frac{240}{r}$$
$$280r = 240(r + 10)$$
$$280r = 240r + 2400$$
$$40r = 2400$$
$$r = 60$$
$$r + 10 = 60 + 10 = 70$$

The motorcycle's speed was 60 miles per hour and the car's speed was 70 miles per hour.

65. Let x be the time in minutes that it takes for the faster pump to fill the tank, so it fills $\dfrac{1}{x}$ of the tank in 1 minute. It takes the slower pump $3x$ minutes to fill the tank, so the slower pump fills $\dfrac{1}{3x}$ of the tank in 1 minute. Together, the pumps fill $\dfrac{1}{21}$ of the tank in 1 minute.

$$\frac{1}{x} + \frac{1}{3x} = \frac{1}{21}$$
$$21x\left(\frac{1}{x} + \frac{1}{3x}\right) = 21x\left(\frac{1}{21}\right)$$
$$21 + 7 = x$$
$$28 = x$$
$$3x = 3(28) = 84$$

The faster pump fills the tank in 28 minutes, while the slower pump takes 84 minutes.

69. $\dfrac{14}{7} = \dfrac{7}{y}$

$$14y = 7(7)$$
$$14y = 49$$
$$y = \frac{49}{14}$$
$$y = \frac{7}{2}$$

The missing length is $y = \dfrac{7}{2} = 3.5$.

73. $\dfrac{\dfrac{1}{4}+\dfrac{5}{4}}{\dfrac{3}{8}+\dfrac{7}{8}} = \dfrac{\dfrac{6}{4}}{\dfrac{3}{8}+\dfrac{7}{8}}$

$= \dfrac{\dfrac{6}{4}}{\dfrac{10}{8}}$

$= \dfrac{6}{4} \div \dfrac{10}{8}$

$= \dfrac{6}{4} \cdot \dfrac{8}{10}$

$= \dfrac{2\cdot3\cdot2\cdot4}{4\cdot2\cdot5}$

$= \dfrac{6}{5}$

77. $D = RT$

$\dfrac{D}{T} = \dfrac{RT}{T}$

$\dfrac{D}{T} = R \text{ or } R = \dfrac{D}{T}$

Exercise Set 5.7

1. $\dfrac{\dfrac{1}{2}}{\dfrac{3}{4}} = \dfrac{1}{2} \div \dfrac{3}{4} = \dfrac{1}{2} \cdot \dfrac{4}{3} = \dfrac{1\cdot4}{2\cdot3} = \dfrac{2}{3}$

5. $\dfrac{\dfrac{1+x}{6}}{\dfrac{1+x}{3}} = \dfrac{1+x}{6} \div \dfrac{1+x}{3} = \dfrac{1+x}{6} \cdot \dfrac{3}{1+x} = \dfrac{3(1+x)}{6(1+x)} = \dfrac{1}{2}$

9. $\dfrac{2+\dfrac{7}{10}}{1+\dfrac{3}{5}} = \dfrac{10\left(2+\dfrac{7}{10}\right)}{10\left(1+\dfrac{3}{5}\right)} = \dfrac{20+7}{10+6} = \dfrac{27}{16}$

13. $\dfrac{-\dfrac{2}{9}}{-\dfrac{14}{3}} = -\dfrac{2}{9} \div \left(-\dfrac{14}{3}\right) = -\dfrac{2}{9} \cdot \left(-\dfrac{3}{14}\right) = \dfrac{2\cdot3}{9\cdot14} = \dfrac{1}{21}$

17. $\dfrac{\dfrac{m}{n}-1}{\dfrac{m}{n}+1} = \dfrac{n\left(\dfrac{m}{n}-1\right)}{n\left(\dfrac{m}{n}+1\right)} = \dfrac{m-n}{m+n}$

21. $\dfrac{1+\dfrac{1}{y-2}}{y+\dfrac{1}{y-2}} = \dfrac{(y-2)\left(1+\dfrac{1}{y-2}\right)}{(y-2)\left(y+\dfrac{1}{y-2}\right)}$

$= \dfrac{y-2+1}{(y-2)y+1}$

$= \dfrac{y-1}{y^2-2y+1}$

$= \dfrac{y-1}{(y-1)^2}$

$= \dfrac{1}{y-1}$

25. $\dfrac{\dfrac{x}{y}+1}{\dfrac{x}{y}-1} = \dfrac{y\left(\dfrac{x}{y}+1\right)}{y\left(\dfrac{x}{y}-1\right)} = \dfrac{x+y}{x-y}$

29. $\dfrac{\dfrac{ax+ab}{x^2-b^2}}{\dfrac{x+b}{x-b}} = \dfrac{ax+ab}{x^2-b^2} \div \dfrac{x+b}{x-b}$

$= \dfrac{ax+ab}{x^2-b^2} \cdot \dfrac{x-b}{x+b}$

$= \dfrac{a(x+b)}{(x+b)(x-b)} \cdot \dfrac{x-b}{x+b}$

$= \dfrac{a}{x+b}$

33. $\dfrac{3+\dfrac{12}{x}}{1-\dfrac{16}{x^2}} = \dfrac{x^2\left(3+\dfrac{12}{x}\right)}{x^2\left(1-\dfrac{16}{x^2}\right)}$

$= \dfrac{3x^2+12x}{x^2-16}$

$= \dfrac{3x(x+4)}{(x+4)(x-4)}$

$= \dfrac{3x}{x-4}$

37. $\dfrac{\dfrac{s}{r}+\dfrac{r}{s}}{\dfrac{s}{r}-\dfrac{r}{s}} = \dfrac{rs\left(\dfrac{s}{r}+\dfrac{r}{s}\right)}{rs\left(\dfrac{s}{r}-\dfrac{r}{s}\right)} = \dfrac{s^2+r^2}{s^2-r^2}$

41. The longest bar corresponds to Steffi Graf, so Steffi Graf has won the most prize money in her career.

45. answers may vary

49. $\dfrac{1}{\dfrac{1}{R_1}+\dfrac{1}{R_2}} = \dfrac{R_1R_2(1)}{R_1R_2\left(\dfrac{1}{R_1}+\dfrac{1}{R_2}\right)} = \dfrac{R_1R_2}{R_2+R_1}$

53. $\dfrac{y^{-2}}{1-y^{-2}} = \dfrac{\dfrac{1}{y^2}}{1-\dfrac{1}{y^2}} = \dfrac{y^2\left(\dfrac{1}{y^2}\right)}{y^2\left(1-\dfrac{1}{y^2}\right)} = \dfrac{1}{y^2-1}$

Chapter 5 Test

1. $x^2+4x+3=0$

$(x+1)(x+3)=0$

$x+1=0 \quad \text{ or } \quad x+3=0$

$x=-1 \qquad\qquad x=-3$

The expression $\dfrac{x+5}{x^2+4x+3}$ is undefined for $x=-1$ and $x=-3$.

5. $\dfrac{7-x}{x-7} = \dfrac{-(x-7)}{x-7} = -1$

9. $\dfrac{x^2-13x+42}{x^2+10x+21} \div \dfrac{x^2-4}{x^2+x-6}$

$= \dfrac{x^2-13x+42}{x^2+10x+21} \cdot \dfrac{x^2+x-6}{x^2-4}$

$= \dfrac{(x-6)(x-7)}{(x+3)(x+7)} \cdot \dfrac{(x+3)(x-2)}{(x+2)(x-2)}$

$= \dfrac{(x-6)(x-7)}{(x+7)(x+2)}$

13. $\dfrac{5a}{a^2 - a - 6} - \dfrac{2}{a - 3}$

$= \dfrac{5a}{(a - 3)(a + 2)} - \dfrac{2}{a - 3}$

$= \dfrac{5a}{(a - 3)(a + 2)} - \dfrac{2(a + 2)}{(a - 3)(a + 2)}$

$= \dfrac{5a - 2(a + 2)}{(a - 3)(a + 2)}$

$= \dfrac{5a - 2a - 4}{(a - 3)(a + 2)}$

$= \dfrac{3a - 4}{(a - 3)(a + 2)}$

17. $\dfrac{4y}{y^2 + 6y + 5} - \dfrac{3}{y^2 + 5y + 4}$

$= \dfrac{4y}{(y + 5)(y + 1)} - \dfrac{3}{(y + 1)(y + 4)}$

$= \dfrac{4y(y + 4) - 3(y + 5)}{(y + 1)(y + 5)(y + 4)}$

$= \dfrac{4y^2 + 16y - 3y - 15}{(y + 1)(y + 5)(y + 4)}$

$= \dfrac{4y^2 + 13y - 15}{(y + 1)(y + 5)(y + 4)}$

21. The LCD is $x^2 - 25 = (x + 5)(x - 5)$.

$$\dfrac{10}{x^2 - 25} = \dfrac{3}{x + 5} + \dfrac{1}{x - 5}$$

$$(x^2 - 25)\left(\dfrac{10}{x^2 - 25}\right) = (x + 5)(x - 5)\left(\dfrac{3}{x + 5} + \dfrac{1}{x - 5}\right)$$

$$10 = 3(x - 5) + 1(x + 5)$$

$$10 = 3x - 15 + x + 5$$

$$10 = 4x - 10$$

$$20 = 4x$$

$$5 = x$$

Since $x = 5$ causes the denominators $x^2 - 25$ and $x - 5$ to be 0, the equation has no solution.

25. $\dfrac{5 - \dfrac{1}{y^2}}{\dfrac{1}{y} + \dfrac{2}{y^2}} = \dfrac{y^2\left(5 - \dfrac{1}{y^2}\right)}{y^2\left(\dfrac{1}{y} + \dfrac{2}{y^2}\right)} = \dfrac{5y^2 - 1}{y + 2}$

29. Let x be the time in hours that it takes for both inlet pipes together to fill the tank.

The first pipe fills $\dfrac{1}{12}$ of the tank in 1 hour, the second pipe fills $\dfrac{1}{15}$ of the tank in 1 hour, and the two pipes fill $\dfrac{1}{x}$ of the tank in 1 hour.

$$\dfrac{1}{12} + \dfrac{1}{15} = \dfrac{1}{x}$$

The LCD is $60x$.

$$60x\left(\dfrac{1}{12} + \dfrac{1}{15}\right) = 60x\left(\dfrac{1}{x}\right)$$

$$5x + 4x = 60$$

$$9x = 60$$

$$x = \dfrac{60}{9}$$

$$x = \dfrac{20}{3}$$

It takes both pipes $\dfrac{20}{3} = 6\dfrac{2}{3}$ hours to fill the tank.

CHAPTER 6

Exercise Set 6.1

1. The tallest bar corresponds to France, so France is the most popular tourist destination.

5. The height of the bar is near 40, so approximately 40 million tourists go to Italy each year.

9. The highest point corresponds to 1999, and is at a height of about 74,800.

13. The greatest decrease was from 1984 to 1986.

17.

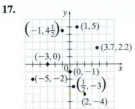

$(1, 5)$ and $(3.7, 2.2)$ are in quadrant I.

$\left(-1, 4\dfrac{1}{2}\right)$ is in quadrant II.

$(-5, -2)$ is in quadrant III.

$(2, -4)$ and $\left(\dfrac{1}{2}, -3\right)$ are in quadrant IV.

$(-3, 0)$ lies on the x-axis.

$(0, -1)$ lies on the y-axis.

21. Point C is 3 units right and 2 units up from the origin, so its coordinates are $(3, 2)$.

25. Point G is 2 units right and 1 unit down from the origin so its coordinates are $(2, -1)$.

29. Point D is 1 unit right and 3 units up from the origin, so its coordinates are $(1, 3)$.

33. a. The ordered pairs are $(2001, 28.5)$, $(2002, 29.5)$, $(2003, 32.4)$, and $(2004, 34.3)$.

b. The ordered pair $(2004, 34.3)$ indicates that in 2004, $34.3 billion was spent on pet-related expenditure.

c.

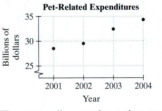

d. The scatter diagram shows that pet-related expenditures increased every year.

37. $x - 4y = 4$

In $(\ , -2)$, the y-coordinate is -2.

$$x - 4(-2) = 4$$

$$x + 8 = 4$$

$$x = -4$$

In $(4, \ )$, the x-coordinate is 4.

$$4 - 4y = 4$$

$$-4y = 0$$

$$y = 0$$

The completed coordinates are $(-4, -2)$ and $(4, 0)$.

41. $y = -7x$

$-\dfrac{1}{7}y = x$

$x = -\dfrac{1}{7}y$	$y = -7x$
0	$-7(0) = 0$
-1	$-7(-1) = 7$
$-\dfrac{1}{7}(2) = -\dfrac{2}{7}$	2

45. $y = \dfrac{1}{2}x$

$2y = x$

$x = 2y$	$y = \dfrac{1}{2}x$
0	$\dfrac{1}{2}(0) = 0$
-6	$\dfrac{1}{2}(-6) = -3$
$2(1) = 2$	1

49. $y = 2x - 12$

$y + 12 = 2x$

$\dfrac{1}{2}(y + 12) = x$

$x = \dfrac{1}{2}(y + 12)$	$y = 2x - 12$
0	$2(0) - 12 = 0 - 12 = -12$
$\dfrac{1}{2}(-2 + 12) = \dfrac{1}{2}(10) = 5$	-2
3	$2(3) - 12 = 6 - 12 = -6$

53. $x = -5y$

$y = 0: x = -5(0) = 0$

$y = 1: x = -5(1) = -5$

$x = 10: 10 = -5y$

$-2 = y$

The ordered pairs are $(0, 0)$, $(-5, 1)$, and $(10, -2)$.

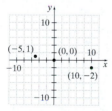

57. a.

x	100	200	300
$y = 80x + 5000$	$80(100) + 5000$ $= 8000 + 5000$ $= 13,000$	$80(200) + 5000$ $= 16,000 + 5000$ $= 21,000$	$80(300) + 5000$ $= 24,000 + 5000$ $= 29,000$

b. Find x when $y = 8600$.

$8600 = 80x + 5000$

$3600 = 80x$

$45 = x$

45 computer desks can be produced for $8600.

61. $x + y = 5$

$y = 5 - x$

65. $10x = -5y$

$-2x = y$

$y = -2x$

69. true

73. Points in quadrant IV are to the right and down from the origin, so the x-coordinate is positive and the y-coordinate is negative. (positive, negative) corresponds to quadrant IV.

77. If the x-coordinate of a point is 0, the point is neither to the left nor to the right of the origin, so it is on the y-axis.

81. answers may vary

85. The distance between $(-1, 5)$ and $(3, 5)$ and also points $(3, -4)$ and $(-1, -4)$ is each 4 units. The distance between $(3, 5)$ and $(3, -4)$ and also points $(-1, -4)$ and $(-1, 5)$ is each 9 units. The perimeter is then $9 + 9 + 4 + 4 = 26$ units.

89. answers may vary

Exercise Set 6.2

1. $x - y = 6$

$y = 0: x - 0 = 6$

$x = 6$

$x = 4: 4 - y = 6$

$-y = 2$

$y = -2$

$y = -1: x - (-1) = 6$

$x + 1 = 6$

$x = 5$

The ordered pairs are $(6, 0)$, $(4, -2)$, and $(5, -1)$.

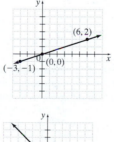

5. $y = \dfrac{1}{3}x$

$x = 0: y = \dfrac{1}{3}(0) = 0$

$x = 6: y = \dfrac{1}{3}(6) = 2$

$x = -3: y = \dfrac{1}{3}(-3) = -1$

The ordered pairs are $(0, 0)$, $(6, 2)$, and $(-3, -1)$.

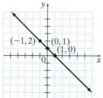

9.

13.

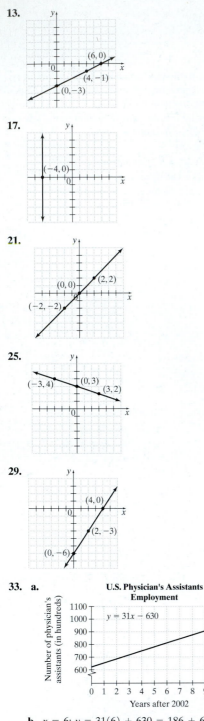

17.

21.

25.

29.

33. a.

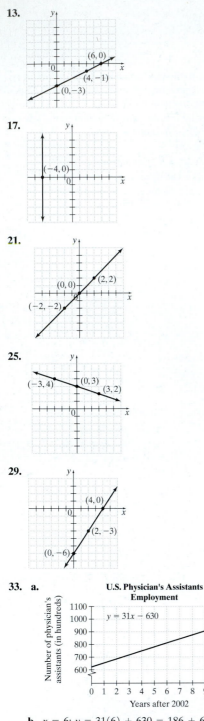

U.S. Physician's Assistants
Employment

$y = 31x - 630$

Years after 2002

b. $x = 6$: $y = 31(6) + 630 = 186 + 630 = 816$ Yes, the point
(6, 816) lies on the line.

37. The fourth vertex is the bottom-right corner of the rectangle. The
x-coordinate must line the point up with the top-right corner, and
the y-coordinate must line the point up with the bottom-left corner. The coordinates are $(4, -1)$.

41. $y = 2x$

$x = 0$: $y = 2(0) = 0$

$y = 0$: $0 = 2x$

$\quad 0 = x$

x	y
0	0
0	0

45.

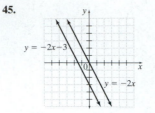

$y = -2x - 3$

$y = -2x$

49. The perimeter is the distance around.

$x + 5 + y + 5 = 22$

$x + y + 10 = 22$

$x + y = 12$

$x = 3$: $3 + y = 12$

$y = 9$

If x is 3 centimeters, then y is 9 centimeters.

Exercise Set 6.3

1. x-intercept: $(-1, 0)$
y-intercept: $(0, 1)$

5. x-intercepts: $(-2, 0), (1, 0), (3, 0)$
y-intercept: $(0, 3)$

9. $x - y = 3$

$y = 0$: $x - 0 = 0$

$\quad x = 3$

x-intercept: $(3, 0)$

$x = 0$: $0 - y = 3$

$\quad -y = 3$

$\quad y = -3$

y-intercept: $(0, -3)$

13. $-x + 2y = 6$

$y = 0$: $-x + 2(0) = 6$

$\quad -x = 6$

$\quad x = -6$

x-intercept: $(-6, 0)$

$x = 0$: $-0 + 2y = 6$

$\quad 2y = 6$

$\quad y = 3$

y-intercept: $(0, 3)$

17. $2x - y = 0$

$y = 0$: $2x - 0 = 0$

$\quad 2x = 0$

$\quad x = 0$

x-intercept: $(0, 0)$

$(0, 0)$ is also the y-intercept. Let $x = 1$ to find a second point.

$x = 1$: $2(1) - y = 0$

$\quad 2 - y = 0$

$$-y = -2$$
$$y = 2$$
$(1, 2)$ is another point on the line.

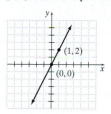

21. The graph of $x = -1$ is a vertical line with x-intercept $(-1, 0)$.

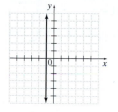

25. $y + 7 = 0$
$$y = -7$$
The graph of $y = -7$ is a horizontal line with y-intercept $(0, -7)$.

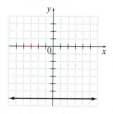

29. $x = y$
$$y = 0: x = 0$$
x-intercept: $(0, 0)$
$(0, 0)$ is also the y-intercept. Let $x = 3$ to find a second point.
$$x = 3: 3 = y$$
$(3, 3)$ is another point on the line.

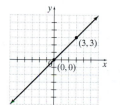

33. $5 = 6x - y$
$$y = 0: 5 = 6x - 0$$
$$5 = 6x$$
$$\frac{5}{6} = x$$
x-intercept: $\left(\frac{5}{6}, 0\right)$
$$x = 0: \quad 5 = 6(0) - y$$
$$5 = -y$$
$$-5 = y$$
y-intercept: $(0, -5)$

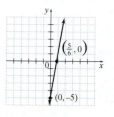

37. $x = -4\frac{1}{2}$
This is a vertical line with x-intercept
$\left(-4\frac{1}{2}, 0\right)$.

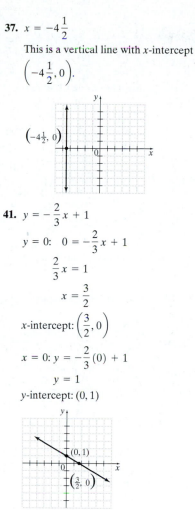

41. $y = -\frac{2}{3}x + 1$
$$y = 0: \quad 0 = -\frac{2}{3}x + 1$$
$$\frac{2}{3}x = 1$$
$$x = \frac{3}{2}$$
x-intercept: $\left(\frac{3}{2}, 0\right)$
$$x = 0: y = -\frac{2}{3}(0) + 1$$
$$y = 1$$
y-intercept: $(0, 1)$

45. $\dfrac{-6 - 3}{2 - 8} = \dfrac{-9}{-6} = \dfrac{3}{2}$

49. $\dfrac{0 - 6}{5 - 0} = \dfrac{-6}{5} = -\dfrac{6}{5}$

53. The graph of $x = 3$ is a vertical line with x-intercept $(3, 0)$. This is graph a.

57. A circle can have no x- and y-intercepts. That is, it does not have to intersect the axes.

61. a. $3x + 6y = 1200$
$$x = 0: 3(0) + 6y = 1200$$
$$6y = 1200$$
$$y = 200$$
The ordered pair $(0, 200)$ corresponds to manufacturing 0 chairs and 200 desks.

b. $3x + 6y = 1200$
$$y = 0: 3x + 6(0) = 1200$$
$$3x = 1200$$
$$x = 400$$
The ordered pair $(400, 0)$ corresponds to manufacturing 400 chairs and 0 desks.

c. Manufacturing 50 desks corresponds to $y = 50$.
$$3x + 6y = 1200$$
$$y = 50: 3x + 6(50) = 1200$$
$$3x + 300 = 1200$$
$$3x = 900$$
$$x = 300$$
When 50 desks are manufactured, 300 chairs can be manufactured.

65. a. $y = 29.2x + 919$

$x = 0: y = 29.2(0) + 919$

$y = 919$

The y-intercept is $(0, 919)$.

b. The y-intercept of $(0, 919)$ means that there were 919 stores in 1999 (0 years after 1999).

Exercise Set 6.4

1. $m = \dfrac{y_2 - y_1}{x_2 - x_1} = \dfrac{5 - (-2)}{-1 - 6} = \dfrac{7}{-7} = -1$

5. $m = \dfrac{y_2 - y_1}{x_2 - x_1} = \dfrac{1 - 1}{-2 - 5} = \dfrac{0}{-7} = 0$

9. $(x_1, y_1) = (-1, 2), (x_2, y_2) = (2, -2)$

$m = \dfrac{y_2 - y_1}{x_2 - x_1} = \dfrac{-2 - 2}{2 - (-1)} = \dfrac{-4}{3} = -\dfrac{4}{3}$

13. Line 1 has positive slope and line 2 has negative slope. Thus, line 1 has the greater slope.

17. $y = 5x - 2$

The slope is $m = 5$.

21. $2x + y = 7$

$y = -2x + 7$

The slope is $m = -2$.

25. $2x - 3y = 10$

$-3y = -2x + 10$

$y = \dfrac{2}{3}x - \dfrac{10}{3}$

The slope is $m = \dfrac{2}{3}$.

29. $x = 2y$

$\dfrac{1}{2}x = y$

$y = \dfrac{1}{2}x$

The slope is $m = \dfrac{1}{2}$.

33. $-3x - 4y = 6$

$-4y = 3x + 6$

$y = -\dfrac{3}{4}x - \dfrac{3}{2}$

The slope is $m = -\dfrac{3}{4}$.

37. $y = \dfrac{2}{9}x + 3$

$y = -\dfrac{2}{9}x$

$\dfrac{2}{9} \neq -\dfrac{2}{9}$ so the lines are not parallel.

$\left(\dfrac{2}{9}\right)\left(-\dfrac{2}{9}\right) = -\dfrac{4}{81} \neq -1$, so the lines are not perpendicular.

The lines are neither parallel nor perpendicular.

41. $6x = 5y + 1$

$-5y = -6x + 1$

$y = \dfrac{6}{5}x - \dfrac{1}{5}$

$-12x + 10y = 1$

$10y = 12x + 1$

$y = \dfrac{6}{5}x + \dfrac{1}{10}$

Both lines have slope $\dfrac{6}{5}$ and the y-intercepts are different, so they are parallel.

45. $m = \dfrac{y_2 - y_1}{x_2 - x_1} = \dfrac{0 - (-3)}{0 - (-3)} = \dfrac{3}{3} = 1$

a. The slope of a parallel line is 1.

b. The slope of a perpendicular line is $-\dfrac{1}{1} = -1$.

49. slope $= \dfrac{\text{rise}}{\text{run}} = \dfrac{6 \text{ feet}}{10 \text{ feet}} = \dfrac{3}{5}$

The pitch of the roof is $\dfrac{3}{5}$.

53. grade $= \dfrac{\text{rise}}{\text{run}} = \dfrac{2580 \text{ meters}}{6450 \text{ meters}} = 0.40 = 40\%$

The grade of the track is 40%.

57. $m = \dfrac{y_2 - y_1}{x_2 - x_1} = \dfrac{86 - 74}{2006 - 2002} = \dfrac{12}{4} = 3 = \dfrac{3}{1}$

Every 1 year, there are/should be 3 million more U.S. households with personal computers.

61. $y - (-6) = 2(x - 4)$

$y + 6 = 2x - 8$

$y = 2x - 14$

65. $(x_1, y_1) = (0, 0), (x_2, y_2) = (1, 1)$

$m = \dfrac{y_2 - y_1}{x_2 - x_1} = \dfrac{1 - 0}{1 - 0} = \dfrac{1}{1} = 1$

The slope is $m = 1$; d.

69. $(x_1, y_1) = (2, 0), (x_2, y_2) = (4, -1)$

$m = \dfrac{y_2 - y_1}{x_2 - x_1} = \dfrac{-1 - 0}{4 - 2} = \dfrac{-1}{2} = -\dfrac{1}{2}$

The slope is $m = -\dfrac{1}{2}$; e.

73. answers may vary

77. The lowest points on the graph correspond to 1994 and 2000. The average fuel economy for those years was 28.1 miles per gallon.

81. The run is $\dfrac{1}{2}x$.

slope $= \dfrac{\text{rise}}{\text{run}}$

$\dfrac{2}{5} = \dfrac{4}{\dfrac{x}{2}}$

$2\left(\dfrac{x}{2}\right) = 5(4)$

$x = 20$

85. One side is formed by the points $(1, 3)$ and $(2, 1)$. The slope is

$m = \dfrac{y_2 - y_1}{x_2 - x_1} = \dfrac{1 - 3}{2 - 1} = \dfrac{-2}{1} = -2$.

The opposite side is formed by the points $(-4, 0)$ and $(-3, -2)$.

The slope is $m = \dfrac{y_2 - y_1}{x_2 - x_1} = \dfrac{-2 - 0}{-3 - (-4)} = \dfrac{-2}{1} = -2$.

Since the slopes are the same, the sides are parallel.

Another side is formed by the points $(1, 3)$ and $(-4, 0)$. The slope is

$m = \dfrac{y_2 - y_1}{x_2 - x_1} = \dfrac{0 - 3}{-4 - 1} = \dfrac{-3}{-5} = \dfrac{3}{5}$.

The opposite side is formed by the points $(2, 1)$ and $(-3, -2)$. The slope is

$m = \dfrac{y_2 - y_1}{x_2 - x_1} = \dfrac{-2 - 1}{-3 - 2} = \dfrac{-3}{-5} = \dfrac{3}{5}$

Since the slopes are the same, the sides are parallel.

The opposite sides of the quadrilateral are parallel, so the quadrilateral is a parallelogram.

89. $m = \dfrac{y_2 - y_1}{x_2 - x_1} = \dfrac{-2.9 - (-10.1)}{9.8 - 14.3} = \dfrac{7.2}{-4.5} = -1.6$

Exercise Set 6.5

1. $y = mx + b$
$y = 5x + 3$

5. $y = mx + b$
$y = \dfrac{2}{3}x + 0$
$y = \dfrac{2}{3}x$

9. $y = mx + b$
$y = -\dfrac{1}{5}x + \dfrac{1}{9}$

13. From $y = \dfrac{2}{3}x + 5$, the y-intercept is $(0, 5)$. The slope is $\dfrac{2}{3}$, so another point on the graph is $(0 + 3, 5 + 2)$ or $(3, 7)$.

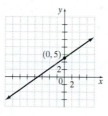

17. $4x + y = 6$
$y = -4x + 6$

the slope is $-4 = \dfrac{-4}{1}$ and the y-intercept is $(0, 6)$. Another point on the graph is $(0 + 1, 6 - 4)$ or $(1, 2)$.

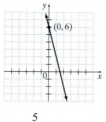

21. $x = \dfrac{5}{4}y$
$\dfrac{4}{5}x = y$
$y = \dfrac{4}{5}x + 0$

The slope is $\dfrac{4}{5}$ and the y-intercept is $(0, 0)$. Another point on the line is $(0 + 5, 0 + 4)$ or $(5, 4)$.

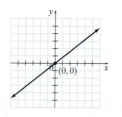

25. $y - y_1 = m(x - x_1)$
$y - (-5) = -8[x - (-1)]$
$y + 5 = -8(x + 1)$
$y + 5 = -8x - 8$
$y = -8x - 13$
$8x + y = -13$

29. $y - y_1 = m(x - x_1)$
$y - 0 = -\dfrac{1}{2}[x - (-3)]$
$y = -\dfrac{1}{2}(x + 3)$
$2y = -1(x + 3)$
$2y = -x - 3$
$x + 2y = -3$

33. $m = \dfrac{y_2 - y_1}{x_2 - x_1} = \dfrac{-5 - 3}{-2 - (-1)} = \dfrac{-8}{-1} = 8$
$y - y_1 = m(x - x_1)$
$y - 3 = 8[x - (-1)]$
$y - 3 = 8(x + 1)$
$y - 3 = 8x + 8$
$y - 11 = 8x$
$-11 = 8x - y$
$8x - y = -11$

37. $m = \dfrac{y_2 - y_1}{x_2 - x_1} = \dfrac{\dfrac{1}{13} - 0}{-\dfrac{1}{8} - 0} = \dfrac{\dfrac{1}{13}}{-\dfrac{1}{8}} = \dfrac{1}{13}\left(-\dfrac{8}{1}\right) = -\dfrac{8}{13}$
$y - y_1 = m(x - x_1)$
$y - 0 = -\dfrac{8}{13}(x - 0)$
$y = -\dfrac{8}{13}x$
$13y = -8x$
$8x + 13y = 0$

41. $m = \dfrac{y_2 - y_1}{x_2 - x_1} = \dfrac{10 - 7}{7 - 10} = \dfrac{3}{-3} = -1$
$y - y_1 = m(x - x_1)$
$y - 7 = -1(x - 10)$
$y - 7 = -x + 10$
$y = -x + 17$

45. $y - y_1 = m(x - x_1)$
$y - 9 = 1[x - (-7)]$
$y - 9 = x + 7$
$y = x + 16$

49. A line parallel to the x-axis is a horizontal line. $y = 7$

53. A line perpendicular to the y-axis is a horizontal line.
$y = -3$

57. a. The ordered pairs are $(1, 32)$ and $(3, 96)$.
$m = \dfrac{s_2 - s_1}{t_2 - t_1} = \dfrac{96 - 32}{3 - 1} = \dfrac{64}{2} = 32$
$s - s_1 = m(t - t_1)$
$s - 32 = 32(t - 1)$
$s - 32 = 32t - 32$
$s = 32t$

b. $t = 4$: $s = 32(4)$
$s = 128$

The speed of the rock 4 seconds after it was dropped is 128 feet per second.

61. a. The ordered pairs are $(4, 5700)$ and $(0, 7032)$.
$m = \dfrac{y_2 - y_1}{x_2 - x_1} = \dfrac{7032 - 5700}{0 - 4} = \dfrac{1332}{-4} = -333$
$y - y_1 = m(x - x_1)$

$y - 7032 = -333(x - 0)$

$y - 7032 = -333x$

$y = -333x + 7032$

b. The year 2007 is 8 years past 1999, so it corresponds to $x = 8$.

$x = 8: y = -333(8) + 7032$

$y = -2664 + 7032$

$y = 4368$

4368 cinema sites are predicted for 2007.

65. a. The ordered pairs are $(3, 10,000)$ and $(5, 8000)$.

$m = \dfrac{S_2 - S_1}{p_2 - p_1}$

$= \dfrac{8000 - 10,000}{5 - 3}$

$= \dfrac{-2000}{2}$

$= -1000$

$S - S_1 = m(p - p_1)$

$S - 10,000 = -1000(p - 3)$

$S - 10,000 = -1000p + 3000$

$S = -1000p + 13,000$

b. $p = 3.50: S = -1000(3.50) + 13,000$

$S = -3500 + 13,000$

$S = 9500$

9500 Fun Noodles will be sold when the price is $3.50 each.

69. $x = -1: x^2 - 3x + 1 = (-1)^2 - 3(-1) + 1$

$= 1 + 3 + 1$

$= 5$

73. The graph of $y = -3x - 2$ has slope $m = -3$ and y-intercept $(0, -2)$. This is graph d.

77. a. A line parallel to the line $y = 3x - 1$ will have slope 3.

$y - y_1 = m(x - x_1)$

$y - 2 = 3[x - (-1)]$

$y - 2 = 3(x + 1)$

$y - 2 = 3x + 3$

$y = 3x + 5$

$-5 = 3x - y$

$3x - y = -5$

b. A line perpendicular to the line $y = 3x - 1$ will have slope $-\dfrac{1}{3}$.

$y - y_1 = m(x - x_1)$

$y - 2 = -\dfrac{1}{3}[x - (-1)]$

$y - 2 = -\dfrac{1}{3}(x + 1)$

$3y - 6 = -1(x + 1)$

$3y - 6 = -x - 1$

$3y = -x + 5$

$x + 3y = 5$

Exercise Set 6.6

1. The domain is the set of x-coordinates:

$\{-7, 0, 2, 10\}$.

The range is the set of y-coordinates:

$\{-7, 0, 4, 10\}$.

5. Each x-value is only assigned to one y-value, so the relation is a function.

9. The vertical line $x = 1$ will intersect the graph in two points, so the graph is not the graph of a function.

13. No vertical line will intersect the graph more than once, so the graph is the graph of a function.

17. On June 1, the graph shows sunset to be at approximately 9:30 P.M.

21. The graph passes the vertical line test, so it is the graph of a function.

25. According to the graph, the minimum wage increased to over $5.75 in 2005.

29. $f(x) = 2x - 5$

$f(-2) = 2(-2) - 5 = -4 - 5 = -9$

$f(0) = 2(0) - 5 = 0 - 5 = -5$

$f(3) = 2(3) - 5 = 6 - 5 = 1$

33. $f(x) = 3x$

$f(-2) = 3(-2) = -6$

$f(0) = 3(0) = 0$

$f(3) = 3(3) = 9$

37. $h(x) = -5x$

$h(-1) = -5(-1) = 5$

$h(0) = -5(0) = 0$

$h(4) = -5(4) = -20$

41. The ordered-pair solution corresponding to $f(3) = 6$ is $(3, 6)$.

45. When $x = 0$, $y = -1$, so $f(0) = -1$.

49. $2x + 5 < 7$

$2x < 2$

$x < 1$

53. $\dfrac{3}{x} + \dfrac{3}{2x} + \dfrac{5}{x} = \dfrac{3 \cdot 2}{x \cdot 2} + \dfrac{3}{2x} + \dfrac{5 \cdot 2}{x \cdot 2}$

$= \dfrac{6}{2x} + \dfrac{3}{2x} + \dfrac{10}{2x}$

$= \dfrac{6 + 3 + 10}{2x}$

$= \dfrac{19}{2x}$

The perimeter is $\dfrac{19}{2x}$ meters.

57. answers may vary

61. $f(x) = \dfrac{136}{25}x$

a. $f(35) = \dfrac{136}{25}(35) = \dfrac{4760}{25} = \dfrac{952}{5} = 190.4$

The proper dosage for a 35-pound dog is 190.4 milligrams.

b. $f(70) = \dfrac{136}{25}(70) = \dfrac{9520}{25} = \dfrac{1904}{5} = 380.8$

The proper dosage for a 70-pound dog is 380.8 milligrams.

Exercise Set 6.7

1. $x - y > 3$

$(0, 3): 0 - 3 > 3$

$-3 > 3$ False

$(0, 3)$ is not a solution of the inequality.

$(2, -1): 2 - (-1) > 3$

$2 + 1 > 3$

$3 > 3$ False

$(2, -1)$ is not a solution of the inequality.

5. $x < -y$

$(0, 2): 0 < -2$ False

$(0, 2)$ is not a solution of the inequality.

$(-5, 1): -5 < -1$ True

$(-5, 1)$ is a solution of the inequality.

9. Graph the boundary line, $2x - y = -4$, with a dashed line.

Test $(0, 0)$: $2x - y > -4$

$2(0) - 0 > -4$

$0 - 0 > -4$

$0 > -4$ True

Shade the half-plane containing $(0, 0)$.

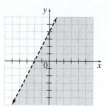

13. Graph the boundary line, $x = -3y$, with a solid line.

Test $(1, 1)$: $x \leq -3y$

$$1 \leq -3(1)$$
$$1 \leq -3 \quad \text{False}$$

Shade the half-plane not containing $(1, 1)$.

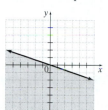

17. Graph the boundary line, $y = 4$, with a dashed line.

Test $(0, 0)$: $y < 4$

$$0 < 4 \quad \text{True}$$

Shade the half-plane containing $(0, 0)$.

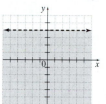

21. Graph the boundary line, $5x + 2y = 10$, with a solid line.

Test $(0, 0)$: $5x + 2y \leq 10$

$$5(0) + 2(0) \leq 10$$
$$0 + 0 \leq 10$$
$$0 \leq 10 \quad \text{True}$$

Shade the half-plane containing $(0, 0)$.

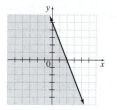

25. Graph the boundary line, $x - y = 6$, with a solid line.

Test $(0, 0)$: $x - y \leq 6$

$$0 - 0 \leq 6$$
$$0 \leq 6 \quad \text{True}$$

Shade the half-plane containing $(0, 0)$.

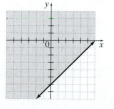

29. Shade the boundary line, $2x + 7y \, 1 = 5$, with a dashed line.

Test $(0, 0)$: $2x + 7y > 5$

$$2(0) + 7(0) > 5$$
$$0 + 0 > 5$$
$$0 > 5 \quad \text{False}$$

Shade the half-plane not containing $(0, 0)$.

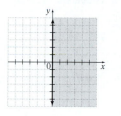

33. The point of intersection appears to be $(-2, 1)$.

37. The graph is the half-plane with boundary line $x = 2$; a.

41. answers may vary

45. Test $(1, 1)$: $y \geq -\dfrac{1}{2}x$

$$1 \geq -\frac{1}{2}(1)$$
$$1 \geq -\frac{1}{2} \quad \text{True}$$

$(1, 1)$ is included in the graph of $y \geq -\dfrac{1}{2}x$.

Exercise Set 6.8

1. $y = kx$

$$3 = k(6)$$
$$\frac{1}{2} = k$$
$$y = \frac{1}{2}x$$

5. $m = \dfrac{y_2 - y_1}{x_2 - x_1} = \dfrac{3 - 0}{1 - 0} = \dfrac{3}{1} = 3$

$$y = 3x$$

9. $y = \dfrac{k}{x}$

$$7 = \frac{k}{1}$$
$$7 = k$$
$$y = \frac{7}{x}$$

13. y varies directly as x is written as $y = kx$.

17. z varies directly as x^2 is written as $z = kx^2$.

21. x varies inversely as $\sqrt{y}$ is written as $x = \dfrac{k}{\sqrt{y}}$.

25. $y = \dfrac{k}{x}$

$y = 5$ when $x = 60$: $\quad 5 = \dfrac{k}{60}$

$$300 = k$$
$$y = \frac{300}{x}$$

$x = 100$: $y = \dfrac{300}{100} = 3$

$y = 3$ when $x = 100$.

29. $a = \dfrac{k}{b^3}$

$a = \dfrac{3}{2}$ when $b = 2$: $\quad \dfrac{3}{2} = \dfrac{k}{2^3}$

$$\frac{3}{2} = \frac{k}{8}$$
$$12 = k$$

$$a = \frac{12}{b^3}$$

$$b = 3: a = \frac{12}{3^3} = \frac{12}{27} = \frac{4}{9}$$

$a = \dfrac{4}{9}$ when $b = 3$.

33. Let c be the cost per headphone when h headphones are manufactured.

$$c = \frac{k}{h}$$

$c = 9$ when $h = 5000$: $9 = \dfrac{k}{5000}$

$$45{,}000 = k$$

$$c = \frac{45{,}000}{h}$$

$h = 7500: c = \dfrac{45{,}000}{7500} = 6$

The cost to manufacture 7500 headphones is \$6 per headphone.

37. Let w be the weight of an object when it is d miles from the center of the Earth.

$$w = \frac{k}{d^2}$$

$w = 180$ when $d = 4000$:

$$180 = \frac{k}{4000^2}$$

$$180 = \frac{k}{16{,}000{,}000}$$

$$2{,}880{,}000{,}000 = k$$

$$w = \frac{2{,}880{,}000{,}000}{d^2}$$

$$w = \frac{2{,}880{,}000{,}000}{4010^2}$$

$d = 4010$: $= \dfrac{2{,}880{,}000{,}000}{16{,}080{,}100}$

$$\approx 179.1$$

The man will weigh about 179.1 pounds when he is 10 miles above the surface of the Earth.

41. $-3x + 4y = 7$

$\underline{3x - 2y = 9}$

$2y = 16$

45. If y varies directly as x, then $y = kx$. If x is tripled, to become $3x$, then $y = k(3x) = 3(kx)$, and y is multiplied by 3.

Chapter 6 Test

1. $12y - 7x = 5$

$x = 1: 12y - 7(1) = 5$

$12y - 7 = 5$

$12y = 12$

$y = 1$

The ordered pair is $(1, 1)$.

5. $m = \dfrac{y_2 - y_1}{x_2 - x_1} = \dfrac{2 - (-5)}{-1 - 6} = \dfrac{7}{-7} = -1$

9. $2x + y = 8$

$y = 0: 2x + 0 = 8$

$2x = 8$

$x = 4$

x-intercept: $(4, 0)$

$x = 0: 2(0) + y = 8$

$y = 8$

y-intercept: $(0, 8)$

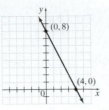

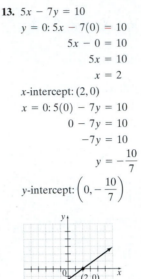

13. $5x - 7y = 10$

$y = 0: 5x - 7(0) = 10$

$5x - 0 = 10$

$5x = 10$

$x = 2$

x-intercept: $(2, 0)$

$x = 0: 5(0) - 7y = 10$

$0 - 7y = 10$

$-7y = 10$

$y = -\dfrac{10}{7}$

y-intercept: $\left(0, -\dfrac{10}{7}\right)$

17. $y = 2x - 6$

The slope is $m = 2$ and the y-intercept is $(0, -6)$.

$-4x = 2y$

$-2x = y$

$y = -2x + 0$

The slope is $m = -2$ and the y-intercept is $(0, 0)$. Since the slopes are different, the lines are not parallel. Since $2(-2) = -4 \neq -1$, the lines are not perpendicular. The lines are neither parallel nor perpendicular.

21. $m = \dfrac{1}{8}; b = 12$

$$y = \frac{1}{8}x + 12$$

$$8y = x + 8(12)$$

$$8y = x + 96$$

$$x - 8y = -96$$

25. No vertical line will intersect the graph more than once, so the graph is the graph of a function.

29. a. The ordered pairs are $(2000, 69.3)$, $(2001, 70.0)$, $(2002, 69.9)$, $(2003, 70.1)$, $(2004, 70.3)$, $(2005, 70.5)$.

b.

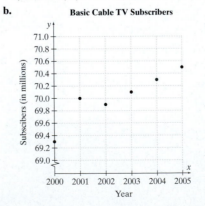

Basic Cable TV Subscribers

CHAPTER 7

Exercise Set 7.1

1. a. First equation:

$x + y = 8$

$2 + 4 \stackrel{?}{=} 8$

$6 = 8$ False

$(2, 4)$ is not a solution of the first equation, so it is not a solution of system.

b. First equation:

$x + y = 8$

$5 + 3 \stackrel{?}{=} 8$

$8 = 8$ True

Second equation:

$3x + 2y = 21$

$3(5) + 2(3) \stackrel{?}{=} 21$

$15 + 6 \stackrel{?}{=} 21$

$21 = 21$ True

Since $(5, 3)$ is a solution of both equations, it is a solution of the system.

5. a. First equation:

$2y = 4x + 6$

$2(-3) \stackrel{?}{=} 4(-3) + 6$

$-6 \stackrel{?}{=} -12 + 6$

$-6 = -6$ True

Second equation:

$2x - y = -3$

$2(-3) - (-3) \stackrel{?}{=} -3$

$-6 + 3 \stackrel{?}{=} -3$

$-3 = -3$ True

Since $(-3, -3)$ is a solution of both equations, it is a solution of the system.

b. First equation:

$2y = 4x + 6$

$2(3) \stackrel{?}{=} 4(0) + 6$

$6 \stackrel{?}{=} 0 + 6$

$6 = 6$ True

Second equation:

$2x - y = 0$

$2(0) - 3 \stackrel{?}{=} 0$

$0 - 3 \stackrel{?}{=} 0$

$-3 = 0$ False

$(0, 3)$ is not a solution of the second equation, so it is not a solution of the system.

9.

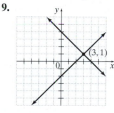

First equation:

$x + y = 4$

$3 + 1 \stackrel{?}{=} 4$

$4 = 4$ True

Second equation:

$x - y = 2$

$3 - 1 \stackrel{?}{=} 2$

$2 = 2$ True

The solution of the system is $(3, 1)$.

13.

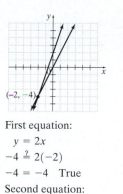

First equation:

$y = 2x$

$-4 \stackrel{?}{=} 2(-2)$

$-4 = -4$ True

Second equation:

$3x - y = -2$

$3(-2) - (-4) \stackrel{?}{=} -2$

$-6 + 4 \stackrel{?}{=} -2$

$-2 = -2$ True

The solution of the system is $(-2, -4)$.

17.

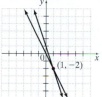

First equation:

$2x + y = 0$

$2(1) + (-2) \stackrel{?}{=} 0$

$2 - 2 \stackrel{?}{=} 0$

$0 = 0$ True

Second equation:

$3x + y = 1$

$3(1) + (-2) \stackrel{?}{=} 1$

$3 - 2 \stackrel{?}{=} 1$

$1 = 1$ True

The solution of the system is $(1, -2)$.

21.

First equation:

$x + y = 5$

$y = -x + 5$

Second equation:

$x + y = 6$

$y = -x + 6$

The lines have the same slope, but different y-intercepts, so they are parallel. The system has no solution.

25.

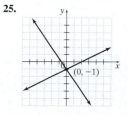

First equation:

$x - 2y = 2$

$0 - 2(-1) \stackrel{?}{=} 2$

$0 + 2 \stackrel{?}{=} 2$

$2 = 2$ True

Second equation:

$3x + 2y = -2$

$3(0) + 2(-1) \stackrel{?}{=} -2$

$0 - 2 \stackrel{?}{=} -2$

$-2 = -2$ True

The solution of the system is $(0, -1)$.

29.

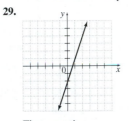

First equation:

$y - 3x = -2$

$y = 3x - 2$

Second equation:

$6x - 2y = 4$

$-2y = -6x + 4$

$y = 3x - 2$

The graphs of the equations are the same line, so the system has an infinite number of solutions.

33.

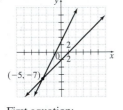

First equation:

$y = x - 2$

$-7 \stackrel{?}{=} -5 - 2$

$-7 = -7$ True

Second equation:

$y = 2x + 3$

$-7 \stackrel{?}{=} 2(-5) + 3$

$-7 \stackrel{?}{=} -10 + 3$

$-7 = -7$ True

The solution of the system is $(-5, -7)$.

37.

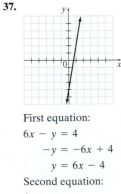

First equation:

$6x - y = 4$

$-y = -6x + 4$

$y = 6x - 4$

Second equation:

$\dfrac{1}{2}y = -2 + x$

$y = -4 + 6x$

$y = 6x - 4$

The graphs of the equations are the same line, so the system has an infinite number of solutions.

41. $4\left(\dfrac{y + 1}{2}\right) + 3y = 0$

$2(y + 1) + 3y = 0$

$2y + 2 + 3y = 0$

$5y + 2 = 0$

$5y = -2$

$y = -\dfrac{2}{5}$

45. answers may vary

49. The lines cross at points corresponding to the years 1984 and 1988. The number of pounds of imported fishery products was equal to the number of pounds of domestic catch in 1984 and 1988.

53. answers may vary

57. answers may vary

Exercise Set 7.2

1. $\begin{cases} x + y = 3 \\ x = 2y \end{cases}$

Substitute $2y$ for x in the first equation and solve for y.

$x + y = 3$

$2y + y = 3$

$3y = 3$

$y = 1$

Now solve for x.

$x = 2y = 2(1) = 2$

The solution of the system is $(2, 1)$.

5. $\begin{cases} y = 3x + 1 \\ 4y - 8x = 12 \end{cases}$

Substitute $3x + 1$ for y in the second equation and solve for x.

$4y - 8x = 12$

$4(3x + 1) - 8x = 12$

$12x + 4 - 8x = 12$

$4x + 4 = 12$

$4x = 8$

$x = 2$

Now solve for y.

$y = 3x + 1 = 3(2) + 1 = 6 + 1 = 7$

The solution of the system is $(2, 7)$.

9. $\begin{cases} 3x - 4y = 10 \\ y = x - 3 \end{cases}$

Substitute $x - 3$ for y in the first equation and solve for x.

$3x - 4y = 10$

$3x - 4(x - 3) = 10$

$3x - 4x + 12 = 10$

$-x + 12 = 10$

$-x = -2$

$x = 2$

Now solve for y.

$y = x - 3 = 2 - 3 = -1$

The solution of the system is $(2, -1)$.

13. $\begin{cases} 3x + 2y = 16 \\ x = 3y - 2 \end{cases}$

Substitute $3y - 2$ for x in the first equation and solve for y.

$3x + 2y = 16$

$3(3y - 2) + 2y = 16$

$9y - 6 + 2y = 16$

$11y - 6 = 16$

$11y = 22$

$y = 2$

Now solve for x.

$x = 3y - 2 = 3(2) - 2 = 6 - 2 = 4$

The solution of the system is $(4, 2)$.

17. $\begin{cases} 4x + 2y = 5 \\ -2x = y + 4 \end{cases}$

Solve the second equation for x.

$-2x = y + 4$

$x = -\dfrac{1}{2}y - 2$

Substitute $-\dfrac{1}{2}y - 2$ for x in the first equation and solve for y.

$4x + 2y = 5$

$4\left(-\dfrac{1}{2}y - 2\right) + 2y = 5$

$-2y - 8 + 2y = 5$

$-8 = 5 \quad$ False

Since the statement $-8 = 5$ is false, the system has no solution.

21. $\begin{cases} x + 2y + 5 = -4 + 5y - x \\ 2x + x = y + 4 \end{cases}$

Simplify each equation.

$\begin{cases} 2x + 9 = 3y \\ 3x = y + 4 \end{cases}$

Solve the second simplified equation for y.

$3x = y + 4$

$3x - 4 = y$

Substitute $3x - 4$ for y in the first simplified equation and solve for x.

$2x + 9 = 3y$

$2x + 9 = 3(3x - 4)$

$2x + 9 = 9x - 12$

$2x + 21 = 9x$

$21 = 7x$

$3 = x$

Now solve for y.

$y = 3x - 4 = 3(3) - 4 = 9 - 4 = 5$

The solution of the system is $(3, 5)$.

25. $\begin{cases} 3x - y = 1 \\ 2x - 3y = 10 \end{cases}$

Solve the first equation for y.

$3x - y = 1$

$-y = -3x + 1$

$y = 3x - 1$

Substitute $3x - 1$ for y in the second equation and solve for x.

$2x - 3y = 10$

$2x - 3(3x - 1) = 10$

$2x - 9x + 3 = 10$

$-7x + 3 = 10$

$-7x = 7$

$x = -1$

Now solve for y.

$y = 3x - 1 = 3(-1) - 1 = -3 - 1 = -4$

The solution of the system is $(-1, -4)$.

29. $\begin{cases} 5x + 10y = 20 \\ 2x + 6y = 10 \end{cases}$

Solve the first equation for x. (Note that the second equation could also be easily solved for x.)

$5x + 10y = 20$

$5x = -10y + 20$

$x = -2y + 4$

Substitute $-2y + 4$ for x in the second equation and solve for y.

$2x + 6y = 10$

$2(-2y + 4) + 6y = 10$

$-4y + 8 + 6y = 10$

$2y + 8 = 10$

$2y = 2$

$y = 1$

Now solve for x.

$x = -2y + 4 = -2(1) + 4 = -2 + 4 = 2$

The solution of the system is $(2, 1)$.

33. $\begin{cases} \dfrac{1}{3}x - y = 2 \\ x - 3y = 6 \end{cases}$

Solve the second equation for x.

$x - 3y = 6$

$x = 3y + 6$

Substitute $3y + 6$ for x in the first equation and solve for y.

$\dfrac{1}{3}x - y = 2$

$\dfrac{1}{3}(3y + 6) - y = 2$

$y + 2 - y = 2$

$2 = 2$

Since $2 = 2$ is a true statement, the two equations in the original system are equivalent. The system has an infinite number of solutions.

37. $\quad 3x + 2y = 6$

$-2(3x + 2y) = -2(6)$

$\overline{\quad -6x - 4y = -12}$

41. $\quad 3n + 6m$

$\underline{+2n - 6m}$

$\quad 5n$

45. $\begin{cases} -5y + 6y = 3x + 2(x - 5) - 3x + 5 \\ 4(x + y) - x + y = -12 \end{cases}$

Simplify each equation.

$\begin{cases} y = 2x - 5 \\ 3x + 5y = -12 \end{cases}$

Substitute $2x - 5$ for y in the second simplified equation and solve for x.

$3x + 5y = -12$

$3x + 5(2x - 5) = -12$

$3x + 10x - 25 = -12$

$13x - 25 = -12$

$13x = 13$

$x = 1$

Now solve for y.

$y = 2x - 5 = 2(1) - 5 = 2 - 5 = -3$

The solution of the system is $(1, -3)$.

49. no; answers may vary

53. Using a graphing calculator, the solution of the system is $(-2.6, 1.3)$.

57. a. $\begin{cases} y = -0.50x + 21.92 \\ y = 0.71x + 11.03 \end{cases}$

Substitute $-0.50x + 21.92$ for y in the second equation and solve for x.

$y = 0.71x + 11.03$

$-0.50x + 21.92 = 0.71x + 11.03$

$-0.50x + 10.89 = 0.71x$

$10.89 = 1.21x$

$9 = x$

Now solve for y.

$y = -0.50x + 21.92$

$= -0.50(9) + 21.92$

$= -4.50 + 21.92 = 17.42$

Rounded to the nearest whole numbers, the solution of the system is $(9, 17)$.

b. In $1973 + 9 = 1982$, the percent of households that used fuel oil and electricity was the same, 17%.

c.

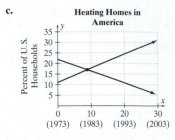

Heating Homes in America

Exercise Set 7.3

1. $\begin{cases} 3x + y = 5 \\ 6x - y = 4 \end{cases}$

Add the equations to eliminate y, then solve for x.

$3x + y = 5$

$\underline{6x - y = 4}$

$9x \quad\;\; = 9$

$\quad\;\; x = 1$

Now solve for y.

$3x + y = 5$

$3(1) + y = 5$

$3 + y = 5$

$y = 2$

The solution of the system is $(1, 2)$.

5. $\begin{cases} 3x + y = -11 \\ 6x - 2y = -2 \end{cases}$

Multiply the first equation by 2.

$\begin{cases} 2(3x + y) = 2(-11) \\ 6x - 2y = -2 \end{cases} \rightarrow \begin{cases} 6x + 2y = -22 \\ 6x - 2y = -2 \end{cases}$

Add the equations to eliminate y, then solve for x.

$6x + 2y = -22$

$\underline{6x - 2y = -2}$

$12x \quad\;\;\; = -24$

$\quad\;\; x = -2$

Now solve for y.

$3x + y = -11$

$3(-2) + y = -11$

$-6 + y = -11$

$y = -5$

The solution of the system is $(-2, -5)$.

9. $\begin{cases} x + 5y = 18 \\ 3x + 2y = -11 \end{cases}$

Multiply the first equation by -3.

$\begin{cases} -3(x + 5y) = -3(18) \\ 3x + 2y = -11 \end{cases} \rightarrow \begin{cases} -3x - 15y = -54 \\ 3x + 2y = -11 \end{cases}$

Add the equations to eliminate x, then solve for y.

$-3x - 15y = -54$

$\underline{3x + 2y = -11}$

$-13y = -65$

$y = 5$

Now solve for x.

$x + 5y = 18$

$x + 5(5) = 18$

$x + 25 = 18$

$x = -7$

The solution of the system is $(-7, 5)$.

13. $\begin{cases} 2x + 3y = 0 \\ 4x + 6y = 3 \end{cases}$

Multiply the first equation by -2.

$\begin{cases} -2(2x + 3y) = -2(0) \\ 4x + 6y = 3 \end{cases} \rightarrow \begin{cases} -4x - 6y = 0 \\ 4x + 6y = 3 \end{cases}$

Add the equations to eliminate x.

$-4x - 6y = 0$

$\underline{4x + 6y = 3}$

$0 = 3$

Since the statement $0 = 3$ is false, the system has no solution.

17. $\begin{cases} 3x - 2y = 7 \\ 5x + 4y = 8 \end{cases}$

Multiply the first equation by 2.

$\begin{cases} 2(3x - 2y) = 2(7) \\ 5x + 4y = 8 \end{cases} \rightarrow \begin{cases} 6x - 4y = 14 \\ 5x + 4y = 8 \end{cases}$

Add the equations to eliminate y, then solve for x.

$6x - 4y = 14$

$\underline{5x + 4y = 8}$

$11x \quad\;\;\; = 22$

$\quad\;\; x = 2$

Now solve for y.

$3x - 2y = 7$

$3(2) - 2y = 7$

$6 - 2y = 7$

$-2y = 1$

$y = -\dfrac{1}{2}$

The solution of the system is $\left(2, -\dfrac{1}{2}\right)$.

21. $\begin{cases} 4x - 3y = 7 \\ 7x + 5y = 2 \end{cases}$

Multiply the first equation by 5 and the second equation by 3.

$\begin{cases} 5(4x - 3y) = 5(7) \\ 3(7x + 5y) = 3(2) \end{cases} \rightarrow \begin{cases} 20x - 15y = 35 \\ 21x + 15y = 6 \end{cases}$

Add the equations to eliminate y, then solve for x.

$20x - 15y = 35$

$\underline{21x + 15y = 6}$

$41x \quad\;\;\;\; = 41$

$\quad\;\; x = 1$

Now solve for y.

$4x - 3y = 7$

$4(1) - 3y = 7$

$4 - 3y = 7$

$-3y = 3$

$y = -1$

The solution of the system is $(1, -1)$.

25. $\begin{cases} 2x - 5y = 4 \\ 3x - 2y = 4 \end{cases}$

Multiply the first equation by -3 and the second equation by 2.

$\begin{cases} -3(2x - 5y) = -3(4) \\ 2(3x - 2y) = 2(4) \end{cases} \rightarrow \begin{cases} -6x + 15y = -12 \\ 6x - 4y = 8 \end{cases}$

Add the equations to eliminate x, then solve for y.

$-6x + 15y = -12$

$\underline{6x - 4y = 8}$

$11y = -4$

$y = -\dfrac{4}{11}$

Multiply the first original equation by -2 and the second original equation by 5.

$$\begin{cases} -2(2x - 5y) = -2(4) \\ 5(3x - 2y) = 5(4) \end{cases} \rightarrow \begin{cases} -4x + 10y = -8 \\ 15x - 10y = 20 \end{cases}$$

Add the equations to eliminate y, then solve for x.

$$\begin{array}{r} -4x + 10y = -8 \\ 15x - 10y = 20 \\ \hline 11x \qquad\quad = 12 \end{array}$$

$$x = \frac{12}{11}$$

The solution of the system is $\left(\dfrac{12}{11}, -\dfrac{4}{11}\right)$.

29. $\begin{cases} \dfrac{10}{3}x + 4y = -4 \\ 5x + 6y = -6 \end{cases}$

Multiply the first equation by 3 and the second equation by -2.

$$\begin{cases} 3\left(\dfrac{10}{3}x + 4y\right) = 3(-4) \\ -2(5x + 6y) = -2(-6) \end{cases} \rightarrow \begin{cases} 10x + 12y = -12 \\ -10x - 12y = 12 \end{cases}$$

Add the equations to eliminate x.

$$\begin{array}{r} 10x + 12y = -12 \\ -10x - 12y = 12 \\ \hline 0 = 0 \end{array}$$

Since the statement $0 = 0$ is true, the system has an infinite number of solutions.

33. $\begin{cases} -4(x + 2) = 3y \\ 2x - 2y = 3 \end{cases}$

Rewrite the first equation.

$$-4(x + 2) = 3y$$
$$-4x - 8 = 3y$$
$$-4x = 3y + 8$$
$$-4x - 3y = 8$$

$$\begin{cases} -4x - 3y = 8 \\ 2x - 2y = 3 \end{cases}$$

Multiply the second equation by 2.

$$\begin{cases} -4x - 3y = 8 \\ 2(2x - 2y) = 2(3) \end{cases} \rightarrow \begin{cases} -4x - 3y = 8 \\ 4x - 4y = 6 \end{cases}$$

Add the equations to eliminate x, then solve for y.

$$\begin{array}{r} -4x - 3y = 8 \\ 4x - 4y = 6 \\ \hline -7y = 14 \\ y = -2 \end{array}$$

Now solve for x.

$$2x - 2y = 3$$
$$2x - 2(-2) = 3$$
$$2x + 4 = 3$$
$$2x = -1$$
$$x = -\frac{1}{2}$$

The solution of the system is $\left(-\dfrac{1}{2}, -2\right)$.

37. $\begin{cases} \dfrac{3}{5}x - y = -\dfrac{4}{5} \\ 3x + \dfrac{y}{2} = -\dfrac{9}{5} \end{cases}$

Multiply the first equation by 5 and the second equation by 10 to eliminate fractions.

$$\begin{cases} 5\left(\dfrac{3}{5}x - y\right) = 5\left(-\dfrac{4}{5}\right) \\ 10\left(3x + \dfrac{y}{2}\right) = 10\left(-\dfrac{9}{5}\right) \end{cases} \rightarrow \begin{cases} 3x - 5y = -4 \\ 30x + 5y = -18 \end{cases}$$

Add the equations to eliminate y, then solve for x.

$$\begin{array}{r} 3x - 5y = -4 \\ 30x + 5y = -18 \\ \hline 33x \qquad\quad = -22 \end{array}$$

$$x = -\frac{22}{33} = -\frac{2}{3}$$

Now use the equation $3x - 5y = -4$ to solve for y.

$$3x - 5y = -4$$
$$3\left(-\frac{2}{3}\right) - 5y = -4$$
$$-2 - 5y = -4$$
$$-5y = -2$$
$$y = \frac{2}{5}$$

The solution of the system is $\left(-\dfrac{2}{3}, \dfrac{2}{5}\right)$.

41. $\begin{cases} 0.02x + 0.04y = 0.09 \\ -0.1x + 0.3y = 0.8 \end{cases}$

Multiply the first equation by 100 and the second equation by 10 to eliminate decimals.

$$\begin{cases} 100(0.02x + 0.04y) = 100(0.09) \\ 10(-0.1x + 0.3y) = 10(0.8) \end{cases}$$
$$\rightarrow \begin{cases} 2x + 4y = 9 \\ -x + 3y = 8 \end{cases}$$

Multiply the second equation by 2.

$$\begin{cases} 2x + 4y = 9 \\ 2(-x + 3y) = 2(8) \end{cases} \rightarrow \begin{cases} 2x + 4y = 9 \\ -2x + 6y = 16 \end{cases}$$

Add the equations to eliminate x, then solve for y.

$$\begin{array}{r} 2x + 4y = 9 \\ -2x + 6y = 16 \\ \hline 10y = 25 \\ y = 2.5 \end{array}$$

Use the equation $-x + 3y = 8$ to solve for x.

$$-x + 3y = 8$$
$$-x + 3(2.5) = 8$$
$$-x + 7.5 = 8$$
$$-x = 0.5$$
$$x = -0.5$$

The solution of the system is $(-0.5, 2.5)$.

45. Three times a number, subtracted from 20, is 2 is written as $20 - 3x = 2$.

49. To eliminate the variable y, multiply the second equation by 2.

$$3x - y = -12$$
$$2(3x - y) = 2(-12)$$
$$6x - 2y = -24$$

53. answers may vary

57. $\begin{cases} 2x + 3y = 14 \\ 3x - 4y = -69.1 \end{cases}$

Multiply the first equation by -3 and the second equation by 2.

$$\begin{cases} -3(2x + 3y) = -3(14) \\ 2(3x - 4y) = 2(-69.1) \end{cases}$$
$$\rightarrow \begin{cases} -6x - 9y = -42 \\ 6x - 8y = -138.2 \end{cases}$$

Add the equations to eliminate x, then solve for y.

$$-6x - 9y = -42$$
$$\underline{6x - 8y = -138.2}$$
$$-17y = -180.2$$
$$y = 10.6$$

Now solve for x.

$$2x + 3y = 14$$
$$2x + 3(10.6) = 14$$
$$2x + 31.8 = 14$$
$$2x = -17.8$$
$$x = -8.9$$

The solution of the system is $(-8.9, 10.6)$.

Exercise Set 7.4

1. In choice b, the length is not 3 feet longer than the width. In choice a, the perimeter is $2(8 + 5) = 2(13) = 26$ feet, not 30 feet. Choice c gives the solution, since $9 = 6 + 3$ and $2(9 + 6) = 2(15) = 30$.

5. In choice b, the total number of coins is $20 + 44 = 64$, not 100. In choice c, the total value of the coins is
$60(0.10) + 40(0.25) = 6.00 + 10.00 = \$16.00$, not $\$13.00$.
Choice a gives the solution, since $80 + 20 = 100$ and
$80(0.10) + 20(0.25) = 8.00 + 5.00 = 13.00$.

9. Let x be the amount in the larger account, and y be the amount in the smaller account.
$$\begin{cases} x + y = 6500 \\ x = y + 800 \end{cases}$$

13. Let x be the first number and y the second.
$$\begin{cases} x + 2y = 8 \\ 2x + y = 25 \end{cases}$$
Solve the first equation for x.
$$x + 2y = 8$$
$$x = 8 - 2y$$
Substitute $8 - 2y$ for x in the second equation, and solve for y.
$$2x + y = 25$$
$$2(8 - 2y) + y = 25$$
$$16 - 4y + y = 25$$
$$16 - 3y = 25$$
$$-3y = 9$$
$$y = -3$$
Now solve for x.
$$x = 8 - 2y = 8 - 2(-3) = 8 + 6 = 14$$
The numbers are 14 and -3.

17. Let a be the price of an adult's ticket and c be the price of a child's ticket.
$$\begin{cases} 3a + 4c = 159 \\ 2a + 3c = 112 \end{cases}$$
Multiply the first equation by -2 and the second equation by 3.
$$\begin{cases} -2(3a + 4c) = -2(159) \\ 3(2a + 3c) = 3(112) \end{cases} \rightarrow \begin{cases} -6a - 8c = -318 \\ 6a + 9c = 336 \end{cases}$$
Add the equations to eliminate a and solve for c.
$$-6a - 8c = -318$$
$$\underline{6a + 9c = 336}$$
$$c = 18$$
Now solve for a.
$$2a + 3c = 112$$
$$2a + 3(18) = 112$$
$$2a + 54 = 112$$
$$2a = 58$$
$$a = 29$$
The price of an adult's ticket is $\$29$ and the price of a child's ticket is $\$18$.

21. Let x be the value of one McDonald's share and let y be the value of one Ohio Art Company share.
$$\begin{cases} 35x + 69y = 1551 \\ x = y + 25 \end{cases}$$
Substitute $y + 25$ for x in the first equation and solve for y.
$$35x + 69y = 1551$$
$$35(y + 25) + 69y = 1551$$
$$35y + 875 + 69y = 1551$$
$$875 + 104y = 1551$$
$$104y = 676$$
$$y = 6.5$$
Now solve for x.
$$x = y + 25 = 6.5 + 25 = 31.5$$
On that day, the closing price of the McDonald's stock was $\$31.5$ per share and the closing price of The Ohio Art Company stock was $\$6.50$ per share.

25. The distance downstream is the same as the distance upstream and it is 18 miles. The time downstream is 2 hours and the time upstream is $4\frac{1}{2}$ hours.
$$\begin{cases} 18 = 2(x + y) \\ 18 = \dfrac{9}{2}(x - y) \end{cases}$$
Multiply the first equation by $\frac{1}{2}$ and the second equation by $\frac{2}{9}$.
$$\begin{cases} \dfrac{1}{2}(18) = \dfrac{1}{2}[2(x + y)] \\ \dfrac{2}{9}(18) = \dfrac{2}{9}\left[\dfrac{2}{9}(x - y)\right] \end{cases} \rightarrow \begin{cases} 9 = x + y \\ 4 = x - y \end{cases}$$
Add the equations to eliminate y, then solve for x.
$$9 = x + y$$
$$\underline{4 = x - y}$$
$$13 = 2x$$
$$6.5 = x$$
Now solve for y.
$$9 = x + y$$
$$9 = 6.5 + y$$
$$2.5 = y$$
The rate that Pratap can row in still water is 6.5 miles per hour and the rate of the current is 2.5 miles per hour.

29. Let x be the number of hours that Jim spent on his bicycle and y be the number of hours he spent walking. Then the distance he rode was $40x$ miles and the distance he walked was $4y$ miles.
$$\begin{cases} x + y = 6 \\ 40x + 4y = 186 \end{cases}$$
Solve the first equation for y.
$$x + y = 6$$
$$y = 6 - x$$
Substitute $6 - x$ for y in the second equation and solve for x.
$$40x + 4y = 186$$
$$40x + 4(6 - x) = 186$$
$$40x + 24 - 4x = 186$$
$$36x + 24 = 186$$
$$36x = 162$$
$$x = 4.5$$
Jim spent 4.5 hours on his bike.

33. Let x be the number of pounds of high-quality coffee and let y be the number of pounds of the cheaper coffee.
$$\begin{cases} x + y = 200 \\ 4.95x + 2.65y = 200(3.95) \end{cases}$$
Solve the first equation for x.
$$x + y = 200$$
$$x = 200 - y$$

Substitute $200 - y$ for x in the second equation, and solve for y.

$$4.95x + 2.65y = 200(3.95)$$
$$4.95(200 - y) + 2.65y = 790$$
$$990 - 4.95y + 2.65y = 790$$
$$990 - 2.30y = 790$$
$$-2.30y = -200$$
$$y = \frac{200}{2.30}$$
$$y \approx 87$$

$$x = 200 - y \approx 200 - 87 = 113$$

Wayne should blend 113 pounds of the coffee that sells for $4.95 per pound with 87 pounds of the cheaper coffee.

37. Let x be the measure of one angle and y be the measure of the other.

$$\begin{cases} x + y = 90 \\ x = 10 + 3y \end{cases}$$

Substitute $10 + 3y$ for x in the first equation and solve for y.

$$x + y = 90$$
$$10 + 3y + y = 90$$
$$10 + 4y = 90$$
$$4y = 80$$
$$y = 20$$

$$x = 10 + 3y = 10 + 3(20) = 10 + 60 = 70$$

The angles measure $20°$ and $70°$.

41. Let x be the rate of the faster group and y be the rate of the slower group.

$$\begin{cases} 240x + 240y = 1200 \\ y = x - \dfrac{1}{2} \end{cases}$$

Substitute $x - \dfrac{1}{2}$ for y in the first equation and solve for x.

$$240x + 240y = 1200$$
$$240x + 240\left(x - \frac{1}{2}\right) = 1200$$
$$240x + 240x - 120 = 1200$$
$$480x - 120 = 1200$$
$$480x = 1320$$
$$x = 2.75$$

$$y = x - \frac{1}{2} = x - 0.5 = 2.75 - 0.5 = 2.25$$

The hiking rates are $2.75 = 2\dfrac{3}{4}$ miles per hour and $2.25 = 2\dfrac{1}{4}$ miles per hour.

45. Let x be the length and y the width.

$$\begin{cases} 2(x + y) = 144 \\ x = y + 12 \end{cases}$$

Substitute $y + 12$ for x in the first equation and solve for y.

$$2(x + y) = 144$$
$$2(y + 12 + y) = 144$$
$$2(2y + 12) = 144$$
$$4y + 24 = 144$$
$$4y = 120$$
$$y = 30$$

$$x = y + 12 = 30 + 12 = 42$$

The length is 42 inches and the width is 30 inches.

49. $(6x)^2 = (6x)(6x) = 36x^2$

53. The price of the result must be between $0.49 and $0.65, so choice a is the only possibility.

Chapter 7 Test

1. False; a system of two linear equations can have no solutions, exactly one solution, or infinitely many solutions.

5. First equation:

$$2x - 3y = 5$$
$$2(1) - 3(-1) \stackrel{?}{=} 5$$
$$2 + 3 \stackrel{?}{=} 5$$
$$5 = 5 \quad \text{True}$$

Second equation:

$$6x + y = 1$$
$$6(1) + (-1) \stackrel{?}{=} 1$$
$$6 - 1 \stackrel{?}{=} 1$$
$$5 = 1 \quad \text{False}$$

Since the statement $5 = 1$ is false, $(1, -1)$ is not a solution of the system.

9. $\begin{cases} 3x - 2y = -14 \\ y = x + 5 \end{cases}$

Substitute $x + 5$ for y in the first equation and solve for x.

$$3x - 2y = -14$$
$$3x - 2(x + 5) = -14$$
$$3x - 2x - 10 = -14$$
$$x - 10 = -14$$
$$x = -4$$

$$y = x + 5 = -4 + 5 = 1$$

The solution of the system is $(-4, 1)$.

13. $\begin{cases} 3x + y = 7 \\ 4x + 3y = 1 \end{cases}$

Solve the first equation for y.

$$3x + y = 7$$
$$y = 7 - 3x$$

Substitute $7 - 3x$ for y in the second equation and solve for x.

$$4x + 3y = 1$$
$$4x + 3(7 - 3x) = 1$$
$$4x + 21 - 9x = 1$$
$$21 - 5x = 1$$
$$-5x = -20$$
$$x = 4$$

$$y = 7 - 3x = 7 - 3(4) = 7 - 12 = -5$$

The solution of the system is $(4, -5)$.

17. $\begin{cases} 0.01x - 0.06y = -0.23 \\ 0.2x + 0.4y = 0.2 \end{cases}$

Multiply the first equation by 100 and the second equation by 10 to eliminate decimals.

$$\begin{cases} 100(0.01x - 0.06y) = 100(-0.23) \\ 10(0.2x + 0.4y) = 10(0.2) \end{cases} \rightarrow \begin{cases} x - 6y = -23 \\ 2x + 4y = 2 \end{cases}$$

Multiply the first equation by -2.

$$\begin{cases} -2(x - 6y) = -2(-23) \\ 2x + 4y = 2 \end{cases} \rightarrow \begin{cases} -2x + 12y = 46 \\ 2x + 4y = 2 \end{cases}$$

Add the equations to eliminate x, then solve for y.

$$-2x + 12y = 46$$
$$\underline{2x + 4y = 2}$$
$$16y = 48$$
$$y = 3$$

Now solve for x.

$$x - 6y = -23$$
$$x - 6(3) = -23$$
$$x - 18 = -23$$
$$x = -5$$

The solution of the system is $(-5, 3)$.

21. Let t be the number of farms in Texas and let m be the number of farms in Missouri.

$$\begin{cases} t + m = 336 \\ t = m + 116 \end{cases}$$

Substitute $m + 116$ for t in the first equation and solve for m.

$$t + m = 336$$
$$m + 116 + m = 336$$
$$2m + 116 = 336$$
$$2m = 220$$
$$m = 110$$
$$t = m + 116 = 110 + 116 = 226$$

There are 226 thousand farms in Texas and 110 thousand in Missouri.

CHAPTER 8

Exercise Set 8.1

1. $\sqrt{16} = 4$, because $4^2 = 16$ and 4 is positive.

5. $-\sqrt{100} = -10$. The negative sign indicates the negative square root of 100.

9. $-\sqrt{121} = -11$. The negative sign indicates the negative square root of 121.

13. $\sqrt{900} = 30$, because $30^2 = 900$ and 30 is positive.

17. $\sqrt{\dfrac{1}{100}} = \dfrac{1}{10}$, because $\left(\dfrac{1}{10}\right)^2 = \dfrac{1}{100}$ and $\dfrac{1}{10}$ is positive.

21. $\sqrt[3]{125} = 5$, because $5^3 = 125$.

25. $-\sqrt[3]{8} = -2$, because $2^3 = 8$.

29. $\sqrt[3]{-125} = -5$, because $(-5)^3 = -125$.

33. $\sqrt{81} = 9$, because $9^2 = 81$ and 9 is positive.

37. $\sqrt[3]{-\dfrac{27}{64}} = -\dfrac{3}{4}$, because $\left(-\dfrac{3}{4}\right)^3 = -\dfrac{27}{64}$.

41. $\sqrt[6]{1} = 1$, because $1^6 = 1$.

45. $\sqrt{37} \approx 6.083$

49. $\sqrt{2} \approx 1.41$

$$90\sqrt{2} \approx 90(1.41) = 126.90$$

The distance from home plate to second base is approximately 126.90 feet.

53. $\sqrt{x^4} = x^2$, because $(x^2)^2 = x^4$.

57. $\sqrt{81x^2} = 9x$ because $(9x)^2 = 81x^2$.

61. $\sqrt{16a^6b^4} = 4a^3b^2$, because $(4a^3b^2)^2 = 16a^6b^4$

65. $\sqrt[3]{-8x^3y^{27}} = -2xy^9$, because $(-2xy^9)^3 = -8x^3y^{27}$

69. $32 = 16 \cdot 2$ or
$$32 = 4 \cdot 8$$

73. $27 = 9 \cdot 3$

77. The length of the side is $\sqrt{49}$. Since $7^2 = 49$, $\sqrt{49} = 7$ and the sides of the square have length 7 miles.

81. $\sqrt{\sqrt{81}} = \sqrt{9} = 3$, since $3^2 = 9$ and $9^2 = 81$.

85. $T = 2\pi\sqrt{\dfrac{L}{g}} = 2\pi\sqrt{\dfrac{30}{32}} \approx 2(3.14)(0.968) \approx 6.1$

The period of the pendulum is 6.1 seconds.

89.

x	$y = \sqrt{x}$
0	$\sqrt{0} = 0$
1	$\sqrt{1} = 1$
3	$\sqrt{3} \approx 1.7$
4	$\sqrt{4} = 2$
9	$\sqrt{9} = 3$

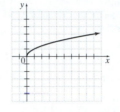

93. The graph of $y = \sqrt{x + 4}$ "starts" at $(-4, 0)$.

Exercise Set 8.2

1. $\sqrt{20} = \sqrt{4 \cdot 5} = \sqrt{4}\sqrt{5} = 2\sqrt{5}$

5. $\sqrt{33}$ is in simplest form.

9. $\sqrt{60} = \sqrt{4 \cdot 15} = \sqrt{4}\sqrt{15} = 2\sqrt{15}$

13. $\sqrt{52} = \sqrt{4 \cdot 13} = \sqrt{4}\sqrt{13} = 2\sqrt{13}$

17. $7\sqrt{63} = 7\sqrt{9 \cdot 7} = 7\sqrt{9}\sqrt{7} = 7 \cdot 3\sqrt{7} = 21\sqrt{7}$

21. $\sqrt{\dfrac{8}{25}} = \dfrac{\sqrt{8}}{\sqrt{25}} = \dfrac{\sqrt{4 \cdot 2}}{5} = \dfrac{\sqrt{4}\sqrt{2}}{5} = \dfrac{2\sqrt{2}}{5}$

25. $\sqrt{\dfrac{9}{4}} = \dfrac{\sqrt{9}}{\sqrt{4}} = \dfrac{3}{2}$

29. $\sqrt{\dfrac{11}{36}} = \dfrac{\sqrt{11}}{\sqrt{36}} = \dfrac{\sqrt{11}}{6}$

33. $\sqrt{x^7} = \sqrt{x^6 \cdot x} = \sqrt{x^6}\sqrt{x} = x^3\sqrt{x}$

37. $\sqrt{36a^3} = \sqrt{36a^2 \cdot a} = \sqrt{36a^2}\sqrt{a} = 6a\sqrt{a}$

41. $\sqrt{\dfrac{12}{m^2}} = \dfrac{\sqrt{12}}{\sqrt{m^2}} = \dfrac{\sqrt{4 \cdot 3}}{m} = \dfrac{\sqrt{4}\sqrt{3}}{m} = \dfrac{2\sqrt{3}}{m}$

45. $\sqrt{\dfrac{88}{x^{12}}} = \dfrac{\sqrt{88}}{\sqrt{x^{12}}} = \dfrac{\sqrt{4 \cdot 22}}{x^6} = \dfrac{\sqrt{4}\sqrt{22}}{x^6} = \dfrac{2\sqrt{22}}{x^6}$

49. $\sqrt{\dfrac{36}{121}} = \dfrac{\sqrt{36}}{\sqrt{121}} = \dfrac{6}{11}$

53. $\sqrt{\dfrac{20}{9}} = \dfrac{\sqrt{20}}{\sqrt{9}} = \dfrac{\sqrt{4 \cdot 5}}{3} = \dfrac{\sqrt{4}\sqrt{5}}{3} = \dfrac{2\sqrt{5}}{3}$

57. $\sqrt{\dfrac{23y^3}{4x^6}} = \dfrac{\sqrt{23y^3}}{\sqrt{4x^6}}$

$$= \dfrac{\sqrt{y^2 \cdot 23y}}{2x^3}$$

$$= \dfrac{\sqrt{y^2}\sqrt{23y}}{2x^3}$$

$$= \dfrac{y\sqrt{23y}}{2x^3}$$

61. $\sqrt[3]{250} = \sqrt[3]{125 \cdot 2} = \sqrt[3]{125} \cdot \sqrt[3]{2} = 5\sqrt[3]{2}$

65. $\sqrt[3]{\dfrac{23}{8}} = \dfrac{\sqrt[3]{23}}{\sqrt[3]{8}} = \dfrac{\sqrt[3]{23}}{2}$

69. $\sqrt[3]{80} = \sqrt[3]{8 \cdot 10} = \sqrt[3]{8}\sqrt[3]{10} = 2\sqrt[3]{10}$

73. $(2x + 3)(x - 5) = 2x^2 - 10x + 3x - 15$
$$= 2x^2 - 7x - 15$$

77. $\sqrt{x^6 y^3} = \sqrt{x^6 y^2 y} = \sqrt{x^6}\sqrt{y^2}\sqrt{y} = x^3 y\sqrt{y}$

81. $\sqrt[3]{-8x^6} = \sqrt[3]{-8}\sqrt[3]{x^6} = -2x^3$

85. answers may vary

89. Use $\dfrac{\sqrt{6A}}{6}$ with $A = 30.375$.

$$\frac{\sqrt{6 \cdot 30.375}}{6} = \frac{\sqrt{182.25}}{6}$$
$$= \frac{13.5}{6}$$
$$= 2.25$$

Each edge of a Rubik's cube is 2.25 inches long.

93. $h = 169$ and $w = 64$.

$$B = \sqrt{\frac{hw}{3600}}$$
$$= \sqrt{\frac{169 \cdot 64}{3600}}$$
$$= \frac{\sqrt{169 \cdot 64}}{\sqrt{3600}}$$
$$= \frac{\sqrt{169}\sqrt{64}}{60}$$
$$= \frac{13 \cdot 8}{60}$$
$$= \frac{104}{60}$$
$$= \frac{26}{15} \approx 1.7$$

The body surface area is about 1.7 square meters.

Exercise Set 8.3

1. $4\sqrt{3} - 8\sqrt{3} = (4 - 8)\sqrt{3} = -4\sqrt{3}$

5. $6\sqrt{5} - 5\sqrt{5} + \sqrt{2} = (6 - 5)\sqrt{5} + \sqrt{2} = \sqrt{5} + \sqrt{2}$

9. $2\sqrt{2} - 7\sqrt{2} - 6 = (2 - 7)\sqrt{2} - 6 = -5\sqrt{2} - 6$

13. $\sqrt{45} + 3\sqrt{20} = \sqrt{9 \cdot 5} + 3\sqrt{4 \cdot 5}$
$$= \sqrt{9}\sqrt{5} + 3\sqrt{4}\sqrt{5}$$
$$= 3\sqrt{5} + 3 \cdot 2\sqrt{5}$$
$$= 3\sqrt{5} + 6\sqrt{5}$$
$$= (3 + 6)\sqrt{5}$$
$$= 9\sqrt{5}$$

17. $4x - 3\sqrt{x^2} + \sqrt{x} = 4x - 3x + \sqrt{x} = x + \sqrt{x}$

21. $\sqrt{\dfrac{5}{9}} + \sqrt{\dfrac{5}{81}} = \dfrac{\sqrt{5}}{\sqrt{9}} + \dfrac{\sqrt{5}}{\sqrt{81}}$
$$= \frac{\sqrt{5}}{3} + \frac{\sqrt{5}}{9}$$
$$= \frac{3\sqrt{5}}{9} + \frac{\sqrt{5}}{9}$$
$$= \left(\frac{3}{9} + \frac{1}{9}\right)\sqrt{5}$$
$$= \frac{4}{9}\sqrt{5}$$
$$= \frac{4\sqrt{5}}{9}$$

25. $12\sqrt{5} - \sqrt{5} - 4\sqrt{5} = (12 - 1 - 4)\sqrt{5} = 7\sqrt{5}$

29. $\sqrt{5} + \sqrt{15}$ is in simplest form.

33. $\sqrt{8} + \sqrt{9} + \sqrt{18} + \sqrt{81} = \sqrt{4 \cdot 2} + 3 + \sqrt{9 \cdot 2} + 9$
$$= \sqrt{4}\sqrt{2} + 3 + \sqrt{9}\sqrt{2} + 9$$
$$= 2\sqrt{2} + 3 + 3\sqrt{2} + 9$$
$$= (2 + 3)\sqrt{2} + 3 + 9$$
$$= 5\sqrt{2} + 12$$

37. $2\sqrt{45} - 2\sqrt{20} = 2\sqrt{9 \cdot 5} - 2\sqrt{4 \cdot 5}$
$$= 2\sqrt{9}\sqrt{5} - 2\sqrt{4}\sqrt{5}$$
$$= 2 \cdot 3\sqrt{5} - 2 \cdot 2\sqrt{5}$$
$$= 6\sqrt{5} - 4\sqrt{5}$$
$$= (6 - 4)\sqrt{5}$$
$$= 2\sqrt{5}$$

41. $6 - 2\sqrt{3} - \sqrt{3} = 6 + (-2 - 1)\sqrt{3} = 6 - 3\sqrt{3}$

45. $\sqrt{9x^2} + \sqrt{81x^2} - 11\sqrt{x} = 3x + 9x - 11\sqrt{x}$
$$= 12x - 11\sqrt{x}$$

49. $\sqrt{32x^2} + \sqrt{32x^2} + \sqrt{4x^2}$
$$= \sqrt{16x^2 \cdot 2} + \sqrt{16x^2 \cdot 2} + 2x$$
$$= \sqrt{16x^2}\sqrt{2} + \sqrt{16x^2}\sqrt{2} + 2x$$
$$= 4x\sqrt{2} + 4x\sqrt{2} + 2x$$
$$= (4x + 4x)\sqrt{2} + 2x$$
$$= 8x\sqrt{2} + 2x$$

53. $2\sqrt[3]{9} + 5\sqrt[3]{9} - \sqrt[3]{25} = (2 + 5)\sqrt[3]{9} - \sqrt[3]{25}$
$$= 7\sqrt[3]{9} - \sqrt[3]{25}$$

57. $\sqrt[3]{81} + \sqrt[3]{24} = \sqrt[3]{27 \cdot 3} + \sqrt[3]{8 \cdot 3}$
$$= \sqrt[3]{27}\sqrt[3]{3} + \sqrt[3]{8}\sqrt[3]{3}$$
$$= 3\sqrt[3]{3} + 2\sqrt[3]{3}$$
$$= (3 + 2)\sqrt[3]{3}$$
$$= 5\sqrt[3]{3}$$

61. $\sqrt{40x} + x\sqrt[3]{40} - 2\sqrt{10x} - x\sqrt[3]{5}$
$$= \sqrt{4 \cdot 10x} + x\sqrt[3]{8 \cdot 5} - 2\sqrt{10x} - x\sqrt[3]{5}$$
$$= \sqrt{4}\sqrt{10x} + x\sqrt[3]{8}\sqrt[3]{5} - 2\sqrt{10x} - x\sqrt[3]{5}$$
$$= 2\sqrt{10x} + 2x\sqrt[3]{5} - 2\sqrt{10x} - x\sqrt[3]{5}$$
$$= (2 - 2)\sqrt{10x} + (2x - x)\sqrt[3]{5}$$
$$= x\sqrt[3]{5}$$

65. $(x + 6)^2 = (x)^2 + 2(x)(6) + (6)^2 = x^2 + 12x + 36$

69. answers may vary

73. Two triangular end pieces and two rectangular side panels are needed. Each side panel has area $8 \cdot 3 = 24$ square feet.

$$2 \cdot 24 + 2 \cdot \frac{3\sqrt{27}}{4} = 48 + \frac{3\sqrt{9 \cdot 3}}{2}$$
$$= 48 + \frac{3\sqrt{9}\sqrt{3}}{2}$$
$$= 48 + \frac{3 \cdot 3\sqrt{3}}{2}$$
$$= 48 + \frac{9\sqrt{3}}{2}$$

The total area of wood needed is $\left(48 + \dfrac{9\sqrt{3}}{2}\right)$ square feet.

77. The expression $6 + 7\sqrt{6}$ cannot be simplified.

81. $\sqrt{\dfrac{x^3}{16}} - x\sqrt{\dfrac{9x}{25}} + \dfrac{\sqrt{81x^3}}{2}$

$= \dfrac{\sqrt{x^3}}{\sqrt{16}} - x\dfrac{\sqrt{9x}}{\sqrt{25}} + \dfrac{\sqrt{81x^3}}{2}$

$= \dfrac{\sqrt{x^2 \cdot x}}{4} - x\dfrac{\sqrt{9 \cdot x}}{5} + \dfrac{\sqrt{81x^2 \cdot x}}{2}$

$= \dfrac{\sqrt{x^2}\sqrt{x}}{4} - x\dfrac{\sqrt{9}\sqrt{x}}{5} + \dfrac{\sqrt{81x^2}\sqrt{x}}{2}$

$= \dfrac{x\sqrt{x}}{4} - \dfrac{3x\sqrt{x}}{5} + \dfrac{9x\sqrt{x}}{2}$

$= \left(\dfrac{1}{4} - \dfrac{3}{5} + \dfrac{9}{2}\right)x\sqrt{x}$

$= \left(\dfrac{5}{20} - \dfrac{12}{20} + \dfrac{90}{20}\right)x\sqrt{x}$

$= \dfrac{83}{20}x\sqrt{x}$

$= \dfrac{83x\sqrt{x}}{20}$

Exercise Set 8.4

1. $\sqrt{8} \cdot \sqrt{2} = \sqrt{16} = 4$

5. $\left(\sqrt{6}\right)^2 = 6$

9. $\left(2\sqrt{5}\right)^2 = \left(2\sqrt{5}\right)\left(2\sqrt{5}\right) = 4\left(\sqrt{5}\right)^2 = 4 \cdot 5 = 20$

13. $\sqrt{3x^5} \cdot \sqrt{6x} = \sqrt{3x^5 \cdot 6x}$

$= \sqrt{18x^6}$

$= \sqrt{9x^6 \cdot 2}$

$= \sqrt{9x^6}\sqrt{2}$

$= 3x^3\sqrt{2}$

17. $\sqrt{6}\left(\sqrt{5} + \sqrt{7}\right) = \sqrt{6} \cdot \sqrt{5} + \sqrt{6} \cdot \sqrt{7} = \sqrt{30} + \sqrt{42}$

21. $\sqrt{7y}\left(\sqrt{y} - 2\sqrt{7}\right) = \sqrt{7y} \cdot \sqrt{y} - \sqrt{7y} \cdot 2\sqrt{7}$

$= \sqrt{7y \cdot y} - 2\sqrt{7y \cdot 7}$

$= \sqrt{7y^2} - 2\sqrt{49y}$

$= \sqrt{y^2 \cdot 7} - 2\sqrt{49 \cdot y}$

$= \sqrt{y^2}\sqrt{7} - 2\sqrt{49}\sqrt{y}$

$= y\sqrt{7} - 2 \cdot 7\sqrt{y}$

$= y\sqrt{7} - 14\sqrt{y}$

25. $\left(\sqrt{3} + \sqrt{5}\right)\left(\sqrt{2} - \sqrt{5}\right)$

$= \sqrt{3} \cdot \sqrt{2} - \sqrt{3}\sqrt{5} + \sqrt{5} \cdot \sqrt{2} - \sqrt{5} \cdot \sqrt{5}$

$= \sqrt{6} - \sqrt{15} + \sqrt{10} - \sqrt{25}$

$= \sqrt{6} - \sqrt{15} + \sqrt{10} - 5$

29. $\left(\sqrt{x} + 6\right)\left(\sqrt{x} - 6\right) = \left(\sqrt{x}\right)^2 - (6)^2 = x - 36$

33. $\left(\sqrt{6y} + 1\right)^2 = \left(\sqrt{6y}\right)^2 + 2\left(\sqrt{6y}\right)(1) + (1)^2$

$= 6y + 2\sqrt{6y} + 1$

37. $\dfrac{\sqrt{21}}{\sqrt{3}} = \sqrt{\dfrac{21}{3}} = \sqrt{7}$

41. $\dfrac{\sqrt{75y^5}}{\sqrt{3y}} = \sqrt{\dfrac{75y^5}{3y}} = \sqrt{25y^4} = 5y^2$

45. $\dfrac{\sqrt{72y^5}}{\sqrt{3y^3}} = \sqrt{\dfrac{72y^5}{3y^3}}$

$= \sqrt{24y^2}$

$= \sqrt{4y^2 \cdot 6}$

$= \sqrt{4y^2}\sqrt{6}$

$= 2y\sqrt{6}$

49. $\dfrac{\sqrt{3}}{\sqrt{5}} = \dfrac{\sqrt{3}}{\sqrt{5}} \cdot \dfrac{\sqrt{5}}{\sqrt{5}} = \dfrac{\sqrt{15}}{5}$

53. $\dfrac{1}{\sqrt{6y}} = \dfrac{1}{\sqrt{6y}} \cdot \dfrac{\sqrt{6y}}{\sqrt{6y}} = \dfrac{\sqrt{6y}}{6y}$

57. $\sqrt{\dfrac{3}{x}} = \dfrac{\sqrt{3}}{\sqrt{x}} = \dfrac{\sqrt{3}}{\sqrt{x}} \cdot \dfrac{\sqrt{x}}{\sqrt{x}} = \dfrac{\sqrt{3x}}{x}$

61. $\sqrt{\dfrac{2}{15}} = \dfrac{\sqrt{2}}{\sqrt{15}} = \dfrac{\sqrt{2}}{\sqrt{15}} \cdot \dfrac{\sqrt{15}}{\sqrt{15}} = \dfrac{\sqrt{30}}{15}$

65. $\dfrac{3x}{\sqrt{2x}} = \dfrac{3x}{\sqrt{2x}} \cdot \dfrac{\sqrt{2x}}{\sqrt{2x}} = \dfrac{3x\sqrt{2x}}{2x} = \dfrac{3\sqrt{2x}}{2}$

69. $\sqrt{\dfrac{y}{12x}} = \dfrac{\sqrt{y}}{\sqrt{12x}}$

$= \dfrac{\sqrt{y}}{\sqrt{4}\sqrt{3x}}$

$= \dfrac{\sqrt{y}}{2\sqrt{3x}}$

$= \dfrac{\sqrt{y}}{2\sqrt{3x}} \cdot \dfrac{\sqrt{3x}}{\sqrt{3x}}$

$= \dfrac{\sqrt{3xy}}{2 \cdot 3x}$

$= \dfrac{\sqrt{3xy}}{6x}$

73. $\dfrac{4}{2 - \sqrt{5}} = \dfrac{4}{2 - \sqrt{5}} \cdot \dfrac{2 + \sqrt{5}}{2 + \sqrt{5}}$

$= \dfrac{4\left(2 + \sqrt{5}\right)}{2^2 - \left(\sqrt{5}\right)^2}$

$= \dfrac{8 + 4\sqrt{5}}{4 - 5}$

$= \dfrac{8 + 4\sqrt{5}}{-1}$

$= -8 - 4\sqrt{5}$

77. $\dfrac{\sqrt{3} + 1}{\sqrt{2} - 1} = \dfrac{\sqrt{3} + 1}{\sqrt{2} - 1} \cdot \dfrac{\sqrt{2} + 1}{\sqrt{2} + 1}$

$= \dfrac{\left(\sqrt{3} + 1\right)\left(\sqrt{2} + 1\right)}{\left(\sqrt{2}\right)^2 - 1}$

$= \dfrac{\sqrt{3}\sqrt{2} + \sqrt{3} \cdot 1 + 1 \cdot \sqrt{2} + 1^2}{2 - 1}$

$= \dfrac{\sqrt{6} + \sqrt{3} + \sqrt{2} + 1}{1}$

$= \sqrt{6} + \sqrt{3} + \sqrt{2} + 1$

81. $\dfrac{3}{\sqrt{x} - 4} = \dfrac{3}{\sqrt{x} - 4} \cdot \dfrac{\sqrt{x} + 4}{\sqrt{x} + 4}$

$= \dfrac{3\left(\sqrt{x} + 4\right)}{\left(\sqrt{x}\right)^2 - (4)^2}$

$= \dfrac{3\sqrt{x} + 12}{x - 16}$

85. $4z^2 + 6z - 12 = (2z)^2$

$4z^2 + 6z - 12 = 4z^2$

$6z - 12 = 0$

$6z = 12$

$z = 2$

89. Area = (length)(width)

$$13\sqrt{2}\cdot 5\sqrt{6} = 13\cdot 5\cdot \sqrt{2}\cdot \sqrt{6}$$
$$= 65\sqrt{12}$$
$$= 65\sqrt{4}\sqrt{3}$$
$$= 65\cdot 2\sqrt{3}$$
$$= 130\sqrt{3}$$

The area is $130\sqrt{3}$ square meters.

93. $\sqrt{5}\cdot\sqrt{5} = \left(\sqrt{5}\right)^2 = 5$

The statement is true.

97. $\sqrt{11} + \sqrt{2}$ cannot be simplified because the radicands are different. The statement is false.

101. answers may vary

Exercise Set 8.5

1. $\sqrt{x} = 9$
$$(\sqrt{x})^2 = 9^2$$
$$x = 81$$

5. $\sqrt{x} - 2 = 5$
$$\sqrt{x} = 7$$
$$(\sqrt{x})^2 = 7^2$$
$$x = 49$$

9. $\sqrt{x} = \sqrt{3x - 8}$
$$(\sqrt{x})^2 = \left(\sqrt{3x - 8}\right)^2$$
$$x = 3x - 8$$
$$-2x = -8$$
$$x = 4$$

13. $\sqrt{9x^2 + 2x - 4} = 3x$
$$\left(\sqrt{9x^2 + 2x - 4}\right)^2 = (3x)^2$$
$$9x^2 + 2x - 4 = 9x^2$$
$$2x - 4 = 0$$
$$2x = 4$$
$$x = 2$$

17. $\sqrt{x + 7} = x + 5$
$$\left(\sqrt{x + 7}\right)^2 = (x + 5)^2$$
$$x + 7 = x^2 + 10x + 25$$
$$0 = x^2 + 9x + 18$$
$$0 = (x + 6)(x + 3)$$
$$x + 6 = 0 \quad\text{or}\quad x + 3 = 0$$
$$x = -6 \qquad\qquad x = -3$$

$x = -6$ does not check, so the solution is $x = -3$.

21. $\sqrt{16x^2 + 2x + 2} = 4x$
$$\left(\sqrt{16x^2 + 2x + 2}\right)^2 = (4x)^2$$
$$16x^2 + 2x + 2 = 16x^2$$
$$2x + 2 = 0$$
$$2x = -2$$
$$x = -1$$

$x = -1$ does not check, so the equation has no solution.

25. $\sqrt{x - 7} = \sqrt{x} - 1$
$$\left(\sqrt{x - 7}\right)^2 = \left(\sqrt{x} - 1\right)^2$$
$$x - 7 = x - 2\sqrt{x} + 1$$
$$2\sqrt{x} = 8$$
$$\sqrt{x} = 4$$
$$(\sqrt{x})^2 = 4^2$$
$$x = 16$$

29. $\sqrt{x + 8} = \sqrt{x} + 2$
$$\left(\sqrt{x + 8}\right)^2 = \left(\sqrt{x} + 2\right)^2$$
$$x + 8 = x + 4\sqrt{x} + 4$$
$$4 = 4\sqrt{x}$$
$$1 = \sqrt{x}$$
$$1^2 = (\sqrt{x})^2$$
$$1 = x$$

33. $\sqrt{x + 6} + 1 = 3$
$$\sqrt{x + 6} = 2$$
$$\left(\sqrt{x + 6}\right)^2 = 2^2$$
$$x + 6 = 4$$
$$x = -2$$

37. $\sqrt{16x^2 - 3x + 6} = 4x$
$$\left(\sqrt{16x^2 - 3x + 6}\right)^2 = (4x)^2$$
$$16x^2 - 3x + 6 = 16x^2$$
$$-3x + 6 = 0$$
$$-2x = -6$$
$$x = 2$$

41. $\sqrt{x + 9} = \sqrt{x} - 3$
$$\left(\sqrt{x + 9}\right)^2 = \left(\sqrt{x} - 3\right)^2$$
$$x + 9 = x - 6\sqrt{x} + 9$$
$$0 = -6\sqrt{x}$$
$$0 = \sqrt{x}$$
$$0^2 = (\sqrt{x})^2$$
$$0 = x$$

$x = 0$ does not check, so the equation has no solution.

45. $\sqrt{x} + 3 = 7$
$$\sqrt{x} = 4$$
$$(\sqrt{x})^2 = 4^2$$
$$x = 16$$

49. $\sqrt{2x + 1} = x - 7$
$$\left(\sqrt{2x + 1}\right)^2 = (x - 7)^2$$
$$2x + 1 = x^2 - 14x + 49$$
$$0 = x^2 - 16x + 48$$
$$0 = (x - 4)(x - 12)$$
$$x - 4 = 0 \text{ or } x - 12 = 0$$
$$x = 4 \qquad\qquad x = 12$$

$x = 4$ does not check, so the only solution is $x = 12$.

53. $\sqrt{1 - 8x} - x = 4$
$$\sqrt{1 - 8x} = x + 4$$
$$\left(\sqrt{1 - 8x}\right)^2 = (x + 4)^2$$
$$1 - 8x = x^2 + 8x + 16$$
$$0 = x^2 + 16x + 15$$
$$0 = (x + 15)(x + 1)$$
$$x + 15 = 0 \quad\text{or}\quad x + 1 = 0$$
$$x = -15 \qquad\qquad x = -1$$

$x = -15$ does not check, so the solution is $x = -1$.

57. Let x be the width of the rectangle, then the length is $2x$.
$$2(2x + x) = 24$$
$$2(3x) = 24$$
$$6x = 24$$
$$x = 4$$
$$2x = 2(4) = 8$$

The length of the rectangle is 8 inches.

61. answers may vary

65.

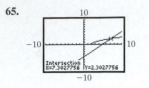

The solution of $\sqrt{x-2} = x - 5$ is $x \approx 7.30$.

Exercise Set 8.6

1. $a^2 + b^2 = c^2$

$2^2 + 3^2 = c^2$

$4 + 9 = c^2$

$13 = c^2$

$\sqrt{13} = \sqrt{c^2}$

$\sqrt{13} = c$

$c = \sqrt{13} \approx 3.61$

5. $a^2 + b^2 = c^2$

$7^2 + 24^2 = c^2$

$49 + 576 = c^2$

$625 = c^2$

$\sqrt{625} = \sqrt{c^2}$

$25 = c$

9. $a^2 + b^2 = c^2$

$4^2 + b^2 = 13^2$

$16 + b^2 = 169$

$b^2 = 153$

$\sqrt{b^2} = \sqrt{153}$

$b = 3\sqrt{17}$

$b = 3\sqrt{17} \approx 12.37$

13. $a^2 + b^2 = c^2$

$a^2 + 2^2 = 6^2$

$a^2 + 4 = 36$

$a^2 = 32$

$\sqrt{a^2} = \sqrt{32}$

$a = 4\sqrt{2}$

$a = 4\sqrt{2} \approx 5.66$

17. The pole, wire, and ground form a right triangle with legs of 5 feet and 20 feet.

$a^2 + b^2 = c^2$

$5^2 + 20^2 = c^2$

$25 + 400 = c^2$

$425 = c^2$

$\sqrt{425} = \sqrt{c^2}$

$\sqrt{425} = c$

$c = \sqrt{425} \approx 20.6$

The length of the wire is 20.6 feet.

21. $b = \sqrt{\dfrac{3V}{h}}$

$6 = \sqrt{\dfrac{3V}{2}}$

$6^2 = \left(\sqrt{\dfrac{3V}{2}}\right)^2$

$36 = \dfrac{3V}{2}$

$72 = 3V$

$24 = V$

The volume is 24 cubic feet.

25. $v = \sqrt{2.5r}$

$v = \sqrt{2.5(300)}$

$v = \sqrt{750}$

$v \approx 27.4$

The maximum safe speed is 27 miles per hour.

29. $d = 3.5\sqrt{h}$

$d = 35\sqrt{295.7}$

$d \approx 60.2$

You can see a distance of 60.2 kilometers.

33. $\sqrt{100} = 10$ and $-\sqrt{100} = -10$, so -10 and 10 are numbers whose square is 100.

37. To find y,

$3^2 + y^2 = 7^2$

$9 + y^2 = 49$

$y^2 = 40$

$y = \sqrt{40} = 2\sqrt{10}$

Let b be the second leg of the right triangle with hypotenuse 5. To find x, first find b.

$a^2 + b^2 = c^2$

$3^2 + b^2 = 5^2$

$9 + b^2 = 25$

$b^2 = 16$

$\sqrt{b^2} = \sqrt{16}$

$b = 4$

Thus,

$x = y - 4$ or

$x = 2\sqrt{10} - 4$

41. answers may vary

Chapter 8 Test

1. $\sqrt{16} = 4$, because $4^2 = 16$ and 4 is positive.

5. $\sqrt[4]{-81}$ is not a real number since the index 4 is even and the radicand -81 is negative.

9. $\sqrt{y^7} = \sqrt{y^6 \cdot y} = \sqrt{y^6}\sqrt{y} = y^3\sqrt{y}$

13. $\sqrt{\dfrac{5}{16}} = \dfrac{\sqrt{5}}{\sqrt{16}} = \dfrac{\sqrt{5}}{4}$

17. $\sqrt{\dfrac{3}{4}} + \sqrt{\dfrac{3}{25}} = \dfrac{\sqrt{3}}{\sqrt{4}} + \dfrac{\sqrt{3}}{\sqrt{25}}$

$= \dfrac{\sqrt{3}}{2} + \dfrac{\sqrt{3}}{5}$

$= \dfrac{5\sqrt{3}}{10} + \dfrac{2\sqrt{3}}{10}$

$= \dfrac{5\sqrt{3} + 2\sqrt{3}}{10}$

$= \dfrac{(5+2)\sqrt{3}}{10}$

$= \dfrac{7\sqrt{3}}{10}$

21. $\dfrac{\sqrt{50}}{\sqrt{10}} = \sqrt{\dfrac{50}{10}} = \sqrt{5}$

25. $\dfrac{8}{\sqrt{6}+2} = \dfrac{8}{\sqrt{6}+2} \cdot \dfrac{\sqrt{6}-2}{\sqrt{6}-2}$

$= \dfrac{8(\sqrt{6}-2)}{(\sqrt{6})^2 - 2^2}$

$= \dfrac{8(\sqrt{6}-2)}{6-4}$

$= \dfrac{8(\sqrt{6}-2)}{2}$

$= 4(\sqrt{6}-2)$

$= 4\sqrt{6} - 8$

29. $\sqrt{2x-2} = x - 5$

$(\sqrt{2x-2})^2 = (x-5)^2$

$2x - 2 = x^2 - 10x + 25$

$0 = x^2 - 12x + 27$

$0 = (x-3)(x-9)$

$x - 3 = 0 \quad \text{or} \quad x - 9 = 0$

$\quad x = 3 \qquad\qquad x = 9$

$x = 3$ does not check, so the solution is $x = 9$.

CHAPTER 9

Exercise Set 9.1

1. $\quad k^2 - 49 = 0$

$(k+7)(k-7) = 0$

$k + 7 = 0 \quad \text{or} \; k - 7 = 0$

$\quad k = -7 \qquad\quad k = 7$

The solutions are $k = -7$ and $k = 7$.

5. $\quad 2x^2 - 32 = 0$

$2(x^2 - 16) = 0$

$2(x+4)(x-4) = 0$

$x + 4 = 0 \quad \text{or} \; x - 4 = 0$

$\quad x = -4 \qquad\quad x = 4$

The solutions are $x = -4$ and $x = 4$.

9. $\quad x^2 + 7x = -10$

$x^2 + 7x + 10 = 0$

$(x+2)(x+5) = 0$

$x + 2 = 0 \quad \text{or} \; x + 5 = 0$

$\quad x = -2 \qquad\quad x = -5$

The solutions are $x = -2$ and $x = -5$.

13. $x^2 = 21$

$x = \sqrt{21}$ or $x = -\sqrt{21}$

The solutions are $x = \pm\sqrt{21}$.

17. $x^2 = -4$ has no real solution because the square root of -4 is not a real number.

21. $7x^2 = 4$

$x^2 = \dfrac{4}{7}$

$x = \sqrt{\dfrac{4}{7}} \quad$ or $\quad x = -\sqrt{\dfrac{4}{7}}$

$x = \dfrac{\sqrt{4}}{\sqrt{7}} \cdot \dfrac{\sqrt{7}}{\sqrt{7}} \quad x = -\dfrac{\sqrt{4}}{\sqrt{7}} \cdot \dfrac{\sqrt{7}}{\sqrt{7}}$

$x = \dfrac{2\sqrt{7}}{7} \qquad x = -\dfrac{2\sqrt{7}}{7}$

The solutions are $x = \pm\dfrac{2\sqrt{7}}{7}$.

25. $(x-5)^2 = 49$

$x - 5 = \sqrt{49}$ or $x - 5 = -\sqrt{49}$

$x - 5 = 7 \qquad\quad x - 5 = -7$

$\quad x = 12 \qquad\qquad x = -2$

The solutions are $x = -2$ and $x = 12$.

29. $\left(m - \dfrac{1}{2}\right)^2 = \dfrac{1}{4}$

$m - \dfrac{1}{2} = \sqrt{\dfrac{1}{4}}$ or $m - \dfrac{1}{2} = -\sqrt{\dfrac{1}{4}}$

$m - \dfrac{1}{2} = \dfrac{1}{2} \qquad m - \dfrac{1}{2} = -\dfrac{1}{2}$

$\quad m = 1 \qquad\qquad m = 0$

The solutions are $m = 0$ and $m = 1$.

33. $(3y + 2)^2 = 100$

$3y + 2 = \sqrt{100}$ or $3y + 2 = -\sqrt{100}$

$3y + 2 = 10 \qquad\quad 3y + 2 = -10$

$\quad 3y = 8 \qquad\qquad 3y = -12$

$\quad y = \dfrac{8}{3} \qquad\qquad y = -4$

The solutions are $y = -4$ and $y = \dfrac{8}{3}$.

37. $(2x - 11)^2 = 50$

$2x - 11 = \sqrt{50} \qquad$ or $\quad 2x - 11 = -\sqrt{50}$

$2x - 11 = 5\sqrt{2} \qquad\qquad 2x - 11 = -5\sqrt{2}$

$2x = 11 + 5\sqrt{2} \qquad\qquad 2x = 11 - 5\sqrt{2}$

$x = \dfrac{11 + 5\sqrt{2}}{2} \qquad\qquad x = \dfrac{11 - 5\sqrt{2}}{2}$

The solutions are $x = \dfrac{11 \pm 5\sqrt{2}}{2}$.

41. $x^2 - 2 = 0$

$x^2 = 2$

$x = \sqrt{2}$ or $x = -\sqrt{2}$

The solutions are $x = \pm\sqrt{2}$.

45. $\dfrac{1}{2} n^2 = 5$

$n^2 = 10$

$n = \sqrt{10}$ or $n = -\sqrt{10}$

The solutions are $n = \pm\sqrt{10}$.

49. $3z^2 = 36$

$z^2 = 12$

$z = \sqrt{12}$ or $z = -\sqrt{12}$

$z = 2\sqrt{3} \qquad z = -2\sqrt{3}$

The solutions are $z = \pm 2\sqrt{3}$.

53. $\quad h = 16t^2$

$87.6 = 16t^2$

$\dfrac{87.6}{16} = t^2$

$5.475 = t^2$

$\sqrt{5.475} = t$ or $-\sqrt{5.475} = 5$

$2.3 \approx t \qquad\qquad -2.3 \approx 5$

Since the time of a dive is not a negative number, reject the solution -2.3. The dive lasted approximately 2.3 seconds.

57. $A = s^2$

$20 = s^2$

$\sqrt{20} = s$ or $-\sqrt{20} = s$

$2\sqrt{5} = s \qquad -2\sqrt{5} = s$

Since the length of a side is not a negative number, reject the solution $-2\sqrt{5}$. The sides have length $2\sqrt{5} \approx 4.47$ inches.

61. $x^2 + 6x + 9 = (x)^2 + 2(x)(3) + (3)^2$
$= (x + 3)^2$

65. answers may vary

69. $A = \pi r^2$
$36\pi = \pi r^2$
$36 = r^2$
$\sqrt{36} = r \text{ or } -\sqrt{36} = r$
$6 = r \qquad -6 = r$

Since the radius is not a negative number, reject the solution -6.
The radius of the circle is 6 inches.

73. $y = -0.07(x - 192.5)^2 + 3135$
$727 = -0.07(x - 192.5)^2 + 3135$
$-2408 = -0.07(x - 192.5)^2$
$\dfrac{-2408}{-0.07} = (x - 192.5)^2$
$34{,}400 = (x - 192.5)^2$
$\sqrt{34{,}400} = x - 192.5$
$20\sqrt{86} = x - 192.5$
$192.5 + 20\sqrt{86} = x$
$378 \approx x$
or $-\sqrt{34{,}400} = x - 192.5$
$-20\sqrt{86} = x - 192.5$
$192.5 - 20\sqrt{86} = x$
$7 \approx x$

There will be 727 Barnes and Noble Booksellers open for business when $x = 7$, which is 7 years after 1999, or 2006.

Exercise Set 9.2

1. $x^2 + 8x = -12$
$x^2 + 8x + \left(\dfrac{8}{2}\right)^2 = -12 + \left(\dfrac{8}{2}\right)^2$
$x^2 + 8x + 4^2 = -12 + 4^2$
$x^2 + 8x + 16 = -12 + 16$
$(x + 4)^2 = 4$
$x + 4 = \sqrt{4} \quad \text{or } x + 4 = -\sqrt{4}$
$x = -4 + 2 \qquad x = -4 - 2$
$x = -2 \qquad\qquad x = -6$
The solutions are $x = -6$ and $x = -2$.

5. $x^2 - 6x = 0$
$x^2 - 6x + \left(\dfrac{-6}{2}\right)^2 = 0 + \left(\dfrac{-6}{2}\right)^2$
$x^2 - 6x + (-3)^2 = 0 + (-3)^2$
$x^2 - 6x + 9 = 9$
$(x - 3)^2 = 9$
$x - 3 = \sqrt{9} \text{ or } x - 3 = -\sqrt{9}$
$x = 3 + 3 \qquad x = 3 - 3$
$x = 6 \qquad\qquad x = 0$
The solutions are $x = 0$ and $x = 6$.

9. $x^2 - 2x - 1 = 0$
$x^2 - 2x = 1$
$x^2 - 2x + \left(\dfrac{-2}{2}\right)^2 = 1 + \left(\dfrac{-2}{2}\right)^2$
$x^2 - 2x + (-1)^2 = 1 + (-1)^2$
$x^2 - 2x + 1 = 1 + 1$
$(x - 1)^2 = 2$
$x - 1 = \sqrt{2} \text{ or } x - 1 = -\sqrt{2}$
$x = 1 + \sqrt{2} \qquad x = 1 - \sqrt{2}$
The solutions are $x = 1 \pm \sqrt{2}$.

13. $3x^2 - 6x = 24$
$x^2 - 2x = 8$
$x^2 - 2x + \left(\dfrac{-2}{2}\right)^2 = 8 + \left(\dfrac{-2}{2}\right)^2$
$x^2 - 2x + (-1)^2 = 8 + (-1)^2$
$x^2 - 2x + 1 = 8 + 1$
$(x - 1)^2 = 9$
$x - 1 = \sqrt{9} \text{ or } x - 1 = -\sqrt{9}$
$x = 1 + 3 \qquad x = 1 - 3$
$x = 4 \qquad\qquad x = -2$
The solutions are $x = -2$ and $x = 4$.

17. $2x^2 = 6x + 5$
$2x^2 - 6x = 5$
$x^2 - 3x = \dfrac{5}{2}$
$x^2 - 3x + \left(\dfrac{-3}{2}\right)^2 = \dfrac{5}{2} + \left(\dfrac{-3}{2}\right)^2$
$x^2 - 3x + \dfrac{9}{4} = \dfrac{5}{2} + \dfrac{9}{4}$
$\left(x - \dfrac{3}{2}\right)^2 = \dfrac{10}{4} + \dfrac{9}{4}$
$\left(x - \dfrac{3}{2}\right)^2 = \dfrac{19}{4}$
$x - \dfrac{3}{2} = \sqrt{\dfrac{19}{4}} \quad \text{or } x - \dfrac{3}{2} = -\sqrt{\dfrac{19}{4}}$
$x = \dfrac{3}{2} + \dfrac{\sqrt{19}}{2} \qquad x = \dfrac{3}{2} - \dfrac{\sqrt{19}}{2}$
$x = \dfrac{3 + \sqrt{19}}{2} \qquad x = \dfrac{3 - \sqrt{19}}{2}$
The solutions are $x = \dfrac{3 \pm \sqrt{19}}{2}$.

21. $x^2 + 6x - 25 = 0$
$x^2 + 6x = 25$
$x^2 + 6x + \left(\dfrac{6}{2}\right)^2 = 25 + \left(\dfrac{6}{2}\right)^2$
$x^2 + 6x + 3^2 = 25 + 3^2$
$x^2 + 6x + 9 = 25 + 9$
$(x + 3)^2 = 34$
$x + 3 = \sqrt{34} \qquad \text{or } x + 3 = -\sqrt{34}$
$x = -3 + \sqrt{34} \qquad x = -3 - \sqrt{34}$
The solutions are $x = -3 \pm \sqrt{34}$.

25. $2y^2 - 3y + 1 = 0$
$2y^2 - 3y = -1$
$y^2 - \dfrac{3}{2}y = -\dfrac{1}{2}$
$y^2 - \dfrac{3}{2}y + \left(\dfrac{-\frac{3}{2}}{2}\right)^2 = -\dfrac{1}{2} + \left(\dfrac{-\frac{3}{2}}{2}\right)^2$
$y^2 - \dfrac{3}{2}y + \left(-\dfrac{3}{4}\right)^2 = -\dfrac{1}{2} + \left(-\dfrac{3}{4}\right)^2$
$y^2 - \dfrac{3}{2}y + \dfrac{9}{16} = -\dfrac{1}{2} + \dfrac{9}{16}$
$\left(y - \dfrac{3}{4}\right)^2 = \dfrac{1}{16}$
$y - \dfrac{3}{4} = \sqrt{\dfrac{1}{16}} \quad \text{or } y - \dfrac{3}{4} = -\sqrt{\dfrac{1}{16}}$
$y = \dfrac{3}{4} + \dfrac{1}{4} \qquad y = \dfrac{3}{4} - \dfrac{1}{4}$
$y = 1 \qquad\qquad y = \dfrac{1}{2}$
The solutions are $y = \dfrac{1}{2}$ and $y = 1$.

29. $3z^2 + 6z + 4 = 0$

$3z^2 + 6z = -4$

$z^2 + 2z = -\dfrac{4}{3}$

$z^2 + 2z + \left(\dfrac{2}{2}\right)^2 = -\dfrac{4}{3} + \left(\dfrac{2}{2}\right)^2$

$z^2 + 2z + 1^2 = -\dfrac{4}{3} + 1^2$

$z^2 + 2z + 1 = -\dfrac{4}{3} + 1$

$(z + 1)^2 = -\dfrac{1}{3}$

The equation has no solution, since the square root of $-\dfrac{1}{3}$ is not a real number.

33. $\dfrac{3}{4} - \sqrt{\dfrac{25}{16}} = \dfrac{3}{4} - \dfrac{\sqrt{25}}{\sqrt{16}}$

$= \dfrac{3}{4} - \dfrac{5}{4}$

$= \dfrac{3-5}{4}$

$= \dfrac{-2}{4}$

$= -\dfrac{1}{2}$

37. $\dfrac{6 + 4\sqrt{5}}{2} = \dfrac{6}{2} + \dfrac{4\sqrt{5}}{2} = 3 + 2\sqrt{5}$

41. answers may vary

45. $x^2 + kx + \left(\dfrac{k}{2}\right)^2$ is a perfect square trinomial. If $x^2 + kx + 16$ is

a perfect square trinomial, then $\left(\dfrac{k}{2}\right)^2 = 16$

$\dfrac{k}{2} = \sqrt{16}$ or $\dfrac{k}{2} = -\sqrt{16}$

$k = 2 \cdot 4 \qquad k = 2(-4)$

$k = 8 \qquad\quad k = -8$

$x^2 + kx + 16$ is a perfect square trinomial
when $k = 8$ or $k = -8$.

49. The solutions are $x = -6$ and $x = -2$.

Exercise Set 9.3

1. $x^2 - 3x + 2 = 0$

$a = 1, b = -3, c = 2$

$x = \dfrac{-b \pm \sqrt{b^2 - 4ac}}{2a}$

$x = \dfrac{-(-3) \pm \sqrt{(-3)^2 - 4(1)(2)}}{2(1)}$

$= \dfrac{3 \pm \sqrt{9 - 8}}{2}$

$= \dfrac{3 \pm \sqrt{1}}{2}$

$= \dfrac{3 \pm 1}{2}$

$x = \dfrac{3 + 1}{2} = 2$ or $x = \dfrac{3 - 1}{2} = 1$

The solutions are $x = 2$ and $x = 1$.

5. $4x^2 - 3 = 0$

$4x^2 + 0x - 3 = 0$

$a = 4, b = 0, c = -3$

$x = \dfrac{-b \pm \sqrt{b^2 - 4ac}}{2a}$

$x = \dfrac{-0 \pm \sqrt{0^2 - 4(4)(-3)}}{2(4)}$

$= \dfrac{0 \pm \sqrt{0 + 48}}{8}$

$= \dfrac{\pm\sqrt{48}}{8}$

$= \pm\dfrac{4\sqrt{3}}{8}$

$= \pm\dfrac{\sqrt{3}}{2}$

The solutions are $x = \pm\dfrac{\sqrt{3}}{2}$.

9. $y^2 = 7y + 30$

$y^2 - 7y - 30 = 0$

$a = 1, b = -7, c = -30$

$y = \dfrac{-b \pm \sqrt{b^2 - 4ac}}{2a}$

$y = \dfrac{-(-7) \pm \sqrt{(-7)^2 - 4(1)(-30)}}{2(1)}$

$= \dfrac{7 \pm \sqrt{49 + 120}}{2}$

$= \dfrac{7 \pm \sqrt{169}}{2}$

$= \dfrac{7 \pm 13}{2}$

$y = \dfrac{7 + 13}{2} = 10$ or $y = \dfrac{7 - 13}{2} = -3$

The solutions are $y = 10$ and $y = -3$.

13. $m^2 - 12 = m$

$m^2 - m - 12 = 0$

$a = 1, b = -1, c = -12$

$m = \dfrac{-b \pm \sqrt{b^2 - 4ac}}{2a}$

$m = \dfrac{-(-1) \pm \sqrt{(-1)^2 - 4(1)(-12)}}{2(1)}$

$= \dfrac{1 \pm \sqrt{1 + 48}}{2}$

$= \dfrac{1 \pm \sqrt{49}}{2}$

$= \dfrac{1 \pm 7}{2}$

$m = \dfrac{1 + 7}{2} = 4$ or $m = \dfrac{1 - 7}{2} = -3$

The solutions are $m = 4$ and $m = -3$.

17. $6x^2 + 9x = 2$

$6x^2 + 9x - 2 = 0$

$a = 6, b = 9, c = -2$

$x = \dfrac{-b \pm \sqrt{b^2 - 4ac}}{2a}$

$$x = \frac{-9 \pm \sqrt{9^2 - 4(6)(-2)}}{2(6)}$$

$$= \frac{-9 \pm \sqrt{81 + 48}}{12}$$

$$= \frac{-9 \pm \sqrt{129}}{12}$$

The solutions are $x = \dfrac{-9 \pm \sqrt{129}}{12}$.

21. $x^2 - 6x + 2 = 0$

$a = 1, b = -6, c = 2$

$$x = \frac{-b \pm \sqrt{b^2 - 4ac}}{2a}$$

$$x = \frac{-(-6) \pm \sqrt{(-6)^2 - 4(1)(2)}}{2(1)}$$

$$= \frac{6 \pm \sqrt{36 - 8}}{2}$$

$$= \frac{6 \pm \sqrt{28}}{2}$$

$$= \frac{6 \pm 2\sqrt{7}}{2}$$

$$= \frac{2\left(3 \pm \sqrt{7}\right)}{2}$$

$$= 3 \pm \sqrt{7}$$

The solutions are $x = 3 \pm \sqrt{7}$.

25. $3x^2 = 1 - 2x$

$3x^2 + 2x - 1 = 0$

$a = 3, b = 2, c = -1$

$$x = \frac{-b \pm \sqrt{b^2 - 4ac}}{2a}$$

$$x = \frac{-2 \pm \sqrt{2^2 - 4(3)(-1)}}{2(3)}$$

$$= \frac{-2 \pm \sqrt{4 + 12}}{6}$$

$$= \frac{-2 \pm \sqrt{16}}{6}$$

$$= \frac{-2 \pm 4}{6}$$

$$x = \frac{-2 + 4}{6} = \frac{1}{3} \text{ or } x = \frac{-2 - 4}{6} = -1$$

The solutions are $x = \dfrac{1}{3}$ and $x = -1$.

29. $20y^2 = 3 - 11y$

$20y^2 + 11y - 3 = 0$

$a = 20, b = 11, c = -3$

$$y = \frac{-b \pm \sqrt{b^2 - 4ac}}{2a}$$

$$y = \frac{-11 \pm \sqrt{11^2 - 4(20)(-3)}}{2(20)}$$

$$= \frac{-11 \pm \sqrt{121 + 240}}{40}$$

$$= \frac{-11 \pm \sqrt{361}}{40}$$

$$= \frac{-11 \pm 19}{40}$$

$$y = \frac{-11 + 19}{40} = \frac{1}{5} \text{ or } y = \frac{-11 - 19}{40} = -\frac{3}{4}$$

The solutions are $y = \dfrac{1}{5}$ and $y = -\dfrac{3}{4}$.

33. $\dfrac{m^2}{2} = m + \dfrac{1}{2}$

$$\frac{m^2}{2} - m - \frac{1}{2} = 0$$

$$m^2 - 2m - 1 = 0$$

$a = 1, b = -2, c = -1$

$$m = \frac{-b \pm \sqrt{b^2 - 4ac}}{2a}$$

$$m = \frac{-(-2) \pm \sqrt{(-2)^2 - 4(1)(-1)}}{2(1)}$$

$$= \frac{2 \pm \sqrt{4 + 4}}{2}$$

$$= \frac{2 \pm \sqrt{8}}{2}$$

$$= \frac{2 \pm 2\sqrt{2}}{2}$$

$$= \frac{2\left(1 \pm \sqrt{2}\right)}{2}$$

$$= 1 \pm \sqrt{2}$$

The solutions are $m = 1 \pm \sqrt{2}$.

37. $4p^2 + \dfrac{3}{2} = -5p$

$$4p^2 + 5p + \frac{3}{2} = 0$$

$$8p^2 + 10p + 3 = 0$$

$a = 8, b = 10, c = 3$

$$p = \frac{-b \pm \sqrt{b^2 - 4ac}}{2a}$$

$$p = \frac{-10 \pm \sqrt{10^2 - 4(8)(3)}}{2(8)}$$

$$= \frac{-10 \pm \sqrt{100 - 96}}{16}$$

$$= \frac{-10 \pm \sqrt{4}}{16}$$

$$= \frac{-10 \pm 2}{16}$$

$$p = \frac{-10 + 2}{16} = -\frac{1}{2} \text{ or } p = \frac{-10 - 2}{16} = -\frac{3}{4}$$

The solutions are $p = -\dfrac{1}{2}$ and $p = -\dfrac{3}{4}$.

41. $x^2 - \dfrac{11}{2}x - \dfrac{1}{2} = 0$

$$2x^2 - 11x - 1 = 0$$

$a = 2, b = -11, c = -1$

$$x = \frac{-b \pm \sqrt{b^2 - 4ac}}{2a}$$

$$x = \frac{-(-11) \pm \sqrt{(-11)^2 - 4(2)(-1)}}{2(2)}$$

$$= \frac{11 \pm \sqrt{121 + 8}}{4}$$

$$= \frac{11 \pm \sqrt{129}}{4}$$

The solutions are $x = \dfrac{11 \pm \sqrt{129}}{4}$.

45. $3x^2 = 21$

$$3x^2 - 21 = 0$$

$$3x^2 + 0x - 21 = 0$$

$a = 3, b = 0, c = -21$

$$x = \frac{-b \pm \sqrt{b^2 - 4ac}}{2a}$$

$$x = \frac{-0 \pm \sqrt{0^2 - 4(3)(-21)}}{2(3)}$$

$$= \frac{\pm\sqrt{0 + 252}}{6}$$

$$= \frac{\pm\sqrt{252}}{6}$$

$$= \frac{\pm 6\sqrt{7}}{6}$$

$$= \pm\sqrt{7}$$

The solutions are $x = \pm\sqrt{7}$ or $x \approx -2.6$ and $x \approx 2.6$.

49.
$$x^2 = 9x + 4$$
$$x^2 - 9x - 4 = 0$$
$$a = 1, b = -9, c = -4$$

$$x = \frac{-b \pm \sqrt{b^2 - 4ac}}{2a}$$

$$x = \frac{-(-9) \pm \sqrt{(-9)^2 - 4(1)(-4)}}{2(1)}$$

$$= \frac{9 \pm \sqrt{81 + 16}}{2}$$

$$= \frac{9 \pm \sqrt{97}}{2}$$

The solutions are $x = \dfrac{9 \pm \sqrt{97}}{2}$ or $x \approx 9.4$ and $x \approx -0.4$.

53. $y = -3$ is a horizontal line with y-intercept $(0, -3)$.

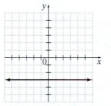

57.
$$5x^2 + 2 = x$$
$$5x^2 - x + 2 = 0$$
$$b = -1, \text{ which is choice c.}$$

61. Let x be the width of the chocolate bar. Then the length is $2x + 0.5$.

Area = (length)(width)
$$50.8 = (2x + 0.5)(x)$$
$$50.8 = 2x^2 + 0.5x$$
$$0 = 2x^2 + 0.5x - 50.8$$
$$0 = 20x^2 + 5x - 508$$
$$a = 20, b = 5, c = -508$$

$$x = \frac{-b \pm \sqrt{b^2 - 4ac}}{2a}$$

$$x = \frac{-5 \pm \sqrt{5^2 - 4(20)(-508)}}{2(20)}$$

$$= \frac{-5 \pm \sqrt{25 + 40{,}640}}{40}$$

$$= \frac{-5 \pm \sqrt{40{,}665}}{40}$$

$$x = \frac{-5 + \sqrt{40{,}665}}{40} \approx 4.9 \text{ or}$$

$$x = \frac{-5 - \sqrt{40{,}665}}{40} \approx -5.2$$

Since the width is not a negative number, discard the solution $x \approx -5.2$.

$$2x + 0.5 = 2(4.9) + 0.5 = 9.8 + 0.5 = 10.3$$

The width was approximately 4.9 feet and the length was approximately 10.3 feet.

65. answers may vary

69.
$$h = -16t^2 + 120t + 80$$
$$0 = -16t^2 + 120t + 80$$
$$a = -16, b = 120, c = 80$$

$$t = \frac{-b \pm \sqrt{b^2 - 4ac}}{2a}$$

$$t = \frac{-120 \pm \sqrt{120^2 - 4(-16)(80)}}{2(-16)}$$

$$= \frac{-120 \pm \sqrt{14{,}400 + 5120}}{-32}$$

$$= \frac{-120 \pm \sqrt{19{,}520}}{-32}$$

$$t = \frac{-120 + \sqrt{19{,}520}}{-32} \approx -0.6 \text{ or}$$

$$t = \frac{-120 - \sqrt{19{,}520}}{-32} \approx 8.1$$

Since the time of the flight is not a negative number, discard the solution -0.6. The rocket will strike the ground approximately 8.1 seconds after it is launched.

Exercise Set 9.4

1.

x	$y = 2x^2$
-2	$2(-2)^2 = 2(4) = 8$
-1	$2(-1)^2 = 2(1) = 2$
0	$2(0)^2 = 2(0) = 0$
1	$2(1)^2 = 2(1) = 2$
2	$2(2)^2 = 2(4) = 8$

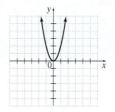

5. $y = x^2 - 1$

$$y = 0: \quad 0 = x^2 - 1$$
$$1 = x^2$$
$$\pm\sqrt{1} = x$$
$$\pm 1 = x$$

x-intercepts: $(-1, 0), (1, 0)$

$x = 0: y = 0^2 - 1 = -1$

y-intercept: $(0, -1)$

$$y = x^2 + 0x - 1$$
$$a = 1, b = 0, c = 1$$

$$\frac{-b}{2a} = \frac{-0}{2(1)} = \frac{0}{2} = 0$$

$x = 0$: $y = 0^2 - 1 = -1$
The vertex is $(0, -1)$.

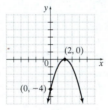

9. $y = -x^2 + 4x - 4$
$y = 0$: $0 = -x^2 + 4x - 4$
$\quad\quad\quad 0 = x^2 - 4x + 4$
$\quad\quad\quad 0 = (x - 2)^2$
$\quad\quad\quad 0 = x - 2$
$\quad\quad\quad 2 = x$
The x-intercept is $(2, 0)$.
$x = 0$: $y = -0^2 + 4(0) - 4 = -4$
The y-intercept is $(0, -4)$.
$y = -x^2 + 4x - 4$
$a = -1, b = 4, c = -4$
$\dfrac{-b}{2a} = \dfrac{-4}{2(-1)} = \dfrac{-4}{-2} = 2$
$x = 2$: $y = -(2)^2 + 4(2) - 4 = -4 + 8 - 4 = 0$
The vertex is $(2, 0)$.

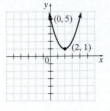

13. $y = x^2 - 4x + 5$
$y = 0$: $\quad 0 = x^2 - 4x + 5$
$a = 1, b = -4, c = 5$
$y = \dfrac{-(-4) \pm \sqrt{(-4)^2 - 4(1)(5)}}{2(1)}$
$\quad = \dfrac{4 \pm \sqrt{16 - 20}}{2}$
$\quad = \dfrac{4 \pm \sqrt{-4}}{2}$
There are no x-intercepts, since $\sqrt{-4}$ is not a real number.
$x = 0$: $y = 0^2 - 4(0) + 5 = 5$
y-intercept: $(0, 5)$
$y = x^2 - 4x + 5$
$a = 1, b = -4, c = 5$
$\dfrac{-b}{2a} = \dfrac{-(-4)}{2(1)} = \dfrac{4}{2} = 2$
$x = 2$: $\quad y = 2^2 - 4(2) + 5 = 4 - 8 + 5 = 1$
vertex: $(2, 1)$

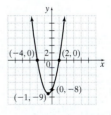

17. $y = \dfrac{1}{3} x^2$
$y = 0$: $0 = \dfrac{1}{3} x^2$
$\quad\quad\quad 0 = x^2$
$\quad\quad\quad 0 = x$
x-intercept: $(0, 0)$
The y-intercept is also $(0, 0)$.
$y = \dfrac{1}{3} x^2 + 0x + 0$
$a = \dfrac{1}{3}, b = 0, c = 0$
$\dfrac{-b}{2a} = \dfrac{-0}{2\left(\dfrac{1}{3}\right)} = \dfrac{0}{\dfrac{2}{3}} = 0$
$x = 0$: $\quad y = \dfrac{1}{3}(0)^2 = \dfrac{1}{3}(0) = 0$
vertex: $(0, 0)$
For additional points, use $x = \pm 3$.
$x = -3$: $y = \dfrac{1}{3}(-3)^2 = \dfrac{1}{3}(9) = 3$
$x = 3$: $y = \dfrac{1}{3}(3)^2 = \dfrac{1}{3}(9) = 3$
$(-3, 3)$ and $(3, 3)$ are also on the graph.

21. $y = x^2 + 2x - 8$
$y = 0$: $0 = x^2 + 2x - 8$
$\quad\quad\quad 0 = (x + 4)(x - 2)$
$\quad\quad\quad 0 = x + 4 \text{ or } 0 = x - 2$
$\quad\quad\quad -4 = x \quad\quad\quad 2 = x$
x-intercepts: $(-4, 0), (2, 0)$
$x = 0$: $y = 0^2 + 2(0) - 8 = -8$
y-intercept: $(0, -8)$
$y = x^2 + 2x - 8$
$a = 1, b = 2, c = -8$
$\dfrac{-b}{2a} = \dfrac{-2}{2(1)} = \dfrac{-2}{2} = -1$
$x = -1$: $y = (-1)^2 + 2(-1) - 8 = 1 - 2 - 8 = -9$
vertex: $(-1, -9)$

25. $y = 2x^2 - 11x + 5$
$y = 0$: $\quad 0 = 2x^2 - 11x + 5$
$\quad\quad\quad 0 = (2x - 1)(x - 5)$
$\quad\quad\quad 0 = 2x - 1 \quad \text{or } 0 = x - 5$
$\quad\quad\quad 1 = 2x \quad\quad\quad\quad 5 = x$
$\quad\quad\quad \dfrac{1}{2} = x$

x-intercepts: $\left(\frac{1}{2}, 0\right), (5, 0)$

$x = 0$: $y = 2(0)^2 - 11(0) + 5 = 5.$

$y = $ intercept: $(0, 5)$

$y = 2x^2 - 11x + 5$

$a = 2, b = -11, c = 5$

$\frac{-b}{2a} = \frac{-(-11)}{2(2)} = \frac{11}{4}$

$x = \frac{11}{4}$: $y = 2\left(\frac{11}{4}\right)^2 - 11\left(\frac{11}{4}\right) + 5$

$= 2\left(\frac{121}{16}\right) - \frac{121}{4} + 5$

$= \frac{121}{8} - \frac{121}{4} + 5$

$= \frac{121}{8} - \frac{242}{8} + \frac{40}{8}$

$= -\frac{81}{8}$

vertex: $\left(\frac{11}{4}, -\frac{81}{8}\right)$

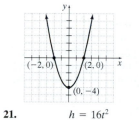

29. $\dfrac{\frac{1}{7}}{\frac{2}{5}} = \frac{1}{7} \div \frac{2}{5} = \frac{1}{7} \cdot \frac{5}{2} = \frac{5}{14}$

33. $\dfrac{2x}{1 - \frac{1}{x}} = \dfrac{x(2x)}{x\left(1 - \frac{1}{x}\right)} = \dfrac{2x^2}{x - 1}$

37. a. The maximum height appears to be about 256 feet.
b. The fireball appears to reach its maximum height when $t = 4$ seconds.
c. The fireball appears to return to the ground when $t = 8$ seconds.

41. With $a < 0$, the parabola opens downward. A downward-opening parabola that does not touch or cross the x-axis (no x-intercept) is graph D.

Chapter 9 Test

1. $x^2 - 400 = 0$

$(x + 20)(x - 20) = 0$

$x + 20 = 0$ or $x - 20 = 0$

$\qquad x = -20 \qquad\qquad x = 20$

The solutions are $x = \pm 20.$

5. $x^2 - 26x + 160 = 0$

$x^2 - 26x = -160$

$x^2 - 26x + \left(\frac{-26}{2}\right)^2 = -160 + \left(\frac{-26}{2}\right)^2$

$x^2 - 26x + (-13)^2 = -160 + (-13)^2$

$x^2 - 26x + 169 = -160 + 169$

$(x - 13)^2 = 9$

$x - 13 = \sqrt{9}$ or $x - 13 = -\sqrt{9}$

$x - 13 = 3 \qquad\qquad x - 13 = -3$

$\qquad x = 16 \qquad\qquad\quad x = 10$

The solutions are $x = 10$ and $x = 16.$

9. $(3x - 5)(x + 2) = -6$

$3x^2 + 6x - 5x - 10 = -6$

$3x^2 + x - 10 = -6$

$3x^2 + x - 4 = 0$

$a = 3, b = 1, c = -4$

$x = \dfrac{-b \pm \sqrt{b^2 - 4ac}}{2a}$

$x = \dfrac{-1 \pm \sqrt{1^2 - 4(3)(-4)}}{2(3)}$

$= \dfrac{-1 \pm \sqrt{1 + 48}}{6}$

$= \dfrac{-1 \pm \sqrt{49}}{6}$

$= \dfrac{-1 \pm 7}{6}$

$x = \dfrac{-1 + 7}{6} = 1$ or $x = \dfrac{-1 - 7}{6} = -\dfrac{4}{3}$

The solutions are $x = 1$ and $x = -\dfrac{4}{3}.$

13. $3x^2 - 7x + 2 = 0$

$a = 3, b = -7, c = 2$

$x = \dfrac{-b \pm \sqrt{b^2 - 4ac}}{2a}$

$x = \dfrac{-(-7) \pm \sqrt{(-7)^2 - 4(3)(2)}}{2(3)}$

$= \dfrac{7 \pm \sqrt{49 - 24}}{6}$

$= \dfrac{7 \pm \sqrt{25}}{6}$

$= \dfrac{7 \pm 5}{6}$

$x = \dfrac{7 + 5}{6} = 2$ or $x = \dfrac{7 - 5}{6} = \dfrac{1}{3}$

The solutions are $x = \dfrac{1}{3}$ and $x = 2.$

17. $y = x^2 - 4$

$y = 0$: $\quad 0 = x^2 - 4$

$\qquad\qquad 0 = (x + 2)(x - 2)$

$\qquad\qquad 0 = x + 2$ or $\quad 0 = x - 2$

$\qquad\quad -2 = x \qquad\qquad\quad 2 = x$

x-intercepts: $(-2, 0), (2, 0)$

$x = 0$: $\quad y = 0^2 - 4 = -4$

y-intercept: $(0, -4)$

$y = x^2 + 0x - 4$

$a = 1, b = 0, c = -4$

$\dfrac{-b}{2a} = \dfrac{-0}{2(1)} = \dfrac{0}{2} = 0$

$x = 0$: $y = 0^2 - 4 = -4$

vertex: $(0, -4)$

21. $\qquad\qquad h = 16t^2$

$120.75 = 16t^2$

$\dfrac{120.75}{16} = t^2$

$7.546875 = t^2$

$\sqrt{7.546875} = t$ or $-\sqrt{7.546875} = t$

$\qquad 2.7 \approx t \qquad\qquad -2.7 \approx t$

Since the time of the dive is not a negative number, discard the solution $t \approx -2.7$. The dive took approximately 2.7 seconds.

APPENDIX

Exercise Set Appendix B

1. $a^3 + 27 = a^3 + 3^3$
$= (a + 3)[a^2 - (a)(3) + 3^2]$
$= (a + 3)(a^2 - 3a + 9)$

5. $5k^3 + 40 = 5(k^3 + 8)$
$= 5(k^3 + 2^3)$
$= 5(k + 2)[k^2 - (k)(2) + 2^2]$
$= 5(k + 2)(k^2 - 2k + 4)$

9. $x^3 + 125 = x^3 + 5^3$
$= (x + 5)[x^2 - (x)(5) + 5^2]$
$= (x + 5)(x^2 - 5x + 25)$

13. $27 - t^3 = 3^3 - t^3$
$= (3 - t)[3^2 + (3)(t) + t^2]$
$= (3 - t)(9 + 3t + t^2)$

17. $t^3 - 343 = t^3 - 7^3$
$= (t - 7)[t^2 + (t)(7) + 7^2]$
$= (t - 7)(t^2 + 7t + 49)$

Exercise Set Appendix C

1. mean $= \dfrac{21 + 28 + 16 + 42 + 38}{5} = \dfrac{145}{5} = 29$

$16, 21, \underline{28}, 38, 42$
median $= 28$
no mode

5. mean
$= \dfrac{0.2 + 0.3 + 0.5 + 0.6 + 0.6 + 0.9 + 0.2 + 0.7 + 1.1}{9}$
$= \dfrac{5.1}{9}$
≈ 0.6
$0.2, 0.2, 0.3, 0.5, \underline{0.6}, 0.6, 0.7, 0.9, 1.1$
median $= 0.6$
mode: 0.2 and 0.6

9. $\dfrac{1454 + 1250 + 1136 + 1127 + 1107}{5} = \dfrac{6074}{5}$
$= 1214.8$

The mean height of the five tallest buildings is 1214.8 feet.

13. $\dfrac{7.8 + 6.9 + 7.5 + 4.7 + 6.9 + 7.0}{6} = \dfrac{40.8}{6} = 6.8$

The mean time was 6.8 seconds.

17. $74, 77, \underline{85, 86}, 91, 95$

median $= \dfrac{85 + 86}{2} = 85.5$

The median test score was 85.5.

21. The values 70 and 71 both occur twice while other values only occur once, so the mode of the pulse rates is 70 and 71.

25. Since the mode is 21, the value 21 must occur at least twice in the set of numbers. Since there are an odd number of numbers in the set, the median is the middle number. That is, the median of 20 is one of the numbers in the set.
The missing numbers are 21, 21, and 20.

Exercise Set Appendix D

1. The set of negative integers from -10 to -5 in roster form is $\{-9, -8, -7, -6\}$.

5. The set of whole numbers in roster form is $\{0, 1, 2, 3, 4, \dots\}$.

9. Since 3 is a listed element of the set $\{1, 3, 5, 7, 9\}$, the statement $3 \in \{1, 3, 5, 7, 9\}$ is true.

13. $\{a, e, i, o, u\} \subseteq \{a, e, i, o, u\}$ since every element of the left-hand set is also an element of the right-hand set.

17. Since 9 is not an even number, it is not an element of the set $\{x | x \text{ is an even number}\}$. Thus the statement $9 \notin \{x | x \text{ is an even number}\}$ is true.

21. $A \cup B = \{1, 2, 3, 4, 5, 6\} \cup \{2, 4, 6\}$
$= \{1, 2, 3, 4, 5, 6\}$

25. $C \cup D = \{1, 3, 5\} \cup \{7\} = \{1, 3, 5, 7\}$

29. Since every element of $B = \{2, 4, 6\}$ is also an element of $A = \{1, 2, 3, 4, 5, 6\}$, the statement $B \subseteq A$ is true.

33. Since 2 is an element of $A = \{1, 2, 3, 4, 5, 6\}$, but not of $C = \{1, 3, 5\}$, the statement $A \subseteq C$ is false.

37. Since the empty set is a subset of every set, the statement $\varnothing \subseteq A$ is true.

41. Since the union of a set with the empty set is the original set, the statement that $\{a, b, c\} \cup \{ \ \}$ is $\{a, b, c\}$ is true.

Exercise Set Appendix E

1. $90° - 19° = 71°$
The complement of a 19° angle is a 71° angle.

5. $90° - 11\frac{1}{4}° = 78\frac{3}{4}°$

The complement of an $11\frac{1}{4}°$ angle is a $78\frac{3}{4}°$ angle.

9. $180° - 30.2° = 149.8°$
The supplement of a 30.2° angle is a 149.8° angle.

13. $\angle 1$ and the angle marked 110° are vertical angles, so $m\angle 1 = 110°$.

$\angle 2$ and the angle marked 110° are supplementary angles, so $m\angle 2 = 180° - 110° = 70°$.

$\angle 3$ and the angle marked 110° are supplementary angles, so $m\angle 3 = 180° - 110° = 70°$.

$\angle 4$ and $\angle 3$ are alternate interior angles, so $m\angle 4 = m\angle 3 = 70°$.

$\angle 5$ and the angle marked 110° are alternate interior angles, so $m\angle 5 = 110°$

$\angle 6$ and $\angle 5$ are supplementary angles, so $m\angle 6 = 180° - m\angle 5 = 180° - 110° = 70°$.

$\angle 7$ and the angle marked 110° are corresponding angles, so $m\angle 7 = 110°$.

17. $180° - 25° - 65° = 90°$
The third angle measures 90°.

21. Since the triangle is a right triangle, one angle measures 90°.
$180° - 45° - 90° = 45°$
The other two angles of the triangle measure 45° and 90°.

25. Since the triangle is a right triangle, one angle measures 90°.

$180° - 39\frac{3}{4}° - 90° = 50\frac{1}{4}°$

The other two angles of the triangle measure $50\frac{1}{4}°$ and 90°.

29. $\dfrac{6}{9} = \dfrac{3}{x}$

$6x = 27$

$x = \dfrac{27}{6}$

$x = 4.5$

33. $a^2 + b^2 = c^2$
$a^2 + 5^2 = 13^2$
$a^2 + 25 = 169$
$a^2 = 144$

Since a represents a length, we assume that a is positive. Since $a^2 = 144$, $a = 12$. The other leg of the right triangle has length 12.

INDEX

Photo Credits

CHAPTER TEST PREP VIDEO CD, INTRODUCTORY ALGEBRA 3E, ELAYN MARTIN-GAY

0-13-186838-1
© 2007 Pearson Education, Inc.
Pearson Prentice Hall
Pearson Education, Inc.
Upper Saddle River, NJ 07458
Pearson Prentice Hall™ is a trademark of Pearson Education, Inc.

YOU SHOULD CAREFULLY READ THE TERMS AND CONDITIONS BEFORE USING THE CD-ROM PACKAGE. USING THIS CD-ROM PACKAGE INDICATES YOUR ACCEPTANCE OF THESE TERMS AND CONDITIONS.

Pearson Education, Inc. provides this program and licenses its use. You assume responsibility for the selection of the program to achieve your intended results, and for the installation, use, and results obtained from the program. This license extends only to use of the program in the United States or countries in which the program is marketed by authorized distributors.

LICENSE GRANT

You hereby accept a nonexclusive, nontransferable, permanent license to install and use the program ON A SINGLE COMPUTER at any given time. You may copy the program solely for backup or archival purposes in support of your use of the program on the single computer. You may not modify, translate, disassemble, decompile, or reverse engineer the program, in whole or in part.

TERM

The License is effective until terminated. Pearson Education, Inc. reserves the right to terminate this License automatically if any provision of the License is violated. You may terminate the License at any time. To terminate this License, you must return the program, including documentation, along with a written warranty stating that all copies in your possession have been returned or destroyed.

LIMITED WARRANTY

THE PROGRAM IS PROVIDED "AS IS" WITHOUT WARRANTY OF ANY KIND, EITHER EXPRESSED OR IMPLIED, INCLUDING, BUT NOT LIMITED TO, THE IMPLIED WARRANTIES OF MERCHANTABILITY AND FITNESS FOR A PARTICULAR PURPOSE. THE ENTIRE RISK AS TO THE QUALITY AND PERFORMANCE OF THE PROGRAM IS WITH YOU. SHOULD THE PROGRAM PROVE DEFECTIVE, YOU (AND NOT PEARSON EDUCATION, INC. OR ANY AUTHORIZED DEALER) ASSUME THE ENTIRE COST OF ALL NECESSARY SERVICING, REPAIR, OR CORRECTION. NO ORAL OR WRITTEN INFORMATION OR ADVICE GIVEN BY PEARSON EDUCATION, INC., ITS DEALERS, DISTRIBUTORS, OR AGENTS SHALL CREATE A WARRANTY OR INCREASE THE SCOPE OF THIS WARRANTY.

SOME STATES DO NOT ALLOW THE EXCLUSION OF IMPLIED WARRANTIES, SO THE ABOVE EXCLUSION MAY NOT APPLY TO YOU. THIS WARRANTY GIVES YOU SPECIFIC LEGAL RIGHTS AND YOU MAY ALSO HAVE OTHER LEGAL RIGHTS THAT VARY FROM STATE TO STATE.

Pearson Education, Inc. does not warrant that the functions contained in the program will meet your requirements or that the operation of the program will be uninterrupted or error-free. However, Pearson Education, Inc. warrants the CD-ROM(s) on which the program is furnished to be free from defects in material and workmanship under normal use for a period of ninety (90) days from the date of delivery to you as evidenced by a copy of your receipt. The program should not be relied on as the sole basis to solve a problem whose incorrect solution could result in injury to person or property. If the program is employed in such a manner, it is at the user's own risk and Pearson Education, Inc. explicitly disclaims all liability for such misuse.

LIMITATION OF REMEDIES

Pearson Education, Inc.'s entire liability and your exclusive remedy shall be:
1. the replacement of any CD-ROM not meeting Pearson Education, Inc.'s "LIMITED WARRANTY" and that is returned to Pearson Education, or
2. if Pearson Education is unable to deliver a replacement CD-ROM that is free of defects in materials or workmanship, you may terminate this agreement by returning the program.

IN NO EVENT WILL PEARSON EDUCATION, INC. BE LIABLE TO YOU FOR ANY DAMAGES, INCLUDING ANY LOST PROFITS, LOST SAVINGS, OR OTHER INCIDENTAL OR CONSEQUENTIAL DAMAGES ARISING OUT OF THE USE OR INABILITY TO USE SUCH PROGRAM EVEN IF PEARSON EDUCATION, INC. OR AN AUTHORIZED DISTRIBUTOR HAS BEEN ADVISED OF THE POSSIBILITY OF SUCH DAMAGES, OR FOR ANY CLAIM BY ANY OTHER PARTY.

SOME STATES DO NOT ALLOW FOR THE LIMITATION OR EXCLUSION OF LIABILITY FOR INCIDENTAL OR CONSEQUENTIAL DAMAGES, SO THE ABOVE LIMITATION OR EXCLUSION MAY NOT APPLY TO YOU.

GENERAL

You may not sublicense, assign, or transfer the license of the program. Any attempt to sublicense, assign or transfer any of the rights, duties, or obligations hereunder is void.

This Agreement will be governed by the laws of the State of New York.

Should you have any questions concerning this Agreement, you may contact Pearson Education, Inc. by writing to:
ESM Media Development
Higher Education Division
Pearson Education, Inc.
1 Lake Street
Upper Saddle River, NJ 07458

Should you have any questions concerning technical support, you may write to:
New Media Production
Higher Education Division
Pearson Education, Inc.
1 Lake Street
Upper Saddle River, NJ 07458

YOU ACKNOWLEDGE THAT YOU HAVE READ THIS AGREEMENT, UNDERSTAND IT, AND AGREE TO BE BOUND BY ITS TERMS AND CONDITIONS. YOU FURTHER AGREE THAT IT IS THE COMPLETE AND EXCLUSIVE STATEMENT OF THE AGREEMENT BETWEEN US THAT SUPERSEDES ANY PROPOSAL OR PRIOR AGREEMENT, ORAL OR WRITTEN, AND ANY OTHER COMMUNICATIONS BETWEEN US RELATING TO THE SUBJECT MATTER OF THIS AGREEMENT.

System Requirements

Windows
Pentium II 300 MHz processor
Windows 98, NT, 2000, ME, or XP
64 MB RAM (128 MB RAM required for Windows XP)
4.3 MB available hard drive space (optional-for minimum QuickTime installation)
800 x 600 resolution
8x or faster CD-ROM drive
QuickTime 6.x
Sound card

Macintosh
PowerPC G3 233 MHz or better
Mac OS 9.x or 10.x
64 MB RAM
10 MB available hard drive space for Mac OS 9, 19 MB on OS X (optional—if QuickTime installation is needed)
800 x 600 resolution
8x or faster CD-ROM drive
QuickTime 6.x

Support Information

If you are having problems with this software, call (800) 677-6337 between 8:00 a.m. and 8:00 p.m. EST, Monday through Friday, and 5:00 p.m. through 12:00 a.m. EST on Sundays. You can also get support by filling out the web form located at: http://247.prenhall.com/mediaform

Our technical staff will need to know certain things about your system in order to help us solve your problems more quickly and efficiently. If possible, please be at your computer when you call for support. You should have the following information ready:
- Textbook ISBN
- CD-ROM ISBN
- corresponding product and title
- computer make and model
- Operating System (Windows or Macintosh) and Version
- RAM available
- hard disk space available
- Sound card? Yes or No
- printer make and model
- network connection
- detailed description of the problem, including the exact wording of any error messages.

NOTE: Pearson does not support and/or assist with the following:
- third-party software (i.e. Microsoft including Microsoft Office suite, Apple, Borland, etc.)
- homework assistance
- Textbooks and CD-ROMs purchased used are not supported and are non-replaceable. To purchase a new CD-ROM, contact Pearson Individual Order Copies at 1-800-282-0693.